PROCESSING AND IMPACT ON ACTIVE COMPONENTS IN FOOD

PROCESSING AND IMPACT ON ACTIVE COMPONENTS IN FOOD

Edited by

VICTOR PREEDY

Department of Nutrition and Dietetics,
Diabetes & Nutritional Sciences Division, School of Medicine,
King's College London, UK

ELSEVIER

AMSTERDAM • BOSTON • HEIDELBERG • LONDON
NEW YORK • OXFORD • PARIS • SAN DIEGO
SAN FRANCISCO • SINGAPORE • SYDNEY • TOKYO
Academic Press is an imprint of Elsevier

Academic Press is an imprint of Elsevier
32 Jamestown Road, London NW1 7BY, UK
225 Wyman Street, Waltham, MA 02451, USA
525 B Street, Suite 1800, San Diego, CA 92101-4495, USA

First edition 2015

British Library Cataloguing-in-Publication Data
A catalogue record for this book is available from the British Library

Library of Congress Cataloging-in-Publication Data
A catalog record for this book is available from the Library of Congress

ISBN: 978-0-12-404699-3

For information on all Academic Press publications
visit our website at elsevierdirect.com

Typeset by TNQ Books and Journals
www.tnq.co.in

Contents

9. Carotenoids in Pumpkin and Impact of Processing Treatments and Storage

JOÃO GUSTAVO PROVESI, EDNA REGINA AMANTE

10. Cassava Production and Processing and Impact on Biological Compounds

LEON BRIMER

11. Radiation Processing for Sprout Inhibition of Stored Potatoes and Mitigation of Acrylamide in Fries and Chips

JYOTI TRIPATHI, PRASAD S. VARIYAR, REKHA S. SINGHAL, ARUN SHARMA

12. Production of Jerusalem Artichoke (*Helianthus tuberosus* L.) and Impact on Inulin and Phenolic Compounds

VIBE BACH, MORTEN RAHR CLAUSEN, MERETE EDELENBOS

13. Bioactive Compounds in Asparagus and Impact of Storage and Processing

JOSÉ MARÍA FUENTES ALVENTOSA, JOSÉ MANUEL MORENO ROJAS

14. Bioactive Components in Potatoes as Influenced by Thermal Processing

CARLA S.P. SANTOS, SARA CUNHA, SUSANA CASAL

15. Processing of Fructans and Oligosaccharides from Agave Plants

CLARITA OLVERA CARRANZA, ANGELA ÁVILA FERNANDEZ, GUSTAVO R. BUSTILLO ARMENDÁRIZ, AGUSTÍN LÓPEZ-MUNGUÍA

Section 2
FRUIT

16. Effects of 1-Methylcyclopropene on Active Composition in Fruits

JUNFENG GUAN, MUQIANG HU, CHENGGUO SHEN, SHUO ZHOU, YUDOU CHENG, JINGANG HE

17. How Cultivars Influence Fruit Composition: Total Phenols, Flavonoids Contents, and Antioxidant Activity in the Pulp of Selected Asian Pears

JUNFENG GUAN, JINGANG HE, CHENGGUO SHEN, LIMEI LI, YANXIA WANG, YUDOU CHENG

27. Formation of Biogenic Amines in Cheese

TUBA ŞANLI, EBRU ŞENEL

28. Changes in Volatile Compounds of Cheese

SEVAL ANDIÇ, YUSUF TUNÇTÜRK, GÖKHAN BORAN

Section 4

OILS FATS SPREADS

29. Oxidation Products of Corn Oil at Room Temperature

ENCARNACIÓN GOICOECHEA, MARÍA D. GUILLÉN

30. Aldehydes after Prolonged Heating at Frying Temperature

ANDREA MARTÍNEZ-YUSTA, ENCARNACIÓN GOICOECHEA, MARIA D. GUILLÉN

31. Influence of Filtration on Composition of Olive Oils

KAROLINA BRKIć BUBOLA, OLIVERA KOPRIVNJAK

32. Virgin Olive Oil: Losses of Antioxidant Polar Phenolic Compounds due to Storage, Packaging, and Culinary Uses

ANASTASIOS KOIDIS, DIMITRIOS BOSKOU

Section 5

MEATS

33. Volatile Compounds in High-Pressure-Processed Pork Meat Products

ANA RIVAS-CAÑEDO, MARIA TERESA DÍAZ, ANTONIA PICÓN, ESTRELLA FERNÁNDEZ-GARCÍA, MANUEL NÚÑEZ

34. Imidazole Dipeptides in Meat from Different Animal Species and Effect of Cooking Method on their Contents in Beef and Turkey Meat

PIER GIORGIO PEIRETTI, CLAUDIO MEDANA

35. Cantonese Sausage, Processing, Storage and Composition

WEIZHENG SUN, FEIBAI ZHOU, MOUMING ZHAO

Section 6

GRAIN, BEANS, PULSES, NUTS AND SEEDS

56. Fish Roe Lipids: Composition and Changes During Processing and Storage

P.G. PRABHAKARA RAO, K. BALASWAMY, T. JYOTHIRMAYI, M.S.L. KARUNA, R.B.N. PRASAD

Section 8
BEVERAGES

57. Use of Oak Wood to Enrich Wine with Volatile Compounds

PEDRO M. PÉREZ-JUAN, MARÍA DOLORES LUQUE DE CASTRO

58. Influence of Diammonium Phosphate Addition to Fermentation on Wine Biologicals

MAR VILANOVA, ISAK S. PRETORIUS, PAUL A. HENSCHKE

59. The Application of Foliar Urea on the Concentration of Amines in Wine

CARMEN ANCÍN-AZPILICUETA, RODRIGO NIETO-ROJO, JULIO GÓMEZ-CORDÓN

60. Effect of Wine Production Techniques on Wine Resveratrol and Total Phenolics

JELENA CVEJIĆ, MILICA ATANACKOVIĆ

61. Arsenic in Wines and Beers from European Markets: Alert of Arsenic Species in Response to Processing

J.-H. HUANG, KAN-NIAN HU, JÖRG ILGEN, GUNTER ILGEN, CHRISTINE ALEWELL

62. Aflatoxin B1 in Beer at Different Stages of Production

ANTONIO RUIZ-MEDINA, Mª LUISA FERNÁNDEZ-DE CÓRDOVA

63. Behavior of Triazole Fungicide Residues from Barley to Beer

SIMÓN NAVARRO, GABRIEL PÉREZ-LUCAS, NURIA VELA, GINÉS NAVARRO

64. Effects of Processing Stages on the Profile of Phenolic Compounds in Beer

HAIFENG ZHAO

65. Furan and Other Furanic Compounds in Coffee: Occurrence, Mitigation Strategies, and Importance of Processing

MONICA ANESE

66. Mineral Composition Variability of Coffees: A Result of Processing and Production

REBECA CRUZ, SIMONE MORAIS, SUSANA CASAL

67. Free Radical Processes in Coffee I—Solid Samples

BERNARD A. GOODMAN, CHAHAN YERETZIAN

68. Free Radical Processes in Coffee II—Liquids

BERNARD A. GOODMAN, CHAHAN YERETZIAN

69. Acrylamide in Coffee: Influence of Processing

CRISTINA M.D. SOARES, RITA C. ALVES, M. BEATRIZ P.P. OLIVEIRA

70. Use of Microwaves to Extract Chlorogenic Acids from Green Coffee Beans

LINGAMALLU JAGAN MOHAN RAO, KULATHOORAN RAMALAKSHMI

71. Tea Leaf Age, Shade and Characteristic Levels of L-Theanine, Caffeine, (−)-Epigallocatechin Gallate (EGCG), (−)-Epigallocatechin (EGC), (−)-Epicatechin (EC), and (−)-Epicatechin Gallate (ECG)

KEHAU A. HAGIWARA, ANTHONY D. WRIGHT

72. The Impact of Variety, Environment and Agricultural Practices on Catechins and Caffeine in Plucked Tea Leaves

HAO CHENG, KANG WEI, LIYUAN WANG

73. Cocoa Processing and Impact on Composition

JASMINKA GIACOMETTI, SLAVICA MAZOR JOLIĆ, DJURO JOSIĆ

74. Tagatose Stability in Beverages as Impacted by Composition and Thermal Processing

LEONARD N. BELL

75. High-Pressure Processing, Strawberry Beverages, and Composition of 'Bioactives'

RAVI KIRAN TADAPANENI, INDIKA EDIRISINGHE, BRITT BURTON-FREEMAN

76. Processing Pomegranates for Juice and Impact on Bioactive Components

NALLELY NUNCIO-JÁUREGUI, ÁNGEL CALÍN-SÁNCHEZ, LAURA VÁZQUEZ-ARAÚJO, ANTONIO JOSÉ PÉREZ-LÓPEZ, MARÍA J. FRUTOS-FERNÁNDEZ, ÁNGEL A. CARBONELL-BARRACHINA

77. Carotenoids in Nonthermally Treated Fruit Juices

I. ODRIOZOLA-SERRANO, G. OMS-OLIU, R. SOLIVA-FORTUNY, O. MARTÍN-BELLOSO

Section 9

HERBS AND SPICES AND MISCELLANEOUS VEGETATION

78. Dehydration of Basil Leaves and Impact of Processing Composition

RAFFAELLA BOGGIA, PAOLA ZUNIN, VILMA HYSENAJ, ALDO BOTTINO, ANTONIO COMITE

79. Variations in Polyphenols and Lipid Soluble Vitamins in *Moringa oleifera*

JULIA P. COPPIN, H. RODOLFO JULIANI, QINGLI WU, JAMES E. SIMON

Section 10

CONFECTIONARY AND OTHER FOOD ITEMS

80. Effect of Processing on Dark Chocolate Composition: A Focus on Allergens

SEFAT E. KHUDA, KRISTINA M. WILLIAMS

81. Honey Production and Compositional Parameters

JONATHAN M. STEPHENS, DAVID R. GREENWOOD, LIAM FEARNLEY, JESSIE BONG, RALF C. SCHLOTHAUER, KERRY M. LOOMES

82. Jam Processing and Impact on Composition of Active Compounds

TAHA M. RABABAH, MUHAMMAD H. AL-U'DATT, SUSAN BREWER

List of Contributors

Kjersti Aaby Norwegian Institute of Food, Fisheries, and Aquaculture Research, Aas, Norway

Nissreen Abu-Ghannam School of Food Science and Environmental Health, Dublin Institute of Technology, Dublin, Ireland

Alfredo Aires CITAB/UTAD-Centre for Research and Technology for Agro-Environment and Biological Sciences, University of Trás-os-Montes e Alto Douro, Vila Real, Portugal

Christine Alewell Institute of Environmental Geosciences, University of Basel, Basel, Switzerland

Domingos P.F. Almeida Instituto Superior de Agronomia, Universidade de Lisboa, Lisbon, Portugal

Sergio Almonacid Department of Chemical and Environmental Engineering, Universidad Técnica Federico Santa María, Valparaíso, Chile; Centro Regional para el Estudio de Alimentos Saludables (CREAS), Valparaíso, Chile

Muhammad H. Al-u'datt Department of Nutrition and Food Technology, Jordan University of Science and Technology, Irbid, Jordan

José María Fuentes Alventosa Tecnología, Postcosecha e Industria Agroalimentaria. Instituto de Investigación y Formación Agraria y Pesquera (IFAPA) Córdoba, Spain

Rita C. Alves REQUIMTE, Departamento de Ciência Químicas, Faculdade de Farmácia da Universidade do Porto, Porto, Portugal; REQUIMTE, Instituto Superior de Engenharia do Porto, Instituto Politécnico do Porto, Porto, Portugal

Edna Regina Amante Laboratory of Fruits and Vegetables, Department of Food Science and Technology, Federal University of Santa Catarina, Florianópolis, Brazil

Ana L. Amaro Centro de Biotecnologia e Química Fina, Escola Superior de Biotecnologia, Centro Regional do Porto da Universidade Católica Portuguesa, Porto, Portugal

Carmen Ancín-Azpilicueta Department of Applied Chemistry, Universidad Pública de Navarra, Pamplona, Spain

Seval Andiç Department of Food Engineering, Faculty of Engineering and Architecture, Yüzüncü Yıl University, Van, Turkey

Monica Anese Dipartimento di Scienze degli Alimenti, University of Udine, Udine, Italy

Metin Atamer Ankara University, Faculty of Agriculture, Department of Dairy Technology, Ankara, Turkey

Milica Atanacković Department of Pharmacy, Faculty of Medicine, University of Novi Sad, Novi Sad, Serbia

Vibe Bach Department of Food Science, Aarhus University, Kirstinebjergvej, Denmark

K. Balaswamy CSIR-Central Food Technological Research Institute (CFTRI), Resource Centre, Habshiguda, Hyderabad, India

Gustavo Barbosa-Cánovas Department of Biological Systems Engineering, Washington State University, Pullman, WA, USA

Leonard N. Bell Auburn University, Auburn, USA

Nacer Bellaloui USDA-ARS, Crop Genetics Research Unit, Jackson, TN, USA

M. Benlloch-Tinoco Food Technology Department, Food Investigation and Innovation Group, Universitat Politècnica de Valencia, Valencia, Spain

Daniela Bermúdez-Aguirre Department of Biological Systems Engineering, Washington State University, Pullman, WA, USA

Chiranjib Bhattacharjee Jadavpur University, Kolkata, India

Mandana Bimakr Department of Food Science and Technology, Faculty of Agriculture, University of Zanjan, Zanjan, Iran; Department of Food Technology, Faculty of Food Science and Technology, Universiti Putra Malaysia, Selangor, Malaysia

Gry Aletta Bjørlykke Institute of Marine Research, Bergen, Norway

Raffaella Boggia University of Genoa, Genoa, Italy

Jessie Bong School of Biological Sciences, University of Auckland, New Zealand

Gökhan Boran Department of Food Engineering, Faculty of Engineering and Architecture, Yüzüncü Yıl University, Van, Turkey

Dimitrios Boskou Department of Chemistry, Aristotle University of Thessaloniki, Thessaloniki, Greece

Aldo Bottino University of Genoa, Genoa, Italy

Joyce Irene Boye Food Research and Development Centre, Agriculture and Agri-Food Canada, Quebec, Canada

Susan Brewer Department of Food Science and Human Nutrition, University of Illinois at Urbana-Champaign, Urbana, USA

Leon Brimer Department of Veterinary Disease Biology, University of Copenhagen, Denmark

Nigel Brunton School of Agriculture and Food Science, University College Dublin, Dublin, Ireland

Karolina Brkić Bubola Department of Agriculture and Nutrition, Institute of Agriculture and Tourism, Poreč, Croatia

Sandy Van Buggenhout Laboratory of Food Technology and Leuven Food Science and Nutrition Research Centre (LFoRCe), Department of Microbial and Molecular Systems (M^2S), Katholieke Universiteit Leuven, Leuven, Belgium

Britt Burton-Freeman Institute for Food Safety and Health, Illinois Institute of Technology, Illinois, USA

Joselyn Bustamante Department of Chemical and Environmental Engineering, Universidad Técnica Federico Santa María, Valparaíso, Chile

Gustavo R. Bustillo Armendáriz Bustar Alimentos, SAPI de CV, Guadalajara, México

Mónica Calderón-Santiago Department of Analytical Chemistry, Annex Marie Curie Building, Campus of Rabanales, University of Córdoba, Córdoba, Spain

Ángel Calín-Sánchez Miguel Hernández University, Alicante, Spain

M.M. Camacho Food Technology Department, Food Investigation and Innovation Group, Universitat Politècnica de Valencia, Valencia, Spain

Ángel A. Carbonell-Barrachina Miguel Hernández University, Alicante, Spain

Clarita Olvera Carranza Instituto de Biotecnología, Universidad Nacional Autónoma de México, México

J. Carranza-Concha Universidad Autónoma de Zacatecas, Zacatecas, México

Susana Casal REQUIMTE, Faculdade de Farmácia, Universidade do Porto, Portugal

Anoma Chandrasekara Department of Applied Nutrition, Wayamba University of Sri Lanka, Gonawila, Sri Lanka

Yudou Cheng Institute of Genetics and Physiology, Hebei Academy of Agricultural and Forestry Sciences, Shijiazhuang, PR China

Hao Cheng Tea Research Institute Chinese Academy of Agricultural Sciences (TRICAAS), Zhejiang, China

Morten Rahr Clausen Department of Food Science, Aarhus University, Kirstinebjergvej, Denmark

Antonio Comite University of Genoa, Genoa, Italy

Anna Concollato Institute of Marine Research, Matre Research Station, Matredal, Norway

Julia P. Coppin New Use Agriculture and Natural Plant Products Program, Rutgers University, New Jersey, USA

Mª Luisa Fernández-de Córdova Department of Physical and Analytical Chemistry, University of Jaén, Jaén, Spain

Rebeca Cruz REQUIMTE, Serviço de Bromatologia e Hidrologia, Faculdade de Farmácia, Universidade do Porto, Porto, Portugal

Raimondo Cubadda DiAAA, Universitá degli Studi del Molise, Campobasso, Italy

Sara Cunha REQUIMTE, Faculdade de Farmácia, Universidade do Porto, Portugal

Jelena Cvejić Department of Pharmacy, Faculty of Medicine, University of Novi Sad, Novi Sad, Serbia

Malgorzata R. Cyran Plant Breeding and Acclimatization Institute, National Research Institute (IHAR-PIB), Radzikow, Poland

Ranjana Das Jadavpur University, Kolkata, India

Bidyut C. Deka ICAR Research Complex, Nagaland Centre, Nagaland, India

Maria Teresa Díaz Departamento de Tecnología de Alimentos, Instituto Nacional de Investigación y Tecnología Agraria y Alimentaria (INIA), Madrid, Spain

Stephen O. Duke USDA-ARS, Natural Products Utilization Research Unit, MS, USA

Merete Edelenbos Department of Food Science, Aarhus University, Kirstinebjergvej, Denmark

Indika Edirisinghe Institute for Food Safety and Health, Illinois Institute of Technology, Illinois, USA

Ines Eichholz Humboldt-Universität zu Berlin, Division Urban Plant Ecophysiology, Section Quality Dynamics/ Postharvest Physiology, Berlin, Germany

Liam Fearnley School of Biological Sciences, University of Auckland, New Zealand

Katarzyna Felisiak Department of Food Science and Technology, Faculty of Food Science and Fisheries, West Pomeranian University of Technology, Poland

Angela Ávila Fernandez Centro de Investigación, División Académica de Ciencias de la Salud, Universidad Juárez Autónoma de Tabasco, México

Estrella Fernández-García Departamento de Tecnología de Alimentos, Instituto Nacional de Investigación y Tecnología Agraria y Alimentaria (INIA), Madrid, Spain

Alessandra Fratianni DiAAA, Universitá degli Studi del Molise, Campobasso, Italy

María J. Frutos-Fernández Miguel Hernández University, Alicante, Spain

Ali Ganjloo Department of Food Science and Technology, Faculty of Agriculture, University of Zanjan, Zanjan, Iran

Hugo S. Garcia UNIDA-Instituto Tecnologico de Veracruz, M.A. de Quevedo 2779, Col. Formando Hogar, Veracruz, Ver. 91897, México

Jasminka Giacometti University of Rijeka, Rijeka, Croatia

Encarnación Goicoechea Department of Food Technology, Faculty of Pharmacy, Lascaray Research Center, University of the Basque Country (UPV/EHU), Vitoria, Spain

Julio Gómez-Cordón Avanzare Innovación Tecnológica, Logroño, Spain

Bernard A. Goodman State Key Laboratory for Conservation and Utilization of Subtropical Agro-Bioresources, Guangxi University, Nanning, Guangxi, China

David R. Greenwood School of Biological Sciences, University of Auckland, New Zealand

Junfeng Guan Institute of Genetics and Physiology, Hebei Academy of Agricultural and Forestry Sciences, Shijiazhuang, PR China

Manisha Guha Department of Grain Science and Technology, Central Food Technological Research Institute, Council of Scientific and Industrial Research (CSIR), Mysore, India

María D. Guillén Department of Food Technology, Faculty of Pharmacy, Lascaray Research Center, University of the Basque Country (UPV/EHU), Vitoria, Spain

Kehau A. Hagiwara Daniel K. Inouye College of Pharmacy, Department of Pharmaceutical Sciences, University of Hawaii at Hilo, Hawaii, USA

Jingang He Institute of Genetics and Physiology, Hebei Academy of Agricultural and Forestry Sciences, Shijiazhuang, PR China

Marc Hendrickx Laboratory of Food Technology and Leuven Food Science and Nutrition Research Centre (LFoRCe), Department of Microbial and Molecular Systems (M^2S), Katholieke Universiteit Leuven, Leuven, Belgium

Paul A. Henschke The Australian Wine Research Institute, Glen Osmond, Australia

Kan-Nian Hu Presentation Road, Brighton, MA, USA

Muqiang Hu Institute of Genetics and Physiology, Hebei Academy of Agricultural and Forestry Sciences, Shijiazhuang, PR China

J.-H. Huang Institute of Environmental Geosciences, University of Basel, Basel, Switzerland

Susanne Huyskens-Keil Humboldt-Universität zu Berlin, Division Urban Plant Ecophysiology, Section Quality Dynamics/Postharvest Physiology, Berlin, Germany

Vilma Hysenaj University of Genoa, Genoa, Italy

Jörg Ilgen Wein-Bastion Ulm, Ulm, Germany

Gunter Ilgen Central Analytic, Bayreuth Center for Ecology and Environmental Research (BayCEER), University of Bayreuth, Bayreuth, Germany

Jacek Jaczynski West Virginia University, Division of Animal and Nutritional Sciences, Morgantown, WV, USA

Amit Kumar Jaiswal School of Food Science and Environmental Health, Dublin Institute of Technology, Dublin, Ireland

Slavica Mazor Jolić Kraš,d.d., Zagreb, Croatia

Djuro Josić University of Rijeka, Rijeka, Croatia; Brown University, Rhode Island, USA

H. Rodolfo Juliani New Use Agriculture and Natural Plant Products Program, Rutgers University, New Jersey, USA

T. Jyothirmayi CSIR-Central Food Technological Research Institute (CFTRI), Resource Centre, Habshiguda, Hyderabad, India

Dipankar Kalita Tezpur University, Assam, India

M.S.L. Karuna Centre for Lipid Research, CSIR-Indian Institute of Chemical Technology (IICT), Tarnaka, Hyderabad, India

Sefat E. Khuda US Food and Drug Administration, Maryland, USA

Griet Knockaert Laboratory of Food Technology and Leuven Food Science and Nutrition Research Centre (LFoRCe), Department of Microbial and Molecular Systems (M^2S), Katholieke Universiteit Leuven, Leuven, Belgium

Anastasios Koidis Institute for Global Food Security, Queen's University, Belfast, UK

Olivera Koprivnjak Department of Food Technology and Control, School of Medicine, University of Rijeka, Croatia

Bjørn Olav Kvamme Institute of Marine Research, Bergen, Norway

Lien Lemmens Laboratory of Food Technology and Leuven Food Science and Nutrition Research Centre (LFoRCe), Department of Microbial and Molecular Systems (M^2S), Katholieke Universiteit Leuven, Leuven, Belgium

Jørgen Lerfall Department of Food Technology, Sør-Trøndelag University College, Trondheim, Norway

Limei Li Institute of Genetics and Physiology, Hebei Academy of Agricultural and Forestry Sciences, Shijiazhuang, PR China

Ann Van Loey Laboratory of Food Technology and Leuven Food Science and Nutrition Research Centre (LFoRCe), Department of Microbial and Molecular Systems (M^2S), Katholieke Universiteit Leuven, Leuven, Belgium

Kerry M. Loomes School of Biological Sciences, University of Auckland, New Zealand

Leticia X. Lopez-Martinez Universidad Autónoma del Estado de México, Facultad de Química. Paseo Colón y Paseo Tollocan S/N, Toluca, Estado de México 50120, México

Agustín López-Munguía Instituto de Biotecnología, Universidad Nacional Autónoma de México, México

M.D. Luque de Castro Departamento de Química Analítica, Universidad de Córdoba, Córdoba, Spain

Zhen Ma Food Research and Development Centre, Agriculture and Agri-Food Canada, Quebec, Canada

Charu Lata Mahanta Tezpur University, Assam, India

N.G. Malleshi Department of Grain Science and Technology, Central Food Technological Research Institute, Council of Scientific and Industrial Research (CSIR), Mysore, India

S. Mandal Division of Agricultural Engineering, ICAR Research Complex, Umiam, Meghalaya, India

O. Martín-Belloso Department of Food Technology, University of Lleida, Lleida, Spain

N. Martínez-Navarrete Food Technology Department, Food Investigation and Innovation Group, Universitat Politècnica de Valencia, Valencia, Spain

Claudio Medana Department of Molecular Biotechnology and Health Sciences, University of Torino, Torino, Italy

Alemu Mengistu USDA-ARS, Crop Genetics Research Unit, Jackson, TN, USA

Seyed Hossein Mirdehghan Department Horticultural Sciences, College of Agriculture, Vali-e-Asr University of Rafsanjan, Kerman, Iran

Simone Morais REQUIMTE, Instituto Superior de Engenharia do Porto, Instituto Politécnico do Porto, Porto, Portugal

J.M. Moreno-Rojas Área de Tecnología, Poscosecha e Industria Agroalimentaria, Córdoba, Spain; Instituto de Investigación y Formación Agraria y Pesquera (IFAPA), Córdoba, Spain

Bożena Muszyńska Department of Pharmaceutical Botany, Jagiellonian University Collegium Medicum, Medyczna street 9, 30-688 Kraków, Poland

A. Nath Division of Agricultural Engineering, ICAR Research Complex, Umiam, Meghalaya, India

Simón Navarro Departamento de Química Agrícola, Geología y Edafología, Facultad de Química, Universidad de Murcia, Campus Universitario de Espinardo, Murcia, Spain

Ginés Navarro Departamento de Química Agrícola, Geología y Edafología, Facultad de Química, Universidad de Murcia, Campus Universitario de Espinardo, Murcia, Spain

S.V. Ngachan Division of Agricultural Engineering, ICAR Research Complex, Umiam, Meghalaya, India

Rodrigo Nieto-Rojo Department of Applied Chemistry, Universidad Pública de Navarra, Pamplona, Spain

Nallely Nuncio-Jáuregui Miguel Hernández University, Alicante, Spain

Manuel Núñez Departamento de Tecnología de Alimentos, Instituto Nacional de Investigación y Tecnología Agraria y Alimentaria (INIA), Madrid, Spain

I. Odriozola-Serrano Department of Food Technology, University of Lleida, Lleida, Spain

Ana Oliveira Centro de Biotecnologia e Química Fina, Escola Superior de Biotecnologia, Centro Regional do Porto da Universidade Católica Portuguesa, Porto, Portugal

M. Beatriz P.P. Oliveira REQUIMTE, Departamento de Ciência Químicas, Faculdade de Farmácia da Universidade do Porto, Porto, Portugal

Rolf Erik Olsen Institute of Marine Research, Matre Research Station, Matredal, Norway

G. Oms-Oliu Department of Food Technology, University of Lleida, Lleida, Spain

Gianfranco Panfili DiAAA, Universitá degli Studi del Molise, Campobasso, Italy

Violeta Pardío-Sedas Facultad de Medicina Veterinaria y Zootecnia, Universidad Veracruzana, Colonia Unidad Veracruzana, Veracruz, México

Pedro M Pérez-Juan Bodega y Viñedos Vallebravo, Puente de Génave, Pol. Ind. La Vicaría, Jaén, Spain

Pier Giorgio Peiretti Institute of Science of Food Production, National Research Council, Grugliasco, Italy

Antonio José Pérez-López Catholic University of San Antonio, Murcia, Spain

Gabriel Pérez-Lucas Departamento de Química Agrícola, Geología y Edafología, Facultad de Ciencias Sociosanitarias, Avenida de las Fuerzas Armadas, Lorca, Murcia, Spain

Antonia Picón Departamento de Tecnología de Alimentos, Instituto Nacional de Investigación y Tecnología Agraria y Alimentaria (INIA), Madrid, Spain

Marlene Pinto Department of Chemical and Environmental Engineering, Universidad Técnica Federico Santa María, Valparaíso, Chile

P.G. Prabhakara Rao CSIR-Central Food Technological Research Institute (CFTRI), Resource Centre, Habshiguda, Hyderabad, India

I. Pradas-Baena Área de Tecnología, Poscosecha e Industria Agroalimentaria, Córdoba, Spain; Instituto de Investigación y Formación Agraria y Pesquera (IFAPA), Córdoba, Spain

R.B.N. Prasad Centre for Lipid Research, CSIR-Indian Institute of Chemical Technology (IICT), Tarnaka, Hyderabad, India

Isak S. Pretorius Macquarie University, Sydney, Australia

João Gustavo Provesi Department of Food Science and Technology, Federal Institute of Education, Science and Technology of Santa Catarina, Urupema, Brazil

Taha M. Rababah Department of Nutrition and Food Technology, Jordan University of Science and Technology, Irbid, Jordan

Russly A. Rahman Halal Product Research Institute, Universiti Putra Malaysia, Selangor, Malaysia; Department of Process and Food Engineering, Faculty of Engineering, Universiti Putra Malaysia, Selangor, Malaysia; Department of Food Technology, Faculty of Food Science and Technology, Universiti Putra Malaysia, Selangor, Malaysia

Prasad Rallabhandi Center for Food Safety and Applied Nutrition, U.S. Food and Drug Administration, Laurel, Maryland, USA

Kulathooran Ramalakshmi Plantation Products, Spices and Flavour Technology Department, Central Food Technological Research Institute, Mysore, India

Lingamallu Jagan Mohan Rao Plantation Products, Spices and Flavour Technology Department, Central Food Technological Research Institute, Mysore, India

Ashish Rawson Indian Institute of Crop Processing Technology, Thanjavur, India

Krishna N. Reddy USDA-ARS, Crop Production Systems Research Unit, Jackson, TN, USA

Siv Fagertun Remberg Department of Plant and Environmental Sciences, Norwegian University of Life Sciences, Aas, Norway

Isabel Revilla Area de Tecnología de Alimentos, Universidad de Salamanca, Spain

Ana Rivas-Cañedo Departamento de Tecnología de Alimentos, Instituto Nacional de Investigación y Tecnología Agraria y Alimentaria (INIA), Madrid, Spain

Sascha Rohn University of Hamburg, Hamburg School of Food Science, Institute of Food Chemistry, Hamburg, Germany

José Manuel Moreno Rojas Tecnología, Postcosecha e Industria Agroalimentaria. Instituto de Investigación y Formación Agraria y Pesquera (IFAPA) Córdoba, Spain

Antonio Ruiz-Medina Department of Physical and Analytical Chemistry, University of Jaén, Jaén, Spain

Shyam S. Sablani Biological Systems Engineering Department, Washington State University, Pullman, USA

Tuba Şanlı Ankara University, Faculty of Agriculture, Department of Dairy Technology, Ankara, Turkey

Carla S.P. Santos REQUIMTE, Faculdade de Farmácia, Universidade do Porto, Portugal

Mohammad Sayyari Department Horticultural Sciences, College of Agriculture, Bu-Ali Sina University, Hamedan, Iran

Ralf C. Schlothauer Comvita Innovations, Institute for Innovation in Biotechnology, University of Auckland, Auckland, New Zealand

Ebru Şenel Ankara University, Faculty of Agriculture, Department of Dairy Technology, Ankara, Turkey

María Serrano Department of Applied Biology, University Miguel Hernández, Alicante, Spain

Fereidoon Shahidi Department of Biochemistry, Memorial University of Newfoundland, St. John's, Canada

Arun Sharma Food Technology Division, Bhabha Atomic Research Centre, Mumbai, India

Girdhari M. Sharma Center for Food Safety and Applied Nutrition, U.S. Food and Drug Administration, Laurel, Maryland, USA

Chengguo Shen Institute of Genetics and Physiology, Hebei Academy of Agricultural and Forestry Sciences, Shijiazhuang, PR China

James E. Simon New Use Agriculture and Natural Plant Products Program, Rutgers University, New Jersey, USA

Ricardo Simpson Department of Chemical and Environmental Engineering, Universidad Técnica Federico Santa María, Valparaíso, Chile; Centro Regional para el Estudio de Alimentos Saludables (CREAS), Valparaíso, Chile

R.K. Singh Division of Agricultural Engineering, ICAR Research Complex, Umiam, Meghalaya, India

Rekha S. Singhal Food Engineering and Technology Department, Institute of Chemical Technology, Mumbai, India

Erik Slinde Institute of Marine Research, Bergen, Norway; Department of Chemistry, Biotechnology and Food Science, Norwegian University of Life Science, Ås, Norway

Cristina M.D. Soares REQUIMTE, Departamento de Química e Bioquímica, Faculdade de Ciências da Universidade do Porto, Portugal

R. Soliva-Fortuny Department of Food Technology, University of Lleida, Lleida, Spain

Shangxin Song Key Laboratory of Meat Processing and Quality Control, Nanjing Agricultural University, Nanjing, China

Oddvin Sørheim Nofima, Ås, Norway

Yadahally N. Sreerama Department of Grain Science and Technology, Central Food Technological Research Institute, Council of Scientific and Industrial Research (CSIR), Mysore, India

Jonathan M. Stephens Comvita Innovations, Institute for Innovation in Biotechnology, University of Auckland, Auckland, New Zealand

Katarzyna Sułkowska-Ziaja Department of Pharmaceutical Botany, Jagiellonian University Collegium Medicum, Kraków, Poland

Weizheng Sun College of Light Industry and Food Sciences, South China University of Technology, Guangzhou, China

Mariusz Szymczak Department of Food Science and Technology, Faculty of Food Science and Fisheries, West Pomeranian University of Technology, Poland

Ravi Kiran Tadapaneni Institute for Food Safety and Health, Illinois Institute of Technology, Illinois, USA

Reza Tahergorabi North Carolina A&T State University, Greensboro, NC, USA

Grzegorz Tokarczyk Department of Food Science and Technology, Faculty of Food Science and Fisheries, West Pomeranian University of Technology, Poland

Jyoti Tripathi Food Technology Division, Bhabha Atomic Research Centre, Mumbai, India

Apollinaire Tsopmo Food Science and Nutrition Program, Carleton University, Ottawa, Canada

Yusuf Tunçtürk Department of Food Engineering, Faculty of Engineering and Architecture, Yüzüncü Yıl University, Van, Turkey

Daniel Valero Department of Food Technology, University Miguel Hernández, Alicante, Spain

Prasad S. Variyar Food Technology Division, Bhabha Atomic Research Centre, Mumbai, India

Laura Vázquez-Araújo University of Vigo, Ourense, Spain

Nuria Vela Facultad de Enfermería, Universidad Católica San Antonio de Murcia, Campus de Los Jerónimos, Guadalupe, Murcia, Spain

Mar Vilanova Misión Biológica de Galicia, CSIC, Pontevedra, Spain

Liyuan Wang Tea Research Institute Chinese Academy of Agricultural Sciences (TRICAAS), Zhejiang, China

Yanxia Wang Institute of Genetics and Physiology, Hebei Academy of Agricultural and Forestry Sciences, Shijiazhuang, PR China

Kang Wei Tea Research Institute Chinese Academy of Agricultural Sciences (TRICAAS), Zhejiang, China

Kristina M. Williams US Food and Drug Administration, Maryland, USA

Anthony D. Wright Daniel K. Inouye College of Pharmacy, Department of Pharmaceutical Sciences, University of Hawaii at Hilo, Hawaii, USA

Qingli Wu New Use Agriculture and Natural Plant Products Program, Rutgers University, New Jersey, USA

Hongmei Xiao Key Laboratory of Meat Processing and Quality Control, Nanjing Agricultural University, Nanjing, China

Chahan Yeretzian Zurich University of Applied Sciences, Institute of Chemistry and Biological Chemistry, Wädenswil, Switzerland

Andrea Martínez-Yusta Department of Food Technology, Faculty of Pharmacy, Lascaray Research Center, University of the Basque Country (UPV/EHU), Vitoria, Spain

Haifeng Zhao College of Light Industry and Food Sciences, South China University of Technology, Guangzhou, P. R. China

Mouming Zhao College of Light Industry and Food Sciences, South China University of Technology, Guangzhou, China

Feibai Zhou College of Light Industry and Food Sciences, South China University of Technology, Guangzhou, China

Shuo Zhou Institute of Genetics and Physiology, Hebei Academy of Agricultural and Forestry Sciences, Shijiazhuang, PR China

Paola Zunin University of Genoa, Genoa, Italy

Preface

The foods that people eat have extensive diversity. This relates not only to the nature of the individual items of food, but their components as well. Moreover, the qualitative and quantitative nature of these components can vary considerably due to the way the food has been processed. In its simplest definition, production and processing includes cultivars or varieties, small scale or large industrial processes, stabilization, storage, chilling, freezing, freeze-drying, bottling, canning, shaping and reforming as well as other processes. Thus, in the end, food from shops and markets reflects the net sum of all processing, packaging and storage effects. On the other hand, some foods such as fruit and vegetables need little post-harvest processing. Nevertheless, there are significant changes in the amount and profile of biologically active compounds during ripening and storage up to the point of consumption. These biologically active components may be either micronutrients as well as nutraceuticals which have effects on health. Some foods have a role in health due to their macromolecular profiles: these include meats, grains and dairy products. Occasionally, the way the food is processed will affect the amounts or formation of harmful components. Some components will contribute to the sensory profile of the food.

The science behind the post-harvest compositional changes in food is transferable to other foods. However, finding all this information in a coherent book has been problematical as hitherto no publication has attempted to marshal together all the relevant information. This is addressed in **Processing and Impact on Active Components in Food** where the focus is on components that have a biological role in the body.

Its unique features are the 3 sections within each contribution:

How composition is altered
Analytical techniques
Summary points

There are 10 main sections covering vegetables and root crops, fruit, dairy and eggs, oils, meats, grains, beans, pulses, nuts and seeds, marine foods, beverages, herbs and other vegetation, confectionary and other food items. There are detailed descriptions on carrots, broccoli, cabbage, parsnips, mushrooms, peas, pumpkin, cassava, potatoes, Jerusalem artichoke, asparagus, agave plants, pears, strawberries, melons, blueberries, raisins, pomegranates, milk, cheese, butter, corn oil, olive oil, pork, beef, turkey, sausages, wheat, maize, rye, rice, barley, oats, faba beans, soybean, sesame seeds, jackfruit seeds, melon seeds, fish including herring and roe, shellfish including oysters, wine, beer, coffee, tea, fruit juices, basil, Moringa oleifera, chocolate, honey and jam, as well as many other food items.

The book is designed for food scientists, technologist, food industry workers, dietitians and nutritionists, as well as research scientists. Contributions are from leading national and international experts including those from world renowned institutions.

Professor Victor R. Preedy,
King's College London

Biography

Victor R. Preedy BSc, PhD, DSc, FSB, FRSH, FRIPH, FRSPH, FRCPath, FRSC is a senior member of King's College London (Professor of Nutritional Biochemistry) and King's College Hospital (Professor of Clinical Biochemistry: Hon). He is attached to both the Diabetes and Nutritional Sciences Division and the Department of Nutrition and Dietetics. He is also Director of the Genomics Centre and a member of the School of Medicine. Professor Preedy graduated in 1974 with an Honours Degree in Biology and Physiology with Pharmacology. He gained his University of London PhD in 1981. In 1992, he received his Membership of the Royal College of Pathologists and in 1993 he gained his second doctoral degree, for his outstanding contribution to protein metabolism in health and disease. Professor Preedy was elected as a Fellow to the Institute of Biology in 1995 and to the Royal College of Pathologists in 2000. Since then he has been elected as a Fellow to the Royal Society for the Promotion of Health (2004) and The Royal Institute of Public Health (2004). In 2009, Professor Preedy became a Fellow of the Royal Society for Public Health and in 2012 a Fellow of the Royal Society of Chemistry. In his career Professor Preedy has carried out research at Imperial College London (National Heart Hospital) and the MRC Centre at Northwick Park Hospital. He is a leading expert on the science of health. He has lectured nationally and internationally. To his credit, Professor Preedy has published over 570 articles, which includes 165 peer-reviewed manuscripts based on original research, 100 reviews and over 50 books and volumes.

SECTION 1

VEGETABLES AND ROOT CROPS

CHAPTER

1

Effect of Processing on Active Compounds in Fresh-Cut Vegetables

I. Pradas-Baena[*,†], *J.M. Moreno-Rojas*[*,†], *M.D. Luque de Castro*[‡]

[*]Área de Tecnología, Poscosecha e Industria Agroalimentaria, Córdoba, Spain, [†]Instituto de Investigación y Formación Agraria y Pesquera (IFAPA), Córdoba, Spain, [‡]Departamento de Química Analítica, Universidad de Córdoba, Córdoba, Spain

CHAPTER POINTS

- The value of fresh-cut products lies in their typical freshness, convenience and health properties.
- Freshcut products with a high content of bioactive compounds can only be obtained from raw materials with an also high bioactive content.
- The contents of bioactive compounds in fresh-cut products can be affected by various processing steps.
- Wounding increases respiration and ethylene production, thus leading to faster loss of certain vitamins.
- Wounding can also promote the synthesis of new phenolic compounds and the oxidation of endogenous phenols. The amount of phenols present after wounding is the result of a balance between phenol synthesis and oxidation.
- Carotenoid contents are maintained virtually unchanged after fresh-cut processing.
- The modified atmosphere packaging and low temperature used during storage of fresh-cut products help maintain the levels of bioactive compounds over longer periods.
- Controlled abiotic stresses such as ultraviolet radiation or hyperoxia have been shown to increase the contents of bioactive compounds.
- Extraction of antioxidants is a key step in the analysis of active compounds. Combining two cycles with different solvents is recommended for more efficient extraction.
- High-performance liquid chromatography is the most useful, popular, and reliable separation technique available for the analysis of specific classes of phenolic compounds, carotenoids, and vitamins prior to detection.

INTRODUCTION

The International Fresh-cut Produce Association (IFPA) defines fresh-cut products as fruits or vegetables that have been trimmed and/or cut into a 100% usable product which is packaged to offer consumers high nutrition, convenience, and flavor while still maintaining freshness (Rico *et al.*, 2007).

Over the last few decades, nutrition policies have strongly promoted the consumption of a diet containing a minimum of 400 g of fruit and vegetables per day for the prevention of chronic diseases such as heart disease, cancer, diabetes, and obesity, and also to prevent and alleviate various micronutrient deficiencies. These new healthy eating campaigns, which have been designed to increase fruit and vegetable consumption, and to favor lifestyle modifications (e.g., taking the time needed for purchasing and cooking fresh products), together with the increase in single-person households, have boosted the demand for ready-to-eat vegetables.

Fresh-cut fruit and vegetables have emerged to fulfil new consumers' demands for healthy, palatable, easily prepared plant foods (Allende *et al.*, 2006). These products constitute a major, rapidly expanding food segment of interest to growers, processors, retailers, and consumers. Today, the fresh-cut industry is expanding faster than any other

Processing and Impact on Active Components in Food
http://dx.doi.org/10.1016/B978-0-12-404699-3.00001-9

segment of the fruit and vegetable market due to its supplying of both the food service industry and retail outlets, and expanding production to new markets around the world.

On a worldwide scale, the USA is the largest consumer of fresh-cut products. In 2005, *per capita* consumption in the USA was 30 kg as compared to only 6 kg in France and 1.5–2.0 kg in Spain; in Europe the United Kingdom tops the table—a result of its deeply rooted tradition of consuming ready-to-eat products. France has for many years been the second largest European consumer, but was recently surpassed by Italy.

Fresh-cut products currently available at markets include the following:

- Lettuce (cleaned, chopped, shredded)
- Spinach and leafy greens (washed and trimmed)
- Broccoli and cauliflower (florets)
- Cabbage (shredded)
- Carrots (baby, sticks, shredded)
- Celery (sticks)
- Onions (whole peeled, sliced, diced)
- Potatoes and other roots (peeled, sliced)
- Mushrooms (sliced)
- Jicama, zucchini, cucumber (sliced, diced)
- Garlic (fresh peeled, sliced)
- Tomato and pepper (sliced).

HOW COMPOSITION IS ALTERED

Fresh-cut processing results in stress on vegetable tissues that leads to phytochemical accumulation or loss via an increased or reduced activity in key enzymes of secondary metabolic pathways.

Fresh-cut processing causes many cells to be broken and intracellular products such as oxidizing enzymes to be released, thereby accelerating product decay. In addition to other enzymes involved in decay processes and causing softening [polygalacturonase (PG) and pectinmethylesterase (PME)] or off-flavors [lipoxygenase (LOX)], for example, the concentration of active compounds during fresh-cut processing is largely modified by polyphenol oxidase (PPO), peroxidase (POD), and phenylalanine ammonia-lyase (PAL), which play different roles in the process.

Thus, in the presence of oxygen, PPO transforms phenolic compounds into *o*-quinones which undergo further polymerization to give dark, insoluble polymers known as "melanins."

POD is another enzyme with polyphenol oxidase activity that catalyses the single-electron oxidation of a wide variety of compounds in the presence of hydrogen peroxide. PPO might act as promoter of POD activity by producing hydrogen peroxide during the oxidation of phenolic compounds in PPO-catalysed reactions (Tomás-Barberán and Espin, 2001).

PAL plays a crucial role at the interface between plant primary and secondary metabolism by catalyzing deamination of L-phenylalanine to *trans*-cinnamic acid and free ammonium ion. This reaction is the first step in the biosynthesis of a broad range of phenylpropanoid-derived secondary products in plants.

In summary, the enzymes PPO and POD reduce the levels of secondary metabolites, whereas PAL increases them.

Fresh-cut vegetable processing usually involves washing, peeling, slicing, and/or shredding before packaging and storage at a low temperature (Figure 1.1). All these operations affect the nutrient contents of fresh-cut products.

HARVEST

The quality of the raw materials is one of the most essential factors governing that of fresh-cut products, harvesting of which should therefore be carried out as carefully as possible to avoid overstressing plant tissues.

RECEPTION

The first step on receipt of fresh-cut raw materials is quality control, a mandatory step in ensuring standard product quality. Thus, obtaining a high-quality fresh-cut product entails starting from a good raw material with

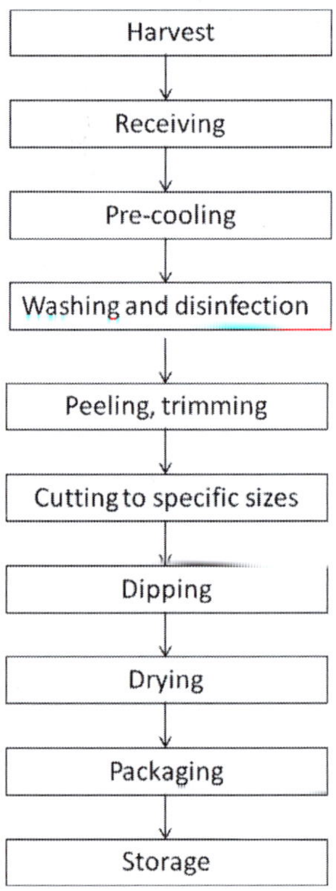

FIGURE 1.1 Flow-chart of freshcut vegetable processing. The typical steps in fresh-cut vegetable processing are shown.

a high content in active compounds. In turn, the quality of the raw material is influenced by diverse factors such as the particular cultivar (genetics), climatic conditions, cultivation practices, growing location, or maturity at harvest.

PRE-COOLING

Fresh-cut products should be pre-cooled as soon as possible after harvesting in order to extend their potential shelf-life. Temperature is the most important factor influencing the quality of fresh products. A low storage temperature reduces respiration and ethylene production rates, thereby delaying nutritional decay.

PEELING

Some vegetables (e.g., potatoes, carrots) require peeling. There are several available peeling methods; at an industrial scale, however, vegetables are peeled mechanically, chemically, or with high-pressure steam peelers. In each case, the peeling operation should be as gentle as possible. The ideal method is hand peeling with a sharp knife (Jongen, 2002). Industrial peeling affects active compounds in fresh-cut vegetables. Barry-Ryan and O'Beirne (1999) found increased levels of vitamin C retained in hand-prepared shredded iceberg lettuce relative to machine-peeled samples. Also, Ewald *et al.* (1999) observed losses of 39% of quercetin in onions after peeling, trimming, and chopping. Such high losses of this flavonoid resulted from the removal of the first and second layers, which contain more than 90% of all quercetin in onions.

CUTTING

The cutting operations used to prepare fresh-cut vegetables are known to disrupt plant cells and induce physiological changes such as increased rates of respiration and browning that, in turn, trigger nutritional changes and quality losses (Venneria *et al.*, 2012).

Mechanical injury (wounding) of living plant tissues increases respiration and ethylene production. It also induces other metabolic pathways producing secondary metabolites. Wounding, together with ethylene production, boost the activity of PAL, a key enzyme for phenol biosynthesis. Phenols have a protective action against pathogen attack and water loss. Moreover, phenolic compounds are substrates for PPO enzyme transformation, so their accumulation can lead to increased browning (Figure 1.2).

Cutting boosts ethylene production; this increases respiration and senescence, and leads to even faster loss of some vitamins (Figure 1.2). Fernando Reyes *et al.* (2007) found ascorbic acid contents to be decreased by wounding in fresh-cut celery, carrot, parsnip, zucchini, potato, and white cabbage (Table 1.1); by contrast, vitamin C contents were scarcely affected by cutting in fresh-cut sweet potato, radish, red cabbage (Fernando Reyes *et al.*, 2007), and tomato (Odriozola-Serrano *et al.*, 2008) (Table 1.1).

Phenol accumulation in response to wounding, which has a two-fold effect on phenol metabolism, is one of the most widely studied phenomena. One effect is oxidation of endogenous phenols as a result of cell membrane rupturing; this allows phenols to mix with oxidative enzyme systems, from which they are normally isolated by membranes (Figure 1.2). This response reduces the amount of phenols. The other effect of wounding is stimulation of

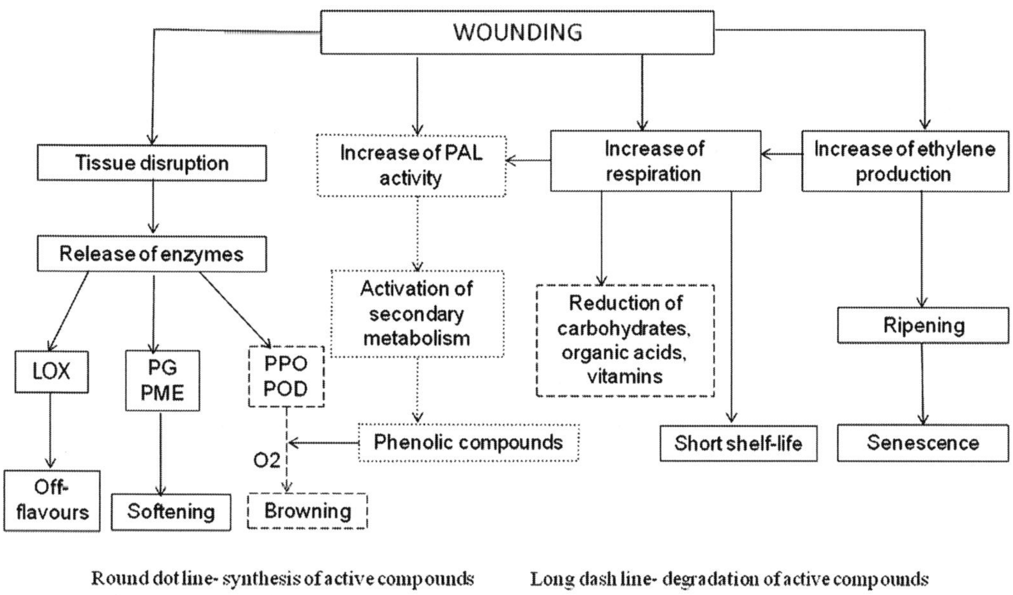

FIGURE 1.2 Effect of wounding on freshcut products. The physiological processes in vegetables after wounding are shown.

TABLE 1.1 Changes in Active Compounds and Antioxidant Capacity in Various Fresh-Cut Products After Wounding

Crop	Phenols Content (%)	Antioxidant Capacity (%)	Ascorbic Acid Content (%)	Reference
Iceberg lettuce	↑81	↑233	n.d.	Fernando Reyes *et al.*, 2007
Iceberg lettuce	↑280	↑140	—	Kang and Saltveit, 2002
Iceberg lettuce	↑55	↑345	—	Heredia and Cisneros-Zevallos, 2009a
Romaine lettuce	↑290	↑255	—	Kang and Saltveit, 2002
Bell pepper	n.s.	n.s.	—	Heredia and Cisneros-Zevallos, 2009a
Celery	↑30	↑442	↓53	Fernando Reyes *et al.*, 2007
Celery	↑39	↑57	—	Heredia and Cisneros-Zevallos, 2009a
Carrot	↑191	↑77	↓82	Fernando Reyes *et al.*, 2007
Carrot	↑287	↑240	—	Jacobo-Velázquez et al., 2011
Carrot	↑103	↑375	—	Heredia and Cisneros-Zevallos, 2009a
Parsnips	↑13	↑12	↓76	Fernando Reyes *et al.*, 2007
Asparagus	n.s.	n.s.	—	Heredia and Cisneros-Zevallos, 2009a
Sweet potato	↑17	↑12	n.s.	Fernando Reyes *et al.*, 2007
Zucchini	↓26	↓21	↓53	Fernando Reyes *et al.*, 2007
Radish	↓7	n.s.	n.s.	Fernando Reyes *et al.*, 2007
Radish	n.s.	↓13	—	Heredia and Cisneros-Zevallos, 2009a
Potato (cv. Russet)	↓15	↓51	↓32	Fernando Reyes *et al.*, 2007
Potato (cv. all Blue)	↑60	↑85	—	Reyes and Cisneros-Zevallos, 2003
White potato	n.s.	n.s.	—	Heredia and Cisneros-Zevallos, 2009a
Tomato	n.s.	n.s.	n.s.	Odriozola-Serrano et al., 2008
Tomato	n.s.	n.s.	—	Heredia and Cisneros-Zevallos, 2009a
Red cabbage	↓9	↓9	n.s.	Fernando Reyes *et al.*, 2007
Red cabbage	n.s.	n.s.	—	Heredia and Cisneros-Zevallos, 2009a
White cabbage	n.s.	↑17	↓11	Fernando Reyes *et al.*, 2007
Green cabbage	n.s.	↓11	—	Heredia and Cisneros-Zevallos, 2009a
Red onion	↑18	↑12	—	Heredia and Cisneros-Zevallos, 2009a
White onion	↑20	↑45	—	Heredia and Cisneros-Zevallos, 2009a
Jicama	↑18	↑19	—	Heredia and Cisneros-Zevallos, 2009a

↑ increase; ↓ decrease; n.s., non-significant differences; n.d., not detected; —, not measured.
The table shows the changes in phenols content, ascorbic acid and antioxidant capacity in various fresh-cut products after wounding. The changes are expressed as percentages.

cells adjacent to the injury zone to synthesise more phenols in order to repair damage. This response is caused by changes in PAL activity, which is the key metabolic enzyme in the phenylpropanoid pathway leading to an increase in phenol contents (Figure 1.2).

Phenol contents after wounding are thus the result of a rate balance between phenol synthesis and degradation (see Table 1.1). A number of authors have found an increase in phenols after wounding in fresh-cut lettuce (Kang and Saltveit, 2002; Fernando Reyes *et al.*, 2007; Heredia and Cisneros-Zevallos, 2009a), carrot (Fernando Reyes *et al.*, 2007; Heredia and Cisneros-Zevallos, 2009a; Jacobo-Velázquez *et al.*, 2011), potato (Reyes and Cisneros-Zevallos, 2003), celery (Fernando Reyes *et al.*, 2007; Heredia and Cisneros-Zevallos, 2009a), sweet potato, parsnips (Fernando Reyes *et al.*, 2007), red and white

TABLE 1.2 The Effect of Different Washing Solutions on the Total Polyphenols of Two Different Fresh-Cut Vegetables

	Washing Solutions								
Crops	**Water**	**Chlorine**	**Ozonated Water (10 mg/l)**	**Ozonated Water (20 mg/l)**	**Ozonated Water (10 mg/l active by UV-C)**	**Lactic Acid (20 ml/l)**	**Acidified Sodium Chlorite (250 mg/l)**	**Peroxyacetic Acid (300 ml/l)**	**Reference**
Shredded iceberg lettuce	23.3 n.s.	21.1 n.s.	23.8 n.s.	20.9 n.s.	19.8 n.s.	—	—	—	Beltran *et al.*, 2005
Rocket leaves (*Diplotaxis tenuifolia*)	131.8 n.s.	113.4 n.s.	135.7 n.s.	—	—	115.7 n.s.	120.6 n.s.	139.0 n.s.	Martinez-Sanchez *et al.*, 2006

—, Not measured; n.s. non-significant differences (between washing solutions).
The table shows the content of total polyphenols (mg/100g of fresh weight) of shredded iceberg lettuce and rocket leaves after washing with different washing solutions. Means of three replicates compared using the least significant differences (LSD).

onions, and jicama (Heredia and Cisneros-Zevallos, 2009a). The increase may have resulted from a high synthesis rate of phenols relative to their degradation rate. The trend observed in fresh-cut white, red, and green cabbage (Fernando Reyes *et al.*, 2007; Heredia and Cisneros-Zevallos, 2009a), tomato (Odriozola-Serrano *et al.*, 2008; Heredia and Cisneros-Zevallos, 2009a), bell pepper, asparagus, and radish (Heredia and Cisneros-Zevallos, 2009a) can be ascribed to the similarly important kinetics of phenolic synthesis and degradation. Conversely, the reduction in phenolic contents exhibited by zucchini, radish, potato, and red cabbage (Fernando Reyes *et al.*, 2007) may be related to the degradation rate of phenols exceeding their synthesis rate.

In addition to the previous reasoning, the response to vegetable wounds depends on various factors such as the type of tissue, cultivar, maturity, storage temperature, and cutting method. For example, the differences in phenol contents among wounded potatoes (Table 1.1) can be ascribed to differences in cultivar, among other factors.

Generally, the synthesis of phenolic compounds increases with wounding severity. Thus, a greater increase in phenol content and PAL enzyme activity was observed in shredded carrots as compared to either sliced or whole carrots (Heredia and Cisneros-Zevallos, 2009b; Jacobo-Velázquez *et al.*, 2011). The specific phenolic compounds found to be increased after wounding in carrots were 3-*O*-caffeoylquinic (neochlorogenic), ferulic, 3,5-dicaffeoylquinic (isochlorogenic), and 4,5-dicaffeoylquinic acids.

The antioxidant capacity of fresh-cut vegetables depends on a wide variety of compounds such as phenols, vitamins, and carotenoids. Based on the data of Table 1.1, antioxidant capacity changed similarly to phenolic compounds. This suggests that phenols are the main contributors to changes in antioxidant activity in fresh-cut vegetables.

Other compounds such as lycopene and carotenoids exhibit insubstantial changes after fresh-cut processing. Odriozola-Serrano *et al.* (2008) showed lycopene contents to be well maintained after fresh-cut processing of tomatoes. Similarly, other studies have shown that the total carotenoid content of carrots and sweet potatoes are not significantly decreased by wounding (Fernando Reyes *et al.*, 2007).

WASHING

At reception, vegetables are usually covered with sand, mud, and earth and must therefore be washed before processing. A second wash is usually applied after peeling and/or cutting.

Antioxidant constituents are susceptible to degradation when exposed to an acid pH such as that conferred by hypochlorite used for sanitation. The interaction of these constituents with enzymes such as PPO, POD, or ascorbate oxidase promotes degradation of active compounds.

Washing in cold chlorinated water is necessary to remove the dirt and reduce microbial charge before cutting. Chlorine-based treatments have been used for sanitation in food processing for several decades and are the most widely used choice in the food industry. New sanitizers have been introduced over the past few years in response to increasing concern about the products of chlorine degradation by organic matter with formation of potentially harmful substances. One of the alternatives to hypochlorite is ozone, which is considered to be a safe, effective sanitizing agent for use in the fresh-cut product industry. The effects on substrates of ozone (Beltran *et al.*, 2005; Martinez-Sanchez *et al.*, 2006), and ozone and chlorine (Zhang *et al.*, 2005), as sanitizers in washing water have been studied. Beltran *et al.* (2005) and Martinez-Sanchez *et al.* (2006) found no effect of different washing treatment on the final phenolic content of fresh-cut iceberg lettuce and rocket leaves, respectively (Table 1.2). A different effect was reported by Zhang *et al.*

(2005) in fresh-cut celery. Thus, they found fresh-cut celery treated with ozonated water to exhibit higher retention of vitamin C content than those washed with tap water alone. This effect may have resulted from inhibition of PPO activity by ozonated water.

DRYING

Fresh-cut vegetables should be dried prior to packaging. This process is a good method for increasing the shelf-life of fresh-cut products—an excess of water or tissue fluids provides an excellent medium for microbial growth. Also, moisture accelerates some enzymatic reactions leading to rapid degradation of bioactive compounds.

OTHER WAYS IN WHICH COMPOSITION IS ALTERED

The storage conditions of fresh-cut vegetables can modify their nutritional composition. The most important parameters influencing preservation during storage are temperature and atmospheric composition. Using appropriate modified atmosphere packaging (MAP) in combination with refrigeration can help maintain the compositional quality of fresh-cut products.

The most common components of MAP are low oxygen and high carbon dioxide concentrations to maintain the quality and to increase the shelf-life of fresh-cut products. The use of low levels of oxygen in combination with refrigeration helps preserve vitamin C, which is easily degraded by an oxygen-rich atmosphere and/or a high temperature. Thus, Fonseca *et al.* (2005) found degradation of ascorbic acid in shredded Galega kale to be more marked in air than in the low-oxygen atmosphere of packaging systems. Interestingly, Gil *et al.* (1999) found the vitamin C contents of fresh-cut spinach to be better maintained under MAP than in air; however, the equilibrium between the two vitamin forms (ascorbic acid and dehydroascorbic acid) was shifted to the latter—which is thus the predominant form of the vitamin during storage.

Because oxidative reactions require oxygen, MAP with low oxygen levels can help avoid the oxidation of active compounds. Odriozola-Serrano *et al.* (2008) reported good retention of lycopene in fresh-cut tomatoes stored in MAP with low oxygen levels.

Some recent studies have examined the effect of applying controlled post-harvest abiotic stresses to fruits and vegetables in order to enhance their nutraceutical content. Martínez-Hernández *et al.* (2011) concluded that an ultraviolet C type radiation (UV-C) pretreatment increases the contents in bioactive compounds of fresh-cut products such as broccoli. This was also the case with fresh-cut carrots subjected to hyperoxia stress, which had increased concentrations of phenolic compounds relative to air-treated samples (Jacobo-Velázquez *et al.*, 2011).

Controlled abiotic stress has proved useful to the fresh produce industry with a view to obtaining healthier products by enhancing the nutraceutical content of fresh-cut or whole products (Cisneros-Zevallos, 2003).

ANALYTICAL TECHNIQUES

Antioxidant Capacity

The antioxidant capacity of vegetables is derived from the synergistic action of a wide variety of antioxidants such as phenolic compounds, vitamins, and/or carotenoids. Antioxidant capacity can be assessed with a number of methods that fall into either of two general categories, namely:

- assays based on an electron transfer reaction that is monitored via the color change in the oxidant as it is reduced; and
- assays based on a hydrogen atom transfer reaction where the antioxidant and the substrate compete for free radicals.

The former group (electron transfer assays) includes the Trolox Equivalent Antioxidant Capacity (TEAC) assay, the ferric reducing ability of plasma (FRAP) assay and the 2,2-diphenyl1picrylhydrazyl (DPPH) radical scavenging capacity assay. The latter group consists mainly of the oxygen radical absorbance capacity (ORAC) assay (Tabart *et al.*, 2009).

One key step in assessing antioxidant capacity is the extraction of antioxidants (usually with methanol, acetone or ethanol in water mixtures). However, no single solvent can efficiently extract all antioxidants present in a food. As a result, extraction residues contain substantial amounts of antioxidant. Combining at least two extraction cycles with water–organic solvents of different polarity is thus recommended to ensure efficient extraction of antioxidants with dissimilar chemical properties (e.g., hydrophilic and lipophilic) (Pérez-Jiménez *et al.*, 2008).

Phenolic Compounds

The most commonly used method for determining the total content of phenols is that of Folin–Ciocalteu (F-C), a simple reproducible method, the results of which are generally well correlated with antioxidant activity as measured by the DPPH, ORAC, TEAC, or FRAP tests. The F-C method is a colorimetric method based on the transfer of electrons from phenols and other reductants to molybdenum in an alkaline medium, the resulting blue complexes being measured at 750–765 nm. This is a non-specific method because the reagent reacts with other reducing substances in addition to phenols.

High-performance liquid chromatography (HPLC, from now on LC) is currently the most useful, popular, and reliable technique available for determining the phenolic profiles of foods. C18 reverse-phase columns are commonly used for individual separations, and acetonitrile and methanol the most widely used mobile phases—acidified with acetic, formic, or phosphoric acid in many cases. The most common detectors used in combination with LC include UV/Vis molecular absorption spectrophotometers, photodiode array (DAD) instruments, and fluorescence. DAD is the most frequent choice because it allows real-time scanning of UV/Vis spectral data as the analytes are eluted from the chromatographic column. Although phenols can be determined from their retention times and by comparison of their UV/Vis spectra to appropriate standards (Tsao, 2010), they are best identified by mass spectrometry (MS) because the MS technique allows users to confirm structures and obtain information about phenolic molecular masses from fragmentation patterns.

The emergence of ultra-high-performance liquid chromatography (UHPLC or UPLC) has significantly enhanced separation performance and constitutes an advantageous alternative to traditional LC as it affords the separation of phenols with significantly improved efficiency, resolution and sensitivity, in addition to substantially reduced overall analysis times (Stalikas, 2007).

Vitamin C

Although the main active form of vitamin C is ascorbic acid (AA), its oxidized form, dehydroascorbic acid (DHA), also possesses biological activity.

Vitamin C is usually extracted from vegetables by using pure water or acid solutions. Ascorbic acid is easily oxidized under alkaline conditions, so it requires an acid extractant to prevent oxidation. The solvents typically used for this purpose include metaphosphoric or oxalic acid, whether alone or in combination with other acids and/or organic solvents (e.g., methanol), to which an antioxidant such as EDTA or BHT can also be added (Campos *et al.*, 2009).

Ascorbic acid can be determined by using various methods based on titration or direct photometric measurements. Detection of this compound is most often preceded by enzymatic reaction or chromatographic separation. Complex matrices require a chromatographic (usually LC) step in which the pH of the mobile phase must be adjusted to 5 or lower in order to preserve the stability of vitamin C during analysis.

The detection and quantitation of ascorbic acid by UV absorption spectroscopy should be performed at wavelengths over the range 245–255 nm, which is where it absorbs maximally. Because of its weak molar absorptivity, DHA is usually determined indirectly after reduction to AA. The DHA concentration is obtained as the difference between total AA after DHA reduction and AA in the original sample.

Another UV-based method for DHA determination uses absorbance measurements at 348 nm after pre-column derivatization with 1,2-phenylenediamine (OPDA) and LC separation (Zapata and Dufour, 1992).

Carotenoids

Carotenoids are hydrophobic and soluble in organic solvents. Tetrahydrofuran, methanol, ethyl acetate and acetone are some of the solvents used for carotenoid extraction, which can be assisted by saponification.

A great variety of carotenoid pigment isomers have been determined with LC separation (usually with a C18 or C30 reverse-phase column) and conventional UV/Vis, but also DAD or MS detection (Gayosso-García Sancho *et al.*, 2011). Thin layer chromatography has also been used to separate carotenoids, albeit less frequently than LC (Chedea *et al.*, 2010).

References

Allende, A., Tomás-Barberán, F.A., Gil, M.I., 2006. Minimal processing for healthy traditional foods. Trends Food Sci. Technol. 17, 513–519.

Barry-Ryan, C., O'Beirne, D., 1999. Ascorbic acid retention in Shredded Iceberg lettuce as affected by minimal processing. J. Food Sci. 64, 498–500.

Beltran, D., Selma, M.V., Marin, A., Gil, M.I., 2005. Ozonated water extends the shelf life of fresh-cut lettuce. J. Agric. Food Chem. 53, 5654–5663.

Campos, F.M., Ribeiro, S.M.R., Della Lucia, C.M., Pinheiro-Sant'Ana, H.M., Stringheta, P.C., 2009. Optimization of methodology to analyze ascorbic and dehydroascorbic acid in vegetables. Quim. Nova. 32, 87–91.

Chedea, V.S., Kefalas, P., Socaciu, C., 2010. Patterns of carotenoid pigments extracted from two orange peel wastes (Valencia and Navel var.). J. Food Biochem. 34, 101–110.

Cisneros-Zevallos, L., 2003. The use of controlled postharvest abiotic stresses as a tool for enhancing the nutraceutical content and adding-value of fresh fruits and vegetables. J. Food Sci. 68, 1560–1564.

Ewald, C., Fjelkner-Modig, S., Johansson, K., Sjöholm, I., Åkesson, B., 1999. Effect of processing on major flavonoids in processed onions, green beans, and peas. Food Chem. 64, 231–235.

Fernando Reyes, L., Emilio Villarreal, J., Cisneros-Zevallos, L., 2007. The increase in antioxidant capacity after wounding depends on the type of fruit or vegetable tissue. Food Chem. 101, 1254–1262.

Fonseca, S.C., Oliveira, F.A.R., Brecht, J.K., Chau, K.V., 2005. Influence of low oxygen and high carbon dioxide on shredded Galega kale quality for development of modified atmosphere packages. Postharvest. Biol. Technol. 35, 279–292.

Gayosso-García Sancho, L.E., Yahia, E.M., González-Aguilar, G.A., 2011. Identification and quantification of phenols, carotenoids, and vitamin C from papaya (*Carica papaya* L., cv. Maradol) fruit determined by HPLC-DAD-MS/MS-ESI. Food Res. Int. 44, 1284–1291.

Gil, M.I., Ferreres, F., Tomás-Barberán, F.A., 1999. Effect of postharvest storage and processing on the antioxidant constituents (flavonoids and vitamin C) of fresh-cut spinach. J. Agric. Food Chem. 47, 2213–2217.

Heredia, J.B., Cisneros-Zevallos, L., 2009a. The effects of exogenous ethylene and methyl jasmonate on the accumulation of phenolic antioxidants in selected whole and wounded fresh produce. Food Chem. 115, 1500–1508.

Heredia, J.B., Cisneros-Zevallos, L., 2009b. The effect of exogenous ethylene and methyl jasmonate on pal activity, phenolic profiles and antioxidant capacity of carrots (*Daucus carota*) under different wounding intensities. Postharvest. Biol. Technol. 51, 242–249.

Jacobo-Velázquez, D.A., Martínez-Hernández, G.B., del, C., Rodríguez, S., Cao, C.-M., Cisneros-Zevallos, L., 2011. Plants as biofactories: physiological role of reactive oxygen species on the accumulation of phenolic antioxidants in carrot tissue under wounding and hyperoxia stress. J. Agric. Food Chem. 59, 6583–6593.

Jongen, W., 2002. Fruit and Vegetable Processing—Improving Quality. Woodhead Publishing, Cambridge, England.

Kang, H.M., Saltveit, M.E., 2002. Antioxidant capacity of lettuce leaf tissue increases after wounding. J. Agric. Food Chem. 50, 7536–7541.

Martínez-Hernández, G.B., Gómez, P.A., Pradas, I., Artés, F., Artés-Hernández, F., 2011. Moderate UV-C pretreatment as a quality enhancement tool in fresh-cut Bimi® broccoli. Postharvest. Biol. Technol. 62, 327–337.

Martinez-Sanchez, A., Allende, A., Bennett, R.N., Ferreres, F., Gil, M.I., 2006. Microbial, nutritional and sensory quality of rocket leaves as affected by different sanitizers. Postharvest. Biol. Technol. 42, 86–97.

Odriozola-Serrano, I., Soliva-Fortuny, R., Martín-Belloso, O., 2008. Effect of minimal processing on bioactive compounds and color attributes of fresh-cut tomatoes. LWT - Food Sci. Technol. 41, 217–226.

Pérez-Jiménez, J., Arranz, S., Tabernero, M., Diaz-Rubio, M.E., Serrano, J., Goni, I., Saura-Calixto, F., 2008. Updated methodology to determine antioxidant capacity in plant foods, oils and beverages: extraction, measurement and expression of results. Food Res. Int. 41, 274–285.

Reyes, L.F., Cisneros-Zevallos, L., 2003. Wounding stress increases the phenolic content and antioxidant capacity of purple-flesh potatoes (*Solanum tuberosum* L.). J. Agric. Food Chem. 51, 5296–5300.

Rico, D., Martin-Diana, A.B., Barat, J.M., Barry-Ryan, C., 2007. Extending and measuring the quality of fresh-cut fruit and vegetables: a review. Trends Food Sci. Technol. 18, 373–386.

Stalikas, C.D., 2007. Extraction, separation, and detection methods for phenolic acids and flavonoids. J. Separation Sci. 30, 3268–3295.

Tabart, J., Kevers, C., Pincemail, J., Defraigne, J.O., Dommes, J., 2009. Comparative antioxidant capacities of phenolic compounds measured by various tests. Food Chem. 113, 1226–1233.

Tomás-Barberán, F., Espin, J.C., 2001. Phenolic compounds and related enzymes as determinants of quality in fruits and vegetables. J. Sci. Food Agric. 81, 853–876.

Tsao, R., 2010. Chemistry and biochemistry of dietary polyphenols. Nutrients 2, 1231–1246.

Venneria, E., Marinelli, L., Intorre, F., Foddalai, M.S., Aurigemma, C., Durazzo, A., Maiani, G., De Giusti, M., 2012. Effect of harvest time and minimal processing on nutritional and microbiological quality of three leaf crops. J. Agri. Biodiversity Res. 1, 11–17.

Zapata, S., Dufour, J.P., 1992. Ascorbic, dehydroascorbic and isoascorbic acid simultaneous determinations by reverse phase ion interaction HPLC. J. Food Sci. 57, 506–511.

Zhang, L., Lu, Z., Yu, Z., Gao, X., 2005. Preservation of fresh-cut celery by treatment of ozonated water. Food Control 16, 279–283.

CHAPTER

2

Changes in β-Carotene During Processing of Carrots

Griet Knockaert, Lien Lemmens, Sandy Van Buggenhout, Marc Hendrickx, Ann Van Loey

Laboratory of Food Technology and Leuven Food Science and Nutrition Research Centre (LFoRCe), Department of Microbial and Molecular Systems (M[2]S), Katholieke Universiteit Leuven, Leuven, Belgium

CHAPTER POINTS

- Thermal processing induces β-carotene isomerization and degradation, especially in intense processing conditions.
- Thermal processing mostly increases β-carotene bioaccessibility.
- Addition of oil during thermal processing has a positive effect on β-carotene bioaccessibility, which can possibly compensate for the increase in β-carotene degradation and isomerization.
- High-pressure processing generally does not affect total β-carotene concentration in a negative way and induces less β-carotene isomerization compared to thermal processing.
- In most cases, HP processing can not improve β-carotene bioaccessibility, not even in the presence of oil.

INTRODUCTION

β-Carotene, belonging to the group of carotenoids, is an important micronutrient to which particular health related properties have been attributed. As a result of its specific molecular structure consisting of a polyene chain with 11 conjugated double bonds and a β-ring at each end of the chain, β-carotene has provitamin A and antioxidant activity (Rodriguez-Amaya, 2001). β-Carotene is present in various plant-based food products. In the human diet, carrots (*Daucus carota*) are a common source of β-carotene and they are often consumed as processed products (e.g., soups, juices, purees, etc.). One major purpose of food processing is food preservation. The shelf-life of food products may be enhanced by destruction of pathogenic microorganisms and inactivation of spoilage microorganisms and enzymes (Ramaswamy and Marcotte, 2006). Processing might also affect β-carotene in different ways. In this chapter, the effect of processing of carrot-based products on β-carotene stability on the one hand, and on β-carotene bioaccessibility on the other hand, is discussed.

EFFECT OF PROCESSING ON β-CAROTENE STABILITY

Due to its conjugated system of double bonds, β-carotene is susceptible to isomerization and degradation during processing. Unfortunately, β-carotene *cis*-isomers have a decreased provitamin A activity and an altered antioxidant activity (Schieber and Carle, 2005). Furthermore, although some of the first oxidation products of β-carotene (e.g., apocarotenals) are suggested to be biologically active as anticancer agents (Sharoni *et al.*, 2012), complete degradation of β-carotene results in a complete loss of its health-related properties (Rodriguez-Amaya, 2001). In nature, β-carotene is mainly present as all-*trans*-β-carotene which is thermodynamically the most stable form. Nevertheless, 9-*cis*-, 13-*cis*-, and 15-*cis*-β-carotene have also been detected in small amounts in untreated carrot products (Rodriguez-Amaya, 2001).

Thermal processing, which is the conventional method for food pasteurization and sterilization, has been shown to induce β-carotene degradation and isomerization. As β-carotene in carrot

Processing and Impact on Active Components in Food
http://dx.doi.org/10.1016/B978-0-12-404699-3.00002-0

TABLE 2.1 Kinetic Parameters ± Standard Error (T_{ref} = 110°C) for the Modeling of Total β-Carotene Degradation and of the Formation of Different β-Carotene *Cis*-Isomers During Thermal Processing of Carrot Puree or Olive Oil/Carrot Emulsion, Estimated by One-Step Regression Analysis Using a Fractional Conversion Model

	Carrot Puree		Olive Oil/Carrot Emulsion	
	E_a (kJ/mol)	k_{ref} (per min)	E_a (kJ/mol)	k_{ref} (per min)
Total β-carotene degradation	nd	nd	45.0 ± 8.6	0.10 ± 0.01
Total *cis*-isomer formation	11.5 ± 1.7	0.038 ± 0.000089	70.5 ± 7.4	0.099 ± 0.01
9-*Cis*-β-carotene formation	14.1 ± 3.1	0.023 ± 0.0011	nd	nd
13-*Cis*-β-carotene formation	11.7 ± 2.0	0.066 ± 0.0035	120 ± 8.6	0.39 ± 0.05
15-*Cis*-β-carotene formation	nd	nd	98.4 ± 8.7	0.11 ± 0.01

nd, not determined.

Adapted with permission from Lemmens et al., *β-Carotene isomerization kinetics during thermal treatments of carrot puree,* Journal of Agricultural and Food Chemistry, *58, 6816–6824, ©2010 American Chemical Society and adapted with permission from Knockaert* et al., *Carrot β-carotene degradation and isomerization kinetics during thermal processing in the presence of oil,* Journal of Agricultural and Food Chemistry, *60, 10312–10319, ©2012 American Chemical Society.*

products might be protected by the food matrix, intense process conditions which also cause severe destruction of the food matrix are necessary to induce β-carotene degradation and isomerization. In carrot pieces, only sterilization processes resulted in significantly higher percentages of *cis*-isomers as compared to raw carrot pieces. β-Carotene degradation, however, was still limited at sterilization conditions (Knockaert *et al.*, 2011; Lemmens *et al.*, 2011a). The formation of 13-*cis*-β-carotene was more pronounced in thermally sterilized compared to thermally pasteurized carrot juice (Marx *et al.*, 2003). During thermal processing of carrot products, 13-*cis*-β-carotene is the predominantly formed cis-isomer, followed by 15-*cis*- and 9-*cis*-β-carotene, which is mainly formed at more severe treatment conditions (Chen *et al.*, 1995; Marx *et al.*, 2003). Furthermore, the formation of 13,15-di-*cis*-β-carotene has been reported after UHT processing of carrot juice (Chen *et al.*, 1995). Kinetic studies of β-carotene degradation and isomerization are useful tools to predict β-carotene changes during processing and to identify process conditions resulting in safe carrot products with a high nutritional value. Isomerization of all-*trans*-β-carotene and formation of 13-*cis*- and 9-*cis*-β-carotene in plain carrot puree during thermal processing (80–150°C) has been described by a fractional conversion model, which is characterized by a plateau value after prolonged heating (Lemmens *et al.*, 2010). As β-carotene isomerization is a reversible reaction (Dugave, 2006), the isomerization reaction from all-*trans*-β-carotene to the *cis*-isomers is in equilibrium with the reverse reaction at the moment the plateau is reached. In Table 2.1, the obtained reaction rate constants and activation energies are shown. The highest reaction rate constant for the formation of 13-*cis*-β-carotene confirms that this isomer is the most easily formed *cis*-isomer during thermal processing of carrot puree. Using the kinetic model, it can be predicted that β-carotene isomerization in carrot puree is quite limited during industrially relevant process conditions (see Figure 2.1).

Next to high temperature, several other factors might influence β-carotene stability. Lipids, for example, are important to consider as they are often added to foods during processing and addition of oil has already been shown to enhance β-carotene isomerization. Addition of 5% olive oil to a carrot puree during thermal pasteurization increased the concentration of β-carotene *cis*-isomers 2.5 times (up to 19%) (Knockaert *et al.*, 2012a). During thermal sterilization of carrot juice, the concentration of 13-*cis*-β-carotene was increased from 6% to 18.8% after addition of 1% grape seed oil (Marx *et al.*, 2003). Solubilization of β-carotene crystals in the oil droplets, which makes β-carotene more susceptible to high temperatures, might explain this effect. The presence of oil also affects the kinetics of β-carotene isomerization during thermal processing. In Table 2.1, a comparison is made between the kinetic parameters for thermal isomerization of β-carotene in a carrot puree without oil (Lemmens *et al.*, 2010) and in an olive oil/carrot emulsion (= carrot puree + 5% olive oil) (Knockaert *et al.*, 2012b). β-Carotene isomerization during thermal processing (85°C–130°C) was largely accelerated by the presence of oil (expressed by the higher reaction rate constants). Furthermore, β-carotene solubilization probably also resulted in a higher temperature dependence of the reaction rate con stants and thus a higher activation energy compared to the carrot puree without added oil.

The presence of oil also influences β-carotene degradation kinetics. Whereas no or only very limited β-carotene degradation was observed during thermal processing (80°C–150°C) of the carrot puree without oil (Lemmens *et al.*, 2010), β-carotene degradation was already observed at the lowest temperature studied (85°C) during thermal processing of the oil/carrot emulsion (Knockaert *et al.*, 2012b). Oil is reported to act as

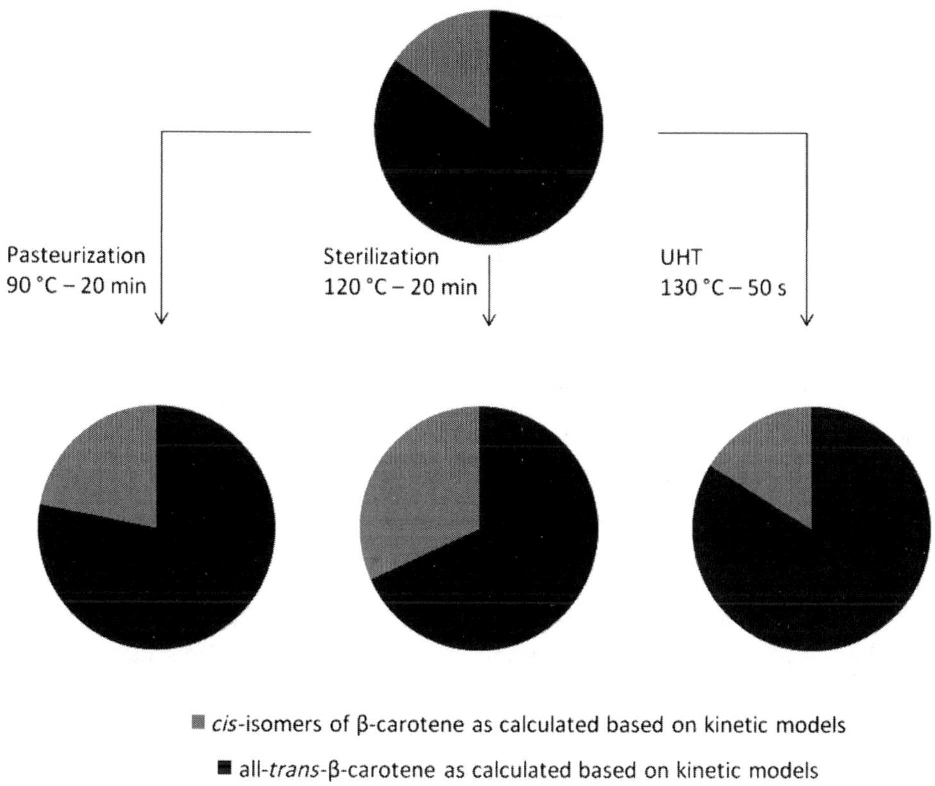

FIGURE 2.1 Isomerization of all-*trans*-β-carotene to its *cis*-isomers in carrot puree during industrially relevant processes, estimated based on the kinetic study of Lemmens *et al.* (2010). *Reprinted with permission from Lemmens* et al., *β-Carotene isomerization kinetics during thermal treatments of carrot puree,* Journal of Agricultural and Food Chemistry, *58, 6816–6824, ©2010 American Chemical Society.*

a prooxidant and its presence might enhance β-carotene degradation which is mainly attributed to oxidation (Bonnie and Choo, 1999). β-Carotene degradation in the oil/carrot emulsion could be described by a fractional conversion model. The obtained kinetic parameters are given in Table 2.1.

From the previous section, it is clear that thermal processing has an adverse effect on β-carotene stability in carrot products. Nowadays however, consumers demand high-quality products which are also safe and they appreciate the fresh appearance of minimally processed food products (Rastogi *et al.*, 2007). In this context, high pressure (HP) processing has been developed as a promising alternative to thermal preservation. In industry, HP processing is already used for food pasteurization (400–600 MPa/temperature ≤45°C) as vegetative microorganisms can be inactivated by HP (Balasubramaniam *et al.*, 2008). HP sterilization, however, is still at the research stage. As some bacterial spores can tolerate pressures up to 1700 MPa at room temperature, a combination of HP (600 MPa or higher) and an elevated initial temperature (60–90°C) might be used. As a result of adiabatic heating during pressure build up, process temperatures can be reached substantially faster compared to thermal sterilization, which results in shorter process times (Barbosa-Cánovas and Juliano, 2008). HP is not expected to negatively affect low-molecular-weight components such as β-carotene because covalent bonds are not directly disrupted by HP at room temperature (McInerney *et al.*, 2007). In some studies, indeed, no or only a very small negative effect of HP processing (400–600 MPa) at room temperature or slightly elevated temperature (≤75°C) on total β-carotene concentration in carrot puree or carrot pieces has been reported (Tauscher, 1998; McInerney *et al.*, 2007; Knockaert *et al.*, 2011). On the contrary, in the study of Knockaert *et al.* (2012a), β-carotene loss up to 18% was observed after HP pasteurization (600 MPa, 45°C, 20 min) of homogenized carrot puree. In the context of HP sterilization, only one study reports on the effect on total β-carotene concentration of carrot products, until now. HP processing during 5–10 min at 600 MPa and 117°C did not affect the total β-carotene concentration in carrot pieces (Knockaert *et al.*, 2011).

In contrast to β-carotene degradation, limited information on the effect of HP processing on β-carotene isomerization is available. Knockaert *et al.* (2012c) investigated the effect of combined thermal/HP processing at different pressure and temperature combinations on β-carotene isomerization in presence of oil. In Figure 2.2, an example of the formation of total *cis*-isomers in an olive oil/carrot emulsion during combined thermal/HP treatments at 100°C and different pressure levels is given. During the combined thermal/HP treatments, clearly less isomerization occurred, compared to thermal processing, and the degree of isomerization was independent on

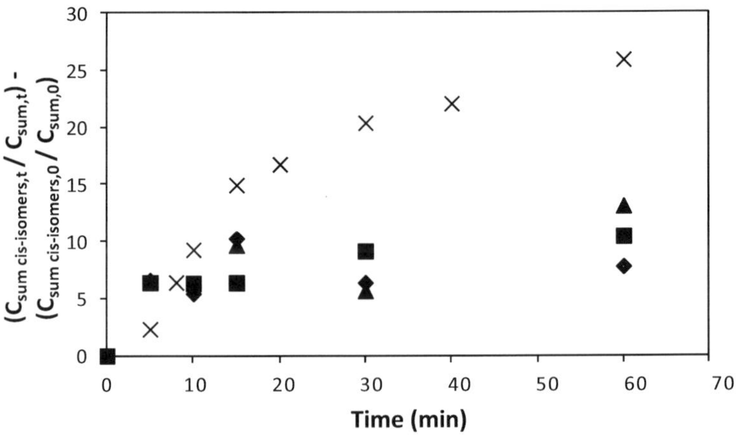

FIGURE 2.2 Formation of total *cis*-isomers in an olive oil/carrot emulsion during combined thermal/HP treatments at 100°C and 300 MPa (♦), 500 MPa (■), and 700 MPa (▲), and during a thermal treatment at 100°C (×). *Reprinted from* Food Chemistry, *Knockaert* et al., *Isomerization of carrot β-carotene in presence of oil during thermal and combined thermal/high pressure processing, 138, 1515–1520, ©2012, with permission from Elsevier.*

the pressure level in case HP was applied. It was hypothesized that the transfer of β-carotene from the carrot cells to the oil phase, which is suggested to be a prerequisite for isomerization, was hindered during combined thermal/HP processing, thereby limiting the possibility of β-carotene to isomerize. Strengthening of the carrot cell walls by changes in pectin during combined thermal/HP processing are believed to be responsible for the observed effect.

EFFECT OF PROCESSING ON β-CAROTENE BIOACCESSIBILITY

The previous section clearly showed that different processing techniques may have diverse effects on β-carotene concentration and stability. Nevertheless, it is only useful to produce carrot products with a high all-*trans*-β-carotene concentration and a low degree of isomerization if β-carotene can effectively be absorbed in the human body during digestion. A prerequisite for absorption is that β-carotene should be bioaccessible, i.e., it needs to be released from the carrot matrix during digestion and made accessible for absorption into mucosa (=incorporation in mixed micelles) (Hedrén *et al.*, 2002). For the determination of β-carotene bioaccessibility, human digestion is imitated by applying *in vitro* digestion procedures and both the release from the carrot matrix and the incorporation in the mixed micelles may be considered. A number of factors, described by the term SLAMENGHI, can affect the bioaccessibility of carotenoids (van het Hof *et al.*, 2000).

In the context of processing of carrot products, the carrot matrix (M in SLAMENGHI) plays an important role for β-carotene bioaccessibility. The intactness of the carrot matrix might be changed during processing, which could influence the release of β-carotene from the matrix. In most studies, thermal processing of carrot products has been shown to improve β-carotene bioaccessibility (both release from the carrot matrix and micellarization), with the increase being more pronounced at more intense process conditions (Hedrén *et al.*, 2002; Hornero-Méndez and Mínguez-Mosquera, 2007; Knockaert *et al.*, 2011). In a pilot scale study on carrot pieces, the release of β-carotene from the carrot matrix during *in vitro* digestion and the incorporation in the micelles was significantly improved after thermal processes at different sterilization intensities, whereas thermal pasteurization processes were not intense enough to have a significant effect (Lemmens *et al.*, 2011a). Lemmens *et al.* (2011b) could model the increase in β-carotene bioaccessibility (release from the matrix) during thermal processing (90–120°C) of carrot particles (500–4000 μm) as a function of time by a fractional conversion model. Here also, the increase was the largest at the highest process temperatures. In the context of improved β-carotene bioaccessibility in thermally processed carrot products, β-eliminative depolymerization of pectin should be considered. It results in softening and solubilization of the cell walls and is accelerated at higher temperatures. Consequently, digestive enzymes may have more access to their substrates, which results in a better release of β-carotene for absorption. An inverse correlation between β-carotene bioaccessibility and hardness of thermally treated carrot pieces was found by Lemmens *et al.* (2009) (see Figure 2.3) and it was shown that pectin structural changes as a result of β-eliminative depolymerization are a key factor for β-carotene bioaccessibility. The inverse correlation between hardness and β-carotene bioaccessibility of thermally treated carrot pieces was confirmed in the previously described pilot scale study on carrot pieces (Lemmens *et al.*, 2011a). Conversely, Tydeman *et al.* (2010) stated that particle size reduction is more important than heat treatment to enhance β-carotene bioaccessibility.

In contrast to thermal processing, HP processing cannot in most cases improve β-carotene bioaccessibility. In homogenized carrot puree for example, an additional

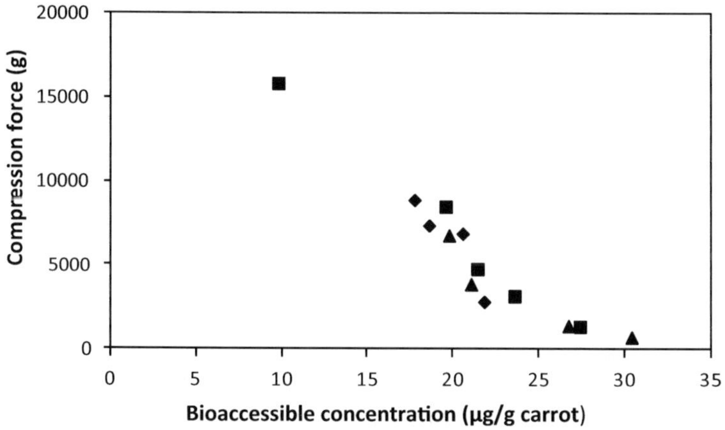

FIGURE 2.3 Relation between the hardness of carrot pieces and the bioaccessible β-carotene concentration of thermally treated carrot pieces at 90°C (♦), 100°C (■) and 110°C (▲). *Adapted from* Food Research International, *42, Lemmens* et al., *Towards a better understanding of the relationship between the β-carotene* in vitro *bio-accessibility and pectin structural changes: a case study on carrots, 1323 1330, ©2009, with permission from Elsevier.*

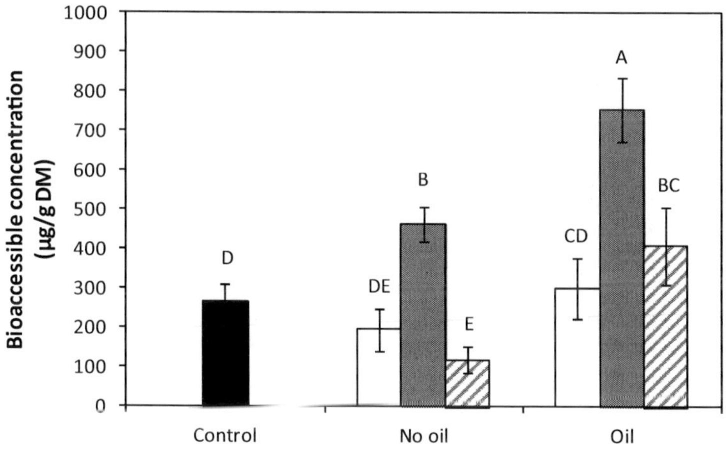

FIGURE 2.4 Mean bioaccessible β-carotene concentration (± standard deviation) of a non-HP homogenized carrot puree without oil (■) and of carrot purees homogenized at 10 MPa with or without 5% olive oil, followed by no treatment (□), followed by a thermal pasteurization process (■) or followed by a HP pasteurization process (striped). Results indicated with the same letter are not significantly different. *Adapted from* Food Chemistry, *133, Knockaert* et al., *Changes in β-carotene bioaccessibility and concentration during processing of carrot puree, 60–67, ©2012, with permission from Elsevier.*

HP pasteurization process (600 MPa, 45°C, 20 min) could not improve the amount of β-carotene released from the matrix during *in vitro* digestion, as illustrated in Figure 2.4 (Knockaert *et al.*, 2012a). McInerney *et al.* (2007) also did not find a positive effect of HP pasteurization (400 or 600 MPa, 25°C, 2 min) of carrot pieces on β-carotene bioaccessibility. During HP pasteurization, the temperature is not high enough for β-eliminative depolymerization of pectin to take place. Only in one study (Knockaert *et al.*, 2011) was a small positive effect of HP pasteurization of carrot pieces on β-carotene bioaccessibility observed. HP sterilization of carrot pieces resulted in a lower amount of bioaccessible β-carotene compared to the raw carrot pieces, despite the strong hardness loss (Knockaert *et al.*, 2011). This might be explained by the strengthening effect of combined thermal/HP processing on the carrot cell walls (as described above for isomerization), thereby limiting the release of β-carotene from the carrot matrix during *in vitro* digestion.

In order to improve β-carotene bioaccessibility during processing of carrot products, the addition of lipids [=Effector of absorption (E in SLAMENGHI)] should also be considered. The positive effect of addition of oil during thermal processing of carrot puree has been shown. After thermal pasteurization (Knockaert *et al.*, 2012a; see Figure 2.4) or boiling (Hedrén *et al.*, 2002) of carrot puree in the presence of oil, the amount of β carotene released from the carrot matrix during *in vitro* digestion was significantly higher compared to thermally treated carrot puree without added oil. In the study of Hornero-Méndez and Mínguez-Mosquera (2007), however, no effect of oil addition during thermal processing of carrot puree on the amount of β-carotene released from the matrix was observed, whereas it had a positive effect on the amount of β-carotene incorporated in the micelles. Similarly, Veda *et al.* (2006) described a positive effect of addition of oil during thermal processing on β-carotene micellarization during *in vitro* digestion.

In contrast to thermal processing, the presence of oil during HP pasteurization of homogenized carrot puree could not increase the amount of β-carotene released from the matrix during *in vitro* digestion, as illustrated in Figure 2.4 (Knockaert *et al.*, 2012a). Under certain pressure–temperature conditions, oil can crystallize which might hinder the solubilization of β-carotene in the oil.

References

Balasubramaniam, V.M., Farkas, D., Turek, E.J., 2008. Preserving foods through high-pressure processing. Food Technol. 62, 32–38.

Barbosa-Cánovas, G.V., Juliano, P., 2008. Food sterilization by combining high pressure and thermal energy. In: Gutiérrez-López, G.F., Barbosa-Cánovas, G.V., Welti-Chanes, J., Parada-Arias, E. (Eds.), Food Engineering: Integrated Approaches. Springer, New York, pp. 9–46.

Bonnie, T.P., Choo, Y.M., 1999. Oxidation and thermal degradation of carotenoids. J. Oil Palm Res. 2, 62–78.

Chen, B.H., Peng, H.Y., Chen, H.E., 1995. Changes of carotenoids, color, and vitamin A contents during processing of carrot juice. J. Agric. Food Chem. 43, 1912–1918.

Dugave, C., 2006. Cis-trans Isomerization in Biochemistry. Wiley-VCH Verlag, Germany.

Hedrén, E., Diaz, V., Svanberg, U., 2002. Estimation of carotenoid accessibility from carrots determined by an *in vitro* digestion method. Eur. J. Clin. Nutr. 56, 425–430.

Hornero-Méndez, D., Mínguez-Mosquera, M.I., 2007. Bioaccessibility of carotenes from carrots: effect of cooking and addition of oil. Innovative Food Sci. Emer. Technol. 8, 407–412.

Knockaert, G., De Roeck, A., Lemmens, L., Van Buggenhout, S., Hendrickx, M., Van Loey, A., 2011. Effect of thermal and high pressure processes on structural and health-related properties of carrots (*Daucus carota*). Food Chem. 125, 903–912.

Knockaert, G., Lemmens, L., Van Buggenhout, S., Hendrickx, M., Van Loey, A., 2012a. Changes in β-carotene bioaccessibility and concentration during processing of carrot puree. Food Chem. 133, 60–67.

Knockaert, G., Sudheer, K.P., Lemmens, L., Van Buggenhout, S., Hendrickx, M., Van Loey, A., 2012b. Carrot β-carotene degradation and isomerization kinetics during thermal processing in the presence of oil. J. Agric. Food Chem. 60, 10312–10319.

Knockaert, G., Sudheer, K.P., Lemmens, L., Van Buggenhout, S., Hendrickx, M., Van Loey, A., 2012c. Isomerization of carrot β-carotene in presence of oil during thermal and combined thermal/high pressure processing. Food Chem. 138, 1515–1520.

Lemmens, L., Colle, I., Knockaert, G., Van Buggenhout, S., Van Loey, A., Hendrickx, M., 2011a. Influence of pilot scale in pack pasteurization and sterilization treatments on nutritional and textural characteristics of carrot pieces. Food Res. Int. 50, 526–533.

Lemmens, L., Colle, I.J.P., Van Buggenhout, S., Van Loey, A.M., Hendrickx, M.E., 2011b. Quantifying the influence of thermal process parameters on in vitro β-carotene bioaccessibility: a case study on carrots. J. Agric. Food Chem. 59, 3162–3167.

Lemmens, L., De Vleeschouwer, K., Moelants, K.R.N., Colle, I.J.P., Van Loey, A.M., Hendrickx, M.E., 2010. β-Carotene isomerization kinetics during thermal treatments of carrot puree. J. Agric. Food Chem. 58, 6816–6824.

Lemmens, L., Van Buggenhout, S., Oey, I., Van Loey, A., Hendrickx, M., 2009. Towards a better understanding of the relationship between the β-carotene *in vitro* bio-accessibility and pectin structural changes: a case study on carrots. Food Res. Int. 42, 1323–1330.

Marx, M., Stuparic, M., Schieber, A., Carle, R., 2003. Effects of thermal processing on *trans-cis*-isomerization of β-carotene in carrot juices and carotene containing preparations. Food Chem. 83, 609–617.

McInerney, J.K., Seccafien, C.A., Stewart, C.M., Bird, A.R., 2007. Effects of high pressure processing on antioxidant activity, and total carotenoid content and availability, in vegetables. Innovative Food Sci. Emer. Technol. 8, 543–548.

Ramaswamy, H., Marcotte, M., 2006. Food Processing—Principles and Applications. CRC Press, Boca Raton.

Rastogi, N.K., Raghavarao, K.S.M.S., Balasubramaniam, V.M., Niranjan, K., Knorr, D., 2007. Opportunities and challenges in high pressure processing of foods. Critical Rev. Food Sci. Nutr. 47, 69–112.

Rodriguez-Amaya, D.B., 2001. A Guide to Carotenoid Analysis in Foods. International Life Sciences Institute Press, Washington.

Schieber, A., Carle, R., 2005. Occurrence of carotenoid *cis*-isomers in food: technological, analytical, and nutritional implications. Trends Food Sci. Technol. 16, 416–422.

Sharoni, Y., Linnewiel-Hermoni, K., Khanin, M., Salman, H., Veprik, A., Danilenko, M., Levy, J., 2012. Carotenoids and apocarotenoids in cellular signaling related to cancer: a review. Molecular Nutr. Food Res. 56, 259–269.

Tauscher, B., 1998. Effect of high pressure treatment to nutritive substances and natural pigments. In: Autio, K. (Ed.), Fresh Novel Foods by High Pressure. Technical Research Centre of Finland, Finland, pp. 83–95.

Tydeman, E.A., Parker, M.L., Wickham, M.S.J., Rich, G.T., Faulks, R.M., Gidley, M.J., Fillery-Travis, A., Waldron, K.W., 2010. Effect of carrot (*Daucus carota*) microstructure on carotene bioaccessibility in the upper gastrointestinal tract. 1. In vitro simulations of carrot digestion. J. Agric. Food Chem. 58, 9847–9854.

van het Hof, K.H., West, C.E., Weststrate, J.A., Hautvast, J.G.A.J., 2000. Dietary factors that affect the bioavailability of carotenoids. J. Nutr. 130, 503–506.

Veda, S., Kamath, A., Platel, K., Begum, K., Srinivasan, K., 2006. Determination of bioaccessibility of β-carotene in vegetables by *in vitro* methods. Molecular Nutr. Food Res. 50, 1047–1052.

CHAPTER

3

Brassica Composition and Food Processing

Alfredo Aires

CITAB/UTAD-Centre for Research and Technology for Agro-Environment and Biological Sciences, University of Trás-os-Montes e Alto Douro, Vila Real, Portugal

CHAPTER POINTS

- *Brassica* vegetables.
- Nutriments and nutriments.
- Phtochemicals and human health.
- Pre-harvest and post-harvest factors affecting phytochemical composition.
- Food processing effects on phytochemical composition.

INTRODUCTION

The *Brassicaceae* family (formerly *Cruciferae*) comprises a large number of plant species with economic importance, including many familiar vegetables (e.g., broccoli, Brussels sprouts, cabbage, cauliflower, kale, Savoy cabbage, and Chinese kale), oil crops (oil-seed rape), and ornamental plants (wallflower, alyssum) (Gomez-Campo and Prakash, 1999). Many of those with economic significance are extensively consumed as commodities and used in the industry worldwide. They are found all over the world, but most species occur in the north temperate region, Mediterranean areas, and a few in the southern hemisphere. The family *Brassicaceae* is a large group of plants, having around 3000 species grouped into 350 genera, including several types of edible plant. They are mostly cultivated as annual or perennial herbaceous plants (Gomez-Campo and Prakash, 1999). The genus of *Brassica* belongs to this family and from an economical and agronomical point of view is the most important genus. According to Gomez-Campo and Prakash (1999), the principal genus *Brassica* is the most important in the *Brassicaceae* family, and includes vegetables and forage forms such as *Brassica oleracea* L., *Brassica napus* L., and *Brassica rapa* L. The same author stated that the *acephala* (kale, collard greens), *botrytis* (cauliflower, Romanesco broccoli, broccoflower), *capitata* (white cabbage), *capitata rubra* (red cabbage), *costata* (tronchuda kale, Portuguese cabbage), *gemífera* (brussels sprouts), *gongylodes* (kohlrabi), *italic* (broccoli), *sabauda* (Savoy cabbage), and *viridis* (collards, tree kale) are the most important cultivars of *Brassica oleracea* in western-hemisphere countries. *Brassica* vegetables represent an important part of the human diet worldwide. They are consumed by people all over the world and are considered important food crops in China, Japan, India, and Europe (Cartea and Velasco, 2008).

The interest in *Brassica* plants exceeds the information on proteins, lipids, carbohydrates, vitamins, amino acids, and minerals since other compounds exist that can explain their protective mechanisms in human health. Due to their agricultural importance, *Brassica* plants have been the subject of much scientific interest, particularly the *Brassica oleracea* species which have assumed an important role in human nutrition, as they are the predominant dietary source of glucosinolates (Cartea and Velasco, 2008) but also have high contents of phenolics and other antioxidant compounds (Ferreres *et al.* 2007).

PHYTOCHEMICAL COMPOSTION OF *BRASSICA* VEGETABLES

Polyphenols

The nutritional interest of *Brassica* crops is partially related to their polyphenol composition. The polyphenol content of different *Brassica* plants has been described, revealing distinct qualitative and quantitative profiles. A few years ago, Podsedek (2007) carried out an extensive review on polyphenol profiles in different *Brassica*

Processing and Impact on Active Components in Food
http://dx.doi.org/10.1016/B978-0-12-404699-3.00003-2

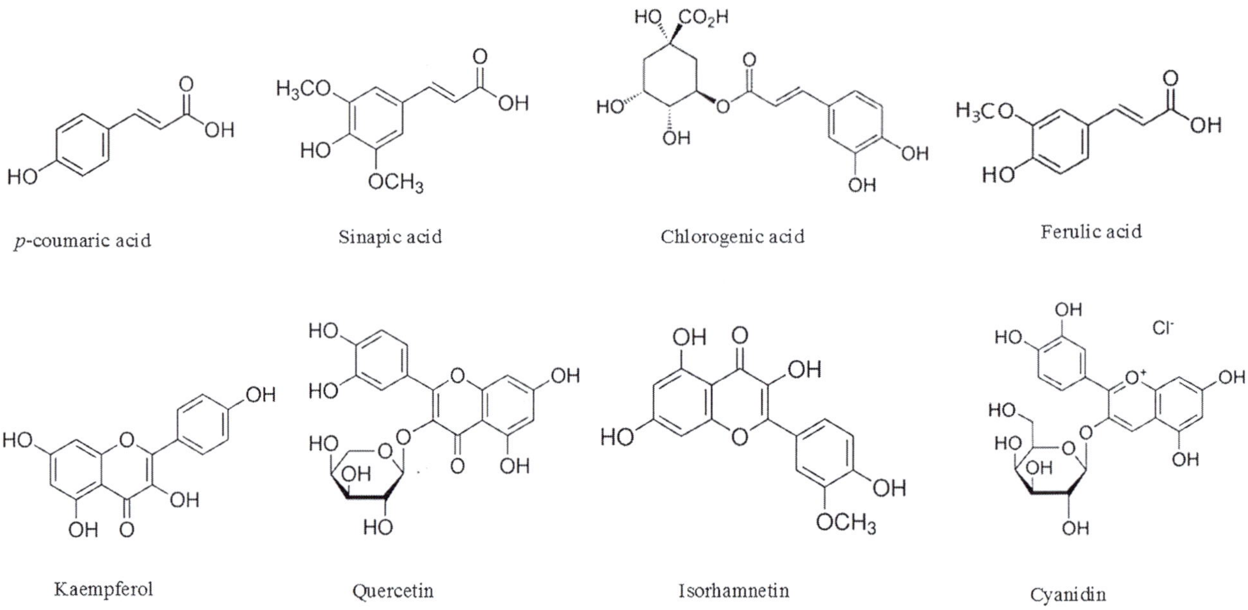

FIGURE 3.1 Hydroxycinnamic acids and flavonoid aglycones most common in *Brassica* crops.

species. Several authors have also reported the presence of different classes of polyphenols in different *Brassica* crops. In general, in *Brassica*, the preponderant classes of polyphenols are quercetin derivatives, kaempferol derivatives, hydroxycinamic acids, and isorhamnetin derivatives (Lin and Harnly, 2009). Anthocyanins have also been identified in *Brassica* (Moreno *et al.*, 2010). The pigmentation found in red cabbage and in broccoli sprouts is caused by anthocyanins. The major anthocyanins found in *Brassica* crops are mainly cyanidin 3-*O*-(sinapoyl)(feruloyl)diglucoside-5-*O*-glucoside and cyanidin 3-*O*-(sinapoyl)(sinapoyl) diglucoside-5-*O*-glucoside, with significant differences among species, cultivars, and crops within the same species (Moreno *et al.*, 2010) (Figure 3.1).

Glucosinolates

Another important group of phytochemicals, or even the most important group of compounds normally present in *Brassicacea* plants, are the glucosinolates. The glucosinolates are a group of organic anions containing *β-D-thioglucose* and sulfonated oxime moieties derived from amino-acids (Figure 3.2). This class of phytochemicals comprises at least 120 compounds with well-defined chemical structures (Halkier and Gershenzon, 2006).

The species *Brassica* and other cruciferous plants in which glucosinolates occur include: broccoli, rapeseed, white cabbage, red cabbage, cauliflower, kale, Brussel sprouts, Swede/turnip, turnip greens/tops, leaf rape,

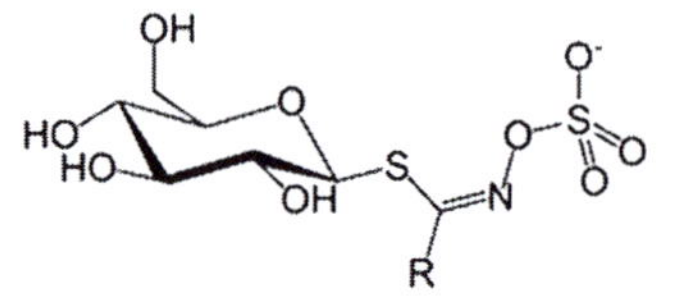

FIGURE 3.2 Glucosinolates structure; side group R varies. *Adapted from Fahey* et al. *(2001) and Halkier and Gershenzon (2006).*

Chinese cabbage, radish, mustard seeds, horseradish, cress, and mustard greens. They intervene in plant defense mechanisms against insects, fungi, nematodes, bacteria, and viruses, depending on their profile and concentration (Fahey *et al.*, 2001). Glucosinolates occur in plants together with the enzyme myrosinase (*β*-thioglucosidase-EC 3.2.1-3.2.3) which is responsible, after tissue disruption (cut, ground, or chewed) for their degradation in several biologically active compounds such as isothiocyanates, thiocyanates, nitriles, epithionitrile, and different indole compounds (Halkier and Gershenzon, 2006). The type of compounds produced (Table 3.1) is specific to the respective glucosinolate present in plant tissue and it always depends on the conditions under which the hydrolysis occurs (Halkier and Gershenzon, 2006).

Differences in chemical properties and biological activity between glucosinolates and their respective hydrolysis products are largely determined by side chain structure (Fahey *et al.*, 2001). Among the various hydrolysis products, the isothiocyanates are recognized as one of the major inhibitors of microbial activity but also have anticarcinogenic properties (Kuroiwa *et al.*, 2006).

TABLE 3.1 Examples of Some of the Most Important Glucosinolates and their Hydrolysis Products in *Brassica* Vegetables Crops

Glucosinolates	Isothiocyanates and Others	Nitriles, Cianides
Sinigrin (2-Propenil glucosinolate)	Allyl-isothiocyanate (AITC)	Allyl-nitrile, Allyl-cyanide (ACN)
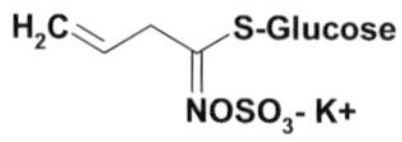	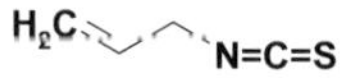	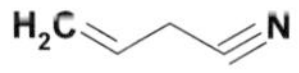
Glucotropaeolin (Benzyl glucosinolate)	Benzyl-isothiocyanate (BITC)	Benzyl-nitrile, Benzyl-cyanide (BCN)
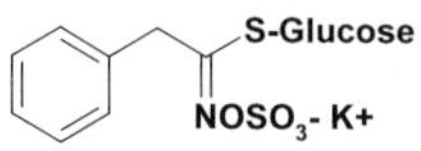	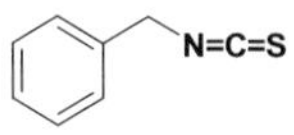	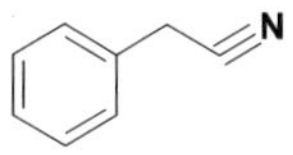
Gluconasturtiin (2-Phenylethyl glucosinolate)	2-Phenylethyl-isotiocyanate (PEITC)	2-Phenylethyl-nitrile, 2-Phenylethyl-cyanide (PCN)
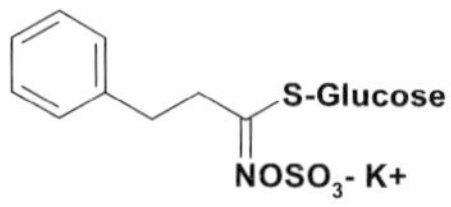	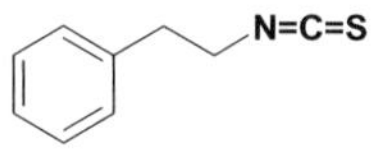	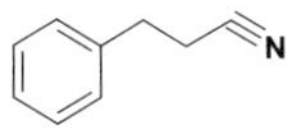
Glucobrassicin (3-Indolylmethyl glucosinolate)	Indole-3-Carbinol (I_3C)	Indole-3-acetonitrile (IAN)
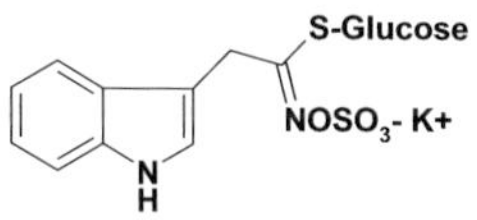	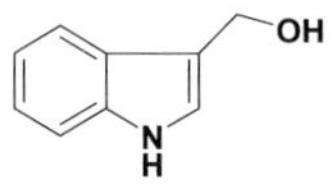	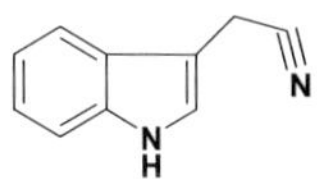
Glucoraphanin (4-Methylsulfinylbutyl glucosinolate	Sulforaphane (SFN)	Sulphorafane-nitrile (SF-CN)
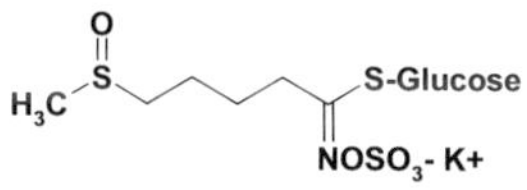	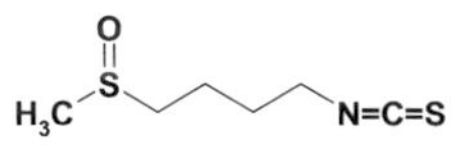	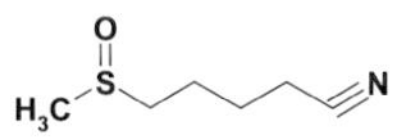

Adapted from Fahey et al. *(2001).*

Vitamins and Carotenoids

The literature shows that the major antioxidant compounds, and thus the major protective dietary antioxidants, are vitamins C and E, carotenoids, and polyphenols, especially flavonoids. It is well accepted now that these antioxidant compounds scavenge radicals and contribute both to the first and second defense lines against oxidative stress (Posdedek, 2007). As a result, they protect cells against oxidative damage, and may therefore prevent chronic diseases such as cancer, cardiovascular disease, and diabetes. *Brassica*, like other vegetables, contains a considerable amount of such compounds. Bernhardt and Schlich (2006), Korus and Lisiewska (2007), and Singh *et al.* (2007) presented studies which reported the richness of such compounds in various cultivars of *Brassica* including cabbage, cauliflower, Brussel sprouts, Chinese cabbage, broccoli, and kale cultivars. In general all authors concluded that *Brassica* and other cruciferous vegetables are a relatively good source of abundant antioxidants, and there is a substantial and significant variation, both within and between the species and cultivars. All authors reported that broccoli inflorescences, followed by Brussel sprouts, and kale are amongst the *Brassica* vegetables with the highest content of vitamim C, β-carotene, lutein and DL-α-tocopherol. Posdedek (2007) recently presented an extensive review on the importance of such compounds in *Brassica* vegetables.

Antioxidant Activity

Brassica vegetable extracts have been screened for antioxidant activity using various methods. Human health benefits associated with *Brassica* vegetable consumption can be explained by the presence of antioxidants. Thus, Brassica vegetables are the focus of intense research based on their content of secondary metabolites (Podsedek, 2007). The antioxidant potential of *Brassica oleracea* vegetables is high when compared to other vegetables. The broccoli, kale, red cabbage, and Brussel sprouts are, among various vegetables, the most studied

Brassica. Wu *et al.* (2004) studying the antioxidant activity of different *Brassica* crops with linoleic acid emulsion and liposomal phospholipid suspension antioxidant systems found the following ranking: broccoli > cauliflower > cabbage > Chinese cabbage. In a previous study, Azuma *et al.* (1999) had found a similar order in antioxidant potential of *Brassica*: red cabbage > broccoli > cabbage > Chinese cabbage.

The order of antioxidant activities depended on the sample extraction method, the antioxidant activity assay used, the type of the reactive species in the reaction mixture, and mainly of course the antioxidant levels and their composition (Podsedek, 2007). Therefore, according to Podsedek (2007), in order to determine the potential antioxidant capacity of *Brassica* crops, we must look for the chemical composition and the activity of water and lipid-soluble antioxidants must always be considered. Several studies have been conducted to evaluate the antioxidant properties of *Brassica* in different solvent extractions. For example, Ferreres *et al.* (2007), De Pascale (2007), Kusznierewicz *et al.* (2008), Watchel-Galor *et al.* (2008), and Volden *et al.* (2009), extracted hydrophilic antioxidants from tronchuda, turnip, cabbage, broccoli, cabbage, Choy sum (Chinese-cabbage), and cauliflower. The same procedure was used recently by Šamec *et al.* (2011) on Cabbage and Chinese cabbage, whilst Roy *et al.* (2009) used organic solvents such as acetone or n-hexane to extract lipid-soluble antioxidants from broccoli.

The antioxidant properties of *Brassica* vegetables, as for other important biological and beneficial properties for human health, may depend not only on the genotype but also on different parameters, such as environmental and agronomical conditions in which plants are grown, and the way these vegetables are prepared and processed. All these factors must be taken into consideration when the objective is to maximize or optimize the antioxidant properties of such vegetables.

FACTORS AFFECTING PHYTOCHEMICAL CONTENT OF *BRASSICA*

As with the primary nutrients, the phytochemical content of *Brassica* is affected by several factors including genetic intrinsic characters (species, cultivar and plant part), environmental conditions, production season, soil fertility, and soil composition (Cartea *et al.*, 2008). The agronomical practices and food processing (pre- and post-harvest conditions) can highly affect the average levels of bioactive components in *Brassica* (López-Berenguer *et al.*, 2007; De Pascale *et al*, 2007; Francisco *et al.*, 2010; Šamec *et al.*, 2011; Aires *et al.*, 2012).

Pre- and Post-Harvest Factors

The level of bioactive compounds in *Brassica* crops as in other vegetables varies according to pre-harvest and post-harvest factors. Pre-harvest factors include genetic, environmental, or climate conditions (location, soil moisture, soil temperature, air temperature, and rainfall), growing locations, and agronomic factors (fertilizers, sowing date, and water irrigation, production system—organic or conventional). Whilst post-harvest factors include food-processing operations and food product storage conditions. There is also an additional factor that could be very incisive in affecting the levels of bioactive compounds—the harvest period (Figure 3.3).

Several studies showed that genotype (Kushad *et al.*, 1999), plant parts (Šamec *et al.*, 2011), maturity stages, environmental conditions (Cartea *et al.*, 2008), agronomical practices (De Pascale *et al.*, 2007), harvesting time, transportation, storage time, and temperature (Korus *et al.*, 2011) could play a key factor in influencing the levels of phytochemicals in harvested crops. Wacthchel-Galor *et al.* (2008) and Korus and Lisiewska (2011) showed that despite pre-harvest factors being important in the variation of phytochemicals, factors such as food-processing operations, either at domestic or industrial level, must be considered, since they can interfere with the availability of phytochemicals and antioxidants. Recently, Aires *et al.* (2012) showed that culinary treatment is fundamental to optimizing the levels of phytochemicals such as glucosinolates in different *Brassica* vegetables. Vallejo *et al.* (2003) reported that cooling conditions, transport, type of atmospheric control used in storage, and shelf-life period can interfere with the quality and quantity of phytochemicals. Korus *et al.* (2011) stated that harvesting, transportation, storage, and refrigeration are more important in decreasing or increasing of phytochemicals and other important nutritional compounds.

Apart from the pre-harvest factors described above, various post-harvest stages, including food-processing operations, also have a major influence on the levels and profile of phytochemicals in *Brassica crops* and their products.

Post-Harvest Factor—Food Processing

The phytochemical composition of unprepared vegetables has been much studied. Since a large part of ingested vegetables, in this particular case *Brassica*, is processed prior to consumption, it is important to discuss how processing operations affect the levels of these beneficial compounds.

Food processing operations are fundamental for food processors to optimize levels of phytochemicals. In general food processing includes domestic, conventional, or industrial processing (e.g., freezing, blanching, pasteurization, high-pressure processing, boiling, steaming,

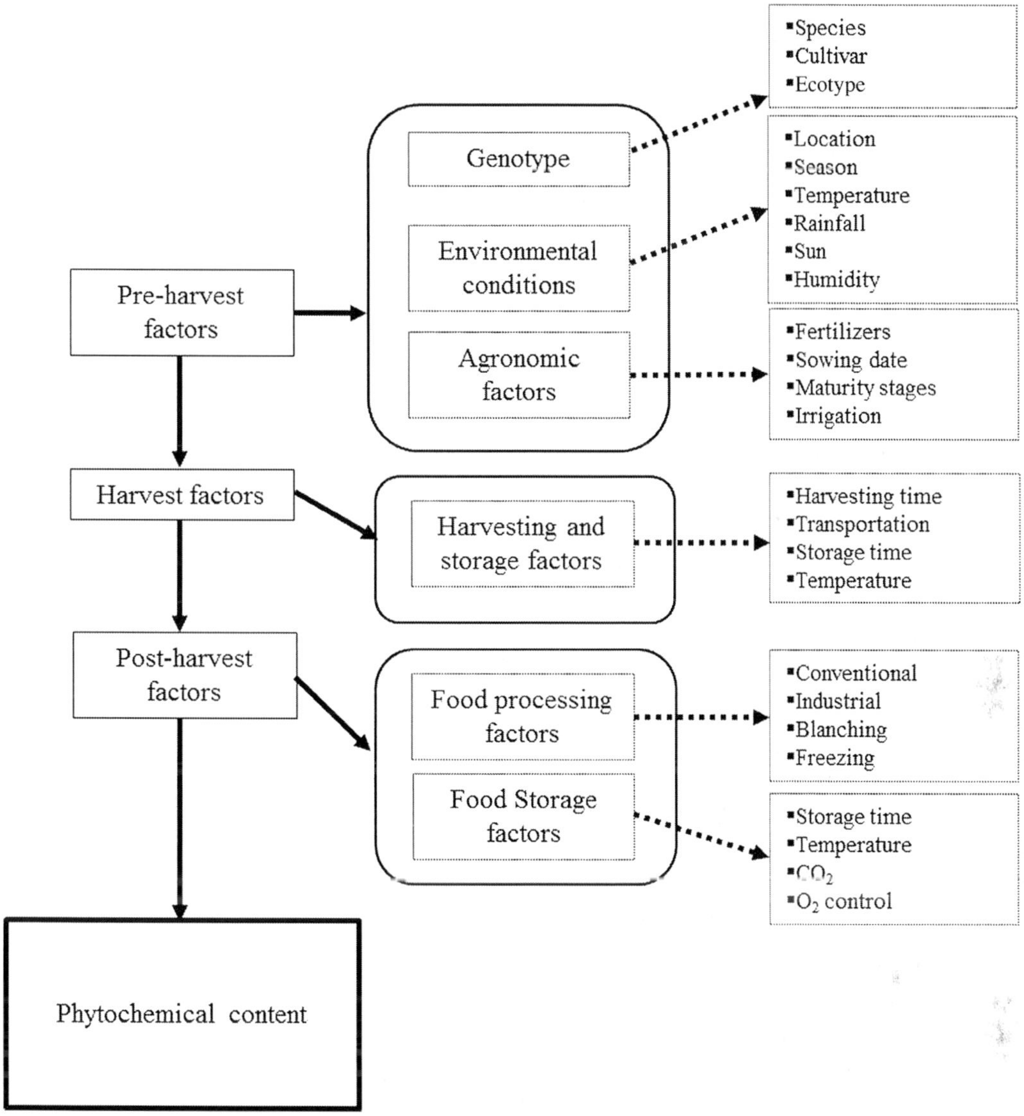

FIGURE 3.3 Factors affecting the chemical composition of vegetables.

stewing, and microwaving). Minimal processing operations (e.g., thawing, shredding, chopping, and cutting) at domestic or industrial level are also common prior to consumption, and may influence stability, and thereby, the levels of phytochemicals, and also the bioavailability of antioxidants.

Processing methods are generally believed to be responsible for a depletion of naturally occurring phytochemicals in *Brassica*. However, in some cases, there is no change in their content and sometimes processing can even lead to the formation of novel compounds with biological activity (Turkmen *et al.*, 2005; Volden *et al.*, 2009). Significant effects of various processing methods on phytochemical contents, such as glucosinolates, polyphenols (phenolics and flavonoids), antioxidant vitamins such as C, E, β-carotene, α-tocoperols, and related compounds have been reported. In Table 3.2 some of the many effects of various food processing operations on phytochemicals in *Brassica* crops are summarized, presented by several authors.

The literature indicates that the level of phytochemicals in vegetables is dependent on several post-harvest stages of the production chain. Thus, the food processor chain must be optimized in order to restrict the loss of phytochemicals. Recent data showed a consistent trend for the effects of processing on the total antioxidant activity, phenolics, vitamins, and glucosinolates in vegetable *Brassica* crops. The data reported suggest that the effect of processing operations is different in different products. Moreover, differences in processing methods may have different effects on the content of distinct phytochemicals (see Table 3.2). For example, Zhang and Hamauzu (2004) studied the changes in phenolics, ascorbic acid, and carotenoids during conventional (5 min) boiling and

TABLE 3.2 Effects of Some Food Processing (Domestic or Industrial) Operations on the Levels of Phytochemicals in *Brassica* Vegetables

Processing Operation	Processing Conditions	*Brassica* Product	Effects	References
Freezing	−24°C during 12 months	Cauliflower	No losses were detected in total aliphatic and indole glucosinolates	Volden *et al.* (2009)
	Between −20°C and −30°C during 12 months	Kale leaves	Loss between 48 and 61% of L-ascorbic acid levels, 58–75% of total phenolics, and 47% of total antioxidant activity	Korus and Lisiewska (2011)
Blanching	2.5 min at 96–98°C	Kale leaves	34% decrease in the content of vitamin C, 51% decrease in polyphenols, 33% decrease in antioxidative activity.	Korus and Lisiewska (2011)
Canning vegetables	Canned and stored during 12 months	Kale	Loss of 83% of L-ascorbic acid levels, 76% of total phenolics and 75% of total antioxidant activity	Korus and Lisiewska (2011)
	40 min at 120°C	Red cabbage	Loss of 73% of total glucosinolates	Oerlemans *et al.* (2006)
Boiling	5 min (100°C)	Broccoli	Loss of 19.2–65.9% of ascorbic acid	Zhang and Hamauzu (2004)
	15 min (100°C)	Turnip	Loss of 81% of total glucosinolates	Aires *et al.* (2012)
Steaming	7.5 min (95°C)	Broccoli	Gain of 18% of total phenolics	Turkmen *et al.* (2005)
	10 min (95°C)	Cauliflower	Gain of 13% of total antioxidant activity	Wachtel-Galor *et al.* (2008)
Stewing	8 min	Broccoli	Gain in the levels of β-carotene and α-tocopherol	Bernhardt and Schlich (2006)
Microwaving	1.5 min	Broccoli	Gain of 25% of total phenolics	Turkmen *et al.* (2005)
	2.0 min	Broccoli	Gain of 16–18% of lutein and loss of 22.7% of carotenoids	Zhang and Hamauzu (2004)
Fermentation	14 days of fermentation	White cabbage	The sauerkraut product collected after 14 days of fermentation exhibits about seven-times higher in DPPH assay relatively to initial value determined for fresh cabbage material.	Kusznierewicz *et al.* (2008)
Shredding	Internal leaves shredded in to 72 pieces.	Red cabbage	Decrease in total soluble phenolics content around 9%, and 14% of total total anthocyanin content.	Reyes *et al.* (2007)

microwave cooking of broccoli florets and stems. They observed a loss of about 71.9% and 42.2% of total phenolics in the fresh florets and stems, respectively for both conventional and microwave processing, but also reported a slight increase in the level of lutein during conventional and microwave cooking. The opposite tendency was noted for compounds. The authors observed that the time of cooking had a higher influence on antioxidant levels such as ascorbic acid, lutein, β-carotene, and violaxanthin than the cooking methods. Turkmen *et al.* (2005) reported losses of 62% of total phenolics when broccoli florets were submitted to boiling for 5 min. Similar results were found recently by Francisco *et al.* (2010) who reported a loss of 65–80% of individual phenolics under conventional and high-pressure cooking in turnip tops. However, losses were reduced to 20–30% by steam cooking, showing that this is the ideal method to preserve secondary metabolites in *Brassica* crops. Roy *et al.* (2009), evaluating the effect of steaming on the nutritional value of broccoli by measuring the total phenolics and antioxidant activity of both raw and steamed broccoli, found that total phenolics and total antioxidant activity of steamed broccoli was higher due to higher extractability of antioxidant compounds with this food processing method. According to these authors, steaming may increment significantly the extractability of bioactive compounds and the antioxidant potential. Similar results were obtained by Bernhardt and Schlich (2006) for β-carotene and α-tocopherol contents in fresh broccoli inflorescences. They noted that steaming leads to a softening of the plant tissue and the denaturation of proteins so that the carotenoids can be extracted a lot more easily and thus be more available for consumers. These findings contradict the assumption that processed vegetables have lower nutrition quality and value than their equivalent unprocessed parts. Also, Wachtel-Galor *et al.* (2008) reported that the total phenolic content of broccoli, cauliflower, cabbage, and Chinese cabbage was higher in steamed > boiled > microwaved, but their concentration decreased with longer cooking time, regardless of method. Wachtel-Galor *et al.* (2008), concluded that the effect of cooking on phytochemicals content of different vegetables may differ. It seems that antioxidant changes during cooking largely depend on the crop analyzed. However, previously Czarniecka-Skubina (2002) reported that retention of vitamin C in Brussel sprouts was strongly dependent on the cooking method used. They reported that high retention of this vitamin was found when cooking in a microwave oven, whilst pressure cooking in steam and acuthermal pots caused a loss of vitamin C of 3.7% to 10.6%. Sultana *et al.* (2008) reported the effects of different cooking methods (boiling, frying, and microwave cooking) on the antioxidant activity and phytochemical composition of various selected vegetables (cabbage, cauliflower, yellow turnip, and white turnip) and they concluded that all the cooking methods affected the phytochemical profiles with respect to content. As most research studies conclude, cooking with steam seems to be the best food cooking process due to a positive effect on phytochemical levels of *Brassica* vegetables, keeping or increasing their antioxidant potential, thus incrementing their beneficial properties for human health.

Brassica vegetables usually undergo domestic processing before cooking. Some of them such as broccoli, cauliflower, and turnip are cut; others such as kale and cabbage (Savoy cabbage, white cabbage, or red cabbage) are chopped. Operations such as cutting and shredding or slicing may cause a rapid enzymatic depletion of several naturally occurring antioxidants as a result of the cellular disruption, which allows the contact of substrates and enzymes (myrosinase enzyme) affecting the levels of phytochemicals. Reyes *et al.* (2007) reported the decrease of several glucosinolates and polyphenols as a consequence of thawing, cutting, and shredding of broccoli and red cabbage, respectively.

Industrial processing such as blanching, canning, sterilizing, and freezing, as well as cooking methods are expected to affect the yield, chemical composition and bioavailability of phytochemicals or antioxidants. Blanching is necessary and fundamental as a pre-treatment to a freezing procedure, in order to inactivate enzymes responsible for the degradation of food products. After this procedure, the vegetables are frozen and preserved at a constant temperature, normally below 0°C. Volden *et al.* (2009) found for cauliflower reduction in total phenolics, total aliphactic and indole glucosinolates when the blanching method was used, but with no freezing. However, when cauliflower heads were submitted to freezing after blanching, during 12 months at −24°C, no major losses were detected (Volden *et al.*, 2009). A similar tendency was found by Korus and Lisiewska (2011) for kale leaves. Bernhardt and Schlich (2006) found a similar tendency for all-β-carotene when fresh broccoli was frozen after being blanched. For these authors the freezing process does not cause a decrease in antioxidant compounds, on the contrary, the freezing is an effective way of preserving such compounds and others important to human health. They conclude that blanching prior to freezing helps to slow or stop the enzyme activity that can cause undesirable changes in food composition, flavor, and texture, and helps to protect the product's phytochemicals, vitamins, color, and other food attributes. With regard to canning methods, the results reported by various authors are conclusive; canning methods seem to be the worst preservation methods for *Brassica* vegetables, since they cause high losses of phytochemicals and antioxidant compounds. Czarniecka-Skubina (2002) reported a decrease of 66% of vitamin C in Brussel sprouts when the canning method was used, which was about two-fold higher than in the case of blanching and freezing. Korus and Lisiewska (2011) reported similar losses on kale.

Based on the results shown above, it is normally accepted that natural antioxidants and phytochemicals found in *Brassica* vegetables are significantly lost during food processing. However, it was recently demonstrated that processed vegetables may retain their antioxidant activity and phytochemicals. The recent report by Kusznierewicz *et al.* (2008) showed that fermentation processing as well as heat treatment, both short and prolonged, greatly improves the antioxidant properties of cabbage. For Kusznierewicz *et al.* (2008), heat processing during fermentation of cabbage seemed to compensate for the loss of natural antioxidants by the formation of non-nutrient antioxidants such as Maillard reactions products (MRPs). Their results showed that fermented cabbage increases its antioxidative potential upon heating. They noticed that fermentation of cabbage (sauerkraut) by lactic bacteria gradually increases the antioxidative activity and the sauerkraut product collected after 14 days of fermentation exhibits values about seven times higher in a DPPH assay relative to initial values determined for fresh cabbage material. The same authors concluded that in the case of fermented cabbage, thermal treatment can be the reason for the formation of compounds with antioxidant activity. Fermentation processes as well as heat treatment increased the initial values of antioxidant activity of fresh cabbage.

FINAL CONSIDERATIONS

Brassica vegetables are consumed throughout the year as the ingredients of different salads or after cooking of raw and frozen vegetables. The contribution of *Brassica* vegetables to health improvement can be related to their phytochemical composition. The vast literature indicates that the level of phytochemicals and related compounds are always dependent on both pre-harvest and post-harvest stages of the production chain. Therefore, it is fundamental to know how to act, either at domestic or at industrial levels, in order to preserve these beneficial compounds. The different stages of the production chain must be optimized in order to minimize the losses of beneficial nutraceutical and bioactive compounds. Research should be focused on the relationship between the content of beneficial compounds in *Brassica* vegetables and their stability.

References

Aires, A., Carvalho, R., Rosa, E., 2012. Glucosinolate composition of Brassica is affected by Postharvest, Food processing and myrosinase activity. J. Food Processing Preservation 36, 214–224.

Azuma, K., Ippoushi, K., Ito, H., Higashio, H., Terao, J., 1999. Evaluation of antioxidative activity of vegetable extracts in linoleic acid emulsion and phospholipid bilayers. J. Sci. Food Agric. 79, 2010–2016.

Bernhardt, S., Schlich, E., 2006. Impact of different cooking methods on food quality: retention of lipophilic vitamins in fresh and frozen vegetables. J. Food Eng. 77, 327–333.

Cartea, M.E., Velasco, P., Obregón, S., Padilla, G., de Haro, A., 2008. Seasonal variation in glucosinolate content in *Brassica oleracea* crops grown in northwestern Spain. Phytochemistry 68, 403–410.

Czarniecka-Skubina, E., 2002. Effect of the material form, storage and cooking methods on the quality of Brussels sprouts. Polish J. Food Nutr. Sci. 11/52, 75–82.

De Pascale, S., Maggio, A., Pernice, R., Fogliano, V., Barbieri, G., 2007. Sulphur fertilization may improve the nutritional value of *Brassica rapa* L. subsp. *Sylvestris*. Eur. J. Agronomy 26, 418–424.

Fahey, J.W., Zalcmann, A.T., Talalay, P., 2001. The chemical diversity and distribution of glucosinolates and isothiocyanates among plants. Phytochemistry 56, 5–51.

Ferreres, F., Sousa, C., Valentão, P., Seabra, R.M., Pereira, J.A., Andrade, P.B., 2007. Tronchuda cabbage (*Brassica oleracea* L. var. *costata* DC) seeds: Phytochemical characterization and antioxidant potential. Food Chem. 101, 549–558.

Francisco, M., Velasco, P., Moreno, D.A., García-Viguera, C., Cartea, M.E., 2010. Cooking methods of *Brassica rapa* affect the preservation of glucosinolates, phenolics and vitamin C. Food Res. Int. 43, 1455–1463.

Gómez-Campo, C., Prakash, S., 1999. Origin and Domestication. In: Gómez-Campo, C. (Ed.), Biology of *Brassica* Coenospecies, Developments in plant genetics and breeding. fourth ed. Elsevier Inc., Amesterdam, pp. 33–58.

Halkier, B.A., Gershenzon, J., 2006. Biology and biochemistry of glucosinolates. Annu. Rev. Plant Biol. 57, 303–333.

Korus, A., Lisiewska, Z., 2011. Effect of preliminary processing and method of preservation on the content of selected antioxidative compounds in kale (*Brassica oleracea* L. var. *acephala*) leaves. Food Chem. 129, 149–154.

Korus, A., Lisiewska, Z., Słupski, J., 2011. Retention of oxalates in frozen products of three Brassica species depending on the methods of freezing and preparation for consumption. Int. J. Refrigeration 34, 1527–1534.

Kuroiwa, Y., Nishikawa, A., Kitamura, Y., Kanki, K., Ishii, Y., Umemura, T., Hirose, M., 2006. Protective effects of benzyl isothiocyanate and sulforaphane but not resveratrol against initiation of pancreatic carcinogenesis in hamsters. Cancer. Lett. 241, 275–280.

Kushad, M.M., Brown, A.F.B., Kurilich, A.C., Juvik, J.A., Klein, B.P., Wallig, M.A., Jeffery, E.H., 1999. Variation of glucosinolates in vegetable crops of *Brassica oleracea*. J. Agric. Food. Chem. 47, 1541–1548.

Kusznierewicz, B., Śmiechowska, A., Bartosek, A., Namieśnik, J., 2008. The effect of heating and fermentation on antioxidant properties of white cabbage. Food Chem. 108, 853–861.

Lin, L.Z., Harnly, J.M., 2009. Identification of the phenolic components of collard greens, kale, and Chinese broccoli. J. Agric. Food Chem. 57, 7401–7408.

López-Berenguer, C., Carvajal, M., Moreno, D.A., García-Viguera, C., 2007. Effects of microwave cooking conditions on bioactive compounds present in broccoli inflorescences. J. Agric. Food Chem. 55, 10001–10007.

Moreno, D.A., Perez-Balibrea, S., Ferreres, F., Gil-Izquierdo, A., García-Viguera, C., 2010. Acylated anthocyanins in broccoli sprouts. Food Chem. 123, 358–363.

Oerlemans, K., Barret, D.M., Suades, C.B., Verkerk, R., Dekker, M., 2006. Thermal degradation of glucosinolates in red cabbage. Food Chem. 95, 19–29.

Podsedek, A., 2007. Natural antioxidants and antioxidant capacity of *Brassica* vegetables: A review. LWT- Food Sci. Technol. 40, 1–11.

Reyes, L.F., Villarreal, J.E., Cisneros-Zevallos, L., 2007. The increase in antioxidant capacity after wounding depends on the type of fruit or vegetable tissue. Food Chem. 101, 1254–1262.

Roy, M.K., Juneja, L.R., Isobe, S., Tsushida, T., 2009. Steam processed broccoli (*Brassica oleracea*) has higher antioxidant activity in chemical and cellular assay systems. Food chem. 114, 263–269.

Šamec, D., Piljac-Žegarac, J., Bogović, M., Habjanič, K., Grúz, J., 2011. Antioxidant potency of white (*Brassica oleracea* L. var. *capitata*) and Chinese (*Brassica rapa* L. var. *pekinensis* (Lour.)) cabbage: The influence of development stage, cultivar choice and seed selection. Sci. Hortic. 128, 78–83.

Singh, J., Upadhyay, A.K., Prasad, K., Bahadur, A., Rai, M., 2007. Variability of carotenes, vitamin C, E and phenolics in *Brassica* vegetables. J. Food Comp. Anal. 20, 106–112.

Sultana, B., Anwar, F., Iqbal, S., 2008. Effect of different cooking methods on the antioxidant activity of some vegetables from Pakistan. Int. J. Food Sci. Technol. 43, 560–567.

Turkmen, N., Sari, F., Velioglu, Y.S., 2005. The effect of cooking methods on total phenolics and antioxidant activity of selected green vegetables. Food Chem. 93, 713–718.

Vallejo, F., Tomás-Barberán, F.A., Garcia-Viguera, C., 2003. Phenolic compound contents in edible parts of broccoli inflorescences after domestic cooking. J. Sci. Food Agric. 83, 1511–1516.

Volden, J., Bengtsson, G.B., Wicklund, T., 2009. Glucosinolates, L-ascorbic acid, total phenols, anthocyanins, antioxidant capacities and colour in cauliflower (*Brassica oleracea* L. ssp. *botrytis*); effects of long-term freezer storage. Food Chem. 112, 967–976.

Wachtel-Galor, S., Wong, K.W., Benzie, I.F.F., 2008. The effect of cooking on *Brassica* vegetables. Food Chemistry 110, 706–710.

Wu, X., Beecher, G.R., Holden, J.M., Haytowitz, D.B., Gebhardt, S.E., Prior, R.L., 2004. Lipophilic and hydrophilic antioxidant capacities of common foods in the United States. J. Agric. Food. Chem. 52, 4026–4037.

Zhang, D., Hamauzu, Y., 2004. Phenolics, ascorbic acid, carotenoids and antioxidant activity of broccoli and their changes during conventional and microwave cooking. Food Chem. 88, 503–509.

CHAPTER

4

Ascorbic Acid, β-Carotene and Antioxidant Activity of Broccoli During Short-Term Refrigerated Storage

A. Nath*, S. Mandal*, R.K. Singh*, Bidyut C. Deka†, S.V. Ngachan*

*Division of Agricultural Engineering, ICAR Research Complex, Umiam, Meghalaya, India, †ICAR Research Complex, Nagaland Centre, Nagaland, India

CHAPTER POINTS

- Shelf-life of broccoli is very short.
- It undergoes rapid changes in bioactive components during storage.
- It provides antioxidant protection against chronic diseases, including chronic heart disease, arthritis, and cancer.
- Different aspects of ascorbic acid, β-carotene, and total antioxidant content of broccoli florets during post-harvest storage experience rapid changes.
- In preventing deterioration of broccoli during storage, temperature and packaging materials play a significant role in the retention of these phytochemicals.

INTRODUCTION

Interest in food composition has extended past nutrients to include bioactive components that may prevent chronic disease. Some nutrients, such as the antioxidant vitamins, carotenoids, tocopherols, and ascorbic acid, appear to play a dual role in metabolism. These are required for normal growth and development, and they appear to provide antioxidant protection against chronic diseases, including chronic heart disease, arthritis, and cancer (Krinsky *et al.*, 2000). Cruciferous vegetables are an excellent dietary source of antioxidants and vitamins, precursors to a group of isothiocyanates shown to be anticarcinogenic (Jeffery and Jarrell, 2001).

Broccoli (*Brassica oleracea*) has been described as a vegetable with a high nutritional value due to its important content of vitamins, antioxidants, anti-carcinogenic compounds (Nestle, 1998) and health-promoting phytochemicals (Chun *et al.*, 2005). Broccoli inflorescences are harvested while they are totally immature, which implies severe changes in nutrient, water, and hormonal status. Epidemiological studies have shown an inverse association between the consumption of Brassica vegetables and the risk of cancer (Day *et al.*, 1994). Of the case-controlled studies, 56% demonstrate a strong association between increased broccoli consumption and protection against cancer (Verhoeven *et al.*, 1996). This protective effect has largely been attributed to the complement of phytochemicals in broccoli, which include the vitamins C and E, the flavonols quercetin and kaempferol, the carotenoids β-carotene and lutein, and the glucosinolates (Podsedek, 2007).

Broccoli is generally a high-priced green vegetable compared to other vegetables locally available and is a highly valued vegetable due to its richness in phytochemicals. This vegetable reaches the retail market at least 1–2 days after harvest. Most of the time fresh-looking, green color florets are preferred for consumption. This crop is generally sold in the retail market either without any packaging or sometimes as a packed form in polyethylene bags of 250–500 g. Most of the time, after purchase consumers of this vegetable keep it at home for 2–3 days in either open, packed, or refrigerated condition. Though this vegetable is purchased at a high price due to its higher phytochemical properties, its degradation during storage in such conditions is not known. Various literature is available on the

Processing and Impact on Active Components in Food
http://dx.doi.org/10.1016/B978-0-12-404699-3.00004-4

degradation of chlorophyll content during controlled atmosphere (CA) storage and also at specified storage temperatures under modified atmosphere packaging (MAP). When broccoli was stored at 0°C and 95% relative humidity, its green color was maintained for much longer than when unrefrigerated (Wang and Hruschka, 1977). Lebermann *et al.* (1968) found that total chlorophyll content of broccoli heads decreased after 16 days of storage and the loss was much greater at 7°C than at 1°C. Deschene *et al.* (1991) reported that florets from freshly cut heads of broccoli rapidly senesced when stored in air at 23 or 10°C. Chlorophyll levels declined by 80–90% within 4 days at 23°C and within 10 days at 10°C. Takeda *et al.* (1993) found that chlorophyll levels in broccoli decreased during storage at 20 or 23°C, whereas at 2°C chlorophyll levels remained nearly constant. Storage, transport and processing have been determined to significantly influence the levels of ascorbic acid, glucosinolates and flavonoids in broccoli (Leja *et al.*, 2001; Vallejo *et al.*, 2003). For example, Vallejo *et al.* (2003) demonstrated major losses of total glucosinolates (71–80%) and total flavonoids (62–59%) but not vitamin C in broccoli inflorescences stored for 7 days at 1°C to mimic cold storage followed by a 3-day period at 15°C to mimic the retail period. Conversely, Winkler *et al.* (2007) showed that extended storage (1 or 4°C for up to 28 days) followed by a 3-day period did not influence glucosinolates or flavonoids in broccoli inflorescences. Forney and Jordan (1999) reported that the O_2 concentration at 1–2% and CO_2 concentration at 5–10%, in a temperature range of 0–5°C was desirable for MAP of broccoli to avoid odor development due to the presence of sulphur-containing volatiles.

Leja *et al.* (2001) studied the broccoli of Lord cultivar stored at 20°C and at 5°C for 3 and 7 days, respectively, either non-packaged or packaged in polymeric film samples. They reported the influence of packaging and temperature on various components of broccoli flower buds during short-term storage (Table 4.1). Short-term storage at room temperature induced accumulation of total phenols, especially in non-packaged broccoli. With low-temperature treatment, phenol content rose only after 7-day storage of non-packaged heads. Both low temperature and application of polymeric foil stopped losses of ascorbic acid. Total antioxidant activity increased considerably during storage in all treatments. Changes of fatty acids were manifested as a slight decrease in saturated fatty acids in cold storage and increase of polyunsaturated fatty acids in most treatments.

The aim of this chapter is to discuss some metabolic changes, namely ascorbic acid, β-carotene, and antioxidant activity in broccoli flower buds during short-term refrigerated storage of broccoli heads under commonly applied conditions. Nath *et al.* (2011) studied the changes in ascorbic acid content, β-carotene and antioxidant activity of broccoli (cv. 'Pushpa', very common and suitable for hill regions of India) during refrigerated storage. In their studies, one set of samples, having total weight of 200–250 g (eight to 10 florets), were kept in plastic perforated trays both in open ambient storage (BTAT) conditions (15 ± 1°C and 55 ± 2% RH) and laboratory refrigerated storage (BTRS) conditions (4 ± 0.5°C and 50 ± 2% RH). In another set, eight to 10 florets/bag were packaged using a commercial polypropylene (PP) film with 10 pin holes and stored at ambient (BPPAT) conditions (15 ± 1°C and 55 ± 2% RH) and laboratory refrigerated storage (BPPRT) conditions (4 ± 0.5°C and 50 ± 2% RH). Broccoli florets were sampled at 0-, 48-, 96-, and 144-h intervals for further analysis.

CHANGES IN ASCORBIC ACID

Changes in ascorbic acid content of broccoli during refrigerated storage were determined by using the 2,6-dichlorophenol-indophenol dye method of Freed (1966). Broccoli floret samples of 2.5 g were ground with about 25 ml of 4% oxalic acid and filtered through Whatman no. 4 filter paper. The filtrate was collected in a 50-ml volumetric flask and the volume was made up with 4% oxalic acid and titrated against the standard dye to a pink point. The amount of ascorbic acid was calculated using Eq. (1) given below and expressed as mg/100 g on a fresh-weight (FW) basis.

$$\text{Ascorbic acid (mg/100g)} = \frac{\text{Titre value} \times \text{Dye factor} \times \text{Volume made ups}}{\text{Aliquot of extract taken for estimation} \times \text{Wt. or vol. of the sample taken for estimation}} \quad (1)$$

Nath *et al.* (2011) studied the changes in ascorbic acid content of broccoli during refrigerated storage and they reported that the initial ascorbic acid content of fresh broccoli florets was 130 mg/100 g which decreased linearly during storage under different treatments (Figure 4.1). However, broccoli florets packed in micro-perforated PP and stored in refrigerated conditions (4°C) showed no significant change in ascorbic acid content by the end of storage (144 h). The ascorbic acid content decreased rapidly in the florets kept in open ambient plastic trays compared to the florets kept in ambient PP micro-perforated packets and open plastic tray refrigerated samples (r^2 = 0.989). By 144 h

TABLE 4.1 Influence of Packaging and Temperature on Various Components of Broccoli Flower Buds During Short-Term Storage

	Days of Storage[a]					
	0 time	**1 day**		**3 days**		**7 days**
Components		**20°C**	**5°C**	**20°C**	**5°C**	**5°C**
Total Phenols (per 100 mg FW)[b]						
Non-packaged	56.2[c]	76.0[f]	57.9[c]	78.9[g]	57.8[c]	71.2[e]
Packaged		65.22[d]	53.25[b]	73.18[e]	50.47[a]	56.08[c]
Ascorbic Acid (per 100 mg FW)						
Non-packaged	60.1[c]	69.3[f]	68.3[ef]	44.7[a]	71.0[g]	75.1[h]
Packaged		71.13[g]	67.61[e]	61.60[d]	68.05[ef]	58.37[b]
Total Antioxidant Activity (%)						
Non-packaged	4.8[a]	30.1[c]	26.6[c]	41.8[d]	30.0[c]	69.5[f]
Packaged		11.7[ab]	14.4[b]	30.5[c]	23.7[c]	59.4[e]
Saturated Fatty Acids (%)						
Non-packaged	36.1[d]	35.4[d]	28.3[bc]	38.6[d]	28.2[bc]	25.1[ab]
Packaged		24.6[ab]	30.8[c]	30.5[c]	27.0[abc]	23.5[a]
Polyunsaturated Fatty Acids (%)						
Non-packaged	55.5[ab]	61.7[bc]	70.0[e]	54.2[a]	59.7[abc]	70.3[e]
Packaged		70.27[e]	69.20[de]	63.28[cd]	63.24[cd]	69.87[e]
Malondialdehyde (μM/g FW)						
Non-packaged	13.8[d]	13.0[cd]	12.7[bcd]	17.9[e]	13.6[d]	10.8[b]
Packaged		11.0[bc]	11.2[bc]	14.4[d]	13.0[d]	7.07[a]
Monodialdehydes (μM/g FW)						
Non-packaged	239[f]	179[cde]	197.6[e]	189.7[de]	163[bcd]	207[e]
Packaged		153[abc]	182[de]	132[a]	138[ab]	125[a]

[a] *Means followed by the same superscript letters are not significantly different.*
[b] *Fresh weight.*
Different letters (a h) within columns are significantly different at $p \leq 0.05$ according to Duncan Multiple Range Test for separation of means.
Source: Leja et al. *(2001).*

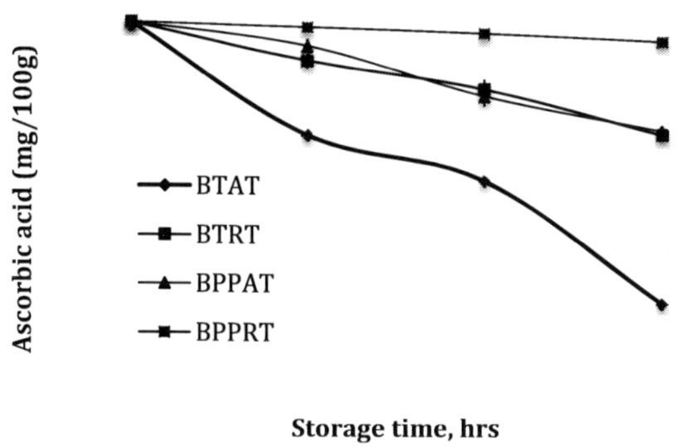

FIGURE 4.1 Effect of storage atmosphere on ascorbic acid content (mg/100g FW) of broccoli florets during short-term storage (means ± SE, n = 5). *Source: Nath* et al., *2011.*

storage, ascorbic acid content decreased from 130 mg/100 g to 35.3 mg/100 g FW, a 72.86% reduction in the open ambient plastic tray samples, while the same decreased from 130 to 92 mg/100 g FW, a 29.23% decline, in the open refrigerated samples of broccoli florets in plastic trays. During storage, the samples in open ambience in plastic trays had significantly lower levels of ascorbic acid content compared to the PP micro-perforated packed refrigerated samples (P<0.01), open refrigerated in plastic tray and ambient PP micro-perforated stored samples (P<0.05) at 96 h (Figure 4.1). At the end of storage (144 h), the open ambient in plastic tray stored broccoli contained significantly lower levels of ascorbic acid content than in the PP micro-perforated packed refrigerated stored samples (P<0.01). These results were in agreement with the earlier report of MAP of broccoli in similar packaging material (Rai *et al.*, 2008). Wang

FIGURE 4.2 Effect of amount of broccoli wrapped on vitamin C retention in broccoli florets during storage at 1°C over 15 days. *Source: Carvalho and Clemente, 2004.*

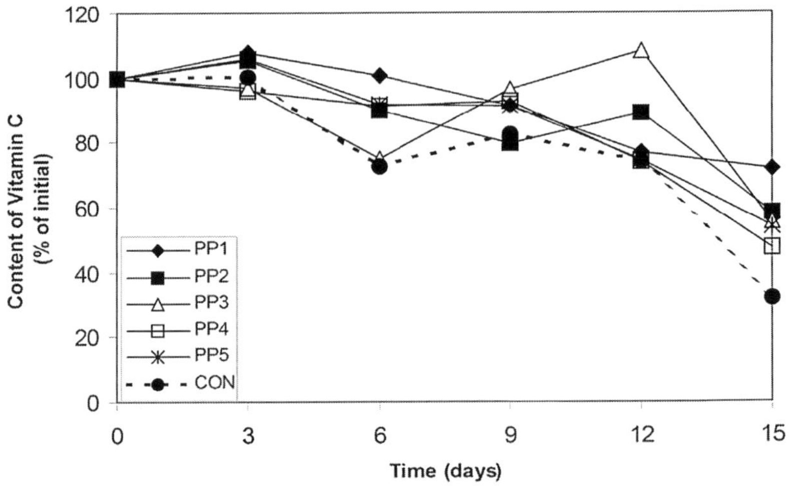

and Win (2002) and Leja *et al.* (2001) also reported a rapid decrease in ascorbic acid content of broccoli and other vegetables during storage at 15°C. Higher ascorbic acid (71.9–95.1%) retention in PP micro-perforated packed broccoli might be due to a limited level of atmospheric oxygen available for oxidation. During storage, oxidizing enzymes like ascorbic acid oxidase, peroxodase, catalase, and polyphenol oxidase might help in reducing the ascorbic acid of fruits and vegetables (Mapson, 1970).

Carvalho and Clemente (2004) studied the influence of broccoli (*Brassica oleracea* var. *itálica*) fill weight on post-harvest quality. In their experiments, they harvested fresh broccoli heads and packed them in PP bi-oriented (ITAP/BEMIS, Cambe-PR, BRA) (thickness = 40 μm; $PO_2 > 2.000\,cc.m^2.dia^{-1}$) in different levels of weights (PP = 240.0 g packaging; PP2 = 270.0 g packaging; PP3 = 300.0 g packaging; PP4 = 330.0 g packaging; PP5 = 360.0 g packaging; Control = 300.0 g no packaging) and analyzed at an intervals of 3 days up to 15 days starting at time zero. They observed that in all treatments, a decrease in vitamin C levels (Figure 4.2) was observed along the time, compared to the zero time, with more accentuated decrease registered on the 9th and 12th days. However, in some treatments such as the PP3, PP2 and Control sample, increases in vitamin levels on the 6th and 9th days were observed. This notable increase occurred due to the tendency of some vegetable textures to present increases in their vitamin C levels due to ripening (Lorient and Linden, 1996). This may indicate that an increase in ascorbic acid rates occurs because of vegetable stress. The harvest, manipulation, cutting, and cooling (1°C) can promote stress conditions, so the plant increases ascorbic acid synthesis as a protection mechanism (Rozek *et al.*, 1994). At the end of the experiment, the control sample presented the lowest residual vitamin C levels, as expected and observed by other authors (Barth *et al.*, 1993). The greater loss of humidity observed for this treatment explains the intensive losses of vitamin C. It occurred also due to the effect of atmosphere change inside the package which promoted vitamin C retention as a function of the CO_2 increment as the O_2 reduction (Barth and Zhuang, 1996). In the packaged treatments, PP3, PP4, and PP5 showed the lowest levels of residual vitamin C as the intensive changes in atmosphere caused injuries, promoting a lower retention of the vitamin. Generally, high CO_2 concentrations cause vitamin C degradation (Lee and Kader, 2000).

CHANGES IN β-CAROTENE

A change in β-carotene content of broccoli during refrigerated storage was determined by using the colorimetric method of Srivastava and Kumar (2002). A 5-g fresh sample was crushed in 10–15 ml acetone by adding a few crystals of anhydrous Na_2SO_4. The supernatant was decanted into a beaker and the process was repeated twice. The combined supernatant was transferred to a separating funnel and 10–15 ml petroleum ether was added and mixed thoroughly. After discarding the lower layer, the upper layer was collected in a 100-ml volumetric flask and volume was made up with petroleum ether. The OD was recorded at 452 nm and β-carotene was expressed as mg/100 g on a FW basis by Eq. (2), as follows:

$$\beta\text{-Carotene (mg/100 g)} = \frac{\text{OD} \times 13.9 \times 10^4 \times 100}{\text{Weight of the sample} \times 560 \times 1000}$$

Nath *et al.* (2011) studied the changes in β-carotene content of broccoli florets under different storage condition (Figure 4.3). The recorded initial β-carotene content of broccoli was 26.5 mg/100 g and it decreased with the advancement of storage time, irrespective of treatments (Figure 4.3). However, broccoli florets packed in PP micro-perforated and stored under refrigerated conditions

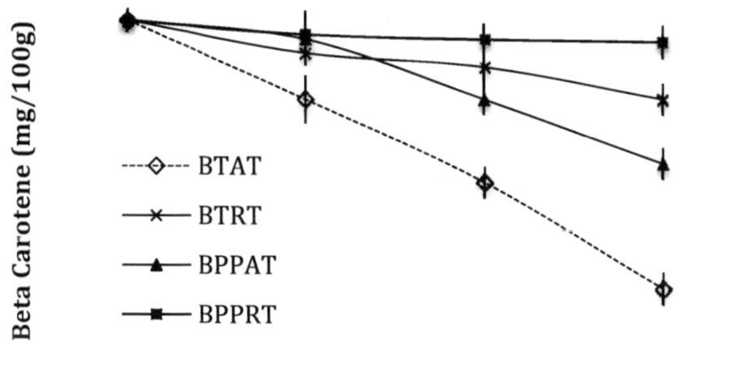

FIGURE 4.3 Effect of storage atmosphere on β-carotene content (mg/100g FW) of broccoli florets during short-term storage (means ± SE, $n = 5$). *Source: Nath* et al. *2011.*

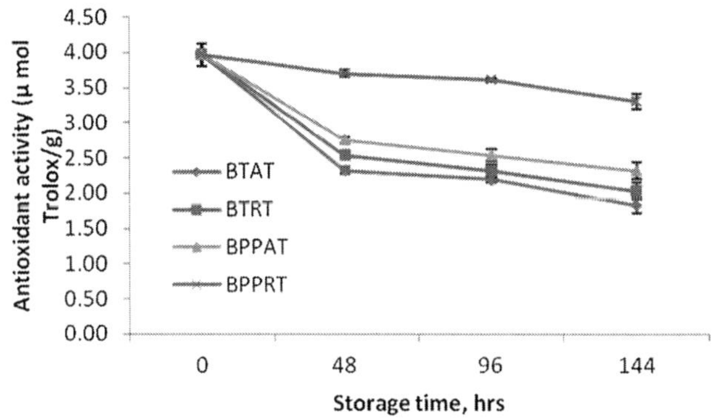

FIGURE 4.4 Effect of storage atmosphere on total antioxidant activity (μmol Trolox/g) of broccoli florets during short-term storage (means ± SE, $n = 5$). *Source: Nath* et al. *2011.*

(4°C) showed no significant change in β-carotene content by the end of storage (144h). β-carotene content decreased rapidly in the florets of open ambient plastic tray samples compared to ambient PP micro-perforated packed florets and open plastic tray refrigerated samples ($r^2 = 0.995$). At 144h storage, β-carotene content decreased from 26.5mg/100g to 7.62mg/100g FW, a 71.25% reduction in the open ambient plastic tray samples, while β-carotene content decreased from 26.5 to 16.19mg/100g FW, a 38.91% decline in the open refrigerated samples in plastic tray. During storage, the samples in open ambient plastic trays had significantly lower levels of β-carotene content compared to the PP micro-perforated packed refrigerated samples ($P < 0.01$), open refrigerated in plastic tray and ambient PP micro-perforated stored samples ($P < 0.05$) at 96h, (Figure 4.3). The open ambient in plastic tray stored broccoli florets recorded significantly lower levels of β-carotene content than in the PP micro-perforated packed refrigerated stored samples ($P < 0.01$) at the end of storage (144h). These results were in agreement with the earlier report on MAP of broccoli (Kidmose and Hansen, 1999). Hussein *et al.* (2000) also reported decrease in β-carotene content of broccoli during storage at higher temperatures. Higher β-carotene retention in PP micro-perforated packed broccoli might be due to limited levels of atmospheric oxygen available for oxidation and, therefore, oxidizing enzymes are less active for reduction of β-carotene of the broccoli during storage. During storage, oxidizing enzymes like superoxide dismutase (SOD), catalase (CAT), ascorbate peroxidase (APX), and peroxidase (POX) might have helped in reducing the β-carotene of the broccoli (Li *et al.*, 2004).

CHANGES IN ANTIOXIDANT ACTIVITY

Changes in antioxidant activity of of broccoli during refrigerated storage were determined using the Ferric Reducing Antioxidant Power (FRAP) assay of Benzic and Strain (1996) as a measure of "Antioxidant Power." The working FRAP reagent was prepared by mixing 10ml of 300-mmol/l acetate buffer, 1ml of 10-mmol/l TPTZ (2,4,6- tripyridyl-s- triazine) in 40- mmol/l hydrochloric acid and 1ml of 20-mmol/l Ferric chloride. A 100-μl (0.1ml) extract was added to 3.9ml of freshly prepared FRAP reagent and mixed thoroughly, and a reading was taken at 593nm in an UV-VIS Spectrophotometer. A standard curve was prepared using different concentrations of Trolox. The results were expressed in micromole Trolox equivalents per gram FW.

Nath *et al.* (2011) studied the changes in antioxidant activities of fresh harvested broccoli florets (Figure 4.4) and they reported that the antioxidant content of fresh harvested broccoli florets before storage was 3.96μmol Trolox/g. From their results, the changes in antioxidant activities of broccoli florets during storage under different treatments showed a linear decrease, irrespective of treatments with the advancement of storage time (Figure 4.4). However, broccoli florets packed in PP micro-perforated and stored under refrigerated conditions (4°C) showed no significant change in antioxidant activities by the end of storage (144h). Antioxidant activities decreased rapidly in the florets of open ambient plastic tray samples compared to ambient PP micro-perforated packed florets and open plastic tray refrigerated samples ($r^2 = 0.964$). At 144h storage, antioxidant activities decreased from 3.96 to 1.83μmol Trolox/g FW, a 53.79% reduction in the open ambient plastic tray samples while, antioxidant activities decreased from 3.96 to 2.00μmol Trolox/g FW, a 49.49% decline in the open refrigerated samples of broccoli florets in plastic trays. During storage, the open ambient in plastic tray treated samples had significantly lower levels of antioxidant activities compared with the PP micro-perforated packed refrigerated samples ($P < 0.01$), open refrigerated in plastic tray and ambient PP micro-perforated stored samples ($P < 0.05$) at 96h (Figure 4.4). At the end of storage (144h), the open ambient in plastic tray stored broccoli had significantly lower levels of antioxidant activities than the PP micro-perforated packed refrigerated stored samples ($P<0.01$). Costa *et al.* (2005) reported similar results in hot-air-treated broccoli florets during storage at 20°C. Conversely, Leja *et al.* (2001) reported an increase in antioxidant activity of broccoli (cv. Lord) flower buds during short-term storage at 20°C and at 5°C for 3 and 7 days, respectively, either non-packaged or packaged in polymeric film samples. Higher antioxidant activities retention in PP micro-perforated packed broccoli might be due to the limited level of atmospheric oxygen available for oxidation.

Xu *et al.* (2012) studied the effect of ethanol treatment on quality and antioxidant activity in post-harvest broccoli florets. In their studies, they used fresh broccoli heads (*Brassica oleracea* L. var. *Italica*, cv. Chaoda No. 1) treated with ethanol concentrations of 100, 250, 500, 750, and 1000 µl/l and stored in plastic trays, covered with polyethylene film to diminish water loss and stored at 20°C with 95% relative humidity in darkness. The samples were analyzed for antioxidant activities using peroxidase (POD, EC 1.11.1.7), catalase (CAT, EC 1.11.1.6), and superoxide dismutase (SOD, EC 1.15.1.1) activities according to the method of Toivonen and Sweeney (1998) with some modifications, while ascorbate peroxidase (APX, EC 1.11.1.11) activity was measured as described by Vicente *et al.* (2006). The 1-1-diphenyl-2-picrylhydrazyl (DPPH) radical scavenging activity was estimated following the method of Larrauri *et al.* (1998). They reported that ethanol-treated samples inhibited the decrease of CAT, APX, and SOD activities and the increase of POD activity,

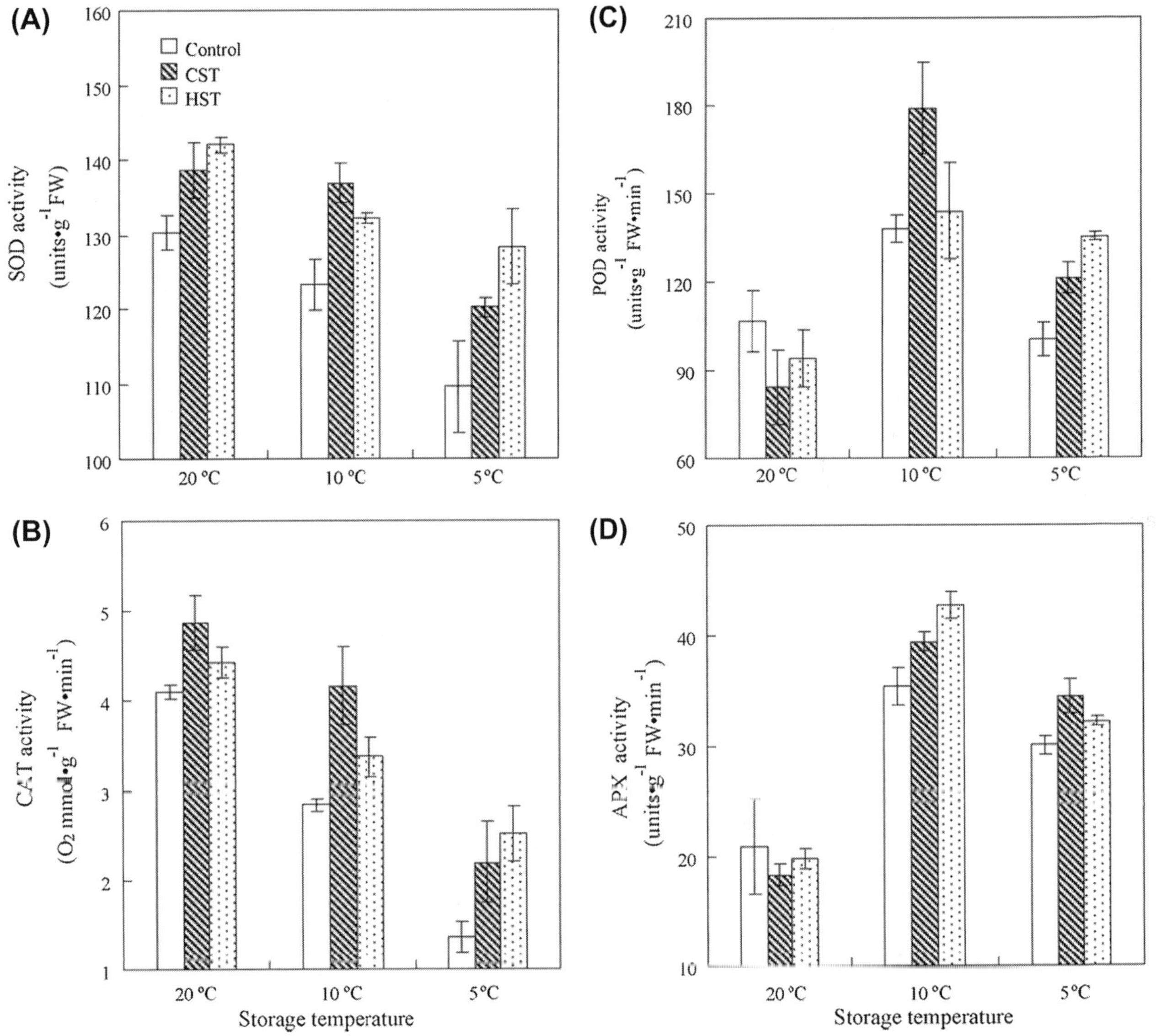

FIGURE 4.5 Changes in activities of superoxide dismutase (A), catalase (B), peroxidase (C), and ascorbate peroxidase (D) in broccoli with cold-shock and heat-shock treatment during storage at 20, 10, and 5°C. CST, cold-shock treatment (0°C, 1 h); HST, heat-shock treatment (55°C, 30 s). The vertical bars indicate SE ($n=5$). *Source: Zhang* et al. *2009.*

which resulted in higher CAT, APX, and SOD activities and lower POD activity in treatment than in control during storage. Treatment with ethanol significantly inhibited the yellowing of broccoli and maintained higher activities of CAT, APX, and SOD. These results suggest that the effect of ethanol in delaying the yellowing was correlated to enhanced antioxidant enzyme activity. The scavenging activities against DPPH in both ethanol-treated and control broccoli florets exhibited an increase at the first 3 days and then decreased during further storage. Ethanol treatment improved the DPPH radical scavenging activity, which was significantly higher ($P < 0.05$) in ethanol-treated broccoli florets than in the control during the storage.

Zhang *et al.*, (2009) studied the comparison of the antioxidant enzymes of broccoli after cold or heat-shock treatment at different storage temperatures. In their experiments, they treated broccoli samples both in HST (hot water-based), and CST (ice water-based) and stored at 5, 10, or 20°C in the dark and measured the enzyme activities such as superoxide dismutase (SOD), catalase (CAT), peroxidase (POD), and ascorbate peroxidase (APX). They reported that the lowering of storage temperature, SOD activity, and CAT activity decreased in broccoli (Figure 4.5A and 4.5B). There was no difference between CST and HST in SOD activity of broccoli stored at 20°C, and both treatments increased SOD activity in comparison with the controls during storage at 20°C. The SOD activity of CST was higher than that of HST, but the latter was higher than the controls at 10°C storage. However, SOD activity of HST was the highest and that of the controls was the lowest in broccoli stored at 5°C. The POD activity of broccoli stored at 10°C was higher than in broccoli stored at 20 or 5°C regardless of treatment (Figure 4.5C). Again, under storage at 10°C, a significant increase in APX activity of broccoli was observed, and highest APX activity was recorded in HST, while the APX activity of CST was also higher than in the controls. In contrast, at the 5°C storage temperature, the highest APX activity was observed in CST, followed by HST, and the lowest was recorded in controls (Figure 4.5D).

References

Barth, M.M., Kerbel, E.L., Perry, A.K., Schmidt, S.J., 1993. Modified atmosphere packaging affects ascorbic acid, enzyme activity and market quality of broccoli. J. Food Sci. 58, 140–143.

Barth, M.M., Zhuang, H., 1996. Packaging design affects antioxidant vitamin retention and quality of broccoli florets during postharvest storage. Postharvest Biol. Technol. 9, 141–150.

Benzie, I.E.F., Strain, J.J., 1996. The Ferric Reducing Antioxidant of Plasma (FRAP) as a measure of Antioxidant power: The FRAP assay,. Anal. Biochem. 239, 70–76.

Carvalho, Paulo de Tarso, Clemente, Edmar, 2004. The influence of the broccoli (*Brassica oleracea* var. *italica*) fill weight on postharvest quality. Cienc. Tecnol. Aliment. Campinas 24 (4), 646–651.

Chun, O.K., Kim, D.O., Smith, N., Schroeder, D., Han, J.T., Lee, C.Y., 2005. Daily consumption of phenolics and total antioxidant capacity from fruit and vegetables in the American diet. J. Sci. Food Agric. 85, 1715–1724.

Costa, M.L., Civello, P.M., Chaves, A.R., Martinez, G.A., 2005. Effect of hot air treatments on senescence and quality parameters of harvested broccoli (*Brassica oleracea* L. var. *italica*) heads. J. Sci. Food Agric. 85 (7), 1154–1160.

Day, G.L., Shore, R.E., Blot, W.J., McLaughlin, J.K., Austin, D.F., Greenberg, R.S., Liff, J.M., Preston-Martin, S., Sarkar, S., Schoenberg, J.B., Fraumeni Jr., J.F., 1994. Dietary factors and second primary cancers: a follow-up of oral and pharyngeal cancer patients. Nutr. Cancer. 21, 223–232.

Deschene, A., Paliyath, G., Lougheed, E.C., Dumbroff, E.B., Thompson, J.E., 1991. Membrane deterioration during postharvest senescence of broccoli tiorets: modification by temperature and controlled atmosphere storage. Postharvest Biol. Technol. 1, 19–31.

Forney, C.F., Jordan, M.A., 1999. Anaerobic production of methanethiol and other compounds by *Brassica* vegetables. HortScience 34 (4), 696–699.

Freed, M., 1966. Method of vitamin assay. Interscience Publication Inc., New York.

Hussein, A., Odumeru, J.A., Ayanbadejo, T., Faulkner, H., McNab, W.B., Hager, H., Szijarto, L., 2000. Effects of processing and packaging on vitamin C and beta-carotene content of ready-to-use (RTU) vegetables. Food Res. Int. 33 (2), 131–136.

Jeffery, E.H., Jarrell, V., 2001. Cruciferous vegetables and cancer prevention. In: Wildman, R.E.C. (Ed.), Handbook of nutraceuticals and functional foods. CRC Press, Boca Raton. FL, pp. 169–191.

Kidmose, U., Hansen, M., 1999. The influence of postharvest storage, temperature and duration on quality of cooked broccoli florets. J. Food Qual. 22 (2), 135–146.

Krinsky, N.I., Beecher, G.R., Burk, R.F., Chan, A.C., Erdman, J.W., Jacob, R.A., Jialal, I., Kolonel, L.N., Marshall, J.R., Taylor Mayne, S., Prentice, R.L., Schwarz, K.B., Steinberg, D., Traber, M.G., 2000. Dietary reference intakes for vitamin C, vitamin E, selenium and carotenoids. National Academy Press, Washington, DC.

Larrauri, J.A., Sanchez-Moreno, C., Saura-Calixto, F., 1998. Effect of temperature on the free radical scavenging capacity of extracts from red and white grape pomace peels. J. Agric. Food Chem. 46, 2694–2697.

Lebermann, K.W., Nelson, A.L., Steinberg, M.P., 1968. Postharvest changes of broccoli stored in modified atmospheres. I. Respiration of shoots and color of flouer heads. Food Technol. 22, 487–490.

Lee, S.K., Kader, A.A., 2000. Preharvest and postharvest factors influencing vitamin C content of horticultural crops. Postharvest Biol. Technol. 20 (3), 207–220.

Leja, M., Mareczek, A., Starzynska, A., Rozek, S., 2001. Antioxidant ability of broccoli flower buds during short-term storage. Food Chem. 72 (2), 219–222.

Li, JinCai, Maezawa, S., Nakano, K., 2004. Correlations between antioxidative enzyme activities and antioxidative substrates and senescence in broccoli (*Brassica oleracea* L.) flower buds at different storage temperatures. J. Japanese Soci. Horticultural Sci. 73 (4), 399–403.

Lorient, D., Linden, G., 1996. Bioquímica Agroindustrial. Acribia, Zaragoza.

Mapson, C.W., 1970. Vitamins in fruits: Stability of L-ascorbic acid. In: Biochemistry of fruits and their products. Academic Press, London, pp. 376–387.

Nath, A., Bagchi, B., Misra, L.K., Deka Bidyut, C., 2011. Changes in post-harvest phytochemical qualities of broccoli florets during ambient and refrigerated storage. Food Chem. 127, 1510–1514.

Nestle, M., 1998. Broccoli sprouts in cancer prevention. Nutr. Rev. 56, 127–130.

Podsedek, A., 2007. Natural antioxidants and antioxidant capacity of *Brassica* vegetables: a review. LWT- Food Sci. Technol. 40, 1–11.

Rai, D.R., Tyagi, S.K., Jha, S.N., Mohan, S., 2008. Qualitative changes in the broccoli (*Brassica oleracea italica*) under modified atmosphere packaging in perforated polymeric film. J. Food Sci.Technol. 45 (3), 247–250.

Rozek, S., Leja, m., Myczkowski, J., Mareczek, A., 1994. The effect of fertilization with different forms of nitrogen on greenhouse lettuce quality and its changes during storage. I Content of certains nutritive compounds. Folia Horticulturae. 6 (1), 41–45.

Srivastava, R.P., Kumar, S., 2002. Fruits and vegetables preservation—Principles and Practices, International book Distributing Co. pp 353–363.

Takeda, Y., Yoza, K.I., Nogata, Y., Ohta, H., 1993. Effects of storage temperatures on polyamine content of some leafy vegetables. J. Japanese Soci. Horticultural Sci. 62, 425–430.

Toivonen, P.M.A., Sweeney, M., 1998. Differences in chlorophyll loss at 13° C for two broccoli (*Brassica oleracea* L.) cultivars associated with antioxidant enzyme activities. J. Agric. Food Chem. 46, 20–24.

Vallejo, F., Tomas-Barberan, F.A., Garcia-Viguera, C., 2003. Health-promoting compounds in broccoli as influenced by refrigerated transport and retail sale period. J. Agric. Food Chem. 51, 3029–3034.

Verhoeven, D.T.H., Goldbohm, R.A., Van Poppel, G., Verhagen, H., Van den Brandt, P.A., 1996. Epidemiological studies on *Brassica* vegetables and cancer risk. Cancer Epidemiology, Biomarkers & Prevention 5, 733–748.

Vicente, A.R., Martinez, G.A., Chaves, A.R., Civello, P.M., 2006. Effect of heat treatment on strawberry fruit damage and oxidative metabolism during storage. Postharvest Biol. Technol. 40,116–122.

Wang, C.Y., Hruschka, H.W., 1977. Quality maintenance in polyethylene-packaged broccoli. Marketing Res. Rept, USDA, Washington, DC.

Wang, Q., Win, K., 2002. Effect of 1-MCP on storage life, quality and antioxidant enzyme activities of broccoli. J. Zhejiang Univ. Agric. Life Sci. 28 (5), 507–512.

Winkler, S., Faragher, J., Franz, P., Imsic, M., Jones, R., 2007. Glucoraphanin and flavonoid levels remain stable during simulated transport and marketing of broccoli (*Brassica oleracea* var. *italica*) heads. Postharvest Biol. Technol. 43, 89–94.

Xu, F., Chen, X., Jin, P., Wang, X., Wang, J., Zheng, Y., 2012. Effect of ethanol treatment on quality and antioxidant activity in postharvest broccoli florets. Eur. Food Res. Technol 235, 793–800.

Zhang, Z., Nakano, K., Maezawa, S., 2009. Comparison of the antioxidant enzymes of broccoli after cold or heat shock treatment at different storage temperatures. Postharvest Biol. Technol. 54, 101–105.

CHAPTER

5

Blanching as a Treatment Process: Effect on Polyphenol and Antioxidant Capacity of Cabbage

Nissreen Abu-Ghannam, Amit Kumar Jaiswal

School of Food Science and Environmental Health, Dublin Institute of Technology, Dublin, Ireland

CHAPTER POINTS

- Cabbage is grown in temperate climate areas and is available throughout the year.
- Cabbage is an excellent source of several phytochemicals such as, phenolic acids, flavonoids, and glucosinolates, with more than 20 phenolic acids being identified. Phytochemicals are characterized as being rich sources of antioxidants.
- The photochemical composition in cabbage varies with cultivar type, climatic conditions, maturity at harvest, and storage conditions.
- Blanching is an essential step in cabbage processing in order to ensure both safety and quality requirements for either immediate consumption or further storage. In addition, blanching of cabbage renders this vegetable suitable for human consumption with respect to color, texture, and flavor.
- Blanching methods rely on the use of either conventional hot water or steam blanchers and utilize temperatures in the range of 80–100°C for periods of time not exceeding 15 min. Time–temperature combinations and blanching methods employed are controlled to a great extent by the variety of cabbage. Conventional hot water blanching continues to be the most preferred industrial method due to low capital costs and blanching uniformity of the product.
- Cabbage blanching causes significant losses in its photochemical content leading to substantial reductions in its antioxidant potential. Losses could be reduced by using steam blanchers; however, losses in conventional blanching could be minimized by reducing the water to cabbage ratio.
- Blanching time is a greater factor in determining antioxidant losses than temperature, with up to 70% losses in the antioxidant capacity of cabbage taking place within the initial phase of blanching (first 6 min), regardless of the applied blanching temperature.
- There is a need for standardization of antioxidant capacity methods in order to provide meaningful and useful comparisons for the effects of blanching on the content of polyphenols in cabbage and their respective antioxidant activities.
- While the attainment of specified safety and quality parameters in food processing are of paramount importance, it is strongly suggested that specific nutritional components of significant relevance to a product (in this case cabbage being an excellent source of antioxidants) should also be included in the processing criteria parameters.

Processing and Impact on Active Components in Food
http://dx.doi.org/10.1016/B978-0-12-404699-3.00005-6

INTRODUCTION

The family Brassicaceae or Cruciferae is a large group with 3500 species and lies within 350 genera including Arabidopsis, *Brassica, Camelina, Crambe,* Descurainia, *Glaucocarpum, Sinapis,* and *Thlaspi.* The genus Brassica is the most important within the Brassicaceae family which includes some crops and species of valuable economic importance in the human diet worldwide such as *Brassica oleracea* L., *Brassica napus* L., and *Brassica rapa* L. *B. oleracea* includes several types of edible crops such as cauliflower, broccoli, kohlrabi, kale, cabbage, and Brussel sprouts. Among these, cabbage is one of the oldest vegetables and has been used in culinary applications in Europe for over 4000 years.

Cabbage is grown in most major temperate vegetable growing areas and is available year-round. It can be classified as green (York Cabbage, Round Cabbage, Savoy Cabbage), white (Coleslaw cabbage), and red or purple, although color could range from nearly white to reddish-purple. The cabbage shape is broadly divided into two categories as round or conical. York cabbage has mainly conical heads and the edible head is a terminal bud composed of many tightly packed overlapping leaves. White cabbage is light- to dark-green in color and has thick leaves with round, tightly wrapped heads and with a smooth plain-leaved "ball head."

Epidemiological and clinical investigations thus far have demonstrated that a high consumption of fruits and vegetables, including cabbage, is positively associated with the prevention of cardiovascular diseases, cancer, aging, diabetes, osteoporosis, hypertension, and stroke (Maritess *et al.*, 2005). Recent research indicates that the health protective effects associated with Brassica vegetables such as cabbage could be attributed to the presence of several types of antioxidants.

ANTIOXIDANTS IN CABBAGE

Cabbage is rich in several nutritive and non-nutritive bioactive compounds, which are well recognized for their antioxidant properties and potential health benefits (Jaiswal *et al.*, 2012a,b). These include polyphenols such as flavonoids, hydroxycinnamic acid derivatives, hydroxybenzoic acid derivatives, carotenoids, and vitamins such as ascorbic acid and α-tocopherol (Jaiswal *et al.*, 2011; Podsedek *et al.*, 2006). The antioxidant properties of polyphenols are suggested to play a positive role in the alleviation of oxidative stress and the prevention of free-radical mediated diseases (Halvorsen *et al.*, 2002). In addition, polyphenols have been associated with a wide range of health promoting properties, including anti-obesity, anti-diabetic, anti-hypertensive, anti-mutagenic, and anti-cancer properties (Pandey and Rizvi, 2009; Varoni *et al.*, 2012). It is well established that the content of antioxidants in cabbage varies with cultivar type, climatic conditions, maturity at harvest, and storage conditions (Kusznierewicz *et al.*, 2008; Šamec *et al.*, 2011; Singh *et al.*, 2007).

Cabbage as a Source of Polyphenols

Polyphenols are a large group of secondary metabolites that are widespread in the plant kingdom, including cabbage, and are well known for their free radicals scavenging capacity, metal chelating capacity and lipid peroxidation properties (Jaiswal *et al.*, 2011; Jaiswal *et al.*, 2012a,b). Polyphenols from various cabbage varieties have been characterized in a number of studies. Nielsen *et al.* (1998) reported that white cabbage leaves contain more than 20 types of phenolic compounds, including glucosides of kaempferol and quercetin with/without further acylation with hydrocinnamic acid. York cabbage (*Brassica oleracea* var. *capitata* alba subvar. *conica*) and white cabbage (*Brassica oleracea* var. *capitata*) have significant amounts of phenolic acids such as hydroxycinnamic acid derivatives and hydroxybenzoic acid derivatives; however, the concentration of these compounds are cultivar dependent (Jaiswal *et al.*, 2011). Seventeen phenolic compounds were identified in Tronchuda cabbage (*Brassica oleracea* L. var. *costata* DC) mainly belonging to derivatives of quercetin and kaempferol while anthocyanin derivatives were associated with colored varieties of cabbage such as red cabbage (Ferreres *et al.*, 2006; McDougall *et al.*, 2007; Scalzo *et al.*, 2008).

Antioxidant Properties of Cabbage

Free radicals play an important role in causation and generation of several diseases such as cardiovascular diseases, inflammatory diseases, neurodegenerative disorders, cancer, and aging because they are highly reactive to biomolecules and thus cause damage to DNA, proteins, carbohydrates, and lipids (Gutteridge and Halliwell, 1993). Diets rich in foods containing antioxidants such as cabbage can exert health benefits by inhibiting these free radicals. Antioxidants are defined as "any substance that, when present at low concentrations compared with that of an oxidizable substrate, significantly delays or inhibits oxidation of that substrate" (Halliwell *et al.*, 1995). Several methods have been recommended for antioxidant capacity estimation such as 2,2-diphenyl-1-picrylhydrazyl (DPPH) radical scavenging activity, hydroxyl ($OH^{\bullet}$) radical scavenging activity, hydrogen peroxide (H_2O_2) radical scavenging activity, ferric reducing antioxidant power (FRAP), oxygen radical absorbance capacity (ORAC), 2,2-azinobis-3- ethylbenzothiazoline-6-sulfonic acid (ABTS) radical scavenging activity, superoxide anion radical (O_2^-), peroxyl radical ($ROO^{\bullet}$) scavenging capacity, and lipid peroxides inhibitory ability. Variations within these methods are attributed to the reaction mechanisms and type of free radicals being targeted.

Numerous studies have reported on the antioxidant potential of cabbage (Amin and Lee, 2005; Jacob *et al.*, 2011; Jaiswal *et al.*, 2011; Jaiswal *et al.*, 2012a,b; Kusznierewicz *et al.*, 2008; Šamec *et al.*, 2011). Red cabbage exhibited 2.2- to 5-fold higher antioxidant capacity than Savoy and white cabbage (Podsedek *et al.*, 2006). Jacob *et al.* (2011) reported that red cabbage contains a higher antioxidant capacity as compared to its green counterpart in different antioxidant systems; similarly green cabbage showed a higher antioxidant capacity than white cabbage (Jaiswal *et al.*, 2011). In general, red cabbage have the highest antioxidant activity, followed by green cabbage, mustard cabbage, Chinese cabbage and Chinese white cabbage (Amin and Lee, 2005) which could be attributed to the presence of different antioxidant components such as phenolic compounds and also being influenced by the genotype, environment, and growth conditions (Singh *et al.*, 2007).

Typically, most cabbage varieties undergo some form of heat treatment for a certain time period prior to consumption mainly to attain an acceptable sensory texture. Antioxidants are widely known for their heat labile properties suggesting potential losses in their bioactivity and associated health-promotion properties upon the application of heat treatments. It is relevant then to evaluate the effects of heat processing conditions on the antioxidant content and capacity of cabbage and to propose suggestions to minimize heating effects. Such findings would be of importance not only for the food industry, but also from a public health perspective as health professionals strive to enhance population health using practical and feasible approaches.

Blanching

Blanching is a short heat treatment that is typically applied to vegetables prior to further processing with the aim of enhancing both safety and quality attributes. It imparts benefits such as the destruction of surface microflora on vegetables and the enhancement of the color, texture and also the keeping quality of vegetable products (Jaiswal *et al.*, 2012c). In addition, blanching is essential for vegetable products destined for further storage such as freezing or drying in order to inactivate certain enzymes including lipoxygenase, polyphenoloxidase, polygalacturonase, and chlorophyllase which are associated with losses in quality and nutritional properties. Apart from blanching, other processing methodologies of vegetables, including drying and freezing, are insufficient to inactivate these enzymes thus leading to deterioration in texture, color, and flavor during storage.

The quality of blanched products depends significantly on the time–temperature combinations of blanching and also on the vegetable type. Under-blanching speeds up the activity of enzymes and is worse than no blanching (Jaiswal *et al.*, 2012c). Over-blanching causes losses in texture, color, phytochemicals and minerals. Typically, industrial blanching processes utilize temperatures ranging from 70 to 95°C and times are usually within 10 min (Morales-Blancas *et al.*, 2006); whereas for domestic purposes vegetables are generally blanched, or given an extended blanching period which ultimately leads to cooking, for 10–15 min in water at temperatures ranging from 95–100°C.

Generally, blanching is carried out by the application of a wet medium such as steam or hot water in order to provide uniform heating and a high-heat transfer rate. Both in domestic and industrial processing, several blanching methods may be employed such as conventional water blanching, microwave, or steam blanching; the regime being dictated by the nature of the raw material and the desired properties of the final product. Traditionally, blanching is carried out either at a low temperature (55–75°C) for long-time, typically referred to as LTLT or high-temperature short-time (80–100°C) for less than 10 min, referred to as HTST depending upon the type of vegetable (Abu-Ghannam and Crowley, 2006). Conventional water blanching usually imparts more uniform processing. However, prolonged water blanching results in considerable losses in phytochemicals and antioxidant properties (Jaiswal *et al.*, 2012c).

Microwave heating is three- to five-times faster than conventional heating and relies on the application of dielectric heating. This is accomplished by using microwave radiation to heat water and other polarized molecules within the food, leading to heat generation in the entire volume at the same rate due to internal thermal dissipation of water molecule vibrations in the food (Kamel and Stauffer, 1993). It has advantages over conventional heating methods such as precision timing, speed, and energy saving.

Steam blanching is generally carried out in a steam blancher where the vegetable product is exposed directly to a food-grade steam typically at a temperature close to 100°C. Steam blanching results in minimum losses in phytochemicals and antioxidant capacity (Faller and Fialho, 2009; Podsędek *et al.*, 2008; Turkmen *et al.*, 2005; Wachtel-Galor *et al.*, 2008); furthermore, it requires less time than conventional blanching because the heat transfer coefficient of condensing steam is greater than that of hot water and it is proven to be comparatively economical as it saves energy (De Corcuera *et al.*, 2004).

Effects of Blanching on the Antioxidant Content of Cabbage

A number of studies have investigated the effect of blanching on the polyphenolic content of cabbage (Table 5.1). Different effects have been reported depending on differences in process conditions and systems applied in

TABLE 5.1 Summary of Literature Findings on the Effects of Blanching on Losses of Cabbage Polyphenol Content

Cabbage Type	Mode of Blanching	Conditions	Reported Outcomes	Reference
York cabbage	Conventional blanching	14 min (80–100°C)	75–80% reduction in total phenolic content	(Jaiswal *et al.*, 2012c)
Red cabbage	Conventional blanching	1 min (94–96°C)	43% reduction in total phenolic content	(Volden *et al.*, 2008)
Red cabbage	Conventional blanching	15 min	37% reduction in total phenolic content	(Amin and Lee, 2005)
Chinese white cabbage	Conventional blanching	15 min	82% decrease in total phenolic content	(Amin and Lee, 2005)
Mustard cabbage	Conventional blanching	15 min	54% reduction in total phenolic content	(Amin and Lee, 2005)
Chinese Cabbage	Conventional blanching	15 min	23% reduction in total phenolic content	(Amin and Lee, 2005)
York cabbage	Conventional blanching	14 min (80–100°C)	74–78% reduction in total flavonoid content	(Jaiswal *et al.*, 2012c)
Red cabbage (Koda and Kissendrup)	Conventional blanching	20 min	57–71% reduction in Anthocyanins	(Podsędek *et al.*, 2008)
Red cabbage (Koda and Kissendrup)	Conventional blanching	20 min	60–70% reduction in hydroxybenzoic acids	(Podsędek *et al.*, 2008)
Red cabbage (Koda and Kissendrup)	Conventional blanching	20 min	20–27% reduction in hydroxycinnamic acids	(Podsędek *et al.*, 2008)
Red cabbage	Conventional blanching	1 min (94–96°C)	59% reduction in total monomeric anthocyanin content	(Volden *et al.*, 2008)
Red cabbage	Boiling	10 min	16% reduction in total phenolic content	(Volden *et al.*, 2008)
Organic and conventional retail vegetables (white cabbage)	Boiling	5 min	0–14 % reduction in soluble polyphenols	(Faller and Fialho, 2009)
Organic and conventional retail vegetables (white cabbage)	Boiling	5 min	60–65% reduction in hydrolyzable polyphenols	(Faller and Fialho, 2009)
Red cabbage	Boiling	10 min	49% reduction in total monomeric anthocyanin content	(Volden *et al.*, 2008)
Red cabbage	Steaming	10 min	No reduction in total phenolic content	(Volden *et al.*, 2008)
Organic and conventional retail vegetables (white cabbage)	Steaming	10 min	50–64% reduction in soluble polyphenols	(Faller and Fialho, 2009)
Organic and conventional retail vegetables (white cabbage)	Steaming	10 min	0–63% reduction in hydrolyzable polyphenols	(Faller and Fialho, 2009)
Red cabbage (Koda and Kissendrup)	Steaming	20 min	25–26% reduction in Anthocyanins	(Podsędek *et al.*, 2008)
Red cabbage (Koda and Kissendrup)	Steaming	20 min	35–40% reduction in hydroxybenzoic acids	(Podsędek *et al.*, 2008)
Red cabbage (Koda and Kissendrup)	Steaming	20 min	10–13% increase in hydroxycinnamic acids	(Podsędek *et al.*, 2008)
Red cabbage	Steaming	10 min	21% reduction in total monomeric anthocyanin content	(Volden *et al.*, 2008)
Organic and conventional retail vegetables (white cabbage)	Microwave	3.5 min at 2450 W	39–41% reduction in soluble polyphenols	(Faller and Fialho, 2009)
Organic and conventional retail vegetables (white cabbage)	Microwave	3.5 min at 2450 W	44–45% reduction in hydrolyzable polyphenols	(Faller and Fialho, 2009)

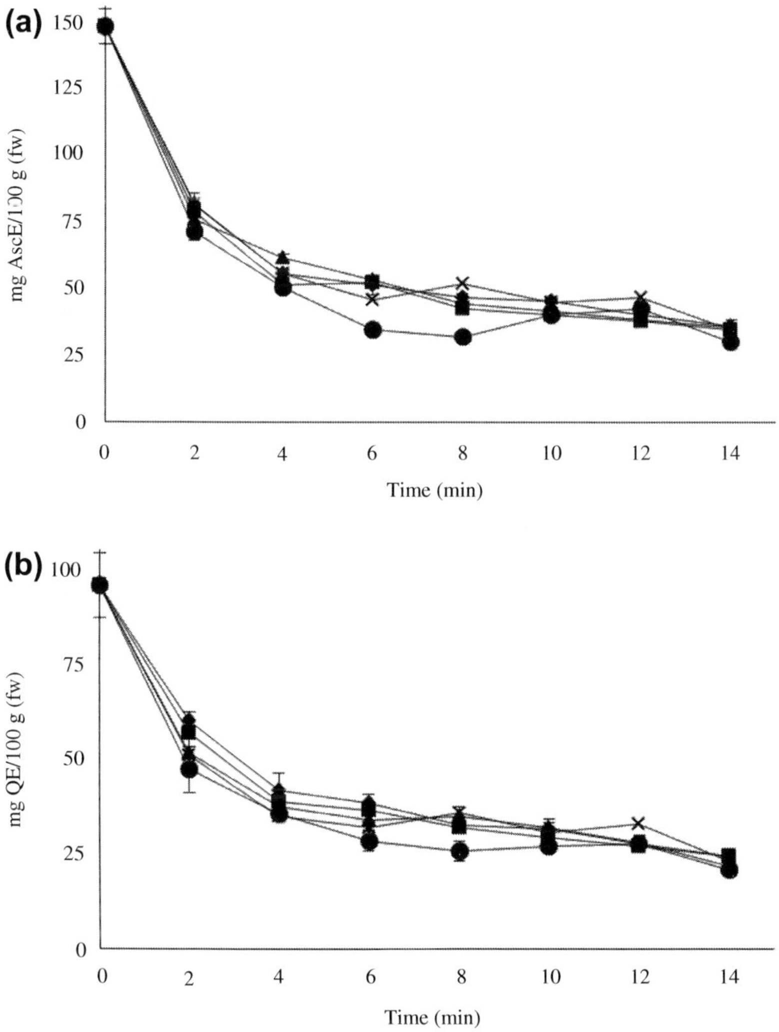

FIGURE 5.1 Effect of blanching temperature [80 (◆), 85(■), 90 (▲), 95 (✕), and 100°C (●)] and time combination on (a) total phenolic content and (b) total flavonoid content of York cabbage. *Adapted from Jaiswal* et al. *(2012c).*

the blanching process. A detailed study was carried out by Jaiswal *et al.* (2012c) to evaluate the effect of conventional water blanching on the polyphenols of York cabbage. A range of blanching temperatures (80, 85, 90, 95, and 100°C) for 0–14 min with a 2 min interval was studied with the aim of including both low and high blanching temperatures (Figure 5.1). Reductions in the total phenolic content up to 45% were reported at the lower blanching temperatures (80–90°C) within 2 min of blanching and reached up to 50% at high blanching temperatures (95–100°C) for the same blanching period of 2 min. Degradation in total phenolic content continued up to 76% within 6 min at high blanching temperatures; however, no significant reductions in total phenolic content were reported after this for up to 14 min, a stage at which the cabbage was considered to be completely processed. (Figure 5.1a). Jaiswal *et al.* (2012c) also reported similar observations for the effects of blanching time on the degradation of total flavonoid content of York cabbage (Figure 5.1b). These findings highlight the deleterious effects of blanching in compromising the significance of cabbage in being one of the richest vegetable sources of antioxidants.

Amin and Lee (2005) applied conventional blanching methods for red, green, mustard, Chinese, and Chinese white cabbage for 5, 10, and 15 min. A significant reduction in the total phenolic content was observed irrespective of cabbage type. Mustard cabbage showed a steady trend in phenolic contents loss, with up to 25, 51, and 54% reported to be lost after blanching for 5, 10, and 15 min, respectively. Similarly, Chinese white cabbage was blanched for 5, 10, and 15 min and correspondingly 58, 63, and 82% reductions in total phenolic content were reported (Amin and Lee, 2005). The previous studies indicate that blanching has a detrimental effect on the phenolic content of cabbage with losses in general exceeding 50% of the initial content in raw cabbage. Losses in polyphenol content are attributed to the disruption of the plant tissue due to the heating effect, leading to polyphenols leaching out into the blanching water environment (Gonçalves *et al.*, 2010). Furthermore, the reciprocal interconversion of insoluble phenolics into more soluble forms can also occur, which may lead to additional losses in polyphenols.

Podsędek *et al.* (2008) examined the effects of water to cabbage ratio and blanching time on the polyphenol content of Kissendrup and Koda red cabbage varieties upon the application of conventional water blanching. Reductions of up to 50% in water volume led to better retention of phenolics as compared to when blanching was carried out using a 2:1 ratio of water to cabbage. Conversely, shortening the conventional blanching time from 20 to 10 min increased the retention of phenolics by only 3.8–6.7%. This suggests that polyphenol losses upon blanching could be attributed to their respective solubility and stability, which would be highly influenced by the type of the blanching environment (steam or hot water) and also its corresponding volume to facilitate the leaching and transfer of phenolic compounds from the cell interior to the blanching environment. Also it is worth noting that free polyphenols leach out faster in the water when compared with bound polyphenols. As blanching temperatures increase, a higher loss in phenolic compounds is expected; however, the effect of blanching time does not necessarily follow a similar trend. It seems that for a given blanching treatment, losses in phenolic compounds initially increase as blanching time proceeds, but only up to a certain point, after which a plateau state is attained and longer blanching times do not correspond to higher losses of phenolic compounds. This highlights the role of heat in degrading cabbage cellular material leading to the leaching of polyphenols. A higher level of phenolic losses would require the application of a higher temperature rather than a longer blanching time for a given blanching treatment, thus emphasizing the role of heat in causing phenolic mass transfer. Conversely, blanching time profiling with respect to polyphenol losses would be of significant importance in identifying those times that correspond to minimum polyphenol losses while also attaining appropriate quality attributes such as color and texture. At a domestic or industrial level, the first 6–8 min of blanching are detrimental with respect to a number of quality parameters including phenolic content, and there is a need for more detailed studies at both processing and consumer levels to examine whether the polyphenols, with their established health benefits, are being largely compromised by perceived consumer quality attributes for attaining certain textures and colors of vegetables upon blanching.

Effects of Blanching on the Antioxidant Capacity of Cabbage

A summary of the effects of blanching on the antioxidant capacity of a range of cabbage varieties is illustrated in Table 5.2. There were variations in the applied antioxidant methodologies in those reported studies suggesting that specific inferences regarding the antioxidant capacity, for a given blanching treatment and a cabbage variety, might require more in depth considerations. In general, conventional blanching of cabbage can result in high reductions in antioxidant properties.

Jaiswal *et al.* (2012) studied the effects of a range of blanching times and temperatures on the antioxidant capacity of York cabbage by reporting on the H_2O_2 radical scavenging capacity (Figure 5.2a), lipid peroxidation ability (Figure 5.2b), ferric reducing antioxidant potential, and DPPH free radical scavenging capacity, which collectively are methods employed to characterize properties of antioxidants. The study highlighted that maximum losses of up to 70% in antioxidant capacity were observed in the initial period of blanching (mainly within 6 min) with the blanching temperature playing a minor effect. For example, lower blanching temperatures in the range of 80–90°C resulted in up to 60% losses in H_2O_2 scavenging capacity within 6 min as compared to 68% for the same period when blanching was carried out at the higher temperatures of 95–100°C. Only minimal losses were reported when blanching time was extended beyond the first 6 min as seen in Figure 5.2a.

A similar pattern was also observed with respect to reductions in the lipid peroxidation inhibitory ability of blanched York cabbage with losses ranging between 60 and 67%, being slightly high for the higher blanching temperatures but without a significant difference. The same trend was also observed for the other measured antioxidant systems such as DPPH free radical scavenging capacity and ferric reducing antioxidant potential.

The extent of antioxidant losses during cabbage blanching can also be attributed to the cabbage variety. Cabbage varieties have different leaf morphology with respect to color, texture, and thickness. It is expected then, that there will be variations in the effects of blanching on cellular disruptions and the consequences of polyphenol losses and reductions in antioxidant capacity. Amin *et al.* (2005) studied the effects of blanching at 98°C for red cabbage, Chinese cabbage, mustard cabbage, and Chinese white cabbage. Losses in β-carotene bleaching activity were reported to be highest in Chinese cabbage (40%) within 15 min of blanching, followed by Chinese white cabbage (19%), mustard cabbage (9%), and red cabbage (4%). However, in the case of DPPH scavenging capacity, red cabbage showed maximum activity losses (40%), followed by Chinese cabbage (38%), and mustard cabbage (23%).

Overall, there is an obvious correlation between antioxidant content and antioxidant capacity or activity of blanched cabbage regardless of the variety. The general trend observed is that losses are more pronounced within the initial phase of blanching, and the factors influencing such losses are blanching methodology, cabbage variety,

TABLE 5.2 Summary of Literature Findings on the Effect of Blanching on the Antioxidant Properties of Cabbage

Cabbage Type	Mode of Blanching	Conditions	Consequence	Reference
York cabbage	Conventional blanching	14 min (80–100°C)	80–82% reduction in DPPH scavenging capacity	(Jaiswal *et al.*, 2012c)
Red cabbage	Conventional blanching	15 min	40% reduction in DPPH scavenging capacity	(Amin and Lee, 2005)
Chinese cabbage	Conventional blanching	15 min	38% reduction in DPPH scavenging capacity	(Amin and Lee, 2005)
Mustard cabbage	Conventional blanching	15 min	23% reduction in DPPH scavenging capacity	(Amin and Lee, 2005)
Chinese white Cabbage	Conventional blanching	15 min	11% reduction in DPPH scavenging capacity	(Amin and Lee, 2005)
York cabbage	Conventional blanching	14 min (80–100°C)	73–75% reduction in H_2O_2 scavenging capacity	(Jaiswal *et al.*, 2012c)
York cabbage	Conventional blanching	14 min (80–100°C)	72–78% reduction in lipid peroxidation (LPO) in a haemoglobin-induced linoleic acid system	(Jaiswal *et al.*, 2012c)
Red cabbage	Conventional blanching	1 min (94–96°C)	42% reduction in ferric reducing antioxidant potential	(Volden *et al.*, 2008)
York cabbage	Conventional blanching	14 min (80–100°C)	67–75% reduction in ferric reducing antioxidant potential	(Jaiswal *et al.*, 2012c)
Red cabbage	Conventional blanching	1 min (94–96°C)	51% reduction in oxygen radical antioxidant capacity	(Volden *et al.*, 2008)
Red cabbage (Koda and Kissendrup)	Conventional blanching	20 min	38–42% reduction in ABTS radical scavenging activity	(Podsędek *et al.*, 2008)
Red cabbage	Conventional blanching	15 min	4% reduction in β-carotene bleaching activity	(Amin and Lee, 2005)
Chinese cabbage	Conventional blanching	15 min	40% reduction in β-carotene bleaching activity	(Amin and Lee, 2005)
Mustard cabbage	Conventional blanching	15 min	9% reduction in β-carotene bleaching activity	(Amin and Lee, 2005)
Chinese white Cabbage	Conventional blanching	15 min	19% reduction in β-carotene bleaching activity	(Amin and Lee, 2005)
Organic and conventional retail vegetables (white cabbage)	Boiling	5 min	34–69% reduction in DPPH scavenging capacity	(Faller and Fialho, 2009)
Red cabbage	Boiling	10 min	17% reduction in ferric reducing antioxidant potential	(Volden *et al.*, 2008)
Red cabbage	Boiling	10 min	19% reduction in oxygen radical antioxidant capacity	(Volden *et al.*, 2008)
Organic and conventional retail vegetables (white cabbage)	Steaming	10 min	45–81% reduction in DPPH scavenging capacity	(Faller and Fialho, 2009)
Red cabbage	Steaming	10 min	2% reduction in ferric reducing antioxidant potential	(Volden *et al.*, 2008)
Red cabbage	Steaming	10 min	13% reduction in oxygen radical antioxidant capacity	(Volden *et al.*, 2008)
Red cabbage (Koda and Kissendrup)	Steaming	20 min	12–20% reduction in ABTS radical scavenging activity	(Podsędek *et al.*, 2008)
Organic and conventional retail vegetables (white cabbage)	Microwave	3.5 min at 2450 W	8–47% reduction in DPPH scavenging capacity	(Faller and Fialho, 2009)

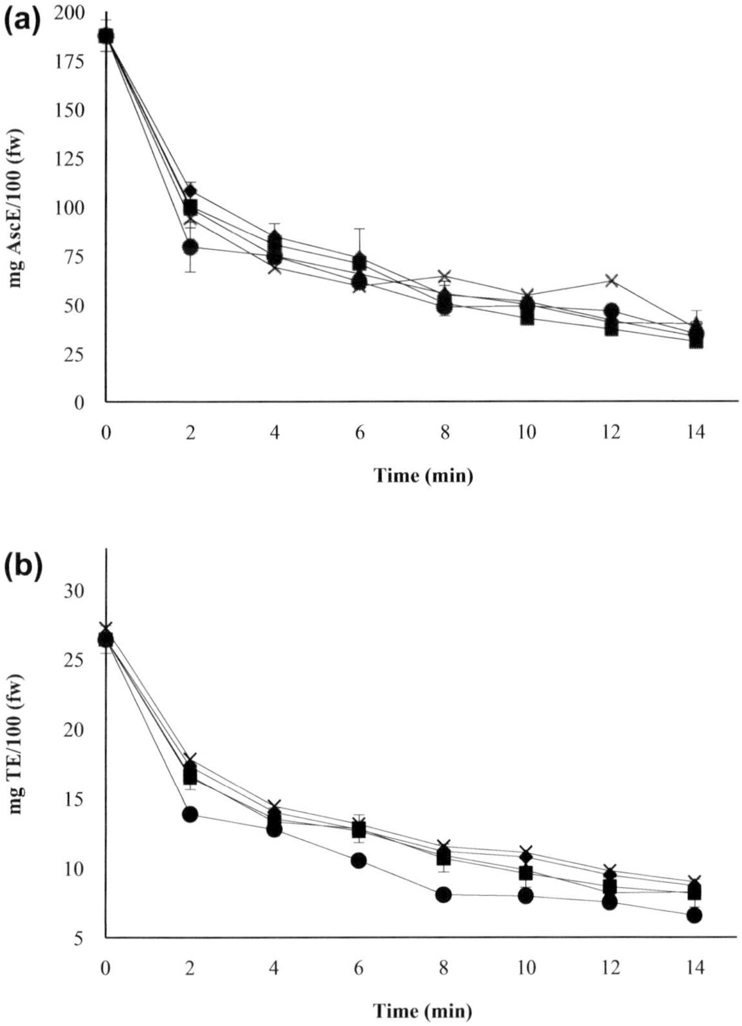

FIGURE 5.2 Effect of blanching temperature [80 (-◆-), 85(-■-), 90 (-▲-), 95 (-✳-), and 100°C (-●-)] and time combination on antioxidant capacity (expressed as relative standard equivalent) on (a) H_2O_2 radical scavenging capacity (b) Lipid peroxidation inhibitory ability of York cabbage. *Adapted from Jaiswal* et al. *(2012c).*

and blanching time. The blanching temperature effect is not of significant influence, particularly if the temperature applied is within the range of 80–100°C. From the perspective of food safety and quality retention requirements, higher blanching temperatures would be recommended in order to eliminate surface microflora and to inactivate enzymes associated with quality deterioration during further storage such as freezing.

Steam blanching produces smaller volumes of waste and lower disposal charges than water blanchers; however, uneven blanching can occur if the vegetable product is piled too high in addition to possibilities of limited cleaning of vegetable surfaces which is essential for the elimination of surface microflora. Conventional blanching using hot water operates at lower capital costs and better energy efficiency as compared to steam blanching. Conversely, the cost of large volumes of potable water and dilute effluent treatment are drawbacks to this processing method.

As cabbage, with its wide available varieties, is considered a rich source of antioxidants, it is thus important to provide accurate quantifications of antioxidant capacity losses during blanching. However, it should be noted that the methodologies and procedures employed to estimate antioxidant capacities vary considerably in the reported literature and there is a need for standardization in order for comparisons to be meaningful and useful to both the consumer and the food industry.

The rationale for the industrial treatments adopted for vegetable blanching has been mainly based on achieving a high-quality product with acceptable sensory attributes from a consumer perspective. The findings presented in this chapter regarding cabbage blanching highlight the importance of including the nutritional aspect, specifically antioxidant content and activity, when specifying the blanching methodologies and time–temperature combinations employed. This is particularly important as the majority of cabbage varieties require only blanching as a processing step prior to consumption. Due to the substantial losses in antioxidant capacity that occur as very early stages of blanching, it is strongly suggested that a review of the sensory attributes resulting from blanching should be in parallel consideration with antioxidant losses.

References

Abu-Ghannam, N., Crowley, H., 2006. The effect of low temperature blanching on the texture of whole processed new potatoes. J. Food Eng. 74, 335–344.

Amin, I., Lee, W.Y., 2005. Effect of different blanching times on antioxidant properties in selected cruciferous vegetables. J. Sci. Food Agr 85, 2314–2320.

De Corcuera, J.I.R., Cavalieri, R.P., Powers, J.R., 2004. Blanching of Foods. In: Heldman, D. (Ed.), Encyclopedia of Agricultural, Food, and Biological Engineering. Marcel Dekker, Inc., New York, pp. 1–5.

Faller, A., Fialho, E., 2009. The antioxidant capacity and polyphenol content of organic and conventional retail vegetables after domestic cooking. Food Res. Int. 42, 210–215.

Ferreres, F., Sousa, C., Vrchovská, V., Valentão, P., Pereira, J.A., Seabra, R.M., Andrade, P.B., 2006. Chemical composition and antioxidant activity of Tronchuda cabbage internal leaves. Eur. Food Res. Technol. 222, 88–98.

Gonçalves, E.M., Pinheiro, J., Abreu, M., Brandão, T.R.S., Silva, C.L.M., 2010. Carrot (*Daucus carota* L.) peroxidase inactivation, phenolic content and physical changes kinetics due to blanching. J. Food Eng. 97, 574–581.

Gutteridge, J.M.C., Halliwell, B., 1993. Free radicals in disease processes: A compilation of cause and consequence. Free Radical Res. 19, 141–158.

Halliwell, B., Gutteridge, J.M.C., Guohua, C., Cutler, R., 1995. The definition and measurement of antioxidants in biological systems. Free Radical Biol. Med. 18, 125–126.

Halvorsen, B.L., Holte, K., Myhrstad, M.C.W., Barikmo, I., Hvattum, E., Remberg, S.F., Wold, A.B., Haffner, K., Baugerød, H., Andersen, L.F., 2002. A systematic screening of total antioxidants in dietary plants. J. Nutr. 132, 461–471.

Jacob, J.A., Mahal, H., Mukherjee, T., Kapoor, S., 2011. Free radical reactions with the extract of Brassica family. Food Chem. 129, 1132–1138.

Jaiswal, A.K., Abu-Ghannam, N., Gupta, S., 2012a. A comparative study on the polyphenolic content, antibacterial activity and antioxidant capacity of different solvent extracts of *Brassica oleracia* vegetables. Int. J. Food Sci. Technol. 47, 223–231.

Jaiswal, A.K., Rajauria, G., Abu-Ghannam, N., Gupta, S., 2012b. Effect of different solvents on polyphenolic content, antioxidant capacity and antibacterial activity of Irish York cabbage. J. Food Biochem. 36, 344–358.

Jaiswal, A.K., Gupta, S., Abu-Ghannam, N., 2012c. Kinetic evaluation of color, texture, polyphenols and antioxidant capacity of Irish York cabbage after blanching treatment. Food Chem. 131, 63–72.

Jaiswal, A.K., Rajauria, G., Abu-Ghannam, N., Gupta, S., 2011. Phenolic composition, antioxidant capacity and antibacterial activity of selected Irish Brassica vegetables. Nat. Prod. Commun. 6, 1299–1304.

Kamel, B.S., Stauffer, C.E., 1993. Advances in baking technology. Blackie Academic and Professional., New York.

Kusznierewicz, B., Bartoszek, A., Wolska, L., Drzewiecki, J., Gorinstein, S., Namieśnik, J., 2008. Partial characterization of white cabbages (*Brassica oleracea var. capitata f. alba*) from different regions by glucosinolates, bioactive compounds, total antioxidant activities and proteins. LWT-Food Sci. Technol. 41, 1–9.

Maritess, C., Small, S., Waltz-Hill, M., 2005. Alternative nutrition therapies in cancer patients. Sem. Oncol. Nurs. 21, 173–176.

McDougall, G.J., Fyffe, S., Dobson, P., Stewart, D., 2007. Anthocyanins from red cabbage — stability to simulated gastrointestinal digestion. Phytochem. 68, 1285–1294.

Morales-Blancas, E., Chandia, V., Cisneros-Zevallos, L., 2006. Thermal inactivation kinetics of peroxidase and lipoxygenase from broccoli, green asparagus and carrots. J. Food Sci. 67, 146–154.

Nielsen, J.K., Nørbæk, R., Olsen, C.E., 1998. Kaempferol tetraglucosides from cabbage leaves. Phytochem. 49, 2171.

Pandey, K.B., Rizvi, S.I., 2009. Plant polyphenols as dietary antioxidants in human health and disease. Oxid. Med. Cell. Longev. 2, 270–278.

Podsedek, A., Sosnowska, D., Redzynia, M., Anders, B., 2006. Antioxidant capacity and content of *Brassica oleracea* dietary antioxidants. Int. J. Food Sci. Technol. 41, 49–58.

Podsędek, A., Sosnowska, D., Redzynia, M., Koziołkiewicz, M., 2008. Effect of domestic cooking on the red cabbage hydrophilic antioxidants. Int. J. Food Sci. Technol. 43, 1770–1777.

Šamec, D., Piljac-Žegarac, J., Bogović, M., Habjanič, K., Grúz, J., 2011. Antioxidant potency of white (*Brassica oleracea L. var. capitata*) and Chinese (*Brassica rapa L. var. pekinensis* (Lour.)) cabbage: The influence of development stage, cultivar choice and seed selection. Sci. Hort. 128, 78–83.

Scalzo, R.L., Genna, A., Branca, F., Chedin, M., Chassaigne, H., 2008. Anthocyanin composition of cauliflower (*Brassica oleracea L. var. botrytis*) and cabbage (*B. oleracea L. var. capitata*) and its stability in relation to thermal treatments. Food Chem. 107, 136–144.

Singh, J., Upadhyay, A., Prasad, K., Bahadur, A., Rai, M., 2007. Variability of carotenes, vitamin C, E and phenolics in Brassica vegetables. J. Food Comp. Anal. 20, 106–112.

Turkmen, N., Sari, F., Velioglu, Y.S., 2005. The effect of cooking methods on total phenolics and antioxidant activity of selected green vegetables. Food Chem. 93, 713–718.

Varoni, E.M., Lodi, G., Sardella, A., Carrassi, A., Iriti, M., 2012. Plant polyphenols and oral health: old phytochemicals for new fields. Curr. Med. Chem. 19, 1706–1720.

Volden, J., Borge, G.I.A., Bengtsson, G.B., Hansen, M., Thygesen, I.E., Wicklund, T., 2008. Effect of thermal treatment on glucosinolates and antioxidant-related parameters in red cabbage (*Brassica oleracea L. ssp. capitata f. rubra*). Food Chem. 109, 595–605.

Wachtel-Galor, S., Wong, K.W., Benzie, I.F.F., 2008. The effect of cooking on *Brassica* vegetables. Food Chem. 110, 706–710.

CHAPTER

6

Effect of Different Types of Processing and Storage on the Polyacetylene Profile of Carrots and Parsnips

*Anastasios Koidis**, *Ashish Rawson*†, *Nigel Brunton*‡

*Institute for Global Food Security, Queen's University, Belfast, UK, †Indian Institute of Crop Processing Technology, Thanjavur, India, ‡School of Agriculture and Food Science, University College Dublin, Dublin, Ireland

CHAPTER POINTS

- FaOH, FaDOH, and FaDOAc are the three major polyacetylenes naturally occurring in carrots and, to some extent, in parsnips at total levels ranging from 100 to 300 mg/kg fresh weight.
- Their retention during minimal and conventional vegetable processing is important because of increased evidence of their bioactivity and association with health benefits.
- Post-harvest storage and peeling have been shown to reduce polyacetylene content
- Conventional thermal processing such as boiling and blanching significantly reduce polyacetylene levels. Severity of thermal treatment is directly proportional to degradation of polyacetylene compounds.
- High pressure and *sous-vide* are two promising novel technologies that induce better retention of polyacetylenes although limited information is available.
- Blanching (thermal or non-thermal, using ultrasound) prior to storage retains the polyacetylene content.
- Short-time storage does not affect polyacetylene content but longer storage, especially thermally treatment carrot products has reduced PA content.
- Falcarinol-type polyacetylenes generally tend to be stable as long as they are in the plant matrix.
- Polyacetylene levels are individually determined using C18 RP-HPLC coupled with mass spectrometer or other detector after appropriate solvent extraction.
- In a domestic environment, roasting significantly reduces polyacetylene content, and whilst losses are less dramatic for boiling, levels are much lower than for fresh vegetables.

TYPES, OCCURRENCE, AND BIOACTIVE PROPERTIES OF FALCARINOL TYPE C17-POLYACETYLENES

Carrots and parsnips are root vegetables belonging to the *Apiaceae* family. In common with most vegetables, they contain, in small quantities, secondary metabolites with specific physiological and disease-protection roles in the plant. There is also some evidence to suggest that some secondary metabolites could play a role in conferring protection against some diseases on the consumer. The possible beneficial effect of β-carotene from carrots has been particularly well studied. However, in recent times the focus of studies on the beneficial effects of carrot-rich diets observed in epidemiology has shifted to another group of bioactive compounds present in carrots and parsnips, known as polyacetylenes (PAs). The physiological role of this metabolite group in the plant is to provide resistance again insects, fungi, and other microorganisms (Hansen and Boll, 1986). Although the PA group is quite wide (more than 2000 types reported),

Processing and Impact on Active Components in Food
http://dx.doi.org/10.1016/B978-0-12-404699-3.00006-8

the particular variants found in carrots are lipophilic-C17 acetylenes of the FaOH type (Figure 6.1). C17 acetylenes are found in the edible parts of *Apiaceae* vegetables including carrots, parsnips, celery, and parsley. Apart from the *Apiaceae* (and to a lesser extent, *Araliaceae*), they are less common in food plants of other plant families, although they have been found in tomatoes and aubergines of the *Solanaceae* family and even in marine organisms (sponges).

Most studies have identified three C17 PAs in carrots: **falcarinol** (FaOH), **falcarindiol** (FaDOH) and **falcarindiol 3-acetate** (FaDOAc), as shown in Figure 6.1. The levels range from 20–300 mg/kg fresh weight (FW) in extracts and they are particularly cultivar dependant (Hansen *et al.*, 2003; Christensen and Kreutzmann, 2007). In parsnip and celery extracts, FaDOAc is absent; however, the presence of both FaOH and FaDOH has been reported at similar levels to those in carrot extracts (Lund and White, 1990; Zidorn *et al.*, 2005). The most commonly occurring PA is FaOH (Figure 6.1A) although it is not always the most abundant. Some hydrolyzed or oxidized forms of these PAs have also been reported.

In terms of the anatomical distribution of PAs in roots, their levels are generally found to be higher in the carrot epidermis (i.e., the peel) than in the inner root parts (phloem and xylem). Studies suggested that FaDOH (FaDOH) is present mostly in the phloem and in the upper part of the carrot root whereas FaOH is more homogeneously distributed (Czepa and Hofmann, 2004; Baranska *et al.*, 2005). PAs of the FaOH type have been linked to perceived bitterness of carrot puree, which is one of the main reasons for low preference scores in sensory evaluations (Hansen *et al.*, 2003). Studies involving both instrumental and sensory analysis showed that amongst all PAs, FaDOH makes the strongest contribution to carrot bitterness (Czepa and Hoffman, 2003). Therefore, by peeling carrots rich in PAs, the bitter off-taste can be reduced while a high FaOH content is maintained (Kreutzmann *et al.*, 2008).

Mild toxicity and skin irritation means that, historically, C17 PAs were considered undesirable. However, more recent studies have shown that PAs in small amounts exhibit anti-inflammatory and anti-platelet, anti-fungal and anti-viral, and anti-bacterial properties, in addition to cytotoxicity and anti-cancer activity observed in various pre-clinical studies (Kobæk-Larsen *et al.*, 2005; Hansen *et al.*, 2003). FaOH has emerged as the most active PA in carrots in terms of cytotoxicity against cancer cell lines.

ANALYSIS AND QUANTIFICATION OF POLYACETYLENES

Extraction and Purification Methods

Falcarinol-type PAs are lipophilic and readily soluble in non-polar solvents. Therefore they can be extracted from freeze-dried plant material using conventional solid–liquid extraction with ethyl acetate, acetonitrile,

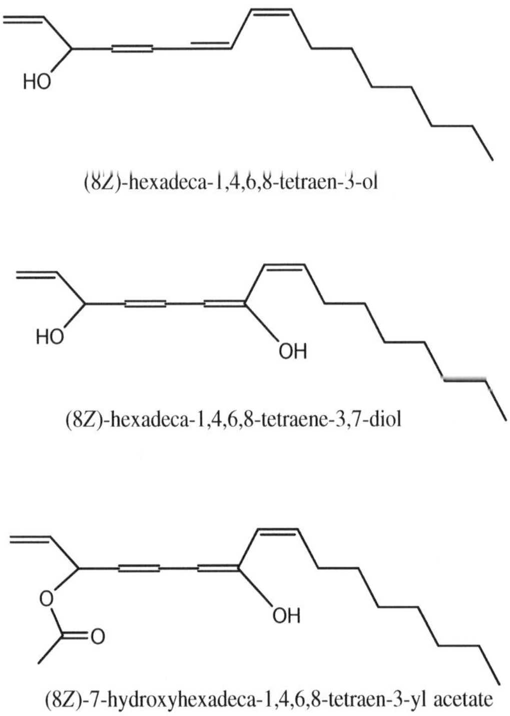

Falcarinol (FaOH)

Falcarindiol (FaDOH)

Falcarindiol-3-acetate (FADOAc)

FIGURE 6.1 The chemical structures of the three most important PAs in carrots and parsnips.

diethyl ether, and methylene chloride. Apart from solid–liquid extraction, pressurized liquid extraction of FAOH type PAs has also been reported (Pferschy-Wenzig *et al.*, 2009; Rawson *et al.*, 2010a) which resulted in improvements in efficiency (extraction yield) and reproducibility. PAs are considered thermally unstable and may undergo photodecomposition if exposed to daylight (Bohlmann *et al.*, 1973); thus, they are usually extracted in low light using relatively low temperatures.

Isolation and purification of PAs from their natural sources (e.g., carrots) is essential for the next steps of the analysis because commercial standards of pure PAs are not available. The most common method involves initial bulk extraction with an appropriate solvent (e.g., ethyl acetate), bulk silica bonded column chromatography to obtain fractions of different polarity, where PA-rich fractions are separated from the carotenoid ones. After appropriate concentration, the PA fraction is usually isolated with preparative high-pressure chromatography in three different fractions: FaOH, FaDOH, and FaDOAc (Hansen *et al.*, 2003; Czepa and Hoffmann, 2004; Christensen and Kreutzmann, 2007). Preliminary identification of the purified PAs can be achieved by comparing their characteristic ultraviolet (UV) spectra with published spectroscopic data. However, for accurate structural confirmation both nuclear magnetic resonance (NMR) spectroscopy and mass spectrometry analyses are necessary.

Determination of Polyacetylene Levels in Vegetable Extracts

Separation of PAs by analytical reversed-phase HPLC has been typically performed on C18 columns and a gradient consisting of methanol/water or acetonitrile/water is usually used for their separation (Christensen and Kreutzmann, 2007). C17- PAs have very characteristic UV-spectra due to their conjugated triple bonds and, hence, are easily identified in extracts by single-wavelength UV detector or diode array detection. As referred to earlier, specific PAs are identified using compounds isolated and purified in house from the source material (Hansen *et al.*, 2003; Kidmose *et al.*, 2004). Typically a wavelength of 205 nm is used for UV detection; however, selectivity at this wavelength is low as other compounds absorb in this region. Moreover, chromatographic runs tend to be very long (>90 min) as all carotenoids need to be removed from the column.

Given the aforementioned lack of selectivity of UV detectors it is not surprising that Liquid chromatography mass spectrometry (LC-MS) methodologies are often used to characterize purified PAs and to determine the PA levels in *Apiaceae* food plants (Zidorn *et al.*, 2005). For example, Pferschy-Wenzig *et al.* (2009) used ultra-performance liquid chromatography (UPLC) coupled with time-of-flight tandem mass spectrometry (TOF-MS) to analyse carrot extracts. FaOH was determined by extracting the main ion species generated in the ESI-positive mode, m/z 268 $[M+H-H_2O+MeCN]^+$, from the full MS chromatogram. Rai and co-workers (2011) further optimized the LC separation method achieving significantly shorter times (40% less) and detected all PAs using negative ion tandem mass spectrometry. A typical chromatograph is shown in Figure 6.2.

Spectroscopic techniques have been found to be useful in the qualitative and quantitative analysis of PAs of the FaOH-type. For example, NIR-FT (Near Infrared–Fourier transform) has been used for analyzing PAs of carrots and parsnips. The main advantage of NIR-FT over other methods is the fact that it can be carried out *in situ* on carrot disks with no requirement for the preparation of extracts. Baranska *et al.*, (2005) and Schulz and Baranska (2007) reported that the PA spectra show strong and polarized bands due to the presence of triple bonds in the region of 2200 cm^{-1}. They suggested that the number of triple bonds influence the frequency of the –C≡C– stretching modes. Thus, NIR-FT is a valuable supporting technique for the identification and quantification of PAs.

COMPOSITIONAL CHANGES IN THE CONTENTS AND PROFILE OF POLYACETYLENES DURING PROCESSING

Carrots are consumed in both raw and processed forms, but most other *Apiaceae* vegetables are often processed prior to consumption. Apart from minimally processed vegetables discussed in the next paragraph, conventional heat processing (water immersion, sous-vide processing, and canning, etc.) is usually applied to fresh carrots and parsnips to extend shelf-life. Culinary practices used in domestic preparation of carrots and parsnips also include roasting, grilling, or boiling.

Thermal and Non-Thermal Processing and the Effect of Minimal Unit Operations

In general, PAs are considered to be heat-sensitive compounds due to their acetylenic structure (Hansen *et al.*, 2003; Kidmose *et al.*, 2004). For example, Hansen *et al.* (2003) demonstrated that steam blanching of carrot cubes or shreds at 90°C for 2 min resulted in losses between 25 and 50% of all three major PAs, which is in line with the loss of FaOH and FaDOH observed in parsnip disks after blanching (90 s, 95°C) in the study of Rawson *et al.* (2010b). Boiling may reduce PA content further, i.e., 70% loss in FaOH content of carrot pieces boiled in water for 12 min (Hansen *et al.*, 2003). The reduction

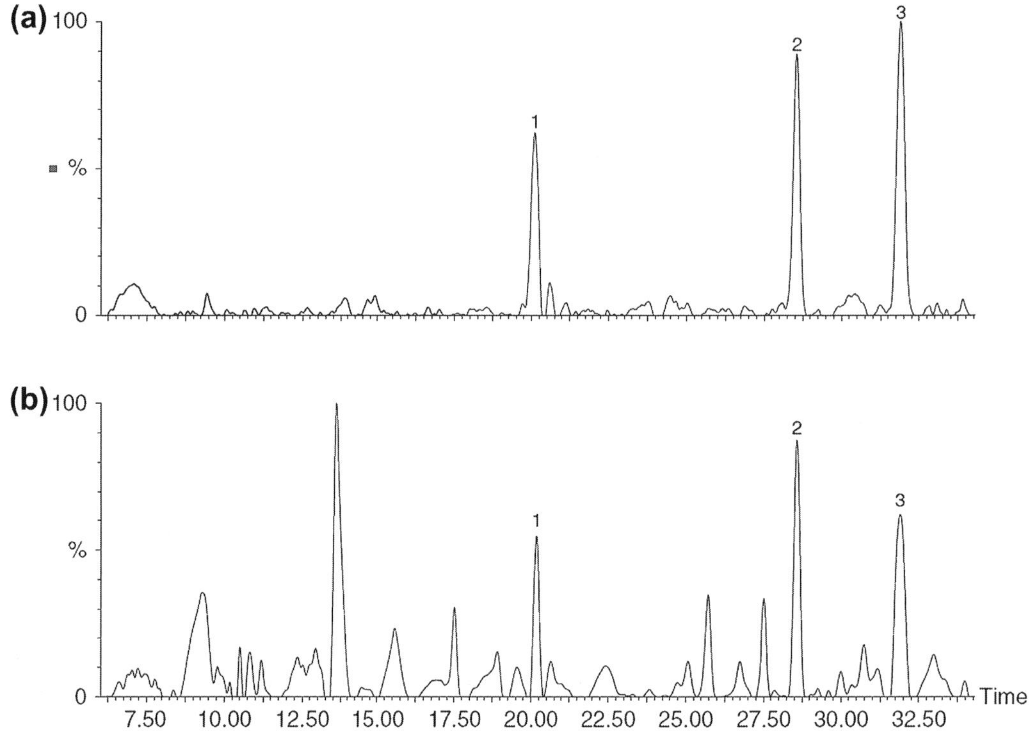

FIGURE 6.2 Typical LC-MS total ion chromatograms of carrot extracts using C18-reversed phase column showing FaDOH (1), FaDOAc (2), and FaOH (3) in: (a) a mixture of purified PAs, and (b) the crude ethyl acetate extract of carrots (Rai *et al.*, 2011).

in the content of FaOH and FaDOH during boiling has been reported to be due to successive dehydrogenation and oxidation reactions resulting in their conversion to degradation products (Rawson *et al.*, 2010b). In contrast, Rawson *et al.* (2010a) observed that, during hot water immersion of carrot disks at high temperatures (70–90°C) and long immersion times (5–60 min), an increase in the level of PAs was observed. In particular, the content of FaOH increased with increasing processing temperatures. Similar effects were observed by Rawson *et al.* (2009) when carrots were processed under full-scale commercial thermal treatments (canning to two sterilization values) for FaOH; however, they observed that FaDOH and FaDOAc levels were decreased. This finding is mainly attributed to the leaching of sugars and other solids into the water during severe thermal conditions that influence the PA content when expressed on a dry-weight basis. It is also possible that thermal processing resulted in increased extractability of compounds as thermal treatment in water may lead to the solubilization of the intercellular cementing pectin.

Sous-vide or vacuum-cooked foods are cooked under controlled conditions of temperature and time inside heat-stable vacuumized pouches. *Sous-vide* products are typically heated at relatively mild temperatures (65–95°C) for a long period of time. After heating, the products are quickly cooled and kept in chilled storage (1–4°C) for up to 21 days. There is limited published research and a lack of comparable data on the PA retention of *sous-vide* cooked vegetables. However, in one study conducted to compare the effect of *sous-vide* processing (blanching at 95°C followed by thermal processing at 90°C for 10 min) with water immersion processing (70°C for 2 min), Rawson *et al.* (2010b) demonstrated that the first step, blanching, had the greatest influence on the retention of PAs in *sous-vide*-processed parsnip disks. It resulted in significant decreases of 24.5 and 24.0% of FaOH and FaDOH, respectively. Subsequent *sous-vide* processing did not result in additional losses in PAs compared to blanched samples.

Ultrasound assisted drying is a non-thermal processing method which generates a biocidal effect through physical (cavitation, mechanical effects) and/or chemical (formation of free radicals due to sonochemical reaction) principles. To minimize enzyme induced quality losses, drying of vegetables is usually preceded by blanching. Ultrasound drying has shown some potential to be a replacement for blanching techniques. During blanching of carrots, the integrity of some plant cells is disrupted and the tissues are heat-humidified. However, subsequent ultrasound waves can have a positive effect on the stability of phytochemicals present in the food material, irrespective of ultrasound conditions. Rawson *et al.* (2011) studied the effect of ultrasound-assisted hot air drying (UPHD) of carrot disks and compared it with conventional blanched hot-air-dried (BHD) carrot disks for the retention of PA s. They reported that BHD led to significant decreases in the levels of FaDOH, FaDOAc, and FaOH compared with control samples, which was attributed to the relatively high temperatures required for blanching treatment (80°C, 3 min) leading to oxidative and thermal degradation. Conversely, UPHD samples sonicated at the highest treatment conditions had higher PA retention levels. They concluded that higher ultrasound amplitudes and longer holding times may lead to improved retention compared to hot-air-dried samples with no prior treatment. It is possible that ultrasound treatment may increase the extractability of PAs by providing the activation energy required to dissociate PAs from pectin-rich cell walls.

The technology to produce **high hydrostatic pressure** (HPP) processed products is not new; however, use of this non-thermal method of processing has only become common recently. HPP can be used to improve microbiological quality, thus extending shelf-life during refrigeration and, in some cases, at room temperature storage (Yuste *et al.*, 2001). Because of the pressure resistance of spores and some food-deteriorating enzymes at room temperature, a combination of HPP with other treatments (e.g., thermal) is sometimes needed to obtain sterile and shelf-stable food products. Therefore, HPP processing may be carried out at low temperatures or used in conjunction with heat depending on the required shelf-life of the product. Rawson *et al.* (2012a) investigated the pressure stability (400, 500 MPa) of PAs at elevated temperatures (50, 60, 70°C) in carrot disks. They reported that the rate of degradation for FaDOAc was the highest of the three PAs investigated when temperature was held constant and pressure increased, indicating that FaDOAc may be barosensitive. In relation to overall temperature sensitivity, FaDOH was found to be most sensitive followed by FaDOAc and FaOH. The temperature instability of PAs has also been noted in other studies (Hansen *et al.*, 2003). These observations suggest that synergistic application of HPP-temperature may result in lower retentions of PAs, depending on the severity of the treatment conditions. Similar results for degradation of other lipophilic bioactive compounds such as carotenoids, vitamins, and chlorophyll at high pressure and elevated temperature have been reported (De Ancos *et al.*, 2000).

Ready-to-eat food (RTE) uncooked, minimally processed products with fresh-like characteristics are becoming popular amongst consumers. Carrot and parsnip are core ingredients in many RTE foods such as chilled freshly prepared salads. **Minimal processing** such as cutting, peeling, and washing can have significant effects on the levels of many phytochemicals in vegetables including PAs. For example, Czepa and Hofmann (2004) observed that minimal processing, and especially steam peeling, significantly reduced the FaDOH content in carrots. In another study, Koidis and co-authors (2012) investigated the effect of industrial-scale unit operations (peeling, washing, cutting in various shapes, shredding) on the levels of PAs in carrots and parsnips and observed that the initial total PA level of unprocessed carrots and parsnips was significantly decreased at the end of processing. This decrease was more evident when they were cut in disks, cubes, or batons and less when shredded. Among the different operations, peeling resulted in greatest losses which suggests that PAs are susceptible to degradation or losses following various unit operations.

C17-PAs are generally known to be heat and light unstable. In a study conducted by Leonti *et al.* (2010), it was observed that pure isolated FaOH degraded rapidly when it was stored in solvents such as DMSO, n-hexane, or ethanol. Storage at −80°C in an atmosphere of argon and protection from light retarded the degradation process but did not inhibit it entirely. In addition, Leonti *et al.* (2010) observed that at room temperature FaOH decomposed rapidly in DMSO with an apparent half-life of less than 24 h. The various by-products formed during the degradation of FaOH were FaDOH, 4,5-dihydrofalcarinol, falcarinone, *E*-heptadeca-1, 8-diene-4, 6-diyne-3, 10-diol and dihydroseselidiol. While these products were formed when PA degraded in its pure form, it would not be accurate to infer that they were also formed inside a food matrix. In a recent study, Rawson *et al.* (2010b) reported the effect of thermal processing on PA content of parsnip disks with particular attention on the presence of potential degradation products. The authors reported that, following thermal processing (*sous-vide* processing), various oxidizing and hydrolyzed forms of FaOH such as falcarindione, dehydrofalcarinol, or dehydrofalcarinone (Figure 6.4) were detected, which were not present in the unprocessed product. Interestingly, most of the products formed retained the bisacetylenic moiety of the parent PAs suggesting probable degradation reactions involve oxidation, dehydrogenation, or dehydroxylation reactions (Figure 6.3).

Packaging and Storage

Table 6.1 summarizes the studies to date on the response of PAs to storage in minimally and fully processed plant foods. The majority of studies to date have been carried out on carrots and parsnips and a variety of responses have been reported. Generally, in thermally processed samples, a reduction in PA levels during storage has been reported. Rawson *et al.* (2010b) reported that chill storage of *sous-vide* stored parsnip disks resulted in a reduction in FaOH levels following 21 days of anaerobic storage. In the same study the authors reported that water immersion cooking followed by aerobic chill storage for 5 days resulted in a 70% reduction in FaDOH levels. Prior to freezing, carrots and parsnips are often

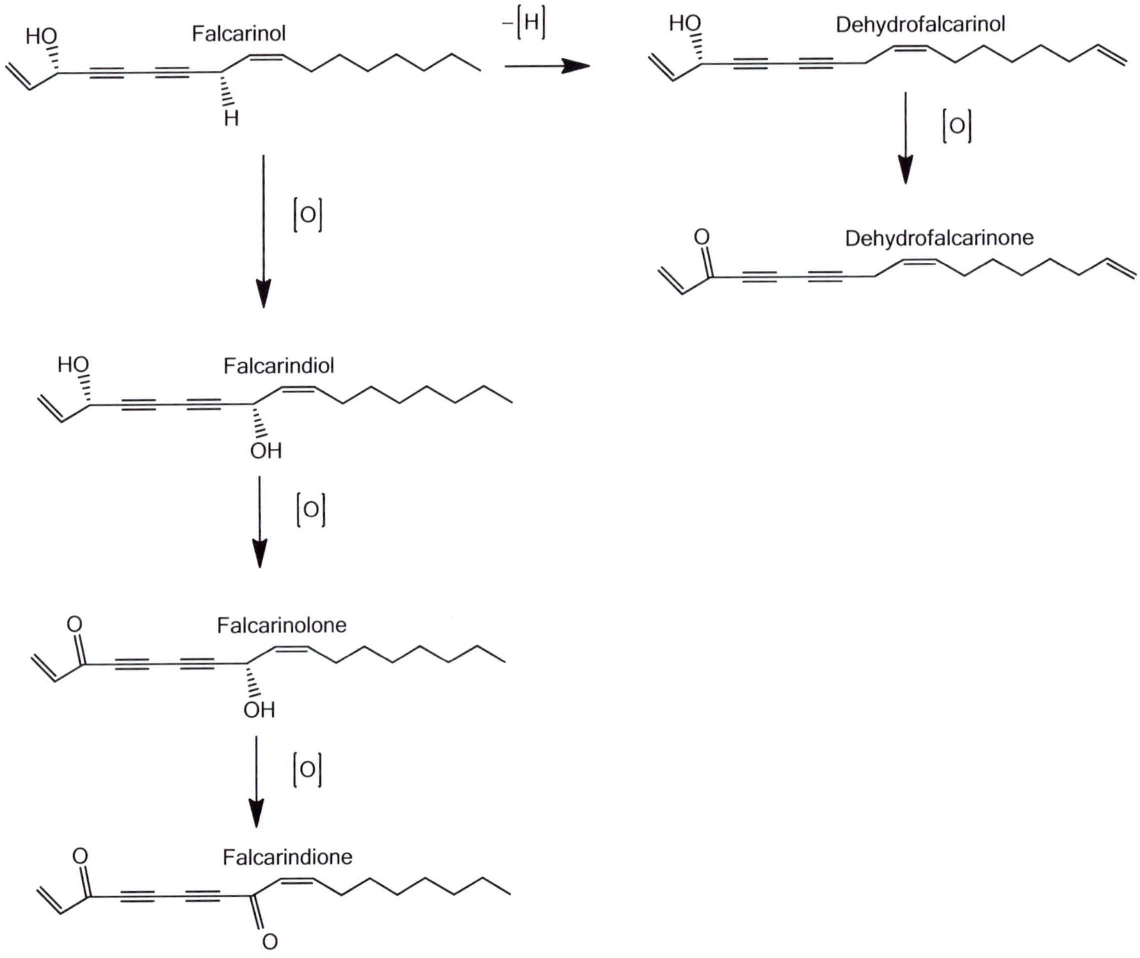

FIGURE 6.3 Suggested mechanism of degradation of PAs following thermal processing present in carrot and parsnip. *Adapted from Rawson et al., 2010b.*

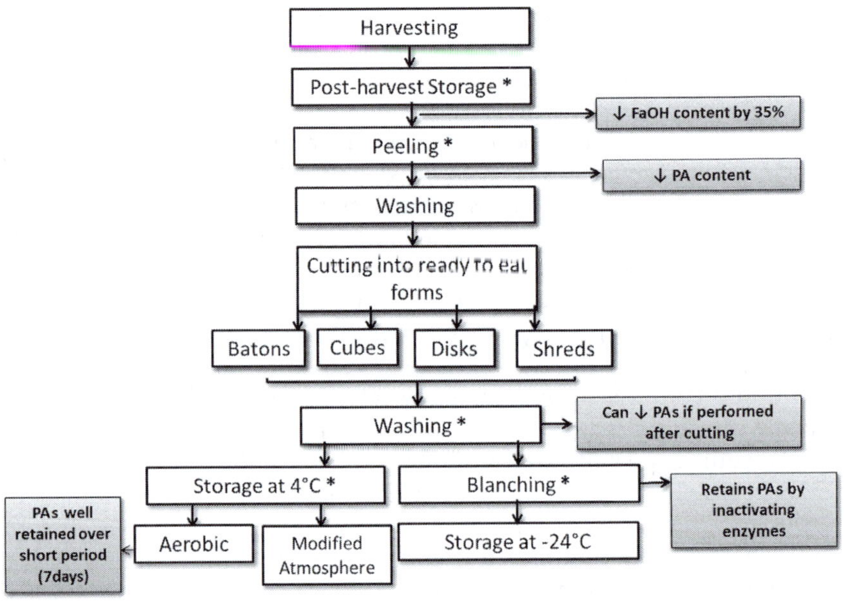

FIGURE 6.4 Flow diagram illustrating the influence of minimal-processing unit operations on PA retention in plant foods. Processing steps with asterisks indicate that notable decreases in PA levels occur.

TABLE 6.1 The Effect of Storage on the Polyacetylene Profile of Minimally and Fully Processed Carrots and Parsnips

Vegetables	Processing Step	Storage Conditions	Reaction to Storage	Reference
Carrot	Peel removed shortly after harvest	At 1°C after exposure to ultraviolet light (UV-C)	Level of PAs increased in peel	Mercier *et al.* (1993)
Carrot	Not processed	At 1°C and an RH of 98%	After 120 days of storage the FaOH content was reduced by almost 35%	Hansen *et al.* (2003)
Carrot	Blanched	Storage at −24°C for 120 days	No significant reduction in FaOH	Hansen *et al.* (2003)
Carrot	Not processed	120 days at 1°C and −24°C	FaOH, FaDOH, and FaDOAc significantly higher in carrots that were stored for 4 months at 1°C than at −24°C	Kidmose *et al.* (2004)
Parsnip disks	Sous-vide (90°C for 10 min)—(initial blanching step was applied)	Anaerobically stored at 4°C for 21 days	Significant decrease in FaOH levels, no change in FaDOH	Rawson *et al.* (2010b)
Carrot	Not processed	At 1°C and an RH of 98%	At (FaOH) below 200 mg/kg, DW levels of PAs ↑ above that level PAs ↓	Kjellenberg *et al.* (2010, 2012)
Parsnip disks	Water Immersion (70°C for 2 min)	Aerobically stored at 4°C	FaDOH losses of up to 70% occurring after 5 days' storage	Rawson *et al.* (2010a)
Carrot disks	Minimally processed	Stored at 4°C for 6 days Gas 1) 5% CO_2, 5% O_2, and 90% N_2; and, Gas 2) 10% CO_2, 80% O_2, and 10% N_2 as gas mixtures. Two packaging films: (1) polyester–polyethylene (PP); and (2) alert breathable film (Abf)	Gas 1 showed a better retention of FaOH compared to Gas 2; The Abf film (polyamide-based breathable material) showed better retention of FaOH than PP	Valverde *et al.* (2011)
Carrot disks	Blanched or not blanched	Stored at −20°C for 60 days	Better PA retention in samples blanched prior to freezing	Rawson *et al.* (2012a)

blanched to prevent enzyme-induced quality deterioration. As outlined above, whilst blanching often results in a net loss of PAs it may deliver benefits with respect to PA retention prior to long-term frozen storage. For example, Rawson *et al.* (2012a) reported better retention of PAs in carrot disks that were blanched prior to freezing than in those that were not blanched. Hansen *et al.* (2003) reported no reduction in FaOH levels of carrots stored at −24°C for 4 months.

Blanching will influence the retention of PA content in storage as it will serve to inactivate enzymes such as desaturase, acetylenase, dehydrogenase, and lipooxygenase, responsible for their degradation (Minto and Blacklock, 2008). In both of the studies above, blanching was carried out at 90°C for 75 s; however, Hansen *et al.* (2003), used steam blanching rather than water immersion as used by Rawson *et al.* (2012b); therefore, the greater retention of PA in the former study could be attributed to more complete enzyme inactivation in steam-blanched samples. Degradative enzymes will also play a critical role in the retention of PAs in minimally processed plant foods as in general the tissue is still respiring, especially in short-shelf-life products such as minimally processed carrots. However, this may not always result in a loss of PAs during storage as the enzymes responsible for synthesis of these compounds may also be active. For example, Kidmose *et al.* (2004) reported that the content of FaOH, FaDOH, and FaDOAc was significantly higher in whole carrots that were stored for 4 months at refrigerated rather than frozen temperatures. According to the authors, this may indicate that there was a net production of PAs during post-harvest storage or a smaller degradation than during frozen storage. Kjellenberg *et al.* (2012) reported that during post-harvest storage at 1°C the amount of PAs levelled off, increasing in samples initially low and decreasing in samples initially high in PAs. Interestingly the authors (Kjellenberg *et al.*, 2012) reported in a later study that post-harvest stored carrots exhibited an inverse correlation between PAs and the amount of soluble sugar.

Koidis *et al.* (2012) conducted storage studies on minimally processed carrots and parsnips in order to investigate the fate of PAs during 7 days of storage. They reported that the PA content in RTE carrots and parsnips cut in disks, cubes, and batons was well retained suggesting that the FaOH type PAs are relatively stable in the food plant matrix during storage for a short duration under refrigeration. Modified atmosphere packaging is often used to retain product quality in minimally processed carrots although it does affect the retention of PAs in carrots (Valverde *et al.*, 2011).

RECOMMENDATIONS TO ENSURE BIOACTIVE RETENTION DURING PROCESSING AND STORAGE

The bulk of practices used to process PA-containing vegetables can be classified as either minimal or full processing. Carrots and parsnips are minimally processed to convert them into convenient RTE or cook forms for sale to consumers from chilled retail display cabinets. This relatively new manufacturing practice is represented in simplified form in Figure 6.4. This figure also provides a schematic representation of the most important factors that have been found to affect the concentration of PAs in processed plant foods. For example, both post-harvest storage and peeling have been shown to reduce PA content. The latter practice however cannot be omitted if manufacturers want to produce RTE foods in forms suitable for consumers (cubes, shreds or batons). In addition, washing in a chlorinated solution can also modestly reduce PA content if performed after conversion in RTE forms. Storage in chill cabinets (4°C), either aerobically or in a modified atmosphere, has relatively less effect on PA content provided it is done for a relatively short period (7 days). If PA-containing vegetables are to be stored for long periods at freezing temperatures, it is essential that a blanching step is performed prior to frozen storage as this will serve to activate PA degrading enzymes which otherwise would remain active.

With regard to full-scale processing, whilst boiling results in PA losses both following storage and during subsequent chill storage, the degree of loss is not as marked as for more severe processing methods (e.g., canning). In contrast, HPP at ambient temperature results in increases in levels of PAs in comparison to fresh samples for both carrot and parsnip slices. Therefore, this technique may be an option to processors interested in producing long-shelf-life products with enhanced levels of PAs. Keepability can be improved further if HPP is used in combination with heat at moderate temperatures (50–70°C). Blanching results in the marked degradation of PAs. For example the inclusion of a blanching step prior to sous-vide processing results in a significant decrease in levels of FaOH and FaDOH in parsnip disks. Subsequent *sous-vide* processing had little effect on levels of PA. In contrast, chill storage for up 20 days resulted in significant decreases in these compounds. Practically, exclusion of the blanching step may not be an option for processors when there is a lag between peeling, slicing of products and application of the heat treatment.

More in-depth studies are required to provide recommendations to consumers on the best practices for retaining PAs in cooked vegetables. As a culinary practice, roasting results in significant losses of PAs, and whilst losses are less dramatic as compared to boiling, PA levels of roasted vegetables are much lower than for fresh vegetables. Processors interested in retaining PAs in their products may examine the feasibility of the non-thermal processing methods available especially if more research studies are conducted to support initial promising findings.

References

Baranska, M., Schulz, H., Baranski, R., Nothnagel, T., Christensen, L.P., 2005. *In situ* simultaneous analysis of polyacetylenes, carotenoids and polysaccharides in carrot roots. J. Agri. Food Chem. 53, 6565–6571.

Bohlmann, F., Burkhardt, F., Zdero, C., 1973. Naturally Occurring Acetylenes. Academic Press, London.

Christensen, L.P., Kreutzmann, S., 2007. Determination of polyacetylenes in carrot roots (*DaucuscarotaL.*) by high-performance liquid chromatography coupled with diode array detection. J. Separation Sci. 30, 483–490.

Czepa, A., Hofmann, T., 2003. Structural and sensory characterization of compounds contributing to the bitter off-taste of carrots (*Daucus carota L.*) and carrot puree. J. Agri. Food Chem. 51, 3865–3873.

Czepa, A., Hofmann, T., 2004. Quantitative studies and sensory analyses on the influence of cultivar, spatial tissue distribution, and industrial processing on the bitter off-taste of carrots *(Daucus Carota L.)* and carrot products. J. Agri. Food Chem. 52, 4508–4514.

De Ancos, B., Gonzalez, E., Pilar Cano, M., 2000. Effect of high pressure treatment on the carotenoid composition and the radical scavenging activity of persimmon fruit purees. J. Agri. Food Chem. 48, 3542–3548.

Hansen, L., Boll, P.M., 1986. Polyacetylenes in *Araliaceae*: Their chemistry, biosynthesis and biological significance. Phytochemistry 25, 285–293.

Hansen, S., Purup, S., Christensen, L., 2003. Bioactivity of FaOH and the influence of processing and storage on its content in carrots (*Daucus carota* L). J. Sci. Food Agri. 83, 1010–1017.

Kidmose, U., Hansen, S.L., Christensen, L.P., Edelenbos, M., Larsen, E., Norbaek, R., 2004. Effects of genotype, root size, storage, and processing on bioactive con pounds in organically grown carrots (*Daucus carota L.*). J. Food Sci. 69 (9), S388–S394.

Kjellenberg, L., Johansson, E., Gustavsson, K.E., Olsson, M.E., 2012. Polyacetylenes in fresh and stored carrots (*Daucus carota*): relations to root morphology and sugar content. J. Sci. Food Agri. 92, 1748–1754.

Kjellenberg, L., Johansson, E., Gustavsson, K.E., Olsson, M.E., 2010. Effects of Harvesting Date and Storage on the Amounts of Polyacetylenes in Carrots, *Daucus carota*. J. Agri. Food Chem. 58, 11703–11708.

Kobæk-Larsen, M., Christiansen, L.P., Vach, W., Ritskes-Hoitinga, J., Brandt, K., 2005. Inhibitory effects of feeding with carrots or (+)-FaOH on development of azoxymethane-induced preneoplastic lesions in the rat colon. J. Agri. Food Chem. 53, 1823–1827.

Koidis, A., Rawson, A., Tuohy, M., Brunton, N., 2012. Influence of unit operations on the levels of polyacetylenes in minimally processed carrots and parsnips: An industrial trial. Food Chem. 132, 1406–1412.

Kreutzmann, S., Christensen, L.P., Edelenbos, M., 2008. Investigation of bitterness in carrots *(Daucus carota L.)* based on quantitative chemical and sensory analyses. LWT — Food Sci. Tech. 41, 193–205.

Leonti, M., Casu, L., Raduner, S., Cottiglia, F., Floris, C., Altmann, K. H., Gertsch, J. 2010. Falcarinol is a covalent cannabinoid CB1 receptor antagonist and induces pro-allergic effects in skin. Biochem. Pharma. 79, 1815–1826.

Lund, E.D., White, J.M., 1990. Polyacetylenes in normal and water stressed Orlando Gold Carrots (*Daucus carota*). J. Sci. Food Agri. 51, 507–516.

Mercier, J., Ponnampalam, R., Berard, L.S., Arul, J., 1993. Polyacetylene content and UV-induced 6-methoxymellein accumulation in carrot cultivars. J. Sci. Food Agri. 63, 313–317.

Minto, R.E., Blacklock, B.J., 2008. Biosynthesis and function of polyacetylenes and allied natural products. Prog. Lipid Res. 47, 233–306.

Pferschy-Wenzig, E.M., Getzinger, V., Kunert, O., Woelkart, K., Zahrl, J., Bauer, R., 2009. Determination of FaOH in carrot (*Daucus carota* L.) genotypes using liquid chromatography/mass spectrometry. Food Chem. 114, 1083–1090.

Rai, D.K., Brunton, N., Koidis, A., Rawson, A., McLouglin, P., Griffith, W., 2011. Characterisation of Polyacetylenes isolated from *carrot (Daucus Carota)* extracts by Negative Ion Tandem Mass Spectrometry. Rapid Comm. Mass Spec. 25, 2231–2239.

Rawson, A., Koidis, A., Tuohy, M.G., Brunton, N., 2009. Effect of thermal processing on levels of polyacetylenes in carrot disks (*Daucus carota*, cv Nerac). In: Institute of Food Technologists Annual Meeting, California, USA, June 6 –10, 2009.

Rawson, A., Koidis, A., Patras, A., Tuohy, M., Brunton, N., 2010a. Modelling the effect of thermal processing on polyacetylene levels in carrot disks. Food Chem. 121, 62–68.

Rawson, A., Koidis, A., Rai, D.K., Tuohy, M., Brunton, N., 2010b. Influence of sous vide and water immersion processing on polyacetylene content and instrumental color of parsnip (*Pastinaca sativa*) disks. J. Agri. Food Chem. 58, 7740–7747.

Rawson, A., Tiwari, B.K., Tuohy, M., O'Donnell, C.P., Brunton, N., 2011. Effect of ultrasound on polyacetylenes in carrot. Ultrason. Sonochem. 18, 1172–1179.

Rawson, A., Tiwari, B.K., Tuohy, M., Brunton, N., 2012a. Impact of frozen storage on polyacetylene content, texture and colour in carrots disks. J. Food Eng. 108 (4), 563–569.

Rawson, A., Brunton, N., Tuohy, M., 2012b. Degradation kinetics of polyacetylenes in carrots as affected by combined treatment of pressure and temperature compared to sous-vide processing. Food Chem. 133, 15–20.

Schulz, H., Baranska, M., 2007. Identification and quantification of valuable plant substances by IR and Raman spectroscopy. Vibrational Spectroscopy. 43 (1), 13–25.

Valverde, J., Rawson, A., Quesada, L., Brunton, N., 2011. Bioactives in minimally processed carrots. TResearch 6, 20–21.

Yuste, J., Capellas, M., Pla, R., Fung, D.Y.C., Mor-Mur, M., 2001. High pressure processing for food safety and preservation: A Review. J. Rapid Meth. Autom. Microbiol. 9, 1–10.

Zidorn, C., Jöhrer, K., Ganzera, M., Schubert, B., Sigmund, E.M., Mader, J., Greil, R., Ellmerer, E.P., Stuppner, H., 2005. Polyacetyl*enes from the Apiaceae* vegetables carrot, celery, fennel, parsley, and parsnip and their cytotoxic activities. J. Agri. Food Chem. 53, 2518–2523.

CHAPTER

7

Impact of Food Processing on Non-Hallucinogenic Indole Derivatives in Edible Mushrooms

Bożena Muszyńska, Katarzyna Sułkowska-Ziaja

Department of Pharmaceutical Botany, Jagiellonian University Collegium Medicum, Kraków, Poland

CHAPTER POINTS

- All the edible mushrooms examined are a source of bioactive indole compounds.
- The temperature during the preparation of mushroom dishes affects the composition and amount of indole compounds because of their sensitivity to elevated temperature.
- An increase in temperature causes a partial decrease in their amounts, but most of the heat-treated fruiting bodies of mushrooms contain, on average, about half of the total amount of indole compounds in comparison with the unprocessed ones.
- The analysis indicated that thermally processed fruiting bodies of *Cantharellus cibarius, Boletus badius, Lactarius deliciosus, Macrolepiota procera, Pleurotus ostreatus,* and *Suillus bovinus* are still good sources of bioactive indole compounds from mushrooms.
- Among indole compounds, the most temperature resistant is L-tryptophan. Research indicates that mushrooms are an alternative to animal meat, because they are a good source of L-tryptophan, which, as an exogenous amino acid, is important in the daily human diet.

INTRODUCTION

Mushrooms are widely distributed around the globe and play an important role both in their natural habitats as well as in human economy. They have been a component of the human diet from the prehistoric times. In ancient Rome, people appreciated dishes made from mushrooms. Even today, in many European countries, mushrooms are highly prized for their taste and aroma. Nowadays they may be derived not only from their natural habitats and commercial farms, but also from *in vitro* cultures conducted under appropriate conditions. Moreover, in the Far East mushrooms have been used for their medicinal properties from time immemorial (Wasser and Weiss, 1999).

Mushrooms are able to accumulate both primary and secondary metabolites. Some of these compounds may play an antioxidant role, e.g., phenolic and indole compounds, flavonoids, terpenoids, sterols, ascorbic acid, ergothioneine, statins, carotenoids. Mushrooms also accumulate trace elements, e.g., selenium (Barros *et al.*, 2007; 2008; Ferreira *et al.*, 2007). Medicinal and antioxidant properties of mushrooms are an excellent combination that constitute their dietary value. Recent studies have shown that edible mushroom species contain non-hallucinogenic indole compounds and their derivatives (Muszyńska *et al.*, 2007; 2009; 2011a,b,c, 2012, 2013). The indole skeleton is the basis of the substances serving important functions in the human body, such as L-tryptophan, serotonin, and melatonin. These compounds fulfil the role of neurotransmitters or their precursors, exhibit antioxidant, anticancer, anti-ageing activity, regulate the diurnal cycle in humans and participate in blood coagulation. These compounds and their derivatives are also anti-inflammatory and analgesic therapeutics (Kaneez and Arshad, 2007; Martin *et al.*, 1972; Singhal *et al.*, 2012).

Most of the human body's total serotonin is located in the enterochromaffin cells in the gut, where it is used to regulate intestinal movements (Keszthelyi *et al.*, 2009). In the central nervous system, serotonin takes

Processing and Impact on Active Components in Food
http://dx.doi.org/10.1016/B978-0-12-404699-3.00007-X

part in neuron to neuron communication and appears to enhance the perception of well-being and modulate the intensity of emotional states. Serotonin has also some cognitive functions, including memory and learning. Tryptophan and 5-hydroxytryptophan are precursors of serotonin. 5-Hydroxytryptophan, which is a direct precursor of serotonin (Birdsall, 1998), when ingested, easily crosses the blood–brain barrier into the central nervous system, where it is efficiently converted to serotonin. This metabolite, usually derived from the seeds of *Griffonia simplicifolia*, is sold in some countries (Canada, United Kingdom, USA) as a dietary supplement for use as an antidepressant and sleep aid in a typical dose of 50–100 mg (Turner *et al.*, 2006).

Melatonin is a strong antioxidant. It delays the ageing processes, scavenging the free radicals, such as the hydroxyl radical, so it may reduce damage to the nervous system in some types of Parkinson's disease. There are also studies proving its efficacy in the treatment of Alzheimer's disease. In neurodegenerative diseases, melatonin plays a neuroprotective role (Reiter *et al.*, 2001; Singhal *et al.*, 2012; Wang, 2009) and also shows immunomodulatory and anti-inflammatory activity (Esposito and Cuzzocera, 2010).

The qualitative and quantitative analyses of indole compounds in methanolic extracts from the fruiting bodies of selected edible mushroom species belonging to the Basidiomycota taxon collected from natural habitats and of commercial origin were described by Muszyńska (Muszyńska *et al.*, 2007, 2009, 2011a,b,c, 2012, 2013). Eleven edible mushroom species were chosen for the study, not only due to their popularity among consumers but most of all because of their medicinal and dietary value. In addition, all those species contain numerous primary and secondary metabolites exhibiting significant biological activity (for example: polysaccharides, unsaturated fatty acids, water- and fat-soluble vitamins, a wide spectrum of proteins, flavonoids, terpenoids, sterols, carotenoids, phenolic compounds, and trace elements, e.g., selenium) (Barros *et al.*, 2007, 2008; Ferreira *et al.*, 2007).

HOW COMPOSITION IS ALTERED

Different conditions, especially temperature during the preparation of dishes, are responsible for the degradation of indole compounds. For these reasons, the thermal processing of mushrooms was designed to simulate the cooking conditions in which mushroom dishes (like sauces and soups) are normally prepared, since only a few edible mushroom species can be eaten raw. Because not all indole compounds are stable at higher temperatures (the most stable compound among them is L-tryptophan because, for example, it is resistant to temperatures up to 90°C), so it is appropriate to determine the actual amounts of these compounds in dishes in order to assess whether mushrooms are a good source of these compounds to consumers (Cuq and Cheftel, 1983). The amounts of indole compounds in these species were analyzed in selected fruiting bodies of mushrooms before and after thermal processing by using high-pressure liquid chromatography (HPLC) methods.

The materials for analysis were fresh fruiting bodies of edible mushrooms: *Armillaria mellea* (Honey mushroom) (Muszyńska *et al.*, 2011c), *Boletus badius* (Bay Bolete) (Muszyńska *et al.*, 2009, 2011a), *Boletus edulis* (King Bolete) (Muszyńska *et al.*, 2011a,b), *Cantharellus cibarius* (Chanterelle), *Lactarius deliciosus* (Saffron milk-cap), *Leccinum scabrum* Bull. Gray (the birch bolete), *Macrolepiota procera* (Scop.) Singer (the parasol mushroom), *Suillus bovinus* (L.: Fr.) O. Kuntze (the Jersey cow mushroom) (Muszyńska *et al.*, 2013), which were collected between 2007 and 2011 from selected regions of southern Poland (mixed forests near Kraków). The materials for analysis also included fresh fruiting bodies of cultivated mushrooms *Lentinula edodes* (Berk.), Pegler (the shiitake), *Auricularia polytricha* (Mont.) Sacc (the black, or cloud ear, fungus) (Muszyńska *et al.*, 2011b), and *Pleurotus ostreatus* (Oyster mushroom) (Muszyńska *et al.*, 2011a,b), which were of commercial origin. Fresh mushrooms (50 g of each species) were cleaned of plant debris and after taxonomic identification according to Knudsen and Vesterholt (2008) and online keys (http://www.mycokey.com/) were frozen and immediately dried by lyophilization (Freezone 4.5, Labconco; temperature: −40°C).

The method of extraction applied to the fruiting bodies of mushrooms and the HPLC analysis proved to meet the optimal conditions for the qualitative and quantitative determination of indole compounds in the test material. The studies showed that all the species contained bioactive indole compounds, but differences were observed in qualitative composition and quantity between the thermally processed mushrooms and the unprocessed ones, which were likely caused by elevated temperature. The temperature during the preparation of mushroom dishes affects the composition and amount of indole compounds. The effect of elevated temperature on the levels of indole compounds is ambiguous. Both a decrease and an increase in the amount of indole compounds can be observed. The decrease is probably due to the decomposition of these compounds under the influence of temperature because, as it is known, indole compounds are thermolabile. In turn, the increase in the amount of some compounds (indole, 5-hydroxytryptophan, L-tryptophan, tryptamine, 5-methyltryptamine) may be associated with thermal decomposition of other compounds. Most of the heat-treated fruiting bodies contained on average about half of the total amount of indole compounds in comparison with the unprocessed ones (Figure 7.1).

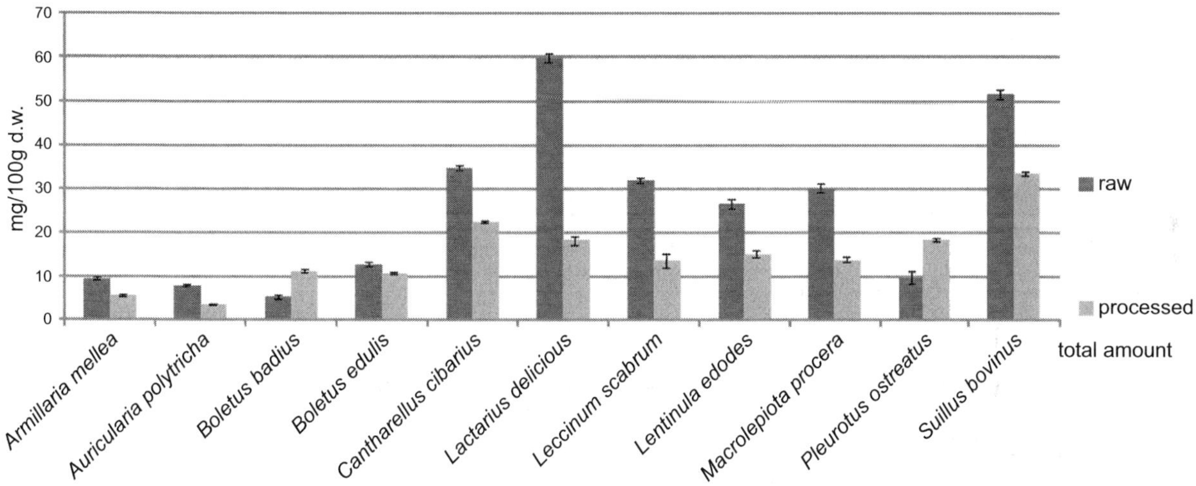

FIGURE 7.1 Total amounts of indole compounds in the fruiting bodies of edible mushrooms before and after thermal processing. *Muszyńska* et al. *2011a,b,c, 2012, 2013.*

The highest total amounts of indole compounds were found in the fruiting bodies of *S. bovinus*, both in the unprocessed as well as thermally processed ones. They were 51.42 and 33.44 mg/100 g dry weight (d.w.), respectively. In turn, the lowest amounts of indole compounds were found in the extracts from the fruiting bodies of *A. polytricha* (10.44 mg/100 g d.w. in the unprocessed fruiting bodies and 3.59 in the cooked ones) (Muszyńska *et al.*, 2013).

The greatest variety of indole compounds was found in the extract from the thermally unprocessed fruiting bodies of *C. cibarius* (Table 7.1). The extract contained seven out of the 11 indole compounds present in the tested mushrooms. The least number of indole compounds was determined in the extracts from the processed fruiting bodies of *A. mellea* (L-tryptophan and tryptamine). The compounds found in the largest amounts in the tested mushrooms were serotonin, L-tryptophan and 5-hydroxytryptophan. Among the indole compounds, serotonin was determined in the highest amounts in the unprocessed fruiting bodies of *C. cibarius* (29.61 mg/100 g d.w.).

After thermal processing, a lack of serotonin in the extract from this species was observed. In the cases of *A. mellea*, *B. badius*, *B. edulis*, *L. deliciosus*, *P. ostreatus* serotonin was found in the unprocessed fruiting bodies, but did not appear in the extracts from the fruiting bodies subjected to thermal processing. In unprocessed *L. scabrum* the amount of this metabolite was 13.09 mg/100 g d.w., which decreased to 2.07 in the cooked sample (Table 7.2).

Conversely, serotonin was determined in the extracts from the processed fruiting bodies of *L. edodes* and *S. bovinus*, whereas in the extracts from the unprocessed fruiting bodies of these mushroom species its presence was not revealed. Serotonin, when ingested, does not cross the blood–brain barrier into the central nervous system but can regulate intestinal movements (Birdsall, 1998).

L-Tryptophan is an amino acid exogenous to the human body and, therefore, it must be supplied to the body via food. It was investigated in each of the tested mushrooms and its amounts ranged from 0.01 to 25.90 mg/100 g d.w. L-Tryptophan is a substrate for the synthesis of such crucial substances for the human organism as neurotransmitters (dopamine, melatonin, serotonin, adrenaline) and vitamins (niacin) (Kaneez and Arshad, 2007). The highest amount of L-tryptophan was determined in the extracts from the thermally unprocessed fruiting bodies of *S. bovinus* (25.90 mg/100 g d.w.). Among the extracts from the heat-treated fruiting bodies, the extract from the fruiting bodies of *S. bovinus* also contained the highest amount of L-tryptophan (17.71 mg/100 g d.w.). It can be noted that processed edible mushrooms, especially *S. bovinus*, can constitute a vital source of that amino acid and thus an alternative to other foods that contain L-tryptophan (Muszyńska *et al.*, 2013).

The smallest amount of L-tryptophan was found in the fruiting bodies of *C. cibarius* (0.01 mg/100 g d.w.). In the case of *B. edulis*, the amount of L-tryptophan was higher in the processed material than in raw fruiting bodies (Muszyńska *et al.*, 2011a, 2012, 2013). The reason for this may be the degradation of indole compounds such as serotonin, 5-hydroxytryptophan (to tryptophan), which indicated high levels in unprocessed *B. edulis*.

5-Hydroxytryptophan, a direct precursor of serotonin and melatonin, was found in both fresh and processed fruiting bodies, but in the largest amount in the unprocessed mushrooms of the species tested. The highest amounts of this metabolite were found in the extracts from the unprocessed fruiting bodies of *L. edodes* (24.83 mg/100 g d.w.), *M. procera* (22.94), *S. bovinus* (15.83) and among the heat-treated fruiting

TABLE 7.1 Amounts (mg/100 g d.w.) of Indole Compounds in Selected Edible Basidiomycota mushrooms

	L-Tryptophan	5-CH_3-Tryptophan	5-OH-Tryptophan	Tryptamine	5-CH_3-Tryptamine	Melatonin	Serotonin	Indole	Kynurenine Sulphate	Indoleacetonitryle	Indoleacetic Acid
Armilaria mellea	4.47 ± 0.010	a	a	2.74 ± 0.010	a	a	2.21 ± 0.010	a	a	a	a
Auricularia polytricha	0.16 ± 0.050	a	7.35 ± 0.200	2.77 ± 0.050	0.19 ± 0.010	a	a	a	a	a	a
Boletus badius	0.68 ± 0.047	a	a	0.47 ± 0.045	a	a	0.52 ± 0.042	a	3.53 ± 0.157	a	a
Boletus edulis	0.39 ± 0.020	a	0.18 ± 0.001	1.17 ± 0.003	a	0.68 ± 0.006	10.14 ± 0.270	a	a	0.04 ± 0.001	a
Cantharellus cibarius	0.01 ± 0.002	a	0.02 ± 0.003	0.01 ± 0.002	a	0.14 ± 0.011	29.61 ± 0.177	a	4.81 ± 0.290	0.01 ± 0.002	a
Lactarius delicious	a	a	0.25 ± 0.017	a	a	1.29 ± 0.077	18.42 ± 0.930	a	39.20 ± 1.960	0.62 ± 0.045	2.04 ± 0.130
Leccinum scabrum	9.56 ± 0.040	8.32 ± 0.050	a	a	a	a	a	a	a	a	a
Lentinula edodes	0.58 ± 0.020	a	24.83 ± 0.300	0.04 ± 0.010	a	0.13 ± 0.010	1.03 ± 0.020	a	a	a	a
Macrolepiota procera	3.47 ± 0.050	a	22.94 ± 0.500	0.92 ± 0.040	2.54 ± 0.050	a	a	0.02 ± 0.003	a	a	a
Pleurotus ostreatus	a	a	2.08 ± 0.003	0.91 ± 0.002	a	a	6.52 ± 1.770	a	a	0.01 ± 0.002	0.21 ± 0.020
Suillus bovinus	25.90 ± 0.200	a	15.83 ± 0.500	3.15 ± 0.050	6.51 ± 0.300	a	a	0.03 ± 0.010	a	a	a

[a] *Content lower than 0.001 mg/100 g dry weight.*
Data are presented as the mean ± SD of three series. Statistical significance was tested at $p \leq 0.05$.
Muszyńska et al. *(2011a,b c, 2012, 2013).*

TABLE 7.2 Amounts (mg/100 g d.w.) of Indole Compounds in Selected Edible Basidiomycota Mushrooms after Thermal Processing

	L-Tryptophan	5-CH_3-Tryptophan	5-OH-Tryptophan	Tryptamine	5-CH_3-Tryptamine	Melatonin	Serotonin	Indole	Kynurenine sulphate	Indoleacetonitryle	Indoleacetic acid
Armilaria mellea	2.78 ± 0.050	a	a	2.77 ± 0.050	a	a	a	a	a	a	a
Auricularia polytricha	a	a	3.52 ± 0.050	a	0.06 ± 0.010	a	a	0.01 ± 0.005	a	a	a
Boletus badius	1.74 ± 0.040	6.55 ± 0.200	a	a	a	0.71 ± 0.020	a	a	a	2.07 ± 0.040	a
Boletus edulis	8.35 ± 0.200	a	a	a	a	a	a	1.6 ± 0.040	a	0.73 ± 0.020	a
Cantharellus cibarius	4.17 ± 0.050	3.75 ± 0.050	a	3.30 ± 0.040	a	4.40 ± 0.050	a	1.96 ± 0.030	a	4.94 ± 0.060	a
Lactarius delicious	8.76 ± 0.200	3.83 ± 0.050	a	3.15 ± 0.050	a	a	a	1.65 ± 0.040	a	0.83 ± 0.020	a
Leccinum scabrum	2.74 ± 0.040	2.88 ± 0.050	5.65 ± 0.200	0.36 ± 0.020	a	a	2.07 ± 0.040	a	a	a	a
Lentinula edodes	0.63 ± 0.040	a	a	a	a	0.04 ± 0.005	14.33 ± 0.400	a	a	a	a
Macrolepiota procera	0.03 ± 0.005	a	10.11 ± 0.500	2.56 ± 0.040	0.93 ± 0.050	a	a	0.02 ± 0.005	a	a	a
Pleurotus ostreatus	8.92 ± 0.200	3.79 ± 0.050	a	3.21 ± 0.050	a	a	a	1.64 ± 0.040	a	0.84 ± 0.020	a
Suillus bovinus	17.71 ± 0.200	a	a	2.50 ± 0.050	0.12 ± 0.010	a	13.11 ± 0.200	a	a	a	a

[a] *Content lower than 0.001 mg/100 g dry weight.*
Data are presented as the mean ± SD of three series. Statistical significance was tested at $p \leq 0.05$.
Data from Muszyńska et al. *(2012, 2013).*

bodies in the extracts from *M. procera* (10.11 mg/100 g d.w.) (Muszyńska *et al.*, 2013).

5-Methyltryptophan was determined among the unprocessed mushrooms in only *L. scabrum*, but in the largest amount (8.32 mg/100 g d.w.). 5-Methyltryptophan had been determined in the processed fruiting bodies of four species: *B. edulis*, *C. cibarius*, *L. deliciosus*, and *P. ostreatus*, in amounts comparable to those found in *L. scabrum* (Muszyńska *et al.*, 2011a, 2012).

Tryptamine is found in trace amounts in the brains of mammals and plays a role as a neuromodulator or neurotransmitter (Jones *et al.*, 1982; Juorio and Durden, 1984; Martin *et al.*, 1972). It was found both in the processed and in the unprocessed mushrooms. The amounts of tryptamine, the second metabolite common to all the studied species, did not vary very much, ranging from 0.01 to 3.15 mg/100 g d.w. in the unprocessed fruiting bodies, and were similar to those in the processed samples: 0.36–3.15 mg/100 g d.w. (Muszyńska *et al.*, 2009, 2011a,b,c, 2012, 2013).

5-Methyltryptamine was determined in three species of mushrooms: *S. bovinus*, *M. procera* and *A. polytricha* (6.51, 2.54, 0.19 mg/100 g d.w., respectively), and also in the heat-treated samples of the same species, but only in small amounts (0.12, 0.93, 0.06, respectively) (Muszyńska *et al.*, 2013). An indole compound also rarely found in the tested mushrooms was melatonin. It was present in small amounts in the extracts from the fruiting bodies of *B. edulis*, *C. cibarius*, *L. deliciosus*, *L. edodes*, and *M. procera* (ranging from 0.07 to 1.29 mg/100 g d.w., respectively) (Muszyńska *et al.*, 2011, 2013). In the case of *C. cibarius*, an increase in melatonin content was observed after processing (4.40 mg/100 g d.w.) (Muszyńska *et al.*, 2013). The presence of indoleacetic acid was revealed only in two species: *L. deliciosus* and *P. ostreatus* (2.04, 0.21 mg/100 g d.w., respectively). The fruiting bodies of *L. deliciosus*, *C. cibarius*, and *B. badius* contained large amounts of tryptophan degradation product: kynurenine sulfate (39.20, 4.81, and 3.53 mg/100 g d.w., respectively) (Muszyńska *et al.*, 2009; 2011a). In the case of kynurenine sulfate and indoleacetic acid, elevated temperature leads to the decomposition of these compounds. The highest concentration of indoleacetonitrile was observed in the extracts from the processed fruiting bodies of *C. cibarius* (4.94 mg/100 g d.w.) and *B. edulis* (2.07 mg/100 g d.w.) (Muszyńska *et al.*, 2012).

In conclusion, it can be suggested that edible mushrooms can be a valuable dietary source of physiologically active indole compounds, but the preparation of mushroom dishes and dishes containing mushrooms influences the levels of those compounds because of their sensitivity to elevated temperature. An increase in temperature causes a partial decrease in their amounts, but cooked mushrooms still remain an excellent source of these metabolites, especially the precursors of all the indole derivatives possessing antidepressant and antioxidant activity like L-tryptophan, 5-hydroxytryptotbe taken into account that serotonin does not cross the blood–brain barrier, while L-tryptophan and 5-hydroxytryptophan have this ability and are the precursors of serotonin, and therefore have an antidepressant action. L-Tryptophan is a substrate of dopamine, melatonin, adrenaline, as well as vitamins (niacin). Consumption of dishes which contain mushrooms is safe and beneficial because of good assimilability of the nutrients that protect against civilization diseases and have vitalizing potential for the human organism. In addition, the present analysis indicates that thermally processed fruiting bodies of *Cantharellus cibarius*, *Boletus badius*, *Lactarius deliciosus*, *Macrolepiota procera*, *Pleurotus ostreatus*, and *Suillus bovinus* are the richest sources of bioactive indole compounds.

OTHER WAYS IN WHICH COMPOSITION IS ALTERED

Not only temperature, but other different conditions such as UV irradiation, pH, free radicals, the presence of phenyloxidase in mushrooms, and storage conditions, are responsible for the degradation of indole compounds. Out of all indole compounds, the most stable during processing and storage is L-tryptophan. A few reports have indicated that L-tryptophan is sensitive to oxidation by hydrogen peroxide and large tryptophan losses were found on reaction of protein with oxidizing lipids (Nielsen *et al.*, 1985). L-Tryptophan destruction is associated with lack of flavors in food. Indole compounds, especially serotonin and melatonin, are light-sensitive. Aqueous solutions of these compounds when stored in sterile vials under vacuum are stable for at least 6 months. Melatonin has influence at the level of tryptophan and its direct derivative 5-hydroxytryptophan. Melatonin inhibits the tryptophan 2,3-dioxygenase enzyme, which destroys tryptophan by enhancing its catabolism. Inhibition of 2,3-dioxygenase by melatonin leads to an increase in the amount of tryptophan (Walsh *et al.*, 1994).

ANALYTICAL TECHNIQUES

The lyophilized fruiting bodies (Freezone 4.5, Labconco; temperature: −40°C) of mushrooms were weighed (5 g of each species), ground in a mortar and then subjected to extraction with petroleum ether in percolators in order to remove the lipid fraction according to the procedure developed in our laboratory (Muszyńska *et al.*, 2007). After the extraction with ether, the material was dried and subjected to extraction with methanol placed in a percolator (which was kept in the dark). These oil fractions were discarded and the remaining biomass was dried and again subjected to extraction with methanol in a percolator for 24h. The extracts were concentrated by distillation in a vacuum evaporator (Buchi, Germany) under reduced pressure (200mBa) at 40°C. To remove the remaining lipids, the concentrated extracts were frozen. Polysaccharides were removed using the Chihara method. The residues were quantitatively dissolved in methanol, filtered through Whatman No. 3 paper and purified by thin-layer chromatography (TLC). For the purification of the extracts, we used TLC on aluminium-backed silica gel 60 plates (Merck, Darmstadt, Germany, Art. No. 1.055540001), onto which the methanol extracts were loaded. Chromatograms were developed in a mobile phase which was found to be optimal for the separation of indole compounds: n-propanol/ethyl acetate/water (7:1:2 v/v/v). Spots containing indole compounds were identified under a UV lamp at $\lambda = 254$ nm. The obtained fractions were extracted from the chromatograms with methanol, filtered through a syringe-driven filter unit (Millex, Millipore Corporation, USA) and then concentrated by distillation in a vacuum evaporator under reduced pressure at 40°C. The extracts, quantitatively dissolved in 1.5ml of methanol, were subjected to HPLC analysis.

The method of sample preparation by thermal processing was as follows: dry materials after lyophilization were weighed (5g of each species) and ground in a mortar; 80ml of distilled water was added to each sample and the mixtures were boiled in a thermostated water bath (Labart Sp. z o.o., Poland) at 100°C for 60min in a Soxhlet apparatus. After that time, the whole mixtures containing the fruiting bodies with the aqueous extracts from mushrooms were frozen and lyophilized again. After lyophilization, the crushed dry mass (fruiting bodies with aqueous extracts) was treated with the same procedure as the unprocessed samples (Muszyńska *et al.*, 2012).

The extracts obtained from the unprocessed and thermally processed mushrooms were analyzed for indole compounds: L-tryptophan, 5-hydroxytryptophan, 5-methyltryptophan, serotonin, melatonin, tryptamine, 5-methyltryptamine, kynurenine sulfate, indoleacetic acid, indoleacetonitrile, indole and indoleacetamide (standards from Sigma-Aldrich St Louis, MO, USA, all of HPLC grade) by the HPLC method. The analyses were carried out according to the procedure developed by Kysilka and Wurst (1985) with our modifications (Muszyńska *et al.*, 2009). HPLC analyses were performed using a Hitachi apparatus (Merck, Japan) with pump: L-7100; column: Purospher® RP-18 (4×200mm, 5μm) thermostated at 25°C. The solvent system used was: methanol/water/ammonium acetate (15:14:1 v/v/v); flow rate: 1ml/min. Detection was carried out in a UV detector at $\lambda = 280$ nm. The identification of indole compounds was made by comparing the retention times of sample peaks with those of the standards. An example chromatogram is presented in Figure 7.2.

The results were expressed in mg/100g d.w., calculated by internal normalization of the chromatographic peak area. The statistical analyses were carried out using Student's test. The results were expressed as mean values with the standard deviation (SD). All the analyses were conducted using Statistica 10 (StatSoft, Poland). Statistical significance was tested at $p \leq 0.05$.

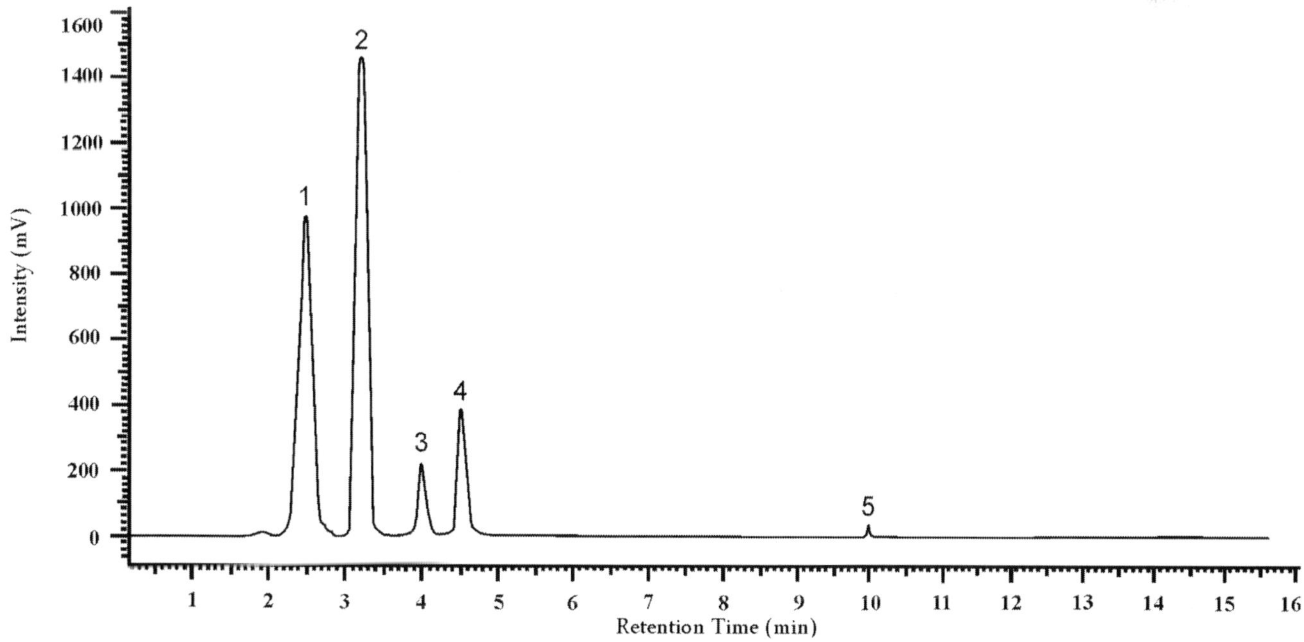

FIGURE 7.2 HPLC chromatogram of extract from fruiting bodies of *Suillus bovinus*: (1) 5-hydroxytryptophan; (2) L-tryptophan; (3) tryptamine; (4) 5-methyltryptamine; (5) indol.

References

Barros, L., Baptista, P., Correia, D.M., Casal, S., Oliveira, B., Ferreira, I.C.F.R., 2007. Fatty acid and sugar compositions and nutritional value of five wild edible mushrooms from Northeast Portugal. Food Chem. 105, 140–145.

Barros, L., Cruz, T., Baptista, P., Estevinho, L.M., Ferreira, I.C.F.R., 2008. Wild and commercial mushrooms as source of nutrients and nutraceuticals. Food Chem. Toxicol. 46, 2742–2747.

Birdsall, T.C., 1998. 5-Hydroxytryptophan: A Clinically-Effective Serotonin Precursor. Altern. Med. Rev. 3, 271–280.

Cuq, J.C., Cheftel, J.C., 1983. Tryptophan degradation during heat treatments: Part 1—The degradation of free tryptophan. Food Chem. 12, 1–14.

Esposito, E., Cuzzocrea, S., 2010. Antiinflammatory Activity of Melatonin in Central Nervous System. Curr. Neuropharmacol. 8, 228–242.

Ferreira, I.C.F.R., Baptista, P., Vilas-Boas, M., Barros, L., 2007. Free-radical scavenging capacity and reducing power of wild edible mushrooms from northeast Portugal: Individual cap and stipe activity. Food Chem. 100, 1511–1516.

Jones, R.S., 1982. Tryptamine: a neuromodulator or neurotransmitter in mammalian brain? Prog. Neurobiol. 19, 117–139.

Juorio, A.V., Durden, D.A., 1984. The distribution and turnover of tryptamine in the brain and spinal cord. Neurochem. Res. 9, 1283–1293.

Kaneez, F.S., Arshad, S.S., 2007. The metabolism of serotonin in neuronal cells in culture and platelets. Exp. Brain Res. 183, 411–416.

Keszthelyi, D., Troost, F.J., Masclee, A.M., 2009. Understanding the role of tryptophan and serotonin metabolism in gastrointestinal function. Neurogastroenterology And Motility: The Official Journal of the European Gastrointestinal Motility Society 21, 1239–1249.

Knudsen, H., Vesterholt, J., 2008. Funga Nordica: agaricoid, boletoid and cyphelloid genera. Nordsvamp, Copenhagen.

Kysilka, R., Wurst, M., 1985. High Performance Liquid Chromatographic Determination of Hallucinogenic Indoleamins with Simultaneous UV Photometric and Voltametric Detection. J. Chromatogr. 320, 414–420.

Martin, W.R., Sloan, J.W., Christian, S.T., Clements, T.H., 1972. Brain levels of tryptamine. Psychopharmacologia (Berl.). 24, 331–346.

Muszyńska, B., Maślanka, A., Sułkowska-Ziaja, K., Krzek, J., 2007. Analysis of indole compounds and nitric bases in fruiting bodies of *Lactarius deterrimus* by TLC-UV. J. Planar Chromatogr. Mod. TLC 20, 55–58.

Muszyńska, B., Sułkowska-Ziaja, K., Ekiert, H., 2009. Indole compounds in fruiting bodies of some selected Macromycetes species and their culture in vitro. Pharmazie 64, 479–480.

Muszyńska, B., Sułkowska-Ziaja, K., Ekiert, H., 2011a. Indole compounds in fruiting bodies of some edible Basidiomycota species. Food Chem. 125, 1306–1308.

Muszyńska, B., Sułkowska-Ziaja, K., Ekiert, H., 2011b. Indole compounds in edible *Basidiomycota* species: *Boletus edulis* (Bull. ex Fr.), *Suillus luteus* (L. ex Fr.) and *Pleurotus ostreatus* (Jacq.ex Fr.) Kumm. Int. J. Med. Mushrooms 13, 449–454.

Muszyńska, B., Maślanka, A., Sułkowska-Ziaja, K., Ekiert, H., 2011c. Analysis of indole compounds in *Armillaria mellea* fruiting bodies. Acta. Pol. Pharm. 68, 93–97.

Muszyńska, B., Sułkowska-Ziaja, K., 2012. Analysis of indole compounds in edible Basidiomycota species after thermal processing. Food Chem. 132, 455–459.

Muszyńska, B., Sułkowska-Ziaja, K., Wójcik, A., 2013. Levels of physiological active indole derivatives in the fruiting bodies of some edible mushrooms (Basidiomycota) before and after thermal processing. Mycoscience 54, 321–326.

Nielsen, H.K., De Weck, D., Finot, P.A., Liardon, R., Hurrell, R.F., 1985. Stability of tryptophan during food processing and storage. Brit. J. Nutr. 53, 281–292.

Reiter, R.J., Tan, D.X., Burkhardt, S., Manchester, L.C., 2001. Melatonin in Plants. Nutr. Rev. 59, 286–290.

Singhal, N.K., Srivastava, G., Agrawal, S., Jain, S.K., Singh, M.P., 2012. Melatonin as a Neuroprotective Agent in the Rodent Models of Parkinson's Disease: Is it All Set to Irrefutable Clinical Translation? Mol. Neurobiol. 45, 186–199.

Turner, E.H., Loftis, J.M., Blackwell, A.D., 2006. Serotonin a la carte: Supplementation with the serotonin precursor 5-hydroxytryptophan. Pharmacol. Therapeut. 109, 325–338.

Walsh, H.A., Daya, S., Whiteley, C.G., 1994. Mode of inhibition of rat liver tryptophan 2,3-dioxygenase by melatonin. J. Pineal Res. 16, 188–192.

Wang, X., 2009. The anti-apoptotic activity of melatonin in neurodegenerative diseases. CNS Neurosci. Ther. 15, 345–357.

Wasser, S.P., Weis, A.L., 1999. Medicinal Properties of Substances Occurring in Higher Basidiomycetes Mushrooms: Current Perspectives. Int. J. Med. Mushrooms 1, 47–50.

CHAPTER

8

Impact of Processing on Bioactive Compounds of Field Peas

Joyce Irene Boye, Zhen Ma

Food Research and Development Centre, Agriculture and Agri-Food Canada, Quebec, Canada

CHAPTER POINTS

- Pulse ingredients including field peas have remained under exploited, partially due to the presence of undesirable beany flavors, as well as the presence of antinutritional components.
- A variety of processing treatments can be used to inactivate or remove antinutritional compounds in peas while improving their organoleptic acceptability.
- Thermal treatments including boiling, autoclaving, and microwave cooking and treatments such as germination and fermentation can be explored to reduce the amount of trypsin inhibitor, phytic acid, and phenolics in field peas.
- In general, protein content and protein and starch digestibility, as well as mineral bioactivity are significantly improved after soaking, wet cooking, and fermentation.
- There is a need for research to further identify and design efficient processing treatments to improve the sensory palatability of pea seeds while minimizing undesirable impacts on nutritional quality.

INTRODUCTION

The field pea (*Pisum sativum*) belongs to the family of cool season legume crops (*Leguminosae*) commonly referred to as pulses. The five major types of pea grown worldwide are green pea, maple pea, Austrian winter pea, marrowfat pea, and yellow pea (Roy *et al.*, 2010). Field peas are high in energy, protein, dietary fiber and mineral content and are a good complement to cereal grains.

The major producing countries of fresh/green field peas are China (with 9808 billion tons in production, valued at US$ 3.2 billion), followed by India, France, and the United Kingdom (Food and Agriculture Organization of the United Nations, 2010). Canada is the leading producer of dry field peas, with a production of 2862 billion tons, valued at US$ 0.47 billion, followed by the USA, India, Russia, France, and Australia.

The health and nutritional properties of pulse legumes (e.g., pea, lentil, chickpea, bean, etc.) are subjects of growing interest. Various studies have indicated that frequent consumption of pulses such as peas can reduce the risk of cardiovascular disease (Flight and Clifton, 2006), and diabetes, and may have protective effects against numerous types of cancers such as breast cancer, colon and rectal cancer (Campos-Vega *et al.*, 2010). Specific components responsible for these beneficial effects are still under investigation. Studies also suggest that some components in pulses traditionally considered to be antinutritional may have positive health benefits. For example, phytic acid has antioxidant and DNA protective effects (Campos-Vega *et al.*, 2010); lectins may play a key role in preventing certain cancers and may be used for controlling obesity (González *et al.*, 2005); some protease inhibitors have been shown to have anti-inflammatory properties (Roy *et al.*, 2010); phenolic compounds including phenolic acids and flavonoids exhibit antioxidant properties (Martine, 2002); and resistant starch, and indigestible oligosaccharides may exert prebiotic effects (Martínez-Villaluenga *et al.*, 2008). Several studies have also shown that protein hydrolysates from pulses possess angiotensin I converting enzyme (ACE) inhibitory properties and have bile acid binding properties and may help to prevent cardiovascular disease (Li and Aluko, 2010).

Processing and Impact on Active Components in Food
http://dx.doi.org/10.1016/B978-0-12-404699-3.00008-1

HOW COMPOSITION IS ALTERED

The composition of the field pea is affected by genotype, environment (i.e., location, biotic and abiotic stresses, etc.), and processing. Tables 8.1A and 8.1B present the macronutrient and micronutrient (minerals) composition of various field pea varieties. Starch is the major component and its content in raw peas ranges from 33.4% to 49.4%. Total dietary fiber content ranges from 14.3% to 16.8%, and the level of resistant starch varies from 8.1% to 58%, depending on variety (Hoover *et al.*, 2010). Peas also contain high amounts of protein ranging from 20.2% to 30.9%. The major proteins in peas are albumins and globulins representing 15–25% and 50–60% of the total protein, respectively. Ash content of peas varies from 1.34% to 4.06% (Table 8.1A). The composition of antinutritional factors (ANFs) in peas is presented in Table 8.2.

Figure 8.1 presents a schematic showing the different ways peas are processed and utilized. As shown below, processing can have an impact on the macronutrients, micronutrients, and antinutritional composition of peas.

Antinutritional Factors

Enzyme Inhibitors

Protease inhibitors are heat-labile proteins that are sensitive to both dry and wet heating, thus thermal treatment can be used as an effective method to decrease their activity. As shown in Table 8.2, boiling, roasting, microwave cooking, and autoclaving can markedly decrease or completely remove the trypsin inhibitors in field peas. Trypsin inhibitor (TI) content decreased by 17.0–19.9% after soaking (i.e., trypsin inhibitor unit (TIU) levels decreased from 1135.27–1297.22 per g to 942.31–1039.71 per g). Dehulling and soaking also caused decreases in chymotrypsin inhibitory activity (Alonso *et al.*, 1998). Protein isolates and concentrates obtained after fractionation of pea flours also displayed reduced trypsin inhibitor activity compared to the pea seeds (Fernández-Quintela *et al.*, 1997). Alonso *et al.* (1998) reported a 100% reduction of α-amylase inhibitory activity (AIA) after extrusion.

Lectins

Some lectins are heat-labile, while others may be resistant to mild-heat treatment. In addition to the type of heat treatment used (i.e., dry or wet), the haemagglutinating activity (HA) of pulses will thus depend on the processing temperature applied. Unfortunately, there is very little information in the literature on the effect of different processing treatments on HA in different pea varieties. Alonso *et al.* (1998) reported a decrease of HA from 5.1–6.2 to 0.1–0.1 (units of HA, per mg) after extrusion of pea flours which represented 98–100% reduction. Microwave cooking of soaked seeds also decreased lectin content.

Phytic Acid

Soaking and thermal treatment have all been shown to decrease the levels of phytic acid in peas. As shown in Table 8.2, soaking reduced phytic acid levels in pea

TABLE 8.1A Macronutrient Composition of Field Peas as Affected by Variety and Processing

Pea Variety/Type	Moisture (%)	Ash (%)	Protein (%)	Fat (%)	Starch (%)	Fiber (%)	Reference
Field pea (raw, Greece)	NA	3.05–4.06 (db)	24.3–32.5 (db)	0.76–3.95 (db)	33.4–47.5 (db)	NA	Nikolopoulou *et al.* (2007)
Field pea (raw)	NA	2.8–3.3	21.0–25.7	1.2–1.7	45.3	6.6–7.2 (CF)	Hanbury *et al.* (2000)
Field pea (raw, Canadian)	8.11	2.94	21.62	2.43	NA	4.66 (CF)	Khattab *et al.* (2009)
Field pea (raw, different varieties)[a]	NA	1.34–3.39	20.16–26.66	1.0–1.69	41.6–47.5	NA	Wang and Daun (2004a)
Field pea (raw, different cultivars)[b]	NA	2.53–2.72	22.90–24.50	NA	45.70–49.45	14.26–16.83 (DF)	Wang *et al.* (2008)
Field pea (soaked, different cultivars)[b]	NA	2.39–2.52	23.50–25.20	NA	46.45–50.35	15.75–17.28 (DF)	Wang *et al.* (2008)
Field pea (cooked, different cultivars)[b]	NA	1.70–2.09	23.75–25.75	NA	46.80–51.00	17.72–20.91 (DF)	Wang *et al.* (2008)
Field pea (dehulled, different cultivars)[b]	NA	2.54–2.75	24.55–26.70	NA	50.45–53.80	9.03–9.69 (DF)	Wang *et al.* (2008)

CF, crude fiber; DF, dietary fiber; db, dry basis; NA, not available.

[a] *Values are the range for different varieties of field peas including Alfetta, Carneval, Keoma, and Majoret.*

[b] *Values are the range for different varieties of field peas including Nitouche, Keoma, SW Parade, Eclipse, Delta, and CDC Mozart.*

by 40.1–48.5% (Khattab and Arntfield, 2009). Phytic acid exists in dried pulses in its water-soluble salt form, thus phytate ions can leach into the soak water which would explain the reduction. Further reduction in phytic acid content can be obtained by prolonging the soaking time (Bishnoi *et al.*, 1994). Boiling pea seeds also reduced phytic acid levels by 51.0–58.5% (Table 8.2). Furthermore, combining soaking and cooking provides greater reduction in phytic acid content compared with cooking of unsoaked pea seeds (Bishnoi *et al.*, 1994). Autoclaving with water decreased phytic acid content in field peas by 67.5–70.2% (Table 8.2). However, dry heat treatment such as micronization and roasting are less effective in reducing phytic acid content.

Phenolics

Soaking of pea seeds in water for 12–18h resulted in a reduction in polyphenolic compounds by 50–58% whereas pressure cooking of soaked pea seeds decreased polyphenolic content by 89%. However, ordinary and pressure cooking of unsoaked pea seeds had only marginal effects (8–12%) on reducing polyphenolic content of different pea cultivars. In general, boiling resulted in the highest reduction in tannin content of pea followed by autoclaving and microwave cooking (Khattab and Arntfield, 2009). Extrusion cooking also markedly reduced (83.6%) tannin content, likely due to the destruction of condensed tannins at high temperature (Alonso *et al.*, 1998). Roasting seeds, however, provided the least reduction of tannin content (Khattab and Arntfield, 2009).

Macronutrients

Protein Content and Digestibility

Soaking, cooking, and dehulling increased determined protein content in different varieties of field peas by 2.6–5.0%, 3.0–6.7%, and 5.4–10.4%, respectively (Wang *et al.*, 2008). Peas generally have lower protein digestibility amino acid scores (PDCAAS) compared with other legumes such as soybean (Table 8.3), however its protein quality can be improved when used in mixed diets. According to Khattab *et al.* (2009), *in vitro* protein digestibility (IVPD) significantly improves after soaking, boiling, microwave cooking, and autoclaving. Dry and wet heat treatments have a significant impact on improving IVPD either by inactivating the protease inhibitors or by denaturing the proteins which exposes the interior parts of the protein structure allowing greater access of gastrointestinal enzymes for hydrolysis. Roasting and micronization, however, reduced the IVPD of peas.

Starch Content and Digestibility

Starch content and digestibility are affected by processing. Soaking, cooking, and dehulling increased the determined starch content of field peas by 0.8–2.3%, 2.4–6.7%, and 7.6–11.8%, respectively (Wang *et al.*, 2008). Processing of pea may lead to a loss of soluble solids in the pea seeds which would explain this increase.

An important property of starch in terms of nutritional value is starch digestibility. Pressure cooking was the most effective method for increasing starch digestibility,

TABLE 8.1B Mineral Content of Field Peas as Affected by Variety and Processing

Minerals	Different Field Pea Varieties[a] (mg/100g, db)	Field Peas (*P. sativum*)	Canadian Field Peas (mg/100g, db)	Different Field Pea Varieties (raw)[b] (mg/100g)	Different Field Pea Varieties (soaked)[b] (mg/100g)	Different Field Pea Varieties (cooked)[b] (mg/100g)	Different Field Pea Varieties (dehulled)[b] (mg/100g)
Cu	0.4–0.9	4.8 (mg/kg)	0.5–106.9	0.55–0.84	0.60–0.98	0.63–1.09	0.50–0.78
Fe	4.3–7.9	53 (mg/kg)	4.1–7.9	4.21–5.20	4.09–5.11	3.56–4.96	3.63–4.80
Zn	2.5–5.2	30 (mg/kg)	2.5–6.4	2.42–3.74	2.35–3.68	2.18–3.16	2.35–3.70
Ca	59.6–106.9	0.07 (%)	59.6–106.9	75.0–91.0	80.3–96.9	81.6–108.8	37.5–49.7
K	876.1–1463.9	0.91 (%)	687.4–1473.2	888.0–953.0	794.5–909.5	440.0–599.0	920.5–985.5
Mg	130.4–172.3	0.12 (%)	115.4–172.3	117.0–128.0	117.5–130.8	109.5–120.5	96.5–110.5
P	270.3–950.5	0.34 (%)	226.5–950.5	297.8–318.8	300.0–335.0	317.0–365.5	322.0–380.5
Mn	0.8–2.4	14 (mg/kg)	0.2–0.4	0.89–1.33	0.90–1.36	0.94–1.38	0.77–1.24
References	Wang and Daun (2004b)	Hanbury *et al.* (2000)	Wang and Daun (2004a)	Wang *et al.* (2008)	Wang *et al.* (2008)	Wang *et al.* (2008)	Wang *et al.* (2008)

db, dry basis.

[a] *Different field pea varieties include Alfetta, Carneval, Keoma, and Majoret.*

[b] *Different field pea varieties include Nitouche, Keoma, SW Parade, Eclipse, Delta, and CDC Mozart.*

TABLE 8.2 Impact of Processing on Antinutrients in Field Peas

Varieties and Treatment	Enzyme Inhibitors	Phytic Acid	Tannin and Total Phenolics	Oligosaccharides[a]	References
Canadian and Egyptian field peas (raw)	1135.27–1297.22 (TIU/g)	0.815–0.89 (g/100g)	27.80–30.93 (tannin, mg/g)	5.63–5.73 (g/100g)	Khattab and Arntfield (2009)
Canadian and Egyptian field peas (soaked)	942.31–1039.71 (TIU/g)	0.458–0.488 (g/100g)	3.72–4.14 (tannin, mg/g)	3.61–3.64 (g/100g)	Khattab and Arntfield (2009)
Canadian and Egyptian field peas (boiled)	0 (TIU/g)	0.369–0.399 (g/100g)	1.67–2.94 (tannin, mg/g)	2.00–2.05 (g/100g)	Khattab and Arntfield (2009)
Canadian and Egyptian field peas (roasted)	0 (TIU/g)	0.503–0.563 (g/100g)	4.00–4.74 (tannin, mg/g)	4.26–4.36 (g/100g)	Khattab and Arntfield (2009)
Canadian and Egyptian field peas (microwaved)	0 (TIU/g)	0.295–0.354 (g/100g)	1.27–5.06 (tannin, mg/g)	2.22–2.29 (g/100g)	Khattab and Arntfield (2009)
Canadian and Egyptian field peas (autoclaved)	0 (TIU/g)	0.265 (g/100g)	2.57–2.91 (tannin, mg/g)	1.27–1.29 (g/100g)	Khattab and Arntfield (2009)
Canadian and Egyptian field peas (fermented)	669.76–758.31 (TIU/g)	0.265–0.295 (g/100g)	1.19–2.26 (tannin, mg/g)	1.60–1.63 (g/100g)	Khattab and Arntfield (2009)
Canadian and Egyptian field peas (micronized)	73.35–90.78 (TIU/g)	0.555–0.592 (g/100g)	3.16–5.40 (tannin, mg/g)	4.49–4.60 (g/100g)	Khattab and Arntfield (2009)
Australian field peas (raw)	0.40–2.10 (TIA, mg/g, db)	0.22–0.99 (g/100g, db)	0.01–1.10 (tannin, g/100g, db)	2.4–5.8 (g/100g, db)	Wang and Daun (2004a)
Canadian yellow field peas (raw)	2.05–3.32 (TIA, g/kg, db)	NA	NA	50.4–72.6 (g/kg, db)	Wang *et al.* (2003)
Canadian yellow field peas (soaked for 24h)	1.96–3.03 (TIA, g/kg, db)	NA	NA	45.4–65.0 (g/kg, db)	Wang *et al.* (2003)
Canadian yellow field peas (cooked for 30min)	0.34–0.41 (TIA, g/kg, db)	NA	NA	28.2–41.3 (g/kg, db)	Wang *et al.* (2003)
Field peas (different cultivars, Polish)[c]	1.28–7.32 (TIU/mg)	6.32–13.36 (mg/g)	NA	52.8–75.8 (g/kg, db)	Zdunczyk *et al.* (1997)
Field peas	5.4–7.8 (TIU per mg, db); 4.4–12.5 (TIA, mg/g, db); 740–10240 (CTIA, per g, db); 14–80 (AIA, per g, db)	0.2–1.3 (%, db)	0.0–1.3 (tannin, %, db) 0.25 (total phenolics, %, db)	NA	Campos-Vega *et al.* (2010) and references therein

db, dry basis; CTIA, chymotrypsin inhibitor activity.
[a] *Oligosaccharide content is the total of verbascose, stachyose, and raffinose.*

followed by ordinary cooking, sprouting, dehulling, and soaking (Bishnoi and Khetarpaul, 1993). Enhanced starch digestibility of cooked peas is likely due to the swelling and rupture of starch granules which takes place during moist cooking, which may contribute to a more randomized configuration facilitating hydrolysis by α-amylase. Uncooked samples, however, are ungelatinized and are less readily hydrolyzed.

Soluble Fiber

The major soluble sugars found in peas are sucrose and the oligosaccharides raffinose, starchyose, and verbascose (i.e., soluble fiber). The effect of various processing treatments (e.g., soaking, boiling, roasting, microwaving, autoclaving, and micronization) on the oligosaccharide content of peas is presented in Table 8.2. Different processing treatments modify the physicochemical structure of peas which can affect polysaccharide composition including fiber. Soaking significantly reduces sucrose (16.4–26.7%) and stachyose content (3.6–13.6%), likely due to leaching of the sugars into the soak water. Cooking also causes marked reductions in oligosaccharide content possibly due to heat hydrolysis of raffinose, stachyose, and verbascose, leaching of the sugars into the cooking water, or due to the formation of other complex compounds during cooking. In addition to soluble fiber, peas contain high amounts of insoluble fiber (e.g., hemicellulose, cellulose, and lignin) (Stoughton-Ens *et al.*, 2010). Soaking and cooking increased the total dietary fiber content in peas by 10.4–21.2% and 24.3–46.6%, respectively, whereas dehulling decreased it from 14.26–16.83% to 9.03–9.69% (Table 8.1A) (Wang *et al.*, 2008).

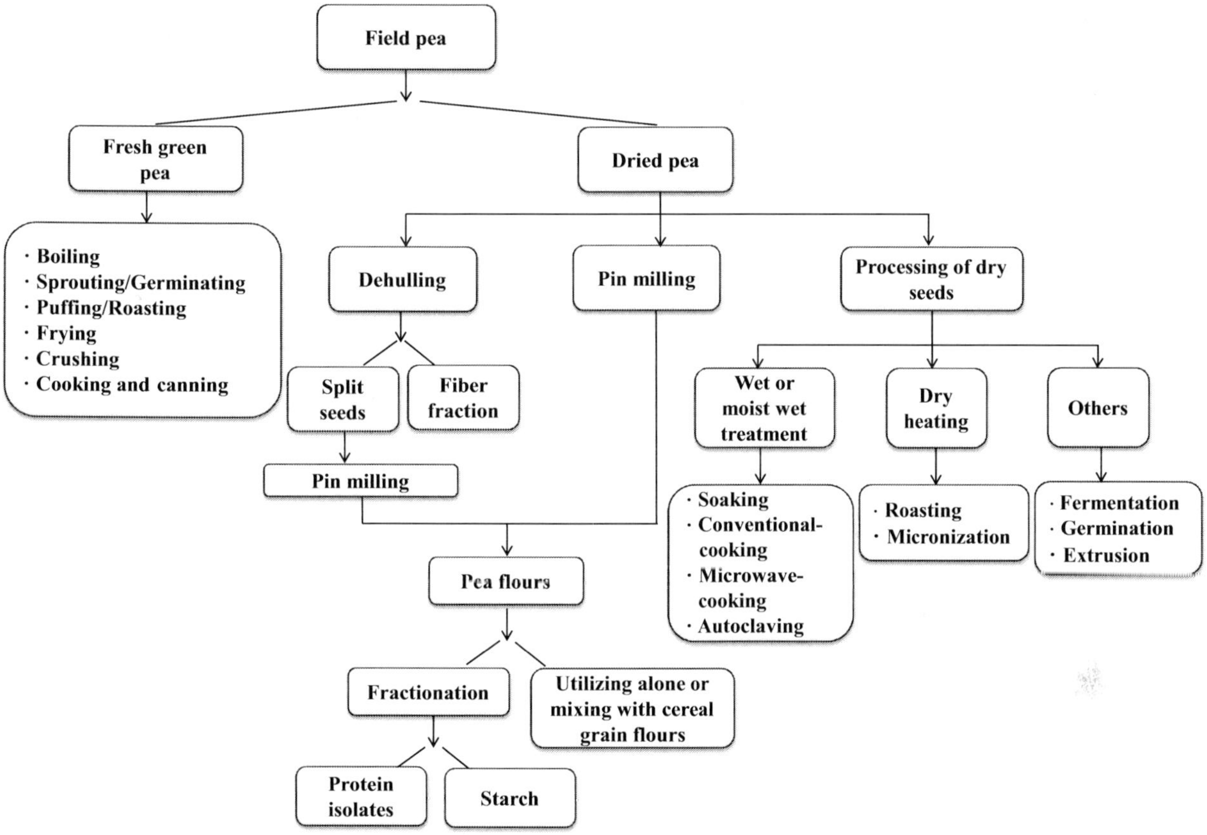

FIGURE 8.1 Schematic Showing the Processing and Utilization of Field Pea

TABLE 8.3 Protein Digestibility Corrected Amino Acid Scores (PDCAAS) for Field Peas

Field Pea Type	Treatment	Total Protein, g/100 g Edible Portion, Unless Indicated	Protein Digestibility (%)	PDCAAS (%) Calculated Using Reference Pattern for 1- to 2-Year-Old Child, and LAA	PDCAAS (%) Calculated Using Reference Pattern for 3- to 10-Year-Old Child, and LAA	Methods Used (Reference Pattern and Digestion Method)	References
Pea (Organic cultivation, 2002)	Cooked (and freeze-dried)	25.9 (DM)	90.3	82 Met+Cys		2- to 5-year-old child *in vivo* (rat)	Jørgensen *et al.* (2008)
Pea, Canadian	Soaked 18h. Dried overnight at 55°C and ground	21.6	83.7	44 Met+Cys	48 Met+Cys	FAO/WHO (1973). *In vitro*	Khattab *et al.* (2009)
Pea, Canadian	Cooked in microwave oven for 15min, with water. Dried overnight at 55°C and ground	21.6	89.1	84 Met+Cys	91 Met+Cys	FAO/WHO (1973). *In vitro*	Khattab *et al.* (2009)
Pea, var. Nebraska, (Egyptian)	Roasted for 15min at 180°C, in sand bath	23.1	75.0	44 Met+Cys	48 Met+Cys	FAO/WHO (1973). *In vitro*	Khattab *et al.* (2009)
Pea, var. Nebraska, (Egyptian)	Cooked in microwave oven for 15min, with water. Dried overnight at 55°C and ground	23.1	90.9	92 Ile	92 Ile	FAO/WHO (1973). *In vitro*	Khattab *et al.* (2009)

LAA, limiting amino acid.
Adapted from Boye et al., *2012.*

MICRONUTRIENTS

The relative low bioavailability of minerals in peas can be attributed to the presence of phytates and iron-binding polyphenols. Processing such as soaking and hydrothermal processing can increase phytate and polyphenol degradation and thereby increase mineral bioactivity. Addition of ascorbic acid which enhances iron bioavailability may also counteract the deleterious effects of phytate and polyphenols (Siegenberg *et al.*, 1991). Differences in the mineral composition of several field pea varieties as affected by processing are provided in Table 8.1B. Cooking can have negative impacts on some micronutrients. As reported by Dang *et al.* (2000), cooking peas in boiling water reduces the folate content by 55% suggesting that folates may leach into the cooking water.

OTHER WAYS IN WHICH COMPOSITION IS ALTERED

In addition to dehulling, soaking, and wet and dry thermal treatments, germination and fermentation can positively impact the composition of field peas by improving their nutritional quality while maintaining their acceptable sensory properties.

TABLE 8.4 Brief Description of Analytical Methods used to Evaluate the Composition of some Bioactive Compounds in Field Peas

Bioactive Compounds	Sample Preparation	Method/Instrument of Analysis
ANTINUTRITIONAL FACTORS		
Trypsin inhibitors	Extraction with NaOH, using benzoyl-DL-arginine-p-nitroanalide hydrochloric as substrate	Colorimetric method using a spectrophotometer at 410 nm
Phytic acid	Extraction and separation by ion-exchange chromatography	Quantified colorimetrically using a spectrophotometer at 500 nm (using Wade reagent)
Lectin	Measured as haemagglutination activity, trypsin treated rat erythrocytes used	One unit of haemagglutinating (lectin) activity (HU) is defined as the amount of sample in the last dilution which causes agglutination of 50% of the blood cells
Oligosaccharides	Extraction with ethanol (lactose is used as an internal standard) or with 2-propanol (arabinose used as an internal standard)	High-performance anion exchange chromatography (HPAE) with pulsed amperometric detection (PAD) or by HPLC with a differential refractometer detector
Tannins	Preparation of tannin extract using HCl (heating)	Gravimetric determination of tannin based on binding with insoluble polyvinylpyrrolidone (PVP) or by using Prussian blue photometric method
Total phenolic compounds	Extraction with methanol containing HCl, mixing of extract with Folin–Ciocalteu reagent	Colorimetric method using a spectrophotometer at 725 nm
DIETARY PROTEIN AND STARCH QUALITY		
Total amino acids	Hydrolysis of sample with HCl or alkali (cysteine and cystine are measured as cysteic acid and methionine as methionine sulphone after performic acid oxidation and HCl hydrolysis; tryptophan analysis is performed on hydrolysates obtained by alkaline hydrolysis of the samples)	High performance liquid chromatography (gas chromatography can also be used)
In vitro protein digestibility (IVPD)	Incubation of samples with a mixture of multienzymes (chymotrypsin, trypsin, and peptidase) at 37°C, by adjusting the pH to 8.0	The pH(x) of the sample is recorded at 10/15 min after adding the multienzyme solution, IVPD is calculated as follows: $Y = 210.464 - 18.103\, pH(x)$
Amino acid score (AAS)		AAS (dmnl) = mg of limiting amino acid in 1 g test protein/mg of limiting amino acid in a reference protein or requirement pattern
Protein digestibility corrected amino acid score (PDCAAS)	Relies on determination of protein contents, amino acid profile and protein digestibility	PDCAAS (% or dmnl) = [(mg of limiting amino acid in 1 g of test protein)/(mg of same amino acid in 1 g of reference protein or requirement pattern)] × true fecal digestibility (%) × 100
Starch digestibility	Incubation with α-amylase	Measurement of glucose content using a glucose oxidase-peroxidase (GOPOD) kit; or measuring maltose liberated by using a spectrophotometer at a wavelength of 550 nm

dmnl, dimensionless.

Germination increases the determined protein content in pea seeds due to the utilization of carbohydrates as energy sources for developing sprouts. Alonso *et al.* (1998) reported a significant increase in determined protein content from 18.8–21.6% to 20.2–22.8% after germination for 72h, and successive increases were noticed with increases in germination time. Fermentation decreased the phytic acid and trypsin inhibitor contents in field peas by 63.8–70.2% and 41.0–41.5%, respectively (Table 8.2). Germination similarly results in a reduction of phytic acid in field and vegetable peas. Germination of peas for 24h reduced the polyphenolic content of pea by 21.6–62.0%. Reduction levels in total polyphenols ranging between 37.8 and 51.3%, and 48.6 and 74.0% for samples germinated for 48h and 72h, respectively, have been reported (Alonso *et al.*, 1998). The extent of reduction in the total polyphenol content of germinated seeds thus appears to depend on the time allowed for germination.

Starch digestibility of peas significantly increased after 48h germination due to the pre-digestion of starch by amylolytic enzymes. Similar to thermal treatments, germination and fermentation can also increase mineral bioactivity by increasing phytate and polyphenol degradation. Germinated pea (2 or 4 days with or without light) increased levels of vitamin B_1 (Thiamine) and B_2 (Riboflavin) (Urbano *et al.*, 2005).

ANALYTICAL TECHNIQUES

Understanding the impact of processing on the bioactive compounds of field peas requires the use of reliable and reproducible quantitative methods. TI levels in foods are expressed either as TIU or TI activity (TIA). The analytical method used to measure TI levels is achieved indirectly by inhibiting the activity of trypsin. A synthetic substrate (benzyme-DL-arginen-para-nitroanilide) is subjected to hydrolysis by trypsin to produce yellow-colored p-nitroanilide, the degree of inhibition by the sample extract of the yellow-color production is then measured at 410nm using a spectrophotometer (Table 8.4). Lectin content is frequently measured by determining the HA as indicated in Table 8.4. Total phenolic compounds in peas can be measured by colorimetric methods after mixing of extract with Folin–Ciocalteau reagent. High-pressure liquid chromatrography-mass spectrometry (HPLC-MS) can also be used for the determination of total phenolic compounds in field peas.

Amino acid analysis is often based on separation of free amino acids by ion-exchange chromatography followed by post-column derivatization after hydrolyzing protein into individual amino acids. PDCAAS is a measure of protein quality. The PDCAAS is a function of the content of indispensable amino acid and protein digestibility. Protein digestibility can be measured using both *in vivo* and *in vitro* techniques. One of the frequently used methods for evaluating *in vitro* protein digestibility (IVPD) is by measuring the pH drop after incubating the sample with a mixture of multienzymes for specific periods (Table 8.4). A summary of analytical techniques commonly used in previous studies for measuring other bioactive compounds such as phytic acid, oligosaccharides and tannins is presented in Table 8.4.

References

Alonso, R., Orue, E., Marzo, F., 1998. Effects of extrusion and conventional processing methods on protein and antinutritional factor contents in pea seeds. Food Chem. 63, 505–512.

Bishnoi, S., Khetarpaul, N., Yadav, R.K., 1994. Effect of domestic processing and cooking methods on phytic acid and polyphenol contents of pea cultivars (*Pisum sativum*). Plant Foods Hum. Nutr. 45, 381–388.

Bishnoi, S., Khetarpaul, N., 1993. Effect of domestic processing and cooking methods on in-vitro starch digestibility of different pea cultivars (*Pisum sativum*). Food Chem. 47, 177–182.

Boye, J., Wijesinha-Bettoni, R., Burlingame, B., 2012. Protein quality evaluation twenty years after the introduction of the protein digestibility corrected amino acid score method. Br. J. Nutr. 108, S183–S211.

Campos-Vega, R., Loarca-Piña, G., Oomah, B.D., 2010. Minor components of pulses and their potential impact on human health. Food Res. Int. 43, 461–482.

Dang, J., Arcot, J., Shrestha, A., 2000. Folate retention in selected processed legumes. Food Chem. 68, 295–298.

Fernández-Quintela, A., Macarulla, M.T., Del Barrio, A.S., Martínez, J.A., 1997. Composition and functional properties of protein isolates obtained from commercial legumes grown in northern Spain. Plant Foods Hum. Nutr. 51, 331–341.

Flight, I., Clifton, P., 2006. Cereal grains and legumes in the prevention of coronary heart disease and stroke: a review of the literature. Eur. J. Clin. Nutr. 60, 1145–1159.

González De Mejía, E., Prisecaru, V.I., 2005. Lectins as bioactive plant proteins: A potential in cancer treatment. Crit. Rev. Food Sci. Nutr. 45, 425–445.

Hanbury, C.D., White, C.L., Mullan, B.P., Siddique, K.H.M., 2000. A review of the potential of *Lathyrus sativus* L. and *L. cicera* L. grain for use as animal feed. Anim. Feed Sci. Technol. 87, 1–27.

Hoover, R., Hughes, T., Chung, H.J., Liu, Q., 2010. Composition, molecular structure, properties, and modification of pulse starches: a review. Food Res. Int. 43, 399–413.

Jørgensen, H., Brandt, K., Lauridsen, C., 2008. Year rather than farming system influences protein utilization and energy value of vegetables when measured in a rat model. Nutr. Res. 28, 866–878.

Khattab, R.Y., Arntfield, S.D., Nyachoti, C.M., 2009. Nutritional quality of legume seeds as affected by some physical treatments, Part 1: Protein quality evaluation. LWT-Food Sci. Technol. 42, 1107–1112.

Khattab, R.Y., Arntfield, S.D., 2009. Nutritional quality of legume seeds as affected by some physical treatments 2. Antinutritional factors. LWT—Food Sci. Technol. 42, 1113–1118.

Li, H., Aluko, R.E., 2010. Identification and inhibitory properties of multifunctional peptides from pea protein hydrolysate. J. Agric. Food Chem. 58, 11471–11476.

Martine, M.J., 2002. Non-nutrient bioactive substances of pulses. Br. J. Nutr. 88, S307–S319.

Martínez-Villaluenga, C., Frias, J., Vidal-Valverde, C., 2008. Alpha-galactosides: antinutritional factors or functional ingredients? Crit. Rev. Food Sci. Nutr. 48, 301–316.

Nikolopoulou, D., Grigorakis, K., Stasini, M., Alexis, M.N., Iliadis, K., 2007. Differences in chemical composition of field pea (*Pisum sativum*) cultivars: Effects of cultivation area and year. Food Chem. 103, 847–852.

Roy, F., Boye, J.I., Simpson, B.K., 2010. Bioactive proteins and peptides in pulse crops: Pea, chickpea and lentil. Food Res. Int. 43, 432–442.

Siegenberg, D., Baynes, R.D., Bothwell, T.H., Macfarlane, B.J., Lamparelli, R.D., Car, N.G., MacPhail, P., Schmidt, U., Tal, A., Mayet, F., 1991. Ascorbic acid prevents the dose-dependent inhibitory effects of polyphenols and phytates on nonheme-iron absorption. Am. J. Clin. Nutr. 53, 537–541.

Stoughton-Ens, M.D., Hatcher, D.W., Wang, N., Warkentin, T.D., 2010. Influence of genotype and environment on the dietary fiber content of field pea (*Pisum sativum* L.) grown in Canada. Food Res. Int. 43, 547–552.

Urbano, G., Aranda, P., Vílchez, A., Aranda, C., Cabrera, L., Porres, J.M., López-Jurado, M., 2005. Effects of germination on the composition and nutritive value of proteins in *Pisum sativum*, L. Food Chem. 93, 671–679.

Wang, N., Daun, J.K., 2004a. The Chemical composition and nutritive value of Canadian pulses. Canadian Grain Commission (CGC), 19–29.

Wang, N., Daun, J.K., 2004b. Effect of variety and crude protein content on nutrients and certain antinutrients in field peas (*Pisum sativum*). J. Sci. Food Agric. 84, 1021–1029.

Wang, N., Daun, J.K., Malcolmson, L.J., 2003. Relationship between physicochemical and cooking properties, and effects of cooking on antinutrients, of yellow field peas (*Pisum sativum*). J. Sci. Food Agric. 83, 1228–1237.

Wang, N., Hatcher, D.W., Gawalko, E.J., 2008. Effect of variety and processing on nutrients and certain anti-nutrients in field peas (*Pisum sativum*). Food Chem. 111, 132–138.

Zdunczyk, Z., Godycka, I., Amarowicz, R., 1997. Chemical composition and content of antinutritional factors in Polish cultivars of peas. Plant Foods Hum. Nutr. 50, 37–45.

CHAPTER

9

Carotenoids in Pumpkin and Impact of Processing Treatments and Storage

*João Gustavo Provesi**, *Edna Regina Amante*†

*Department of Food Science and Technology, Federal Institute of Education, Science and Technology of Santa Catarina, Urupema, Brazil, †Laboratory of Fruits and Vegetables, Department of Food Science and Technology, Federal University of Santa Catarina, Florianópolis, Brazil

CHAPTER POINTS

- In general, there is a predominance of one to four carotenoids in pumpkins. β-Carotene is the principal carotenoid in most species.
- There are significant variations in the concentrations of carotenoids in pumpkins, even within the same species or variety.
- Typically, cultivation in regions with lower average temperatures tends to cause a decrease in concentrations of carotenoids. The ripening process is accompanied by an increase in carotenogenesis.
- During processing and storage of pumpkin products, carotenoids may undergo degradation reactions such as oxidation and isomerization, depending mainly on the temperature and contact with oxygen and exposure to light.
- Home-prepared pumpkin (prepared in water, either boiled or steamed pumpkin), have higher retention than when prepared in oil, such as fried or stir-fried pumpkin. However, studies reported a higher bioavailability of carotenoids in preparations with oil.
- The traditional hot air-drying method causes greater loss of carotenoids during dehydration of pumpkins than methods where the temperature or contact with oxygen are lower, such as freeze-drying or vacuum methods.
- The losses of carotenes during industrial processing of pumpkin puree are small, because of the low concentration of oxygen in the product. However, the impact on xanthophylls is higher.

INTRODUCTION

The pumpkin is a member of the Cucurbitaceae family and is native to tropical and subtropical America. The term 'pumpkin' is often applied to fruits of different species of the genus *Cucurbita*, such as *C. pepo* L., *C. maxima* Duchesne, *C. moschata* Duchesne, *C. argyrosperma* Huber, and *C. ficifolia* Bouché (Ferriol and Picó, 2008). Besides the different species of pumpkins, there are also several varieties, which differ in shape, color, and chemical composition.

Pumpkin seeds are an excellent source of unsaturated fatty acids, fibers, and minerals. Pectin is a polysaccharide that is widely used in the food industry and can be extracted from pumpkin pulp. In addition, recent studies have also reported the presence of polysaccharides that have biological activity, such as the capacity to increase serum levels of insulin and reduce blood glucose, showing potential use in the control of diabetes (Caili *et al.*, 2006).

Therefore, the main feature of pumpkin pulp is its concentration of carotenoids. These natural pigments, usually C_{40} tetraterpenoids with a long conjugated chain of double bonds, give flowers and fruits a coloration that ranges from yellow to red. Some carotenoids, such as β-carotene and α-carotene, are precursors of vitamin A, a fat-soluble vitamin which is important for normal growth and development of the human body. Moreover, because of their antioxidant activity, consumption of carotenoids has been related to risk reduction of atherosclerosis and other cardiovascular diseases, protection against macular degeneration, and prevention of certain types of carcinomas, such as those of skin and prostate (Rao and Rao, 2007).

Considering their nutritional value and low production costs, pumpkins have a great potential for

Processing and Impact on Active Components in Food
http://dx.doi.org/10.1016/B978-0-12-404699-3.00009-3

industrialization. There is a range of products that can be developed from pumpkins, offering consumers foods that have long shelf-life and good nutritional value. However, it is necessary to consider the stability of carotenoids during processing and storage of pumpkin products because these compounds have an unsaturated chemical structure, and thus they are susceptible to reactions such as oxidation and isomerization.

Therefore, the goals of this chapter are: (1) to present the carotenoid composition of different species and varieties of pumpkins; (2) to elucidate how processing may affect the composition of carotenoids in pumpkins and processed products made of pumpkin.

CAROTENOIDS IN PUMPKINS: COMPOSITION AND CHANGES

In general, one to four carotenoids are predominant in the pumpkin species, with several other carotenoids at low concentrations or traces. β-carotene is the major carotenoid in most cases. Other carotenoids are also commonly noted, such as α-carotene, lutein, zeaxanthin, and violaxanthin (Figure 9.1).

The number of different compounds may be higher considering the minor carotenoids and esters forms. Arima and Rodriguez-Amaya (1988, 1990) identified 11 carotenoids in *C. maxima* 'Jerimum Caboclo', 14 in *C. moschata* 'Menina Brasileira', 15 in *C. maxima* 'Exposição', and 19 carotenoids in *C. moschata* 'Baianinha'.

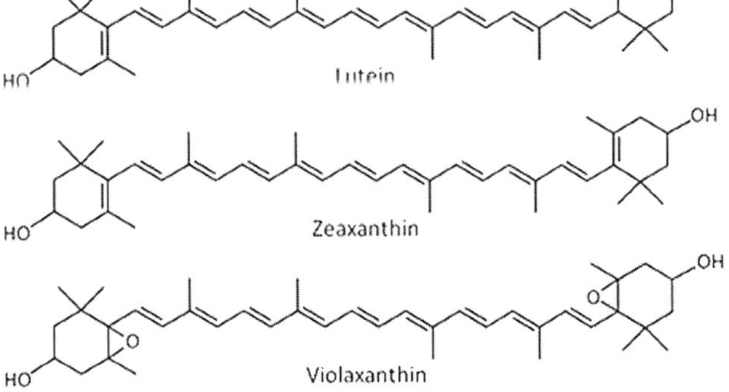

FIGURE 9.1 The chemical structure of the principal carotenoids found in pumpkins.

Table 9.1 lists the concentrations noted for these carotenoids in different species and varieties of pumpkins.

Similarly to that which occurs in other vegetables, the composition of carotenoids in pumpkins differs very much, with significant qualitative and quantitative variations. The concentrations noted for β-carotene range from 0.6 μg/g to values greater than 70 μg/g, higher than those noted in fruits such as papaya, mango, and guava (Wilberg and Rodriguez-Amaya, 1995). α-Carotene is not detected in some varieties, but its concentration is close to 60 μg/g in a sample of *C. moschata* 'Long Island Cheese'. In several varieties, especially in the *C.maxima* species, lutein is the major carotenoid instead of β-carotene.

In general, the *C. moschata* varieties have the same pattern and their major carotenoids are β-carotene and α-carotene. The concentration of these carotenoids varies the most in *C. maxima*, where lutein is the major carotenoid in many of its varieties (Azevedo-Meleiro and Rodriguez-Amaya, 2007).

The composition of carotenoids in pumpkins and other vegetables may be affected by some factors such as maturation stage, differences in growing environment, edaphoclimatic conditions, harvesting and post-harvesting treatment, and analytical variations, among other factors (Figure 9.2) (Rodriguez-Amaya *et al.*, 2008; Provesi *et al.*, 2012). Since there is interaction among species and varieties, the influence of such factors on the carotenoids in each species and variety is not always obvious.

Climate conditions significantly affect carotenoid content in pumpkins. Typically, the biosynthesis of carotenoids decreases in lower temperatures. For example, the species *C. moschata* 'Menina Brasileira' and *C. maxima* 'Exposição' cultivated in the southern region of Brazil, where the average temperatures are lower than throughout the rest of the country, showed lower concentrations of carotenoids in relation to the same varieties harvested in the northeast and southeast regions of Brazil (Godoy and Rodriguez-Amaya, 1998; Azevedo-Meleiro and Rodriguez-Amaya, 2007; Provesi *et al.*, 2011).

Iacuzzo and Dalla Costa (2009) studied the variation in total carotenoids in three different harvests for two varieties of pumpkins widely cultivated in Italy. Changes were smaller in the variety 'Violina', from 21 μg/g in the first year, to 18 μg/g in the second year, and to 26 μg/g in the third year. However, for the variety 'Tetsukabuto' concentrations of 62 μg/g in the first year decreased to 37 μg/g in the second year, returning to higher levels (80 μg/g) in the third year.

Ripening of fruits is generally accompanied by enhanced carotenogenesis, the carotenoids increasing markedly in type and quantity. The concentration of total carotenoids in *C. moschata* 'Menina Brasileira' increased from 5.4 to 79.6 μg/g during ripening (Arima

TABLE 9.1 Carotenoid Concentration (μg/g) in Different Varieties of Pumpkins

Pumpkin	Variety	Origin	Year	Carotenoids (μg/g)[a]			Method	Reference
				β-Carotene	α-Carotene	Lutein		
C. pepo	Acorn Table	Graz, Austria	2000	21	1.5	18[b]	HPLC	Murkovic *et al.* (2002)
	Acorn Tay Bell	Graz, Austria	2000	9.4	1.7	3.7[b]	HPLC	Murkovic *et al.* (2002)
	Carneval di Venezia	Graz, Austria	2000	0.6	0.3	—	HPLC	Murkovic *et al.* (2002)
	Melonette Jaspée Vende	Graz, Austria	2000	13	0.5	4.3[b]	HPLC	Murkovic *et al.* (2002)
	Mogango	Campinas, SP, Brazil	2001–2002	5.4 ± 1.6	nd	9.8 ± 2.9	HPLC	Azevedo-Meleiro and Rodriguez-Amaya (2007)
	Sweet Lightning	Stuttgart, Germany	2006	7.00 ± 0.17		1.31 ± 0.06	HPLC	Kurz *et al.* (2008)
	Tonda Padana	Graz, Austria	2000	23	1.2	15[b]	HPLC	Murkovic *et al.* (2002)
C. maxima	Amazonka	Warsaw, Poland	2005	65 ± 24.6	nq	nq	Spectrophotometric	Konopacka *et al.* (2010)
	Autumn Cup	Graz, Austria	2000	52	8	27[b]	HPLC	Murkovic *et al.* (2002)
	Bambino	Warsaw, Poland	2005	17 ± 6.9	nq	nq	Spectrophotometric	Konopacka *et al.* (2010)
	Bischofsmutze	Stuttgart, Germany	2006	4.73 ± 1.8		0.32 ± 0.01	HPLC	Kurz *et al.* (2008)
	Buen Gusto	Graz, Austria	2000	33	10	63[b]	HPLC	Murkovic *et al.* (2002)
	Exposição	Campinas, SP, Brazil	2001–2002	15.4 ± 4.2	nd	10.7 ± 3.9	HPLC	Azevedo-Meleiro and Rodriguez-Amaya (2007)
	Exposição	Curitibanos, SC, Brazil	2010	13.38 ± 2.25	0.43 ± 0.05	10.43 ± 0.13	HPLC	Provesi *et al.* (2011)
	Exposição	Campinas, SP, Brazil	1983–1987	17 (3.1-28)	0.1 (nd-0.2)	10 (7.2-25)	Open-column chromatography	Godoy and Rodriguez-Amaya (1998)
	Flat White Boer	Graz, Austria	2000	62	75	75[b]	HPLC	Murkovic *et al.* (2002)
	Gelber Zentner	Graz, Austria	2000	22	—	8[b]	HPLC	Murkovic *et al.* (2002)

(Continued)

TABLE 9.1 Carotenoid Concentration (μg/g) in Different Varieties of Pumpkins—cont'd

				Carotenoids (μg/g)[a]				
Pumpkin	**Variety**	**Origin**	**Year**	**β-Carotene**	**α-Carotene**	**Lutein**	**Method**	**Reference**
	Golden Nuggets	Stuttgart, Germany	2005	18.64 ± 0.01		22.75 ± 0.23	HPLC	Kurz *et al.* (2008)
	Halloween	Stuttgart, Germany	2006	8.13 ± 0.84		8.13 ± 0.60	HPLC	Kurz *et al.* (2008)
	Hokkaido I	Stuttgart, Germany	2005	27.14 ± 0.92		29.74 ± 1.56	HPLC	Kurz *et al.* (2008)
	Hokkaido II	Stuttgart, Germany	2006	70.92 ± 4.24		38.67 ± 3.28	HPLC	Kurz *et al.* (2008)
	Hyvita	Graz, Austria	2000	25	9.9	170[b]	HPLC	Murkovic *et al.* (2002)
	Imperial Elite	Graz, Austria	2000	74	11	71[b]	HPLC	Murkovic *et al.* (2002)
	Japan 117	Graz, Austria	2000	72	10	18[b]	HPLC	Murkovic *et al.* (2002)
	Justynka F_1	Warsaw, Poland	2005	70 ± 13.9	nq	nq	Spectrophotometric	Konopacka *et al.* (2010)
	Karowita	Warsaw, Poland	2005	20 ± 5.4	nq	nq	Spectrophotometric	Konopacka *et al.* (2010)
C. maxima	Otylia F_1	Warsaw, Poland	2005	6 ± 1.8	nq	nq	Spectrophotometric	Konopacka *et al.* (2010)
	Snow Delite	Graz, Austria	2000	64	15	16[b]	HPLC	Murkovic *et al.* (2002)
	Uchiki Kuri	Graz, Austria	2000	25	14	30[b]	HPLC	Murkovic *et al.* (2002)
	Umber Cup	Graz, Austria	2000	37	7.9	110[b]	HPLC	Murkovic *et al.* (2002)
	Walfish	Graz, Austria	2000	43	9	39[b]	HPLC	Murkovic *et al.* (2002)

C. moschata	Burpee Butterbush	Graz, Austria	2000	31	9.8	0.8[b]	HPLC	Murkovic *et al.* (2002)
	Butternuts	Stuttgart, Germany	2006	11.40 ± 0.06	0.58 ± 0.01	1.40 ± 0.08	HPLC	Kurz *et al.* (2008)
	Goianinha	Campinas, SP, Brazil	2001–2002	56.7 ± 7.6	23.8 ± 3.3	18.3 ± 5.0	HPLC	Azevedo-Meleiro and Rodriguez-Amaya (2007)
	Long Island Cheese	Graz, Austria	2000	7	59	1.4[b]	HPLC	Murkovic *et al.* (2002)
	Martinica	Graz, Austria	2000	54	16	4.1[b]	HPLC	Murkovic *et al.* (2002)
	Menina Brasileira	Campinas, SP, Brazil	2001–2002	66.7 ± 9.1	26.8 ± 5.1	17.4 ± 3.5	HPLC	Azevedo-Meleiro and Rodriguez-Amaya (2007)
	Menina Brasileira	Curitibanos, SC, Brazil	2010	19.45 ± 2.55	12.60 ± 1.56	0.59 ± 0.18	HPLC	Provesi *et al.* (2011)
	Mousquée de Provence	Graz, Austria	2000	49	28	11[b]	HPLC	Murkovic *et al.* (2002)
	Muscade de Provance	Stuttgart, Germany	2005	8.97 ± 0.69	10.60 ± 0.42		HPLC	Kurz *et al.* (2008)
C. maxima × *C. moschata*	Tetsuka Buto	Graz, Austria	2000	35	24	120[b]	HPLC	Murkovic *et al.* (2002)
	Tetsuka Buto	Campinas, SP, Brazil	2001–2002	30.5 ± 5.4	Tr	56.6 ± 9.7	HPLC	Azevedo-Meleiro and Rodriguez-Amaya (2007)
	Tetsuka Buto	Campinas, SP, Brazil	1983–1987	13 (3.7–18)	0.2 (nd–0.5)	19 (3.5–34)	Open-column chromatography	Godoy and Rodriguez-Amaya (1998)

[a] *In some studies, the values are shown as mean and standard deviation, while in others studies only the mean value is available. Values in parentheses refer to the concentration ranges of carotenoids. nd, not detected; nq, not quantified.*

[b] *In this study, lutein and zeaxanthin were not separated. For this reason, the concentration refers to the sum of these carotenoids.*

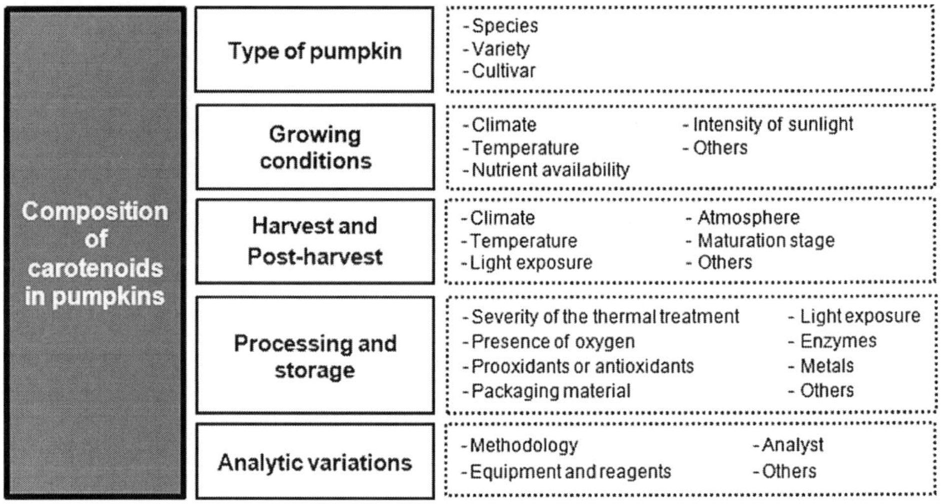

FIGURE 9.2 The main factors that affect the composition of carotenoids in different species and varieties of pumpkins.

and Rodriguez-Amaya, 1988). The concentrations of β-carotene in pumpkins obtained at a local market in Thailand and stored at room temperature increased from 6.5 μg/g to 16.3 μg/g after 4 weeks, 19.7 μg/g at 8 weeks, 51.1 μg/g at 10 weeks and 73.3 μg/g after 12 weeks (Chavasit *et al.*, 2002).

In addition to all the aforementioned factors, there is the possibility of analytical variations, which involve methodology, equipment, and the analyst. These factors may significantly affect the results and must always be taken into consideration when comparing different studies (Rodriguez-Amaya, 1999).

EFFECTS OF PROCESSING ON CAROTENOIDS IN PUMPKIN PRODUCTS

It is clear that raw pumpkin is an excellent source of carotenoids for human consumption. However, due to the presence of double bonds in their structure, carotenoids in pumpkin products are susceptible to many degradation reactions during preparation, processing, and storage.

The major reactions that may occur in these compounds are oxidation and isomerization. Oxidation may occur either by autoxidation, a spontaneous free radicals chain reaction in the presence of oxygen, or by photoxidation, involving singlet oxygen. In the case of isomerization, *trans*-isomers are more common and stable in foods while *cis*-isomers are usually formed during food processing. Besides the changes in the color of the products, these reactions may also lead to a significant decrease in the biological activities of carotenoids. The occurrence of these reactions depends directly on several factors such as the concentration of oxygen, metals, enzymes, unsaturated lipids, pro-oxidant and antioxidant compounds, exposure to light, severity of heat treatment, packaging, storage conditions, among other factors (Rodriguez-Amaya, 1999; Zepka and Mercadante, 2009).

The study and discussion on how processing affects the concentrations of carotenoids is extremely important in order to ensure higher nutritional value in food products. However, investigation on the real effect of processing on these compounds is not so simple to carry out. There are several changes that may occur in the food that should be considered when calculating the retention of these compounds to avoid either under- or overestimated results (see more in the *Analytical Techniques* section).

Table 9.2 shows some studies involving the effects of processing on carotenoids in pumpkin products.

Home-prepared pumpkin is probably still the most common form of consumption. Pumpkins can be consumed cooked, baked, fried, or roasted, or used in the preparation of other foods, such as sauces and jams. Time–temperature is an important factor to be considered in all types of preparations, although other aspects of the food product and method of preparation also need to be considered.

In general, foods prepared in water, such as boiled pumpkin, have higher retention than foods prepared in oil, such as fried or stir-fried pumpkin (Azizah *et al.*, 2009; Veda *et al.*, 2010). Naturally, pumpkin was exposed to a lower temperature during cooking in water than in the frying process. Furthermore, carotenoids are fat-soluble compounds, and thus the frying process also causes a higher loss by lixiviation of these compounds.

Steaming is a good method to reduce the loss of fat-soluble nutrients during preparation of vegetables. Low losses in the concentration of β-carotene (close to 15%) were noted when steaming pumpkins for 40 min (Sungpuag *et al.*, 1999).

However, preparations with oil have better bioavailability of carotenoids after ingestion, because it facilitates their absorption and utilization by the human body (Veda *et al*, 2010).

TABLE 9.2 Retention of Carotenoids after Preparation or Processing of Pumpkin Products

			Retention Index (approximate)			
Processing	**Species/Variety**	**Processing Conditions**	**Total carotenoids**	**β-Carotene**	**α-Carotene**	**Reference**
Preparation at home	*C. moschata* 'Menina Brasileira'	Boiled for 10 min	—	88%	94%	Godoy and Rodriguez-Amaya (1988)
		Stir-fried for 10 min	—	88%	105%	Godoy and Rodriguez-Amaya (1988)
		Boiled for 20 min (pieces)	—	83%	82%	Provesi *et al.* (2011)
	C. maxima Variety not specified	Boiled for 10 min	—	67%	—	Veda *et al.* (2010)
		Stir-fried for 10 min	—	46%	—	Veda *et al.* (2010)
		Pressure-cooking for 10 min	—	44%	—	Veda *et al.* (2010)
		Steaming for 40 min	—	85%	—	Sungpuag *et al.* (1999)
	C. maxima 'Exposição'	Boiled for 20 min (pieces)	—	84%	—	Provesi *et al.* (2011)
Dehydration	*C. maxima* 'Amazonka'	Crisps: 1.2-mm-thick slices dried convectively (85°C, 4 m/s)	31%	39%	—	Konopacka *et al.* (2010)
	C. maxima 'Bambino'		19%	27%	—	Konopacka *et al.* (2010)
	C. maxima 'Justynka F_1'		67%	55%	—	Konopacka *et al.* (2010)
	C. maxima 'Karowita'		67%	56%	—	Konopacka *et al.* (2010)
	C. maxima 'Amazonka'	*Porous cubes*: 10 × 10 mm cubes dried convectively (70°C, 2 m/s)	57%	68%	—	Konopacka *et al.* (2010)
	C. maxima 'Bambino'		52%	91%	—	Konopacka *et al.* (2010)
	C. maxima 'Justynka F_1'		74%	64%	—	Konopacka *et al.* (2010)
	C. maxima 'Karowita'		89%	89%	—	Konopacka *et al.* (2010)
Pumpkin puree	*C. moschata* 'Menina Brasileira'	Commercial sterilization (120°C for 20 min) of pumpkin puree	—	83%	83%	Provesi *et al.* (2011)
	C. maxima 'Exposição'		—	79%	—	Provesi *et al.* (2011)

Regarding industrial processing, the two most common processes for pumpkins are dehydration, one of the oldest methods of food preservation, and thermal treatment of its pulp for production of pumpkin puree.

Pumpkin flour is one of the main processed pumpkin products. It is added in the processing of pasta, breads, and cookies. Pumpkin crisps are also an interesting food product because of their appealing shape and taste. Dehydration of these products is usually carried out by hot air-drying, which is a less costly method. However, the higher temperatures and contact with oxygen during the process may accelerate the oxidation of carotenoids.

In studies involving the drying of 12 different varieties of the pumpkins *C. maxima* and *C. pepo*, higher retentions were obtained with dehydration by freeze drying, followed by vacuum-microwave, only vacuum, and lastly the traditional convective method (Nawirska *et al.*, 2009). Pumpkin flour obtained by freeze drying also showed a better color than that obtained by hot-air drying, possibly because of a lower degradation of carotenoids (Que *et al.*, 2008).

Whenever drying is carried out with larger pieces, to produce crisps or flakes, for example, the area of exposure to oxygen is an important factor in the retention of carotenoids. Konopacka *et al.* (2010) investigated the effect of drying on the carotenoids in the production of ready-to-eat dried crisps (1.2-mm-thick slices) and crunchy cubes (10 × 10 mm) made from different cultivars of pumpkin. It is interesting to note that in most cases the pore size of the cubes are characterized by higher dietary values than the crisps, approximately double on average.

Pumpkin puree is obtained through the processing of the pulp and is already available to consumers in many countries, where it can be easily incorporated into pastas, sauces, and jams.

In the temperature range of 60–100°C for 0–2 h, the degradation of β-carotene in pumpkin puree followed first-order reaction kinetics. While after 2 h at 60°C, the concentrations of carotenoids were close to 14 μg/g, after the same period of time at 100°C, the concentrations decreased to slightly over 4 μg/g (Dutta *et al.*, 2006).

Therefore, a product such as the pumpkin puree, whose pH is higher than 4.5, requires care in relation to microbiological aspects. Thus, a more intense (> 100°C) thermal treatment, such as commercial sterilization, is recommended to ensure safety of the food product (Provesi *et al.*, 2012).

Provesi *et al.* (2011) proposed an industrial process to obtain pumpkin puree of *C. moschata* 'Menina Brasileira' and *C. maxima* 'Exposição', where the effects of two steps of processing (cooking and sterilization) on the principal carotenoids were evaluated. Contrary to the expected outcome, the major losses occurred during the cooking stage, despite the temperature being lower than that used in commercial sterilization. The reason for this is that cooking in water is carried out in open systems, with direct contact with oxygen causing lixiviation of nutrients, while in commercial sterilization the purees are already bottled, in closed systems, with a lower concentration of dissolved oxygen.

Moreover, the higher losses noted were in xanthophylls, a group of carotenoids which has one or more oxygen groups in their structure, and therefore has lower oxidation stability. The content of lutein in the pumpkin 'Exposição' decreased almost 82% after cooking. Carotenes, such as α-carotene and β-carotene, which are compounds that have no oxygen in their structure, obtained retention index relatively higher after processing (> 75%). This difference in stability between carotenes and xanthophylls was also observed in other foods (De Sá and Rodriguez-Amaya, 2004; Zepka and Mercadante, 2009).

During storage of pumpkin products, factors such as temperature, light exposure, and contact with oxygen may also cause loss of carotenoids. The type of product, its conservation, and the packaging used play an important role in this context.

During the storage of pumpkin purees for 180 days, Provesi *et al.* (2011) observed no changes in the concentrations of carotene. Inactivation of enzymes during processing, low concentration of dissolved oxygen in the product, storage at temperatures lower than 30°C, and protection against light contributed to this stability.

The food matrix and physical state of carotenoids in the food product also influence the behavior of carotenoids during storage. For example, in crystalline form, as in carrot juice, carotenoids have higher stability than when dissolved in oil drops (Marx *et al.*, 2003).

In summary, during preparation, processing, and storage of pumpkin products there are several conditions for the degradation of carotenoids, which may cause a decrease in the biological activity of these compounds or color changes in the products. The use of new alternative technologies instead of the conventional thermal treatment and the utilization of packaging that renders better protection to food products are measures that can increase the retention of carotenoids, thereby improving the nutritional and sensorial quality in pumpkin products.

ANALYTICAL TECHNIQUES

The determination of carotenoids by spectrophotometry is still commonly used because it is a simple and inexpensive method. However, high-performance liquid chromatography (HPLC) has significantly greater power of resolution, reproducibility, and speed of separation of these compounds (Meléndez-Matínez *et al.*, 2007).

The extraction usually occurs with organic solvents such as acetone, methanol, among others, or combinations of these. After extraction, it is possible to perform the saponification step, whose aim is to release the carotenoids in esterified form through treatment with an alkaline solution (Oliver and Palou, 2000). For species or varieties of pumpkins that have a low content of xanthophylls, saponification may not be required.

For chromatographic analysis, reversed-phase C_{18} and C_{30} columns have been widely used. In the mobile phase, the methods with gradient allow the separation of a larger number of compounds, especially when evaluation of the profile of carotenoids is desired (Meléndez-Matínez *et al.*, 2007). For species and varieties of pumpkins with high content of carotenes, isocratic methods are sufficient since the separation is relatively easy.

The detector that is mostly applied in the identification of carotenoids by HPLC is the photodiode array detector. This system makes it possible to obtain the absorption spectrum in the UV-visible region, providing data about the wavelengths of maximum absorption and fine structure; both are important factors for identification. It is also possible to use mass spectrometry, which allows differentiation of compounds that have identical absorption spectra (Kurz *et al.*, 2008).

Procedures for analysis of carotenoids in pumpkins can be found in the works of Murkovic *et al.* (2002), Azevedo-Meleiro and Rodriguez-Amaya (2007), and Kurz *et al.* (2008). However, there is no methodology that is applied to any species or variety of pumpkins. The methodology to be used depends on the profile of carotenoids in pumpkins and on the purpose of the analysis.

When the purpose is to investigate the effects of processing on carotenoids, it is important to know that a simple comparison between the concentrations of carotenoids before and after processing may generate significant distortions in the results.

It is relatively common to find studies reporting retention indices over 100% in processed foods. Therefore, it is very important to make it clear that there is no way for carotenoids to be biosynthesized during food processing (Rodriguez-Amaya, 1999). The causes of this fact may stem from several factors. Thermal treatment can weaken or break the main cell structures, which then facilitates the extraction of carotenoids during analysis. However, this can not be considered a real increase in carotenoids. A good extraction process must ensure the same effectiveness in raw and processed foods. Moreover, there is the possibility of an enzymatic degradation of carotenoids in raw foods, where enzymes have not yet been inactivated by heat, particularly when they are not analyzed immediately after processing.

Consequently, it is strongly recommended that studies to evaluate the effects of processing on carotenoids take into account the gain or loss of weight during processing through incorporation or loss of water or water-soluble solids. A simple way to ensure that is to calculate the concentration of carotenoids in samples based on the mass of raw food, multiplying the concentration obtained for the sample by the ratio of the food mass after processing and the food mass before processing.

Thus, the retention index (RI) can be calculated as proposed by Murphy *et al.* (1975) cited by De Sá and Rodriguez-Amaya (2004), as follows: %RI = 100 × (nutrient content per g of processed food × g of food after processing) / (nutrient content per g of raw food × g of food before processing).

The analysis of carotenoids in pumpkins is not complex, but requires attention. A trained analyst, who understands each step of the analysis and its risks, can minimize the errors and obtain more reliable results.

References

Arima, H.K., Rodriguez-Amaya, D.B., 1988. Carotenoid composition and vitamin A value of commercial Brazilian squashes and pumpkins. J. Micronutrient Anal. 4, 177–191.

Arima, H.K., Rodriguez-Amaya, D.B., 1990. Carotenoid composition and vitamin A value of a squash and pumpkin from Northeastern Brasil. Archivos Lationamericanos de Nutrición 40, 284–292.

Azevedo-Meleiro, C.H., Rodriguez-Amaya, D.B., 2007. Qualitative and quantitative differences in carotenoid composition among *Cucurbita moschata, Cucurbita maxima*, and Cucurbita pepo. J. Agric. Food. Chem. 55, 4027–4033.

Azizah, A.H., Wee, K.C., Azizah, O., Azizah, M., 2009. Effect of boiling and stir frying on total phenolics, carotenoids and radical scavenging activity of pumpkin (*Cucurbita moschata*). Int. Food Res. J. 16, 45–51.

Caili, F., Huan, S., Quanhong, L., 2006. A review on pharmacological activities and utilization technologies of pumpkin. Plant Food Hum. Nutr. 61, 73–80.

Chavasit, V., Pisaphab, R., Sungpuag, P., Jittinandana, S., Wasantwisut, E., 2002. Changes in β-carotene and Vitamin A contents of vitamin A-rich foods in Thailand during preservation and storage. J. Food Sci. 67, 375–379.

De Sá, M.C., Rodriguez-Amaya, D.B., 2004. Optimization of HPLC quantification of carotenoids in cooked green vegetables — Comparison of analytical and calculated data. J. Food Composition Anal. 17, 37–51.

Dutta, D., Dutta, A., Raychaudhuri, U., Chakraborty, R., 2006. Rheological characteristics and thermal degradation kinetics of beta-carotene in pumpkin puree. J. Food Eng. 76, 538–546.

Ferriol, M., Picó, B., 2008. Pumpkin and Winter Squash. In: Prohens, J., Nuez, F. (Eds.), Vegetables I. Springer, New York, pp. 317–349.

Godoy, H.T., Rodriguez-Amaya, D.B., 1998. Occurrence of *cis* isomers of provitamins A in Brazilian vegetables. J. Agric. Food Chem. 46, 3081–3086.

Iacuzzo, F., Dalla Costa, L., 2009. Yield performance, quality characteristics and fruit storability of winter squash cultivars in sub-humid areas. Sci. Hortic. 120, 330–335.

Konopacka, D., Seroczynka, A., Korzeniewska, A., Jesionkowska, K., Niemirowicz-Szczytt, K., Plocharski, W., 2010. Studies on the usefulness of *Cucurbita maxima* for the production of ready-to-eat dried vegetables snacks with a high carotenoid content. LWT—Food Sci. Technol. 43, 302–309.

Kurz, C., Carle, R., Schieber, A., 2008. HPLC-DAD-MS characterization of carotenoids from apricots and pumpkins for the evaluation of fruit product authenticity. Food Chem. 110, 522–530.

Marx, C., Stuparic, M., Schieber, A., Carle, R., 2003. Effects of thermal processing on *trans-cis*-isomerization of β-carotene in carrot juices and carotene-containing preparations. Food Chem. 83, 609–617.

Meléndez-Martínez, A.J., Vicario, I.M., Heredia, F.J., 2007. Review: Analysis of carotenoids in orange juice. J. Food Composition Anal. 20, 638–649.

Murkovic, M., Mulleder, U., Neunteufl, H., 2002. Carotenoid content in different varieties of pumpkins. J. Food Composition Anal. 15, 633–638.

Nawirska, A., Figiel, A., Kucharska, A.Z., Sokól-Letowska, A., Biesiada, A., 2009. Drying kinetics and quality parameters of pumpkin slices dehydrated using different methods. J. Food Eng. 94, 14–20.

Oliver, J., Palou, A., 2000. Chromatographic determination of carotenoids in foods. J. Chromatogr. A 881, 543–555.

Provesi, J.G., Dias, C.O., Amante, E.R., 2011. Changes in carotenoids during processing and storage of pumpkin puree. Food Chem. 128, 195–202.

Provesi, J.G., Dias, C.O., Amboni, R.D.M.C., Amante, E.R., 2012. Characterisation and stability of quality indices on storage of pumpkin (*Cucurbita moschata* and *Cucurbita maxima*) purees. Int. J. Food Sci. Technol. 47, 67–74.

Que, F., Mao, L., Fang, X., Wu, T., 2008. Comparison of hot air drying and freeze-drying on the physicochemical properties and antioxidant activities of pumpkin (*Cucurbita moschata* Duch.) flours. Int. J. Food Sci. Technol. 43, 1195–1201.

Rao, A.V., Rao, L.G., 2007. Carotenoids and human health. Pharm. Res. 55, 207–216.

Rodriguez-Amaya, D.B., Kimura, M., Godoy, H.T., Amaya-Farfan, J., 2008. Updated Brazilian database on food carotenoids: Factors affecting carotenoid composition. J. Food Composition Anal. 21, 445–463.

Rodriguez-Amaya, D.B., 1999. A Guide to Carotenoid Analysis in Foods. International Life Sciences Institute (ILSI) Press, Washington, DC.

Sungpuag, P., Tangchitpianvit, S., Chittchang, U., Wasantwisut, E., 1999. Retinol and beta carotene content of indigenous raw and home-prepared foods in Northeast Thailand. Food Chem. 64, 163–167.

Veda, S., Platel, K., Srinivasan, K., 2010. Enhanced bioaccessibility of β–carotene from yellow–orange vegetables and green leafy vegetables by domestic heat processing. Int. J. Food Sci. Technol. 45, 2201–2207.

Wilberg, V.C., Rodriguez-Amaya, D.B., 1995. Quantification of major carotenoids of fresh and processed guava, mango and papaya. LWT—Food Sci. Technol. 28, 474–480.

Zepka, L.Q., Mercadante, A.Z., 2009. Degradation compounds of carotenoids formed during heating of a simulated cashew apple juice. Food Chem. 117, 28–34.

CHAPTER

10

Cassava Production and Processing and Impact on Biological Compounds

Leon Brimer

Department of Veterinary Disease Biology, University of Copenhagen, Denmark

CHAPTER POINTS

- The majority of studies concerning the fate of cassava (root or leaf) material constituents during processing are concerned with cyanogenic glycosides and their degradation products.
- Grating (of roots) in most cases results in hydrolysis of more than 90% of the glycosides within hours.
- Monitored removal of the cyanohydrins formed is essential for chemical food safety.
- Vitamin D active compounds in yellow roots are relatively well retained during cooking.
- Pounding/cooking effectively removes most cyanogens from leaves.

List of Abbreviations

DW dry weight
FAO Food and Agricultural Organisation (of the UN)
FW fresh weight
HCN hydrogen cyanide
NPK Nitrogen Phosphorus Potassium (fertilizer)
OECD The Organization for Economic Co-operation and Development
ROS Reactive Oxygen Species
w/w (%) Weight -weight percentage
w/v (%) Weight –volume percentage

INTRODUCTION

Cassava (*Manihot esculenta* Crantz) a perennial shrub originating in South America spread to Africa, India, South East Asia, and the Pacific. In 2010 world production of its bulky storage roots was 230 million tons. Nigeria was the largest producer, with 37 million tons, followed by Brazil, Indonesia, and Thailand. The water content of fresh roots varies around 65% (w/w). The dry matter mostly is starch, composed of amylose (~20%) and amylopectine (~80%). Clones showing some free sugar accumulation are cultivated in the Amazonas (Carvalho *et al.*, 2004), known as sugary cassavas or "mandiocabas" (Vieira *et al.*, 2011).

Protein content in the roots is low, i.e., in the interval 1.5 to 4.7 % (DW) for the mean values of crude protein found in six investigations (OECD, 2009), and may only be slightly enhanced by fermentation (Tivana *et al.*, 2007).

The leaves are used in salads and in the preparation of relishes. In Congo and Central Africa leaves are fermented to produce the food product *Ntoba mbodi* (Mokemiabeka *et al.*, 2011). The range of mean values from five investigations of the crude protein content is 14.7–36.4% (DW) (OECD, 2009).

The root protein has less than the recommended minimum of all essential amino acids except for tryptophan. In contrast the leaf levels of all of the essential amino acids except methionine exceed the FAO's recommendations. The nitrogen-to-protein conversion factor based on nitrogen recovered from total amino acid analysis including ammonia ranges from 4.75 to 5.87; the traditional factor of 6.25 being non-valid for cassava root (Yeoh and Truong, 1996). In the leaves the content of the amino acids glutamic acid and leucine is the highest, with values above 4 g/100 g true protein, followed by aspartic acid, arginine, and alanine (OECD, 2009).

Roots with yellow parenchyma have elevated contents of carotenoids; some with vitamin A effect. Total carotene may vary between 1 and 18 μg/g (FW) (Carvalho *et al.*, 2012). Vitamin levels in mature cassava roots and leaves are low; the young leaves normally eaten by humans are richer in ascorbic acid, thiamine and β-carotene (OECD, 2009).

Processing and Impact on Active Components in Food
http://dx.doi.org/10.1016/B978-0-12-404699-3.00010-X

The mineral content of the edible cassava root parenchyma is low as compared to cereals, the contents of most elements being comparable to those of edible root parenchyma and of leaves. On a dry weight basis the levels in leaves are much higher, the use of NPK fertilizer significantly increasing the concentrations of N, P, K, S, Mn, and Cu, while the concentrations of Mg and Ca may decrease (OECD, 2009). Significant differences in the content of iron and zinc between different cultivars have been shown for both roots and leaves (Burns *et al.*, 2012).

All parts of the cassava plant contain cyanogenic glycosides, linamarin being dominant. Highest concentrations are in the cortex (peel) of the root and in the leaves (Brimer, 2010). Upon processing these are hydrolyzed to release cyanohydrins, which in turn release the toxic HCN (Figure 10.1). Antinutrients and other bioactive agents in roots/leaves include coumarins, nitrate, oxalate, phytic acid, saponins, tannins and trypsin inhibitors (OECD, 2009).

HOW COMPOSITION IS ALTERED

Processing may decrease the concentration of a compound by removing (extracting, evaporating) it from the remaining food matrix; i.e., the final meal or an intermediate product. Alternatively, processing may facilitate degradation of a compound by oxidation, reduction, or hydrolysis; or by reaction with other constituents forming chemical bonds.

Shelf-life of Fresh Cassava Roots

Cassava roots are perishable with a shelf-life of 24 to no more than 72 h after harvest if no special measures are taken. Deterioration involves a primary physiological deterioration resulting in internal discoloration followed by bacterial deterioration. Methods to prolong shelf-life have been developed; however, most have proven complicated or expensive in many of the cassava producing countries (Westby, 2002). The earliest biochemical event during the initiation of post-harvest physiological deterioration (PPD) is a rapid burst of reactive oxygen species (ROS) accumulation. Transgenic plants showing a reduction in ROS accumulation showed a delayed onset of PPD up to around 14 to 21 days after harvest (Zidenga *et al.*, 2012). Evidence exists for a causal link between cyanogenesis and the onset of this oxidative burst. Due to deterioration, cassava roots, if not immediately used, must be processed.

The General Fate of Cyanogenic Glycosides During Processing of Cassava

Cassava is one of very few food plants for which a number of toxic varieties are deliberately grown. Toxic in this context means that the harvested material has to be processed; if for no other reason than to reduce the content of the cyanogenic glycosides present. An international trade standard defines bitter cassava roots (roots that need processing) as roots with a total cyanogenic potential of more than 50 mg/kg fresh weight (FW) releasable HCN (Codex Alimentarius, 2011). A large number of scientific studies are available concerning the fate of cyanogenic glycosides during various types of processing.

Linamarin (≥90%) and lotaustralin (≤10%) are the important constituents of the edible root and leaves of parenchyma. In both compounds the sugar part is β-D-glucose; the compounds are cyanogenic *glucosides*. As such they are not very toxic, since they are (to a great extent) absorbed in the gastrointestinal tract and excreted unchanged in urine. However, during chewing and processing, the glucosides will normally be degraded in a reaction cascade starting with hydrolysis of the glycosidic bond, followed by cleavage of the cyanohydrin formed; one of the end products being the toxic HCN (hydrogen cyanide). Hydrolysis is catalyzed by a specific β-glucosidase (linamarase) present in the plant tissue separated from the glucosides. Upon tissue damage the compartmentation is broken and the hydrolysis starts. The cyanohydrins formed are stable at acidic pH while they are spontaneously cleaved under alkaline conditions (Brimer, 2010; OECD, 2009). Upon ingestion they are sources of HCN, since they will either be degraded in the gastrointestinal tract or absorbed and then degraded to release HCN.

Some degree of glucoside hydrolysis cannot be avoided during processing of cassava tubers or leaves. Hydrolysis may also occur during the process of eating

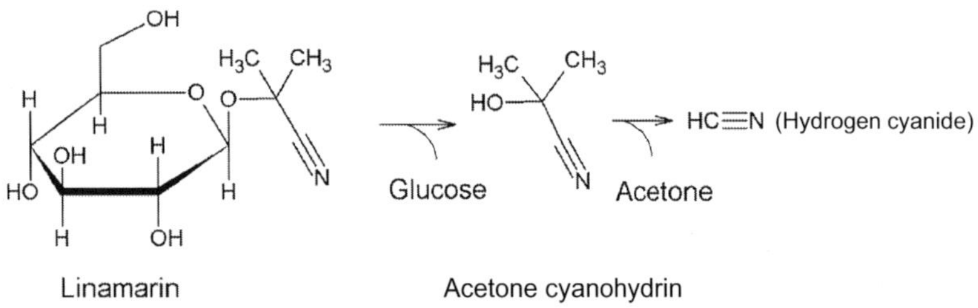

FIGURE 10.1 The degradation of the cyanogenic glucoside linamarin to release hydrogen cyanide.

the food; if the linamarase has not been denaturated. Investigations also show that hydrolysis in the gastrointestinal tract cannot be totally ruled out (Carlson *et al.*, 1999). Therefore, processing of toxic cassava varieties must, in addition to steps to improve shelf-life, include steps that ensure proper removal of the cyanogens (i.e., the glucosides, the cyanohydrins and the HCN) present. The importance of keeping a focus on this was recently illustrated by the finding of relatively high concentrations of total cyanogens in two investigations of traded products in Zambia/Malawi (Nyirenda *et al.*, 2011) and in Australia (Miles *et al.*, 2011), respectively. The latter analysis was of imported ready-to-eat chips.

Products and Processing—Cassava Root Tubers

A huge number of food products based on processing of cassava root parenchyma exist worldwide. Some are restricted to specific geographical regions, others are spread wider. In many cases the same product is named differently in different countries/regions. An example is "gari," a fermented and roasted product with a long shelf-life and multiple uses. Gari is produced commercially—and in households—in major parts of West Africa, south of Sahel (Nigeria and Ghana). The same type of product is produced (mainly household production) in parts of South East Africa (Mozambique) where it is called "rale."

Products may, however, also have more or less identical names while still being relatively different. An example is the Brazilian product "farinha" ("farinha de mandioca"). Most productions include the following steps: washing, peeling, grating, pressing, sieving, and roasting. However, in the Amazon region the whole roots may first be retted ("farinha de aqua") or the pulp may be fermented (Chuzel, 2001).

A description of the process interrelationships of different cassava products as identified through investigations in 233 villages in six African countries covers most unit operations used in the processing of sweet and bitter cassava roots (Westby, 2002). The only type of products not included are the fermented "sour starches" from Brazil and bordering countries, and the different cassava root based drinks or sauces from the Amazon region and the West Indies. These include "mingao" (fermented starch dissolved in boiling water and flavored with fruits) and "cassareep" or "tucupay" (juice pressed from tubers, with added spices and concentrated) (Balagopalan, 2002), and the fermented alcoholic beverage "carixi" made from a mixture of cassava and sweet potatoes by the Juruna people of Brazil (Santos *et al.*, 2012). An overview of cassava root based food products world-wide may be gained from various reviews (Balagopalan, 2002; Chuzel, 2001; Gnonlonfin *et al.*, 2012; Moorthy and Mathew, 1998; Westby, 2002).

Returning to the process interrelationships it becomes clear (Westby, 2002) that fresh peeled (maybe slightly subdivided) roots may be subjected to one of the following unit processes:

1. drying;
2. soaking (retting, steeping);
3. cooking;
4. frying;
5. size reduction (grating); or
6. fermentation (dry in heaps or wet—by extended soaking or after grating).

Most cassava root based food products, whether dishes or products, such as flour, will have undergone more than one unit process; i.e., being a result of a "unit process chain." Also, some products will be a result of additional unit processes such as: pressing/dewatering (of grated root material), steaming (e.g., of size reduced and pounded root material), roasting, or smoking.

Unit Processes and their Influence on the Composition of Cassava Root-Based Food Products

Drying

Drying of fresh (non-presoaked) peeled cassava root—or more often "chips" (pieces)—results in a decrease in the total content of cyanogens. The percentage loss depends on the size of the chips, temperature, and drying time. In general, thicker chips show a higher decrease. This phenomenon is caused by the faster drying of thin chips, which reduces the contact time between the linamarase and the glucosides. A number of drying studies are reviewed by Padmaja (1995).

A recent study showed a strong increase in the content of the coumarin compound scopoletin (7-hydroxy-6-methoxychromen-2-one) after peeling, cutting into 2 cm thick slices (chipping) and drying for 6 days. The mean content in fresh roots from four varieties varied between 4.1 to 11.1 mg/kg, while the content after 6 days rose to 242 mg/kg for the cultivar showing the highest increase (Gnonlonfin *et al.*, 2011).

Soaking

During the soaking of roots in water the texture undergoes noticeable changes and the roots are rendered soft. Attempts have been made to quantify softening and use the result as an index to monitor the extent of fermentation. Softening of tissues can be attributed to the action of pectinolytic and cellulytic enzymes produced by microorganisms. Softening can be helpful in subsequent size reduction operations such as crushing and grinding. Softening studies have been made in plain water (allowing the influence of the natural microbial flora) and after the addition of microbial inoculums. The influence of soaking on the texture was reviewed comprehensively by Moorthy and Mathew (1998), while the corresponding reduction in cyanogens are overviewed by Padmaja (1995). The latter quotes a study where Bourdoux *et al.* found that soaking roots in water for 1 day decreased the total cyanogenic potential from 108 to 60 mg/kg, while a 5-day treatment resulted in a final content of 3 mg/kg (Padmaja, 1995).

Recently, the first study to disclose which microorganisms are the most effective in softening cassava root material during soaking/fermentation was published. A strain of *Saccharomyces cerevisiae* proved to be the best (Padonou *et al.*, 2010).

Cooking (Baking, Frying, and Steaming)

Cooking of cassava root material cut into pieces, as seen in the preparation of "cassava salad" in eastern Nigeria and as part of food preparation in Sri Lanka and South America reduces the content of total cyanogens. The percentage reduction depends on parameters, such as boiling time, volume of water as related to volume of cassava material, and the tuber piece size. Cooking where the tuber pieces are placed in cold water and heated to the boiling point is more effective than if the material is added to boiling water. This is due to a faster heat inactivation of the linamarase in the latter case. Most often the reduction in the total cyanogenic potential will not be higher than around 50% (maximum 70%) (Padmaja, 1995). An investigation using three different varieties of cassava showed that boiling in general is more effective in reducing the total cyanogenic potential than baking, frying and steaming (Nambisan and Sundaresan, 1985).

Cultivars containing elevated concentrations of pro-vitamin A carotenoids that can be metabolized into vitamin A (retinol) in humans have been spread to improve the vitamin A nutritional status in northeast Brazil. The all-trans (all-*E*) form of β-carotene has the highest pro-vitamin A activity. Cooking can extract/degrade the content of carotenoids and promote isomerization in β-carotene, increasing the concentration of Z isomers with low pro-vitamin A activity. Investigations have been made concerning the content of carotenoids and the influence of various home cooking processes on the retention of β-carotene in yellow sweet cassava (Carvalho *et al.*, 2012). Using seven different cultivars with a mean total β-carotene content varying between 1.99 and 8.11 μg/g (FW) three different 30 min cooking processes were tested. The percentage retention of β-carotene varied with both the cultivar and cooking method. However, most cultivars and methods showed retention in the range from around 65–95% (Carvalho *et al.*, 2012); as presented in Table 10.1 for three of the cultivars.

Size Reduction (Grating)

If not processed into chips, most processing of cassava roots initially involves a size reduction of peeled fresh or soaked roots by grating, resulting in wet pulp. Grating is also a necessary part of starch extraction.

In West Africa, pulp made from fresh roots was reported early on (Hahn, 1988, among others) to be processed directly by molding it into balls of around 3–5 cm diameter and put into boiling water (Gnonlonfin *et al.*, 2012).

TABLE 10.1 Total Carotenoids, β-Carotene and All-*E*-β-Carotene in Raw Sweet Yellow Cassava Roots from Three Different Cultivars, and the Retention Percentages (%) of Total Carotenoids and β-Carotene, Respectively, After Submission to Three Different Home Cooking Methods[a]

Cultivar	Raw Roots Total Carotenoids (μg/g)	Raw Roots Total β-carotene (μg/g)	Raw Roots All-*E*-β-carotene (μg/g)	Cooking Method[b]	Total Carotenoids Retention %	Total β-carotene Retention %
Híbrido 2003 14 08 begin_of_the_skype_highlighting	10.32 ± 0.94	8.11 ± 0.06	7.27 ± 0.05	1	89.2	90.84
				2	89.2	71.25
				3	62.97	32.06
Híbrido 2003 14 11 begin_of_the_skype_highlighting	7.03 ± 0.11	5.37 ± 0.51	4.23 ± 0.47	1	77.9	94.31
				2	73.83	76.97
				3	80.9	75.76
1721-Aipim cacau	4.43 ± 0.46	2.84 ± 0.35	1.58 ± 0.16	1	65.42	65.42
				2	69.38	69.38
				3	65.23	65.23

[a] *Results from Carvalho* et al., *(2012).*

[b] *Cooking methods: (1) complete immersion in water (1:1.5 w/v) and boiling in a pot without lid; (2) complete immersion in water (1:1.5 w/v) and boiling in a pot with lid on; (3) Half-immersion in water (1:1 w/v) and boiling in a pot with lid on. All cooking was for 30 min.*

Wet pulp from fresh roots may also be fermented as is the case in the production of gari (Onabolu *et al.*, 2002a) and agbelima (Amoa-Awua *et al.*, 1996), the first depending on the naturally available flora, the latter being added an inoculum.

Wet pulp produced by soaking peeled roots for about 3 days or unpeeled for up to 5 days, squeezing the roots to remove water, and then pounding of the material, have been described from various regions of Zaire. After some kind of drying (usually in the form of balls), the dried material is finally pounded again to form flour (Banea *et al.*, 1992).

The literature is scarce concerning investigations showing the changes in the total cyanogenic potential and/or the changes in the relative concentrations of linamarin to cyanogens and free HCN, respectively. What can be said in general is that the finer the grating, the more cell disruption, which immediately initiates the degradation of linamarin to the corresponding cyanohydrin.

Fermentation (Dry in Heaps or Wet—by Extended Soaking or After Grating)

Heap fermentation of cassava root material was reported from among others Uganda, Congo, and Mozambique. A recent Mozambican study described the process as follows: roots are peeled and the root ends removed. Roots longer than 50 cm are divided by cutting transversely. Thus prepared, roots are piled into heaps. Piles are placed in the shade, and covered, e.g., by peel material (Tivana *et al.*, 2007). In the Mozambique study, the average cyanogenic content of the peeled fresh root material was around 260 mg HCN equivalents/kg. Flour made from material heap fermented for 4 days by sun drying had a total cyanogenic potential of 17 mg HCN equivalents/kg as compared to flour made from directly sun dried material, showing a content of 158 mg HCN equivalents/kg (Tivana *et al.*, 2007).

Changes in cyanogens during grating and natural lactic fermentation of cassava roots at 30°C to form gari were studied by Vasconcelos *et al.*, 1990. More than 95% of the initial linamarin content was hydrolyzed within 3 h of grating. Significant concentrations of cyanohydrin (and free HCN) were left in the mash. The cyanohydrins were stable in the wet mash under the acidic conditions formed as a result of the fermentation. After 100 h the total cyanogenic potential thus only had decreased from around 170 to approximately 75 mg HCN equivalents/kg (dry weight, DW) (Vasconcelos *et al.* as cited in Westby, 2002). Further changes in the content of cyanohydrins in a cassava mash and in gari (after roasting) were studied in a model production of gari in Nigeria. Continuous dewatering of the mash by pressure during the fermentation resulted in higher contents than dewatering after fermentation (Onabolu *et al.*, 2002a).

Most of the few existing studies of changes in total protein percentage, mineral content, vitamin content, and amino acid composition during fermentation of cassava roots have been reviewed by Moorthy and Mathew (1998). In general a loss of approximately 50% or more is seen for minerals (Ca, Mg, K, Fe, P, Na, and Mn) as calculated on a dry weight basis. Loss of zinc seems to be higher while Cu was retained. Protein content may be slightly enhanced as a result of fermentation, however, considerable loss (up to around 50%) may also be seen (Moorthy and Mathew, 1998). A large reduction in ascorbic acid from around 80 to 7.5 mg/kg (FW) during gari fermentation was attributed to leaching out of the ascorbic acid into the steep water. The content of riboflavin was reported to remain more or less unchanged (Moorthy and Mathew, 1998).

Unit Processes and their Influence on the Composition of Cassava Leaf Based Food Products

A study of leaf meals prepared from five different cassava cultivars harvesting the leafs from the plants at three different plant ages (12, 15, and 17 months) showed some consistent variation in the concentrations of crude protein, β-carotene, Fe, Zn, Mn, Cu, Mg, Ca, P, K, and S as a function of leaf age. At 12 months crude protein, β-carotene, iron, phosphorus, and sulfur were at their highest. However, the leaf meal from the 17-month old plants had the highest vitamin C, zinc, and calcium levels (Wobeto *et al.*, 2006).

The effect of cooking on the content of the different cyanogens and on the content of free amino acids has been studied recently. Thus, Ngudi *et al.* cooked aliquots of 100 g of raw pounded cassava leaves (obtained commercially from exotic shops in Ghent, Belgium) in 250 ml of water and added 10 ml of palm oil. After 30 min of cooking, the samples were analyzed and the results compared to those obtained from the uncooked material.

The initial total cyanogens before cooking ranged from 36–107 mg HCN equivalents/kg (DW), the residual being significantly reduced to below around 2 mg HCN equivalents/kg (DW). The total free amino acids varied from around 11–38 g/kg (DW) in the raw and from 7.4–25.6 in the cooked leaf material (Ngudi *et al.*, 2003).

Cassava leaves were also reported to contain low concentrations of such antinutrients as oxalate, nitrate trypsine inhibitors, saponins, and polyphenolic substances (OECD, 2009). The contents were reported to vary with cultivar and plant age (Wobeto *et al.*, 2007). No studies on the influence of processing seems to exist.

Other Ways in which Composition is Altered

With the intensified industrialization of the processing of cassava root and leaf material to products that can be exported, the question concerning composition changes during long-term storage and as a result of, for example, freezing and thawing arises. Except for a single study on the effect of 4 weeks of storage on the content of the different classes of cyanogens in prepared gari, no others studies seem available. In this study 15 samples of gari, nine roasted from cassava mash fermented for at least 4 days (type A) and six roasted from mash fermented for only 1 day (type B), were stored for 4 weeks. Storage was on a shelf at room temperature (Nigeria) to simulate home settings. Free cyanide, linamarin, and cyanohydrin were determined weekly. Free cyanide was absent in all samples. Loss of both cyanohydrin and linamarin did occur over the 4 weeks with a loss of around 50 to 70% for both types of cyanogens (Onabolu *et al.*, 2002b). Handling frozen cassava material must be anticipated to cause a loss in cyanogens as a result of (repeated) freezing and thawing, due to the fact that the cellular barriers that keep the linamarase and the cyanogenic glucosides separated in the fresh material will be damaged.

ANALYTICAL TECHNIQUES

Cyanogenic glycosides are known to be biosynthesized by more than 2000 higher plants. Analytical methods to disclose the genuine qualitative and quantitative content in a given plant material have been developed, the most advanced being reverse-phase high-pressure liquid chromatography with post-column enzymatic hydrolysis and electrochemical detection of the cyanide formed (Brimer and Dalgaard, 1984). For the analysis of food or feed material of known botanic origin; i.e., with a knowledge about the cyanogenic glycosides expected to be present, simpler methods are, however, satisfactory in order to judge the potential toxicity upon ingestion. Analyses traditionally have focused on the total amount of cyanide that can be formed (the so-called *total cyanogenic potential*, TCP or *cyanogenic potential*, CNp). A range of assays has been developed for this purpose. These are based on: (i) hydrolysis of the glycosides; (ii) degradation of cyanohydrins to cyanide; and (iii) measurement of total cyanide. Assays have been further developed to allow: (i) determination of cyanide coming from the non-glycosidic cyanogenic compounds (NGC) (=cyanohydrins) plus the cyanide already present; or (ii) from the cyanide alone. These developments, in turn, have made it possible to calculate the cyanogenic potential due to the glycoside(s) denoted GLY. The assays rely on the complete hydrolysis of all glycoside(s) present, as well as the degradation of the cyanohydrin(s) to form cyanide. The major challenge in such analytical procedures is the reliable and complete hydrolysis of the glycosides. A general overview of methods and their performance are found in EFSA, 2007.

Most analytical methods used in investigations of the fate of cyanogens during processing of cassava material include the use of either picrate (Bradbury *et al.*, 1999) or isonicotinate/1,3-dimethyl barbiturate (Essers *et al.*, 1993) as chromogens.

References

Amoa-Awua, W.K.A., Appoh, F.E., Jakobsen, M., 1996. Lactic acid fermentation of cassava dough into agbelima. Int. J. Food Microbiol. 31, 87–98.

Balogopalan, 2002. Cassava utilization in food, feed and industry. In: Hillocks, R.J., Thresh, J.M., Bellotti, A. (Eds.), Cassava: Biology, Production and Utilization. CABI, Wallingford. (chapter 15). pp. 301–318.

Banea, M., Poulter, N.H., Rosling, H., 1992. Shortcuts in cassava processing and risk of dietary cyanide exposure in Zaire. Food Nutr. Bull 14, 137–143.

Bradbury, M.G., Egan, S.V., Bradbury, J.H., 1999. Picrate paper kits for determination of total cyanogens in cassava roots and all forms of cyanogens in cassava products. J. Sci. Food Agric. 79, 593–601.

Brimer, L. 2010. Cyanogenic Glycosides. In the Proceedings of the Symposium "Bioactive compounds in plants—benefits and risks for man and animals", Norwegian Academy of Sciences and Letter (13–14 November 2008, Oslo, Norway). Editor Bernhoft, A. The Norwegian Academy of Sciences and Letters/Novus:Oslo (ISBN 978-82-7099-583-7), pp 125–143.

Brimer, L., Dalgaard, L., 1984. Cyanogenic glycosides and cyanohydrins in plant tissues; qualitative and quantitative determination by enzymatic post-column cleavage and electrochemical detection, after separation by high-performance liquid chromatography. J. Chromatogr. 303, 77–88.

Burns, A.E., Gleadow, R.M., Zacarias, A.M., Cuambe, C.E., Miller, R.E., Cavagnaro, T.R., 2012. Variations in the chemical composition of cassava (*Manihot exculenta* Crantz) leaves and roots as affected by genotype and environmental variation. J. Agric. Food Chem. 60, 4946–4956.

Carlsson, L., Mlingi, N., Juma, A., Ronquist, G., Rosling, H., 1999. Metabolic fates in humans of linamarin in cassava flour ingested as stiff porridge. Food Chem. Toxicol. 37, 307–312.

Carvalho, L.J.C.B., de Souza, C.R.B., de Mattos Cascardo, J.C., Junior, C.B., Campos, L., 2004. Identification and characterization of a novel cassava (*Manihot esculenta* Crantz) clone with high free sugar content and novel starch. Plant Mol. Biol. 56, 643–659.

Carvalho, L.M.J., Oliveira, A.R.G., Godoy, R.L.O., Pacheco, S., Nutti, M.R., de Carvalho, J.L.V., Pereira, E.J., Fukuda, W.G., 2012. Retention of total carotenoid and β-carotene in yellow sweet cassava (*Manihot exculenta* Crantz) after domestic cooking. Food Nutr. Res. 56 (15788—DOI: 10.3402/fnr.56i0.15788).

Chuzel, G., 2001. The cassava processing industry in Brazil: traditional techniques, technological developments, innovations and new markets. Afr. J. Food Nutr. Sec 1 (1), 46–59.

Codex Alimentarius Commission, 2011. Codex standard for bitter cassava. (Codex Stan 300-2010; amended 2011), p.4.

EFSA (European Food Safety Authority), 2007. Opinion of the scientific panel on contaminants in the food chain on a request from the commission related to cyanogenic compounds as undesirable substances in animal feed. EFSA J. 434, 1–67.

Essers, S.A.J.A., Bosveld, M., van der Grift, R.M., Voragen, A.G.J., 1993. Studies on the quantification of specific cyanogens in cassava products and introduction of a new chromogen. J. Sc. Food Agric. 63, 287–296.

Gnonlonfin, B.G.J., Gbaguidi, F., Gbenou, J.D., Sanni, A., Brimer, L., 2011. Changes in scopoletin concentration in cassava chips from four varieties during storage. J. Sci. Food Agric. 91, 2344–2347.

Gnonlonfin, B.G., Sanni, A., Brimer, L., 2012. Preservation of cassava (*Manihot esculenta* Crantz): a major crop to nourish people worldwide. In: Bhat, R., Alias, A.K., Paliyath, G. (Eds.), Progress in Food Preservation. Wiley-Blackwell, Hoboken, pp. 331–342.

Hahn, S.K., 1998. An overview of traditional processing and utilization of cassava in Africa. In: Hahn, S.K., Reynolds, L., Egbunike, G.N. (Eds.), "Cassava as Livestock Feed in Africa", Proceedings of the IITA/ILCA /University of Ibadan Workshop on the potential utilization of cassava as livestock feed in, Africa. pp. 16–27. Available at http://www.fao.org/wairdocs/ilri/x5458e/x5458e05.htm.

Miles, D., Jansson, E., Mai, M.C., Azer, M., Day, P., Shadbolt, C., Stitt, V., Kiermeier, A., Szabo, E., 2011. A survey of total hydrocyanic acid content in ready-to-eat cassava-based chips obtained in the Australian market in 2008. J. Food Protec. 74, 980–985.

Mokemiabeka, S., Dhellot, J., Kobawila, S.C., Diakabana, P., Ntietie Loukombo, R.N., Nyanga-Koumou, A.G., Louembe, D., 2011. Softening and mineral content of cassava (*Manihot esculenta* Crantz) leaves during the fermentation to produce Ntoba mbodi. Adv. J. Food Sci. Technol. 3, 418–423.

Moorthy, S.N., Mathew, G., 1998. Cassava fermentation and associated changes in physicochemical and functional properties. Cri. Rev. Food Sci. Nutr. 38 (2), 73–121.

Nambisan, B., Sundaresan, S., 1985. Effect of processing on the cyanoglucoside content of cassava. J. Sci. Food Agric. 36 1197–????.

Ngudi, D.D., Kuo, Y.-H., Lambein, F., 2003. Cassav cyanogens and free amino acids in raw and cooked leaves. Food Chem. Toxicol. 41, 1193–1197.

Nyirenda, D.B., Chiwona-Karltun, L., Chitundu, M., Haggblade, S., Brimer, L., 2011. Chemical food safety of cassava products in regions adopting cassava production and processing—experience from Southern Africa. Food Chem. Toxicol. 49, 607–612.

OECD, 2009. Consensus document on compositional considerations for new varieties of cassava (*Manihot esculenta* Crantz): Key food and feed nutrients, antinutrients, toxicants and allergens. Series on the Safety of Novel Foods and Feeds No. 18, 50 pages (ENV/JM/MONO(2009)44).

Onabolu, A.O., Oluwole, O.S.A., Rosling, H., Bokanga, M., 2002a. Processing factors affecting the level of residual cyanohydrins in gari. J. Sci. Food Agric. 82, 966–969.

Onabolu, A.O., Oluwole, O.S.A., Bokanga, M., 2002b. Loss of residual cyanogens in a cassava food during short-term storage. Int. J. Food Sci. Nutr. 53, 343–349.

Padmaja, G., 1995. Cyanide detoxification in cassava for food and feed uses. Cri. Rev. Food Sci. Nutr. 35, 299–339.

Padonou, S.W., Nielsen, D.S., Akissoe, N.H., Hounhouigan, J.D., Nago, M.C., Jakobsen, M., 2010. Development of starter culture for improved processing of Lafun, an African fermented cassava food product. J. Appl. Microbiol. 109, 1402–1410.

Santos, C.C.A.A., de Almeida, E.G., de Melo, G.V.P., Schwan, R.F., 2012. Microbiological and physicochemical characterization of carixi, and alcoholic beverage produced by the indigenous Juruna people of Brazil. Int. J. Food Microbiol. 156, 112–121.

Tivana, L.D., Bvochora, J.M., Mutukumira, A.N., Owens, J.D., 2007. A Study of heap fermentation process of cassava roots in Nampula Province, Mozambique. J. Root Crops 33, 119–128.

Vasconcelos, A.T., Twiddy, D.R., Westby, A., Reilly, P.J.A., 1990. Detoxification of cassava during gari preparation. Int. J. Food Sci. Technol. 25, 198–203.

Vieira, E.A., de Freitas Fialho, J., Faleiro, F.G., Bellon, G., da Fonceca, K.G., Carvalho, L.J.C.B., 2011. Molecular characterization of sugary and non-sugary cassava accessions. Ciênc. Agrotec. Lavras 35, 455–461.

Westby, A., 2002. Cassava utilization, storage and small-scale processing. In: Hillocks, R.J., Thresh, J.M., Bellotti, A. (Eds.), Cassava: Biology, Production and Utilization. CABI, Wallingford. (chapter 14). pp. 281-300.

Wobeto, C., Correa, A.D., de Abreu, C.M.P., dos Santos, C.D., de Abreu, J.R., 2006. Nutrients in the cassava (*Manihot esculenta* Crantz) leaf meal at three ages of the plant. Ciênc. Technol. Aliment., Campinas 26, 865–869.

Wobeto, C., Correa, A.D., de Abreu, C.M.P., dos Santos, C.D., Pereira, H.V., 2007. Antinutrients in cassava (*Manihot esculenta* Crantz) leaf poder at three age of the plant. Ciênc. Technol. Aliment., Campinas 27, 108–112.

Yeoh, H.-H., Truong, V.-D., 1996. Protein contents, amino acid compositions and nitrogen-to-protein conversion factors for cassava roots. J. Sci. Food Agric. 70, 51–54.

Zidenga, T., Leyva-Guerrero, E., Moon, H., Siritunga, D., Sayre, R., 2012. Extending Cassava Root Shelf-life via Reduction of Reactive Oxygen Species Production. Plant Physiol. 159 (4), 1396–1407.

CHAPTER

11

Radiation Processing for Sprout Inhibition of Stored Potatoes and Mitigation of Acrylamide in Fries and Chips

Jyoti Tripathi, Prasad S. Variyar*, Rekha S. Singhal†, Arun Sharma**

*Food Technology Division, Bhabha Atomic Research Centre, Mumbai, India †Food Engineering and Technology Department, Institute of Chemical Technology, Mumbai, India

INTRODUCTION

Acrylamide (2-propenamide) is an organic chemical used commercially for the production of polyacrylamide that finds application as a flocculant in water treatment, in the manufacture of paper, in ore processing, and in gel electrophoresis. The compound has been classified as "a probable human carcinogen" by the International Agency on Research in Cancer (IARC, 1994) and long-term exposure to high levels of this compound has been shown to damage the nervous system (JECFA. 2011). In recent years, its substantial presence in a wide range of cooked and fried carbohydrate-rich foods has caused an international health alarm. Information on the occurrence, mechanism of formation, analytical methodologies, exposure to different population groups world over, and methods for its reduction, have been extensively investigated and reported. Current epidemiological and toxicological evidences, however, do not provide sufficient data to determine if amounts of acrylamide consumed in normal diet are likely to have an adverse impact on human health, particularly carcinogenicity. Thus, an acrylamide minimization concept encouraging reduction in acrylamide, to the extent possible, in food has been proposed to reduce the harmful effects of this compound.

MECHANISM OF FORMATION

Acrylamide has been demonstrated to be mainly formed via the Maillard reaction between free asparagine and a carbonyl source, principally reducing sugars. The mechanism involves the formation of a Schiffs base (e.g., *N*-(D-gluco-1-yl)-3′-aminopropionamide) as a key intermediate from *N*-glucosyl conjugate in aqueous systems (Figure 11.1). The Schiff base then undergoes decarboxylation to azomethine ylide or is converted to an Amadori compound via the intermediary oxazolidine-5-one. Acrylamide is then released either directly from azomethine ylide/decarboxylated Amadori compound or by loss of ammonia from 3′-aminopropionamide derived from azomethine ylide (Figure 11.1). Formation of acrylamide via azomethine ylide was demonstrated to be the dominant pathway. Other amino acids producing low amounts of acrylamide include aspartic acid, arginine, alanine, cysteine, threonine, valine, methionine, and glutamine. Several mono- and disaccharides as well as non-reducing sugars such as sucrose that release reducing sugars are efficient reactants. Formation of acrylamide from acrylic acid, produced from acrolein via lipid oxidation has also been suggested as a minor pathway (Figure 11.1).

INFLUENCE OF DIFFERENT PARAMETERS ON ACRYLAMIDE FORMATION

Initial studies on formation of acrylamide performed in model systems focused parameters such as heating time, temperature, and the ratio of amino acids and reducing sugars. Acrylamide formation is generally temperature dependent with a substantial increase between 120 and 170°C and a decrease thereafter. Moderate amounts (5–50 μg/kg) were detected in heated protein rich foods such as meat while higher levels (150–4000 μg/kg) have been found in carbohydrate rich

Processing and Impact on Active Components in Food
http://dx.doi.org/10.1016/B978-0-12-404699-3.00011-1

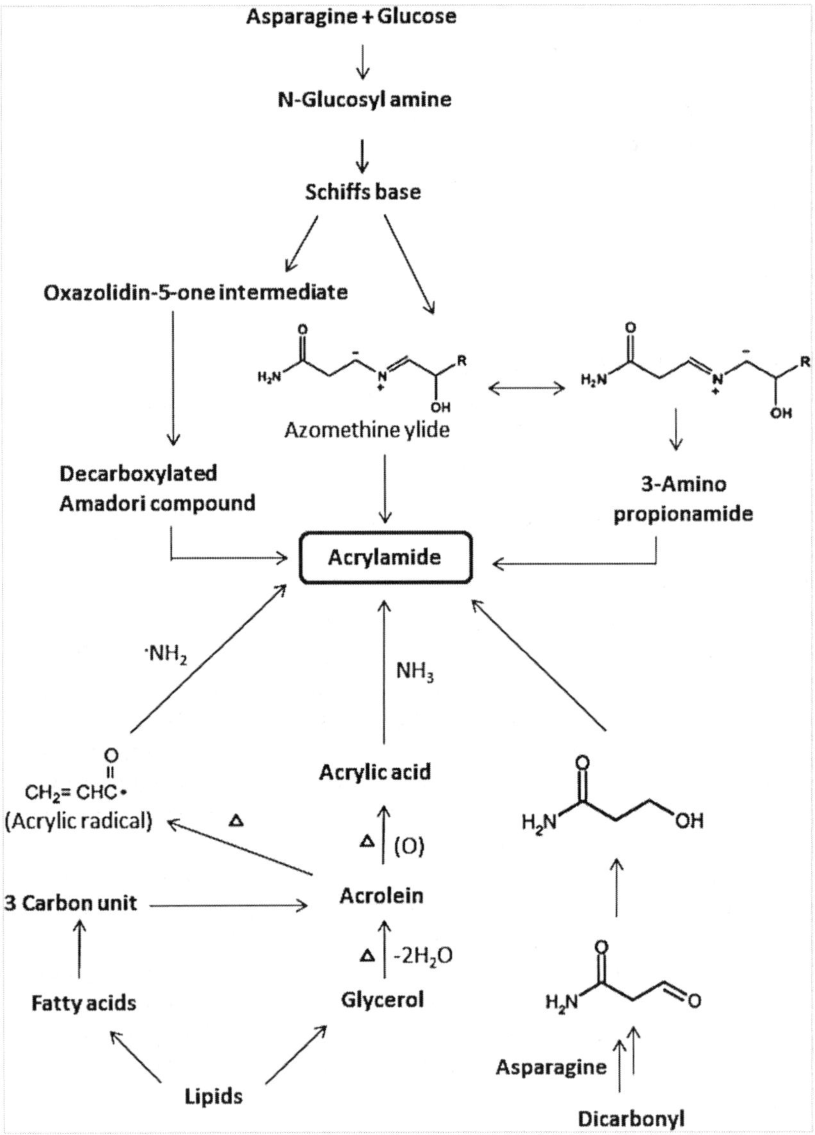

FIGURE 11.1 Mechanism of acrylamide formation via different pathways.

foods such as potatoes. Basic pH conditions also favor release of acrylamide during post-harvest storage. Agronomic factors such as type of fertilizer, harvest season and climatic conditions, variety, and processing methods are other factors that affect acrylamide formation. Favorable light and temperature during plant cultivation can enhance amino acid and protein content and thus favor formation of acrylamide. Sulfur deficiency results in an enhancement in asparagine and a consequent increase in acrylamide. Selection of varieties with low levels of asparagine and reducing sugar can help in reducing acrylamide. Acrylamide level in potato chips prepared from potatoes stored at low temperature were found to be far higher than those stored at higher temperature. At low frying temperatures ($<160°C$), temperature of frying had a greater influence on acrylamide formation. On the other hand, time of frying had a greater impact on the formation of the compound at higher temperatures ($>170°C$). Deep frying was shown to induce less acrylamide than shallow frying. Use of low temperature vacuum frying and short time heating and limited frying by microwave can also reduce acrylamide content. Formation of acrylamide is also dependent on the type of food matrix. In actual food matrices, unlike in model systems, formation of this compound via various heat processing techniques such as baking, frying and roasting is more complex. Other related factors include water content in raw materials, precursor levels and the type of pretreatment. Activation energy of acrylamide formation was found to be larger at low moisture content thus requiring more heat for its formation. During high heat processing, formation of the compound increased with higher water activities. At intermediate water activities, an increased rate of browning of potato products via Maillard browning was, however, noted. Thus, the formation and elimination of Maillard reaction products and acrylamide is influenced differently by water activity. Controlling water activity is therefore useful in obtaining desired browning without an increase in acrylamide.

Potato tubers contain a significant amount of precursors such as free asparagine, glucose, and fructose resulting in detection of considerable amounts of acrylamide in potato products such as French fries. Among the bakery products, bread and cookies contain a substantial amount of acrylamide due to the processing conditions such as favorable baking time, temperature, and presence of comfortable precursor levels. Amount of acrylamide, however, reduces with long baking time. Cereal derived foods are another major source for acrylamide and are formed during heat processing. Wheat and rye are the two important products in this class. A direct relation between asparagine content and acrylamide levels was demonstrated in baked/toasted wheat and rye breads. Reducing sugars such as glucose and fructose are also important contributors. Selection based on low asparagine, fructose, and glucose content can thus aid in reduction of acrylamide. Coffee is a commercially important food product that has been shown to contain acrylamide in good amounts. This is mainly due to the high temperatures employed during roasting. Increasing the degree of roasting was shown to reduce the acrylamide content. The main factors that decide the content of acrylamide in coffee include the time, degree of roasting and storage conditions, with a clear reduction at ambient temperature. A high amount of acrylamide (260–1530 μg) has also been reported in almonds which are known to contain considerable amounts of asparagine, glucose, fructose, and sucrose. Acrylamide content, however, decreased in roasted almonds during storage at room temperature.

ANALYTICAL TECHNIQUES FOR ESTIMATION OF ACRYLAMIDE IN FOOD MATRIX

Isolation of Acrylamide from Food Matrix

Water as well as methanol has been widely used for the extraction of acrylamide from various food matrices. Higher recovery is reported with aqueous sodium chloride as a result of reduced emulsification during extraction (Young *et al.*, 2004). Use of accelerated solvent extraction (ASE) wherein elevated temperature and pressure resulted in improved extraction efficiency has also been suggested (Cavalli *et al.*, 2003). A defatting step using non-polar solvents such as hexane or petroleum ether in case of high fat containing foods, or a deproteinating procedure employing high level of methanol, acetonitrile or saline solution for high protein foods are sometimes used before or in combination with extraction step. A good chromatographic quantification requires an effective sample cleanup. This usually involves the use of a solid-phase extraction (SPE) step. In this regard an oasis HLB cartridge, a water-wettable reversed-phase sorbent with hydrophilic-lipophilic balance, is the one commonly used.

Analysis

Gas chromatography (GC)—mass spectrometry (GC/MS) and liquid chromatography tandem mass spectrometry (LC-MS/MS) are the most commonly used techniques for the detection and estimation of acrylamide (Zhang *et al.*, 2005). GC-based methods require prior derivatization to its bromo derivative (Wenzl *et al.*, 2003; Ahn *et al.*, 2002). Polar carbowax columns and chemical ionization mass spectrometry in the selected ion mode (SIM) are employed for obtaining good reproducibility and high sensitivity of detection. LC-MS/MS methods comprising of porous graphitic carbon (Hypercarb) as the stationary phase and water as mobile phase in LC and use of either electrospray ionization (ESI) or atmospheric pressure chemical ionization (APCI) as the ionization mode in the MS resulted in efficient separation and quantification of acrylamide. A LC-MS/MS method for simultaneous analysis of acrylamide as well as glucose, fructose and sucrose and asparagine in bread with low limit of detection was recently proposed (Nielsen *et al.*, 2006). Some of the newer analytical methods that have been validated include the use of capillary zone electrophoresis (CZE), micellar electrokinetic capillary chromatography in combination with MS, LC or GC techniques combined with time-of-flight mass spectrometry (TOF-MS), near infrared (NIR) spectrometry in combination with multiple regression analysis, and computer vision based image analysis coupled to a semiautomatic segmentation algorithm.

While MS is a method of choice for detection of acrylamide, it has the drawback that the presence of coextractives having the same magnitude of m/z as acrylamide or its fragments makes the method non-specific. Therefore, some non-MS methods such as LC-UV (Paleologos and Kontominas, 2005) and LC-diode array detection (DAD) with the detectors set at 210 and 225, pulsed electrochemical detection and GC-ECD (electron capture) have been developed and optimized (Casella *et al.*, 2006). Thin-layer chromatography (TLC) with fluorescence detection after derivatization with dansulfinic acid has also been proposed (Alpmann and Morlock 2008). Besides, MS/MS, proton transfer reaction mass spectrometry (PTR-MS), pyrolysis (Py) - GC-MS and FT-IR spectrometry have also been considered to be very useful techniques for monitoring real time changes in acrylamide content.

STRATEGIES FOR REDUCTION OF ACRYLAMIDE

The possible methods that can be adopted to reduce the content of acrylamide in food can be broadly divided into: (i) modifying reaction conditions to reduce formation or transformation of key precursors; (ii) controlling conditions so as to divert pathway towards formation of some other small; and (iii) inhibiting key steps in Maillard pathway. Control of precursors namely asparagine and reducing sugars can be achieved by selection of cultivars with a reduced concentration of these precursors, avoiding low temperature storage and avoiding the addition of reducing sugars during further treatment/processing. Besides acrylamide, three other final products namely maleimide, succinimide, and niacinamide are also formed in asparagine pathway (Yaylayan *et al.*, 2003). A shift towards the formation of these compounds can aid in reducing acrylamide. Appropriate heating temperature and avoiding heating for long periods can help in reducing acrylamide by inhibiting some of the steps in Maillard pathway. Other methods to reduce or change precursors include modification of raw materials, optimization of heat processing methods, parameters, and change in storage temperature. Modification of pretreatments such as additional immersion/blanching procedures before frying can result in significant reduction in levels of acrylamide by modifying precursor concentrations. Vacuum frying and microwave precooking steps can effectively alleviate acrylamide content. Besides, the use of asparaginase, coating with egg/breadcrumbs and lactic acid fermentation can also help in eliminating acrylamide. New and improved mitigation strategies involve the use of exogenous non-toxic food additives without affecting the sensory quality of the processed food. pH modifiers such as citric acid and acetic acid act by lowering pH and protonating α-amino group of asparagine thereby preventing nucleophilic addition reaction to the carbonyl group (Jung *et al.*, 2003). Addition of proteins/amino acids results in a marked reduction in acrylamide. For example, glutathione acts on acrylamide via the decomposition of glycine fragments of GSH that degrades via Strecker degradation to aldehydes such as formaldehyde. Formaldehyde reacts with acrylamide leading to its decomposition to smaller molecules. Soy proteins are also used to reduce acrylamide formation as they compete with acrylamide for key reaction intermediates. Addition of bicarbonates such as $NaHCO_3$ and $KHCO_3$, mono- and divalent cations (Na^+ and Ca^{2+}) and antioxidant plant extracts from spices, olive, green tea, and bamboo leaves in food have also been demonstrated to efficiently reduce acrylamide content (Zhang *et al.*, 2009).

A major criteria in the selection of various methods for reducing acrylamide, is their impact on the sensory attributes of food. For example low pH can inhibit Maillard reaction resulting in undesirable flavors and sour product taste thus impacting sensory quality (Mestdagh *et al.*, 2008). Similarly, addition of sulfur containing amino acids can generate unpleasant odors while use of $CaCl_2$ can produce bitter aftertaste despite improved texture. Genetic modification techniques and gamma-irradiation as effective alternative approaches for mitigating acrylamide content could be exploited. Fan *et al.* (2006) have reported the use of radiation processing for reducing acrylamide. A dose of 1.5 kGy resulted in degradation of most of the acrylamide in water but had a limited effect on acrylamide levels in oil and potato chips even up to a dose of 10 kGy. While the use of gamma irradiation has so far been restricted, the use of this process has the advantage that the sensory properties are not compromised as it is a cold process. Thus there is a need to further explore such technologies for reducing the negative effects of acrylamide without affecting sensory acceptability.

EFFECT OF POST-HARVEST STORAGE ON ACRYLAMIDE CONTENT IN POTATO

Acrylamide content has been shown to vary widely in different foods and even within a given food product (Table 11.1). Furthermore, contribution to the acrylamide intake depends on the extent of intake of a particular food by a given population. Hence the current focus of public health measures for reducing acrylamide in diet has been on reducing this compound in the highest potential impact foods rather than on reducing its mean intake by the population.

TABLE 11.1 Range of Acrylamide Content Reported in Different Food Categories

Sample no.	Product Category	Acrylamide Content (μg/kg)
1	French fries	413–600
2	Potato crisps	466–1000
3	Breads	31–150
4	Breakfast cereals	400–738
5	Biscuits	500–657
6	Roast coffee	230–450
7	Baby foods	40–120
8	Processed cereal based foods for infants	100–203

Potato-derived foods constitute a major source of dietary acrylamide contributing anywhere from 30%–70% of the intake in adult population. Lowering the major precursors such as asparagine and reducing sugars present in high amounts in potato can result in reduced acrylamide formation. In this regard, agronomic factors that could be manipulated and proposed currently include breeding or suppressing genes that encode the biosynthesis of free asparagine, selection of varieties with lower levels of asparagine, and reducing sugars (Table 11.2), and growing in soil with high sulfur content.

The content of major sugars namely starch, sucrose, glucose, and fructose in potato can vary with

TABLE 11.2 Concentrations of Asparagine and Reducing Sugars in Different Potato Cultivars

Sample no.	Potato Cultivar	Asparagine (mg/kg)	Reducing Sugars (mg/kg)
1	Kufri chipsona-2		681
2	Kufri badshah		3368
3	Kufri surya	3184	
4	Kufri lavkar	2074	
5	Kufri anand		
6	Kufri satlaj		
7	Kufri bahar		
8	Kufri jyoti		
9	Agria	2547	478
10	Appell	2030	2120
11	Bintje	3280	680
12	Charlotte	2942	1877
13	Desiree	3750	2080
14	Eba	2771	1056
15	Erntestolz	3214	220
16	Hermes	4030	395
17	Lady Claire	4250	144
18	Lady Rosetta	2378	169
19	Markies	2570	130
20	Marlene	3323	130
21	Naturella	2010	4050
22	Nicola	3684	2849
23	Panda	3139	308
24	Santana	3520	1230
25	Saturna	3520	293

post-harvest treatment . At storage temperatures above 10°C, starch and other sugars remain in balance with free sugars either being utilized in other reactions or converted to starch. However, below 10°C, reducing sugars begin to accumulate. This occurs as a result of so-called "cold sweetening" by induced hydrolysis of sugar and epimerization. Significant reduction in reducing sugars can be achieved if potato is reconditioned at room temperature following cold storage. Studies reported on the distribution of acrylamide precursors in potatoes grown in different parts of the world have clearly revealed that their composition can vary widely depending on the conditions of storage. On a dry weight basis, the content of acryl amide in French fries was found to be >10,000 μg/kg when the product was prepared from potatoes stored at 4°C (Fiselier *et al.*, 2004). A high content of reducing sugars was the result of storage at low temperature which was shown to be responsible for such high content of the compound. Acrylamide levels were highly correlated with the content of glucose and fructose in the tuber. Among the sugars, fructose was found to be twice as effective as glucose in generating acrylamide. In a recent study by the authors on six varieties of potatoes stored at 4°C and 14°C, no significant differences were observed in the asparagine content at both temperatures (Mulla *et al.*, 2011). However, reducing sugar content of potatoes increased from 657.5 to 1531.9 mg/kg and 1870.9 mg/kg when stored at these temperatures respectively. Among the varieties screened, *Kufri chipsona*-2 had the lowest content of reducing sugars (680.68 ± 56.50 mg/Kg) and also the lowest content of acrylamide (416.75 ± 15.25 mg/kg) was found in the chips prepared from this variety. Highest level of reducing sugars (3367.87 ± 174.50) and a consequent higher acrylamide (1902.55 ± 10.50 mg/kg) concentration were found in the variety *Kufri badshah*.

RADIATION PROCESSING FOR MITIGATION OF ACRYLAMIDE CONTENT IN POTATO CHIPS

Radiation Processing

Among the currently available methods of food preservation, radiation processing is an effective technology that involves exposure of food to high energy ionizing radiations. Ionizing radiation used in food irradiation includes gamma rays from radionuclides (cobalt-60 or cesium-137), X-rays generated from machine sources operated at or below an energy level of 5 MeV and electron beams generated from machine sources at or below an energy level of 10 MeV. Both gamma and X-rays are part of electromagnetic spectrum and have identical physical properties and effects on matter. Gamma rays are emitted from radio isotopes in a discontinuous spectrum of energies while X-rays are machine generated and exhibit a wide continuous spectrum of radiation. The SI unit for measuring radiation doses given to foods is Gray (Gy). When 1 Joule of energy is deposited in 1 kg of food material by means of irradiation the amount of dose is said to be 1 Gy. In radiation processing of food, the dose is generally measured in terms of kGy (1000 Gy). While dose rate of gamma ray sources do not exceed 10 Gy/s, that of electron accelerators vary between 10^4–10^9 Gy/s.

Major functions achieved by gamma irradiation are insect disinfestation, sprout inhibition in bulbs and tubers (Figure 11.2), delay in ripening of fruits and elimination of spoilage and pathogenic microorganisms and parasites (Table 11.3). Advantages of radiation processing over other treatments like heat are that it is a cold process, environment friendly and can be used for agricultural commodities without changing their fresh-like character. No adverse effects on nutritional and textural properties are noted. Even solanin, a toxic glycoalkaloid, is reported to be reduced in radiation treated potatoes (Nair *et al.*, 1981). Unlike fumigation, no toxic residues are left behind in products so treated. Their highly penetrative nature results in a high degree of homogeneity and pre-packed food stuffs can be treated for hygienization and improvement in shelf-life. Radiation processing of food has been thoroughly and extensively studied in order to ensure its toxicological, nutritional, and microbiological safety (Sharma, 2004). Worldwide research over the last five decades has adequately demonstrated the safety of irradiated food.

On the basis of radiation dose, application of the method can be classified into low dose (<1 kGy), medium dose (1–10 kGy) and high dose (>10 kGy). Low doses are mainly employed to control insect infestation, delay ripening, and inhibit sprouting, while the medium doses are mainly used for reducing microorganisms and extending shelf-life. High doses are employed for achieving sterilization. At the doses recommended, no perceptible changes with respect to chemical, physical, and organoleptic quality have been observed. No significant changes in macronutrients (proteins, carbohydrates, fats) or micronutrients (such as vitamins) and thus in the nutritional quality have been observed. The sensory quality has also been demonstrated to be insignificantly

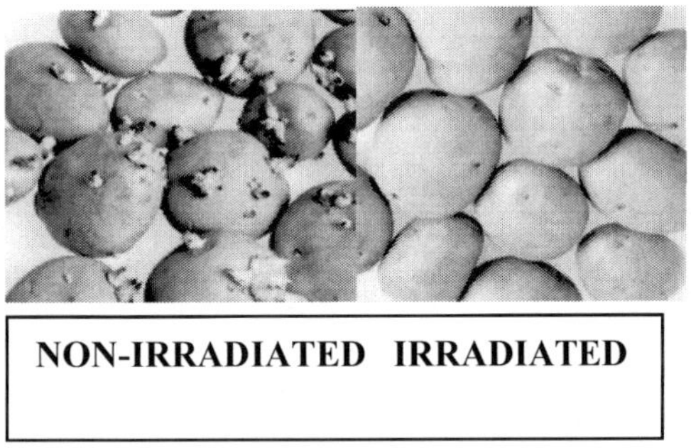

FIGURE 11.2 Sprout inhibition in radiation processed potatoes.

TABLE 11.3 Items Permitted for Irradiation Under Food Safety and Standards Act of India (FSSAI)

Sample no.	Food Item	Irradiation Dose (kGy) Minimum	Maximum	Purpose
1	Onion	0.03	0.09	Sprout inhibition
2	Potato	0.06	0.15	Sprout inhibition
3	Shallots (small onion), garlic, ginger	0.03	0.15	Sprout inhibition
4	Rice	0.25	1.00	Insect disinfestation
5	Semolina, wheat flour, refined flour	0.25	1.00	Insect disinfestation
6	Pulses	0.25	1.00	Insect disinfestation
7	Dried sea food	0.25	1.00	Insect disinfestation
8	Raisins, figs and dried dates	0.25	1.00	Insect disinfestation
9	Mango	0.25	0.75	Shelf-life extension and quarantine treatment
10	Meat and meat products including chicken	2.50	4.00	Shelf-life extension and quarantine treatment
11	Fresh sea food	1.00	3.00	Shelf-life extension
12	Frozen sea food	4.00	6.00	Microbial pathogen control
13	Spices	6.00	14.00	Microbial pathogen control

affected. The changes observed in most cases were similar to other processing techniques like pasteurization, canning or even cold storage.

In 1980, the Joint FAO/IAEA/WHO Expert Committee on Wholesomeness of Irradiated Food (JECFI) stated "Irradiation of any food commodity up to an overall average dose of 10 kGy introduces no toxicological hazard; hence, toxicological testing of food/herb so treated is no longer required." In 1997, the FAO/IAEA/WHO joint study group further evaluated the data on wholesomeness of food irradiated with doses above 10 kGy and concluded "Food irradiated with any dose to achieve technical objectives is safe and nutritionally adequate. No upper limit therefore needs to be imposed as long as food irradiation is based on prevailing good manufacturing practices." No transmittable genetic defects such as teratogenicity or oncogenicity have been observed or attributed to the consumption of irradiated diets. It may however be noted that irradiation is not a replacement for comprehensive product safety. In particular, irradiation cannot reverse spoilage process. Irradiated products need to be handled and stored in the same way as non-irradiated products. Presently, radiation processing is practiced in several countries for over 100 food items. International trade in radiation processed food stuffs has increased considerably in the last two decades.

APPLICATION FOR SPROUT INHIBITION IN BULBS AND TUBERS

Among the various physiological factors that affect storage of bulbs and tubers, sprouting is the most apparent manifestation of deterioration. There is a time lag after harvest, the dormant period, before the bulbs start sprouting. The dormancy period depends on several factors such as variety, climactic conditions during growth, microbial infections, mechanical damage, maturity at time of harvest, and storage environment, particularly temperature (Thomas, 1986). In potatoes the dormancy breaks when tubers are kept at temperatures above 5°C. Sprouting can have detrimental effects on their nutritive value and marketability. Hence there is a need for inhibiting sprouting.

The methods recommended for inhibiting sprouting in potatoes are storage at low temperatures and use of chemicals (e.g., maleic hydrazide and isopropyl-*N*-(3-chlorophenyl) carbamate) to prolong dormancy. A temperature of 4–5°C and a relative humidity of 92–95% have been shown to be ideal for prolonged storage. Gamma irradiation in the range of 0.03 to 0.15 kGy has been shown to inhibit sprouting in tubers, bulbs, rhizomes, and corms. Ionizing radiation suppresses nucleic acid synthesis (DNA and RNA) in the meristem (bud) which prevents germination and consequently results in sprout inhibition as a function of the absorbed radiation dose (Farkas, 2006). Doses exceeding 0.15–0.20 kGy result in increased darkening or browning, decreased wound healing ability, increased storage rot, spoilage, sweetening, a decrease in vitamin content, and changes in chemical composition which do not disappear during subsequent storage.

Effect on Acrylamide Content

Reducing sugars are known to increase gradually during cold (4–5°C) storage and are the main precursors responsible for the formation of acrylamide in processed potato products like chips and crisps. Unlike non-irradiated potatoes, that require cold storage to prevent sprouting, irradiated potatoes can be stored at a higher temperature of 15°C. This reduces the production of sugars and has the advantage that the precursors of acrylamide, that are mainly reducing sugars, are kept at low levels. In our studies on the effect of radiation processing on the acrylamide content in six varieties of potatoes, the levels of reducing sugars were found to be lower in the irradiated tubers compared to their corresponding non-irradiated controls after a storage period of six months at both 4°C and 14°C. Such results were also observed earlier in five Indian potato cultivars (Joshi *et al.*, 1990). Gokeman *et al.* (2007), in their studies on the controlled atmosphere storage and low dose irradiation on two potato cultivars, found a decreased rate of sweetening of the tubers at low doses. Our studies on the acrylamide content in potato chips prepared from irradiated potato kept for six months at 4°C and 14°C demonstrated a decreased formation of this compound compared to the chips prepared from their corresponding non-irradiated controls. Interaction between the temperature and time of storage was found to be significant for acrylamide formation in chips prepared from non-irradiated potatoes ($P = 0.0386$) but was non-significant for the product prepared from irradiated tubers. Color of the french fries prepared was also lighter and acceptable. Radiation processing therefore can be advantageous to the manufacturers of processed potato products by providing desired lighter color to fries and chips with lower acrylamide content. Thus irradiation of potatoes at sprout inhibition doses results in its suitability for processing to chips with retention of quality factors such as color and a lower acrylamide content, thereby enhancing consumer safety. In a parallel work by the authors (unpublished), the content of acrylamide in *pooris* prepared from wheat (var. Lokwan) irradiated (0.2 –1.00 kGy) and stored over a period of six months was investigated. A lower content of acrylamide and better color retention in *pooris* prepared with irradiated samples was observed compared to the non-irradiated controls. Thus, use of irradiation as a general mitigation strategy for reducing acrylamide in food can be proposed. This, however, needs further confirmation in other food stuffs.

References

Ahn, J.S., Castle, L., Clarke, D.B., Lloyd, A.S., Philo, M.R., Speck, D.R., 2002. Verification of the findings of acrylamide in heated foods. Food. Addit. Contam. 19, 1116–1124.

Alpmann, A., Morlock, G., 2008. Rapid and sensitive determination of acrylamide in drinking water by planar chromatography and fluorescence detection after derivatization with dansulfinic acid. J. Separation Sci. 31, 71–77.

Casella, I.G., Pierri, M., Contursi, M., 2006. Determination of acrylamide and acrylic acid by isocratic liquid chromatography with pulsed electrochemical detection. J. Chromatogr. A. 1107, 198–203.

Cavalli, S., Maurer, R., Höfler, F., 2003. LC-GC Europe. In: The Applications Book. Thermo Electron Corporation Press, San Jose, CA. April 2003, p 1.

Fan, X.T., Mastovska, K., 2006. Effectiveness of Ionizing Radiation in Reducing Furan and Acrylamide Levels in Foods. J. Agric. Food. Chem. 54, 8266–8270.

Farkas, J., 2006. Irradiation for better foods. Trends Food Sci. Technol. 17, 148–152.

Fiselier, K., Gama-Baumgartner, F., Fiscalini, A., Biedermann, M., Grob, K., Imhof, D., Beer, M., 2004. Good manufacturing practice (GMP) for French fries low in acrylamide: results of a pilot project. Mitteilungen aus Lebensmitteluntersuchung und Hygiene 95, 127–134.

Gökmen, V., Akbudak, B., Serpen, A., Acar, J., Turan, Z.M., Eris, A., 2007. Effects of controlled atmosphere storage and low-dose irradiation on potato tuber components affecting acrylamide and colour formations upon frying. Eur. Food Res. Technol. 224, 681–687.

IARC. 1994. Acrylamide. In IARC Monographs on the evaluation of the Carcinogenic Risks to Humans, Vol. 60, Some Industrial Chemicals, pp. 389–433. Lyon, France: Int. Agency Res. Cancer. http://monographs.Iarc.fr?ENG/Monographs/vol60/mono60-16.pdf

JECFA. 2011. Safety evaluation of certain contaminants in food. Acrylamide. 72nd meeting of the Joint FAO/WHO Expert Committee on Food Additives (JECFA), FAO JECFA Monograph 8, pp. 1–151. WHO Food Additive Series 63. http://whqlibdoc.who.int/publications/2011/9789241660631_eng.pdf

Joshi, M.R., Srirangarajan, A.N., Thomas, P., 1990. Effects of gamma irradiation and temperature on sugar and vitamin C changes in five Indian potato cultivars during storage. Food Chem. 35, 209–216.

Jung, M.Y., Choi, D.S., Ju, J.W., 2003. A Novel Technique for Limitation of Acrylamide Formation in Fried and Baked Corn Chips and in French Fries. J. Food. Sci. 68, 1287–1290.

Mestdagh, F., de Wilde, T., Delporte, K., van Peteghem, C., de Meulenaer, B., 2008. Impact of chemical pre-treatments on the acrylamide formation and sensorial quality of potato crisps. Food Chem. 106, 914–922.

Mulla, M.,Z., Bharadwaj, V.,R., Annapure, U.,S., Variyar, P.,S., Sharma, A., Singhal, R.,S., 2011. Acrylamide content in fried chips prepared from irradiated and non-irradiated stored potatoes. Food Chem. 127, 1668–1672.

Nair, P.M., Behere, A.G., Ramaswamy, N.K., 1981. Glycoalkaloids of Solanum tuberosum Linn. J. Sci. Ind. Res. 40, 529–535.

Nielsen, N.J., Granby, K., Hedegaard, R.V., Skibsted, L.H., 2006. A liquid chromatography — tandem mass spectrometry method for simultaneous analysis of acrylamide and the precursors, asparagine and reducing sugars in bread. Anal. Chim. Acta. 557, 211–220.

Paleologos, E.K., Kontominas, M.G., 2005. Determination of acrylamide and methacrylamide by normal phase high performance liquid chromatography and UV detection. J. Chromatogr. A. 1077, 128–135.

Sharma, A., 2004. Post-harvest processing of fruits and vegetables by ionizing radiation. In: Dris, R., Mohan Jain, S. (Eds.), Production Practices and Quality Assessment of Food Crops: Post-harvest treatment and technology. Kluwer Academic publishers, Dordrecht, USA. Volume 4, pp. 261–295.

Thomas, P., 1986. Radiation preservation of foods of plant origin. Part 3. Tropical fruits: bananas, mangoes, and papayas. CRC Crtit. Rev. Food Sci. Nutr. 23, 147.

Wenzl, T., de la Calle, M.,B., Anklam, E., 2003. Analytical methods for the determination of acrylamide in food products: a review. 20, 885–902.

Yaylayan, V.A., Wnorowski, A., Locas, C.P., 2003. Why Asparagine Needs Carbohydrates To Generate Acrylamide. J. Agric. Food Chem. 51, 1753–1757.

Young, M.S., Jenkins, K.M., Mallet, C.R., 2004. Solid-Phase Extraction and Cleanup Procedures for Determination of Acrylamide in Fried Potato Products by Liquid Chromatography/Mass Spectrometry. J. AOAC Int. 87, 961–964.

Zhang, Y., Ren, Y., Zhang, Y., 2009. New Research Developments on Acrylamide: Analytical Chemistry, Formation Mechanism, and Mitigation Recipes. Chem. Rev. 109, 4375–4397.

Zhang, Y., Zhang, G.Y., Zhang, Y., 2005. Occurrence and analytical methods of acrylamide in heat-treated foods. J. Chromatogr. A 1075, 1–21.

CHAPTER

12

Production of Jerusalem Artichoke (*Helianthus tuberosus* L.) and Impact on Inulin and Phenolic Compounds

Vibe Bach, Morten Rahr Clausen, Merete Edelenbos

Department of Food Science, Aarhus University, Kirstinebjergvej, Denmark

CHAPTER POINTS

- Jerusalem artichoke tubers contain health-promoting compounds such as inulin and phenolic compounds.
- Inulin is a fructose polymer with a degree of polymerization of 2–50 in Jerusalem artichoke tubers.
- The degree of polymerization and content of inulin in Jerusalem artichoke tubers depends on the maturity of the tubers at harvest.
- Inulin in Jerusalem artichoke tubers is degraded to shorter polymers during storage in cold rooms or in the field.
- The phenolic profile of Jerusalem artichoke tubers is mainly composed of phenolic acids.
- The content of phenolic acids reaches a maximum as the tubers mature, but decreases during storage.
- The content of phenolic compounds in Jerusalem artichoke is higher in the peel than in the tuber flesh.

INTRODUCTION

Jerusalem artichoke (JA) (*Helianthus tuberosus* L.) is a perennial, temperate zone plant of the Asteraceae family, native to North America. The plants can grow up to 3–4 m in height, and they produce yellow flowers and underground tubers. JA is grown for its edible tubers, as a forage crop, for biofuel production, and for the extraction of inulin and production of fructose. The nutritional information of raw JA tubers is seen in Table 12.1. JA stores energy in the form of the fructose polymer inulin. JA tubers have a low caloric value, contain virtually no fat and have a high content of the minerals potassium, calcium, and iron (Table 12.1). The tubers also contain several secondary metabolites such as coumarins (Cabello-Hurtado *et al.*, 1998), terpenes (Bach *et al.*, 2012; Macleod *et al.*, 1982) and phenolic acids (Mattila and Hellström, 2007; Tchoné *et al.*, 2006). The scope of this chapter covers inulin and phenolic compounds, their structure, health effects, and concentrations in raw JA tubers used for human consumption.

HOW COMPOSITION IS ALTERED

Inulin

Inulin is a linear fructose polymer built from 2–60 D-fructose units connected through β(2→1) glycosidic linkages with a terminal α(1→2) bonded D-glucose (Nair *et al.*, 2010; Niness, 1999) (Figure 12.1). Inulin without terminally bonded glucose is called inulo-*n*-ose (*n* = number of fructose units). Inulin polymers with less than 10 fructose units are often referred to as fructooligosaccharides (FOS) (Niness, 1999). The inulin in JA tubers is a mixture of polymers with different degrees of polymerization (DP) containing 2–50 fructose units (Marx *et al.*, 1997; Yildiz, 2011).

Health Effects of Inulin

Inulin is a prebiotic, soluble dietary fiber, i.e., it is not degraded by enzymes in the human digestive system,

Processing and Impact on Active Components in Food
http://dx.doi.org/10.1016/B978-0-12-404699-3.00012-3

TABLE 12.1 Constituents of Raw Jerusalem Artichoke Tubers and their Contents per 100g Fresh Weight (FW)

Constituents	Content/100 g FW
Energy (kJ)	158–317[a,b]
Protein (g)	2.0–3.6[b,c]
Fat (g)	0.1–0.6[a,b]
Total carbohydrate (g)	10.5–17.6[a,b]
Total sugar (g)	1.6–2.2[d,e]
Total dietary fibers:	
Inulin (g)	7.4–12.0[d,e]
Others than inulin (g)	0.6–4.0[a,b]
Moisture (g)	77.2–86.7[d,e]
Potassium (mg)	278–1250[b,c,f]
Calcium (mg)	28–352[b,c,f]
Iron (mg)	0.5–2[b,c,f]

[a] *Kays and Nottingham (2008)*
[b] *Saxholt* et al. *(2008)*
[c] *Seiler (1990).*
[d] *Bach* et al. *(2012).*
[e] *Bach* et al. *(2013b).*
[f] *Somda* et al. *(1999).*

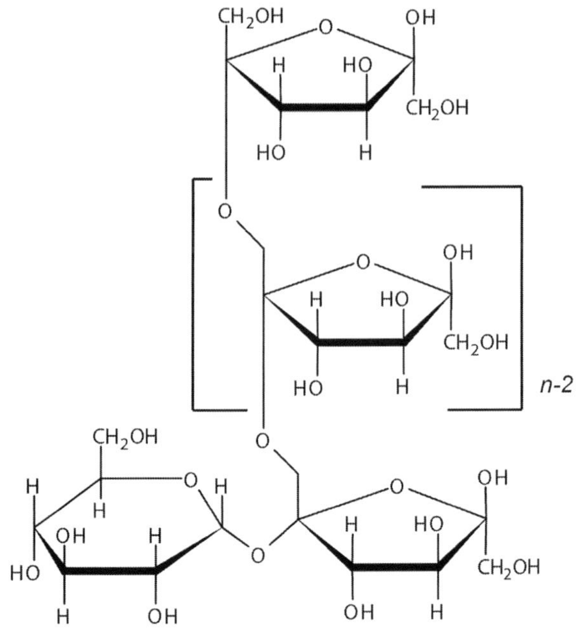

FIGURE 12.1 Chemical structure of inulin (n = 2–60). Fructooligosaccharides have n ≤ 8.

but fermented selectively by beneficial bacteria in the gut (Nair *et al.*, 2010; Niness, 1999). JA tubers are suitable constituents in a diet for diabetics as the glucose content is low (Figure 12.2A), and as inulin increases the blood glucose level is less than with other carbohydrates (Niness, 1999). Inulin is also characterized by a range of other health-promoting effects, which make JA tubers a valuable part of a human diet. Inulin promotes good digestive health, influences lipid metabolism by reducing concentration of triacylglycerides in the blood and enhances mineral bioavailability of calcium, magnesium, and iron, thereby decreasing the risk of osteoporosis (Nair *et al.*, 2010). Inulin has also been associated with a reduced risk of developing colon cancer and inhibition of tumor growth (Nair *et al.*, 2010). Inulo-*n*-ose has the same health benefits as inulin as it contains the non-degradable β (2→1) bonds although it has no end-glucose.

Concentration of Inulin

JA tubers contain 13–23 g dry matter (DM) per 100 g fresh weight (FW) of which 7.4–12.0 g is inulin, and 1.6–2.2 g is fructose, glucose and sucrose making up the total sugars (Table 12.1). The main sugar in JA tubers is sucrose, whereas tubers contain very little fructose and hardly any glucose (Figure 12.2A). This is true for most of the season except for just before sprouting where a dramatic increase in fructose content has been observed (Marx *et al.*, 1997).

Changes in Degree of Polymerization of Inulin During Production

The content and DP of inulin is dependent on cultivar, growing conditions, harvest time, and maturity of tubers at harvest. While only minor variation in total inulin content is seen around termination of tuber growth and during storage in the field (Figure 12.2B), significant changes may occur in the DP of the inulin molecules. Saengthongpinit and Sajjaanantakul (2005) determined the polymerization of inulin in raw JA tubers 16–20 weeks after planting by quantifying the polymers in groups based on their DP. They found that DP 3–10 comprised 47%, DP 11–20 comprised 27–30%, DP 21–30 comprised 10%, and DP > 30 comprised 4–5% of the total content of inulin. The content of inulin and the DP reaches a maximum as the tubers mature during the fall. The time of peak maturity depends on the cultivar (Kocsis *et al.*, 2007; Slimestad *et al.*, 2010). Marx *et al.* (1997) found that JA tubers contained inulin with a DP up to 50 in the early autumn, but no inulin with DP above 20 in the spring of the following year. Kocsis *et al.* (2007) reported that the average DP (DP*n*) in early cultivars of JA was 11–13 when analyzed 14 weeks after planting, but fell to 5–7 when analyzed 47 weeks after planting. In comparison, the DP*n* in late cultivars was 13–18 after 14 weeks falling to 6–8 after 47 weeks. This result is in line with

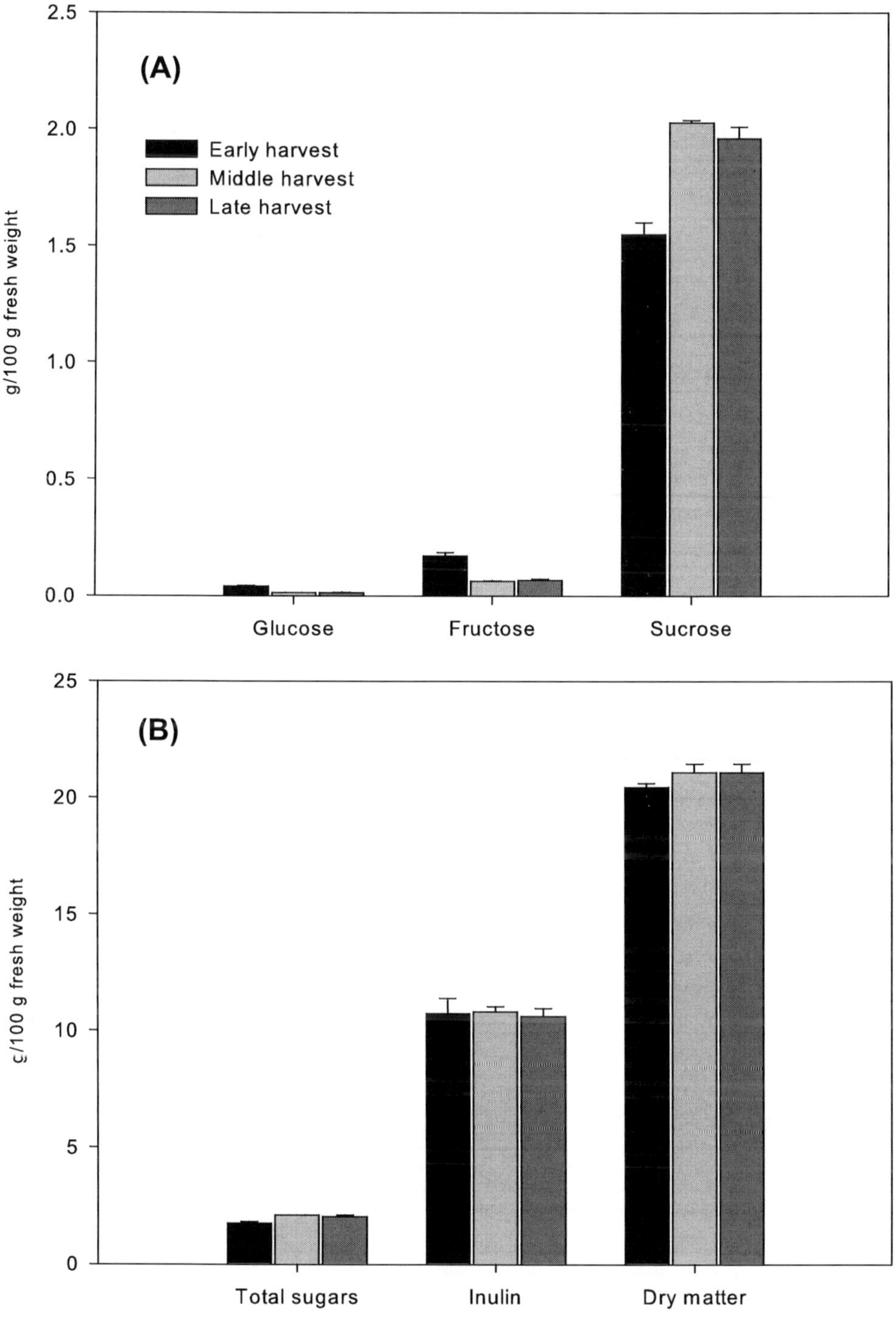

FIGURE 12.2 Carbohydrates and dry matter in Jerusalem artichoke tubers. The bar plot shows the concentrations of (A) glucose, fructose, sucrose, and (B) total sugars, inulin, dry matter in Jerusalem artichoke tubers at an early (October), a middle (November), and a late harvest (February the following year). Concentrations are calculated as mean across three cultivars using data from Bach *et al.* (2012).

the findings of Schorr-Galindo and Guiraud (1997) who reported that early cultivars reached a maximum DP*n* of 11 in September and fell to 4 before the end of November, while very late cultivars had a DP*n* of 8 in September, then rose and finally fell to 3–4 in December–January.

JA tubers are frost tolerant down to approximately −5°C (Uemura and Yoshida, 1986) and may therefore be stored in the field during winter, either under straw, uncovered, or in cold rooms. During storage the inulin molecules are metabolized into FOS and sucrose (Cabezas *et al.*, 2002; Saengthongpinit and Sajjaanantakul, 2005), which are used for osmotic adjustment, prevention of water crystallization, and protection of cell structures during water deficit caused by low temperatures (Marx *et al.*, 1997; Portes *et al.*, 2008). However, inulin may also be degraded into inulo-*n*-ose (Edelman and Jefford, 1968; Saengthongpinit and Sajjaanantakul, 2005). Modler *et al.* (1993) investigated the DP of inulin in JA tubers stored for 18 weeks at 5°C and 2°C. In both cases the content of inulin with DP above 10 decreased, while the content of inulin with DP below 10 increased. This is in agreement with Saengthongpinit and Sajjaanantakul (2005) who reported a gradual increase in DP 3–10 and a decrease in DP > 10 in JA tubers stored for 10 weeks at 5°C and 2°C, especially after 4 weeks storage at 5°C. Inulin in JA tubers is stable at −18°C for 10 weeks of storage, probably because all metabolic activities in the tubers stop as such at low freezing temperature (Cabezas *et al.*, 2002; Saengthongpinit and Sajjaanantakul, 2005).

OTHER WAYS IN WHICH COMPOSITION IS ALTERED

Structure and Health Benefits of Phenolic Compounds

JA tubers contain secondary metabolites like phenolic and polyphenolic compounds. Phenolic compounds have positive physiological functions such as antioxidant, antimutagenic and antitumor activities. Tchoné *et al.* (2006) identified 22 phenolic compounds in JA tubers and found that the phenolic profile was dominated by phenolic acids, mainly salicylic and chlorogenic acid. Mattila and Hellström (2007) reported chlorogenic acid, caffeic acid, and an unidentified phenolic acid as the major phenolic acids in JA tubers. Bach *et al.* (2013a) identified chlorogenic acid and gallic acid as the major phenolic compounds in JA tubers. The structure of the major phenolic acids in JA tubers is shown in Figure 12.3.

Concentration of Phenolic Compounds and Changes During Production

The content of total phenolic compounds in JA tubers depends on the cultivar and on the analyzed part of the tuber. The peel of JA tubers contains two to 12 times higher content of total phenolic compounds than the internal flesh although phenolic compounds are present throughout the tuber (Seljåsen and Slimestad, 2007; Takeuchi and Nagashima, 2011; Tchoné *et al.*, 2006). Seljåsen and Slimestad (2007) and Takeuchi and Nagashima (2011) reported between 39 and 129 mg gallic acid equivalents (GAE)/100 g FW in the peel and between 4.9 and 31 mg GAE/100 g FW in the tuber flesh. The concentrations of major phenolic acids in peel and flesh of JA tubers is shown in Figure 12.3.

As with inulin, the content of phenolic compounds in JA tubers is affected by the maturity of the tubers at harvest. During growth, the content of phenolic acids in the tubers rises and reaches a maximum and then decreases at termination of growth (Tchoné *et al.*, 2006). Bach *et al.* (2013a) reported the concentration of total phenolic compounds to be 46.6 mg GAE/100 g FW in the autumn and 14.0 mg GAE/100 g FW in the following spring. Seljåsen and Slimestad (2007) investigated the antioxidant capacity of JA tubers as measured by 2,2-diphenyl-1-picrylhydrazyl (DPPH) assay and found antioxidant activity in the peel, but not in the flesh of the tubers. In order to get a high content of phenolic compounds from JA tubers, they should be eaten unpeeled and preferably in the autumn before overwintering in the fields or storage in cold rooms.

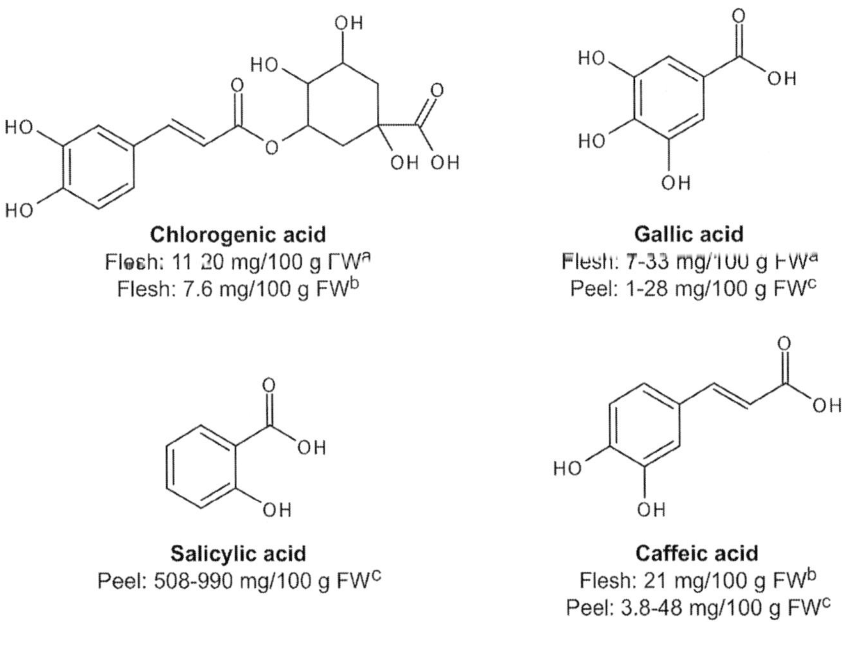

FIGURE 12.3 Chemical structure of phenolic compounds in Jerusalem artichoke tubers. Below the name of the individual compounds are indicated their concentration as reported in the literature. [a]Bach *et al.* (2013a); [b]Mattila and Hellström (2007); [c]Tchoné *et al.* (2006).

ANALYTICAL TECHNIQUES

The content of inulin in JA tubers is most often analyzed using high performance anion exchange chromatography with pulsed amperometric detection (HPAEC-PAD). The method makes it possible to separate the individual polymers according to their DP, but quantification of the individual polymers is challenging due to lack of pure standards (Böhm *et al.*, 2005; Saengthongpinit and Sajjaanantakul, 2005; Slimestad *et al.*, 2010). Inulin standards normally contain an unknown mixture of inulin polymers, and only the total content of inulin is known. The content of individual inulin polymers or polymers grouped according to DP is normally stated as relative content.

The total inulin content can be quantified using acidic or enzymatic hydrolysis of the inulin polymers. First the concentrations of glucose, fructose and sucrose in the sample are determined, then the inulin is hydrolyzed, and the concentrations of individual sugars are determined again. The total content of inulin can then be calculated, and the DP*n* can be determined from the ratio between fructose and glucose (Kocsis *et al.*, 2007; Praznik *et al.*, 2002; Schorr-Galindo and Guiraud, 1997; Slimestad *et al.*, 2010). Praznik *et al.* (2002) reported that the DP*n* of inulin in JA tubers was 9.8 in the autumn and 5.9 the following spring, when inulin was acid-hydrolyzed and sugars determined by spectrophotometric enzymatic assay and almost similar contents were found (9.9 in the autumn and 4.9 in the spring) by size-exclusion chromatography of pure extracts.

Clausen *et al.* (2012) investigated the inulin and sugar content in JA tubers by ^1H nuclear magnetic resonance (NMR) spectroscopy. They found that the content of inulin decreased, but also that the number of inulin molecules increased from early harvest in October to late harvest in February the following year, corresponding to a degradation of inulin to shorter chain polymers (Figure 12.4). By using signals specific to the glucose and fructose units, it was possible to determine that the DP*n* of inulin was 4.8 in October, 3.7 in November, and 3.8 in February the following year. For comparison, the DP*n* results obtained by HPAEC-PAD on the same samples were 4.9, 3.3, and 3.3, respectively (unpublished results).

The content of phenolic compounds in JA tubers has mainly been assessed using Folin-Ciocalteu (FC) type assays, which only gives information on the total content of phenolic compounds. The method is a non-specific assay, which also measures some non-phenolic compounds like aromatic amines, sulphur dioxide, Fe^{2+}, tryptophan, tyrosine, guanine, xanthine, hydrogen sulphide, reducing sugars, and ascorbic acid. The results of such measurements are normally denoted as total phenolic compounds, but in reality the FC results depict an expression of the total reducing capacity of the sample. For more detailed information on the content of phenolic compounds in JA tubers, HPLC was used by Bach *et al.* (2013a) and Mattila and Hellström (2007) while GC-MS and LC-MS was used by Tchoné *et al.*

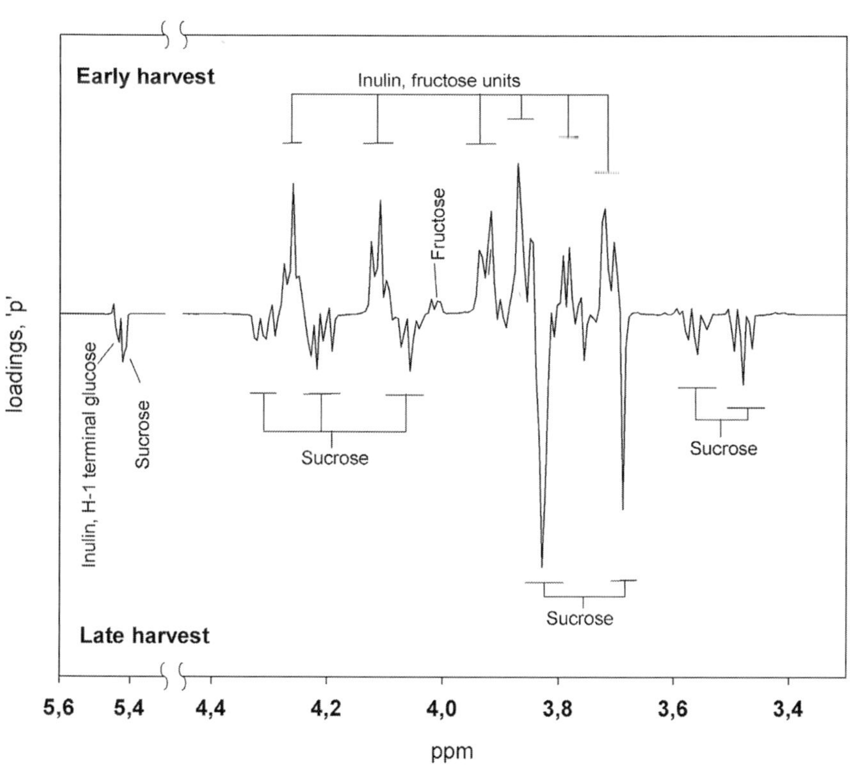

FIGURE 12.4 Effect of harvest time on the carbohydrate composition of Jerusalem artichoke tubers analyzed by ^1H NMR spectroscopy and multivariate data analysis (partial least squares-discriminant analysis, PLS-DA). The positive inulin loadings show that total inulin content decreases from early to late harvest, while the negative inulin loadings, corresponding to H-1 of the terminal glucose unit, show that the number of inulin molecules increases from early to late harvest. Furthermore, the results show that sucrose and fructose content increase and decrease, respectively, from early to late harvest. *Data from Clausen et al. (2012).*

(2006). Most studies on phenolic compounds in JA tubers have been performed on dried tuber material, except for Tchoné *et al.* (2006), who extracted phenolic compounds from fresh tubers. Kim *et al.* (2010) extracted up to 73.8mg GAE/100g FW from fermented JA tubers pressurized with enzyme mixtures for increased liquefication at a hydrostatic pressure of 150MPa but only 35.5mg GAE/100g FW from non-fermented tubers using non-pressurized hot-water extraction.

References

Bach, V., Jensen, S., Clausen, M.R., Bertram, H.C., Edelenbos, M., 2013a. Enzymatic browning and after-cooking darkening of Jerusalem artichoke tubers (*Helianthus tuberosus* L.). Food Chem. 141, 1445–1450.

Bach, V., Kidmose, U., Kjeldsen, G.B., Edelenbos, M., 2012. Effects of harvest time and variety on sensory quality and chemical composition of Jerusalem artichoke (*Helianthus tuberosus* L.) tubers. Food Chem. 133, 82–89.

Bach, V., Thybo, A.K., Kidmose, U., Edelenbos, M., 2013b. Sensory quality and appropriateness of raw and boiled Jerusalem artichoke tubers (*Helianthus tuberosus* L.). J. Sci. Food Agric. 93, 1211–1218.

Böhm, A., Kaiser, I., Trebstein, A., Henle, T., 2005. Heat-induced degradation of inulin. Eur. Food Res. Technol. 220, 466–471.

Cabello-Hurtado, F., Durst, F., Jorrín, J.V., Werck-Reichhart, D., 1998. Coumarins in *Helianthus tuberosus*: Characterization, induced accumulation and biosynthesis. Phytochemistry 49, 1029–1036.

Cabezas, M.J., Rabert, C., Bravo, S., Shene, C., 2002. Inulin and sugar contents in *Helianthus tuberosus* and *Cichorium intybus* tubers: Effect of postharvest storage temperature. J. Food Sci. 67, 2860–2865.

Clausen, M.R., Bach, V., Edelenbos, M., Bertram, H.C., 2012. Metabolomics reveals drastic compositional changes during overwintering of Jerusalem artichoke (*Helianthus tuberosus*) tubers. J. Agric. Food. Chem. 60, 9495–9501.

Edelman, J., Jefford, T.G., 1968. The mechanism of fructosan metabolism in higher plants as exemplified in Helianthus tuberosus. New Phytol. 67, 517–531.

Kim, D., Fan, J.P., Chung, H.C., Han, G.D., 2010. Changes in extractability and antioxidant activity of Jerusalem artichoke (*Helianthus tuberosus* L.) tubers by various high hydrostatic pressure treatments. Food Sci. Biotechnol. 19, 1365–1371.

Kays, S.J., Nottingham, S.F., 2008. Biology and Chemistry of Jerusalem Artichoke CRC Press. Taylor and Francis Group. Boca Raton, Florida, USA.

Kocsis, L., Liebhard, P., Praznik, W., 2007. Effect of seasonal changes on content and profile of soluble carbohydrates in tubers of different varieties of Jerusalem artichoke (*Helianthus tuberosus* L.). J. Agric. Food Chem. 55, 9401–9408.

Macleod, A.J., Pieris, N.M., de Troconis, N.G., 1982. Aroma volatiles of *Cynara scolymus* and *Helianthus tuberosus*. Phytochemistry 21, 1647–1651.

Marx, S.P., Nösberger, J., Frehner, M., 1997. Seasonal variation of fructan-β-fructosidase (FEH) activity and characterization of a β-(2-1)-linkage specific FEH from tubers of Jerusalem artichoke (*Helianthus tuberosus*). New Phytol. 135, 267–277.

Mattila, P., Hellström, J., 2007. Phenolic acids in potatoes, vegetables, and some of their products. J. Food Comp. Anal. 20, 152–160.

Modler, H.W., Jones, J.D., Mazza, G., 1993. Observations on long-term storage and processing of Jerusalem artichoke tubers (*Helinathus tuberosus*). Food Chem. 48, 279–284.

Nair, K.K., Kharb, S., Thompkinson, D.K., 2010. Inulin dietary fiber with functional and health attributes—a review. Food Rev. Int. 26, 189–203.

Niness, K.R., 1999. Inulin and oligofructose: What are they? J. Nutr. 129, 1402S–1406S.

Portes, M.T., Figueiredo-Ribeiro, R.d.C.L., de Carvalho, M.A.M., 2008. Low temperature and defoliation affect fructan-metabolizing enzymes in different regions of the rhizophores of *Vernonia herbacea*. J. Plant Physiol. 165, 1572–1581.

Praznik, W., Cieslik, E., Filipiak-Florkiewicz, A., 2002. Soluble dietary fibres in Jerusalem artichoke powders: Composition and application in bread. Nahrung/Food 46, 151–157.

Saengthongpinit, W., Sajjaanantakul, T., 2005. Influence of harvest time and storage temperature on characteristics of inulin from Jerusalem artichoke (*Helianthus tuberosus* L.) tubers. Postharvest Biol. Technol. 37, 93–100.

Saxholt, E, Christensen, A.T., Møller, A., Hartkopp, H.B., Ygil, K.H., and Hels, O.H., 2008. Danish Food Composition Databank, revision 7. Department of Nutrition. National Food Institute, Technical University of Denmark.

Schorr-Galindo, S., Guiraud, J.P., 1997. Sugar potential of different Jerusalem artichoke cultivars according to harvest. Bioresour. Technol. 60, 15–20.

Seiler, G.J., 1990. Protein and mineral concentrations in tubers of selected genotypes of wild and cultivated Jerusalem-artichoke (*Helianthus tuberosus*, Asteraceae). Econ. Bot. 44, 322–335.

Seljåsen, R., Slimestad, R., 2007. Fructooligosaccharides and phenolics in flesh and peel of spring harvested *Helianthus tuberosus*. In: Desjardins, Y. (Ed.), Proceedings of the 1st International Symposium on Human Health Effects of Fruits and Vegetables. Vol. 744, pp. 447–450, .

Slimestad, R., Seljaasen, R., Meijer, K., Skar, S.L., 2010. Norwegian-grown Jerusalem artichoke (*Helianthus tuberosus* L.): morphology and content of sugars and fructo-oligosaccharides in stems and tubers. J. Sci. Food Agric. 90, 956–964.

Somda, Z.C., McLaurin, W.J., Kays, S.J., 1999. Jerusalem artichoke growth, development, and field storage. II. Carbon and nutrient element allocation and redistribution. J. Plant Nutr. 22, 1315–1334.

Takeuchi, J., Nagashima, T., 2011. Preparation of dried chips from Jerusalem artichoke (*Helianthus tuberosus*) tubers and analysis of their functional properties. Food Chem. 126, 922–926.

Tchoné, M., Bärwald, G., Annemüller, G., Fleischer, L.G., 2006. Separation and identification of phenolic compounds in Jerusalem artichoke (*Helianthus tuberosus* L.). Sci. Aliments 26, 394–408.

Uemura, M., Yoshida, S., 1986. Studies on freezing injury in plant cells. II. Protein and lipid changes in the plasma membranes of Jerusalem artichoke tubers during a lethal freezing *in vivo*. Plant Physiol. 80, 187–195.

Yildiz, S., 2011. The metabolism of fructooligosaccharides and fructooligosaccharide-related compounds in plants. Food Rev. Int. 27, 16–50.

CHAPTER

13

Bioactive Compounds in Asparagus and Impact of Storage and Processing

José María Fuentes Alventosa, José Manuel Moreno Rojas

Tecnología, Postcosecha e Industria Agroalimentaria. Instituto de Investigación y Formación Agraria y Pesquera (IFAPA) Córdoba, Spain

CHAPTER POINTS

- The most important nutraceuticals of asparagus include phenols (flavonoids and hydroxycinnamates), saponins, fructooligosaccharides (FOS), and vitamin C (ascorbic acid).
- Phenols: flavonoid profile is determined by sample location and variety. High temperature season increases the amount of phenols. Ferulic acid concentration and its derivatives are higher during the post-harvest toughening of asparagus spears, with a lower increase under refrigeration conditions and the absence of oxygen in storage. There are alternative sterilization methods in which phenols are more stable.
- Saponins: their profile is determined by sample location and variety, but their content is higher with lower soil temperatures. High losses have been reported among them in commercially canned asparagus. Frozen storage does not affect their concentration in the spears.
- Fructooligosaccharides: post-harvest storage does not affect their stability.
- Ascorbic acid: samples harvested during high temperature season showed lower concentrations of this vitamin in spears. The use of a drying process with a low temperature for a short period of time may result in relatively high retention of this vitamin. MAP conditions or coating techniques used for the storage of the asparagus spears contribute to their preservation.
- Minerals: their content in asparagus is influenced by varieties. An increase or decrease of each mineral related to spear peeling occurs during industrial processing. They do not seem to be affected by frozen storage.

INTRODUCTION

Asparagus (*Asparagus officinalis* L.) is a member of the Liliaceae family whose origin is believed to be the eastern Mediterranean and further east, in the Caucasus Mountains. It is a perennial and monocotyledonous plant that is grown for its edible stems (spears). Spears are consumed as a seasonal vegetable and they are highly appreciated for their delicious taste, low energy content, and nutritional quality (Table 13.1).

Asparagus are divided into two main groups: green asparagus and white asparagus. From a botanical point of view, white and green asparagus are the same plant. The difference between them is the way their spears grow. Edible stems are white while they grow below the soil, but when they emerge and are exposed to direct sunlight, they acquire a green hue through the chlorophyll function.

Asparagus are available in the market as fresh, frozen, and canned products. Several steps comprise product (frozen and canned asparagus) processing, including cleaning of the fresh spears, washing, cutting to separate the edible portions of the spears, and blanching to slow down enzyme action. In canned processing, after being sealed in cans the spears are heated with a mixture of brine and a small quantity of citric acid at 120°C for

Processing and Impact on Active Components in Food
http://dx.doi.org/10.1016/B978-0-12-404699-3.00013-5

TABLE 13.1 Nutritional Composition of Fresh Asparagus

	White Asparagus	Green Asparagus
Moisture[a]	93–94	92–93
Proteins[a]	2.8	1.9
Sugars[a]	4	3
Lipids	Traces	Traces
Calories[b]	25	22
Fiber[a]	0.9	0.9
Phosphorus[c]	60	75
Calcium[c]	26	40
Magnesium	16	21
Sodium[c]	4	4
Potassium[c]	360	370
Cooper[c]	0.02	2.5
Iron[c]	0.03–0.2	1.2–1.9
Manganese[c]	0.01	0.3–0.5
Zinc[c]	0.04	0.7
Chrome[c]	0.007	0.007
Vitamin A[d]	50	980
Vitamin B_1[c]	0.11	0.23
Vitamin B_2[c]	0.08	0.15
Vitamin B_3[c]	1.1	2.2
Vitamin C[c]	28	48

[a] *Expressed as a percentage.*
[b] *Expressed as calories/100 g of product.*
[c] *Expressed as mg/100 g of product.*
[d] *Expressed as International units/100 g of product.*
The table shows the differences in nutritional composition between white and green asparagus.
Sources: Espejo J.A., González-Fernández J.L. (2000) En: El arte culinario del esparrago de Huetor-Tajar. Consejo Regulador de la Denominación Específica "Espárrago de Huétor-Tájar". *Howard F.D., Mac Gillivray J.H., Ymaguchi M. (1962) Nutrient composition of fresh California-grown vegetables.* California Agr. Exp. Sta. Bull. *738: 43. United States Department of Agriculture (1981) Science and Education Administration Nutritive Value of Foods 52–53. Hartmann H.D (1989) Spargel. Grundlagen für den Anbau.* Verlag Eugen Ulmer, Stuttgart. *Amaro M.A. (1993) Variaciones del contenido de elementos minerales en espárrago* (Asparagus officinalis, *L.). Doctoral thesis, Faculty of Veterinary Medicine, University of Cordoba, Spain.*

15 min to avoid spoilage. After that a cooling step is necessary before storage (Figure 13.1).

Also, asparagus extracts have been traditionally used as diuretics, laxatives, antitussives, etc. In addition, recent studies have shown additional biological activities of these extracts, e.g., their antioxidant and antitumoral activities (Chin *et al.*, 2002; Rodríguez *et al.*, 2005). For this reason, besides their organoleptic and nutritional quality, asparagus are valuable vegetables for their composition of bioactive compounds, among which we find phenols, saponins, minerals, and vitamins.

On that basis, consumers are increasingly demanding a healthy diet and including the health-enhancing characteristics of food in their purchase decisions.

The nutraceutical composition of asparagus may vary according to several factors, such as cultivars and varieties, origin, extraction methods, storage, packaging, and other post-harvest protocols. Indeed, handling, processing, and the usual preservation methods—canning, pickling, freezing, and drying—may affect the phytochemical content, causing irreversible modifications.

HOW COMPOSITION IS ALTERED

This section has been structured according to the main nutraceuticals cited above (phenols, saponins, fructooligosaccharides, and ascorbic acid), with the aim of easier understanding.

Phenols

Influence of Post-Harvest Storage on Ferulic Acid and its Derivatives Content

Ferulic acid and its derivatives are the main phenolic compounds contained in asparagus cell walls. Post-harvest toughening of asparagus spears is associated with a large increase in the monomeric and diferulic acids in stem tissue cell walls.

White and green asparagus tissues represent highly different systems from the point of view of their mechanical properties. There are several differences between the distribution and the quantity of these phenols in white asparagus and those in green asparagus after post-harvest storage. First, the total amount of phenols is higher in white asparagus than in green asparagus. Regarding distribution, their accumulation in white asparagus mainly occurs in the latter two portions (middle and lower sections) of the spears, while equivalent amounts are found in the apical, middle, and lower sections of green asparagus.

Storage conditions have been evaluated on white asparagus (Jaramillo *et al.*, 2007) in order to assess the evolution of ferulic acid and its derivatives contained in asparagus cell walls. Aerobic conditions and anaerobic conditions have been tested, showing a direct relationship between phenol accumulation and increases in texture. It was observed that temperature and storage atmosphere significantly affected phenolics in cell walls.

Losses of Flavonoids in Pickled Asparagus Spears

The pickling process of green asparagus has been shown to greatly influence flavonid concentrations (Fuleki, 1999), evincing important losses of rutin, the most

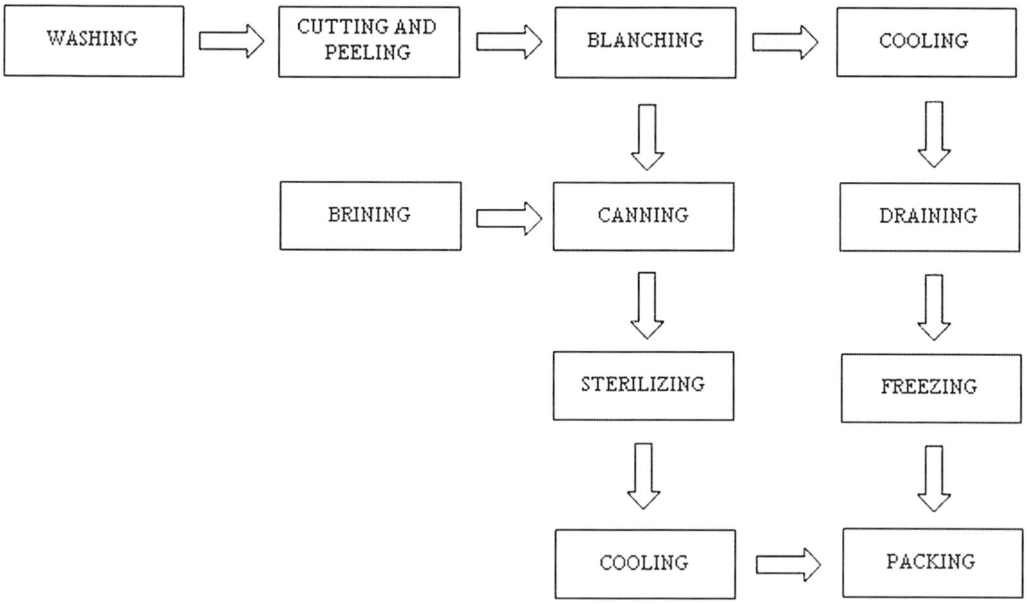

FIGURE 13.1 Flow-chart of asparagus processing. The figure shows the typical steps of asparagus manufacturing process.

important flavonoid in asparagus (0.02 to 0.1%). These losses were identified as highly visible white and greyish deposits that consisted mainly of needle shaped crystals of rutin after 6 months of storage. The precipitation of this bioactive compound is due to the low pH (3.6–4.3) of the pickled product. A different behaviour is found in canned green asparagus owing to a higher pH (5.3–5.8).

Influence of Microwave-Circulated Water Combination and Conventional Sterilization on Flavonoid Content

The various sterilization methods used to treat asparagus also influence the flavonoid content of the final product (Sun *et al.* 2007). Common methods used for sterilization of green asparagus spears include microwave-circulated water combination (MCWC), heating systems, pressured hot-water heating, and steam-heating in retort systems. The reported values regarding rutin content showed a relative heat-stability of this flavonoid, in relation to other antioxidants such as ascorbic acid, not decreasing significantly after sterilization and with no significant differences among above-mentioned sterilization methods.

Saponins

Saponins are, together with flavonoids, the most important group of phytochemicals in white and green asparagus.

The stability of saponins in white asparagus spears in deep freeze storage is over 20 weeks, as reported by Schwarzbach *et al.* (2006). In contrast, the loss of total saponins amounted to 40% after 28 weeks, probably owing to enzymatic or microbiological slowed-down activities at low temperatures or plant tissue senescence.

Moreover, losses of approximately 30% of total saponins (Schwarzbach, 2004) were reported in samples of canned asparagus.

Fructooligosaccharides

Not much information is available regarding fructooligosaccharides (FOS) composition in different varieties of asparagus. Despite this, their biological activity is well established. FOS are well known for their health benefits, as they act as prebiotics in the human body.

Differences in FOS content have been observed during storage. Different concentrations were reported, depending on preservation temperature and spear portion. Shiomi *et al.* (2007) reported that at higher temperatures (20°C), the level of 1-kestose increased rapidly and to a great extent, in relation to the lower temperatures (4 and 10°C), and significant differences were found among spear portions during the last week of storage, which emphasizes the lowest content in the bottom portion. Another FOS, neokestone, was analyzed, and the variation was similar to 1-kestose but with lower content and a subsequent increase. A significant increase was observed at the top portion during storage at higher temperatures (20°C), while there were no significant differences among the different portions at lower temperatures during storage. Regarding nystose (FOS), the only changes in content were detected in the top and middle portions, which significantly increased FOS quantity during the last 3 days of storage at higher temperatures (20°C).

Therefore, although it is well established that the sugar content of asparagus spears decreases during storage, FOS do not seem to be affected, and an increase at higher temperatures was even reported.

Vitamin C

It is well known that asparagus are a good source of vitamin C, but its concentration decreases dramatically during short storage periods because it is very sensitive to heating, oxygen, and light exposure.

Influence of Different Packaging Conditions on Vitamin C Content

Modified atmosphere packaging (MAP) conditions and their influence on vitamin C content in fresh green asparagus have been studied (Villanueva *et al.*, 2005). Storage of the hybrid UC-157-F1 under normal atmospheric conditions, the typical conditions recommended for commercialization (MAP and refrigeration at 2°C) and transport, and retailing conditions (MAP at 2°C for the first 5 days and then at 10°C), were all tested. It was observed that vitamin C content behaved differently in each of the three different storage conditions. Although vitamin C decreased quickly in all of them, conservation was better and decreases were lower in samples stored under MAP conditions, both at 2°C and 10°C. In this way, 40.3%, 58.3%, and 49.1% of ascorbic acid retention were reported after 14 days of storage for the samples stored under normal atmospheric conditions, under MAP conditions at 2°C and under MAP conditions at 10°C, respectively.

Further interesting research has been conducted by testing the packaging system. The use of MAP and ordinary and microperforated oriented polypropylene (OPP) films were tested under the following conditions: 10 days at 15°C and 75% relative humidity (Saito *et al.*, 2000). It was reported that vitamin C was affected to a large degree in microperforated film packages at all of the oxygen levels (1–6%), but approximately 50% remained in ordinary OPP.

Additional studies on packaging evaluated the effect of silver nanoparticles-PVP coating on ascorbic acid in green asparagus spears stored at 2°C and 10°C (An *et al.*, 2008). They showed that vitamin C content in samples under controlled conditions was reduced to an extent greater than in those preserved under the coated packaging.

Influence of Drying Methods on Vitamin C Content

Drying methods affect vitamin C content as well. Nindo *et al.* (2003) studied the influence of five drying methods, namely, tray drying, spouted bed drying, combined microwave and spouted bed drying, refractance window drying, and freeze drying. Refractance window drying was noted as the best method for vitamin C retention, owing to a moderate time–temperature combination of less than 5 min and 70–80°C product temperatures. Refractance window drying was followed by freeze drying, combined microwave and spouted bed drying, spouted bed drying, and tray drying was the least effective treatment for vitamin C. These results confirm that the application of a drying process with low temperatures for a short period of time can result in relatively high ascorbic acid retention, avoiding its sensitivity to heating processes.

Minerals

Vegetative Cycle of Asparagus

The vegetative cycle of asparagus (alterations in the colouring of the turion) affects essential elements—copper, iron, zinc, and manganese content—as reported by Amaro-López *et al.* (1995). Thus, the concentrations of these minerals increased during the vegetative cycle, with the highest levels in green spears in relation to both white and purple ones, which is likely to be due to the initiation of photosynthesis and the creation of new vegetative organs. The effect of the portion of asparagus on the micronutrient content was also shown, with the highest concentration in the apical portion vis-á-vis the basal portion. The increase in their levels during the vegetative cycle of the asparagus was higher in the tip portion in relation to minor variations in the basal one. Amaro-López *et al.* (1999a) bore out this greater ability of the apical part of the spear to concentrate minerals in fresh green asparagus, where 11 mineral elements (copper, iron, zinc, manganese, nickel, chromium, calcium, magnesium, potassium, and phosphorus) were found in higher concentration in the apical zone spear, chiefly in the tip, except sodium.

Industrial Processing

Several changes in the mineral contents of asparagus during industrial processing have been reported. In this sense, Amaro-López *et al.* (1999b) studied concentration of essential trace elements following asparagus processing consisting of washing, peeling, and blanching operations. Lower levels of iron and manganese but higher levels of copper and zinc were observed throughout processing, and they noted that these modifications could be related to the peeling of the spears (Table 13.2). In this way, there was a higher concentration of iron and manganese in the asparagus peel and, therefore, its removal decreased the levels of these essential trace elements in the peeled spears, mainly as far as iron is concerned. On the other hand, the increase of copper and zinc content in peeled asparagus, compared with that in the same

TABLE 13.2 Mineral Content after Industrial Processing

	Cu	Fe	Zn	Mn	Ca	Mg	Na	K	P
Fresh spear[a]	18.7	185.3	58.1	25.8	4040	1650	550	37,000	7350
Washed spear[a]	19.9	111.1	60.6	22.2	3100	1650	560	37,000	7380
Peeled spear[a]	23.0	79.7	68.3	20.8	3410	1860	100	38,100	8320
Blanched spear[a]	23.3	79.5	66.6	20.1	3450	1780	520	36,600	8270
Frozen Storage[b]—0 days	20.3	48.9	68.2	18.9	3396	1565	382	27,155	4386
Frozen Storage[b]—45 days	20.1	49.3	66.2	18.5	3403	1534	381	27,190	4459
Frozen Storage[b]—90 days	19.4	44.4	66.7	17.8	3375	1530	391	24,284	4523

[a] *Asparagus officinalis, L., var. DESTO and SUR.*
[b] *Asparagus officinalis, L., var.* Ciprés.
The table shows the changes in the mineral contents of asparagus during industrial processing, including frozen storage at different temperatures (mg/kg dry weight).
Sources: Amaro-López, M.A., Moreno-Rojas, R., Zurera-Cosano, G., Sánchez-Segarra, P.J. (1999) Nutritional changes in the essential trace elements content of asparagus during industrial processing. Food Res. Int. *32, 479–486. Amaro, M.A., Zurera, G., Moreno, R., Cañal, C. (1997) Effect of industrial processing on the mineral content of white asparagus.* Int. J. Food Sci. Tech. *32, 401–406. Amaro, M.A., Moreno, R., Zurera, G. (1998) Nutritional estimation of changes in mineral content during frozen storage of white asparagus.* J. Food Qual. *21, 445–458.*

unpeeled spear, during processing (Table 13.2) was due to significantly smaller concentrations of both elements in the asparagus peel in relation to the peeled spear. The effect of processing and varieties in the mineral content in asparagus was also reported.

Canned asparagus showed similar results, a decrease of iron and manganese levels and a significant increase of copper, zinc, magnesium, potassium and phosphorus after peeling.

Amaro *et al.* (1997) also studied the effect of industrial processing (involving washing, peeling, blanching, and sterilization of the spears) on the calcium, magnesium, sodium, potassium, and phosphorus content of white asparagus. A loss of calcium and sodium was reported, but so was a higher concentration of magnesium, potassium, and phosphorus; all these variations being due to peeling (Table 13.2). Differences among varieties and different fluctuations during processing in fresh asparagus mineral content were reported. The tip and middle portions of the spears evinced a higher concentration of these minerals, except sodium.

Frozen Storage

Frozen storage does not seem to affect the mineral content of asparagus to a large extent. In this sense, the essential elements—copper, iron, zinc, manganese, calcium, magnesium, sodium, potassium, and phosphorus—have been analyzed in white asparagus stored at −18°C after 45 and 90 days (Amaro *et al.*, 1998). The freezing process was performed upon washing, peeling, and blanching of the spears. After 90 days of frozen storage a decrease of copper, iron, manganese, and potassium was reported, but not so after 45 days (Table 13.2). A deterioration of the vegetable tissue and cells during the freezing storage period was noted, with a high loss of exuded liquid during defrosting being the cause of this decrease. However, there were no significant losses in the content of zinc, calcium, magnesium, sodium, and phosphorus during storage time. The main concentrations of minerals (except for sodium) were found in the apical part, more specifically in the tip.

OTHER WAYS IN WHICH COMPOSITION IS ALTERED

In this case, harvesting date, cultivars and the different parts of the asparagus spear have been studied, especially with respect to some nutraceuticals as we describe below.

Influence of Cultivars and Part of the Spear on Flavonoid Content

Flavonoids are one of the most abundant phytochemicals in green asparagus, and they are related to the antioxidant activity of ethanolic extracts of green asparagus (Fuentes-Alventosa *et al.*, 2007).

Sample origin and variety determine flavonoid profile. However, genetic factors seem to be more determinant on the composition of flavonoids than environmental conditions. In this sense, in 2008 the flavonoid content (HPLC) of 32 commercial hybrids of green asparagus, developed in recent years by major international asparagus programs and cultivated in the same place, was published (Fuentes-Alventosa *et al.*, 2008). The results varied between 259 and 763 mg/kg of fresh weight. Moreover, the flavonoid content of 65 different native lines of the characteristic *triguero* asparagus from the Hué torTájar population variety (D.O.) was

TABLE 13.3 Flavonid and Saponin Content in Green and Triguero Asparagus

Commercial Hybrids	Flavonoid[a]	Saponin[b]	Genotypes of Triguero Asparagus	Flavonoid[a]	Saponin[b]
Apollo	457	45	HT-1	528	258
Backlim	236	46	HT-2	405	165
Ercole	356	48	HT-3	237	317
G. Welph	247	45	HT-4	248	233
Jiuc-1	358	45	HT-5	475	211
Nj 956	267	50	HT-6	530	188
Plaverd	275	45	HT-7	347	189
Ravel	182	28	HT-8	655	236
Solar	315	49	HT-9	401	160
Supreme	336	50	HT-10	568	231

[a] *Flavonoids expressed as milligrams of rutin per kilogram of fresh weight.*
[b] *Saponins expressed as milligrams of diosgenin per kilogram of fresh weight.*
The table shows the flavonoid and saponin content in 10 commercial hybrids of green asparagus and 10 genotypes of triguero asparagus from the Huétor-Tájar population variety.
Source: Fuentes-Alventosa, J.M. (2007) Caracterización de flavonoides y saponinas en el espárrago triguero de HuétorTájar (Granada) [Flavonoids and Saponins profiles in Triguero asparagus]. Master in Agrofood, University of Cordoba, Spain.

analyzed, ranging between 285 and 804 mg/kg of fresh weight. It must be emphasized that there were several genotypes of *triguero* asparagus with a flavonoid composition which was very distinct and statistically different from the majority of commercial hybrids. Hence, the choice of the appropriate genotype may be an important issue from a nutraceutical point of view (Table 13.3; Fuentes-Alventosa, 2007).

On the other hand, flavonoids are more concentrated toward the tip of the spear, *vis-á-vis* the middle and bottom sections, although the differences are not dramatic. Wang *et al.* (2003) reported that the upper youngest shoot tissue had the highest concentration of rutin (the main flavonoid in asparagus), at levels ranging from 0.03 to 0.06% fresh sample weight.

Influence of Harvesting Date on Phenols Content

Harvesting date influences the quality of asparagus. In general, there are one or two main harvest seasons. In the latter case, different climate conditions affect spear quality. In this sense, seasonal variations have been shown to influence nutritional components of green asparagus (*Asparagus officinalis* L. cv. UC157 F1) (Shou *et al.*, 2007). It was observed that the highest amount of rutin and total soluble phenol in asparagus fresh spears were found in high temperature season (June and July), compared with those in cooler months (Table 13.4).

TABLE 13.4 Seasonal Variations in Phenols and Vitamin C Content

	Spring	Summer	Autumn
Ascorbic acid (mg/kg fresh weight)	0.37	0.28	0.35
Rutin (g/kg dry weight)	1.77	2.95	2.28
Total soluble phenol (mg gallic acid/g fresh weight)	0.72	1.62	1.29

The table shows the comparison of ascorbic acid, rutin, and total soluble phenol composition of asparagus spears harvested in different months.
Source: Shou S., Lu G., Huang X. (2007) Seasonal variations in nutritional components of green asparagus using the mother fern cultivation. Scientia Horticulturae *112, 251–257.*

Influence of Cultivars, Climate Conditions, and Part of the Spear on Saponin Content

As with flavonoids, genetic factors seem to be more determinant in the composition of this phytochemical. In this regard, several studies have been conducted into both white and green asparagus.

The saponin contents of six white asparagus cultivars grown in Germany were measured by High Performance Thin Layer Chromatography (HPTLC) and ranged between 0.14 and 0.80 g/kg of fresh weight (Schwarzbach *et al.*, 2006). All cultivars grew under equal pre-harvest conditions, but the level of saponins was very different, and a probable correlation between saponin contents and bitter taste as well as pathogenic resistance was noted.

In the case of green asparagus, a study into 10 commercial hybrids of green asparagus developed in recent years by major international asparagus programs and 10 different native lines of *triguero* asparagus from the Hué tor-Tájar population variety quantified total saponin contents by a colorimetric method (Fuentes-Alventosa, 2007). Saponin content ranged between 30 and 50 mg/kg of fresh weight for commercial hybrids and 160–320 mg/kg for *triguero* asparagus (Table 13.3).

As occurred with flavonoids, *triguero* asparagus genotypes had a composition which was very different from that of commercial hybrids, with different types of saponins not found in commercial hybrids and, therefore, with probable specific functional properties.

Moreover, it was observed that climate conditions exert an important influence on this parameter, showing an increase of saponin content under lower soil temperatures (Schwarzbach, 2006).

Additionally, the different parts of asparagus spears show differences in saponin levels. Wang *et al.* (2003) reported a protodioscin content (the main saponin in asparagus) that ranged between 0.0024 and 0.25 g/kg of fresh weight at the top and bottom sections, respectively.

Influence of Harvesting Date on Vitamin C Content

Shou *et al.* (2007) studied seasonal variations in nutritional components of green asparagus and did not detect significant differences between spears produced in spring and those produced in autumn in terms of their ascorbic acid content, while lower vitamin levels were found in spears harvested in June and July (Table 13.4).

ANALYTICAL TECHNIQUES

Various analytical techniques have been evaluated for the purposes of the study of the bioactive compounds described in this chapter.

The Folin-Ciocalteu spectrophotometric method described by Vinson et al. (1998) is applied to quantify total soluble phenols, with absorbance at 760 nm and total phenolic content expressed as gallic acid equivalent (mg/g).

High-performance liquid chromatography with diode array detection (HPLCDAD) is often used for the identification and quantification of flavonoids and liquid chromatography-mass spectrometry (LCMS) under identical HPLC conditions to confirm the structures. The range for the spectra from all peaks is 190–600 nm (Fuentes-Alventosa *et al.*, 2007; Wang *et al.*, 2003; Fuleki, 1999). Chromatograms are acquired at 255 nm (Wang *et al.*, 2003; Shou *et al.*, 2007), 280 nm (Sun *et al.*, 2007) and 360 nm (Fuentes-Alventosa *et al.*, 2007, Sun et al., 2007). Ionization energies of 50 and 100 eV (negative mode) and 50 eV (positive mode), with MS scans from m/z 100 to 1000, are used in order to obtain electrospray ionization mass spectra.

Ferulic derivatives are detected and quantified by HPLCDAD and the chromatograms are acquired at 280 nm.

Flavonoids and ferulic derivatives quantification is performed through integration of peak areas, using calibrations made with known amounts of pure compounds.

Regarding saponins, several methods are described in order to be analyzed—High Performance Thin Layer Chromatography (HPTLC) (Schwarzbach *et al.*, 2006), LC/MS with selected ion monitoring (SIM), LC/MS/MS with selected daughter ion monitoring under positive ion mode and negative ion mode (Wang *et al.*, 2003), and a colorimetric method based on reactive anisaldehyde-sulphuric acid-acetic acid for TLC staining (Wang *et al.*, 2007), with absorbance at 630 nm and using diosgenin as a reference standard (Fuentes-Alventosa, 2007). In all cases identification was achieved by mass-spectrometric analyses, using an electrospray ion (ESI-MS) source that operated under positive and negative mode with MS scans from m/z = 200 to m/z = 1400.

Fructooligosaccharides (FOS) are analyzed by high performance anion exchange chromatography (HPAEC) and separated on an HPLCcarbohydrate column PA1. FOS are quantified with reference to standard peaks.

Vitamin C (Shou *et al.*, 2007) could be determined by the 2,6dichlorophenolindophenol method (Albrechi, 1993) and the Brubacher, MüllerMulot, and Southgate methods (1985) in a fluorimeter at 350 nm (excitation) and 430 nm (emission) (Villanueva *et al.*, 2005), using a HPLC method (Saito *et al.*, 2000) or the AOAC's official titrimetric method (Nindo *et al.*, 2003; An *et al.*, 2008).

Analytical determinations of mineral elements copper, iron, zinc, manganese, nickel, chromium, calcium, and magnesium are performed with an atomic absorption spectrophotometer equipped with a standard air-acetylene flame, a spoiler nebulizer (except for magnesium determination), and single element hollow cathode lamps for elemental analyses. In the case of manganese, deuterium background correction is used. Sodium and potassium analysis is performed by flame atomic emission spectrophotometry. Phosphorus is analyzed by visible-ultraviolet spectrophotometry, using the ammonium vanadatemolybdate colorimetric method at a wavelength of 400 nm (AOAC, 1991).

References

Albrechi, J.A., 1993. Ascorbic and retention in lettuce. J. Food Qual. 16, 311–316.

Amaro, M.A., Moreno, R., Zurera, G., 1998. Nutritional estimation of changes in mineral content during frozen storage of white asparagus. J. Food Qual. 21, 445–458.

Amaro, M.A., Zurera, G., Moreno, R., Cañal, C., 1997. Effect of industrial processing on the mineral content of white asparagus. Int. J. Food Sci. Tech. 32, 401–406.

Amaro-López, M.A., Moreno-Rojas, R., Zurera-Cosano, G., Sánchez-Segarra, P.J., 1999b. Nutritional changes in the essential trace elements content of asparagus during industrial processing. Food Res. Int. 32, 479–486.

Amaro-López, M.A., Zurera-Cosano, G., Moreno-Rojas, R., 1999a. Nutritional evaluation of mineral content changes in fresh green asparagus as a function of the spear portions. J. Sci. Food Agric. 79, 900–906.

Amaro-López, M.A., Zurera-Cosano, G., Moreno-Rojas, R., Garcia-Gimeno, R.M., 1995. Influence of vegetative cycle of asparagus (*Asparagus officinalis* L.) on copper, iron, zinc and manganese content. Plant Foods Hum. Nutr. 47, 349–355.

An, J., Zhang, M., Wang, S., Tang, J., 2008. Physical, chemical and microbiological changes in stored green asparagus spears as affected by coating of silver nanoparticles-PVP. LWT 41, 1100–1107.

AOAC, 1991. Official Methods of Analysis, (15th ed, 2nd supplement 991.25), Association of Official Analytical Chemists, Arlington, Virginia.

Brubacher, G., Müller-Mulot, W., Southgate, D. (Eds.), 1985. Methods for the determination of vitamins in foods. Elsevier, London, New York. Recomended by Cost 91, pp. 76–84.

Chin, C.K., Garrison, S.A., Shao, Y., Wang, M., Simon, J., Ho, C.T., Huang, M.T., 2002. Functional elements from asparagus for human health. Acta Hortic. (ISHS) 589, 233–241.

Fuentes-Alventosa, J.M., Jaramillo, S., Rodríguez-Gutiérrez, G., Cermeño, P., Espejo, J.A., Jiménez-Araujo, A., 2008. Flavonoid profile of green asparagus genotypes. J. Agric. Food Chem. 56, 6977–6984.

Fuentes-Alventosa, J.M., Rodríguez, G., Cermeño, P., Jiménez, A., Guillén, R., Fernandez-Bolaños, J., Rodríguez-Arcos, R., 2007. Identification of flavonoid diglycosides in several genotypes of asparagus from the Huetor-Tajar population variety. J. Agric. Food Chem. 55, 10028–10035.

Fuentes-Alventosa, J.M., 2007. Caracterización de flavonoides y saponinas en el espárrago triguero de Hué torTájar (Granada). Master in Agrofood. University of Cordoba (Spain).

Fuleki, T., 1999. Rutin, the main component of surface deposits on pickled green asparagus. J. Food Sci. 64, 252–254.

Jaramillo, S., Rodríguez, R., Jiménez, A., Guillén, R., Fernandez-Bolaños, J., Heredia, A., 2007. Effects of storage conditions on the accumulation of ferulic acid derivatives in white asparagus cell wall. J. Sci. Food Agric. 87, 286–296.

Nindo, C.I., Sun, T., Wang, S.W., Tang, J., Powers, J.R., 2003. Evaluation of drying technologies for retention of physical quality and antioxidants in asparagus. Eur. Food Res. Technol. 36, 507–516.

Rodríguez, R., Jaramillo, S., Rodríguez, G., Espejo, J.A., Guillén, R., Fernández-Bolaños, J., Heredia, A., Jiménez, A., 2005. Antioxidant activity of ethanolic extracts from several asparagus cultivars. J. Agric. Food Chem. 53, 5212–5217.

Saito, M., Rai, D.R., Masuda, R., 2000. Effect of modified atmosphere packaging on glutathione and ascorbic acid content of asparagus spears. J. Food Process. Preservation 24, 243–251.

Schwarzbach, A., 2004. Gemüse 8, 30–31.

Schwarzbach, A., Schreiner, M., Knorr, D., 2006. Effect of cultivars and deep freeze storage on saponin content of white asparagus spears (*Asparagus officinalis* L.). Eur. Food Res. Technol. 222, 32–35.

Shiomi, N., Benkeblia, N., Onodera, S., Omori, T., Takahashi, T., Fujishima, M., Yoshihira, T., Kosaka, S., 2007. Saccharide and fructooligosaccharide contents, and invertase, 1-KHE, 1-SST, 1-FFT and 6G-FFT activities in green asparagus spears during storage: Effects of temperature and spear portion. J. Appl. Glycosci. 54, 187–194.

Shou, S., Lu, G., Huang, X., 2007. Seasonal variations in nutritional components of green asparagus using the mother fern cultivation. Sci. Horticulturae 112, 251–257.

Sun, T., Tang, J., Powers, J.R., 2007. Antioxidant activity and quality of asparagus affected by microwave-circulated water combination and conventional sterilization. Food Chem. 100, 813–819.

Villanueva, M.J., Tenorio, M.D., Sagardoy, M., Redondo, A., Saco, M.D., 2005. Physical, chemical, histological and microbiological changes in fresh green asparagus (*Asparagus officinalis*, L.) stored in modified atmosphere packaging. Food Chem. 91, 609–619.

Vinson, J.A., Hao, Y., Su, X., Zubik, L., 1998. Phenol antioxidant quantity and quality in foods vegetables. J. Agric. Food Chem. 46, 3630–3634.

Wang, M., Tadmor, Y., Wu, Q.-L., Chin, C.-K., Garrison, St.A., Simon, J.E., 2003. Quantification of Protodioscin and Rutin in Asparagus Shoots by LC/MS and HPLC Methods. J. Agric. Food Chem. 51, 6132–6136.

Wang, T.-S., Lii, C.-K., Chang, J.-Y., Kuo, C.-I., Chen, H.-P., 2007. Anti-Fenton reaction activity of three taxa of water yam (*Dioscorea alata* L.). Int. J. Food Sci. Technol. 42, 1107–1113.

CHAPTER

14

Bioactive Components in Potatoes as Influenced by Thermal Processing

Carla S.P. Santos, Sara Cunha, Susana Casal
REQUIMTE, Faculdade de Farmácia, Universidade do Porto, Portugal

CHAPTER POINTS

- Potatoes are an important source of bioactive components in the diet due to the elevated amounts consumed worldwide.
- Raw potato varieties have significantly different amounts of bioactive components, this being the main contributing factor for the amounts ingested after cooking, independently of the thermal method used.
- Potato bioactivity is affected by thermal processing, with boiling inducing generally higher losses than steam cooking, frying, baking or microwave cooking.
- Removal of the peel before or after cooking appears to influence the bioactive compounds in cooked potatoes, particularly in boiled potatoes.
- Before analysis, a cautious choice of the analytical methodologies to be used should be discussed in order to avoid unnecessary costs or loss of valuable information.

INTRODUCTION

The potato plant is a perennial herb belonging to the family *Solanaceae*, species *Solanum tuberosum* L., which produces one of the most important crops for human consumption, the potato (Singh and Kaur, 2009). Originating from South America, in particular in the Andes where it has been cultivated for more than 10,000 years (Camire *et al.*, 2009), it was unknown to the rest of the world until 500 years ago. With more than 4000 varieties described, and being cultivated in more than 160 countries, it is now the fourth most important food crop in the world after maize, wheat, and rice, with a production exceeding 300 million tons per year (Singh and Kaur, 2009; Camire *et al.*, 2009).

Figure 14.1 presents the latest statistics available for the largest potato producers (FAOSTAT, 2010), with China attaining the highest figures, followed by India and Russia. The annual *per capita* potato consumption in 2005 achieved almost 90 kg in Europe, 60 kg in North America, being lower in Asia/Oceania (24 kg), Latin America (21 kg) and Africa (13 kg) (FAO, 2008).

Potatoes are usually eaten cooked, mostly because of the indigestibility of its ungelatinized starch, with great improvement in flavor, texture, and appearance. The most popular cooking methods include boiling, baking, steaming, roasting, deep-frying, and microwaving, depending mostly on local cultural habits. The major industrially processed potato products are potato chips and crisps, followed by some frozen products, and dehydrated products such as potato flakes (Singh and Kaur, 2009).

Potatoes are rich in carbohydrates (Table 14.1) but, in comparison with other carbohydrate staple foods, potatoes' nutrient density is low because of their high water content, almost 80%. Potatoes are very low in fat (Table 14.1) and, although not regarded as an important source of protein [0.6–2.1% fresh weight (FW)], they have a high biological value (90–100) (Camire *et al.*, 2009). Potatoes are also rich in essential minerals, and recent findings highlight that potatoes are an important dietary source of bioactive compounds, secondary plant metabolites that may delay the onset of chronic diseases, such as arthritis, atherosclerosis, cancer, cardiovascular diseases, diabetes, gastrointestinal disorders, and neurodegenerative dysfunctions (Andre *et al.*, 2007; Camire *et al.*, 2009). These potato components include vitamin C and E, carotenoids, phenolic compounds (phenolic acids, flavonoids, anthocyanins), among others

Processing and Impact on Active Components in Food
http://dx.doi.org/10.1016/B978-0-12-404699-3.00014-7

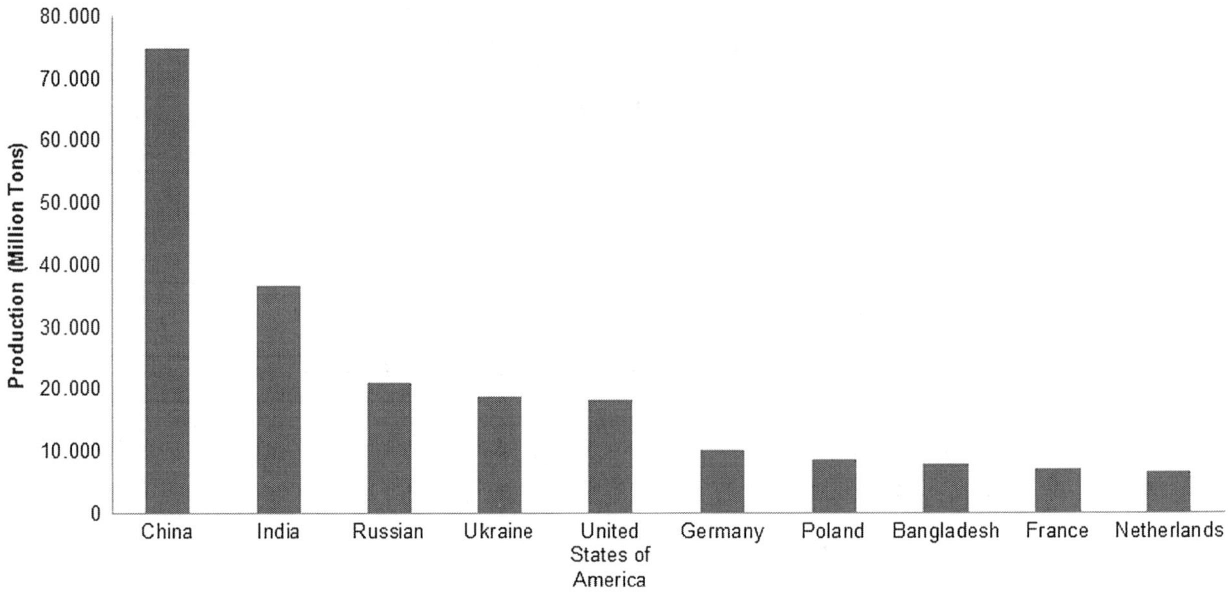

FIGURE 14.1 Top 10 potato producers. *Adapted from FAOSTAT, 2010.*

TABLE 14.1 Nutritional Composition of Raw Potato (per 100 g)*

Nutrients	FW	DW	Bioactive Compounds	FW	DW
Moisture	72–85 g	305–360 g	Phenolic compounds	137–965 mg	596–4196 mg
Carbohydrates	13–18 g	55–79 g	Phenolic acids	58–303 mg	252–1319 mg
Dietary fibre	1.0–2.0 g	4.3–8.7 g	Flavonoids	0.03–0.06 mg	0.13–0.26 mg
Protein	0.6–2.1 g	2.6–9.1 g	Anthocyanins	58–542 mg	253–2357 mg
Lipid	0.1–0.2 g	0.3–0.9 g	Vitamin C	5–54 mg	20–235 mg
Potassium	280–564 mg	1217–2452 mg	Vitamin E	0.02–0.1 mg	0.07–0.4 mg
Phosphorus	30–60 mg	130–261 mg	Carotenoids	0.1–5.8 mg	0.2–23.3 mg
Magnesium	14–18 mg	61–78 mg	Glycoalkaloids	1–20 mg	4–87 mg

Note: contents originally reported in another unit than mg/100 g FW were recalculated using an average moisture content of 77%. FW, fresh weight; DW, dry weight.
* *Adapted from Haase and Weber, 2003; Han et al., 2004; Kalogeropoulos et al., 2007; Mulinacci et al., 2008; Singh and Kaur, 2009; Navarre et al., 2010; Burgos et al., 2012 and 2013; Evers and Deußer, 2012; Elzbieta, 2012; Perla et al., 2012; Tajner-Czopek et al., 2012.*

(Andre *et al.*, 2007; Singh and Kaur, 2009). Therefore, although being implicated in type-2 diabetes and obesity, due to the high glycemic index of its available starch after cooking, a balanced ingestion might have an essential role in human health, particularly in some populations. However, potatoes also contain glycoalkaloids, a group of potentially toxic bitter compounds, whose ingestion should be avoided.

The compositional variability of raw potato is condensed in Table 14.1, expressed on a FW and dry weight (DW) basis. Apart from agronomic practices, genotype, or postharvest storage methods, which are known to influence the concentration of these constituents, subsequent thermal processing determines the amounts truly available to consumers. Therefore, this work will focus on a detailed discussion of the influence of thermal processing on potato nutrients, particularly its bioactive compounds, taking into account the specificities of each component.

HOW POTATOES BIOACTIVE COMPOUNDS ARE INFLUENCED BY THERMAL PROCESSING

Thermal processing is among the most common methods of food preservation and is often simultaneously a condition for consumption, as in the case of potatoes, which need to be cooked to become palatable. Therefore, while contributing to food safety, it is responsible for the occurrence of physicochemical changes in the product, determinant for consumer acceptability and, above all, its nutritional quality (Stadler and Lineback, 2009).

Diverse thermal processing techniques are used, using different temperature/time combinations, heat sources, direct/indirect food contact, etc. Besides the effect of temperature increase, the heating medium is of particular importance, as in the case of water during boiling or oil during frying, where marked exchanges are expected to occur. Also, forced air heating can also induce severe alterations, particularly oxidation of sensitive nutrients.

From the consumer point of view, the physical changes are more visible, namely color and texture, conditioning its perception of quality. Cooked potatoes' color can be the result of chemical browning, due to formation of new compounds during thermal processing, or previously influenced by enzymatic browning achieved before thermal inactivation of the enzymes. While the former is determined by the availability of reducing sugars, free amino acids, the extension of the thermal treatment, etc., the latter is mostly conditioned by raw potato pre-treatments such as blanching (Singh and Kaur, 2009; Pedreschi, 2012). The textural changes are also a visible result of heat and mass transfer, together with chemical reactions, particularly starch gelatinization, being evaluated by attributes such as firmness, hardness, softness, and adhesiveness (Singh and Kaur, 2009).

The chemical changes will also regulate the digestibility and availability of nutrients in the final potato product: starch gelatinization, for instance, increases carbohydrate availability, while the formation of resistant starch reduces the efficiency of endogenous digestive enzymes; protein denaturation increases its digestibility and simultaneously destroys anti-nutritional factors such as trypsin inhibitors; minerals might be leached during water cooking and can react with other dietary components becoming less or even unavailable (Stadler and Lineback, 2009). All these effects should be taken into account when each single food is evaluated for nutritional purposes. More than the chemical composition of raw products, the influence of thermal processing on its content and availability is the main determinant.

The present chapter summarizes the effect of thermal processing on the main bioactive components of potatoes, aiming to understand the mechanisms involved and suggest strategies for increased bioactivity of cooked potatoes.

Phenolic compounds are recognized as increasingly important in the diet due to their antioxidant activity. Therefore, a particular attention will be given to this class, including phenolic acids, flavonoids, and in particular, anthocyanins. In addition, vitamin C is the major vitamin in raw potato, with highly variable amounts (Table 14.1) depending on potato variety, environmental, and storage conditions (Singh and Kaur, 2009). Despite not being among the richest sources of vitamin C, the fact that it is consumed in large quantities makes it an important source of vitamin C in human diet, a fact not always recognized by consumers. Therefore, the effect of thermal processing on this vitamin is particularly important, as it will determine the ingested amounts. Vitamin E will also be discussed, based on its potential enrichment after processing with vegetable oils. Carotenoids, being particularly sensitive molecules, are also included in this review. Finally, the cooking effect on potentially toxic glycoalkaloids is discussed.

Phenolic Compounds

Phenolic compounds, whose functions in plants include defending against invading pathogens (bacteria, fungi, and viruses), also participate in enzyme-catalyzed browning reactions, contributing to the color, taste, and nutritional quality of potatoes (Singh and Kaur, 2009; Perla *et al.*, 2012). In humans, the ingestion of phenolic compounds is increasingly associated with a wealth of health promotion effects (Singh and Kaur, 2009). Potatoes, although not being classified as a rich source of phenolic compounds, are the third daily source of these antioxidants in the American population (Chun *et al.*, 2005). It is therefore worth acquiring an in-depth knowledge of the possible effects of thermal treatments on their contents.

Under the designation of phenolic compounds, two important classes are included: the phenolic acids and flavonoids. A confusing variety of data, however, are found in the literature, mostly because of the analytical techniques used. While most studies make only a global overestimation by using unspecific spectrophotometric methods, few evaluate in detail individual phenolic acids and flavonoids. Therefore, in this chapter we will discuss all phenolic compounds simultaneously.

Some reported results for total phenolic compound variations with domestic cooking are presented in Table 14.2. Generally, boiling is responsible for a higher loss of phenolic compounds, particularly in peeled potatoes, probably due to leaching. The phenolic compounds were apparently preserved after baking (Navarre *et al.*, 2010) but only if processed with the peel (Perla *et al.*, 2012). The same can be shown for microwaving, with comparatively smaller losses if processed without peel (Table 14.2). The highest source of variation however, is the initial amount in raw potato, which is directly dependent on the potato variety used.

When the phenolic acids are analyzed individually, chlorogenic acid is clearly the main compound, followed by caffeic acid (Mattila and Hellström, 2007). However, their changes during thermal processing follow the same pattern observed for total phenols: boiled peeled potatoes present lower values than unpeeled ones (Mattila and Hellström, 2007); baking exhibited higher preservation (Im *et al.*, 2008), as did frying and microwave cooking (Blessington *et al.*, 2010); and the total amounts are higher in pigmented cultivars in comparison with white

TABLE 14.2 Results of Total Phenolic Compounds by Potato Varieties *versus* Thermal Processing, in mg/100g (DW)

	Thermal Processing					
Potatoes Varieties	**Raw**	**Boiled**	**Baked**	**Microwave**	**Steamed**	**References**
Bintje	167	286	367	219	323	Navarre *et al.* (2010) (with skin)
Piccolo	170	230	278	174	249	
Purple Majesty	416	529	620	435	481	
	285	108	756	84	—	Perla *et al.* (2012)
Mesa Russet	143	71	74	75	—	
Silverton Russet	155	71	55	60	—	
Colorado Rose	138	74	75	54	—	
Russet Nugget	94	76	58	58	—	
Dakota Pear	92	80	91	87	—	Xu *et al.* (2009) (with skin)
Goldrosh	235	224	203	190	—	
Nordonna	206	173	171	125	—	
Sangre	147	125	139	133	—	

DW, dry weight.

ones (Mulinacci *et al.*, 2008). Boiling with 3% salt induced higher losses of phenolic acids (Im *et al.*, 2008).

The main flavonoids in potato are anthocyanins, catechin, and epicatechin. In opposition to the observation for phenolic acids, boiling preserved the flavonoids better in comparison with baking or microwaving (Perla *et al.*, 2012), providing that they are processed with the peel. Blessington *et al.* (2010) observed a general increase in all cooked samples, particularly epicatechin after microwave cooking. Only quercetin dihydrate was reduced in all methods, in comparison with the raw potatoes, but these results should be interpreted with caution as they were compared on a FW basis.

Anthocyanins are responsible for many of the blue, red and violet colors in plants, including potatoes, with a marked influence in sensorial acceptance (Singh and Kaur, 2009). In plants they contribute to self-protection and help in chemotaxonomic characterization. To humans, the potential health benefits derive from their antioxidant activity, which plays an important role in the prevention of neuronal and cardiovascular diseases, cancer, and diabetes (Lachman *et al.*, 2012). When observed on a FW basis, as reported by Mulinacci *et al.* (2008), anthocyanins decreased after processing by boiling and microwaving, particularly in the latter. On a DW basis, and for a specific purple variety, Burgos *et al.* (2013) reported reduced alterations by boiling. In a detailed study, Lachman *et al.* (2012) reported a generalized increase after unpeeled cooking, by boiling, steaming, microwaving and roasting. In this case, however, microwaving losses were comparatively higher.

After thermal processing, on average, the anthocyanins content was highest in red and purple varieties, and lowest in white and light purple varieties (Mulinacci *et al.*, 2008; Kaspar *et al.*, 2012; Lachman *et al.*, 2012; Burgos *et al.*, 2013).

Vitamin C

Ascorbic acid (AA) is an important organic acid in plants, with a noticeable role as detoxifier of reactive oxygen species, particularly within photosynthesis. In humans it is crucial for several biological functions, being involved in collagen biosynthesis, it improves the bioavailability of several micronutrients, including iron, and is a cofactor for numerous enzymes, etc. (Singh and Kaur, 2009). Dehydroascorbic acid (DHA), AA's main oxidation product, has also some biological activity, with both being designated as Vitamin C. However, during storage and processing, their hydrolysis to diketogulonic acid should be reduced as much as possible, as this is an inactive compound.

A summary of several studies on AA changes with different thermal processing methods is present in Table 14.3. Unfortunately, most studies do not quantify DHA, therefore giving reductive information on total vitamin C potential activity. When processing methods are compared independently of the variety, the highest losses are observed with boiling (Han *et al.*, 2004). Interestingly, when boiled with salt (3%) the losses are apparently reduced (Han *et al.*, 2004), in contrast to the observations made for the phenolic compounds. These losses are probably direct consequences of leaching

TABLE 14.3 Thermal Processing Influence on Potatoes Ascorbic Acid, in mg/100 g (DW)*

	Potato Varieties													
Thermal Process	**Dejima**	**Sumi**	**Chaju**	**Purple Majesty#**	**Piccolo#**	**Bintje#**	**Undefined**		**EE-2057#**	**Puma Maqui#**	**Morar Nayra#**	**Morada Taruna#**	**Rucuma o Lucuma#**	**Maria Cruz#**
Raw	56	116	140	44	58	55	95	88	18	28	32	33	38	76
Blanched	—	—	—	—	—	—	70	—	—	—	—	—	—	—
Boiled	7	31	32	52	73	71	—	—	10	21	31	22	33	56
Boiled (1% salt)	15	30	30	—	—	—	—	—	—	—	—	—	—	—
Boiled (3% salt)	24	53	58	—	—	—	—	—	—	—	—	—	—	—
Pressure-cooking	33	58	57	—	—	—	—	—	—	—	—	—	—	—
Fried	10	57	52	—	—	—	—	66	—	—	—	—	—	—
Sautéd	35	37	51	—	—	—	—	-	—	—	—	—	—	—
Braised	30	61	58	57	70	72	—	—	—	—	—	—	—	—
Baked	48	81	94	68	71	77	—	—	8	5	21	2	19	24
Microwaved	37	93	104	52	67	65	—	—	7	3	13	2	9	18

Note: contents originally reported in another unit than mg/100 g DW were recalculated using an average moisture content of 77%.
With peel.
* *Adapted from Han* et al., *2004; Navarre* et al., *2010; Haase and Weber, 2003; Burgos* et al., *2009.*

because they are reduced when boiled unpeeled (Burgos *et al.*, 2009; Navarre *et al.*, 2010).

Frying induces lower vitamin C losses than blanching. The absence of water leaching in the former and the rapid inactivation of ascorbate oxidase enzymes contribute to this observation (Haase and Weber, 2003). The thermal processes that preserved AA the most were baking (Navarre *et al.*, 2010) and microwaving (Han *et al.*, 2004), similar to those achieved when potatoes are boiled with peel, but the results are variable when comparing different authors. This observation suggests that leaching, rather than temperature, should be the main detrimental factor for AA loss.

Vitamin E

Vitamin E, the most powerful lipid-soluble antioxidant known, is the global designation given to a group of eight tocochromanols, derived from the same isoprenoid precursors as carotenoids, with important antioxidant and antiradical scavenging activities. While in photosynthetic plants they seem to have a protective effect in the chloroplast machinery and protect plant lipids from oxidation; a variety of positive effects have been attributed in humans, particularly to its most active compound, α-tocopherol (Singh and Kaur, 2009), the main tocochromanol found in potatoes (Chun *et al.*, 2006). However, vitamin E amounts in raw potatoes are reduced (Table 14.1), as might be expected from a very low-fat food. Therefore, studies on thermal processing effects on potato native vitamin E content are scarce. Chun *et al.* (2006) reported no alteration after boiling. Several processing methods use vegetable oils, these being among the richest sources of vitamin E. Therefore, an increase in potato vitamin E content is usually reported after frying, or for other methods with addition of vegetable oils. As an example, after shallow frying in virgin olive oil, Kalogeropoulos *et al.* (2007) reported a significant enrichment (18%) in α-tocopherol comparing with the raw potatoes (Table 14.1).

Carotenoids

Carotenoids are lipid-soluble pigments naturally present in reduced amounts in potatoes, but are mainly responsible for the flesh color (Brown *et al.*, 2008) and therefore its natural division into white (low carotenoid content), yellow (moderate content), or orange (high content) potatoes. The carotenoids family encompasses multiple compounds, chemically divided into xanthophylls and carotenes, based on the presence or absence of oxygen in their molecules. The most abundant carotenoids in potatoes are the non-provitamin A xanthophylls, lutein and zeaxanthin, which are particularly important for eye health, with a reduced amount of β-carotene (Evers and Deußer, 2012). Still, knowing that carotenoids are sensitive molecules, in particular to oxygen, light, and heat, thermal processing should have a great influence on the amounts present after cooking.

A detailed study on the effect of boiling was performed by Burgos *et al.* (2012), using potatoes from diversely colored varieties. Independently of the initial carotenoids content in the raw potatoes, the total amount of carotenoids preserved was clearly correlated with the pulp yellow intensity. This was mostly a consequence of a reduction in violoxanthin and antheraxanthin contents with a simultaneous increase in lutein and zeaxanthin, all on a FW basis. Still, these authors used unpeeled potatoes, contributing certainly to carotenoid protection. Blessington *et al.* (2010) compared different domestic processing methods, including boiling, frying, baking, and microwave cooking, using diced unpeeled potato cubes from different varieties. Except for boiling, again with lower amounts in some samples, no other cooking methods affected the final carotenoid content significantly.

As for industrial processing, Kaspar *et al.* (2012) compared different drying technologies for the production of potatoes flakes. Generally, all processes induced higher carotenoid losses than the domestic processes, and these were more evident in drum-drying than freeze-drying, clearly due to the temperatures applied.

Glycoalkaloids

The steroidal glycoalkaloids are toxic to microorganisms, insects, animals, and, in high levels, even to humans (Camire *et al.*, 2009), their ingestion usually being avoided. Recent findings, however, have shown some beneficial effects, including reduction of blood cholesterol, protection against infection with *Salmonella typhimurium* and chemoprevention (Singh and Kaur, 2009), requiring further studies to define effects and amounts for sustained consumer advice. Potatoes, in particular, have two glycoalkaloids, α-solanine and α-chaconine (Singh and Kaur, 2009). Upon thermal processing, all studies verified a slight decrease, more visible with boiling (Mulinacci *et al.*, 2008), and frying (Tajner-Czopek *et al.*, 2012). These compounds are known to decompose only above 170°C (Takagi *et al.*, 1990). Therefore, while during frying, some destruction can occur, during boiling only leaching can justify the losses.

Unpeeled potatoes had higher glycoalkaloid amounts (Tajner-Czopek *et al.*, 2012). Indeed, during the industrial preparation of dried potato granules, the highest decrease in glycoalkaloids was observed after peeling and blanching (Elzbieta, 2012).

INFLUENCE OF NEW PROCESSING TECHNIQUES

Thermal processing is generally regarded as providing a high level of food safety, but it also contributes to the degradation of various bioactive compounds and therefore potential nutritional attributes. Thus, to mitigate these undesirable thermal effects on food, a number of new processing techniques have emerged in recent years, aiming to ensure the inactivation of microorganisms, while significantly reducing or eliminating the amount of heat or time required for the process.

Microwave heating is among these "new" processing techniques but it has already achieved a stable position in domestic heating. In comparison with other cooking methods, the effects on bioactive compounds are similar and cooking is achieved in a shorter time, which is of great importance to consumers. At the industrial level several techniques are being tentatively implemented. Microwave and radiofrequency methods were tested for pasteurization and sterilization of mashed potatoes (Guan *et al.*, 2004). High hydrostatic pressure was tested in the substitution of blanching for enzyme inactivation, particularly polyphenoloxidase and ascorbate oxidase (Al-Khuseibi *et al.*, 2005), while frozen potatoes with enhanced textural characteristics were tested by application of high-pressure freezing (Luscher *et al.*, 2005). For drying purposes, the combination of vacuum with microwave heating was also investigated (Bondaruk *et al.*, 2007), as well as pulsed electric fields (Lebovka *et al.*, 2004). Most of these techniques, however, are expensive and not yet fully implemented.

TABLE 14.4 Analytical Techniques used for the Quantification of Bioactive Compounds in Potatoes

Bioactive Compounds	Analytical Techniques	References
Vitamin C	Colorimetric (commercial kit)	Haase and Weber (2003)
	HPLC/UV-Vis	Han *et al.* (2004)
	HPLC/DAD/MS	Navarre *et al.* (2010)
Carotenoids	Spectrophotometry and HPLC/DAD	Burgos *et al.* (2012)
		Kaspar *et al.* (2012)
	HPLC / DAD	Blessington *et al.* (2010)
Vitamin E	HPLC /FLD	Chun *et al.* (2006)
		Kalogeropoulos *et al.* (2007)
Phenolic compounds	Spectrophotometry	Kaspar *et al.* (2012)
		Perla *et al.* (2012)
		Burgos *et al.* (2013)
Phenolic acids	HPLC/DAD	Blessington *et al.* (2010)
		Burgos *et al.* (2013)
		Mattila and Hellström (2007)
	HPLC/DAD/MS	Im *et al.* (2008)
		Mulinacci *et al.* (2008)
		Navarre *et al.* (2010)
Flavonoids	Spectrophotometry	Perla *et al.* (2012)
	HPLC/DAD	Blessington *et al.* (2010)
Anthocyanins	Spectrophotometry	Kaspar *et al.* (2012)
		Burgos *et al.* (2013)
	Spectrophotometry + HPLC/DAD	Lachman *et al.* (2012)
	HPLC/DAD/MS	Mulinacci *et al.* (2008)
Glycoalkaloids	HPLC/UV-Vis	Elzbieta (2012)
		Tajner-Czopek *et al.* (2012)
	HPLC/DAD/MS	Mulinacci *et al.* (2008)

Also, most studies on their comparative effect with conventional techniques are still focused mostly on the textural and sensorial effects, while a detailed study on bioactive compounds will also be needed before they can be fully implemented.

ANALYTICAL TECHNIQUES

Analytical techniques in the field of micronutrients analyses have had a huge increase in development in recent decades. The increased awareness of the importance of these compounds in disease prevention pushed forward analytical chemistry to the inevitable goal of identifying compounds and providing sensitive and fast analytical techniques for their quantification. Still, most records on bioactive compounds analysis in food, potatoes in particular, still use older unspecific techniques. Nevertheless, when one intends only to compare different processing methods, as in this case, using similar samples, the results might be correctly interpreted on a comparative basis. However, when comparing with other matrices, or attempting to compare with authors using more specific methodologies, the results are frequently overestimated.

The analytical techniques used for qualitative and/or quantitative determination of the bioactive compounds in this chapter are condensed in Table 14.4, and include nonspecific techniques such as spectrophotometry methods and more specific ones such as high-performance liquid chromatography (HPLC) with diverse detection methods from simple UV-Vis to mass spectrometers (MS).

In particular, the results presented for ascorbic acid in this chapter are derived either from colorimetric reactions, as well as after HPLC analysis with UV-Vis, or even MS detection systems. While other compounds might give false positives by reacting with the colorimetric reagents, HPLC-MS can provide the true identity of each individual compound present in each single sample. The same is true for total phenolic compounds estimation using the Folin-Ciocalteu reaction. Fortunately, an increasing number of authors are individually quantifying phenolic acids and flavonoids, allowing study of each single variation with the thermal processes. The same applies increasingly for xanthophylls. Still, most of these chromatographic techniques require sophisticated and expensive equipment, as well as experienced technicians and regular maintenance. Therefore, one should balance correctly the purpose of the study before deciding to use one of these techniques. As previously mentioned, when a single matrix is under comparison and the objective is only a direct comparison of a single variable, a simple spectrophotometric method might give the results necessary to proceed with the experiment, in a faster and cheaper way. Thus, before analyzing a bioactive compound, it is crucial to choose among the analytical techniques available, particularly with regard to the level and precision of information needed.

References

Al-Khuseibi, M.K., Sablani, S.S., Perera, C.O., 2005. Comparison of Water Blanching and High Hydrostatic Pressure Effects on Drying Kinetics and Quality of Potato, Drying Technol. An Int. J. 23, 2449–2461.

Andre, C.M., Ghislain, M., Bertin, P., Oufir, M., Herrera, M.R., Hoffmann, L., et al., 2007. Andean Potato Cultivars (*Solanum tuberosum* L.) as a Source of Antioxidant and Mineral Micronutrients. J. Agric. Food Chem. 55, 366–378.

Blessington, T., Nzaramba, M.N., Scheuring, D.C., Hale, A.L., Reddivari, L., Miller Jr., J.C., et al., 2010. Cooking Methods and Storage Treatments of Potato: Effects on Carotenoids, Antioxidant Activity, and Phenolics. Am. J. Potato Res., 87, 479–491.

Bondaruk, J., Markowski, M., Błaszczak, W., 2007. Effect of drying conditions on the quality of vacuum-microwave dried potato cubes. J. Food Eng. 81, 306–312.

Brown, C.R., Durst, R.W., Wrolstad, R., De Jong, W., 2008. Variability of Phytonutrient Content of Potato in Relation to Growing Location and Cooking Method. Potato Res. 51, 259–270.

Burgos, G., Auqui, S., Amoros, W., Salas, E., Bonierbale, M., 2009. Ascorbic acid concentration of native Andean potato varieties as affected by environment, cooking and storage. J. Food Compos. Anal. 22, 533–538.

Burgos, G., Amoros, W., Salas, E., Muñoa, L., Sosa, P., Díaz, C., et al., 2012. Carotenoid concentrations of native Andean potatoes as affected by cooking. Food Chem. 133, 1131–1137.

Burgos, G., Amoros, W., Muñoa, L., Sosa, P., Cayhualla, E., Sanchez, C., et al., 2013. Total phenolic, total anthocyanin and phenolic acid concentrations and antioxidant activity of purple-fleshed potatoes as affected by boiling. J. Food Compos. Anal. 30, 6–12.

Camire, M.E., Kubow, S., Donnelly, D.J., 2009. Potatoes and Human Health. Crit. Rev. Food Sci. Nutr. 49, 823–840.

Chun, O., Kim, D., Smith, N., Schroeder, D., Han, J., Lee, C., 2005. Daily consumption of phenolics and total antioxidant capacity from fruit and vegetables in the American diet. J. Sci. Food Agric. 85, 1715–1724.

Chun, J., Lee, J., Ye, L., Exler, J., Eitenmiller, R.R., 2006. Tocopherol and tocotrienol contents of raw and processed fruits and vegetables in the United States diet. J. Food Comp. Anal. 19, 196–204.

Elzbieta, R., 2012. The effect of industrial potato processing on the concentrations of glycoalkaloids and nitrates in potato granules. Food Control 28, 380–384.

Evers, D., Deuβer, H., 2012. Potato Antioxidant Compounds: Impact of Cultivation Methods and Relevance for Diet and Health. In: Bouayed, J., Bohn, T. (Eds.), Well-Being and Health Nutrition. InTech, Croatia, pp. 95–118.

FAO, 2008. Potato world: Production and Consumption — International Year of the Potato 2008. http://www.potato2008.org/en/world/index.html (Last accessed 24.01.13.)..

FAOSTAT, 2010. Food and Agricultural commodities production. http://faostat.fao.org/site/339/default.aspx Last (accessed 05.01.13.).

Guan, D., Cheng, M., Wang, Y., Tang, J., 2004. Dielectric Properties of Mashed Potatoes Relevant to Microwave and Radiofrequency Pasteurization and Sterilization Processes. J. Food Sci. 69, 30–37.

Haase, N.U., Weber, L., 2003. Ascorbic acid losses during processing of French fries and potato chips. J. Food Eng. 56, 207–209.

Han, J.-S., Kozukue, N., Young, K.-S., Lee, K.-R., Friedman, M., 2004. Distribution of Ascorbic Acid in Potato Tubers and in Home-Processed and Commercial Potato Foods. J. Agric. Food Chem. 52, 6516–6521.

Im, H.W., Suh, B.-S., Lee, S.-U., Kozukue, N., Ohnisi-Kameyama, M., Levin, C.E., et al., 2008. Analysis of Phenolic Compounds by High-Performance Liquid Chromatography and Liquid Chromatography/Mass Spectrometry in Potato Plant Flowers, Leaves, Stems, and Tubers and in Home-Processed Potatoes. J. Agric. Food Chem. 56, 3341–3349.

Kalogeropoulos, N., Mylona, A., Chiou, A., Ioannou, M.S., Andrikopoulos, N.K., 2007. Retention and distribution of natural antioxidants (a-tocopherol, polyphenols and terpenic acids) after shallow frying of vegetables in virgin olive oil. LWT — Food Sci. Technol. 40, 1008–1017.

Kaspar, K.L., Park, J.S., Mathison, B.D., Brown, C.R., Massimino, S., Chew, B.P., 2012. Processing of pigmented-flesh potatoes (*Solanum tuberosum L.*) on the retention of bioactive compounds. Int. J. Food Sci. Technol. 47, 376–382.

Lachman, J., Hamouz, K., Orsák, M., Pivec, V., Hejtmánková, K., Pazderu, K., et al., 2012. Impact of selected factors — Cultivar, storage, cooking and baking on the content of anthocyanins in coloured-flesh potatoes. Food Chem. 133, 1107–1116.

Lebovka, N.I., Praporscica, I., Vorobieva, E., 2004. Effect of moderate thermal and pulsed electric field treatments on textural properties of carrots, potatoes and apples. Innovat. Food Sci. Emerg. Technol. 5, 9–16.

Luscher, C., Schlüter, O., Knorr, D., 2005. High pressure–low temperature processing of foods: impact on cell membranes, texture, color and visual appearance of potato tissue. Innovat. Food Sci. Emerg. Technol. 6, 59–71.

Mattila, P., Hellström, J., 2007. Phenolic acids in potatoes, vegetables, and some of their products. J. Food Compos. Anal. 20, 152–160.

Mulinacci, N., Leri, F., Giaccherini, C., Innocenti, M., Andrenelli, L., Canova, G., et al., 2008. Effect of Cooking on the Anthocyanins, Phenolic Acids, Glycoalkaloids, and Resistant Starch Content in Two Pigmented Cultivars of *Solanum tuberosum* L. J. Agric. Food Chem. 56, 11830–11837.

Navarre, D.A., Shakya, R., Holden, J., Kumar, S., 2010. The Effect of Different Cooking Methods on Phenolics and Vitamin C in Developmentally Young Potato Tubers. Am. J. Potato Res. 87, 350–359.

Pedreschi, F., 2012. Frying of Potatoes: Physical, Chemical, and Microstructural Changes. Dry Technol. 30, 707–725.

Perla, V., Holm, D.G., Jayanty, S.S., 2012. Effects of cooking methods on polyphenols, pigments and antioxidant activity in potato tubers. LWT — Food Sci. Technol. 45, 161–171.

Singh, J., Kaur, L., 2009. In: Advances in potato chemistry and technology. First ed. Academic Press, USA.

Stadler, R.H., Lineback, D.R., 2009. Process-induced food toxicants — occurrence, formation, mitigation and health risks. Wiley, USA.

Tajner-Czopek, A., Rytel, E., Kita, A., Pęksa, A., Hamouz, K., 2012. The influence of thermal process of coloured potatoes on the content of glycoalkaloids in the potato products. Food Chem. 133, 1117–1122.

Takagi, K., Toyoda, M., Fujiyama, Y., Saito, Y., 1990. Title Effect of cooking on contents of α-chaconine and α-solanine in potatoes. J. Food Hyg. Soc. Jpn. 31, 67–73.

Xu, X., Li, W., Lu, Z., Beta, T., Hydamaka, A.W., 2009. Phenolic Content, Composition, Antioxidant Activity, and Their Changes during Domestic Cooking of Potatoes. J. Agric. Food Chem. 57, 10231–10238.

CHAPTER

15

Processing of Fructans and Oligosaccharides from Agave Plants

*Clarita Olvera Carranza**, *Angela Ávila Fernandez*†, *Gustavo R. Bustillo Armendáriz*‡, *Agustín López-Munguía**

*Instituto de Biotecnología, Universidad Nacional Autónoma de México, México, †Centro de Investigación, División Académica de Ciencias de la Salud, Universidad Juárez Autónoma de Tabasco, México, ‡Bustar Alimentos, SAPI de CV, Guadalajara, México

CHAPTER POINTS

- The *Agave* genus, endemic of America is a source of fructans.
- *Agave* fructans have a unique complex structure, also known as *agavin*.
- Distilled alcoholic beverages such as *tequila* and *mezcal* are traditional products of *Agave* industrialization.
- *Agave* fructans have soluble fiber and prebiotic properties and may be considered nutraceuticals.
- *Aguamiel* and *pulque*, traditional products of *Agave* have nutritional properties associated to *agavin*.
- *Agavin* may be fractionated by acid or enzymatic hydrolysis to produce *Agave* FOS.
- *Agavin* may be separated into large and low molecular weight fructans by ultrafiltration or ion exchange operations.
- There is a potential market for *Agave* fructans provided their nutraceutical properties are fully recognized.

INTRODUCTION

The *Agavaceae* family is endemic of America; it is found in arid and semi-arid zones of Central and North America and includes nine genera and 330 species, all native to the Americas. Its center of origin and major diversity is located in México where the nine genera, represented by 251 species can be found. The largest genus is *Agave* which includes 200 species, 75% of which are distributed in México (Garcia-Mendoza, 2007). *Agave* contains fructans as the main carbohydrate reserve, with several physiological implications for the plant.

Evidence of the use of *Agave* plants dates back to Mesoamerican civilizations, which consumed these plants as food, drink, and medicine, and incorporated them in religious rituals. After 20 centuries, the most common agave application in the food industry became the production of distilled alcoholic beverages derived from fermented agave sugars, now a successful global industry. This is the case with *tequila* and *mezcal*, both beverages commercially protected by origin certification conferred by the World Intellectual Property Organization (WIPO) since 1978 and 1995, respectively. Nevertheless, traditional fermentation processes are still preserved in various regions of México where local products may be found. In all of these cases, fructans are transformed to fructose losing all their functional properties. A very important exception is *pulque*, a fermented and non-distilled beverage obtained from *aguamiel*, a sweet and nutritional rich sap extracted from agave pines by an extremely delicate process. This procedure was probably developed by the *Otomi* culture c. 2000 BC, and later inherited by almost all Mesoamerican cultures; the process has resisted industrialization until recent times. *Pulque* has several nutritional advantages derived both from the residual content of complex fructan oligosaccharides in *aguamiel*, including microbial glucans and

Processing and Impact on Active Components in Food
http://dx.doi.org/10.1016/B978-0-12-404699-3.00015-9

fructans, as well as by the natural microbiota present in the fermented beverage.

The presence of fructans in *agave* was reported in 1888 (Suzuki *et al.*, 1993) Initially, described as "inulins" because of their similarity to β2-1 fructans, due to their structural complexity and heterogeneity, it has now been proposed that they be named *agavin* (Lopez *et al.*, 2003).

The elucidation of *agavin* structure, as well as the scientific recognition of fructans and oligofructans as soluble fiber and prebiotics (Gomez *et al.*, 2010) has opened a potential market for agave fructans. There is also now wide evidence explaining why *pulque*, in spite of its alcoholic content, has retained through centuries the tradition of nutritional beverage. This evidence provides a new driving force to review these traditional processes and to diversify agave applications. In this chapter we review *agavin* properties, as well as processes related to its extraction and transformation.

AGAVE FRUCTANS STRUCTURE

Fructans are fructose polymers found as reserve carbohydrates in 15% of flowering plants (Hendry, 1993), representing, after starch, the most common energy storage source. Fructans are classified as inulin, when bearing mainly β2–1 linkages among fructose molecules, or levan when fructose molecules are joined by β2–6 links; inulin or levan neoseries, contain a glucose moiety between two chains of either inulin or levan extended by the corresponding β2–1 or β2–6 linkages. More complex structures derive from the possibility of branching, as in graminan, which are mixed fructans containing a terminal glucose and *agavin* which is a complex and highly branched molecule with both β2–1 and β2–6 linkages in which an internal or an external glucose molecule may be present. Fructan structures are illustrated in Figure 15.1.

The complexity and heterogeneity of fructan structures in *Agave* plants is due to the presence of glycosyl transferase enzymes with a given specificity and activity, resulting in a particular composition and distribution in each agave species as reported by Mancilla-Margalli *et al.* (2006). Around 60 to 85% of soluble carbohydrates in *Agave* are fructans with a wide polymerization degree ranging from three to 35 fructose molecules in their structures (Lopez *et al.*, 2003). Furthermore, β2-6 fructose linkages may constitute up to 20% of total linkages in some *Agave* species as shown in Figure 15.2 (Mancilla-Margalli *et al.*, 2006). Glucose, fructose, and sucrose are also part of the residual carbohydrates. More than 20% and up to 30% of pine are fructans on a wet basis, but this may vary among species and particularly with pine age. This is the case of *A. tequilana*, *A. americana*, *A. angustifolia*, *A. potatorum*, *A. salmiana*, *A. fourcroydes*, amongst others. Recent studies demonstrate that during vegetative development of *A. tequilana*, fructan content and structure varies during the first 5 years, the most abundant carbohydrates stored being fructooligosaccharides, while 6- to 7-year-old plants, contain mainly long chain highly branched *agavin*. In conclusion, *agavin* becomes larger and more complex with plant age, but reaches maturity after 5 years (Figure 15.3). This is very interesting in terms of fructan productivity as in the traditional tequila production process agave plants are usually harvested after 8 years of growth (Arrizon *et al.*, 2010; Mellado-Mojica and Lopez, 2012).

The basic fructan biosynthetic pathway proposed by Vijn and Smeekens (1999) is described in Figure 15.4. A sucrose: sucrose 1-fructosyltransferase (1-SST) initiates fructan synthesis catalyzing the transfer of a fructosyl residue from sucrose to another sucrose molecule resulting in the formation of the trisaccharide 1-kestose (Ávila-Fernández et al, 2007).; fructan: fructan 1-fructosyltransferase (1-FFT), elongates the fructosyl chain; fructan: fructan 6G-fructosyltransferase (6-GFT), is implicated in inulin neoseries synthesis, while sucrose: fructan 6-fructosyltransferase (6-SFT), catalyzes the formation and extension of β2–6 linked fructans.

AGAVE FRUCTANS IN TRADITIONAL PRODUCTS

As already stated *tequila* and *mezcal*, as well as *sisal*, and *bacanora* (amongst others), are distilled alcoholic beverages produced from the fermentation and distillation of specific *Agave* species. A particular review of these processes may be found in Narvaez-Zapata and Sanchez-Teyer (2009). In these processes, agave plants are harvested, "jimadas", which means elimination of the leaves, and the pines cooked in ovens or autoclaves to hydrolyze *agavin*, liberating fructose and producing compounds that will later contribute to the aroma and flavor of the distilled product. Cooking also softens the *Agave* pine, facilitating pine milling and the extraction operation. Traditionally, the milling process was carried out in rudimentary stone mills that have now been substituted by modern milling equipment, in which the cooked pines are cut and shredded to facilitate heat and mass transfer during the aqueous extraction of the pine components carried out in diffusers.

The extracted, cooked, and hydrolyzed fructans are then fermented and distilled to produce *tequila* (*A. tequilana* Weber var. *azul*) and *mezcal* (*A. angustifolia*, *A. korwinskii* and *A. salmiana*), two widely known agave beverages. However other less known distilled products such as *raicilla* (*A. maximiliana*), *sisal* (*A. sisalana*), and *bacanora* (*A. angustifolia* Haw) are also locally produced.

In order to diversify the *Agave* agroindustry, a new product was developed: a fructose syrup, referred to as "Agave syrup" or "Agave honey" (*miel de Agave*),

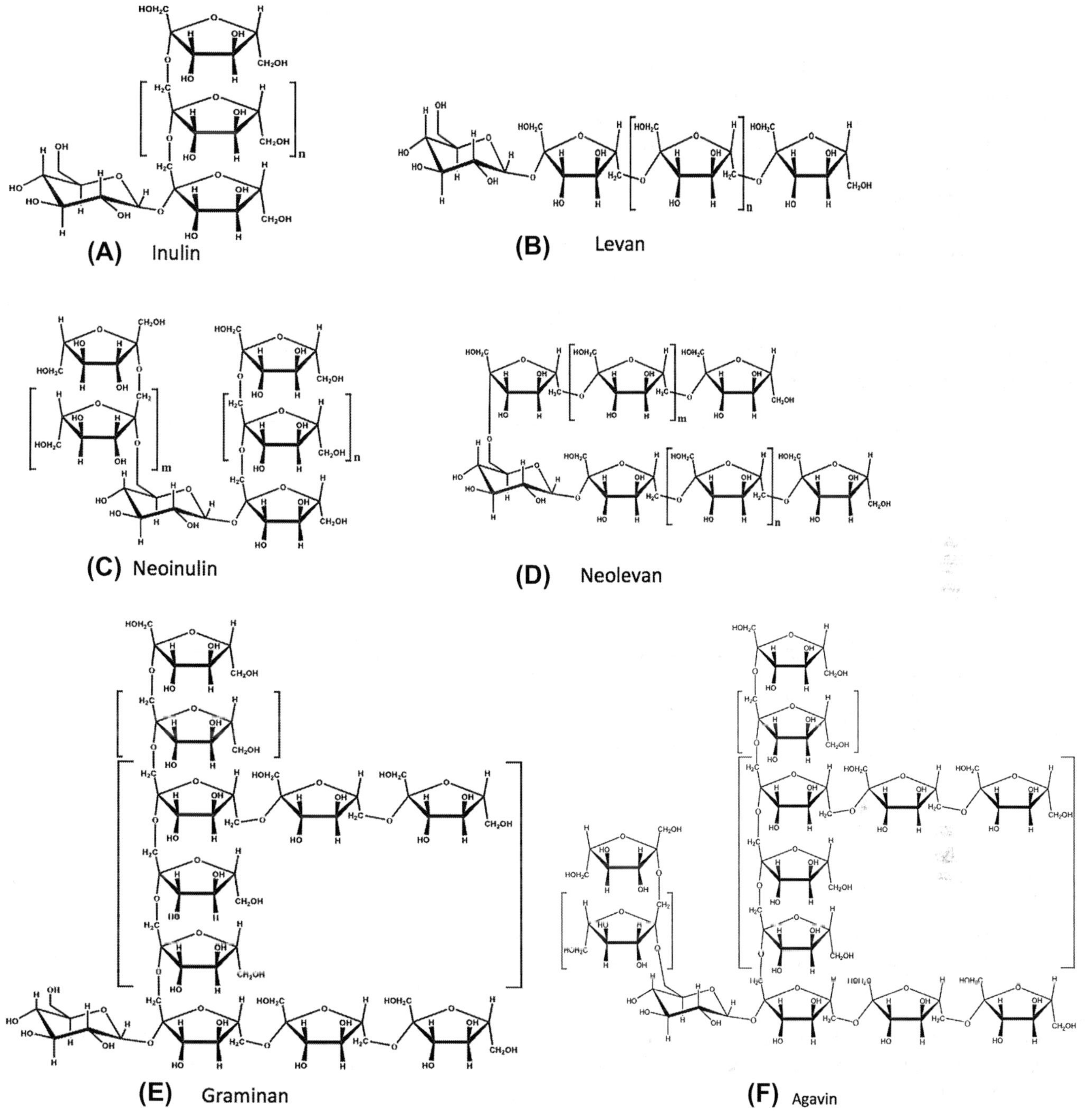

FIGURE 15.1 Fructan structures. (A) Inulin, (B) levan, (C) inulin neoseries, (D) levan neoseries, (E) graminans, and (F) agavin from *A. tequilana* Weber var. *azul* structure. *Source: Muñoz Gutierrez, et al., 2009.*

which is obtained after hydrolysis of the extracted *agavin*, avoiding fermentation. Although this sweetener has found an important market, particularly in the USA and in the European Union, it is only a "more natural" alternative source of fructose to the controversial HFCS obtained from corn starch. Nevertheless, *Agave* syrups rapidly became the second largest *Agave* industrial product. Enzymes (inulinases) are an alternative for this process, both in the tequila and the agave syrup market (Ávila-Fernández, 2009). There is a sweetener market demand for a clear and tasteless *Agave* syrup ingredient that does not affect the color or taste of the product. Therefore, enzymatic hydrolysis is a clean alternative in the process, since no flavor and color, characteristic of the acid and/or thermal hydrolysis, is produced.

Considering the growing demand for functional fructans, thermal hydrolysis may also be avoided in order to extract intact fructans, as will be described later.

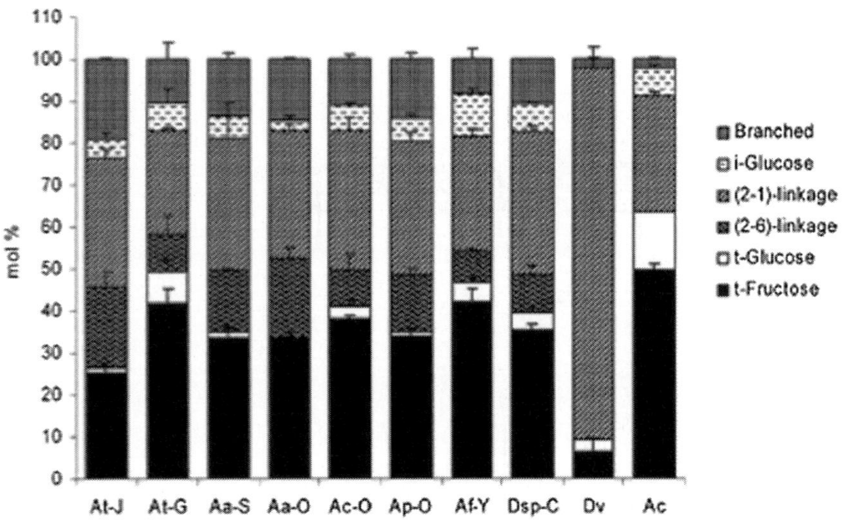

FIGURE 15.2 Glycosyl linkage compositions in mole percentage of fructans from Agave and Dasylirion species. Abbreviations: *A. tequilana* from Jalisco (At-J); *A. tequilana* from Guanajuato (At-G); *A. angustifolia* from Sonora (Aa-S); *A. angustifolia* from Oaxaca (Aa-O); *A. potatorum* from Oaxaca (Ap-O); *A. cantala* from Oaxaca (Ac-O); *A. fourcroydes* from Yucatan (Af-Y); *Dasylirion* spp. from Chihuahua (Dsp-C); *D. variabilis* (DV), and *A. cepa* (Ac). Bars represent the standard deviation of three independent determinations. *Source: Mancilla-Margalli and Lopez, 2006.*

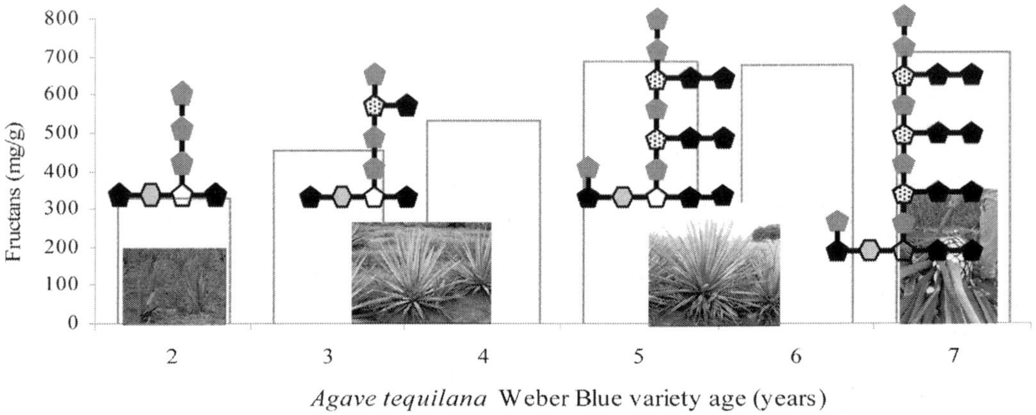

FIGURE 15.3 Fructan contents and their proposed molecular structure relationships in *A. tequilana* Weber *azul* variety along its life cycle in the field. *Source: Mellado-Mojica and Lopez, 2012.*

Agavin has a different fate in *pulque* production. A detailed description of the process was recently reviewed (Escalante *et al.*, 2013). The process is based on the laborious extraction of *aguamiel*, which is a sap obtained from *Agave* "pulquero," a term that includes *Agave* species such as *A. salmiana* var. *salmiana* (green or meek agave), *A. atrovirens* (white agave), *A. mapisaga* (Mexican or long hand agave), *A. atrovirens*, *A. americana*, *A. ferox*, *A. lehmannii*, and *A. altisima*, which are all found in the central states of México.

Aguamiel is a milky white sap collected daily for 3–6 months from a hole made in the pine heart when the plant is ready to flower (around 8–10 years old), an operation known as "capado" (castration). Using a sharp spoon, the cavity is scraped daily, promoting sap flow and its accumulation in the cavity. Eventually, a single agave plant may have produced up to 1000 l of *aguamiel*. When it dies it is removed and substituted by an offspring.

Although *aguamiel* composition varies with agave species, season of the year, weather conditions, and several other factors, it does not seem to be affected through the 3–4 months harvesting period (Ortiz-Basurto *et al.*, 2008). On a dry basis, *aguamiel* is around 90% sugar, mainly sucrose, fructose, and glucose, but also around 10% *agavin*-fructooligosaccharides (*Agave*-FOS). These sugars are the result of fructosyl hydrolase digestion during the storage time between collections. When *aguamiel* is later fermented and the simple sugars transformed in microbial products, *Agave*-FOS are retained, constituting an important prebiotic value of *pulque*. *Aguamiel* also contains essential amino acids, proteins, vitamins, gums, and minerals, and constitutes a water substitute or an alternative food in places where water is scarce or where protein is not available. However, due to raw collection, transport, and storage conditions, it is difficult to prevent fermentation before its arrival to traditional wood barrels, vessels or, more recently, to modern recipients where fermentation takes place (or continues) by the addition of freshly fermented product. Fermentation time varies from a few hours to overnight, until hexose exhaustion. A wide variety of microorganisms have been isolated from

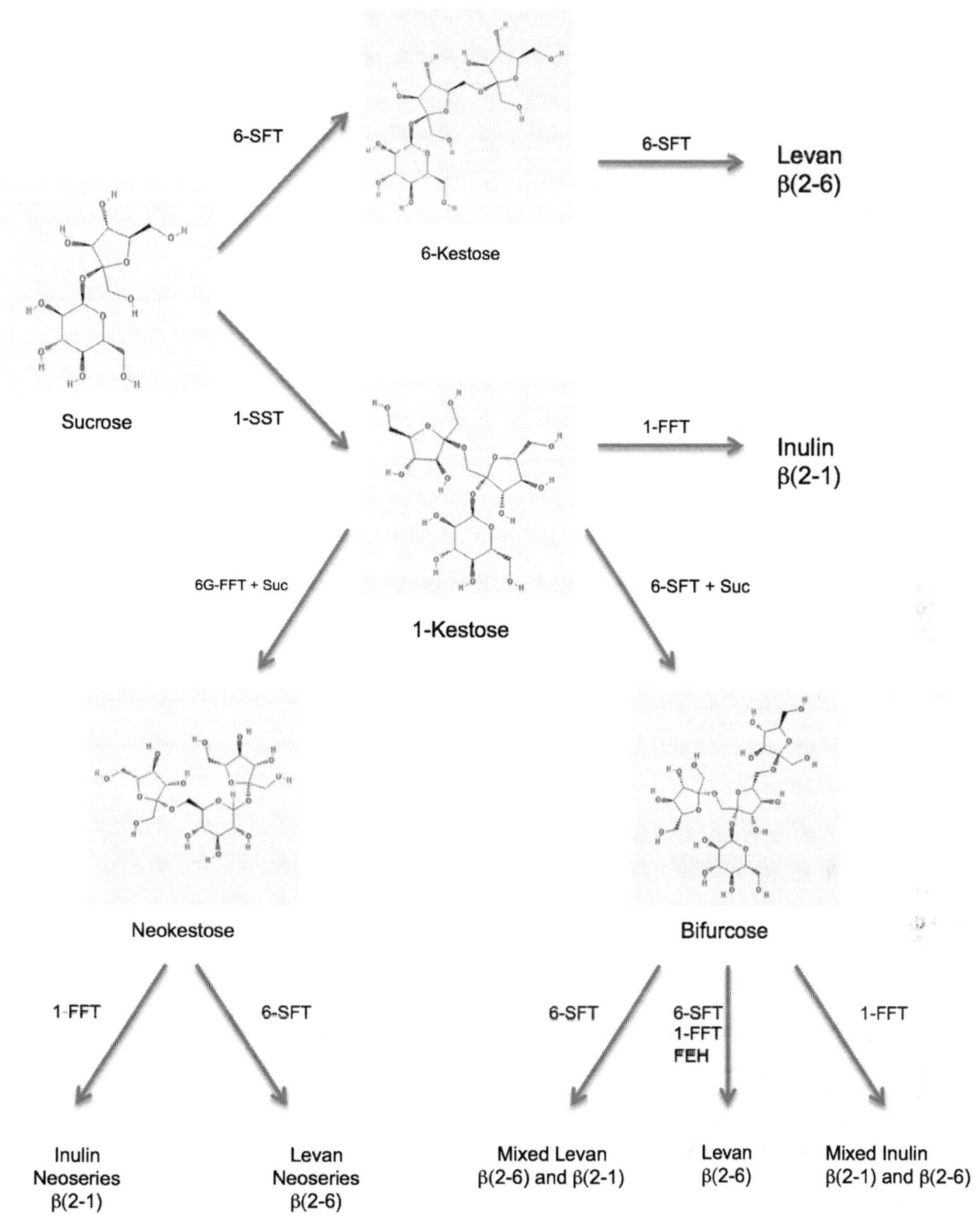

FIGURE 15.4 Model of fructan biosynthesis in plants. Starting from sucrose, structurally different fructan molecules can be produced by the concerted action of different fructosyltransferases. FEH, Fructan exohydrolase.

pulque dominated by *Lactobacillus* species (around 80% of the microbial biodiversity) as demonstrated by 16s RNA sequencing (Escalante *et al.*, 2004). Nevertheless, three main fermentation activities may be distinguished in this process: an alcoholic fermentation (5 to 6%) carried out by yeasts like *Saccharomyces cereviciae* and by *Zymomonas mobilis*, a bacteria originally isolated from *pulque*, a lactic acid fermentation, performed by LAB, such as *Lactobacillus acidophilus, L. plantarum, L. buchneri, L. kefir, L. acetotolerans,* among others, and a viscous fermentation, which is the result of exo-polysaccharide (EPS) synthesis, actually a useful criteria to determine the degree of fermentation and to distinguish between fresh or mature *pulque*. These EPS include dextran, levan, and inulin from lactic acid bacteria, among them *Leuconostoc mesenteroides* and *L. citreum* or *Zymononas mobilis,* which is also a levan producer.

Today, the majority of *pulque* consumed in México is still produced under non-aseptic conditions. The final product is stored in wood barrels and distributed for immediate consumption without the addition of preservatives. Nevertheless, there are an increasing number of companies introducing aseptic production conditions for industrialization and canning, developing modern *aguamiel* mechanical extraction processes, and selective key microorganisms for the fermentation process, including those with probiotic properties such as *Lactobacillus acidophilus*, *L. mesenteroides* (Escalante *et al.*, 2004), and *Z. mobilis*, which has an antagonist activity against pathogenic bacteria and fungal species (Wuanick, 1970; Gonçalves de Lima, 1978). One of the main nutritional interests lies in *aguamiel* and *pulque* fructans and fructooligosaccharides as soluble fiber and prebiotics. The modulation of colonic microflora is achieved by prebiotics promoting the growth of microorganisms such as *bifidobacteria* and *lactobacilli* and inhibiting pathogens such as *Clostridium difficile*. Beneficial microflora produce short-chain fatty acids (SCFA) that trigger a set of processes that benefit the host (Kolida, 2002). This is the case with *agavin* and *Agave*-FOS as will be described later.

Many therapeutic properties have been attributed to *pulque*, particularly in the control of diseases such as dyspepsia, stomach pain, diarrhea, anemia, anorexia, and several infections. Actually, archeological evidence suggests that *pulque* has been used since pre-Hispanic times as an enema. Besides the beneficial effects due to pre and probiotic properties, some of the reported effects may also be associated with *Agave* saponins, which are sterols with hypocholesterolemic, anti-inflammatory, and antibiotic activity (Güçlü-Üstündag and Mazza, 2007).

In conclusion, *aguamiel* and *pulque*—its fermented product—are now considered a natural source of nutraceuticals and an example of transformation schemes by which society may benefit from agave biodiversity.

AGAVE INULIN AND AGAVE FOS AS NUTRACEUTICALS: HEALTH IMPLICATIONS

As a consequence of the saturation of the tequila market as well as the scientific evidence concerning the functional properties of fructans and FOS, the *Agave* agroindustry started diversification a few decades ago, focusing on developing processes to obtain alternative products to alcoholic beverages, based first in the fructose syrup market, but more recently, in the nutraceutical properties of *Agave* fructans and/or *Agave* FOS.

Direct evidence of the prebiotic activity of *Agave* fructans (*A. tequilana)* has been provided *in vitro*, demonstrating that a large increase in the number of *bifidobacteria* and *lactobacilli* is observed as a consequence of the differential impact of fructans from agave on the microbial ecology of the human digestive tract, as well as selective changes in colonic metabolism as determined by the short chain fatty acids profile induced. These experiments were performed in anaerobic batch cultures with human fecal slurries inoculated with a media containing *agavin* (Predilife, a commercial product from the Mexican company Bustar Alimentos, SAPI de CV). Interestingly, the effect is similar to the one observed with inulin from chicory, with no effect observed in the experiment where cellulose was used as control (Gomez *et al.*, 2010).

In experiments where the effect of supplementation with 10% agave fructans from *A. tequilana* and *Dasylirion* spp on glucose and lipid metabolism in mice was studied, it was found that supplementation with *Agave* fructans induced a higher concentration of GLP-1 and its precursor, proglucagon mRNA, in the different colonic segments, suggesting that fermentable fructans from different botanical origin and chemical structure are able to promote the production of satietogenic/incretin peptides in the lower part of the gut, with promising effects on glucose metabolism, body weight, and fat mass development (Urıas-Silvas *et al.*, 2008).

The same group later demonstrated that the presence of mixtures of short-DP fructans (DP< 10) from *A. angustifolia* significantly influenced the rate of fermentation by the probiotic *bifidobacteria* and *lactobacilli* strains more efficiently than commercial chicory inulin (Santiago-Garcia and Lopez, 2009).

Furthermore, low molecular weight (DP) agave fructans contain branching linkages and are highly soluble, an important property in the design of functional products. The health benefits of the individual components of *agavin* have yet to be demonstrated.

Several research groups are performing clinical trials with *Agave* fructans such as the application of an *Agave* syrup to treat nocturnal cough and sleep quality in infants and toddlers with upper respiratory infections, and the incorporation of prebiotic *Agave* products in infant formula.

The research group CIATEJ in Guadalajara is carrying out various projects where the main goal is to verify the effects of oligofructans from *A. tequilana* on the human gut microbiota health through metagenomics with emphasis on the treatment of obesity, evaluating also the immunological profile of pediatric patients, while in INNCM-Salvador Zubirán a research group is developing at least five clinical research protocols in various non dairy formulae to evaluate the prebiotic effect on volunteer patients, particularly patients with cardiovascular problems, cirrhosis, and obesity.

Nevertheless, in order to classify *Agave* fructans as prebiotics, multiple *in vivo* studies are still required to demonstrate their resistance to digestion and the low dose induction of changes in the composition and activity of the gut microbiota.

AGAVE FRUCTANS PRODUCTION PROCESS

As a radical difference to the traditional agave processes, this new market alternative requires preserving the structural integrity of the complex *agavin*, removing compounds that confer undesirable flavors but also mono- and di-saccharides, particularly glucose, fructose, and sucrose, that increase the glycemic index of the final product. The process involves milling or slicing of the pine, aqueous extraction, purification, concentration, and drying. The unit operations are outlined in Figure 15.5, and briefly described in the following lines.

Due to their high solubility, *Agave* fructans are easily extracted with water at 60–90°C. The extraction proceeds to 99% efficiency through aqueous countercurrent

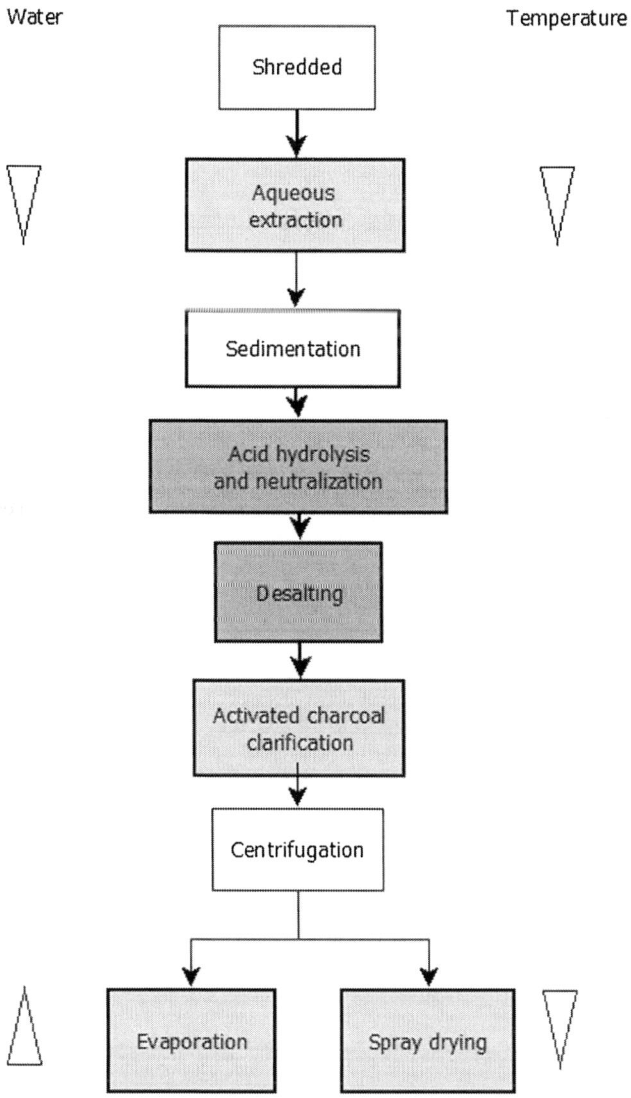

FIGURE 15.5 *Agavin* and *agavin* FOS extraction and production process. Boxes in light gray indicate the key operations in the process. Boxes in dark gray are introduced to transform *agavin* into FOS-*agavin*.

extraction in a diffuser (Diaz Compain, 1968), in which the pines are fed after being shredded in order to facilitate an extraction rate and efficiency (Ávila-Fernández *et al.*, 2009) that is very much higher than that obtained in traditional autoclaves. One important process parameter is pH, which must remain near neutrality to avoid acid hydrolysis that may increase the free fructose content in the final product. Although fructose does not increase the glycemic index, it decreases the final product quality and reduces the *agavin* and FOS extraction efficiency.

After extraction, fiber residues are settled and the *agavin* extract recovered to continue the process. Some compounds such as *saponins* responsible for undesirable flavors and color may be eliminated by an activated charcoal filtration. However, this is an important decision, as evidence regarding the anticancer activity of agave *saponins* has resulted in a patent for a process particularly devoted to obtaining a *saponin* rich agave extract (Gutierrez Uribe and Serna Saldivar, 2009). A drying step may be applied, usually in a spray dryer, if a solid product is required. All operational conditions, sequence, and processing time must be defined bearing in mind the risk of hydrolysis or an eventual contamination.

When mono and di-saccharides are abundant, an ion exchange operation in a cationic resin column may be included in the process. This procedure is commonly used in the food industry to separate mono-saccharides from sugar mixtures, as in the HFCS production process.

AGAVE FRUCTOOLIGOSACCHARIDES PRODUCTION

Chicory inulin FOS are the most common prebiotics but not necessarily the most potent, as FOS have different prebiotic behavior and efficiency depending on their size (Santiago-Garcia and Lopez, 2009). Actually, Mexican regulation clearly differentiates *Agave* inulin from *Agave* FOS in terms of size. Therefore, methods to separate *agave* fructans according to their size by ultrafiltration, precipitation, or ion exchange chromatography (Leenheer *et al.*, 2011; Stewart, 2002; Leenheer and Smits, 1998; Engels *et al.*, 2004) or to transform large *agavin* into FOS through chemical or enzymatic hydrolysis have potential application in the functional food industry. Actually, FOS with a degree of polymerization between three and 10 are also alternative sweeteners for diabetic patients (Franck *et al.*, 2002).

However, although readily transformed to fructose by exo-inulinase (Muñoz-Gutierrez *et al.*, 2009; Montañez-Soto *et al.*, 2011), *agavin* is not a substrate for endo-inulinases. Nevertheless, it has been shown that *agavin* FOS can be obtained by a controlled acid hydrolysis. For this case, a Fructose Equivalent (FE)—analogous to the Dextrose Equivalent used in the starch industry—was introduced to characterize the hydrolysis, following the evolution of the process using HPAEC-PAD. It was found that in both a direct acid hydrolysis or through ion exchange resins, the hydrolysis proceeds through a combination of exo and endo reaction mechanisms releasing fructose and FOS simultaneously. In this context, when the hydrolysis reaches an FE of 30, a maximum proportion of *agavin* FOS with respect to fructose is obtained (Ávila-Fernández *et al.*, 2011). Actually, the hydrolysis reaction may proceed with the natural acidity of the *agavin* extract increasing the temperature, or adding acid to increase the hydrolysis rate. The complete process is outlined in Figure 15.5. The resulting product may either be applied as an agave sweetener with prebiotic properties, or be subjected to further processing to separate the fructose released in a cation exchange column.

ACTUAL AND POTENTIAL MARKET OF AGAVE DERIVED PRODUCTS

The regulation or official norm in México (Norma Mexicana NMX-F-591-SCFI-2010) defines as *Agave* fructans (*fructanos de Agave*) all fructans extracted from *Agave* plants. This term refers to the original polymer extracted from agave pines without any type of treatment, neither physical, chemical, or enzymatic, containing 90% fructan consisting of oligosaccharides with a DP higher than three (see below). The same norm defines *Agave* inulin (*inulina de Agave*) as the product after a purification or fractionation process that reduces the oligosaccharide content with DP <10 to a maximum of 5% retaining a minimum of 85% of fructans with a DP >10. Finally the so called *agavin* or agave FOS are those products that after a purification or fractionation process consist of a minimum of 80% of FOS with a DP <10 and no more than 5% of fructans with a DP >10. At this time, this norm is not yet compulsory and applies only in México.

In theory, the potential market for *Agave* fructans is estimated as the same as inulin from other sources as well as that of soluble fibers (around 100,000 tons of inulin and 200,000 tons of FOS worldwide in 2009); however, in spite of *in vitro* and *in vivo* research already performed in animal models, none of the agave products has been classified as prebiotic. It is expected that once sufficient experimental and clinical studies accumulate and the technological properties of the already existing products are defined, the real market for *Agave* inulin will be defined. With this situation in mind, in México alone, the potential market is estimated at 10,000 tons per year. This estimation considers that almost all fructans consumed in the country are imported mainly from Chile where the company Orafti produces inulin from chicory.

As far as exports is concerned, the volume is still limited compared to the installed capacity of the seven companies devoted to the production of *Agave* syrup (fructose) commercialized at a much lower price than the corresponding price of the original fructans, if they are recognized as prebiotics.

From 2007 to 2012, México exported a total of 2665 tons of a product classified as *Agave* inulin, with exports of more than 1000 tons during the last 2 years at an average selling price of $3.53 USD/Kg. In the same period (2007–2012) México imported 14,758 tons of inulin, more than 1000 tons in 2007, 2000 tons in 2009 and 3000 tons in 2011. The accumulated value of this market is $47.5 million USD, corresponding to an average of $10 million USD per year, as reported in a database of the Jalisco state.

It is clear that agave fructans are natural ingredients with functional and technological properties of enormous potential, still requiring fulfillment of all quality premises, acceptability, and convenience that are explicit through certification. It is also important to increase the knowledge behind the many variables affecting the properties and composition of products derived from the wide biodiversity of *Agave* plants.

The annual increase in inulin imports show that, until now, *Agave* fructans have had a marginal presence in the national and international functional ingredients market, certainly due to the lack of characterization, certification, as well as the incipient basic and applied research. This is a bottleneck that has to be solved by concerted efforts between the public and private sectors in order to promote the *Agave* agroindustry through the recognition of the functional properties of the molecules on which ancient traditions are based.

For this purpose, the medical, functional and technological properties have to be demonstrated, based on complete chemical and biological characterization, as well as clinical studies, so that health claims in functional products can be made.

References

Arrizon, J., Morel, S., Gschaedler, A., Monsan, P., 2010. Comparison of the water-soluble carbohydrate composition and fructan structures of *Agave tequilana* plants of different ages. Food Chem. 122 (1), 123–130.

Ávila-Fernández, A., Olvera-Carranza, C., Rudiño-Piñera, E., Cassab, G.I., Nieto-Sotelo, J., López-Munguía, A., 2007. Molecular characterization of sucrose: sucrose 1-fructosyltransferase (1-SST) from *Agave tequilana* Weber var. azul. Plant. Sci. 173 (4), 478–486.

Ávila-Fernandez, A., Rendon-Poujol, X., Olvera, C., Gonzalez, F., Capella, S., Pena-Alvarez, A., et al., 2009. Enzymatic Hydrolysis of Fructans in the Tequila Production Process. J. Agric. Food Chem. 57, 5578–5585.

Ávila-Fernández, A., Galicia-Lagunas, N., Rodríguez-Alegría, M.E., Olvera, C., López-Munguía, A., 2011. Production of functional oligosaccharides through limited acid hydrolysis of agave fructans. Food Chem. 129 (2), 380–386.

Diaz-Compain, J. E. Apparatus for diffusion and lixiviation of sugar cane, beet-sugar or sweet. US Patent 3554799. 15th May 1968.

Engels, D., Begli, A.H., Kunz, M., Mattes, R., Munir, M., Vogel, M. Method for the production of polyfructans. US Patent 2004/0086902 A1. 6th May 2004.
Escalante, A., Rodríguez, M.E., Martínez, A., López-Munguía, A., Bolívar, F., Gosset, G., 2004. Characterization of bacterial diversity in Pulque, a traditional Mexican alcoholic fermented beverage, as determined by 16S rDNA analysis. FEMS. Microbiol. Lett. 235, 273–279.
Escalante, A., Giles-Gómez, M., Esquivel-Flores, G., Matus Acuña, V., Moreno-Terrazas, R., López-Munguía, A., et al., 2013. Pulque Fermentation. In: Hui, Y.H. (Ed.), Handbook of Plant-Based Fermented Food and Beverage Technology. second ed. CRC Press, Boca Raton, FL, USA, pp. 691–706.
Franck, A., 2002. Technological functionality of inulin and oligofructose. Br. J. Nutr. 87 (Suppl. 2), S287–S291.
García-Mendoza, A.J., 2007. Los agaves de México. Ciencias 87, 14–23.
Gomez, E., Tuohy, K.M., Gibson, G.R., Klinder, A., Costabile, A., 2010. In vitro evaluation of the fermentation properties and potential prebiotic activity of *Agave* fructans. J. Appl. Microbiol. 108 (6), 2114–2121.
Gonçalves de Lima, O., 1978. El maguey y el pulque en los códices mexicanos. pp 278. Fondo de Cultura Económica. México D.F.
Güçlü-Üstündag, Ö., Mazza, G., 2007. Saponins: Properties, applications and processing. Crit. Rev. Food Sci. Nutr. 47, 231–258.
Gutierrez Uribe, J. Serna Saldivar S. Agave syrup extracts having anticancer activity. US patent 2009/0124685 A1.14th May 2009.
Hendry, A.F., 1993. Evolutionary Origins and Natural Functions of Fructans—A Climatological, Biogeographic and Mechanistic Appraisal. New. Phytol. 123 (1), 3–14.
Kolida, S., Tuohy, K., Gibson, G.R., 2002. Prebiotic Effects of Inulin and Oligofructose. Br. J. Nutr. 87 (suppl. 2), S193–S197.
Leenheer, L., Smits G. Process for the manufacture of chicory inulin, hydrolysates and derivatives of inulin, and improved chicory inulin products, hydrolysates and derivatives. EP 0930317 A1. 13th January 1998.
Leenheer, L., Frooninckx, K. M. J., Heroufosse, F. A. Fructooligosaccharide composition, process for its production and use. US Patent 2011/0081449. 7th April 2011.
Lopez, M.G., Mancilla-Margalli, N.A., Mendoza-Díaz, G., 2003. Molecular structures of fructans from *Agave tequilana* Weber var. azul. J. Agric. Food Chem. 51, 7835–7840.
Mancilla-Margalli, N.A., Lopez, M.G., 2006. Water-soluble carbohydrates and fructan structure patterns from Agave and Dasylirion species. J. Agric. Food Chem. 54, 7832–7839.
Mellado-Mojica, E., López, M.G., 2012. Fructan Metabolism in *A. tequilana* Weber Blue Variety along Its Developmental Cycle in the Field. J. Agric. Food Chem. 60 (47), 11704–11713.
Montanez-Soto, J.L., Venegas-González, J., Bernardino-Nicanor, A., Ramos-Ramírez, E.G., 2011. Enzymatic production of high fructose syrup from *Agave tequilana* fructans and its physicochemical characterization. Afr. J. Biotechnol. 10 (82), 19137–19143.
Muñoz-Gutierrez, I., Rodriguez-Alegría, M.E., López-Munguía, A., 2009. Kinetic behaviour and specificity of β-fructosidases in the hydrolysis of plant and microbial fructans. Process Biochem. 44, 891–898.
Narvaez-Zapata, J.A., Sanchez-Teyer, L.F., 2009. Agaves as a raw material: recent technologies and applications. Recent. Pat. Biotechnol. 3 (3), 185–191.
Ortiz-Basurto, R., Pourcelly, G., Doco, T., Williams, P., Dornier, M., Belleville, M.P., 2008. Analysis of the main components of the aguamiel produced by the maguey- pulquero (*Agave mapisaga*) throughout the harvest period. J. Agric. Food Chem. 56, 3682–3687.
Santiago-Garcia, P.A., Lopez, M., 2009. Prebiotic effect of agave fructans and mixtures of different degrees of polymerization from *Agave angustifolia* Haw. Dynamic Biochem. Process Biotechnol. Mol. Biol. (3 Special Issue), 52–58 Global Science Books.
Stewart B.S. Inulin fractions. US Patent 6419978. 16th July 2002.
Suzuki, M., 1993. History of fructan research: Rose to Edelman. In: Suzuki, M., Chatterton, N.J. (Eds.), Science and technology of fructans. CRC Press, Boca Raton, FL, USA, pp. 21–39.
Urias-Silvas, J.E., Cani, P.D., Delmee, E., Neyrinck, A., Lopez, M.G., Delzenne, N.M., 2008. Physiological effects of dietary fructans extracted from *Agave tequilana* Gto. and *Dasylirion* spp. Br. J. Nutr. 99, 254–261.
Vijn, I., Smeekens, S., 1999. Fructan: more than a reserve carbohydrate? Plant. Physiol. 120 (2), 351–360.
Wuanick, M.C., 1970. Cura de vaginitis de etiologia variada pelo emprego cultura de *Zymomonas mobilis* (Lindner). Rev. Inst. Antibiot. Rec. 10, 47–50.

SECTION 2

FRUIT

CHAPTER

16

Effects of 1-Methylcyclopropene on Active Composition in Fruits

Junfeng Guan, Muqiang Hu, Chengguo Shen, Shuo Zhou, Yudou Cheng, Jingang He

Institute of Genetics and Physiology, Hebei Academy of Agricultural and Forestry Sciences, Shijiazhuang, PR China

CHAPTER POINTS

- The changes of sugar and organic acid contents in 1-MCP-treated fruits varied depending on the species, cultivars, and the storage conditions.
- Most volatile compounds were strongly reduced by 1-MCP in many fruits.
- In most cases, 1-MCP-treated fruit retained a higher antioxidant content.
- 1-MCP altered the oil content and fatty acid profiles in avocado.
- 1-MCP could induce accumulation of some essential amino acids in apple.

INTRODUCTION

1-Methylcyclopropene (1-MCP), a new ethylene action inhibitor, has significant effects on extending shelf-life and maintaining the quality of plant products including fruits, vegetables, and floriculture crops. Not only does the use of 1-MCP in commercial use improve product quality after harvest, but it is also a useful compound in the laboratory for further studying the role of ethylene in ripening and senescence.

1-MCP is a gas with a formula of C_4H_6 and a molecular weight of 54 at standard temperature and pressure. 1-MCP is thought to compete with ethylene to bind with ethylene receptors and thereby prevent ethylene-dependent responses. Compared with ethylene, 1-MCP is active at much lower concentrations because the affinity of 1-MCP for the ethylene receptor is approximately 10 times greater than that of ethylene (Sisler and Serek, 1997).

Edward Sisler and Sylvia Blankenship jointly hold the patent on the use of cyclopropenes to inhibit ethylene action since 1996. Commercialization of 1-MCP for ornamental crops, named EthylBloc, was undertaken by Floralife, Inc. The product was approved by the United States Environmental Protection Agency (EPA) in 1999. Subsequently, commercialization of 1-MCP for edible crops, named SmartFresh, was launched by AgroFresh, Inc. 1-MCP is applied at very low concentration with a non-toxic mode of action and negligible residue (Blankenship *et al.*, 2003).

1-MCP has been widely used in quality maintenance of fruits during storage for its significant effects on delaying fruit softening, inhibiting browning, preventing decay, deferring peel yellowing, etc. (Watkins, 2006; Cheng *et al.*, 2012). The effects of 1-MCP are closely related to the storage temperature and time. The effects are usually not significant at low temperature over a short period, while they are significant during long-term storage. Therefore, it is essential to analyse the effects of 1-MCP on the changes of fruit active compounds during storage in order to optimize its application under different scenarios.

In this paper, we mainly describe the effects of 1-MCP on fruit active components including sugars, organic acids, volatile compounds, antioxidant, fatty acids, and amino acids during storage.

SUGARS

Carbohydrate metabolism is an important process in fruit during storage. Total soluble solids (TSS), which are primarily carbohydrates, are influenced by 1-MCP treatment but the effects were variable from an increase

Processing and Impact on Active Components in Food
http://dx.doi.org/10.1016/B978-0-12-404699-3.00016-0

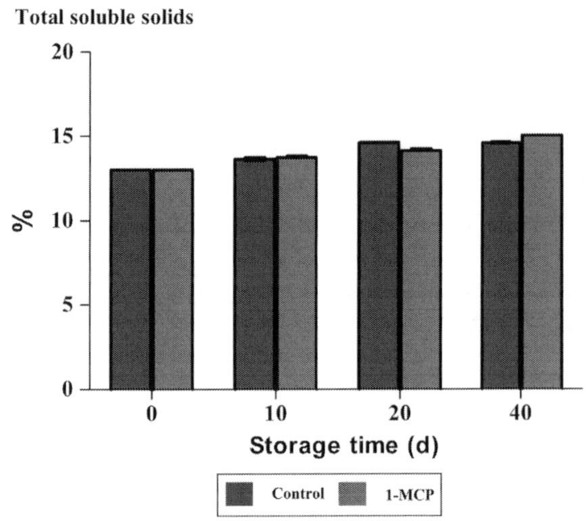

FIGURE 16.1 Total soluble solids content in 'Orin' apple fruit with 1-MCP treatment and control during ambient storage at 25°C. Values are the means ±SE (n=3). Asterisk represents significant difference (P<0.05).

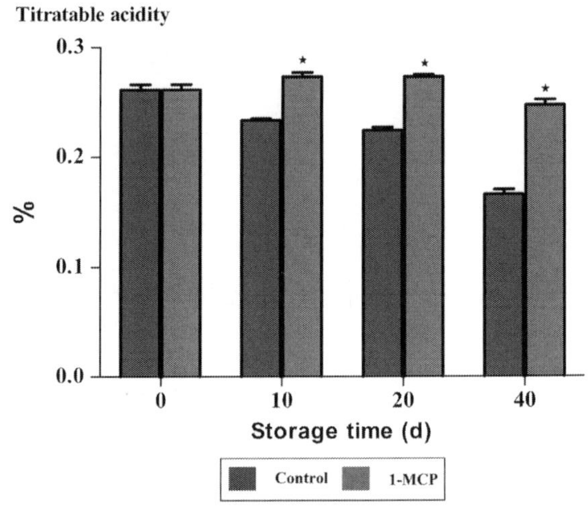

FIGURE 16.2 Titratable acid content in "Orin" apple fruit with 1-MCP treatment and control during ambient storage at 25°C. Values are the means±SE (n=3). Asterisk represents significant difference (P<0.05).

to a decrease depending on the species, cultivars, and the storage conditions. TSS was higher in 1-MCP-treated 'Patharnakh' pear (Mahajan *et al.*, 2010), 'Gala' and 'Fuji' apple (Bai *et al.*, 2005), 'Williams' banana (Golding *et al.*, 1998), while lower in 1-MCP-treated 'Nanicão' banana (Nascimento *et al.*, 2006), cut and intact kiwi fruit (Mao *et al.*, 2007), and 'Shatangju' mandarin (Li *et al.*, 2012) . However, other reports showed that 1-MCP had no effects on TSS in some apple cultivars (Zanella *et al.*, 2003; DeLong *et al.*, 2004) and 'Summit' sweet cherry (Wei *et al.*, 2011).

We analyzed TSS in 'Orin' apple fruits with 1-MCP-treatment and control during cold storage prior to and at 10, 20, and 40 days after treatment. TSS was shown to give a slight increase in 1-MCP-treated fruits and controls. However, there were no significant differences between 1-MCP-treated fruits and controls (Figure 16.1).

The analysis of individual sugar showed that 1-MCP delayed the decline in mannoheptulose and perseitol in early season avocado fruits during cold storage and shelf-life. However, in late season fruits, the same trend was observed in perseitol but not in mannoheptulose. Overall 1-MCP-treated avocado fruits had significantly less sucrose than controls during storage and shelf-life (Meyer *et al.*, 2010). 1-MCP inhibited the sucrose lost and the accumulation of glucose and fructose during cold storage in 'Gold Nijisseiki' and 'Hosui' pear (Itai *et al.*, 2008). 1-MCP treatment did not alter the levels of glucose, fructose, and sucrose in 'Empire' apples during storage, but enhanced the levels of intermediate products in the metabolism pathway of carbohydrates, such as fructose 6-phosphate, glucose 6-phosphate, and sorbitol (Lee *et al.*, 2011).

ORGANIC ACIDS

Organic acids affect the taste and their abundance represents the quality of fruits. In fruits, the organic acids were multitudinous, including malic acid, succinic acid, citric acid, fumaric acid, etc. Some work showed that 1-MCP had different effects on the level of organic acids due to the diversity of fruit cultivars and different storage conditions. 1-MCP effectively retarded the decrease in titratable acid (TA) of which the major component was malic acid in 'Gala', 'Ginger gold,' 'Granny Smith,' 'Law Rome,' 'Empire,' and 'Fuji' apples during air storage (Zanella *et al.*, 2003; Bai *et al.*, 2005), and in 'Patharnakh' pear (Mahajan *et al.*, 2010). However, in the controlled atmosphere (CA) storage, the level of TA was not consistently affected by 1-MCP treatment in 'Law Rome' apples (Bai *et al.*, 2005) and 'Summit' sweet cherry (Wei *et al.*, 2011).

Our data shows a TA loss in 1-MCP-treated 'Orin' apple fruits had been effectively prevented during ambient storage at 25°C, while TA declined significantly in non-treated fruits (Figure 16.2).

Further analysis of individual organic acid indicated that 1-MCP treatment did not delay the decrease in malic acid, but reduced the level of mucic acid of 'Empire' apples (Lee *et al.*, 2011). 1-MCP prevented the degradation of ascorbic acid, malic acid, and citric acid, but not succinic acid in arazá fruits (Carrillo *et al.*, 2011).

VOLATILE COMPOUNDS

The content of volatile compounds is another indicator that reflects the quality of fruits. The metabolism of the volatile compounds is an ethylene dependent process.

Previous reports indicated that the total of volatile compounds are usually reduced by 1-MCP in 'Fuji,' 'Anna,' 'Gala,' 'Delicious,' and 'Granny Smith' apples (Lurie *et al.*, 2002; Bai *et al.*, 2005). The analysis of an individual volatile compound indicated that 1-MCP significantly reduced the levels of alcohols, aldehydes and esters in 'Empire' apple (Lee *et al.*, 2011) such as acetate esters, ethyl butanoate, ethanol, propanol, butanol, acetaldehyde, and propanal but increased 3-methylbutyl 2-methylbutanoate content in 'Conference' pear fruits (Rizzolo *et al.*, 2005).

ANTIOXIDANTS

Antioxidants are an important nutritional attribute of fruits for a human's healthy lifestyle due to their functions of reducing the risk of diseases. 1-MCP is widely used in apples to maintain the content of phenolic compounds during storage (Mattheis, 2008). Further studies show that the effect of 1-MCP on total phenolic contents is different among apple cultivars. In the 'Cripps Pink' apple, 1-MCP reduced the total phenolic contents in the flesh, and the levels of chlorogenic acid and caffeic acid decreased (Hoang *et al.*, 2011). In 'Empire' apples, 1-MCP treatment could keep a higher level of total phenolics than the control in the peel, and lower level in flesh (Fawbush *et al.*, 2009). Moreover, 1-MCP could retain a higher total phenol content in kiwifruit while it demonstrated significant drops during storage period (Jhalegar *et al.*, 2011).

The evidence for the effects of 1-MCP on phenolic compounds in pears is scarcely documented, therefore we have analysed the contents of arbutin, chlorogenic acid, and gallic acid in 'Yali' pear in 1-MCP-treated fruits and control during storage. Arbutin content declined slightly after 60 days storage, while a higher content was retained in 1-MCP fruits than the control (Figure 16.3). A significant decrease was observed in the chlorogenic acid content in control fruits during storage, while 1-MCP could effectively prevent such a loss (Figure 16.4). However, 1-MCP-treated fruits had less gallic acid content than the control at late stage of cold storage (Figure 16.5).

The effects of 1-MCP on total flavonoids content during storage also varied among different apple cultivars. In 'Fuji' apples, 1-MCP could effectively retain the amount of total flavonoid (Lu *et al.*, 2012), while no effect of 1-MCP was detected in 'Empire' apples (Fawbush *et al.*, 2009).

Research has demonstrated that 1-MCP may affect anthocyanin levels in fruits. 1-MCP could transiently inhibit anthocyanin accumulation in grape skins compared with the control before harvest, although the suppression level varied in the different treatment period (Chervin *et al.*, 2004), and it lowered the increases in anthocyanin and phenolic contents in strawberry fruit (Jiang *et al.*, 2001). However, it had no significant effects

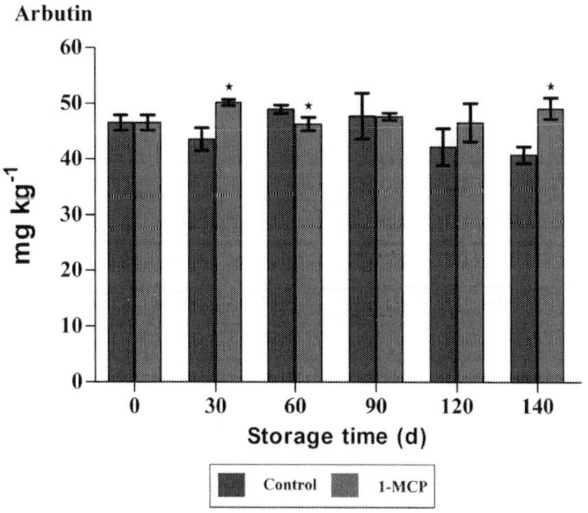

FIGURE 16.3 Arbutin content of "Yali" pear fruit with 1-MCP treatment and control during cold storage at 0°C. Values are the means ± SE (n = 3). Asterisk represents significant difference ($P < 0.05$).

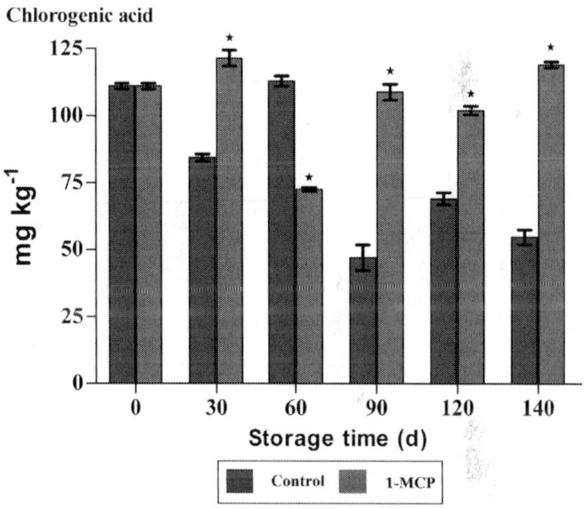

FIGURE 16.4 Chlorogenic acid content of "Yali" pear fruit with 1-MCP treatment and control during cold storage at 0°C. Values are the means ± SE (n = 3). Asterisk represents significant difference ($P < 0.05$).

on anthocyanins in sweet cherry (Mozetič *et al.* 2006) and blueberry (Chiabrando and Giacalone, 2011).

Lycopene, a carotenoid found in tomatoes, grapefruit and additional red fruits, is a proven antioxidant that may lower the risk of certain disorders such as heart disease and cancer. 1-MCP delayed the accumulation of lycopene (Choi and Huber, 2008; Xie and Luo, 2011; Mostofi *et al.*, 2003). However, at the end of ripening, there was no significant difference of the lycopene content between the fruit with and without 1-MCP treatment (Mostofi *et al.*, 2003).

Ascorbic acid is essential to human nutrition because of its well-known disease prevention benefits, and is also considered to be an important indicator of fruit quality. 1-MCP could significantly keep a higher level of ascorbic acid in 1-MCP-treated fruits compared with controls

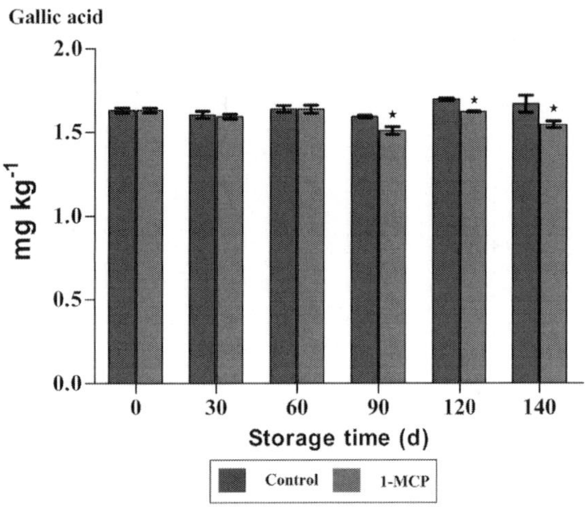

FIGURE 16.5 Gallic acid content of "Yali" pear fruit with 1-MCP treatment and control during cold storage at 0°C. Values are the means ± SE (n = 3). Asterisk represents significant difference ($P < 0.05$).

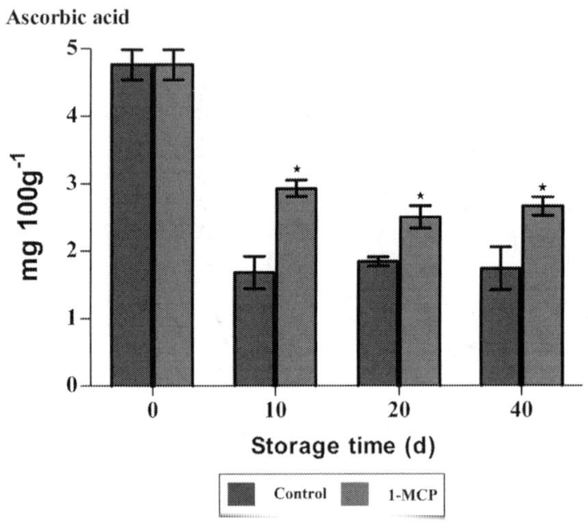

FIGURE 16.6 Ascorbic acid content in "Orin" apple fruits with 1-MCP treatment and control during ambient storage at 25°C. Values are the means ± SE (n = 3). Asterisk represents significant difference ($P < 0.05$).

in 'Fuji' apples (Lu *et al.*, 2012), 'Shatangju' mandarin (Li *et al.*, 2012), pineapples (Selvarajah *et al.*, 2001), and 'McLean's Red' litchi (Sivakumar *et al.*, 2010), but there was no effect on ascorbic acid content in 'Empire' apples (Fawbush *et al.*, 2009).

In 'Orin' apples, a significant decrease was observed in ascorbic acid content during storage, while 1-MCP could significantly keep the fruits with a higher level of ascorbic acid in ambient storage (Figure 16.6).

FATTY ACIDS

Oil content is often associated with horticultural maturity in avocado. During cold storage and shelf-life period, a slight, yet significant increase in oil yield is found in ripening avocados.

In the fatty acid profiles of avocado fruits, oleic acid constitutively represents the main component followed by palmitic, linoleic, palmitoleic, and linolenic acids in order of their abundance. In the early season, the fatty acid composition shows a decrease in palmitic and palmitoleic acids and an increase in linolenic acid during cold storage. 1-MCP treatment results in higher palmitic acid content than the control. During subsequent ripening during shelf-life period, a decrease in palmitic acid and increase in polyunsaturated fatty acids was observed. Again, palmitic acid content was significantly higher for 1-MCP-treated fruit over 3 days of shelf-life. However, those changes were not observed in late season avocado fruit (Meyer and Terry, 2010).

AMINO ACIDS

Studies on the effect of 1-MCP on amino acid metabolism in fruits were rarely seen. Recently, Lee *et al.* (2011) reported that 1-MCP could induce accumulation of some essential amino acids, such as threonine, isoleucine, valine, leucine, and phenylalanine. However, the mechanism of amino acid accumulation induced by 1-MCP remains unknown.

ANALYTICAL TECHNIQUES

'Orin' apple and 'Yali' pear were harvested at commercial mature stage and were transported to the laboratory within 4 h and selected by the uniformity of weight, shape, and color without any visual defects, and then randomly divided into two. The first half were exposed to 1.0 μl/l 1-MCP for 24 h at 25°C. Whilst the second half were sealed with air as control. After treatment, the apples and pears were stored at 25°C and 0°C, respectively.

For measurement of TSS, TA and ascorbic acid in 'Orin' apple, 15 fruits of each replicate were taken at 10, 20, and 40 days during ambient storage. Each treatment has three replicates. The measurement of TSS, TA was carried out according to the method of Bai *et al.* (2005) and the measurement of ascorbic acid according to the method of Lu *et al.* (2012).

For measurement of arbutin, chlorogenic acid and gallic acid, fifteen fruits of each replicate were taken at 0, 30, 60, 90, 120, and 140 days during storage. Each treatment has three replicates. The extraction and HPLC assays of the phenolic compounds were carried out according to the method of Galvis Sánchez *et al.* (2003).

References

Bai, J., Baldwin, E.A., Goodner, K.L., Mattheis, J.P., Brecht, J.K., 2005. Response of four apple cultivars to 1-methylcyclopropene treatment and controlled atmosphere storage. Hortscience 40, 1534–1538.

Blankenship, S.M., Dole, J.M., 2003. 1-Methylcyclopropene: a review. Postharvest Biol. Technol. 28, 1–25.

Carrillo, M.P., Hernández, M.S., Barrera, J., Martínez, O., Fernández-Trujillo, J.P., 2011. 1-Methylcyclopropene delays arazá ripening and improves post-harvest fruit quality. LWT-Food Sci. Technol. 44, 250–255.

Cheng, Y.D., Dong, Y., Yan, H.B., Ge, W.Y., Shen, C.G., Guan, J.F., et al., 2012. Effects of 1-MCP on chlorophyll degradation pathway-associated genes expression and chloroplast ultrastructure during the peel yellowing of Chinese pear fruits in storage. Food Chem. 135, 415–422.

Chervin, C., El-Kereamy, A., Roustan, J.P., Latché, A., Lamon, J., Bouzayen, M., 2004. Ethylene seems required for the berry development and ripening in grape, a non-climacteric fruit. Plant Sci. 167, 1301–1305.

Chiabrando, V., Giacalone, G., 2011. Shelf-life extension of highbush blueberry using 1-methylcyclopropene stored under air and controlled atmosphere. Food Chem. 126, 1812–1816.

Choi, S.T., Huber, D.J., 2008. Influence of aqueous 1-methylcyclopropene concentration, immersion duration, and solution longevity on the post-harvest ripening of breaker-turning tomato (*Solanum lycopersicum* L.) fruit. Postharvest Biol. Technol. 49, 147–154.

DeLong, J.M., Prange, R.K., Harrison, P.A., 2004. The influence of 1-Methylcyclopropene on "Cortland" and "McIntosh" apple quality following long-term storage. HortScience 39, 1062–1065.

Fawbush, F., Nock, J.F., Watkins, C.B., 2009. Antioxidant contents and activity of 1-methylcyclopropene (1-MCP)-treated "Empire" apples in air and controlled atmosphere storage. Postharvest Biol. Technol. 52, 30–37.

Galvis Sánchez, A.C., Gil-Izquierdo, A., Gil, M.I., 2003. Comparative study of six pear cultivars in terms of their phenolic and vitamin C contents and antioxidant capacity. J. Sci. Food Agriculture 83, 995–1003.

Golding, J.B., Shearer, D., Wyllie, S.G., McGlasson, W.B., 1998. Application of 1-MCP and propylene to identify ethylene-dependent ripening processes in mature banana fruit. Postharvest Biol. Technol. 14, 87–98.

Hoang, N.T., Golding, J.B., Wilkes, M.A., 2011. The effect of postharvest 1-MCP treatment and storage atmosphere on "Cripps Pink" apple phenolics and antioxidant activity. Food Chem. 127, 1249–1256.

Itai, A., Tanahashi, T., 2008. Inhibition of sucrose loss during cold storage in Japanese pear (*Pyrus pyrifolia* Nakai) by 1-MCP. Postharvest Biol. Technol. 48, 355–363.

Jhalegar, M.D., Sharma, R., Pal, R., Arora, A., Dahuja, A., 2011. Analysis of physiological and biochemical changes in kiwifruit (*Actinidia deliciosa* cv. Allison) after post-harvest treatment with 1-Methylcyclopropene. J. Plant Biochem. Biotechnol. 20, 205–210.

Jiang, Y.M., Joyce, D.C., Terry, L.A., 2001. 1-Methylcyclopropene treatment affects strawberry fruit decay. Postharvest Biol. Technol. 23, 227–232.

Nascimento, J.R.O.D., Júnior, A.V., Bassinello, P.Z., Cordenunsi, B.R., Mainardi, J.A., Purgatto, E., et al., 2006. Beta-amylase expression and starch degradation during banana ripening. Postharvest Biol. Technol. 40, 41–47.

Lee, J., Rudell, D.R., Davies, P.J., Watkins, C.B., 2011. Metabolic changes in 1-methylcyclopropene (1-MCP)-treated "Empire" apple fruit during storage. Metabolomics 8, 742–753.

Li, Q., Wu, F.W., Li, T.T., Su, X.G., Jiang, G.Q., Qu, H.X., et al., 2012. 1-Methylcyclopropene extends the shelf-life of "Shatangju" mandarin (*Citrus reticulate* Blanco) fruit with attached leaves. Postharvest Biol. Technol. 67, 92–95.

Lu, X., Ma, Y., Liu, X., 2012. Effects of maturity and 1-MCP treatment on postharvest quality and antioxidant properties of "Fuji" apples during long-term cold storage. Horticulture Environ. Biotechnol. 53, 378–386.

Lurie, S., Pre-Aymard, C., Ravid, U., Larkov, O., Fallik, E., 2002. Effect of 1-methylcyclopropene on volatile emission and aroma in cv. Anna apples. J. Agricultural Food Chem. 50, 4251–4256.

Mahajan, B.V.C., Singh, K., Dhillon, W.S., 2010. Effect of 1-methylcyclopropene (1-MCP) on storage life and quality of pear fruits. J. Food Sci. Technol. 47, 351–354.

Mao, L.C., Wang, G.Z., Que, F., 2007. Application of 1-methylcyclopropene prior to cutting reduces wound responses and maintains quality in cut kiwifruit. J. Food Eng. 78, 361–365.

Mattheis, J.P., 2008. How 1-methylcyclopropene has altered the Washington state apple industry. HortScience 43, 99–101.

Meyer, M.D., Terry, L.A., 2010. Fatty acid and sugar composition of avocado, cv. Hass, in response to treatment with an ethylene scavenger or 1-methylcyclopropene to extend storage life. Food Chem 121, 1203–1210.

Mozetič, B., Simčič, M., Trebše, P., 2006. Anthocyanins and hydroxycinnamic acids of Lambert Compact cherries (*Prunus avium* L.) after cold storage and 1-methylcyclopropene treatment. Food Chem., 97, 302–309.

Mostofi, Y., Toivonen, P.M.A., Lessnai, H., Babalar, M., Lu, C., 2003. Effects of 1-methylcyclopropene on ripening of greenhouse tomatoes at three storage temperatures. Postharvest Biol. Technol. 27, 285–292.

Rizzolo, A., Cambiaghi, P., Grassi, M., Zerbini, P.E., 2005. Influence of 1-Methylcyclopropene and Storage Atmosphere on Changes in Volatile Compounds and Fruit Quality of Conference Pears. J. Agricultural Food Chem. 53, 9781–9789.

Sisler, E., Serek, M., 1997. Inhibitors of ethylene responses in plants at the receptor level: recent developments. Physiologia Plantarum 100, 577–582.

Sivakumar, D., Korsten, L., 2010. Fruit quality and physiological responses of litchi cultivar McLean's Red to 1-methylcyclopropene pre-treatment and controlled atmosphere storage conditions. LWT - Food Sci. Technol. 43, 942–948.

Watkins, C.B., 2006. The use of 1-methylcyclopropene (1-MCP) on fruits and vegetables. Biotechnol. Adv. 24, 389–409.

Wei, J.M., Qi, X.D., Guan, J.F., Zhu, X.Q., 2011. Effect of cold storage and 1-MCP treatment on post-harvest changes of fruit quality and cell wall metabolism in sweet cherry. J. Food Agriculture Environ. 9, 118–122.

Xie, J., Luo, Z.S., 2011. Effect of 1-methylcyclopropene on chilling injury of tomato fruit. Acta Horticulturae Sinica 38, 281–287 (In Chinese).

Zanella, A., 2003. Control of apple superficial scald and ripening—a comparison between 1-methylcyclopropene and diphenylamine postharvest treatments, initial low oxygen stress and ultra low oxygen storage. Postharvest Biol. Technol. 27, 69–78.

CHAPTER

17

How Cultivars Influence Fruit Composition: Total Phenols, Flavonoids Contents, and Antioxidant Activity in the Pulp of Selected Asian Pears

Junfeng Guan, Jingang He, Chengguo Shen, Limei Li, Yanxia Wang, Yudou Cheng

Institute of Genetics and Physiology, Hebei Academy of Agricultural and Forestry Sciences, Shijiazhuang, PR China

CHAPTER POINTS

- The average total phenols content was 0.2237 μg quercetin/g fresh weight (FW) among the 22 selected Asian pears, while the average value of flavonoids content was 0.17803 μg quercetin/g FW. The α,α-diphenyl-b-picrazylhydrazyl radical (DPPH·) scavenging effect ranged from 7.08% to 31.5%, with an average value of 18.46%.
- The antioxidant capacity that was measured by ferric reducing antioxidant potential assay ranged from 14.39% to 48.80% μmol trolox equivalent (TE)/100 g FW, and the average value was 29.63% μmol TE/100 g FW.
- The high phenolic content and flavonoids content were significantly correlated with high antioxidant capacity.
- In all of the cultivars, 'Xuehua' pear showed the highest total content for phenols and flavonoids, therefore presented the strongest antioxidant capacity.

INTRODUCTION

The high involvement of fruit and vegetables in the human diet, because of their ability to neutralize reactive oxygen species (ROS) that are hazardous for health, is essential to human life. The antioxidant capacity of plant tissues is closely associated with activity of free radical scavenging enzymes and with the contents of antioxidant substances, mainly phenolic compounds, carotenoids, tocopherol, and ascorbic acid (Bartosz, 1997).

Recently, antioxidant activity has been determined in many species of fruit, vegetables, herbs, and cereals (Kahkonen *et al.*, 1999). Particular attention has been paid to fruit as rich sources of antioxidant compounds (Kalt *et al.*, 1999; Wang and Lin, 2000). Oxidative stress contributes to aging and age-related diseases such as cancer, cardiovascular disease, chronic inflammation, and neurodegenerative diseases (Thomas *et al.*, 2007). Further study and research in epidemiology shows that the intake of fruit is negatively correlated with the incidence of cardiovascular diseases and tumors, which are related to the oxidant damage caused by free radicals (Steinmetz and Potter, 1996). Many researchers assume that it is the antioxidant capacity that gives fruit the ability to prevent the age-related diseases (Lise *et al.*, 1994; Gaziano and Manson, 1996; Hertog *et al.*, 1993).

Pear is regarded as a fruit which is beneficial to the human respiratory system and can even restrain infections of the respiratory system to some degree according to traditional Chinese medicine. The evaluation of the total phenolic and flavonoids contents, and antioxidant activities of 22 Asian pear cultivars was accomplished to offer new information for antioxidant research in pear.

Processing and Impact on Active Components in Food
http://dx.doi.org/10.1016/B978-0-12-404699-3.00017-2

DETERMINATION OF TOTAL PHENOLS AND ANTIOXIDANT ACTIVITY IN THE PULP OF ASIAN PEARS

Total phenols and flavonoids were extracted and total phenolic content was determined using the Folin–Ciocalteu reagent assay according to the method adopted by Bao *et al.* (2005) and the results were expressed as mg of quercetin equivalents per 100 g fresh weight (FW). The flavonoids content was determined by the spectrophotometrical method and the content was calculated in terms of mg/100 g FW of quercertin equivalents (Jia *et al.*, 1999). The antioxidant activity of pears was measured by the FRAP (ferric reducing antioxidant potential) assay (Deighton *et al.*, 2000), the antioxidant capacity was calculated as μmol/100 g FW of Trolox equivalents (TE). The antioxidant activity was determined by the DPPH (α,α-diphenyl-b-picrazylhydrazyl radical) method (Bao *et al.*, 2005).

COMPARISON OF TOTAL PHENOLIC CONTENT, FLAVONOIDS CONTENT AND ANTIOXIDANT ACTIVITY IN THE PULP OF ASIAN PEARS

According to our survey, the total phenolic contents, which ranged from 45.92 to 103.14 μg quercetin/g FW, varied among the 22 cultivars tested (Figure 17.1). The results showed that the 'Xuehua' pear presented the highest total phenolic content. In contrast, the 'Wasekisui' pear presented the lowest. The 22 cultivars exhibited different flavonoids contents, ranging from 23.10~104.28 μg quercetin/g FW (Figure 17.2). 'Xuehua' pear showed the highest flavonoids content and 'Wonhuwang' pear showed the lowest.

The antioxidant activity of each of the 22 Asian pear cultivars was different. The DPPH· scavenging effects fall within a wide range from 7.08% to 31.5%, with an average value of 18.46%, and 'Xuehua' pear presented the strongest scavenging capacity, while 'Wasekisui' pear presented the lowest (Figure 17.3). The antioxidant capacity that was measured by FRAP assay ranged from 14.39 to 48.80 μmol TE/100 g FW, the average value was 29.63 μmol TE/100 g FW. The highest value was found in 'Xuehua' pear and the lowest was 'Lvbaoshi' pear (Figure 17.4). In conclusion, 'Xuehua' pear possessed the strongest antioxidant capacity among the 22 pear cultivars that we detected.

ANTIOXIDANT ACTIVITY IN FRUIT

According to recent research (Guo *et al.*, 2003) on the free radical scavenging effect of several common fruit, such as kiwi fruit, purple mulberry, strawberry, grape, peach, apricot, and 'Yali' pear, the average FRAP value obtained by using $FeSO_4$ as standard solution in pulp

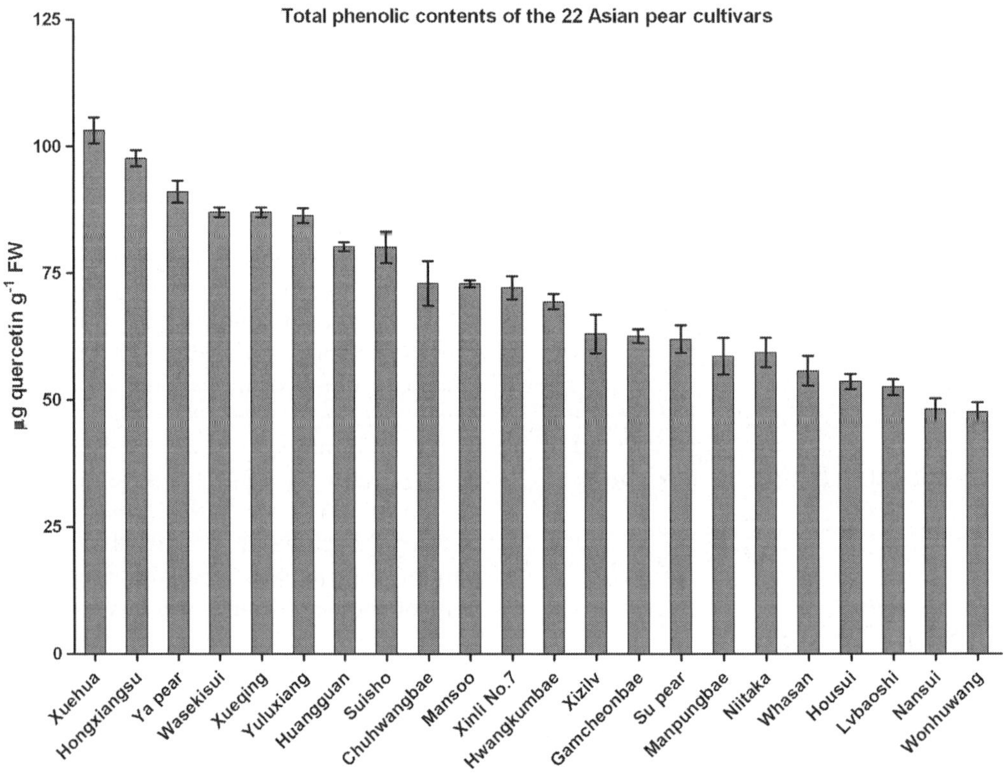

FIGURE 17.1 Total phenolic contents of the 22 pear cultivars.

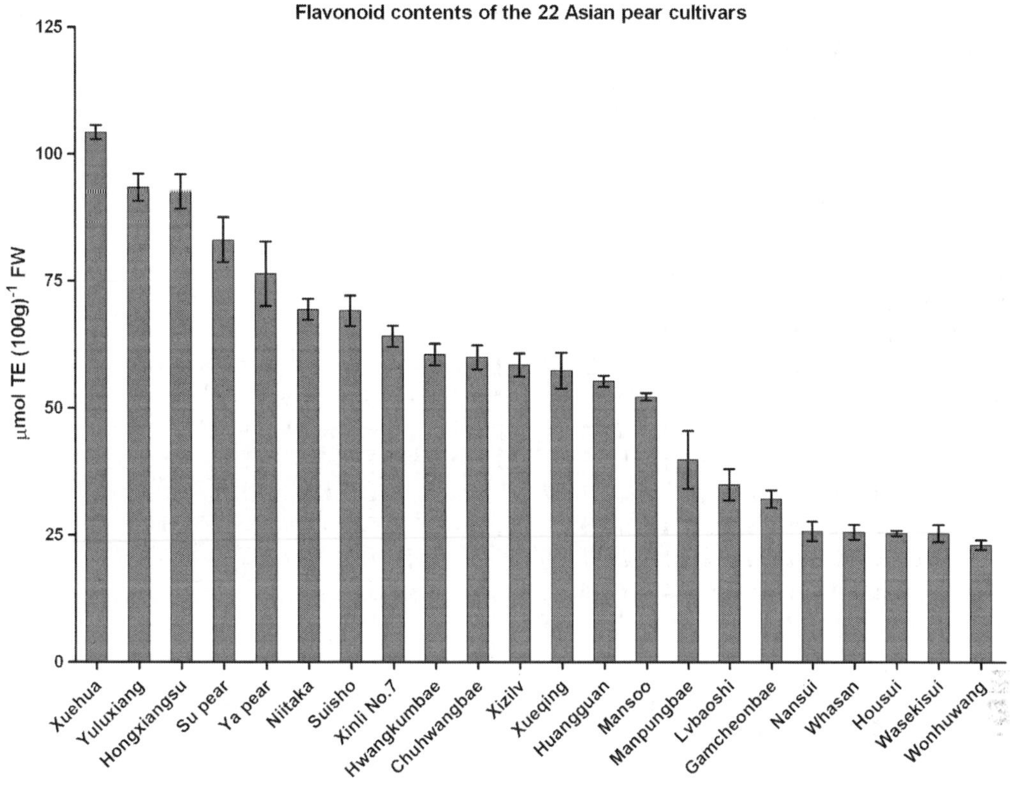

FIGURE 17.2 Flavonoids contents of the 22 pear cultivars.

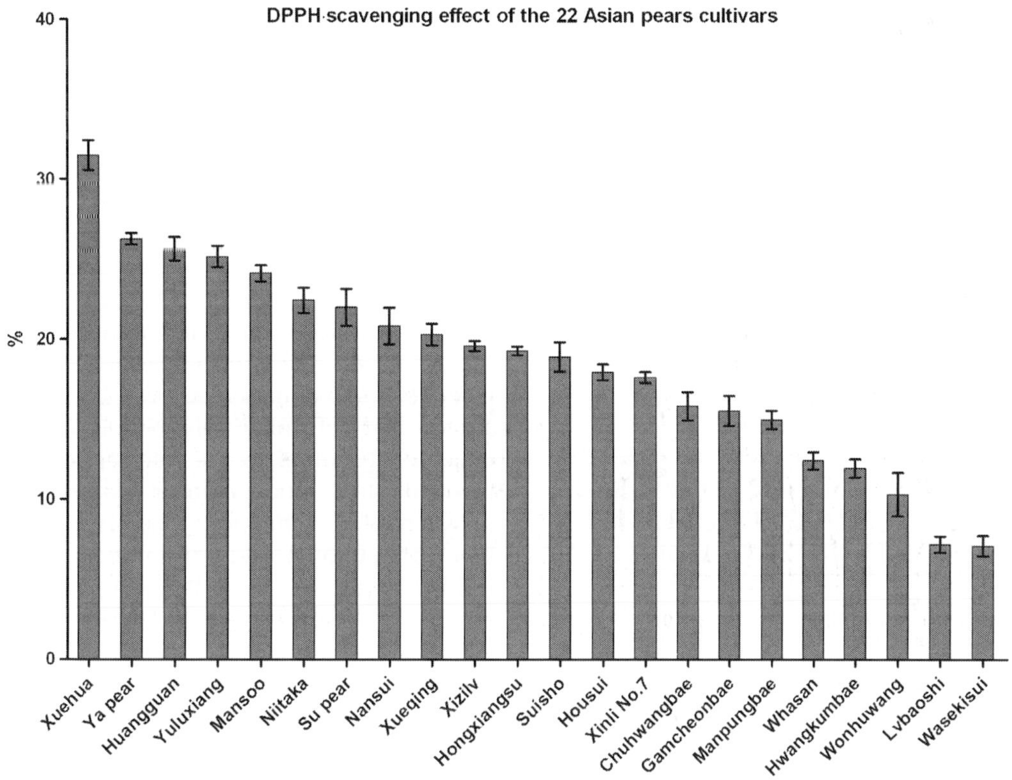

FIGURE 17.3 The DPPH· scavenging effect of the 22 Asian pear cultivars.

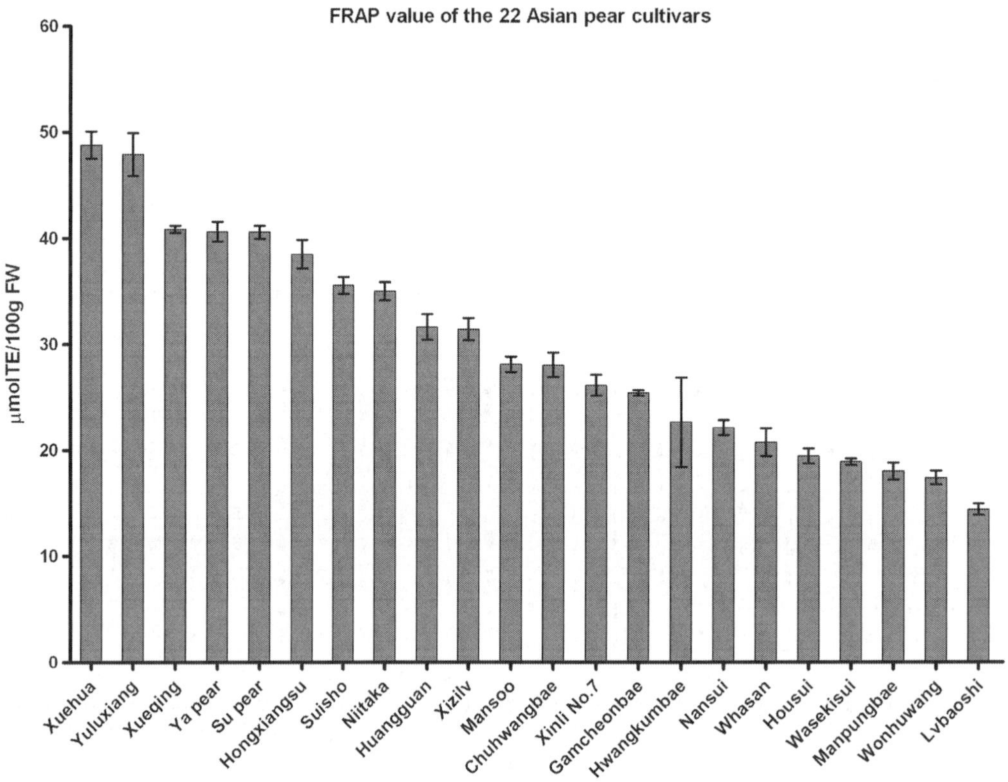

FIGURE 17.4 The FRAP value of the 22 Asian pear cultivars.

was about 1.88 mmol/100 g FW. Carbonaro and Mattera (2001) measured the polyphenols content of a pear (*Pyrus. communis* L., cv. Williams), and found that the contents were 48.2 and 56.1 mg tannic acid/100 g FW, conventionally and organically grown, respectively. Pears ranked relatively low among fruit regarding antioxidant activity and concentration of phenolics (Campanella *et al.*, 2003; García-Alonso *et al.*, 2004), but had higher antioxidant activity than many common vegetables (Höner and Cervellati, 2002). We measured 22 Asian pear cultivars, and found that the DPPH· scavenging effect ranged from 7.08% to 31.5%; the average value was 18.46%. The antioxidant capacity measured by FRAP assay ranged from 14.39 to 48.80 µmol TE/100 g FW, and the average value was 29.63 µmol TE/100 g FW. 'Xuehua' pear possesses the strongest antioxidant power.

RELATIONSHIP BETWEEN PHENOLICS, FLAVONOIDS, AND ANTIOXIDANT ACTIVITY IN FRUIT

Increased fruit and vegetable consumption is associated with a decreased incidence of cardiovascular diseases, cancer, and other chronic diseases. Phenolics, which are abundant in fruit and vegetables, are efficient in protecting low-density lipoprotein from oxidation driven by peroxyl and ferril radicals (Serafini *et al.*, 2000). The beneficial health effects of fruit and vegetables had been attributed, in part, to antioxidant flavonoids present in these foods. Flavonoids played a significant role as antioxidants in reducing chronic disease risk (Silvina and Balz, 2006).

The correlation between antioxidant power and polyphenolics content has been studied by some, but the results vary greatly. In Alothman and colleagues' study (2009), a high level of phenol content was significantly correlated with high antioxidant capacity in several tropical fruit from Malaysia. García-Alonso *et al.* (2004) measured 28 different fruit, and concluded that the antioxidant activity of the analyzed fruit can not only be attributed to their flavonols content, but whether they can be attributed to the result of the action of different antioxidant compounds present in the fruit and to possible synergic and antagonist effects still remain unknown. Du *et al.* (2009) demonstrated that both total polyphenols and vitamin C were major contributors to the total antioxidant capacity in kiwi fruit. In addition, a linear relationship was observed between FRAP values and total phenolics ($r = 0.981$) in 10 blueberry cultivars (Ilkay and Bulent, 2009).

In this current study, the correlation analysis to the relation of total phenols, flavonoids, FRAP value, and DPPH· scavenging effect showed that these four values were positively correlated and were highly significant ($P < 0.01$) (Table 17.1). The correlation coefficients between FRAP value and total phenols and total flavonoids were higher than that between DPPH· scavenging effect and both of the above, but the correlation values

TABLE 17.1 Correlation of Antioxidant Activity (FRAP Value), DPPH Scavenging Effect and the Contents of Total Phenols and Flavonoids in Pear Fruit

	Total Phenolic Contents	Flavonoid Contents	FRAP Value	DPPH· Scavenging Effect
Total phenolic contents	1	0.940**	0.911**	0.755**
Flavonoid contents	0.940**	1	0.898**	0.702**
FRAP value	0.911**	0.898**	01	0.817**
DPPH scavenging effect	0.755**	0.702**	0.817**	1

** *Significant correlation* ($p<0.01$).

of total phenolic content, FRAP value, and DPPH· scavenging effect were higher than that of flavonoids content and the antioxidant capacity detected by the two methods. Therefore, the flavonoids content in the pear pulp presented a high correlation with the antioxidant capacity, and it was confirmed that the phenolic compounds were the main resources of antioxidant capacity in pear pulp.

GENETIC BACKGROUND DETERMINES THE ANTIOXIDANT POTENTIAL OF FRUIT

With the development of modern nutriology, more and more people recognize that phenolic compounds contribute greatly to human health, especially emphasizing that the aim of fruit intake is to gain phenolic compounds. Polyphenols are known to have strong antioxidant potential and there is a significant correlation between phenolic concentration and free radical scavenging activity. Kelly *et al.* (2003) reported that phenolic compounds in the apple pomace can inhibit the growth of cancer cells. In citrus fruit, flavedo extracts mainly containing total phenolic, flavonoids, and Vitamin C exhibit good DNA protection abilities (Deena *et al.*, 2010). The fruit of the strawberry tree (*Arbutus unego* L.) are thought to be a very good source of antioxidants precisely because they have a high flavonoids content (32.37 mg/100 g edible portion) (Pallauf *et al.*, 2008). Hanna *et al.* (2002) found phenolic compounds in pears improved lipid metabolism and increase the plasma antioxidant capacity in rats fed with added cholesterol.

There are many factors that could influence the antioxidant components and capacity such as varieties (cultivar/subspecies) and agronomic and environmental factors (climate, soils, and light exposure) (Emmy *et al.*, 2009). Genetic background (species and cultivars) also plays an important role in determining the antioxidant potential of fruit. It was proved that the high antioxidant capacity of strawberries is strongly influenced by species or cultivar (Jessica *et al.*, 2005) (Table 17.2). Even the effect of the

TABLE 17.2 TEAC Values Obtained by Hydrophilic, Lipophilic, and Total Components of Different Genotypes of Strawberries*

		TEAC		
Species	Variety	Hydrophilic	Lipophilic	Total
Cultivated strawberry	Don	9.76±1.57[c]	0.82±0.25[a]	10.58±1.81[c]
	Idea	11.06±0.79[c]	0.84±0.27[a]	11.9±1.04[c]
	Camarosa	11.49±0.56[bc]	1.02±0.17[a]	12.52±0.44[bc]
	Onda	12.02±0.23[bc]	0.78±0.33[a]	13.81±0.43[bc]
	Sveva	13.56±0.64[ab]	1.17±0.42[a]	14.73±0.91[ab]
	Patty	14.58±0.37[a]	1.43±0.06[a]	16.03±0.36[a]
Apple	Fuji C	0.79±0.14[c]	0.10±0.01[a]	0.89±0.14[d]
	Golden Delicious	0.84±0.09[c]	0.09±0.04[a]	0.94±0.09[cd]
	Rosa Annurca	1.41±0.22b[c]	0.11±0.02[a]	1.52±0.24[bc]
	Mela rasa	1.95±0.40[ab]	0.14±0.01[a]	2.09±0.41[ab]
	AB89.4.1.211	2.45±0.22[a]	0.13±0.02[a]	2.58±0.24[a]

* *Values are mean±standard error of the mean. Values with the same superscript letter are not statistically different at the 5% level. For each TEAC contribution, analysis refers to differences in genotype per species.*

TEAC, antioxidant capacity measured as micromoles of 6-hydroxy-2,5,7,8-tetramethylchromane-2-carboxylic acid equivalent per gram FW.

TABLE 17.3 TEAC Values Obtained by Hydrophilic, Lipophilic, and Total Components of Different Genotypes of the Same Apricot and Peach Varieties Grafted with Five Different Rootstocks*

		TEAC		
Species	**Variety**	**Hydrophilic**	**Lipophilic**	**Total**
Apricot	Citation	0.77 ± 0.08^{c}	0.09 ± 0.01^{a}	0.86 ± 0.08^{c}
	Julior	1.10 ± 0.08^{b}	0.19 ± 0.02^{a}	1.29 ± 0.09^{b}
	Isthara	1.42 ± 0.15^{a}	0.13 ± 0.02^{a}	1.55 ± 0.16^{ab}
	Mir 29C	1.42 ± 0.13^{a}	0.21 ± 0.02^{a}	1.63 ± 0.13^{a}
	Rubira	1.41 ± 0.16^{a}	0.22 ± 0.02^{a}	1.62 ± 0.16^{a}
Peach	Barrier 1	0.81 ± 0.05^{d}	0.10 ± 0.01^{a}	0.91 ± 0.05^{d}
	Citation	0.92 ± 0.06^{cd}	0.16 ± 0.01^{a}	1.08 ± 0.06^{c}
	Isthara	0.96 ± 0.04^{c}	0.21 ± 0.01^{a}	1.18 ± 0.05^{c}
	GF 677	1.12 ± 0.05^{b}	0.21 ± 0.01^{a}	1.33 ± 0.04^{b}
	Julior	1.34 ± 0.06^{a}	0.26 ± 0.01^{a}	1.60 ± 0.06^{a}

* *Values are mean ± standard error of the mean. Values with the same superscript letter are not statistically different at the 5% level. For each TEAC contribution, analysis refers to differences in genotype per specie.*

TEAC, antioxidant capacity measured as micromoles of 6-hydroxy-2,5,7,8-tetramethylchromane-2-carboxylic acid equivalent per gram FW.

rootstock was confirmed in changing the total antioxidant capacity of fruit (Jessica *et al.*, 2005) (Table 17.3). The contents of phenols and flavonoids in pear fruit vary considerably among different cultivars, thus more choices were provided for pear utilization. Considering the fact that the antioxidant power was significantly correlated with the total phenolic content and the flavonoids content, as well as the positive correlation between fruit browning and total phenolic content, more selections can be ensured for pear production and commercialization as an antioxidant functional food and a source for low-browning by-product.

Acknowledgments

This research was supported by the Earmarked Fund for China Agriculture Research System (CARS-29-20) and the Research Fund for Projects on science and technology studies of Hebei Academy of Agricultural and Forestry Sciences (A2012110302).

References

Alothman, M., Rajeev, B., Karim, A.A., 2009. Antioxidant capacity and phenolic content of selected tropical fruits from Malaysia, extracted with different solvents. Food Chem. 115, 785–788.

Bao, J.S., Cai, Y.Z., Sun, M., Wang, G.Y., Harold, C., 2005. Anthocyanins, flavonols and free radical scavenging activity of Chinese Bayberry (*Myrica rubra*) extracts and their color properties and stability. J. Agric. Food Chem. 53, 2327–2332.

Bartosz, G., 1997. Oxidative stress in plants. Acta. Physiol. Plant 19, 47–64.

Campanella, L., Bonanni, A., Favero, G., TomassettiO, M., 2003. Determination of antioxidant properties of aromatic herbs, olives and fresh fruit using an enzymatic sensor. Anal. Bioanal. Chem. 375, 1011–1016.

Carbonaro, M., Mattera, M., 2001. Polyphenoloxidase activity and polyphenol levels in organically and conventionally grown peach (*Prunus persica* L., cv. Regina bianca) and pear (*Pyrus communis* L., cv. Williams). Food Chem. 72, 419–424.

Deena, R., Theeshan, B., Emmanuel, B., Evelyne, T., Okezie, I.A., 2010. Bioactive phenolics and antioxidant propensity of flavedo extracts of Mauritian citrus fruits: Potential prophylactic ingredients for functional foods application. Toxicology 278, 75–87.

Deighton, N., Brennan, R., Finn, C., Davies, H.V., 2000. Antioxidant properties of domesticated and wild Rubus species. J. Sci. Food Agric. 80, 1307–1313.

Du, G.R., Li, M.J., Ma, F.W., Liang, D., 2009. Antioxidant capacity and the relationship with polyphenol and Vitamin C in *Actinidia* fruits. Food Chem. 113, 557–562.

Emmy, H.,K.,I., Khoo, H.,E., Abbe, M.,M.,J., Amin, I.,S.,I., Azrina, A., Halimatul, S.,M.,N., Norzatol, A.,M.,D., Ruzaidi, A.,M.,M., 2009. Antioxidant capacity and total phenolic content of Malaysian underutilized fruits. J. Food Composition Anal. 22, 388–393.

García-Alonso, M., Pascual-Teresa, S., Santos-Buelga, C., Rivas-Gozalo, J.C., 2004. Evaluation of the antioxidant properties of fruits. Food Chem. 84, 13–18.

Gaziano, J.M., Manson, J.E., 1996. Diet and heart disease: the role of fat, alcohol and antioxidants. Cardiol. Clin. 14, 69–83.

Guo, C.J., Yang, J.J., Wei, J.Y., Li, Y.F., Xu, J., Jiang, Y.G., 2003. Antioxidant activities of peel, pulp and seed fractions of common fruits as determined by FRAP assay. Nutr. Res. 23, 1719–1726.

Hanna, L., Shela, G., Antonin, L., 2002. Comparative contents of some bioactive compounds in apples, peaches and pears and their influence on lipids and antioxidant capacity in rats. J. Nutr. Biochem. 13 (10), 603–610.

Hertog, M.G., Feskens, E.J., Hollman, P.C., Katan, M.B., 1993. Dietary antioxidant flavonoids and risk of coronary heart disease: the Zutphen Elderly Study. The Lancet. 342, 1007–1011.

Höner, K., Cervellati, R., 2002. Measurements of the antioxidant capacity of fruits and vegetables using the BR reaction method. Eur. Food Res. Technol. 215, 437–442.

Ilkay, K., Bulent, K., 2009. Antioxidant properties of blackberry and blueberry fruits grown in the Black Sea Region of Turkey. Sci. Hortic. 121, 447–450.

Jessica, S., Alessia, P., Nicoletta, P., Bruno, M., Maurizio, B., 2005. Plant genotype affects total antioxidant capacity and phenolic contents in fruit. Nutrition 21, 207–213.

Jia, Z., Tang, M., Wu, J., 1999. The determination of flavonoid contents in mulberry and their scavenging effects on superoxide radicals. Food Chem. 64, 555–559.

Kahkonen, M.P., Hopia, A.I., Vuorela, H.J., Raucha, J.P., Pihlaja, K., Kujala, T.S., Heinonen, M., 1999. Antioxidant activity of plant extracts containing phenolic compounds. J. Agric. Food Chem. 47, 3954–3962.

Kalt, W., Forney, C.F., Martin, A., Prior, R., 1999. Antioxidant capacity, vitamin C, phenolics and anthocyanins after fresh storage of small fruits. J. Agric. Food Chem. 47, 4638–4644.

Kelly, W., Wu, X.Z., Liu, R.H., 2003. Antioxidant activity of apple peels. J. Agric. Food Chem. 51, 609–614.

Lise, S., Suresh, I.S.R., Brian, F.C.C., 1994. Testing garlic for possible anti-ageing effects on long-term growth characteristics, morphology and macromolecular synthesis of human fibroblasts in culture. J. Ethnopharmacol. 43, 125–133.

Pallauf, K., Rivas-Gonzalo, J.C., Castillo, del M.D., Cano, M.P., Pascual-Teresa, de S., 2008. Characterization of the antioxidant composition of strawberry tree (*Arbutus unedo* L.) fruits. J. Food Composition Anal. 21, 273–281.

Serafini, M., Laranjinha, J.A.N., Almeida, L.M., Maiani, G., 2000. Inhibition of human LDL lipid peroxidation by phenol-rich beverages and their impact on plasma total antioxidant capacity in humans. J. Nutr. Biochem. 11, 585–590.

Silvina, B.L., Balz, F., 2006. Consumption of flavonoid-rich foods and increased plasma antioxidant capacity in humans: Cause, consequence, or epiphenomenon? Free. Radic. Biol. Med. 41, 1727–1746.

Steinmetz, K.A., Potter, J.D., 1996. Vegetables, fruit, and cancer prevention: a review. J. Am. Diet. Assoc. 96, 1027–1039.

Thomas, W.K., Nobunao, W., Shyam, B., 2007. Cell survival responses to environmental stresses via the Keap1-Nrf2-ARE pathway. Annu. Rev. Pharmacol. Toxicol. 47, 89–116.

Triantis, T., Stelakis, A., Dimotikall, D., Papadopoulos, K., 2005. Investigations on the antioxidant activity of fruit and vegetable aqueous extracts on superoxide radical anion using chemiluminescence techniques. Anal. Chim. Acta. 536, 101–105.

Wang, S.Y., Lin, H.S., 2000. Antioxidant activity in fruits and leaves of blackberry, raspberry, and strawberry varies with cultivar and developmental stage. J. Agric. Food. Chem. 48, 140–146.

CHAPTER

18

Freezing of Fruits and Impact on Anthocyanins

Shyam S. Sablani

Biological Systems Engineering Department, Washington State University, Pullman, USA

CHAPTER POINTS

- The maturity level and cultivar of fruit significantly affect the total anthocyanin content and total antioxidant activity.
- In comparison to fresh fruit, the anthocyanin content either increased or did not change significantly after the freezing process.
- Different measurement methods yield slightly different results; there are 3–9% differences between total anthocyanin contents obtained with the pH-differential and HPLC methods.
- Freezing and frozen storage allow a better extraction of total anthocyanins.
- During the initial 4–6 months of frozen storage, there is no significant difference or a slight decrease in content of total anthocyanins, after which the rate of degradation increases.
- Long-term frozen storage has a significant impact on the total anthocyanins and total antioxidant capacity of fruits, depending on the fruit variety and storage time.
- Anthocyanins make a greater contribution to the antioxidant activity of fruits than other phenolic compounds.

INTRODUCTION

Anthocyanins are polyphenolic compounds present in a variety of fruits, responsible for their attractive color. Anthocyanins are glycosides of polyhydroxy and polymethoxy derivatives of 2-phenylbenzopyrylium salt. Concentrations and types of anthocyanins in fruits vary significantly (Table 18.1). Some factors, such as the types and cultivars of fruits (food matrix), as well as growth conditions, weather at the growing site, maturity, material preparation, and analysis methods could create differences in total anthocyanins in fruits. Growth conditions and environmental stress, such as high exposure to UV light and temperature, are important factors influencing the levels of bioactive content as a result of plant defense response (Leong and Oey, 2012). Maturity level also significantly contributes to the total anthocyanin content of fruits.

The anthocyanin content in fruits is at its highest during the ripening stage, in which the biosynthesis rate is accelerated due to the action of the ripening hormone (ethylene), triggering the activation of many enzymes involved in anthocyanin biosynthesis, and eventually declining at the end of maturation stage. Since anthocyanins are synthesized at an increasing rate during maturation, the total anthocyanin content quantified here may serve as the index of maturity and an important quality parameter. The level of total anthocyanins also depends upon the cultivar. De Ancos *et al.* (2000) found higher total anthocyanin contents in the late red raspberry cultivars, Zeva (116 mg/g of fruit) and Rubi (96.08 mg/g of fruit) compared to the early cultivars, Heritage and Autumn Bliss (31.13 and 37.04 mg/g of fruit, respectively), which showed less than half of the late cultivars' concentration. The anthocyanins are significant because of their nutraceutical benefits, antioxidant, and anticarcinogenic properties. Anthocyanins can also be used as natural food colorants in the food industry. However, anthocyanins are labile in nature, and therefore are susceptible to deterioration during processing and storage (Syamaladevi *et al.*, 2011).

Freezing is one of the most common methods of preservation of fruits for long-term storage. Frozen fruits are used as ingredients in many food formulations such as jams, jellies, sauces, purees, toppings, syrups, juice concentrates, as well as bakery and dairy products. The freezing process and frozen storage may change the anthocyanin content of fruits, thereby affecting the antioxidant capacity and possible health benefits of the fruit.

Processing and Impact on Active Components in Food
http://dx.doi.org/10.1016/B978-0-12-404699-3.00018-4

TABLE 18.1 Effect of Frozen Storage Temperature on Stability of Total Anthocyanins

Fruit	Initial Value	Storage Conditions	Comments	Reference
Cherry	46.1–547 mg/g DW	Just after freezing −20°C	48% increase	Leong and Oey (2012)
Nectar	10.9–17.1 mg/g DW		22% increase	
Peach	7.7–16.2 mg/g DW		69% increase	
Plum			108% increase	
Pomegranate		334 days at −23°C		Turfani *et al.* (2012)
Unclarified juice	1091* and 1005** mg/kg FW		7% decrease	
Clarified juice	863* and 839** mg/kg FW		No loss	
Strawberry		24 h at −20°C thawing at 20°C/8 h		Holzwarth *et al.* (2012)
Senga Sengana	559 mg/100 g DW		7% decrease	
Candonga	333 mg/100 g DW		19% decrease	
Sabrosa	325 mg/100 g DW		14% decrease	
Black carrot juice	3747* and 4123** mg/kg juice	319 days at −23°C	0* and 2%** decrease	Turkyılmaz and Ozkan (2012)
Date		1 month at −20°C		Allaith *et al.* (2012)
Khalas	1.17 mg/100 g FW		69% decrease	
Khunaizi	1.27 mg/100 g FW		47% decrease	
Red raspberry	0.78 mg/g DW	378 days		Syamaladevi *et al.* (2011)
		at −20°C	21% increase	
		at −35°C	29% increase	
		at −80°C	16% decrease	
Serviceberry	53.2 mg/100 g FW	10 months at −20°C	13% increase	Michalczyk and Macura (2010)
Blackberry *Tupy*	141 mg/100 g fruit	6 months at −10°C	57% decrease	Jacques *et al.*, (2010)
		6 months at −18°C	50% decrease	
		6 months at −80°C	31% decrease	
Strawberry	31.0 mg/100 g FW	10 months at −18°C	25% decrease	Poiana *et al.* (2010a)
Sweet cherry	49.9 mg/100 g FW		15% decrease	
Sour cherry	93.2 mg/100 g FW		12% decrease	
Blueberry	206 mg/100 g FW	10 months at −18°C	13% decrease	Poiana *et al.* (2010b)
Red raspberry	39.7 mg/100 g FW		15% decrease	
Blackberry	194 mg/100 g FW		8% decrease	
Blackberry		12 months at −18°C		Kopjar *et al.* (2009)
Thornfree	1306 mg/l puree #		31% decrease	
Cacanska bestrna	1201 mg/l puree #		30% decrease	
Strawberry		6 months at −20°C		Oszmianski *et al.* (2009)
Elkat	682 mg/kg FW		7% increase	
Kent	299.4 mg/kg FW		10% decrease	
Myrtle berries	2307 mg/l	12 months at −20°C	18% decrease	Tuberoso *et al.* (2008)

TABLE 18.1 Effect of Frozen Storage Temperature on Stability of Total Anthocyanins—cont'd

Fruit	Initial Value	Storage Conditions	Comments	Reference
Blackberry	248 mg/100 g FW	6 months at −20°C	No significant change	Hager *et al.* (2008a)
Black raspberry	1113 mg/100 g FW	6 months at −20°C	No significant change	Hager *et al.* (2008b)
Blackcurrant	0.9 g/100 g FW	Collected from supermarket	28% lower than fresh	Hollands *et al.* (2008)
Blueberry *Tifblue* (extract)	115 (33.6) mg/100 g FW	2 months at −20°C	(26% decrease)	Srivastava *et al.* (2007)
Powderblue (extract)	121 (36.1) mg/100 g FW		(25% decrease)	
Strawberry (six different genotypes)	37.1–122.3 mg/100 g FW	1 month at −23°C	8% decrease (Totem genotype)	Ngo *et al.* (2007)
Red Raspberry *Heritage*	33.0 mg/100 g FW	12 months at −18°C	No significant change	Sousa *et al.* (2005)
Blueberry	7.2 mg/g DW	3 months at −20°C	No significant change	Lohachoompol *et al.* (2004)
Red raspberry *Ample*	770 nmol/g FW	After freezing at −80°C	No significant change	Mullen *et al.* (2002)
Red raspberry		12 months at −20°C		De Ancos *et al.* (2000)
Autumn Bliss	31.1 mg/100 g FW		5% increase	
Heritage	37.0 mg/100 g FW		17% increase	
Zeva	116.3 mg/100 g FW		18% decrease	
Rubi	96.1 mg/100 g FW		4% decrease	
Strawberry	n.a.	12 months at −20°C	23% decrease	Garcia-Viguera *et al.* (1999)
Strawberry	130 mg/g FW	2 months at 20°C	42% decrease	Larsen and Poll (1995)
Strawberry	n.a	6 months −20°C to −80°C		Deng and Ueda (1993)
Hoko-wase			57–67% decrease	
Toyonaka			73–80% decrease	
Nyoho			66–71% decrease	
Sour cherry		Season 1: 12 months at −20°C	150–350% increase depending upon variety 30–45% decrease depending upon the variety	Urbanyi and Horti (1992)
Fardi	541, 1321 mg/1000 g dry matter			
Pandi	1101, 1039 mg/1000 g dry matter			
Ujfehertoi	793, 1413 mg/1000 g dry matter			

* *pH differential method.*
** *HPLC,*
#*5 weeks of storage.*

Therefore, it is vital to understand the stability of anthocyanins during frozen storage. For frozen fruits, the retention of anthocyanins depends on the freezing rate, composition, pH, cultivar, temperature, and the presence/absence of oxygen (Wrolstad *et al.*, 1970; Mazza and Miniati, 1993).

The freezing of fruits involves the conversion of water into ice, which includes two successive processes: the formation of ice crystals (*nucleation*) and the subsequent increase in crystal size (*growth*) (Zaritzky, 2012). Figure 18.1 shows a typical plot of time–temperatures for the freezing of pure water (Figure 18.1A) and food (Figure 18.1B). The cooling of pure water involves the removal of sensible heat (point 1 to point 2), followed by the removal of latent heat at the freezing point the water is converted to ice (point 3). Nucleation is necessary to initiate freezing, and the temperature can fall below 0°C without the formation of ice crystals. Point 2 indicates the supercooling of water before crystallization begins. The heat of solidification is liberated after

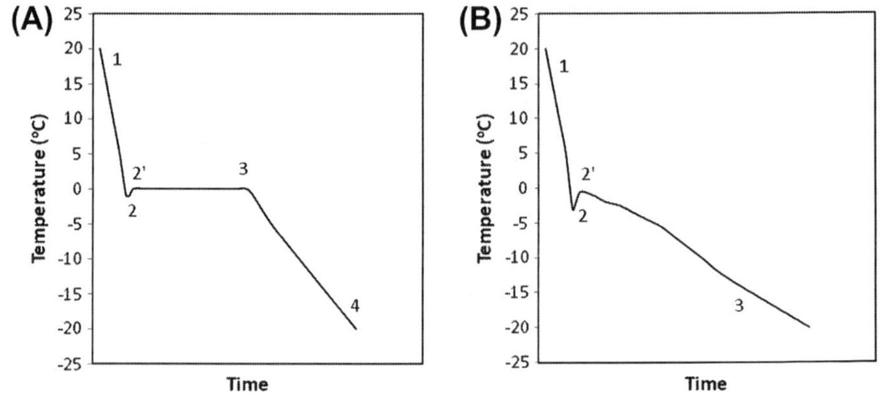

FIGURE 18.1 Typical plot of time–temperatures for the freezing of (A) pure water and (B) food.

initial supercooling, thus increasing the temperature from point 2 to point 2′ (Figure 18.1A), which represents the onset of ice crystallization. Further removal of heat reduces the temperature to point 4. The freezing of foods is more complex than the freezing of pure water. The initial cooling behavior from point 1 to point 2 and point 2′ is similar to that of pure water. However, point 2′, which represents the initial freezing point of the solution, is lower than that of pure water. Further cooling from point 2′ to point 3 represents the growth of ice crystals and ice formation. As freezing continues from point 2′ to point 3, water is separated in the form of ice, which increases the solute concentration and further suppresses the freezing point (Zaritzky, 2012).

This chapter examines the influence of the freezing process and frozen storage on the stability of anthocyanins and antioxidant capacity in fruits, and also explores other factors that affect the stability of anthocyanins in fruits. An understanding of these processes is integral to the development of food processing techniques that preserve important nutrients and aesthetic qualities, maximizing benefits to the consumer.

MECHANISMS OF ANTHOCYANIN DEGRADATION

The mechanisms of anthocyanin degradation during processing and storage are well-known. Anthocyanins are unique among the flavonoids because they carry a positive charge associated with the C-ring in the flavylium ion form (Hollands *et al.*, 2008). In most plant tissues, anthocyanins are in the intensely colored flavylium form, often due to copigmentation which enhances the anthocyanin color and stability. However, when the plant cells are ruptured and anthocyanins are exposed to a higher pH (near neutral), they can form a carbinol pseudo-base, quinoidal-base, or a chalcone; this process is associated with a loss of color. The rate of color is dependent on the pH and temperature, light conditions, the presence/concentrations of metal ions, oxygen, ascorbic acid, enzymes, and anthocyanin concentration. In addition, anthocyanin instability can lead to the formation of polymeric forms, which is associated with a change in color to a browner and less desirable shade (Mazza and Miniati, 1993).

Attempts have been made to explain the stability of anthocyanins in frozen fruits using the molecular mobility concept. At sufficiently low temperatures, fruit products form a maximum-freeze-concentrated matrix which is described by two temperatures, i.e., the glass transition temperatures of the maximum-freeze-concentrated matrices (T_g'') and the onset of ice melting temperature (*Tm″*) (Syamaladevi *et al.*, 2011). The transition from the reversible liquid/rubber state to the glassy state starts in the fruit matrix, below the temperature corresponding to the onset of ice melting temperature (T_m'). According to the glass transition concept, foods are most stable in the glassy state, i.e., at temperatures below their glass transition temperature. Below T_g', viscosity becomes great enough to considerably slow down the rates of chemical reactions. The physical and chemical degradation reactions of frozen food systems may be related to molecular mobility, and thus the T_g' (Torreggiani *et al.*, 1999; Syamaladevi *et al.*, 2011). However, a recent study by Syamaladevi *et al.* (2011) found that storage temperature does not significantly influence the degradation of anthocyanins in red raspberries. Rizzolo *et al* (2003) report no significant difference in total anthocyanin content of blueberry juices after 6 months of storage when frozen at −10, −20 and −30°C. Others found that the glass transition and storage temperatures had no effect on the degradation of anthocyanins during frozen storage of blueberry juice, with or without added sugars (Rizzolo *et al.*, 2003). Torreggiani *et al.* (1999) reported a significant loss of strawberry anthocyanins at −10°C during 4 months of storage, but no direct relationship between anthocyanin loss and T_g'. Several studies also suggest no evidence that the degradation of anthocyanins in frozen raspberries is diffusion-limited or dependent on molecular mobility (Syamaladevi *et al.*, 2011).

EFFECT OF FREEZING ON ANTHOCYANINS

Studies show either no major difference or an increase in the concentration of anthocyanins immediately after freezing compared to fresh fruits (Table 18.1: Leong and Oey, 2012; Syamaladevi *et al.*, 2011; De Ancos *et al.*, 2000). Leong and Oey (2012) found a 22–108% increase in total anthocyanins after cherries, nectarines, peaches, and plums were frozen. This increase was dependent on the fruit matrix, with a 22% increase for nectarines and a 108% increase for plums. For other fruits, this change is much smaller. For example, Poiana *et al.* (2010a) found a slight increase in total anthocyanins in sweet cherries and strawberries immediately after freezing (1.6–2.9%), while Syamaladevi *et al.* (2011) and De Ancos *et al.* (2000) observed no significant change in total anthocyanins of red raspberries after freezing. The greater quantity of anthocyanins in fruits just after freezing has been attributed to a higher extraction efficiency of anthocyanins from frozen fruits compared to that of fresh fruits, due to cellular disruption during freezing and thawing.

The rate of freezing and the type of freezing technique may also influence the stability of anthocyanins in fruits. Freezing induces the formation of ice crystals, which favors localized concentrations of solutes such as anthocyanins, as well as the reallocation of water molecules in the cell structure. However, cell damage from the growth of ice crystals due to temperature fluctuation and turgor loss often leads to the softening of fruit texture. The rate of freezing has been found to influence ice crystal formation that expands the separation between cells in the fruit structure. When fruits are rapidly frozen, smaller ice crystals form, which reduce cell structure disruption, while fruits that are frozen slowly form large intercellular ice crystals that cause more damage (Poina *et al.*, 2010a). Studies examining the influence of different freezing and thawing procedures on anthocyanin retention in fruits found that the effects of differing freezing technologies is minor; however, different thawing regimes significantly affect anthocyanin retention after thawing. Holzwarth *et al.* (2012) observed higher retention of anthocyanins in strawberries when they were thawed at 20°C in a microwave oven compared to thawing at 4°C and 37°C.

De Ancos *et al.* (2000) observed that the two early cultivars of red raspberries, Heritage and Autumn Bliss, which have low total anthocyanin and cyanidin 3-glucoside concentrations, show no degradation after freezing and, indeed, a better extraction of total anthocyanin content due to cellular disruption caused by the freezing process. Rubi and Zeva, the two late cultivars of red raspberry with high total anthocyanin and cyanidin 3-glucoside concentrations, showed a more evident degradation of total anthocyanins caused by the freezing process. This degradative effect could be due to the high content of the more reactive anthocyanin compound cyanidin 3-glucoside, or perhaps cellular disruption caused by the freezing process, producing a release of the oxidoreductase enzymatic system (PPO). De Ancos *et al.* (2000) report that their previous work showed more polyphenol oxidase (PPO) enzyme activity in Rubi and Autumn Bliss raspberry tissues than in Zeva and Heritage tissues. Therefore, the degradative enzymatic reactions could be one of the main reasons for the total anthocyanin concentration losses in Rubi but not in Zeva. Physicochemical characteristic differences have also been found between early and late cultivars: Rubi and Zeva show lower pH and higher °Brix values than the early cultivars, Heritage and Autumn Bliss. These results suggest that the stability of anthocyanins during freezing mainly depends on the pH value, organic acid content, sugar concentration, initial concentration, and initial cyanidin 3-glucoside content.

EFFECT OF FROZEN STORAGE ON ANTHOCYANINS

Frozen storage affects the stability of total anthocyanin content of fruits in different ways (Table 18.1). Several studies show a decreasing trend in total fruit anthocyanins during frozen storage, e.g., in unclarified pomegranate juice (Turfani *et al.*, 2012); strawberries (Holzwarth *et al.*, 2012, Poiana *et al.*, 2010a; Oszmianski *et al.*, 2009; Ngo *et al.*, 2007; Garcia-Viguera *et al.*, 1999; Larsen and Poll, 1995; Deng and Ueda, 1993); black carrot juice (Turkyimaz and Ozkan, 2012); dates (Allaith *et al.*, 2012); blackberries (Jacques *et al.*, 2010; Poiana *et al.*, 2010b; Kopjar *et al.*, 2009); sweet cherries (Poiana *et al.*, 2010a); blueberries (Poiana *et al.*, 2010b; Srivastava *et al.*, 2007); red raspberries (Poiana *et al.*, 2010b; De Ancos *et al.*, 2000); Myrtle berries (Tuberosco *et al.*, 2008); blackcurrants (Hollands *et al.*, 2008); and sour cherries (Urbanyi and Horti, 1992).

A few studies found an increase in total anthocyanins during frozen storage, including red raspberries (Syamaladevi *et al.*, 2011; De Ancos *et al.*, 2000); serviceberries (Michalczyk and Macura, 2010); strawberries (Oszmianski *et al.*, 2009); sour cherries (Urbanyi and Horti, 1992). This has been attributed to a concentration effect from moisture loss or enhanced extraction of anthocyanins due to tissue softening (Hager *et al.*, 2008b). Some studies show no significant change in total anthocyanins during frozen storage of fruits, e.g., clarified pomegranate juice (Turfani *et al.*, 2012), black carrot juice (Turkyimaz and Ozkan, 2012), blackberries (Hager *et al.*, 2008a), black raspberries (Hager *et al.*, 2008b), red raspberries (Sousa *et al.*, 2005; Mullen *et al.*, 2002), and blueberries (Lohachoompol *et al.*, 2004).

It is not feasible to conduct a complete comparison between studies, since the fruit variety, maturity level, initial pH, storage temperature, and time differ. De Ancos *et al.* (2000) observed that frozen storage affects total anthocyanin content and individual anthocyanin distribution in different ways, depending on the cultivar. Total anthocyanins in Heritage and Autumn Bliss raspberries were found to increase slightly, at 17% and 5%, respectively, after frozen storage for 360 days. The relative percentage of individual anthocyanins in the early raspberry cultivars, Heritage and Autumn Bliss, showed small changes during frozen storage, apparently due to increased extraction of the main pigments, cyanidin 3-sophoroside and cyanidin 3-rutinoside, during frozen storage. The total value of anthocyanins in late cultivars was significantly decreased (4–17%) in both Rubi and Zeva cultivars during frozen storage. These results could be explained by the different chemical compositions found between early cultivars (Heritage and Autumn Bliss) and late cultivars (Rubi and Zeva). Raspberry cultivars with low pH, high soluble solids content (°Brix), and high total anthocyanin content retained the initial anthocyanin concentration better during processing.

Many studies have shown that long-term frozen storage has a significant impact on anthocyanins depending on the type of fruit. However, during the initial period of frozen storage (4 to 6 months), there is little or no significant decrease in total anthocyanins, with the rate of degradation increasing after this period (Poiana *et al.*, 2010a,b; Chaovanalikit and Wrolstad, 2004; Sahari *et al.*, 2004). Woodward *et al.* (2009) studied the effect of freeze–thaw cycles on anthocyanin stability. To determine the storage stability of various anthocyanins, acidified buffer solutions (10mMNa/K phosphate buffer in 2% HCl) were spiked with 150 μM delphinidin-3-glucoside, cyanidin-3-glucoside, pelargonidin-3-glucoside, delphinidin, cyanidin, or pelargonidin (individually). Samples were reanalyzed sequentially for six freeze–thaw cycles. One freeze–thaw cycle consisted of 24-h storage at −80°C, followed by a 20-min defrosting at ambient room temperature prior to HPLC analysis. Results showed that anthocyanin glucosides were stable during storage and the freeze–thaw treatment, with no significant losses. Of the three aglycone species, both pelargonidins and cyanidins showed a significant loss following freeze–thaw cycling, while delphinidins remained stable. Pelargonidins showed a linear rate of degradation, with significant losses demonstrated at four freeze–thaw cycles and a total reduction of 11% at six freeze–thaw cycles. Cyanidins also showed a linear rate of degradation, with a maximum reduction of 6% at six freeze–thaw cycles.

Urbanyi and Horti (1992) found that total anthocyanin content increased linearly during the first year of storage. In studies of shorter storage times in the second year, this change was reversed. They attributed this behavior to various phases of storage time and different processes taking place at different levels of fruit maturity. In the less ripe fruits/1st year, the biosynthesis of anthocyanins continue, according to the results found by several researchers. In the case of riper fruits, this goes much slower or does not occur at all.

OTHER FACTORS AFFECTING ANTHOCYANIN DEGRADATION

Home-scale freezers take a longer time to freeze fruits compared to commercial freezers hence anthocyanins susceptible to oxidation suffer damages (Poiana *et al.*, 2010a). This explains the higher loss of anthocyanins in domestic freezing of strawberries and cherries. Hollands *et al.* (2008) found that anthocyanin levels in thermally processed blackcurrant are extremely low compared to the high levels in frozen blackcurrant fruits. Adding sucrose to fruits during frozen storage has been shown to protect anthocyanins and also to retard browning and polymeric color formation (Wrolstad *et al.*, 1990). Kopjar *et al.* (2009) also observed a high retention of anthocyanin content in two varieties of blackberries, Thornfree and Cacanska bestrna, after 12 months of storage at −18°C when samples were coated with glucose. This effect was explained by the fact that the addition of sugar reduces water activity.

In the juice production process, compounds such as pectin, fibers, semi-fibers, starch, and proteins must be clarified, which may affect the stability of anthocyanins in juice during storage. Turfani *et al.* (2012) found that total anthocyanin content of pomegranate juice decreased substantially after clarification. The loss of total anthocyanins was more than 20% higher in clarified juice compared to unclarified juice. The apparent decrease in anthocyanin content is caused by the interaction between anthocyanin and gelatin-tannin flocks formed during clarification. Similarly, studies show an 18% decrease in anthocyanin contents of blackberry juices after clarification (Rommel *et al.*, 1992).

EFFECT OF FROZEN STORAGE ON ANTIOXIDANT CAPACITY

The health benefits of anthocyanins come from their antioxidant activity. Antioxidants are substances that, if present in low concentrations, significantly prevent the oxidation of a substrate. The human body produces reactive carbon, sulfur, nitrogen, and oxygen species as a result of interaction with ionizing radiation and physiological processing, the most damaging of which are the reactive oxygen species, such as superoxide, hydrogen peroxide, and hydroxyl radicals. Many studies of anthocyanin content changes in fruits after freezing and during frozen storage have also evaluated the total antioxidant capacity of fruits. The presence of other phenolic compounds such as flavonoids, phenolic acids, and vitamins C and E also contribute to the total antioxidant activity of fruits. However, Bof *et al.* (2012) found a higher correlation between total anthocyanins and antioxidant capacity in comparison to total phenolics and antioxidant capacity, concluding that anthocyanins make a greater contribution to antioxidant activity than other phenolic compounds. Michalczyk and Macura (2010) also observed that the antioxidant properties of serviceberries are affected mainly by total anthocyanins.

The total antioxidant capacity (TAC) of fruits can be affected by the freezing process and during frozen storage. Similar to total anthocyanins, frozen storage affects the total antioxidant capacity of fruits in different ways (Table 18.2). Several studies report a decrease in total antioxidant capacity during frozen storage, e.g., *Khunaizi* dates (Allaith *et al.*, 2012), guava (Bof *et al.*, 2012), strawberries (Bof *et al.*, 2012; Poiana *et al.*, 2010a), pears (Bof *et al.*, 2012), cherries (Poiana *et al.*, 2010a),

TABLE 18.2 Effect of Freezing and Frozen Storage on Antioxidant Activity

Fruit	Initial Value	Storage Conditions	Comments	Reference
Date	mmol/100g FW	1 month		Allaith *et al.* (2012)
Khalas	2.69	at −20°C	88% increase	
Khunaizi	3.48		24% decrease	
Pulp:	TEAC(μmol/g)	3 months		Bof *et al.* (2012)
Guava	27.0	at −15°C	26% reduction	
Grape	22.0		No significant change	
Fig	5.1		No significant change	
Strawberry	16.0		18% reduction	
Apple	11.1		No significant change	
Pear	7.3		45% reduction	
	mM Fe^{2+}/kg FW	10 months		Poiana *et al.* (2010a)
Strawberry	24.4	at −18°C	42% decrease	
Sweet cherry	13.5		39% decrease	
Sour cherry	43.1		35% decrease	
	mM Fe^{2+}/kg FW	10 months		Poiana *et al.* (2010b)
Blueberry	58.3	at −18°C	23% decrease	
Red raspberry	40.2		38% decrease	
Blackberry	49.6		35% decrease	
	g FW/g DPPH	10 months		Michalczyk and Macura
Serviceberry	21.0	at −20°C	25% decrease	(2010)
		10 days	No significant change	Polinati *et al.* (2010)
Apple	n.a	at −18°C		
Orange	n.a	at −70°C		
Blackberry	n.a.	12 months	8% decrease	Kopjar *et al.* (2009)
Thornfree		at −18°C	8% decrease	
Cacanska bestrna				
Blackberry	(μmol of TE/g of FW) 97.2	6 months at −20°C	No significant change	Hager *et al.* (2008a)
Black raspberry	μmol of TE/g of FW) 192	6 months at −18°C	18% increase	Hager *et al.* (2008b)
Blueberry	(μm/g FW)	2 months	(No significant change)	Srivastava *et al.* (2007)
Tifblue	26.1	at −20°C	(6% decrease)	
(extract)	(17.0)			
Powder blue	27.3			
(extract)	(17.5)			
Red raspberry *Ample*	(number of Fermy's radical × 10^{16} reduced) 406	After freezing at −80°C	No significant change	Mullen *et al.* (2002)

blueberries (Poiana *et al.*, 2010b; Srivastava *et al.*, 2007), blackberries (Poiana *et al.*, 2010b; Kopjar *et al.*, 2009), red raspberries (Poiana *et al.*, 2010b), and serviceberries (Michalczyk and Macura, 2010). This decrease in total antioxidant capacity ranged from 8% to 45% depending upon the fruit, storage temperature, and time. For example, Poina *et al.* (2010a) found that total antioxidant capacity of cherries and strawberries decreased during frozen storage. In the first 4 months of storage, there was a relatively small decrease in antioxidant capacity, followed by a significant decline in following months. At 10 months, antioxidant capacity had decreased by up to 35% for sour cherries, up to 38% for sweet cherries, and up to about 42% for strawberries. Also, at the end of 4 months of frozen storage, the loss of antioxidant activity of sour cherries ranged about 15% for sour cherries, about 19% for sweet cherries, and 23% for strawberries (Poina *et al.*, 2010a).

A few studies noticed no significant change in TAC during the frozen storage of fruits, e.g., grapes, figs, and apples (Bof *et al.*, 2012); apples and oranges (Polinati *et al.*, 2010); blackberries (Hager *et al.*, 2008a); blueberries (Srivastava *et al.*, 2007); and red raspberries (Mullen *et al.*, 2002). However, a few studies reported an increasing trend in TAC during frozen storage, e.g., *Khalas* dates (Allaith *et al.*, 2012) and black raspberries (Hager *et al.*, 2008b).

ANALYTICAL TECHNIQUES

The most common methods to quantify anthocyanin content of fruits are pH differential method/Spectrophotometric method and high pressure liquid chromatography (HPLC) (Table 18.3). The pH differential method is used to determine total anthocyanins, while

TABLE 18.3 Analytical Techniques Used to Quantify TA and TAC

Fruit	TA and TAC Measurement Methods	Reference
Guava, grape, fig, strawberry, apple, pear	TAC: TEAC	Bof *et al.* (2012)
Pomegranate	TA: pH differential and HPLC	Turfani *et al.* (2012)
Strawberry	TA: HPLC-DAD-MS	Holzwarth *et al.* (2012)
Black carrot juice	TA: pH differential method and HPLC-DAD-MS	Turkyılmaz and Ozkan (2012)
Date	TA: pH differential method TAC: FRAP and DPPH assays	Allaith *et al.* (2012)
Cherry, nectar, peach, plum	TA: pH differential method/ spectrophotometer (700 nm)	Leong and Oey (2012)
Red raspberry	TA: Spectrophotometric method	Syamaladevi *et al.* (2011)
Blackberry	TA: Spectrophotometric method	Jacques *et al.*, (2010)
Serviceberry	TA: pH differential methods TAA: DPPH assay	Michalczyk and Macura (2010)
Strawberry, sweet cherry, sour cherry	TA: Spectrophotometric methods (520 nm and 700 nm) TAC: FRAP	Poiana *et al.* (2010a)
Blueberry, Red raspberry, Blackberry	TA: Spectrophotometric methods (520 nm and 700 nm) TAC: FRAP	Poiana *et al.* (2010b)
Apple and orange	TAC: DPPH assay	Polinati *et al.* (2010)
Blackberry	TA: Spectrophotometric method TAC: DPPH	Kopjar *et al.* (2009)
Strawberry	TA: HPLC	Oszmianski *et al.* (2009)
Strawberry	TA: HPLC	Gössinger *et al.* (2009)
Myrtle berries	TA: Spectrophotometric method (520 nm)	Tuberoso *et al.* (2008)
Blackcurrant	TA: HPLC	Hollands *et al.* (2008)

TABLE 18.3 Analytical Techniques Used to Quantify TA and TAC—cont'd

Fruit	TA and TAC Measurement Methods	Reference
Blackberry	TAA: ORAC	Hager *et al.* (2008a)
Black raspberry	TA: HPLC TAC: ORAC	Hager *et al.* (2008b)
Strawberry	TA: Spectrophotometric method	Ngo *et al.* (2007)
Blueberry	TA: Spectrophotometric method TAC: TEAC	Srivastava *et al.* (2007)
Blueberry	TA: Spectrophotometric method	Lohachoompol *et al.* (2004)
Red raspberry	TA: Spectrophotometric method TAC: ESR spectra of Fermy's radical	Mullen *et al.* (2002)
Red raspberry	TA: Spectrophotometric method and HPLC	De Ancos *et al.* (2000)
Strawberry	TA: HPLC	Garcia-Viguera *et al.* (1999)
Strawberry	TA: pH differential method/ spectrophotometer	Larsen and Poll (1995)
Strawberry	TA: Spectrophotometric method	Deng and Ueda (1993)
Sour cherries	TA: Spectrophotometric method (530 nm)	Urbanyi and Horti (1992)
Strawberry	TA: Spectrophotometric method	Wrolstad *et al.* (1990)

the HPLC method is used to quantify individual anthocyanins. Although the pH differential method is fast and easy to perform, critics have noted that this method is not suitable for identifying the pattern of individual glycoside substitution of anthocyanins with sugar compounds. Other fruit substances such as pectin, proteins, lipids, and polyphenol compounds are known to interfere with anthocyanin measurements (Leong and Oey, 2012). Turfani *et al.* (2012) determined the total anthocyanin content of pomegranate juice with both pH differential (spectrophotometric) and HPLC methods. They found differences between total anthocyanin contents obtained by both methods; however, there was high correlation between the amounts of anthocyanins found in pomegranate juice with both methods. Similarly, Lee *et al.* (2002) found that total anthocyanin contents of blueberry juices obtained with the pH differential method differed from those obtained with HPLC. These differences are attributed to several factors: (i) different solvent systems utilized for HPLC and pH differential methods; (ii) different wavelengths, i.e., 520 and 512 nm utilized by HPLC and pH differential methods, respectively; and (iii) the interference of polymeric pigments during anthocyanin analyses. For example, polymeric pigments may be retained in the HPLC column and not included in HPLC measurements, whereas they may have contributed to the results from the pH differential method (Lee *et al.*, 2002). Turkyılmaz and Ozkan (2012) found that in comparison to the HPLC method, the spectrophotometric method determined a 9% lower value of total anthocyanins in black carrot juice, while Turfani *et al.* (2012) observed an opposite trend, with a 3–8% lower value of anthocyanins in pomegranate juice using the HPLC method.

References

Allaith, A.A., Ahmed, S.H., Jafer, F., 2012. Effect of different thermal treatments and freezing on the antioxidant constituents and activity of two Bahraini date cultivars (*Phoenix dactylifera L.*). Int. J. Food Sci. Technol. 47, 783–792.

Bof, C.M.J., Fontana, R.C., Piemolini-Barreto, L.T., Sandri, I.G., 2012. Effect of freezing and processing technologies on the antioxidant capacity of fruit pulp and jelly. Braz. Arch. Biol. Technol. 55 (1), 107–114.

Chaovanalikit, A., Wrolstad, R.E., 2004. Anthocyanin and polyphenolic composition of fresh and processed cherries. J. Food Sci. 69 (1), 73–83.

De Ancos, B., Ibanez, E., Reglero, G., Pilar Cano, M., 2000. Frozen storage effects on anthocyanins and volatile compounds of raspberry fruit. J. Agric. Food Chem. 48, 873–879.

Deng, H., Ueda, Y., 1993. Effects of freezing methods and storage temperature on flavor stability and ester contents of frozen strawberries. J. Japanese Soc. Horticultural Sci. 62 (3), 633–639.

Garcia-Viguera, C., Zafrilla, P., Tomas-Barberan, F.A., 1999. Influence of processing and storage conditions in strawberry jam color. Food Sci. Technol. Int. 5 (6), 487–492.

Gössinger, M., Moritz, S., Hermes, M., Wendelin, S., Scherbichler, H., Halbwirth, H., Stich, K., Berghofer, E., 2009. Effects of processing parameters on color stability of strawberry nectar from puree. J. Food Eng. 90, 171–178.

Hager, T.J., Howard, L.R., Prior, R.L., 2008a. Processing and storage effects on monomeric anthocyanins, percent polymeric color, and antioxidant capacity of processed blackberry products. J. Agric. Food Chem. 56, 689–695.

Hager, A., Howard, L.R., Prior, R.L., Brownmiller, C., 2008b. Anthocyanins, percent polymeric color, and antioxidant capacity of processed black raspberry products. J. Food Sci. 73 (6), H134–H140.

Hollands, W., Brett a, G.M., Radreau, P., Saha, S., Teucher, B., Bennett, R.N., Kroon, P.A., 2008. Processing blackcurrants dramatically reduces the content and does not enhance the urinary yield of anthocyanins in human subjects. Food Chem. 108, 869–878.

Holzwarth, M., Korhummel, S., Carle, R., Kammerer, D.R., 2012. Evaluation of the effects of different freezing and thawing methods on color, polyphenol and ascorbic acid retention in strawberries (Fragaria×ananassa Duch.). Food Res. Int. 48, 241–248.

Jacques, A.C., Pertuzatti, P.B., Barcia, M.T., Zambiazi, R.C., Chim, J.F., 2010. Stability of bioactive compounds in frozen pulp of blackberry (*Rubus fructicosus*) cv. Tupy, Quim. Nova 33 (8), 1720–1725.

Kopjar, M., Tiban, N.N., Pilizota, V., Babic, J., 2009. Stability of anthocyanins, phenols and free radical scavenging activity through sugar addition during frozen storage of blackberries. J. Food process. Preservation 33, 1–11.

Larsen, M., Poll, L., 1995. Changes in the composition of aromatic compounds and other quality parameters of strawberries during freezing and thawing. Z. Lebensm. Unters. Forsch. 201, 275–277.

Lee, J., Durst, R.W., Wrolstad, R.E., 2002. Impact of juice processing on blueberry anthocyanins and polyphenolics: Comparison of two pretreatments. J. Food Sci. 67, 1660–1667.

Leong, S.Y., Oey, I., 2012. Effects of processing on anthocyanins, carotenoids and vitamin C in summer fruits and vegetables. Food Chem. 133, 1577–1587.

Lohachoompol, V., Srzednicki, G., Craske, J., 2004. The change of total anthocyanins in blueberries and their antioxidant effect after drying and freezing. J. Biomed. Biotechnol. 5, 248–252.

Mazza, G., Miniati, E., 1993. Introduction. In: Mazza, G. (Ed.), Anthocyanins in Fruits, Vegetables and Grains. CRC Press, Boca Raton, pp. 85–87.

Michalczyk, M., Macura, R., 2010. Effect of processing and storage on the antioxidant activity of frozen and pasteurized shadblow serviceberry (*Amelanchier canadensis*). Int. J. Food Properties 13, 1225–1233.

Mullen, W., Stewart, A.J., Lean, M.E.J., Gardner, P., Duthie, G.G., Crozier, A., 2002. Effect of freezing and storage on the phenolics, ellagitannins, flavonoids, and antioxidant capacity of red raspberries. J. Agric. Food Chem. 50, 5197–5201.

Ngo, T., Wrolstad, R.E., Zhao, Y., 2007. Color quality of Oregon strawberries— impact of genotype, composition, and processing. J. Food Sci. 72 (1), c25–c32.

Oszmianski, J., Wojdylo, A., Kolniak, J., 2009. Effect of L–ascorbic acid, sugar, pectin and freeze–thaw treatment on polyphenol content of frozen strawberries. LWT- Food Sci Technol. 42, 581–586.

Poiana, M.A., Moigradean, D., Alexa, E., 2010a. Influence of home-scale freezing and storage of antioxidant properties and color quality of different garden fruits. Bulgarian J. Agri. Sci. 16 (2), 163–171.

Poiana, M.A., Moigradean, D., Raba, D., Alda, L.-M., Popa, M., 2010b. The effect of long-term frozen storage on the nutraceutical compounds, antioxidant properties and color indices of different kinds of berries. J. Food, Agri. Environ. 8 (1), 54–58.

Polinati, R.M., Faller, A.L.K., Fialho, E., 2010. The effect of freezing at −18°C and −70°C with and without ascorbic acid on the stability of antioxidant in extracts of apple and orange fruits. Int. J. Food Sci. Technol. 45, 1814–1820.

Rizzolo, A., Nani, R.C., Viscardi, D., Bertolo, G., Torreggiani, D., 2003. Modification of glass transition temperature through carbohydrates addition and anthocyanin and soluble phenol stability of frozen blueberry juices. J. Food Eng. 56 (2–3), 229–231.

Rommel, A., Wrolstad, R.E., Heatherbell, D.A., 1992. Blackberry juice and wine: Processing and storage effects on anthocyanin composition, color and appearance. J. Food Sci. 57, 385–391.

Sahari, M.A., Mohsen Boostani, F., Zohreh, H.E., 2004. Effect of low temperature on the ascorbic acid content and quality characteristics of frozen strawberry. Food Chem. 86, 357–363.

Sousa, M.B., Canet, W., Alvarez, M.D., Tortosa, M.E., 2005. The effect of the pre-treatments and the long and short-term frozen storage on the quality of raspberry (cv. *Heritage*). Eur. Food Res. Technol. 221, 132–144.

Srivastava, A., Akoh, C.C., Yi, W., Fischer, J., Krewer, G., 2007. Effect of storage conditions on the biological activity of phenolic compounds of blueberry extract packed in glass bottles. J. Agric. Food Chem. 55, 2705–2713.

Syamaladevi, R.M., Sablani, S.S., Tang, J., Powers, J., Swanson, B.G., 2011. Stability of anthocyanins in frozen and freeze-dried raspberries during long-term storage—In relation to glass transition. J. Food Sci. 76, E414–E421.

Tuberoso, C.I.G., Barra, A., Cabras, P., 2008. Effect of different technological processes on the chemical composition of myrtle (*Myrtus communis* L.) alcoholic extracts. Eur. Food Res. Technol. 226, 801–808.

Torreggiani, D., Forni, E., Guercilena, I., Maestrelli, A., Bertolo, G., Archer, G.P., Kennedy, C.J., Bone, S., Blond, G., Contreras-Lopez, E., Champion, D., 1999. Modification of glass transition temperature through carbohydrates additions: effect upon color and anthocyanin pigment stability in frozen strawberry juices. Food Res. Int. 32, 441–446.

Turfani, O., Turkyilmaz, M., Yemis, O., Özkan, M., 2012. Effects of clarification and storage on anthocyanins and color of pomegranate juice concentrates. J. Food Qual. 35, 272–282.

Turkyılmaz, M., Ozkan, M., 2012. Kinetics of anthocyanin degradation and polymeric color formation in black carrot juice concentrates during storage. Int. J. Food Sci. Technol. 47, 2273–2281.

Urbanyi, G., Horti, K., 1992. Changes of surface color of the fruit and of the anthocyanin content of sour cherries during frozen storage. Acta Alimentaria 21 (3–4), 307–323.

Woodward, G., Kroon, P., Cassidy, A., Kay, C., 2009. Anthocyanin stability and recovery: implications for the analysis of clinical and experimental samples. J. Agric. Food Chem. 57, 5271–5278.

Wrolstad, R.E., Skrede, G., Lea, P., Enersen, G., 1990. Influence of sugar on anthocyanin pigment stability in frozen strawberries. J. Food Sci. 55 (4), 1064–1072.

Wrolstad, R.E., Putman, T.P., Varseveld, G.W., 1970. Color quality of frozen strawberries: effect of anthocyanin, pH, total acidity and ascorbic acid variability. J. Food Sci. 35, 448–452.

Zaritzky, N., 2012. Physical-Chemical Principles in Freezing, In Handbook of Frozen Food Processing and Packaging. 2nd Edition CRC Press, Boca Raton, FL, p. 3–37.

CHAPTER

19

Strawberry Phenolics and Impact of Ripening

*Kjersti Aaby**, *Siv Fagertun Remberg*†

*Norwegian Institute of Food, Fisheries, and Aquaculture Research, Aas, Norway, †Department of Plant and Environmental Sciences, Norwegian University of Life Sciences, Aas, Norway

CHAPTER POINTS

- Strawberry fruits contain various phenolic compounds, i.e., flavonoids comprising anthocyanins, flavonols, and flavan-3-ols including proanthocyanidins, phenolic acids, and hydrolysable tannins (ellagitannins).
- During ripening from green to red ripe fruits extensive changes in gene expressions, enzyme activities, and synthesis of phenolic metabolites occur.
- Ellagic acid containing compounds and flavan-3-ols, including proanthocyanidins, are present in the highest concentrations in green fruits and do not change significantly in the late developmental stages, i.e., from white to red fruits.
- Differing results are reported for synthesis of flavonols during ripening.
- The concentrations of anthocyanins and cinnamic acid derivatives increase significantly in the late developmental stages, i.e., from turning to red fruits.
- The concentration of anthocyanins continues to increase during post-harvest storage.

List of Abbreviations

ANR Anthocyanidin reductase
ANS anthocyanidin synthase
C4H cinnamate 4-hydroxylate
C3H cinnamate 3-hydroxylate
CHI chalcone isomerase
CHS chalcone synthase
CGT cinnamate glycosyl transferase
COMT-1 caffeic acid *O*-methyltransferase
4CL *p*-coumarate:CoA ligase
DFR dihydroflavonol 4-reductase
FGT flavonoid glycosyltransferase
(F3′H) flavonoid 3′-hydroxylase
(FHT) flavanone 3-hydroxylase
(FLS) flavonol synthase
(GT) galloyltransferase
(GTs) glucosyltransferases
(HPLC) high performance liquid chromatography
(LAR) leucoanthocyanidin reductase
(MS) mass spectrometry
(PAL) phenylalanine ammonia lyase
(UV-vis) ultraviolet-visible light

INTRODUCTION

The fruits of strawberry (*Fragaria* × *ananassa*) are popular both as fresh fruits and after processing and are produced worldwide in high quantities (4.36 million tonnes in 2010 (Food and Agricultural Organization of the United Nations, 2010). In the last decades the interest in strawberries has risen partly due to the increased focus on health effects associated with consumption of fruits and berries. Promising effects on inflammation, proliferation, enzyme inhibition relevant to glycemic control, cardiovascular related diseases, and cognitive behavior of fruits and fruit extracts and their constituents are shown both *in vitro* and *in vivo* (Chong *et al.*, 2010; Clifford and Brown, 2006; Devore *et al.*, 2012). The group of compounds in fruits which has gained special attention is the polyphenols. Polyphenols are responsible both for the bright, red color (anthocyanins) and astringent taste (tannins) in strawberry fruits, in addition to possible health effects. In the plant, phenolic compounds function as protective agents against UV light, herbivores, and pathogen attack, as attractants for pollinators and seed-dispersing animals, and as signal molecules (Crozier *et al.*, 2006). The biosynthesis of different phenolic compounds is thus dependent on the needs of the plant during development, and the concentration in strawberry fruits will thus vary during ripening (Carbone *et al.*, 2009; Fait *et al.*, 2008; Lunkenbein *et al.*, 2006). Other factors influencing the phenolic are genotype (Aaby *et al.*,

Processing and Impact on Active Components in Food
http://dx.doi.org/10.1016/B978-0-12-404699-3.00019-6

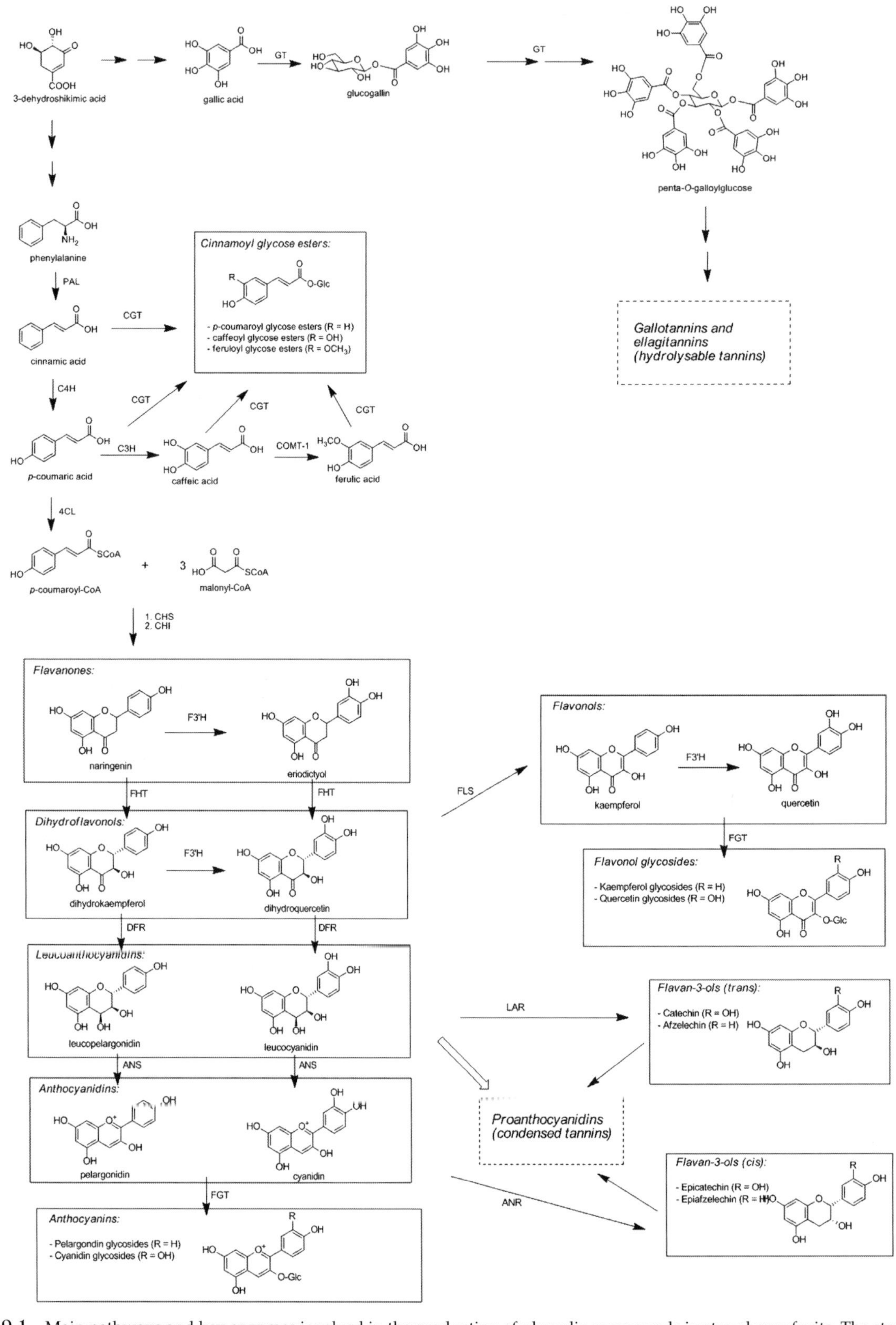

FIGURE 19.1 Main pathways and key enzymes involved in the production of phenolic compounds in strawberry fruits. The structures of the most important metabolites are shown. ANR, anthocyanidin reductase; ANS, anthocyanidin synthase; C3H, cinnamate 3-hydroxylate; CGT, cinnamate glycosyl transferase; COMT-1, caffeic acid *O*-methyltransferase; FGT, flavonoid glycosyltransferase; F3′H, flavonoid 3′-hydroxylase; GT, galloyltransferase; LAR, leucoanthocyanidin reductase.

TABLE 19.1 Concentration of Phenolic Compounds in Red Ripe Fruits of 27 Cultivars of Strawberry (*Fragaria* × *Ananassa*)

Compounds	Concentration (mg/100 g)	
	Range	*Mean*
Anthocyanins		
Cyanidin-3-glucoside	0.1–2.9	1.1
Pelargonidin-3-glucoside	6.9–47.7	25.3
Pelargonidin-3-rutinoside	0.0–2.6	1.1
Cyanidin-3-malonyl-glucoside	0.0–0.8	0.4
Pelargonidin-3-malonyl-glucoside	0.0–20.8	6.0
Pelargonidin-3-acetylgluco-side	0.0–0.5	0.1
Total	8.5–63.7	33.9
Flavonols		
Quercetin glycosides	0.4–5.2	1.8
Kaempferol glycosides	0.0–2.1	0.8
Flavan-3-ols		
(+)-catechin	2.0–8.1	4.5
Procyanidin dimers	4.3–16.1	9.1
Procyanidin trimers	3.0–18.8	7.9
Cinnamic acid derivatives		
Cinnamoyl glucose	0.6–24.9	5.0
P coumaroyl hexoses	1.7–13.6	5.4
Ellagic acid containing compounds		
Ellagic acid	0.2–0.9	0.5
Ellagic acid glycosides	0.1–1.8	0.6
Agrimoniin	5.8–13.8	8.8
Other ellagitannins	1.5–5.1	2.8

Adapted from Aaby et al. *(2012).*

2012; Anttonen *et al.*, 2006; Buendia *et al.*, 2010) and growing conditions, including climate and cultivation practices (Anttonen *et al.*, 2006; Wang *et al.*, 2002). In addition, the phenolic composition will change during post-harvest storage, depending on various factors (Cordenunsi *et al.*, 2005; Nunes *et al.*, 2006; Olsson *et al.*, 2004).

In this chapter, the impact of ripening on phenolics in strawberry fruits is reviewed. The effect of post-harvest storage is also addressed.

HOW COMPOSITION IS ALTERED

Phenolic Compounds in Strawberry Fruits

Phenolic compounds in strawberry fruits are flavonoids, comprising anthocyanins, flavonols and flavan-3-ols including proanthocyanidins, phenolic acids, and ellagitannins (Aaby *et al.*, 2012; Buendia *et al.*, 2010; Määttä-Riihinen *et al.*, 2004) (Figure 19.1 and Table 19.1). The anthocyanins, responsible for the red color of the fruits, mainly consist of pelargonidin glycosides and major compounds are pelargonidin-3-glucoside, contributing 60–95% of the total anthocyanin content in ripe fruits, pelargonidin-3-malonylglucoside (0–33%), pelargonidin-3-rutinoside (0–7%), and cyanidin-3-glucose (1–6%) (Aaby *et al.*, 2012; Buendia *et al.*, 2010; Määttä-Riihinen *et al.*, 2004; Yoshida *et al.*, 2002). The main flavonols are glycosides and glucuronides of kaempferol and quercetin (Aaby *et al.*, 2012; Buendia *et al.*, 2010; Fait *et al.*, 2008). (+)-Catechin is the only flavan-3-ol monomer in strawberry fruits, while proanthocyanidins (condensed tannins), with main degree of polymerization of 3.4–6.3, are abundant (Buendia *et al.*, 2010; Gu *et al.*, 2003). The phenolic acids present are mainly cinnamic acid derivatives, i.e., esters and glycosides of *p*-coumaric acid, cinnamic acid and ferulic acid (Aaby *et al.*, 2012; Buendia *et al.*, 2010; Lunkenbein *et al.*, 2006). The most abundant ellagitannin (hydrolysable tannin) in strawberry is agrimoniin (Aaby *et al.*, 2012).

Biosynthesis of Phenolic Compounds in Strawberry Fruits

The main pathways and key enzymes involved in the production of strawberry phenolics are shown in Figure 19.1. Phenolic compounds are synthesized in the shikimate, phenylpropanoic and flavonoid pathways (Crozier *et al.*, 2006). Ellagitannins stem directly from the shikimic acid pathway via gallic acid and penta-*O*-galloyl-glucose, which is further galloylated and oxidized. Flavonoids and cinnamic acids are derived from phenylalanine, which is converted to cinnamic acid by phenylalanine ammonia lyase (PAL) in the phenylpropanoic pathway. Cinnamic acid is hydroxylated to *p*-coumaric acid by cinnamate 4-hydroxylase (C4H), and through further hydroxylation and methyltransferation, caffeic, ferulic and syringic acid are formed. The final step in the general phenylpropanoic metabolism is the carboxyl activation of *p*-coumaric acid catalyzed by 4-hydroxycinnamate:CoA ligase (4CL). The flavonoid pathway starts with the condensation of *p*-coumaroyl-CoA with three malonyl-CoA residues in a reaction catalyzed by chalcone synthase (CHS). The chalcone formed is isomerized by chalcone isomerase (CHI) to naringenin. Further reactions in strawberries involve flavonol 3-hydroxylase (FHT) to form dihydroflavonols which are converted to flavonols by flavonol synthase (FLS) or are reduced by dihydroflavonol 4-reductase (DFR) to form leucocyanidins, the precursor of anthocyanidins and flavan-3-ols. Different glycosyltransferases (GTs) transform cinnamic acids, anthocyanidins, and flavonols to their glycosylated end-products.

The biosynthetic pathways are regulated by developmental, genetic and environmental signals (Almeida *et al.*, 2007; Carbone *et al.*, 2009).

Impact of Ripening on Strawberry Phenolics

The ripening process in strawberry fruits involves various complex changes in gene expression. During ripening from green to red ripe fruits, three patterns of gene expression were observed (Almeida *et al.*, 2007; Carbone *et al.*, 2009). Key genes in the phenylpropanoid (*PAL*, *C4H*, and *4CL*) and flavonoid (*CHS*, *CHI*, *FHT*, *ANS*, and *LAR*) pathways showed a two-phase expression pattern with peaks early after anthesis and in the final ripening phase, with the highest expression in the last stage. *DFR* and *GTs* were upregulated in turning and ripe fruits, while some genes (*F3′H*, *FLS*, and *ANR*) diminished throughout fruit ripening (Carbone *et al.*, 2009) (Figure 19.2). Overall, an upregulation of gene expression gave a corresponding increase in enzyme activity and synthesis of phenolic metabolites (Carbone *et al.*, 2009; Griesser *et al.*, 2008; Lunkenbein *et al.*, 2006). Different transcription factor patterns suggest biosynthesis of different phenolic compounds at different stages during fruit development. Consequently, flavan-3-ols, including proanthocyanidins, and ellagic acids had a peak in early green fruits, probably to protect immature

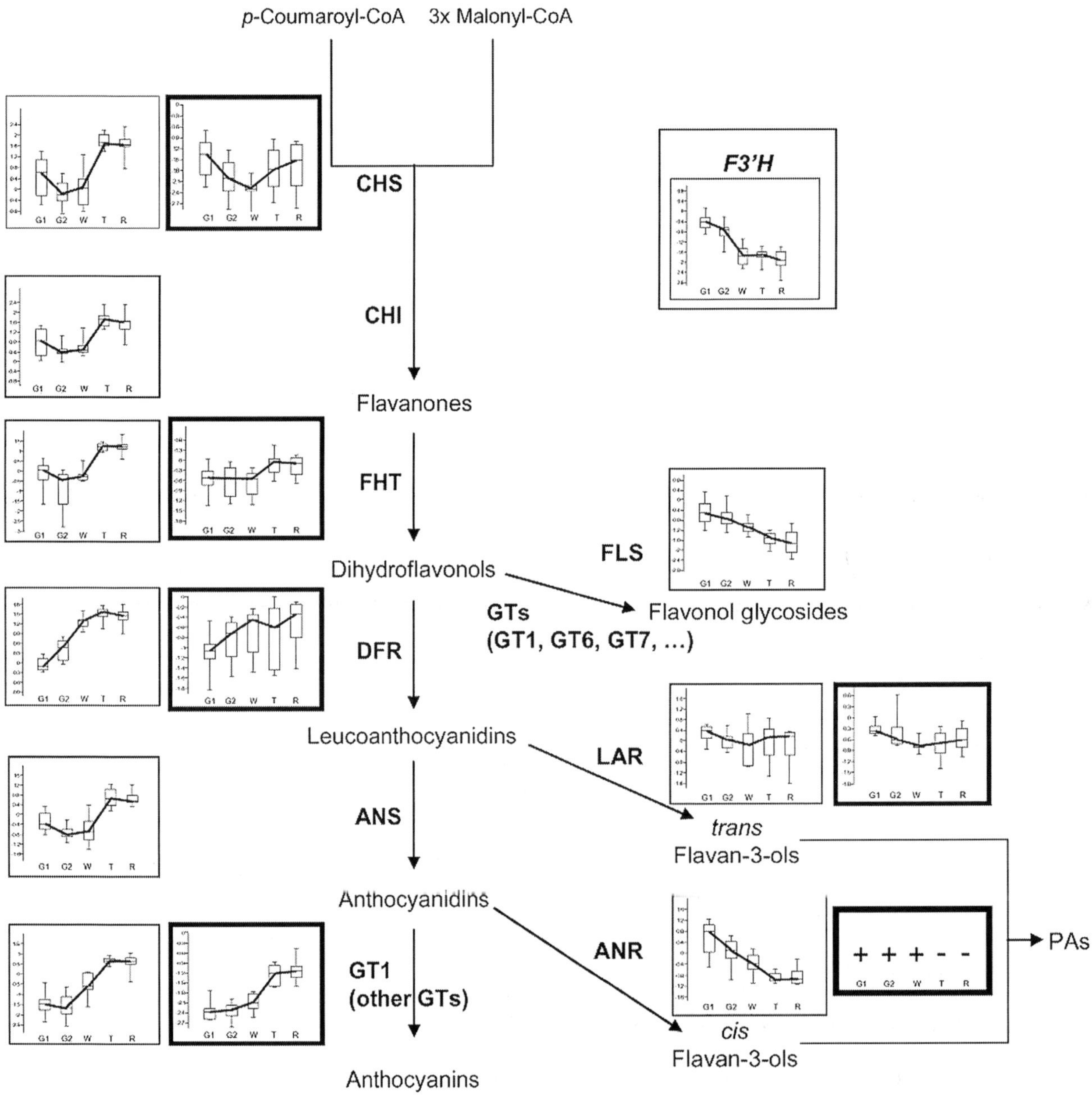

FIGURE 19.2 The flavonoid biosynthetic pathway in strawberry fruits. For each enzymatic step, box plot charts with quantitative gene expression (thin line border) and enzyme activity (thick line border) are given. The *F3′H* chart is in a standalone box (see Figure 19.1 for steps involving F3′H in the pathway). ANR, anthocyanidin reductase; ANS, anthocyanidin synthase; C3H, cinnamate 3-hydroxylate; CGT, cinnamate glycosyl transferase; COMT-1, caffeic acid *O*-methyltransferase; FGT, flavonoid glycosyltransferase; F3′H, flavonoid 3′-hydroxylase; GT, galloyltransferase; LAR, leucoanthocyanidin reductase. *Reprinted from Carbone* et al. *(2009) with permission from John Wiley and Sons.*

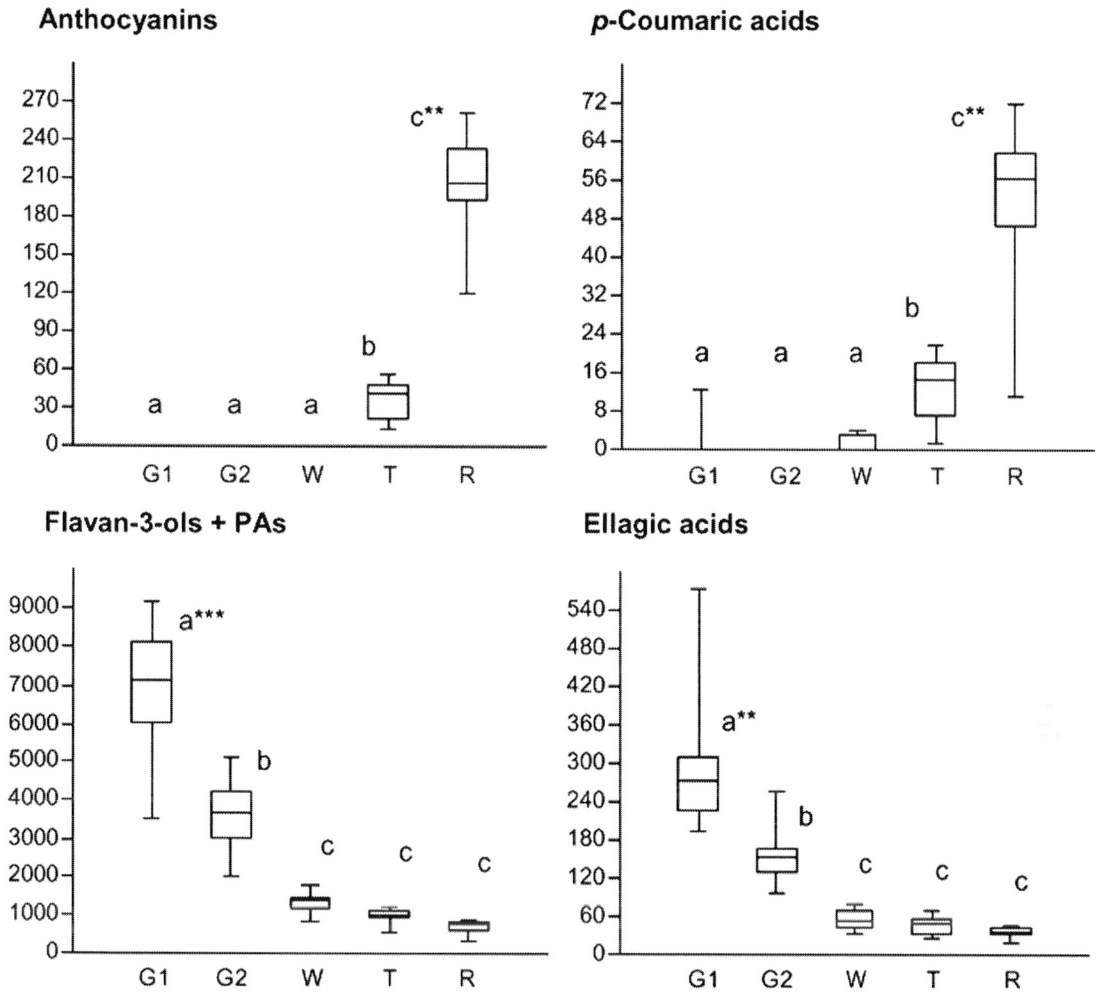

FIGURE 19.3 Total accumulation of phenolic compounds as a function of developmental stages in strawberry fruits. G1 = early green [7–10 daa (day after anthesis)], G2 = intermediate green (12–14 daa), W = white (ca. 20 daa), T = turning (ca. 25 daa), and R = ripe. PA = proanthocyanidins. Box plots show values of the 10 genotypes studied. Metabolite levels are expresses as μg/g FW. Different letters indicate significant differences among mean values (** $p \leq 0.01$; *** $p \leq 0.001$). *Reprinted from Carbone* et al. *(2009) with permission from John Wiley and Sons.*

fruits from feeding and pathogenic attack, and decreased during ripening (Carbone *et al.*, 2009; Fait *et al.*, 2008; Halbwirth *et al.*, 2006; Williner *et al.*, 2003), while anthocyanins acting as attractants, and cinnamic acid derivatives, increased in the later developmental stages (Carbone *et al.*, 2009; Fait *et al.*, 2008; Halbwirth *et al.*, 2006; Latza *et al.*, 1996) (Figure 19.3). Latza *et al.* (1996) showed that accumulation of cinnamoyl glucose succeeded fruit growth and seemed to precede anthocyanin formation with a sharp rise in concentration 21 days after anthesis (Figure 19.4).

The concentration of pelargonidin-3-glucoside increased the most during ripening of fruits of 'Nyoho', while the relative contribution of cyanidin-3-glucoside was found to be higher at early stages of color development (Yoshida *et al.*, 2002). Also, no increase in concentration of cyanidin-3-glucoside was found from green to ripe fruits of 'Osogrande', while an eight-fold increase was found for 'Camino Real' (Pineli *et al.*, 2011). The ellagitannins and galloylglucoses found in the highest concentrations in the achenes were present in the highest concentrations in early developmental stages, while ellagitannins present in the highest concentrations in the receptacle were present in the highest concentrations in big green and white fruits (Fait *et al.*, 2008). Differing results are reported for synthesis of flavonols. Halbwirth *et al.* (2006) found an increase in late stages. Fait *et al.* (2008) found the highest concentration of most flavonols in medium green to white fruits, while no clear dependence on ripening were found by others (Aaby *et al.*, 2012; Carbone *et al.*, 2009; Griesser *et al.*, 2008). The differences in development profile reported for flavonols may be caused by environmental effects as observed by Carbone *et al.* (2009). In this study, significant differences among cultivars at each development stage were also found.

Many studies have shown changes in phenolic content during fruit development from small green to red ripe fruits, but the differences in content of phenolic compounds in fruits in degrees of ripening intended for consumption are hardly studied. As previously shown (Figure 19.3), the concentration of flavan-3-ols and ellagic

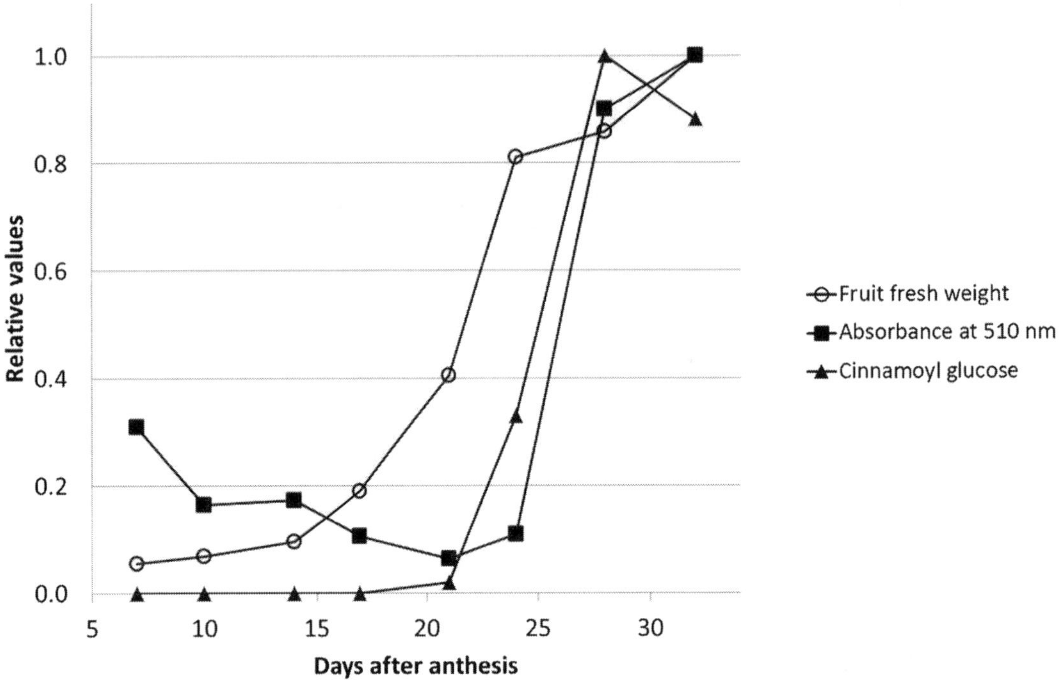

FIGURE 19.4 Strawberry fruit weights, contents of anthocyanins and cinnamoyl glucose during fruit development. Average fresh fruit weights (circles), absorbance at 510 nm, i.e., anthocyanins (squares) and content of cinnamoyl glucose (triangles) in strawberry fruits given as relative values as a function of days after anthesis. *Adapted from Latza* et al. *(1996).*

acids decreased during ripening from early green to white fruits, but there were no significant differences in content in white, turning, and ripe fruits (Carbone *et al.*, 2009). The concentrations of anthocyanins and *p*-coumaric acids, however, increased significantly from white to turning, and from turning to red ripe fruits. The trend was basically the same in three strawberry cultivars at three ripening stages all suitable for consumption (Aaby *et al.*, 2012). The concentration of ellagic acid derivatives and agrimoniin decreased during ripening, but the changes were minor (12–39%), while the concentrations of anthocyanins and *p*-coumaroyl hexoses increased in average 2.5-fold from light red to dark red fruits, and cinnamoyl glucose increased a considerable 4.3-fold. Flavonols and (+)-catechin were not affected by ripening, except for fruits of 'Blink,' where a decrease in concentration of (+)-catechin was found. In agreement with these results, the concentration of ellagic acid did not change from red to dark red fruits of 'Honeoye' and 'Senga Sengana' (Olsson *et al.*, 2004). The concentration of cinnamic acid derivatives increased during ripening of 'Honeoye,' but not 'Senga Sengana.' The effect of ripening on strawberry phenolics seemed thus to some degree to be dependent on genotype. However, developmental effects were found to dominate over genetic and environmental effects on flavonoid gene expression and metabolite levels in strawberry fruits during ripening (Carbone *et al.*, 2009).

OTHER WAYS IN WHICH COMPOSITION IS ALTERED

Effects of Post-Harvest Storage

Various post-harvest factors have been reported to affect both content and stability of phenolic compounds in fruits and berries. In strawberry, as for most horticultural products, temperature is the most important factor affecting physiochemical changes and quality during storage, which also includes alteration in secondary metabolites like phenolic compounds. Olsson *et al.* (2004) observed relatively small changes in antioxidant capacity during storage of 'Honeoye' at 4°C, although a decrease in chlorogenic acid and quercetin was observed after 3 days. Regardless of the initial color stage at harvest, a significant increase in total anthocyanin content in field-ripened strawberries was observed during storage (Nunes *et al.*, 2006). The same trend has also been reported by others, e.g., Cordenunsi *et al.* (2005) who observed an increase in anthocyanin content in strawberries during storage for up to 6 days. The anthocyanin level correlated positively with storage temperature, with lower anthocyanin content found in berries stored at lower (6°C) than higher (16 and 25°C) temperatures. The same trend has been found by Kalt *et al.* (1999) who found increased levels of anthocyanins in strawberries stored at higher temperatures. During 8 days of storage, a 6.8-fold increase in anthocyanin content was found in strawberries stored at 30°C, compared to a 1.7-fold increase when stored at 0°C. Antioxidant capacity also increased, but had the highest increase (1.5-fold) in strawberries stored at 10 and 20°C.

ANALYTICAL TECHNIQUES

Sample Preparation

Phenolic compounds must be released from the fruits before analysis. Homogenization of the plant material followed by solvent extraction with methanol or acetone, often in combination with water and addition of an organic acid, is the most common technique for extraction of phenolic compounds (Aaby *et al.*, 2012; Buendia *et al.*, 2010; Kim and Lee, 2002; Määttä-Riihinen *et al.*, 2004). Acetone is shown to be the best extraction solvent for most phenolic compounds in strawberries (Buendia *et al.*, 2010). The less hydrophilic proanthocyanidins may be insufficiently extracted by aqueous-organic solvents, and direct hydrolysis and derivatization without prior extraction can be utilized (Buendia *et al.*, 2010).

Acid hydrolysis is used to cleave glycosidic bonds to form aglycons. Alkaline hydrolysis cleaves ester bonds, and can be used to confirm the presence of acylated compounds. Hydrolysis is used both for confirmation of identity of parent compounds and for quantification of aglycons. Acid-catalyzed reaction of proanthocyanidins with nucleophilic agents, such as phloroglucinol or thiol, give cleavage products that are used to determine proanthocyanidin subunit composition and degree of polymerization (Buendia *et al.*, 2010; Gu *et al.*, 2003).

Solid phase extraction, usually with C-18 as the solid phase, are used to remove interfering compounds from the crude extracts and to separate the phenolic compounds in an anthocyanin and a non-anthocyanin fraction (Kim and Lee, 2002).

Analysis of Individual Compounds by HPLC

Native proanthocyanidins are separated on normal phase columns by high-performance liquid chromatography (HPLC), while other phenolic compounds and proanthocyanidin phloroglucinol adducts are separated on reversed phase columns (Aaby *et al.*, 2012; Buendia *et al.*, 2010; Määttä-Riihinen *et al.*, 2004). Since the elution of compounds is dependent on their polarity, the elution pattern gives information about substitution on the phenolic ring and conjugation with sugars and acylation, especially within a class of phenolic compounds.

Polyphenols absorb in the UV-region and have traditionally been analyzed by UV detectors. The ultraviolet-visible light (UV-vis) spectrum, obtained by a diode array detector, indicates which class the phenolic compound belongs to, e.g., flavan-3-ols have λ_{max} at 280 nm and anthocyanins at 500–530 nm. Flavan-3-ols, including proanthocyanidins, have low molar absorptivity and improved sensitivity is obtained by using fluorescence detection. In the last decades HPLC with mass spectrometry (MS) detection has become widespread. HPLC-MS^n is predominately used for structural elucidation, but increasingly also used for quantification.

Quantification

Ideally, quantification should be performed by use of authentic standards of compounds of interest. However, due to both uncertainty in identification and lack of commercial standards of many of the complex compounds in strawberries, the phenolic compounds are quantified as equivalents of other compounds, preferable compounds with similar structure and molecular weight, e.g., all anthocyanins in a sample are quantified as pelargonidin-3-glucoside. Also, phenolic conjugates can be analyzed and quantified after acid hydrolysis, e.g., ellagitannins are quantified as ellagic acid. Different methods of extraction, separation, detection, and quantification lead to considerable differences in concentration of some phenolic compounds, most pronounced for proanthocyanidins and ellagitannins. In addition, contents are reported as concentrations of individual compounds or concentrations of groups of compounds. Consequently, it can be difficult to compare results between studies.

References

Aaby, K., Mazur, S., Nes, A., Skrede, G., 2012. Phenolic compounds in strawberry (*Fragaria* x *ananassa* Duch.) fruits: Composition in 27 cultivars and changes during ripening. Food Chem. 132, 86–97.

Almeida, J.R.M., D'Amico, E., Preuss, A., Carbone, F., de Vos, C.H.R., Deiml, B., Mourgues, F., Perrotta, G., Fischer, T.C., Bovy, A.G., Martens, S., Rosati, C., 2007. Characterization of major enzymes and genes involved in flavonoid and proanthocyanidin biosynthesis during fruit development in strawberry (*Fragaria* x *ananassa*). Arch. Biochem. Biophys. 465, 61–71.

Anttonen, M.J., Hoppula, K.I., Nestby, R., Verheul, M.J., Karjalainen, R.O., 2006. Influence of fertilization, mulch color, early forcing, fruit order, planting date, shading, growing environment, and genotype on the contents of selected phenolics in strawberry (*Fragaria* x *ananassa* Duch.) fruits. J. Agric. Food Chem. 54, 2614–2620.

Buendia, B., Gil, M.I., Tudela, J.A., Gady, A.L., Medina, J.J., Soria, C., Lopez, J.M., Tomas-Barberan, F.A., 2010. HPLC-MS analysis of proanthocyanidin oligomers and other phenolics in 15 strawberry cultivars. J. Agric. Food Chem. 58, 3916–3926.

Carbone, F., Preuss, A., De Vos, R.C.H., D'Amico, E., Perrotta, G., Bovy, A.G., Martens, S., Rosati, C., 2009. Developmental, genetic and environmental factors affect the expression of flavonoid genes, enzymes and metabolites in strawberry fruits*. Plant Cell Environ. 32, 1117–1131.

Chong, M.F.F., Macdonald, R., Lovegrove, J.A., 2010. Fruit polyphenols and CVD risk: a review of human intervention studies. Brit. J. Nutr. 104, S28–S39.

Clifford, M.N., Brown, J.E., 2006. Dietary flavonoids and health—broadening the perspective. In: Andersen, Ø.M., Markham, K.R. (Eds.), Flavonoids. Chemistry, biochemistry and applications. CRC Press, Boca Raton, FL. 319–370.

Cordenunsi, B.R., Genovese, M.I., do Nascimento, J.R.O., Hassimotto, N.M.A., dos Santos, R.J., Lajolo, F.M., 2005. Effects of temperature on the chemical composition and antioxidant activity of three strawberry cultivars. Food Chem. 91, 113–121.

Crozier, A., Jaganath, I.B., Clifford, M.N., 2006. Phenols, polyphenols and tannins: An overview. In: Crozier, A., Clifford, M.N., Ashihara, H. (Eds.), Plant secondary metabolites. Occurence, structure and role in the human diet. Blackwell Publishing Ltd, Oxford, UK. 1–24.

Devore, E.E., Kang, J.H., Breteler, M.M.B., Grodstein, F., 2012. Dietary intakes of berries and flavonoids in relation to cognitive decline. Ann. Neurol. 72, 135–143.

Fait, A., Hanhineva, K., Beleggia, R., Dai, N., Rogachev, I., Nikiforova, V.J., Fernie, A.R., Aharoni, A., 2008. Reconfiguration of the achene and receptacle metabolic networks during strawberry fruit development. Plant Physiol. 148, 730–750.

Food and Agriculture Organization of the United Nations, 2010. Available from: http://faostat.fao.org/site/567/ (retrieved 10.01.2013).

Griesser, M., Hoffmann, T., Bellido, M.L., Rosati, C., Fink, B., Kurtzer, R., Aharoni, A., Munoz-Blanco, J., Schwab, W., 2008. Redirection of flavonoid biosynthesis through the down-regulation of an anthocyanidin glucosyltransferase in ripening strawberry fruit. Plant Physiol. 146, 1528–1539.

Gu, L., Kelm, M.A., Hammerstone, J.F., Beecher, G., Holden, J., Haytowitz, D., Prior, R.L., 2003. Screening of foods containing proanthocyanidins and their structural characterization using LC-MS/MS and thiolytic degradation. J. Agric. Food Chem. 51, 7513–7521.

Halbwirth, H., Puhl, I., Haas, U., Jezik, K., Treutter, D., Stich, K., 2006. Two-phase flavonoid formation in developing strawberry (*Fragaria* x *ananassa*) fruit. J. Agric. Food Chem. 54, 1479–1485.

Kalt, W., Forney, C.F., Martin, A., Prior, R.L., 1999. Antioxidant capacity, vitamin C, phenolics, and anthocyanins after fresh storage of small fruits. J. Agric. Food Chem. 47, 4638–4644.

Kim, D.O., Lee, C.Y., 2002. Extraction and isolation of polyphenolics. In: Wrolstad, R.E. (Ed.), Current Protocols in Food Analytical Chemistry. John Wiley & Sons, Inc., New York. 11.12.11–11.12-12.

Latza, S., Gansser, D., Berger, R.G., 1996. Identification and accumulation of 1-*O*-*trans*-cinnamoyl-beta-D-glucopyranose in developing strawberry fruit (*Fragaria ananassa* Duch cv. Kent). J. Agric. Food Chem. 44, 1367–1370.

Lunkenbein, S., Bellido, M., Aharoni, A., Salentijn, E.M.J., Kaldenhoff, R., Coiner, H.A., Munoz-Blanco, J., Schwab, W., 2006. Cinnamate metabolism in ripening fruit. Characterization of a UDP-glucose: Cinnamate glucosyltransferase from strawberry. Plant Physiol. 140, 1047–1058.

Määttä-Riihinen, K.R., Kamal-Eldin, A., Törrönen, A.R., 2004. Identification and quantification of phenolic compounds in berries of *Fragaria* and *Rubus* species (Family Rosaceae). J. Agric. Food Chem. 52, 6178–6187.

Nunes, M.C.N., Brecht, J.K., Morais, A., Sargent, S.A., 2006. Physicochemical changes during strawberry development in the field compared to those that occur in harvested fruit during storage. J. Sci. Food Agric. 86, 180–190.

Olsson, M.E., Ekvall, J., Gustavsson, K.E., Nilsson, J., Pillai, D., Sjöholm, I., Svensson, U., Åkesson, B., Nyman, M.G.L., 2004. Antioxidants, low molecular weight carbohydrates, and total antioxidant capacity in strawberries (*Fragaria* x *ananassa*): Effects of cultivar, ripening, and storage. J. Agric. Food Chem. 52, 2490–2498.

Pineli, L.d.L.d.O., Moretti, C.L., dos Santos, M.S., Campos, A.B., Brasileiro, A.V., Cordova, A.C., Chiarello, M.D., 2011. Antioxidants and other chemical and physical characteristics of two strawberry cultivars at different ripeness stages. J. Food Comp. Anal. 24, 11–16.

Wang, S.Y., Zheng, W., Galletta, G.J., 2002. Cultural system affects fruit quality and antioxidant capacity in strawberries. J. Agric. Food Chem. 50, 6534–6542.

Williner, M.R., Pirovani, M.E., Guemes, D.R., 2003. Ellagic acid content in strawberries of different cultivars and ripening stages. J. Sci. Food Agric. 83, 842–845.

Yoshida, Y., Koyama, N., Tamura, H., 2002. Color and anthocyanin composition of strawberry fruit: Changes during fruit development and differences among cultivars, with special reference to the occurrence of pelargonidin 3-malonylglucoside. J. Japan. Soc. Hort. Sci. 71, 355–361.

CHAPTER

20

Biologically Active Compounds in Melon: Modulation by Preharvest, Post-harvest, and Processing Factors

Ana L. Amaro, Ana Oliveira*, Domingos P.F. Almeida*†

*Centro de Biotecnologia e Química Fina, Escola Superior de Biotecnologia, Centro Regional do Porto da Universidade Católica Portuguesa, Porto, Portugal, †Instituto Superior de Agronomia, Universidade de Lisboa, Lisbon Portugal

CHAPTER POINTS

- Levels of biologically active compounds in melon are affected by preharvest, post-harvest, and processing factors.
- Interactions between genotypic and environmental main effects and storage and processing conditions make it difficult to generalize about the final melon content in bioactive compounds.
- The levels of ascorbic acid, the best documented bioactive compound in melon, vary widely but are generally reduced by processing methods.
- β-Carotene concentration varies widely with genotype and is affected by processing conditions.
- Phenolics, as a group, are generally preserved during storage and processing of raw materials but can be enhanced by roasting.
- Little is known about the effects of processing conditions on folic acid content in melon.
- Cucurbitacin levels are negligible in commercial cultivars but may become important in pharmacological or ethnobotanical applications.

INTRODUCTION

Melon (*Cucumis melo* L.), a species belonging to the *Cucurbitaceae* family, is one of the most widely cultivated crops and consumed commodities worldwide. The average composition of melon fruit is summarized in Table 20.1. Water and sugars account for *ca.* 98% of the melon edible fresh weight. However, melon also contains biologically active compounds, present in small quantities as naturally occurring substances and with an effect on human health. Biologically active compounds can be essential nutrients (e.g., vitamins), non-essential phytochemicals (e.g., polyphenols), and toxic compounds (e.g., cucurbitacins). The most relevant vitamins in melon fruit are vitamin C, pro-vitamin A, namely β-carotene, and folic acid (Lester and Hodges, 2008; Laur and Tian, 2011).

Vitamin C is an essential water-soluble vitamin, constituted by the reduced and oxidized forms of ascorbic acid, ascorbate and dehydroascorbate, respectively. Ascorbic acid is sensitive to degradation during the post-harvest period and processing, being a good marker for abuse from processing to storage conditions.

Carotenoids are a large class of lipid-soluble isoprenoid compounds located in plant plastids and belong to two chemical classes: the oxygen-free carotenes (orange and red pigments) and the oxygen-containing xanthophylls (yellow pigments). The main carotenoids in melons are β-carotene (precursor of vitamin A) and lutein (Laur and Tian, 2011).

Phenolic compounds are secondary plant metabolites commonly found in plants, which have multiple biological effects, including antioxidant activity. Even though common melon fruit does not rank particularly highly in phenolic content, with values ranging from 109.6 mg/100 g in cantaloupe to 59.3 mg/100 g in honeydew (INRA, 2009), phenolics are one the most important classes of biologically active plant phytochemicals. The

Processing and Impact on Active Components in Food
http://dx.doi.org/10.1016/B978-0-12-404699-3.00020-2

TABLE 20.1 Composition of Melons from the *Cantaloupensis* and *Inodorus* Groups

Component (Units per 100 g Edible Weight)	Cantaloupe		Honeydew	
	Average	SE	Average	SE
Water (g)	90.15	0.312	89.82	0.478
Energy (kcal)	34		36	
Energy (kJ)	141		150	
Protein (g)	0.84	0.030	0.54	0.054
Total lipids (g)	0.19	0.036	0.14	0.037
Ash (g)	0.65	0.039	0.41	0.021
Carbohydrates (g)	8.16		9.09	
Total fiber (g)	0.9	0.112	0.8	0.140
Total sugars (g)	7.86		8.12	0.925
Sucrose (g)	4.35	0.666	2.48	0.574
Glucose (g)	1.54	0.322	2.68	0.251
Frutose (g)	1.87	0.304	2.96	0.156
Maltose (g)	0.04	0.020		
Galactose (g)	0.06	0.004		
Starch (g)	0.03	0.025		
Selenium (μg)	0.4	0.055	0.7	
Vitamin C (mg)	36.7	1.384	18.0	1.637
Thiamin (mg)	0.041	0.009	0.038	0.013
Riboflavin (mg)	0.019		0.012	0.002
Niacin (mg)	0.734	0.040	0.418	0.061
Pantothenic acid (mg)	0.105	0.020	0.155	0.068
Vitamin B-6 (mg)	0.072	0.028	0.088	0.012
Folate (μg)	21	7.073	19	3.066
Vitamin A (μg)	169		3	0.037
Vitamin A (IU)	3382		50	0.735
β-Carotene (μg)	2020	252.90	30	0.441
α-Carotene (μg)	16	10.021		
β-Cryptoxanthin (μg)	1	1.170		
Lutein + zeaxanthin (μg)	26	9.999	27	3.810
Vitamin E (α-tocopherol) (mg)	0.05	0.007	0.02	0.006
γ-Tocopherol (mg)	0.11	0.013	0.03	0.003
Vitamin K (μg)	2.5	0.304	2.9	0.130
Total phenols (mg)	109.59	81.53	59.33	36.70

Source: USDA National Nutrient Database for Standard Reference, Release 23.

antioxidant activity of phenolics is due to their redox properties, which allow them to act as reducing agents, hydrogen donators, and singlet oxygen quenchers. The main phenolics in melon are: benzoic, vanillic, and *trans*-cinnamic acids. Others, like *p*-coumaric acid, ferulic acid, gallic acid, *p*-hydroxybenzoic acid, syringic acid, epicatechin, protocatechuic acid, chlorogenic acid, quercetin, and rutin are found in small amounts (Kolayli *et al.*, 2010).

Cucurbitacins are oxygenated tetracyclic triterpenoids with a bitter taste and toxic effects, naturally occurring in the *Cucurbitaceae* family. Fruit of cultivated *Cucurbitaceae*—cucumber, squash, melon, and watermelon—have been selected to reduce or eliminate cucurbitacin content. However, recent findings have demonstrated the potential of these compounds for treating pathologies such as inflammation, cancer, or autoimmune diseases. Therefore, both positive and toxic pharmacological effects to human health must be attributed to this class of biologically active compounds from the *Cucurbitaceae* (Devendra *et al.*, 2011).

Consumers can not directly assess the nutritional or functional composition of melons or melon products, in the same way they evaluate external attributes (appearance and textural quality) or eating quality. Instead, they rely on information available in specialized publications and databases or conveyed in outreach documents or in labeling. Factors that affect melon composition at harvest and loss of nutritional quality post-harvest or upon processing may result in the ingestion of nutritionally inferior fruit. This chapter surveys the pre- and post-harvest factors that affect melon biologically active compounds and emphasize how processing technologies impact these compounds.

HOW COMPOSITION IS ALTERED

Building Melon Composition: The Effect of Pre-harvest Factors

Compositional quality (nutritional and phytochemical) is built during melon production. The content of biologically active compounds results, in the end, from the interaction between a genotype and environmental factors, modulated by the horticultural techniques. A general understanding of how melon bioactive compounds and antioxidant activity are affected by pre-harvest factors provides valuable information for quality building during production and its levels in processed products.

Genotype Effects

Composition is largely affected by genotypic variation. Genotypic variation in *Cucumis melo* can be considered at two different levels. First and foremost is the cultivar group (often attributed the taxonomic level of botanical subspecies or variety). Six or more cultivar groups are considered within *Cucumis melo*, depending on the author (Münger and Robinson, 1991). The most widely produced and consumed melons in Europe and America belong to the Cantaloupensis and Inodorus cultivar groups. Non-sweet melons belonging to the Momordica group are common in Asia and those from Flexuosus and Conomon groups find specific applications in certain regions of the world (Münger and Robinson, 1991). Bitter melon (*Cucumis melo* Momordica group) is a rich source of phenolic compounds and has biological functions including antidiabetic, anticancer, anti-inflammatory, antiviral, and antioxidant activities, but reported adverse effects of bitter melon include hypoglycemia and convulsions in children, reduced fertility in mice, a favism-like syndrome, and headaches (Basch *et al.*, 2003). Cultivar groups differ in fruit morphology and physiology, with considerable differences in composition.

Within each horticultural cultivar group there are several horticultural types and a wide range of cultivars. Cultivar accounts for most of the variation in ascorbic and folic acids, and β-carotene content in melon (Table 10.2; Lester and Hodges, 2008). Ascorbic acid and folic acid contents are also significantly affected by fruit size (Lester and Crosby, 2002). There is also considerable variation in melon phytochemical concentration across fruit mesocarp regions (Lester, 2008). Folic acid and β-carotene concentrations increase from the sub-peel to inner tissues while total ascorbic acid is higher in the outer sub-peel tissues than in the mesocarp tissues closer to the seed cavity (Lester, 2008).

Environmental Effects: Location and Season

Abiotic environmental factors, soil and climate related, vary with the location and season of production. The genotype interacts differentially with abiotic conditions defined by soil type, temperature, moisture, and light, which vary markedly from year to year, affecting melon composition and the content of bioactive compounds. However, cultivar nutritional composition has seldom been characterized under changing environmental conditions. The melon-growing region is becoming an important criterion for consumers' decisions and affects fruit composition. For example, β-carotene and lutein contents range from 0.63 to 38.61 μg/g and 0.07 to 0.33 μg/g, respectively, depending on cultivar and geographical origin (Laur and Tian, 2011). Ascorbic acid and β-carotene content also vary widely among muskmelon cultivars grown in different locations and soil types (Lester and Crosby, 2002). Production season affects total ascorbic acid, β-carotene, and folic acid in green- and orange-fleshed inodorus cultivars (Lester and Crosby, 2002). Melons produced in autumn had higher levels of ascorbic acid and β-carotene than fruits produced in the spring (Lester and Hodges, 2008; Lester *et al.*, 2010). Fruit grown under excess light, resulting in photoinhibition, and high temperatures, resulting in higher respiration rates, have lower sugar, ascorbic acid, and carotenoid levels (Lester and Hodges, 2008). Interactions between genotype and environmental conditions exist. For instance, 'Honey Gold,' 'Orange Delight,' and 'Orange Dew' had higher levels of ascorbic acid when grown in the fall compared to spring-grown fruit, while 'Temptation' had lower ascorbic acid levels in the fall (Lester *et al.*, 2007). Folic acid concentrations in orange-fleshed honeydew fruit were higher (often nearly two-fold) in the spring than in autumn.

Soil type also affects melon composition. Total ascorbic acid and folic acid content are higher when green-fleshed honeydew melons are grown on clay loam than on sandy loam soil (Lester and Crosby, 2002). The content of cucurbitacins increases with temperature and also varies with season, as demonstrated in *Trichosanthes cucumerina* (Devendra *et al.*, 2011).

Cultural Practices

Cropping systems affect the canopy microclimate and modulate plant growth, development, and fruit compositional quality. Fertilization is a key tool to modulate

TABLE 20.2 Comparison of Orange-Fleshed Honey Dew Breeding Line (SVR-03935152) and Commercial Cultivars for Levels of Ascorbate, β-Carotene and 5-Methyltetrahydrofolate (Folic Acid)

	Ascorbate (mg/kg)		β-Carotene (mg/kg)		Folic Acid (μg/kg)	
	Harvest		Harvest		Harvest	
Genotype	*Autumn*	*Spring*	*Autumn*	*Spring*	*Autumn*	*Spring*
Honey Gold	171[BC]	150[B]	12.5[B]	10.2[B]	446[A]	573[A]
Orange Delight	192[AB]	180[A]	18.9[A]	12.6[A]	398[A]	674[A]
Orange Dew	224[A]	181[A]	18.2[A]	15.5[A]	393[A]	695[A]
SVR-03935152	139[C]	161[B]	8.1[C]	10.6[B]	291[B]	573[A]
Temptation	163[BC]	182[A]	12.3[B]	9.9[B]	356[AB]	633[A]

Fruit were harvested and maintained for 3 days at 20°C to simulate retail display. All data are on a fresh-weight basis. Uppercase superscript letters within a column indicate significant differences (LSMEANS 5% level) among cultivars within a storage period.
Adapted from Lester and Hodges (2008).

fruit composition. Ascorbic acid and β-carotene content respond positively to foliar potassium applications (Lester *et al.*, 2010). Organic production of 10 melon cultivars resulted in significantly higher concentrations of ascorbic acid than conventional production, whereas total phenolics content was not consistently different in the two systems (Salandanan *et al.*, 2009). However, most of the variation in antioxidant properties is explained by genotype and not by the production system, conventional or organic.

Grafting affects compositional quality of melon fruit, but the effects of this technique are not consistent. Ascorbic acid, total phenolics, and the content of phenolic acids accumulated at lower levels in melons grafted on *Cucurbita moschata* (Kolayli *et al.*, 2010). Ascorbic acid content was not affected by grafting melon into two interspecific hybrids of *Cucurbita maxima* × *C. moschata* (Zhao *et al.*, 2011).

Maturity at Harvest

Maturity at harvest influences the compositional quality of melon. The level of total phenols is higher in the premature stage, decreases during maturation and increases again during ripening. In contrast, total antioxidant activity and vitamin C content in melon peak at the pre-mature stage and carotenoids peak at the mature stage (Menon and Rao, 2012). Total antioxidant activity does not seem to change during melon ripening or during the normal harvest period (Menon and Rao, 2012).

Total carotenoids have been shown to decrease during the harvest period in one cultivar and to increase in another cultivar with ascorbic acid levels varying in the opposite way (Maietti *et al.*, 2012). This example highlights the difficulty in predicting the outcome of the various genotypic and environmental factors at play in determining the levels of bioactive compounds in melon fruit.

Cucurbitacin levels, when present, reach their highest concentration in mature fruit (Gry *et al.*, 2006).

Post-Harvest Factors: Changes between Harvest and Consumption or Processing

Refrigeration

Temperature is the most important environmental factor limiting post-harvest life of fruit and vegetables. Recommended storage temperature for inodorus melons is 7–10°C and for netted melons is 2–7°C with *ca.* 95% relative humidity (RH). Exposure to lower temperatures may result in chilling injury.

Storage temperature (5 or 10°C) had virtually no effect on total hydrophilic and lipophilic antioxidant activities in orange-fleshed honeydew, but ascorbate, β-carotene, and folic acid contents were differentially affected by storage temperature in different cultivars (Lester and Hodges, 2008). 'Honey Gold,' 'Orange Delight,' and 'Orange Dew' had higher levels of ascorbate and β-carotene content when stored at 5°C while folic acid content was better maintained during storage at 10°C. 'Temptation' had lower ascorbate and β-carotene content during storage at 5°C than at 10°C (Lester and Hodges, 2008).

Storage Time

Time is a key variable requiring management in all kinetic processes and determines the final concentrations of nutritional and bioactive compounds in foods. Genotype effects are relevant in the post-harvest changes of melon bioactive compounds. Total ascorbic acid content declined during cold storage in a non-netted green and in an orange-fleshed muskmelon, but not in a netted muskmelon cultivar. β-Carotene content has been reported to decreases during storage at 7 °C, with the greatest reduction occurring in the initial 2 weeks and little variation thereafter, or to remain constant during storage. Significant interactions between cultivar and storage conditions are reported for the loss of β-carotene. Phenolics have been reported to increase or remain constant (Gil *et al.*, 2006) during melon storage.

Controlled or modified atmosphere

Controlled or modified atmospheres can be used as a supplement to refrigeration to preserve whole cantaloupe quality. The recommended gas levels are 3 to 5% O_2 and 10 to 20% CO_2. These atmospheric conditions reduce ripening and ethylene production and the high CO_2 help prevent fungal decay. The potential benefit, however, is considered limited. O_2 levels below 1% and CO_2 levels higher than 20% are injurious for cantaloupes. The effect of atmospheric composition on the preservation of whole melon phytochemicals has not received systematic attention.

Mild heat treatments

Hot water treatment of whole cantaloupes has been reported to have no effect or to reduce ascorbic acid content during subsequent storage. A hot water treatment of 53 °C for 1 min did not affect vitamin C content of whole Galia melons (Lalaguna, 1998). Despite the inconsistencies reported for ascorbic acid, heat-conditioning treatments affect the enzymic antioxidant system of whole melon. Galia melons treated at 59 °C for 30 s had increased catalase, and reduced glutathione reductase activity (Fogelman *et al.*, 2011), with a potential benefit for chilling tolerance.

Radiation-based treatments

Treatment with gamma radiation at the levels ≤ 1 kGy had no adverse effect on quality attributes and chemical composition, namely vitamin C and soluble solids content of melons (Lalaguna, 1998). Irradiation of whole cantaloupes with electron an beam accelerator of 1.0 kGy did not affect the fruit compositional quality attributes, including carotene content, during subsequently storage at 10 °C (Castell-Perez *et al.*, 2004).

OTHER WAYS IN WHICH COMPOSITION IS ALTERED

Processing Technologies

Melons are presented in the market as fresh whole fruit or minimally processed as ready-to-eat fresh-cut fruit. In addition, melon can be used as raw material for a number of processed food and beverage products, although the volumes of processed melon are low compared with those of other widely consumed fruit.

Fresh-Cut Processing

The relatively large size of the fruit and the need for preparation and waste disposal can make melon consumption inconvenient for the modern consumer. Convenience of consumption can be improved with fresh-cut processing. The value of fresh-cut processing is part of the reason for the popularity of fresh-cut melons within the fresh-cut fruit market. Fresh-cut processing in melon includes fruit washing and sanitation, rind removal, slicing and cubing, and packaging. Additionally, the use of additives, such as calcium salts, and coatings has been studied, although their commercial usage is negligible.

The loss of cellular compartmentation at the cut surface and the consequent interaction between oxidative enzymes and their substrate and the tissue exposure to oxygen are major factors in the loss of melon bioactive compounds (Oms-Oliu *et al.*, 2008a). Vitamin C is slightly better preserved in whole cantaloupes than in fresh-cut cubes. The initial vitamin C content in cantaloupe includes ascorbic acid and dehydroascorbic acid in the same proportions; the reduced form was well preserved after processing, whereas the levels of dehydroascorbic acid decreased (Gil *et al.*, 2006). The total carotenoid content in fresh-cut melon decreases after cutting and during the initial days in storage but remains steady toward the end of storage (Gil *et al.*, 2006). Wounding may stimulate accumulation of phenolic compounds via an increase in phenylalanine ammonia lyase activity; however, there are no significant changes in the concentration of total phenolics during storage of whole and fresh-cut cantaloupe (Gil *et al.*, 2006, Amaro *et al.*, 2013).

The recommended temperature for fresh-cut melon is 0–5°C. Cold storage at 3°C reduces the degradation rate of ascorbic acid in fresh-cut cantaloupe, consistent with Van't Hoff's rule. The activity of ascorbate oxidase and ascorbate peroxidase are reduced at low temperature and enzymic changes to ascorbic acid levels are minor under cold storage. An interruption in the cold-chain results in a significant increase in enzyme activities and ascorbic acid degradation rate.

Fresh-cut melons are thought to benefit from modified atmosphere packaging, with recommended gas levels of 3–5% O_2 + 6–15% CO_2 and 2% O_2 + 10% CO_2 for cantaloupe and honeydew, respectively. Oxygen availability in the package headspace enhances the degradation of vitamin C. An atmosphere with 2.5% O_2 + 7% CO_2 induced a higher production of phenolic compounds compared to atmosphere conditions with higher O_2 levels, which may be related to an enhanced oxidative stress induced by hypoxia or high CO_2 concentration inside packages (Oms-Oliu *et al.*, 2008a). Consistently, ascorbic acid oxidation was greatly promoted by the presence of oxygen and thus, a marked decrease in vitamin C content was observed in cantaloupe samples under 10% O_2 + 7% CO_2, 21, 30, or 70% O_2 than under 2.5% O_2. An atmosphere with 2.5% O_2 + 7% CO_2 induced a higher accumulation of phenolic compounds compared to atmosphere conditions with higher oxygen levels. There are no evidences of CO_2 effect on melon bioactive compounds.

The inhibition of ethylene action in fresh-cut cantaloupe cubes, with a post-cutting treatment with 1-methylcyclopropene, does not affect total phenolics, total carotenoids, and β-carotene content during subsequent storage (Amaro *et al.*, 2013).

The effect of alginate, pectin, and gellan-based edible coatings on the shelf-life of fresh-cut 'Piel de Sapo' melon was investigated during 15 days at 4°C. Vitamin C retention was significantly affected by the coating composition with gellan-based edible coatings significantly reducing the loss of vitamin C and enhancing phenolic accumulation and antioxidant activity (Oms-Oliu *et al.*, 2008b). The effect of these coatings was attributed to the putative modification of internal fruit atmosphere, but elicitor effects may play a role.

Thermal Processing Methods

Thermal technologies are widely used in processed fruit. Blanching is a processing step to ensure enzyme inactivation in frozen or dehydrated products, pasteurization ensures microbial stability in acid fruit products (pH < 4.5) but low acidity or neutral food matrices require sterilization for microbial safety and stability. Melon pH values range from 5.5 to 7.1, advising intensive heat treatments to suppress sporulated bacteria. Pasteurization of melon juice at 90°C for 60 s resulted in a decrease in juice β-carotene and ascorbic acid contents (Chen *et al.*, 2009) while a rapid pasteurization method using saturated water vapor at 100°C for 45 s increased ascorbic acid content of melon cubes (Tirawat *et al.*, 2010).

Melons are generally not suitable for canning, since the intense heat treatments extensively degrade the texture. However, as in other matrices, it is expected that intense heat treatments degrade heat-sensitive ascorbic acid and carotenoids.

Freezing is effective in preserving the composition of low-fat fruit.

Ultrasound Processing

Ultrasound processing, also called sonication, is a non-thermal technology to inactivate microorganisms and enzymes at room temperature. Cantaloupe juice processed at a constant ultrasound frequency of 19 kHz and power intensity of 376 W/cm^2 for 10 min reduced the levels of phenolics by 30% (Fonteles *et al.*, 2012).

High-Pressure Processing

High-pressure processing (HPP) is reported to inactivate microorganisms and denature several enzymes with minimal impacts on the bioactive compounds of fruit. HPP applied to cut melon at 600 MPa for 10 min and subsequently stored at 4°C reduced vitamin C content comparing with untreated controls, in a cultivar dependent manner (Wolbang *et al.*, 2008). HPP-treated melon pieces had higher levels of β-carotene than fresh melon due to enhanced extraction resulting from the denaturing of the protein–carotenoid complex (Wolbang *et al.*, 2008).

Dense-Phase Carbon Dioxide Processing

Dense-phase carbon dioxide processing (DP-CO_2) is a non-thermal alternative to pasteurization for juice. DP-CO_2-processed juices are subjected to elevated pressures of CO_2 (8–35 MPa) for a period of time at mild temperatures (35–65°C). Processing with DP-CO_2 requires 5–8% CO_2 by weight, and the pressures used are at least one order of magnitude less than those typically used in HPP. Because pH of liquid foods is reduced, oxygen is removed from the environment and the temperatures used are moderate, antioxidant activity and vitamins are better preserved by DP-CO_2. DP-CO_2 treatment of Hami melon juice preserves ascorbic acid and β-carotene levels better than a high-temperature/short-time heat treatment (Chen *et al.*, 2009).

Dehydration and Roasting

Osmotic dehydration involves immersion of the fruit tissue in aqueous solutions of high concentrations of sugars or salts. Osmotic infusion of fruits can also be used as a pretreatment step prior to processing operations such as convection drying or freezing thereby reducing the overall energy requirements in the dehydration process (Jacob and Paliyath, 2012).

The process of osmotic dehydration of melon tissue followed by air drying has been studied using salted sucrose, glucose or mannitol solutions, at different concentrations and ratios, under vacuum and mild heat (65°C) and air-dried (60°C). It is not clear how these processes affect melon biologically active compounds but the loss of water-soluble compounds is anticipated: e.g., loss of anthocyanins was observed during osmotic infusion of blueberry and cherry while 2,2-diphenyl-1-picrylhydrazyl (DPPH) radical scavenging activity of fresh and infused fruit extracts did not differ significantly suggesting that the antioxidant activity of infused fruits is not impaired by the process (Jacob and Paliyath, 2012).

Dehydration by spray drying is used in the wide range of liquid products to yield dry powders and agglomerates. Freeze-drying is a process whereby water is removed by sublimation of ice in the materials. It is generally recommended for drying of materials containing heat-sensitive components such as tocopherols, ascorbic acid, carotenoids and phenolics. In muskmelon cubes, freeze-drying preserved the levels of total phenolic compounds, ascorbic acid, and β-carotene (Shofian *et al.*, 2011).

Roasting can increase the antioxidant activity, as demonstrated in bitter melon (Choi *et al.*, 2012). Although flavanols decreased following roasting, total phenolic content and flavan-3-ol and phenolic acid contents increased, leading to a general increase in antioxidant activities after roasting (Choi *et al.*, 2012).

Analytical Techniques

The quantification of total phenolics is made by the Folin–Ciocalteu method and the determination of total antioxidant activity by the reaction of melon extracts with ABTS, DPPH, and ferric-reducing antioxidant power (FRAP). Melon fruit extracts are obtained after tissue homogenization with water/methanol, followed by centrifugation and filtration (Amaro *et al.*, 2013; Fonteles *et al.*, 2012; Oms-Oliu *et al.*, 2008a; Shofian *et al.*, 2011). Ascorbic acid and carotenoid contents are determined through chromatographic techniques using diode array detector (HPLC-DAD) and HPLC with electrospray tandem mass spectrometry detection (HPLC-MS-MS ESI).

Extracts for vitamin C determination are prepared with citric acid, EDTA disodium salt, methanol, and NaF (Gil *et al.*, 2006; Silveira *et al.*, 2013), or metaphosforic acid and DL-1, 4-dithiothreitol (Oms-Oliu *et al.*, 2008). The pH of the filtrate is adjusted to 2.35–2.40 flushed through a Sep-Pak C18 cartridge and then filtered. 1,2-phenylenediamine dihydrochloride solution is added to the extract for dehydroascorbic acid derivatization into the fluorophore 3-(1,2-dihydroxyethyl) furo [3,4-b]quinoxaline-1-one and after 37 min in darkness, samples are analyzed by HPLC (Gil *et al.*, 2006; Silveira *et al.*, 2013; Oms-Oliu *et al.*, 2008; Wolbang *et al.*, 2008). A 0.01% solution of sulphuric acid adjusted to pH 2.6 (Gil *et al.*, 2006) or a mixture of methanol and water with hexadecyl trimethyl ammonium bromide and potassium di-hydrogen phosphate (Silveira *et al.*, 2013) are used as solvents for HPLC analyses.

Carotenoids extraction is performed using methanol and hexane containing BHT (Gil *et al.*, 2006), heptane (Hodges and Lester, 2006), hexane, ethanol and acetate (Amaro *et al.*, 2013) or tetrahydrofuran with methanol and ascorbic acid (Wolfbang *et al.*, 2010). After extractions, aliquots of the samples are injected onto a 5-μm

C18 spherical particles, packed in a 250×4.6-mm (Gil *et al.*, 2006; Oms-Oliu *et al.*, 2008) or 150×4.6-mm chromatographic column (Hodges and Lester, 2006). Primary solvents used for carotenoids were acetonitrile, methanol and methylene chloride, containing 0.1% butylated hydroxytoluene, 0.05M ammonium acetate and 0.05% triethylamine (Gil *et al.*, 2006). Other studies use a mixture of acetonitrile, methanol, dichloromethane, hexane and ammonium acetate (Amaro *et al.*, 2013; Wolbang *et al.*, 2008). Also a ternary mixture of acetonitrile-methanol-ethyl acetate was used (Shofian *et al.*, 2011).

References

Amaro, A.L., Fundo, J.F., Oliveira, A., Beaulieu, J.C., Fernández-Trujillo, J.P., Almeida, D.P.F., 2013. 1-Methylcyclopropene effects on temporal changes of aroma volatiles and phytochemicals of fresh-cut cantaloupe. J. Sci. Food Agric. 93, 828–837.

Basch, W.E., Gabardi, S., Ulbricht, C., 2003. Bitter melon (*Momordica charantia*): A review of efficacy and safety. Am. J. Health-Syst. Ph. 60, 356–359.

Castell-Perez, E., Moreno, M., Rodriguez, O., Moreira, R.G., 2004. Electron beam irradiation treatment of cantaloupes: effect on product quality. Food Sci. Technol. Int. 10, 383–390.

Chen, J., Zhang, J., Feng, Z., Song, L., Wu, J., Hu, X., 2009. Influence of thermal and dense-phase carbon dioxide pasteurization on physicochemical properties and flavor compounds in Hami melon juice. J. Agric. Food Chem. 57, 5805–5808.

Choi, J.S., Kim, H.Y., Seo, W.T., Lee, J.H., Cho, K.M., 2012. Roasting enhances antioxidant effect of bitter melon (*Momordica charantia* L.) increasing in flavan-3-ol and phenolic acid contents. Food Sci. Biotechnol. 21, 19–26.

Devendra, N.K., Attard, E.G., Raghunandan, D., Seetharam, Y.N., 2011. Study on seasonal variation on the content of cucurbitacin of various vegetative parts of *Trichosanthes cucumerina* L. var. *cucumerina*. Int. J. Plant Res. 1, 25–28.

Fogelman, E., Kaplan, A., Tanami, Z., Ginzberg, I., 2011. Antioxidative activity associated with chilling injury tolerance of muskmelon (*Cucumis melo* L.) rind. Sci. Hortic. 128, 267–273.

Fonteles, T.V., Costa, M.G.M., Jesus, A.L.T., Miranda, M.R.A., Fernandes, F.A.N., Rodrigues, R., 2012. Power ultrasound processing of cantaloupe melon juice: Effects on quality parameters. Food Res. Int. 48, 41–48.

Gil, M.I., Aguayo, E., Kader, A.A., 2006. Quality changes and nutrient retention in fresh-cut versus whole fruits during storage. J. Agric. Food Chem. 54, 4284–4296.

Gry, J., Soborn, I., Andersson, H.C., 2006. Cucurbitacins in plant food. Nordic Council of Ministers, Ekspressen Tryk and Kopicenter, Copenhagen.

Hodges, D.M., Lester, D.M., 2006. Comparisons between orange- and green-fleshed non-netted and orange-fleshed netted muskmelons: Antioxidant changes following different harvest and storage periods. J. Am. Soc. Hortic. Sci. 131, 110–117.

INRA, 2009. INRA (Institut National de la Recherche Agronomique)-Phenol Explorer Database. Available at: http://www.phenol-explorer.eu/.

Jacob, J.K., Paliyath, G., 2012. Infusion of fruits with nutraceuticals and health regulatory components for enhanced functionality. Food Res. Int. 45, 93–102.

Kolayli, S., Kara, M., Tezcan, F., Erim, F.B., Sahin, H., Ulusoy, E., Aliyazicioglu, R., 2010. Comparative study of chemical and biochemical properties of different melon cultivars: standard, hybrid, and grafted melon. J. Agric. Food Chem. 58, 9764–9769.

Lalaguna, F., 1998. Response of 'Galia' muskmelons to irradiation as a quarantine treatment. HortScience 33, 118–120.

Laur, L.M., Tian, L., 2011. Provitamin A and vitamin C contents in selected California-grown cantaloupe and honeydew melons and imported melons. J. Food Comp. Anal. 24,194–201.

Lester, G.E., Crosby, K., 2002. Ascorbic acid, folic acid and potassium content in postharvest green-fleshed honeydew muskmelons: influence of cultivar, fruit size, soil type, and year. J. Am. Soc. Hortic. Sci. 127, 843–847.

Lester, G.E., Saftner, R.A., Hodges, D.M., 2007. Market quality attributes of orange-fleshed, non-netted honeydew melon genotypes following different growing seasons and storage temperature durations. HortTechnology 17, 346–352.

Lester, G.E., 2008. Antioxidant, sugar, mineral, and phytonutrient concentrations across edible fruit tissues of orange-fleshed honeydew melon (*Cucumis melo* L.). J. Agric. Food Chem. 56, 3694–3698.

Lester, G.E., Hodges, D.M., 2008. Antioxidants associated with fruit senescence and human health: Novel orange-fleshed non-netted honey dew melon genotype comparisons following different seasonal productions and cold storage durations. Postharvest Biol. Technol. 48, 347–354.

Lester, G.E., Jifon, J.L., Makus, D.J., 2010. Impact of potassium nutrition on postharvest fruit quality: Melon (*Cucumis melo* L.) case study. Plant Soil 335, 117–131.

Maietti, A., Tedeschi, P., Stagno, C., Bordiga, M., Travaglia, F., Locatelli, M., Arlorio, M., Brandolini, V., 2012. Analytical traceability of melon (*Cucumis melo var reticulatus*): proximate composition, bioactive compounds, and antioxidant capacity in relation to cultivar, plant physiology state, and seasonal variability. J. Food Sci. 77, 646–652.

Menon, S.V., Rao, T.V.R., 2012. Nutritional quality of muskmelon fruit as revealed by its biochemical properties during different rates of ripening. Int. Food Res. J. 19, 1621–1628.

Münger, H.M., Robinson, R.W., 1991. Nomenclature of *Cucumis melo* L. Cucurbit Genet. Cooperative Rep. 14, 43–44.

Oms-Oliu, G., Odriozola-Serrano, I., Soliva-Fortuny, R., Martín-Belloso, O., 2008a. The role of peroxidase on the antioxidant potential of fresh-cut 'Piel de Sapo' melon packaged under different modified atmospheres. Food Chem. 106, 1085–1092.

Oms-Oliu, G., Soliva-Fortuny, R., Martín-Belloso, O., 2008b. Using polysaccharide-based edible coatings to enhance quality and antioxidant properties of fresh-cut melon. LWT—Food Sci. Technol. 41, 1862–1870.

Salandanan, K., Bunning, M., Stonaker, F., Külen, O., Kendall, P., Stushnoff, C., 2009. Comparative analysis of antioxidant properties and fruit quality attributes of organically and conventionally grown melons (*Cucumis melo* L.). HortScience 44, 1825–1832.

Shofian, N.M., Hamid, A.A., Osman, A., Saari, N., Anwar, F., Dek, N.S.P., Hairuddin, M.R., 2011. Effect of freeze-drying on the antioxidant compounds and antioxidant activity of selected tropical fruits. Int. J. Mol. Sci. 12, 4678–4692.

Silveira, A.C., Aguayo, E., Artés, F., 2013. Shelf-life and quality attributes in fresh-cut Galia melon combined with fruit juices. LWT—Food Sci. Technol. 50, 343–348.

Tirawat, D., Meno, A., Fujiwara, H., Higo, K., Noma, S., Igura, N., Shimoda, M., 2010. Development of rapid hygrothermal pasteurization using saturated water vapor. Innov. Food Sci. Emerg. Technol. 11, 458–463.

Wolbang, C.M., Fitos, J.L., Treeby, M.T., 2008. The effect of high pressure processing on nutritional value and quality attributes of *Cucumis melo* L. Innov. Food Sci. Emerg. Technol. 9, 196–200.

Zhao, X., Guo, Y., Huber, D.J., Lee, J., 2011. Grafting effects on postharvest ripening and quality of 1-methylcyclopropene-treated muskmelon fruit. Sci. Hortic. 130, 581–587.

CHAPTER

21

Blueberry Phenolic Compounds: Fruit Maturation, Ripening and Post-Harvest Effects

Ines Eichholz, Susanne Huyskens-Keil*, Sascha Rohn†*

*Humboldt-Universität zu Berlin, Division Urban Plant Ecophysiology, Section Quality Dynamics/Postharvest Physiology, Berlin, Germany, †University of Hamburg, Hamburg School of Food Science, Institute of Food Chemistry, Hamburg, Germany

CHAPTER POINTS

- Blueberries are a good source of phytochemicals such as phenolic compounds.
- Processes of maturation and ripening influence the phenolic profile.
- Fruit maturation and ripening are characterized by a decrease in flavonol and hydroxycinnamic acid contents at early maturation and throughout ripening, while an accumulation of anthocyanins occurs.
- Due to the fact that phenolic compounds are mainly present in the berry skin and outer layer of the pulp, the increase of anthocyanin contents during successive harvests is influenced by the diminishing berry weight.
- In stored full-ripe berry fruits, all phenolic compounds are degraded.
- Berries harvested before the full-ripe stage continue maturation in post-harvest that include an increase of anthocyanins and a decrease of minor complex phenolic compounds (hydroxycinnamic acids, flavonols).

INTRODUCTION

In recent years, the research interest has been increasingly directed towards problems of the physiological functionality of food constituents arising from plant secondary metabolism, in particular polyphenols. Phenolic compounds include a wide range of different chemical classes with a high diversity in chemical structures and functions. They play important roles in the interaction of plants with their environment, e.g., protection from UV radiation, as feeding deterrents, pollination attractants, protective compounds against pathogens or various abiotic stresses, antioxidants or signaling molecules (Winkel-Shirley, 2001; Treutter, 2010).

Among berry fruits, blueberries are considered to be a rich source of phenolic compounds and are praised for their high antioxidant activity scores (Prior *et al.*, 1998). From the existing research results, it has to be considered that they comprise not only the main blueberries cultivated, i.e., the Northern and Southern highbush blueberries, but also the lowbush (or wild blueberries) and rabbiteye blueberries. Between these blueberry types as well as between different varieties and within other *Vaccinium* species, there are significant differences in the phenolic content and antioxidant activity (Prior *et al.*, 1998; Taruscio *et al.*, 2004).

Due to the increasing health awareness of the consumers and the apparent relationship of phytochemicals in plant foods with the prevention of degenerative diseases (Watzl and Leitzmann 2005), the content of phenolic compounds in blueberries has been studied. As a result, phenolic compound profile and quantitative composition of blueberries is well documented (e.g., Kalt *et al.*, 1999; Howell *et al.*, 2001; Moyer *et al.*, 2002).

Even though literature on blueberry issues is vast, expanding from different culture practices to potential health effects of blueberry extracts in animal studies, little is reported on the development of phenolic compounds during the process of fruit maturation and ripening pre- and post-harvest. It is known that the content of polyphenols in berries is not only affected by genetic differences,

Processing and Impact on Active Components in Food
http://dx.doi.org/10.1016/B978-0-12-404699-3.00021-4

pre-harvest environmental conditions, but also by physiological processes during maturation, ripening, and post-harvest (Zadernowski *et al.*, 2005). During fruit maturation and ripening, as well as after harvest, an array of physiological changes takes place in berry fruits. Therefore, the understanding of these processes and the identification of suitable maturity stages are important pre-requisites for crop exploitation and for the development of appropriate physiological adapted post-harvest technologies.

HOW COMPOSITION IS ALTERED

Fruit Maturation and Ripening of Blueberries

Fruit development is a dynamic process commencing with pollination and fertilization, followed by intensive cell division and expansion, with concurrent seed development as well as maturation (Vvedenskaya and Vorsa 2004). The growth and development of blueberry fruit during a plant's annual cycle is presented in Figure 21.1. Many complex biochemical changes occur within the single fruit during this time, including those that result in softening, heightened pigmentation, sweetening, and cell enlargement (Gough, 1993).

The development and ripening of blueberry fruits can last from 8 to 16 weeks, depending on cultivar and climatic conditions (Ebert, 2005) and was classified in three stages by Shoemaker (1948): (I) an accelerated growth rate for about 29 days after pollination; (II) a retarded pericarp growth of 5–56 days (the embryos and seeds develop and mature); and (III) a second period of rapid development that continues to fruit maturity (rapid increase in volume caused by cell enlargement).

Maturity may be defined in terms of either physiological maturity or horticultural maturity of plant food crops and is based on the measurement of various qualitative and quantitative parameters (Thompson and Thompson 2003). Physiological maturity of a crop was described by Watada *et al.* (1984) as "the stage of development when a plant or plant part will continue ontogeny even if detached," whereas horticultural commercial maturity is defined as "the stage of development when a plant or plant part possesses the prerequisites for the utilization by consumers for a particular purpose."

Maturation of blueberries occurs during the period between stage III of berry growth and ripening (Gough, 1993). Blueberries, growing on the same bush, do not all ripen at the same time and therefore, berries were harvested weekly over a period of three to four weeks. During maturation, blueberry fruits suffer from physiological changes that include softening of the flesh, a decrease in chlorophyll content with a subsequent increase in anthocyanin concentration (i.e., the berry turns from green to blue), and an increase in size (Gough, 1993). Moreover, blueberries are considered climacteric fruits showing two rises in respiration rate during their development, initially with the beginning of coloration in transition from the green–pink to the blue–pink stage, and secondly during the ripe stage towards over-ripening (Ebert, 2005).

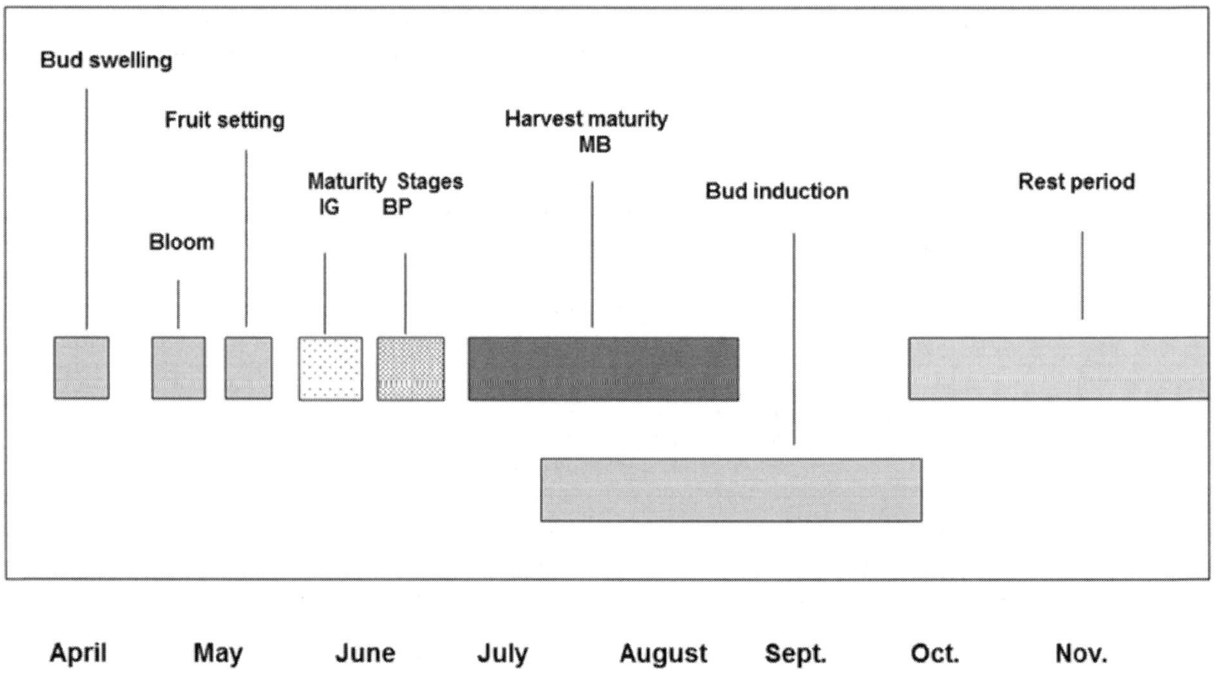

FIGURE 21.1 Growth and development of blueberry fruit during a plant's annual cycle. The figure shows the growth and development of blueberry fruit at three specific maturity stages: IG, Immature Green; BP, Blue Pink; MB, Mature. *Adapted with permission from Ebert (2005).*

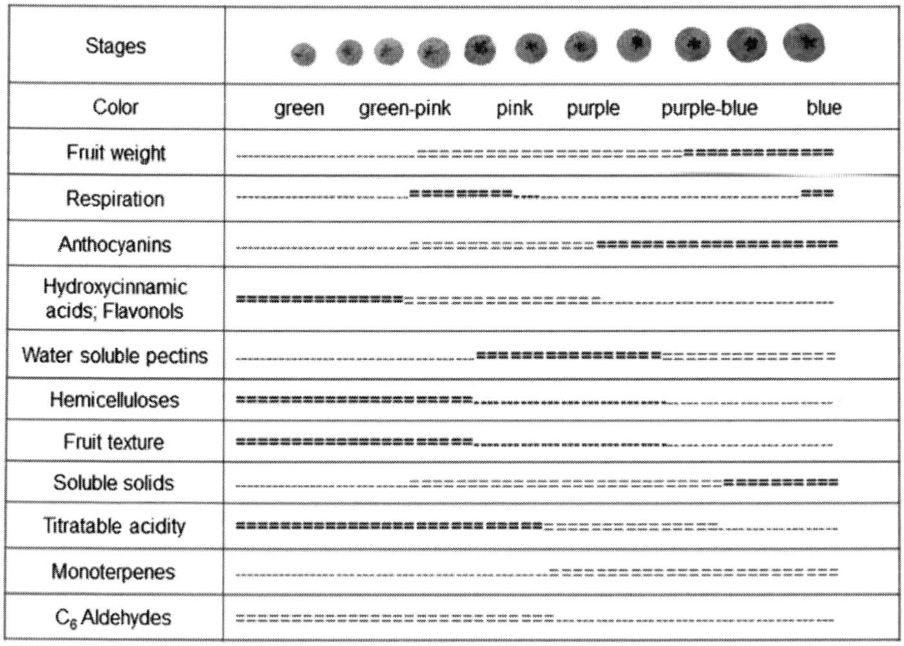

FIGURE 21.2 Physiological and chemical changes during ripening of blueberries. Quality attributes and chemical composition of blueberry cv. Bluecrop during maturation and ripening. Gray lines indicate minor values, dark lines indicate major values. *Source: Eichholz* et al. *(2008).*

All biochemical and physiological changes that take place during fruit ripening are driven by a cascade of molecular events starting with the activation of signalling pathways that stimulate the coordinated expression of fruit ripening-related genes (McManus, 2012). These correspond to visually verifiable changes such as fruit growth and color changes, textural changes such as fruit softening resulting from changes in structural carbohydrate pattern, as well as sensory changes including aroma composition and changes in sweetness and acidity (see Figure 21.2).

Phenolic Composition During Blueberry Fruit Maturation and Ripening

Color changes during fruit ripening implicate both synthesis and degradation of the plant pigments including chlorophyll, carotenoids, as well as flavonoids. Flavonoids and non-flavonoids are the two major classes of phenolic compounds common to small fruits (Howell *et al.*, 2001). The predominant flavonoids found in berries and red grapes are anthocyanins and flavonols, which are almost exclusively present in their glycosylated forms. Wu and Prior (2005) identified and characterized anthocyanins from 25 different fruits. Anthocyanins were found in 14 fruits ranging from two different anthocyanin structures in peaches (*Prunus persica*) and nectarines (*Prunus persica* var. *nucipersica*) to 31 different anthocyanin structures in grapes (*Vitis vinifera*). Blueberry cultivars were found to contain 20–27 different anthocyanins. Phenol profiling of blueberries showed that flavonoids (40%) and phenolic acids (59%) are the dominant phenolic compound fractions (Törrönen *et al.*, 1997). Major flavonoids comprise of several anthocyanins (Moyer *et al.*, 2002); however, they also include flavonols such as quercetin (Taruscio *et al.*, 2004). The most common flavonols in fruits and vegetables are quercetin and kaempferol with their corresponding glycosides. Bilyk and Sapers (1986) quantified the quercetin contents of the ripe highbush blueberries 'Earlible', 'Weymouth,' 'Coville,' and 'Bluetta' (24–29 mg/kg fresh weight), but neither kaempferol nor myricetin were found.

Not to be underestimated is the presence of phenolic acids, which were also found in diverse *Vaccinium* species (Taruscio *et al.*, 2004), including the phenolic acids caffeic, chlorogenic, p coumaric, ferulic, as well as p-hydroxybenzoic acid. Among these phenolic compounds, chlorogenic acid is the colorless phenolic compound to be found dominantly in both highbush and lowbush blueberries in concentrations ranging between 5 and 10 mg/kg fresh fruit (Gao and Mazza 1994).

The influence of cultivar and maturity on the content of polyphenols and antioxidant activity has been reported for berry fruits such as blackberry (*Rubus fructicosus*) (Siriwoharn *et al.*, 2004) and strawberry (*Fragaria* × *ananassa*) (Kosar *et al.*, 2004). Maturity stage is also an important factor affecting the content of phenolic compounds in blueberries. Mainland and Tucker (2000) reported that the anthocyanin and total phenolic content as well as the antioxidant activity (measured by *Oxygen Radical Absorbance Capacity*, ORAC) of five blueberry cultivars increased with progressing maturity. Here, the samples were picked from berries almost ripe with a blue color and red around the stem scar, in fully ripe stage with a blue color ready for marketing, and overripe beginning

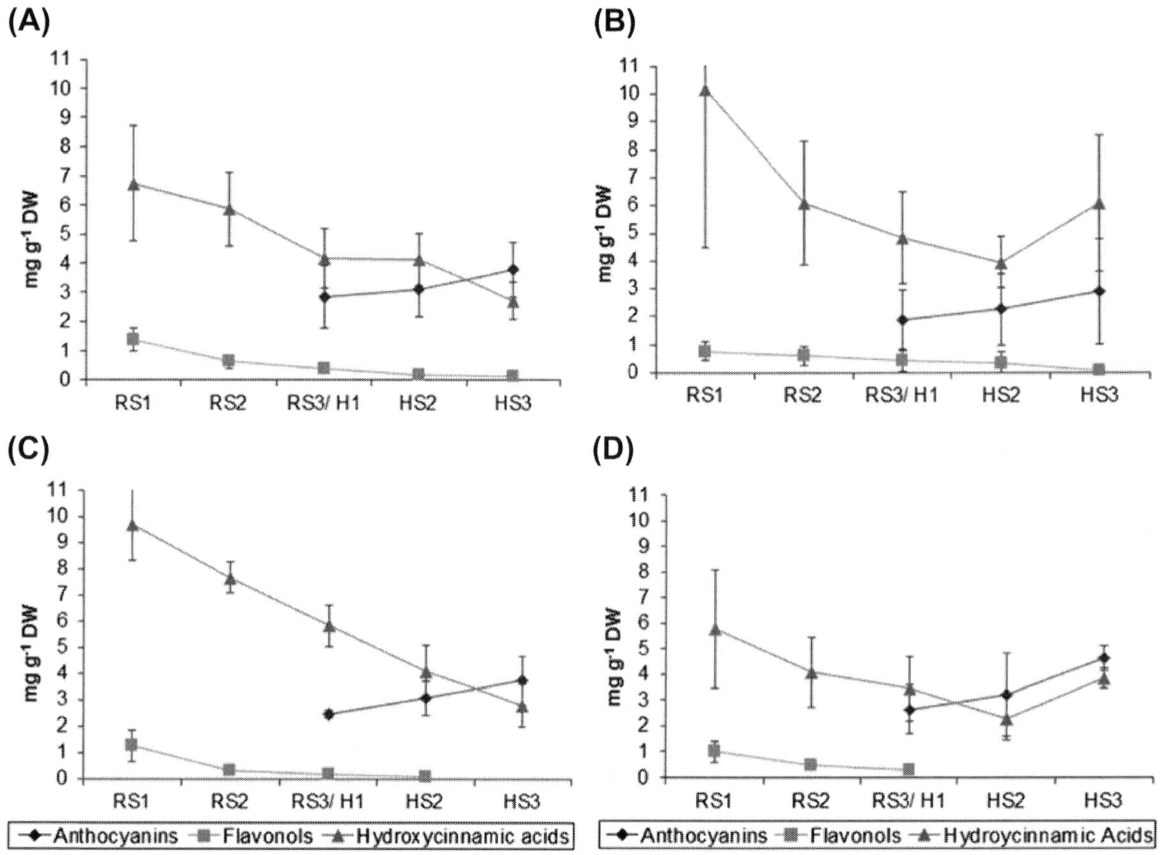

FIGURE 21.3 Content of phenolic subclasses of different blueberry cultivars during fruit development at selected maturity stages. The figure shows the mean ± SD of phenolic subclasses of blueberries. The corresponding phenolic compound is equivalent as follows: Anthocyanins = cyanidin-3-glucoside; Hydroxycinnamic acids = chlorogenic acid; Flavonols = rutin; A) Reka, B) Bluecrop, C) Puru, and D) Berkeley. RS1 = Ripening stage 1 = unripe green; RS2 = Ripening stage 2 = unripe purple; RS 3/H1 = Ripening stage 3/Harvest 1 = 50–70% ripe blue; HS2 = Harvest stage 2 = 20–40% ripe blue; HS3 = Harvest stage 3 ≤ 15% ripe blue. *With permission from Rodarte-Castrejon* et al. *(2008).*

already to soften. However, Rodarte-Castrejon *et al.* (2008) determined different phenolic compound classes and the antioxidant activity [measured by Electronic Paramagnetic Resonance spectroscopy (EPR) and Trolox Equivalent Antioxidant Capacity assay (TEAC) in four blueberry cultivars at five different stages (unripe green, unripe purple, 1.harvest, 2.harvest, and 3.harvest)]. They reported an increase in anthocyanins during maturation and ripening of berries (Figure 21.3). Here, their antioxidant activity was higher in early maturation and during initial pigmentation than at ripeness. Authors concluded that the latter finding was attributed to the higher concentrations of hydroxycinnamic acids and flavonols found prior to ripening; whereas lower antioxidant activity of mature berries might suggest that anthocyanins have less antioxidant capacities than the other phenolic compound classes.

Anthocyanins are not expected to be present in unripe green berries. This was confirmed by Kalt *et al.* (2003). The synthesis of anthocyanins in fruit tissues, as studied in bilberry (*Vaccinium myrtillus*) (Jaakola *et al.*, 2002), correlates with the expression of the flavonoid pathway genes during berry fruit ripening. At the time of harvest, anthocyanins are the second-most prevalent phenolic compounds in blueberry fruits and concentrations are higher in extracts of successive picking dates (Rodarte-Castrejon *et al.*, 2008).

The highest concentration of red pigments in blueberries occurs in the skin and outer layer of the pulp. Moreover, Gao and Mazza (1994) noted the relationship of anthocyanin accumulation with berry size in different blueberry cultivars. The small berries of lowbush blueberry cultivars and hybrids had higher anthocyanin contents in comparison to the larger berries of highbush blueberry cv. Bluecrop. Moyer *et al.* (2002) found that fruit size highly correlated ($r = 0.84$) with anthocyanins within *V. corymbosum*, but did not correlate with anthocyanins across eight other *Vaccinium* species, or within 27 blackberry hybrids. There was also no relationship between fruit weight and anthocyanin content among 135 lowbush clones and the 80 highbush clones studied by Howell *et al.* (2001). However, Rodarte-Castrejon *et al.* (2008) detected anthocyanins only at the ripe stages correlating significantly with fruit weight ($r = -0.521$,

$p \geq 0.01$). Here, fruit size increased progressively until full ripening and, thereafter, decreased with each successive picking date. Therefore, authors explained the correlation by an anthocyanin increase with successive harvests and, conversely, a fruit weight decrease.

The decrease of flavonols during the ripening process occurs not only in blueberries as presented by Rodarte-Castrejon *et al.* (2008), but also in other fruits, e.g., strawberries and cranberries (*Vaccinium macrocarpon* Ait.) (Stöhr and Herrmann, 1975; Celik *et al.*, 2008). Already in 1976, it had been observed that the elevated synthesis of anthocyanins during maturation and ripening was not associated with an accumulation of flavonol glycosides (Starke and Herrmann, 1976). When comparing the total content of flavonols with the total anthocyanins of blackberries, no correlation was found between the accumulation of flavonols and anthocyanins (Bilyk and Sapers, 1986). There is evidence that flavonoid biosynthesis is strongly regulated depending on the developmental stage of the fruit. The enzyme activities are regulated in response to different developmental and environmental cues (Jaakola *et al.*, 2002). In bilberry, a coordinated expression of flavonoid biosynthetic genes in relation to the accumulation of anthocyanins, proanthocyanidins, and flavonols was demonstrated in developing fruits. In strawberry, it was detected that flavonoid biosynthesis has two distinct key flavonoid enzymes activity peaks during fruit ripening at an early and late developmental stage (Halbwirth *et al.*, 2006). If this applies to other berry fruits such as blueberries, Rodarte-Castrejon *et al.* (2008) hypothesized from their results that during maturation and ripening the first flavonoid biosynthesis peak corresponds to flavonols and hydroxycinnamic acids, while the second peak relates to anthocyanin accumulation. Kalt *et al.* (2003) also reported an increase of anthocyanin content and a decrease of phenols during fruit ripening. They suggested that during ripening of highbush blueberry there is a shift in the pool of total phenols towards anthocyanin synthesis, and an overall decline in the content of further phenolic compounds.

Post-Harvest Effects on Phenolic Composition of Blueberry at Different Ripening Stages

Post-harvest handling comprises all steps from the producer to the consumer including harvesting, handling, post-harvest treatment, packaging, cooling, storing, shipping, wholesaling, and retailing. However, losses in post-harvest from harvest until the product has reached the consumer are comparatively high, specifically for highly perishable fruits that are very susceptible to decay and deterioration (Huyskens-Keil and Schreiner, 2004). Recently, strong emphasis has been placed on the development and improvement of new post-harvest treatments in berry fruits ranging from the improvement of new film packaging to the application of post-harvest elicitors accelerating the synthesis of stress compounds in the products, e.g., heat-shock proteins, phenolic compounds, leading to an extended shelf-life and improved product quality (Huyskens-Keil and Schreiner, 2004; Schreiner *et al.*, 2012).

Film packaging is widely used for berry fruits in order to prevent mechanical damage and maintain fruit quality. Studies on environmentally friendly film packaging material for berry fruits are still ongoing (e.g., Netzker *et al.*, 2005). For the blueberry cv. Bluecrop, it was found that biodegradable film packaging (based on corn starch) induced an increase in total phenols and antioxidant activity in fruits stored at 2°C for 3 weeks, indicating maturation processes that, however, did not differ to fruits packed in conventional polypropylene packaging material. In conclusion, a substitution of biodegradable film packaging for PVC or PP-packaging material still needs to be investigated more in detail.

Temperature is known to affect anthocyanin content in berry fruits, e.g., in blueberries (Kalt *et al.*, 1999) and strawberries (Miszczak *et al.*, 1995) during storage. Storing different types of blueberries at temperatures of 0–2°C induced phenolic synthesis and increased the fruit total antioxidant capacity (Kalt *et al.*, 1999; Netzker *et al.*, 2005). Blueberries harvested before the full-ripe stage continued maturation during storage being associated with an increase of anthocyanins and a decrease of minor complex phenolic compounds (hydroxycinnamic acids, flavonols) (unpublished data). In contrast, for stored full-ripe blueberry fruits it was found that anthocyanins and flavonols decreased while phenolic acids (hydroxycinnamic acids and hydroxybenzoic acids) remained almost constant during storage (Figure 21.4, unpublished data).

Controlled atmosphere (CA) storage (2.5 kPa O_2 + 15 kPa CO_2) for up to 6 weeks increased total phenolic content of blueberry cultivars ('Centurion' and 'Maru'). However, a negative relationship existed between antioxidant activity and phenol content during storage in CA storage compared to regular atmospheric storage conditions at low temperature. It is assumed that prolonged CA storage may lead to the oxidation of the phenolic compounds (Schotsmanns *et al.*, 2007). Zheng *et al.* (2005) reported that antioxidant capacity is correlated with total phenolic and anthocyanin contents and, thus, high-oxygen treatments may improve the antioxidant capacity of blueberry fruit, however, only to a limited storage period as these effects were found to diminish with prolonged storage duration.

Post-harvest strategies that included enriched CO_2 atmosphere and/or SO_2 fumigation were found to maintain polyphenol content and total antioxidant activity of blueberry fruit, but varying significantly among cultivars. SO_2 fumigation followed by controlled atmosphere

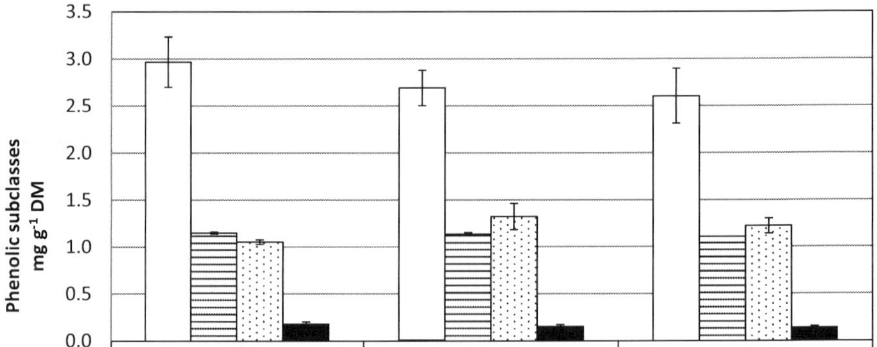

	control/ 0 days		7 days		14 days	
□ AY	2.97	a	2.69	a	2.60	a
⊟ HYZ	1.15	b	1.14	b	1.11	a
⊡ HYB	1.05	a	1.32	b	1.22	ab
■ FLAV	0.19	b	0.16	ab	0.14	a

Storage time

FIGURE 21.4 Change of phenolic subclasses profile in blueberry cv. Bluecrop during storage. Contents of phenolic subclasses (mean ± SD) in blueberry during 14 days of storage under simulated retail condition (12 h at 20°C and 12 h at 5°C, 80% rH) are shown. The corresponding phenolic compound is equivalent as follows: AY = anthocyanins = cyanidin-3-glucoside; HYZ = hydroxycinnamicacids = chlorogenicacid; HYB = hydrobenzoic acids = gallic acid; FLAV = flavonols = rutin. *Source: Authors' unpublished data.*

storage (3% O_2 + 6 or 12% CO_2) is reported to be a promising post-harvest strategy for fresh blueberries to reduce decay, extend market life, and maintain high nutritional value (Cantin *et al.*, 2012).

Elicitor applications also comprise the targeted UV-C and UV-B treatment in post-harvest for meeting hygienic requirements on the one hand and for increasing health-promoting phytochemical compounds of fruits and vegetables on the other hand (Schreiner *et al.*, 2012). Flavonoids have various physiological functions such as antioxidant as well as UV protective effects (Treutter, 2010). Especially anthocyanins and quercetin are known as protective pigments against UV irradiation and thus, increases in fruits upon post-harvest UV-irradiation (Schreiner *et al.*, 2012). Wang *et al.* (2009) indicated that post-harvest application of UV-C (2.15 and 4.30 kJ/m^2) is effective in stimulating the antioxidant capacity and flavonoid content of blueberries, however, differing in dependence on the original flavonoid profile. Time-course measurements of the effect of UV-C revealed that the strongest responses of fruit to UV-C treatment occurred instantly after the illumination and the effects diminished with storage time.

Post-harvest UV-B treatment of ripe black currant (*Ribes nigrum*) induced the increase of total phenol content and phenolic composition (flavonols, anthocyanins, hydroxycinnamic, and hydroxybenzoic acids) to a large extent (Huyskens-Keil *et al.*, 2007). Anthocyanins were concluded to absorb UV radiation within a short time; meanwhile flavonols and phenolic acids were assumed to have an impact on antioxidant protection of UV-B-mediated tissue damage. Recent studies reported an increase of phenolic compounds and corresponding antioxidant activity in blueberry cv. Bluecrop resulting from an UV-B post-harvest mediated stress response (Eichholz *et al.*, 2010).

OTHER WAYS IN WHICH COMPOSITION IS ALTERED- FURTHER IMPACTS ON PHENOLIC COMPOUNDS IN BLUEBERRY FRUITS

It is well accepted that maturity at harvest plays an important role for quality and storability of fruits and vegetables (Shewfelt, 1993). However, various factors in pre- and post-harvest influence the composition and quality of fruits. These include: genetic factors (selection of cultivars and rootstocks), pre-harvest climatic conditions and cultural practices, maturity at harvest and harvesting method, post-harvest handling procedure, post-harvest treatments, and processing methods (Kader, 2002). These factors also affect the phenolic profiles of plant tissues including the flavonoid biosynthesis being tissue specific, developmentally regulated and induced by a variety of environmental factors, e.g., light, UV-radiation, fungal infection, interaction with microorganisms, or wounding (Winkel-Shirley, 2001; Huyskens-Keil *et al.*, 2007). Most of these factors imply stress in plants and, thus, result in an accumulation of phenolic compounds as a plant defense response. For example, excessive fertilization as well as a shortage of macro- and microelements indicates a change in phenol concentration as a defense reaction in plants. However, here the element boron gains special attention. Due to its structure, boron forms stable phenol–borate complexes and, therefore, high boron availability inactivates soluble phenolic compounds in plants (Brown *et al.*, 2008). In this case, higher but not excessive concentrations of boron in fruits, applied by foliar spraying, led to lower total phenolic contents in blueberries, as it was reported by Eichholz *et al.* (2011). Moreover, it has been assumed that under boron deficiency, polyphenols concentrate via the stimulation of the enzyme phenylalanine-ammonium lyase (PAL) (Cakmak *et al.*, 1995), the key enzyme that induces the flavonoid metabolism in plants (Winkel-Shirley, 2001). Nevertheless, it is noticeable that most of the pre-harvest factors accelerating the synthesis of phenolic compounds also influence other (undesired) metabolic processes, e.g., leading to a pronounced loss of biomass and yield of blueberries (Eichholz, 2008).

ANALYTICAL TECHNIQUES

Primarily, when investigating phenolic compounds, total phenolic content (often abbreviated as TPC) is estimated spectrophotometrically in a high variety of different approaches. One of the most cited works in this context is Singleton *et al.* (1999), who described the use of the Folin–Ciocalteu reagent for an easy-to-use assay for all kinds of plant extracts.

Traditionally, as already mentioned for the analysis of the profile of phenolic compounds, high-performance liquid chromatography (HPLC) is used with various detection systems. Detection with UV light is the common one, but also its use over a selected wavelength range (diode array detection), the use of electrochemical detection systems, or mostly preferred coupling to mass spectrometry is highly appreciated. Similarly, as for the estimation of the total phenolic content, also in this case many different operating procedures are applied, differing in the selection of extraction solvents (e.g., 70% methanol *versus* 80% ethanol), use of solid-phase extraction cartridges (e.g., polyamide *versus* C18 material), chromatographic conditions, especially solvent gradients, etc. With regard to the analysis in blueberries, from the mid 1990s, the groups of Kalt *et al.* (1999, 2003), or the US and Scandinavian groups (Heinonen *et al.*, 1998; Prior *et al.*, 1998, 2001; Häkkinen *et al.*, 1999, 2000) presented some approaches that are still the basis for other research groups in this field. At this point, the reader is referred to these and further publications for detailed methodological descriptions.

References

Bilyk, A., Sapers, G.M., 1986. Varietal differences in the quercetin, kaempferol, and myricetin contents of highbush blueberry, cranberry, and thornless blackberry fruits. J. Agric. Food Chem. 34, 585–588.

Brown, P.H., Bellaloui, N., Wimmer, M.A., Bassil, E.S., Ruiz, J., Hu, H., Pfeffer, H., Dannel, F., Römheld, V., 2008. Boron in plant biology. Plant Biol. 4, 205–223.

Cakmak, I., Kurz, H., Marschner, H., 1995. Short-term effects of boron, germanium and high light intensity on membrane permeability in boron deficient leaves of sunflower. Physiol. Plant 95, 11–18.

Cantin, C.M., Minas, I.S., Goulas, V., Jiménez, M., Manganaris, G.A., Michailides, T.J., Crisosto, C.H., 2012. Sulfur dioxide fumigation alone or in combination with CO_2-enriched atmosphere extends the market life of highbush blueberry fruit. Postharvest Biol. Technol. 67, 84–91.

Celik, H., Ögen, M., Serce, S., Kaya, C., 2008. Phytochemical accumulation and antioxidant capacity at four maturity stages of cranberry fruit. Sci. Hort. 117, 345–348.

Ebert, G., 2005. Anbau von Heidelbeeren und Cranberries. Eugen Ulmer KG, Stuttgart, p.104.

Eichholz, I., 2008. Veränderung bioaktiver Inhaltsstoffe in Kulturheidelbeeren (*Vaccinium corymbosum* L.) in Abhängigkeit von der Mineralstoffverfügbarkeit. Verlag im Internet GmbH, Humboldt-Universität zu Berlin, dissertation.de, p. 174.

Eichholz, I., Huyskens-Keil, S., Keller, A., Ulrich, D., Kroh, L.W., Rohn, S., 2010. UV-B-induced changes of volatile metabolites and phenolic compounds in blueberries (*Vaccinium corymbosum* L.). Food Chem. 126, 60–64.

Eichholz, I., Huyskens-Keil, S., Kroh, L.W., Rohn, S., 2011. Phenolic compounds, pectin and antioxidant activity in blueberries (*Vaccinium corymbosum* L.) influenced by boron and mulch cover. J. Appl. Bot. Food Qual. 84, 26–31.

Gao, L., Mazza, G., 1994. Quantitation and distribution of simple and acylated anthocyanins and other phenolics in blueberries. J. Food Sci. 59, 1057–1059.

Gough, R.E., 1993. The highbush blueberry and its management. The Haworth Press- Food Products Press, New York, p. 288.

Halbwirth, H., Puhl, I., Haas, U., Jezik, K., Treutter, D., Stich, K., 2006. Two-phase flavonoid formation in developing strawberry (Fragaria x ananassa) fruit. J. Agric. Food Chem. 54, 1479–1485.

Heinonen, I.M., Meyer, A.S., Frankel, E.N., 1998. Antioxidant activity of berry phenolics on human low-density lipoprotein and liposome oxidation. J. Agric. Food Chem. 46, 4107–4112.

Häkkinen, S., Heinonen, M., Karenlampi, S., Mykännen, H., Ruuskanen, J., Törrönen, R., 1999. Screening of selected flavonoids and phenolic acids in 19 berries. Food Res. Int. 32, 345–353.

Häkkinen, S.H., Törrönen, A.R., 2000. Content of flavonols and selected phenolic acids in strawberries and *Vaccinium* species: influence of cultivar, cultivation site and technique. Food Res. Int. 33, 517–524.

Howell, A., Kalt, W., Duy, J.C., Forney, C.F., McDonald, J.E., 2001. Horticultural factors affecting antioxidant capacity of blueberries and other small fruit. HortTechnol.11, 523–528.

Huyskens-Keil, S., Schreiner, M., 2004. Quality dynamics and quality assurance of fresh fruit and vegetable products in pre- and postharvest. In: Dris, R., Jain, S.M. (Eds.), Production practices and quality assessement of food crops, Vol. 3. Kluwer Academic Publishers, Dordrecht, Boston, London, pp. 401–449.

Huyskens-Keil, S., Eichholz, I., Kroh, L.W., Rohn, S., 2007. UV-B induced changes of phenol composition and antioxidant activity in black currant fruit (*Ribes nigrum* L.). J. Appl. Bot. Food Qual. 81, 140–144.

Jaakola, L., Määttä, K., Pirttilä, A.M., Törrönen, R., Karenlampi, S., Hohtola, A., 2002. Expression of genes involved in anthocyanin biosynthesis in relation to anthocyanin, proanthocyanidin, and flavonol levels during bilberry fruit development. Plant Physiol. 130, 729–739.

Kader, A.A., 2002. Fruits in the global market. In: Knee, M. (Ed.), Fruit quality and its biological basis. Sheffield Academic Press, Sheffied, p. 279.

Kalt, W., Forney, C.F., Martin, A., Prior, R.L., 1999. Antioxidant capacity, vitamin C, phenolics, and anthocyanins after fresh storage of small fruits. J. Agric. Food Chem. 47, 4638–4644.

Kalt, W., Lawand, C., Ryan, D.A.J., McDonald, J.E., Donner, H., Forney, C.F., 2003. Oxygen radical absorbing capacity, anthocyanin and phenolic content of highbush blueberries (*Vaccinium corymbosum* L.) during ripening and storage. J. Am. Soc. Hort. Sci. 128, 917–923.

Kosar, M., Kafkas, E., Paydas, S., Baser, K.H.C., 2004. Phenolic composition of strawberry genotypes at different maturation stages. J. Agric. Food Chem. 52, 1586–1589.

Mainland, C.M., Tucker, J.W., 2000. Blueberry health information-some new mostly review. Acta Hort. 574, 39–43.

McManus, M.T., 2012. Annual Plant Reviews, The plant hormone ethylene. Wiley-Blackwell. , p. 416.

Miszczak, A., Forney, C.F., Prange, R.K., 1995. Development of aroma volatiles and colour during postharvest ripening of 'Kent' strawberries. J. Amer. Soc. Hort. Sci. 120, 650–655.

Moyer, R.A., Hummer, K.E., Finn, C.E., Frei, B., Wrolstad, R.E., 2002. Anthocyanins, phenolics, and antioxidant capacity in diverse small fruits: *Vaccinium*, *Rubus*. Ribes. J. Agric. Food Chem. 50, 519–525.

Netzker, U., Eichholz, I., Huyskens-Keil, S., Rohn, S., Kroh, L.W., 2005. Einfluss von biologisch abbaubaren Verpackungen auf die Nacherntequalität von Kulturheidelbeeren (*Vaccinium corymbosum* L.) unter besonderer Berücksichtigung bioaktiver Inhaltsstoffe. In: Albinus, B. (Ed.), Chancen und Grenzen der nicht-invasiven Qualitätsanalytik im Gartenbau – Wie ist Produktqualität messbar ?. Gartenbauwissenschaftliche Tagung Geisenheim. BGHL, 42, p. 110.

Prior, R.L., Cao, G., Martin, A., Sofic, E., McEwen, J., O'Brien, C., Lischner, N., Ehlenfeldt, M., Kalt, W., Krewer, G., 1998. Antioxidant capacity as influenced by total phenolic and anthocyanin content, maturity, and variety of *Vaccinium* species. J. Agric. Food Chem. 46, 2686–2693.

Prior, R.L., Lazarus, S.A., Cao, G.H., Muccitelli, H., Hammerstone, J.F., 2001. Identification of procyanidins and anthocyanins in blueberries and cranberries (*Vaccinium* spp.) using high-performance liquid chromatography/mass spectrometry. J. Agric. Food Chem. 49, 1270–1276.

Rodarte-Castrejon, A.D., Eichholz, I., Rohn, S., Kroh, L.W., Huyskens-Keil, S., 2008. Phenolic profile and antioxidant activity of highbush blueberry (*Vaccinium corymbosum* L.) during fruit maturation and ripening. Food Chem. 109, 564–572.

Schotsmans, W., Molan, A., MacKay, B., 2007. Controlled atmosphere storage of rabbiteye blueberries enhances postharvest quality aspects. Postharvest Biol. Technol. 44, 277–285.

Schreiner, M., Mewis, I., Huyskens-Keil, S., Jansen, M.A.K., Zrenner, R., Winkler, J.B., O'Brian, N., Krumbein, A., 2012. UV-B induced secondary plant metabolites - potential benefits for plant and human health. Crit. Rev. Plant Sci. 31, 229–240.

Shewfelt, R.L., 1993. Measuring quality and maturity. In: Shewfelt, R.L., Prussia, S.E. (Eds.), Postharvest handling: a systems approach. Academic Press, San Diego, CA, pp. 100–124.

Shoemaker, J.S., 1948. Small fruit culture- a text for instruction and reference Work and a guide for field practice, third ed. McGraw–Hill Book Company, New York, p.433.

Singleton, V.L., Orthofer, R., Lamuela-Raventos, R.M., 1999. Analysis of total phenols and other oxidation substrates and antioxidants by means of folin-ciocalteu reagent. In: Packer, L. (Ed.), Methods in Enzymology. Academic Press, New York. Vol. 299, pp. 152–178.

Siriwoharn, T., Wrolstad, R.E., Finn, C.E., Pereira, C.B., 2004. Influence of cultivar, maturity, and sampling on blackberry (*Rubus* L. Hybrids) anthocyanins, polyphenolics, and antioxidant properties. J. Agric. Food Chem. 52, 8021–8030.

Starke, H., Herrmann, K., 1976. The phenolics of fruits VIII- changes in flavonol concentrations during fruit development. Z. Lebensm. Unters Forsch. 161, 131–135.

Stöhr, H., Herrmann, K., 1975. The phenolics of strawberries and their changes during development and ripeness of the fruits. Z. Lebensm. Unters Forsch. 158, 341–348.

Taruscio, T.G., Barney, D.L., Exon, J., 2004. Content and profile of flavanoid and phenolic acid compounds in conjunction with the antioxidant capacity for a variety of Northwest *Vaccinium* berries. J. Agric. Food Chem. 52, 3169–3176.

Thompson, K., Thompson, A.K., 2003. Fruit and vegetables: harvesting, handling and storage. Blackwell Science Ltd, Oxford, p. 459.

Törrönen, R., Häkkinen, S., Kärenlampi, S., Mykkänen, H., 1997. Flavonoids and phenolic acids in selected berries. Cancer Lett. 114, 191–192.

Treutter, D., 2010. Managing phenol contents in crop plants by phytochemical farming and breeding—visions and constraints. Int. J. Mol. Sci. 11, 807–857.

Vvedenskaya, I.O., Vorsa, N., 2004. Flavonoid composition over fruit development and maturation in American cranberry *Vaccinium macrocarpon* Ait. Plant Sci. 167, 1043–1054.

Wang, C.Y., Chen, C.-T., Wang, S.Y., 2009. Changes of flavonoid content and antioxidant capacity in blueberries after illumination with UV-C. Food Chem. 1, 426–431.

Watada, A.E., Herner, R.C., Kader, A.A., Romani, R.J., Staby, G.L., 1984. Terminology for the description of developmental stages of horticultural crops. HortSci. 19, 20–22.

Watzl, B., Leitzmann, C., 2005. Bioaktive Substanzen in Lebensmitteln. Hippokrates Verlag, Stuttgart, p. 254.

Winkel-Shirley, B., 2001. Flavonoid biosynthesis. A colorful model for genetics, biochemistry, cell biology, and biotechnology. Plant Physiol. 126, 485–493.

Wu, X., Prior, R.L., 2005. Systematic identification and characterization of anthocyanins by HPLC-ESI-MS/MS in common foods in the United States: fruits and berries. J. Agric. Food Chem. 53, 2589–2599.

Zadernowski, R., Naczk, M., Nesterowicz, J., 2005. Phenolic acid profiles in some small berries. J. Agric. Food Chem. 53, 2118–2124.

Zheng, Y., Wang, C.Y., Wang, S.Y., Zheng, W., 2005. Effect of Superatmospheric oxygen on anthocyanins, phenolics and antioxidant activity of blueberries and strawberries. Acta Hort. 857, 475–482.

CHAPTER

22

Production of Raisins and its Impact on Active Compounds

M. Benlloch-Tinoco[*], *J. Carranza-Concha*[†], *M.M. Camacho*[*], *N. Martínez-Navarrete*[*]

[*]Food Technology Department, Food Investigation and Innovation Group, Universitat Politècnica de Valencia, Valencia, Spain, [†]Universidad Autónoma de Zacatecas, Zacatecas, México

CHAPTER POINTS

- Although pre-treating the grapes with NaOH and using microwaves to obtain raisins both lead to a loss in ascorbic acid and minerals, they bring about an increase in the total phenol content.
- Whereas microwave treatment increases the antioxidant activity, NaOH pre-treatment causes it to decrease.
- From these results, microwaves may be recommended as a means of obtaining raisins. An NaOH pre-treatment should only be applied if it is necessary to shorten the process time.
- Gentle heating treatments, such as blanching, do not cause a significant effect either on the bioactive compounds of grapes or on the antioxidant capacity of the fruit.

INTRODUCTION

Grapes are among the world's most important horticultural products. According to the International Organisation of Vine and Wine, the world production of this fruit was around 69.2 million tons in 2011. Wild grapes existed as far back as 35,000 BC and their consumption dates back to prehistoric times (Williamson and Carughi, 2010). Their excellent nutritional and sensory properties are the main factors responsible for the great popularity and commercial interest of this fruit (Jairaj *et al.*, 2009; Pangavhane and Sawhney, 2002). Grapes can be served in different ways, not only as fresh fruit, but also in the form of raisins (Pangavhane and Sawhney, 2002), jams (Esmaiilia *et al.*, 2007), juices or wine (Gardé-Cerdán *et al.*, 2007).

Of the different ways in which grapes are processed, their preservation as raisins is relevant in terms of production in different countries like the United States, Turkey, China, Iran, Chile, South Africa, Greece, Australia, and Uzbekistan (Williamson and Carughi, 2010). They are principally obtained from the Thompson Seedless variety, which represents about 95% of the total production, followed by Fiesta (3%) and Black Corinth (1.5%) (Breska *et al.*, 2010). Raisins are products with exceptional nutritional properties that provide carbohydrates, iron, vitamins, and minerals (Ghrairi *et al.*, 2013) and also important quantities of other bioactive compounds, mainly phenols (Hogan *et al.*, 2009). However, many transformation processes involve grapes being exposed to severe conditions that may cause losses of those compounds responsible for the potential benefits of the fresh fruit (Breksa *et al.*, 2010; Dávalos *et al.*, 2005; Ivanova *et al.*, 2011).

HOW COMPOSITION IS ALTERED

Grapes are a valuable source of fiber, ascorbic and tartaric acid, sugars (glucose, fructose, and sucrose) and minerals (potassium, calcium, magnesium, and phosphorus). The composition of two white grape varieties commonly used to produce raisins is shown in Table 22.1. These compounds are essential for a wide range of metabolic and physiologic processes in the human body (Williams, 2005). Some of their associated benefits are described as follows. Fibers have been recognized to promote laxation as well as blood cholesterol and glucose attenuation (Llobrera and Calleñas, 2007); ascorbic acid can reduce

Processing and Impact on Active Components in Food
http://dx.doi.org/10.1016/B978-0-12-404699-3.00022-6

TABLE 22.1 Micronutrients Analyzed in Thompson Seedless and Imperial Seedless Grape Varieties

mg/100 g	Thompson Seedless	Imperial Seedless
Glucose	9 (2)	6.70 (0.07)
Fructose	10 (2)	8.30 (0.09)
Sucrose	0.6 (0.4)	0.55 (0.05)
TA	476 (124)	459 (49)
AA	3.0 (0.9)	3.4 (0.6)
Mg	7.3 (0.9)	9.2 (0.3)
Ca	9 (2)	10.18 (0.12)
K	233 (2)	149 (3)
P	26 (2)	7.2 (0.6)
TP	315 (57)	415 (40)
TPh	62 (14)	47 (2)
AOA	31 (5)	29.4 (1.1)

Mean value and standard deviation of three sets of analysis in brackets. TA, tartaric acid; AA, ascorbic acid; Mg, magnessium; Ca, calcium; K, potassium; TP, total pectin content (mg galacturonic acid/100g); TPh, total phenol content (mg gallic acid/100g); AOA, antioxidant capacity (mg Trolox/100g).

cancer cases or heart disease rates (Yilmaz and Toledo, 2004); potassium lowers the blood pressure, cardiovascular diseases, morbidity, and mortality; calcium decreases the risk of osteoporosis; and magnesium and phosphorus are involved in functions crucial to all living cells (Ghariari *et al.*, 2013; Loghman-Adham, 1997).

Despite the relevance of all these compounds to human health, the recognized health-promoting effects associated with the consumption of grapes are believed to be mainly linked to their content in phenolic compounds (Dávalos *et al.*, 2005; Hogan *et al.*, 2009; Gurak *et al.*, 2010). Phenols are a large and complex group of compounds which have been confirmed to possess a diversity of favorable effects. Although the exact nature of their protective effect is yet to be established, their role in preventing cardiovascular diseases has been widely demonstrated, as has their anti-inflammatory, antimicrobial, and anti-aging effects (Dávalos *et al.*, 2005; Ghrairi *et al.*, 2013). The phenolic composition of grapes depends on multiple factors, including climate, degree of ripeness, berry size, and grape variety (Dávalos *et al.*, 2005; Soyer *et al.*, 2003). The amount of total polyphenols in white grape varieties tends to be lower than that of red grapes because of their lack of anthocyanins (Ivanova *et al.*, 2011). However, white grapes must still be considered a good source of non-anthocyanic phenols (Rodriguez-Montealegre *et al.*, 2006).

Raisins are derived from the dehydration of grapes, which enables the product's shelf-life to be extended by slowing down reaction rates and avoiding the growth of microorganisms (Jairaj *et al.*, 2009). To date, two methods of dehydration have been basically employed to produce raisins: sun drying and convective drying. Sun drying principally consists of spreading grapes over a platform in a thin layer directly exposed to the sun. This process may take between 2 and 3 weeks. However, hot air drying (HA) involves placing grapes in a dehydration tunnel for at least 24h, during which time they are exposed to an air flow with controlled temperature (>70°C) and velocity (Williamson and Carughi, 2010). However, current research suggests that drying processes may be improved by accelerating dehydration through the combination of various drying technologies or pre-treating the grapes (Contreras *et al.*, 2005; Gowen *et al.*, 2006). In this context, the combination of microwave energy with hot air drying, which is commonly known as microwave-assisted hot-air drying (MW-HA), as well as performing a pre-treatment based on dipping the berries into a NaOH solution (0.03%) as a step prior to convective drying (HA+NaOH), leads to a reduction in the time required to process grapes into raisins of 85 and 32%, respectively. Additionally, if both microwave-assisted hot air drying and NaOH pre-treatment are combined (MW-HA+NaOH), processing times could be reduced by up to 91%. The drying methodology selected to carry out the process may have a substantial repercussion on the product composition. In fact, the use of alternative techniques of reduced drying time, such as those mentioned above (MW-HA, MW-HA+NaOH, or HA+NaOH), offers the opportunity to minimize the impact of drying on the nutritional and functional quality of the product. These are the processes that have been used in this work. Grapes selected from the Thompson seedless variety were subjected to MW-HA (100g at 0.2W/g, 60°C and v=1.5m/s for 7.5h) and HA (450g at 60°C, v=10m/s for 50h) treatments. However, Imperial seedless grapes were processed using MW-HA+NaOH and HA+NaOH treatments under the same conditions as described before, but previously dipping the berries in an NaOH solution (0.03%) at 95°C for 45s. In those cases, the treatment times were 4.5 and 34h for MW-HA+NaOH and HA+NaOH processes, respectively. Table 22.2 shows the major nutritional and functional composition of raisins obtained from grapes subjected to these drying methods. By observing the compositional data shown in Table 22.2, it can be deduced that all these alternative drying processes are suitable for the production of raisins with a nutritional, functional, and physicochemical composition close to that commonly observed for commercial raisins, which means that these products might fully meet consumer expectations.

Any kind of dehydration has been widely recognized to cause damage to the texture, color, taste, and nutritional value of grapes. When the data shown in Tables 22.1 and 22.2 are referred to dry basis, an analysis of the impact caused by using drying to obtain raisins can be performed. The drying processes carried out in this work caused losses of ascorbic acid of up to 20%, mainly due to both hydrolysis and the high

TABLE 22.2 Water Content, Water Activity, Soluble Solids and Micronutrients Analyzed in Raisins

	Thompson Seedless		Imperial Seedless		Commercial Product
	HA	MW-HA	MW-HA + NaOH	HA + NaOH	
x_w (g/100 g)	31 (5)	24 (7)	37 (6)	26 (2)	16.2 (0.5)
a_w	0.854 (0.003)	0.915 (0.003)	0.793 (0.008)	0.820 (0.003)	0.542 (0.009)
⁰Brix	69 (3)	75 (5)	60 (5)	60.03 (0.08)	84.50 (0.05)
Glucose[(1)]	22 (2)	34 (2)	27.7 (1.3)	26 (1)	40 (6)
Fructose[(1)]	34 (2)	41.3 (2.2)	28.7 (5.4)	28.2 (0.2)	43 (7)
Sucrose[(1)]	ND	ND	ND	ND	ND
TA[(1)]	974 (134)	1847 (94)	1506 (97)	1598 (20)	1742 (46)
AA[(1)]	11.54 (0.04)	8 (0.1)	13.34 (0.05)	9.84 (0.03)	5.9 (0.6)
Mg[(1)]	18.6 (1.5)	15 (2)	28 (2)	17 (1)	35 (6)
Ca[(1)]	18.4 (1)	13.9 (1.3)	22 (3)	70 (12)	62 (11)
K[(1)]	687 (3)	702 (11)	500 (12)	602 (10)	821 (90)
P[(1)]	130 (9)	50 (5)	14 (4)	24 (2)	67 (4)
TP	770 (43)	923 (48)	1941 (158)	1248 (21)	944 (200)
TPh	103 (5)	47.3 (1.6)	299 (20)	506 (12)	341 (15)
AOA	103 (5)	30.1 (1.1)	129 (10)	76 (6)	145 (9)

TA, tartaric acid; AA, ascorbic acid; Mg, magnessium; Ca, calcium; K, potassium; TP, total pectin content (mg galacturonic acid/100 g); TPh, total phenol content (mg gallic acid/100 g); AOA, antioxidant capacity (mg Trolox/100 g). ND, not detected. Mean value and standard deviation of three sets of analysis in brackets. MW-HA, microwave-assisted hot air drying; HA, hot air drying; MW-HA + NaOH, microwave-assisted hot air drying combined with NaOH pre-treatment, and HA + NaOH, hot air drying combined with NaOH pre-treatment.
[(1)] *mg/100 g.*

temperatures reached in the product. In the case of tartaric acid, its content can decrease by between 10 and 30%. As regards the behavior of sugars, sucrose was the most seriously affected, becoming completely hydrolyzed after drying, while losses of glucose and fructose of up to 50 and 20%, respectively, were observed after processing. However, dehydration leads to the alteration of cellular tissues due to the high temperatures reached in the product, a fact that markedly affects the mineral and pectin composition of grapes. The mineral content was reduced, which may be related to the diffusion of these micronutrients into intercellular spaces. This phenomenon can be promoted by the loss of cellular structure. In this context, calcium was apparently the most seriously affected mineral, showing losses of up to 40% after processing. As regards pectin content, a reduction of both the water-soluble and oxalate soluble fractions took place during the drying step, which led to a global decrease in pectin content (10–30%).

Figure 22.1 shows the total phenol content and the antioxidant capacity of raisins, expressed as the content of each compound in the raisins referred to the corresponding grape weight used to obtain it, before and after the different drying processes applied in this study. The corresponding gain or loss due to treatments can be calculated from these data and those shown in Table 22.1, as 100 × (Table 22.1 values – Figure 22.1 values)/Table 22.1 values. The results obtained for total phenol content were –27.39%, –8.48%, –66.42%, –198.92% for MW, HA, HA + NAOH, MW + NaOH treatments, respectively and –16.81, –1.83, 14.01, –9.27 for the antioxidant activity.

Both microwave-assisted hot air drying and hot air drying processes have an analogous effect on the variations in the total phenol content. In both cases, an increase in the total phenol content was observed. Pre-treating grapes with NaOH leads to a greater increase in these compounds when they were obtained by employing both drying technologies (MW-HA and HA). However, the MW-HA process should be considered as the most favorable, regardless of the application or not of the NaOH pre-treatment, as the raisins showed a greater augmented antioxidant capacity than those obtained by means of the hot air drying process. As far as is currently known, the most abundant phenols in raisins have been found to be querecitin and kaempferol, as flavonols, and caftaric and coumaric acid, as phenolic acids. Phenolic acids exhibit higher thermolability than flavonols and they show losses of up to 90 and 60%, respectively when hot air is used to obtain raisins. Procyanidins and flavan-3-ols are the most seriously affected

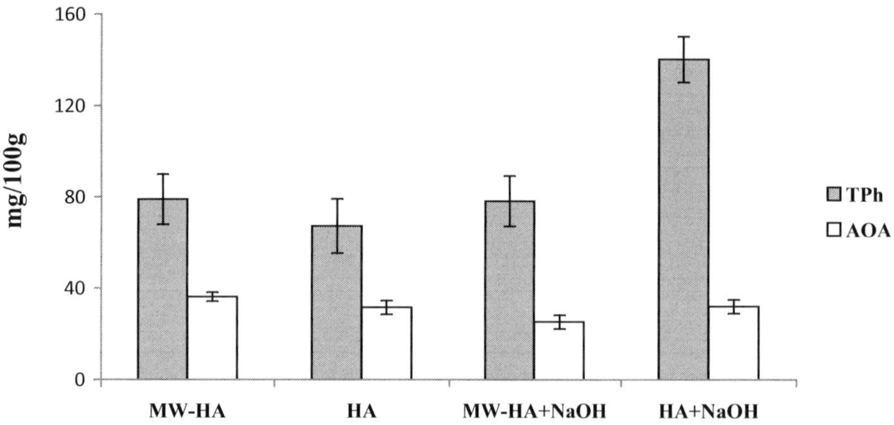

FIGURE 22.1 Total phenol content (mg of GAE/100 g grape) and antioxidant capacity (mg Trolox/100g grape) of raisin. Raisin contents are expressed as the content of each compound in the raisins referred to the corresponding grape weight used to obtain it. The histograms represent the mean value and standard deviation of three sets of analysis of total phenol content (TPh) and antioxidant capacity (AOA).

polyphenols, and are completely degraded at the end of the process (Karadeniz *et al.*, 2000). According to Williamson and Carughi (2010), despite the fact that some phenolic compounds originally present in grapes are lost as a consequence of the drying process, others may be transformed. However, the total phenol content can remain unchangeable or even increase after processing. This increase in phenols could be explained by the fact that these compounds are easier to extract as a result of the structural alteration that takes place during processing. According to literature, the application of thermal treatments can also be associated with the dissociation of conjugated forms into free phenolic acid compounds (Jing *et al.*, 2010). However, the available information on the phenolic content of raisins is far from complete. Therefore, this is considered as an active area of research which is important for establishing associated health benefits (Williamson and Carughi, 2010). As regards antioxidant capacity, it is not usually reduced after the drying treatment. However, this property did not experience the same changes as those observed in phenols, which is linked with the low correlation between the total phenolic content and antioxidant activity in white grapes, due to the lack of anthocyanins (Kuskoski *et al.*, 2005).

OTHER WAYS IN WHICH THE COMPOSITION IS ALTERED

The variations brought about in the most relevant nutrients and functional compounds of grapes after drying can be used as a measure of the quality lost due to processing. Taking into account the recognized relevant contribution of phenolic compounds and other antioxidant compounds to the health-promoting effects of grapes, together with their sensitivity to drying processes (Tarhan, 2006), special attention has to be paid to the effect processing has on them. Hence, both the total phenol content and the antioxidant capacity of grape products may be selected as quality indicators in order to carry out a comparison between different processing methodologies.

Enzyme activity and microbial growth are considered to be the major causes behind the alteration of food products. The most commonly used enzyme inactivation technique is blanching, carried out by placing the product in hot or boiling water (85–100°C), in acid or basic hot or boiling solutions, by steaming or by microwave heating with immersion in water or solutions for a few seconds or minutes (Kidmose and Martens, 1999; Ponne *et al.*, 1991; Ramaswany and Van de Voort, 1990; Severini *et al.*, 2001). This manipulation does not constitute a preservation method and is usually applied when preparing the raw material before other conservation operations are performed, such as sterilization, dehydration, and freezing. The heat received during blanching inevitably causes some changes in the sensory and nutritional characteristics of the product.

In this section, the impact of traditional blanching (TB) and microwave blanching (MWB) on the total phenols and the antioxidant activity of grapes (Festival Seedless white grapes, *Vitis vinifera*) has been compared. Tartaric acid, the major organic acid present in grapes, and total pectin content have also been compared. For TB, the grape (100 g) was submerged in a bath of distilled water at 100°C for 150 s. For MWB, the grape (100 g) was heated while immersed in distilled water (300 g) for 170 s at 900 W. The hot water was then eliminated. In both cases, grapes were rapidly cooled in ice water for 3–5 min.

The microwave and traditional blanching treatments carried out lead to a pectinmethylesterase (PME), peroxidase (POD), and polyphenol oxidase (PPO) inactivation in the range of 63.5 and 70.8, 81.7 and 86.7, and 99.90 and 99.80%, respectively. POD enzyme activity is critical in grapes and this enzyme exhibited the greatest thermal resistance. The enzymatic inactivation results obtained by the two blanching treatments confirmed that while the traditional method had a significantly ($\alpha\{\tau\sigma\} < 0.05$) greater effect than the microwave method on PME and POD, the rates of PPO inactivation remained equal ($\alpha\{\tau\sigma\} < 0.05$) for both. However, the mean value of the water and soluble solid content of the obtained grapes was in the range of 82.0–83.6 g water/100 g sample and

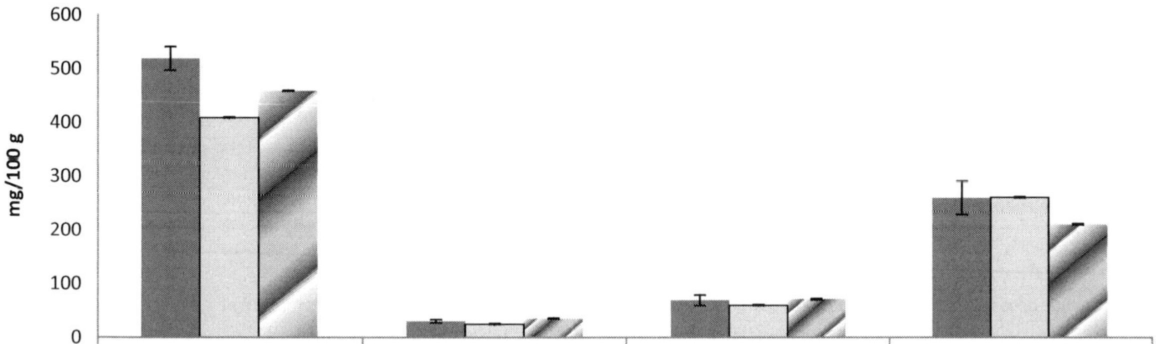

	TA (mg/100 g)	AOA (mg Trolox/100 g)	TPh (mg GAE/100 g)	TP (mg AGU/100 g)
FG	519	30	69	260
MWB	409	25	60	261
TB	459	35	71	211

FIGURE 22.2 Tartaric acid, antioxidant activity, total phenol content and total pectin content of fresh, blanched by microwave and traditional blanched grapes. The histograms represent the mean value and standard deviation of three sets of analysis. TA, tartaric acid; AOA, antioxidant activity; TPh, total phenol content; TP, total pectin content; FG, fresh grape; MWB, blanched by microwave grape; TB, traditional blanched grape.

16.4–15.4 g soluble solids/100 g sample. The water activity of the samples showed no significant differences, this being 0.973. Figure 22.2 shows the other analyzed compounds and the antioxidant activity of grapes. No significant differences between the blanched and fresh samples were observed, except for tartaric acid. This is not surprising given the mild nature of these blanching treatments and the greater sensitivity of organic acids to temperature. If the blanching and drying treatments are compared, the latter ones seem to be more desirable. Despite the fact that a loss of ascorbic acid is detected in raisins, they have a greater total phenolic content. When comparing the antioxidant capacity of blanched grapes and raisins, it was observed to increase in the raisins obtained with any of the drying processes applied, except when a pre-treatment with NaOH and hot air drying were combined. In other works, the polyphenolic content, color, and antioxidant activity of red grape pomace peels were not significantly affected when dried at 60°C (Larrauri *et al.*, 1997). For similar processes, Nwanguma and Eze (1996) found that an increase of mashing temperature from 65 to 75°C led to a significant reduction in the polyphenolic content of wort. Makkar and Singh (1991) reported a decreased content of total proanthocyanidins in cassava and Leucaena leaves (10.1 and 21.4%, respectively) when heated at 90°C for 24 h.

ANALYTICAL TECHNIQUES

Currently, plenty of analytical methods concerning how to determine not only the phenolic compound content of diverse foodstuffs, but also their antioxidant capacity are available. Consequently, selecting the most appropriate method for each case is essential. By consulting the literature, it can be seen that the existing methods for phenol quantification can be classified mainly in function of their cost, specificity, sensitivity and ability to quantify (Williamson and Carughi, 2010). If the priorities are the sensitivity and specificity of the obtained results, high-performance liquid chromatography (HPLC) with mass spectrometric detection should be considered the best option. However, it leads to particularly high associated costs. On the contrary, spectrophotometric methods, such as that of Folin–Ciocalteu, are considered to be of low sensitivity and generate very low specific results, since they can be exclusively used to obtain data concerning the total phenol content. Nevertheless, it is cheaper and easier to use, which are the main reasons why it is one of the most common ways to determine phenols in food products. However, the use of an HPLC-diode array, the cost of which is in between those of HPLC with mass spectrometric detection and spectrophotometric methods, can be considered as an intermediate option since it offers the possibility of getting results with both high sensitivity and specificity. As regards the determination of the antioxidant capacity, assays basically consist of different mechanisms that involve the use of diverse radical or oxidant sources (Du *et al.*, 2009). The most widely known methods are DPPH$^{\cdot}$ and ABTS^{+}, both of which are based on the measurement of free radical scavenging of the antioxidant compounds using a radical and a radical cation, respectively. Also commonly used are ORAC, which consists of determining the antioxidant capacity of food products by measuring their oxygen radical absorbance capacity, and FRAP, that analyzes the product's ferric-reducing antioxidant power. Although it is accepted that all of them accurately describe the antioxidant capacity of food products, they do not tend to generate similar results.

Focusing more closely on the present work, several analytical techniques have been employed to evaluate the effect of each of the above-mentioned preservation processes (dehydration and blanching) on the grape quality. In this context, all the compositional data shown, either throughout the text or included in the different tables and figures, were determined by analyzing grapes before and after any kind of processing. The methodologies corresponding to each one of the analyses performed are described as follows. The moisture content was determined by drying the sample until constant weight was reached at 60°C in a vacuum oven (AOAC 20.013, 1997). The soluble solids of the liquid phase were measured in a previously homogenized sample using an Atago NAR-3T refractometer (Tokyo, Japan). The water activity was analyzed by using a dew point hygrometer (GBX FA-st lab, France). Total acidity was determined by titration with NaOH (0.1 N). It was expressed in mg of the major acid present in grapes (tartaric acid). Ascorbic acid was determined by titration according to AOAC 985.33 (1997). The total pectin content was analyzed by quantifying the galacturonic acid residues following the procedure described by Yu *et al.* (1996). Phosphorus was determined by means of spectrophotometry at 600 nm (Thermo Spectronic UV1, Madison, USA). The rest of the minerals (calcium, potassium, and magnesium) and sugars (glucose, fructose, and sucrose) were measured by high-performance anion-exchange chromatography (HPAEC) using a Metrohm chromatograph (Herisau, Switzerland). In the case of minerals, both a mobile phase, consisting of tartaric acid (4 mmol/l) and dipicolinic acid (0.75 mmol/l), and a Metrosep C2-150 column (4 × 150 mm) with a particle size of 7 μm were employed. However, NaOH (0.1 N) was used as a mobile phase for sugars and the column employed in this case was Metrosep Carb 1 (4.6 × 250 mm), with a particle size of 5 μm. Total phenols were extracted following the methodology described by Peiró *et al.* (2006) and quantified using the Folin–Ciocalteu method (Li *et al.*, 2006). The antioxidant capacity was determined by a modification of the spectrophotometric technique developed by Re *et al.* (1999), using the $ABTS^+$ radical (Sigma) generated by 2.45 mM potassium persulfate ($K_2S_2O_8$) and the results were expressed as the antioxidant capacity equivalent to the mg of Trolox (TEAC) in 100 g of fresh sample.

References

AOAC, 1997. Association of official analytical chemis. Official methods of analysis, 20.013 and 985.33. Washington, DC.

Breska, A.P., Takeoka, G.R., Hidalgo, M.B., Vilches, A., Vasse, J., Ramming, D.W., 2010. Antioxidant activity and phenolic content of 16 raisin grape (*Vitis vinifera* L.) cultivars and selections. Food Chem. 121, 740–745.

Contreras, C., Martin, M.E., Martínez-Navarrete, N., Chiralt, A., 2005. Effect of vacuum impregnation and microwave application on structural changes which occurred during air-drying of apple. LWT—Food Sci. Technol. 38, 471–477.

Dávalos, A., Bartolomé, B., Gómez-Cordovés, C., 2005. Antioxidant properties of comercial grape juices and vinegars. Food Chem. 93, 325–330.

Du, G., Li, M., Ma, F., Liang, D., 2009. Antioxidant capacity and the relationship with polyphenol and Vitamin C in *Actinidia* fruits. Food Chem. 113, 557–562.

Esmaiilia, M., Sotudeh-Gharebaghb, R., Croninc, K., Mousavia, M.A.E., Rezazadehd, G., 2007. Grape Drying: A Review. Food Rev. Int. 23 (3), 257–280.

Gardé-Cerdán, T., Arias-Gil, M., Marsellés-Fontanet, A.R., Ancín-Azpilicueta, C., Martín-Belloso, O., 2007. Effects of thermal and non-thermal processing treatments on fatty acids and free amino acids of grape juice. Food Control 18, 473–479.

Ghrairi, F., Lahouar, L., Amira, E.A., Brahmi, F., Ferchichi, A., Achour, F., et al., 2013. Physicochemical composition of different varieties of raisins (*Vitis vinífera* L.) from Tunisia. Ind. Crops Prod. 43, 73–77.

Gowen, A.A., Abu-Ghannam, N., Frias, J.M., Barat, J.M., Andres, A., Oliveira, J.C., 2006. Comparative study of quality changes occurring on dehydration and rehydration of cooked chickpeas (*Cicer arietinum* L.) subjected to combined microwave-convective and convective hot air dehydration. J. Food Sci. 71, 282–289.

Gurak, P.D., Cabral, L.M.C., Rocha-Leão, M.H.M., Matta, V.M., Freitas, S.P., 2010. Quality evaluation of grape juice concentrated by reverse osmosis. J. Food Eng. 96, 421–426.

Hogan, S., Zhang, L., Li, J., Zoecklein, B., Zhou, K., 2009. Antioxidant properties and bioactive components of Norton (*Vitis aestivalis*) and Cabernet Fran (*Vitis vinifera*) wine grapes. LWT—Food Sci. Technol. 42, 1269–1274.

Ivanova, V., Stefova, M., Vojnoski, B., Dörnyei, A., Márk, L., Dimovska, V., et al., 2011. Identificacion of polyphenolic compounds in red and white grape varieties grown in R. Macedonia and changes of their content during ripening. Food Res. Int. 44, 2851–2860.

Jairaj, K.S., Singh, S.P., Srikant, K.A., 2009. review of solar dryers developed for grape drying. Solar Energy 83, 1698–1712.

Jing, Y., Jing-Feng, C., Yu-Ying, Z., Lin-Chun, M., 2010. Effects of drying processes on the antioxidant properties in sweet potatoes. Agric. Sci. China 9 (10), 1522–1529.

Karadeniz, F., Durst, R.W., Wrolstad, R.E., 2000. Polyphenolic composition of raisins. J. Agric. Food Chem. 48, 5343–5350.

Kidmose, U., Martens, H.J., 1999. Changes in texture, microstructure and nutritional quality of carrot slices during blanching and freezing. J. Sci. Food Agric. 79 (12), 1747–1753.

Kuskoski, M.E., Asuero, A.G., Troncoso, A.M., Manzini-Filho, J., Fett, R., 2005. Aplicación de diversos métodos químicos para determinar actividad antioxidante en pulpa de frutos. Cien. y Tecnología de Alimentos 25 (4), 726–732.

Larrauri, J.A., Rupérez, P., Saura-Calixto, F., 1997. Effect of Drying Temperature on the Stability of Polyphenols and Antioxidant Activity of Red Grape Pomace Peels. J. Agric. Food Chem. 45, 1390–1393.

Li, B.B., Smith, B., Hossain, M.M., 2006. Extraction of phenolics from citrus peels: I. Solvent extraction method. Separation Purif. Technol. 48, 182–188.

Llobrera, A., Cañellas, J., 2007. Dietary fibre content and antioxidatn activity of Mango Negro red grape (*Vitis vinifery*): pomace and stem. Food Chem. 101, 656–666.

Loghman-Adham, M., 1997. Adaptation to changes in dietary phosphorus intake in health and in renal failure. J. Lab. Clin. Med. 129 (2), 176–188.

Makkar, H.P.S., Singh, B., 1991. Effect of drying conditions on tannin, fibre and lignin levels in mature Oak (*Quercus incana*) leaves. J. Sci. Food Agric. 54, 323–328.

Nwanguma, B.C., Eze, O.M., 1996. Changes in the concentrations of the polyphenolic constituents of sorghum during malting and mashing. J. Sci. Food Agric. 70, 162–166.

Pangavhane, D.R., Sawhney, R.L., 2002. Review of research and development work on solar dryers for grape drying. Energy Conversion Manage. 43, 45–61.

Peiró, R., Díaz, V.M.C., Camacho, M.M., Martínez-Navarrete, N., 2006. Micronutrient flow to the osmotic solution during the grapefruit osmotic dehydration. J. Food Eng. 74 (3), 299–307.

Ponne, C.T., Van Remmen, H.H.J., Bartels, P.V., 1991. Application of electromagnetic energy in processing of whole potatoes. Voedingsmiddelentechnologie. 24 (21), 44–46.

Ramaswany, H., Van de Voort, F.R., 1990. Microwave application in food processing. Can. Inst. Food Sci. Tech. J. 23 (1), 17–23.

Rodríguez-Montealegre, R., Romero-Peces, R., Chacón-Vozmediano, J.L., Martínez-Gascueña, J., García-Romero, E., 2006. Phenolic compounds in skins and seeds of ten grape *Vitis vinifera* varieties grown in a warm climate. J. Food Composition Anal. 19, 687–693.

Severini, C., De Pilli, T., Baiano, A., Mastrocola, D., Massini, R., 2001. Study of preventing enzymatic browning of potato by microwave blanching. Sci. Des Alim. 21 (2), 149–160.

Soyer, Y., Koca, N., Karadeniz, F., 2003. Organic acid profile of Turkish White grapes and grape juices. J. Food Composition Anal. 16, 629–636.

Re, R., Pellegrini, N., Proteggente, A., Pannala, N., Yang, M., Rice-Evans, C., 1999. Antioxidant activity applying an improved ABTS radical cation decolorization assay. Free Radic. Biol. Med. 26, 1231–1237.

Tarhan, S., 2006. Selection of chemical and thermal pretreatment combination for plum drying at low moderate drying air temperatures. J. Food Eng. 79, 255–260.

Williams, M.H., 2005. Dietary supplements and sports performance: minerals. J. Int. Soc. Sports Nutr. 2 (1), 43–49.

Williamson, G., Carughi, A., 2010. Polyphenol content and health benefits of raisins. Nutr. Res. 30, 511–519.

Yilmaz, Y., Toledo, R.T., 2004. Health aspects of functional grape seed constituents. Trends Food Sci. Tech. 15, 422–433.

Yu, L., Reitmeieir, C.A., Love, M.H., 1996. Strawberry texture and pectin content as affected by electron beam irradiation. J. Food Sci. 61 (4), 844–846.

CHAPTER

23

Vapor Treatments, Chilling, Storage, and Antioxidants in Pomegranates

Daniel Valero, Seyed Hossein Mirdehghan†, Mohammad Sayyari‡, María Serrano§*

*Department of Food Technology, University Miguel Hernández, Alicante, Spain, †Department Horticultural Sciences, College of Agriculture, Vali-e-Asr University of Rafsanjan, Kerman, Iran, ‡Department Horticultural Sciences, College of Agriculture, Bu-Ali Sina University, Hamedan, Iran, §Department of Applied Biology, University Miguel Hernández, Alicante, Spain

CHAPTER POINTS

- Pomegranate is considered as a very sensitive fruit to low temperature storage with occurrence of chilling injury symptoms.
- Chilling injury of pomegranates are manifested by skin browning, surface pitting and desiccation and accompanied by increases in electrolyte leakage.
- Pomegranate fruit is rich in bioactive compounds such as phenolics and anthocyanins with proved antioxidant activity *in vitro* and *in vivo*, with important differences between cultivars.
- Post-harvest treatments with hot water, putrescine, spermidine, oxalic acid, salicylic acid, and acetyl salicylic acid maintain organoleptic and functional quality of pomegranate.
- Innovative post-harvest vapour treatments with methyl jasmonate and methyl salicylate reduced chilling injury and enhanced the content of bioactive compounds and the total antioxidant activity of the arils along storage.

INTRODUCTION

Pomegranate (*Punica granatum* L.) is considered one of the oldest known edible fruit and probably originated in a region of Iran where it was first cultivated in about 2000 B.C., but spread to the Mediterranean countries and northern Turkey at a very early date (Ward, 2003). Pomegranates are recorded in several ancient Mesopotamian cuneiform sources, from the second half of the 3rd millennium B.C. It is without doubt cited in the Bible under the Hebrew names *rimmôn*, or the Greek *roa*, *roia*, or *kodon* and also in The Holy Quran as the name of *Romman in Al-Anam* and *Ar-Rahman surah*. In Greece, between the 5th and 3rd centuries B.C., Hippocrates recorded numerous medicinal uses of pomegranate to treat heart pain, leucorrhoea, sterility, uterus, pleuritis and lung diseases, phthisis, and as an astringent, vulnerary, febrifuge, antiemetic (a panacea for women in gynaecology) and for cholera. Later the pomegranate medicinal uses are treated by Dioscorides (1st century A.D.), the Armenian Ananias of Shirak (7th century A.D.), Al-Biruni, and Avicenna (Rivera *et al.*, 2012).

Fruit quality depends largely on sugar and acid content of the juice. A high-quality pomegranate should also have an attractive skin, small seeds in the aril, and should be free from sunburn, growth cracks, cuts, bruises, and decay. Skin color and smoothness are other quality indices. Sour and sour–sweet pomegranates have reddish skin, in contrast to sweet pomegranate, which have yellowish-green skin. The edible portion of pomegranate are the arils, which can be consumed fresh but they are also used in the preparation of juices, canned beverages, jellies, jams, and flavorings and colorings for drinks (Melgarejo and Martínez, 1992). The arils are the edible part of the fruit, which consist of around 80% juice and 20% seed. Arils constitute 52% of total fruit (w/w), and the fresh juice contains 85.4% moisture and considerable amounts of total soluble solids (TSS),

Processing and Impact on Active Components in Food
http://dx.doi.org/10.1016/B978-0-12-404699-3.00023-8

total sugars, reducing sugars, anthocyanins, phenolics, organic acids, ascorbic acid, vitamins, polysaccharides, proteins, and essential minerals (Mirdehghan, 2006; Sayyari, 2009).

According to statistics (Table 23.1), the main countries producing pomegranates are Iran and India (more than 500,000 t) followed by China and the U.S.A., while Spain is the greatest European producer of pomegranates and its production exceeded 37,000 t in 2007 (Holland and Bar-Ya'akov, 2008). Most of this production takes place in the south-east of Spain, where the pomegranate tree is adapted to the poor and saline soils of this semi-arid area, contributing to the reduction in the risk of desertification. 'Mollar de Elche' and 'Valenciana' are the most widely spread cultivars in the local area, and display very attractive sensorial characteristics for Spanish consumers; nevertheless, they are almost unknown to other markets and different to 'Wonderful', the varietal group widely grown in the rest of the world. In addition to the high concentration of sugars, 'Mollar de Elche' pomegranate has an additional advantage, that is the very small and soft seeds which are barely discernible in the mouth and thus the arils from this cultivar are easily eaten.

TABLE 23.1 Estimated World Pomegranate Production and Export Based on Research Carried Out by Yael Kachel, Department of Market Research, Israeli Ministry of Agriculture*

Country	Planted Area (ha)	Production (t)	Export (t)
Iran	65,000	600,000	60,000
India	54,750	500,000	22,0000
China	Unknown	260,000	Unknown
USA	6,070	110,000	17,000
Turkey	7,600	90,000	Unknown
Spain	2,400	37,000	14,000
Tunisia	2,600	25,000	2,000
Israel	1,500	17,000	4,000

* *Adapted from Holland and Bar-Ya'akov (2008).*

POMEGRANATE PHYTOCHEMICALS

Although pomegranates have been used broadly in the folk medicine of many cultures, other properties have been recently claimed, such as antioxidant activity and beneficial effects against several diseases. In fact, in recent years the number of scientific publications related to health-beneficial properties of pomegranate has considerably increased providing the scientific basis for some uses of the pomegranate in traditional medicine. In this sense, pomegranate consumption shows biological action in atherosclerosis, in the prevention of several inflammatory- and infective-mediated diseases, and cancer (Faria and Calhau, 2011; Ismail *et al.*, 2012). Among the large number of phytochemicals present in pomegranate fruit, the main ones responsible for these health-beneficial effects are polyphenolic in nature, including ellagic acid, ellagitannins (punicalagins),

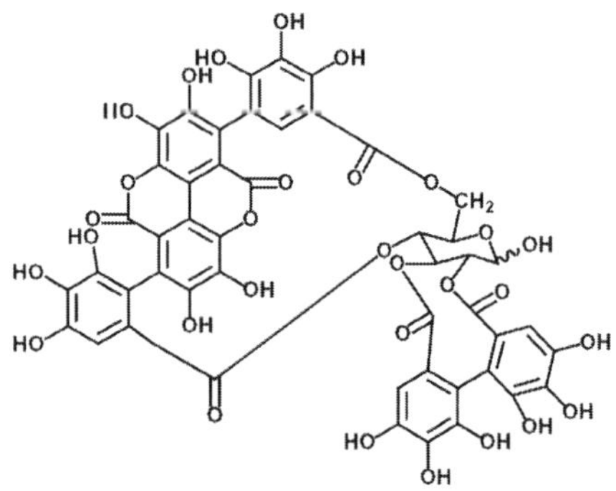

Punicalagin

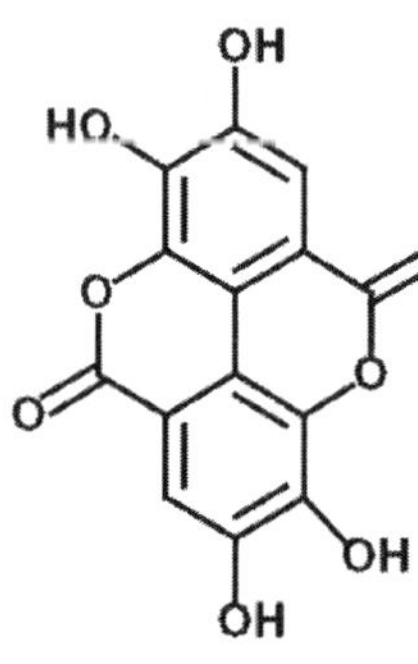

Ellagic acid

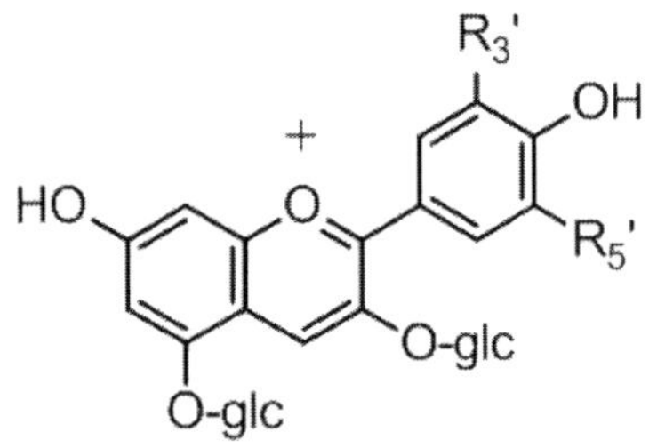

Anthocyanin

R3'	R5'	Anthocyanin
H	H	Pelargonidin
OH	H	Cyanidin
OH	OH	Delphinidin

FIGURE 23.1 Chemical structures for the main phenolic compounds in pomegranate. Punicalagin is the main ellagitannis present in pomegranate, while ellagic acid is a representative of the tannins. Anthocyanins are the main polyphenols responsible for the red color of pomegranates.

punicic acid, anthocyanins, flavonols, flavan-3-ols, and flavones (Figure 23.1).

Free ellagic acid and its glycosidic combinations with several sugars (glucose, rhamnose, and arabinose) are present in pomegranates and concentrations differ between cultivars (Mena *et al.*, 2011). Ellagitannins are the main polyphenols in pomegranates, along with punicalagin and punicalin.

Anthocyanins are the pigments responsible for the typical color of the pomegranate fruit, including delphinidin-, cyanidin-, and pelargonidin- 3-glucosides, and 3,5-diglucosides. Differences in both anthocyanin type and concentration exist between cultivars, the major anthocyanin being cyanidin-3-glucoside for 'Mollar de Elche' sweet cultivar, the most important pomegranate in Spain, while cyanidin-3,5-diglucoside has been found to be the major anthocyanin in the sour cultivars (Alighourchi *et al.*, 2008; Sayyari *et al.*, 2010).

The potent antioxidant and antiatherogenic properties of pomegranates have been attributed to their high content of polyphenols, including ellagic acid and punicalagin. In fact, Seeram *et al.* (2004) reported for the first time direct evidence of the absorption of ellagic acid in humans from a food source such as pomegranate juice. Further epidemiological studies provide evidence that ellagic acid and ellagitannins could protect against several chronic diseases in spite of the discrepancy between *in vitro* and *in vivo* studies due to the low bioavailability of these compounds. Ellagitannins are hydrolyzed *in vivo* to ellagic acid, and then metabolized by the gut microbiota to urolithin derivatives, which are considered to have a direct effect on reducing colon cancer development by inhibiting cell proliferation and inducing cell apoptosis (Landete, 2011).

The antioxidant activity measured from hydrophilic extracts (H-TAA) on a wide range of pomegranates was much higher than that found in other fruits typical of the Mediterranean diet, such as green and red pepper, orange, and kiwi fruit (Figure 23.2). However differences existed between the pomegranates, with the sweet cultivars (Mollares and Valencianas groups) showing higher H-TAA than sour types. In addition, total phenolic compounds were also present at high concentrations in the arils for all pomegranate cultivars ranging from 30 to 120 mg gallic acid eq./100 g depending on cultivars. In fact, a positive high correlation ($r^2=0.810$) was found between the content of phenolics and H-TAA (Figure 23.3). When lipophilic total antioxidant activity (L-TAA) was determined, there were also differences between cultivars although L-TAA was lower than H-TAA with values ranging from 15 to 21 mg Trolox/100 g, which were highly correlated ($r^2=0.690$) to the content of total carotenoids (Figure 23.3). These results show that phenolic compounds are the main compounds responsible for the H-TAA, while carotenoids are the main contributors to L-TAA, as has been observed in other fruits (Valero and Serrano, 2010). Thus, for future epidemiological studies, those cultivars having the highest TAA, both H-TAA and L-TAA, *in vitro* should be used since higher *in vivo* biological effects could be expected.

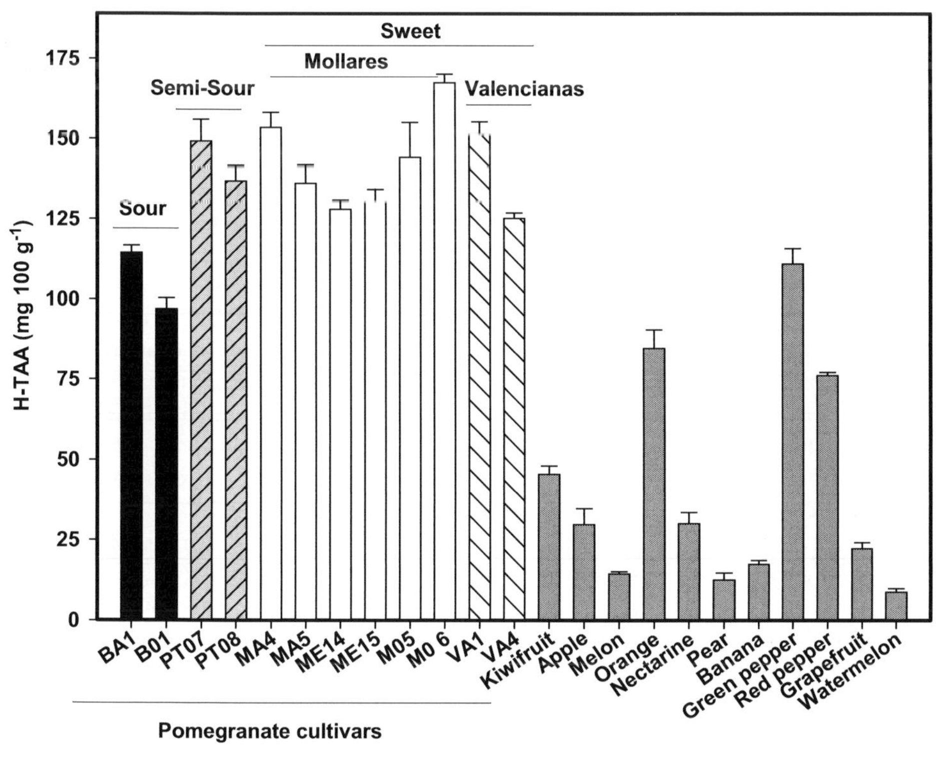

FIGURE 23.2 Hydrophilic total antioxidant activity (H-TAA) in the arils of 12 different pomegranate cultivars and in some characteristic fruits of the Mediterranean diet. Histograms show the mean ± SEM of determinations made in 10 individual fruits (unpublished data).

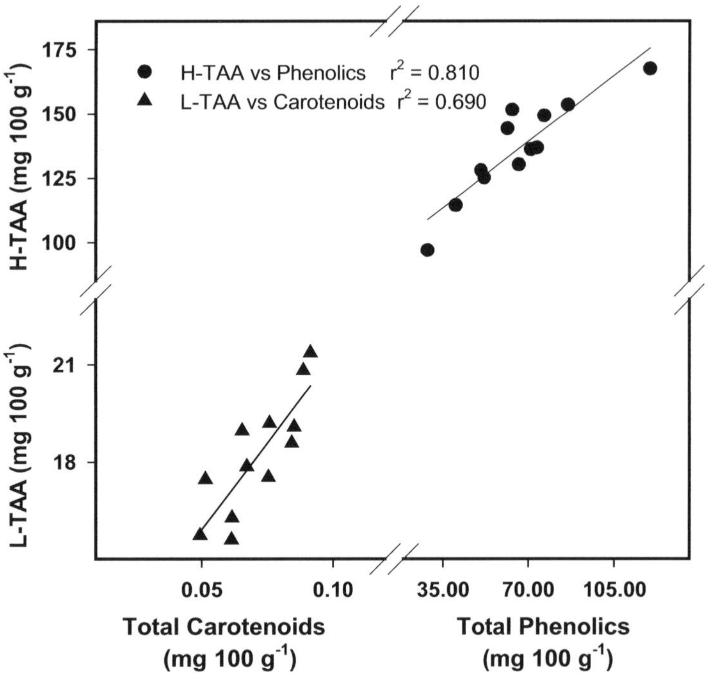

FIGURE 23.3 Correlations between hydrophilic total antioxidant activity (H-TAA) and total phenolics, and between lipophilic total antioxidant activity (L-TAA) and carotenoids in the arils of 12 pomegranate cultivars. Correlations were made taking into account data for 12 pomegranate cultivars named in Figure 23.2 (unpublished data).

COLD STORAGE, CHILLING INJURY, AND POMEGRANATE BIOACTIVE COMPOUNDS

During post-harvest storage, pomegranate fruits exhibit important losses in quality parameters due to several physiological and enzymatic disorders, the major storage problem being water loss and browning symptoms in both peel and arils. However, to avoid excessive desiccation and decay, storage at low temperatures is necessary, although pomegranate is very sensitive to cold temperature storage (below 5°C) and develops a wide range of alterations known as chilling injury (CI) symptoms, manifested as skin browning, surface pitting and higher susceptibility to decay (Valero and Serrano, 2010). Most of these symptoms could reach the arils leading to reduction also in the internal fruit quality by affecting firmness, aril color, vitamin C and acidity contents.

To avoid CI incidence in pomegranate, several post-harvest treatments have been carried out with satisfactory results such as polyamine [putrescine (Put) and spermidine (Spd) at 1 mM] treatments (Mirdehghan *et al.*, 2007a), hot water dips at 45°C (Mirdehghan *et al.*, 2006; 2007b), salicylic acid (SA) at 0.7, 1.4 and 2.0 mM (Sayyari *et al.*, 2009), oxalic acid (OA) at 2, 4, and 6 mM (Sayyari *et al.*, 2010), acetyl salicylic acid (ASA) at 0.1, 0.5, and 1 mM (Sayyari *et al.*, 2011). Figure 23.4 shows the CI index for 'Mollar de Elche' pomegranate stored at chilling temperature (2°C) for 28, 56, and 84 days, plus 4 days at 20°C. It is clear that all treatments showed beneficial effects on alleviating CI during storage. In the case of hot water, Put, and Spd treatments, the reduction of CI was accompanied by lower percentages of electrolyte leakage, higher ratios of unsaturated/saturated fatty acids and increases in the endogenous concentrations of putrescine and spermidine in the husk. Thus, both treatments were able to maintain membrane fluidity at low temperature storage which is considered as an acclimation mechanism of CI-sensitive fruits to cold storage (Mirdehghan *et al.*, 2007a,b). Similarly, SA, OA, and ASA diminished electrolyte leakage and thus maintained membrane integrity. The mechanism by which ASA improved CI symptoms is still unknown. However, there are evidences that ASA is converted to SA in plant tissues after exogenous application, and SA was also effective in reducing CI in the Iranian pomegranate 'Malas Saveh' (Sayyari *et al.*, 2009).

Apart from the beneficial effect of the above treatments on maintaining pomegranate quality during post-harvest storage, there is some evidence that processing pomegranates through appropriate post-harvest strategies could induce some modulation in the content of bioactive compounds such as ascorbic acid, total phenolics, total anthocyanins, and TAA measured separately in both hydrophilic and lipophilic extracts.

In this sense, during post-harvest storage of the sour pomegranate 'Malas Saveh', a reduction (≅25%) of ascorbic acid (vitamin C) was observed, and the treatments with SA were effective in delaying the vitamin C losses in a concentration-dependent manner (Sayyari *et al.*, 2009). Similar effects on maintaining ascorbic acid concentration during storage were obtained for the sweet

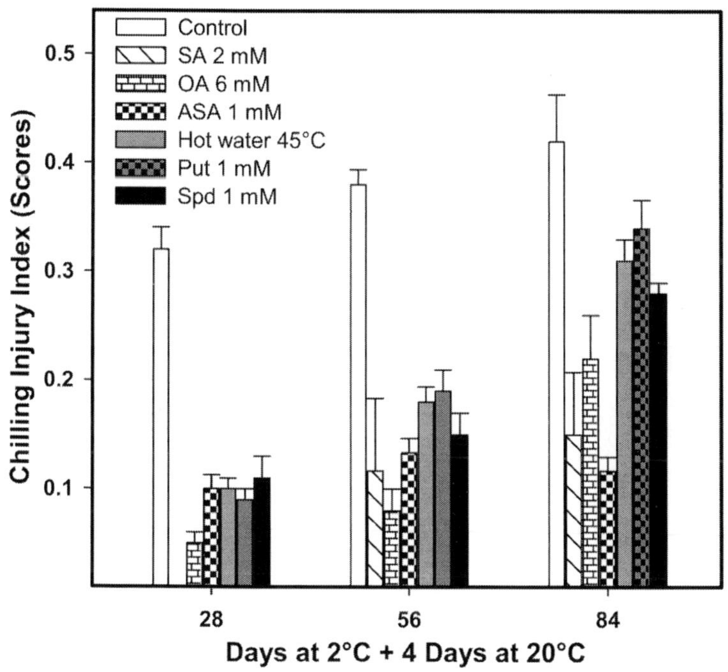

FIGURE 23.4 Chilling injury index in control and treated pomegranates with salicylic acid (SA), oxalic acid (OA), acetyl salicylic acid (ASA), hot water (45°C), putrescine (Put), and spermidine (Spd) along cold storage plus 4 days at 20°C. Chilling injury was determined according to the method described in *Analytical Determinations*. Histograms show the mean ± SEM of determinations made in 15 individual fruits. *Adapted from the PhD Theses of Mirdehgan (2006) and Sayyari (2009).*

pomegranate 'Mollar de Elche' after ASA or OA treatments (Sayyari *et al.*, 2010, 2011).

Some evidence exists regarding the changes in other bioactive compounds and antioxidant activity during cold storage of fruits, although no general tendency is found. In the case of pomegranate, losses of total phenolics have been reported and thus loss of health-beneficial compounds occurred. However, the post-harvest application of the above treatments (SA, OA, ASA, hot water, Put, and Spd) were effective in maintaining higher phenolic content after a prolonged storage (3 months; Figure 23.5).

Other bioactive compounds with antioxidant activity are anthocyanins, which are also responsible for the red color of pomegranate arils. Generally, anthocyanins increase with storage time, and this has been correlated with the advancement of the ripening process. However, the application of these treatments led to higher concentrations of total anthocyanins during storage, as can be shown in Figure 23.5 after 3 months. Since total phenolics and anthocyanins contribute to H-TAA, pomegranates treated with these compounds had higher H-TAA at the end of storage and thus would provide fruits with enhanced health-beneficial properties (Mirdehghan *et al.*, 2006, 2007c; Sayyari *et al.*, 2010, 2011).

When dealing with TAA, most of the research has been focused on those moieties contributing to hydrophilic TAA, such as phenolics and anthocyanins. However, a recent paper described for the first time that pomegranate arils also exhibit TAA derived from lipophilic compounds (L-TAA). Thus, oxalic acid induced maintenance of L-TAA after 84 days of cold storage while decreases were observed in control arils (Sayyari *et al.*, 2010). The mechanism by which oxalic acid retained L-TAA is still unclear, but there is evidence supporting the fact that oxalic acid is a natural antioxidant compound by reducing lipid peroxidation *in vitro* (Kayashima and Katayama, 2002).

METHYL JASMONATE AND METHYL SALICYLATE VAPOR TREATMENTS AND POMEGRANATE ANTIOXIDANT PROPERTIES

Methyl jasmonate (MeJA) and its free-acid jasmonic acid, collectively referred to as jasmonates, are important cellular regulators involved in diverse developmental processes, such as seed germination, root growth, fertility, fruit ripening, and senescence (Wasternack and Parthier, 1997). Methyl salicylate (MeSA), like MeJA, is a volatile compound which is synthetized from SA and plays important roles in plant development since it can be converted back to SA (Hayat and Ahmed, 2007).

In addition, both MeJA and MeSA activate plant defence mechanisms in response to several environmental stresses including CI induced by low temperature. Treatment with MeJA was effective in reducing the development of CI symptoms in a number of horticultural crops, including loquat and peach fruit (Meng *et al.*, 2009; Cao *et al.*, 2010). Thus, MeJA has a potential application in post-harvest treatments for alleviating CI and maintaining fruit quality. However, the mode of action of MeJA in reducing CI and

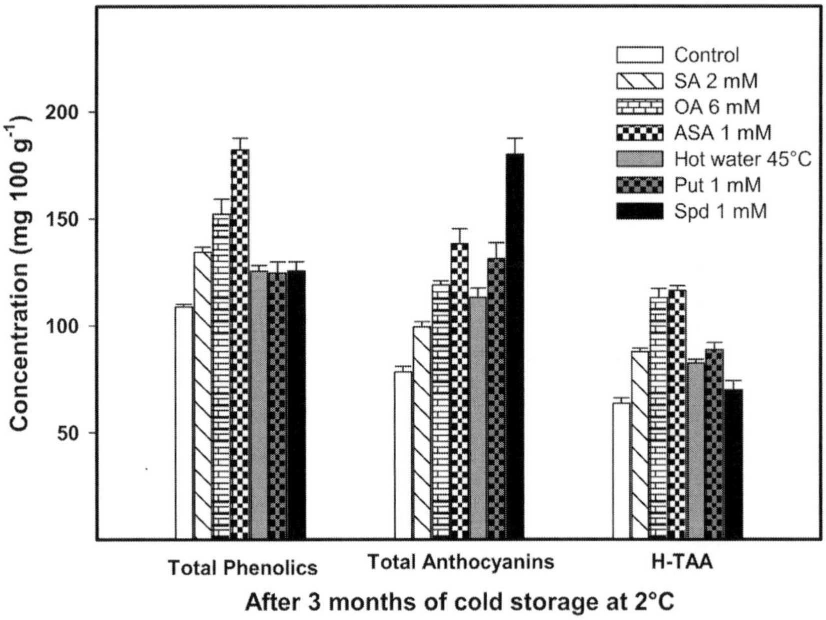

FIGURE 23.5 Concentration of total phenolics, total anthocyanins, and hydrophilic total antioxidant activity (H-TAA) in 'Mollar de Elche' pomegranate cultivar after 3 months of cold storage as affected by treatment: control, salicylic acid (SA), oxalic acid (OA), acetyl salicylic acid (ASA), hot water (45°C), putrescine (Put), and spermidine (Spd). Histograms show the mean ± SEM of determinations made in 15 individual fruits. *Adapted from the PhD Theses of Mirdehgan (2006) and Sayyari (2009).*

TABLE 23.2 Effect of Post-harvest Treatments with MeSA or (MeJA) at 0.1 mM On Chilling Injury Index, Total Phenolics, Total Anthocyanins, and Total Antioxidant Activity in the Hydrophilic Fraction (H-TAA) During 'Mollar De Elche' Pomegranate Storage*

Days	Control	MeSA	MeJa
	Chilling Injury Index		
0	0	0	0
42	0.28 ± 0.01 Aa	0.12 ± 0.01 Ab	0.12 ± 0.01 Ab
84	0.36 ± 0.03 Ba	0.15 ± 0.02 Ab	0.17 ± 0.04 Ab
	Total Phenolics (mg gallic acid eq./100 g)		
0	233 ± 8 A		
42	266 ± 5 Ba	258 ± 11 Ba	296 ± 16 Bb
84	284 ± 7 Ca	353 ± 12 Cb	381 ± 15 Cc
	Total Anthocyanins (mg cyanidin 3-glucoside eq./100 g)		
0	59 ± 4 A		
42	82 ± 5 Ba	106 ± 9 Bb	118 ± 5 Bb
84	115 ± 4 Ca	134 ± 7 Cb	131 ± 4 Cb
	H-TAA (mg Trolox eq./100 g)		
0	82 ± 6 A		
42	86 ± 5 Aa	76 ± 4 Aa	84 ± 3 Aa
84	61 ± 3 Ba	115 ± 3 Bb	115 ± 7 Bb

* *For each parameter, different capital letters show significant differences (P < 0.05) during storage time while small letters show significant differences (P < 0.05) between treatments.*

Adapted from the PhD Thesis of Sayyari (2009).

quality deterioration has not been clearly elucidated, although it has been proposed to occur throughout an increase in both free putrescine and proline in tomato (Zhang *et al.*, 2012). Conversely, the effect of MeSA on delaying CI has been attributed to the biosynthesis of heat shock proteins and to the increase of antioxidant enzyme activities, which could protect membrane deterioration caused by the lipid peroxidation (Asghari and Soleimani-Aghdam, 2010).

In pomegranate, post-harvest treatment with MeJA and MeSA reduced significantly the CI symptoms (Table 23.2) manifested by skin browning, surface pitting, and desiccation. The primary cause of CI is thought to be damage to plant cell membranes, including disorganization of mitochondria and chloroplast, which sets off a cascade of secondary reactions, including ethylene production, increased respiration, reduced photosynthesis and interference with energy production, accumulation of toxic compounds, such as ethanol and acetaldehyde, and altered cellular structure (Valero and Serrano, 2010). Thus, both treatments were able to maintain membrane structure and selective permeability leading to lower values of electrolyte leakage. In addition, other parameters related to fruit quality, such as fruit firmness, TSS and total acidity (TA) were also maintained while significant losses occurred in control pomegranates (Sayyari *et al.*, 2011). Overall, these results would indicate that both MeJA and MeSA could increase the content of endogenous polyamines (Put and Spd), and thus polyamines may be involved in reducing CI due to their ability to preserve membrane integrity, both by lowering the membrane phase transition temperature fluidity and by retarding lipid peroxidation, resulting in increased cell viability, due to their membrane-binding capacity and/or antioxidant properties. This hypothesis is supported by the fact that exogenous polyamine treatments after harvest but before cold storage decreased CI in chilling sensitive fruits, such as pomegranate (Mirdehghan *et al.*, 2007a,c).

Total phenolics in the arils increased along storage, although these increases were significantly higher in MeJA- and MeSA-treated than in control pomegranates (50, 63, and 22% increases, respectively), with respect to those levels observed at harvest (Table 23.2). A similar evolution was found for total anthocyanins, which also increased with storage time, but concentrations were significantly higher in treated pomegranates. It is important to point out that H-TAA decreased in control fruits, probably due to losses in ascorbic acid (Sayyari *et al.*, 2009), while in those treated with MeJA or MeSA, a significant increase was obtained. For comparative purposes, literature regarding the effect of MeJA on the bioactive compounds and antioxidant activity in fruits is limited, and only few examples are found in small berries, blackberries, bayberries, strawberries, and raspberries, while no scientific information is available on the relationship between MeSA and phytochemicals. In these berry reports MeJA generally led to significant increases in total phenolics, total anthocyanins, and total antioxidant activity. Specifically, in blackberries the enhancement of total phenolics including anthocyanins was correlated to higher levels of phenylalanineammonialyase activity, the key enzyme in the first step of the phenylpropanoid pathway and directly involved in the biosynthesis of phenolic compounds (Wang *et al.*, 2009). However, further studies are necessary to understand the mechanism of action by which MeJA and MeSA enhance the phytochemicals in pomegranate. The results would suggest that both MeJA and MeSA have potential post-harvest applications for reducing CI, maintaining quality, and improving the health benefits of pomegranate fruit consumption by increasing its antioxidant capacity.

CONCLUDING REMARKS

Although pomegranate fruit is sensitive to developing CI symptoms, several post-harvest technologies, such as polyamines, heat treatments, salicylic acid, acetyl salicylic acid, oxalic acid, and vapor treatments with MeJA and MeSA effectively alleviated this physiological disorder and maintained fruit quality during cold storage. Interestingly, the application of these post-harvest treatments considered as natural compounds and to be environmentally friendly, had an additional benefit in terms of enhanced content of bioactive compounds and antioxidant activity.

ANALYTICAL TECHNIQUES

CI index of pomegranate is usually calculated with a four-point hedonic scale based on the percentage of husk surface affected by CI symptoms (dehydration, browning, and pitting): 0 (no symptoms), 1 (1–25% of damaged area), 2 (26–50% of damaged area), and 3 (>51% of the damaged area) using the following formula:

CI = Σ(value of hedonic scale) × (number of fruit with the corresponding scale number) / (4 × total number of fruit in the sample).

Total antioxidant activity (TAA) was quantified according to Arnao *et al.* (2001) which enables us to determine TAA due to both hydrophilic and lipophilic compounds in the same extraction. Briefly, five grams of arils were homogenized in 5 ml of 50 mM phosphate buffer pH7.8 and 3 ml of ethyl acetate, then centrifuged at 10,000 × *g* for 15 min at 4°C. The upper fraction was used for total antioxidant activity due to lipophilic compounds (L-TAA) and the lower for total antioxidant activity due to hydrophilic compounds (H-TAA). In both cases, TAA was determined using the enzymatic system composed of the chromophore 2,2′-azino-bis-(3-ethylbenzothiazoline-6-sulfonic acid) diammonium salt (ABTS), the horseradish peroxidase enzyme and its oxidant substrate (hydrogen peroxide), in which $ABTS^{\bullet +}$ radicals are generated and monitored at 730 nm. The decrease in absorbance after adding the extract was proportional to the TAA of the sample. A calibration curve was performed with Trolox ((R)-(+)-6-hydroxy-2,5,7,8-tetramethyl-croman-2-carboxylic acid) (0–20 nmol) from Sigma (Madrid, Spain), and results are expressed as mg of Trolox equivalent/100 g.

Total phenolics were extracted according to the Tomás-Barberán *et al.* (2001) protocol using water : methanol (2 : 8) containing 2 mM NaF and quantified using the Folin–Ciocalteu reagent and results were expressed as mg gallic acid equivalent/100 g fresh weight. Total anthocyanins were determined according to Sayyari *et al.* (2010) and calculated as cyanidin 3-glucoside equivalent (molar absorption coefficient of 23900 l/cm per mol and molecular weight of 449.2 g/mol) and results expressed as mg/100 g fresh weight.

Acknowledgments

The authors greatly appreciate the technical assistance of Dr. Juan Miguel Valverde and Dr. Huertas M. Díaz-Mula in laboratory analytical determinations.

References

Alighourchi, H., Barzegar, M., Abbasi, S., 2008. Anthocyanins characterization of 15 Iranian pomegranate (*Punica granatum* L.) varieties and their variation after cold storage and pasteurization. Eur. Food Res. Technol. 2227, 881–887.

Arnao, M.B., Cano, A., Acosta, M., 2001. The hydrophilic and lipophilic contribution to total antioxidant activity. Food Chem. 73, 239–244.

Asghari, M., Soleimani-Aghdam, M., 2010. Impact of salicylic acid on post-harvest physiology of horticultural crops. Trends Food Sci. Technol. 21, 502–509.

Cao, S.F., Zheng, Y.H., Wang, K.T., Rui, H.J., Tang, S.S., 2010. Effect of methyl jasmonate on cell wall modification of loquat fruit in relation to chilling injury after harvest. Food Chem. 118, 641–647.

Faria, A., Calhau, C., 2011. The bioactivity of pomegranate: Impact on health and disease. Crit. Rev. Food Sci. Nutr. 51, 626–634.

Hayat, S., Ahmed, A., 2007. Salicylic Acid: A Plant Hormone. Springer, Dordrecht.

Holland, D., Bar-Ya'akov, I., 2008. The pomegranate: new interest in an ancient fruit. Chron. Horticult. 48 (3), 12–15.

Ismail, T., Sestili, P., Akhtar, S., 2012. Pomegranate peel and fruit extracts: A review of potential anti-inflammatory and anti-infective effects. J. Etnopharmacol. 143, 397–405.

Landete, J.M., 2011. Ellagitannins, ellagic acid and their derived metabolites: A review about source, metabolism, functions and health. Food Res. Int. 44, 1150–1160.

Kayashima, T., Katayama, T., 2002. Oxalic acid is available as a natural antioxidant in some systems. Biochem. Biophys. Acta 1573, 1–3.

Melgarejo, P., and Martínez, R. (1992). El Granado. Ed. Mundi-Prensa, Madrid.

Mena, P., García-Viguera, C., Navarro-Rico, J., Moreno, D.A., Bartual, J., Saura, D., et al., 2011. Phytochemical characterisation for industrial use of pomegranate (*Punica granatum* L.) cultivars grown in Spain. J. Sci. Food Agric. 91, 1893–1906.

Meng, X.H., Han, J., Wang, Q., Tian, S.P., 2009. Changes in physiology and quality of peach fruits treated by methyl jasmonate under low temperature stress. Food Chem. 114, 1028–1035.

Mirdehghan, S.H., 2006. Physiological and biochemical changes during growth and storage of pomegranate fruit in relation to chilling injury: role of heat treatment. PhD Thesis, Shiraz University, Shiraz.

Mirdehghan, S.H., Rahemi, M., Castillo, S., Martínez-Romero, D., Serrano, M., Valero, D., 2007a. Pre-storage application of polyamines by pressure or immersion improves shelf life of pomegranate stored at chilling temperature by increasing endogenous polyamine levels. Postharvest Biol. Technol. 44, 26–33.

Mirdehghan, S.H., Rahemi, M., Martínez-Romero, D., Guillén, F., Valverde, J.M., Zapata, P.J., et al., 2007b. Reduction of pomegranate chilling injury during storage after heat treatment: role of polyamines. Postharvest Biol. Technol. 44, 19–25.

Mirdehghan, S.H., Rahemi, M., Serrano, M., Guillén, F., Martínez-Romero, D., Valero, D., 2007c. The application of polyamines by pressure or immersion as a tool to maintain functional properties in stored pomegranate arils. J. Agric. Food Chem. 55, 755–760.

Mirdehghan, S.H., Rahemi, M., Serrano, M., Guillén, F., Martínez-Romero, D., Valero, D., 2006. Prestorage heat treatment to maintain nutritive and functional properties during postharvest cold storage of pomegranate. J. Agric. Food Chem. 54, 8495–8500.

Rivera, D., Matilla, G., Obón, C., Alcaraz, F., 2012. Plants and Humans in the Near East and The Caucasus. Editum—Plants and Humans, Murcia.

Sayyari, M., 2009. Effect of chemical and non-chemical factors on chilling and scald resistance of stored pomegranate cv. Malas Saveh. PhD Thesis, Tehran University, Tehran.

Sayyari, M., Babalar, M., Kalantari, S., Serrano, M., Valero, D., 2009. Effect of salicylic acid treatment on reducing chilling injury in stored pomegranates. Postharvest Biol. Technol. 53, 152–154.

Sayyari, M., Castillo, S., Valero, D., Díaz-Mula, H.M., Serrano, M., 2011. Acetyl salicylic acid alleviates chilling injury and maintains nutritive and bioactive compounds and antioxidant activity during postharvest storage of pomegranates. Postharvest Biol. Technol. 60, 136–142.

Sayyari, M., Valero, D., Babalar, M., Kalantari, S., Zapata, P.J., Serrano, M., 2010. Prestorage oxalic acid treatment maintained visual quality, bioactive compounds, and antioxidant potential of pomegranate after long-term storage at 2°C. J. Agric. Food Chem. 58, 6804–6808.

Seeram, N.P., Lee, R., Heber, D., 2004. Bioavailability of ellagic acid in human plasma after consumption of ellagitannins from pomegranate (*Punica granatum* L.) juice. Clin. Chim. Acta. 348, 63–68.

Tomás-Barberán, F.A., Gil, M.I., Cremin, P., Waterhouse, A.L., Hess-Pierce, B., Kader, A.A., 2001. HPLC-DAD-ESIMS analysis of phenolic compounds in nectarines, peaches, and plums. J. Agric. Food Chem. 49, 4748–4760.

Valero, D., Serrano, M., 2010. Postharvest Biology and Technology for Preserving Fruit Quality. CRC Press-Taylor & Francis, Boca Raton, USA.

Wang, K., Jin, P., Cao, S., Shang, H., Yang, Z., Zheng, Y., 2009. Methyl jasmonate reduces decay and enhances antioxidant capacity in Chinese bayberries. J. Agric. Food Chem. 57, 5809–5815.

Ward, C., 2003. Pomegranates in eastern Mediterranean contexts during the Late Bronze age. World Archaeol 34, 529–541.

Wasternack, C., Parthier, B., 1997. Jasmonate-signalled plant gene expression. Trends Plant Sci. 2, 302–307.

Zhang, X., Sheng, J., Li, F., Meng, D., Shen, L., 2012. Methyl jasmonate alters arginine catabolism and improves postharvest chilling tolerance in cherry tomato fruit. Postharvest Biol. Technol. 64, 160–167.

SECTION 3

DAIRY AND EGGS

CHAPTER

24

Milk Pasteurization, Curdling and Salting: Steps which can Modify Omega-3-Fortified Cheese

Daniela Bermúdez-Aguirre, Gustavo Barbosa-Cánovas
Department of Biological Systems Engineering, Washington State University, Pullman, WA, USA

CHAPTER POINTS

- *Queso fresco* and goat cheese have shown higher retention of omega-3 after milk pasteurization.
- Microencapsulated fish oil has shown the highest retention in cheddar when it is added at the end of the process.
- Mozzarella has been the most challenged product to fortify with omega-3 due to the conventional process that must be followed.
- Some non-thermal approaches can be used to modify the structure of milk fat globules and bond the omega-3 molecules to the fat–casein matrix.
- Feeding ruminants with grass and nutritional supplements to produce milk rich in omega-3 do not show important results in cheese fortification.

INTRODUCTION

An important increase of omega-3- and omega-6-fortified food products is currently taking place in the market. Eicosapentaenoic (EPA; C20:5 n-3), docosahexanoic (DHA C22:6 n-3) and α-linolenic acid (ALA; C18:3 n-3) acids are known as polyunsaturated fatty acids (PUFA) or long-chain n-3 PUFA (LC PUFA), and it is becoming more frequent to find them in the composition of some processed and fortified products. These are polyunsaturated acids with 3, 5, or 6 double bonds in the carbon chain with 18, 20, and 22 carbons, respectively. Humans are not able to synthesize PUFA from their food; it is necessary to ingest at least 650 mg of LC n-3 PUFA per day, according to ISSFAL (2004) to cover nutrition requirements. Linolenic acid (LA; C18:2 n-6) and α-linolenic acid (ALA; C18:3 n-3) can be metabolized to a longer-chain fatty acid; for example, it is believed that 10–15% of ALA is converted to EPA, although the synthesis of DHA is limited (Trautwein, 2001). These PUFA are well known for their health benefits, such as reduction of cardiovascular diseases, reduction of blood pressure, anti-inflammatory and anti-allergic effects, protection against cancer, and development and function of the brain, nervous system, and retina.

PUFA are naturally found in high concentration in some products such as fish oil, flaxseed, canola and soybean oils, algae, and fresh grass, among others (Table 24.1). These fatty acids are also found in other products such as eggs, red meat, and turkey, but in much lower concentrations, and are very often oxidized during food processing due to the common unit operations of heating, drying, salting, and mixing, among others, or during storage, considerably reducing the concentration of PUFA. ALA is found in some green leafy vegetables, in the chloroplast, but because of the low fat content of those vegetables, they are not considered as a good source of omega-3. For example, spinach has about 0.13 g ALA/100 g, but the fat content of spinach is about 0.30 g/100 g (Trautwein, 2001).

One food product that is becoming more important in the PUFA market is cheese. Different types of this dairy product have been fortified with animal and vegetable sources of omega-3 and omega-6; however, cheese production still faces several challenges. First, omega-3 should be added to the cheese in such a way that flavor interaction is not noticed after fortification. Secondly,

Processing and Impact on Active Components in Food
http://dx.doi.org/10.1016/B978-0-12-404699-3.00024-X

TABLE 24.1 Some Food Products Considered Natural Rich Sources of Omega-3

Vegetable Sources	g/tbsp (Reported as ALA)	Animal Sources	g/85 g Serving Size (Edible Portion) (Reported as DHA + EPA)
Olive oil	0.1	Tuna (fresh)	0.24–1.28
Walnuts	0.7	Sardines	0.98–1.70
Soybean oil	0.9	Salmon (Atlantic farmed)	1.09–1.83
Canola oil	1.3	Mackerel	0.34–1.57
Walnut oil	1.4	Herring (Pacific)	1.81
Flaxseeds	2.2	Trout Rainbow (Farmed)	0.98
Flaxseed (linseed oil)	8.5	Oyster (Pacific)	1.17
		Lobster	0.07–0.41
		Crab (Alaskan, King)	0.35
		Shrimp (Mixed species)	0.27
		Clam	0.24

Adapted from Kris-Etherton et al. *(2002).*

and not less important, the concentration of omega-3 needs to remains at an effective level after the processing steps of cheese-making have been undertaken.

This chapter describes how the concentration of omega-3 changes during the cheese-making process and how the food industry can enhance the retention of this important nutrient in cheese without flavor modification.

HOW THE COMPOSITION IS ALTERED

The cheese-making process is well known for the different stages required to create a cheese with specific characteristics. There are very specific processes for cheese-making which only require the addition of starter cultures or enzymes to obtain the final product, but there are also more complex processes that require the use of temperature, pressure, and storage under specific conditions to develop flavor compounds, for example. During these processes, some nutrients can be affected, depending on processing conditions (high temperature, oxygen concentration, interaction between compounds, pressure, etc.). The fortification of cheese thus represents a challenge in terms of retaining the nutrient content at desirable concentration without any change after processing or during storage.

The Cheese-Making Process for Different Types of Cheese

It is not possible to summarize when during the processing steps for cheese-making the composition of omega-3 is affected, due to the different compositions of cheeses, the variable stages during each unique cheese process, and other external factors such as the source of omega-3. Therefore, a brief summary for some common types of cheese and how omega-3 is altered during processing will be described here, but again, because of the wide variety of cheese, it is not possible to affirm, for example, that all soft cheeses will have the same behavior in terms of omega-3, such as *queso fresco,* a soft-Hispanic cheese.

Soft Cheese

QUESO FRESCO

This soft and Hispanic style cheese requires a single process, compared to other semi-hard or hard cheeses. Pasteurized milk is curded with some enzymes (mainly rennet), and then the curds are pressed, salted, and molded for further storage under refrigerated conditions. One of the main challenges during the incorporation of omega-3 in this cheese is related with the high moisture content of the product (about 50%), which can "wash off" the nutrient from the product. The incorporation of omega-3 in *queso fresco* has been tested experimentally during different stages of the cheese-making process, such as after milk pasteurization, during curdling, and during salting (Bermúdez-Aguirre and Barbosa-Cánovas, 2012). For this cheese, the highest retention of the nutrient was observed after milk pasteurization using microencapsulated omega-3 (8.49 mg/g), followed by the addition of omega-3 during salting (8.07 mg/g), as shown by Bermúdez-Aguirre and Barbosa-Cánovas (2012). The microcapsules of omega-3 remain attached to milk particles, such as fat globules and casein micelles, after pasteurization allowing retention of the nutrient during processing. However, when microcapsules are added to the cheese during curdling, together with the specific enzymes required for the curd (mainly rennet), the nutrient is lost in a high concentration. The reason for that could be the enzymatic activity over the coating material of the microcapsules of omega-3, releasing the content to the medium and reacting with the nutrient. Finally, when the omega-3 is added during salting, no further steps are required in this kind of cheese, and the powdered nutrient is maintained as part of the cheese matrix, and only a small decrease in concentration occurs during the overnight draining of the cheese.

SPREADABLE FRESH CHEESE

Another highly consumed cheese is the spreadable, commonly known as 'Philadelphia' type. In an interesting research, this cheese (not flavored, 370 g fat/kg and with garlic flavor 350 g fat/kg) was fortified with fish oil rich in omega-3 to study the sensory acceptability of the final product. Incorporation of omega-3 was conducted only by blending the fish oil with the cheese from 0 to 80 g/kg at room temperature. High levels of fortification were achieved in this cheese, as shown in Table 24.2, with a final concentration of about 6 mg omega-3/g of cheese. However, as the concentration of omega-3 was higher in this cheese, the sensory quality decreased as fishy flavor was detected (Kolanowski and Weißbrodt, 2007). In this example, there were no processing conditions that drastically affected the concentration of omega-3 in the product. Omega-3 (fish oil) was only blended with the cheese until a homogeneous mix was achieved. But the main drawback is the undesirable effect on the sensory quality of the product. However, these authors also showed that when a specific flavor is added to the product (such as garlic), the fish oil taste can be masked to be detected less often by panelists.

GOAT CHEESE

Goat cheese is also a product with an important market because of its specific sensory characteristics. Hughes *et al.* (2012) studied the fortification of soft goat cheese with omega-3 using fish oil in concentrations from 60 to 100 g oil/ 833 g cheese. Fish oil was added after milk pasteurization (62°C, 30 min), using a manual blender to achieve a full emulsion. After that, the culture and rennet was added and the typical protocol for goat cheese was followed. The final concentration of EPA and DHA was about 4.53 mg omega-3/g cheese (127 mg/28-g serving), and this is reported as the maximum concentration that the cheese can retain regardless of the use of higher concentration of fish oil, such as 80 or 100 g. It is interesting to see how the incorporation of fish oil before curd formation can achieve a higher retention for this kind of cheese, similar to the one reported for *queso fresco* in the previous section of this chapter. Conversely, for goat cheese, omega-3 was added in liquid form, which makes it more difficult to retain the nutrient in the matrix. However, after the incorporation of the nutrient, the starter culture and the rennet were added, allowing 24 h to set. This time is considerably longer compared with other soft cheeses such as *queso fresco,* which usually takes less than an hour to show curd formation. Indeed, this longer period of time for curd formation for goat cheese allows for the holding of molecules of omega-3 into the fat–casein matrix during cheese-making. However, the study also shows a saturation point at which no more fish oil can be retained, and it is lost in the whey portion (Hughes *et al.*, 2012).

Semi-Hard Cheese

CHEDDAR CHEESE

Cheddar cheese is one of the most popular cheeses in the United States, and is highly consumed alone or as a part of other foods. Cheese-making for this product is not complicated, but every step must be conducted at the right moment to achieve the typical characteristics of cheddar.

In a study to process fortified reduced fat (50%) cheddar cheese, incorporation of fish oil was conducted using microencapsulated oil. The typical cheddar cheese process was followed; before pressing the curd, omega-3 powder was mixed with the ground curd. The authors reported a final concentration of omega-3 in cheddar cheese of about 2.53 mg/g. There were no changes in the omega-3 concentration during aging of the cheese (about 6 months); the level of oxidation was minimal in affecting the final concentration of the nutrient (Martini *et al.*, 2009). This is a clear example in which fortification of cheese was achieved at the end of the cheese-making process and the correct source of omega-3 was used (microencapsulated), which provided the nutrient with better stability even during aging of the cheese. However, it is important to highlight that cheddar cheese has a specific process that includes ripening, cutting, cooking, and washing, among others, and the omega-3 was added at the final stage of the process before pressing the curd, which at a certain point protected the nutrient from the action of temperature, or the enzymatic and microbial activity.

In another report about the fortification of cheddar cheese, using fish oil high in omega-3, researchers replaced up to 75% of the milk fat using fish oil. No important changes in the physicochemical characteristics of the cheese using different omega-3 ratios were observed, and only a slight change of flavor for the cheese without omega-3 was detected (Aryana, 2007). However, the quantification of omega-3 at the end of the cheese making was not reported for this study, nor was the initial concentration of oil rich omega-3.

TABLE 24.2 Examples of Cheese Enriched with Different Sources of Omega-3[a]

Type of Cheese	Processing Conditions	Final Omega-3 Concentration (mg/g)	Reference
Alpine	Using regular milk from cows fed with alpine grass	0.39[b]	Hauswirth *et al.* (2004)
Alpine—Silage	Using regular milk from cows fed with alpine grass and silage	0.24[b]	Hauswirth *et al.* (2004)
Alpine—Linseed supplemented	Using regular milk from cows fed with alpine grass and supplemented with linseed	0.23[b]	Hauswirth *et al.* (2004)
Cheddar (50% reduced fat)	Regular cheese-making process for cheddar, grinding the curd and adding microencapsulated omega-3 (fish oil)	2.53	Martini *et al.* (2009)
Cheddar	Using regular milk from cows fed with alpine grass	0.16[b]	Hauswirth *et al.* (2004)
Cheddar	Adding flaxseed oil and microencapsulated fish oil during salting	5.08 and 8.69, respectively	Bermúdez-Aguirre and Barbosa-Cánovas (2011)
Emmental	Using regular milk from cows fed with alpine grass	0.20[b]	Hauswirth *et al.* (2004)
Mozzarella	Adding flaxseed oil and microencapsulated fish oil during salting	0.88 and 2.69, respectively	Bermúdez-Aguirre and Barbosa-Cánovas (2011)
Mozzarella	Vegetable (flaxseed) oil added to milk pasteurized with high hydrostatic pressure (483 MPa, 40°C and 5 min)	6.64	Bermúdez-Aguirre and Barbosa-Cánovas (2012)
Processed cheese	Purified fish oil blended with the product	12	Kolanowski and Weißbrodt (2007)
Processed cheese slices	Processed cheese model based on rennet casein and soya oil; fortified cheese was done replacing the soya oil by tuna fish oil	0.495–3.96	Ye *et al.* (2009)
Queso fresco	Flaxseed oil and microencapsulated (fish oil) omega-3 added to milk after pasteurization	1.62 and 8.49, respectively	Bermúdez-Aguirre and Barbosa-Cánovas (2011)
Queso fresco	Vegetable (flaxseed) oil added to milk pasteurized with Pulsed electric fields (30 kV, 40°C, 30 pulses, 2 μs each)	5.20	Bermúdez-Aguirre and Barbosa-Cánovas (2012)
Queso fresco	Vegetable (flaxseed) oil added to milk pasteurized with ultrasound (63°C, 120 μm, 5 min)	5.12	Bermúdez-Aguirre and Barbosa-Cánovas (2012)
Soft goat cheese	Fish oil blended with milk after pasteurization (60–100 g fish oil/833 g cheese)	4.53	Hughes *et al.* (2012)
Spreadable fresh cheese (Philadelphia type)	Purified fish oil blended with the product (20 g/kg)	6	Kolanowski and Weißbrodt (2007)

[a] *Recommended daily intake LC n-3 PUFA 650 mg/d (ISSFAL, 2004).*
[b] *Reported as EPA.*

Finally, in a study to fortify cheddar cheese using different sources of omega-3 such as flaxseed oil, liquid fish oil, microencapsulated fish oil, and flaxseed powder, researchers tested three different stages during the cheese-making process (after milk pasteurization, during curdling and during salting) to add the nutrient and find the highest retention of the product. Results showed that microencapsulated fish oil had the highest retention during salting, followed by the flaxseed oil (Bermúdez-Aguirre and Barbosa-Cánovas, 2011). As no further processing steps are required, omega-3 remains in the cheese matrix and only minor loss is observed on the whey fraction of the product. However, very low concentration of the nutrient was observed in the same study when omega-3 was added as oil, regardless of the processing step (after milk pasteurization or during curdling). It is evident that the cheese-making process that involves heating, ripening, cutting, mixing, again heating and milling of the curd, affects the stability of the lipids, given the action of the temperature, oxygen, microbial, and enzymatic activity.

PROCESSED CHEESE

Several varieties of processed cheese can be mentioned; however, some of the most common types are processed cheese spread, processed cheese slices, and blended cheese. Processed cheese slices are similar to processed cheddar cheese (Guinee *et al.*, 2004). Processed cheese slices can then be included as part of the semi-hard cheese group. Ye *et al.* (2009) used a model for processed cheese slices, based mainly on rennet casein and soya oil; about half of the percentage of the cheese composition was water. For fortified cheese, tuna fish oil (99 mg n-3 PUFA/g) was used, and depending on the required final concentration, soya oil was substituted by the nutrient at the beginning of the process. Processing of this cheese included mixing ingredients manually, hydrating for 1 h, heating at 90°C (10 min), shaping and sealing in a plastic bag, and cooling and storing (4°C). Peroxide values (PV) and thiobarbituric acid reactive substances (TBARS) were quantified to detect the stability of the fish oil. PV showed an increase in lipid oxidation during the first 17 days of storage at 30°C, mainly for the fish oil, and minor oxidation for encapsulated fish oil. A similar trend was observed for Propanal concentration in cheese (secondary lipid oxidation products). TBARS values showed a fast increase during storage at 30°C for both fish oil and microencapsulated fish oil, being slightly lower for the encapsulated form. When cheese was stored at refrigerated conditions TBARS did not show important changes. The authors concluded that during the heating (90°C) and the blending of the ingredients, temperature and oxygen led to oxidation of omega-3. However, it is clear that using microencapsulated fish oil considerably reduces lipid oxidation during the processing steps.

Hard Cheese

MOZZARELLA CHEESE

One of the most laborious cheese-making processes is that required for mozzarella cheese. Several steps are required to achieve the typical characteristics of this cheese, such as flavor, odor, and melting properties, among others. Several mild thermal treatments are required during the process to allow the growth of the thermophilic culture, to cook the whey and the curd, and to stretch the curd. During these stages of processing, not only could the omega-3 be affected, but also other nutrients present in milk.

Few studies have been reported in the fortification of mozzarella cheese with omega-3, possibly due to the complexity of the process and the low retention of the nutrient for the above-mentioned conditions. A similar study to that reported for cheddar cheese was conducted for mozzarella, using different sources of omega-3 during three steps of the cheese-making process. However, in this case, all sources showed very low retention regardless of the processing step in which the omega-3 was added. Highest retention was observed for microencapsulated fish oil with a value of 2.69 mg/g added during curdling; meanwhile, the flaxseed oil showed the highest concentration (1.40 mg/g) when added during salting (Bermúdez-Aguirre and Barbosa-Cánovas, 2012). It is well known that lipid oxidation can occur easily when the product is heated, exposed to light, exposed to oxygen, etc. So, the fortification of mozzarella cheese using omega-3 is still a challenge for food scientists because of the rapid oxidation and loss of the nutrient, even when a microencapsulated form is used.

Other Ways in which Composition is Altered

Some innovative resources that have been researched to increase the concentration of omega-3 in cheese include the use of non-thermal technologies. These novel approaches have illuminated the modification of some components in milk, such as the unfolding of proteins due to the use of high hydrostatic pressure, the decrease of fat globule size and agglomeration with casein molecules using ultrasound, or clustering of fat globules with the use of pulsed electric fields. Based on this research, the use of these technologies was explored in order to take advantage of the new configuration of milk after non-thermal technologies and to study the possibility for better bonding of the omega-3 molecules before the cheese-making process.

Meanwhile, other researchers believe that a better way to achieve this is to include one of the ingredients, in this specific case, milk, with high concentration of omega-3. Some studies have been conducted directly on the countryside, feeding ruminants with sources high in omega-3, with the expectance of producing milk with high levels of omega-3.

High Hydrostatic Pressure (HHP)

This non-thermal technology was used to improve the retention of omega-3 in cheese because the internal modification of milk components. Bermúdez-Aguirre and Barbosa-Cánovas (2012) illustrated the use of pressure about 483 MPa, 40°C during 5 min applied in milk containing flaxseed omega 3. The main idea was to use the modification of proteins and other components such as fat globules in milk because of the pressure to retain more omega-3 molecules. However, for *queso fresco* and cheddar, there was no important enhancement in nutrient retention. In fact, for *queso fresco,* HHP had an antagonist effect, resulting in a final concentration of omega-3 lower than the control sample (milk thermally pasteurized), because pressurized fat globules are not very close together (Figure 24.1A), thus allowing the retention of water molecules in the cheese matrix and then retaining very low concentration of the hydrophobic oil in the same cheese matrix (O'Reilly *et al.*, 2001). For mozzarella cheese, there was an increase in the final concentration of omega-3 of about 0.7 mg/g higher for the pressurized cheese than for the control sample.

Similar to HHP, the use of high-pressure homogenization (HPH) has been explored as an alternative during

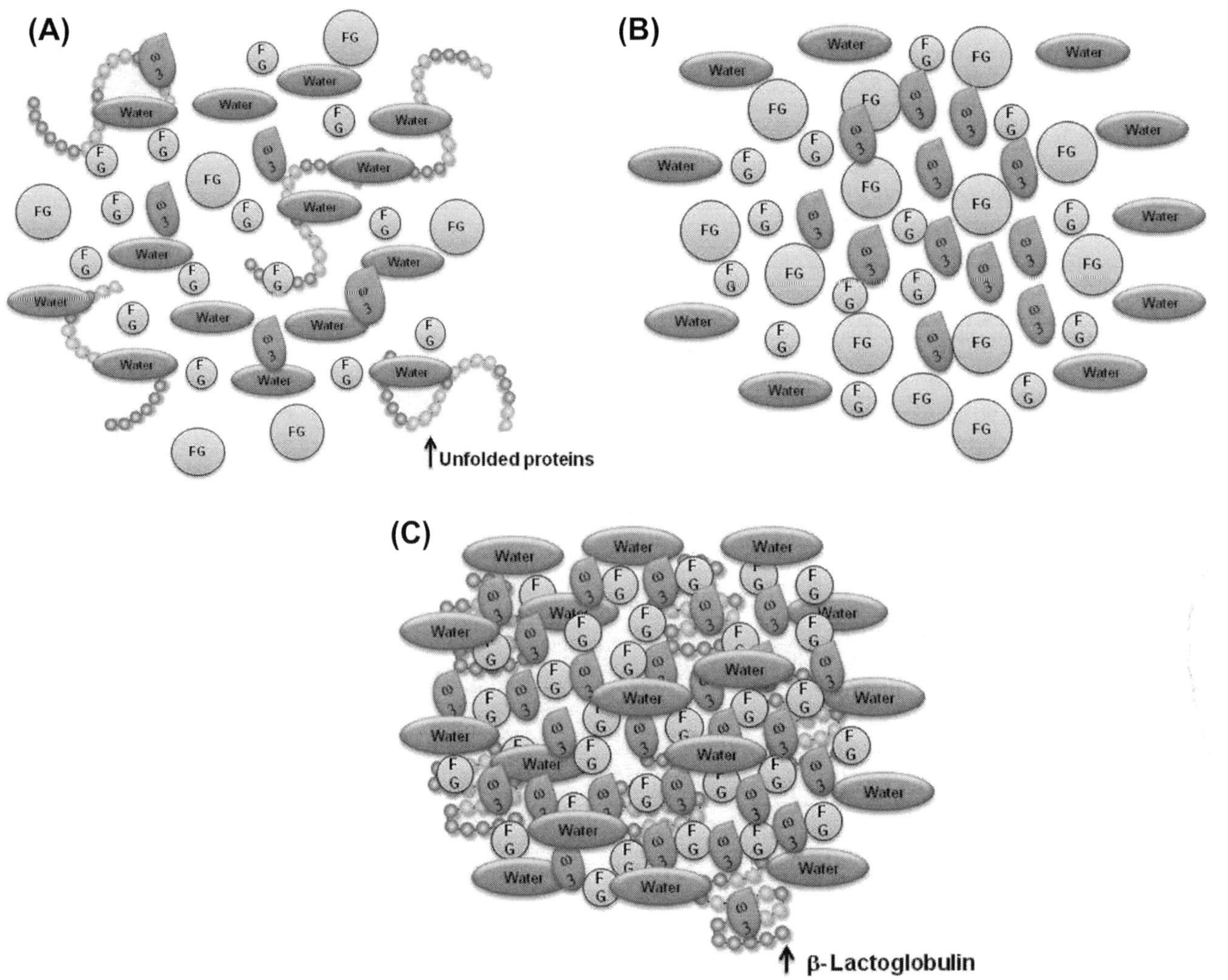

FIGURE 24.1 Schematic view of the cheese matrix fortified with omega-3 after (A) high hydrostatic pressure, (B) pulsed electric fields, (C) ultrasound. FG, Fat globule (hydrophobic molecule); ω3, drops of omega-3 (oil, hydrophobic molecule).

cheese fortification. HPH uses pressure lower than typical HHP; the effect on milk components then is minor. Calligaris *et al.* (2013) used HPH (20–100 MPa) to incorporate fish and vegetable (flaxseed) oil during the production of *queso fresco.* It is believed that HPH modifies and unfolds the proteins of milk, changing the emulsifying capacity and allowing more fat to be trapped in the cheese matrix. However, results of the work done by Calligaris *et al.* (2013) did not show important differences in fat content between fortified and unfortified cheese, probably due to the poor incorporation of the oil in milk during the HPH.

Pulsed Electric Fields (PEF)

This novel technology has previously been used to pasteurize some liquids foods and to extract certain compounds from matrices. However, because of the positive results in microbial inactivation (pathogens and spoilage microorganisms) in milk, further studies have been conducted. PEF has also shown positive results in the rennet coagulation of milk (Yu *et al.*, 2009), speeding up the process for the formation of curd, but also ensuring the microbial quality of milk. In addition, during the PEF processing of milk, some structural changes of milk components have been observed that, similar to HHP, could allow the retention of, in this case, omega-3 molecules, together with the fat/casein matrix. Bermúdez-Aguirre and Barbosa-Cánovas (2012) used PEF to pasteurize milk that was used for cheese-making. PEF treatment was carried out at 30 kV, 40°C, and 30 pulses (2 μs each) in raw whole milk, and cheese fortification was conducted with flaxseed oil. Three kinds of cheese were processed: *queso fresco*, cheddar, and mozzarella. Results showed a good retention of omega-3 for the three cheeses, with cheddar having the lowest concentration (4.93 mg/g) and mozzarella with the highest concentration (6.19 mg/g). The reason for that could be the better integration of fat globules after milk processing with PEF. It is well known that PEF can agglomerate fat globules in milk (Figure 24.1B), which could keep omega-3 molecules together because of the hydrophobic behavior of both, and at some point protecting them from environmental factors such as temperature.

Ultrasound (US)

Finally, the third non-thermal technology included in this chapter is ultrasound, which has promising results for the dairy industry. US is able to create perfect emulsions, to reduce syneresis in some dairy products such as yogurt, to increase yield, and to improve the quality of some cheeses, and to eliminate biofilms in industry equipment, among others uses.

When ultrasound is applied to milk, the size of fat globules is reduced and smaller globules are able to agglomerate with casein molecules to form more complex systems. The milk fat lipid globule membrane (MFLG) is disrupted because of the action of the US, and depending on the intensity of the treatment and time, can be partially or totally disrupted, providing the fat globules with a granular surface (Bermúdez-Aguirre *et al.*, 2008). Again, similar to PEF, the hydrophobic character of the fat globules can be used to hold other hydrophobic molecules (such as oil) and keep them together (Figure 24.1C). Also, proteins such as beta-lactoglobulin (β-Lg) have affinity for hydrophobic molecules such as fatty acids, so omega-3 could be attached to these proteins and form perfect emulsions (Zimet and Livney, 2009).

The use of US was tested for fortification of *queso fresco* with omega-3; three processing times were chosen for milk pasteurization (5, 10, and 30 min) and flaxseed was used as a source of omega-3. The highest concentration of omega-3 (5.12 mg/g) was observed after 5 min of treatment; meanwhile, the concentration decreased in accordance with the processing time up to 4.11 mg/g (Bermúdez-Aguirre and Barbosa-Cánovas, 2012). The reason for the decrease in the concentration of omega-3 during sonication could be attributed to oxidation problems. Since US generates heat during processing, even when the temperature is controlled by an external source, oxidation could be promoted for the instant generation of heat (hot spots), but also because of the intense cavitation that takes place during sonication which generates implosions and explosions in the milk with the constant presence of oxygen. Additionally, external sources, such as the material of some sonotrodes, could lead to lipid oxidation. However, the omega-3 concentration found in *queso fresco* after sonication is good enough to be claimed as a source of the nutrient.

Enrichment of Milk with Omega-3 before Cheese-Making

One strategy that researchers are exploring in order to increase the content of omega-3 in cheese is the feeding of cows, dairy ewes, sheep, and similar ruminants using sources rich in conjugated linoleic acid (CLA) such as linseed (oil or seeds). Reports in the literature show how the use of linseed during the feeding of ruminants together with the regular pasture can enhance the content of CLA in milk and cheese. Pecorino cheese was processed with milk previously enriched by feeding sheep with a diet containing a concentrate with 30% extruded linseed, showing an increase up to 250% of alpha-linolenic acid, which can be claimed as "omega-3" dietary source because of the final concentration in cheese (Mele *et al.*, 2011). Other authors have also compared the concentration of omega-3 in cheese produced with milk from ruminants fed with alpine grass (which has a high concentration of ALA) with ruminants fed with an enriched diet based on linseed (Hauswirth *et al.*, 2004). Results showed that alpine grass represents the best option for high concentration of ALA, EPA, and CLA on cheese, rather than the enriched diet. However, the concentration of omega-3 in alpine, emmental or cheddar cheese (regardless of the diet of cows) is far below the concentration found in cheese enriched with this nutrient during processing, as shown in Table 24.2.

Similar results were found by Avramis *et al.* (2003) in a study conducted with Holstein cows fed with a fish meal containing a supplement of herring (60%) to produce cheddar cheese. Even though the final concentration of PUFA was not reported, some properties of the milk changed undesirably, such as the fat content that decreased about 1.71%. However, some properties of cheddar cheese were improved, such as the reduction of ripening, better texture and stronger flavor.

Based on these few results in which researchers have tried to increase the concentration of PUFA in milk, feeding ruminants with fish supplements or vegetable sources of omega-3, it has been observed that final concentration of the nutrient in cheese is not very high, and several grams of cheese should be consumed per day to satisfy the daily requirements of omega-3 in the diet. According to ISSFAL (2004), the recommended daily intake of LC n-3 PUFA is about 650 mg/d. Thus, an average individual eating an alpine cheese should consume more than 1.5 kg of cheese to fulfill this requirement; meanwhile, eating one of the fortified cheeses (average 5 mg/g), an individual should consume about 130 g of cheese per day, if this is the only source of omega-3 in the diet.

ANALYTICAL TECHNIQUES

One of the most explored areas in the last few years has been the analytical techniques to quantify dietary supplements in the correct way. Several techniques are still being explored, in an effort to find an official method that meets three important stages: method development, single-laboratory validation (SLV) and collaborative study. The omega-3 official method was reported under a collaborative study stage in 2008 due to related problems because of microencapsulation that very well protects the components, but makes it difficult to extract and recover those components during analytical measurements (AOAC, 2008).

According to Pike (1998), there are several options to quantify PUFAs in food items; the American Oil Chemists' Society (AOCS) proposes method Cd15-78; the Association of Official Analytical Chemists (AOAC) refers to method number 969.33; and the International Union of Pure and Applied Chemists (IUPAC) proposes method 2.301. Basically, all of these methods require an extraction of the fat components and further analysis of those components using gas chromatography/mass spectroscopy (GC-MS) or high-performance liquid chromatography (HPLC).

In the literature, authors have reported the quantification of omega-3, mainly using gas chromatography (GC), which is able to separate fatty acid methyl esters and identify each peak, thus identifying the exact concentration of each compound (Bermúdez-Aguirre and Barbosa-Cánovas, 2012; Gómez-Cortés *et al.*, 2009; Martini *et al.*, 2009; Hughes *et al.*, 2012; Hauswirth *et al.*, 2004). Some variations to the methodology have been proposed and validate the official method of facilitating extraction and quantification, for example, the one proposed by O'Fallon *et al.* (2007). Other authors prefer the use of HPLC to quantify the fatty acids, such as Mele *et al.* (2011).

Other authors report omega-3 in foods in terms of values that provide the degree of lipid oxidation, such as PV, thiobarbituric acid reactive substances (TBARS), and propanal values (Ye *et al.*, 2009; Calligaris *et al.*, 2013). However, it is important to highlight that these values include all of the lipid species in the food product rather than only omega-3 fatty acids and might not be very accurate.

References

A.O.A.C, 2008. Methods for Dietary Supplements: Successful conclusion of a 5-year effort. Inside Lab. Manage. AOAC Int., 26–31.

Aryana, K.J., 2007. Cheddar cheese manufactured with oil high in omega-3 fatty acids. Milchwissenschaft 62 (2), 167–171.

Avramis, C.A., Wang, H., McBride, B.W., Wright, T.C., Hill, A.R., 2003. Physical and processing properties of milk, butter, and cheddar cheese from cows fed supplemental fish meal. J. Dairy Sci. 86, 2568–2576.

Bermúdez-Aguirre, D., Barbosa-Cánovas, G.V., 2011. Quality of selected cheeses fortified with vegetable and animal sources of omega-3. LWT—Food Sci. Technol. 44, 1577–1584.

Bermúdez-Aguirre, D., Barbosa-Cánovas, G.V., 2012. Fortification of *queso fresco*, cheddar and mozzarella using selected sources of omega-3 and some nonthermal approaches. Food Chem. 133, 787–797.

Bermúdez-Aguirre, D., Mawson, R., Barbosa-Cánovas, G.V., 2008. Microstructure of fat globules in whole milk after thermo-sonication treatment. J. Food Sci. 73 (7), E325–E332.

Calligaris, S., Gulotta, A., Ignat, A., Bermúdez-Aguirre, D., Barbosa-Cánovas, G.V., Nicoli, M.C., 2013. Milk pre-treatment by high pressure homogenization in the manufacturing of "*queso fresco*" fortified with omega-3 fatty acids. LWT—Food Sci. Technol. 50, 626–633.

Gómez-Cortés, P., Bach, A., Luna, P., Juárez, M., de la Fuente, M.A., 2009. Effects of extruded linseed supplementation on n-3 fatty acids and conjugated linoleic acid in milk and cheese from ewes. J. Dairy Sci. 92, 4122–4134.

Guinee, T.P., Carič, M., Kaláb, M., 2004. Pasteurized processed cheese and substitute/imitation cheese products. In: Fox, P.F., McSweeney, P.L.H., Cogan, T.M., Guinee, T.P. (Eds.), Cheese: Chemistry, Physics and Microbiology, vol. 2. Elsevier, Academic Press, London, pp. 349–394.

Hauswirth, C.B., Scheeder, M.R.L., Beer, J.H., 2004. High ω-3 Fatty acid content in alpine cheese: The basis for an alpine paradox. Circulation 109, 103–107.

Hughes, B.H., Perkins, L.B., Calder, B.L., Skonberg, D.I., 2012. Fish oil fortification of soft goat cheese. J. Food Sci. 77 (2), S128–S133.

ISSFAL, 2004. Report of the sub-committee on recommendations for the daily intake of polyunsaturated fatty acids in healthy adults. U. K. Int. Soc. Stud. Fatty Acids Lipids.

Kolanowski, W., Weißbrodt, J., 2007. Sensory quality of dairy products fortified with fish oil. Int. Dairy J. 17, 1248–1253.

Kris-Etherton, P.M., Harris, W.S., Appel, L.J., 2002. Fish Consumption, Fish Oil, Omega-3 Fatty Acids, and Cardiovascular Disease. Circulation 106, 2747–2757.

Martini, S., Thurgood, J.E., Brothersen, C., Ward, R., McMahon, D.J., 2009. Fortification of reduced-fat Cheddar cheese with n-3 fatty acids: Effect on off-flavor generation. J. Dairy Sci. 92, 1876–1884.

Mele, M., Contarini, G., Cercaci, L., Serra, A., Buccioni, A., Povolo, M., Conte, G., Funaro, A., Banni, S., Lercker, G., Secchiari, P., 2011. Enrichment of Pecorino cheese with conjugated linoleic acid by feeding dairy ewes with extruded linseed: Effect on fatty acid and triglycerides composition and on oxidative stability. Int. Dairy J. 21, 365–372.

O'Fallon, J.V., Busboom, J.R., Nelson, M.L., Gaskins, C.T., 2007. A direct method for fatty acid methyl ester synthesis: Application to wet meat tissues, oils and feedstuffs. J. Anim. Sci. 85, 1511–1521.

O'Reilly, C.E., Kelly, A.L., Murphy, P.M., Beresford, T.P., 2001. High pressure treatment: Applications in cheese manufacture and ripening. Trends Food Sci. Technol. 12, 51–59.

Pike, O.A., 1998. Fat Characterization. In: Nielsen, S. (Ed.), Food Analysis. Aspen Publication, Maryland, pp. 217–235.

Trautwein, E.A., 2001. n-3 Fatty acids—physiological and technical aspects for their use in food. Eur J. Lipid Sci. Technol. 103, 45–55.

Ye, A., Cui, J., Taneja, A., Zhu, X., Singh, H., 2009. Evaluation of processed cheese fortified with fish oil emulsion. Food Res. Int. 42, 1093–1098.

Yu, L.J., Ngadi, M., Raghavan, G.S.V., 2009. Effect of temperature and pulsed electric field treatment on rennet coagulation properties of milk. J. Food Eng. 95 (1), 115–118.

Zimet, P., Livney, Y.D., 2009. Beta-lactoglobulin and its nanocomplexes with pectin as vehicles for ω-3 polyunsaturated fatty acids. Food Hydrocolloids 23, 1120–1126.

CHAPTER

25

Use of *Lactobacillus* spp to Degrade Pesticides in Milk

Mónica Calderón-Santiago, María Dolores Luque de Castro

Department of Analytical Chemistry, Annex Marie Curie Building, Campus of Rabanales, University of Córdoba, Córdoba, Spain

CHAPTER POINTS

- Many foods can be contaminated by pesticides, with milk as the main source of pesticide residues in human diets.
- Microorganisms present in cheese can degrade pesticides during storage.
- Pesticide residues and some nutrients present in milk can be degraded by heating for a long time.
- A strategy to degrade pesticides in milk by use of microorganisms was recently proposed.
- An organophosphorus hydrolase exhibiting high activity on pesticides was obtained by cloning a *Lactobacillus brevis* gen.

INTRODUCTION

The growing demand for agricultural products has raised the need to use pesticides to prevent pest and insect growth. Pesticides are generally deemed safe for crops and animals by virtue of their relatively fast degradation. However, the local environmental conditions can alter their degradation rate and extend their persistence over long periods (Ragnarsdottir, 2000). Although a number of foods can be contaminated by pesticides, those of animal origin are the most difficult to free from these substances; this is especially the case with milk, which is the main source of pesticide residues in human diet owing to the high complexity of its matrix (Kampire *et al.*, 2011). Pesticides can reach milk-producing animals through water, forage, and the environment. Animals can metabolize and detoxify most pesticides except chlorinated hydrocarbons, which are stored in body fat and contaminate meat and milk for human consumption as a result (Li and Bradley, 1969). In fact, pesticides have been found in milk from cows, buffalo, and sheep (Bulut *et al.*, 2011), and even from women, where placenta and milk are the preferential excretory routes of lipophilic pesticides (Siddiqui and Saxena, 1985). Organophosphorus pesticides are the most frequently encountered in milk as a result of their widespread use as herbicides and for cow ectoparasite control; however, they are more rapidly degraded than other pesticides. Figure 25.1 shows the organochlorine and organophosphorus pesticides most frequently found in milk.

Although pesticides concentrate mainly in the fatty fraction of milk, they rarely disappear when the fat is removed by cooking or other processing steps. All steps by which milk is transformed into dairy products or made suitable for marketing have been studied in depth in order to find ways to efficiently remove pesticides. Also, various strategies for complete pesticide removal from milk have been proposed.

The presence of pesticides in food must be avoided or at least controlled because most of them are mutagenic and teratogenic, and their long-term, low-dose exposure is linked to effects on human health such as immunosuppression, hormone disruption, diminished intelligence, reproductive abnormalities, and cancer (Aktar *et al.*, 2009). For these reasons, national and transnational governments have set maximum residue levels of pesticides in or on food and feed of plant and animal origin intended for human or animal consumption. Since pesticide use can not be suppressed altogether without placing food production at risk, a number of strategies have been developed to remove pesticides from contaminated food.

Processing and Impact on Active Components in Food
http://dx.doi.org/10.1016/B978-0-12-404699-3.00025-1

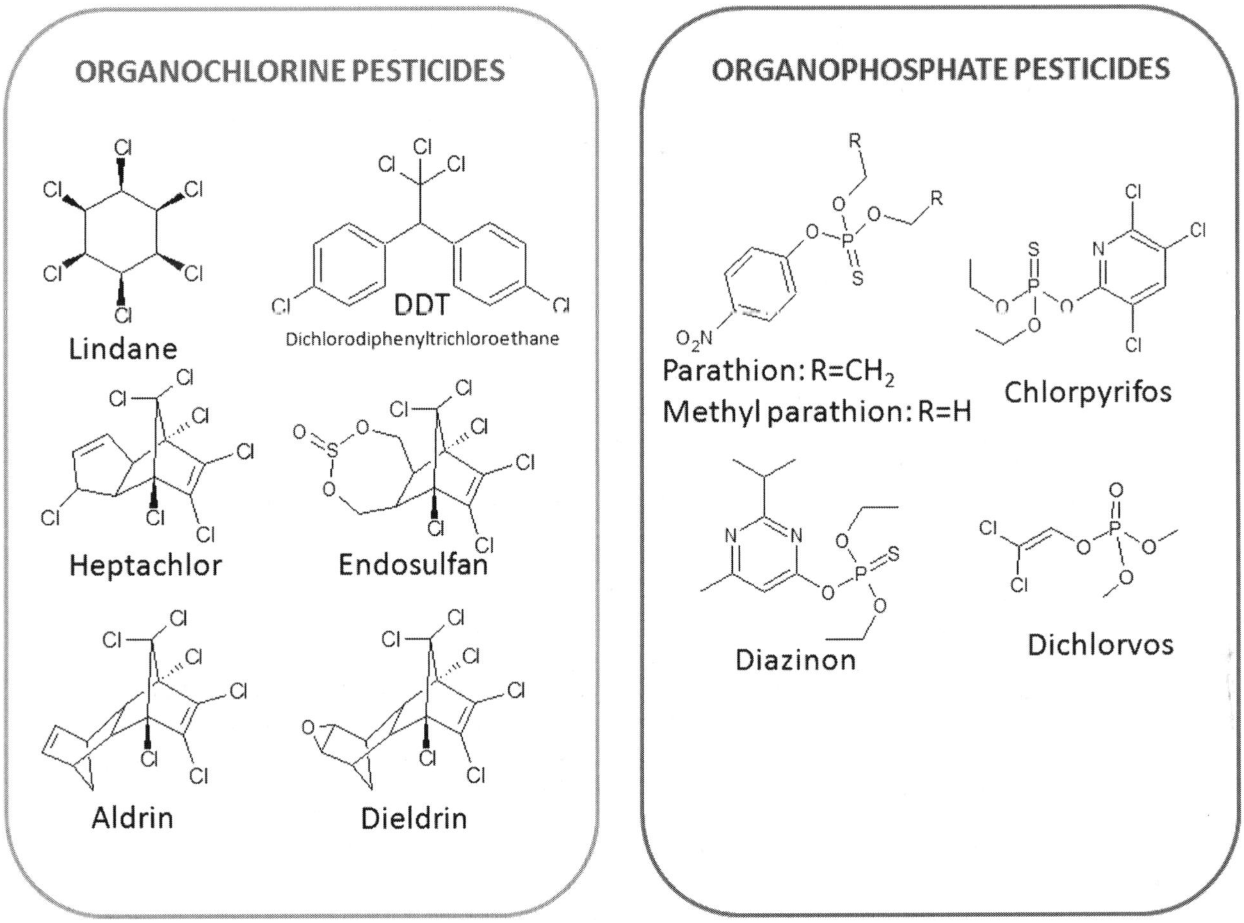

FIGURE 25.1 Pesticides most commonly encountered in milk.

One key aspect in dealing with pesticides in food is the way in which they are analyzed. A number of sensitive, selective methods for milk analysis exist to ensure compliance with legal limits and facilitate the detection of pesticides and their removal. However, existing methods fail to identify the nature of the products formed by pesticides in response to milk treatments.

HOW COMPOSITION IS ALTERED

Influence of Milk Processing and Storage on Pesticide Removal Efficiency

Safety concerns have raised the need to examine the behavior of pesticides in milk during its processing and storage. Milk processing typically involves pasteurization, boiling, sterilization and/or skimming. Butter, cheese, and yogurt production are also included here as they are these dairy products are among the most popular ingredients of human diet.

Butter

Butter production is not a useful step for pesticide removal. In fact, some pesticides such as DDT and lindane have been found to undergo preconcentration during butter production due to their high affinity for the fatty fraction of milk (Li *et al.*, 1970). Diminishing butter contamination by pesticides therefore requires an additional step to remove them.

Cheese

Pesticide degradation during cheese production occurs mainly at the heating and salting stages. For example, heating and salting were found to remove 98% of leptophos (an organophosphorus pesticide) present in milk used for cheese production (Abu-Elamayem *et al.*, 1979). Pesticide degradation is also influenced by the presence of microorganisms, which act mainly during storage. Thus, lindane and DDT levels were found to fall by 37 and 70–60%, respectively, after six months of storage of cheese from cow milk (Abou-Arab, 1999). Conversely, lindane levels in feta cheese from ovine milk underwent no change during cheese ripening and storage for up to 240 days, possibly because the pesticide was bound to other cheese components (Mallatou *et al.*, 2002).

Yogurt

Bo *et al.* (2011) studied the degradation of the organophosphorus pesticides dimethoate, fenthion, malathion, methyl parathion, monocrotophos, phorate, and trichlorphon in bovine milk during yogurt processing and identified time and temperature as the most influential factors on the pesticide concentrations. The nature of the fermentation starter proved highly influential as well. Pesticide degradation was ascribed to the action of microorganisms used to obtain yogurt.

Milk

Kampire *et al.* (2011) showed heat treatments such as pasteurization to efficiently degrade residual organochlorine pesticides. Bo and Zhao (2010) studied pesticide degradation by heating at a variable temperature and concluded that increased processing temperatures led to faster degradation. At 100°C, the half-lives of denthion, dimethoate, malathion, methyl parathion, monocrotophos, phorate, and trichlorfon ranged from 4 to 6h.

OTHER WAYS IN WHICH COMPOSITION IS ALTERED

Available Methods for the Removal or Degradation of Pesticides in Milk

Available methods for pesticide removal are mainly based on their physical and chemical properties (especially volatility and polarity). Bills and Sloan (1967) found molecular distillation to remove 95–99% of chlorinated pesticides such as DDT, DDD, or DDE. Their method uses a temperature of 200°C and a pressure of 5×10^{-4} torr, and is similar to one previously proposed by Kroger (1968) called 'steam deodorization'. Kroger's method, which uses a temperature of 180–195°C and a pressure of 0.01–0.5mmHg, completely removes heptachlor epoxide and dieldrin from milk.

A number of polarity-based methods use a sorbent to remove pesticides from food; most, however, have only been applied to water. The main problem with milk is that its large fatty fraction can be co-extracted with the pesticides. Thus, pinus bark is able to adsorb 97% of heptachlor, aldrin, endrin, dieldrin, DDD, DDT, and DDE—but only 37% of lindane—and activated carbon efficiently removes 97% of all these pesticides (Brás *et al.*, 1999), but both are useless for the removal of pesticides from milk. This is also the case with the acid-treated date stones used by Bakouri *et al.* (2009), who conducted a comprehensive physico-chemical study of the adsorption of aldrin, dieldrin, and endrin on this sorbent, albeit in aqueous solutions of the pesticides rather than milk.

As regards pesticide degradation, Basfar and Mohamed (2012) have patented a method using ionizing radiation that is applicable to aqueous solutions, fruits, and vegetables, and also to milk. Pesticide degradation by exposure to non-ionizing radiation (ultraviolet light) is also possible. Thus, Li and Bradley (1969) used UV energy to destroy organophosphate and organochlorine pesticides, which are photosensitive, in milk and butter oil. UV light was used to degrade six chlorinated hydrocarbon pesticides of which only methoxychlor was significantly destroyed. The main disadvantage of using UV energy for this purpose is that it can also degrade some nutrients such as vitamins and diminish food quality as a result.

A more effective strategy here may be provided by some microorganisms (especially fungi and bacteria), which have for some time been used to degrade pesticides in food, water, and soil. As stated above, microorganisms are responsible for pesticide degradation during cheese ripening and storage, and also during yogurt fermentation. Therefore, they may be useful to degrade pesticides in milk as well. Langlois *et al.* (1970) studied the degradation of DDT in skimmed milk in the presence of three different species (*Bacillus*, *Escherichia coli*, and *Enterobacter aerogenes*), none of which, however, was capable of degrading DDT in this matrix. Abou-Arab (2002) found *Lactobacillus plantarum* to degrade lindane and DDT in fermented sausage and tryptone soya broth. More recently, Zhao and Wang (2012) assessed the degradation potential of *Lactobacillus delbruekii* ssp. *bulgarius*, *Lactobacillus paracasei*, and *Lactobacillus plantarum* on seven pesticides spiked to skimmed milk (at a 0.5-mg/kg level for monocrotophos and phorate, and 1.2mg/kg for the other five pesticides, all in a 12% w/w solution) incubated at 42°C. As can be seen from Table 25.1, dimethoate and methyl parathion were the most stable pesticides, whereas malathion was the most labile. *Lactobacillus bulgarius* proved especially active on dimethoate, fenthion, and monocrotophos, whereas *L. plantarum* was more effective on malathion, methyl parathion, and trichlorphon. As can also be seen from Table 25.1, the pesticide half-lives ranged from 16 to 45h with *Lactobacillus*; the effect was not significant, however, since heating at 42°C without microorganism inoculation led to half-lives of 32–50h. In any case, the results showed that each *Lactobacillus* species was effective at removing some pesticides and not others, so a combination of microorganisms may be needed to efficiently degrade a mixture of pesticides.

In vitro studies can be useful to assess the pesticide degradation efficiency of microorganisms, identify the resulting degradation products, and evaluate their

TABLE 25.1 Residual Concentration (Mean ± SD, mg/kg) and Degradation Kinetic Parameters of Seven Organophosphorus Pesticides in Skimmed Milk Cultured with *Lactobacillus* spp. at 42°C for Variable Lengths of Time

Pesticide (Amount Spiked)	Stain Inoculated	Residual Concentration at Different Times (h)			Degradation Kinetic Parameters		
		8	16	24	k (per h)	R^2	$t_{1/2}$ (h)
Dimethoate (1.2 mg/kg)	*L. bulgaricus*	1.100 ± 0.025	0.909 ± 0.046	0.808 ± 0.039	0.0184	0.975	37.7
	L. paracasei	1.178 ± 0.042	0.986 ± 0.031	0.897 ± 0.024	0.0169	0.974	41.0
	L. plantarum	0.990 ± 0.017	0.841 ± 0.048	0.735 ± 0.032	0.0168	0.933	41.2
	Control	1.050 ± 0.021	0.908 ± 0.035	0.808 ± 0.037	0.0165	0.996	42.0
Fenthion (1.2 mg/kg)	*L. bulgaricus*	0.996 ± 0.036	0.872 ± 0.046	0.679 ± 0.044	0.0247	0.972	28.1
	L. paracasei	0.992 ± 0.035	0.843 ± 0.048	0.711 ± 0.013	0.0226	0.974	30.7
	L. plantarum	1.019 ± 0.033	0.804 ± 0.043	0.698 ± 0.017	0.0236	0.959	29.4
	Control	0.934 ± 0.051	0.800 ± 0.024	0.667 ± 0.037	0.0210	0.990	33.0
Malathion (1.2 mg/kg)	*L. bulgaricus*	0.987 ± 0.026	0.853 ± 0.042	0.677 ± 0.046	0.0233	0.986	29.7
	L. paracasei	0.991 ± 0.034	0.728 ± 0.032	0.614 ± 0.041	0.0279	0.964	24.8
	L. plantarum	0.912 ± 0.044	0.663 ± 0.036	0.484 ± 0.033	0.0420	0.979	16.5
	Control	0.986 ± 0.046	0.821 ± 0.034	0.699 ± 0.056	0.0218	0.998	31.8
Methyl parathion (1.2 mg/kg)	*L. bulgaricus*	0.919 ± 0.056	0.838 ± 0.035	0.683 ± 0.051	0.0204	0.844	34.0
	L. paracasei	1.096 ± 0.026	0.977 ± 0.034	0.867 ± 0.038	0.0153	0.992	45.3
	L. plantarum	1.163 ± 0.027	0.942 ± 0.045	0.807 ± 0.043	0.0213	0.955	32.5
	Control	1.051 ± 0.033	0.919 ± 0.028	0.822 ± 0.049	0.0153	0.998	45.3
Monocrotophos (0.5 mg/kg)	*L. bulgaricus*	0.423 ± 0.036	0.388 ± 0.012	0.303 ± 0.034	0.0220	0.937	31.5
	L. paracasei	0.438 ± 0.013	0.348 ± 0.031	0.326 ± 0.345	0.0188	0.994	36.9
	L. plantarum	0.432 ± 0.044	0.377 ± 0.031	0.311 ± 0.029	0.0199	0.986	34.8
	Control	0.439 ± 0.047	0.353 ± 0.030	0.320 ± 0.027	0.0196	0.962	35.4
Phorate (0.5 mg/kg)	*L. bulgaricus*	0.422 ± 0.032	0.369 ± 0.029	0.295 ± 0.047	0.0200	0.916	34.6
	L. paracasei	0.411 ± 0.029	0.346 ± 0.042	0.291 ± 0.037	0.0230	0.984	30.1
	L. plantarum	0.412 ± 0.039	0.342 ± 0.016	0.297 ± 0.029	0.0204	0.988	34.0
	Control	0.392 ± 0.035	0.341 ± 0.034	0.288 ± 0.032	0.0185	0.964	37.4
Trichlorphon (1.2 mg/kg)	*L. bulgaricus*	1.210 ± 0.041	0.988 ± 0.033	0.877 ± 0.053	0.0229	0.926	30.3
	L. paracasei	1.136 ± 0.025	0.900 ± 0.011	0.778 ± 0.019	0.0233	0.977	30.1
	L. plantarum	1.230 ± 0.033	0.990 ± 0.014	0.847 ± 0.012	0.0242	0.979	28.6
	Control	1.036 ± 0.030	0.892 ± 0.034	Not detected	0.0140	0.874	49.5

k, degradation rate constant, R^2, regression coefficient, $t_{1/2}$ half-life.
Adapted with permission from Zhao and Wang (2012).

toxicity in relation to their precursors. Harishankar *et al.* (2012) studied the *in vitro* degradation of the organophosphorus pesticide chlorpyrifos by five intestinal bacteria (viz. *Lactobacillus lactis, Lactobacillus plantarum, Lactobacillus fermentum, Escherichia coli*, and *Enterococcus faecalis*). As can be seen from Figure 25.2, *L. fermentum* degraded chlorpyrifos to 3,5,6-trichloro-2-pyridol by 70% and *L. lactis* degraded chlorpyrifos to chlorpyrifos oxon by 61%. The study, which used liquid chromatography–mass spectrometry (LC–MS) with ion trapping and atomic pressure chemical ionization, revealed differences in degradation (metabolite) profiles between bacterial strains, which clearly shows the need to accurately identify metabolites in order assess toxicity and the metabolic pathways behind each microorganism.

Cho *et al.* (2009) studied the microorganisms potentially responsible for chlorpyrifos degradation during fermentation of 'kimchi', a vegetable-based fermented food. They identified four organisms as responsible for chlorpyrifos degradation, namely: *Leuconostoc mesenteroides* WCP907, *Lactobacillus brevis* WCP902, *Lactobacillus plantarum* WCP931, and *Lactobacillus sakei* WCP904. These microorganisms degraded chlorpyrifos by over 80% in 3 days and 100% in 9 days. Islam *et al.* (2010) cloned the genes of *Lactobacillus brevis* and obtained an organophosphorus hydrolase potentially useful for decontaminating raw food materials after testing its degrading effect on 'kimchi' production. The enzyme exhibited high hydrolase activity against methyl parathion, diazinon, chlorpyrifos, coumaphos, and parathion.

One other important subject to be studied in using bacteria to degrade pesticides is their nutritional requirements. Barragán-Huerta and Rodríguez-Vázquez (2010) used green bean coffee as nutrient source for *Pseudomonas aeruginosa* isolated from this natural product and found the microorganism to remove 51% of endosulfan and 67% of DDT after 7 days of incubation.

As can be seen, pesticide-degrading bacteria are useful to decontaminate not only milk and dairy products, but also other fermented foods or even soil and water, where pesticide accumulation can lead to animal and vegetable contamination. Contamination in this context can be assessed by metabolomics and fate analysis. The former can help identify the pathways involved in pesticide degradation and determine the physico-chemical properties of the resulting metabolites, and the latter to predict the fate of pesticides under specific environmental conditions.

ANALYTICAL TECHNIQUES

Methods for Determining Pesticides in Milk

As stated above, the most common pesticides in milk belong to the organochlorine or organophosphorus families; therefore, most pesticide determination methods focus on these two groups. MS in combination with capillary electrophoresis, gas chromatography, or LC for separation is the preferred choice for pesticide determination; some methods, however, use an electron capture detector instead. In any case, the sample treatment is complex and similar, whichever the choice. Figure 25.3 shows the most common methods for determining pesticides in milk and depicts their sample treatment process. For example, Armendáriz *et al.* (2004) developed a method to determine up to 15 organochlorine pesticides in milk by gas chromatography coupled to electron capture detection and involving fat separation, solid-phase extraction, and purification as sample

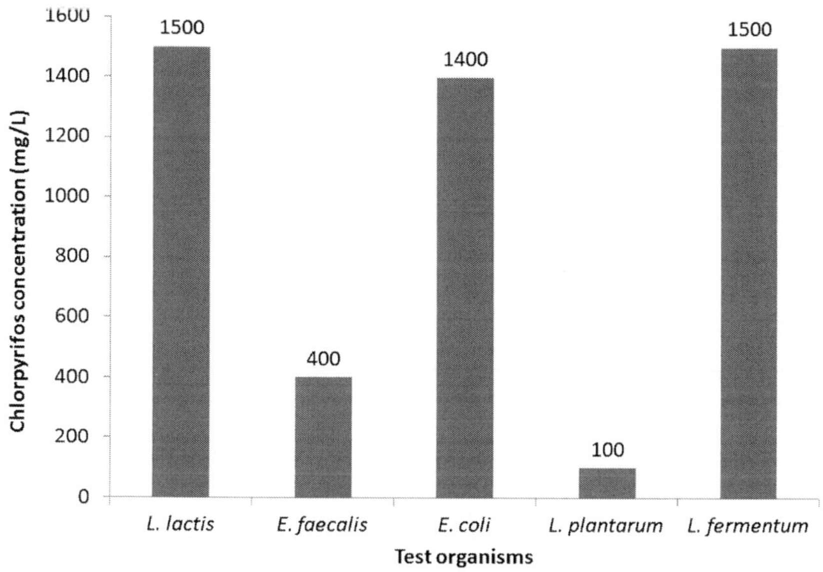

FIGURE 25.2 Bar graph showing the chlorpyrifos concentration tolerated by various bacterial strains. *With permission from Harishankar* et al. *(2012).*

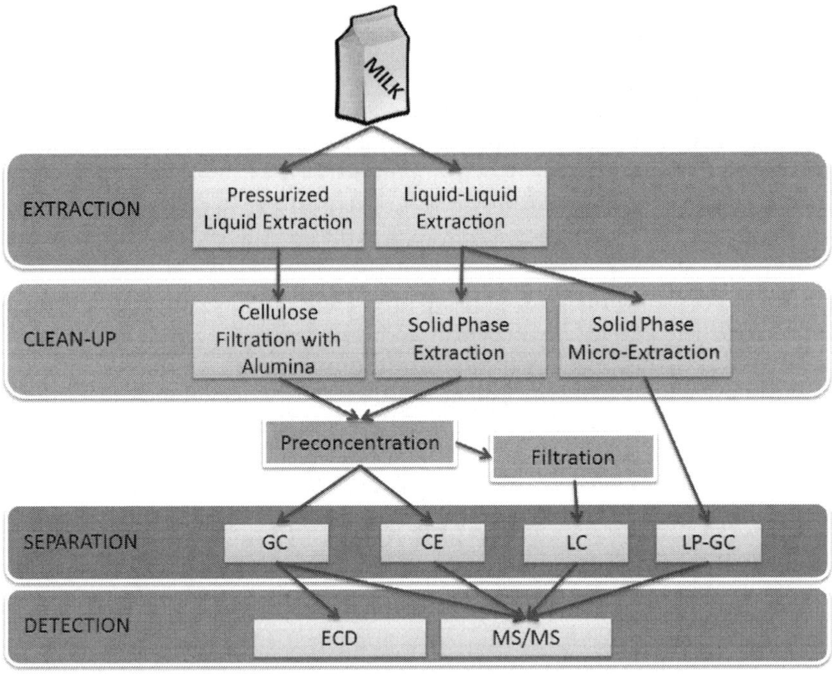

FIGURE 25.3 Most common strategies for determining pesticides in milk.

treatment. More recently, Zheng *et al.* (2009) developed a method for the determination of 128 pesticides in milk by LC coupled to MS. The former method requires removing fat by centrifugation prior to extraction of the pesticides on a glass column; by contrast, the latter requires no fat separation but liquid–liquid extraction with acetonitrile prior to clean-up by passage through solid-phase extraction cartridges. Both use a concentration step before analysis.

Acknowledgments

Funding of this research by Spain's Ministry of Economy and Competitiveness and the FEDER Program through project CTQ2012-37428 is gratefully acknowledged. M. Calderón Santiago is also grateful to Spain's Ministry of Science and Innovation for award of an FPU scholarship.

References

Abou-Arab, A.A.K., 1999. Effects of processing and storage of dairy products on lindane residues and metabolites. Food Chem. 64, 467–473.

Abou-Arab, A.A.K., 2002. Degradation of organochlorine pesticides by meat starter in liquid media and fermented sausage. Food Chem. Toxicol. 40 (1), 33–41.

Abu-Elamayem, M.M., Abdel-All, A., Tantawy, G.A., Marei, A.S.M., 1979. Fate of leptophos in milk and wheat during the processing steps. Alexandria J. Agric. Res. 27 (3), 659–663.

Aktar, M.W., Sengupta, D., Chowdhury, A., 2009. Impact of pesticides use in agriculture: their benefits and hazards. Interdiscip. Toxicol. 2 (1), 1–12.

Armendáriz, C., Pérez de Ciriza, J.A., Farré, R., 2004. Gas chromatographic determination of organochlorine pesticides in cow milk. Int. J. Food Sci. Nutr. 55 (3), 215–221.

Bakouri, H.E., Usero, J., Morillo, J., Rojas, R., Ouassini, A., 2009. Drin pesticides removal from aqueous solutions using acid-treated date stones. Bioresour. Technol. 100, 2676–2684.

Barragán-Huerta, B.E., Rodríguez-Vázquez, R., 2010. Green bean coffee as nutrient source for pesticide-degrading bacteria. In: Méndez-Vilas, A. (Ed.), Current Research, Technology and Education Topics in Applied Microbiology and Microbial Biotechnology, vol. 2. Formatex, Badajoz, Spain, pp. 1322–1327.

Basfar, A.A., Mohamed, K.A., 2012. Removal of pesticide residues in food by ionizing radiation. U. S. Patent Appl. 20120021103.

Bills, D.D., Sloan, J.L., 1967. Removal of chlorinated insecticide residues from milk fat by molecular distillation. J. Agric. Food Chem. 15 (4), 676–678.

Bo, L.Y., Zhang, Y.H., Zhao, X.H., 2011. Degradation kinetics of seven organophosphorus pesticides in milk during yoghurt processing. J. Serb. Chem. Soc. 76 (3), 353–362.

Bo, L.Y., Zhao, X.H., 2010. Preliminary study on the degradation of seven organophosphorus pesticides in bovine milk during lactic acid fermentation or heat treatment. Afr. J. Microbiol. Res. 4 (11), 1171–1179.

Brás, I.P., Santos, L., Alves, A., 1999. Organochlorine pesticides removal by pinus bark sorption. Environ. Sci. Technol. 33 (4), 631–634.

Bulut, S., Akkaya, L., Gök, V., Konuk, M., 2011. Organochlorine pesticide (OCP) residues in cow's, buffalo's, and sheep's milk from Afyonkarahisar region. Turkey. Environ. Monit. Assess. 181, 555–562.

Cho, K.M., Math, R.K., Islam, S.M.A., Lim, W.J., Hong, S.Y., Kim, J.M., Yun, M.G., Cho, J.J., Yun, H.D., 2009. Biodegradation of chlorpyrifos by lactic acid bacteria during kimchi fermentation. J. Agric. Food Chem. 57 (5), 1882–1889.

Harishankar, M.K., Sasikala, C., Ramya, M., 2012. Efficiency of the intestinal bacteria in the degradation of the toxic pesticide chlorpyrifos. 3 Biotech. http://dx.doi.org/10.1007/s13205-012-0078-0.

Islam, S.M., Math, R.K., Cho, K.M., Lim, W.J., Hong, S.Y., Kim, J.M., Yun, M.G., Cho, J.J., Yun, H.D., 2010. Organophosphorus hydrolase (OpdB) of *Lactobacillus brevis* WCP902 from Kimchi is able to degrade organophosphorus pesticides. J. Agric. Food Chem. 58, 5380–5386.

Kampire, E., Kiremire, B.T., Nyanzi, S.A., Kishimba, M., 2011. Organochlorine pesticide in fresh and pasteurized cow's milk from Kampala markets. Chemosphere 84, 923–927.

Kroger, M., 1968. Effect of various physical treatments on certain organochlorine hydrocarbon insecticides found in milk fat. J. Dairy Sci. 51 (2), 196–198.

Langlois, B.E., Collins, J.A., Sides, K.G., 1970. Some factors affecting degradation of organochlorine pesticides by bacteria. J. Dairy Sci. 53 (12), 1671–1675.

Li, C., Bradley Jr., R.L., Schultz, L.H., 1970. Fate of organochlorine pesticides during processing of milk into dairy products. J. Assoc. Off. Agric. Food Chem. 53 (1), 127–139.

Li, C.F., Bradley, R.L., 1969. Degradation of chlorinated hydrocarbon pesticides in milk and butteroil by ultraviolet energy. J. Dairy Sci. 52 (1), 27–30.

Mallatou, H., Pappas, C.P., Albanis, T.A., 2002. Behaviour of pesticides lindane and methyl parathion during manufacture, ripening and storage of feta cheese. J. Dairy Technol. 55 (4), 211–216.

Ragnarsdottir, K.V., 2000. Environmental fate and toxicology of organophosphate pesticides. J. Geol. Soc. 157 (4), 859–876.

Siddiqui, M.K., Saxena, M.C., 1985. Placenta and milk as excretory routes of lipophilic pesticides in women. Hum. Toxicol. 4 (3), 249–254.

Zhao, X.H., Wang, J., 2012. A brief study on the degradation kinetics of seven organophosphorus pesticides in skimmed milk cultured with *Lactobacillus* spp. at 42 °C. Food Chem. 131 (1), 300–304.

Zheng, J., Pang, G., Fan, C., Wang, M., 2009. Simultaneous determination of 128 pesticide residues in milk by liquid chromatography-tandem electrospray mass spectrometry. Chin. J. Chromatogr. 27 (3), 254–263.

CHAPTER

26

Yayik Butter Profiles from Different Species of Mammals' Milk

Ebru Şenel, Metin Atamer

Ankara University, Faculty of Agriculture, Department of Dairy Technology, Ankara, Turkey

CHAPTER POINTS

- Sheep Yayik butter had the highest lactic content due to the higher dry matter level of raw material.
- Bitterness and rancidity defects were observed in goat Yayik butter having a peroxide value of 3.2 meq O_2/kg oil.
- Aroma and flavor scores of goat Yayik butter were lower than those of cow and sheep Yayik butter at day 30.
- The cloudy yellowish color in goat Yayik butter was criticized as a defect.
- The accumulation of free fatty acids was the highest in goat Yayik butter.
- C_{10}, C_{14}, and C_{16} acids in cow Yayik butter, C_8, C_{14}, C_{16}, and $C_{18:0}$ in sheep Yayik butter, and C_6, C_8, C_{12}, C_{16}, $C_{18:0}$, and $C_{18:1}$ in goat Yayik butter were the predominant free fatty acids.

INTRODUCTION

Yogurt is still being used as raw material in traditional butter manufacturing in Anatolia, Turkey. Butter produced from yogurt is called 'Yayik Butter'. The first record of butter dates back to the Chalcolithic Age. Approximately 7500-year-old potsherds of a special type of jar called a 'churn' were found in the excavations at Yarikkaya/Hattuşaş in Central Anatolia (Sauter *et al.*, 2003). Although efforts have been made to fully characterize this traditional product in the mid-2000s, fairly limited data is available on the properties of Yayik butter. In these studies, some physicochemical and sensorial properties (Hayaloğlu and Konar, 2001; Sağdıç *et al.*, 2004), fatty acids profiles (Sağdıç *et al.*, 2004), and carbonyl compounds of Yayik butter (Atamer *et al.*, 2004a), and isolation and identification of lactic acid bacteria to select the best lactic acid bacteria combination for butter production (Sağdıç *et al.*, 2002) have been investigated. Atamer *et al.* (2004b) determined some technological parameters of traditional Yayik butter production (see Table 26.1).

The production of Yayik butter by traditional methods is somewhat different from that of cream butter. The main difference is the raw material, and yogurt is the starting material for the former product. Yayik butter is manufactured in two different ways in order to increase the yield of the end product, the cream layer of yogurts are collected and churned. Alternatively, yogurts are diluted without removing the cream layer and then churned to make butter. Dilution of yogurt is achieved by mixing potable water with yogurt at rates of 50–150% of total weight of yogurt. The churning temperatures of Yayik butter range from 17 to 24°C (an average of 18°C) and are higher than that of cream butter (8–12°C). In Yayik butter production, hot water (40–50°C) is added to raw material (yogurt) to reach the desired churning temperature. This application is necessary to complete the churning process within approximately 30–40 min. The addition of water to yogurt leads to a reduction in the viscosity of yogurt which facilitates the coalescence of milk fat globules. Our preliminary experiments showed that the churning temperature of 8–10°C recommended for cream butter extended the churning time up to 6–8 h (unpublished data). It is a well-known fact that the formation of yogurt coagulum is mainly due to the interaction of denaturated α-lactalbumin and β-lactoglobulin with κ-casein which gives rise to a three-dimensional network entrapping all milk constituents including fat globules and aqueous phase (Tamime and Robinson, 2007). In other words, fat globules are buried in this network. As a result, the release of fat globules from the network to

Processing and Impact on Active Components in Food
http://dx.doi.org/10.1016/B978-0-12-404699-3.00026-3

TABLE 26.1 Technological Parameters in the Manufacture of Yayik Butter

Farms	Dilution Rate (Yogurt: Water)	Churning pH	Churning Temperature (°C)	Churning Time	Number of Washing
A	1:0.5	3.95	24	30	7
B	1:0.7	3.81	20	64	9
C	1:1.5	3.93	17	48	6
D	1:1.0	3.86	10	40	7
E	1:1.5	3.87	18	55	8
F	1:0.5	4.08	18	17	4

From Atamer et al. *(2004b).*

form the coalescence is expected to occur very slowly. At the end of the churning period, buttermilk (called 'Yayik Ayran') is separated and generally used for manufacturing of some traditional products locally named 'Kurut' and 'Keş'. Remaining butter grains are washed with cold water (7–8°C) to remove residual buttermilk. The number of washings varies from five to nine. The granules are kneaded by hand during washing. Sometimes, a small amount of salt is added to the granules prior to kneading.

Atamer *et al.* (2004b) determined that the major parameters for the optimization of Yayik butter manufacturing were milk species used in the manufacture of yogurt, the fat content of the raw material, the cooling temperature for crystallization, the dilution rate of yogurt, churning pH and temperature, the number of washes of butter granules, and the starter culture employed in yogurt making.

Generally, the typical characteristics of dairy products are determined by many factors, the type of milk used as a raw material being the most important one. Many studies have been conducted on Yayik butter produced from yogurts made by different milk species. Sağdıç *et al.* (2004) determined the chemical compositions, fatty acids profiles, and microbiological characteristics of Yayik butter samples made from cow, sheep, or goat yogurt. A more comprehensive work on Yayik butter made from different types of milk was carried out by Şenel *et al.* (2011). In this study, some properties of Yayik butters such as peroxide value, free fatty acid composition, and lactic acid, and their effects on the lipolytic and oxidative stability of the end products during the storage period of 30 days were investigated. The results obtained from this study are summarized hereafter.

THE PRODUCTION OF YAYIK BUTTER FROM DIFFERENT MAMMALS' MILK

The manufacturing steps of Yayik butter produced from different mammals' milk based on the traditional technique are given in Figure 26.1.

Firstly, yogurt samples were produced for the manufacture of Yayik butters. The raw milks (cow, goat, sheep), whose initial compositions were unchanged, were used for the production of yogurts.

Whole raw milks (cow, sheep, goat) were pasteurized at 85°C for 15 min and then cooled to 45°C. After the starter culture addition (2%, w/w), milks were incubated at 43°C until pH 4.6–4.7 was attained. Yogurts were cooled to ~10°C to control the after- acidification and kept at 4°C for 12 h for physical ripening. After the yogurt production, Yayik butters were produced according to the method described in Figure 26.1 (based on Atamer *et al.*, 2007).

SOME PROPERTIES OF YAYIK BUTTERS

The results of chemical and sensory analysis of Yayik butter produced from different mammals' milk are presented Table 26.2.

The serum pHs of cow, sheep, and goat Yayik butters were 4.10, 3.99, and 4.04, respectively, at the first day of storage and no significant differences between Yayik butter samples were noted ($p > 0.01$). Conversely, significant differences between the butters regarding lactic acid contents were evident ($p < 0.01$). Sheep Yayik butter had the highest lactic acid contents due probably to the higher dry matter level of raw material. There is a general agreement in the literature that an increase in the lactic acid content of fermented milk products is mainly related to the non-fat solid content of raw milk. Atamer *et al.* (1996) noted that the lactic acid content of yogurt samples rose with the increase in total solids content of yogurt.

Oxidation is the main cause of the off-flavor which affects the shelf-life of dairy products containing high fat content. Peroxides are formed by the reaction between the unsaturated fatty acids and the oxygen. Peroxides have no influence on flavor but decompose to yield carbonyl compounds resulting in oxidized flavor defect (O'Conner and O'Brien, 1995). The sensory threshold for peroxide value in cream butter was 2.0 meq O_2/kg oil (Downey, 1975). No data are available in the literature

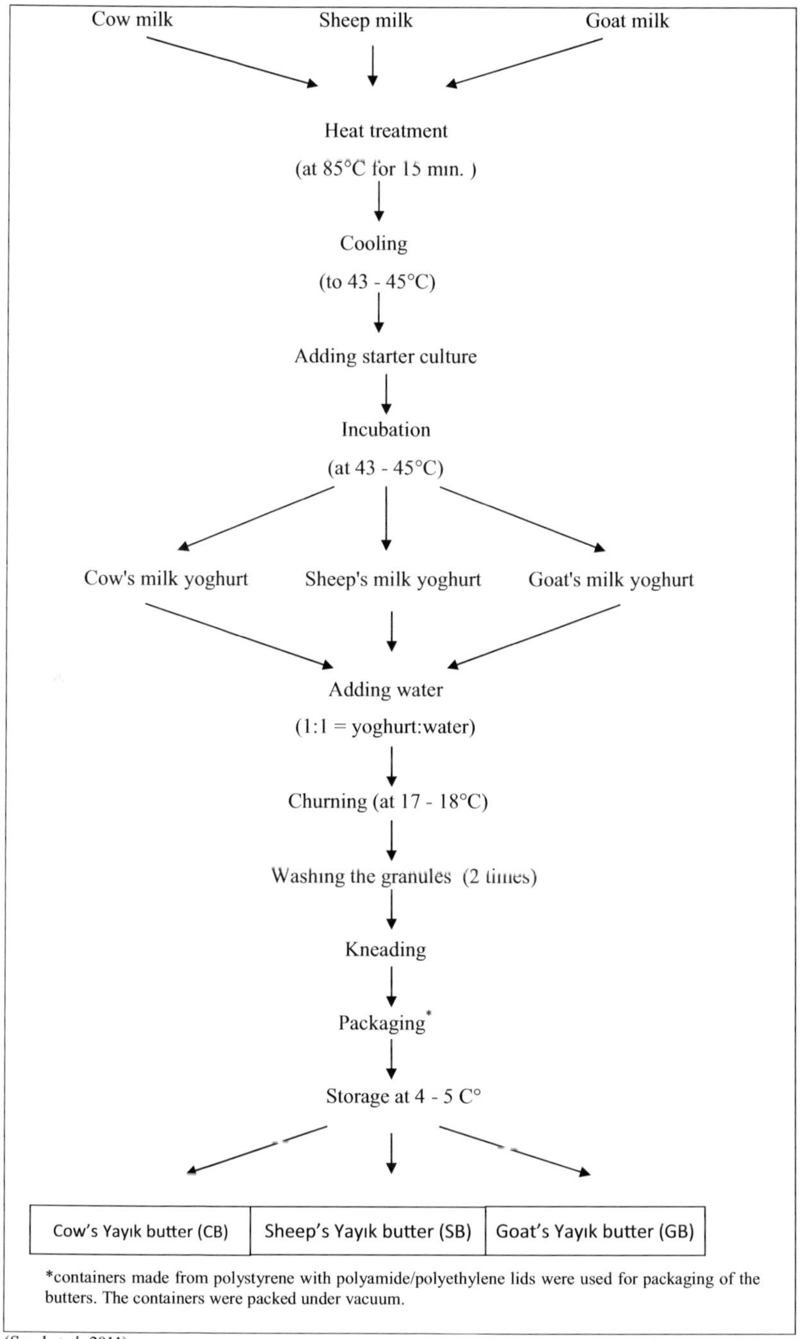

(Şenel *et al.*, 2011)

FIGURE 26.1 The production of Yayik butter samples.

on the threshold level of peroxide in Yayik butter. Oxidized flavor was not perceived in Yayik butter having a peroxide value higher than 2.0 meq O_2/kg oil by the panelists (Şenel *et al.*, 2010). However, in this study, some flavor and aroma defects such as bitterness and rancidity were perceived in goat Yayik butter having peroxide value of 3.22 meq O_2/kg oil on the 30th day of storage. The peroxide values of goat Yayik butter were significantly higher than those of the other samples during the storage ($p<0.01$). This result showed that Yayik butter made from goat milk yogurt had weaker oxidative stability than Yayik butters made from cow and goat milk yogurts. Oxidative stability of dairy products is influenced by many factors such as concentrations of oxygen, metals (Cu^{+2}, Fe^{+3}), antioxidants, water activity, etc. (O'Conner and O'Brien, 1995). Therefore, there is a difficulty in interpreting the results regarding peroxide value. However, it may be attributed to higher levels of free fatty acids in goat Yayik butter than that of cow and sheep Yayik butters. Young (2001) postulated that the existence of large amounts of free fatty acids could facilitate the lipid oxidation.

TABLE 26.2 Variations in Some Properties of Yayik Butter Samples During Storage[n]

Butter Samples	Days	pH in Serum	Lactic Acid (g/100g)	Peroxide Value (meq O_2/kg oil)	Flavor and Aroma (45 point)	Total Sensory Scores (90 points)
CYB	1	4.10±0.00	0.97±0.00	0.61±0.11	43.00	88.00
	30	4.00±0.01	1.24±0.00	0.39±0.02	40.00	84.00
SYB	1	3.99±0.01	1.65±0.00	1.20±0.11	43.00	88.00
	30	4.00±0.00	1.63±0.02	0.99±0.03	37.50	79.00
GYB	1	4.04±0.03	1.19±0.01	1.68±0.10	41.00	86.00
	30	4.02±0.03	1.13±0.01	3.22±0.42	31.25	69.00

[n] *Mean of three replicates.*
CYB, butter produced from cow's yogurt; SYB. butter produced from sheep's yogurt; GYB, butter produced from goat's yogurt.
From Şenel et al. *(2011).*

Sensory analysis revealed that Yayik butter samples received fairly close scores at day 1, but the aroma and flavor scores of goat Yayik butter were lower than those of cow and sheep Yayik butters at day 30. In addition, some flavor defects such as rancidity and bitterness in goat Yayik butter were perceived by the panelists. Results obtained from this study indicated that the higher level of total free fatty acids and peroxide value appeared to be responsible for the development of flavor defects in goat Yayik butter at day 30. Additionally, the cloudy yellowish color of goat Yayik butter was criticized as a defect by the panel group. The occurrence of some color defects in butter owing to oxidation was reported previously (Atamer, 1993). The reason of the discoloration could be attributed to the high level of peroxide of in goat Yayik butter.

The compositions of free fatty acids in Yayik butters from different milk species are given in Table 26.3.

As seen in Table 26.3, capric (C_{10}), mrystic (C_{14}), and palmitic (C_{16}) acids were the predominant free fatty acids in cow Yayik butter and accounted for 56% of total free fatty acids. In sheep Yayik butter, caprylic (C_8), mrystic (C_{14}), palmitic (C_{16}), and stearic ($C_{18:0}$) acids were the dominant free fatty acids and accounted for 61.62% of total free fatty acids.

Six free fatty acids (C_6, C_8, C_{12}, C_{16}, $C_{18:0}$, and $C_{18:1}$) were predominant in goat Yayik butter, and accounted for 90.46% of total free fatty acids. Capric acid in cow Yayik butter, caprylic acid in sheep Yayik butter and goat Yayik butter were the most abundant volatile free fatty acids. The total amount of Σ(C6–C10) free fatty acids varied from 60.78 to 73.96 mg/kg. The total amount of C_{12}–$C_{18:0}$ free fatty acids in all Yayik butters produced from different milk species were significantly higher than those of short-chain and unsaturated free fatty acids. This result could be attributed to the distribution of fatty acids in milk fat globules. Fat is present in the form of globules in milk, which are surrounded by a membrane called milk fat globule membrane. The membrane triglycerides have a predominantly high melting point (Mulder and Walstra, 1974). The arrangement of triglycerides in fat globules is mainly dependent on their melting point. On cooling or at physical ripening (crystallization), glycerides with high melting points form concentric crystalline layers on the innermost side of the fat globules, whereas liquid fat with a low melting point is located at the core of globules (Marshall and Arbuckle, 2000). Therefore, high-melting glycerides are considered to be easily affected by lipases. A similar result was also reported by Şenel (2006) who showed that the accumulation of ΣC_{12}–C_{18} free fatty acids from triglycerides was higher than ΣC_4–C_{10} and $\Sigma C_{18:1}$–$C_{18:2}$ during the storage of Yayik butters. Goat Yayik butter had a higher total amount of free fatty acids than the sheep and cow Yayik butters. This result may be related to a higher concentration (~46%) of lipoprotein lipase (LPL) activity located in the cream phase of the goat milk than in cow or sheep milk (Chillard *et al.*, 1984).

The accumulation of free fatty acids increased during storage. The changes in the amounts of caprylic, myristic, stearic, and oleic acids of all Yayik butters during storage were significant ($p<0.01$). The amounts of myristic acid in cow Yayik butter were highest at days 1 and 30. The level of stearic acid in cow Yayik butter remained fairly constant, whereas it showed a steady increase in sheep and goat Yayik butter ($p<0.01$).

Considering unsaturated free fatty acids (oleic, linoleic acids), the concentration of unsaturated free fatty acids in goat Yayik butter was higher than the others. The total amounts of unsaturated free fatty acids in goat Yayik butter were 610.53 and 624.93 mg/kg at days 1 and 30, respectively.

Generally, the amount of individual free fatty acids in the samples increased during the storage. The increases in the capric, lauric, palmitic, and linoleic acid levels of the samples were significant in all Yayik butters ($p<0.01$). In addition, the increase in the caprylic acid level of goat Yayik butter, the myristic acid levels of cow and sheep Yayik butters, the stearic acid levels of sheep and goat Yayik butter,

TABLE 26.3 Free Fatty Acids Composition of Yayik Butter Samples[n]

Free Fatty Acids (mg/kg)	Days	CYB	SYB	GYB
Caproic (C6)	1	11.97 ± 1.45	9.44 ± 0.55	15.87 ± 0.89
	30	10.89 ± 0.63	9.65 ± 0.45	16.72 ± 0.98
Caprylic (C8)	1	16.10 ± 0.08	38.26 ± 0.56	37.87 ± 1.17
	30	13.80 ± 0.27	37.24 ± 0.66	43.19 ± 1.66
Capric (C10)	1	36.27 ± 0.12	12.50 ± 0.66	16.35 ± 0.31
	30	38.68 ± 0.26	14.46 ± 0.51	17.90 ± 0.28
Σ(C6– C10)	1	64.35 ± 1.43	60.21 ± 1.69	70.10 ± 2.06
	30	63.37 ± 1.06	61.34 ± 1.60	77.80 ± 2.35
Lauric (C12)	1	53.78 ± 1.42	66.18 ± 1.66	86.73 ± 0.28
	30	56.33 ± 0.95	69.89 ± 2.32	89.14 ± 0.17
Myrstic (C14)	1	142.83 ± 0.70	110.82 ± 0.63	74.36 ± 1.27
	30	147.59 ± 0.70	117.17 ± 0.58	76.22 ± 0.16
Palmitic (C16)	1	457.58 ± 1.37	434.46 ± 5.67	390.27 ± 2.32
	30	458.35 ± 1.29	449.44 ± 0.42	394.79 ± 2.03
Stearic (C18:0)	1	106.76 ± 0.74	147.19 ± 0.47	272.61 ± 1.02
	30	106.97 ± 0.17	152.97 ± 0.73	277.24 ± 1.09
Σ(C10– C18:0)	1	760.95 ± 0.70	758.66 ± 5.44	823.97 ± 2.27
	30	769.24 ± 1.14	789.47 ± 2.56	837.38 ± 1.93
Oleic (C18:1)	1	217.10 ± 1.08	328.87 ± 8.30	560.25 ± 2.71
	30	288.87 ± 1.60	334.51 ± 7.54	569.62 ± 2.70
Linoleic (C18:2)	1	42.59 ± 0.26	38.66 ± 1.64	50.28 ± 0.75
	30	45.84 ± 0.56	42.54 ± 1.75	55.32 ± 0.62
Σ(C18.1–C18:2)	1	259.69 ± 0.89	367.52 ± 5.54	610.53 ± 3.18
	30	334.71 ± 1.23	377.05 ± 2.64	624.93 ± 3.51
Total Acids	1	1085.00 ± 2.17	1186.40 ± 11.30	1504.59 ± 3.71
Free Fatty	30	1167.32 ± 1.97	1227.87 ± 4.02	1540.14 ± 3.67

[n] *Mean of three replicates.*
CYB, butter produced from cow's yogurt; SYB, butter produced from sheep's yogurt; GYB, butter produced from goat's yogurt.
From Şenel et al. *(2011).*

and the oleic acid levels of cow Yayik butter during the storage were also statistically significant ($p<0.01$).

THE EFFECTS OF OTHER TECHNOLOGICAL PARAMETERS ON YAYIK BUTTER CHARACTERISTICS

As mentioned above, the manufacturing parameters affecting the Yayik butter profile are the fat content of raw material, churning pH/pH of yogurt, dilution rates of yogurt, and number of washes of butter granules. Therefore, in a study carried out by Şenel (2006), the effects of fat content of raw material and churning pH on some properties of Yayik butter were investigated. The researcher demonstrated that the churning of yogurt at pH4.0 was recommended in order to obtain a well-balanced aroma–flavor and spreadability in the end products. Yayik butter produced from yogurt with 14% fat content was favorable in terms of yield and fat recovery rate.

In another study, the effects of the dilution rate of yogurt and washing number of granules on some properties of Yayik butter were investigated by Öztekin-Öztürk (2010). The author proposed the following production parameters to obtain a product with high oxidative and lipolytic stabilities, sensory properties and fat recovery; 50% dilution rate +3-times washing or 50% dilution rate +6-times washing.

Additionally, threshold levels of some quality parameters including titratable acidity, pH in serum, acid degree, peroxide value, and tyrosine concentration in Yayik butter stored for 180 days at different storage temperatures (4 or 15°C) were determined by Şenel *et al.* (2010). Results of this study revealed that the changes in aroma and flavor of Yayik butter were largely affected by the titratable acidity and acid degree values (lipolysis) of the samples. Aroma and flavor impairment become more pronounced at day 100. Threshold levels of titratable acidity and acid degree values were determined as 6.19 °SH and 1.08 mg KOH/g oil, respectively.

ANALYTICAL TECHNIQUES

The analytical techniques used in the present study are summarized below.

pH in Serum

pHs in serum of Yayik butters were determined according to IDF (1981) using a pH-meter model Mettler Toledo (Analytical, Schwerzenback, Switzerland) fitted with a standard, combined glass electrode.

Lactic Acid

Lactic acid value of Yayik butter was measured spectrophotometrically as per the method of Steinholt and Calbert (1960).

Peroxide Value

In the detection of peroxide value, firstly the extraction of fat from butters was performed according to AOAC (2000). Peroxide value of fats extracted from butter samples were detected by the spectrophotometric method described by Downey (1975).

Free Fatty Acid Composition

Free fatty acids compositions were detected by a GC system (Agilent 6890 series, Agilent Tech. Inc. CA, USA) according to the method of Deeth *et al.* (1983).

Operating conditions for GC were as follows; Detector: FID (Agilent Tech. Inc. CA, USA) at 260°C, column: capillary (30 m × 320 μm id. With 0.25 μm film thickness (HP-FFAP Agilent Tech. Inc. Model 19091F-433), injection mode/volume: Split (1/10) 5 μl at 250°C, flow rates: H_2 : Air : N_2 = 33 : 370 : 30 ml/min, oven temperature: 120°C for 0 min increased to 200°C at a rate of 10°C/min then held at 200°C for 2 min and increased to 205°C at a rate of 10°C/min and held for 2 min, then increased to 210°C at a rate of 10°C/min and held for 2 min then increased to 215°C at a rate of 10°C/min and held for 3 min and increased to 230°C at a rate of 10°C/min held for 3 min.

Sensory Analysis

Sensory evaluation was performed according to the method of Tobias (1976) modified by Şenel (2006). Butters were evaluated by a panel group of 10 experienced staff (consisting of five male and five female staff) working at the Dairy Technology Department.

Acknowledgment

This work was supported by The Scientific and Technological Research Council of Turkey. Project number: TOVAG 105O157.

References

AOAC, 2000. Official methods of analysis of AOAC international, seventeenth ed. AOAC International, Gaitherburg, MD, USA.

Atamer, M., 1993. Butter Technology. Ankara University Publishing, Ankara No: 1313, pp. 90.

Atamer, M., Caric, D.M., Milinovic, S.P., Gavaric, D., 1996. Quality of set yoghurt manufactured from UF milk. Proceedings for Natural Sciences, Matica Srpska NoviSad No. 9119–26.

Atamer, M., Gürsoy, A., Öztekin, Ş., Şenel, E., 2004a. Determination of some carbonyl compounds in Yayik Tereyagi. International Dairy Symposium, Isparta Turkey 174–175.

Atamer, M., Şenel, E., Öztekin, Ş., 2004b. A traditional product: Yayik Tereyağ conventional way of manufacturing and its some properties. International Dairy Symposium, Isparta Turkey 149–152.

Atamer, M., Şenel, E., Öztekin, F.Ş., 2007. A compartive study on some properties of butter produce from yoghurt and cream, Food Industry Milk and Dairy Products. J. Chem. Technolo. Assoc. Belgrad, No: 1–2 18, 3–7.

Chillard, Y., Selset-Attou, G., Bas, P., Morand-Fehr, P., 1984. Characteristic of lipolytic system in goat milk. J. Dairy Sci. 67, 2216–2223.

Deeth, H.C., Fitz-Gerald, C.H., Snow, A.J., 1983. A gas chromatographic method for the quantitative determination of free fatty acids in milk and milk products, New Zealand. J. Dairy Sci. Technol. 18, 13–20.

Downey, W.K., 1975. Butter Quality. Published by An Foras Taluntais 19 Sadyamount Avenue Dublin 4, Dairy Research & Review Series No.7.

Hayaloğlu, A.A., Konar, A.A., 2001. Comparative study on physicochemical and sensorial properties of butter made from yoghurt and cream. Milchwissenschaft 56 (12), 675–677.

IDF, 1981. Determination of The pH of The Serum. IDF Standard 104. Belgium: International Dairy Federation, Brussels.

Marshall, R.T., Arbuckle, W.S., 2000. Ice Cream, fifth ed. Gainthersburg Maryland USA.

Mulder, H., Walstra, P., 1974. The Milk Fat Globule. Emulsion science as applied to milk products and comparable foods. Centre for Agricultural Publishing and Documantation Wageningen, The Netherland.

O'Conner, T.P., O'Brien, N.M., 1995. Lipid oxidation. In: Fox, P.F. (Ed.), *Advanced Dairy Chemistry* Volume 2, Lipids. second ed. Chapman & Hall, UK, pp. 309–333.

Öztekin-Öztürk, F.Ş., 2010. Effects of dilution rate of yoghurt and washing numbers of granules on the properties of Yayik Butter. Ph.D. Thesis. Ankara University, Graduate School of Natural and Applied Sciences, Department of Dairy Technology, Ankara, pp. 147.

Sağdıç, A., Arıcı, M., Şimşek, O., 2002. Selection of starters for a traditional Turkish Yayik butter made from yoghurt. Food Microbiol. 19, 303–312.
Sağdıç, O., Dönmez, M., Demirci, M., 2004. Comparison of characteristic and fatty acid profiles of traditional Turkish Yayik butters produced from goats', ewes' or cows' milk. Food Control 15, 485–490.
Sauter, F., Puchinger, L., Schoop, U., 2003. Studies in organic archaeometry, V^{-1} Fat analysis sheds light on everyday life in prehistoric Anatolia: traces of lipids identified in chalcolithic potsherds excavated near Bogazkale, Central Turkey. Arcivoc (xv), 15–21.
Steinholt, T.K., Calbert, H.E., 1960. A rapid colorimetric for the determination of lactic acid in milk and milk products,. Milchwissenschaft 15, 7–11.
Şenel, E., 2006. The effect of some production parameters on Yayik butter produced from yoghurt. Ankara University Graduate School of Natural and Applied Sciences Department of Dairy Technology, Ph.D. Thesis, Ankara, pp. 168.
Şenel, E., Öztekin, F.Ş., Atamer, M., 2010. Determination of rejection levels for acid, peroxide, tyrosine and titratable acidity in Yayik butter. 1st International Congress on Food Technology, Antalya Turkey 202.
Şenel, E., Atamer, M., Öztekin, F.Ş., 2011. The oxidative and lipolytic stability of Yayik butter produced from different species of mammals' milk (cow, sheep, goat) yoghurt,. Food Chem. 127, 333–339.
Tamime, A.Y., Robinson, R.K., 2007. Tamime and Robinson's Yoghurt, Science and technology, third ed. Woodhead Publishing Limited, Abington Hall, Abington, Cambridge CB21 6AH, England.
Tobias, J., 1976. Organoleptic properties of dairy products. In: Harper, W.J., Hall, C.W. (Eds.), Dairy Technology and Engineering. AVI Publishing Company Inc, 631 Westport, USA, pp. 75–140.
Young, W.P., 2001. Proteolysis and lipolysis of Goat milk cheese. J. Dairy Sci. 84 (Suppl. E), E84–E92.

CHAPTER

27

Formation of Biogenic Amines in Cheese

Tuba Şanlı, Ebru Şenel

Ankara University, Faculty of Agriculture, Department of Dairy Technology, Ankara, Turkey

CHAPTER POINTS

- Cheese is the main fermented milk product that contains potentially harmful levels of biogenic amines (BA), especially tyramine, histamine, and putrecine.
- The use of raw material having poor hygienic quality and length of ripening time can increase the formation of BA in cheese.
- The ripening time can also play an important role the formation of BA as it depends on the degree of proteolysis.
- Lactic acid bacteria have an important role in the production of fermented milk products; however, they are also the most important bacterial group that have amino acid decarboxylase activities.
- *Enterococcus faecalis*, *Enterococcusfaecium* and *Enterococcusdurans* could produce large amounts of tyramine and putrecine, cadaverine, and histamine in cheese.
- Control of the microorganisms producing BA by adequate heat treatment of milk is one of the most important factors for reducing BA accumulation in cheese.
- BA production in cheese has been also related to several environmental factors, particularly pH, temperature and salt concentration.
- HPLC is considered the most suitable technique for determination of BA in cheese.

INTRODUCTION

Biogenic amines (BA) are low-molecular-weight nitrogenous compounds having biological activity resulting primarily from the decarboxylation of amino acids (Silla Santos, 1996; Durlu-Özkaya, 2002; Karovicova and Kohajdova, 2005; Buncova et al., 2010). The chemical structure of BA are aliphatic, alicyclic, or heterocyclic organic bases (Karovicova and Kohajdova, 2005).

In spite of being considered as naturally occurring in some vegetables and fruits, fish, and fish products, BA in other foods are produced by fermentation (e.g., wine, cheese, and beer) as a result of microbial activity (Shalaby, 1996; Bodmer *et al.*, 1999). The most important BA occurring in foods are histamine, putrescine, cadaverine, tyramine, tryptamine, β-phenylethylamine, spermine, products of decarboxylation from histidine, ornithine, lysine, tyrosine, tryptphane, phenylalanine, arginine amino acids, respectively (Shalaby, 1996; Bodmer *et al.*, 1999; Linares *et al.*, 2011).

Biologically active amines are histamine, triptamine, β-phenylethylamine, and tyramine, which have major physiological or vasoactive effects. BA is responsible for many pharmacological effects caused by several types of foodborne diseases such as histamine poising (scombroid poising), and tyramine toxicity (cheese reaction) (Shalaby, 1996). Histamine is the most toxic amine in foods such as fish, cheese, wine, and meat products (Bodmer *et al.*, 1999; Linares *et al.*, 2011). Many foods contain small amounts of histamine which can easily be tolerated. However, the ingestion of food containing unusually high levels of histamine can result in histamine poisoning (Chong *et al.*, 2011). Histamine poisoning is also referred to as scombroid fish poisoning because of its association with consumption of fish belonging to the *Scombroid* family such as tuna, mackerel, and sardines (Chong *et al.*, 2011; Linares *et al.*, 2011). Thus, some outbreaks of histamine poising have occurred after consuming cheese and fish. Histamine poisoning produces the following symptoms: sweating, headache, flushing, urticaria, abdominal cramps, diarrhea, nausea, and vomiting (Bodmer *et al.*, 1999).

Cheese is one of the foods which has the highest content of amines and is the most prevalent product associated with amine poisoning (Durlu-Özkaya, 2002). The

Processing and Impact on Active Components in Food
http://dx.doi.org/10.1016/B978-0-12-404699-3.00027-5

consumption of cheese that has a high level of tyramine may result in dangerous intoxication, known as the 'cheese reaction', which is characterized by an increase in blood pressure. 'Cheese reaction' can cause headaches and hypertension (Shalaby, 1996; Santos *et al.*, 2003). The intake of high levels of tyramine and phenylethylamine can also lead to migraine (Santos *et al.*, 2003).

Normally, foods containing low quantities of BA are metabolized in the human gut to physiologically less active forms. The enzymes monoamino oxidase (MAO) and diamine oxidase (DAO) play an important role in the detoxification process. Inadequate detoxification can be the result of consumption of high concentrations of BA, or it may be because of genetic reasons, or because of the inhibitory effect of some medicines or alcohol, which result in BA accumulation in the body (Bodmer *et al.*, 1999; Spano *et al.*, 2010).

The toxicological level of BA is difficult to determine because it depends on the BA intake concentrations, the presence of other different amines, and individual characteristics of BA (Silla- Santos, 1996). For example, secondary amines such as putrecine and cadaverine contribute to synergistic effects and increase histamine toxicity (Shalaby, 1996).

There is little specific legislation with regard to BA content in foods. For fish, the legal limit is 20 mg histamine/100 g in some countries. However, the upper limit for BA of 100 mg histamine/kg in food, and the legal upper limit for BA of 2 mg histamine per litre of alcoholic beverage have been detected (Silla-Santos, 1996). In addition, a level of 100 mg amine/kg food is very dangerous for healthy individuals. This level is calculated based on the food-borne histamine intoxication. Therefore, the level of toxicity of histamine in food might be changed in the presence or absence of other biogenic amines. The maximum tolerable legal limit for tyramine is 200 mg/kg in cheese in some countries. Generally it is considered that an intake of 5–10 mg of histamine can affect sensitive people and 10 mg histamine in food is a tolerable limit for the body. However, 100 mg is considered to induce a medium toxicity, and 1000 mg is highly toxic (Karavicova and Kohajdova, 2005).

OCCURRENCE OF BIOGENIC AMINES IN CHEESE

Along with fish and wine, cheese can contain high levels of BA (Linares *et al.*, 2011). Therefore, the presence of BA in various types of cheese has been widely investigated by many authors (Vale and Gloria 1998, Durlu-Özkaya *et al.*, 1999, 2000, 2002; Valsamaki *et al.*, 2000; Komprda *et al.*, 2007; Elsanhoty *et al.*, 2009; Yıldız *et al.*, 2010, Şenel *et al.*, 2012).

Cheese is an ideal environment for BA production because it contains proteins, enzymes, cofactors, water, and microorganisms. While occurrence of BA in milk is very low (1 mg/dm^3), the concentration of BA in cheese is 1 g/kg (Karovicova and Kohajdova 2005). BA accumulation in cheese can be influenced firstly by the microbial quality of raw milk, the sanitation procedures adopted, the use of starter cultures, environmental conditions and ripening time (Novella-Rodriguez *et al.*, 2004). The BA detected in different types of cheese are tyramine, histamine, putrecine, cadaverine, tryptamine, and β-phenylethylamine (Linares *et al.*, 2011). The accumulation of BA in different cheeses is variable. Also, it can vary within some types of cheese. Moreover, BA contents not only vary between different cheese types, but also within the same type of cheese. However, BA are often distributed within cheese unevenly and the results of experiments regarding BA distribution in cheese are inconsistent. While tyramine and histamine are the main BA in cheese both quantitatively and toxicologically (Silla-Santos, 1996; Burdychova and Komprda 2007, Komprda *et al.*, 2008), histamine in Edam cheese has not been determined (Bunkova *et al.*, 2010). However, tyramine and putrescine were found as predominant BA in Kaşar cheese (Durlu-Özkaya, 2002) and Edam cheese (Valsamaki *et al.*, 2000).

ROLE OF THE RIPENING PROCESS AND PROTEOLYSIS IN FORMATION OF BA IN CHEESE

Cheese ripening is important to develop the flavor and texture characteristic of the cheese variety (Pinho *et al.*, 2001). However, the well-ripened and/or long-term stored cheese is an ideal environment for BA formation (Pinho *et al.*, 2001; Vale and Gloria 1998). The ripening time can also play an important role in the formation of BA as it depends on the degree of proteolysis (Lineras *et al.*, 2011). The proteolysis which is the most important event in cheese ripening, leads to accumulation of free amino acids which can be substrates for decarboxylase activity (El-Sayed, 1996; Linares *et al.*, 2011). During cheese ripening, casein is degraded by proteolytic enzymes, and as a result free amino acid content increases. BA are produced by degradation of these amino acids resulting from the bacterial decarboxylase activity (Shalaby, 1996). BA, especially histamine, putrecine, and cadaverine, are increased to different levels during ripening. Thus each different type of cheese has its own characteristic free amino acid and BA content, resulting from its specific degradation (Spano *et al.*, 2010; Linares *et al.*, 2011). The gradual increase in the content of some BA in some different types of cheese during the ripening period are presented in Table 27.1.

Several studies have reported on the types of BA and on the amounts formed depending on the length

TABLE 27.1 BA Contents (mg/kg) in Some Cheeses During Ripening Period

Biogenic Amines	Storage Period						References
	1	15	30	60	90	120	
Histamine	2.85	3.87	6.67	22.18	28.99	44.23	Elsanhoty *et al.* (2009)
	2.40	25.9	38.2	47.0	76.40	84.6	Valsamaki *et al.* (2000)
	<0.05	1.19	2.16	3.5	66.34	—	Novella-Rodriguez *et al.* (2004)
	3.05	3.45	5.37	27.99	43.06	—	Novella-Rodriguez *et al.* (2004)
Tyramine	3.96	30.24	69.87	246.12	297.34	365.23	Elsanhoty *et al.* (2009)
	0.00	140	158	152	204	246	Valsamaki *et al.* (2000)
	<0.05	0.96	2.54	7.22	10.90	—	Novella-Rodriguez *et al.* (2004)
	4.49	29.25	93.47	245.32	324.67	—	Novella-Rodriguez *et al.* (2004)
Putrecine	9.62	26.99	38.22	73.18	25.78	92.23	Elsanhoty *et al.* (2009)
	1.62	36.2	53.6	77.7	153	193	Valsamaki *et al.* (2000)
	0.79	3.86	7.83	8.17	14.61	—	Novella-Rodriguez *et al.* (2004)
	10.01	33.97	42.86	74.15	86.40	—	Novella-Rodriguez *et al.* (2004)
Cadaverine	52.47	71.19	84.35	154.12	198.11	223.32	Elsanhoty *et al.* (2009)
	0.27	63.0	67.9	44	60.2	82.8	Valsamaki *et al.* (2000)
	1.29	1.93	5.77	15.52	32.73	—	Novella-Rodriguez *et al.* (2004)
	53.47	72.04	87.15	177.81	196.47	—	Novella-Rodriguez *et al.* (2004)

of the ripening period. Valsamaki *et al.* (2000) observed that the level of the BA was dramatically increased from 3 to 15 days, changed slightly from 15 to 60 days, and almost doubled from 60 to 120 days. Novella-Rodriguez *et al.* (2002) reported that there was a gradual increase in the value of BA throughout ripening. Also, the maximum level of BA in cheese was found to occur after 90 days of storage. Similar results were observed by Bunkova *et al.* (2010). They reported that the tyramine level in Edam cheese exceeds 100 mg/kg after 98 days' storage, this level is considered to be the safety limit, according to Silla-Santos (1996). The same trend was observed for histamine. It is also common to find the highest concentrations of histamine in long-ripened cheeses.

From a qualitative and quantitative point of view, microorganisms in the individual layers of ripening cheese can vary, which can cause different concentration of BA in different parts of cheeses (Komprda *et al.*, 2007; Pinho *et al.*, 2001). The BA contents of different type cheeses are presented in Table 27.2.

FACTORS AFFECTING BIOGENIC AMINES OCCURRENCE IN CHEESE

The occurrence of BA depends on several factors. The most important factors are the presence of microorganisms with amino acid decarboxylation activity, availability of substrate (free amino acids precursors), and environment conditions (pH, amount of salt, and storage temperature).

Microorganisms with Amino Acid Decarboxylase Activity

The main factor contributing to the occurrence of BA in cheese is the presence of microorganisms with decarboxylase activity. Amino acid decarboxylase activity has been described for several groups of microorganisms such as *Enterobacteria*, *Pseudomonas* spp., *Enterococci*, *Micrococci*, and lactic acid bacteria (LAB) (Silla-Santos, 1996; Shalaby 1996; Marino *et al.*, 2000). These microorganisms can be present in the microflora of milk as starter cultures, used to make cheese, or by contamination during manufacturing (Marino *et al.*, 2000; Martuscelli *et al.*, 2005, Burdychova and Komprda, 2007). The amine-producing abilities of various bacteria are different; therefore, a direct correlation could not be found between the BA content in cheese and total bacteria count (Burdychova and Komprda, 2007).

LAB have an important role in the production of fermented milk products; however, they are also the most important bacterial group which may have amino acid decarboxylase activities. Shalaby (1996) reported that the histamine occurred as a result of the decarboxylase activity of starter culture bacteria such as *Streptococcus lactis* and *Lactobacillus helveticus*. Burdychova and Komprda (2007) isolated histamine produced by *Lactobacillus helveticus* strain, a starter culture used for cheese production. Similarly, *Lactobacillus brevis* and *Lactobacillus casei* subsp. *casei* with tyrosine decarboxylase activity and some strains of lactobacilli with histidine decarboxylase activity such as *Lactobacillus curvatus* and *Lactobacillus helveticus* were isolated in Spanish cheeses (Roig-Sagues *et al.*, 2002)

TABLE 27.2 Levels of Biogenic Amines in Different Types of Cheese

Cheese Type	Biogenic Amines (mg/kg)										
	TYM	HIT	PUT	CAD	TRM	PHA	SPD	SPM	AGM	SER	Total BA
Feta[1]											
(fresh)	0.00	2.40	1.62	0.27	4.39	3.51	—	—	—	—	12.2
(120 days)	246	84.6	193	82.8	5.74	4.94	—	—	—	—	617
Kaşar[2]											
(fresh)	109.5	35.2	34.7	42.6	ND	3.7	75.7	ND	—	—	201.4
(aged)	309.6	45.9	88.1	26.5	ND	ND	114.8	ND	—	—	584.9
Urfa[3]	18.0	1.5	1.2	2.1	1.6	3.6	—	—	—	—	28.0
Van herby[3]	5.6	9.6	10.1	4.4	2.1	7.9	—	—	—	—	18.6
Civil[4]	8.8	1.5	0.5	2.2	ND	ND	—	—	—	—	13.1
Dutch[5] (22 weeks)	299.8	1.8	60.8	2.0	ND	54.3	0.3	0,2	—	—	—
Gorgonzola[6]	3.9	13.5	4.1	2.5	9.6	4.3	13.1	2.5	6.4	6.2	66.1
Tilsit[6]	3.1	16.1	11.5	10.6	2.0	5.4	5.1	9.7	0.3	1.7	65.3
Gouda[6]	7.0	28.4	23.3	7.3	6.3	2.5	3.0	2.7	2.4	4.0	82.7
Mozzarella[6]	2.0	15.7	3.5	2.3	0.5	0.8	1.9	2.8	0.1	0.7	30.4
Parmesan[6]	1.7	2.1	2.4	0.7	ND	0.1	0.3	0.3	ND	0.4	7.9
White cheese[7]											
(with starter culture)	57.0	8.7	69.3	30.08	6.9	5.9	—	—	—	—	179
(without starter cultures)	131.0	26.5	112.0	156	11.2	9.3	—	—	—	—	442
Tulum[8]	120	80.2	206	130	—	332	979	—	—	—	—

[1] *Valssamaki* et al. *(2000)*
[2] *Durlu-Özkaya (2002)*
[3] *Şenel* et al. *(2012)*
[4] *Yıldız* et al. *(2010)*
[5] *Komprda* et al. *(2007)*
[6] *Vale and Gloria (1998)*
[7] *Durlu-Özkaya* et al. *(2000)*
[8] *Durlu-Özkaya* et al. *(1999).*
TYM, Tyramine; HIT, Histamine; PUT, Putrescine; CAD, Cadaverine; TRM, Tryptamine; PHA, Phenylethylamine; SPD, Spermidine; SPM, Spermine; AGM, Agmatine; SER, Serotonin; ND, not detected.

The high BA content in cheese has mostly been linked to non-starter bacteria possessing decarboxylase activity (Shalaby, 1996; Bodmer *et al.*, 1999). Novella-Rodriguez *et al.* (2002) determined that the BA in cheese caused by non-starter bacteria in cheese was higher than that of cheese where the starter culture contains *Lactococcus lactis* subsp. *lactis* and *Lactococcus lactis* subsp. *cremoris*. However, during fermentation the occurrence of BA can be controlled by using starter cultures which have less effect on decarboxylation of the amino acids (Chang *et al.*, 2011).

To decrease BA formation, researchers focused on controlling the suitable starter cultures with amine oxidase to degrade the BA. *Brevibacterium lineans* leads to reduced levels of tyramine and histamine during cheese ripening (Karavicova and Kahajdova, 2005).

Gram-positive bacteria, especially lactic acid bacteria, are the main histamine and tyramine producers in cheeses (Novella-Rodriguez *et al.*, 2002; Spano *et al.*, 2010; Linares *et al.*, 2011). There is evidence that some enterococci strains such as *Enterococcus faecalis*, *Enterococcus faecium*, and *Enterococcus durans* could produce large amounts of tyramine and putrecine, cadaverine and histamine in cheese. The production of putrescine and cadaverine in cheese has mainly been linked to most of the Gram-negative bacteria, especially *Enterobacteiaceae* (Linares *et al.*, 2011). Marino *et al.* (2000) found that the large amounts of cadaverine present in cheese can be associated with high counts of *Enterobacteriaceae*, which means that cadaverine may be regarded as an indicator of poor quality or poor hygiene in cheese manufacturing. The presence of tyramine in Cheddar cheese has been associated with *Enterococcus faecalis* (Silla Santos, 1996). In addition, it has been reported that species from other microbiological groups such as *Bacillus macerans* isolated from Italian cheese have the ability to decarboxylase histidine (Rodrigues-Jerez *et al.*, 1994). The BA-producing microorganisms identified from some types of cheese are shown in Table 27.3.

TABLE 27.3 BA-producing Microorganism Identified from Cheeses

Biogenic Amine	Producer Microorganisms		References
Histamine	Gram negative	*Hafnia alvei, Esherichia coli, Klebsiella pneumonia, Serratia* spp., *Enterobacter sakazakii, Esherichia coli, Pseudomonas* spp.	Roig-Sagues *et al.* (2002); Martuscelli *et al.* (2005)
	Gram positive	*Enterococcus casseliflavus, Lactobacillus buchneri, Lactobacillus casei, Lactobacillus delbrueckii, Lactobacillus lactis, Lactobacillus curvatus, Lactobacillus helveticus*	Roig-Sagues *et al.* (2002); Burdychova and Komprda (2007)
	Molds and yeast	*Debaryomyces hansei, Geotrichum candidum*	Linares *et al.* (2012); Roig-Sagues *et al.* (2002)
Tyramine	Gram negative	*Cedecea spp.,Hafnia alvei, Esherichia coli, Citrobacter braakii, Pseudomonas* spp.	Roig-Sagues *et al.* (2002); Martuscelli *et al.* (2005)
	Gram positive	*Enterococcus casseliflavus, Enterococcus durans, Enterococcus faecalis, Enterococcus faecium, Lactobacillus brevis, Lactobacillus delbrueckii, Lactobacillus plantarum, Lactobacillus lactis, Leuconostoc mesenteroides, Pediococcus damnosus, Lactobacillus curvatus, Lactobacillus casei/paracasei*	Roig-Sagues *et al.* (2002); Bunkova *et al.* (2010); Burdychova and Komprda (2007)
	Molds and yeast	*Yarrowia lipolytica*	Linares *et al.* (2012)
Cadaverine	Gram negative	*Citrobacter braakii, Esherichia coli, Enterobacter sakazakii, Pseudomonas* spp., *Enterobacter aerogenes, Enterobacter cloacae, Klebsiella axytoca*	Martuscelli *et al.* (2005); Marino *et al.* (2000)
	Gram positive	*Lactobacillus curvatus, Lactobacillus casei/paracasei*	Bunkova *et al.* (2010)
	Molds and yeast	*Yarrowia lipolytica*	Linares *et al.* (2012)
Putrecine	Gram negative	*Citrobacter braakii, Esherichia coli, Enterobacter sakazakii, Pseudomonas* spp.	Martuscelli *et al.* (2005)
	Gram positive	*Lactobacillus curvatus*	Bunkova *et al.* (2010)
	Molds and yeast	*Debaryomyces hansei, Yarrowia lipolytica*	Linares *et al.* (2012)

Pasteurization

Certainly, the use of good-quality milk and good hygiene practices in cheese manufacturing are important for reducing the risk of BA formation (Chang *et al.*, 2011).The presence of BA in cheese has been used as an indicator of hygienic quality because high levels these amines in cheese could indicate microbial activity in raw milk and contamination during cheese manufacturing (Karovicova and Kohajdova, 2005; Spano *et al.*, 2010). According to Elsanhoty *et al.* (2009), cadeverine, putrecine, and histamine contents are also higher in cheese made from raw milk because of the higher *Enterobacteriaceae* and *Lactobacilli* counts.

Pasteurization is the most common heat treatment used in cheese production for the reduction of pathogenic and spoilage microorganisms (Novella-Rodriguez *et al.*, 2004). Many authors conclude that the BA content of cheese made with pasteurized milk is significantly low as a result of the reduction of its microbial counts. Pasteurization reduces the bacterial load present in milk, including the BA producers. Thereby, cheeses produced with pasteurized milk have lower BA concentrations than those with raw milk (Novella-Rodriguez *et al.*, 2002). The BA contents of different types of cheeses made from pasteurized and raw milk are demonstrated in Table 27.4.

The lower BA content of cheese made with pasteurized milk has been associated with inactivation of some native milk proteases resulting in slower proteolysis and consequently amino acids of BA (Novella-Rodriguez *et al.*, 2004; Linares *et al.*, 2011). Novella-Rodriguez *et al.*, (2004) reported that *Enterobacteriaceae* and *Enteroccoci* counts are also lower in cheese made from pasteurized milk. The high concentrations of tyramine, which is 30-times higher in cheese made from raw milk than from pasteurized milk, associated with the number of *Enteroccoci*. Some BA-producing microorganisms can resist pasteurization or can contaminate the cheese during the subsequent elaboration procedures if the hygienic conditions are not adequate (Linares *et al.*, 2012).

Marino *et al.*, (2008) observed that the Italian cheese made from raw milk contained lower amounts of BA than cheeses made from thermized milk (67°C, for 20 s). This thermal treatment is considered insufficient to significantly reduce the decarboxylase activity of bacteria in cheese.

ENVIRONMENTAL CONDITIONS RELATED TO BA FORMATION IN CHEESE

BA production in cheese has been also related to several environmental factors, particularly pH, temperature, and salt concentration (Shalaby, 1996). One of the key factors in BA formation is the storage temperature (Silla Santos, 1996; Chang *et al.*, 2011; Linares *et al.*, 2011). With respect to the influence of storage temperature on formation of BA, there are reports that storage temperature can affect the content of BA in cheese. The optimal temperature for growing bacteria with decarboxylase activity is between 20 and 37°C reported by Karovica and Kohajdova (2005). A number of studies show that low ripening and storage temperature decrease the accumulation of BA such as histamine, putrecine, and cadaverine. Bunkova *et al.* (2010) observed that a reduction in tyramine, putrecine, and cadaverine content in cheeses which were stored temperature 5°C instead of 10°C. Pinho *et al.* (2001) detected that the concentration of BA in Azeitao cheese is increased at room temperature (25°C) during 4 weeks of storage. The BA activity of *Bacillus macerans* which was isolated from Italian cheese was detected at 30°C storage temperature with maximum histamine content (Rodriguez-Jerez *et al.*, 1994).

The pH of medium is another important factor effecting on amino decarboxylase activity (Linares *et al.*, 2011). Some studies have described that acidic environments cause a more strong amino acid decarboxylase activity and is optimum at acidic pH around 4.0–5.5 (Karovicova and Kohajdova, 2005; Linares *et al.*, 2011). Santos *et al.* (2003) reported that an acidic pH lead to an increase in

TABLE 27.4 BA Contents (mg/kg) of Different Types of Cheeses made from Pasteurized and Raw Milk

		Tyramine	Histamine	Cadaverine	Putrecine	References
Goat cheese	Raw milk	4.49	3.05	53.47	10.01	Novella-Rodriguez *et al.* (2004)
	Pasteurized milk	<0.05	<0.05	1.29	0.79	
Toma (semi hard Italian cheese)	Raw milk	111	23	220	330	Gennaro *et al.* (2003)
	Pasteurized milk	35	Not detected	136	93	
Domiati cheese (made from buffalo milk)	Raw milk	3.96	2.85	52.47	9.62	Elsanhoty *et al.* (2009)
	Pasteurized milk	<0.05	<0.05	1.25	0.81	
Blue cheeses	Raw milk	0–1051.98	0–1041.81	0–756.78	0–875.8	Linares *et al.* (2011)
	Pasteurized milk	0–526.63	0–127.02	40–89.4	0–237.56	

the levels of BA. Josten (1988) observed that the histamine content in Gouda cheese with pH 5.39 at two weeks' ripening was almost twice that of the same cheese with pH 5.19 at the end of a year of storage. Moreover, it is determined that a reduction of salt content in moisture results in an increase in BA content. In spite of the fact that low pH is a key factor which represents a potential risk of BA accumulation in the final product, it is difficult to change the pH, which is inherent to the fermentation process.

The presence of sodium chloride can also control the formation of BA in cheese (Linares *et al.*, 2011). Thus, BA formation can be decreased depending on the increase in the concentration of salt which decreases amine formation. Valsamaki *et al.* (2000) found that the high salt content in Feta cheese decreases BA formation. The decrease in BA formation could be explained by the inhibitory effect of high salt content on the growth rate of BA decarboxylase activity.

ANALYTICAL TECHNIQUES

Determination of BA in Cheese

For the determination of BA in foods different chromatography techniques including thin-layer chromatography (TLC), gas chromatography (GC), and high-performance liquid chromatography (HPLC) have been used (Karovicova and Kohajdova, 2005, Dadakova *et al.*, 2009). The semi-quantitative TLC technique, which is used as an early technique for the determination of BA in foods (Linares *et al.*, 2011), is a useful technique for routine control because of its simplicity, speed, and its applicability to many numbers of samples in minimal time (Dadakova *et al.*, 2009). GC is not often used for the determination of amines. However, HPLC is considered the most suitable technique for the determination of BA in cheese (Moret and Conte, 1996).

The complex sample matrix is a typical problem in the analysis of food for BA (Karovicova and Kohajdova, 2005). To ensure adequate sensitivity for all the methods, an extraction step should be used before detection (Mayer *et al.*, 2010). Perchloric acid, trichloroacetic acid (TCA), hydrocloric acid (HCl), and organic solvents are extraction agents widely used for amine separation from a sample matrix (Shalaby, 1996). The chromatography techniques involve pre- or post-column derivatization with o-phthadehyde, dansyl chloride, and fluorenylmethyl (Karovicova and Kohajdova, 2005). Generally, dansyl chloride is the mostly common used derivatization agent of BA prior to HPLC analysis (Pinho *et al.*, 2001, Martuscelli *et al.*, 2005; Komprda *et al.*, 2008). Pre-separation procedures as well as HPLC and UPLC conditions have been used for the determination of BA in cheese, and are summarized in Table 27.5.

Ultra-performance liquid chromatography (UPLC) is a new category of liquid chromatography. It can dramatically decrease the analysis time by using small particles (<2 μm) and short columns (5 cm). Compared with the

TABLE 27.5 Pre-Separation Procedures as well as HPLC and UPLC Conditions used for Determination of BA in Cheese

Sample Extraction	Derivatization	Column	Mobil Phase	Detection	References
0.1 M HCl	DCl (40°C, 60 min)	C18 Spherisorb S3ODS2	Acetonitrile/Phosfate buffer/H_2O	HPLC/UV 254 nm	Martuscelli *et al.* (2005)
0.1 M HCl	DCl (40°C, 60 min)	C18 (150 mm × 4.6 mm, particle size 5 μm) with guard column Meta Guard ODS-2	Gradient elution, H_2O 35–0%/acetonitrile 65–100%	HPLC/UV 254 nm	Komprda *et al.* (2008)
0.2 M perchloric acid	DCl (70°C, 15 min)	C18 150 × 4.6 mm I.D. Spherisorb ODS column	Linear gradient	HPLC/UV 436 nm	Pinho *et al.* (2001)
5% TCA	—	NOVA PAK C18 (30 g × 150 mm, 5 μm)	Acetonitirle : 0.02 N acetic acid(1 : 9)/0.02 N acetic acid: acetonitrile : methanol (1 : 9 : 9)	HPLC/UV 254 nm	Elsanhoty *et al.* (2009)
0.6 N perchloric acid	Pre-column with AQC (6-aminoquinoly-N-hydroxysuccinimidyl carbamate)	NOVA PAK C18 (3.9 mm × 150 mm,4 μm)	Sodium acetate/methanol	HPLC/ Fluorescence, 395 nm	Mayer *et al.* (2010)
0.6 N perchloric acid	Pre-column with AQC (6-aminoquinoly-N-hydroxysuccinimidyl carbamate)	BEH C18 (1.7 μm, 2.1 mm × 50 mm)	Sodium acetate/methanol	UPLC/ Ultraviolet detector 254 nm	Mayer *et al.* (2010)

HPLC method, UPLC is more advantageous for determination of BA due to the shorter run time, reduced solvent consumption, higher resolution, and sensitivity (Dadakova *et al.*, 2009).

References

Bodmer, S., Imark, C., Kneubühl, M., 1999. Biogenic amines in foods: Histamine and food processing. Inflamm. Res. 48, 296–300.

Bunkova, L., Bunka, F., Mantlova, G., Cablova, A., Sedlacek, I., Svec, P., Pachlova, V., Kracmar, S., 2010. The effect of ripening and storage condition on distribution of tyramine, putracine and cadevarine in Edam-cheese. Food Microbiol. 27, 880–888.

Burdychova, R., Komprda, T., 2007. Biogenic amine-forming microbial communities in cheese. FEMS. Microbiol. Lett. 276, 149–155.

Chong, C.Y., Abu Bakar, F., Russly, A.R., Jamilah, B., Mahyudin, N.A., 2011. The effects of food processing on biogenic amines formation. Int. Food Res. J. 18, 867–876.

Dadakova, e., Krizek, M., Pelikanova, T., 2009. Determination of biogenic amines in foods using ultra-performance liquid chromatography (UPLC). Food Chem. 116, 365–370.

Durlu-Özkaya, F., Alichanidis, E., Litopoulou-tzanetakı, E., Tunail, N., 1999. Determination of biogenic amine content of Beyaz cheese and biogenic amine production ability of some lactic acid bacteria. Milchwissenschaft 54, 680–682.

Durlu-Özkaya, F., Ayhan, K., Özkan, G., 2000. Biogenic amine determination in Tulum cheese by high performance liquid chromatography (HPLC). Milchwissenschaft 55, 27–28.

Durlu-Özkaya, F., 2002. Biogenic amine content of some Turkish cheese. J. Food Process. Preservation 26, 259–265.

Elsanhoty, R.M., Mahrous, H., Ghanaimy, G.A., 2009. Chemical, microbial counts and evaluation of biogenic amines during the ripening of Egyptian soft Domiati cheese made from raw and pasteurized buffaloes milk. Int. J. Dairy Technol. 4 (2), 80–90.

El-Sayed, M.M., 1996. Biogenic amines in processed cheese available in Egypt. Int. Dairy J. 6, 1079–1086.

Gennaro, M.C., Gianotti, V., Marengo, E., Pattono, D., Turi, R.M., 2003. A chemo- metric investigation of the effect of the cheese-making process on contents of biogenic amines in a semi-hard Italian cheese (Toma). Food Chem. 82, 545–551.

Josten, H.M.J., 1988. Conditions allowing the formation of biogenic amines in cheese (3). Factors influencing the amounts formed. Neth. Milk Dairy J. 41, 329–357.

Karovicova, J., Kohajdova, Z., 2005. Biogenic amines in food. Chem. Pap. 59, 70–79.

Komprda,T.,Smela,D.,Novicka,K.,Kalhotka,L.,Sustova,K.,Pechova,P., 2007. Content and distribution of biogenic amines in Dutch-type hard cheese. Food Chem. 10, 129–137.

Komprda, T., Burdychova, R., Dohnal, V., Cwikova, O., Sladkova, P., 2008. Some factors influencing biogenic amines and polyamines content in Dutch-type semi-hard cheese. Euro. Food Res. Technol. 227, 29–36.

Linares, D.M., Martin, M.C., Ladero, V., Alvarez, M.A., Fernandez, M., 2011. Biogenic amines in dairy products. Clin. Rev. Food Sci. Nutr. 51, 691–703.

Linares, D.M., Rio, B., Ladero, V., Martinez, N., Fernandez, M., Martin, M.C., Alvarez, M.A., 2012. Factors influencing biogenic amines accumulation in dairy products. Front. Microbiol. 3, 1–10.

Marino, M., Maifreni, M., Moret, S., Rondinini, G., 2000. The capacity of *Enterobacteriaceae* species to produce biogenic amines in cheese. Lett. Appl. Microbiol. 31, 169–173.

Marino, M., Maifreni, M., Bartolomeoli, I., Rondinini, G., 2008. Evaluation of amino acid decarboxylase microbiota throughout the ripening of an Italian PDO cheese produced using different manufacturing practices. J. Appl. Microbiol. 105, 540–549.

Martuscelli, M., Gardini, F., Torriani, S., Mastrocola, D., Serio, A., Chaves-Lopez, C., Schirone, M., Suzzi, G., 2005. Production of biogenic amines during the ripening of Pecorino Abruzzese cheese. Int. Dairy J. 15, 571–578.

Mayer, H.K., Fiechter, G., Fischer, E., 2010. A new ultra-performance liquid chromatography method for determination of biogenic amines in cheese. J. Chromatogr. A. 1217, 3251–3257.

Moret, S., Conte, L., 1996. High-performance liquid chromatographic evaluation of biogenic amine in foods. An analysis of different methods of sample preparation in relation to food characteristic. J. Chromatogr. A. 729, 363–369.

Novella-Rodriguez, S., Veciana-Nogues, M.T., Trujillo-Mesa, A.J., Vidal-Carou, M.C., 2002. Profile of biogenic amines in goat cheese made from pasteurized and pressurized milks. Food Chem. Toxicol. 67 (8), 2940–2944.

Novella-Rodriguez, S., Veciana-Nogues, M.T., Roig-Sagues, A., Trujillo-Mesa, A.J., Vidal-Carou, M.C., 2004. Evaluation of biogenic amines and microbial counts throughout the ripening of goat cheeses from pasteurized and raw milk. Properties of J. Dairy Res. 71, 245–252.

Pinho, O., Ferreira, I.M.P.L.O., Mendes, E., Oliveira, B.M., Ferreira, M., 2001. Effect of temperature on evolution of free amino acid and biogenic amine contents during storage of Azeitao cheese. Food Chem. 75, 287–291.

Rodriguez-Jerez, J.J., Giaccone, V., Colavita, G., Parisi, E., 1994. *Bacillis macerans*—a new potent histamine production microorganism isolated from Italian cheese. Food Microbiol. 11, 409–415.

Roig-Sagues, A.X., Molina, A.P., Hernandez-Herrero, M.M., 2002. Histamine and tyramine-forming microorganism in Spanish traditional cheeses. Eur. Food Res. Technol. 215, 96–100.

Santos, W.C., Souza, M.R., Cerqueira, M.M.O.P., Gloria, M.B.A., 2003. Bioactive amines formation in milk by *Lactococcus* in the presence or not rennet and NaCl at 20 and 32°C. Food Chem. 81, 595–606.

Shalaby, A.R., 1996. Significance of biogenic amines to food safety and human health. Food Res. Int. 29 (7), 675–690.

Silla-Santos, M.H., 1996. Biogenic amines: their importance in foods. Int. J. Food Microbiol. 29, 213–231.

Spano, G., Russo, P., Lonvaud-Funel, A., Lucas, P., Alexandre, H., Grandvalet, C., Coton, E., Coton, M., Barnavon, L., Bach, B., Rattray, F., Bunte, A., Magni, C., Ladero, V., Alvarez, M., Fernandez, M., Lopez, P., Palencia, P.F., Corbi, A., Trip, H., Lokema, J.S., 2010. Biogenic amines in fermented foods. Eur. J. Clin. Nutr. 64, 95–100.

Şenel, E., Yıldız, F., Yetişemiyen, A., Durlu-Özkaya, F., Öztekin, F.Ş., Şanlı, E., 2012. Evaluation of the biogenic amine content and some chemical and microbiological properties of Urfa and Van herby cheeses. Kafkas Univ. Vet. Fak. Derg. 18 (4), 537–544.

Vale, S., Gloria, B.A., 1998. Biogenic amines in Brazilian cheeses. Food Chem. 63 (3), 343–348.

Valsamaki, K., Michaelidou, A., Polychroniadou, A., 2000. Biogenic amine production in Feta cheese. Food Chem. 71, 259–266.

Yıldız, F., Yetişemiyen, A., Şenel, E., Durlu-Özkaya, F., Öztekin, Ş., Şanlı, E., 2010. Some properties of Civil cheese: a type of traditional Turkish cheese. Int. J. Dairy Technol. 63 (4), 575–580.

CHAPTER

28

Changes in Volatile Compounds of Cheese

Seval Andiç, Yusuf Tunçtürk, Gökhan Boran

Department of Food Engineering, Faculty of Engineering and Architecture, Yüzüncü Yıl University, Van, Turkey

CHAPTER POINTS

- Any factor or measure affecting the chemical composition of milk used in cheese making eventually affects the formation and composition of volatile compounds in cheese.
- Fat catabolism by microbial activity and obtained volatiles seem to shape the main part of the cheese flavor and aroma in the majority of cheese varieties.
- Concentrations of the cheese volatiles are greatly affected by raw material properties, ripening conditions and storage as well as the manufacturing processes.
- The most common cheese volatiles, which are responsible for cheese flavor and aroma, can be grouped as short-chain fatty acids and their esters, aldehydes, alcohols, ketones, and sulfur compounds.

INTRODUCTION

Cheese, as a term, includes a group of fermented milk-based food products. It is one of the most important products of the dairy industry and it is also a rich source of milk nutrients (Fox and McSweeney, 2004). The milk used for cheese making comes from different species such as cow, sheep, and goat. Cheese making involves the conversion of liquid milk into cheese by coagulation.

Using basic techniques, more than a hundred cheese varieties can be produced depending on the raw material and the starter culture used and manufacturing processes employed. In addition, the nature and quality of end product is determined predominantly by the ripening conditions and the storage. Some cheeses, mainly acid-coagulated varieties, are consumed fresh, but most cheese varieties undergo a period of ripening which varies from 2 weeks to 2 years (Fox and McSweeney, 2004).

Cheese is biologically and biochemically dynamic, which is greatly dependent on cheese composition and ripening conditions. Factors including moisture content, salt concentration, pH value, cheese microflora, and storage temperature regulate and control the biochemical changes that occur during ripening, and thus determine the flavor, aroma, and texture of the end product. Flavor and texture development is largely controlled by complex biochemical reactions, with glycolytic, proteolytic, and lipolytic activities being the primary events during cheese ripening (McSweeney, 2004).

The unique flavor of a cheese variety is the result of a complex balance among volatile and non-volatile chemical compounds originating from the metabolism of milk fat, protein, carbohydrates, and citrate (Fox and Wallace, 1997). Nevertheless, non-volatile compounds do not directly determine the flavor in cheese, instead they provide a reservoir of volatile flavor precursors. Thus, non-volatile compounds have a key role in the formation of volatiles in cheese.

HOW COMPOSITION IS ALTERED

General Aspects of Volatile Formation in Cheese

Cheese flavor is primarily an outcome of enzymatic activity. The sources of the enzymes involved in cheese ripening are various and are given in Figure 28.1. Enzymes from various sources degrade macronutrients into amino acids, fatty acids, and lactic acid, which are major precursors of volatile compounds. The catabolism of free amino acids, free fatty acids, and lactic acid, and metabolism of citrate produce volatile aroma compounds which contribute to cheese flavor (Figure 28.2) (McSweeney and Sousa, 2000).

Processing and Impact on Active Components in Food
http://dx.doi.org/10.1016/B978-0-12-404699-3.00028-7

The complete and rapid metabolism of the lactose in cheese curd is essential for the production of good-quality cheese (McSweeney and Fox, 2004). Also, the production of lactic acid is essential for consistent ripening (Califano and Bevilacqua, 2000). During the manufacture of cheese, lactose is converted mainly to L-lactic acid by the starter bacteria. L-lactate is metabolized to different products by microorganisms in cheese (Figure 28.3) (McSweeney and Fox, 2004). Proteolysis is an important biochemical event that occurs during cheese ripening, which influences the taste and aroma of cheese. Proteolytic enzymes from different sources (lactic acid bacteria, the secondary starters, non-starter lactic acid bacteria, milk, and coagulants) play an important role in the degradation of casein to peptides and free amino acids (Irigoyen, *et al.*, 2007). Catabolism of free amino acids (Figure 28.4) may yield volatile aroma compounds such as aldehydes, acids, alcohols, sulfur compounds, and esters (Hemme *et al.*, 1981; McSweeney, 2004). Enzymatic hydrolysis of triglycerides to free fatty acids is another biochemical reaction in cheese ripening, which produces important precursors in the formation of several volatile compounds (Figure 28.5) (Collins *et al.*, 2003; McSweeney, 2004; Hayaloğlu and Özer, 2011).

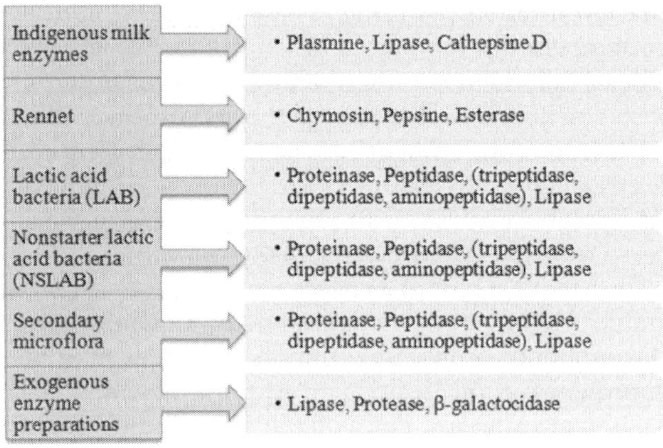

FIGURE 28.1 The sources of enzymes involved in cheese ripening.

Effects of Raw Material

Milk used in cheese manufacturing might have differences in its chemical composition depending on the origin. Some of the cheeses are specifically manufactured from ewes' milk while some are from cows' or goats' milk. Conversely, some of the cheese varieties can be manufactured from milk of any of the above-mentioned species or from their mixtures. The effect of milk source on volatile composition of cheese is obvious as reported by several researchers. For example, the total amount of volatile components in Teleme cheeses were higher in those samples made with ewes' milk compared to that of goats' milk independent from the ripening stage probably due to differences in composition and microflora of milk from different animals (Table 28.1) (Massouras *et al.*, 2006).

The concentration of fat might influence the volatile composition of the end product. In low-fat cheese, the concentration of furanones, phenylethanal, 1-octen-3-one, and free fatty acids were at higher levels compared to that of full fat cheese after 9 months of ripening, not solely due to differences in matrix and flavor release, but also differences in ripening biochemistry, which causes differences in concentrations of flavor-contributing compounds (Table 28.1) (Drake *et al.*, 2010). Kondyli *et al.* (2002) reported that Feta cheese made with low-fat milk had higher content of aldehydes compared to that of full fat Feta cheese. Controversially, alcohols, ketones, and esters were in higher concentrations in full fat Feta cheese due to differences in the fat content of milk (Table 28.1).

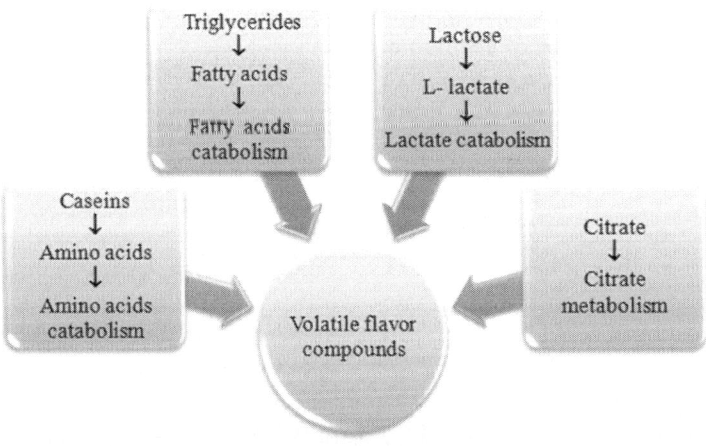

FIGURE 28.2 The sources contributed to the formation of volatile compounds.

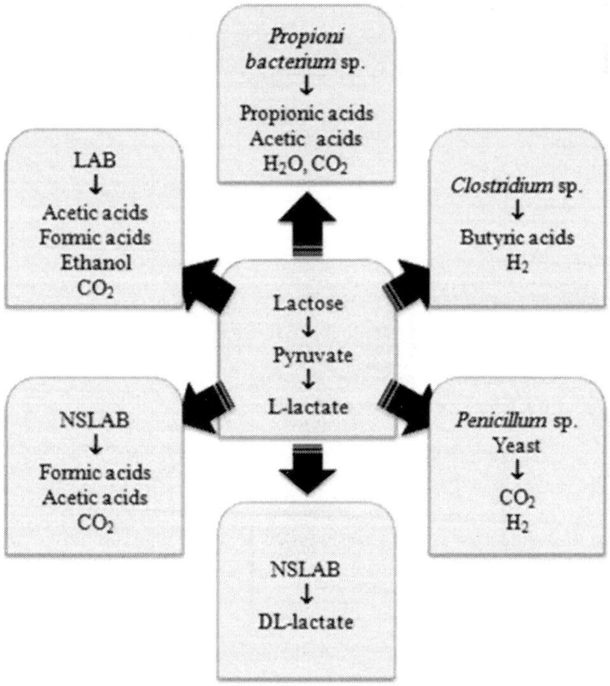

FIGURE 28.3 The groups of microorganisms involved in L-lactate metabolism. LAB, lactic acid bacteria; NSLAB, non-starter lactic acid bacteria.

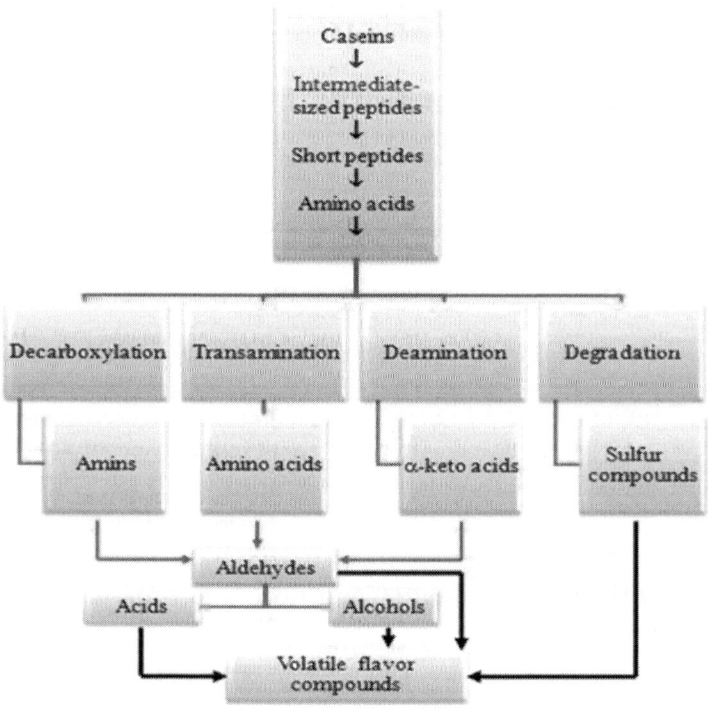

FIGURE 28.4 The formation of volatile compounds by proteolysis and amino acid catabolism.

The chemical composition of milk might differ depending on the season. Therefore, milking season might eventually affect the volatile composition of cheese. It is concluded that the differences in the microbiological and chemical composition of milk, which depend on the seasonal variations of the climatic conditions and the feeding regime, affected the volatile composition and their final concentrations in the cheese studied (Table 28.1) (Fernández-García *et al.*, 2004). Muñoz *et al.* (2003) concluded that the cheese samples manufactured in summer undergo an extensive proteolysis, which resulted in higher concentrations of free

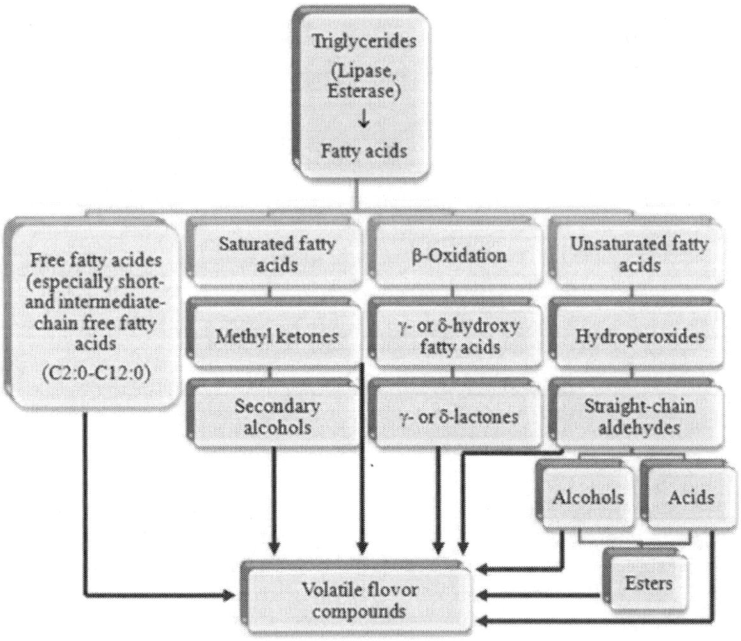

FIGURE 28.5 The formation of volatile compounds by lipolysis and free fatty acid catabolism.

amino acids and volatiles of proteolytic activity. Another raw material related issue affecting the volatile composition of cheese is the breed of milk animal. Ferreira *et al.* (2009) reported the obvious effect of the breed on volatile composition of cheese due to differences in fat content.

Effects of Manufacturing Processes

The microflora contributing to the cheese manufacturing and ripening has an essential role in formation and composition of cheese volatiles. The microflora associated with cheese ripening is extremely diverse; which might be divided into two groups: the starter lactic acid bacteria (LAB) and the secondary microflora. The starter bacteria are the primary group that is responsible for acid production while the secondary microflora does not play any active role during cheese manufacturing, but instead is involved in the ripening process. It is important to note that raw milk flora, in particular, non-starter lactic acid bacteria (NSLAB) increases the diversity of flavor and is involved in the production of typical characteristics of cheese (Steele and Ünlü, 1992). Any treatment affecting the natural microflora and enzymes of milk eventually affect the biochemical changes in cheese, which result in a great diversity of volatiles. As an example, it is reported that raw milk cheeses had higher concentrations of alcohols and esters and consequently their flavor was stronger compared to that of cheese made with pasteurized milk (Table 28.1) (Fernández-García *et al.*, 2002).

In Majorero cheese made with goats' milk and coagulated by hygienized animal rennet paste, higher concentration of volatile components were reported to be due to the addition of adjunct starter culture and lipolytic activity of rennet paste (Castillo *et al.*, 2007). It is also concluded that addition of adjunct culture may be employed in avoiding the unbalanced flavor and the loss of original aroma caused by pasteurization of cheese milk (Table 28.1). Kondyli *et al.* (2002) reported that addition of adjunct starter culture resulted in higher concentrations of acetaldehyde, butan-2-one, 3-methyl butanal and ethanol (Table 28.1). In another study reported by Rodríguez-Alonso *et al.* (2009), free fatty acids, esters, methyl ketones, alcohols and branched chains aldehydes were more abundant in raw milk cheese compared to pasteurized milk cheese due to more intense lipolysis in raw milk cheese (Table 28.1). Cheese samples made with addition of adjunct culture of heterofermentative lactobacilli strains had higher levels of volatiles except aldehydes. Addition of adjunct culture increased cheese proteolysis that was apparent with higher concentrations of free amino acids, which are precursors of volatiles formed during ripening (Table 1) (Irigoyen *et al.*, 2007).

Effects of Ripening Conditions and Storage

The ripening is exclusively carried out by the indigenous bacteria population present in milk and in the dairy environment in case of raw milk cheeses while it is performed by starter microorganisms and the secondary microflora in pasteurized milk cheeses. Microorganisms including bacteria, yeasts, and moulds are present in cheese throughout ripening and contribute to the maturation process either directly through their metabolic activity or indirectly through the release of enzymes by autolysis into the cheese matrix.

TABLE 28.1 Changes in the Concentration of Volatiles in Cheese

Affecting Factors and Volatiles	Samples				Cheese Type	Reference(s)
VARIETY OF MILK	*EWES' MILK CHEESE*		*GOATS' MILK CHEESE*			
Total volatile compounds (ARU × 10^3)	151897		102643		Teleme cheese	Massouras *et al.* (2006)
Fat reduction	**Full-fat cheese**	**Reduced-fat cheese (50% reduced-fat)**	**Low-fat cheese (6% fat containing)**			
Phenylethanal (μg/kg)	380.4	821.5	594.4		Cheddar cheese	Drake *et al.* (2010)
1-Octan-3-one (μg/kg)	ND	26.80	380.07			
Acetic acids (mg/kg)	933.5	1457	4924			
Butyric acids (mg/kg)	3201	756.7	4071			
Fat reduction	**Full-fat cheese**		**Low-fat cheese**			
Butanal (TIC)	61		116		Feta cheese	Kondyli *et al.* (2002)
3-Methyl butanal (TIC)	160		280			
Pentanal (TIC)	157		785			
Hexanal (TIC)	357		821			
Heptanal (TIC)	289		710			
Ethanol (TIC)	26686		13138			
Butan-2-ol (TIC)	6727		5189			
Pentan-2-one (TIC)	630		ND			
3-Hydroxy, butan-2-one (TIC)	1477		1237			
Ethyl acetate (TIC)	507		ND			
Seasonal variation*	**Spring**	**Summer**	**Autumn**	**Winter**		
Aldehydes and branched chain aldehydes (RA)	16.7	10.71	38.43	18.8	Castellano cheese	Fernández-García *et al.* (2004)
Branched chain alcohols (RA)	206.22	219.5	239.26	142.55		
Esters (RA)	55.86	75.517	121.59	65.64		
2-Methyl ketones and diketones (RA)	67.6	55.81	121.84	44.9		
Differences in starter and rennet	**Pasteurized milk, commercial starter and rennet added**	**Raw milk, hygienized rennet paste added**	**Pasteurized milk, specific starter and hygienized rennet paste added**			
Total volatile compounds (μg/100 g)	26760	36715	36715		Majorero cheese	Castillo *et al.* (2007)
Adjunct culture	**Low-fat cheese**	**Low fat cheese with 0.06% adjunct culture added**	**Low fat cheese with 0.09% adjunct culture added**			
Ethanol (TIC)	13138	42081	45886		Feta cheese	Kondyli *et al.* (2002)
Acetaldehyde (TIC)	280	774	781			
Butan-2-one (TIC)	409	20552	19140			
3-Methyl butanal (TIC)	280	565	585			
Heat treatment*	**Raw milk cheese**		**Pasteurized milk cheese**			
Alcohols (RA)	355.4		46.42		Manchego cheese	Fernández-García *et al.* (2002)
Esters (RA)	72.27		7.35			

(Continued)

TABLE 28.1 Changes in the Concentration of Volatiles in Cheese—cont'd

Affecting Factors and Volatiles	Samples				Cheese Type	Reference(s)
VARIETY OF MILK	*EWES' MILK CHEESE*		*GOATS' MILK CHEESE*			
Heat treatment*	**Raw milk cheese**		**Pasteurized milk cheese**			
Fatty acids (RA)	80.6		ND			
Esters (RA)	1805.2		9			
Alcohols (RA)	44597.4		7517.7		Arzúa-Ulloa and Tetilla cheese	Rodríguez-Alonso *et al.* (2009)
Branched chain aldehydes (RA)	120.2		1.2			
Methyl ketones (RA)	140.3		ND			
Adjunct culture*	**Starter culture added cheese**		**Starter and adjunct culture added cheese**			
Hydrocarbons (AU)	30839		32685			
Alcohols (AU)	30591		38516			
Aldehydes (AU)	5774		5245			
Ketones (AU)	59304		80104		Roncal-type ewes' milk cheese	Irigoyen *et al.* (2007)
Acids (AU)	20335		26771			
Esters (AU)	797		1488			
Sulfur-containing compounds (AU)	381		386			
High-pressure (HP) treatment (600 MPa)	**Control (no HP treatment)**	**HP treatment on day 1 in 60 days of ripening**	**HP treatment on day 30 in 60 days of ripening**	**HP treatment on day 50 in 60 days of ripening**		
Alcohols (ARU × 10^4)	325.6	89.3	153.4	224		
Esters (ARU × 10^4)	131.6	30.2	80.1	113.2	Raw goat' milk cheese	Delgado *et al.* (2011a)
Ketones (ARU × 10^4)	170	379.4	277.1	126.4		
Others (ARU × 10^4)	13.1	29.1	30.2	18.8		
Ripening	**Day 1**	**Day 60**	**Day 180**			
Aldehydes (ARU × 10^3)	20224	21905	13480			
Ketones (ARU × 10^3)	5069	20521	49415			
Alcohols (ARU × 10^3)	3964	36878	56845		Ewes' milk cheese	Massouras *et al.* (2006)
Acids (ARU × 10^3)	894	15100	13755			
Total volatiles (ARU × 10^3)	37512	108574	151897			
Ripening*	**Day 1**	**Day 60**	**Day 90**			
Acids (ARU × 10^4)	6303.3	11024.9	10930			
Alcohols (ARU × 10^4)	192.2	940.5	820		Raw goats' milk cheese	Delgado *et al.* (2011c)
Esters (ARU × 10^4)	147.7	321.1	499.1			
Ketones (ARU × 10^4)	70.5	302.4	144.1			
Packaging material	**Wooden barrels**		**Tin vessels**			
Aldehydes (TIC × 10^3)	4551.1		8492.3			
Ketones (TIC × 10^3)	17.685.2		7649.9		Feta cheese	Kondyli *et al.* (2012)
Esters (TIC × 10^3)	77113.1		14965.8			
Alcohols (TIC × 10^3)	110692.2		69306.7			

TABLE 28.1 Changes in the Concentration of Volatiles in Cheese—cont'd

Affecting Factors and Volatiles	Samples		Cheese Type	Reference(s)
VARIETY OF MILK	*EWES' MILK CHEESE*	*GOATS' MILK CHEESE*		
Packaging method*	**Vacuum**	**Non-vacuum**		
Aldehydes (AU)	1.24	1.88	Turkish Motal cheese	Andic *et al.* (2011)
Esters (AU)	57.12	14.94		
Acids (AU)	129.12	54.19		
Ketones (AU)	11.91	307.63		
Alcohols (AU)	9.78	2.71		
Total volatile compounds (AU)	209.17	381.35		
Packaging material*	**Polystyrene container with brine**	**Cryovac container in vacuum**		
Total volatile compounds (ARU)	69.34	55.16	Turkish White cheese	Akalın *et al.* (2011)
Storage temperature*	**+4°C**	**−18°C**		
Aldehydes (AU)	1.88	1.23	Turkish Motal cheese	Andic *et al.* (2011)
Esters (AU)	14.94	26.62		
Acids (AU)	54.19	19.87		
Ketones (AU)	307.63	3.81		
Alcohols (AU)	2.71	5.52		
Total volatile compounds (AU)	381.35	57.05		

ND, not detected; TIC, total ion chromatogram; AU, arbitrary unit; ARU, area units; RA, Relative. abundance.
* *The values calculated from the data reported.*

The total concentration of volatiles typically increases in almost all cheeses during ripening while differing in the profile. Massouras *et al.* (2006) reported that the total concentration of volatiles increased during ripening (Table 28.1). They also reported a decrease in the concentration of aldehydes in the mixture of ewes' and goats' milk cheeses from the beginning of ripening and in goats' and ewes' milk cheeses after 60 days of ripening. Aldehydes are generally accepted as transitory compounds in cheese ripening and they consequently transform into relevant acids and alcohols. Kondyli *et al.* (2012) reported a similar trend in changes of aldehydes in Feta cheese samples stored in wooden barrels (Table 28.1). In the study reported by Delgado *et al.* (2011c), the amount of acids, alcohols, and ketones of raw goats' milk cheese decreased due to the reduced microbial activity during the last month of ripening at refrigerated conditions while the amount of esters increased at every stage of ripening due to esterification of acids and alcohols (Table 28.1) (Delgado *et al.*, 2011c). The reduction of ketones might be due to their further reduction to the secondary alcohols by enzymatic activity (Bergamini *et al.*, 2010). In another study reported by Foda *et al.* (2008) on White Herby cheese, storage duration played key roles in changes in the concentration of these compounds. Some of the aldehydes increased during the storage while some were decreasing due to occurrence of biochemical reactions at different rates, which is greatly dependent on storage conditions. The herbs used in this study reduced the concentration of the main ketones, which are equally responsible for cheese aroma as the aldehydes, compared to the control during the cold storage, possibly due to antimicrobial effects of the herbs used.

The packaging material might also affect the volatile formation and concentration in cheese. Kondyli *et al.* (2012) reported the effect of packaging material on volatile profiles of Feta cheese. According to the results obtained, Feta cheese stored in wooden barrels had higher concentrations of volatile components except aldehydes compared to that of cheese stored in tin vessels (Table 28.1). Andic *et al.* (2011) studied the effects of vacuum packaging and frozen storage on formation of free fatty acids and volatiles in Motal cheese. Storage at −18°C resulted in lower amounts of volatiles compared to that of those stored at +4°C, which clearly confirms the effect of temperature on the rate of biochemical reactions. It is reported that esters and alcohols were in higher concentration in vacuum packed cheeses stored at +4°C. Besides, the concentration of ketones was higher in non-vacuum packed cheeses as a result of higher counts of moulds (Table 28.1). Akalın and Karaman (2011) reported that vacuum-packed cheese samples had higher volatile contents compared to that of cheese ripened in brine as a result of lower salt content and higher proteolysis level in vacuum packed cheese (Table 28. 1).

OTHER WAYS IN WHICH COMPOSITION IS ALTERED

Treatment of high hydrostatic pressure has significant effects on cheese volatiles as reported by Delgado *et al.* (2011a, b). According to the results obtained, pressure treatment had obvious effects on volatile profiles of cheese samples, particularly if applied at the beginning of ripening. Treatment with high hydrostatic pressure at the early stages of ripening decreased the abundance of most volatiles while enhancing the formation of ketones, hydrocarbons, and δ-decalactone (Table 28.1) (Delgado *et al.*, 2011a).

Light exposure may also affect the volatile formation in cheese. Chemical structure and color of a packaging material affects the formation of volatiles when cheese is exposed to the light (Pettersen *et al.*, 2005). Light exposure may cause a significant increment in the concentration of aldehydes while decreasing the concentration of some ketones (Sunesen *et al.*, 2002).

ANALYTICAL TECHNIQUES

There are more than 600 volatile compounds identified in different cheese varieties, many of which have been related to particular flavors of cheese (Curioni and Bosset, 2002). The volatile fractions of cheese constitute an approximate objective image of its odor and aroma. Various quantification techniques have been developed to determine the concentration of volatile compounds in cheese. The commonly used methods for separation and quantification involve GC-MS equipment while there are a variety of methods used for extraction of volatile compounds, including the purge and trap method, headspace techniques, extraction and distillation methods, and the solid-phase micro-extraction method.

The fundamental approach in all these methods is to first heat the samples and free the volatiles and then, adsorb them on to a fiber, and finally desorb volatiles in the injection port of gas chromatography equipment or is to directly inject these freed volatiles into the gas chromatography equipment. After separation of volatiles in a chromatography column, volatile compounds are identified based on comparison with peaks of standard compounds and/or data from library of mass spectrophotometer. The results can be given in arbitrary units, any concentration unit or peak area.

References

Akalın, A.S., Karaman, A.D., 2011. Influence of packaging systems on the biochemical characteristics and volatile compounds of industrially produced Turkish White cheese. J. Food Biochem. 35, 663–680.

Andic, S., Tuncturk, Y., Javidipour, I., 2011. Effects of frozen storage and vacuum packaging on free fatty acid and volatile composition of Turkish Motal cheese. Food Sci. Technol. Int. 17 (4), 375–394.

Bergamini, C.V., Wolf, I.V., Perotti, M.C., Zalazar, C.A., 2010. Characterization of biochemical changes during ripening in Argentinean sheep cheese. Small Ruminant Res. 94, 79–89.

Califano, A.N., Bevilacqua, A.E., 2000. Multivariate analysis of the organic acids content of Gouda-type cheese during ripening. J. Food Composition Anal. 13, 949–960.

Castillo, I., Calvo, M.V., Alonso, L., Juárez, M., Fontecha, J., 2007. Changes in lipolysis and volatile fraction of goat cheese manufactured employing a hygienized rennet paste and defined strain starter. Food Chem. 100, 590–598.

Collins, Y.F., McSweeney, P.L.H., Wilkinson, M.G., 2003. Lipolysis and free fatty acid catabolism in cheese: a review of current knowledge. Int. Dairy J. 13, 841–866.

Curioni, P.M.G., Bosset, J.O., 2002. Key odorants in various cheese types as determined by gas chromatography-olfactometry. Int. Dairy J. 12, 959–984.

Delgado, F.J., Gonzáles-Crespo, J., Cava, R., Ramírez, R., 2011a. Changes in the volatile profile of a raw goat milk cheese treated by hydrostatic high pressure at different stages of maturation. Int. Dairy J. 21, 135–141.

Delgado, F.J., Gonzáles-Crespo, J., Cava, R., Ramírez, R., 2011b. Effect of high-pressure treatment on the volatile profile of a mature raw goat milk cheese with paprika on rind. Innovative Food Sci. Emerg. Technol. 12, 98–103.

Delgado, F.J., Gonzáles-Crespo, J., Cava, R., Ramírez, R., 2011c. Formation of the aroma of a raw goat milk cheese during maturation analyzed by SPME-GC-MS. Food Chem. 129, 1156–1163.

Drake, M.A., Miracle, R.E., McMahon, D.J., 2010. Impact of fat reduction on flavor and flavor chemistry of Cheddar cheeses. J. Dairy Sci. 93, 5069–5081.

Fernández-García, E., Carbonell, M., Nuñez, M., 2002. Volatile fraction and sensory characteristics of Manchego cheese. 1. Comparison of raw and pasteurized milk cheese. J. Dairy Res. 69, 579–593.

Fernández-García, E., Gaya, P., Medina, M., Nuñez, M., 2004. Evolution of the volatile components of raw ewes' milk Castellano cheese: seasonal variation. Int. Dairy J. 14, 39–46.

Ferreira, I.M.P., Pinho, O., Sampaio, P., 2009. Volatile fraction of DOP "Castelo Branco" cheese: Influence of breed. Food Chem. 112, 1053–1059.

Foda, M.I., Sleet, F.L., El-Ghorab, A.H., 2008. Sensory evaluation and related volatile components of White Herby cheese. Int. J. Dairy Sci. 3 (4), 160–169.

Fox, P.F., McSweeney, P.L.H., 2004. Cheese: An Overview, thired ed. In: Fox, P.F., McSweeney, P.L.H., Cogan, T.M., Guinee, T.P. (Eds.), Cheese: Chemistry, physics and microbiology, vol. 1. Elsevier Inc, Amsterdam, pp. 1–18.

Fox, P.F., Wallace, J.M., 1997. Formation of flavour compounds in cheese. Adv. Appl. Microbiol. 45, 17–85.

Hayaloğlu, A., Özer, B., 2011. Peynirde Olgunlaşma. In: Hayaloğlu, A., Özer, B. (Eds.), Peynir Biliminin Temelleri. Sidas Ltd, İzmir, pp. 173–209.

Hemme, D.H., Schmal, V., Auclair, J.E., 1981. Effect of the addition of extracts of thermophilic lactobacilli on acid production by *Streptococcus thermophilus* in milk. J. Dairy Res. 51, 345–351.

Irigoyen, A., Ortigosa, M., Juansaras, I., Oneca, M., Torre, P., 2007. Influence of an adjunct culture of *Lactobacillus* on the free amino acids and volatile compounds in a Roncal-type ewe's-milk cheese. Food Chem. 100, 71–80.

Kondyli, E., Katsiari, M.C., Masouras, T., Voutsinas, L.P., 2002. Free fatty acids and volatile compounds of low-fat Feta-type cheese made with a commercial adjunct culture. Food Chem. 79, 199–205.

Kondyli, E., Pappa, E. C., and Valchou, A. M., 2012. Effect of package type on the composition and volatile compounds of Feta Cheese. Small Ruminant Res. 108, 95–101.

Massouras, T., Pappa, E.C., Mallatou, H., 2006. Headspace analysis of volatile flavor compounds of teleme cheese made from sheep and goat milk. Int. J. Dairy Technol. 59 (4), 250–256.

McSweeney, P.L.H., 2004. Biochemistry of Cheese Ripening: Introduction and Overview. third ed. In: Fox, P.F., McSweeney, P.L.H., Cogan, T.M., Guinee, T.P. (Eds.), Cheese: Chemistry, physics and microbiology, vol. 1. Elsevier Inc, Amsterdam, pp. 347–360.

McSweeney, P.L.H., Sousa, M.J., 2000. Biochemical phatways for the production of flavour compounds in cheese during ripening: a review. Le lait 80 (3), 293–324.

McSweeney, P.L.H., Fox, P.R., 2004. Metabolism of residual lactose and of lactate and citrate, third ed. In: Fox, P.F., McSweeney, P.L.H., Cogan, T.M., Guinee, T.P. (Eds.), Chemistry, physics and microbiology, vol. 1. Elsevier Inc, Amsterdam, pp. 361–371.

Muñoz, N., Ortigosa, M., Torre, P., Izco, J.M., 2003. Free amino acids and volatile compounds in an ewe's milk cheese as affected by seasonal and cheese-making plant variations. Food Chem. 83, 329–338.

Pettersen, M.K., Thomas, E., Nilsson, A., 2005. Oxidative stability of cream cheese stored in thermoformed trays as affected by packaging material, drawing depth and light. Int. Dairy J. 15, 355–362.

Rodríguez- Alonso, P., Centeno, J.A., Garabal, J.I., 2009. Comparison of the volatile profiles of Arzúa-Ulloa and Tetilla cheeses manufactured from raw and pasteurized milk. LWT—Food Sci. Technol. 42, 1722–1728.

Steele, J.L., Ünlü, G., 1992. Impact of lactic acid bacteria on cheese flavour development. Food Technol. 46, 128–135.

Sunesen, L.O., Lund, P., Sorensen, J., Holmer, G., 2002. Development of volatile compounds in processed cheese during storage. Lebensmittel Wissenschaft Technol. 35, 128–134.

SECTION 4

OILS FATS SPREADS

CHAPTER

29

Oxidation Products of Corn Oil at Room Temperature

Encarnación Goicoechea, María D. Guillén

Department of Food Technology, Faculty of Pharmacy, Lascaray Research Center, University of the Basque Country (UPV/EHU), Vitoria, Spain

CHAPTER POINTS

- In corn oil samples stored at room temperature in closed receptacles a great variety of compounds are formed.
- Samples with a low level of oxidation contain hydroperoxides and (*Z*,*E*)-conjugated double bonds.
- Samples with intermediate oxidation levels, in addition to the above-mentioned hydroperoxides, contain hydroxy-derivatives having (*Z*,*E*)-conjugated dienic systems.
- Samples with a high oxidation level contain hydroperoxides having (*E*,*E*)-conjugated double bonds and mono- and di-epoxides, in addition to the previously mentioned compounds.
- Oxidized corn oil also contains aldehydes, principally alkanals, (*E*)-2-alkenals, and 4-hydroxy-(*E*)-2-alkenals, followed by (*E*,*E*)-2,4-alkadienals and 4-hydroperoxy-(*E*)-2-alkenals, with 4,5-epoxy-(*E*)-2-alkenals in very small proportions.
- The headspace of corn with a low oxidation level contains just a few compounds, and that of samples having a high oxidation level contains, in addition to some of the above-mentioned aldehydes, acids, alcohols, ketones, alkylfurans, and lactones, together with toxic mono- and polycyclic aromatic hydrocarbons (PAHs) of low molecular weight.
- The fact that 4-oxo-(*E*)-2-alkenals, detected in the oil headspace by solid phase microextraction–gas chromatography/mass spectrometry, were absent in the oil liquid matrix studied by proton nuclear magnetic resonance, and on the contrary 4-hydroperoxy-(*E*)-2-alkenals, detected in the oil liquid matrix, were absent in the headspace of the oil suggest that the first compounds do not really exist in the oil but are formed in the chromatographic run from the second ones.

INTRODUCTION

The most important cause of oil deterioration is oxidation, a process which provokes a reduction not only of its nutritional and sensorial value, but also of the food product in which it is contained, producing compounds which are potentially toxic for human health. During this process very complex reactions take place as a result of the incorporation of oxygen into the triglyceride structures and the generation of very reactive unstable new compounds. This can cause the breakdown of the fatty acyl group chains, producing volatile or not volatile compounds and bring about the reaction between different acyl group chains to give oligomeric or polymeric structures. The nature and concentration of the compounds generated in this complex process depend on several factors, mainly the proportions of the different acyl groups present in the oil (the higher the unsaturation degree, the lower the oxidative stability), but also on the temperature, time, air/oil contact surface ratio, etc.

Lipid oxidation occurs not only during technological or culinary processes at high temperatures, but also during storage at low temperatures. A large number of studies on lipid oxidation during storage have employed the high-temperature Rancimat, or Oxidative Stability

Processing and Impact on Active Components in Food
http://dx.doi.org/10.1016/B978-0-12-404699-3.00029-9

Index (OSI) tests (100–130°C) (Gordon and Mursi, 1994; Tasan *et al.*, 2011) or the Schaal oven test (60°) (Snyder *et al.*, 1985). However, their results can not be extrapolated because the mechanism of lipid oxidation depends on the conditions under which the process takes place (Frankel, 2005). Food lipid oxidation at room temperature is very slow, which may explain why there are only a few studies on this process (Martin-Polvillo *et al.*, 2004; Guillén and Goicoechea, 2007, 2008a).

In this context, the aim of this chapter is to review current knowledge about the oxidation products that are generated in corn oil when stored at room temperature. According to the Codex Alimentarius the composition, expressed as a percentage of the total content of fatty acids, of corn oils can vary in the ranges 34.0–65.6 of linoleic (L%), 20.0–42.2 of oleic (O%), 8.6–22.5 of saturated (S%), and 0.0–2.0 of linolenic (Ln%) (Codex Stan 210-1999). Despite the fact that this oil is often used for frying and that it is an ingredient in many manufactured products, the changes that occur in its composition during oxidation have been little studied. Furthermore, the oxidation products identified at higher temperatures will also be reviewed, together with the analytical techniques employed.

HOW COMPOSITION IS ALTERED

Different reactions can alter the composition of corn oil submitted to storage for prolonged periods of time in the presence of air. Among these can be cited: reactions leading to the formation of oxygenated functional groups in the acyl group chains of triglycerides accompanied or not by the formation of conjugated dienic systems; reactions leading to the formation of dimers, oligomers, or polymers of triglycerides supporting modified or unmodified acyl groups; and breakdown reactions leading to the formation of compounds of low molecular weight and truncated acyl group chains, these latter supported on monomers, dimers, or polymers of triglycerides.

As a consequence, many different compounds can contribute to the alteration of the corn oil composition, some of which are discussed below.

New Oxygenated Functional Groups Supported on Acyl Group Chains

In relation to these kinds of functional groups it has been proved that hydroperoxides, hydroxides, epoxides, or ketones, associated or not with conjugated dienic systems, can be formed. Figure 29.1 shows some of the possible structures of this kind, derived from linoleic, the main acyl group in corn oil, and identified in the oxidation mechanisms of linoleic acid and methyl linoleate under varied oxidative conditions (Aerts and Jacobs, 2004; Schneider *et al.*, 2004, 2005, 2008; Sun and Salomon, 2004; Zhang *et al.*, 2006).

The presence of some of these oxidation products was detected by ^{1}H NMR in several corn oil samples stored for prolonged periods of time at room temperature in closed receptacles with different air/oil ratios, and with different oxidation levels (Guillén and Goicoechea, 2009). The concentration of these new compounds

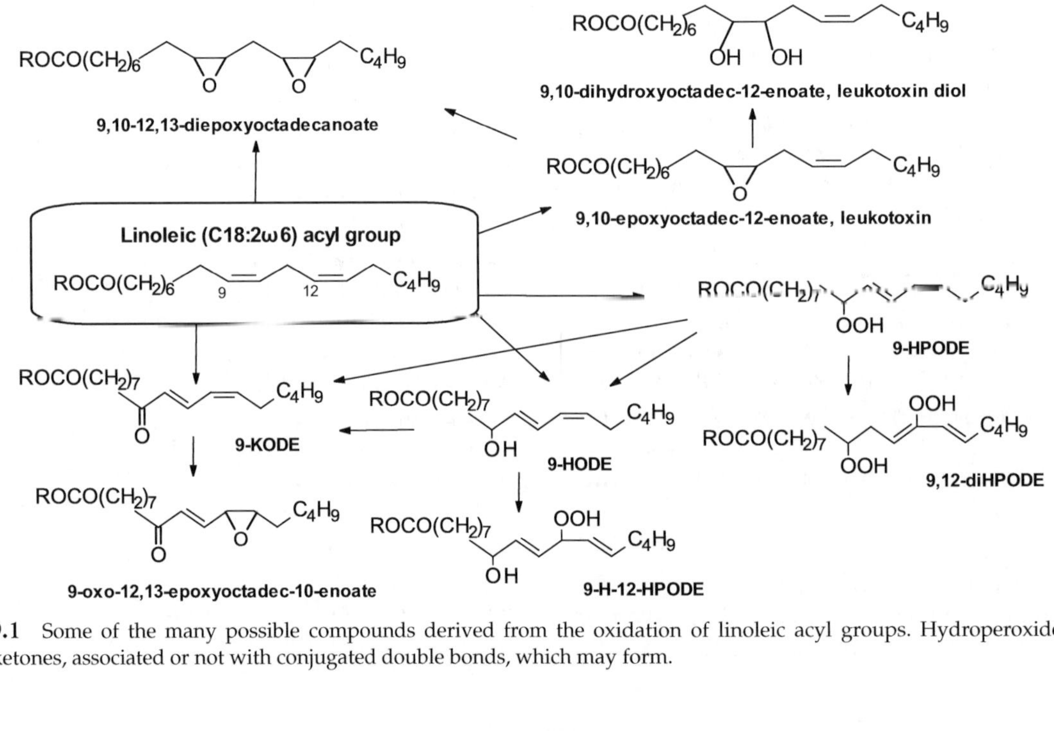

FIGURE 29.1 Some of the many possible compounds derived from the oxidation of linoleic acyl groups. Hydroperoxides, hydroxides, epoxides, or ketones, associated or not with conjugated double bonds, which may form.

rose in line with the decrease in unsaturated acyl groups, mainly linoleic.

In those stored corn oil samples with low oxidation levels *hydroperoxy-(Z,E)-conjugated dienes* were detected by the above-mentioned technique. Figure 29.1 shows, as an example, one of the possible structures of these oxidation products (see 9-HPODE). In corn oil samples with an intermediate oxidation level, in addition to hydroperoxy-(*Z*,*E*)-conjugated dienes, the presence of *hydroxy-(Z,E)-conjugated dienic systems* was detected, which could be associated with the limited amount of oxygen in contact with the oil during oxidation. As far as we know, the formation in vegetable oils of these kinds of products with hydroxy groups has only been described in stored sunflower oils and seeds (Mikolajczak *et al.*, 1968; Guillén and Goicoechea, 2007). An example of one structure of this type is given in Figure 29.1 (see 9-HODE), which could be formed directly from the linoleic acyl group or from the corresponding hydroperoxide (see 9-HPODE). The hydroxides 9-HODE and 13-HODE derived from linoleic groups are considered to be very good markers of lipid peroxidation *in vivo* and are the main compounds generated in the oxidation of human low-density lipoproteins (LDL), which is the precondition for atherosclerosis (Yoshida and Niki, 2010).

Furthermore, in stored corn oil samples with high oxidation levels, as well as the functional groups mentioned, the presence of *hydroperoxy-(E,E)-conjugated dienic systems* was also detected, their intensity increasing along with the oxidation level of the samples. According to other authors, hydroperoxy-(*Z*,*E*)-conjugated dienic systems are considered to be formed under kinetic control at low temperatures (Schneider, 2009). However, with prolonged autoxidation or at higher temperatures, the double-bond configuration of the conjugated hydroperoxides rearranges from (*Z*,*E*) to (*E*,*E*). The apparent driving force for this transformation is the greater thermodynamic stability of the (*E*,*E*)-conjugated double bond in comparison to the (*Z*,*E*). Hydroperoxy-(*E*,*E*)-conjugated dienic systems could be formed at the expense of the (*Z*,*E*), and could therefore be considered to form under thermodynamic control.

Quantitative study of the ^{1}H NMR spectra evidenced that the concentration of hydroperoxy-(*Z*,*E*)-conjugated dienic systems was always higher than that of hydroperoxy-(*E*,*E*)-conjugated dienic systems, and both kinds of hydroperoxy derivatives always showed higher concentrations than those of hydroxy derivatives. Moreover, it was observed that the concentration in all samples of hydroperoxy groups was higher than that of conjugated dienic systems, which indicates either the existence of hydroperoxy groups in acyl chains in which there are no conjugated dienic systems or the existence of acyl chains having two hydroperoxy groups and one conjugated dienic system, in agreement with other authors (Schneider *et al.*, 2004, 2005, 2008; Sun and Salomon, 2004; Zhang *et al.*, 2006).

Furthermore, the presence of chains supporting one and two *epoxy groups* was also observed (Guillén and Goicoechea, 2009; Goicoechea and Guillén, 2010). Among these compounds it is worth noting the presence of *9,10-epoxyoctadec-12-enoate* groups (see Figure 29.1), whose hydrolysis gives *9,10-epoxyoctadec-12-enoic acid* or *leukotoxin* (LTX); this latter compound is so named as it is able to be formed endogenously in cells and tissues, causing degeneration and necrosis of leukocytes, among other toxic effects (Greene and Hammock, 1999).

In addition to these, other oxygenated functional groups can also be supported on truncated acyl groups, which will be dealt with in the following paragraphs. As already mentioned, all these modified chains could be supported on monomers, dimers, oligomers, or polymers of triglycerides.

Oxidation Products Coming from Breakdown of Acyl Chains

Some of the previously described structures can evolve to give other new ones. They can break down, generating compounds of low molecular weight and truncated acyl groups both able to support different functional groups. Among these, aldehydes, ketones, acids, alkylfurans, hydrocarbons, and lactones can be mentioned, some of them related to rancid flavor and/or toxic effects. Figure 29.2 shows some structures of low molecular weight of these types of oxidation products that can be derived from linoleic acyl groups, Figure 29.3 shows those from oleic and Figure 29.4 from linolenic.

Among all these kinds of compounds, *aldehydes* supported on both either small molecules or truncated acyl groups of triglycerides can be determined in stored corn oil by ^{1}H NMR spectroscopy (Guillén and Goicoechea, 2009). So, in oxidized corn oil samples at not very advanced oxidation levels *(E)-2-alkenals, 4-hydroxy-(E)-2-alkenals* and *alkanals* were detected, showing similar concentrations; from this, it can be inferred that the reactions that produce these compounds are closely related. In these samples, *4-hydroperoxy-(E)-2-alkenals* and *(E,E)-2,4-alkadienals* were also detected but in lower concentrations. Nevertheless, in the most oxidized sample *4-hydroxy-(E)-2-alkenals* and *(E)-2-alkenals* showed the highest concentrations and *4,5-epoxy-(E)-2-alkenals* were also identified. It is worth noting the presence of the above-mentioned oxygenated α,β-unsaturated compounds, which are considered as genotoxic and cytotoxic, able to cause cancer, atherosclerosis, Parkinson's or Alzheimer's disease. They are generated in the degradation of omega-3 and omega-6 polyunsaturated fatty acids and esters, both in biological systems and foodstuffs and can be absorbed through the diet (Guillén and Goicoechea, 2008b).

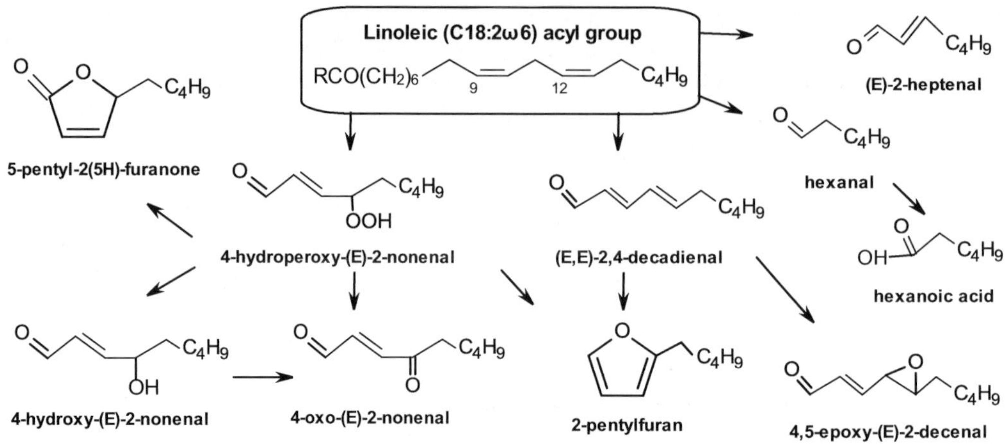

FIGURE 29.2 Some of the many possible compounds, of low molecular weight, which may form during the breakdown of oxidized linoleic acyl groups. Not only aldehydes, but also acids, alkylfurans, and alkylfuranones, among others, can be generated.

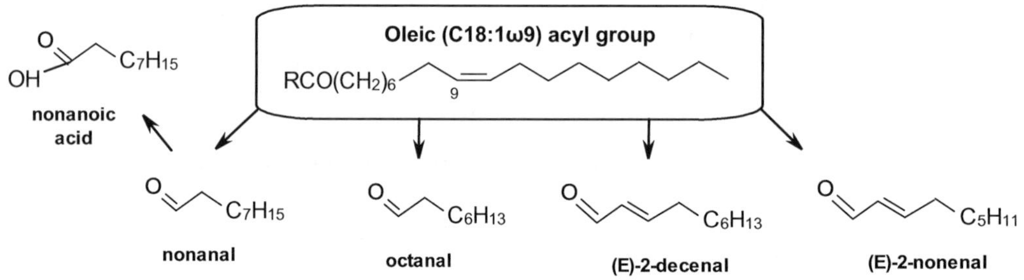

FIGURE 29.3 Some oxidation compounds, of low molecular weight, which may form during the breakdown of oxidized oleic acyl groups. Not only aldehydes, but also acids, among others, can be generated.

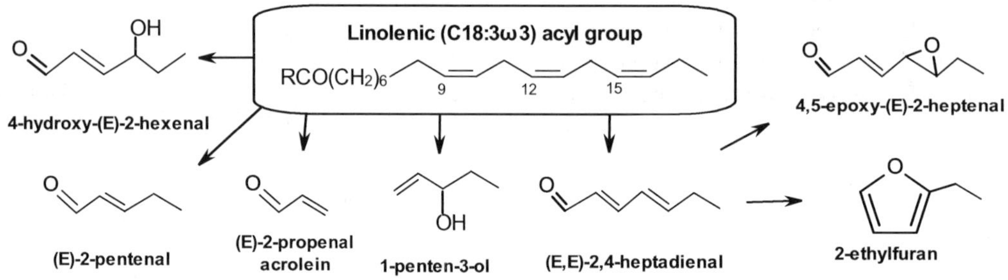

FIGURE 29.4 Some of the many possible oxidation compounds, of low molecular weight, which may form during the breakdown of oxidized linolenic acyl groups. Not only aldehydes, but also alkylfurans and alcohols, among others, can be generated.

The compounds of low molecular weight formed in breakdown reactions can be studied by SPME-GC/MS. Thus, using this technique the compounds present in the headspace of stored corn oil samples for prolonged periods of time were studied, and around 140 volatile oxidation compounds were identified (Goicoechea and Guillén, unpublished results). In samples with a low oxidation level only a few compounds were identified, and in those with a high oxidation level the presence of typical volatile oxidation compounds was evidenced, including *aldehydes*, *acids*, *alcohols*, *ketones*, *hydrocarbons*, *alkylfurans*, and *lactones*. Table 29.1 shows some of the most significant compounds found. In agreement with corn oil composition, those volatiles derived from linoleic acyl groups showed the highest abundances.

It is remarkable that aldehydes were the most numerous group of volatile compounds, but acids showed the highest abundances. This fact can be explained by the storage conditions, which give time for the oxidation of aldehydes into acids and their accumulation inside the receptacle, in agreement with Wu and Chen (1992). Related to this fact, high abundances of lactones were observed, probably coming in part from the cyclization of acids.

According to the results obtained by ^1H NMR, several toxic oxygenated α,β-unsaturated aldehydes were

TABLE 29.1 Some of the Most Significant Volatile Compounds which were Identified by Solid Phase Microextraction–Gas Chromatography/Mass Spectrometry in the Headspace of Corn Oil Samples Stored at Room Temperature and which had High Oxidation Level

Family*	Compound
Acids (10)	Hexanoic Formic Acetic
Alcohols (8)	1-Octen-3-ol 1-Pentanol 1-Penten-3-ol
Esters (3)	Pentylhexanoate
Hydrocarbons (40)	Pentane Octane Naphthalene Methylnaphthalene Toluene
Ketones (12)	2-Heptanone 3-Nonen-2-one
Aldehydes (51)	Hexanal Pentanal Nonanal (*E*)-2-Octenal (*E*)-2-Heptenal (*E*)-2-Decenal (*E*)-2-Undecenal (*E*)-2-Butenal 2,4-Decadienal 2,4-Heptadienal 2,3-Epoxyoctanal 4-Oxononanal 4-Hydroxy-(*E*)-2-nonenal 4,5-Epoxy-(*E*)-2-decenal 4-Oxo-(*E*)-2-nonenal 4-Hydroxy-(*E*)-2-hexenal 4,5-Epoxy-(*E*)-2-heptenal
Lactones (9)	5-Pentyl-2(5H)-furanone
Alkylfurans (7)	2-Pentylfuran 2-Butylfuran 2-Ethylfuran

In samples with a low oxidation level only a few compounds were identified, and in those with a high oxidation level the presence of typical volatile oxidation compounds was evidenced, including aldehydes, acids, alcohols, ketones, hydrocarbons, alkylfurans and lactones.

* *In brackets the total number of volatile compounds of each family.*

From Goicoechea and Guillén (unpublished results).

detected by SPME-GC/MS, such as *4-hydroxy-(E)-2-nonenal* and *4,5-epoxy-2-decenals* resulting from the oxidation of omega-6 polyunsaturated acyl groups (linoleic), and in much lower proportions *4-hydroxy-(E)-2-hexenal* and *4,5-epoxy-2-heptenal* resulting from omega-3 polyunsaturated acyl groups (linolenic). However, *4-hydroperoxy-(E)-2-alkenals* detected by ^{1}H NMR in the liquid matrix of the oxidized oils were not found in their headspaces, and the opposite is true for *4-oxo-(E)-2-alkenals*, the reason for which reason could be that these latter compounds are not present in the oil but are formed by being heated in the chromatographic run, in agreement with results of other authors (Gardner and Grove, 1998). It is also worth noting the detection of another group of toxic compounds in the headspace of stored corn oil samples. These were mono- and poly-cyclic aromatic hydrocarbons (PAHs) of low molecular weight, such as toluene, naphthalene, and fluorene, whose abundance rose in line with the oxidation level of the samples. The formation of these cyclic compounds at room temperature in closed receptacles could be favored by the presence of free unsaturated conjugated radicals in the oil and the pressure in the closed system.

The detection of all these potentially toxic compounds in corn oil samples stored at room temperature helps to explain the results obtained by other authors in previous papers, in which after oral administration to rats of corn oil oxidized at room temperature for 48 months, tumor promotion and initiation was found in all the organs investigated (Perjesi *et al.*, 2002).

OTHER WAYS IN WHICH COMPOSITION IS ALTERED

The composition of corn oil can be altered by the effects of oxygen, heat, and pro-oxidants; however, the modification of its composition has only been studied under certain conditions. Not many studies can be found regarding the oxidation products generated in corn oil at higher temperatures. In the study by GC/MS of the headspace of corn oil submitted to accelerated storage conditions (60°C, 8 days) 28 volatile compounds were identified, mainly hexanal and pentane (Snyder *et al.*, 1985).

Significant differences can be found when comparing the nature and proportions of the above-mentioned oxidation products detected by ^{1}H NMR in stored corn oil samples (Guillén and Goicoechea, 2009) with those found in corn oil samples submitted to 70°C with aeration (Guillén and Ruiz, 2005). Corn oil oxidized at 70°C did not contain hydroxy-(*Z,E*)-conjugated dienic systems; this fact could be related to the limited concentration of air in the stored oil. In addition, in corn oil oxidized at 70°C both hydroperoxy-(*Z,E*)- and -(*E,E*)-conjugated dienic systems were present in all samples, with higher concentrations of hydroperoxy-(*E,E*)- than hydroperoxy-(*Z,E*)-conjugated dienic systems. This fact could be associated with the higher temperature to which these samples were submitted. These differences indicate that both processes evolve in different ways. As for the aldehydes detected by ^{1}H NMR, the main difference is that the concentration of 4-hydroxy-(*E*)-2-alkenals in samples stored at room temperature was always much higher than that of 4-hydroperoxy-(*E*)-2-alkenals. However, in the samples oxidized at 70°C the concentration of 4-hydroperoxy-(*E*)-2-alkenals was initially higher than that of 4-hydroxy-(*E*)-2-alkenals, and as the process evolved the concentration of the former decreased and that of the latter increased. As mentioned, this could be due either to the limited concentration of oxygen in the oxidation at room temperature, which could favor the formation of hydroxy- over that of hydroperoxy-aldehydes, or to the presence of hydroxy-(*Z,E*)-conjugated dienes amongst the oxidation products, which can evolve directly to give 4-hydroxy-(*E*)-2-alkenals, as has been described (Schneider *et al.*, 2004).

Naz *et al.* (2004) studied the oxidative deterioration of stored corn oil (air/light, 30°C, 30 days) and after its use for deep-frying at 180°C. However, no structural identification of the oxidation products was carried out, because the measurements employed were Peroxide Value, Anisidine Value, and Iodine Value. Macku and Shibamoto (1991) identified by GC/MS a total number of 52 volatile compounds in the headspace of corn oil submitted to 180°C for 4h, mainly hexanal and octane.

Great similarities were found by ^{1}H NMR between the composition of corn oil samples submitted to 190°C with aeration in a convection oven (Guillén and Ruiz, 2008) and to 190°C in a microwave oven (Guillén and Ruiz, 2006). In these samples significant concentrations of hydroperoxides and dienic systems were absent, showing that at these high temperatures either they were not generated or if so, they degraded too fast to be detected. Regarding aldehydes, the most abundant were (*E*)-2-alkenals, (*E,E*)-2,4-alkadienals and alkanals, and very low proportions of 4-hydroxy-(*E*)-2-alkenals and 4,5-epoxy-2-alkenals; 4-hydroperoxy-(*E*)-2-alkenals were not detected.

Wu and Chen (1992) submitted corn oil to stir frying (200°C, 3min) and afterwards stored it (55°C, 30 weeks). In the headspace of these samples a total number of 49 volatile compounds were identified by GC/MS, the main one before storage being (*E,E*)-2,4-decadienal and (*E*)-2-heptenal, and after storage hexanoic acid and (*E*)-2-heptenal. Simultaneously the Peroxide Value of the samples was measured, but the authors concluded that these values did not reflect the changes of volatile content in the headspace.

ANALYTICAL TECHNIQUES

In general, most of the studies on oils oxidation are carried out by classical chemical parameters (Wu and Chen, 1992, Naz *et al.*, 2004; Martin-Polvillo *et al.*, 2004), such as: Peroxide Value, which gives the concentration of hydroperoxy groups in the sample; the absorbance at 234nm in ultraviolet spectroscopy, which provides information about the concentration in the oil of conjugated dienes; Iodine Value, which informs about the number of double bonds present in the oil; Anisidine Value, which gives information about the concentration of certain secondary carbonylic compounds; or the 2-thiobarbituric acid (TBA) test which, though initially supposed to react only with the secondary oxidation compound malonaldehyde, has been recently found to also react with many other substances, that is why this method now is called 2-Thiobarbituric Acid Reactive Substances (TBARS) test. All these classical methods inform on the concentration of the compounds monitorable by the corresponding test, giving a partial characterization of the oxidative status of the oil. This is because hydroperoxydes and conjugated dienic systems can be found between primary and secondary oxidation compounds, and in addition because each oil has a different capacity to generate certain compounds. In addition these parameters do not inform on the identity and concentration of the different kinds of compounds having the same functional group (Wu and Chen, 1992; Frankel, 2005).

Chromatographic techniques like GC together with different detectors (MS, Flame Ionization Detector, etc.) are useful for a more detailed description and identification of the oxidation products, because they can provide qualitative and quantitative information (Snyder *et al.*, 1985; Wu and Chen, 1992; Perjesi *et al.*, 2002); however, they require previous extraction techniques. Among the techniques employed for the extraction of volatile compounds it is worth noting the advantages of SPME, which is simple, fast and solvent-free, and is able to extract compounds of very different nature, including the toxic oxygenated α,β-unsaturated aldehydes (Guillén and Goicoechea, 2008a; Guillén and Uriarte, 2012).

In recent years there has also been evidence of the usefulness of spectroscopic techniques, like Fourier transform infrared spectroscopy (FTIR) and ^{1}H NMR, for the characterization of food lipids and the study of their oxidation process under different conditions. These techniques are fast and simple, require no chemical modification of the sample subject of study, and provide information on the oil sample as a whole. By means of FTIR information related to the functional groups is obtained, and ^{1}H NMR gives information about the various types of protons present in the liquid phase of the sample. From these sources, very important information can be obtained regarding oxidation products, as described above (Guillén and Cabo, 1999; Guillén and Ruiz, 2005; Guillén and Goicoechea, 2007, 2009).

LIST OF ABBREVIATIONS

9-HPODE 9-hydroperoxyoctadeca-10,12-dienoate
9-HODE 9-hydroxyoctadeca-10,12-dienoate
9,12-diHPODE 9,12-dihydroperoxyoctadeca-11,13-dienoate

9-H-12-HPODE 9-hydroxy-12-hydroperoxyoctadeca-10,13-dienoate
9-KODE 9-oxooctadeca-10,12-dienoate
GC/MS Gas chromatography/ mass spectrometry
^{1}H NMR Proton nuclear magnetic resonance
SPME Solid phase microextraction

Acknowledgment

This work has been supported by the Spanish Ministry of Science and Innovation (MICINN, AGL2009-09904, AGL2012-36466), the Basque Government (EJ-GV, GIC10/85-IT-463-10), and the University of the Basque Country (UPV/EHU, UFI-11/21).

References

Aerts, H.A.J., Jacobs, P.A., 2004. Epoxide yield determination of oils and fatty acid methyl esters using ^{1}H NMR. J. Am. Oil Chem. Soc. 81, 841–846.

Codex Stan 210-1999. Standard for named vegetable oils. Codex alimentarius.

Frankel, E.N., 2005. In: The Oily Press (Ed.), Lipid Oxidation. second ed. Davis, United States of America.

Gardner, H.W., Grove, M.J., 1998. Soybean lipoxygenase-1 oxidizes 3Z-nonenal A route to 4S-hydroperoxy-2E-nonenal and related products. Plant Physiol. 1359–1366.

Goicoechea, E., Guillén, M.D., 2010. Analysis of hydroperoxides, aldehydes and epoxides by ^{1}H Nuclear Magnetic Resonance in sunflower oil oxidized at 70 and 100 °C. J. Agr. Food Chem. 58 (10), 6234–6245.

Gordon, M.H., Mursi, E.A., 1994. Comparison of oil stability based on the Metrohm Rancimat with storage at 20°C. J. Am. Oil Chem. Soc. 71 (6), 649–651.

Greene, J.F., Hammock, B.D., 1999. Toxicity of linoleic acid metabolites. Adv. Exp. Med. Biol. 469 (4), 471–477.

Guillén, M.D., Cabo, N., 1999. Usefulness of the frequency data of the Fourier transform infrared spectra to evaluate the degree of oxidation of edible oils. J. Agr. Food Chem. 47, 709–719.

Guillén, M.D., Goicoechea, E., 2007. Detection of primary and secondary oxidation products by Fourier Transform infrared spectroscopy (FTIR) and Proton Nuclear Magnetic Resonance (^{1}H NMR) in sun flower oil during the storage. J. Agr. Food Chem. 55, 10729–10736.

Guillén, M.D., Goicoechea, E., 2008a. Formation of oxygenated alpha, beta-unsaturated aldehydes and other toxic compounds in sunflower oil oxidation at room temperature in closed receptacles. Food Chem. 111 (1), 157–164.

Guillén, M.D., Goicoechea, E., 2008b. Toxic oxygenated α,β-unsaturated aldehydes and their study in foods. A review. Crit. Rev. Food Sci. Nutr. 48 (2), 119–136.

Guillén, M.D., Goicoechea, E., 2009. Oxidation of corn oil at room temperature: Primary and secondary oxidation products and determination of their concentration in the oil liquid matrix from ^{1}H nuclear magnetic resonance data. Food Chem. 116 (1), 183–192.

Guillén, M.D., Ruiz, A., 2005. Oxidation process of oils having high content in linoleic acyl groups and formation of toxic hydroperoxy- and hydroxy-alkenals. A study by ^{1}H Nuclear Magnetic Resonance. J. Sci. Food Agr. 85, 2413–2420.

Guillén, M.D., Ruiz, A., 2006. Study by means of ^{1}H nuclear magnetic resonance of the oxidation process undergone by edible oils of different natures submitted to microwave action. Food Chem. 96 (4), 665–674.

Guillén, M.D., Ruiz, A., 2008. Monitoring of heat-induced degradation of edible oils by proton NMR. Eur. J. Lipid Sci. Tech. 110 (1), 52–60.

Guillén, M.D., Uriarte, P.S., 2012. Aldehydes contained in edible oils of a very different nature after prolonged heating at frying temperature: Presence of toxic oxygenated α,β unsaturated aldehydes. Food Chem. 131 (3), 915–926.

Macku, C., Shibamoto, T., 1991. Headspace volatile compounds formed from heated corn oil and corn oil with glycine. J Agr Food Chem. 39 (7), 1265–1269.

Martin-Polvillo, M., Marquez-Ruiz, G., Dobarganes, M.C., 2004. Oxidative stability of sunflower oils differing in unsaturation degree during long-term storage at room temperature. J. Am.Oil Chem. Soc. 81 (6), 577–583.

Mikolajczak, K.L., Freidinger, R.M., Smith Jr., C.R., Wolff, I.A., 1968. Oxygenated fatty acids of oil from sunflower seeds after prolonged storage. Lipids 3 (6), 489–494.

Naz, S., Siddiqi, R., Sheikh, H., Sayeed, S.A., 2004. Volume Date 2005. Deterioration of olive, corn and soybean oils due to air, light, heat and deep-frying. Food Res. Int. 38 (2), 127–134.

Perjesi, P., Pinter, Z., Gyongyi, Z., Ember, I., 2002. Effect of rancid corn oil on some onco/suppressor gene expressions *in vivo*. A short-term study. Anticancer Res. 22 (1A), 225–230.

Schneider, C., Porter, N.A., Brash, A.R., 2004. Autoxidative transformation of chiral omega-6 hydroxy linoleic and arachidonic acids to chiral 4-Hydroxy-2E-nonenal. Chem. Res. Toxicol 17 (7), 937–941.

Schneider, C., Boeglin, W.E., Yin, H., Ste, D.F., Hachey, D.L., Porter, N.A., Brash, A.R., 2005. Synthesis of dihydroperoxides of linoleic and linolenic acids and studies on their transformation to 4-hydroperoxynonenal. Lipids 40 (11), 1155–1162.

Schneider, C., Porter, N.A., Brash, A.R., 2008. Routes to 4-hydroxynonenal: Fundamental issues in the mechanisms of lipid peroxidation. J. Biol. Chem. 283 (23), 15539–15543.

Schneider, C., 2009. An update on products and mechanisms of lipid peroxidation. Mol. Nutr. Food Res. 53 (3), 315–321.

Snyder, J.M., Frankel, E.N., Selke, E., 1985. Capillary gas chromatographic analyses of headspace volatiles from vegetable oils. J. Am. Oil Chem. Soc. 62 (12), 1675–1679.

Sun, M., Salomon, R.G., 2004. Oxidative Fragmentation of Hydroxy Octadecadienoates Generates Biologically Active ganma-Hydroxyalkenals. J. Am. Chem. Soc. 126 (18), 5699–5708.

Tasan, M., Gecgel, U., Demirci, M., 2011. Effects of storage and industrial oilseed extraction methods on the quality and stability characteristics of crude sunflower oil (*Helianthus annuus L.*). Grasas Aceites 62 (4), 389–398.

Wu, C.M., Chen, S.Y., 1992. Volatile compounds in oils after deep frying or stir frying and subsequent storage. J. Am. Oil Chem. Soc. 69 (9), 858–865.

Yoshida, Y., Niki, E., 2010. Hydroxyoctadecadienoic acid (HODE) as a marker of linoleic acid oxidation. In: Aldini, Giancarlo (Ed.), Biomarkers for Antioxidant Defense and Oxidative Damage, pp. 85–97.

Zhang, W., Sun, M., Salomon, R.G., 2006. Preparative singlet oxygenation of linoleate provides doubly allylic dihydroperoxides: putative intermediates in the generation of biologically active aldehydes in vivo. J. Org. Chem. 71 (15), 5607–5615.

CHAPTER

30

Aldehydes after Prolonged Heating at Frying Temperature

Andrea Martínez-Yusta, Encarnación Goicoechea, Maria D. Guillén

Department of Food Technology, Faculty of Pharmacy, Lascaray Research Center, University of the Basque Country (UPV/EHU), Vitoria, Spain

CHAPTER POINTS

- The heating of edible oils at frying temperature provokes the formation of aldehydes from the beginning of the process.
- The aldehydes formed can be alkanals, (*E*)-2-alkenals, and 2,4-alkadienals, oxygenated saturated and α, β-unsaturated aldehydes supported on truncated acyl groups of triglycerides or on small molecules.
- Special attention should be paid to the presence in oils of oxygenated α, β-unsaturated aldehydes, as they are considered potential causal agents of several degenerative diseases.
- The nature and concentration of the aldehydes present in oils after heating at high temperature depend on the composition in acyl groups of the oil and on the conditions under which the heating process takes place.
- Solid phase microextraction-Gas chromatography/mass spectrometry and Proton nuclear magnetic resonance are suitable techniques to determine volatile aldehydes in oil headspace and total aldehydes in oil respectively.

INTRODUCTION

Frying is one of the most widely used cooking methods around the world, not only in the home but also in the food industry, because of its simplicity, speed, and the desirable sensorial properties that fried food develops. Nevertheless, it is widely known that heating oils at frying temperatures provokes oil degradation and the formation of new compounds, some of which are toxic.

For this reason, the thermo-degradation of edible oils and fats is a subject of great interest, as are the complex reactions that take place and the different factors that affect them, such as oil nature, temperature, time, aeration, etc. Furthermore, the compounds formed in this process modify not only the quality of the original oil, in nutritional, sensorial, or technological terms, but also, in some cases, its safety. Aldehydes are considered among the most important of the broad range of compounds which may be formed in this process.

In this context, the aim of this chapter is to review current knowledge concerning the aldehydes present in edible oils after their submission to heating at frying temperatures, paying special attention to how lipid composition is altered to give rise to these compounds, to the factors that affect this formation, their role in the evolution of oil quality and safety, and the analytical techniques employed in their study.

HOW COMPOSITION IS ALTERED

Evolution of the Oil Degradation Process at Frying Temperature

When oils or fats are submitted to high temperature a thermo-degradative process begins and affects their composition. The study, by ^1H NMR, of the evolution of this process indicates that it is very different to that

Processing and Impact on Active Components in Food
http://dx.doi.org/10.1016/B978-0-12-404699-3.00030-5

which occurs at lower temperatures. As Figure 30.1 shows, the first new compounds, formed from triglycerides at high temperatures and detectable by the above mentioned technique, are aldehydes (Guillén and Ruiz, 2006, 2008), and unlike what occurs in degradation at lower temperatures neither intermediate hydroperoxides nor conjugated dienic systems in acyl groups are detected. This can be observed in Figure 30.2, which shows the spectral region of the ^{1}H NMR proton signals of hydroperoxides, conjugated dienic systems and aldehydes of sunflower oil heated at 190°C and 70°C over different periods of time (Guillén and Ruiz, 2005b; Guillén and Uriarte, 2009, 2012c; Goicoechea and Guillén, 2010). During the evolution of this oil at 190°C only the signals of aldehydes appear, whereas when evolving at 70°C, hydroperoxide signals and those of *Z,E* and *E,E* conjugated dienic systems are generated prior to the formation of aldehydes. This fact indicates that hydroperoxides and conjugated double bonds associated to primary oxidation compounds, are either not formed at frying temperature or if formed, degrade very rapidly and are not present in the oil in sufficient concentrations to be detected by ^{1}H NMR. These results have been corroborated in the above mentioned studies and are in agreement with the absence of significant changes in the classical Peroxide Value of the oils submitted to frying temperature (Dobarganes and Pérez-Camino, 1988), and with the observation of other authors concerning the formation of aldehydes in oils heated at high temperature without observing previous induction periods (Silvagni *et al.*, 2010).

Main Factors Influencing the Presence of Aldehydes in Oils Submitted to High Temperatures

The nature and concentration of aldehydes contained in oils after heating at high temperatures depends on many factors, some of the most important being:

1. *The oil composition in acyl groups,* which, as will be commented on later, determines not only the nature of the aldehydes formed, but also their concentration (Belitz *et al.*, 2004; Guillén and Uriarte, 2012a,b,c,d; Juita *et al.*, 2012).
2. *The conditions under which the heating* is carried out (convection, conduction, and infrared or microwave radiation) (Guillén and Ruiz, 2006, 2008; Guillén and Uriarte, 2012b).
3. *The oil volume* involved and *the air/oil contact surface* during the heating process: the first being small and the second being large, both favor the formation of aldehydes (Guillén and Ruiz, 2008; Guillén and Uriarte, 2009, 2012c).
4. *The amount of oxygen present in the proximity* of the oil, which is related to the air/oil contact surface (Pai *et al.*, 1979; Guillén and Ruiz, 2008; Guillén and Uriarte, 2009, 2012c).

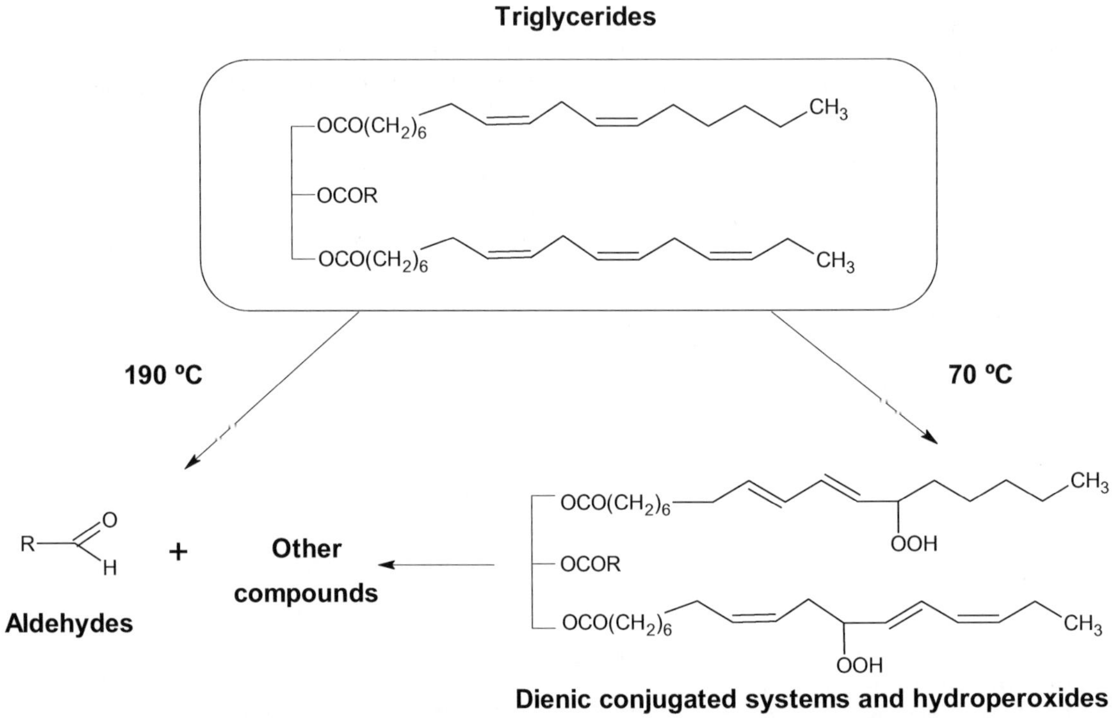

FIGURE 30.1 Scheme of a possible pathway of aldehyde formation during oil degradation at 190°C and 70°C. At 190°C the first new compounds formed from triglycerides and detected by Proton nuclear magnetic resonance are aldehydes. However, at 70°C before aldehydes, intermediate hydroperoxides and conjugated dienic systems are also detected.

5. *The heating temperature* and *the heating time*, influencing not only the amount of aldehydes but also their nature (Song *et al.*, 2011; Guillén and Uriarte, 2012b).
6. *The boiling point* of the aldehydes formed, because this determines whether they escape into the atmosphere or if they remain in the oil (Guillén and Uriarte, 2012b).
7. *The presence either of antioxidants or of prooxidants* (Guillén and Ruiz, 2008).
8. *The presence of foods* in the system, because the aldehydes coming from oil can either react with food components having amine groups through the Maillard reaction, or can be absorbed in certain foods. Furthermore, new aldehydes coming from the food, if this contains lipids, can be added to the oil (Takeoka *et al.*, 1995; Seppanen and Csallany, 2004; Procida, *et al.*, 2009).

Types of Aldehydes that may be Present in Heated Oils

The aldehydes formed in high temperature oil degradation are very numerous and can be classified in different groups:

1. By *size*. This functional group can either be supported on *small molecules*, (examples are all indicated in Figure 30.3a), or *on truncated acyl groups* of triglycerides, these latter being in monomer, dimer, oligomer, or polymer form (an example is indicated in Figure 30.3b).
2. By the *presence of other functional groups* (see Figure 30.3). In addition to *alkanals*, *(E)-2-alkenals*, *alkadienals* and *alkatrienals*, which are the most frequently detected, *oxygenated saturated* and *α,β-unsaturated aldehydes* (OαβUAs) can also be generated. The group of oxygenated saturated aldehydes includes *2,3-epoxyalkanals* and *4-oxoalkanals*, which were described for the first time in the nineties (Takeoka *et al.*, 1995; Buettner and Schieberle, 2001). The group of OαβUAs includes *4,5-epoxy-2-alkenals*, which have long been well known, *4-oxo-(E)-2-alkenals* and *4-hydroxy-(E)-2-alkenals*; some aldehydes of these two latter types have only very recently been found in oxidized and heated oils (Seppanen and Csallany, 2004, 2006; Guillén *et al.*, 2005; Guillén and Ruiz, 2004, 2005a,b,c, 2006, 2008; Guillén and Goicoechea, 2008; Goicoechea and Guillén, 2010; Guillén and Uriarte, 2009, 2012a,b,c,d). All these types of aldehydes could also be supported on truncated acyl groups of triglycerides.

It should be mentioned that not all the aldehydes described above are present in all oils heated at frying temperature. Some aldehydes come from only one or two specific acyl groups and their presence in determined oils after heating depends on the concentration of these acyl groups in the oil. For this reason some aldehydes will never be present in certain heated edible oils if the acyl groups from which they come are absent or in low concentrations in the original oil. Pioneer studies on oil oxidation associated the formation of some alkanals, alkenals, and alkadienals to certain acyl

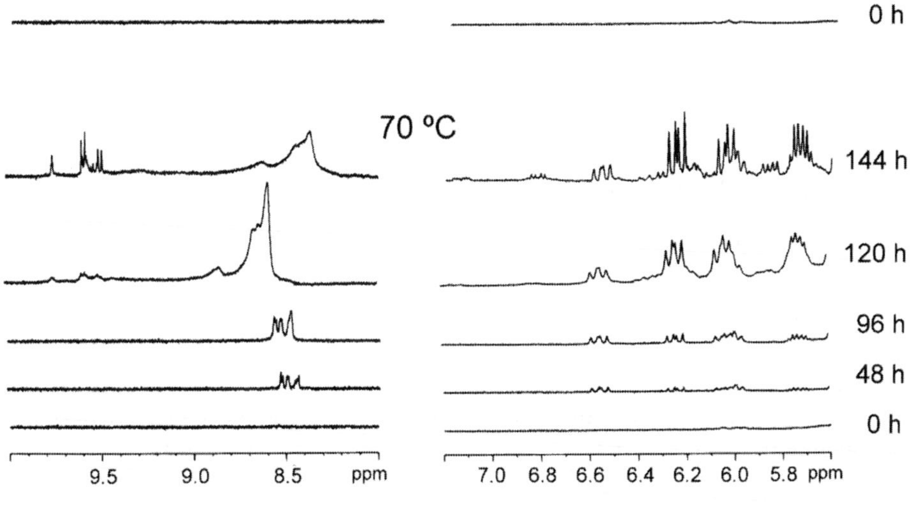

FIGURE 30.2 Two expanded regions of the Proton nuclear magnetic resonance spectra of sunflower oil heated at 190°C and 70°C. Region between 8.0 and 10.0 ppm in which hydroperoxides (8.3–9.0 ppm) and aldehydes (9.3–9.9 ppm) protons signals appear. Region between 5.5 and 7.2 ppm in which protons of dienic conjugated systems appear.

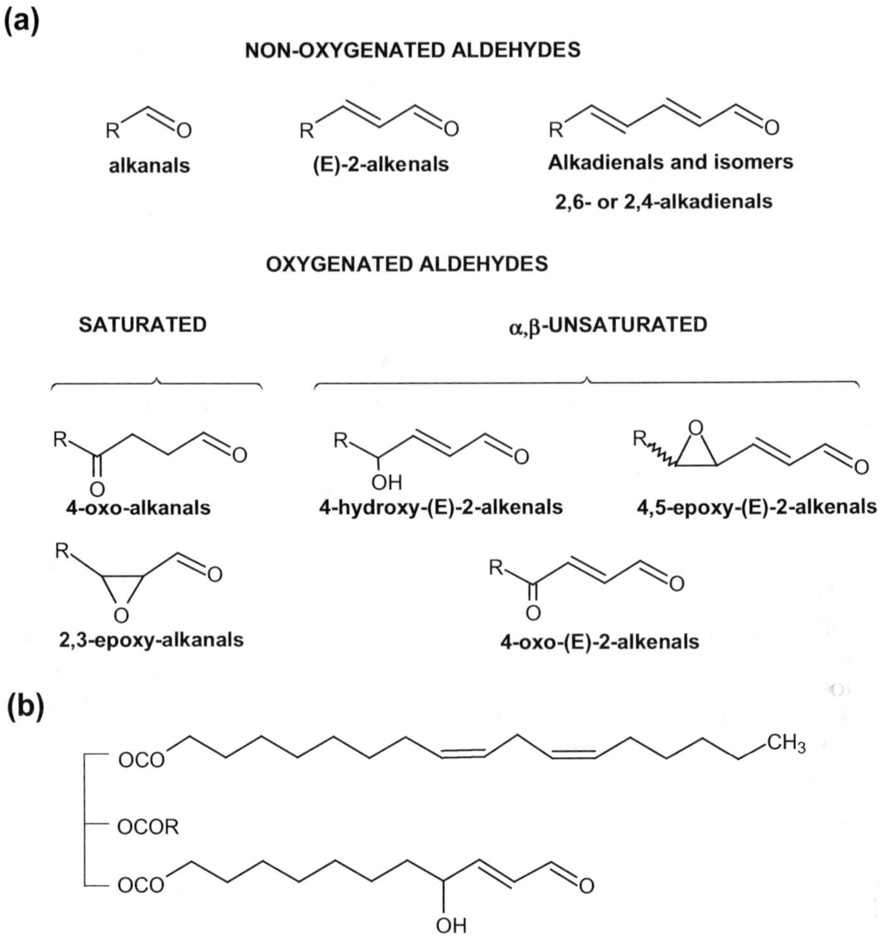

FIGURE 30.3 Types of aldehydes that can be generated in oils at frying temperatures. (a) Examples of aldehydes supported on small molecules. (b) Example of aldehyde supported on a truncated acyl group of triglycerides.

groups (Belitz *et al.*, 2004). Recently, close relationships have been found between abundance of all kinds of aldehydes in the headspace of heated oils and the molar percentages of certain acyl groups in the original oil (Guillén and Uriarte, 2012b), in agreement with other studies (Juita *et al.*, 2012).

Changes in Oil Characteristics due to the Presence of Aldehydes Resulting from Oil Degradation at High Temperature

As a consequence of oil degradation at high temperatures, in addition to aldehydes many other products are formed and together they contribute to the global changes observed in oil properties. Some of the characteristics that can be influenced by the formation of aldehydes are:

1. *Oil sensorial properties.* In general, the odor of most n-alkanals, (*E*)-2-alkenals and 2,3-epoxyalkanals has been described as pleasant; 4,5-epoxy-2-decenal is even used to provide metallic notes to foods. Only some alkadienals have been described as having fishy or rancid notes; to the best of our knowledge, the odor of other oxygenated saturated aldehydes or OαβUAs has not been described yet. In any case, the odor of heated oil, to which not only aldehydes but all compounds formed as consequence of the oil degradation contribute, becomes more unpleasant as the heating time increases.
2. *Oil reactivity.* The presence of aldehydes in oil modifies its global reactivity. It is well known that n-alkanals are less reactive than (*E*)-2-alkenals and these latter less reactive than (*E,E*)-2,4-alkadienals (Sheerin *et al.*, 1997). The most reactive of all are the OαβUAs; their reactivity is due to the presence in the molecule of at least three different functional groups or reactive sites. It is also well known that aldehydes present in the oil can react during frying with amine groups of amino acids or of proteins of foods through the Maillard reaction, which modifies the texture, color, odor, and taste of the fried foods.
3. *Oil safety.* It decreases as the concentration of aldehydes increases, and this is especially important in the case of OαβUAs, which are very reactive and have proven biological activity. These aldehydes

are today considered as potential causal agents of several diseases, such as different types of cancer, atherogenesis, diabetes, Alzheimer's or Parkinson's, among others (Zarkovic, 2003, Guillén and Goicoechea, 2008). Some of them have been found in cells and tissues and can also be formed in the oxidation and thermo-degradation of oils at high temperatures. It is noteworthy that the formation of these aldehydes, when oils rich in omega-6 or in omega-3 are submitted to frying temperature, begins in the first minutes of heating (Guillén and Uriarte, 2012b).

OTHER WAYS IN WHICH COMPOSITION IS ALTERED

In addition to high temperature, aldehydes can also be formed in oils whenever they are in contact with air for a certain period of time or whenever they are heated at any temperature. In these cases the rate of formation of aldehydes will be also a function of the nature of the lipids involved, the air/oil contact surface, the oil volume, the time during which the food lipids are maintained under degradative conditions, the presence or absence of pro- or anti-oxidants, etc. On the other hand, the absence of oxygen, low temperatures, and the presence of antioxidants hinder both the degradation of oils and fats and also the formation of aldehydes in any type of processing and in storage.

However, it should be mentioned that a certain amount of aldehydes is desirable in some foods because they contribute to the food flavor, although their concentration should not overtake certain levels, in order to avoid undesirable flavors and to guarantee the quality and safety of the food.

ANALYTICAL TECHNIQUES

The analytical techniques used to determine the occurrence of aldehydes in heated oils are very varied and their use depends on the aim of the study. They are indicated in Table 30.1 and can be summarized as follow:

1. The study of the *aldehydes supported on truncated acyl groups of triglycerides* requires saponification, methylation and fractionation of the sample and the analysis of the polar fraction of FAMEs by GC/MS. The aldehydes found in unused and used frying oils are saturated C8 and C9, and mono-unsaturated C10, C11, and C12 acyl groups supporting the aldehyde functional group (Kamal-Eldin *et al.*, 1997) (see Table 30.1). The influence of the oil nature on the formation of these aldehydes was not considered and only the first two were detected in all studied oils.
2. The methods used in the determination of *the volatile aldehydes contained in edible oils after being submitted to high temperature for a certain time* require a laborious preparation of the sample that usually involves distillation or distillation-extraction followed by other steps, to finally identify and quantify the aldehydes by GC or GC/MS (see Table 30.1). One of the studies dealt with the determination of *alkanals, alkenals,* and *(E,E)-2,4-alkadienals* in coconut oil submitted to heating for 48 h at 180°C in vacuo and in air (Pai *et al.*, 1979). Another study referred to the volatile *alkanals, alkenals,* and *alkadienals* contained in soybean oil, after heating at 200°C for 1 h, in both deep-frying and stir-frying conditions, and in corn oil and lard under stir-frying conditions (Wu and Chen, 1992).
3. Some studies are only focused on a reduced number of *OαβUAs* contained in heated oil. The formation of these aldehydes in oil degradation has only recently been discovered, for this reason they have rarely been studied. As Table 30.1 shows, the incorporation of the toxic *4-hydroxy-(E)-2-nonenal* into food fried in thermally degraded soybean oil has been studied, as well as the effect of intermittent and continuous heating at frying temperature on the formation of four *4-hydroxy-(E)-2-alkenals* in soybean oil (Seppanen and Csallany, 2004, 2006). More recently, the effect of temperature on the formation of *4-hydroxy-(E)-2-nonenal* in three oils has been analyzed (Han and Csallany, 2008). The method used to determine these aldehydes involves a very laborious methodology that includes derivatization with DNPH, TLC, extraction and analysis by HPLC-UV/Vis.
4. A much simpler process is the study of *the volatile aldehydes contained in the headspace of the oils* after submission to high temperature. In recent years the authors have used SPME for extraction, which requires neither modification of the sample nor the use of solvents, followed by GC/MS. In this case a thorough study of the chromatograms is necessary in order to extract all the information included in them (see Table 30.1). Song *et al.*, (2011) have studied the effect of both heating temperature (100–180°C, 2 h) and heating time (1–6 h, 140°C), on the amount of beef flavour precursors in tallow headspace, such as *alkanals, alkenals,* and *alkadienals.* Likewise, the evolution of the abundance of the volatile aldehydes contained in the headspace of three oils of different nature during heating at frying temperature for prolonged periods have been studied (Guillén and Uriarte, 2012b). In this study not only *alkanals, alkenals,* and *alkadienals* but also *oxygenated saturated aldehydes* and *OαβUAs* were identified and quantified. It is worth noting the identification, for the first time, of some of the latter cited aldehydes as coming from thermal degradation of edible oils.

TABLE 30.1 Summary of Studies on the Occurrence of Aldehydes in Frying Oils

Aldehydes Subject of Study	Oil or Fat	T^a (°C)	Used Receptacle	Air/Oil Contact Surface	Oil Volume	Heating Source	Sample Preparation	Analytical Technique	Reference
NVA in oil	High oleic sunflower, sunflower and partially hydrogenated rapeseed/palm	180	Continuous industrial fryer	$2\,m^2$	4001	Electric resistance	Saponification, Methylation, Extraction	GC/MS	Kamal-Eldin *et al.*, 1997
VA in oil	Coconut	180	Tubes	25.4 mm Ø	25 g	Oil bath	Distillation under vacuum, Fractionation steps, Derivatization with DNPH	GC/MS	Pai *et al.*, 1979
	Soybean Soybean, corn and lard	200	Flask (3 l) Wok (40 cm Ø)	n.d. n.d.	900 g 300 g	Silicon-oil bath, n.d.	Distillation-Extraction, Drying, Concentration	GC/MS	Wu and Chen, 1992
HNE and three other homologues in oil	Soybean	185	Fryer (5 l) Open test tubes	n.d. 12 mm	2600 g 5 g	Electric resistance Sand bath	Derivatization with DNPH, Extraction, Centrifugation, TLC, Extraction	HPLC/UV/VIS	Seppanen and Csallany, 2006
HNE in oil	Soybean	185	Beaker (1 l)	n.d.	300 g	Hotplate	Derivatization with DNPH, Extraction, Centrifugation, TLC, Extraction	HPLC/UV/VIS	Seppanen and Csallany, 2004
	Soybean, corn and butter	190	Open test tubes	25 mm	3 g	Glycerol bath	Derivatization with DNPH, Extraction, Centrifugation, TLC, Extraction	HPLC/UV/VIS	Han and Csallany, 2008
VA in oil headspace	Beef tallow	100–180	Flask (250 ml)	n.d.	100 g	Thermostatic oil bath	SPME	GC/MS	Song *et al.*, 2011
	Extra virgin olive, sunflower and virgin linseed	190	Discontinuous industrial fryer	12 × 30 × 17 cm	41	Electric resistance	SPME	GC/MS	Guillén and Uriarte, 2012b
VA + NVA in oil	Virgin olive, corn and linseed	190	Crystal Petri dishes	80 mm Ø	10 g	Microwave oven	n.r.	^{1}H NMR	Guillén and Ruiz, 2006
	Virgin olive, olive, hazelnut, corn, sunflower, soybean and linseed	190	Crystal Petri dishes	80 mm Ø	10 g	Convection oven	n.r.	^{1}H NMR	Guillén and Ruiz, 2008
	Sunflower	190	Domestic fryer	20.5 × 23.5 × 13.5 cm	21	Electric resistance	n.r.	^{1}H NMR	Guillén and Uriarte, 2009
	Extra virgin olive, sunflower and virgin linseed	190	Discontinuous industrial fryer	12 × 30 × 17 cm	41	Electric resistance	n.r.	^{1}H NMR	Guillén and Uriarte, 2012a,c,d

HNE, 4-hydroxy-(E)-2-nonenal; n.d., no data; n.r., not required; NVA, non-volatile aldehydes; VA, volatile aldehydes. Aldehydes were the subject of study in several heated oils under the conditions indicated, together with the preparation required and the analytical technique used.

TABLE 30.2 Concentration (mmol/l) of Total and of Each Kind of Aldehydes Determined by Proton Nuclear Magnetic Resonance in Virgin Linseed, Sunflower, and Extra Virgin Olive Oil, after 20h of Heating at 190°C in a Discontinuous Industrial Fryer (data from Guillén and Uriarte, 2012d).

Group of Aldehydes	Virgin Linseed Oil	Sunflower Oil	Extra Virgin Olive Oil
Total aldehydes	12.00	20.68	15.06
(*E*,*E*)-2,4-alkadienals	4.52	7.16	1.75
(*E*)-2-alkenals	3.72	4.48	7.08
4-hydroxy-(*E*)-2-alkenals	–	1.45	–
(*Z*,*E*)-2,4-alkadienals	tr	1.54	–
n-Alkanals	3.54	4.05	5.68
4-oxo-alkanals	–	0.89	1.04

The great influence of oil composition not only on the nature, but also on the concentration of the aldehydes contained is noteworthy.

5. Finally, the study of the evolution of the concentration of the different kinds of aldehydes (volatile and non-volatile) in various oils during heating at 190°C for prolonged periods of time, has been carried out by ^{1}H NMR (Guillén and Ruiz, 2006, 2008; Guillén and Uriarte, 2009; 2012a,c,d) (see Table 30.1). This technique takes very few minutes and does not require preparation of the sample, which does not suffer modifications. The groups of aldehydes that can be determined, if their concentration is high enough to be detected by this technique, include *n-alkanals, (E)-2-alkenals, (E,E)-2,4-alkadienals, (Z,E)-2,4-alkadienals, 4,5-epoxy-2-alkenals, 4-hydroxy (E)-2-alkenals, 4-hydroperoxy-(E)-2-alkenals, 4-oxo-alkanals and 4-oxo-(E)-2-alkenals*. An additional advantage is that the evolution of the degradation of the oil acyl groups and the formation of other new compounds can also be simultaneously analyzed. In this way, the influence of temperature, of the oil nature, of the presence of antioxidants, of the heat source, of the heating time, and of the air/oil contact surface have been considered. Table 30.2 shows the concentration of the different groups of aldehydes detected by ^{1}H NMR in oils of different nature after 20h of heating at 190°C in an industrial fryer; the great influence of the oil composition not only on the nature but also on the concentration of the aldehydes contained is noteworthy.

LIST OF ABBREVIATIONS

DNPH 2,4-dinitrophenylhydrazine
FAME Fatty acid methyl ester
GC/MS Gas chromatography/mass spectrometry
HPLC High performance liquid chromatography
^{1}H NMR Proton nuclear magnetic resonance
OαβUAs Oxygenated α,β-unsaturated aldehydes
SPME Solid phase microextraction
TLC Thin layer chromatography
UV/VIS Ultraviolet/visible spectroscopy

Acknowledgments

This work has been supported by the Spanish Ministry of Economy and Competitiveness (MINECO, AGL2012-36466), the Basque Government (EJ-GV, GIC10/IT-463-10), and the UPV/EHU (UFI-11/21). Andrea Martínez-Yusta thanks the EJ-GV for a predoctoral fellowship.

References

Belitz, H.D., Grosch, W., Schieberle, P., 2004. Lipids. In Food Chemistry (3rd revised ed.), Springer-Verlag, Berlin, Germany, pp. 201.

Buettner, A., Schieberle, P., 2001. Aroma Properties of a Homologous Series of 2,3-Epoxyalkanals and trans-4,5-Epoxyalk-2-enals. J. Agr. Food Chem. 49 (8), 3881–3884.

Dobarganes, M.C., Perez-Camino, M.C., 1988. Fatty acid composition: a useful tool for the determination of alteration level in heated fats. Rev. Franç. Corps Gras 35 (2), 67–70.

Guillén, M.D., Cabo, N., Ibargoitia, M.L., Ruiz, A., 2005. Study of both sunflower oil and its headspace throughout the oxidation process. Occurrence in the headspace of toxic oxygenated aldehydes. J. Agr. Food Chem. 53 (4), 1093–1101.

Guillén, M.D., Goicoechea, E., 2008. Toxic Oxygenated α, β-Unsaturated Aldehydes and their Study in Foods: A Review. Crit. Rev. Food Sci. 48 (2), 119–136.

Goicoechea, E., Guillén, M.D., 2010. Analysis of Hydroperoxides, Aldehydes and Epoxides by ^{1}H Nuclear Magnetic Resonance in Sunflower Oil Oxidized at 70 and 100°C. J. Agr. Food Chem. 58 (10), 6234–6245.

Guillén, M.D., Ruiz, A., 2004. Formation of hydroperoxy- and hydroxyalkenals during thermal oxidative degradation of sesame oil monitored by proton NMR. Eur. J. Lipid Sci. Tech 106 (10), 680–687.

Guillén, M.D., Ruiz, A., 2005a. Study by proton nuclear magnetic resonance of the thermal oxidation of oils rich in oleic acyl groups. J. Am. Oil Chem. Soc. 82 (5), 349–355.

Guillén, M.D., Ruiz, A., 2005b. Oxidation process of oils with high content of linoleic acyl groups and formation of toxic hydroperoxy- and hydroxyalkenals. A study by ^{1}H nuclear magnetic resonance. J. Sci. Food Agr. 85 (14), 2413–2420.

Guillén, M.D., Ruiz, A., 2005c. Monitoring the oxidation of unsaturated oils and formation of oxygenated aldehydes by proton NMR. Eur. J. Lipid Sci. Tech 107 (1), 36–47.

Guillén, M.D., Ruiz, A., 2006. Study by means of ^{1}H nuclear magnetic resonance of the oxidation process undergone by edible oils of different natures submitted to microwave action. Food Chem. 96 (4), 665–674.

Guillén, M.D., Ruiz, A., 2008. Monitoring of heat-induced degradation of edible oils by proton NMR. Eur. J. Lipid Sci. Tech 110 (1), 52–60.

Guillén, M.D., Uriarte, P.S., 2009. Contribution to further understanding of the evolution of sunflower oil submitted to frying temperature in a domestic fryer: Study by ^{1}H nuclear magnetic resonance. J. Agr. Food Chem. 57 (17), 7790–7799.

Guillén, M.D., Uriarte, P.S., 2012a. Study by ^{1}H NMR spectroscopy of the evolution of extra virgin olive oil composition submitted to frying temperature in an industrial fryer for a prolonged period of time. Food Chem. 134 (1), 162–172.

Guillén, M.D., Uriarte, P.S., 2012b. Aldehydes contained in edible oils of a very different nature after prolonged heating at frying temperature: Presence of toxic oxygenated α,β unsaturated aldehydes. Food Chem. 131 (3), 915–926.

Guillén, M.D., Uriarte, P.S., 2012c. Simultaneous control of the evolution of the percentage in weight of polar compounds, iodine value, acyl groups proportions and aldehydes concentrations in sunflower oil submitted to frying temperature in an industrial fryer. Food Control 24 (1–2), 50–56.

Guillén, M.D., Uriarte, P.S., 2012d. Monitoring by ^{1}H nuclear magnetic resonance of the changes in the composition of virgin linseed oil heated at frying temperature. Comparison with the evolution of other edible oils. Food Control 28 (1), 59–68.

Han, I.H., Csallany, A.S., 2008. Temperature dependence of HNE formation in vegetable oils and butter oil. J. Am. Oil Chem. Soc. 85 (8), 777–782.

Juita, J., Dlugogorski, B.Z., Kennedy, E.M., Mackie, J.C., 2012. Identification and Quantitation of Volatile Organic Compounds from Oxidation of Linseed Oil. Ind. Eng. Chem. Res. 51 (16), 5645–5652.

Kamal-Eldin, A., Marquez-Ruiz, G., Dobarganes, C., Appelqvist, L.A., 1997. Characterization of aldehydic acids in used and unused frying oils. J. Chromatogr. A 776 (2), 245–254.

Pai, J.S., Lomanno, S.S., Nawar, W.W., 1979. Effect of heat treatments on the volatile composition of coconut oil. J. Am. Oil Chem. Soc. 56 (4), 494–497.

Procida, G., Cichelli, A., Compagnone, D., Maggio, R.M., Cerretani, L., Del Carlo, M., 2009. Influence of chemical composition of olive oil on the development of volatile compounds during frying. Eur. Food Res. Technol. 230 (2), 217–229.

Seppanen, C.M., Csallany, A.S., 2004. Incorporation of the toxic aldehyde 4-hydroxy-2-*trans*-nonenal into food fried in thermally oxidized soybean oil. J. Am. Oil Chem. Soc. 81 (12), 1137–1141.

Seppanen, C.M., Csallany, A.S., 2006. The effect of intermittent and continuous heating of soybean oil at frying temperature on the formation of 4-hydroxy-2-*trans*-nonenal and other α-,β-unsaturated hydroxyaldehydes. J. Am. Oil Chem. Soc. 83 (2), 121–127.

Sheerin, A.N., Silwood, C., Lynch, E., Grootveld, M., 1997. Production of lipid peroxidation products in culinary oils and fats during episodes of thermal stressing: a high field ^{1}H NMR investigation. Biochem. Soc. T. 25 (3), 495S.

Silvagni, A., Franco, L., Bagno, A., Rastrelli, F., 2010. Thermoinduced Lipid Oxidation of a Culinary Oil: A Kinetic Study of the Oxidation Products by Magnetic Resonance Spectroscopies. J. Phys. Chem. A 114 (37), 10059–10065.

Song, S., Zhang, X., Hayat, K., Liu, P., Jia, C., Xia, S., Xiao, Z., Tian, H., Niu, Y., 2011. Formation of the beef flavour precursors and their correlation with chemical parameters during the controlled thermal oxidation of tallow. Food Chem. 124 (1), 203–209.

Takeoka, G.R., Buttery, R.G., Perrino Jr, C.T., 1995. Synthesis and occurrence of oxoaldehydes in used frying oils. J. Agr. Food Chem. 43 (1), 22–26.

Wu, C.M., Chen, S.Y., 1992. Volatile compounds in oils after deep frying or stir frying and subsequent storage. J. Am. Oil Chem. Soc. 69 (9), 858–865.

Zarkovic, N., 2003. 4-Hydroxynonenal as a bioactive marker of pathophysiological processes. Mol. Aspects Med. 24 (4/5), 281–291.

CHAPTER

31

Influence of Filtration on Composition of Olive Oils

*Karolina Brkić Bubola**, *Olivera Koprivnjak*†

*Department of Agriculture and Nutrition, Institute of Agriculture and Tourism, Poreč, Croatia, †Department of Food Technology and Control, School of Medicine, University of Rijeka, Croatia

CHAPTER POINTS

- Composition of active compounds of virgin olive oil (VOO) depends on several agronomic factors, as well as technological aspects such as olive fruit storage, crushing technique, malaxation conditions, extraction method, filtration, and oil storage conditions.
- Filtration of VOO influence antioxidant compounds, especially phenols and pigments, as well as volatile compounds responsible for pleasant VOO aroma.
- Different filter aids have diverse retention power toward different classes of phenol or volatile compounds.
- Filtration decreases chlorophyll and carotenoid concentration in VOO, regardless of type of filter aid.
- Filtration based on a filter bag or inert gas flow has no influence on fatty acids composition.
- Tocopherol content in VOO is not affected by filtration using filter paper or by filtration applying filter bag or inert gas flow.

INTRODUCTION

Virgin olive oil (VOO) is obtained exclusively by mechanical and physical processes which include collecting, washing and crushing of olive fruits, malaxation of olive paste, centrifugation, decantation or filtration, storage, and oil packaging (Figure 31.1). Immediately after extraction from olive fruits, VOO represents a turbid juice which can be consumed directly, without further processing. Turbidity is caused by suspended solid particles of plant tissue and vegetable water emulsified in the oil. According to industrial practice, VOOs are stored in tanks prior to bottling, allowing the vegetation, water, and sediments to settle down by gradual decantation. The 'veiled' VOO can retain its turbidity for several weeks or even months before full deposition of the residue (Gordillo *et al.*, 2011). In the meantime, enzymes and microorganisms from plant tissue particles and vegetable water droplets can deteriorate the quality of VOO by facilitating hydrolysis or promoting oxidation of stored oil (Botino *et al.*, 2004; Lozano-Sánchez *et al.* 2010). Therefore, filtration is becoming a standard step in VOO production that allows shelf life extension and hastens product finalization by giving the oil a clear and brilliant appearance (Lozano-Sánchez *et al.*, 2010).

Several different filtration systems, which can be classified in three groups, are available for VOO (Table 31.1): (i) conventional systems based on use of organic or inorganic filter aids in conjunction with filtration equipment (filter tanks and filter presses); (ii) cross-flow filtration characterized by tangential flow through the polymeric and ceramic membrane; and (iii) two new filtration systems proposed as alternative processes (the polypropylene filter bag system and filtration based on the flow of an inert gas) (Lozano-Sánchez *et al.* 2010). At the industrial scale, the most widespread is the so-called precoat-bodyfeed conventional filtration system, which employs filtration aid and vertical leaf-filter or horizontal filter-press as the filtration apparatus (Masella *et al.*, 2011) (Figure 31.2).

However, during VOO filtering, quantitative and qualitative changes take place, especially on minor components affecting the VOO quality (Lozano-Sánchez *et al.*, 2010), such as antioxidant compounds (phenols, tocopherols, and pigments) and volatile substances responsible

Processing and Impact on Active Components in Food
http://dx.doi.org/10.1016/B978-0-12-404699-3.00031-7

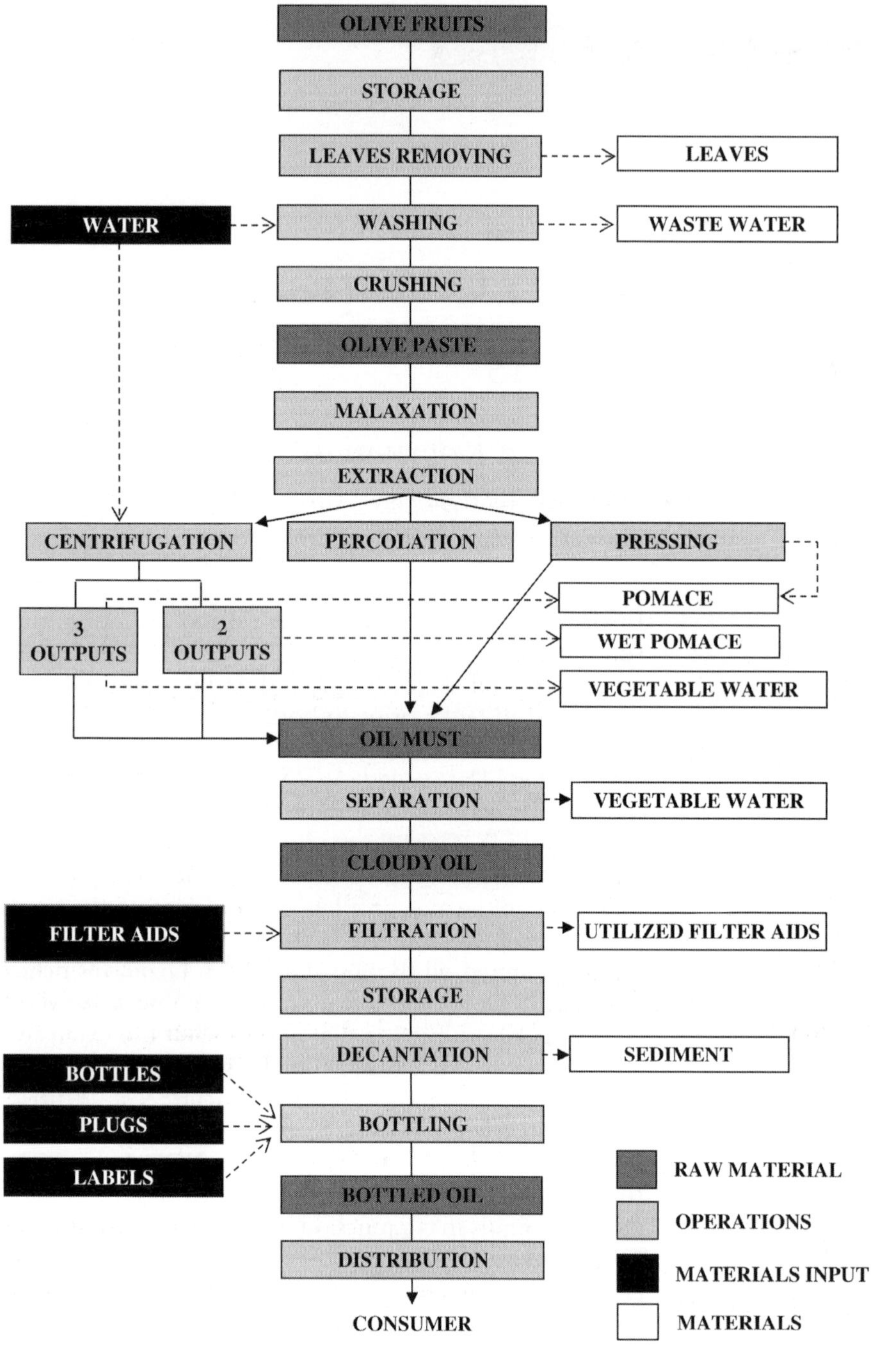

FIGURE 31.1 Flow diagram of virgin olive oil production.

for pleasant VOO aroma. Moreover, filtration could affect the major components, i.e., unsaturated fatty acids. Accumulating evidence suggests that most of the mentioned compounds contribute to the health benefits achievable by regular consumption of VOO, such as reduction of coronary disease risk factors, prevention of several types of cancer and modification of immune and inflammatory responses (Segura-Carretero *et al.*, 2010). In this chapter, changes of these compounds in relation to different filtration systems applied to VOO are discussed.

INFLUENCE OF FILTRATION ON ACTIVE COMPONENTS AND QUALITY OF VOO

Fatty Acids

The nutritional value of VOO can be attributed to its high levels of unsaturated fatty acids, especially oleic acid. VOO habitually comply with EU criteria for stating nutrition and health claims related to this fatty acid (European Union, 2012). However, there is only one report

TABLE 31.1 Virgin Olive Oil Filtration Systems

	Filtration System	Filtration Scale	Reference
Conventional	Filter paper Filter paper + anhydrous Na_2SO_4	Laboratory	Fregapane *et al.* (2006)
	Filter paper	Laboratory	Tsimidou *et al.* (2005)
	Layer of hydrophilic cotton + vacuum pump	Laboratory	Brkić Bubola *et al.* (2012)
	Precoat-bodyfeed filtration	Industrial plant	Masella *et al.* (2011)
	Diatomaceous earth Celulose fibers (Filtracel® 1000, Vitacel® L90) Starch + vacuum pump	Laboratory	Lozano-Sánchez *et al.* (2011)
Cross-flow membrane filtration	Microfiltration (membrane: Carbosep M14) Ultrafiltration (membranes: Kerasept K15, Carbosep M1 and Tami T50)	Pilot plant	Bottino *et al.* (2008)
	Microfiltration (membrane: Carbosep M14, Celgard 2500, Accurel PP1E, K150, K750) Ultrafiltration (membrane: Kerasept K15 and K300, Carbosep M1, Tami T50, T150, T300)	Pilot plant	Bottino *et al.* (2004)
New filtration system	Polypropylene bag Argon gas flow Nitrogen gas flow	Pilot plant	Lozano-Sánchez *et al.* (2012)

FIGURE 31.2 Conventional system for virgin olive oil filtration–horizontal filter-press. *Source: author's personal album.*

on influence of filtration on fatty acids composition (Lozano-Sánchez *et al.*, 2012). According to this research, carried out on a single monovarietal VOO sample, filtration based on a filter bag or inert gas flow (argon and nitrogen) has no influence on fatty acids composition (Lozano-Sánchez *et al.*, 2012).

Pigments

Lipophilic chlorophyll derivatives, beside lutein and β-carotene, are mainly responsible for VOO color. They play an important role in the oxidative stability of VOO due to the antioxidant nature in the dark and pro-oxidant activity in the light (Motilva and Romero, 2010). Both chlorophylls and carotenoids have been widely investigated for a range of beneficial biological activities in humans. However, while lipid matrix contributes highly to the bioavailability of carotenoids, for chlorophylls it is generally assumed that they are unabsorbable by humans (Ferruzzi and Blakeslee, 2007). Most authors reported that filtration of VOO through simple cellulose material, such as a cotton layer or filter paper, led to the decrease of chlorophyll concentration or chromatic coordinate a* corresponding to a green area (Brkić Bubola *et al.*, 2012; Gordillo *et al.*, 2011; Gomez-Caravaca *et al.*, 2007). However, by examining eight samples different in production year, production plant, and storage condition, Gomez-Caravaca *et al.* (2007) have noted that the final result depends on differences in initial turbidity and green color intensity of VOO sample. As regards membrane microfiltration, research carried out on a single VOO sample showed variations of chlorophyll reduction of between 17% and 85%, depending on type and porosity of used membranes (Bottino *et al.*, 2008). New filtration systems based on filter bag and inert gas flow (argon and nitrogen) decreased chlorophyll content in the range of 60–70% (Lozano-Sánchez *et al.*, 2012). When considering carotenoids concentration and chromatic coordinate *b** corresponding to a yellow area, the decreasing trend is less pronounced than for chlorophylls (Lozano-Sánchez *et al.*, 2012; Brkić Bubola *et al.*, 2012; Gordillo *et al.*, 2011), resulting in more apparent contribution of yellow to the color of filtered VOO. Bearing in mind the aforementioned pro-oxidant activity in the light and limited involvement of chlorophylls in nutritional and health properties of VOO, it could be concluded that benefits of VOO filtration in improving oil stability are more important than the disadvantage caused by reduction of pigments content.

Tocopherols

Extra VOOs have relatively high total tocopherol content with more than 97% of α-tocopherol (Beltrán *et al.*, 2010) and could be considered a source of vitamin E (European Union, 2012). As regards influence of filtration on tocopherol concentration in VOO, only a few publications deal with this topic. According to the results of Fregapane *et al.* (2006), filtration of six monovarietal VOO samples through filter paper with or without anhydrous Na_2SO_4 did not affected α-tocopherol content. Likewise, tocopherol content was not affected by applying filter bag or inert gas flow in clarification of VOO (Lozano-Sánchez *et al.*, 2012).

Polar Phenols

Polar phenolic compounds act as natural antioxidants protecting the VOO from oxidation caused by atmospheric oxygen. Also, they are considered potent alternative agents to combat chronic degenerative diseases, cardiovascular diseases, and cancer (Dais and Boskou, 2009). Recently, the European Union has adopted criteria for stating health claims related to polyphenols from VOO and protection of blood lipids from oxidative stress (at least 5 mg of hydroxytyrosol and its derivatives per 20 g of olive oil) (European Union, 2012). In addition, polar phenols are responsible for the characteristic bitter and pungent taste of VOO (Bendini *et al.*, 2007). Results regarding the influence of VOO filtration through cotton layer or filter paper on polar phenol concentration are not consistent. Tsimidou *et al.* (2005) examined six samples, Gomez-Caravaca *et al.* (2007) eight, and Fregapane *et al.* (2006) five samples, finding in some cases reduction and in other cases increase of polar phenol content in filtered VOO. Increase of total polar phenols was also found by Lozano-Sánchez *et al.* (2012) in case of VOO filtration with filter bag or clarification with inert gas flow. Gomez-Caravaca *et al.* (2007) proposed an explanation of increased values that could be linked to the partial elimination of water from VOO samples and to the greater availability of polar phenols for extraction with a polar solvent in analytical procedure. Comparison of filtration through columns built up with different filter aids (cellulose fibres, pregelatinised starch, and diatomaceous earth) showed a decrease of total polar phenol content in filtered VOO in the range of 10–60% depending on type of filter aid and oil sample (Lozano-Sánchez *et al.*, 2011). In the same article it was stressed that different filter aids have diverse retention power toward different classes of polar phenolic compounds, resulting in different phenol profile of filtered VOO.

Volatile Compounds

The delicate and unique olive oil aroma consists of a complex mixture of volatile compounds, which includes mainly aldehydes, alcohols, ketones, and esters. Some of them are potentially toxic, but their toxicity is negligible based on current consumption trends, known volatiles, and known toxicity data (Kalua *et al.*, 2010). The pleasant green and fruity odor of VOO is primarily ascribed to C6 and C5 volatiles, biogenerated from polyunsaturated fatty acids through the lipoxygenase pathway (Kalua *et al.*, 2007). Knowledge about influence of filtration processes on these compounds is quite limited since only two research reports related to this topic have been published so far. Brkić Bubola *et al.* (2012) reported that filtration of two monovarietal VOO samples through hydrophilic cotton layer had no significant influence on the total C6 volatile compounds and total esters concentration. For total C5 volatiles, total aldehydes, alcohols, and ketones, slight although significant changes (either decrease or increase and independently on varietal affiliation) were found. When microfiltration or ultrafiltration of a commercial VOO was performed by using ceramics or polymer inorganic membranes, reduction of carbonyl compounds, determined by high-performance liquid chromatography (HPLC), was 70–80% (C5 compounds), 40–70% (C6 compounds) and 55–70% of total desirable carbonyl compounds (Bottino *et al.*, 2008).

VOO QUALITY

Market quality of VOO is differentiated on the basis of free fatty acid content (FFA), peroxide values (PV), UV absorption indices (K_{232} and K_{270}), and sensory characteristics (EEC, 1991). Filtration through a layer of hydrophilic cotton (Brkić Bubola *et al.*, 2012), filtration using filter paper with or without anhydrous Na_2SO_4 (Fregapane *et al.*, 2006), or by filter bag or inert gas flow (Lozano-Sánchez *et al.*, 2012) showed no significant or very limited changes in FFA. Similar results were found for microfiltration or ultrafiltration performed by using different inorganic membranes (Bottino *et al.*, 2004; Bottino *et al.*, 2008). However, according to Fregapane *et al.* (2006), the rate of triacylglycerol hydrolysis during 8 months of storage at 25°C and 40°C is higher in unfiltered oils, probably because of higher water content. As regards oxidative deterioration of triacylglycerol matrix (PV, K_{232}, and K_{270}) during filtration with cotton layer and filter paper or cross-flow filtration through inorganic membranes, it seems to be negligible or moderate in most cases (Brkić Bubola *et al.* 2012; Fregapane *et al.*, 2006, Tsimidou *et al.*, 2005; Bottino *et al.*, 2004). Fregapane *et al.* (2006) stated that the effect of filtration of five monovarietal VOO samples on the rate of peroxide formation during 8 months of storage is minor regarding differences in oxidation rate due to genetic factors. Interesting results were obtained for filtration using filter bag and argon or nitrogen gas flow where PV was reduced in filtered samples to less than half of the initial value in unfiltered VOO samples (Lozano-Sánchez *et al.*, 2012). However, among trade quality criteria, the sensory characteristics of VOO are the most appreciated by consumers. Fregapane *et al.* (2006) did not observe differences relating to the filtration method for the main positive attributes of olive oil—fruitiness, pungency, and bitterness. The rancid defect, a sign of the autoxidation processes, appeared earlier in the unfiltered than in the filtered and dehydrated oils during storage at 25°C and 40°C. Thus, filtration and dehydration could help to enhance the sensory stability of extra virgin olive oil. Lozano-Sánchez *et al.* (2012) evidenced that filtering by polypropylene filter bag or clarifying by inert gas flow does not lower the intensity of the main organoleptic attributes contributing to the consumer acceptance. Except leaf/grass notes, these treatments produced oils with higher intensity of pleasant flavors and apple notes, as well as bitter taste regarding unfiltered samples. Treatments with argon and nitrogen enhanced fruitiness and pungency, as well. Brkić Bubola *et al.* (2012) noticed only slight changes in sensorial profiles of filtered and unfiltered VOO from two olive varieties using cotton layer as a filter aid. Filtered oils had slightly higher intensities of sensory characteristics apple and grass but without significant differences in sensory scores with regard to unfiltered samples. Bottino *et al.* (2008) found that membrane filtration processes reduce the intensity of the green components of the oil flavor, but could contribute to better acceptance of VOO by decreasing the intensity of some less pleasant flavors, such as eucalyptus.

OTHER WAYS IN WHICH VOO MINOR ACTIVE COMPONENTS ARE ALTERED

Composition and quality of VOO depend on several agronomic factors such as cultivar, fruit ripeness, climatic conditions, and fruit soundness. Polar phenols and tocopherols are two groups of active components that are highly influenced by genetic factors (Beltrán *et al.*, 2010; Škevin *et al.*, 2003). During fruit ripening, many authors reported a decrease of concentration of total phenols (Škevin *et al.*, 2003; Dag *et al.*, 2011), α-tocopherol (Beltrán *et al.*, 2010), chlorophyll and carotenoids (Criado *et al.*, 2007). Environmental factors, such as higher ambient temperature and water deficiency, generate a stress situation that induces the production of phenolics in olive fruits (Romero and Moltiva, 2010). Poor health status of fruits prior to harvesting caused by olive fly infestation is highly negatively correlated to the phenols concentration in VOO (Koprivnjak *et al.*, 2010).

Beside filtration, VOO composition and especially content of minor components are also influenced by other technological aspects such as olive fruit storage before processing, crushing technique, malaxation conditions, extraction method, and oil storage conditions (Figure 31.1). These effects have been evaluated by different authors. Decrease of polar phenols and chlorophylls concentration in VOO obtained from olive fruits stored at ambient temperature is much more expressed than in refrigerator conditions, while there are no significant changes of carotenoids (Ben Youssef *et al.*, 2012). Crushing of olives with mills causing higher tissue fragmentation and higher paste temperature produces VOO with greater content of polar phenols and chlorophylls, as a result of better contact between solid and liquid phase and major solubility at higher temperatures (Inarejos-Garcia *et al.*, 2011). Malaxation represents a slow continuous kneading of crushed olives, aimed at breaking off the emulsion formed during crushing. Longer malaxation times contribute to the higher concentration of chlorophylls and carotenoids in VOO (Ranalli *et al.*, 2003), while the concentration of phenolic compounds remains constant or decreases as a result of the balance between oxidative reactions catalyzed by endogenous oxidoreductase and the distribution of polar phenols between the oil and the water phase (Angerosa *et al.*, 2001, Inarejos-Garcia *et al.*, 2009). Higher malaxation temperature as well as inert atmosphere in contact with the olive paste positively influences the concentration of polar phenols in VOO (Servili *et al.*, 2002). During extraction of oil from olive paste by centrifugation, the amount of water added determines the dilution of the aqueous phase and lowers the concentration of the phenolic substances that are more soluble in vegetable waste water (Di Giovacchino *et al.*, 2001).

ANALYTICAL TECHNIQUES FOR DETERMINATION OF VOO ACTIVE COMPONENTS

Fatty Acids

Determination of VOO fatty acids composition usually is performed by GC separation of their methyl esters which is described by the official method (EEC, 1991).

Pigments

Quantitative determination of total chlorophyll and carotenoid compounds is usually determined following the procedure of Mínguez-Mosquera et al. (1991) and results are expressed as pheophytin α and lutein content, since they are major components of the chlorophyll and carotenoid fractions, respectively. Currently, determination, identification, and quantification of individual pigments in VOO are based on HPLC coupled with UV-Vis spectrophotometric detector (Tsimidou et al., 2005).

Tocopherols

To date, there is no official method for analysis of tocopherols in olive oil. Nevertheless, there are a great number of gas and liquid chromatographic methods suitable for their analysis. Separation and quantitative determination of tocopherols in VOO is commonly performed by normal or reversed phase HPLC coupled with different detectors (ultraviolet, fluorescence, amperometric) (Tsimidou, 2010).

Phenols

Quantitative determination of total polar phenols in olive oil is usually performed according to Folin-Ciocalteu colorimetric method (Gutfinger, 1981). However, this method is not specific and gives no indication of the nature of the phenolic compounds present. Determination of individual phenolic compounds in VOO involves three basic steps: extraction from the oil sample, analytical separation, and identification and quantification. Currently, for the isolation of the polar phenolic fraction of VOO, two basic extraction techniques are used, liquid–liquid extraction and solid-phase extraction methods. The systems not only vary in the solvents and/or solid-phase cartridges used but also in the quantities of sample needed for analysis, volumes of the solvents, and other such details. Identification and quantification of individual phenolic compounds in VOO is based traditionally on HPLC coupled with different detectors (UV, fluorescence, coulometric electrode array detection, amperometric detector), GC-FID, and capillary electrophoresis coupled to different detectors (UV, electrochemical), and can be helped nowadays by the complementary use of mass spectrometry and nuclear magnetic resonance (Segura-Carretero *et al.*, 2010). The most preferred method for analysis of the phenolic fraction of olives and olive oils is reversed-phase HPLC with gradient elution. An usual problem associated with this methodology may be the use of standards for calibration that may not be available in the market (Dais and Boskou, 2009; Segura-Carretero *et al.*, 2010).

References

Angerosa, F., Mostallino, R., Basti, C., Vito, R., 2001. Influence of malaxation temperature and time on the quality of virgin olive oils. Food Chem. 72, 19–28.

Beltrán, G., Jiménez, A., del Rio, C., Sánchez, S., Martínez, L., Uceda, M., Aguilera, M.P., 2010. Variability of vitamin E in virgin olive oil by agronomical and genetic factors. J. Food Compos. Anal. 23, 633–639.

Ben Youssef, N., Youssef, O., Nizar, D., Bechir, B., Chedly, A., Mokhtar, Z., 2012. Effect of olive storage period at two different temperatures on oil quality of two tunisian cultivars of *Olea europea*, Chemlali and Chétoui. Afr. J. Biotechnol. 11, 888–895.

Bendini, A., Cerretani, L., Carrasco-Pancorbo, A., Gómez-Caravaca, A.M., Segura-Carretero, A., Fernández-Gutiérrez, A., 2007. Phenolic molecules in virgin olive oils: A survey of their sensory properties, health effects, antioxidant activity and analytical methods. An overview of the last decade. Molecules 12, 1679–1719.

Bottino, A., Capannelli, G., Comite, A., Ferrari, F., Marotta, F., Mattei, A., Turchini, A., 2004. Application of membrane processes for the filtration of extra virgin olive oil. J. Food Eng. 65, 303–309.

Bottino, A., Capannelli, G., Mattei, A., Rovellini, P., Zunin, P., 2008. Effect of membrane filtration on the flavor of virgin olive oil. Eur. J. Lipid Sci. Technol. 110, 1109–1115.

Brkić Bubola, K., Koprivnjak, O., Sladonja, B., 2012. Influence of filtration on volatile compounds and sensory profile of virgin olive oils. Food Chem. 132, 98–103.

Criado, M.N., Motilva, M.J., Goñi, M., Romero, M.P., 2007. Comparative study of the effect of the maturation process of the olive fruit on the chlorophyll and carotenoid fractions of drupes and virgin oils from Arbequina and Farga cultivars. Food Chem. 100, 748–755.

Dag, A., Kerem, Z., Yogev, N., Zipori, I., Lavee, S., Ben-David, E., 2011. Influence of time of harvest and maturity index on olive oil yield and quality. Scientia Horticulturae 127, 358–366.

Dais, P., Boskou, D., 2009. Detection and Quantification of Phenolic Compounds in Olive Oil, Olives, and Biological Fluids. In: Boskou, D. (Ed.), Olive Oil Minor Constituents and Health. CRC Press, Boca Raton, pp. 56–107.

Di Giovacchino, L., Costantini, N., Serraiocco, A., Surricchio, G., Basti, C., 2001. Natural antioxidants and volatile compounds of virgin olive oils obtained by two or three-phases centrifugal decanters. Eur. J. Lipid Sci. Technol. 103, 279–285.

EEC, 1991. Characteristics of olive oil and olive-residue oil and the relevant methods of analysis. Regulation EEC/2568/91 and latter modifications. Off. J. Eur. Communities: Legis. 248, 1–137.

European Union, 2012. Commission regulation 432/2012 establishing a list of permitted health claims made on foods, other than those referring to the reduction of disease risk and to children's development and health. Off. J. Eur. Union 20: Legis. 136, 21–35.

Ferruzzi, M.G., Blakeslee, J., 2007. Digestion, absorption, and cancer preventative activity of dietary chlorophyll derivatives. Nutr. Res. 27, 1–12.

Fregapane, G., Lavelli, V., Leon, S., Kapuralin, J., Salvador, M.D., 2006. Effect of filtration on virgin olive oil stability during storage. Eur. J. Lipid Sci. Technol. 108, 134–142.

Gomez-Caravaca, A.M., Cerretani, L., Bendini, A., Segura-Carretero, A., Fernandez- Gutierrez, A., Lercker, G., 2007. Effect of filtration systems on the phenolic content in virgin olive oil by HPLC-DADMSD. Am. J. Food Technol. 2, 671–678.

Gordillo, B., Ciaccheri, L., Mignani, A.G., Gonzalez-Miret, M.L., Heredia, F.J., 2011. Influence of Turbidity Grade on Color and Appearance of Virgin Olive Oil. J. Am. Oil Chem. Soc. 88, 1317–1327.

Gutfinger, T., 1981. Phenols in olive oils. J. Am. Oil Chem. Soc. 58, 966–968.

Inarejos-García, A.M., Fregapane, G., Salvador, M.D., 2011. Effect of crushing on olive paste and virgin olive oil minor components. Eur. Food Res. Technol. 232, 441–451.

Inarejos-García, A.M., Gómez-Rico, A., Salvador, M.D., Fregapane, G., 2009. Influence of malaxation conditions on virgin olive oil yield, overall quality and composition. Eur. Food Res. Technol. 228, 671–677.

Kalua, C.M., Allen, M.S., Bedgood Jr., D.R., Bishop, A.G., Prenzler, P.D., Robards, K., 2007. Olive oil volatile compounds, flavour development and quality: A critical review. Food Chem. 100, 273–286.

Kalua, C., Prenzler, P., Ryan, D., Robards, K., 2010. Volatile Compounds in Australian Olive Oils: How Different Are They From Other Oils. In: Preedy, V.R., Watson, R.R. (Eds.), Olives and Olive Oil in Health and Disease Prevention. Elsevier Inc., San Diego, pp. 201–210.

Koprivnjak, O., Dminić, I., Kosić, U., Majetić, V., Godena, S., Valenčič, V., 2010. Dynamics of oil quality parameters changes related to olive fruit fly attack. Eur. J. Lipid Sci. Technol. 112, 1033–1040.

Lozano-Sánchez, J., Cerretani, L., Bendini, A., Gallina-Toschi, T., Segura-Carretero, A., Fernández-Gutiérrez, A., 2012. New Filtration Systems for Extra-Virgin Olive Oil: Effect on Antioxidant Compounds, Oxidative Stability, and Physicochemical and Sensory Properties. J. Agric. Food Chem. 60, 3754–3762.

Lozano-Sánchez, J., Cerretani, L., Bendini, A., Segura-Carretero, A., Fernández-Gutiérrez, A., 2010. Filtration process of extra virgin olive oil: effect on minor components, oxidative stability and sensorial and physicochemical characteristics. Trends Food Sci. Technol. 21, 201–211.

Lozano-Sánchez, J., Segura-Carretero, A., Fernández-Gutiérrez, A., 2011. Characterisation of the phenolic compounds retained in different organic and inorganic filter aids used for filtration of extra virgin olive oil. Food Chem. 124, 1146–1150.

Masella, P., Parenti, A., Spugnoli, P., Baldi, F., Mattei, A., 2011. A Predictive Classification Model for the Management of Virgin Olive Oil Filtration at Industrial Scale. Sep. Sci. Technol. 46, 1709–1715.

Mínguez-Mosquera, M.I., Rejano-Navarro, L., Gandul-Rojas, B., Sánchez-Gòmez, A.H., Garrido-Fernandez, J., 1991. Color-Pigment Correlation in Virgin Olive Oil. J. Am. Oil Chem. Soc. 68, 332–336.

Motilva, M.J., Romero, M.P., 2010. The Effect of the Ripening Process of the Olive Fruit on the Chlorophyll and Carotenoid Fractions of Drupes and Virgin Oils. In: Preedy, V.R., Watson, R.R. (Eds.), Olives and Olive Oil in Health and Disease Prevention. Elsevier Inc., San Diego, pp. 59–68.

Ranalli, A., Pollastri, L., Contento, S., Iannucci, E., Lucera, L., 2003. Effect of olive paste kneading process time on the overall quality of virgin olive oil. Eur. J. Lipid Sci. Technol. 105, 57–67.

Romero, M.P., Motilva, M.J., 2010. Effect of Climatic Conditions on Quality of Virgin Olive Oil. In: Preedy, V.R., Watson, R.R. (Eds.), Olives and Olive Oil in Health and Disease Prevention. Elsevier Inc., San Diego, pp. 43–49.

Segura-Carretero, A., Carrasco-Pancorbo, A., Bendini, A., Cerretani, L., Fernández-Gutiérrez, A., 2010. Analytical Determination of Polyphenols in Olive Oil. In: Preedy, V.R., Watson, R.R. (Eds.), Olives and Olive Oil in Health and Disease Prevention. Elsevier Inc., San Diego, pp. 509–524.

Servili, M., Selvaggini, R., Taticchi, A., Baldioli, M., Montedoro, G.F., 2002. The use of biotechnology means during oil mechanical extraction process: relationship with sensory and nutritional parameters of virgin olive oil quality. Acta Hortic. 586, 557–560.

Škevin, D., Rade, D., Štrucelj, D., Mokrovčak, Ž., Nederal, S., Benčić, Đ., 2003. The influence of variety and harvest time on the bitterness and phenolic compounds of olive oil. Eur. J. Lipid Sci. Technol. 105, 536–541.

Tsimidou, M.Z., 2010. Squalene and Tocopherols in Olive Oil: Importance and Methods of Analysis. In: Preedy, V.R., Watson, R.R. (Eds.), Olives and Olive Oil in Health and Disease Prevention. Elsevier Inc., San Diego, pp. 561–567.

Tsimidou, M.Z., Georgiou, A., Koidis, A., Boskou, D., 2005. Loss of stability of "veiled" (cloudy) virgin olive oils in storage. Food Chem. 93, 377–383.

CHAPTER

32

Virgin Olive Oil: Losses of Antioxidant Polar Phenolic Compounds due to Storage, Packaging, and Culinary Uses

*Anastasios Koidis**, *Dimitrios Boskou*†

*Institute for Global Food Security, Queen's University, Belfast, UK, †Department of Chemistry, Aristotle University of Thessaloniki, Thessaloniki, Greece

CHAPTER POINTS

- Olive oil is a high natural quality product produced by only physical means, naturally rich in bioactive polar phenols.
- Its polar fraction consists of a complex mixture of phenolic acids (such as caffeic acid or syringic acid), simple phenols like tyrosol and hydroxytyrosol, secoiridoid derivatives of the glycosides oleuropein and ligstrodide, lignans, and flavonoids.
- Polar phenols significantly contribute to the oxidative stability of olive oil.
- Between olive oil production and consumption, valuable polar phenols can be lost during various stages, which results in loss of quality.
- These stages are: (i) during production at the mill; (ii) industrial refining—where relevant; (iii) during storage both industrial soon after production and domestic storage; (iv) during transfer; and last but not least (v) the culinary practices in consumers' homes.
- The changes that occur relate to oxidation, including autoxidation and photo-oxidation, enzymic hydrolysis, and high temperature handling during various operations.
- Analysis of polar phenols are realised with chromatographic methods (HPLC), after appropriate extraction (liquid–liquid extraction or SPE). Recent advances included high-end LC-MS methodologies.
- Separation and identification of polar phenols are realised in preparative liquid chromatography systems and LC-NMR.
- Evaluation of antioxidant activity of olive oil or its polar extract is performed by radical scavenging assays. The protective effect of polar phenols in olive oil is measured by other techniques such as Rancimat or PV value.

VIRGIN OLIVE OIL AND ITS POLAR FRACTION

According to accumulated epidemiological evidence, important health benefits are linked with high consumption of olive oil as part of a balanced diet. It has been established that this is due to the presence of both macro-constituents such as fatty acids and minor constituents such as bioactive polar phenols. As with all vegetable oils, virgin olive oil consists of mixed triacylglycerols particularly rich in monounsaturated fatty acids (mainly oleic acid or C18 : 1). They account for more than 98% of the composition of saponified matter. The unsaponified matter of olive oil ranges from 1% to 2% and contains, amongst other compounds, squalene, pigments, tocopherols, waxes, and the polar fraction.

The polar fraction of olive oil consists of a complex mixture of phenolic compounds, phenolic acids (such as caffeic acid or syringic acid), simple phenols like tyrosol and hydroxytyrosol, secoiridoid derivatives of

Processing and Impact on Active Components in Food
http://dx.doi.org/10.1016/B978-0-12-404699-3.00032-9

the glycosides oleuropein and ligstrodide, lignans, and flavonoids (Figure 32.1). This complex mixture is very often erroneously described as 'polyphenols' of olive oil. Due to their wide range and diversity, accurate quantification is difficult and it is often done conventionally. The total phenol content of olive oil is 500–1000 mg/kg although most oils have usually between 100–300 mg/kg (Boskou, 2009b). Levels of individual phenols in olive oil vary because of the natural variability in composition, the olive variety, and the agronomic conditions. Fresh oils contain more complex forms of aglycons deriving from secoiridoids. Free phenols such as tyrosol and hydroxytyrosol are found at higher levels in stored virgin olive oils as a result of enzymic liberation of their respective aglycons. This also contributes to the loss of bitter flavor of fresh olive oil over storage.

Effect on Stability

It is established that polar phenols significantly contribute to the stability of virgin olive oil (Gallina-Toschi *et al.*, 2005). There is a good correlation of stability over shelf-life and total or individual phenol content (Tsimidou *et al.*, 1992). Most of the phenols present in olive oil are antioxidant compounds, but the components that are mainly responsible for the remarkable resistance of olive oil to oxidation are the dialdeydic form of elenolic acid linked to hydroxytyrosol (3,4-DHPEA-EDA), decarboxymethyl oleuropein aglycon, and hydroxytyrosol. Tyrosol, lignans, and ligstroside aglycon are weaker antioxidants (Boskou, 2006). Apart from several simple and complex polar phenols that increase oil stability, positive effect has been observed for other olive oil constituents such as tocopherols and squalene.

Biological Role

Olive oil intake is linked with increased resistance to cardiovascular diseases, and other health benefits. In recent years, attention has been drawn from the unsaturated fatty acid profile to the minor constituents, and especially polar phenols. Phenolic compounds are well known antioxidants acting as radical scavengers, neutralizing reactive oxygen species (ROS) *in vitro*. One of their benefits to health is associated with the radical scavenging of ROS produced in the body due to oxidative stress. The antioxidant effect of particular phenols *in vivo* is deriving from the ability of preventing both the initiation and propagation of low-density lipoprotein oxidation which has demonstrated significant health benefits in *in vitro* and *in vivo* models (Covas *et al.*, 2006.). Several clinical and animal studies have provided evidence that phenolic compounds are absorbed and exert their biological effects in a dose-dependent manner (Visioli *et al.*, 2001). However, concrete scientific proof about their direct influence on human health and their exact role *in vivo* is yet to be demonstrated. Other beneficial effects of polar phenols observed *in vitro* are anti-tumoral and anti-inflammatory effects. Important evidence on bioactivity is reported for some of the phenols such as oleocanthal (Figure 32.1) that claim to have similar anti-inflammatory action to the body as Ibuprofen® due to the ibuprofen-like structure (Beauchaup, 2005).

Other Active Ingredients

Secondary active ingredients in olive oil include *tocopherols* (Vitamin E analogs), *phytosterols*, and *squalene*. This chapter is mainly associated with polar phenols in VOO although some information about the secondary active ingredients will be mentioned where appropriate.

Production and Types

Virgin olive oil is obtained by only mild physical methods, i.e., crushing, extraction, and centrifugation, and is well documented in the literature (Petrakis, 2006).

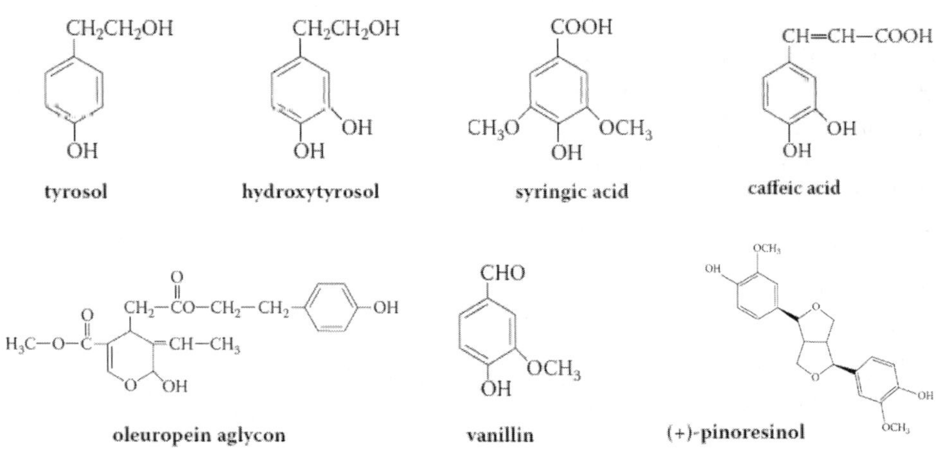

FIGURE 32.1 Chemical structures of phenols (tyrosol, hydroxytyrosol, vanillin), phenolic acids (syringic and caffeic acids), tyrosol derivatives (oleuropein) and lignans/ligstroside/flavonoids (pinoresinol and oleocanthal) present in the polar fraction of olive oil.

The term 'virgin' relates to the quality and authenticity of olive oil, whereas 'extra virgin' is a very high quality and pure unprocessed olive oil. Acidity is one of the important quality parameters that distinguishes 'extra virgin' (>0.8%) from 'virgin' olive oil (0.8–1.5%). Olive oil with high acidity often results from lower quality olive drupes and other agronomical or processing factors. Low quality olive oil is usually industrially processed. Refining includes de-odoring, acidity neutralization, and degumming. During this process part of the chlorophyll is lost, along with other bioactive compounds, which explains the lighter color appearance of refined oils compared to virgin olive oils. The term 'olive oil' is often referred to a mixture of VOO and refined olive oils (EC, 2007).

FACTORS THAT INFLUENCE THE RETENTION OF POLAR PHENOLS

Olive oil composition may vary depending of the geographical location, the agronomical conditions and the type of equipment used for harvesting and extraction. Important active components can be lost during various stages between olive oil production and consumption. These stages are: (i) production at the mill; (ii) industrial refining—where relevant; (iii) storage, both industrial soon after production, and domestic storage; (iv) transfer; and last but not least (v) domestic culinary practices. The changes that occur relate to *oxidation*, including autoxidation and photo-oxidation, enzymic activity, and other changes due to high temperature handling. These stages and the associated risks are summarized in Figure 32.2.

Before and During Production

Agronomical Factors

The importance of agronomic factors, the influence of the olive cultivar, and the effect of maturation on the natural occurrence of phenolic compounds in olive oil have been investigated in many studies. Irrigation of the trees and harvest time were found to influence the maximum retention of active ingredients such as polar phenols, although the optimal times are highly cultivar dependent (Briante *et al.*, 2001; Boskou *et al.*, 2006). Agronomical factors are difficult to control but olive cultivar and harvest time choice might be the two easier to manipulate parameters for the production of high polyphenol content olive oil. Other influential steps and factors are shown below.

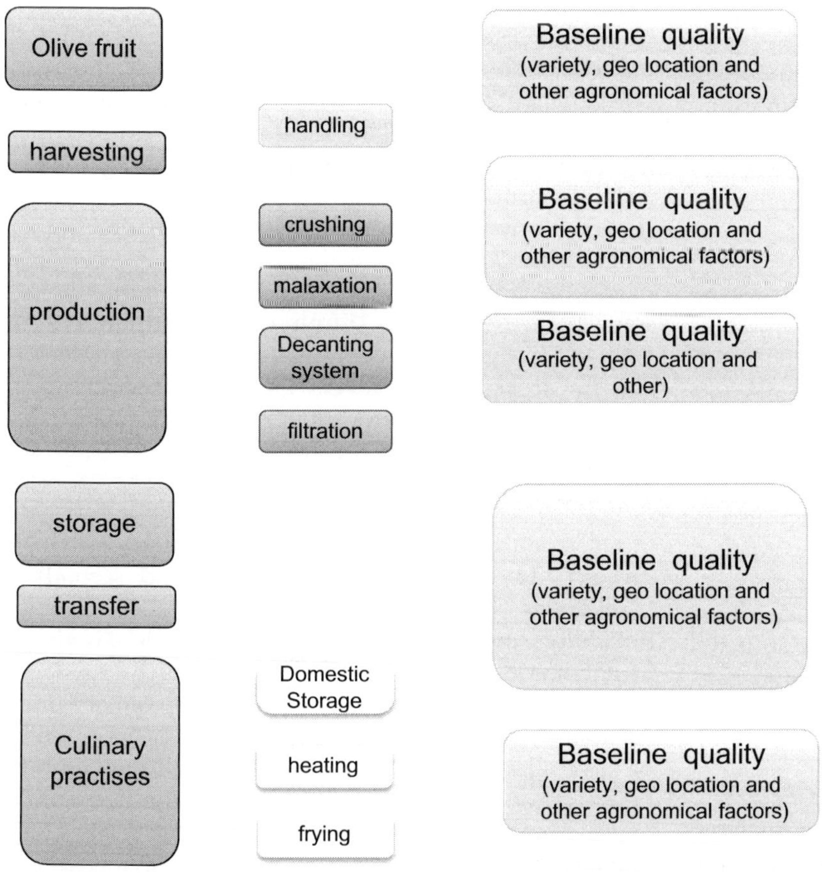

FIGURE 32.2 Various stages from olive oil production to consumption: the quality parameters and associated risks.

Storage of Olives Before Milling

Olives should ideally be processed within 24–48 h after harvest. This is because stocking for longer periods will initiate hydrolytic and oxidative enzymic reactions which result in loss of bioactive costituents, i.e., loss of high quality. In busy periods such as the October–November harvest, where short time stocking is particularly challenging, better mill organisation and improved storage conditions are required.

Extraction Types and Equipment Used

VOO extraction can be achieved by a wide variety of methods and mechanical systems. Extensive literature exists on the effect of the processing system on olive oil quality characteristics and the level of polar phenols (Salvador *et al.*, 2003).

Crushing: both stone crushing and crushing with hammer have been demonstrated suitable for higher retention of polar compounds with the stone mill more efficient in decreasing bitterness and pungency (Caponio *et al.*, 2001). *Malaxation temperature*: in malaxation olive paste undergoes a series of enzymic activities of glycosidases and esterases present in the drupe and the skin that favor the release of hydroxytyrosol. Depending on the temperature employed, the extraction is either cold <30°C) or standard (~50°C) which influences the retention of polar phenols. *Pressing and centrifugation system*: in modern olive oil mills, both two-phase and three-phase systems produce very good quality olive oil and can impact on the retention of phenols. Generally, the two-phase system where no extra water is added, results in oil with higher phenol content and longer induction periods (Petrakis, 1994; 2006). This is because there is no leakage of polar phenols due to increased solubility and polarity of the water layer.

Filtration

In a nutshell, filtration of newly produced cloudy olive oil results in a notable but not significant reduction of polar phenol content (Tsimidou *et al.*, 2005). For detailed information on the retention of polar phenols during filtration, see Chapter 31.

Refining

Olive oil that needs refining often has high acidity derived from low quality olive fruits (defective, infested by diseases) and other quality defects. The high temperatures and other chemical and physical procedures of refining drastically decrease the polar constituents content as well as easily oxidized substances such as tocopherols (Garcia *et al.*, 2006).

Storage Prior to Packaging

In modern olive oil mills, VOO is stored in large stainless steel containers. Due to potential oxidative deterioration, these containers should be made of non-light permeable material such as inox with inner headspace of less than 5%. Some studies showed the advantage of nitrogen headspace atmosphere that further prevents oxidation although this is recommended only in prolonged storage. For the same reason, the product should be stored at room temperature (~20°C). Typical storage is 2–3 months and during this period the sediment in the bottom part of the tank is removed to avoid water accumulation and microbial growth (Koidis *et al.*, 2008).

During In-Pack Storage

There is extensive literature on the changes in the phenol fraction during storage and various conditions have been studied. It is accepted that polar phenol content changes during storage due to hydrolytic and oxidative reactions. This reduction could be controlled under reduced oxygen conditions during storage. *Hydrolysis* can be caused by the enzymic liberation of free fatty acids from the lipid structure which slightly increases acidity and drops oil quality. This is a minor risk and only occurs during long-term storage. There are actually beneficial hydrolytic reactions as discussed above, where polar phenols such as tyrosol and hydroxytyrosol are liberated from the glycosides resulting in a phenol content enrichment and higher antioxidant activity of the oil in the course of storage.

Oxidation and *photo-oxidation* are the main enemies of polar phenols and of olive oil quality. Polar phenols can be damaged by elevated temperatures and light. Due to their structure they are highly susceptible to oxidation to ortho- and para-quinone forms (Shingai *et al.*, 2011). In terms of the effect of the lipid substrate, oxidation results in the production of hydroperoxides, highly reactive species that favor the generation of secondary oxidation products such as aldehydes and ketones with distinct off flavors. Virgin olive oil should not be exposed to light, as it contains endogenous photosensitizers at significant levels (5–40 mg pheophytin-α) (Psomiadou and Tsimidou, 2002).

Culinary Uses

Oil oxidation involves the decomposition of triacylglycerols to primary (hydroperoxides) and secondary breakdown products accompanied by off flavors (rancidity). In comparison to other vegetable oils richer in polyunsaturated fatty acids, VOO is much less structurally altered during high temperature culinary uses such as domestic frying or oven cooking. This stability of olive oil and its resistance to rapid high temperature deterioration are due to its rich monounsaturated profile and the presence of natural antioxidants including polar phenols and other antioxidant and antipolymerization compounds (tocopherols, sterols). They inhibit oxidative deterioration and more specifically polymerization and radical formation (Blekas and Boskou, 1999). However, all studies agree that when virgin olive oil is heated for several frying cycles or other high temperature operations, its polar phenolic fraction content and its antioxidant activity are significantly reduced (Gomez-Alonso *et al.*, 2003). Unlike lipid oxidation, polyphenol oxidation (as discussed earlier) does not produce perceptible off-flavors but only causes significant reduction in the total polar phenol content (Shingai *et al.*, 2011). Brenes *et al.* (2002) investigated the changes occurring in VOO subjected to simulated domestic frying, microwave heating, and boiling with water in a pressure cooker. Heating at elevated temperatures (>180°C) caused a significant loss of tocopherols and hydroxytyrosol derivatives, but lignans (such as pinoresinol) showed notable resistance. Prolonged high temperature heating can also facilitate the formation of aldehydes as discussed in Chapter 30. Microwave heating caused the lower losses of phenolic compounds compared to other methods. Boiling in a pressure cooker caused rapid hydrolysis of the secoiridoid aglycons. In some Mediterranean countries, partially due to the abundance of olive oil, it is common practise to use it as a frying medium. Although olive oil substitutes the water of food products during frying, it largely consists only of monounsaturated lipids since most of its bioactive polar fraction has been diminished in the process. If consumers expect potential health effects from the intake of phytochemicals present in olive oil fried foods , the number of heating operations should be kept to a minimum (Boskou, 2009a).

An attempt to summarize the influence of different treatment on polar phenol content and oil stability (as measured by the stability index in accelerated oxidation tests) is shown in Figure 32.3. It is evident that while filtration and frying in high temperature inevitably lowers phenol content, longer storage (due to the release of polar phenol molecules) and freezing/ thawing cycles keep the polar content of VOO relatively stable.

ANALYSIS OF POLAR PHENOLS AND OTHER BIOACTIVES IN OLIVE OILS

Analysis of olive oil lipid and polar factions created huge scientific interest for various reasons such as investigating bioactivity of minor constituents, characterisation of oils from different geographic location, impact of extraction equipment on bioactive content, bioactive retention during processing handling and storing, authenticity of olive oil, and methods to detect fraudulent adulteration, to name a few.

Separation, Identification, and Qualification of Polar Phenols

In recent years there have been a large number of analytical methods developed for the isolation, separation, structural determination, and quantification of polar phenols in olive oils. As explained earlier, polar phenols exhibit a wide range and diversity which makes their accurate quantification challenging. Appropriate sample preparation is the first important step in the determination of phenolic compounds (Figure 32.4). Extraction of phenolics depends on the nature of the sample. Liquid–liquid extraction (usually aqueous or methanolic) is generally

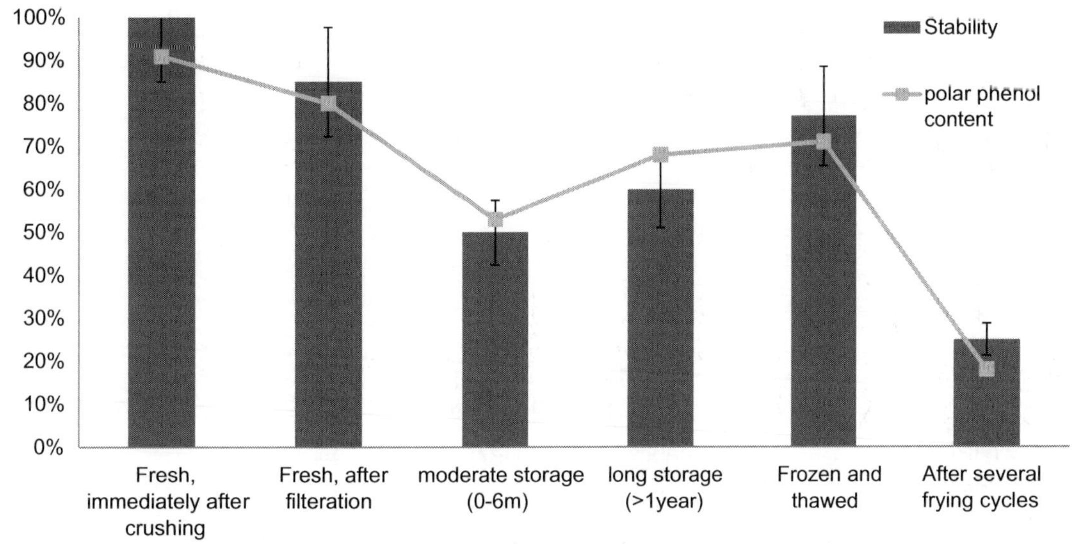

FIGURE 32.3 Analysis of polar phenols extracted from olive oil. Note: The data represent a mixture of author's own experimental evidence and experience as well as cumulative evidence from different sources in the literature. This figure is not based in one study only and should be therefore used with caution.

used for the isolation of phenolic compounds from olive oil. Current methods use Solid phase Extraction (SPE) with C18 column to pre-concentrate the sample in polar compounds (Pellegrini *et al.*, 2001) for better results.

Chromatographic methods offer quantification through separation and analysis of the polar fraction peaks of the oil usually using Reverse Phase High Performance Chromatographic (RP-HPLC) system under typical conditions (C18 column, gradient elution of mobile phase: acetonitrile and water). Most phenolics absorb in the 280-nm range although usually two or three different wavelengths are monitored using an Ultraviolet (UV) or Diode Array Detector (DAD). The on-line coupling of liquid chromatography with mass spectrometer (LC-MS) is a major development in the accurate determination of phenolics as explained in Figure 32.4. This combination, along with the use of various mass spectrometry operation modes, allows the detection of polar non-volatile and thermolabile phenolics present at very low concentrations (Dais and Boskou, 2009). Mass spectrometry provides not only quantification but also structural information of phenolics which is used to further characterize the polar fraction. While one-dimensional ^{1}H and ^{13}C Nuclear Magnetic Resonance (NMR) can also be used to study the structure of phenols of the polar fraction, another coupling, LC-NMR, allows separation and identification/structure elucidation simultaneously. Significantly improved separation of polar phenols is possible with on-line pre-concentration of olive oil extracts using SPE coupled with LC-NMR in a complex LC-SPE-NMR technique (Christophoridou *et al.*, 2005). Preparative HPLC can be used to purify polar phenol extracts and isolate the specific fractions or individual phenols for the enrichment of other lipid substrates with olive antioxidants (Artajo *et al.*, 2006). Gas chromatography is less used due to the extra derivatisation step required and the high column temperature that affect some heat sensitive phenolics during analysis. As an alternative to chromatographic methods, capillary electrophoresis coupled to different detectors was also proposed. (Carrasco-Pancorbo *et al.*, 2004). Bendini *et al.* (2007) summarized in a review paper the analytical method development for the identification and quantification of phenolic compounds in olive oil and olive fruits.

Due to the extreme diversity of the polar phenols, there are not commercial standards for every polar phenol and no standards library built-in to the analytical instruments. Thankfully, the use of mass spectroscopy detectors that provides structural information and the wealth of published data make identification and quantification work less cumbersome.

In cases where screening or estimation of the total phenol content is required, the Folin-Ciocalteau method is often used in the literature. This simple colorimetric method is based on the reaction of the free hydroxyl groups with a complex of phosphomolybdate and phosphotungstate and the spectroscopic detection of the concentration of this yellow complex using a spectrophotometer. The non-specificity of the method derives from the inclusion of all reducing agents (with free OH groups) that may be present in the extract in the result. Caffeic acid is often used as a standard for result expression. Polyphenol-rich olive oil is usually identified if polar phenol levels exceed 300 mg/kg of olive oil. With the development of more accurate and

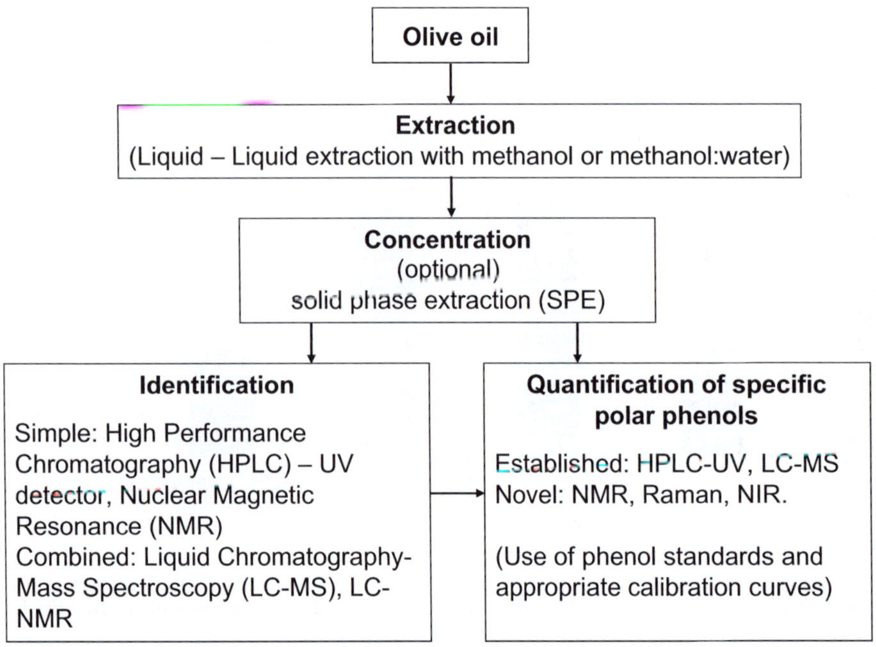

FIGURE 32.4 Relative stability index and polar phenol content losses of virgin olive oil after some different treatments or uses. The data represent a mixture of author's own experimental evidence and experience as well as cumulative evidence from different sources in the literature. This figure is not based in one study only and should therefore be used with caution.

powerful instrumental methods presented earlier, the Folin–Ciocalteau method remains today an useful—although limited—tool for the indication of polar phenol levels in olive oil extracts.

Evaluation of Antioxidant Capacity of Olive Oil and Polar Phenols

Evaluation of antioxidant activity of the total polar phenol fraction or individual phenols is usually based on determinations of the shelf-life of the oil. A standard olive oil may take more than one year to be oxidized and researchers are using accelerated oxidation tests such as Rancimat assay and indexes such as the Peroxide Value and absorbance constants K_{232}, K_{270} (EU, 2007). Methods have been developed to measure the antioxidant activity directly in the oil in addition to methods for the effect of phenolic extracts, pure phenols, or isolated fractions.

The antioxidant capacity of olive oil is determined by ABTS [2,2′-azino-bis(3-ethylbenzothiazoline-6-sulphonic acid)] and DPPH (1,7-diphenyl-2-picrylhydrazyl) radical decolorisation assays. The methods are applied after appropriate extraction or directly in the oil. Losses of antioxidant activity of olive oil due to heating, measured by the same assays were also reported by Pellegrini *et al.* (2001) and Gomez-Alonso *et al.* (2003). Determination of these assays is based on the radical scavenging capacity of polar phenols, i.e., their capacity to protect against oxidation by peroxyl radicals. The results are expressed as caffeic or gallic acid equivalent. Other methods include Oxygen radical absorbance capacity (ORAC), FRAP (Ferric-reducing antioxidant capacity) and Electron spin resonance (ESR) spectroscopy (Ninfali *et al.*, 2001; Gomez-Alonso *et al.*, 2003). Antioxidant behavior is more complex in emulsions than in pure triacylglycerol oil systems or bulk oil because there are more variables involved in lipid oxidation (pH, emulsifiers). Overall, antioxidant methods discussed here provide a good indication for the *in vitro* bioactivity of olive oil polar phenols and the autoxidation protection of olive oil. Their bioavailability and their role in health and nutrition, however, are much more complex.

References

Artajo, L.S., Romero, M.P., Moltiva, M.J., 2006. Enrichment of olive oil: antioxidant capacity functionalized with phenolic compounds. 4th EuroFed Lipid Congress, Madrid, Book of Abstracts, p. 471.

Beauchamp, G.K., Keast, R.S., Morel, D., Lin, J., Pika, J., Han, Q., Lee, C.H., Smith, A.B., Breslin, P.A., 2005. Ibuprofen-like activity in extra-virgin olive oil. Nature 437, 45–46.

Bendini, A., Cerretani, L., Carrasco-Pancorbo, A., Gomez-Caravaca, A.M., Segura-Carretero, A., Fernandez- Gutierrez, A., Lerker, G., 2007. Phenolic molecules in virgin olive oils: a survey of their sensory properties, health effects, antioxidant activity and analytical methods. An overview of the last decade. Molecules 12, 1679–1719.

Blekas, G., Boskou, D., 1999. Phytosterols and stability of frying oils. In: Boskou, D., Elmadfa, I. (Eds.), Frying of Food. Technomic Publishing, Lancaster, PA, pp. 205–222.

Boskou, D., 2009a. Culinary Applications of Olive Oil- Minor Constituents and Cooking. In: Boskou, D. (Ed.), Olive oil: minor constituents and health. CRC Press, FL, pp. 1–4.

Boskou, D., 2009b. Phenolic compounds in olives and olive oil. In: Boskou, D. (Ed.), Olive oil: minor constituents and health. CRC Press, FL, pp. 11–44.

Boskou, D., Tsimidou, M., Blekas, D., 2006. Polar phenolic compounds. In: Boskou, D. (Ed.), Olive Oil, Chemistry and Technology. AOCS Press, Champaign, IL, pp. 73–92.

Brenes, M., Garcia, M.C., Dobarganes, M.C., 2002. Influence of thermal treatments simulating cooking processes and the polyphenol content in virgin olive oil. J. Agric. Food Chem. 50, 5962–5967.

Briante, R., La Cara, F., Tonziello, M.P., Febbrio, F., Nucci, R., 2001. Antioxidant activity of the main bioactive derivatives from oleuropein hydrolysis by hyperthermophilic beta-glycosidase. J. Agric. Food Chem. 49, 3198–3203.

Caponio, F., Gomes, T., 2001. Influence of olive crushing temperature on phenols in olive oil. Eur. Food Res. Technol. 212, 156–159.

Carrasco-Pancorbo, A., Cruces-Blanco, C., Carretero, A.S., Gutierrez, F., 2004. Sensitive determination of phenolic acids in extra virgin olive oil by capillary zone electrophoresis. J. Agric. Food Chem. 52, 6687–6693.

Christophoridou, S., Dais, P., Tseng, L.H., Spraul, M., 2005. Separation and identification of phenolic compounds in olive oil by coupling high-performance liquid chromatography with post column solid phase extraction to nuclear magnetic resonance spectroscopy (LC-SPE-NMR). J. Agric. Food Chem. 53, 4667–4679.

Covas, M.I., de la Torre, K., Farre-Albaladejo, M., Kaikkonen, J., Fito, M., Lopez-Sabater, C., Pujadas-Bastardes, M.A., Joglar, J., Weinbrenner, T., Lamuela-Raventós, R.M., de la Torre, R., 2006. Postprandial LDL phenolic content and LDL oxidation are modulated by olive oil phenolic compounds in humans. Free Rad. Biol. Med. 40, 608–616.

Dais, P., Boskou, D., 2009. Detection and quantification of phenolic compounds, in olive oil, olives, and biological fluids. In: Boskou, D. (Ed.), Olive oil: minor constituents and health. CRC Press, IL, pp. 55–105.

European Commission, 2007. Council Regulation (EC) No. 1234/2007 of 22 October 2007 establishing a common organisation of agricultural markets and on specific provisions for certain agricultural product. Official Journal L 299, 16/11/2007, 1–149.

Gallina-Toschi, T., Cerretani, L., Bendini, A., Bonoli-Carbognin, M., Lercker, G., 2005. Oxidative stability and phenolic content of virgin olive oil: an analytical approach by traditional and high resolution techniques. J. Sep. Sci. 28, 850–870.

Garcia, A., Ruiz-Mendez, M.V., Romero, C., Brenes, M., 2006. Effect of refining on the phenolic composition of crude olive oil. J. Amer. Oil Chem. Soc. 83, 159–164.

Gomez-Alonso, S., Fregapane, G., Salvador Desamparados, M., 2003. Changes in phenolic composition and antioxidant activity of virgin olive oil during frying. J. Agric. Food Chem. 51, 667–672.

Koidis, A., Triantafillou, E., Boskou, D., 2008. Endogenous microflora in turbid virgin olive oils and the physicochemical characteristics of these oils. Eur. J. Lipid Sci. Technol. 110, 164–171.

Ninfali, P., Aluigi, G., Bacchiocca, M., Magnani, M., 2001. Antioxidant capacity of extra-virgin olive oil. J. Amer. Oil Chem. Soc. 78, 243–247.

Pellegrini, N., Visioli, F., Buratti, S., Brighenti, F., 2001. Direct analysis of total antioxidant activity of olive oil and studies on the influence of heating. J. Agric. Food Chem. 49, 2532–2538.

Petrakis, C., 1994. Good Manufacturing Practice (GMP) guidelines for virgin olive oil production. Grasas Aceites 45, 53–54.

Petrakis, C., 2006. Olive oil extraction. In: Boskou, D. (Ed.), Olive Oil, Chemistry and Technology. AOCS Press, Champaign, Illinois, pp. 191–225.

Psomiadou, E., Tsimidou, M., 2002. Stability of olive oil. II. Photo-oxidation studies. J. Agric. Food Chem. 50, 722–727.

Salvador, M., Aranda, F., Gomez-Alonso, S., Fregapane, G., 2003. Influence of extraction system, production year and area on Cornicabra virgin olive oil: a study of five crop seasons. Food Chem. 80, 359–366.

Shingai, Y., Fujimoto, A., Nakamura, M., Masuda, T., 2011. Structure and function of the oxidation products of polyphenols and identification of potent lipoxygenase inhibitors from Fe-catalyzed oxidation of resveratrol. J. Agric. Food Chem. 59, 8180–8186.

Tsimidou, M., Georgiou, A., Koidis, A., Boskou, D., 2005. Loss of stability in cloudy "veiled" olive oil during storage. Food Chem. 93, 377–383.

Tsimidou, M., Papadopoulos, G., Boskou, D., 1992. Phenolic compounds and stability of virgin olive oil. Food Chem. 45, 141–144.

Visioli, F., Caruso, D., Plasmati, E., Patelli, R., Mulinacci, N., Romani, A., 2001. Hydroxytyrosol, as a component of olive mill waste water, is dose-dependently absorbed and increases the antioxidant capacity of rat plasma. Free Radic. Res. 34, 301–305.

SECTION 5

MEATS

CHAPTER

33

Volatile Compounds in High-Pressure-Processed Pork Meat Products

Ana Rivas-Cañedo, Maria Teresa Díaz, Antonia Picón, Estrella Fernández-García, Manuel Núñez

Departamento de Tecnología de Alimentos, Instituto Nacional de Investigación y Tecnología Agraria y Alimentaria (INIA), Madrid, Spain

CHAPTER POINTS

- High-pressure-processing does not cause marked effects on the volatile fraction of pork meat products immediately after treatment. However, pressure-induced changes are perceptible during storage.
- The effect of high-pressure-processing on the volatile fraction greatly depends on the type of pork meat product and on treatment conditions, especially the pressure level.
- Pressurization limits microbial metabolism in cooked pork shoulder leading to a more stable volatile fraction. Pressure level determines the volatile profile after a prolonged storage period.
- The effect of high-pressure-processing on dry-cured products is very moderate. Only severe pressure-treatments enhance lipid oxidation. Processing procedures prior to pressure-treatment might increase the effect of pressurization.
- Formation of lipid oxidation compounds in dry-fermented sausages is enhanced by pressurization. The increase would likely be unnoticeable to consumers due to the spices added to the product.
- No single sampling technique renders a volatile profile identical to that actually present in the sample. The sampling technique should be carefully chosen to meet analytical requirements.

INTRODUCTION

Nowadays, consumers seek safe, long-lasting products with high-quality attributes, free from preservatives and additives. In this respect, mild preservation technologies not only address consumers' demands for minimally processed foods, but also meet other requirements, such as being energy saving and environmentally friendly. Among these technologies, high-pressure processing (HPP), which applies pressures between 300 and 900 MPa to food for a variable time, is one of the most promising for the meat industry. HPP at low temperatures inhibits pathogenic and spoilage microorganisms and inactivates some key food enzymes (Patterson, 2005), yielding a safe product with an extended shelf-life, while maintaining the label of 'preservative-free'. In addition, pressurization generally shows a minimal impact on the nutritional and sensory characteristics of the product, although this effect is matrix-dependent.

Concerning the meat sector, HPP is of particular interest in eliminating microbial cross-contamination during post-processing operations, such as deboning or slicing, which affect both safety and shelf-life. A main drawback when used for fresh meats is the color alteration induced by pressure, leading to a cooked-like aspect (Hugas *et al.*, 2002). This negative aspect of HPP with regard to meat is less pronounced in the case of meat products. For this reason, the most popular applications of HPP within the meat industry are those concerning ready-to-eat, cooked, and dry-cured meat products (Campus, 2010).

Aroma is probably one of the most important quality attributes of meat products, being closely related to their

Processing and Impact on Active Components in Food
http://dx.doi.org/10.1016/B978-0-12-404699-3.00033-0

volatile profile. Manufacturing procedures greatly determine the resulting volatile fraction. The study of the volatile compounds of meat products after treatments helps in identifying those changes that might affect the final flavor. Some chemical families of volatile compounds can be used to follow the extent of some critical reactions, such as the formation of linear aldehydes as an index of lipid oxidation. Literature on the effect of HPP on the volatile fraction of meat products is scarce. The large variation in treatment conditions (e.g., pressure level, temperature and time of exposure) found in the literature makes it difficult to draw an inference about the effect of pressurization on the volatile fraction of meat products.

As might be expected, the effect of HPP on the volatile fraction of meat products greatly depends on their composition and characteristics. Thus, results are here presented considering the type of meat product.

COMPOSITION OF THE VOLATILE FRACTION OF PORK MEAT PRODUCTS AS ALTERED BY HHP TREATMENTS

Cooked Meat Products

The first stages of the manufacturing process for cooked meat products, such as cooked ham or cooked pork shoulder, are the selection and preparation of the raw material, which includes deboning and trimming of connective and adipose tissues. The pieces are then injected with brine containing salt, sugar, nitrate or nitrite, ascorbate, and aromas, among others, and mechanically tumbled to allow the brine to penetrate the full piece. After tumbling, the pieces are cooked at a final core temperature of at least 67–71°C and then are allowed to cool (Toldrá and Flores, 2007). During the cooking stage, protein coagulation takes place, and most of the endogenous enzymes are inactivated. In this sense, although product conditions while temperature rises at the start of cooking are optimal to undergo lipolysis and proteolysis, enzymes responsible for these processes have a short time of action.

The volatile profile of cooked meat products is mainly due to thermally induced reactions, mostly Maillard reaction and lipid degradation (Mottram, 1998). In addition, the composition of the brine injected during the manufacturing process may contribute to the resulting volatile fraction. As an example, methylketones, terpenoids, linear aldehydes C>4, and benzene compounds were the most abundant volatile compounds in cooked pork shoulder, accounting for 21, 19, 11.5, and 8.5 %, respectively, of the volatile fraction (Rivas-Cañedo *et al.*, 2011).

Pressurization itself does not substantially modify the volatile profile of cooked meat products. In a study on sliced cooked pork shoulder pressurized at 400, 500, or 600 MPa for 5 or 10 min, followed by a 28-day storage period at 4°C under vacuum-packaging, minor changes were observed after HPP regardless of pressure level (Rivas-Cañedo *et al.*, 2011). However, the effects of HPP appeared afterwards throughout storage, with pressure level affecting the volatile fraction to a greater extent than time of exposure. Table 33.1 summarizes the main effects of HPP on the volatile profile of cooked pork shoulder. The volatile profile (leaving out ethanol, the major compound, to ease data comprehension) of control and HPP samples at days 0, 14, and 28 is depicted in Figure 33.1. The volatile profile of control samples underwent substantial changes during the storage period whereas that of HPP samples remained fairly stable during the first 14 days. Great differences among treatments were observed during the last two weeks of storage. At day 28, 400-MPa samples showed a volatile profile close to that of control samples, whereas the volatile fraction of 500- and 600-MPa samples looked much alike.

Compounds resulting from lipolysis and lipid oxidation do not seem to be affected by HPP. Only carboxylic acids C_4–C_8 tended to increase throughout storage, with control and 400-MPa samples showing higher levels than 500- and 600-MPa samples at day 28 (Rivas-Cañedo *et al.*, 2011). A more complex effect was recorded for compounds resulting from Strecker degradation. On the one hand, a decrease throughout storage of 2- and 3-methylbutanal in control and 400-MPa samples and an increase from day 14 onwards in 500- and 600-MPa samples was observed, with significantly higher levels in 500- and 600-MPa samples at day 28. On the other hand, the corresponding alcohols and acids tended to increase during refrigerated storage, with control samples showing higher levels than pressurized samples (Rivas-Cañedo *et al.*, 2011). The most marked effect of pressurization is probably that observed on compounds of microbial origin. HPP exerts a lethal effect on microorganisms and an inhibitory effect on the metabolism of surviving bacteria. Compounds such as ethanol, ethyl esters, and acetic acid were at lower levels in 500- and 600-MPa samples at day 28, whereas other compounds, namely ethanal, diacetyl, acetoin, and 2,3-butanediol were at higher levels (Rivas-Cañedo *et al.*, 2011).

Dry-Cured Meat Products

The first stages of the manufacture of dry-cured meat products involve the selection and preparation of the raw materials, which include bleeding and partial skin removal, when applicable. Afterwards, the pieces are rubbed on the surface with curing ingredients and additives (salt, nitrate and/or nitrite, ascorbic acid, etc.) and covered with layers of dry salt at 2–4°C for several days, depending on the piece's weight. The salt is then brushed

TABLE 33.1 Main Effects of HPP on the Volatile Compounds of Sliced Cooked Pork Shoulder

HPP Treatment					
Pressure Level (MPa)	**Exposure Time (min)**	**Temperature (°C)**	**Storage**	**Effects**	**Reference**
400 500 600	5 10	12	Vacuum/4°C/28 days	The pressure level had a more marked influence than the exposure time. A minor effect of HPP was recorded on day 0. HPP samples remained stable for 14 days. The pressure level determined the volatile profile after 14 days of storage. Lipid degradation: minor effect, with an increasing tendency of fatty acids C_4–C_8 throughout storage, less pronounced for 500 and 600 MPa samples. Strecker degradation: uneven effect. Branched-chain aldehydes showed a decreasing tendency throughout storage in control and 400-MPa samples and an increasing tendency in 500- and 600-MPa samples. Branched-chain alcohols and acids showed an increasing tendency throughout storage. Microbial metabolism: marked effect with an increasing tendency throughout storage and accumulation of volatile compounds dependent on the pressure level.	Rivas-Cañedo *et al.* (2011)

This table summarizes HPP and storage conditions together with the main effects of high-pressure on the volatile profile of cooked pork shoulder, as reported in the literature.

or rinsed off and the pieces are held at 0–4°C and 90% relative humidity (RH) for 20–60 days. Afterwards, the pieces are subjected to the ripening–drying stage, in which time, temperature and RH are combined for at least 6–9 months to achieve a final weight loss around 32–36% in relation to the initial weight.

Complex biochemical reactions leading to the final volatile profile take place during the manufacture of these products. Firstly, proteolysis and lipolysis generate peptides, free amino acids and free fatty acids that contribute to the flavor and the aroma of the product. These reactions are mainly due to endogenous enzymes, with a minor contribution of microbial enzymes (Toldrá, 1998), although molds and yeasts growing on the surface are also responsible for proteolysis and lipolysis. Secondly, lipid oxidation and further interactions of the resulting compounds with proteins, peptides and free amino acids, as well as Strecker degradation of free amino acids and Maillard reactions, generate most of the volatile compounds in dry-cured meat products (Toldrá, 1998). Their volatile fraction is composed of numerous compounds, such as aldehydes, ketones, alcohols, hydrocarbons, lactones, and esters (Toldrá, 1998). Aldehydes, benzene compounds and carboxylic acids were the most abundant chemical families among the 22 volatile compounds identified in dry-cured pork loin, accounting for 67% of the volatile fraction which also included methylketones, alcohols, alkanes, and pyrazines (Campus *et al.*, 2008). In the volatile profile of dry-cured Serrano ham, 108 compounds were identified, grouped into alkanes, alkenes, aldehydes, ketones (mostly methylketones), alcohols, esters, benzene compounds, sulfur compounds, alkylfurans, and pyrazines (Rivas-Cañedo *et al.*, 2009a). The most abundant chemical families were linear and branched-chain aldehydes, linear alkanes, secondary alcohols, and sulfur compounds.

Table 33.2 summarizes the main effects of HPP and storage on the volatile fraction of dry-cured meat products. HPP exerts a moderate effect on the volatile fraction of dry-cured meat products, probably because of their

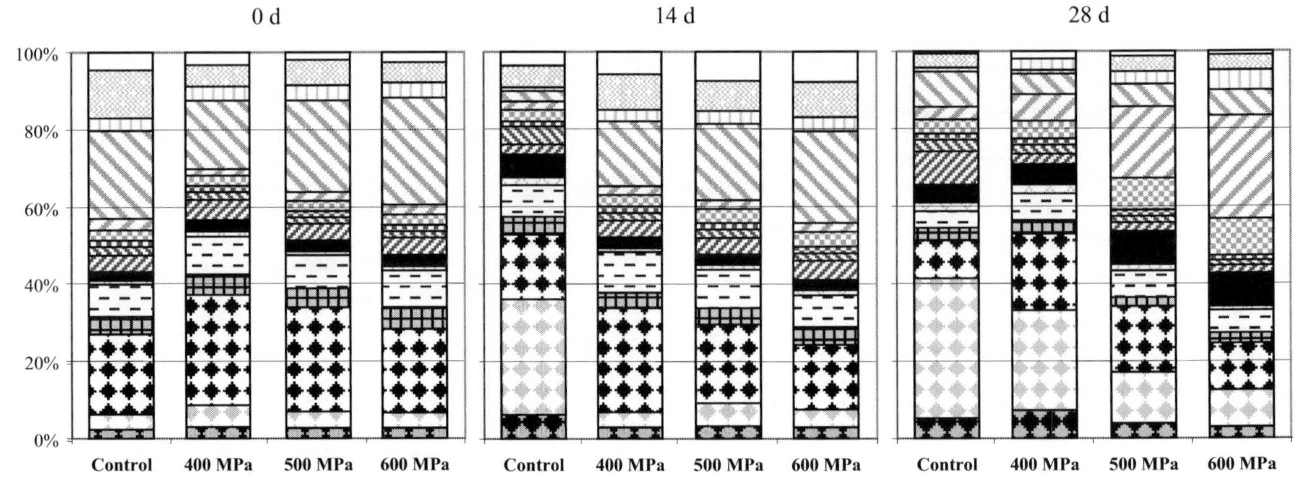

FIGURE 33.1 Volatile profile of control and HPP cooked pork shoulder during refrigerated storage (4°C). Linear alkanes, linear aldehydes C>4, branched-chain aldehydes, methylketones, diacetyl + acetoin, 2,3-butanediol, 1-alcohols C<4, 1-alcohols C>4, 2-alcohols, branched-chain alcohols, esters, benzene compounds, sulfur compounds, terpenoids, acetic acid, carboxylic acids C_4–C_9. The figure shows the relative percentage of the volatile compounds of control and pressurized cooked pork shoulder, grouped into the most relevant chemical families, at days 0, 14, and 28 of refrigerated storage (unpublished data).

low water activity and high stability. Concerning dry-cured pork loin subjected to 300, 350, or 400 MPa for 10 min at 20°C and then stored at 4°C in the dark for up to 45 days, a slight effect on lipo-oxidation volatile compounds was observed (Campus *et al.*, 2008). None of the isolated linear aldehydes, considered to be main secondary products and important indicators of lipid oxidation, were affected by pressurization. Only 2-pentanone was significantly affected by HPP, with an increase after HPP only in 400-MPa samples followed by decrease during refrigerated storage. The effect of HPP was very moderate on compounds associated with Strecker and Maillard reactions, with just a decrease in the levels of trimethylpyrazine in samples treated at pressures below 400 MPa (Campus *et al.*, 2008).

The effect of pressurization on lipo-oxidation compounds in dry-cured ham was dependent on the pressure level. No enhancement of lipid oxidation in sliced HPP (400 MPa, 10 min, 12°C) Serrano dry-cured ham was observed after 3 days of refrigerated storage (Rivas-Cañedo *et al.*, 2009a). Most of the volatile compounds affected by treatment showed lower levels in HPP samples, and only 2-heptanone increased after treatment. These compounds mainly derive from mold growth and metabolism. The lower levels in pressurized samples were related to the inhibition of mold metabolism because the storage period was too short to permit mold growth (Rivas-Cañedo *et al.* (2009a). On the contrary, a more severe HPP treatment (600 MPa, 6 min, 12°C) of Iberian dry-cured ham followed by 30 days of refrigerated storage under 12-h light/dark cycles significantly increased the levels of linear aldehydes and the perception of rancid odor in the sensory analysis (Fuentes *et al.*, 2010).

The composition of the volatile fraction of Iberian ham may be also altered by the type of packaging system. Thus, a marked effect of the packaging system on lipid oxidation was found (Table 33.2) when three different types of vacuum-packaging of samples for HPP treatment were compared (Fuentes *et al.*, 2010). These authors reported higher levels of lipid-derived aldehydes for Iberian ham slices piled up in the package than for stretched out slices or 90-g pieces.

HPP does not exert a marked effect on compounds associated with Maillard reaction and Strecker degradation. In dry-cured ham, Rivas-Cañedo *et al.* (2009a) did not report a significant effect of HPP on any of those compounds whereas Fuentes *et al.* (2010) observed a significant decrease in 2-methylbutanal levels only for one of the types of vacuum-packaged samples.

Dry-Fermented Meat Products

During the manufacture of dry-fermented sausages, chopped meat and pork fat are mixed with salt, sugars, nitrates and/or nitrites, sodium ascorbate, spices, and starter cultures. The mixture is held at 4°C for 24 h to facilitate interaction between its components and then stuffed into casings. Afterwards, sausages are subjected to temperatures of 18–26°C and 90% RH for 48 h in ripening cabinets and then held at 12–14°C and 75–85% RH for the ripening–drying stage (Ordóñez *et al.*, 1999). In some cases, sausages are subjected to low ripening temperatures (below 14°C) and a fast drying stage to limit fermentation. Products thus obtained, such as 'espetec', usually show final pH values of 5.5–6.0.

The volatile profile of dry-fermented products is generated through complex pathways during manufacture.

TABLE 33.2 Main Effects of HPP on the Volatile Compounds of Dry-Cured Meat Products

		HPP Treatment					
Meat Product	**Sample Format**	**Pressure Level (MPa)**	**Exposure Time (min)**	**Temperature (°C)**	**Storage**	**Effects**	**Reference**
Dry-cured pork loin	Slices	300 350 400	10	20	Vacuum/4°C/45 days/dark	Lipid oxidation: minor effect, with most lipo-oxidation volatile compounds unaffected. Strecker and Maillard reactions: negligible effect.	Campus *et al.* (2008)
Serrano dry-cured ham	Slices	400	10	12	Vacuum/4°C/3 days	Lipid oxidation: slight effect, with decreasing levels for most lipo-oxidation volatile compounds. Strecker and Maillard reactions: no effect of HPP.	Rivas-Cañedo *et al.* (2009a)
Iberian dry-cured ham	90-g pieces Stretched out slices Piled up slices	600	6	12	Vacuum/4°C/ 30 days/ 12-h light–dark cycles	Lipid oxidation: marked enhancement by HPP. Influence of the sample format. Strecker degradation: moderate effect, with decreasing levels of 2-methylbutanal in slices stretched out in the package.	Fuentes *et al.* (2010)

This table summarizes HPP and storage conditions together with the main effects of high-pressure on the volatile profile of dry-cured pork meat products, as reported in the literature.

In the fermentation stage, endogenous muscle enzymes and microbial enzymes from starter cultures, which include lactic acid bacteria (LAB), micrococcaceae, yeasts and molds, carry out glycolysis, proteolysis, and lipolysis (Toldrá, 2007). Carbohydrates are metabolized mainly by LAB to generate either lactic acid alone (homofermentative pathway) or lactic acid with variable amounts of volatile end products such as acetate, formate, ethanol or acetoin (heterofermentative pathway). The consequent pH drop favors proteolysis and lipolysis. Free amino acids generated during proteolysis are further metabolized to yield branched-chain aldehydes and secondary products such as acids, alcohols and esters (Toldrá, 2007). Free fatty acids resulting from lipolysis may undergo oxidative phenomena to yield, as secondary products, aliphatic hydrocarbons, alcohols, aldehydes and ketones (Toldrá, 2007). In addition, terpenoids from the spices added during the formulation of these products complete their volatile profile. According to Rivas-Cañedo *et al.* (2009b), the volatile fraction of dry-fermented 'salchichón' was mainly composed of terpenoids, which accounted for more than 80%, followed by alcohols and alkanes.

HPP does not markedly affect the volatile fraction of dry-fermented meat products, though some differences among products are observed. Table 33.3 summarizes the main effects of HPP and storage on two types of dry-fermented sausages. The effect of HPP on compounds derived from microbial metabolism is moderate. In 'salchichón', methylketones and 1-propanol were reported to decrease after HPP (400 MPa, 10 min, 12°C) followed by 3 days of refrigerated storage, whereas higher levels of acetic acid were observed in pressurized samples. Branched-chain aldehydes and toluene, coming from amino acid catabolism and Strecker degradation, were reported to increase after HPP (Rivas-Cañedo *et al.*, 2009b). In 'espetec' subjected to HPP (400 MPa, 10 min, 12°C) followed by 3 days of refrigerated storage, some compounds, namely ethanal, acetoin, and 2,3-butanediol, were found at lower levels in HPP samples whereas 2-propanol and ethyl acetate showed the opposite trend. Among compounds associated with amino acid metabolism, only toluene was affected by HPP, showing higher levels than in control samples.

The most marked effect of HPP on dry-fermented sausages is that concerning lipo-oxidation products, with compound-dependent results. HPP increased the levels of linear aldehydes, 1-alcohols, and enols, as well as of numerous linear alkanes in 'salchichón' (Rivas-Cañedo *et al.*, 2009b). Added spices would mask the effect on its sensory characteristics. On the contrary, compounds such as octane, pentanal, 1-alcohols and 1-octen-3-ol, decreased in HPP 'espetec' (Rivas-Cañedo *et al.*, 2012). It is worth noting that the levels of hexanal, known as an index of lipid oxidation in meat products, decreased in 'espetec' during storage, regardless of pressurization.

ANALYTICAL TECHNIQUES FOR THE STUDY OF THE VOLATILE FRACTION OF PORK MEAT PRODUCTS

The study of the volatile fraction in food products pursues objectives such as the characterization of compounds of sensory relevance or the ascertainment of the effect of manufacturing processes on those compounds. Volatile compounds, as part of a complex mixture, must be separated and identified. In this respect, gas chromatography (GC), which involves the analysis of organic compounds in the vapor phase at the typical GC operating temperatures between 40 and 300°C, is an ideal technique (Wampler, 1997). Volatile compounds are present at low levels (ppm, ppb, or even at ppt) in a complex matrix consisting of non-volatile components such as proteins and lipids. For this reason, isolating and concentrating the compounds by several orders of magnitude is necessary. However, the wide range of polarity, solubility, and pH of volatile compounds, together with the instability of many of them, make extraction and concentration a difficult task (Parliment, 1986). Although there are numerous sampling techniques for volatile compounds, no single procedure gives a complete quantitative or qualitative picture of the volatile profile of the original sample since every technique introduces biases in the resulting isolate (Reineccius, 2007).

Concerning the study of the effect of HPP on the volatile fraction of pork meat products, only two headspace sampling techniques, namely dynamic headspace extraction (DHE) and solid-phase microextraction (SPME), have been used. The advantages of these sampling techniques may have favored their use over others. When a piece of food is placed into a sealed vessel and allowed to stand, the more volatile compounds in the sample matrix will pass into the headspace. The concentration of volatile compound in the headspace depends on factors such as the amount of compound in the original sample, its volatility, the solubility of the compound in the sample matrix, and the temperature and volume of the vessel (Wampler, 1997). A portion of the surrounding headspace can be directly injected in the GC port (static headspace) or the headspace can be continuously swept away by a flow of inert carrier gas to be, afterwards, concentrated in either a cooled trap or suitable adsorbent materials (dynamic headspace sampling). In the latter case, a large amount of the volatile compound is forced to leave the sample matrix and pass to the headspace. Consequently, an increased amount of volatile analyte is obtained which will be afterwards transferred to the GC

TABLE 33.3 Main Effects of HPP on the Volatile Compounds of Fermented Sausages

		HPP treatment					
Meat Product	**Sample Format**	**Pressure Level (MPa)**	**Exposure Time (min)**	**Temperature (°C)**	**Storage**	**Effects**	**Reference**
Dry-fermented 'salchichón' (pH4.6–4.9)	Slices	400	10	12	Vacuum/4°C/3 days	Microbial metabolism: slight effect, with decreasing levels of most volatile compounds after HPP. Strecker degradation: increasing tendency with HPP. Lipid oxidation: significant effect of HPP, with increasing tendency of lipo-oxidation volatile compounds.	Rivas-Cañedo *et al.* (2009b)
Low-acid fermented 'espetec' (pH5.5–5.8)	8-cm pieces	400	10	12	Vacuum/ 4°C/3 days	Microbial metabolism: uneven effect, with decreasing levels for most volatile compounds and increasing levels for others. Strecker degradation: negligible effect. Lipid oxidation: most marked effect of HPP, with decreasing levels of lipo-oxidation volatile compounds in HPP samples.	Rivas-Cañedo *et al.* (2012)

The table summarizes HPP and storage conditions together with the main effects of high-pressure on the volatile profile of fermented sausages, as reported in the literature.

(Wampler, 1997). Numerous sorbents can be used as trapping materials determining the resulting volatile profile (Pillonel *et al.*, 2002). This sampling technique offers many advantages, including simple sample preparation, elimination of solvent peak and increased sensitivity. Main disadvantages are related to the complexity of the instrumentation, i.e, valve leaking (Wampler, 1997).

Although SPME was initially developed for the analysis of water pollutants, it has found a wide range of applications in the food sector (Balasubramanian and Panigrahi, 2010). The SPME device, which can be used either in the direct extraction mode or in the headspace configuration, consists in a modified syringe containing a solid-core fused silica fibre coated with a film of stationary phase. In the headspace application, the food sample is placed in a sealed vial and the FIBRE is exposed to the headspace of the sample for a period of time, during which the volatile compounds are ab/adsorbed onto the stationary phase. Afterwards, the trapped compounds are thermally desorbed in the injection port of the GC and transferred to the column. Headspace extraction is based on the partitioning of analytes among three phases: sample matrix, headspace and the fibre coating. Each compound will behave differently depending on its polarity, volatility, partition coefficient, etc. (Harmon, 1997). Again, different parameters, such as temperature and time of extraction or rate of agitation, affect extraction. Headspace SPME shows many advantages, including elimination of solvent, simplicity, sensitivity and ease of automation.

Rivas-Cañedo *et al.* (2012) compared DHE and SPME in pressurized cooked pork shoulder and 'espetec'. Under the conditions of their study, DHE extracted more efficiently low-boiling alkanes and aldehydes as well as alcohols, because of the constant renewal of the headspace by the carrier gas, whereas SPME extracted more efficiently a higher number of chemical families, including esters, fatty acids, sulfur, and benzene compounds. The resulting volatile profile, therefore, differed depending on the extraction technique. In addition, the effect of pressurization on 'espetec' and cooked pork shoulder was better categorized by DHE and SPME, respectively, thus marking the importance of the choice of the sampling technique.

References

Balasubramanian, S., Panigrahi, S., 2010. Solid-phase microextraction (SPME) techniques for quality characterization of food products: a review. Food Bioprocess Technol. 4, 1–26.

Campus, M., 2010. High pressure processing of meat, meat products and seafood. Food Eng. Rev. 2, 256–273.

Campus, M., Flores, M., Martinez, A., Toldrá, F., 2008. Effect of high pressure treatment on colour, microbial and chemical characteristics of dry cured loin. Meat Sci. 80, 1174–1181.

Fuentes, V., Ventanas, J., Morcuende, D., Estévez, M., Ventanas, S., 2010. Lipid and protein oxidation and sensory properties of vacuum-packaged dry-cured ham subjected to high hydrostatic pressure. Meat Sci. 85, 506–514.

Harmon, A.D., 1997. Solid-phase microextraction for the analysis of flavours. In: Marsili, R. (Ed.), Techniques for Analyzing Food Aroma. Marcel Dekker, Inc., New York, pp. 81–112.

Hugas, M., Garriga, M., Monfort, J.M., 2002. New mild technologies in meat processing: high pressure as a model technology. Meat Sci. 62, 359–371.

Mottram, D.S. 1998. Flavour formation in meat and meat products: a review. Food Chem. 62, 415–424.

Ordóñez, J.A., Hierro, E.M., Bruna, J.M., De la Hoz, L., 1999. Changes in the components of dry-fermented sausages during ripening. Crit. Rev. Food Sci. Nutr. 39, 329–367.

Parliment, T.H., 1986. Sample preparation techniques for gas–liquid chromatographic analysis of biologically derived aromas. In: Parliment, T.H., Croteau, R. (Eds.), Biogeneration of Aromas. American Chemical Society, Washington, DC, pp. 34–52.

Patterson, M.F. 2005. Microbiology of pressure-treated foods. J. App. Microbiol. 98, 1400–1409.

Pillonel, L., Bosset, J.O., Tabacchi, R., 2002. Rapid preconcentration and enrichment techniques for the analysis of food volatile. A review. Lebensmittel-Wiss. Technol. 35, 1–14.

Reineccius, G.A., 2007. Flavour-isolation techniques. In: Berger, R.G. (Ed.), Flavours and Fragances: Chemistry, Bioprocessing and Sustainability. Springer, Berlin, pp. 409–426.

Rivas-Cañedo, A., Fernández-García, E., Nuñez, M., 2009a. Volatile compounds in dry-cured Serrano ham subjected to high pressure processing. Effect of the packaging material. Meat Sci. 82, 162–169.

Rivas-Cañedo, A., Juez-Ojeda, C., Nuñez, M., Fernández-García, E., 2011. Effects of high-pressure processing on the volatile compounds of sliced cooked pork shoulder during refrigerated storage. Food Chem. 124, 749–758.

Rivas-Cañedo, A., Juez-Ojeda, C., Nuñez, M., Fernández-García, E., 2012. Volatile compounds in low-acid fermented sausage "espetec" and sliced cooked pork shoulder subjected to high pressure processing. A comparison of dynamic headspace and solid-phase microextraction. Food Chem. 132, 18–26.

Rivas-Cañedo, A., Nuñez, M., Fernández-García, E., 2009b. Volatile compounds in Spanish dry-fermented sausage 'salchichón' subjected to high pressure processing. Effect of the packaging material. Meat Sci. 83, 620–626.

Toldrá, F., 1998. Proteolysis and lipolysis in flavour development of dry-cured meat products. Meat Sci. 49, S101–S110.

Toldrá, F., 2007. Fermented Meat Production. In: Hui, Y.H. (Ed.), Handbook of Food Products Manufacturing. John Wiley & Sons, Inc., New Jersey, pp. 265–279.

Toldrá, F., Flores, M., 2007. Sensory quality of meat products. In: Hui, Y.H. (Ed.), Handbook of Food Products Manufacturing. John Wiley & Sons, Inc., New Jersey, pp. 303–328.

Wampler, T.P., 1997. Analysis of food volatiles using headspace–gas chromatographic techniques. In: Marsili, R. (Ed.), Techniques for Analyzing Food Aroma. Marcel Dekker, Inc., New York, pp. 27–58.

CHAPTER

34

Imidazole Dipeptides in Meat from Different Animal Species and Effect of Cooking Method on their Contents in Beef and Turkey Meat

*Pier Giorgio Peiretti**, *Claudio Medana*†

*Institute of Science of Food Production, National Research Council, Grugliasco, Italy, †Department of Molecular Biotechnology and Health Sciences, University of Torino, Torino, Italy

CHAPTER POINTS

- This study summarizes the content of two papers on imidazole dipeptide content in meat from different animal species and the effect of cooking method in beef and turkey meat.
- Total imidazole dipeptide content was in decreasing order: turkey > chicken > horse > pork > rabbit > beef.
- Boiling beef caused a 50% loss in carnosine; cooking by microwave/grill led to a reduction in carnosine of 10%.
- The greatest loss of anserine and homocarnosine content in beef meat was observed for the boiling method (70%).
- In turkey meat, the lowest decreases in carnosine, anserine and homocarnosine content were observed for microwave/grilling and for broiling at high temperature.
- This research has found some indications about preparation methods that could guarantee low carnosine and homologue degradation.
- Freeze-dried samples of meat of beef (fillet), horse (fillet), pork (loin), chicken (breast), turkey (breast) and rabbit (hind leg) were analyzed to determine their imidazole dipeptide contents.
- Beef and turkey meat were prepared by commonly used cooking methods and then frozen in liquid nitrogen, freeze-dried and stored at −20°C until analyzed.
- Carnosine, anserine and homocarnosine were identified in the meat on the basis of the preliminary separation/fragmentation study and quantified using calibration curves obtained from the standard carnosine.
- The multistage mass spectrometry study using the Orbitrap technology allowed us to clarify that homocarnosine is a homohistidine derivative.

INTRODUCTION

Carnosine (β-alanyl-L-histidine), anserine (β-alanyl-L-3-methyl-L-histidine) and homocarnosine—whose structure, elucidated in a previous paper of ours (Peiretti *et al.*, 2011), is γ-aminopropyl-L-homohistidine—are endogenous imidazole dipeptides found in mammalian tissues. The exact chemical structure of homologs of carnosine and their relative ratio in meat from different species is the subject of various published works. Ophidine (β-alanyl-2-methyl-L-histidine) has been identified in pig meat and anserine/ophidine ratio has been proposed to distinguish between genuine and imitation hams (Carnegie *et al.*, 1982). Balenine (β-alanyl-1-methyl-L-histidine) was recognized in meat samples by Harris and Milne (1987). Recently, the occurrence of a sixth imidazolic amino acid in addition to histidine, carnosine, anserine, ophidine, and balenine was reported (Yamada *et al.*, 2009). A more complete analytical determination of these isomers is currently in progress in our laboratory (data not shown) to extend the mass spectrometric

Processing and Impact on Active Components in Food
http://dx.doi.org/10.1016/B978-0-12-404699-3.00034-2

characterization with regard to anserine, balenine, and homocarnosine described by Aldini *et al.* (2004). These dipeptides have been extensively studied because of their physiological effects, including antioxidant properties (Yanai *et al.*, 2004) or buffering activity (Abe, 2000) and their potential therapeutic applications (Quinn *et al.*, 1992), such as a possible role in anti-ageing (Hipkiss and Brownson, 2000). Chan and Decker (1994) reported that carnosine and anserine could inhibit lipid oxidation through a combination of free radical scavenging and metal chelation. Wu *et al.* (2003) reported that anserine and carnosine are electron donors and could react with free radicals to convert them to more stable products and terminate radical chain reactions.

HOW COMPOSITION IS ALTERED

Before 2010, the availability of data on carnosine, anserine, and homocarnosine content in meat of different species was limited. Analogously, the effect of cooking methods on imidazole dipeptide content was not so well documented. The objectives of our research reported herein were, first of all, to determine the content of these compounds in raw meats obtained from different species, and, secondly, to measure the changes in concentration in beef and turkey meat samples submitted to different cooking treatments.

Firstly, we found that total imidazole dipeptides were abundant in the following decreasing order: turkey > chicken > horse > pork > rabbit > beef, while the susceptibility of these meats to lipid oxidation increased from chicken to beef, with the exception of turkey meat, which presented a high thiobarbituric acid reactive substance content, even though its total imidazole dipeptide content was the highest (Peiretti *et al.*, 2011). The final w/w percentages of carnosine, anserine and homocarnosine present in each sample are shown in Table 34.1. The mean percentage content of carnosine *vs.* total imidazole dipeptides is lowest in turkey meat (19.9%), followed by rabbit (23.3%) and chicken (30.6%) meat, while beef, pork, and horse meat display a higher carnosine percentage (88.0%, 89.3%, and 98.9%, respectively). As far as the mean percentage content of anserine *versus* total imidazole dipeptides is concerned, an almost inverse trend was observed; the meats with the highest anserine levels were those with the lowest carnosine contents. Turkey, rabbit, and chicken meat were very rich in anserine, with a percentage of between about two- and four-times that of carnosine. Carnosine, by contrast, was present in horse meat in very high amounts compared to anserine (about 10-fold higher) and there was also a greater quantity of carnosine than anserine in pork and beef. The mean percentage content of homocarnosine *versus* total imidazole dipeptides was highest in pork, turkey, and chicken meats, while it was present only in a small percentage in horse meat. Finally, the greatest amount of total imidazole dipeptides was found in turkey meat, while the lowest percentage of meat antioxidants was reported in beef.

Due to the high variable content of antioxidant compounds and in order to clarify whether thermal treatments can affect food protein cross-linking, we describe the detection and quantification of carnosine, anserine, and homocarnosine in turkey and beef meat submitted to different cooking methods (Peiretti *et al.*, 2012). The mean percentages of carnosine and its homologs in cooked meat compared to their initial values in raw beef and turkey meat are reported in Figures 34.1 and 34.2, respectively. In the case of beef, it can be seen that boiling led to a mean loss of 50% of carnosine compared to the corresponding raw meat sample. A probable cause of this depletion is the high water solubility of carnosine ($LogD_{7.4} = -0.23$, as reported by Nielsen *et al.*, 2002) and related compounds. Cooking by broiling at medium and high temperatures led, respectively, to a reduction in carnosine of between 40 and 50% and of 30–40% in raw meat. Cooking with a microwave oven with or without a grill led to a smaller carnosine loss (mean values of 7% and 19%, respectively), thus confirming that they are good cooking methods. As far as anserine and homocarnosine content is concerned, the greatest loss was again observed for boiling (mean values of about 70%); while in increasing order of loss, the heating methods for the latter two molecules followed the sequence: microwave/grill (mean value 35%), microwave irradiation (mean value 43%), broiling at high temperature (mean value 58%), and broiling at medium temperature (mean value 66%). This confirms yet again that boiling led to the maximum antioxidant loss, and that microwave heating is the mildest method and that grilling at a high temperature induces correspondingly less antioxidant degradation. This suggests that the formation of a hard surface layer on the meat prevents degradation of carnosine and related compounds. A similar trend was observed for turkey meat, even though the antioxidant loss was relatively small compared to that observed for beef. Cooking did not cause a significant decrease in homocarnosine content in turkey meat.

Jones *et al.* (2011) analyzed 10 commonly consumed foods within the British diet and found the highest level of carnosine in beef, followed by pork, chicken, turkey, and lamb, while anserine content was highest in turkey and chicken meats, and lowest in lamb, beef, and pork.

Dissimilar results concerning the effect of cooking on carnosine amounts have been found in literature. Whereas some authors have reported that imidazole dipeptides, such as carnosine, are unaffected by cooking (Park *et al.*, 2005; Mora *et al.*, 2008), others obtain relatively small

TABLE 34.1 Carnosine, Anserine, and Homocarnosine Contents (%w/w) in Raw Meats Obtained from Different Species (Triple Extraction, Means ± S.D.)

Meat Cut	Supermarket	Carnosine	Anserine	Homocarnosine
Horse fillet	A	1.98 ± 0.62	0.020 ± 0.0044	0.0065 ± 0.0032
Horse fillet	B	2.54 ± 0.13	0.020 ± 0.0018	0.0040 ± 0.0018
Horse fillet	C	1.73 ± 0.083	0.010 ± 0.0026	0.0037 ± 0.00093
Pork loin	A	1.80 ± 0.15	0.059 ± 0.0093	0.16 ± 0.037
Pork loin	B	1.57 ± 0.39	0.041 ± 0.0084	0.14 ± 0.047
Pork loin	C	1.76 ± 0.015	0.050 ± 0.0031	0.18 ± 0.020
Beef fillet	A	1.49 ± 0.65	0.28 ± 0.14	0.024 ± 0.0085
Beef fillet	B	1.22 ± 0.40	0.098 ± 0.028	0.017 ± 0.0081
Beef fillet	C	1.43 ± 0.18	0.16 ± 0.015	0.014 ± 0.0036
Rabbit hindleg	A	0.46 ± 0.084	1.45 ± 0.18	0.042 ± 0.012
Rabbit hindleg	B	0.36 ± 0.085	1.25 ± 0.41	0.054 ± 0.040
Rabbit hindleg	C	0.39 ± 0.015	1.12 ± 0.12	0.027 ± 0.0016
Chicken breast	A	0.70 ± 0.20	1.70 ± 0.53	0.16 ± 0.056
Chicken breast	B	0.92 ± 0.36	1.67 ± 0.68	0.093 ± 0.059
Chicken breast	C	0.76 ± 0.25	1.46 ± 0.43	0.11 ± 0.059
Turkey breast	A	0.36 ± 0.13	2.11 ± 0.82	0.15 ± 0.032
Turkey breast	B	0.79 ± 0.18	2.46 ± 0.69	0.12 ± 0.039
Turkey breast	C	0.62 ± 0.15	2.19 ± 0.78	0.19 ± 0.062

Table from Peiretti et al. *(2011).*

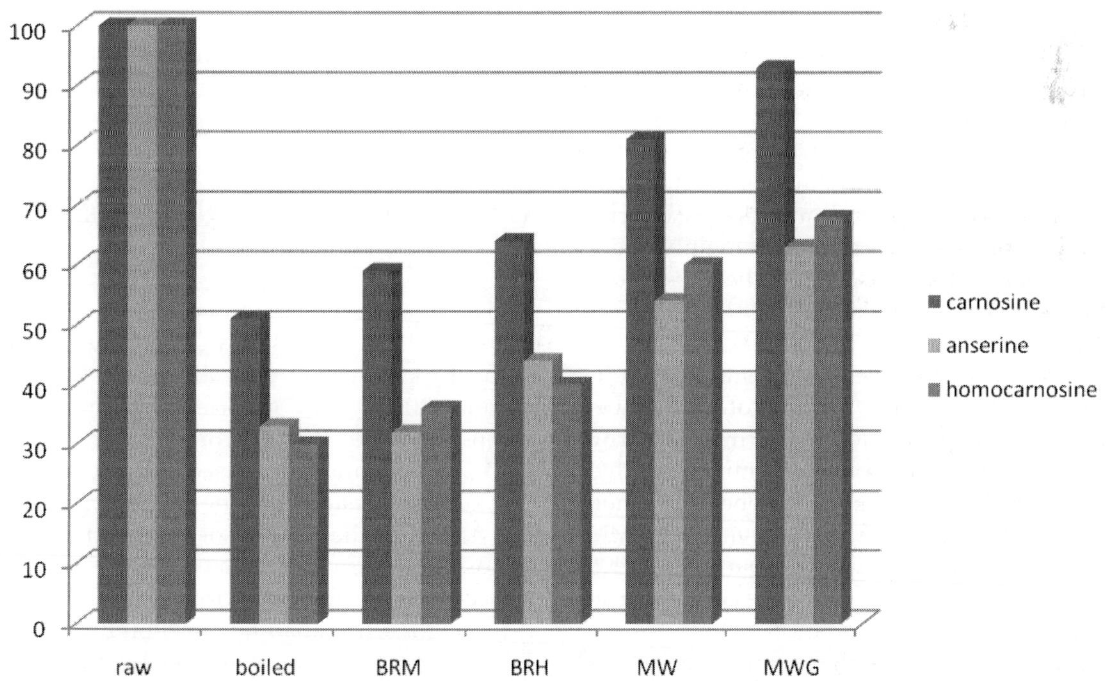

FIGURE 34.1 The mean carnosine, anserine and homocarnosine w/w percentage values measured in the beef samples. Raw meat (not cooked); boiled in water 100°C × 10′; BRM, broiled at a medium temperature 140°C × 10′; BRH, broiled at a high temperature 200°C × 10′; MW, microwave irradiation 100°C × 7′; MWG, microwave irradiation + grill 70°C × 3′ + 7′. *Figure from Peiretti* et al. *(2012).*

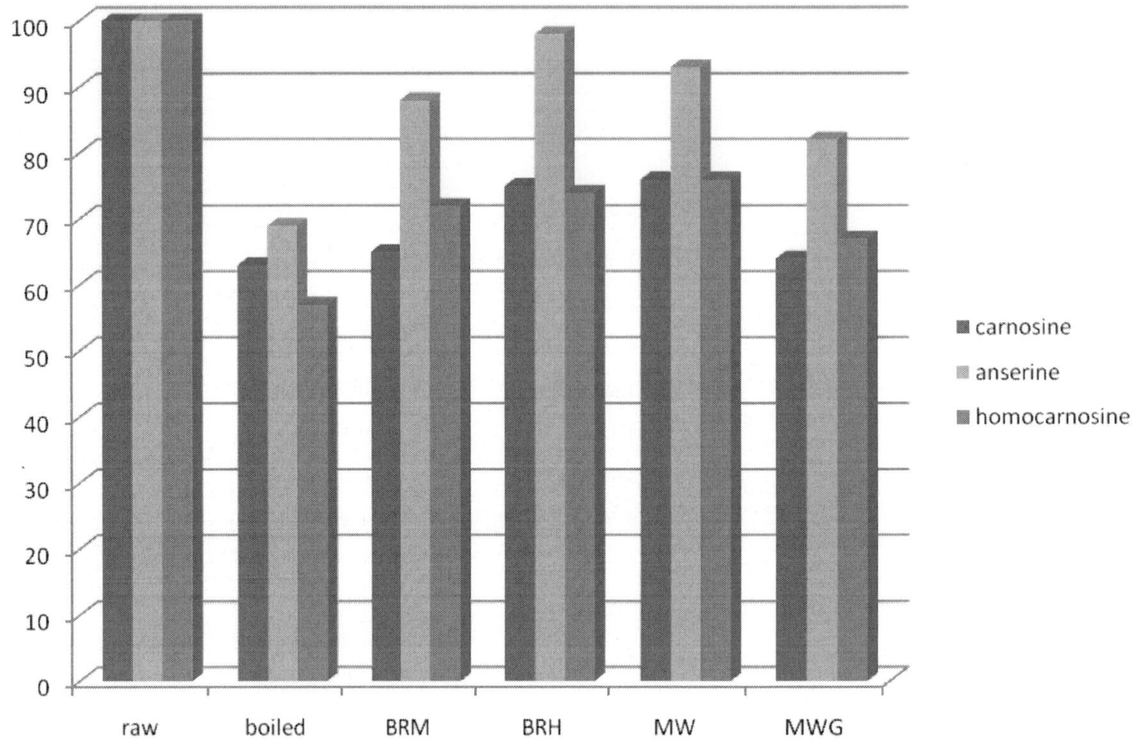

FIGURE 34.2 The mean carnosine, anserine, and homocarnosine w/w percentage values measured in the turkey samples. Raw meat (not cooked); boiled in water 100°C × 10′; BRM, broiled at a medium temperature 140°C × 10′; BRH, broiled at a high temperature 200°C × 10′; MW, Microwave irradiation 100°C × 7′; MWG, microwave irradiation + grill 70°C × 3′ + 7′. *Figure from Peiretti* et al. *(2012).*

differences in carnosine amounts between uncooked and cooked beef meat (Purchas *et al.*, 2004a).

Park *et al.* (2005) found that carnosine concentrations increased from 274 to 277 mg/100 g of meat solids (dry weight basis) for raw and cooked beef, respectively, and suggested that no significant degradation of carnosine occurred during cooking. The authors propose that the lower carnosine concentration could be due to the higher fat content of the beef used in their experiment. A higher fat content could decrease carnosine concentrations, since carnosine is mainly found in the cytosol of skeletal muscle and the presence of adipose fat could decrease skeletal muscle carnosine concentrations.

Furthermore, heat treatments that remove proteins from untreated extracts may cause protein denaturation by aggregation. In previous experiments, carnosine was extracted from many skeletal muscle sources including beef, pork, and chicken using heat treatment (Chan *et al.*, 1993; Gopalakrishnan, *et al.*, 1999; Maikhunthod and Intarapichet, 2005). Chan *et al.* (1993) suggested that the lack of complete recovery of carnosine might be due to the degradation of carnosine or an association of carnosine with proteins precipitated during concentration. Gopalakrishnan *et al.* (1999) extracted carnosine from mechanically separated pork at 60, 70, and 80°C and found that protein content in undried extracts decreased significantly with increasing temperatures. Maikhunthod and Intarapichet (2005) determined carnosine content in the extracts of broiler meat prepared by heating at 60, 80, and 100°C. They found that protein concentrations in the extract decreased while carnosine increased with increasing temperatures.

OTHER WAYS IN WHICH COMPOSITION IS ALTERED

The involvement of dipeptides in the modulation of enzymatic activities and during the heating of beef soup stock solution was discussed by Kuroda and Harada (2000). They reported that levels of histidine and β-alanine in the macromolecular fraction of beef soup stock solution increased during the heating process, and suggested that the formation of this isopeptide occurs in the heating of various foods containing carnosine.

Decker and Faraji (1990) found that heating carnosine at 100°C for 15 min had no effect on its ability to inhibit the catalysis of lipid oxidation by iron, hemoglobin, lipoxidase and singlet oxygen, therefore concluding that carnosine has excellent potential for use as a natural antioxidant in processed foods. Bauchart *et al.* (2006) found

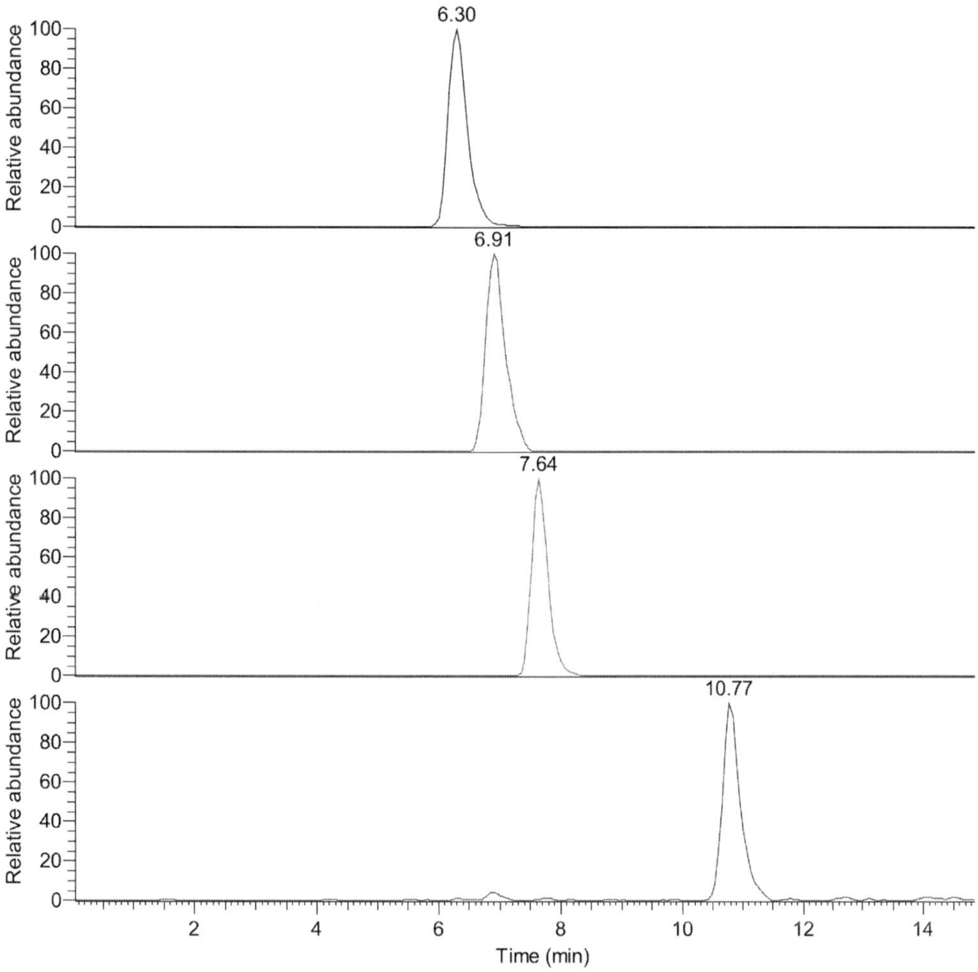

FIGURE 34.3 Carnosine (top line), anserine (second line), homocarnosine (third line), and phenylephrine I.S. (fourth line). MS^2 chromatograms of the microwave cooked turkey samples. A selected reaction monitoring of *m/z* 227–110 transition is reported for carnosine. The difference in the signals due to product ions originated by *m/z* 241 precursor ion is reported for isobaric anserine and homocarnosine: anserine *m/z* 170–224; homocarnosine *m/z* 224–170. Analytes concentrations of this sample were 8.16 (carnosine), 36.34 (anserine), 3.17 (homocarnosine) and 20.0 (phenylephrine) mg/l. *Figure from Peiretti* et al. *(2012).*

that carnosine, anserine and glutathione account for 8–9% of peptide amino acids in fresh muscle and showed good carnosine and anserine stability during beef aging. The levels of these compounds were lower in cooked meats than in fresh muscle and this could be attributed to losses in the cooking juice. At the same time, numerous larger compounds, most probably peptides, were generated in a very reproducible manner during aging and even more during cooking.

A moderate loss of carnosine was described by Purchas *et al.* (2004a) during cooking of beef *semitendinosus* muscle, but the mechanisms remained unclear, probably largely through losses in the cooking juices, while slow cooking (90 min at 70°C) of lamb *longissimus* and *semimembranosus* muscles led to significant reductions in carnosine content. Purchas *et al.* (2004b) found that losses of carnosine were higher in surface samples than in inner samples, but it was not possible to determine the extent to which this was due to losses in exudates from the steaks or to a breakdown of these compounds as a result of the higher temperatures. Carnosine levels decreased by 26% following cooking treatment in a clam cooker, set at 200°C for an internal meat temperature of 71°C (Purchas *et al.*, 2006). It was not possible to determine the extent to which the reductions on cooking were due to losses in cooking juices or to changes in the compounds.

Pereira-Lima *et al.* (2000) studied the effect of heating conditions on the presence of carnosine and anserine in beef broths and found a significant correlation between carnosine and anserine levels and heating temperature and time. They suggest that cooking temperature is more important than heating time in determining the concentrations of these dipeptides. These results could be due to the solubilization and liberation of these compounds at higher cooking temperatures, a variable which has a more pronounced effect than cooking time on the structural changes of meat (Spanier *et al.*, 1990).

ANALYTICAL TECHNIQUES

Meat Sample Preparation

In a first trial (Peiretti *et al.*, 2011), three meat samples of cuts of beef (fillet), horse (fillet), pork (loin), chicken (breast), turkey (breast), and rabbit (hind leg) were purchased directly from three local supermarkets in Turin, Italy, for a total of 18 meat cut samples. In a second trial (Peiretti *et al.*, 2012), commercial samples of beef (fillet) and turkey (breast) (600 g per tray with six slices of meat each) were purchased from two local supermarkets in Turin, Italy. A slice of raw meat (100 g, thickness ~0.4 cm for turkey breast and ~0.6 cm for beef fillet) for each species and from each supermarket was immediately frozen at −20°C and then freeze-dried until analyzed. The other meat slices were prepared for standard cooking with commonly used cooking methods: (1) broiled at 200°C for 10 min (5 min per side) in a meat grill (Tefal model GC3001) until a core temperature of 100°C was reached; (2) broiled at a medium temperature (140°C) for 10 min (5 min per side) until a core temperature of 70°C was reached; (3) cooked by microwave (3 min) and then grilled (7 min), until a core temperature of 70°C was reached; (4) cooked in a domestic microwave oven (Whirlpool model AMW234 WH) with a power setting of 750 W (10 min, 5 min per side), until a core temperature of 100°C was reached; and (5) boiled in water (100°C) for 10 min. The internal temperature was determined using TP 642 thermocouple piercing probes attached to a digital thermometer (HD 8464; Delta OHM, Padova, Italy).

All the cooked meat samples were immediately frozen in liquid nitrogen, freeze-dried and then stored at −20°C until analyzed. The samples were analyzed in triplicate. Each sample was weighed (50 mg) in a 10-ml polyethylene tube, together with the internal standard (phenylephrine hydrochloride) and treated with 0.6 M trichloroacetic acid in order to denature the proteins. After centrifugation (15 min, 9000*g*), 500 µl of supernatant solution was 10-fold diluted using 5 mM heptafluorobutanoic acid and transferred into vials for analysis.

Measurements of Imidazole Dipeptides Content

Standard solutions were prepared, using a starting eluent as a solvent, from water–methanol 1:1 stock solutions in the range 1–200 mg/l (Peiretti *et al.*, 2011, 2012).

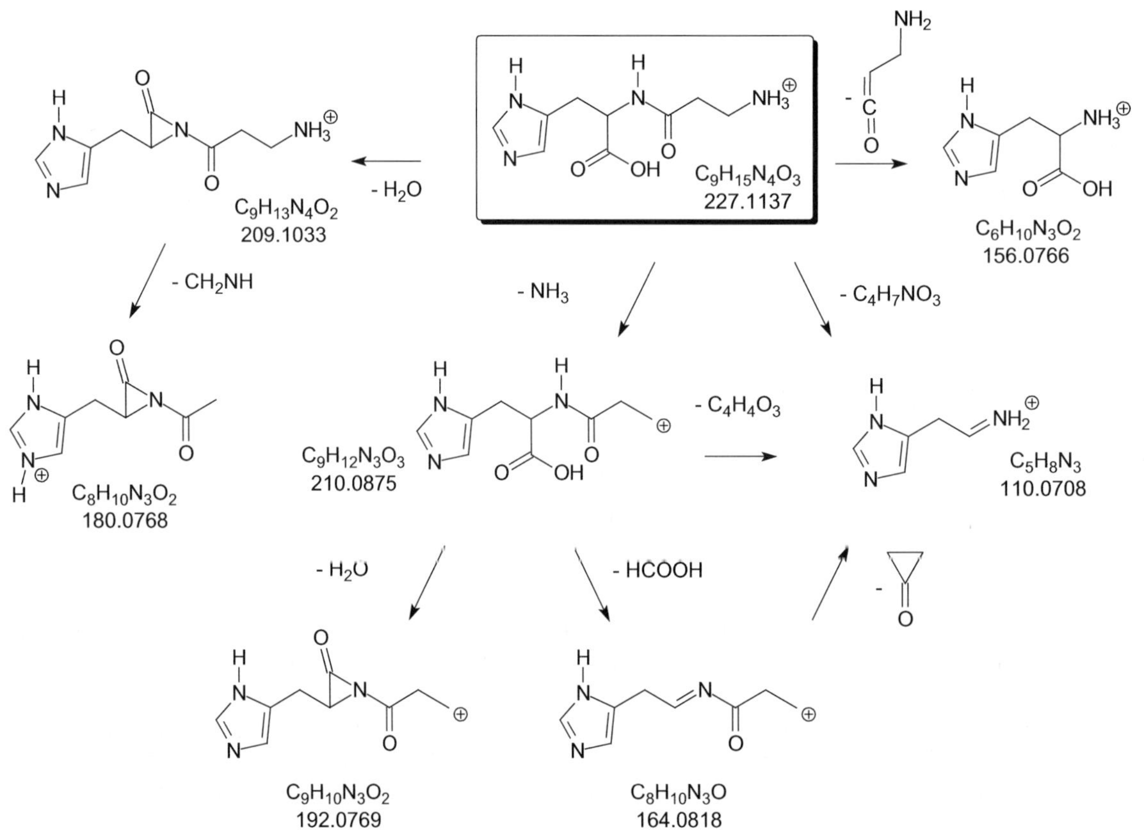

FIGURE 34.4 Carnosine MS² and MS³ fragmentation pathway. Multistage attributions were confirmed through MS³ experiments. *Figure from Peiretti* et al. *(2011).*

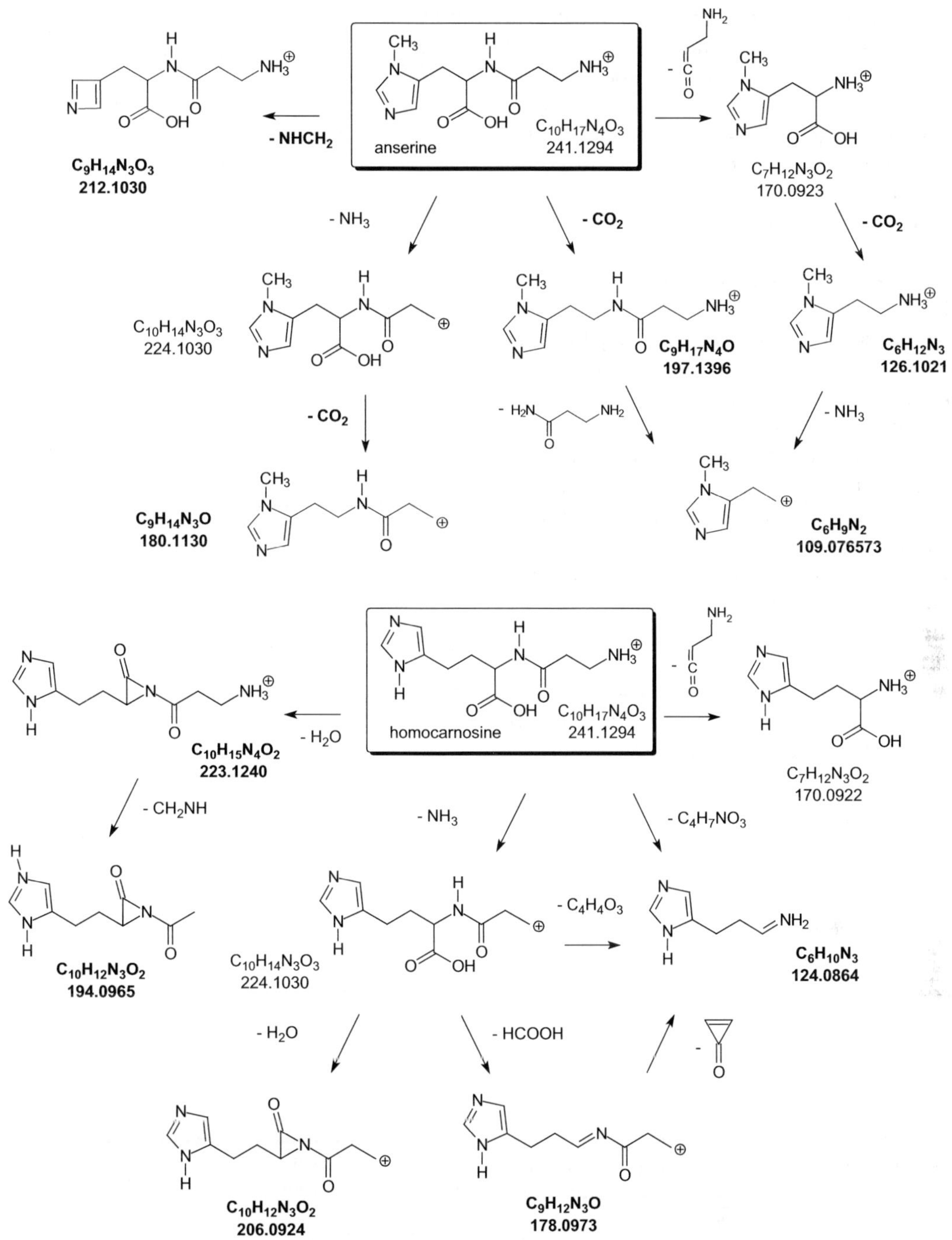

FIGURE 34.5 The main anserine and homocarnosine MS^2 and MS^3 fragmentation pathways. The ions that were useful for characterization are in bold. *Figure from Peiretti* et al. *(2011).*

For each meat type (beef, rabbit, pork, horse, chicken, and turkey), a standard addition calibration curve was built in order to exclude ion suppression effects due to the matrix.

The chromatographic separations were run on an Ultimate 3000 HPLC (Dionex, Milan, Italy) coupled to a high resolving power mass spectrometer LTQ Orbitrap (Thermo Scientific, Rodano, Italy), equipped with an atmospheric pressure interface and an ESI ion source. The samples were analyzed using an RP C18 column (Phenomenex Synergi 150 × 2.1 mm, 3-μm particle size) at a flow rate of 200 μl/min. A gradient mobile phase composition was adopted: 95/5 to 40/60 in 25 min, 5 mM heptafluorobutanoic acid/acetonitrile. The injection volume was 5 μl. Analyses were run using both full MS (50–700 *m/z* range) and MS/MS acquisition in the positive ion mode.

Figure 34.3 shows the MS^2 chromatograms of carnosine (top line), anserine (second line) and homocarnosine (third line) of the microwave-cooked turkey meat sample. The anserine and homocarnosine peaks were difficult to resolve by simple LC-MS analysis and originate a single asymmetrical peak. In order to obtain a good visible discrimination, the anserine and homocarnosine signals were reported as the arithmetic difference between the main product ions originated by the positional isomeric precursor ions at *m/z* 241. Anserine was evidenced by subtracting the intensity of *m/z* 224 from that of *m/z* 170, while the homocarnosine peak rises were shown by subtracting *m/z* 170 − 224. This was possible owing to the different relative intensity between these two ions.

Figure 34.4 shows the relative fragmentation interpretation of carnosine. All of the proposed pathways were confirmed by MS^3 analysis. Ammonia loss is the most intense process; the ion obtained originates three main product ions in successive fragmentations by eliminating formic acid, water and cycloprop-2-en-1-one.

The anserine and homocarnosine isobaric compounds protonate to give two ions at 241.1300 *m/z*, with different structures and spectra. The main differences in the fragmentation pathways are shown in Figure 34.5: the N-methylation that occurs in anserine makes $NHCH_2$ loss and decarboxylation possible, both from the precursor ion and from the first-generation product ions; the homocarnosine spectrum shows the same pattern as carnosine, but shifted by one methylene unit (14.0157 u).

References

Abe, H., 2000. Role of histidine-related compounds as intracellular proton buffering constituents in vertebrate muscle. Biochemistry-Moscow 65, 757–765.

Aldini, G., Orioli, M., Carini, M., Maffei Facino, R., 2004. Profiling histidine-containing dipeptides in rat tissues by liquid chromatography/electrospray ionization tandem mass spectrometry. J. Mass Spec. 39, 1417–1428.

Bauchart, C., Rémond, D., Chambon, C., Patureau Mirand, P., Savary-Auzeloux, I., Reynès, C., Morzel, M., 2006. Small peptides (<5 kDa) found in ready-to-eat beef meat. Meat Sci. 74, 658–666.

Carnegie, P.R., Hee, K.P., Bell, A.W., 1982. Ophidine (β-alanyl-L-3-methylhistidine, 'balenine') and other histidine dipeptides in pig muscles and tinned hams. J. Sci. Food Agr. 33, 795–801.

Chan, K.M., Decker, E.A., Means, W.J., 1993. Extraction and activity of carnosine, a naturally occurring antioxidant in beef muscle. J. Food Sci. 58, 1 4.

Chan, K.M., Decker, E.A., 1994. Endogenous skeletal muscle antioxidants. Crit. Rev. Food Sci. Nutr. 34, 403–426.

Decker, E.A., Faraji, H., 1990. Inhibition of lipid oxidation by carnosine. J. Am. Oil Chem. Soc. 67, 650–652.

Gopalakrishnan, J., Decker, E.A., Means, W.J., 1999. Antioxidant activity of mechanically separated pork extracts. Meat Sci. 52, 101–110.

Harris, C.I., Milne, G., 1987. The identification of the N tau-methyl histidine-containing dipeptide, balenine, in muscle extracts from various mammals and the chicken. Comp. Biochem. Physiol. B 86, 273–279.

Hipkiss, A.R., Brownson, C., 2000. A possible new role for the anti-ageing peptide carnosine. Cell. Mol. Life Sci. 57, 747–753.

Kuroda, M., Harada, T., 2000. Incorporation of histidine and β-alanine into the macromolecular fraction of beef soup stock solution. J. Food Sci. 65, 596–603.

Jones, G., Smith, M., Harris, R., 2011. Imidazole dipeptide content of dietary sources commonly consumed within the British diet. Proc. Nutr. Soc. 70 (OCE6), E363.

Maikhunthod, B., Intarapichet, K.O., 2005. Heat and ultrafiltration extraction of broiler meat carnosine and its antioxidant activity. Meat Sci. 71, 364–374.

Mora, L., Sentandreu, M.A., Toldra, F., 2008. Effect of cooking conditions on creatinine formation in cooked ham. J. Agric. Food Chem. 56, 11279–11284.

Nielsen, C.U., Supuran, C.T., Scozzafava, A., Frokiaer, S., Steffansen, B., Brodin, B., 2002. Transport characteristics of L-carnosine and the anticancer derivative 4-toluenesulfonylureido-carnosine in a human epithelial cell line. Pharm. Res. 19, 1337–1344.

Park, Y.J., Volpe, S.L., Decker, E.A., 2005. Quantitation of carnosine in human plasma after dietary consumption of beef. J. Agric. Food Chem. 53, 4736–4739.

Peiretti, P.G., Medana, C., Visentin, S., Giancotti, V., Zunino, V., Meineri, G., 2011. Determination of carnosine, anserine, homocarnosine, pentosidine and thiobarbituric acid reactive substances contents in meat from different animal species. Food Chem. 126, 1939–1947.

Peiretti, P.G., Medana, C., Visentin, S., Dal Bello, F., Meineri, G., 2012. Effect of cooking method on carnosine and its homologues, pentosidine and thiobarbituric acid-reactive substance contents in beef and turkey meat. Food Chem. 132, 80–85.

Pereira-Lima, C.I., Ordoñez, J.A., García de Fernando, G.D., Cambero, M.I., 2000. Influence of heat treatment on carnosine, anserine and free amino acid composition of beef broth and its role in flavour development. Eur. Food Res. Technol. 210, 165–172.

Purchas, R.W., Rutherfurd, S.M., Pearce, P.D., Vather, R., Wilkinson, B.H.P., 2004a. Concentrations in beef and lamb of taurine, carnosine, coenzyme Q10, and creatine. Meat Sci. 66, 629–637.

Purchas, R.W., Rutherfurd, S.M., Pearce, P.D., Vather, R., Wilkinson, B.H.P., 2004b. Cooking temperature effects on the forms of iron and levels of several other compounds in beef semitendinosus muscle. Meat Sci. 68, 201–207.

Purchas, R.W., Busboom, J.R., Wilkinson, B.H.P., 2006. Changes in the forms of iron and in concentrations of taurine, carnosine, coenzyme Q10, and creatine in beef *longissimus* muscle with cooking and simulated stomach and duodenal digestion. Meat Sci. 74, 443–449.

Quinn, P.J., Boldyrev, A., Formazuyk, V.E., 1992. Carnosine its properties, function and potential therapeutic applications. Mol. Asp. Med. 13, 379–444.

Spanier, A.M., McMillin, K.W., Miller, J.A., 1990. Enzyme activity levels in beef: effect of postmortem aging and end-point cooking temperature. J. Food Sci. 55, 318–326.

Wu, H.-C., Shiau, C.-Y., Chen, H.-M., Chiou, T.-K., 2003. Antioxidant activities of carnosine, anserine, some free amino acids and their combination. J. Food Drug Anal. 11, 148 153.

Yamada, S., Kawashima, K., Baba, K., Oku, T., Ando, S., 2009. Occurrence of a novel acetylated amino acid, Nα-acetylhistidine, in skeletal muscle of freshwater fish and other ectothermic vertebrates. Comp. Biochem. Physiol. B 152, 282–286.

Yanai, M.N., Shiotani, S., Mizuno, M., Nabetani, H., Nakajima, M., 2004. Characteristics of antioxidant activity of carnosine and anserine mixture isolated from chicken extract: comparison with other botanical antioxidants. J. Jpn. Soc. Food Sci. Technol. 51, 238–246.

CHAPTER

35

Cantonese Sausage, Processing, Storage and Composition

Weizheng Sun, Feibai Zhou, Mouming Zhao

College of Light Industry and Food Sciences, South China University of Technology, Guangzhou, China

CHAPTER POINTS

- Composition of Cantonese sausage varies depending on their formula, process and storage characteristics.
- Proteins undergo substantial structural, functional, nutritional, and componential changes during processing of Cantonese sausage.
- Salting, drying condition, proteolysis and protein oxidation account for most of the observed protein structural and functional changes during Cantonese sausage processing.
- The content of the water-soluble protein (sarcoplacmic proteins) and the salt-soluble protein (myofibrillar proteins) decrease markedly due to proteolysis and protein denaturation.
- The rate of protein digestibility *in vitro* decreased during Cantonese sausage processing, indicating a loss of nutritional quality.
- The kind and content of peptides from proteolysis, possessing antioxidant activities, varies widely, depending on process parameters.
- The lipolysis of phospholipids in muscle, especially PC, was the main contributor to the increase of free fatty acids, while in adipose tissue neutral lipid, and phospholipids were equal in their release of free fatty acids.
- Residual nitrite content decreased gradually during processing and storage.

INTRODUCTION

Cantonese sausage is a traditional Chinese meat product. It is one traditional kind of semi-dry sausage made in the region of Guangdong province, in the south of China. It is characterized by its good conservation capacity at room temperature as a low-humidity meat product. It has gained much popularity and acceptance among Chinese sausages and is recognized as a product with low water activity values, a natural wrinkle, a characteristic aroma, a typical red color, and a usefully long life without refrigeration.

Cantonese sausage material is made from a mixture of lean pork and pork back fat (Sun *et al.*, 2011e). The quality level of Cantonese sausage is divided into three grades, namely fancy grade, grade one, and grade two, according to the ratio of lean pork and pork back fat. The ratio of lean pork and pork back fat of the fancy grade, grade one, and grade two are 8:2, 7:3, and 6:4, respectively (Wu *et al.*, 2010). Its ingredients include sugars, salt, nitrite, Chinese liquor (in some references called wine) or soy sauce. It is characterized by the addition of relatively high amount of Chinese liquor (4%) and sugar (8–14%). Water (20%) can be added up to 20% of the meat to facilitate mixing. These raw materials were mixed together using a mixer machine and stuffed into natural pork casing of variable diameters. It hung vertically in drying chambers for three days (approximately 72h). The Cantonese sausage formula does not include culture starters and thus it is a natural fermentation process (Wu *et al.*, 2010). Drying is the terminal processing operation in the production of Cantonese sausages. High temperature (42–60°C) is used during drying, resulting in a unique cured meat flavor.

Cantonese sausage processes involve complex physicochemical and biochemical behaviors which are directly related to the final quality (Sun *et al.*, 2011e). Changes in the majority components (proteins and lipids) and the added ingredients (sugar, nitrite, and condiments) in the preparation of the original mixture are attributed to the above physicochemical and biochemical behaviors (Ordóñez *et al.*, 1999). Proteolysis and lipolysis constitute two of the most important biochemical mechanisms

Processing and Impact on Active Components in Food
http://dx.doi.org/10.1016/B978-0-12-404699-3.00035-4

during the processing of Cantonese sausage with relevant consequences for changes in the components.

Proteolysis during the ripening of Cantonese sausage involves a decrease of high molecular-weight proteins, resulting in the formation of polypeptides and free amino acids (Sun *et al.*, 2009). Polypeptide fragments can be further degraded by peptidyl peptidases and aminopeptidases which will lead to the production of smaller peptides and individual amino acids (Mullen *et al.*, 2000). Bioactive peptides, abundant in animal products, are known to have a wide range of activities, including antihypertension, antioxidant, antimicrobial, and immunomodulatory activities, in addition to nutrient utilization (Elias *et al.*, 2008; Su *et al.*, 2011).

It is well-known that several muscle proteases (cathepsins, calpains, peptidases, and aminopeptidases) and lipases (lysosomal acid lipase, acid phospholipase, and adipose tissue lipase) contribute to the proteolysis and lipolysis during the processing of cured meat products (Toldrá, 2006). Cantonese sausages are dried at 42–60°C for 3 days and this temperature does not result in a total loss of enzyme activity. These enzymes remain active during storage of Cantonese sausage. Thus storage also can alter composition of Cantonese sausage.

Therefore, it is of both scientific and commercial interest to understand how typical processing and storage contribute to the composition of Cantonese sausage.

EFFECT OF PROCESSING AND STORAGE ON PROTEIN COMPOSITION OF CANTONESE SAUSAGE

The meat proteins can be divided into three different groups based on function: myofibrillar, sarcoplasmic, and connective tissue proteins. The myofibrillar proteins constitute between 50 and 55% of the total protein content, while the sarcoplasmic proteins account for approximately 30–34%. The remaining 10–15% of the proteins are the connective tissue proteins (Tornberg, 2005). The meat proteins as the main constituents make up the structure of the meat and meat product (Tornberg, 2005). From the standpoint of nutrition, they are probably the most important. They undergo substantial structural, functional, nutritional and compositional changes during processing of cured meat products (Ordóñez *et al.*, 1999; Sun *et al.*, 2011a, 2011c).

Protein structural change

During Cantonese sausage processing, conditions such as heating, salting, mixing, and degradation may change the structure of proteins. Structural changes of myofibrillar proteins of Cantonese sausage during processing were elucidated using Raman spectroscopy. There was a decrease ($p<0.05$) in α-helix accompanied by an increase ($p<0.05$) in β-sheet structure during 0–18 h and 36–72 h. However, reciprocal changes were found during 18–36 h. Sun *et al.* (2011d) considered that a rebuilding process of secondary structures occurred at the final stage of drying. They used Fourier deconvolution combined with iterative curve fitting and the hierarchical cluster analysis to analyze the Raman spectrum and proved the above rebuilding process. Results from Fourier transform infrared (FTIR) spectroscopy also confirmed this rebuilding process. Sun *et al.* (2011d) found that the obvious bands at 1686/cm appeared and normalized intensities of the I_{1652}/I_{2061} of myofibrillar proteins increased during the rest in the processing period.

With regard to changes in secondary structure of sarcoplasmic proteins (measured by CD) during Cantonese sausage processing, α-helical content markedly decreased from 21.1% to 5.6%, while β-strand increased from 18.7% to 33.1% as processing time proceeded.

Surface hydrophobicity (H_0) is able to monitor subtle changes in chemical and physical states of protein. H_0 can be a suitable parameter to estimate protein changes (Chelh *et al.*, 2006). During Cantonese sausage processing, the H_0 increased at the first stage of drying and thereafter remained constant or even decreased slightly at the rest stage (Sun *et al.*, 2011b, d).

Salting, thermal treatment, proteolysis and protein oxidation during Cantonese sausage processing contribute to the meat protein structural changes. Salting could change the FTIR absorption frequency of the band around 1653/cm of myofibrillar proteins, as the frequency of this band shifted from 1651/cm in unsalted samples to 1654/cm in 9% cured samples (Wu *et al.*, 2006). The muscle fibers are broken after the chopping process. Myofibrillar proteins are extracted due to the action of the salt. The salt facilitates electrostatic processes involved in the formation of the film of protein that surrounds the fatty particles and favors protein–fat and protein–protein interactions (Tornberg, 2005).

It is generally accepted that heat treatment can enhance protein surface hydrophobicity. Protein hydrophobicity had a 2.5-fold increase between 30 and 70°C during thermal denaturation of actomyosin in threadfin bream (Yongsawatdigul and Park, 1999). Chelh *et al.* (2006) have also observed a considerable increase in hydrophobicity of myofibrillar proteins, which was positively correlated to heating time and temperature. Hydrophobicity of meat proteins (including sarcoplasmic proteins and myofibrillar proteins) increased during Cantonese sausage drying between 42°C and 60°C for about three days, which was caused by exposure of the molecules of interior hydrophobic amino acids. Slow heating rate could slowly unravel conformation of meat proteins, exposing more hydrophobic domains at any given temperatures, which would promote aggregation of meat proteins (Yongsawatdigul and Park, 1999).

Protein oxidation can lead to exposure of hydrophobic amino acids, which are buried inside in native status (Sante-Lhoutellier *et al.*, 2007). Protein oxidation is the structural modification induced directly by reactive oxygen species (Roseiro *et al.*) or indirectly by reaction with the by-products of lipid peroxidation (Lund *et al.*, 2011; Shacter, 2000). Total protein carbonyl group content and loss of free sulphydryl groups (SH) are the routine indicators used to evaluate the protein oxidation level. During Cantonese sausage processing, the total carbonyl content was found to significantly increase ($p < 0.05$) in meat proteins, while the SH groups level decreased accompanied by the gradual increases in S-S group level (Sun *et al.*, 2011a).

Protein hydrolysis also results in numerous smaller molecules exposing more hydrophobic regions hitherto concealed in interior proteins (Wu *et al.*, 1998). In a native protein molecule, hydrophobic groups are buried inside the core of the folded structure. After partial hydrolysis, some of these groups are exposed, resulting in increased surface hydrophobicity. Excessive enzyme hydrolysis can lead to a decrease in the surface hydrophobicity of protein due to the fact that small peptides could have fewer hydrophobic binding sites than larger peptides. This could be the reason that H_0 remained constant or even decreased slightly at the rest stage of Cantonese sausage processing.

Protein Compositional Change

In order to evaluate protein change during Cantonese sausage processing, meat proteins were classified into five fractions on the basis of their solubility, namely the non-protein nitrogen (NPN), the water-soluble protein (sarcoplacmic proteins), the salt-soluble protein (myofibrillar proteins), the alkali-soluble fraction, and the alkali-insoluble fraction. As processing time proceeded, changes in the composition of Cantonese sausage proteins (Figure 35.1) were illustrated by the progressive decrease in the water-soluble protein fraction (sarcoplasmic proteins) and the salt-soluble protein fraction (myofibrillar proteins), accompanied by an increase in the alkali-soluble and non-protein fractions (Sun *et al.*, 2011a). Similar phenomena were found in the processing of the Thai-style fermented pork sausage (Nham) (Visessanguan *et al.*, 2004).

Loss in solubility of Cantonese sausage sarcoplasmic proteins and myofibrillar proteins indicated their denaturation or other structural changes during drying due to heat treatment and proteolysis. Protein denaturation and aggregation processes could induce a sharp decrease in native protein molecules. The alkali-soluble fraction is thought to be the denatured myofibrillar fraction due to their electrophoretic pattern was the similar to actomyosin (Visessanguan *et al.*, 2004). An increase in this fraction could also be attributed to protein aggregation, induced by protein oxidation. In order to further estimate changes in protein solubility, the proteins of Cantonese sausage during processing were solubilized in five different solvent systems which can disrupt ionic, hydrogen, and disulfide bonds as well as hydrophobic interactions (Sun *et al.*, 2011a). The results suggested that hydrophobic interactions, hydrogen, ionic, and disulfide bonds played important roles in the formation of a three-dimensional network structure of Cantonese sausage proteins.

Sugar addition levels affected the protein composition of Cantonese sausage during processing. A reduction in sugar level addition induced a decrease in sarcoplasmic and myofibrillar protein content while increasing the NPN and alkaline-soluble fraction proteins (Qiu *et al.*, 2012). A different degree of proteolysis might be

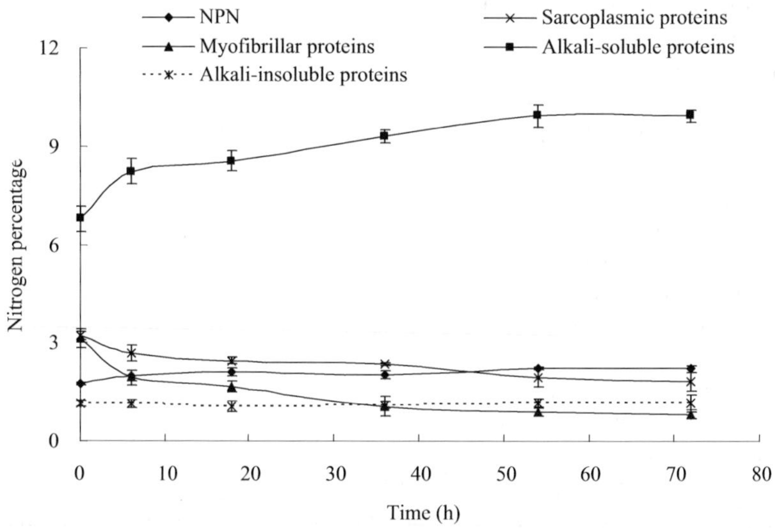

FIGURE 35.1 Changes in the nitrogen percentages of protein fractions in Cantonese sausage during processing. *Adapted from Sun* et al. *(2011a).*

attributable to the above phenomena. Cantonese sausage with a higher sugar concentration may suppress the proteolysis ability of muscular enzyme due to lower a_w and higher osmotic pressure. The protective effect of sugar against protein denaturation also impacts on proteolysis.

Protein Nutritional Change

Meat proteins are important nutritional sources for humans. Proteins must be broken down into amino acids or small peptides before passing through the small intestine wall and entering the bloodstream and, therefore, nutritional quality of meat proteins is largely dependent on their digestibility. Nevertheless, Cantonese sausage processes (salting, drying, sugar addition) can affect the meat proteins structure and, thus, their digestibility. Meat protein oxidation and aggregation have an influence on their degradation by enzymes of the digestive tract (Sante-Lhoutellier *et al.*, 2007).

The effect of the physicochemical changes of sarcoplasmic and myofibrillar proteins, especially oxidation behavior, on their mechanism of action on *in vitro* protein digestibility during Cantonese sausage processing has been evaluated (Sun *et al.*, 2011b, d). For gastric pepsin, protein digestibility and rate of proteolysis of sarcoplasmic proteins increased to 26.77% and 0.0201 over the first 6h drying, while it decreased over 6–72h (Table 35.1). For pancreatic trypsin and α-chymotrypsin, protein digestibility and rate of proteolysis of sarcoplasmic proteins during processing showed significant decrease ($p<0.05$) from 53.26% and 0.0591 to 32.65% and 0.0350, respectively (Table 35.1). For gastric pepsin, the rate of proteolysis of myofibrillar proteins decreased during processing (Figure 35.2). For pancreatic trypsin and α-chymotrypsin, the rate of proteolysis of myofibrillar proteins increased significantly ($p<0.05$) firstly, then decreased ($p<0.05$) at the final stage (Figure 35.2). Results from correlation analysis showed that protein physicochemical changes, especially protein oxidation and aggregation, influenced protein digestibility and thus resulted in a loss of nutritional quality.

EFFECT OF PROCESSING AND STORAGE ON PEPTIDE RELEASE IN CANTONESE SAUSAGE

Proteolysis can release some peptides during Cantonese sausage processing (Sun *et al.*, 2009). The peptides related to the sensory aspects of meat products were studied in depth. In recent years, other bioactive properties, such as antioxidant activity, have been receiving increasing attention. Peptides from Cantonese sausages at different drying periods were divided into two fractions (P1 and P2). As shown in Figure 35.3, the content of P1 remained constant at the initial drying stage and increased significantly ($p<0.05$) at the rest drying stage. The content of P2 increased during the first 6h, and finally remained consistently at approximately 160mg/100g at rest stage (Figure 35.3).

Antioxidant activity was evaluated for each peptide fraction (Figure 35.3). The DPPH radical scavenging activity of P1 increased significantly ($p<0.05$) during processing except between 18 and 36h. The DPPH radical scavenging activity of P2 gradually increased throughout drying ($p<0.05$, 92.27% at 72h). Their amino acid compositions and molecular weight distribution contributing to antioxidant activity were discussed, suggesting that amino acid residues of His, Glu, Gly, and Arg are very important to the radical scavenging activity of peptides (Sun *et al.*, 2009).

Proteolysis is mainly attributed to endogenous proteases and microbial proteases. The activities of the above enzymes significantly depend on the processing conditions such as temperature, time, water activity, redox potential, and salt content (Toldrá *et al.*, 1997). The antioxidant properties of peptides are largely dependent on their nature, such as molecular weight, amino acid composition and their sequences (Chen *et al.*, 1998). As is well known, hydrolysates produced by different enzymes from the same protein source have different antioxidant activities. Therefore, characteristics of proteolysis occurring during the different processing and storage stages, including degree and type of protease action, are extremely important for the content and antioxidant activity of these released peptides. The product with higher salt concentration had low peptide release due to the influence of NaCl on the water binding capacity of the protein structure and proteases (Roseiro *et al.*, 2008).

Thus, to obtain good antioxidant peptides during the ripening of meat products, development of new processes or improvements in existing processes may be useful. In addition, the addition of proteases and an efficient starter may be innovative techniques.

EFFECT OF PROCESSING AND STORAGE ON LIPID COMPOSITION OF CANTONESE SAUSAGE

Lipids (2–40%) are added to Cantonese sausage, which play an important role in the typical flavor development. During Cantonese sausage processing, lipids undergo lipolysis controlled by both lipases and phospholipases, which cause free fatty acids release and thus compositional changes. Lipid composition has great impact on the functionality, oxidative stability and flavor development of meat products. Because of the differences in the ripening condition and materials used in Cantonese sausage,

TABLE 35.1 *In Vitro* Protein Digestibility of Sarcoplasmic Proteins from Raw Muscle and Cantonese Sausage at Different Processing Periods*

	Protein Digestibility	Rate of Proteolysis	Standardized Coefficients
PEPSIN DIGESTIBILITY			
Control	19.76 ± 1.13[ab]	0.0163 ± 0.0005[bc]	0.976 ± 0.007[a]
0 h	22.44 ± 1.68[ab]	0.0173 ± 0.0012[bc]	0.974 ± 0.016[a]
3 h	19.92 ± 1.88[ab]	0.0155 ± 0.0012[ab]	0.948 ± 0.024[a]
6 h	26.7 ± 1.07[c]	0.0201 ± 0.0012[d]	0.946 ± 0.029[a]
18 h	22.16 ± 1.93[ab]	0.0168 ± 0.0007[bc]	0.948 ± 0.018[a]
36 h	23.35 ± 1.77[b]	0.0183 ± 0.0010[cd]	0.957 ± 0.010[a]
54 h	20.39 ± 2.09[ab]	0.0152 ± 0.0017[ab]	0.918 ± 0.033[a]
72 h	18.48 ± 0.77[a]	0.0136 ± 0.0011[a]	0.909 ± 0.036[a]
TRYPSIN AND α-CHYMOTRYPSIN DIGESTIBILITY			
Control	56.07 ± 2.22[d]	0.0599 ± 0.0014[d]	0.854 ± 0.010[a]
0 h	53.26 ± 2.63[d]	0.0591 ± 0.0027[d]	0.862 ± 0.004[a]
3 h	55.07 ± 1.25[d]	0.0602 ± 0.0016[d]	0.830 ± 0.041[a]
6 h	47.65 ± 2.24[c]	0.0518 ± 0.0037[c]	0.808 ± 0.069[a]
18 h	46.88 ± 2.79[c]	0.0521 ± 0.0026[c]	0.832 ± 0.019[a]
36 h	42.03 ± 0.86[b]	0.0455 ± 0.0018[b]	0.788 ± 0.044[a]
54 h	33.90 ± 1.20[a]	0.0368 ± 0.0011[a]	0.802 ± 0.021[a]
72 h	32.65 ± 3.07[a]	0.0350 ± 0.0022[a]	0.821 ± 0.027[a]

**Values in a column followed by different letters are significantly different* ($p < 0.05$)
From Sun et al. *(2011b).*

lipids might undergo unique lipolytic or oxidation patterns during processing, which differ from other dry-cured meat products like hams or fermented sausages.

Qiu *et al.* (2013) profiled fatty acid composition of the three lipid fractions (neutral lipid, phospholipids and free fatty acids) and composition of triacylglycerols and phospholipids of lipid from muscle and adipose tissue of Cantonese sausage during processing. Degradation pattern of triacylglycerols in lipid from muscle and adipose tissue was different.

During processing, both neutral lipid and phospholipids degraded gradually (Table 35.2). In lipid from muscle, phospholipids exhibited the most significant degradation, while in lipid from adipose tissue, degradation of neutral lipids and phospholipids seem to be similar. In lipid from muscle, POL and PSO constantly decreased during processing, which proved that their degradation in triacylglycerol fraction was more significant, and the release of fatty acid was mainly oleic acid. Phosphatidylcholines (PC) and phosphatidylethanolamines (PE) were the main degradation fractions of phospholipids and PE exhibited faster degradation rates than PC during processing. The main release fatty acid of phospholipids was polyunsaturated fatty acids (PUFAs), mainly linoleic (C18:2) and arachidonic (C20:4) acid. In lipid from adipose tissue, PSO and PPP were the main degradation fractions of triacylglycerol fraction and the release of fatty acid was mainly saturated fatty acids. PC and PE were also the main degradation fractions of phospholipids. However, the rate of degradation of phospholipids from adipose tissue was less than muscle. Total free fatty acids, whether in lean pork or in back fat, increased ($p < 0.05$) greatly through the processing of Cantonese sausage. The rate of release of unsaturated fatty acids, especially PUFAs was higher than saturated fatty acids. In muscle, the amount of PUFAs in free fatty acids did not show a marked increase as expected due to the occurrence of oxidation of PUFAs, especially arachidonic acid (C20:4).

The activities of lipase and phospholipases from Cantonese sausages were estimated at different processing periods. These decreased throughout processing, but remain active all along the process. A highly significant correlation was observed between their activities and salt content, water activity, total carbohydrate content, and pH. Sugar addition levels also affected the lipid composition of Cantonese sausage because of its influence on activities of lipase and phospholipases.

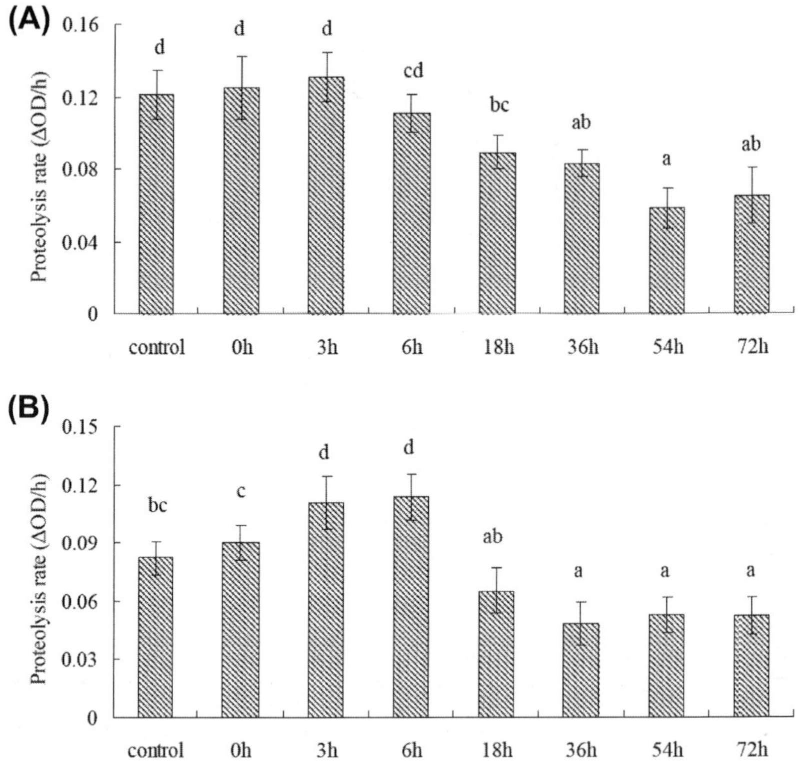

FIGURE 35.2 *In vitro* proteolysis rate of myofibrillar proteins from raw muscle and Cantonese sausage at different processing periods by gastric pepsin (A), pancreatic trypsin + α-chymotrypsin (B), previously treated with pepsin. Values in a column followed by different letters are significantly different ($p < 0.05$). *Adapted from Sun* et al. *(2011d).*

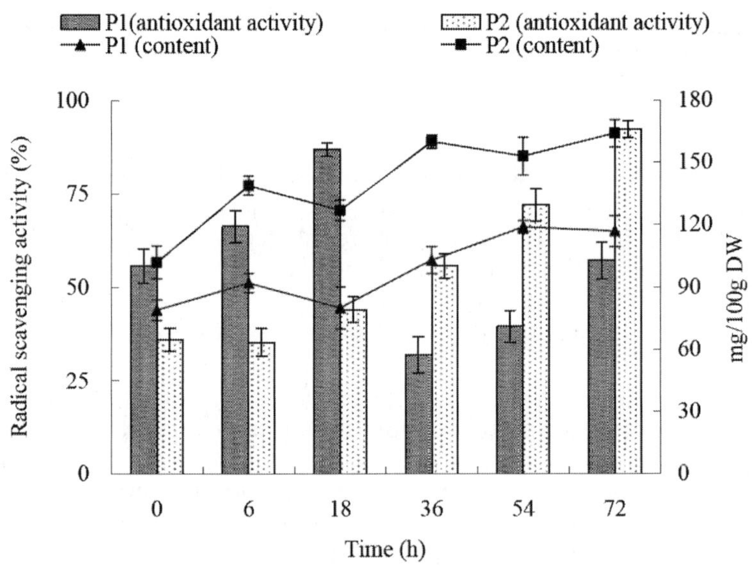

FIGURE 35.3 Changes in the DPPH radical scavenging activity (%) and content during drying of Cantonese sausage for peptides. *Adapted from Sun* et al. *(2009).*

EFFECT OF PROCESSING AND STORAGE ON RESIDUAL NITRITE CONTENT OF CANTONESE SAUSAGE

Nitrite is widely used in meat products due to its indispensable chromogenic function and bacteriostasis. However, the use of nitrite has potential health risks (Majumdar, 2003). Nitrite can form N-nitroso compounds, via reaction with amines and amides, which are related to an increasing risk of gastric, esophageal, nasopharyngeal, and bladder cancers. Nitrite can also oxidize reduced iron (Fe^{2+}) to its oxidized state (Fe^{3+}) in hemoglobin, resulting in a reduction in the total oxygen-carrying capacity of the blood.

Residual nitrite content decreased gradually during processing of Cantonese sausage (Sun *et al.*, 2011e). It also decreases during the storage. In a mild acid environment (pH5.5–6.0), nitrite turned into nitrous acid, which

TABLE 35.2 Changes in Lipid Composition During Cantonese Sausage Processing[1]

	Time					
	0 h	**6 h**	**18 h**	**36 h**	**54 h**	**72 h**
MUSCLE						
Neutral lipid	74.66±2.16[a]	76.51±1.50[ab]	77.14±1.19[ab]	79.02±2.41[ab]	80.89±2.13[cd]	83.37±2.07[d]
Phospholipids	25.23±2.18[d]	23.32±1.48[cd]	22.35±1.26b[cd]	20.46±2.37[bc]	18.53±2.08[ab]	15.90±2.01[a]
Free fatty acids	0.11±0.03[a]	0.27±0.03[b]	0.51±0.07[c]	0.52±0.06[c]	0.58±0.05[c]	0.73±0.06[c]
ADIPOSE TISSUE						
Neutral lipid	95.90±0.23[b]	95.51±0.22[a]	95.54±0.2[1a]	95.98±0.18[b]	96.11±0.22[b]	96.41±0.31[b]
Phospholipids	4.00±0.22[b]	4.25±0.21[c]	4.23±0.19[c]	3.78±0.21[ab]	3.64±0.25[ab]	3.28±0.30[a]
Free fatty acids	0.09±0.02[a]	0.24±0.02[b]	0.23±0.04[b]	0.24±0.03[b]	0.26±0.03[b]	0.31±0.05[b]

[1]*Results are expressed as% of total lipids.*
[a–d]*Values in a row followed by different letters are significantly different* ($p<0.05$).
From Qiu et al. *(2013).*

TABLE 35.3 Changes in Residual Nitrite Content during the Processing of Cantonese Sausage*

	Time (h)/nitrite Nitrite (DW) (mg/kg)					
	0	6	18	36	54	72
Lab	220.6±12.7[a]	213.3±10.8[a]	154.9±7.6[b]	84.2±9.9[c]	53.4±9.7[d]	41.7±5.3[d]
Industry	199.7±12.6[a]	145.5±10.2[b]	58.8±6.6[c]	26.4±3.7[d]	20.7±4.3[d]	13.7±3.6[d]

**Values in the same row and parameter group followed different letters are significantly different* ($P<0.05$).
From Sun et al. *(2011).*

was easily decomposed and consumed in the chromogenic reaction. The fermentation conditions (pH, humidity, temperature, and water content) of meat products affected residual nitrite content due to the chromogenic reaction between them. Sun *et al.* (2011e) found that the residual nitrite level in for laboratorial sausage decrease slower than industrial one (Table 35.3). Some microorganisms, such as *Staphylococcus* strains, were used to initiate color formation during sausage fermentation due to their nitrite and/or nitrate-reductase activities (Gøtterup *et al.*, 2008). Wu *et al.* (2010) collected eight Cantonese sausage samples (grade one) from eight different manufacturers (with same nitrite addition) to monitor their residual nitrite content and found that their values ranged from 26.08 to 46.20 mg/kg. This indicated that processing technology and microbiological status affected residual nitrite content deeply (Wu *et al.*, 2009, 2010).

Acknowledgments

The authors are grateful to the National Natural Science Foundation of China (No.31201387), Natural Science Foundation of Guangdong Province (No. S2012040007533) and the Doctoral Fund of Ministry of Education of China (No. 20120172120017) for their financial support.

References

Chelh, I., Gatellier, P., Santé-Lhoutellier, V., 2006. Technical note: A simplified procedure for myofibril hydrophobicity determination. Meat sci. 74 (4), 681–683.

Chen, H. M., Muramoto, K., Yamauchi, F., Fujimoto, K., Nokihara, K., 1998. Antioxidative properties of histidine-containing peptides designed from peptide fragments found in the digests of a soybean protein. J. Agri. Food Chem. 46 (1), 49–53.

Elias, R.J., Kellerby, S.S., Decker, E.A., 2008. Antioxidant activity of proteins and peptides. Crit. Rev. Food Sci. Nutr. 48 (5), 430–441.

Gøtterup, J., Olsen, K., Knøchel, S., Tjener, K., Stahnke, L.H., Møller, J.K., 2008. Colour formation in fermented sausages by meat-associated *Staphylococci* with different nitrite- and nitrate-reductase activities. Meat sci. 78 (4), 492–501.

Lund, M.N., Heinonen, M., Baron, C.P., Estevez, M., 2011. Protein oxidation in muscle foods: A review. Mol. Nutr. Food Res 55 (1), 83–95.

Majumdar, D., 2003. The blue baby syndrome. Resonance 8 (10), 20–30.

Mullen, A., Stoeva, S., Laib, K., Gruebler, G., Voelter, W., Troy, D., 2000. Preliminary analysis of amino acids at various locations along the *M.longissimus dorsi* in aged beef. Food Chem. 69 (4), 461–465.

Ordóñez, J.A., Hierro, E.M., Bruna, J.M., Hoz, L.d.l., 1999. Changes in the components of dry-fermented sausages during ripening. Crit. Rev. Food. Sci. Nutr. 39 (4), 329–367.

Qiu, C.Y., Sun, W.Z., Cui, C., Zhao, M.M., 2012. Effect of sugar level on physicochemical, biochemical characteristics and proteolysis properties of Cantonese sausage during processing. J. Food Qual. 35 (1), 34–42.

Roseiro, L., Santos, C., Sol, M., Borges, M., Anjos, M., Gonçalves, H., Carvalho, A., 2008. Proteolysis in *Painho de Portalegre* dry fermented sausage in relation to ripening time and salt content. Meat sci. 79 (4), 784–794.

Sante-Lhoutellier, V., Aubry, L., Gatellier, P., 2007. Effect of oxidation on in vitro digestibility of skeletal muscle myofibrillar proteins. J. Agri. Food Chem. 55 (13), 5343–5348.

Shacter, E., 2000. Quantification and significance of protein oxidation in biological samples. Drug Metab. Rev. 32 (3–4), 307–326.

Su, G., Zheng, L., Cui, C., Yang, B., Ren, J., Zhao, M., 2011. Characterization of antioxidant activity and volatile compounds of Maillard reaction products derived from different peptide fractions of peanut hydrolysate. Food Res. Int. 44 (10), 3250–3258.

Sun, W., Cui, C., Zhao, M., Zhao, Q., Yang, B., 2011a. Effects of composition and oxidation of proteins on their solubility, aggregation and proteolytic susceptibility during processing of Cantonese sausage. Food Chem. 124 (1), 336–341.

Sun, W., Zhao, H., Zhao, Q., Zhao, M., Yang, B., Wu, N., Qian, Y., 2009. Structural characteristics of peptides extracted from Cantonese sausage during drying and their antioxidant activities. Innovative Food Sci. Emerg. Technol. 10 (4), 558–563.

Sun, W., Zhao, M., Yang, B., Zhao, H., Cui, C., 2011b. Oxidation of sarcoplasmic proteins during processing of Cantonese sausage in relation to their aggregation behaviour and *in vitro* digestibility. Meat sci. 88 (3), 462–467.

Sun, W., Zhao, Q., Zhao, M., Yang, B., Cui, C., Ren, J., 2011c. Structural evaluation of myofibrillar proteins during processing of Cantonese sausage by Raman spectroscopy. J. Agri. Food Chem. 59 (20), 11070–11077.

Sun, W., Zhou, F., Zhao, M., Yang, B., Cui, C., 2011d. Physicochemical changes of myofibrillar proteins during processing of Cantonese sausage in relation to their aggregation behaviour and *in vitro* digestibility. Food Chem. 129 (2), 472–478.

Sun, W.Z., Zhao, M.M., Zhao, H.F., Zhao, Q.Z., Yang, B., Wu, N., Qian, Y.L., 2011e. Effect of manufacturing level on the biochemical characteristics of Cantonese sausage during processing. J. Food. Biochem. 35 (4), 1015–1033.

Toldrá, F., 2006. The role of muscle enzymes in dry-cured meat products with different drying conditions. Trends Food Sci. Technol. 17 (4), 164–168.

Toldrá, F., Flores, M., Sanz, Y., 1997. Dry-cured ham flavour: enzymatic generation and process influence. Food Chem. 59 (4), 523–530.

Tornberg, E., 2005. Effects of heat on meat proteins—Implications on structure and quality of meat products. Meat sci. 70 (3), 493–508.

Visessanguan, W., Benjakul, S., Riebroy, S., Thepkasikul, P., 2004. Changes in composition and functional properties of proteins and their contributions to Nham characteristics. Meat sci. 66 (3), 579–588.

Wu, W., Hettiarachchy, N., Qi, M., 1998. Hydrophobicity, solubility, and emulsifying properties of soy protein peptides prepared by papain modification and ultrafiltration. J. Am. Oil. Chem. Soc. 75 (7), 845–850.

Wu, Y., Cui, C., Sun, W., Yang, B., Zhao, M., 2009. Effects of Staphylococcus condimenti and Micrococcus caseolyticus on the volatile compounds of Cantonese sausage. J. Food. Process. Eng. 32 (6), 844–854.

Wu, Y., Zhao, M., Yang, B., Sun, W., Cui, C., Mu, L., 2010. Microbial analysis and textural properties of Cantonese sausage. J. Food. Process. Eng. 33 (1), 2–14.

Wu, Z., Bertram, H.C., Kohler, A., Böcker, U., Ofstad, R., Andersen, H.J., 2006. Influence of aging and salting on protein secondary structures and water distribution in uncooked and cooked pork. A combined FT-IR microspectroscopy and ^{1}H NMR relaxometry study. J. Agri. Food Chem. 54 (22), 8589–8597.

Yongsawatdigul, J., Park, J., 1999. Thermal aggregation and dynamic rheological properties of Pacific whiting and cod myosins as affected by heating rate. J. Food Sci. 64 (4), 679–683.

SECTION 6

GRAIN, BEANS, PULSES, NUTS AND SEEDS

CHAPTER

36

The Effects of Processing on Gluten from Wheat, Rye, and Barley, and its Detection in Foods

Girdhari M. Sharma, Prasad Rallabhandi

Center for Food Safety and Applied Nutrition, U.S. Food and Drug Administration, Laurel, Maryland, USA

CHAPTER POINTS

- Gluten (prolamins and glutelins) is a major storage protein in wheat, rye, and barley, which comprises of sulfur-rich, sulfur-poor, and high molecular weight proteins.
- The food quality is dependent up on the structure, proportion, and composition of gluten, which can be affected by thermal and non-thermal processing conditions.
- Gluten aggregates created during thermal processing by the exchange of sulfydryl-disulfide bonds between glutenins at low temperatures (<90°C), and glutenins and gliadins at high temperatures (>90°C) reduces gluten extractability.
- Gluten hydrolysis by non-thermal processing (chemical, enzymatic, or fermentation) yields short gluten polypeptides with improved solubility and altered functionality.
- ELISA is commonly used for gluten detection/quantification, and molecular changes in gluten due to thermal and non-thermal processing affect antigen–antibody interaction contributing to variations in ELISA sensitivity.

INTRODUCTION

Wheat, rye, and barley are widely used cereal grains in human and animal food. Their consumption varies depending upon geographical location and lifestyle. Wheat is a predominantly consumed cereal grain as it is abundantly available and has many food uses. Wheat storage proteins such as gluten play an important role in food preparation because of their unique functional properties. Gluten is present in the endosperm, and acts as a source of nitrogen, sulfur, and carbon during seed germination. Gluten is commonly used as a dough strengthener, thickener, and texturizing agent in various foods. Food processing changes the structural and molecular properties of gluten. These changes, which have a significant effect on the extractability, functionality, and detection of gluten, are discussed below.

GLUTEN PROTEINS AND THEIR PROPERTIES

The propagation programs and the commercial production of grains focus on their high yield, quality and suitability for use in various food products. The quality of a grain is generally defined by kernel hardness and protein content. Among all wheat cultivars, durum wheat grain is the hardest. While the protein content significantly varies depending upon the available nitrogen and moisture levels of soil during the growing period, the grain hardness is strongly affected by abnormal weather conditions during the harvest season (Bushuk, 1998). Figure 36.1 depicts the physical characteristics of grains of various wheat cultivars, rye, and barley. Cereal grains contain about 8–16% protein (Table 36.1), and gluten forms a major part of storage protein in wheat, rye, and barley (Diaz-Amigo and Yeung, 2010). Generally, the harder varieties of wheat contain higher protein content and dough strength, compared to the softer varieties. The end food-use of a grain is dependent upon both its hardness and protein quality, which is determined by

Processing and Impact on Active Components in Food
http://dx.doi.org/10.1016/B978-0-12-404699-3.00036-6

FIGURE 36.1 Different wheat varieties, rye, and barley grains.

TABLE 36.1 Grains, Protein Contents, and their Food Use

Grain	% Protein[a]	Common Food Use
Wheat[b]		
1. Hard Red Spring	15.40	Bread, bulgur, couscous
2. Hard Red Winter	12.61	Bread, French baguette
3. Hard White	11.31	Bread, bulgur, couscous, flat bread, noodles, French baguette
4. Soft Red Winter	10.35	Cake, pastry, cookie, flat bread
5. Soft White	10.69	Cake, pastry, cookie, flat bread
6. Durum	13.68	Pasta products
Rye	10.34	Bread, spirits
Barley	9.91	Animal feed, malt, beer, spirits

[a] *http://ndb.nal.usda.gov/ndb/foods/list accessed Nov 16, 2012.*
[b] *Bushuk (1998).*

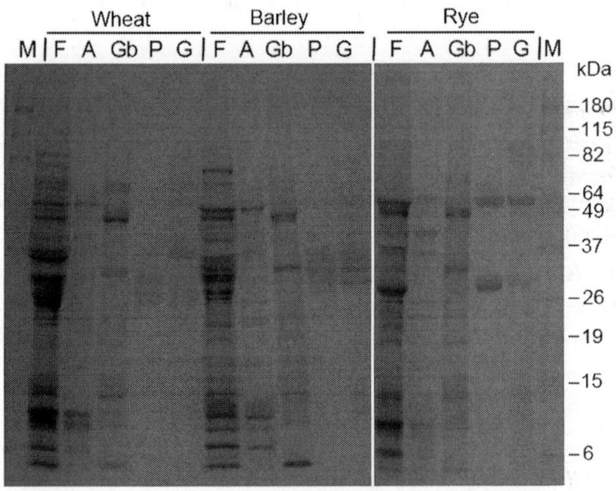

FIGURE 36.2 Coomassie stained protein fractions from wheat, rye, and barley separated by gel electrophoresis under reducing conditions. M, molecular weight markers; F, flour; A, albumin; Gb, globulin; P, prolamin; G, glutelin.

the molecular structure and properties of major proteins, such as gluten. For example, the tetraploid durum wheat is used in pasta products because of its hard vitreous kernel and high protein content, whereas the hexaploid wheat varieties are used for cookies and cakes (soft kernel), and breads (hard kernel). The protein content and common food uses of various commercial wheat cultivars, rye, and barley are listed in Table 36.1.

Cereal grain storage proteins are classified based on their solubility and extractability into albumins/globulins and gluten proteins (prolamins and glutelins). Osborne (1924) developed a method for the isolation of grain proteins based on their solubility in different solvents: albumins in water, globulins in dilute salt solution, prolamins in aqueous alcohol, and glutelins in acid/alkali solution. Figure 36.2 shows various proteins fractionated from the flours of wheat, barley, and rye (Tatham *et al.*, 2000) and separated by gel electrophoresis under reducing conditions. The prolamin fraction of gluten is termed gliadin in wheat, secalin in rye, and hordein in barley. Glutelin in wheat is called glutenin, whereas there is no distinct term for glutelin in rye and barley. Both prolamins and glutelins are proline and glutamine rich in these cereals. Gliadin contributes to the viscosity and extensibility of gluten, whereas glutenin confers elasticity (Wall, 1979). Gluten is also classified into sulfur-rich (S-rich), sulfur-poor (S-poor) and high-molecular-weight (HMW) prolamins, and protein sequence similarity indicates their common evolutionary origin (Shewry and Tatham, 1990; Shewry and Halford, 2002). The cysteine residues in S-rich monomeric proteins are involved in intra-chain disulfide bonds, while those in polymeric proteins are involved in both intra- and inter-chain disulfide bonds. Better baking quality of breads has been correlated to a higher ratio of polymeric to monomeric proteins of gluten in the

TABLE 36.2 Molecular Properties of Gluten Fractions from Wheat, Rye, and Barley

Grain Gluten Proteins	Molecular Weight (kDa)	Proportion of Gluten (%)	Properties[d]
Wheat[a]			
α/β-gliadin	28 – 35	28 – 33	S-rich, monomer
γ-gliadin	31 – 35	23 – 31	S-rich, monomer
ω-gliadin	39 – 55	7 – 13	S-poor, monomer
HMW-GS	67 – 88	7 – 13	S-rich, polymer
LMW-GS	32 – 39	19 – 25	S-rich, polymer
Rye[b]			
γ-40k secalin	40	24	S-rich, monomer
γ-75k secalin	70	46	S-rich, polymer
ω-secalin	50	16	S-poor, monomer
HMW subunits	100	6 – 9	S-rich, polymer
Barley[c]			
A (or γ)-hordein	<30	<5	S-rich, monomer
B hordein	30–50	70–80	S-rich, polymer
C hordein	50–75	10–20	S-poor, monomer
D hordein	>90	<5	S-rich, polymer

[a] *Wieser, H. (2007). Chemistry of gluten proteins.* Food Microbiol. *24, 115–119.*
[b] *Gellrich, C., Schieberle, P., and Wieser, H. (2003).* Cereal Chem. *80, 102–109.*
[c] *Salplachta, J., and Bobalova, J. (2009).* J. Mass Spectrom. *44, 1287–1292.*
[d] *Tatham* et al. *(2000).*
HMW, high molecular weight; LMW, low molecular weight; GS, glutenin subunit

flour (Park *et al.*, 2006). The molecular properties and the proportion of different types of gluten in wheat, rye, and barley are compiled in Table 36.2.

EFFECT OF FOOD PROCESSING ON GLUTEN AND ITS COMPOSITION

Gluten can be present in food as an inherent component of grain or may be added extraneously to improve the functional properties of a food. Any changes in the structure, proportion, or composition of gluten will affect the food quality, resulting in a desirable or undesirable final product. Food processing is carried out for several reasons such as safety, improved food quality, shelf-life, functionality, and storage (Sathe and Sharma, 2009). Broadly, food processing methods are divided into thermal and non-thermal (Figure 36.3). Generally, foods containing gluten undergo a combination of both thermal and non-thermal processing.

Thermal Processing

Thermal processing imparts temperature changes on food and affects various food components, including proteins. The general consequences of thermal processing on gluten and other proteins include the Maillard reaction, denaturation, and aggregation. The Maillard reaction is a chemical reaction, favored at a high temperature, between a reducing sugar and free amino groups in proteins. This reaction is responsible for the desirable color and flavor formation in thermally processed foods. The formation of color compounds increases at 150°C, but does not significantly change up to 120°C, in a wet gluten-glucose model system (Fogliano *et al.*, 1999). However, high temperature may affect gluten functionality by reducing the protein extractability due to polymerization (Schofield *et al.*, 1983). Gluten polymerization can be attributed to the changes in covalent bonds among its fractions. Rombouts *et al.* (2011) observed decreased gluten extractability in the presence of sodium dodecyl sulfate, under both reducing and non-reducing conditions, upon heat treatment of gluten at 130°C for 24 h. This was attributed to the formation of isopeptide bonds at 130°C, as well as disulfide bonds, resulting in the creation of polymerized unextractable gluten. Gluten proteins are generally S-rich, except for ω-gliadin in wheat and its homologous counterparts, ω-secalin in rye and C hordein in barley (Table 36.2). An exchange of sulfydryl-disulfide bonds between the gluten proteins occurs during thermal processing, primarily affecting glutenins at 55–70°C, and gliadins at temperatures exceeding 70°C (Schofield *et al.*, 1983). Lagrain *et al.* (2008) proposed that conformational changes in glutenin at temperatures below 90°C expose free sulfydryl groups, which form disulfide bonds with other glutenins to create aggregates. At temperatures above 90°C, these free sulfydryl groups on glutenin can also be involved in sulfydryl-disulfide exchange between glutenin and gliadin. They also proposed that reducing agents increase the number of free sulfydryl groups, facilitating glutenin-gliadin cross-linking at high temperatures. The significant loss of extractable S-rich α-, β- and γ-gliadins compared to S-poor ω-gliadin observed at high temperature is possibly because the latter cannot participate in sulfydryl-disulfide exchange (Schofield *et al.*, 1983; Lagrain *et al.*, 2008).

The complex biochemical changes that occur in various wheat flour components (e.g., gluten and starch) at different stages of bread making have been reviewed in detail by Goesaert *et al.* (2005). Among thermal processing methods (Figure 36.3), microwave heating is popular in households for ready-to-cook/eat foods and reheating of foods. Microwave drying of wheat did not change the protein content, but caused a loss of dough elasticity and stretchability indicating a change in gluten functionality (Walde *et al.*, 2002). More recently, Yalcin *et al.* (2008) studied the functional properties of 1% gluten prepared in distilled water at pH3.0 after subjecting the gluten to microwave heating at 350, 560, and 706 W power settings for various time intervals (1–3 min). The authors observed a decrease in protein solubility with an increase in the treatment time at all power levels. The emulsion capacity was low for gluten heated at 706 W, but the emulsion stability remained essentially unchanged at all treatment times. Also, foam volume did not change significantly,

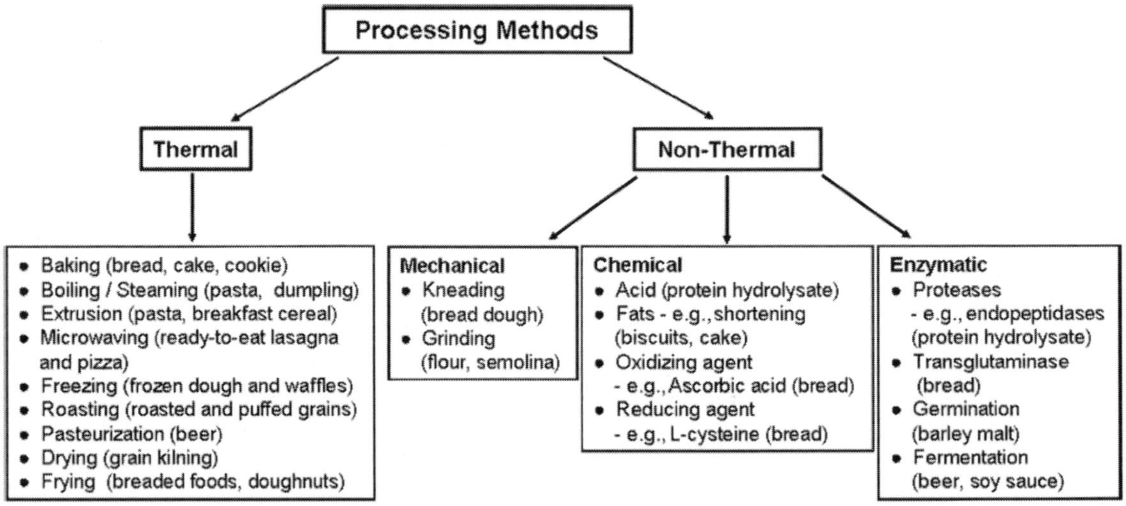

FIGURE 36.3 Various processing methods affecting gluten in foods. Examples of foods produced by respective process are shown in parentheses.

but foam stability increased with an increase in treatment time at any given power level. Further studies on the microwave heating-induced molecular changes in gluten may help better understand the mechanisms responsible for altered protein functionality.

Non-Thermal Processing

The non-thermal processing methods that affect gluten structure can be classified into mechanical, chemical, and enzymatic (Figure 36.3). Development of the gluten network occurs during the dough mixing stage, resulting in increased dough resistance that can be adversely affected by over-mixing (Goesaert *et al.*, 2005). Dough conditioners like oxidizing and reducing agents that control disulfide bonds, and hence the size of the gluten polymer, are often added to improve bread texture.

Hydrolysis of gluten is achieved by acid, proteases, and during fermentation, yielding short gluten polypeptides with improved solubility. Enzymatic hydrolysis results in breakdown of gluten into fragments of variable size with altered functionality. Kong *et al.* (2007) reported a higher degree of hydrolysis of wheat gluten with Alcalase, as compared to pepsin, trypsin, and pancreatin. Moreover, they observed that the gluten hydrolysates generated by Alcalase had increased solubility at a higher degree of hydrolysis. Emulsifying and foaming properties improved at 5%, but were reduced at 10 and 15% degree of hydrolysis. Even though the protein hydrolysates are generally used as an added food ingredient, protein hydrolysis can also occur during processing in some foods (e.g., sourdough bread, barley malt). Fermentation of food generally reduces the pH and increases the enzymatic activity promoting hydrolysis. Sourdough fermented with *Lactobacilli* for 24h showed a reduction in pH from ~6.2 to 3.6 and an increase in the amino nitrogen from 4.1 up to 9.1mmol of glycine per kg of dough (Thiele *et al.*, 2004).

Barley malt is used in making beer, spirits, and malt syrup. Malting, a controlled germination of barley, causes various biochemical changes in the grain. The enzymes produced during malting promote the breakdown of proteins and carbohydrates. Osman *et al.* (2002) showed that the optimum pH and temperature for the activity of barley malt endoproteases is dependent upon the type of gluten substrate. Also, proteolysis increased free amino nitrogen (5- to 7-fold) and soluble proteins (2-fold) in the malted compared to unmalted barley. Moreover, the authors reported high endoprotease activity in the kilned malt, suggesting further hydrolysis of gluten during mashing.

ANALYTICAL TECHNIQUES FOR GLUTEN DETECTION IN FOODS

Gluten may be present unintentionally in foods due to grain contamination (during harvest, transport, or storage), cross-contact during product changeover, and/or poor manufacturing practices, posing a serious risk to sensitive individuals. Gluten sensitivity/intolerance causes severe health problems, like allergies and celiac disease (Rallabhandi, 2012). Since avoidance of dietary gluten is the only option for these patients, a sensitive gluten detection method is warranted. Regulatory agencies from different countries have defined 'gluten-free,' and mandatory food labeling compliance.

Various analytical methods are used to detect trace amounts of gluten for ensuring food safety. These are similar to food allergen detection methods, such as enzyme linked immunosorbent assay (ELISA), lateral flow devices (LFD), polymerase chain reaction (PCR),

TABLE 36.3 Factors Affecting Gluten Detection

Assay Related	Non-Assay Related
Target analyte and specificity	Food matrices (fat, polyphenols)
Reference material	Thermal processing (aggregation, Millard reaction)
Extraction efficiency	Non-thermal processing (hydrolysis, pH)
Limit of detection/ quantification	Storage stability

mass spectrometry (MS), and surface plasmon resonance (SPR). These techniques differ considerably with respect to their cost, sensitivity, and specificity. All methods except PCR detect either whole or part of the gluten proteins, whereas PCR detects the gene encoding gluten. Irrespective of the method, there are assay and non-assay related factors that affect gluten detection in foods (Table 36.3). The extraction efficiency plays an important role as the analytical techniques require gluten to be present in a soluble form to facilitate detection. Since gluten is not soluble in commonly used aqueous buffers, extraction is generally achieved in the presence of 60–80% alcohol solution. Non-assay related factors such as interference by food matrix components and food processing can reduce the extraction efficiency. Food matrix interference can be avoided by removing the interfering component, such as fat, prior to extraction. However, the removal of interfering substances should not result in a partial or complete loss of gluten from the food. Efficient extraction of gliadin from thermally processed foods has been achieved at high extraction temperature (50°C) and by using reducing (e.g., β-mercaptoethanol) and chaotropic (e.g., guanidine hydrochloride) agents (Garcia *et al.*, 2005).

ELISA

ELISA and LFD are commonly used immunoassays for gluten detection. There are various mono- and polyclonal antibody-based commercial ELISA kits available for gluten detection, with a limit of quantification ranging from 0.3 to 5 ppm. These kits vary not only in their antibody type and the targeted epitope, but also differ in other aspects such as extraction buffer, incubation time and temperature, and calibration standards. Hence, gluten estimations can vary significantly among the ELISA kits (Allred and Ritter, 2010; Sharma, 2012). The well-characterized monoclonal antibodies used in commercial ELISA kits include Skerritt (Skerritt and Hill, 1990), R5 (Valdes *et al.*, 2003), and more recently G12 antibody (Moron *et al.*, 2008). Changes in gluten proteins, due to their proportion and amino acid sequence in different cultivars, and/or their structure due to processing have a significant impact on the antigen–antibody interaction in ELISA, and hence can affect gluten detection and quantification. Decrease in gluten solubility due to thermal processing may reduce ELISA sensitivity. ELISA detection and quantification using incurred standards/samples is a practical approach to mimic the industrial scenario in accounting for the unextractable protein due to food processing (Taylor *et al.*, 2009).

Chemical modification of gluten can change its molecular properties in such a way that the epitopes are altered, affecting ELISA detection. For example, Kanerva *et al.* (2011) reported that deamidation of gluten peptides caused a drastic reduction in antibody recognition with both R5 and G12 competitive ELISAs. Conversely, γ-irradiated (up to 12.7 kGy) gliadin and wheat flour have been shown to have higher immunoreactivity with ELISA (Leszczynska *et al.*, 2003). Protein hydrolysis (chemical or enzymatic) may result in partial or complete loss of immunoreactive epitopes. The enzymatic digests of gliadin, secalin, and hordein showed about 70% and 40% response with R5 competitive and sandwich ELISA, respectively, as compared to their intact protein counterparts (Gessendorfer *et al.*, 2009). Competitive ELISA requires the presence of one epitope on the target peptide/protein for an effective detection, while sandwich ELISA needs at least two epitopes. The higher response with the competitive format compared to the sandwich format can be attributed to the differences in the number of epitopes required for detection. Hence, the selection of an appropriate ELISA format (competitive or sandwich) depends upon the treatments employed during food processing.

Acknowledgments

This work in part was supported by an FDA Office of Women's Health grant (to P.R.).

References

Allred, L.K., Ritter, B.W., 2010. Recognition of gliadin and glutenin fractions in four commercial gluten assays. J. AOAC Int. 93, 190–196.
Bushuk, W., 1998. Wheat breeding for end-product use. Euphytica 100, 137–145.
Diaz-Amigo, C., Yeung, J.M., 2010. Critical evaluation of uncertainties of gluten testing: Issues and solutions for food allergen detection. In: Juneja, V.K., Sofos, J.N. (Eds.), Pathogens and Toxins in Foods: Challenges and Interventions. ASM Press, Washington, DC, pp. 286–300.
Fogliano, V., Monti, S.M., Musella, T., Randazzo, G., Ritieni, A., 1999. Formation of coloured Maillard reaction products in a gluten-glucose model system. Food Chem. 66, 293–299.
Garcia, E., Llorente, M., Hernando, A., Kieffer, R., Wieser, H., Mendez, E., 2005. Development of a general procedure for complete extraction of gliadins for heat processed and unheated foods. Eur. J. Gastroenterol. Hepatol. 17, 529–539.
Gessendorfer, B., Koehler, P., Wieser, H., 2009. Preparation and characterization of enzymatically hydrolyzed prolamins from wheat, rye, and barley as references for the immunochemical quantitation of partially hydrolyzed gluten. Anal. Bioanal. Chem. 395, 1721–1728.

Goesaert, H., Brijs, K., Veraverbeke, W.S., Courtin, C.M., Gebruers, K., Delcour, J.A., 2005. Wheat flour constituents: how they impact bread quality, and how to impact their functionality. Trends Food Sci. Technol. 16, 12–30.

Kanerva, P., Brinck, O., Sontag-Strohm, T., Salovaara, H., Loponen, J., 2011. Deamidation of gluten proteins and peptides decreases the antibody affinity in gluten analysis assays. J. Cereal Sci. 53, 335–339.

Kong, X., Zhou, H., Qian, H., 2007. Enzymatic preparation and functional properties of wheat gluten hydrolysates. Food Chem. 101, 615–620.

Lagrain, B., Thewissen, B.G., Brijs, K., Delcour, J.A., 2008. Mechanism of gliadin-glutenin cross-linking during hydrothermal treatment. Food Chem. 107, 753–760.

Leszczynska, J., Lacka, A., Szemraj, J., Lukamowicz, J., Zegota, H., 2003. The influence of gamma irradiation on the immunoreactivity of gliadin and wheat flour. Eur. Food Res. Technol. 217, 143–147.

Moron, B., Bethune, M.T., Comino, I., Manyani, H., Ferragud, M., Lopez, M.C., Cebolla, A., Khosla, C., Sousa, C., 2008. Toward the assessment of food toxicity for celiac patients: Characterization of monoclonal antibodies to a main immunogenic gluten peptide. PLoS ONE 3, e2294.

Osborne, T.B., 1924. The Vegetable Proteins, second ed. Longmans and Co., London, UK.

Osman, A.M., Coverdale, S.M., Cole, N., Hamilton, S.E., de Jersey, J., Inkerman, P.A., 2002. Characterisation and assessment of the role of barley malt endoproteases during malting and mashing. J. Inst. Brew. 108, 62–67.

Park, S.H., Bean, S.R., Chung, O.K., Seib, P.A., 2006. Levels of protein and protein composition in hard winter wheat flours and the relationship to breadmaking. Cereal Chem. 83 (4), 418–423.

Rallabhandi, P., 2012. Gluten and celiac disease—an immunological perspective. J. AOAC Int. 95 (2), 349–355.

Rombouts, I., Lagrain, B., Brunnbauer, M., Koehler, P., Brijs, K., Delcour, J.A., 2011. Identification of isopeptide bonds in heat-treated wheat gluten peptides. J. Agric. Food Chem. 59, 1236–1243.

Sathe, S.K., Sharma, G.M., 2009. Effects of food processing on food allergens. Mol. Nutr. Food Res. 53, 970–978.

Schofield, J.D., Bottomley, R.C., Timms, M.F., Booth, M.R., 1983. The effect of heat on wheat gluten and the involvement of sulphydryl-disulphide interchange reactions. J. Cereal Sci. 1, 241–253.

Sharma, G.M., 2012. Immunoreactivity and detection of wheat proteins by commercial ELISA kits. J. AOAC Int. 95 (2), 364–371.

Shewry, P.R., Halford, N.G., 2002. Cereal seed storage proteins: structures, properties and role in grain utilization. J. Exp. Bot. 53 (370), 947–958.

Shewry, P.R., Tatham, A.S., 1990. The prolamin storage proteins of cereal seeds: structure and evolution. Biochem. J. 267 (1), 1–12.

Skerritt, J.H., Hill, A.S., 1990. Monoclonal antibody sandwich enzyme immunoassays for determination of gluten in foods. J. Agric. Food Chem. 38, 1771–1778.

Tatham, A.S., Gilbert, S.M., Fido, R.J., Shewry, P.R., 2000. Extraction, separation, and purification of wheat gluten proteins and related proteins of barley, rye, and oats. In: Marsh, M.N. (Ed.), Methods in Molecular Medicine, vol. 41. Humana Press Inc., Totowa, NJ, pp. 55–73.

Taylor, S.L., Nordlee, J.A., Niemann, L.M., Lambrecht, D.M., 2009. Allergen immunoassays— considerations for use of naturally incurred standards. Anal. Bioanal. Chem. 395, 83–92.

Thiele, C., Grassl, S., Ganzle, M., 2004. Gluten hydrolysis and depolymerization during sourdough fermentation. J. Agric. Food Chem. 52, 1307–1314.

Valdes, I., Garcia, E., Llorente, M., Mendez, E., 2003. Innovative approach to low-level gluten determination in foods using a novel sandwich enzyme-linked immunosorbent assay protocol. Eur. J. Gastroenterol. Hepatol. 15, 465–474.

Walde, S.G., Balaswamy, K., Velu, V., Rao, D.G., 2002. Microwave drying and grinding characteristics of wheat (*Triticum aestivum*). J. Food Eng. 55, 271–276.

Wall, J.S., 1979. Recent Advances in the Biochemistry of Cereals. In: Laidman, D.L., Wyn Jones, R.G. (Eds.), Proceeding of the Phytochemical Soc. of Europe, vol. 16. Academic Press, London, UK, pp. 275–311.

Yalcin, E., Sakiyan, O., Sumnu, G., Celik, S., Koksel, H., 2008. Functional properties of microwave-treated wheat gluten. Eur. Food Res. Technol. 227, 1411–1417.

CHAPTER

37

Carotenoids, Tocols, and Retinols during the Pasta-Making Process

Alessandra Fratianni, Gianfranco Panfili, Raimondo Cubadda

DiAAA, Universitá degli Studi del Molise, Campobasso, Italy

CHAPTER POINTS

- Cereals are a source of tocols and carotenoids while retinols are provided by the other added ingredients.
- Processing may influence the levels of these compounds in the final products depending on the conditions adopted.
- During the pasta-making process the degradation of tocols and carotenoids is mostly limited to the kneading step.
- An increase in geometric isomerization from the trans to the cis forms of carotenoids and retinols is also reported.
- The bioaccessibility of tocols and carotenoids from durum wheat pasta seems not to be influenced by the pasta-making process.

INTRODUCTION

Growing evidence suggests that cereal and cereal-based products contain notable levels of bioactive compounds with health benefits, among which tocols (Figure 37.1) and carotenoids (Figure 37.2A) have been given particular attention in recent years. In the kernel of various cereals the content of these compounds shows large differences due to different genetic and environmental factors (species, cultivar, agronomical techniques, climatic conditions etc.). As far as cereal-based products are concerned, the level of bioactive compounds can vary significantly according to the other ingredients added and to their quantity. For instance, egg pasta contains discrete amounts of retinols (Figure 37.2B) as well as γ-tocopherol provided by the eggs added to durum wheat semolina or flour. Among the most important cereals, corn contains the highest carotenoid levels, particularly β-cryptoxanthin, zeaxanthin and α + β-carotene. In bread wheat the concentration of carotenoids is low. Conversely, durum wheat can contain, according to the variety, a significant quantity of yellow compounds, mainly lutein (Panfili *et al.*, 2004, Zandomeneghi *et al.*, 2000), which is responsible for the yellow color of semolina and the derived pasta. It is well-known that pasta color is perceived as an important quality trait (Fratianni *et al.*, 2005).

Tocols (tocopherols and tocotrienols) are abundant both in bread wheat and durum wheat (Panfili *et al.*, 2003). Cereals, especially oats, rye, and barley, are good sources of tocotrienols. Panfili *et al.* (2003) reported that the highest tocol levels are present in barley and soft wheat. In hulled and dehulled wheat, the most abundant component is β-tocotrienol (β-T3) (Lampi *et al.*, 2010), whereas γ-tocopherol (γ-T3) predominates in corn and α-tocotrienol (α-T3) in oats and barley (Panfili *et al.*, 2003, 2008; Peterson *et al.*, 1995; Tiwari and Cummins, 2009).

The yellow pigments are differently distributed in the kernel: α- and β-carotene are mainly located in the germ, while lutein is equally distributed across the kernel (Panfili *et al.*, 2004). According to Leenhardt *et al.* (2006), 86% of the total kernel carotenoids were recovered in white wheat flour. Borrelli *et al.* (2008) showed that, across the wheat grain, from the central endosperm to the outermost layers, a different distribution pattern for the carotenoids exists, depending on genotypes. In cereal grains, including barley, oat, wheat, and rye, tocopherols are mainly concentrated in the germ (Panfili *et al.*, 2008; Hidalgo and Brandolini, 2008), while tocotrienols are present in the pericarp, aleurone and subaleurone layers, and endosperm fraction (Panfili *et al.*, 2008; Peterson, 1995; Tiwari and Cummins, 2009), where they generally accumulate during grain development (Falk *et al.*, 2004).

Processing and Impact on Active Components in Food
http://dx.doi.org/10.1016/B978-0-12-404699-3.00037-8

Physical removal during technological processes (pearling, dehulling, sieving, air classification. milling) can cause loss of tocols, carotenoids and retinols. Moreover, during processing and storage of foods, alteration or loss may occur due to geometric isomerization and enzymatic or non-enzymatic oxidation (Rodriguez-Amaya, 2002; Rodriguez-Amaya and Kimura, 2004), with consequent alteration of their bioavailability and biological activity. The evaluation of the degree of loss could be of importance in minimizing their degradation during manufacturing and in improving the nutritional and quality properties of the final products.

HOW COMPOSITION IS ALTERED

Impact of Pasta-Making Processes on the Content of Tocols, Carotenoids, and Retinols in Final Products

Pasta processing consists of several stages (Figure 37.3): durum wheat semolina is mixed with water in a paddle mixer and fed into an extrusion worm conveyor, where it is compressed and extruded through a die into a stiff dough. The dough is then cut and dried to achieve a final moisture content that does not exceed 12.5%. The production of high quality pasta depends on the characteristics of semolina and optimal processing conditions.

Being highly unsaturated, carotenoids and retinols are prone to isomerization and, together with tocols, to oxidation during processing and storage of foods. Degradation compounds are numerous, including isomers, epoxides, apocarotenones, apocarotenals, and short-chain cleavage products (Penicaud *et al.*, 2011). Transformation of the trans carotenoids (all E configuration, their usual form in nature, predominant in fresh foods) to the cis-isomers (Z forms) is well-documented. This modification is promoted by acids, heat and light, and results in loss of provitamin A activity and alteration of bioavailability/metabolism. Both E and Z-isomers are subject to oxidation (Rodriguez-Amaya, 2002; Rodriguez-Amaya *et al.*, 2006). The major cause of loss is enzymatic or non-enzymatic oxidation, which depends on the availability of oxygen and on the structure of the compound.

(A)

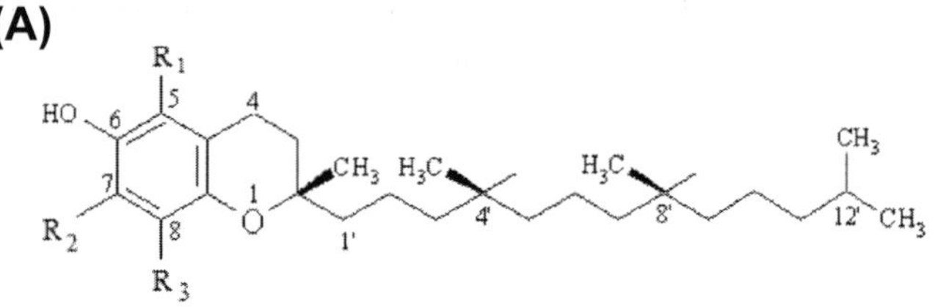

Common name	R_1	R_2	R_3	Chemical name
α-tocopherol	$-CH_3$	$-CH_3$	$-CH_3$	5-7-8-trimethyltocol
β-tocopherol	$-CH_3$	-H	$-CH_3$	5-8-dimethyltocol
γ-tocopherol	-H	$-CH_3$	$-CH_3$	7-8-dimethyltocol
δ-tocopherol	-H	-H	$-CH_3$	8-monomethyltocol

(B)

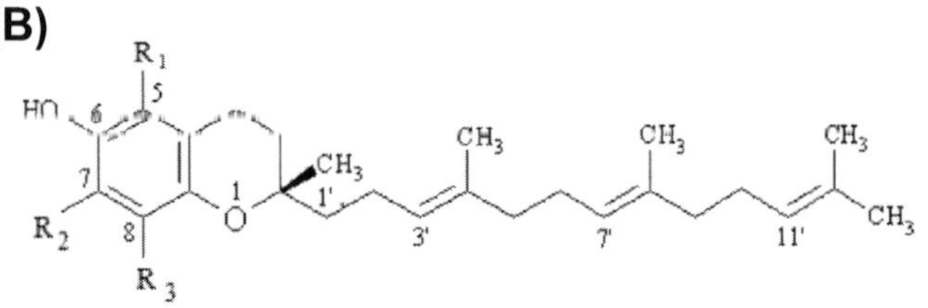

Common name	$-R_1$	$-R_2$	$-R_3$	Chemical name
α-tocotrienol	$-CH_3$	$-CH_3$	$-CH_3$	5-7-8-trimethyltocotrienol
β-tocotrienol	$-CH_3$	-H	$-CH_3$	5-8-dimethyltocotrienol
γ-tocotrienol	-H	$-CH_3$	$-CH_3$	7-8-dimethyltocotrienol
δ-tocotrienol	-H	-H	$-CH_3$	8-monomethyltocotrienol

FIGURE 37.1 Structures and names of tocopherols (A) and tocotrienols (B) of cereals.

It is stimulated by light, heat, metals, enzymes, and peroxides and is inhibited by antioxidants. Oxidative degradation is known to increase with the destruction of the food cellular structure (Rodriguez-Amaya *et al.*, 2006), greater surface area or porosity, length, and severity of the processing conditions (temperatures, time). Compounds could be degraded both by direct and lipoxygenase-mediated oxidation (Leenhardt *et al.*, 2006; Borrelli *et al.*, 2003). This enzyme catalyzes the hydroperoxidation of polyunsaturated fatty acids, such as linoleic acid, producing conjugate hydroperoxides. Radicals from the intermediate steps of this reaction are responsible for oxidative degradation of antioxidant compounds. Enzyme-catalyzed oxidation can occur to a greater extent than thermal degradation. Typically, carotenoid loss in enzymatic oxidation happens rapidly, immediately after tissue disruption, after which carotenoid concentrations stabilize. Non-enzymatic oxidation is usually characterized by a lag phase, followed by rapid decrease of carotenoids, coherent with a free radical mechanism. Enzymatic oxidation can occur prior to heat treatment (Rodriguez-Amaya, 2006).

The effects of pasta processing on carotenoid and tocol amount and color quality of pasta are documented (Table 37.1 and Table 37.2), even if this investigation has been focused only on conventional processing, despite the new tendencies towards the realization of healthy products based on nonconventional flours (Mastromatteo *et al.*, 2012).

Carotenoid losses during pasta processing vary widely with durum wheat cultivar used to produce semolina, depending on the different extent of lipoxygenase activity in the kernel (Laignelet, 1983; Borrelli *et al.*, 2003). In a study by Borrelli *et al.* (2008), the loss of pigments and yellow index after milling and processing were evaluated and correlated with all the parameters involved in the determination of final pasta color. The phase mainly responsible for pigment loss was pasta processing. A decrease of 16.3% in semolina β-carotene content during pasta processing *versus* a 7.9% loss during milling was determined. The isoenzymatic forms LOX-2 and LOX-3, active at the pH of dough, were responsible for the loss of color in pasta products. High temperature, short-time extrusion process, occurrence of oxygen,

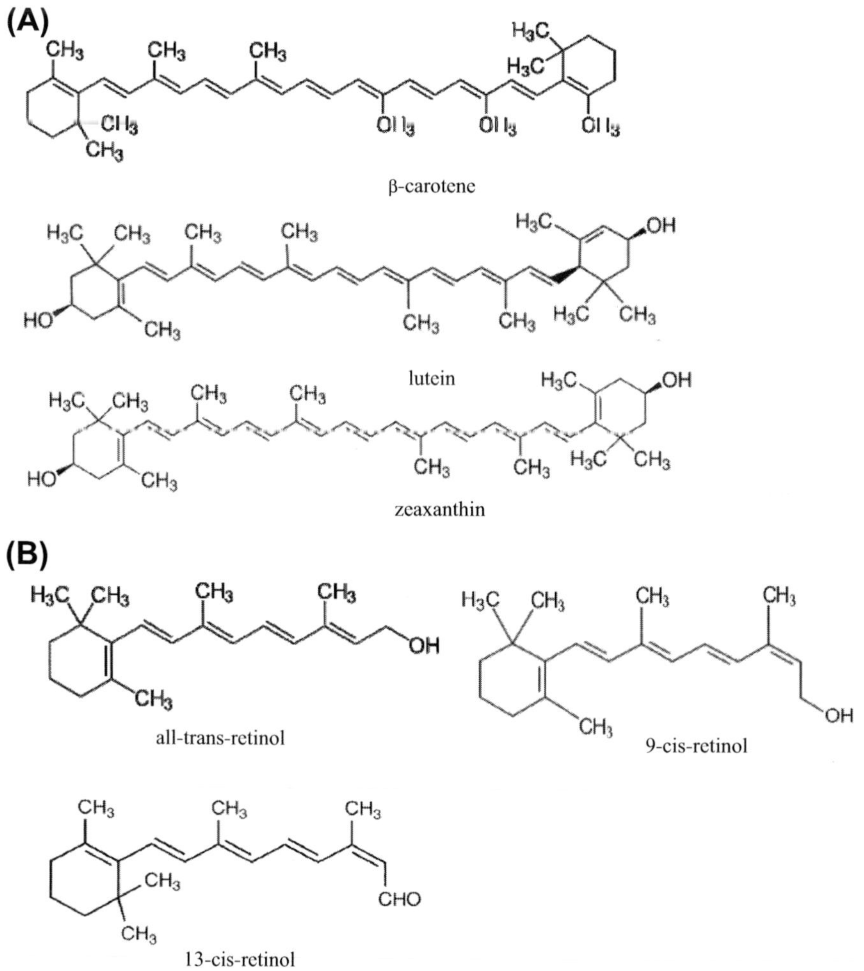

FIGURE 37.2 Structures and names of the main carotenoids (A) in cereals, and retinols (B) in cereal products.

food composition, and presence of antioxidant influence the stability of compounds. A recent paper by Hidalgo *et al.*, 2010, aimed at finding carotenoid evolution during pasta, bread, and water biscuit preparation from einkorn flours, bread flour (from *Triticum aestivum* Blasco) and durum wheat semolina. The longer kneading–extrusion phase determined relevant losses (48%). The drying step did not induce significant changes. The kneading phase leads to the incorporation of water and oxygen in the dough, facilitating the lipoxygenase (LOX)-catalyzed oxidation of polyunsaturated fatty acids. As observed by Borrelli *et al.* (2003) and Leenhardt *et al.* (2006), the LOX reaction, together with mechanical action promoting the substrate-enzyme-O_2 contact, can lead to oxidation of the different antioxidant compounds. During kneading, einkorn consistently showed a lesser percentage degradation (27%), as to 78% of Blasco and 57% of durum semolina, probably because of the lower lipoxygenase activity of its flour (Hidalgo *et al.*, 2010). Hidalgo and Brandolini (2010) made a similar study on tocol stability during pasta processing. Total tocol content decreased throughout the processing stages. The long kneading phase determined high losses (44.2%), while the extrusion step led to an average decrease of 3.7%, under vacuum, and of 29.7%, with no vacuum. Also in this case the drying step did not induce significant changes. Considering the production processes, average tocol losses were 41.0% (under vacuum extrusion) and 55.7% (no vacuum). Einkorn showed a better preservation and supplied more tocols in the end products than durum and bread wheats. In a recent work Fratianni *et al.* (2012) investigated the carotenoid, tocol, and retinol evolution during egg pasta-making. In the final products $\alpha+\beta$-carotene decreased about 25%, lutein about 45%, zeaxanthin 60% and β-criptoxanthin 70%. The overall content of total carotenoids, calculated by summing $\alpha+\beta$-carotene, β-criptoxanthin, lutein and zeaxanthin

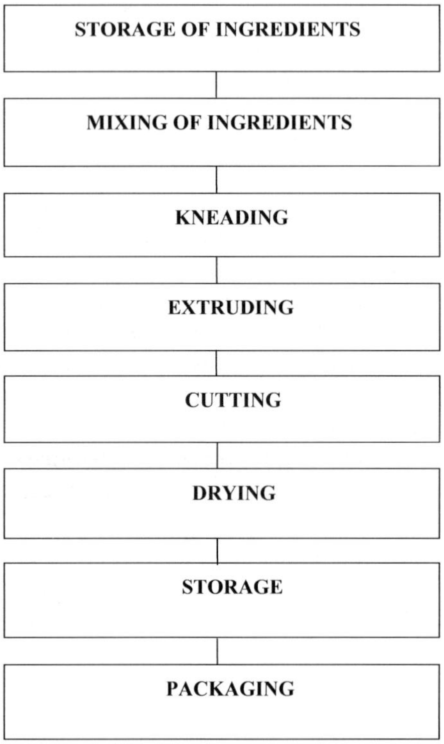

FIGURE 37.3 The pasta-making process.

TABLE 37.1 Tocol Losses (%) During Pasta-Making Processes

Cereal Flour/Semolina	Tocol Loss (%) Tocopherols	Tocotrienols	Total tocols	Extraction and Separation Methods Employed	Reference
T. monococcum flours	—	—	28	Saponification followed by solvent extraction. Normal phase HPLC (Panfili *et al.*, 2004)	Hidalgo *et al.* (2010)
T. aestivum flours	—	—	51		
Durum wheat semolina	—	—	58		
Durum wheat semolina	26	50	40	Saponification followed by solvent extraction. Normal phase HPLC (Panfili *et al.*, 2004)	Fratianni *et al.* (2012)
Durum wheat semolina	50	57	38	Saponification followed by solvent extraction. Normal phase HPLC (Panfili *et al.*, 2004)	Panfili *et al.* (2005)
Emmer semolina*	92	88	89	Solvent extraction (Peterson and Qureshi, 1993[1]) Normal phase HPLC	Fares *et al.* (2008)
Durum wheat semolina*	90	91	91		
Whole durum wheat	30–40	—	—	Solvent extraction. C18 LC-MS (Beleggia *et al.*, 2011)	Beleggia *et al.* (2011)

* *Mean values.*

[1]*Peterson, D.M., Qureshi, A.A. (1993). Genotype and environment effects on tocols of barley and oats. Cereal Chemistry 70, 157–162.*

amounts, decreased about 50% in dried pastas. Regarding single tocol vitamers, β-tocopherol (β-T) showed an average loss of 50%, followed by γ-tocopherol (γ-T) (40%), β-T3 (30%), α-tocopherol (α-T), and α-T3 (about 20%), with an average loss of about 30%. The losses of carotenoid and tocol compounds occurred only during the kneading phase. No losses in carotenoids and tocols were observed during dough extrusion, due to oxygen extraction by the vacuum–extrusion industrial plant and the lower time interval as to the kneading phase. In addition, as observed by Borrelli *et al.* (2003), the progressive decrease of hydroperoxidation activity observed in extruded materials might derive from enzyme destruction or inhibition due to its reaction products. Moreover, physical and chemical modifications during the mixing and the extrusion phases might have irreversibly denatured oxidative enzymes. The drying process did not cause further changes, indicating that the temperature reached during pasta drying inactivated oxidative enzymes but it was not strong enough to cause pigment losses. The thermal resistance of carotenoids and tocols to the temperatures used during drying processes of wheat was also observed by Panfili *et al.* (2005). A significant total carotenoid loss was observed during processing by Beleggia *et al.*, 2011. Overall losses of carotenoids observed from semolina to pasta were between 55% and 80%. The degradation after extrusion went from below 4% to 32%. Further significant losses were found after the drying step, probably due to the high temperature employed in the processes. The broad variation observed was mainly due to the different kneading/extrusion times and drying temperatures, the different characteristics of the semolina samples employed, and enzymatic activities. Regarding tocopherols, a progressive decrease was observed during the pasta steps. In particular, after a significant decrease during the extrusion step (52%), the total content significantly increased in dried pastas. This behavior could be due to an inaccurate storage/manipulation of extruded samples with high activity water, where enzyme-catalyzed oxidations can still occur. The amount of tocols and carotenoids found in the raw material dramatically decreased in semolina and in pasta obtained from selected emmer genotypes, as reported by Fares *et al.* (2008). The decrease in α-tocopherol occurs independently of its content in wholemeal, since the germ, in which this antioxidant is concentrated, is the first portion to be removed during milling. Tocotrienols showed a lesser decrease with respect to tocopherols, varying to 16.8 μg/g of semolina and to 2.04 μg/g in uncooked pasta. After pasta processing and cooking, no β-carotene was found, while the pasta processing caused a strong reduction of lutein. The dramatic effect of processing observed on emmer pasta samples was found also in those from durum and other emmer parents.

The overall behavior of the different compounds observed during the pasta-making process may account for the synergism among the different antioxidants, as reported by Tesoriere *et al.* (1996).

The levels of the (*E*/*Z*)-geometrical isomers of lutein and zeaxanthin were investigated in lasagne and egg noodles by Humphries and Khachik (2003). The all-*E*-isomers of lutein and zeaxanthin in wheat and pasta products were higher than their corresponding *Z*-isomers; the only exception was lasagna in which nearly equal amounts of all-*E*- and 9-*Z*-zeaxanhin were present. The *Z*-isomers of lutein and zeaxanthin in raw fruits and vegetables presumably originate from photochemical isomerization of their all-*E*-isomers. In cooked and processed foods, these

TABLE 37.2 Carotenoid Losses (%) During Pasta-Making Processes

Cereal Flour/Semolina	Carotenoid Loss (%)	Extraction and Separation Methods Employed	Reference
T. monococcum flours	27	Saponification followed by solvent extraction. Normal phase HPLC (Panfili *et al.*, 2004).	Hidalgo *et al.* (2010)
T. aestivum flours	57		
Durum wheat semolina	78		
Durum wheat semolina	50	Saponification followed by solvent extraction. Normal phase HPLC (Panfili *et al.*, 2004).	Fratianni *et al.* (2012)
Durum wheat semolina	30	Saponification followed by solvent extraction. Normal phase HPLC (Panfili *et al.*, 2004).	Panfili *et al.* (2005)
Emmer semolina*	78	Solvent extraction (Fares *et al.*, 2008). Normal phase HPLC.	Fares *et al.* (2008)
Durum wheat semolina*	69		
Whole durum wheat	55–80	Solvent extraction (Digesù *et al.*, 2009[1]). C18 LC-MS (Beleggia *et al.*, 2011).	Beleggia *et al.* (2011)

* *Mean values.*

[1] *Digesù, A. M., Platani, C., Cattivelli, L., Mangini, G., Blanco, A. (2009) Genetic variability in yellow pigment components in cultivated and wild tetraploid wheats. Journal of Cereal Science 50, 210–218.*

geometrical isomers are more likely to form from their all-*E* isomers, as a result of heat treatment. The concentrations of *Z*-isomers of lutein and zeaxanthin in wheat and pasta products appear to be proportional to the concentrations of their all-*E*-isomers. Updike and Schwartz (2003), Aman *et al.* (2005) and De Oliveira and Rodriguez-Amaya (2007) reported higher *Z*-isomers in thermally processed corn compared to the raw material.

Regarding retinols, the only effect of the pasta-making process is reported by Fratianni *et al.* (2012) on egg pasta, in which a geometric isomerization from the *trans* (*E*) to the *cis* (*Z*) forms was observed. In particular, a slight constant increase in 13-*cis* retinol was observed, during the production process, while the amount of the 9-*cis* retinol isomer remained constant. After the kneading step another compound was detected, probably 9-13-di-*cis* retinol (Panfili *et al.*, 2008). Slight decreases in all-trans retinol were found during processing. The 13-*cis*/*trans* ratio increased from about 12% in the mix to 32% in final dried pastas. In milk and dairy products a correlation between the increase in 13-*cis* isomer levels and the severity of heat treatments has been demonstrated (Panfili *et al.*, 1998), together with the influence of microbial evolution during storage (Panfili, *et al.*, 2008). As a result, the ratio between 13-*cis* and all-*trans* retinol, named degree of retinol isomerization, was suggested as a process and product indicator.

All results confirm that the investigated metabolites undergo a transformation during the pasta-making process depending on the processing conditions adopted.

The breeding and technological approaches may play an important role in improving the amount of compounds in food products and in controlling oxidative activity during pasta production. In addition, the study of their kinetic degradation may allow the identification of the fastest reactions, i.e., generally those with the greatest impact, and also the quantification of the effect of the factors which can lower their content. These kinetics have been examined extensively in model systems (Minguez-Mosquera and Jaren-Galan, 1995), but data available for the different losses during the technological processes of cereals is still lacking in the literature.

Impact of Pasta-Making Processes on the Bioavailability of Tocols, Carotenoids, and Retinols

For a nutritional evaluation of a certain foodstuff, as well as for comparison with other foods, not only the amount of nutrients, but also their bioavailability and bioaccessibility have to be considered. Bioavailability can be defined as the fraction of an ingested nutrient that reaches the systemic circulation (Werner and Böhm, 2011). Bioaccessibility describes the fraction of a nutrient that is available for absorption into enterocytes. The absorption of lipophilic micronutrients, such as carotenoids and tocols, is strictly linked to their release from the food matrix, their transfer into emulsified lipid droplets in the stomach, as well as their subsequent incorporation into mixed micelles, along with products of lipolysis in the small intestine.

Carotenoids and tocols represent an example of micronutrients for which processing-induced food structure changes (e.g., matrix disruption or heat treatment) strongly influence their bioavailability and bioaccessibility, by softening or breaking the cell walls and denaturing proteins complexed with compounds, thus facilitating their release from the food matrices (Rodriguez-Amaya *et al.*, 2006).

In recent years, *in vitro* studies have been performed to study the impact of thermal and/or mechanical processing on the carotenoid bioaccessibility (Van Buggenhout *et al.*, 2010), however, only a few papers on the bioaccessibility of tocopherols are available (Reboul *et al.*, 2006). Bioaccessibility of tocotrienols is usually disregarded. Available results indicate that there is still conflicting information regarding the effect of food processing. Relative to pasta, only a recent paper by Werner and Böhm (2011) investigated the bioaccessibility of these lipophilic micronutrients from pure durum wheat pasta and pasta containing 10% eggs. The bioaccessibility of carotenoids and tocols from durum wheat pasta was quite high (71% and 70%, respectively), whereas these micronutrients were considerably less accessible from pasta containing 10% egg (57% and 49%, respectively). The authors attributed this behavior to the *in vitro* digestion method, which was characterized by an insufficient lipolysis. Significant differences among individual tocochromanols, and among (all-*E*)-lutein, (all-*E*)-zeaxanthin, and (all-*E*)-canthaxanthin, or different stereoisomers of lutein were not found.

OTHER WAYS IN WHICH COMPOSITION IS ALTERED

Since lipophilic vitamins are distributed in a tissue specific manner in cereal kernels, removal of the hulls, aleurone layer, and germ during the milling process influences their levels in final products, including starchy endosperm and flour. Milling of cereal grains generally removes the bran and germ layers where most vitamins and provitamins (tocopherols, β-carotene, and zeaxanthin) are located (Panfili *et al.*, 2004), causing significant losses of these compounds in the form of by-products, including hulls and polishes waste. During milling, the tocol content of whole grain becomes concentrated in certain milling fractions such as bran, germ, etc. (Tiwari and Cummins, 2009). In a study by Finocchiaro *et al.* (2007) a 71% reduction in tocol levels during milling of dehulled rice was observed, indicating the greater loss of tocols. Similarly, Ha *et al.* (2006) reported a significant loss of tocols by increasing the degree of milling. Milling process of rye caused decrease in content in tocopherols and tocotrienols, the ratio of tocotrienols to tocopherols (T3/T) was the highest in rye flours with an extraction rate of 100–90%, whereas light flour was the poorest source of tocopherols and tocotrienols (Michalska *et al.*, 2007). Relevant carotenoid losses were reported during milling (Hidalgo and Brandolini, 2008; Leenhardt *et al.*, 2006). In a study of Adom *et al.* (2005), the bran/germ fraction contribution to total lutein, zeaxanthin, and β-cryptoxanthin in whole-wheat flour was 51, 78, and 42%, respectively. Although the bran/germ fraction had higher concentrations of these carotenoids, their final contributions in whole-wheat flour differed greatly, ranging from higher bran/germ contribution for zeaxanthin to higher endosperm contribution for β-cryptoxanthin. Debranned bread and durum wheat flour possessed, on average, 45.5% and 36.4% of total yellow pigments, respectively (Zilic *et al*, 2012).

The presence of tocols and other bioactive compounds, such as β-glucans, in the pericarp, aleurone, and sub-aleurone layers suggests that the application of a pearling technology would facilitate the separation of tocol rich fractions. In the pearling by-products a seven and a five-fold increase was observed for tocopherols and tocotrienols (Panfili *et al.*, 2008). Pearling is proven to be more effective than milling in concentrating total tocols and oil in barley flour (Wang *et al.*, 1993).

Physical processes, such as milling, sieving, air classification, genotypes, or species with high levels of carotenoids and tocols, or with a favorable distribution of these compounds in the kernel, could therefore be used to produce enriched fractions for the realization of functional foods.

ANALYTICAL TECHNIQUES

Several analytical procedures have been developed to evaluate carotenoid, retinol, and tocol concentration in cereals and their products and by-products (Panfili *et al.*, 2003, 2004; Burkhardt and Böhm, 2007; Hentschel *et al.*, 2002). Several factors make this analysis inherently difficult. These compounds are highly sensitive to light, heat, air, and active surfaces and their isolation and analysis may be accompanied by degradation, formation of stereoisomers, structural rearrangements, and other physicochemical reactions. A literature review has shown that procedures for the extraction of analytes from food involve various types of solvent, solvent combinations, and procedures (Rodriguez-Amaya and Kimura, 2004). High-performance liquid chromatography (HPLC) is widely accepted as an accurate and sensitive technique to allow the separation and identification of tocol and carotenoid components and distinguish between similar geometrical structures. Specific HPLC procedures to detect carotenoids in durum grain and end-products have been developed by Panfili *et al.* (2004), Burkhardt and Böhm (2007), Hentschel *et al.* (2002). In particular, use of normal phase HPLC to separate all the eight tocol isomers and, among carotenoids, lutein and zeaxanthin, is advisable. For carotenoids reversed phases are also used, in particular those bounded with C30 material, which are able to distinguish among the different carotenoid isomers.

Several authors had recoveries by using different extractive techniques. Saponification of the matrix, through alkaline hydrolysis, followed by solvent extraction, is considered the elective technique for the extraction of these compounds from cereals. It should be applied in case compounds are present in forms bound to the food matrix, where the application of the sole extraction method would not be sufficient to determine their exact amount and, as a consequence, they are difficult to extract by using sole solvents (Panfili *et al.*, 2003; Peterson, *et al.*, 2007). It is also considered a necessary step in order to simplify separation by removing substances, such as chlorophylls and lipids and other analytical interferences, and by hydrolyzing matrix and carotenoid esters. As reported by Panfili *et al.* (2003, 2004), hot saponification followed by solvent extraction provided significantly higher tocol and carotenoid recoveries. Hidalgo *et al.* (2006) also reported that hot saponification gave a significantly higher extraction of α-tocotrienol fraction. Similarly, Peterson *et al.* (2007) demonstrated that the saponification method is less time-consuming for tocol recovery from oat grains and increased yield by 25%. However, this strategy possesses some drawbacks, since strong conditions are used and losses of carotenoids, as well as isomerization and other reactions, can occur. The most sensitive compounds to alkaline treatments are xanthophylls. Oxidation could be avoided by addition of antioxidants, nitrogen flushing, limited exposure to light and oxygen, and rapid cooling of the reaction mixture after saponification. A suitable extraction procedure, such as that of basic hydrolysis, should be used in case of processed foods, where reported increases in compound content, during cooking or thermal processing, are not likely to be true increases. Processed samples can, in fact, have a greater extractability from the food matrix and comparison with unprocessed samples is difficult if a sole solvent extraction is performed. Moreover, results can be artifacts of the analytical/calculation procedure, can derive from inaccurate sample storage and preparation, where enzymatic activity can cause loss of analytes in samples with elevated water activity, or from unaccounted loss of water and leaking of soluble solids during processing.

References

Adom, K.K., Sorrells, M.E., Liu, R.H., 2005. Phytochemicals and antioxidant activity of milled fractions of different wheat varieties. J. Agric. Food Chem. 53, 2297–2306.

Aman, R., Biehl, J., Carle, R., Conrad, J., Beifuss, U., Schieber, A., 2005. Application of HPLC coupled with DAD, APcI-MS and NMR to the analysis of lutein and zeaxanthin stereoisomers in thermally processed vegetables. Food Chem. 62, 753–763.

Beleggia, R., Platani, C., Papa, R., Di Chio, A., Barro,s, E., Mashaba, C., Wirth, J., Fammartino, A., Sautter, C., Conner, S., Rauscher, J., Stewart, D., Cattivelli, L., 2011. Metabolomics and food processing: from semolina to pasta. J. Agric. Food Chem. 59, 9366–9377.

Borrelli, G.M., De Leonardis, A.M., Fares, C., Platani, C., Di Fonzo, N., 2003. Effects of modified processing conditions on oxidative properties of semolina dough and pasta. Cereal Chem. 80, 225–231.

Borrelli, G.M., De Leonardis, A.M., Platani, C., Troccoli, A., 2008. Distribution along durum wheat kernel of the components involved in semolina colour. J. Cereal Sci. 48, 494–502.

Burkhardt, S., Böhm, V., 2007. Development of a new method for the complete extraction of carotenoids from cereals with special reference to durum wheat (Triticum durum Desf.). J. Agric. Food Chem. 55, 8295–8301.

De Oliveira, G.P.R., Rodriguez-Amaya, D.B., 2007. Processed and prepared corn products as sources of lutein and zeaxanthin: compositional variation in the food chain. J. Food Sci. 72, 79–85.

Falk, J., Krahnstover, A., van der Kooij, T.A., Schlensog, M., Krupinska, K., 2004. Tocopherol and tocotrienol accumulation during development of caryopses from barley (Hordeum vulgare L.). Phytochemistry 65, 2977–2985.

Fares, C., Codianni, P., Nigro, F., Platani, C., Scazzina, F., Pellegrini, N., 2008. Processing and cooking effects on chemical, nutritional and functional properties of pasta obtained from selected emmer genotypes. J. Agric. Food Chem. 88, 2435–2444.

Finocchiaro, F., Ferrari, B., Gianinetti, A., Dall'Asta, C., Galaverna, G., Scazzina, F., Pellegrini, N., 2007. Characterization of antioxidant compounds of red and white rice and changes in total antioxidant capacity during processing. Mol. Nutr. Food Res. 51, 1006–1019.

Fratianni, A., Irano, M., Panfili, G., Acquistucci, R., 2005. Estimation of color of durum wheat. Comparison of WSB, HPLC, and reflectance colorimeter measurements. J. Agric. Food Chem. 53, 2373–2378.

Fratianni, A., Di Criscio, T., Mignogna, R., Panfili, G., 2012. Carotenoids, tocols and retinols evolution during egg pasta - making processes. Food Chem. 131, 590–595.

Ha, T.Y., Kob, S.N., Leeb, S.M., Kim, H.R., Chung, S.H., Kima, S.R., 2006. Changes in nutraceutical lipid components of rice at different degrees of milling. Eur. J. Lipid Sci. Technol. 108, 175–181.

Hentschel, V., Kranl, K., Hollmann, J., Lindhauer, M.G., Bohm, V., Bitsch, R., 2002. Spectrophotometric determination of yellow pigment content and evaluation of carotenoids by high-performance liquid chromatography in durum wheat grain. J. Agric. Food Chem. 50, 6663–6668.

Hidalgo, A., Brandolini, A., Pompei, C., Piscozzi, R., 2006. Carotenoids and tocols of einkorn wheat (Triticum monococcum ssp. monococcum L.). J. Cereal Sci. 44, 182–193.

Hidalgo, A., Brandolini, A., 2008. Protein, ash, lutein and tocols distribution in einkorn (Triticum monococcum L. subsp. monococcum.) seed fractions. Food Chem. 107, 444–448.

Hidalgo, A., Brandolini, A., 2010. Tocols stability during bread, water biscuit and pasta processing from wheat flours. J. Cereal Sci. 52, 254–259.

Hidalgo, A., Brandolini, A., Pompei, C., 2010. Carotenoids evolution during pasta, bread and water biscuit preparation from wheat flours. Food Chem. 121, 746–751.

Humphries, J.M., Khachik, F., 2003. Distribution of lutein, zeaxanthin, and related geometrical isomers in fruits, vegetables, wheat and pasta products. J. Agric. Food Chem. 51, 1322–1327.

Laignelet, B., 1983. Lipid in durum wheat and pasta products. In: Barnes, P.J. (Ed.), Lipids in Cereal Technology. Academic Press, London, pp. 269–286.

Lampi, A.M., Nurmi, T., Piironen, V., 2010. Effects of the environment and genotype on tocopherols and tocotrienols in wheat in the HEALTHGRAIN diversity screen. J. Agric. Food Chem. 58, 9306–9313.

Leenhardt, F., Lyan, B., Rock, E., Boussard, A., Potus, J., Chanliaud, E., Remesy, C., 2006. Wheat lipoxygenase activity induces greater loss of carotenoids than vitamin E during breadmaking. J. Agric. Food Chem. 54, 1710–1715.

Mastromatteo, M., Chillo, S., Civica, V., Iannetti, M., Suriano, N., Del Nobile, M.A., 2012. A multistep optimization approach for the production of healthful pasta based on nonconventional flours. J. Food Process Eng. 35, 601–621.

Michalska, A., Ceglinska, A., Amarowicz, R., Piskula, M.K., Szawara-Nowak, D., Zieliński, H., 2007. Antioxidant contents and antioxidative properties of traditional rye breads. J. Agric. Food Chem. 55, 734–740.

Minguez-Mosquera, M.I., Jaren-Galan, M., 1995. Kinetics of the decolouring of carotenoid pigments. J. Sci. Food Agric. 67, 153–161.

Panfili, G., Manzi, P., Pizzoferrato, 1998. Influence of thermal and other manufacturing stresses on retinol isomerization in milk and dairy products. J. Dairy Res. 65, 253–260.

Panfili, G., Fratianni, A., Irano, M., 2003. Normal phase high performance liquid chromatography method for the determination of tocopherols and tocotrienols in cereals. J. Agric. Food Chem. 51, 3940–3944.

Panfili, G., Fratianni, A., Irano, M., 2004. Improved normal-phase high performance liquid chromatography procedure for the determination of carotenoids in cereals. J. Agric. Food Chem. 52, 6373–6377.

Panfili, G., Fratianni, A., Irano, M., 2005. Change in tocochromanol and carotenoid content during technological processes of cereals. Tecnica Molitoria 5, 493–498.

Panfili, G., Fratianni, A., Di Criscio, T., Gammariello, D., Sorrentino, E., 2008. Influence of microorganisms on retinol isomerization in milk. J. Dairy Res. 75, 37–43.

Penicaud, C., Achir, N., Dhuique-Mayer, C., Dornier, M., Bohuon, P., 2011. Degradation of β-carotene during fruit and vegetable processing or storage: reaction mechanisms and kinetic aspects: a review. Fruits 66, 417–440.

Peterson, D.M., 1995. Oat tocols: concentration and stability in oat products and distribution within the kernel. Cereal Chem. 72, 21–24.

Peterson, D.M., Jensen, C.M., Hoffman, D.L., Mannerstedt-Fogelfors, B., 2007. Oat tocols: saponification vs. direct extraction and analysis in high-oil genotypes. Cereal Chem. 84, 56–60.

Reboul, E., Richelle, M., Perrot, E., Desmoulins-Malezet, C., Pirisi, V., Borel, P., 2006. Bioaccessibility of carotenoids and vitamin E from their main dietary sources. J. Agric. Food Chem. 54, 8749–8755.

Rodriguez-Amaya, D.B., 2002. Effects of processing and storage on food carotenoids. Sight Life Newsletter 3, 25–35.

Rodriguez-Amaya, D.B., Kimura, M., 2004. HarvestPlus handbook for carotenoid analysis. HavestPlus Technical Monographs, vol. 2. Washington, DC, USA and Cali, Colombia. IFPRI and CIAT.

Rodriguez-Amaya, D.B., Rodriguez, E.B., Amaya-Farfan, 2006. Advances in food carotenoid research: chemical and technological aspects, implications in human health. Malays. J. Nutr. 12, 101–121.

Tesoriere, L., Bongiorno, A., Pintaudi, A.M., D'Anna, R., D'Arpa, D., Livrea, M.A., 1996. Synergistic interactions between Vitamin A and Vitamin E against lipid peroxidation in phosphatidylcholine liposomes. Arch. Biochem. Biophys. 396, 57–63.

Tiwari, U., Cummins, E., 2009. Nutritional importance and effect of processing on tocols in cereals. Trends Food Sci. Technol. 20, 511–520.

Updike, A.A., Schwartz, S.J., 2003. Thermal processing of vegetables increases cis isomers of lutein and zeaxanthin. J. Agric. Food Chem. 51, 6184–6190.

Van Buggenhout, S., Alminger, M., Lemmens, L., Colle, I., Knockaert, G., Moelants, K., Van Loey, A., Hendrickx, M., 2010. In vitro approaches to estimate the effect of food processing on carotenoid bioavailability need thorough understanding of process induced microstructural changes. Trends Food Sci. Technol. 21, 607–618.

Wang, L., Xue, Q., Newman, R.K., Newman, C.W., 1993. Enrichment of tocopherols, tocotrienols, and oil in barley fractions by milling and pearling. Cereal Chem. 70, 499–501.

Werner, S., Böhm, V., 2011. Bioaccessibility of carotenoids and Vitamin E from pasta: evaluation of an in vitro digestion model. J. Agric. Food Chem. 59, 1163–1170.

Zandomeneghi, M., Festa, C., Carbonaro, L., Galleschi, L., Lenzi, A., Calucci, L., 2000. Front-surface absorbance spectra of wheat flour: determination of carotenoids. J. Agric. Food Chem. 48, 2216–2221.

Žilić, S., Serpen, A., Akıllıoğlu, G., Janković, M., Gökmen, V., 2012. Distributions of phenolic compounds, yellow pigments and oxidative enzymes in wheat grains and their relation to antioxidant capacity of bran and debranned flour. J. Cereal Sci. 56, 652–658.

CHAPTER

38

Dietary Fiber Arabinoxylans in Processed Rye: Milling- and Breadmaking-Induced Changes

Malgorzata R. Cyran

Plant Breeding and Acclimatization Institute, National Research Institute (IHAR-PIB), Radzikow, Poland

CHAPTER POINTS

- The arabinoxylans with different proportions of water-extractable fraction and degree of branching are present in the endosperm (white) flour and wholemeal produced during rye milling.
- Dough mixing initiates a mechanical solubilization of arabinoxylans.
- The acidic conditions during fermentation are favorable for arabinoxylan hydrolyzing enzymes associated with the grain. Among them the endo-xylanases are the most important.
- Their hydrolytic action results in a partial solubilization of water-unextractable arabinoxylans.
- There is a small difference between total arabinoxylans contents in rye bread and in starting flour. Rye bread generally has an increased level of water-extractable arabinoxylans and decreased that of water-unextractable counterparts.
- The water-extractable arabinoxylans in bread exhibit lower molecular size and their chain structure is modified in comparison to the native form present in the flour.

INTRODUCTION

Rye (*Secale cereale* L.) is a traditional bread cereal grown mainly as a fall-sown winter crop in the cool temperature zones of Europe, spread mainly in its northern, central, and eastern parts. Due to its exceptional winter hardiness and low requirements for soil conditions, in comparison to those of wheat and barley, rye can be successfully cultivated in regions with severe climates or in those with light, sandy, and low-pH soils as well as at high altitudes (Bushuk, 2001). Moreover, unlike other cereals, rye can be grown without herbicides and, in most cases, without fungicides, thus providing the harvest from a more ecological environment.

Cereal products, especially wheat and rye breads, still constitute a staple element of the human diet in many regions of the world, although their consumption has considerably decreased in the last few decades. Hence, they are key products in human nutrition. Cereals are also distinguished by their high concentrations of dietary fiber, the complex mixture of fundamental importance for human health and well-being, which consists of non-starch polysaccharides (arabinoxylans, β-glucans, cellulose, arabinogalactans), oligosaccharides (fructans), lignin, minerals, indigestible proteins (structural and enzymic) and resistant starch. Among them, the arabinoxylans (AX) dominate in rye grain, as well in those of wheat and triticale. They are deposited in the cell walls of the grain during its development (Figure 38.1). The cell wall polysaccharides form a relatively minor population, in contrast to starch, filling the cytoplasm along with the storage proteins. A protein matrix surrounds the starch granules and other minor deposits, such as phytin granules and lipid droplets.

To observe how AX alter at following stages of rye bread production, the starting points must be recognized, i.e., the content and physicochemical and structural characteristics of these polysaccharides in a raw material, the mature rye grain, which depend on genetic background and environmental factors. They are further modified during grain milling and breadmaking stages used for production of a certain type of rye bread.

Processing and Impact on Active Components in Food
http://dx.doi.org/10.1016/B978-0-12-404699-3.00038-X

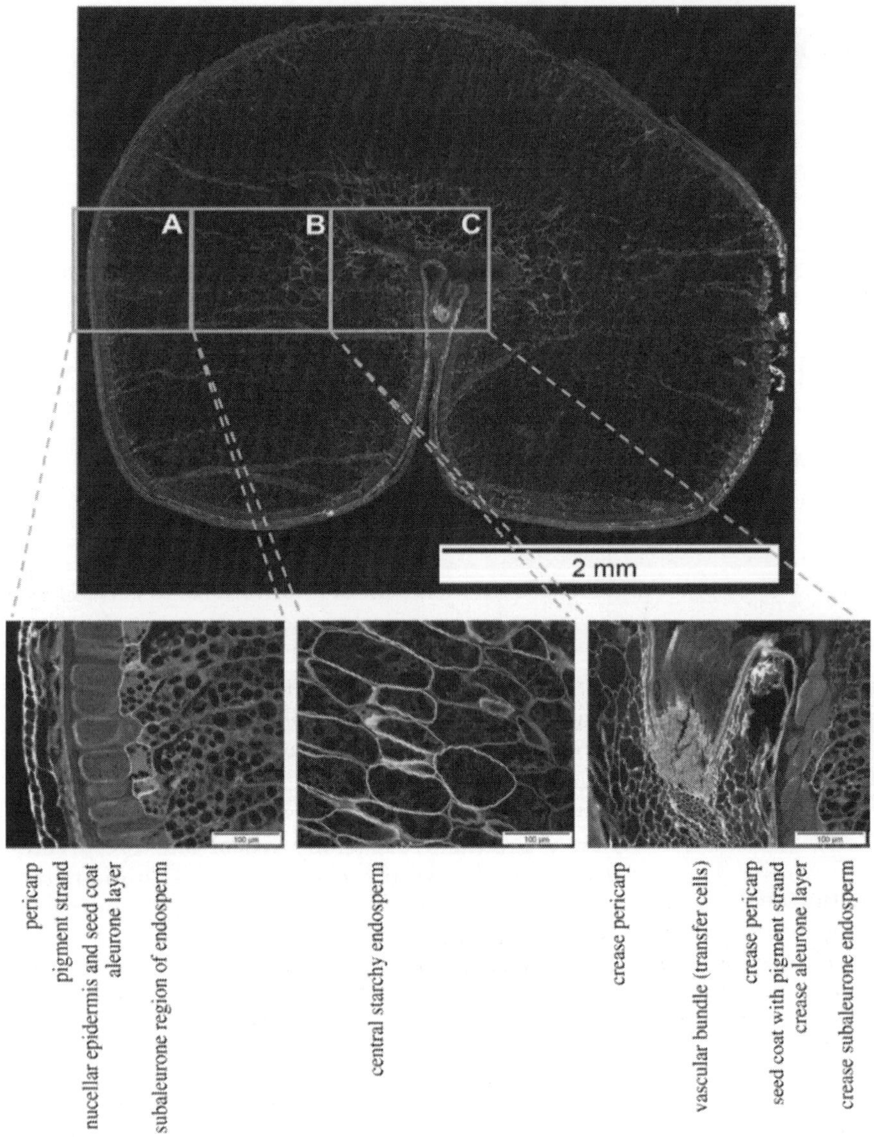

FIGURE 38.1 Cross sections of a rye grain stained with Acid Fuchsin–Calcofluor: 1, outer grain layers; 2, central endosperm; 3, crease region. *Adapted from Dornez* et al., *2011. Reprinted with permission from Elsevier.*

STRUCTURAL FEATURES OF CEREAL ARABINOXYLANS

Although a basic structure of the AX backbone is built of four different β-D-xylopyranosyl residues joined through 1→4 linkages, namely unsubstituted, monosubstituted through *O*-2 or *O*-3 and disubstituted through both *O*-2 and *O*-3 (u-Xyl*p*, 2-Xyl*p*, 3-Xyl*p* and 2,3-Xyl*p*, respectively) with terminal α-L-arabinofuranosyl residues (Ara*f*), the entire AX population is highly heterogeneous with respect to molecular weight, proportion, and sequence of the unsubstituted and differently substituted residues in the backbone (Figure 38.2). They make up long chains with different degrees of polymerization, containing the substituted residues mainly as single and paired, separated by single unsubstituted residues (Goldschmid and Perlin, 1963). However, the "open regions" with two and up to five adjacent unsubstituted residues are present. In addition, the Xyl*p* residues in the backbone might contain the short Ara*f* side chains, terminal Xyl*p* residues as well as the branches with other minor substituents, such as acetyl and α-D-glucuronopyranosyl residues through *O*-2 and feruloyl residues at *O*-5 of Ara*f*. At present, the total AX population is recognized as a mixture of numerous subpopulations with different structural features, occurring at different portions. However, depending on the cereal species, some subpopulations are more abundant than others.

DISTRIBUTION, CONTENT AND WATER-EXTRACTABILITY OF ARABINOXYLANS IN THE GRAIN

The central part of the grain, an endosperm, is enriched in starch granules, as opposed to the outer layers that contain more storage protein and dietary fiber components.

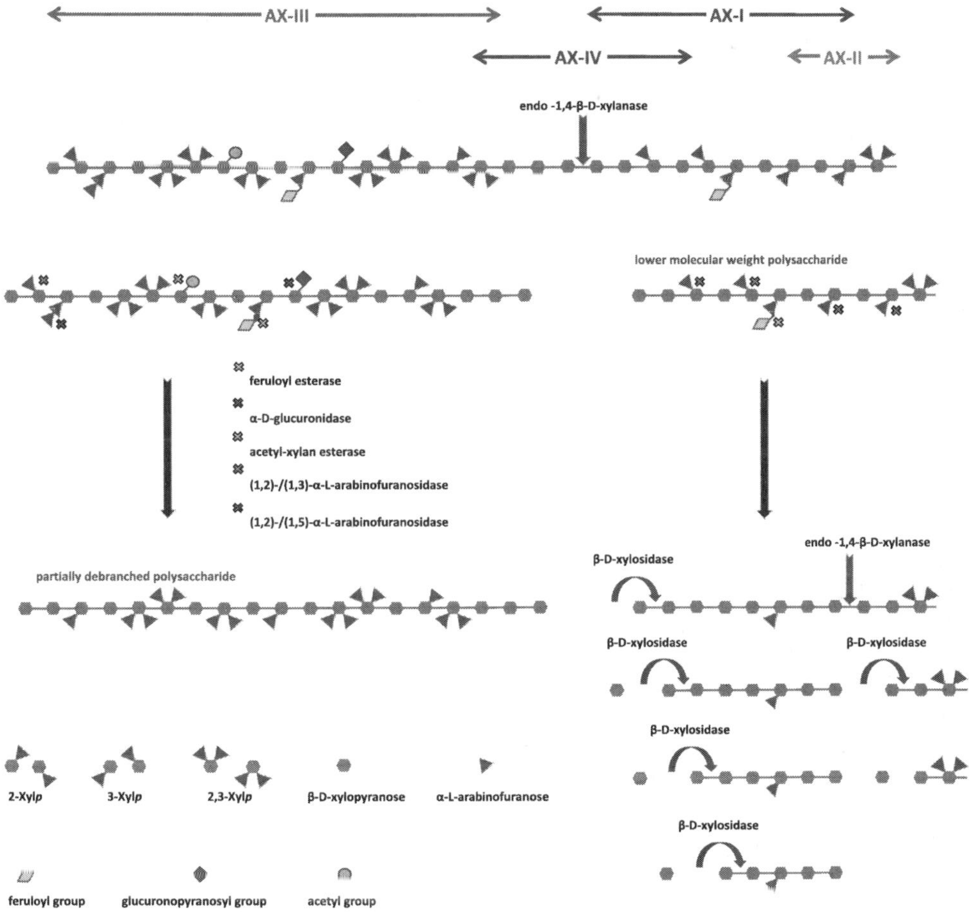

FIGURE 38.2 General structure of rye water-extractable arabinoxylans subfractions, AX-IV, AX-I, AX-II, and AX-III, isolated by sequential ammonium sulfate precipitation at 40, 60, 80, and 100% saturation levels, respectively. A schematic outline of actions of arabinoxylan hydrolyzing enzymes. 2-Xyl*p*, β-D-xylopyranosyl residue substituted at *O*-2 with α-L-arabinofuranosyl residue, 3-Xyl*p*, β-D-xylopyranosyl residue substituted at *O*-3 with α-L-arabinofuranosyl residue, 2,3-Xyl*p*, β-D-xylopyranosyl residue substituted at both *O*-2 and *O*-3 with α-L-arabinofuranosyl residues. The β-D-xylopyranosyl residues are 1→4-linked in arabinoxylan chain.

However, both water-extractable (WE) and water-unextractable (WU) dietary fiber fractions, and consequently both principal types of AX, are differently distributed throughout the grain (Nyman *et al.*, 1984). The WE polysaccharides are almost evenly scattered, while those being WU are concentrated in the outer layers of the grain. For that reason, there are striking differences in the content, proportion of WE- to WU-AX, and their substitution degrees in rye milling fractions with different extraction rates.

In rye grain, the total content of AX ranges from 8.2–11.2% of dry mass (Cyran *et al.*, 1995). They are mainly water-unextractable and a relatively small part of these polysaccharides may be extracted with water. The AX water-extractability, expressed as a proportion of the WE-AX to the total AX content, is temperature-dependent. Besides the evident inter-species differences, those observed within each individual cereal species, are also important. Hot water may extract about 10, 15, 22, and 33% of AX from wheat, hexaploid and tetraploid triticales, and rye wholemeals (WM), respectively (Cyran and Łapiński, 2006). While, only 20–25% of rye AX can be extracted with water at room temperature (Härkönen *et al.*,1997).

The amounts of WE-AX and WU-AX in rye MW vary from 2.4–4.0% and from 5.1–8.4% of dry mass, respectively (Table 38.1). The arabinose-to-xylose (Ara/Xyl) ratio, reflecting the average substitution degree of AX chains with Ara*f*, is usually higher for WE-AX (0.58–0.85) than that of WU polymers (0.52–0.68). While, hot water-extractability of AX present in WM of different rye genotypes may vary from 26–41%.

HOW COMPOSITION IS ALTERED

Before milling, rye grain is usually tempered to 12–14% moisture, to ensure efficient separation of central starchy endosperm from the outer layers of the grain, shorts, and bran. However, dry-milling can also be used to obtain milling fractions enriched in different parts of the grain. Because the content of highly viscous WE-AX in the rye

TABLE 38.1 Content and Arabinose-to-Xylose (Ara/Xyl) Ratio of Water-Extractable and Water-Unextractable Arabinoxylans (WE-AX and WU-AX, respectively) and Arabinoxylans Water Extractability (AX-Extract) in Rye Milling Fractions, Wholemeals, and Endosperm Flours and Resulting Breads

Milling Fraction/Bread	WE-AX (% of dm)	Ara/Xyl of WE-AX	WU-AX (% of dm)	Ara/Xyl of WU-AX	AX-Extract[a] (%)	Reference
Bran (pericarp/testa)	5.5	1.17	34.0	1.02	14	Glitsø and Bach Knudsen (1999)
Bran bread	0.5	1.00	8.4	1.07	6	Glitsø and Bach Knudsen (1999)
Shorts (aleurone)	4.6	0.57	12.5	0.35	27	Glitsø and Bach Knudsen (1999)
Shorts bread	2.8	0.63	7.1	0.37	28	Glitsø and Bach Knudsen (1999)
Endosperm flour	2.9	0.71	1.3	0.83	70	Glitsø and Bach Knudsen (1999)
	2.1–3.0	0.58–0.62	0.7–2.5	0.78–0.86	58–68	Cyran and Cygankiewicz (2004)
Endosperm bread	2.2	0.77	1.1	0.82	66	Glitsø and Bach Knudsen (1999)
	2.2	0.67	1.9	0.75	54	Englyst (1989)
	2.2–3.2	0.58–0.61	0.8–1.4	0.73–0.79	68–77	Cyran and Ceglinska (2011)
Wholemeal	3.7	0.64	5.3	0.63	41	Glitsø and Bach Knudsen (1999)
	2.9	0.63	7.7	0.62	27	Hansen *et al.* (2002)
	2.6–4.0	0.58–0.85	5.3–8.4	0.56–0.68	26–40	Hansen *et al.* (2003)
	2.4	0.58	6.6	0.52	27	Andersson *et al.* (2009)
	2.5	0.68	5.1	0.57	32	Cyran and Łapiński (2006)
Wholemeal bread	2.9	0.73	5.6	0.64	34	Glitsø and Bach Knudsen (1999)
	3.5	0.57	5.0	0.61	41	Hansen *et al.* (2002)
	2.6–2.8	0.53–0.55	5.7–5.9	0.49–0.52	31–33	Andersson *et al.* (2009)
	2.6–3.8	0.55–0.59	4.5–5.1	0.52–0.58	34–43	Cyran and Ceglinska (2011)

[a] *AX-Extract., expressed as per cent WE-AX of total (WE-AX + WU-AX).*

grain is 2–3 times higher than in wheat grain, it is practically not possible to separate the central endosperm from outer layers during rye milling as effectively as in the case of wheat milling. Thus, direct milling of rye grain results in endosperm flour (EF) (white flour) with much lower extraction rate (58–63%) (Cyran *et al.*, 2003) in comparison to that of wheat EF (71–74%) (Cleemput *et al.*, 1997a). The extraction rate of rye WM varies from 95–98%.

Besides the different distribution of WE- and WU-AX throughout the grain, the intact cell walls that occur in the different parts of the grain, such as a seed coat, aleurone layer, subaleurone and central endosperm (Figure 38.1), contain AX with diverse structural characteristics (Table 38.1). Therefore, both types of rye flour, the endosperm and WM, represent a different starting material for processing.

The bran fraction, containing mostly cell walls from the outermost layers of the grain (pericarp/testa), is particularly rich in AX (Table 38.1). There is a clear decrease in the content of total AX observed from the milling fractions of outer layers, bran, and shorts, to that of central endosperm (39.5, 17.1, and 4.2% of total AX content in bran, shorts, and endosperm flour, respectively). Conversely, there is an increase in water-extractability of AX from the cell walls of outer layers to those of central fraction of the grain (14, 27, and 70%, for pericarp/testa, aleurone, and central endosperm cell walls, respectively). Moreover, the AX isolated from rye EF that mostly originate from the cell walls of the central part of the grain, are freely soluble, as an extraction with cold water at 4°C releases 62–66% of the total AX, extracted by hot water (Cyran *et al.*, 2003). The AX present in the pericarp/testa are highly substituted with Ara*f* (Ara/Xyl ratio of 1.02 and 1.17, respectively for WU- and WE-AX). Those from aleurone layer, however, show much lower substitution degrees (Ara/Xyl ratio of 0.35 and 0.57). The intermediate degrees of branching were found for AX of central endosperm (Ara/Xyl ratio of 0.83 and 0.71) (Table 38.1).

OTHER WAYS IN WHICH COMPOSITION IS ALTERED

The intact cell walls of the grain are extensively fragmented during milling. Fragmentation also occurs, to some extent, during mixing and fermentation phases (Autio and Salmenkallio-Marttila, 2001), and it is possible during baking (Brooker, 1993) (Figure 38.3). The degree of cell wall disintegration during mixing of the flour with water and other small ingredients is mainly controlled by the mechanical work input. Nevertheless, it can be supported by the hydrolytic action of the series of cell wall degrading enzymes, including those cleaving the linkages in the AX backbone. Furthermore, dough mixing initiates a mechanical solubilization of AX that is subsequently continued by AX hydrolyzing enzymes associated with the grain.

In the ungerminated grain, the endogenous hydrolytic enzymes are localized in the aleurone and nucellar layers (Jones and Jacobsen, 1991; Beaugrand *et al.*, 2004). The cell walls breakup, water addition, and mixing and fermentation processes provide the appropriate conditions for their action. The grain-associated enzymes also include those of microbial origin, covering the outer surface of the grain. Besides, some hydrolytic activity may derive from minor ingredients of rye dough, such as yeast and starter cultures. Owing to highly heterogeneous structure of AX, a set of endo- and exo-acting enzymes is involved in enzymatic breakdown of these polysaccharides (Figure 38.2). Among them, the endo-(1,4)-β-D-xylanses (endo-xylanases), hydrolyzing an internal 1→4 linkage between Xyl*p* residues in the AX backbone, are the most important, as the cleavage of one such bond in the long chain may cause a drastic reduction in molecular weight, and consequently, a significant decrease in their viscosity enhancing and water-holding capacities. The endo-xylanases act synergistically with other AX-degrading enzymes. The debranched fragments of AX chain are the products of hydrolytic action of α-L-arabinofuranosidases, releasing arabinose from short Ara*f* chains and terminal Ara*f* substituents. Similar products might be obtained by the actions of feruloyl esterases, α-D-glucuronidases and acetyl-xylan esterases, which cleave the bonds between the polysaccharide and its side substituents, feruloyl, glucuronopyranosyl, and acetyl units, respectively. While, the β-D-xylosidases, acting in the exo-mode, make the chains of xylo-oligosaccharides shorter by releasing the xylose from their non-reducing ends. It is assumed that cereal endogenous endo-xylanases hydrolyze the internal 1→4 linkage in the regions of at least five adjacent u-Xyl*p* residues. Nevertheless, the wheat flour endo-xylanase that preferentially hydrolyzed the branched regions of AX chain, has also been reported (Cleemput *et al.*, 1997b).

The bran milling fraction is characterized by the highest level of endo-xylanase activity. It is 12 15-times higher than that of EF and 3–4-times higher than that of WM (Dornez *et al.*, 2009). This can be explained by the fact that the aleurone and nucellar layers, the sites of synthesis and secretion of hydrolyzing enzymes in the grain (Jones and Jacobsen, 1991; Beaugrand *et al.*, 2004), are mostly recovered in bran milling fraction. The level

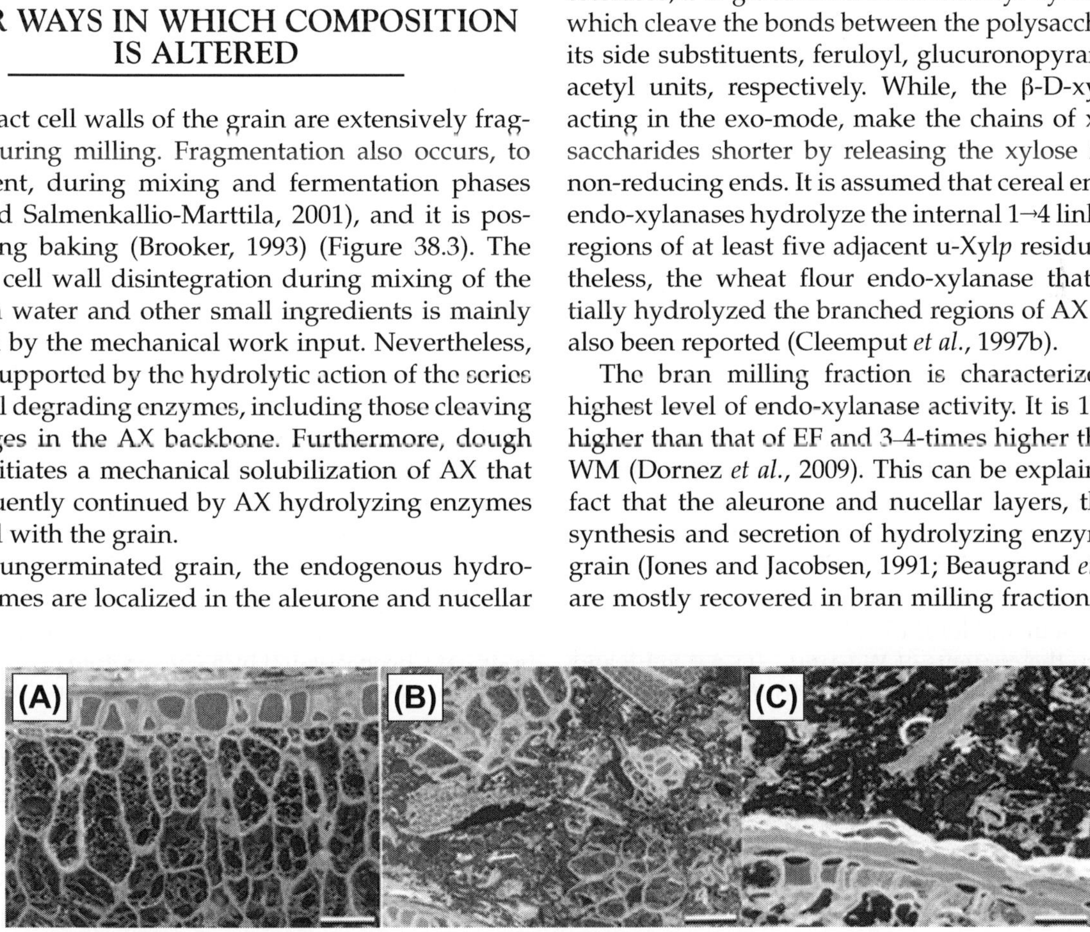

FIGURE 38.3 Embedded cross sections of A, rye grain; B, rye dough and C, rye bread stained with Acid Fuchsin and Calcofluor. The primary cell walls appear blue, protein red and starch black. *Adapted from Autio and Salmenkallio-Marttila, 2001. Reprinted with permission from Elsevier.*

of endo-xylanase activity in rye bran, however, is low in comparison to those of common wheat, oat, barley, and durum wheat (2, 3, 4, and 12-times higher, respectively) (Dornez *et al.*, 2009). The pH optima of endo-xylanase extracted from rye grain is 4.5. It is active in the range from 3.8 to 5.3 pH at 40°C (Rasmussen *et al.*, 2001). These pH and temperature values correspond well with those of typical rye dough fermentation stages used in sourdough and direct rye breadmaking methods, indicating that during the early steps of breadmaking the suitable conditions for endo-xylanases actions are provided. On the other hand, the acidic conditions of rye dough may also result in a partial hydrolysis of acid-labile 1→2 and 1→3 glycosidic bonds between Ara*f* substituents and Xyl*p* residues in the chain, generating the debranched AX regions, similar to those produced by hydrolytic action of α-L-arabinofuranosidases.

The differences in the contents of total AX between starting WM and the corresponding breads are relatively small (Table 38.1). They result mainly from addition of some other ingredients (salt, yeast, acids, starter cultures) to flour during rye dough preparation. Usually, the level of WE-AX observed in the bread is slightly higher than that in the flour used for its production. For yeast-leavened crisp bread, air-leavened crisp bread, and yeast leavened sour-dough breads the amount of WE-AX increased from 2.4 and 2.9% in rye WM to 2.6–2.8 and 3.5% in the resulting WMB, respectively (Andersson *et al.*, 2009; Hansen *et al.*, 2002, Table 38.1). In addition, the AX water-extractability is higher in WMB (31–33% and 41%) than in rye WM (27%). Opposite relationships were reported by Glitsø and Bach Knudsen (1999). Nevertheless, in this case rye bread was produced by wheat bread baking procedure, i.e., with neither direct acid addition nor sourdough preparation. Furthermore, it was supplemented with casein, gluten, soybean oil, lard, and a mixture of vitamins and minerals to provide a diet properly balanced for experimentation in pigs. Hartman *et al.* (2005), however, also reported decreased level of WE-AX in the rye bread obtained by sourdough method in which a deep-frozen starter was used. The authors suggested that some WE-AX populations could become water-unextractable due to formation of cross-links between protein and ferulic acid.

Although the contents of WE- and WU-AX in EF and EB are lower than in WM and WMB, the similar relationships are observed in the levels of AX between flours and breads made from both types of rye flour. Though, the AX water-extractability is much higher in EB than in WMB (58–77% and 31–41%, respectively, Table 38.1).

In contrast to comparatively small differences in the levels of AX in EF and WM and those in corresponding breads, the considerable differences in AX contents were found between bran and shorts milling fractions and resulting breads (Table 38.1). Both types of bread were much poorer in AX than the starting flours. This was caused by a substantial supplementation of bread with wheat starch (1:2.6 and 1:0.5, rye raw material:wheat starch, for bran and shorts breads, respectively) along with other additives that ensured suitable diet for experimentation in pigs.

A slight, but obvious, increase in the level of WE-AX accompanied by a reduction in that of WU-AX in rye bread, despite its "dilution" by other ingredients, clearly indicates that a solubilization of part of WU population by grain-associated enzymes is the major outcome of multisite enzymatic hydrolysis of these polysaccharides, controlled by the endo-xylanases. Similar small changes in AX fractions, caused by hydrolytic actions of endogenous enzymes, were reported during wheat breadmaking (Cleemput *et al.*, 1997). However, in that case the WU-AX solubilization became noticeable during dough mixing, while no such changes could be observed during rye dough mixing (Hansen *et al.*, 2002).

The most evident changes, reflecting the rate of WE-AX depolymerization in rye bread in relation to the counterparts present in the flour, are those perceived in their molecular weight distribution. For high molecular weight AX isolated from rye WM, the weight average molecular weight of about 2×10^6 g/mol was reported. While, much lower values, dependent on the baking method used, were observed for these polymers in rye bread (about 1×10^6 g/mol and 7×10^5 g/mol, respectively for crisp bread and sourdough bread) (Andersson *et al.*, 2009).

The further macromolecular changes in the structure of AX are very difficult to observe, since they are relatively small and as AX are extremely heterogeneous. The overall WE-AX and WU-AX populations include numerous subfractions with diverse structural features. Therefore, structural changes in one subfraction may be covered by changes in the others, resulting in "apparent" no change in Ara/Xyl of the entire fraction. The various techniques used for AX fractionation enable an illustration of the structural adjustment, and help in understanding their behavior during cereal processing. The ammonium sulfate precipitation technique has been used to isolate more homogenous WE-AX subfractions from rye WM, EF, and the resulting breads produced by a direct rye baking method (Table 38.2). The four principal subfractions have been isolated by a consecutive precipitation at 40, 60, 80, and 100% salt saturation, AX-IV, AX-I, AX-II, and AX-III, respectively. The AX-I is the major subfraction, being specific for rye AX. It contains almost exclusively 3-Xyl*p* and u-Xyl*p* residues in the backbone with low substitution degree with Ara*f* (Ara/Xyl of 0.47–0.51) (Figure 38.2, Table 38.2). The AX-II and AX-III subfractions are characterized by the presence of both 3-Xyl*p* and 2,3-Xyl*p* residues as branching sites, however, there is a decrease in a relative proportion of 3-Xyl*p* from AX-I to AX-III and increase in that of 2,3-Xyl*p*, which are

TABLE 38.2 Yield, Relative Proportion, and Structural Characteristics of Water-Extractable Arabinoxylans Isolated from Rye Endosperm Flour and Wholemeal and Resulting Bread by Graded Ammonium Sulfate Precipitation Technique

				Endosperm Flour/Wholemeal						Endosperm Bread/Wholemeal Bread					
	Level of Saturation	Degree of	Major Type	Ara/Xyl		Relative Proportion (%)		Yield (%)		Ara/Xyl		Relative Proportion (%)		Yield (%)	
Class	(%)	Substitution	of Branching	*EF*[a]	*WM*[b]	*EF*[a]	*WM*[b]	*EF*[a]	WM[b]	*EB*[c]	*WMB*[d]	*EB*[c]	*WMB*[d]	*EB*[c]	WMB[d]
AX-IV	40	very low	3-Xylp, 2,3-Xylp	—	0.30–0.31	nd	8–11	—	0.27–0.28	—	—	—	—	—	—
AX-I	60	low	3-Xylp	0.47–0.51	0.49–0.51	69–70	60–62	1.11–1.46	1.60–1.94	0.49–0.50	0.49–0.51	59–66	48–64	1.32–1.36	1.11–1.42
AX-II	80	intermediate	3-Xylp, 2,3-Xylp	0.74–0.91	0.72–0.80	23	20–24	0.37–0.48	0.52–0.76	0.72–0.82	0.66–0.73	24–28	27–40	0.48–0.66	0.58–0.94
AX-III	100	high	2,3-Xylp	0.98–1.28	1.00–1.15	7–8	7–8	0.13–0.14	0.18–0.24	1.08–1.20	0.90–1.05	8–12	7–10	0.17–0.28	0.16–0.24

nd, not detected; Ara/Xyl, arabinose-to-xylose ratio.
[a] *EF, endosperm (white) flour*
[b] *WM, wholemeal*
[c] *EB, endosperm bread*
[d] *WMB, wholemeal bread.*
Source: Cyran et al., *2003; Cyran and Saulnier, 2005; Cyran and Saulnier, 2012*

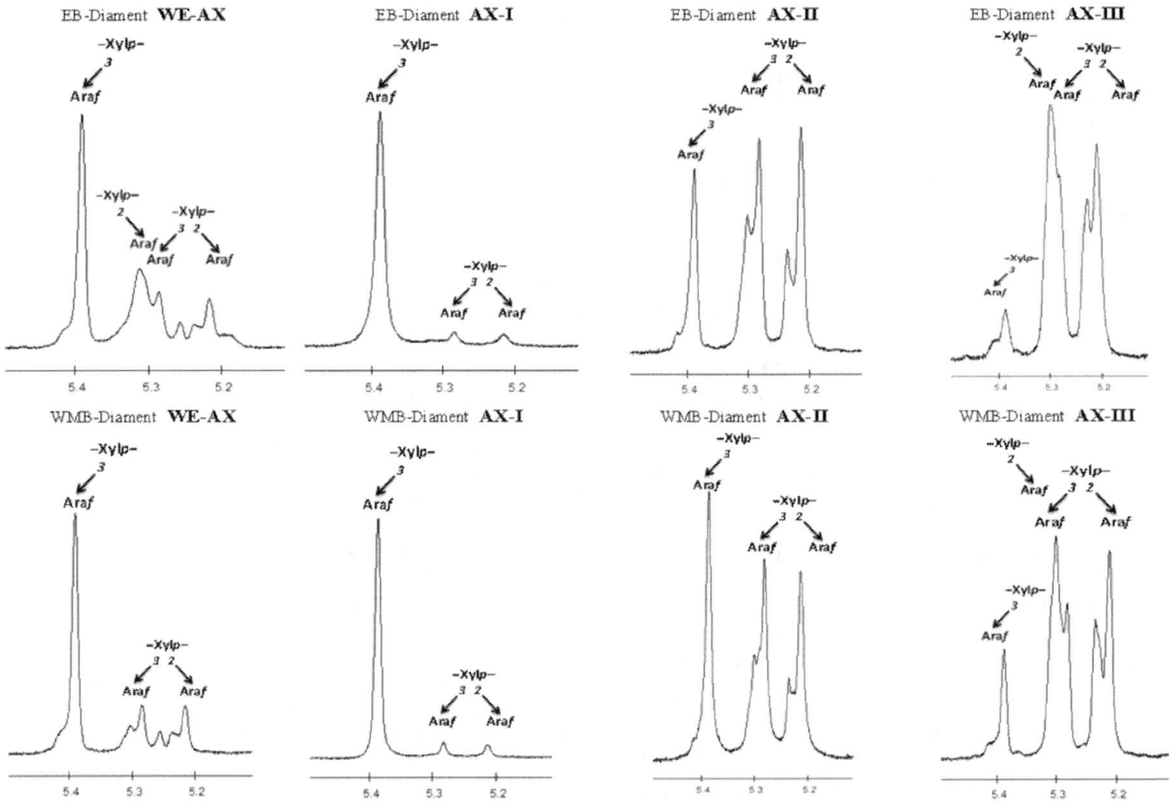

FIGURE 38.4 Schematic structure of β-D-xylopyranosyl (Xyl*p*) residues differently substituted with α-L-arabinofuranosyl (Ara*f*) residues and changes in their relative proportions in endosperm bread (EB) and wholemeal bread (WMB) made from rye cultivar Diament with lower and higher degree of arabinoxylans hydrolysis, respectively, as evidenced by ^{1}H NMR spectroscopy. *Adapted from Cyran and Saulnier, 2012. Reprinted with permission from Elsevier.*

accompanied by increase in their Ara/Xyl ratios. The relative proportions of the branching residues, typical for entire WE-AX population and AX-I, AX-II, and AX-III subfractions, are pictured by their ^{1}H NMR spectra of Ara*f* anomeric protons region (Figure 38.4). Besides the occurrence of 3-Xyl*p* and 2,3-Xyl*p* residues, the highly substituted AX-II and AX-III subfractions also contain 2-Xyl*p*, a very scarce element of rye AX structure, which is usually more abundant in the AX-III subfraction. The AX-IV with both 3-Xyl*p* and 2,3-Xyl*p* branches and very low substitution degree is characteristic of bran milling fraction. It is especially favorable substrate for endo-xylanase, as it contains the long unbranched regions, which are easily accessible.

The entire AX fraction of rye bread, when compared to that of starting flour, exhibits the lower proportion of the major AX-I subfraction and slightly higher proportions of both highly branched subfractions, AX-II and AX-III, which are enriched in 3-Xyl*p* residues, being a product of AX-I degradation during breadmaking (Table 38.2, Figure 38.4). The 2,3-Xyl*p* residues present in the AX backbone can form a steric barrier, limiting the access of endo-xylanase to substrate. Recently, it has been evidenced that a lower degree of AX degradation in rye bread was related not only to the higher proportion of 2,3-Xyl*p* in the highly substituted AX subfractions, but also to their lower ratio of 2,3-Xyl*p* to 2-Xyl*p*, i.e., the di-substituted fragments of the AX chain were concomitantly enriched in 2-Xyl*p* in the subfractions more resistant to endo-xylanase hydrolysis (Cyran and Saulnier, 2012).

There is general agreement that during rye bread baking, heat treatment practically does not cause any significant changes in AX (Meuser and Suckow, 1986). However, some minor changes cannot be excluded, as the increased quantity of xylose reducing ends has been reported after baking of wheat bread made from flour with elevated endo-xylanase activity level (Dornez *et al.*, 2008).

Much larger changes, in AX levels and their structural features, might be expected after addition of microbial endo-xylanases during dough preparation, often applied especially in wheat bread production, to modify dough and bread quality attributes.

ANALYTICAL TECHNIQUES

The analysis of nonstarch polysaccharides (NSP), also denoted cell wall polysaccharides (CWP), is used for determination of a whole spectrum of polysaccharides present in the plant materials, including AX.

The Englyst enzymic-chemical procedure is based on the method of Southgate (1969), which was subsequently optimized for a variety of plant materials (Englyst *et al.*, 1982; Englyst and Cummings, 1984), especially for those with high content of starch (cereals and cereal-based products). It determines the total, WE and WU fractions of NSP. In this analysis, a special attention has been paid to complete digestion of starch by enzymes with no side activity towards cell wall polysaccharides.

A first stage of NSP procedure includes starch gelatinization and digestion, initially with α-amylase (usually thermostable) followed by an incubation with amyloglucosidase. The obtained suspension is centrifuged to separate the WE polymeric material (WE-NSP) and free and low-molecular-weight components (supernatant) from the WU polymeric material (pellet, WU-NSP). The supernatant is mixed with 96% ethanol (1:4, v/v) to precipitate the WE-NSP. They are separated from the ethanol soluble free and low molecular weight components by centrifugation. The precipitated WE-NSP and WU-NSP left in the pellet are hydrolyzed with sulfuric acid. The constituent monosaccharides, released during hydrolysis, are derivatized to alditol acetates and quantified by gas chromatography. The values obtained for arabinose and xylose are converted into that corresponding to polymeric form of AX [AX = (%Arabinose + %Xylose) x 0.88]. Likewise, the constituent hexoses are recalculated to obtain values for their polymeric forms using a conversion factor of 0.9.

At present, besides pancreatic starch degrading enzymes, those of microbial origin are also used for starch digestion in NSP analysis. Moreover, both sulfuric acid and trifluoroacetic acid are utilized in hydrolysis of these polysaccharides. The most commonly applied methods for qualitative and quantitative determination of monosaccharides released during acid hydrolysis are the gas chromatography equipped with flame ionization detector (FID) and high-performance anion-exchange chromatography (HPAEC) with pulsed amperometric detection (PAD).

References

Andersson, R., Fransson, G., Tietjen, M., Åman, P., 2009. Content and molecular-weight distribution of dietary fiber components in whole-grain rye flour and bread. J. Agric. Food Chem. 57, 2004–2008.

Autio, K., Salmenkallio-Marttila, M., 2001. Light microscopic investigations of cereal grains, doughs and breads. Lebensm.-Wiss. u.-Technol. 34, 18–22.

Beaugrand, J., Crônier, D., Thiebeau, P., Schreiber, L., Debeire, P., Chabbert, B., 2004. Structure, chemical composition, and xylanase degradation of external layers isolated from developing wheat grain. J. Agric. Food Chem. 52, 7108–7117.

Brooker, B.E., 1993. The stabilization of air in cake batters–the role of fat. Food Struct. 12, 285–296.

Bushuk, W., 2001. History, world distribution, production, and marketing. In: Bushuk, W. (Ed.), Rye Production, Chemistry and Technology. American Association of Cereal Chemists, Inc., St. Paul, Minnesota, pp. 1–7.

Cleemput, G., Booij, C., Hessing, M., Gruppen, H., Delcour, J.A., 1997a. Solubilisation changes in molecular weight distribution of arabinoxylans and protein in wheat flours during bread-making, and the effects of endogenous arabinoxylans hydrolyzing enzymes. J. Cereal Sci. 26, 55–66.

Cleemput, G., Hessing, M., van Oort, M., Deconynck, M., Delcour, J.A., 1997b. Purification and characterization of a beta-D-xylosidase and endo-xylanase from wheat flour. Plant Physiol. 113, 377–386.

Cyran, M., Rakowska, M., Wasilewko, J., Buraczewska, L., 1995. Degradation of dietary fibre polysaccharides in the intestinal tract of growing pigs used as a model animal for studying digestion in humans. J. Anim. Feed Sci. 4, 217–227.

Cyran, M., Courtin, C.M., Delcour, J.A., 2003. Structural features of arabinoxylans extracted with water at different temperatures from two rye flours of diverse breadmaking quality. J. Agric. Food Chem. 51, 4404–4416.

Cyran, M., Cygankiewicz, A., 2004. Variability in the content of water-extractable and water-unextractable non-starch polysaccharides in rye flour and their relationship to baking quality parameters. Cereal Res. Commun. 32, 143–150.

Cyran, M.R., Saulnier, L., 2005. Cell wall fractions isolated from outer layers of rye grain by sequential treatment with α-amylase and proteinase: Structural investigation of polymers in two ryes with contrasting breadmaking quality. J. Agric. Food Chem. 53, 9213–9224.

Cyran, M., Łapiński, B., 2006. Physico-chemical characteristics of dietary fibre fractions in the grains of tetraploid and hexaploid triticales: a comparison with wheat and rye. Plant Breed. Seed Sci. 54, 77–84.

Cyran, M.R., Ceglinska, A., 2011. Genetic variation in the extract viscosity of rye (*Secale cereale* L.) bread made from endosperm and wholemeal flour: impact of high-molecular-weight arabinoxylan, starch and protein. J. Sci. Food Agric. 91, 469–79.

Cyran, M.R., Saulnier, L., 2012. Macromolecular structure of water-extractable arabinoxylans in endosperm and wholemeal rye breads as factor controlling their extract viscosities. Food Chem. 131, 667–676.

Dornez, E., Cuyvers, S., Gebruers, K., Delcour, J.A., Courtin, C.M., 2008. Contribution of wheat endogenous and wheat kernel associated microbial endoxylanases to changes in the arabinoxylans populations during breadmaking. J. Agric. Food Chem. 56, 2246–2253.

Dornez, E., Gebruers, K., Delcour, J.A., Courtin, C.M., 2009. Grain-associated xylanases: occurrence, variability, and implication for processing. Trends Food Sci. Tech. 20, 495–510.

Dornez, E., Holopainen, U., Cuyvers, S., Poutanen, K., Delcour, J.A., Courtin, C.M., Nordlund, E., 2011. Study of grain cell wall structures by microscopic analysis with four different staining techniques. J. Cereal Sci. 54, 363–373.

Englyst, H., Wiggins, H.S., Cummings, J.H., 1982. Determination of the non-starch polysaccharides in plant foods by gas-liquid chromatography of constituent sugars as alditol acetates. Analyst 107, 307–318.

Englyst, H.N., Cummings, J.H., 1984. Simplified method for the measurement of total non-starch polysaccharides by gas-liquid chromatography of constituent sugars as alditol acetates. Analyst 109, 937–942.

Englyst, H., 1989. Classification and measurement of plant polysaccharides. Anim. Feed Sci. Tech. 23, 27–42.

Glitsø, L.V., Bach Knudsen, K.E., 1999. Milling of whole grain rye to obtain fractions with different dietary fibre characteristics. J. Cereal Sci. 29, 89–97.

Goldschmid, H.R., Perlin, A.S., 1963. Interbranch sequences in the wheat arabino-xylan. Selective enzymolysis studies. Can. J. Chem. 41, 2272–2277.

Hansen, H.B., Andreasen, M.F., Nielsen, M.M., Larsen, L.M., Bach Knudsen, K.E., Meyer, A.S., Christensen, L.P., Hansen, Å., 2002. Changes in dietary fibre, phenolic acids and activity of endogenous enzymes during rye bread-making. Eur. Food Res. Technol. 214, 33–42.

Hansen, H.B., Rasmussen, C.V., Bach Knudsen, K.E., Hansen, Å., 2003. Effects of genotype and harvest year on content and composition of dietary fibre in rye (*Secale cereale* L.) grain. J. Sci. Food Agric. 83, 76–85.

Härkönen, H., Pessa, E., Suortti, T., Poutanen, K., 1997. Distribution and some properties of cell wall polysaccharides in rye milling fractions. J. Cereal Sci. 26, 95–104.

Hartmann, G., Piber, M., Koehler, P., 2005. Isolation and chemical characterization of water-extractable arabinoxylans from wheat and rye during breadmaking. Eur. Food Res. Technol. 221, 487–492.

Jones, R.L., Jacobsen, J.V., 1991. Regulation of synthesis and transport of secreted proteins in cereal aleurone. Int. Rev. Cytol. 126, 49–88.

Meuser, F., Suckow, P., 1986. Non-starch polysaccharides. In: Blanshard, J.M.V., Frazier, P.J., Gilliard, T. (Eds.), Chemistry and Physics of Baking. The Royal Society of Chemistry, London, pp. 42–61.

Nyman, M., Siljeström, M., Pedersen, B., Bach Knudsen, K.E., Asp, N.-G., Johansson, C.-G., Eggum, B.O., 1984. Dietary fiber content and composition in six cereals at different extraction rates. Cereal Chem. 61, 14–19.

Rassmussen, C.V., Hansen, H.B., Hansen, Å., Larsen, L.M., 2001. pH-, temperature- and time-dependent activities of endogenous endo-β-D-xylanase, β-D-xylosidase and α-L-arabinofuranosidase in extracts from ungerminated rye (*Secale cereale* L.) grain. J. Cereal Sci. 34, 49–60.

Southgate, D.A.T., 1969. Determination of carbohydrates in foods. II. Unavailable carbohydrates. J. Sci. Food Agric. 20, 331–335.

CHAPTER

39

Processing of Corn (Maize) and Compositional Features

Leticia X. Lopez-Martinez, Hugo S. Garcia*†

*Universidad Autónoma del Estado de México, Facultad de Química. Paseo Colón y Paseo Tollocan S/N, Toluca, Estado de México 50120, México, †UNIDA-Instituto Tecnologico de Veracruz, M.A. de Quevedo 2779, Col. Formando Hogar, Veracruz, Ver. 91897, México

CHAPTER POINTS

- Corn is a widely diverse grain crop and one of the most extensively cultivated cereals in the world.
- Phenolic compounds appear to be more responsive to environmental factors than carotenoids.
- The stage of maturity can influence directly the content of carotenoids more than phenolic compounds.
- Lutein has the best heat stability of the carotenoids during thermal processing.
- Lime-cooking has the greatest negative impact on the stability of total phenolic compounds and anthocyanins in corn.

INTRODUCTION

The content of bioactive components in grains varies, making quality control and intake recommendations challenging. Variations in these compounds depend on plant factors such as genetics and degree of maturity, as well as on environmental variables including growth conditions, harvest, storage, and processing.

Although processing is often considered to be a negative attribute in nutrition, and some forms reduce the nutritional value of foodstuffs, products such as whole grains are generally not consumed directly by humans but require some degree of processing prior to consumption. Refining reduces the nutrient content of grains, while milling otherwise concentrates desirable components and removes poorly digested compounds and contaminants in grains, whereas cooking is usually accompanied by changes in textural characteristics, increased digestibility of nutrients and some phytochemicals, as well as supporting the importance of processing of grains to enhance their consumption. Overall processing provides shelf stable products that are convenient and good tasting for consumers.

Corn (*Zea mays* L.) is a widely diverse grain crop and one of the most extensively cultivated cereals in the world. Also, maize and milled maize including meal, flour, and bran have been essential dietary components of all socioeconomic classes worldwide, containing several bioactive substances produced by secondary metabolism such as carotenoids and flavonoids. These compounds are pigments with biosynthetic pathways delimited to the plant kingdom and have generated growing interest from food technologists because of their bioactive properties, which are beneficial to human health; furthermore, carotenoids and flavonoids (anthocyanins) are among the most utilized plant colorants in the food industry.

This chapter deals with the production and processing factors that influence the content of the major corn phytochemicals, namely, carotenoids and phenolics.

HOW COMPOSITION IS ALTERED

The Role of Environment in Determining Corn Phytochemicals

Traditionally, organic and sustainable agriculture are the primary cultural practices used in food production; carotenoid and phenolic compound accumulation in

Processing and Impact on Active Components in Food
http://dx.doi.org/10.1016/B978-0-12-404699-3.00039-1

plant tissue appears to be controlled by the physiological, genetic and biochemical attributes of a plant species, and environmental growth factors such as light, temperature and soil fertility.

Carotenoids

Insect and herbivore animal pressures are known to affect nutrient levels in plants. Increased soil nitrogen produces an increase in protein content, whereas low phosphorus soil levels produce increments in sugar content; however, there is very limited information available on the impact that cultural practices could have on the production of carotenoids in maize.

Phenolic Compounds

Asami *et al.* (2003) compared the content of total phenolic compounds in Jubilee Sweet corn grown using conventional, organic and sustainable agricultural practices. The data demonstrated a statistically relevant trend of higher levels of phenolic compounds in organically and sustainably produced corn compared with conventionally produced corn (49, 46, and 27.5 mg/100 g), respectively. Organically and sustainably grown products were produced by cultural methods utilizing very little or no pesticides. Thus pathogenic pressures may explain the elevated content of total phenolic compounds found in organically and sustainably grown maize (Dixon and Pavia, 1995). On the basis of the limited available data, it is not easy to draw conclusions, although it seems that sustainable agriculture or organic farming might have some potential for producing corn enriched with phenolic compounds.

The Role of Genetics

The possible causes for the reported variation in carotenoids and phenolic compounds content across a number of corn varieties might be related to differences in genotype. Also, secondary metabolites in corn are influenced by corn species and varieties.

Carotenoids

The content of carotenoids varies substantially among corn varieties. Among 44 lines of dent and sweet corn, total carotenoids ranged from 0.40 to 33.1 μg/g dry weight. In the same survey, 70-fold differences in zeaxanthin content were reported, whereas lutein content ranged from 0 to 20 μg/g dry weight (Kurlich and Juvik, 1999). De la Parra *et al.* (2007) studied the carotenoids profile of five types of corn (white, yellow, high carotenoid, blue, and red); lutein content was highest in yellow corn (4.05 μg/g dry weight) and the lowest in white corn (0.01 μg/g dry weight). Zeaxanthin content was highest in yellow corn (3.53 μg/g dry weight). As expected, white corn contained the lowest levels of zeaxanthin (0.06 μg/g dry weight).

TABLE 39.1 Main Carotenoid Content (mg/g of d.w.) in Corn Grains at Different Maturation Stages

Sample	Processing	Lutein (mg/kg)	Zeaxanthin (μg/g)	β-carotene (μg/g)
Corn	Fresh	0.92 ± 0.00^{a}	2.80 ± 0.11^{a}	0.95 ± 0.00^{a}
	Blanched	1.79 ± 0.01^{b}	6.34 ± 0.10^{b}	0.96 ± 0.00^{b}
	Dried	1.99 ± 0.01^{b}	7.96 ± 0.12^{b}	0.52 ± 0.11^{b}
	Milled	1.69 ± 0.01^{b}	7.83 ± 0.00^{b}	0.69 ± 0.02^{b}

Means not sharing similar letters within the same column among the methods are significantly different ($P < 0.05$) as determined by one-way ANOVA.
Adapted from Mamatha et al. *(2012).*

In another study Hu and Xu (2011) identified lutein and zeaxanthin in four different maize genotypes (Table 39.1); total carotenoids ranged from 1.24 to 22.5 μg/g dry weight, while the highest content of lutein and zeaxanthin was found in the yellow normal genotype (4.22 μg/g dry weight and 10.44 22 μg/g dry weight, respectively). Yellow corn showed concentrations of 9.65 and 4.14 μg/g dry weight, black corn contained 3.65 and 1.42 μg/g dry weight, while in the white genotype the amounts of lutein and zeaxanthin were the lowest (0.48 and 0.24 μg/g dry weight, respectively). These studies suggest considerable differences in carotenoids according to corn genotype. Genetic variation for carotenoid concentration within species might be advantageous for plant improvement programs.

Phenolic Compounds

Numerous surveys report the range of phenolic contents among corn species. As expected, phenolics differ among corn species and phenotypes within species. López-Martinez *et al.* (2009) determined the content of total phenolic compounds and anthocyanins of eighteen Mexican corn varieties (Table 39.2), some with the same type of pigmentation. Phenolics content in purple samples varied from 465 to 3400 mg/100 g, black samples varied from 475 to 565 mg/100 g, and red samples from 283 to 617 mg/100 g. Anthocyanins content ranged for purple corns from 93 to 851 mg/100 g, for black corns from 76 to 120 mg/100 g, and for red varieties from 85 to 154 mg/100 g. In varieties from China (Zhao *et al.*, 2009) only small differences were detected with reference to the other studies; the black pigmented variety displayed the highest content of anthocyanins (310 mg/100 g), followed by purple (26.5 mg/100 g), red (1.49 mg/100 g), grey (0.29 mg/100 g) and pink (12 mg/100 g). Examination of compositional data from different corn varieties showed that total phenolics content levels varied 10-fold and anthocyanins varied 11-fold.

The observed genetic variability in each study suggests profound differences in potential health benefits among genotypes, and supports the feasibility of developing germoplasms with varieties containing enhanced

TABLE 39.2 Anthocyanins and Total Phenolic Contents of Different Maize Strains and Varieties

Maize Variety	Pigmentation	Total Phenolic Compounds (mg gallic acid/100 g)	Anthocyanins Content (mg Cyanidin 3-Glucoside/100 g)
Black	Black	457 ± 7.4^{h}	76.2 ± 2.2^{j}
Negro Normal OLI04 PV compuesto 1	Black	544 ± 7.3^{f}	120 ± 1.7^{c}
Negro Normal OLIO4 PV compuesto 2	Black	565 ± 9.1^{d}	113 ± 3.8^{f}
Negro IGO4 PV Manuel 3 compuesto 2	Black	472.1 ± 2.9^{g}	113 ± 1.6^{f}
Purple	Purple	465 ± 9.8^{g}	93.2 ± 1.1^{h}
AREQ516540TL	Purple	3400 ± 11^{a}	850 ± 6.2^{a}
Veracruz 43	Purple	1760 ± 10^{b}	389 ± 3.9^{b}
Oaxaca 337	Purple	539 ± 5.2^{f}	131 ± 1.4^{d}
Red	Red	465 ± 4.4^{g}	85.2 ± 2.2^{i}
Pinto	Red	617 ± 7.1^{c}	154 ± 3.3^{c}
Rojo ancho Olinala	Red	456 ± 4.1^{g}	93 ± 3.9^{h}
Manuel 5 IG04 PV Compuesto II	Red	406.76 ± 9.1^{f}	100 ± 2.8^{g}
Mazorca roja-morada	Red	397 ± 2.6^{b}	76.2 ± 2.2^{j}
Rojo Olinala	Red	283 ± 5.1^{j}	62.3 ± 1.1^{k}
Blue	Blue	343 ± 8.6^{h}	99.5 ± 1.8^{g}
Orange	Orange	215 ± 5.1^{j}	30.6 ± 0.9^{l}
Yellow	Yellow	551 ± 3.8^{e}	70.2 ± 0.9^{k}
White	Non/trace	170 ± 1.1^{k}	1.54 ± 0.9^{m}

Means within a column with the same superscript letter are not significantly different ($p > 0.05$) as determined by one-way ANOVA).
Adapted from López-Martínez et al. *(2009).*

levels of these compounds at dosages that could promote health among consumers.

The Importance of Maturity

Development of corn grains is characterized by the occurrence of pericarp color changes varying from creamy white to clean white in white colored varieties, and from creamy yellow to golden in yellow colored varieties. It is believed that color changes reflect fluctuations in coloring compound profiles synthesized in corn grains like carotenoids and anthocyanins. Carotenoid content evolves in different ways during the development and maturation of corn; the decline in lutein and zeaxanthin content may result from their conversion to metabolites because zeaxanthin is an important precursor of abscisic acid and lutein is a precursor for apocarotenoids and some aroma compounds (Toro-Sanchez *et al.*, 2006).

Carotenoids

Immature grains of corn are reported to have the highest concentration of total carotenoids, whereas ripe grains have the lowest. In yellow corn grain the content of lutein increased significantly by 31.7% from 74 days to 93 days post-harvest, and then decreased by 5.7% when fully mature (116 days post-harvest); zeaxanthin content remained unchanged during the maturation period (Xu *et al.*, 2010). Hu and Xu (2011) evaluated the changes of total carotenoids of four cultivars of pigmented corn in three stages of maturity (Table 39.1). The variation of total carotenoids showed significant differences among three stages for each corn type: first carotenoids increased by 21.5% and 10.0% and then decreased by 42.3% and 41.1% for white and black corns, respectively. Total carotenoids of two types of yellow corn decreased by 43% and 36.2% at maturity stage, among different types of colored corns, carotenoids increased when maturity was reached and then declined during further ripening.

Phenolic Compounds

Ripening is typically accompanied by substantial changes in the profile of phenolic compounds. In purple-pigmented corn varieties anthocyanins content increased (Hu and Xu, 2011), the corn was hand-picked at three stages (106 days, 118 days, and complete maturity at 130 days); the anthocyanin content increased continuously and significantly by 3.34-fold. In contrast, anthocyanin content decreased by 90.7, 49.2, and 35.8%, respectively, in white, yellow, and golden yellow varieties of 106 days of maturity. During maturation of a black variety, corn anthocyanins increased continuously by 52.1%, while yellow and golden yellow decreased 77.8 and 42.4% from 106 days to maturity. The levels of anthocyanins decreased for white and yellow corns, contrary to what happened in black corn during maturation, which was similar to the changes observed by Xu *et al.* (2010). Phenolic compounds are produced in all higher plants influenced by different metabolic pathways. The key enzyme for phenolics metabolism is phenylalanine ammonia-lyase; its activity was found to vary greatly with the developmental stage of the plant (Hudec *et al.*, 2007). Additionally, the developmental stage and differences related to species and varieties are major variables affecting the content of phenolic compounds in corn.

Handling and Storage

Whether raw or processed, grains are subjected to handling and storage by producers and consumers, which may affect the integrity of their phytochemicals.

Carotenoids

Drying and storage temperatures are considered significant in the stability of carotenoids. Quackenbusch (1963) analyzed dent corn grains that were oven-dried (38°C) and stored in the dark during 36 months at 3% and 11% moisture, and at 7°C and 25°C. The author reported that total carotenoids levels decreased at 25°C by 61.6% and by 69.2% at 7°C. The losses accounted for 20% of lutein and 25.4% of zeaxanthin. The losses of lutein and zeaxanthin were most rapid between 4 and 8 months of storage in either moisture contents and declined during the late part of the storage period. Using ambient temperature drying under low light, followed by storage in the dark at 4°C and 35% relative humidity during 0, 3, 6, and 18 months, indicated that under these conditions the total carotenoids concentration on the genotypes studied remained fairly stable between 0 and 3 months, but declined by 40% by the sixth month to a level that remained stable over the subsequent 12 months. In a study by Burt *et al.* (2010) using a more rigorous trial with three drying and storage regimes (freeze-drying and storage at −80°C; ambient temperature drying and storage; 90°C drying and room temperature storage), the immediate effects of both 25°C and 90°C compared to freeze-drying showed that total carotenoids decreased *ca.* 20%; none of the genotypes exhibited greater carotenoids losses at 90°C than at 25°C, the stability of total carotenoids concentrations under storage was also assessed with these samples. The freeze-dried material was stored at −80°C and the other two drying treatments were stored at 25°C, in the dark in all cases for 4 months. After the defined storage time, total carotenoid losses ranged from 24% at 90°C to 61% at 25°C, which was consistent with the report by Weber (1987) who found mean carotenoids loss of 46% in four inbred lines of corn after 6 months of storage at room temperature. Freeze-dried samples maintained at −80°C did not show a significant decrease in carotenoids content after four months of storage. These data suggest that the genetic difference is important in the loss of carotenoids during storage.

Phenolic Compounds

Phenolics content may either increase or decrease in grains depending on storage conditions. Zhang *et al.* (2011) studied the content of anthocyanins in purple corn during storage at 4°C and 22°C with a relative humidity of 60–70% for 60 days, and samples withdrawn every 12 days. The loss of anthocyanins was comparable during the first 48 days (34%), but after this time in the treatment at 22°C the loss continued until 52% was reached at day 60. Storage of purple corn at 4°C during 60 days showed losses of only 38% of the initial content of anthocyanins. Zhang *et al.* (2011) tested four drying treatments: solar drying during 9 days at 22–36°C and a relative humidity of 5%–20%; forced air-drying at 60°C for 16 h; microwave drying using a power setup of 80 W for 10 min; and freeze drying for 72 h at −82°C. The forced-air drying showed no influence on the anthocyanin content, while microwave drying lowered the anthocyanin content significantly compared to the control. Each treatment decreased the anthocyanin content by 35%, 5%, 26%, and 23%, respectively; the loss of anthocyanins was primarily a consequence of exposure to high temperatures and moisture content. The content of total phenolic compounds of Supersweet Golden Jubilee corn was analyzed after three post-harvest processing treatments (freezing, freeze-drying, and air-drying) compared to the content in fresh samples that were frozen and stored at −12°C for 6 months. At the end of storage there were no significant differences between frozen and freeze-dried corn samples in terms of total phenolic compounds, in the air-dried treatment the levels decreased by 40% (Asami *et al.*, 2003).

OTHER WAYS IN WHICH COMPOSITION IS ALTERED

The Effect of Processing

Food processing is paramount in determining the bioavailability of bioactive compounds. Thermal food preparation methods applied to cereals generally improve texture, palatability, and nutritive, protein denaturation, increased nutrients availability, and enzyme inhibition (Bark and Gowish, 1991).

Carotenoids

Processing can alter and damage bioactive compounds in grains; heating can result in thermal degradation, leaching and other events that lead to lower levels of phytochemicals in processed food compared with fresh produce; however, in the case of carotenoids, processing can lead to a dissociation from the plant matrix components and increase their content and carotenoids also appear to be less adversely affected by processing compared with other phytochemicals.

Mamatha *et al.* (2012) studied the effect of drying and milling on white corn; the results showed that these processes did not affect lutein and zeaxanthin levels. Fresh corn contained levels of lutein and zeaxanthin of 0.92 and 2.80 µg/g, respectively. The levels increased by 271% and 226% after blanching at 70°C for 5 min. Then the corn was dried at 45°C for 8 h using a cross flow air oven and the levels of lutein and zeaxanthin increased by 11.1% and 27.2% more compared to blanched samples. The milled corn exhibited a decrease in lutein and zeaxanthin by 2.5% and 1.6%, respectively, after blanching and drying (Table 39.3).

Nixtamalization is a process native to Meso-american cultures, known since pre-Hispanic times, that consists of alkaline cooking of corn kernels in a lime or calcium hydroxide solution. In this step, removal of the pericarp occurs. During this process, the maize kernels are lime-cooked at 90–100°C for 0–20 min depending on the type of corn (conventional or pigmented), steeped overnight and the resulting *nixtamal* is washed in preparation for grinding. The ground *nixtamal* or dough (*masa*) is further processed into tortillas by baking at 220°C, or tortilla chips by deep-fat-frying at 175°C for 1 min. De la Parra *et al.* (2007) determined how the processing into *masa*, tortilla and tortilla chips affects the content of carotenoids from five types of corn (white, yellow, high carotenoid, blue, and red). Lime-cooking decreased lutein content in yellow, red, and high carotenoid corns (68.2%, 65%, and 69.5%, respectively), while processing into tortillas and tortilla chips did not affect lutein concentration, except for yellow corn chips (81%). No impact of lime-cooking or further processing was observed in white and blue corns. Zeaxanthin contents were similar among *masa*, tortillas, and tortilla chips for white, red, blue, and high carotenoid corns. Levels of zeaxanthin decreased in yellow corn after lime-cooking (62.5%) as well as after baking (23.6%), and frying (36.3%). Lutein had the best heat stability of the carotenoids.

Phenolic Compounds

Lopez-Martinez *et al.* (2011) studied the effect of processing white, yellow, black, blue, red, and purple corn to *masa*, tortilla, and tortilla chips on changes in total phenolic compounds and anthocyanins (Figure 39.1). The greatest degree of degradation in phenolics was observed when conventional corn varieties (white and yellow) were processed into nixtamal (64.1% and 58%, respectively). Among the blue, black, red, and purple varieties phenolics degradation ranged from 43.1–50.4%. Phenolics losses varied from 51.4% to 57.5% in the conversion of *masa* into tortilla chips for all varieties examined.

TABLE 39.3 Effect of Processing on Carotenoid Composition (Dry Weight Basis) of Corn

Type of Corn	Maturation Stage	Days After Seeding	Lutein (µg/g)	Zeaxanthin (µg/g)	Total Carotenoids (µg/g)
White	M1	86	0.64 ± 0.02^{bD}	0.38 ± 0.02^{bD}	1.77 ± 0.07^{bD}
	M2	98	0.95 ± 0.04^{aC}	0.53 ± 0.02^{aD}	2.15 ± 0.08^{aD}
	Maturity	110	0.48 ± 0.01^{cD}	0.24 ± 0.00^{cD}	1.24 ± 0.05^{cD}
Yellow	M1	96	14.71 ± 0.72^{aB}	7.04 ± 0.10^{aB}	31.81 ± 0.33^{aB}
	M2	108	14.23 ± 0.55^{aA}	7.11 ± 0.14^{aB}	31.10 ± 0.48^{aB}
	Maturity	120	9.65 ± 0.36^{aC}	4.14 ± 0.04^{bA}	20.16 ± 0.16^{bB}
Purple-black	M1	106	5.44 ± 0.25^{aB}	2.59 ± 0.10^{aC}	13.20 ± 0.11^{bC}
	M2	118	5.82 ± 0.28^{bC}	2.35 ± 0.09^{aC}	14.52 ± 0.09^{aC}
	Maturity	130	3.65 ± 0.13^{aA}	1.42 ± 0.04^{bB}	8.55 ± 0.05^{cC}
Yellow normal	M1	109	17.52 ± 0.80^{aA}	10.04 ± 0.12^{aA}	39.61 ± 0.42^{aA}
	M2	124	14.38 ± 0.62^{bA}	8.60 ± 0.10^{bA}	34.82 ± 0.33^{bA}
	Maturity	128	10.44 ± 0.54^{cA}	4.22 ± 0.06^{cA}	22.55 ± 0.27^{cA}

Mean values within a column with different lowercase letters for the same corn type at different maturation stages are significantly different at $P < 0.05$.
Adapted from Hu and Xu (2011).

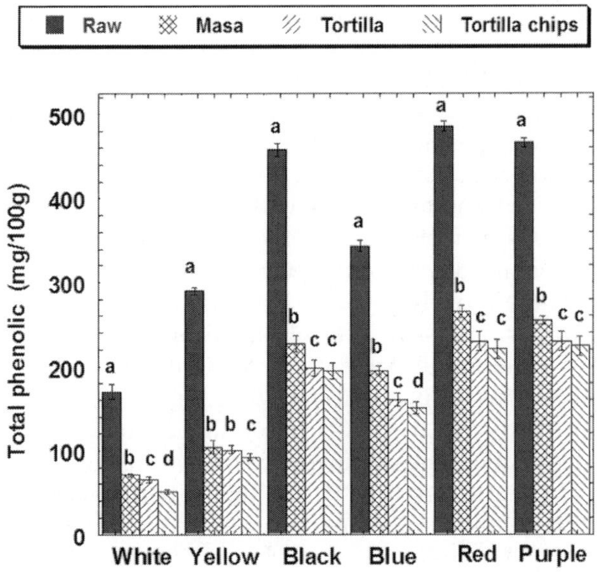

FIGURE 39.1 Total phenolics in raw and processed corn. Means of the same variety within a column with the same superscript are not significantly different (P>0.05). *Adapted from López-Martínez* et al. *(2012).*

Del Pozo-Insfran *et al.* (2007) reported that lime-cooking had the greatest negative impact on the stability of total phenolic compounds and anthocyanins in blue corns of Mexican and American origin. They accounted for losses of 47–50% when corn was processed into *masa*, 56–64% when processed into tortillas and 70–85% when processed into tortilla chips. In summary, thermal processes may release more bound phenolic acid from the breakdown of cellular constituents and deactivate oxidative and hydrolytic enzymes that can degrade some phytochemicals in the grain.

ANALYTICAL TECHNIQUES

Carotenoids

Carotenoid analysis was carried out according to Xu *et al.* (2010). One gram of corn powder was transferred to a 50-ml screw-capped culture tube, and 6.0 ml of methanol containing 0.1 g/l of butylated hydroxytoluene (BHT) was added. Tubes were placed in a water bath at 85°C for 10 min. The tubes were then removed from the water bath, and 120 µl of 80% KOH were added to each tube. Samples were vortexed and returned to the water bath for 10 min more for saponification to occur. After this time the tubes were placed in an ice bath to cool, and then 3.0 ml of cold deionized water was put into each tube, followed by 3.0 ml of hexane. The tubes were vortexed and immediately centrifuged at 4000 g for 10 min. The upper organic phase was collected into a 10 ml disposable tube and evaporated to dryness. The dried extract was reconstituted with 100 µl of methyl tertbutyl-ether followed by 300 µl of methanol. Samples were stored at −20°C under nitrogen until injected into the HPLC column.

TABLE 39.4 Methods Reported to Determine Total Carotenoids, Lutein, and Zeaxanthin

Authors	Section	Determination	Reference
Kurlich and Juvik, (1999) De la Parra *et al.* (2007)	The Role of Genetics	Total carotenoids Lutein Zeaxanthin Lutein Zeaxanthin	Weber (1987) Adom *et al.* (2005)
Xu *et al.* (2010)	The Importance of Maturity	Lutein Zeaxanthin	Li *et al.* (2007)
Quackenbusch (1963) Burt *et al.* (2010)	Handling and Storage	Lutein Zeaxanthin Total carotenoids	Quackenbusch *et al.* (1961) Hentschel *et al.* (2002)
Mamatha *et al.* (2012) De la Parra *et al.* (2007)	The Effect of Processing	Total carotenoids Lutein Zeaxanthin	Raju *et al.* (2007) Adom *et al.* (2005)

The carotenoid extracts were filtered through a 0.45 µm nylon syringe filter prior to HPLC analysis and analyzed in a HPLC system fitted with a C30 carotenoid column using methanol containing 1.0 g of ammonium acetate/L (solvent A) and methyl *tert*-butyl ether (solvent B) as the mobile phase. The following gradient was used: 0–15 min, 100% solvent A; 15–25 min, linear gradient to 10% solvent B; 25–35 min, linear gradient to 30% solvent B; 35–55 min, linear gradient to 50% solvent B. The injection volume was 10 µl, the flow rate was 1.0 ml/min, and the wavelength was 453 nm.

Identification of the main carotenoids was made by comparisons with the retention times and UV spectra of authentic standards; then the quantitative data were calculated from their linear calibration curves under the analytical conditions. The results for the main carotenoids were expressed as micrograms per gram on dry matter basis. The remaining determinations used in this study are depicted in Table 39.4.

Determination of Total Phenolic Compounds and Anthocyanins in Corn

The total soluble phenolics content in the studies realized by Lopez *et al.* (2009, 2011) were analyzed by a procedure adapted from Gao *et al.* (2002). Briefly, extracts of 100 µL of corn were diluted with 500 µl of water. A volume of 700 µl of 0.2 M Folin-Ciocalteu reagent was added and the mixture was allowed to stand for 3 min at room temperature; then 900 µl of Na_2CO_3 1 M were

TABLE 39.5 Methods Reported to Determine Total Phenolics and Anthocyanins

Authors	Section	Determination	Reference
Asami *et al.* (2003) Zhao *et al.* (2009)	The Role of Genetics	Phenolic compounds Anthocyanins	Singlenton and Rossi (1965) Chandra and Li (2001)
Hu and Xu, 2011	The Importance of Maturity	Anthocyanins	Abdel-Aal and Hucl (1999)
Zhang *et al.* (2012) Zhang *et al.* (2011)	Handling and Storage	Anthocyanins	Zhao *et al.* (2008)
Del Pozo-Insfran *et al.* (2007)	The Effect of Processing	Phenolic compounds	Talcott *et al.* (2003)

added and the mixture allowed to stand for 90 min, after which absorbance readings at 765 nm were taken. Acidified methanol was used as the blank. Total phenolics contents were expressed as mg gallic acid/100 g, based on a standard curve of 0–600 mg of gallic acid/mL.

Analysis of total anthocyanins was carried out according to Abdel-Aal and Hucl (1999). One gram of ground whole grain sample was homogenized in a 50 ml centrifuge tube with 25 ml of an acid : ethanol solution (0.225 M HCl in 95 : 5, v/v ethanol : water). The tube was flushed with nitrogen, agitated for 30 min and then centrifuged at 3000 g for 15 min and the supernatants collected. Absorbance readings at 535 nm were taken and corrected for background absorbance at 700 nm (due to turbidity). Anthocyanins were expressed as mg of cyanidin-3 glucoside/100 g. Other determinations used in this study are shown in Table 39.5.

References

Abdel-Aal, E.-S.M., Hucl, P.A., 1999. A rapid method for quantifying total anthocyanins in blue aleurone and purple pericarp wheats. Cereal Chem. 76, 350–354.

Adom, K.K., Sorrells, M.E., Liu, R.H., 2005. Phytochemicals and antioxidant activity of milled fractions of different wheat varieties. J. Agric. Food Chem. 53, 2297–2306.

Asami, D., Hong, Y., Barret, D., Mitchell, A., 2003. Comparison of the total phenolic and ascorbic acid content of freeze-dried and air-dried marionberry, strawberry, and corn grown using conventional, organic, and sustainable agricultural practices. J. Agric. Food Chem. 5, 1237–1241.

Bark, A.A., Gowish, R.A., 1991. Nutritional evaluation and cooking quality of cooked seeds. J. Food Sci. Technol. 28, 312–316.

Burt, A.J., Grainger, C.M., Young, J.C., Shelp, B.J., Lee, E.A., 2010. Impact of postharvest handling on carotenoids concentration and composition in high-carotenoids maize (*Zea mays* L.) kernels. J. Agric. Food Chem. 58, 8286–8292.

Chandra, A., Rana, J., Li, Y.Q., 2001. Separation, identification, quantification, and method validation of anthocyanins in botanical supplement raw materials by HPLC and HPLC-MS. J. Agric. Food Chem. 49, 3515–3521.

Del Pozo-Insfran, D., Bernes, C.H., Serna, S.O., Talcott, S.T., 2006. Polyphenolic and antioxidant content of white and blue corn (*Zea Mays* L.) products. Food Res. Int. 39, 696–703.

De la Parra, C., Serna-Saldivar, S.O., Ruiz, L., 2007. Effect of processing on the phytochemical profiles and antioxidant activity of corn for production of masa, tortillas and tortilla chips. J. Agric. Food Chem. 55, 4177–4183.

Dixon, R.A., Paiva, N.L., 1995. Stress-induced phenylpropanoid metabolism. Plant Cell. 7, 1085–1097.

Gao, L., Oomah, B.D., Mazza, G., 2002. Wheat quality: antioxidant activity wheat millstreams. In: Ng, P., Wrigley, C.W. (Eds.), Wheat quality elucidation. AACC International, ACCC Press, St. Paul, MN. 233 Pp.

Hentschel, V., Kranl, K., Hollmann, J., Lindhauer, M.G., Bohm, V., Bitsch, R., 2002. Spectrophotometric determination of yellow pigment content and evaluation of carotenoids by high-performance liquid chromatography in durum wheat grain. J. Agric. Food Chem. 50, 6663–6668.

Hu, Q.-P., Xu, J.-G., 2011. Profiles of carotenoids, anthocyanins, phenolics, and antioxidant activity of selected color waxy corn grains during maturation. J. Agric. Food Chem. 59, 2026–2033.

Hudec, J., Burdova, M., Kobida, L., Komora, L., Macho, V., Kogan, G., Turianica, I., Kochanova, R., Lozek, O., Haban, M., Chlebo, P., 2007. Antioxidant capacity changes and phenolic profile of *Echinacea purpurea*, nettle (*Urtica dioica* L.), and dandelion (*Taraxacum officinale*) after application of polyamine and phenolic biosynthesis regulators. J. Agric. Food Chem. 55, 5689–5696.

Kurilich, A.C., Juvik, J.A., 1999. Quantification of carotenoid and tocopherol antioxidants in *Zea mays*. J. Agric. Food Chem. 47, 1948–1955.

Li, S., Tayie, F.A., Young, M.F., Rocheford, T., White, W.S., 2007. Retention of Provitamin A Carotenoids in high-carotene maize (*Zea mays*) during traditional African household processing. J. Agric. Food Chem. 55, 10744–10750.

Lopez-Martínez, L.X., Oliart-Ros, R.M., Valerio-Alfaro, G., Lee, C.H., Parkin, K.L., Garcia, H.S., 2009. Antioxidant activity, phenolic compounds and anthocyanins content of eighteen strains of mexican maize. LWT- Food Sci. Technol. 42, 1187–1192.

Lopez-Martinez, L.X., Parkin, K.L., Garcia, H.S., 2011. Phase II-inducing, polyphenols content and antioxidant capacity of corn (Zea mays L.) from phenotypes of white, blue, red and purple colors processed into masa and tortillas. Plant Foods Hum. Nutr. 66, 41–47.

Lopez-Martínez, L.X., Parkin, K.L., Garcia, H.S., 2012. Effect of processing of corn for production of masa, tortillas and tortilla chips on the scavenging capacity of reactive nitrogen species. Int. J. Food Sci. Tech. 47, 1321–1327.

Mamatha, B.S., Arunkumar, R., Baskaran, V., 2012. Effect of processing on major carotenoid levels in corn (*Zea mays*) and selected vegetables: Bioavailability of lutein and zeaxanthin from processed corn in mice. Food Bioproc. Technol. 5, 1355–1363.

Quackenbush, F.W., 1963. Corn carotenoids: effects of temperature and moisture on losses during storage. Cereal Chem. 40, 266–269.

Quackenbush, F.W., Firch, J.G., Rabourn, W.J., McQuistan, M., Petzold, E.N., Kargl, T.E., 1961. Analysis of carotenoids in corn grain. J. Agric. Food Chern. 9, 132–135.

Raju, M., Varakumar, S., Lakshminarayana, R., Krishnakantha, T.P., Baskaran, V., 2007. Carotenoid composition and vitamin A activity of medicinally important green leafy vegetables. Food Chem. 101, 1598–1605.

Singleton, V.L., Rossi, J.A., 1965. Colorimetry of total phenolics with phosphomolybdic-phosphotungstic acid reagents. Am. J. Enol. Viticul. 16, 144–158.

Talcott, S.T., Percival, S.S., Pittet-Moore, J., Celoria, C., 2003. Phytochemical composition and antioxidant stability of fortified yellow passion fruit (*Passiflora edulis*). J. Agric. Food Chem. 51, 935–941.

Toro-Sanchez, D.L., Sanchez, S., Ortiz, A.M., Villanueva, S., Lugo-Cervantes, E., 2006. Generation of aroma compounds from Ditaxis heterantha by *Saccharomyces cerevisiae*. Appl. Microbiol. Biot. 72, 155–162.

Weber, E.J., 1987. Carotenoids and tocols of corn grain determined by HPLC. J. Am. Oil Chem. Soc. 64, 1129–1134.

Xu, J.-G., Hu, Q.-P., Wang, X.-D., Hu, Q.-P., Wang, X.-D., Luo, J.-Y., Tian, C.H.-R., 2010. Changes in the main nutrients, phytochemicals, and antioxidant activity in yellow corn grain during maturation. J. Agric. Food Chem. 58, 5751–5756.

Zhang, C., Ma, Y., Zhao, X., Wang, F., 2011. Effect of drying treatments on anthocyanin, fumonisin B_1, aflatoxin B_1 content of anthocyanin extract from purple corn (*Zea mays* L.) in North China. Adv. Material Res. 361–362, 759–763.

Zhao, X., Zhang, C., Guigas, C., Ma, Y., Corrales, M., Tauscher, B., Hu, X., 2009. Composition, antimicrobial activity, and antiproliferative capacity of anthocyanin extracts of purple corn (*Zea mays* L.) from China. Eur. Food Res. Technol. 228, 759–765.

Zhao, X., Corrales, M., Zhang, C., Hu, X., Ma, Y., Tauscher, B., 2008. Composition and thermal stability of anthocyanins from Chinese purple corn (*Zea mays* L.). J. Agric. Food Chem. 56, 10761–10766.

CHAPTER

40

Impact of Thermal Processing on Faba Bean (*Vicia faba*) Composition

Isabel Revilla

Area de Tecnología de Alimentos, Universidad de Salamanca, Spain

CHAPTER POINTS

- Faba bean seeds are rich in protein, carbohydrates, fiber and vitamins. However, their nutritional value is affected by the presence of antinutritional factors such as trypsin inhibitors, condensed tannins, phytic acid, saponins, lectins, and favism-inducing factors.
- The soaking process positively affects the nutritional profile of faba beans, reducing mainly tannins (48%) and, to a lesser extent, phytic acid, the raffinose family oligosaccharides, and saponin contents.
- Thermal processing (cooking or autoclaving) after soaking strongly reduces the levels of all the antinutritional factors, mainly hemagglutinin activity and trypsin-inhibiting activity, which is almost completely eliminated.
- Autoclaving is more effective at reducing antinutritional factors than cooking at atmospheric pressure, eliciting a higher retention of B-group vitamins and minerals.
- An increase in trypsin-inhibiting activity is observed after dehulling. Nevertheless, *in vitro* protein digestibility increases due to the reduction in other antinutrients such as tannins and polyphenols, this process being the most effective at reducing these compounds.

INTRODUCTION

Faba beans (*Vicia faba* L.), also referred to as broad beans, horse beans, or field beans, have been a staple food for many years in the Mediterranean area, India, Pakistan, and China, owing to their ability to grow in various climatic zones. Two types of varieties are cultivated: 'V. faba major,' or broad beans with large flattened seeds and a 1000-seed weight greater than 1 kg, mainly grown in the southern regions of Europe and used for human food nutrition; and 'V. faba minor,' field beans or horse beans, with medium to relatively small (1000-seed weight less than 500 g), round seeds, used either for human or animal nutrition. Medium-seeded types have been developed throughout the Middle East and North Africa, mainly in Egypt, while small-seeded types are found in Ethiopia.

Faba bean seeds are rich in proteins, carbohydrates, fiber, and vitamins and have a hypocholesterolemic effect (Macarulla *et al.*, 2001). The most common use is as human food in developing countries because they are relatively inexpensive as compared to meat products. However, the biological value of faba beans is negatively affected by the presence of antinutritional factors such as trypsin inhibitors, condensed tannins, phytic acid, saponins, lectins, and favism-inducing factors. Removal of these antinutrients is therefore necessary for effective utilization of food legumes in human nutrition. Certain factors such as trypsin inhibitors, hemagglutinins, and condensed tannins are readily destroyed by heat processing (boiling, cooking, autoclaving or extrusion cooking) or eliminated by pretreatments such as dehulling, soaking, germination, fermentation, and supplementation with various chemical and enzymes. The objective of this work is to review the effects of heating processes and the usual pretreatments undertaken prior to the canning process of faba beans on their nutritional properties.

HOW COMPOSITION IS ALTERED

The steps followed in the processing of dry faba beans for consumption are usually soaking, hydrothermal processing (cooking, autoclaving), and chilling. These

Processing and Impact on Active Components in Food
http://dx.doi.org/10.1016/B978-0-12-404699-3.00040-8

processes are known to reduce antinutrients and to improve legume sensory quality. However, it is also known that certain treatments, such as heat processing may produce physicochemical changes in protein, starch, and other components of legume seeds, affecting the final nutritional properties. In the following sections, the main faba bean components are compiled and briefly described, after which the effect of processing on nutritional characteristics is reviewed.

Raw Faba Bean Composition

The composition of raw faba beans is shown in Table 40.1. Faba bean proteins are formed of two main fractions, albumins (20%) and globulins (80%). The seed albumin fraction includes protease inhibitors and lectins. Lectins cause the agglutination of erythrocytes and bind complex carbohydrates (Sharma and Sehgal, 1992). Trypsin inhibitors inhibit pancreatic serine proteases, thus impairing digestion. However, the trypsin-inhibiting activity of faba beans is very low. On the other hand, protease inhibitors are also considered bioactive substances since they can act as anticarcinogenic agents (Clemente *et al.*, 2004). The carbohydrate fraction of faba beans is composed of soluble sugar, starch (which is the major faba bean component), and dietary fiber. The main soluble sugars are the oligosaccharides of the raffinose family: raffinose, stachyose and verbascose, which are the principal cause of flatulence (Liener, 1980). However, α-galactosides also have many benefits for humans because they are constituents of dietary fiber and can act as a prebiotic. Faba beans have low amounts of fat. The fatty acid composition is characterized by a relatively high proportion of unsaturated fatty acids (21.6–23.4% of saturated fatty acids, 18–18.4% of monounsaturated fatty acids and 31.9–35.1% of polyunsaturated fatty acids) (Grela and Günter, 1995), especially linoleic acid.

Regarding vitamins, the most significant water-soluble vitamins are thiamine, riboflavin, and niacin and their contents range from 0.253–0.640, 0.123–0.190, and 1.52–2.00 mg/g DM (Prodanov *et al.*, 2004, Khalil and Mansour, 1995). Regarding liposoluble vitamins, faba beans show intermediate values of α-tocopherol (15–17 mg/kg) (Grela and Gunter, 1995). Among minerals, whole faba beans are relatively high in iron, copper, zinc, phosphorus, magnesium, potassium, and sodium and relatively low in manganese and calcium.

Condensed tannins are oligomeric and polymeric flavonoids (flavan-3-ols) that are concentrated in the seed coats. They have been considered antinutritional compounds because of their ability to precipitate proteins and reduce the bioavailability of some minerals such as calcium. Nevertheless, they also reduce the LDL fraction and cholesterol in serum (Zdunczyck *et al.*, 2002). Two types of faba beans can be distinguished according to the content of tannins in the teguments: high-tannin (T^+), which are also cultivars with colored flowers, and low-tannin cultivars (T^-), with white flowers (Duc *et al.*, 1999). Vicine and convicine are glucopyranosides concentrated in the cotyledons of faba bean seeds. It has been demonstrated that divicine and isouramil, active derivatives of vicine and convicine, cause favism, an acute hemolysis occurring only in human carriers of a widespread genetic defect [deficiency of the erythrocyte-located glucose-6-phosphate dehydrogenase (G6PD)] (Arese and De Flora, 1990).

Phytic acid (*myo*-inositol 1,2,3,4,5,6,-hexakis(dihydrogen phosphate) can be found in the cotyledons and is the principal storage form of phosphorus in many dry seeds. It decreases the bioavailability of essential minerals such as calcium, zinc, manganese, and iron (Cosgrove, 1980) and may possibly interfere in the utilisation of proteins. However, there are some beneficial effects of phytic acid, such as lowered toxicity of heavy metals (Rimbach and Pallauf, 1997). Saponins are sterol or triterpene high-molecular-weight glycosides. In large amounts they cause gastrointestinal lesions, enter the blood stream and hemolyse red blood cells.

Soaking

Soaking is an integral part of a number of treatments such as cooking, dehulling, canning, or germination. It consists of the hydration of the seeds in water, usually until they reach maximum weight, with or without discarding the soaking medium. The traditional method for home preparation uses tap water and the soaking period (usually overnight, 12 h) is followed by cooking of the rehydrated seeds after discarding the soak water. The results depend on factors such as the faba bean variety, process duration, temperature, pH, and salinity of the soaking media. Thus, the rate of water absorption increases with increasing temperatures while it decreases with increasing concentrations of sodium bicarbonate in soaking solutions (Abdel Kader, 1995).

Although soaking usually is followed by cooking and the effect of both processes are reported together, it is possible to find literature related to the effect of the soaking process alone on nutritional composition. The results show (Figure 40.1) that nearly no changes in lipid contents are seen after soaking. Also, a slight decrease in the protein content of the cotyledons is seen, and the most important reduction is observed for crude fiber content, attributed to the leaching out of these compounds to water. The soaked beans have more or less the same levels of sodium, calcium, and zinc while an important reduction of potassium, copper and magnesium occurs. These results can be attributed to the leaching out of minerals in the medium, reduction depending on the type of soaking medium as well as the time of soaking.

TABLE 40.1 Chemical Composition of Faba Bean Seeds in Dry Matter Basis

Compound	Duc *et al.* (1999) T+	Duc *et al.* (1999) T−	Makkar *et al.* (1997) Color	Makkar *et al.* (1997) White	Khalil and Mansour (1995)	Youssef *et al.* (1987)
Crude protein (g/kg)	310	319	256–289	259–304	292	316
Starch(g/kg)	412	427	406–485	417–476	441	
Crude fiber (g/kg)	99	88	99–142	87–128	50	95
Reducing sugars (g/kg)	38	44	41–80	28–89	72	
Fat (g/kg)	19	20	14–18	18–25	11	12.7
Ash (g/kg)			34–42	33–41	42	33
Na (mg/100g)			3–9	4–33	297	
Ca (mg/100g)			100–155	133–210	220	72.0
K (mg/100g)			1153–1530	1210–1490	748	
Cu (mg/100g)			1.0–1.8	0.9–2.1	2.5	
Zn (mg/100g)			4.1–9.4	4.1–6.3	11.7	
Fe (mg/100g)			5.2–9.1	5.0–6.9	6.6	15.1
Mg (mg/100g)			133–180	22–166	281	8.0
Mn (mg/100g)			1.2–2.3	1.3–1.8	2.3	
Tannins (g/kg)	6.6	0.1	10.3–21.0	0.04–0.23	14.5	
Vicine + convicine (g/kg)	8.3	7.6			8.5	
Phytic acid (g/kg)			13.4–20.0	11.6–19.4	3.90	4.25
Raffinose + Stachiose (g/kg)					18.1	
Tripsin inhibitor activity (TIU/mg)	2.9	2.9	1.9–2.0	2.9–3.5	2.8	
Saponins (g/kg)			25.5–39.3	17.2–20.1		

Table elaborated from data taken and adapted from: Duc et al. *(1999) means of four low-tannin lines (T⁻) compared to the means of their four high-tannin isogenics (T⁺), Makkar* et al. *(1997) means of six color-flowering compared to the means of six white-flowering varieties and Khalil and Mansour (1995), Youssef* et al. *(1987) as means of three replicates of one variety.*

However, an enhanced HCl-extractability of some minerals is observed and a decrease in phytic acid during soaking may partly account for improved extractability (Saharan *et al.*, 2001).

Regarding vitamins, the thiamine content decreases when soaking is carried out in tap water or sodium bicarbonate solution but does not change when soaking is carried out in citric acid solution. The highest vitamin losses observed in basic solutions can be explained in terms of the higher rates of decomposition of thiamine and, at certain levels, riboflavin, due to their chemical instability and higher solubility with increasing pH (Prodanov *et al.*, 2004).

Tannins, polyphenols and phytic acid are significantly reduced by soaking due to the leaching out effect during hydration. Tannin contents undergo a reduction of almost 50% and the longer the soaking time the higher the reduction (Abd El Hadi and Habiba, 2003). Changes in phytic acid contents may be ascribed to the activation of endogenous phytase during the soaking treatment; this hydrolyses phytic acid to *myo*-inositol and free orthophosphate ion. Trypsin inhibitor activity decreases significantly with soaking, which may be due to the water-soluble nature of the inhibitor that permits its migration from seeds into the soaking medium (Abd El-Hadi and Habiba, 2003). This is also the reason for the observed decrease in saponin levels (Sharma and Sehgal, 1992). However, although the leaching out of lectins is observed during soaking, the amount left in seeds is still sufficient to agglutinate rabbit erythrocytes and no significant reduction in hemagglutinin activity is observed (Sharma and Sehgal, 1992; Alonso *et al.*, 2000). Finally, significant losses of the raffinose family oligosaccharides are observed after soaking, with a gradual decrease

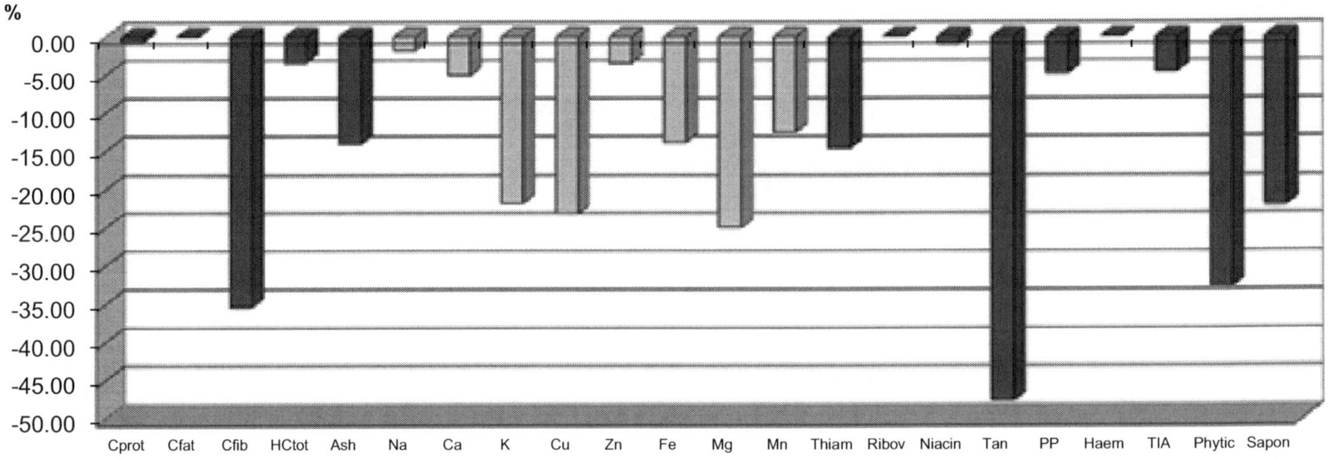

FIGURE 40.1 Effect of soaking (12 h in tap water) on the nutritional composition expressed as % of increase or decrease over raw faba bean values. Values for Cprot (crude protein), Cfat (crude fat), Cfib (crud fiber), CHtot (Total carbohydrates) ash, and minerals are calculated from data from Khalil (2001); Thiam (thiamine), Ribov (riboflavine), and niacin are calculated from data from Prodanov *et al.* (2004), Tan (tanins), PP (total polyphenols), TIA (tripsin inhibitor activity), Phytic acid are calculated from data from Alonso *et al.* (2000) and Sapon (saponins) is calculated from data from Sharma and Sehgal (1992).

in their contents (from 8–27% of losses) as the soaking period progresses. The reduction is higher in sodium bicarbonate solution, possibly because it may tenderize the seed coat as well as the cotyledon and thus enhance the extraction of oligosaccharides (Abdel-Gawad, 1993).

Cooking and Autoclaving

Cooking is probably the oldest treatment for making legumes edible. It usually includes a previous soaking of the seeds and subsequent cooking in boiling water until they become completely soft (between 30 and 90 min). Although cooking methods are known to reduce antinutrient levels and thus improve the nutritive value of legume grains, a long cooking time reduces the nutritive value of legumes as compared with a short treatment. The addition of mineral salts to the cooking media can provide a reduction in cooking time. On the other hand, autoclaving or sterilization processes use higher temperatures (121°C) and shorter times (between 15 and 45 min), enhancing the nutritive value and enabling the shelf-life to be lengthened. Table 40.2 shows the reduction in the initial levels of nutrients and antinutrients in faba beans after different cooking treatments.

The available information indicates that the differences in crude protein between raw and heat-treated faba beans either by cooking or autoclaving are not significant, although the decrease in protein content is slightly higher for autoclaving processes (Khalil and Mansour, 1995). However, faba beans cooked by boiling for a short time have higher contents of total essential amino acids (leucine, tyrosine, threonine, and histidine). Nevertheless, this treatment slightly decreases sulphur-containing amino acids. On the other hand, an improvement in *in vitro* protein digestibility by heat treatments is observed (Khalil and Mansour, 1995). This improvement can be attributed to the denaturation of protein or the destruction of the trypsin inhibitor or a reduction of tannin and phytic acid levels.

The results also show almost no change in lipid contents (1.3 *versus* 1.0 g/100 g; Khalil, 2001), reducing sugars or total carbohydrates after heat treatment. However, hydrothermal processing leads to the retrogradation of starch, changes in the amylase fraction, and the creation of forms with the B-type conformational structure. The forms obtained are characterized by a high resistance to amylolytic enzymes, decreasing starch digestibility (Giczewska and Borowska, 2004). A noticeable reduction of crude fiber due to the prior soaking process is observed; this accounts for a 35% reduction in the initial content (Figure 40.1), but hydrothermal processing may be accompanied by changes, such as the depolymerisation of pectin compounds, which in turn change the insoluble-to-soluble dietary fiber fraction ratio (Giczewska and Borowska, 2004). A slight decrease is also seen in mineral contents due to the heat process, with the exception of sodium, magnesium, and manganese. The most noticeable decreases are observed for potassium and copper, possibly attributable to the leaching of such minerals into the soaking water.

In general, cooking after soaking produces a further decrease in the vitamin content as compared with soaking alone. These losses are higher when previous soaking is carried out in alkaline and neutral solutions and can be explained by the increasing leaching rates, promoted by the high cooking temperature (Prodanov *et al.*, 2004). Moreover, the longer the cooking time, the higher the vitamin losses (Table 40.2) because leaching to the cooking medium of the vitamins studied is more favored. This is the reason why autoclaved faba beans have a

TABLE 40.2 Effect of Soaking and Cooking *versus* Soaking and Autoclaving on the Nutritional Composition Expressed as % Increase (+) or Decrease Over Control Values

	Adapted from Khalil and Mansour (1995)		Adapted from Khalil (2001)[1]; Prodanov *et al.* (2004)[2]; Sharma and Sehgal (1992)[3]; Abdel-Gawad (1993)[4]	
Compound	**Cooking (45 min)**	**Autoclaving (30 min)**	**Cooking**	**Autoclaving**
Crude protein	0.7	5.8	0.9[1]	6.3[1]
Total essential aminoacids	+1.0	0.0	+0.5[1]	1.2[1]
Total non essential aminoacids	0.7	0		
Crude fat	9.1	9.1	23.7[1]	23.7[1]
Total carbohidrates	+4.1	+2.0	+3.4[1]	+1.9[1]
Crude fiber			46.3[1]	40.7[1]
Reducing sugar	16.7	4.2		
Ash	9.5	4.8	22.6[1]	17.0[1]
Na	1.7	0	1.7[1]	0[1]
Ca	5.5	10.0	5.4[1]	10.0[1]
K	37.4	36.6	27.7[1]	15.7[1]
Cu	16.0	16.0	15.4[1]	19.2[1]
Zn	6.0	0	5.2[1]	0.9[1]
Fe	7.6	7.6	12.1[1]	3.4[1]
Mg	3.9	+1.8	0.4[1]	+2.9[1]
Mn	0	0	0.0[1]	4.2[1]
Thiamine	75.0	67.2	24.1[2]	
Riboflavine	42.1	31.6	0.8[2]	
Niacin	94.0	93.0	32.2[2]	
Tannins	55.2	60.0		
Hemagglutinin activity	100.0	100.0	Traces[3]	100.0[3]
Trypsin inhibitor	71.5	84.4		
Phytic acid	30.8	41.0	39.8[1]	48.6[1]
Vicine + convicine	34.7	40.0		
Raffinose + stachyose	47.0	21.0	49.7[4]	56.8[4]
Saponins			35.0[3]	60.0–63.0[3]

[1] *Denotes a cooking treatment of 90 min* versus *an autoclaving time of 45 min*
[2] *denotes a cooking treatment of 30 min*
[3] *denotes a cooking treatment without specifying the time* versus *an autoclaving time of 25 min and*
[4] *denotes a cooking treatment of 60 min* versus *an autoclaving time of 20 min.*
Table calculated from data taken from Khalil and Mansour (1995), Khalil (2001); Prodanov et al. *(2004); and Abdel-Gawad (1993) Sharma and Sehgal (1992).*

higher content and retention of B-group vitamins than cooked faba beans (Khalil and Mansour, 1995). Regarding antinutrients, the tannins and phytic acid of heated beans are significantly reduced and autoclaving has a greater effect on the reduction of both compounds than cooking. The thermal degradation of these molecules, as well as changes in their chemical reactivity or the formation of insoluble complexes, could explain the significant reduction due to thermal processing (Kataria *et al.*, 1989). Despite this, it has been reported that if the autoclaving time is very short (15 min) the levels of phytic acid are higher than after cooking.

Hemagglutinin activity is completely eliminated by autoclaving and a long enough cooking time (Khalil and Mansour, 1995). However, it is possible to detect lectins after soaking and cooking if the heating time is short (Sharma and Sehgal, 1992) because lectin activity is relatively heat-resistant (Armour *et al*, 1998). Trypsin inhibitor activity is significantly reduced by heat treatments autoclaving being more effective than cooking. High-temperature, low-time treatments have been shown to be more effective in destroying protease-inhibiting activity than low-temperature, high-time treatments (Armour *et al.*, 1998).

The total amount of vicine and convicine undergoes an important decrease after heat treatment due to both high temperature and to the leaching out of these compounds to the cooking water (Cardador-Martinez *et al.*, 2012). Regarding the raffinose family oligosaccharides, according to Abdel-Gawad (1992) autoclaving has a greater effect on the elimination of oligosaccharides from legumes than ordinary cooking, although Khalil and Mansour (1995) reported the opposite behavior. Soaking in alkaline medium followed by heating treatment leads to a more marked removal of oligosaccharides than tap water. Finally, the cooking and autoclaving of soaked seeds reduces the saponin content: the longer the autoclaving time, the higher the saponin losses (Sharma and Sehgal, 1992). Losses during cooking may perhaps indicate the thermolabile nature of saponins. This reduction may be due to the formation of a poorly extractable complex between saponins and sugars or amino acids (Khokhar and Chauhan, 1986).

Other Ways in which Composition is Altered

The dehulling process involves physical removal of the seed coat and it can be performed manually or mechanically both with and without prior soaking. After dehulling, two fractions—seed coats and cotyledons—are obtained. The effects of the dehulling process are shown in Figure 40.2, where a reduction in the protein content can be observed while the fat and ash contents remain fairly similar. However, this process leads to a significant decrease in the crude fiber content of the cotyledons and in the contents of iron, manganese, calcium, magnesium, and sodium. In contrast, copper, zinc, and potassium increase. However, as previously mentioned for the soaking process, dehulling improves mineral HCl-extractability between 5 and 12% because of the reduction in phytic acid levels, which range from 4 –13% for soaked and dehulled seeds (Sharma and Sehgal, 1992; Saharan *et al.*, 2001). Despite this, if the dehulling process is carried out without prior soaking an increase in phytic acid levels can be observed (Alonso *et al.*, 2000). Since phytates are mainly located in the cotyledons, the removal of testa by dehulling is responsible for this phenomenon. In the same way, since trypsin inhibitors are linked to proteins, an increase in trypsin inhibiting activity is observed after dehulling. However, *in vitro* protein digestibility increases due to the reduction of other antinutrients such as tannins and polyphenols, which interact with protein to form complexes (Alonso *et al.*, 2000). Major amounts of tannins are located in the seed coat or testa and hence physical removal of the testa strongly decreases tannin and polyphenol levels in cotyledons.

ANALYTICAL TECHNIQUES

Crude protein, crude fat, crude fiber, total carbohydrate, reducing sugars, starch and ash are determined according to the methods of AOAC (1980). Minerals are usually determined in the digested acid solution using atomic absorption and the quantification of amino acid composition is usually carried out using an amino acid analyzer (Makkar *et al.* 1997, Khalil, 2001). Fatty acid composition is determined by GC while tocopherol content is determined by HPLC using a fluorescence spectrometer detector as reported by Grela and Gunter (1995). Thiamin, riboflavin, and niacin are determined by HPLC as

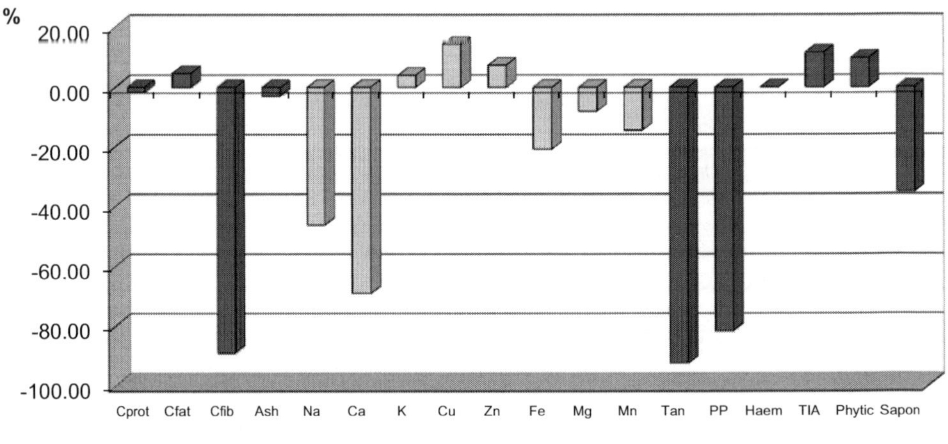

FIGURE 40.2 Effect of dehulling on the nutritional composition expressed as % of increase or decrease over raw faba bean values. Values for Cprot (crude protein), Cfat (crude fat), Cfib (crud fiber), ash, and minerals are calculated from data from Youssef *et al.* (1987); Tan (tannins), PP (total polyphenols), TIA (tripsin inhibitor activity), Phytic acid are calculated from data from Alonso *et al.* (2000) and Sapon (saponins) is calculated from data from Sharma and Sehgal (1992).

described by Prodanov *et al.* (1997) or by microbiological methods (Khalil and Mansour, 1995).

The analysis of antinutrients, which are the most interesting faba bean compounds, marshals the use of a broad number of techniques but here only those most widely used are reviewed briefly. The methods proposed by Wheeler and Ferrel (1971) or Haugh and Lantzsch (1983) are the most widely used to analyse phytic acid. Total polyphenols are analyzed by spectrophotometric methods using the Folin–Ciocalteau or Folin–Denis reagent, while tannins can be determined with the vanillin assay, as described by Alonso *et al.*, (2000). The estimation of lectin contents is carried out by hemagglutination of rabbit erythrocytes, as advocated by Liener (1955). The raffinose family oligosaccharides are analysed with TLC (Khalil and Mansour, 1995) or GC (Abdel-Gawad, 1993). Regarding trypsin inhibitor activity, the method of Kakade *et al.*, (1974) is the most widely reported. Vicine and convicine can be analyzed by HPLC, as described by Cardador-Martinez *et al.*, (2012). Finally, saponins are quantified by means of colorimetric and hemolytic assays, as reported by Makkar *et al.*, (1997).

References

Abd El-Hady, E.A., Habiba, R.A., 2003. Effect of soaking and extrusion conditions on antinutrients and protein digestibility of legume seeds. Lebens-Will. U.-Technol. 36, 285–293.

Abdel Kader, Z.M., 1995. Study of some factors affecting water absorption by faba beans during soaking. Food Chem. 53, 235–238.

Abdel-Gawad, A.S., 1993. Effect of domestic processing on oligosaccharide content of some dry legume seeds. Food Chem. 46, 25–31.

Alonso, R., Aguirre, A., Marzo, F., 2000. Effects of extrusion and traditional processin methods on antinutrientes and in vitro digestibility of protein and starch in faba and kidney beans. Food Chem. 68, 159–165.

Arese, P., De Flora, A., 1990. Pathophysiology of hemolysis in glucose-6-phosphate dehydrogenase deficiency. Semin. Hematol. 27, 1–40.

Armour, J.C., Perera, R.L.C., Buchan, W.C., Grant, G., 1998. Protease inhibitors and lectins in soya beans and effects of aqueous heat-treatment. J. Sci. Food Agric. 78, 225–231.

Cardador-Martinez, A., Maya-Ocaña, K., Ortiz-Moreno, A., Herrera-Cabrera, B., Dávila-Ortiz, G., Múzquiz, M., Martín-Pedrosa, M., Burbano, C., Cuadrado, C., Jiménez-Martínez, C., 2012. Effect of roasting and boiling on the content of vicine, convicine and L-3,4-dihydroxyphenylalanine in *Vicia faba*, L. J. Food Qual. 35, 419–428.

Clemente, A., MacKenzie, D.A., Jonson, I.T., Domoney, C., 2004. Investigation of legume seed protease inhibitors as potential anticarcinogenic proteins. In: Muzquiz, M., Hill, G.D., Cuadrado, C., Pedrosa, M.M., Burbano, C. (Eds.), Proceedings of the fourth international workshop on antinutritional factors in legume seeds and oilseeds. EAA Publications, Wageningen, pp. 137–141.

Cosgrove, D.J., 1980. Inositol phosphates. Their chemistry Biochemistry and Physiology. Elsevier Scientific Publishing Company, Amsterdam.

Duc, G., Marget, P., Esnaulti, R., LeGuen, J., Bastianielli, E., 1999. Genetic variability for feeding value of faba bean seeds (*Vicia faba* L.): comparative chemical composition of isogenic involving zero-tanin and zero-vicine genes. J. Agric. Sci. Camb. 133, 185–196.

Giczewska, A., Borowska, J., 2004. Nutritional value of brad bean seeds. Part 3: Changes of dietary fiber and starch in the production of commercial flours. Nahrung/Food 2, 116–122.

Grela, E.R., Günter, K.D., 1995. Fatty acid composition and tocopherol content of some legume seeds. Animal Feed Sci. Tech. 52, 325–331.

Haugh, G., Lantzsch, W., 1983. Methods for determinatin of phytate of cereal products. J. Sci. Food Agric. 34, 1423–1424.

Kakade, M.L., Rachis, J.J., McGhee, J.E., Puski, G., 1974. Determination of trypsin inhibitor activity of soy products: A collaborative analysis of an improved procedure. Cereal Chem. 51, 376–382.

Kataria, A., Chauhan, B.M., Punia, D., 1989. Antinutrients and protein digestibility (*in vitro*) of mung bean as afected by domestic processing and cooking. Food Chem. 32 (9), 17.

Khalil, A.H., Mansour, E.H., 1995. The effect of cooking, autoclaving and germination on the nutritional quality of faba beans. Food Chem. 54, 177–182.

Khalil, M.M., 2001. Effect of soaking, germination, autoclaving and cooking on chemical and biological value of guar compared with faba bean. Nahrung/Food. 45, 246–280.

Khokhar, S., Chauhan, B.M., 1986. Saponins in food- Their physiological and nutritional significance. Sci. Culture 42 (a), 287–291.

Liener, I.E., 1980. Flatus-producing factors. In: Liener, I.E. (Ed.), Toxic constituents of plant foodstuffs. Academic Press, London, pp. 455–457.

Liener, I.E., 1955. Photometric determination of the haemagglutinating activity of soybean and crude soybean extracts. Arch. Biochem. Biophys. 54, 223–231.

Macarulla, M.T., Medina, C., De Diego, M.A., Chávarri, M., Zulet, M.A., Martínez, J.A., et al., 2001. Effects of the whole seed and a protein isolate of faba bean (*Vicia faba*) on the cholesterol metabolism of hypercholesterolaemic rats. Brit. J. Nutrition 85, 607–614.

Makkar, H.P.S., Becker, K., Abel, Hj, Pawlzik, E., 1997. Nutrient contents, rumen protein degradability and antinutritional factors in some colour–and white–flowering cultivars of *Vicia faba* beans. J. Sci. Food Agric. 75, 511–520.

Prodanov, M., Sierra, I., Vidal-Valverde, C., 2004. Influence of soaking and cooking on the thiamin, riboflavin and niacin contents of legumes. Food Chem. 84, 271–277.

Prodanov, M., Sierra, I., Vidal-Valverde, C., 1997. Effect of germination on the thiamine, riboflavin and niacin contents in legumes. Zeits. Lebens. Unters. Forsch. A 205, 48–52.

Rimbach, G., Pallauf, J., 1997. Cadmium accumulation, zinc status, and mineral bioavailability of growing rats fed diets high in zinc with increasing amounts of phytic acid. Biol. Trace Elem. Res. 57, 59–70.

Saharan, K., Khetarpaul, N., Bishnoi, S., 2001. HCl-extractability of minerals from ricebean and fababean: influence of domestic processing methods. Innovative Food Sci. Emerg. Techn. 2, 323–325.

Sharma, A., Sehgal, S., 1992. Effect of processing and cooking on the antinutritional factos of faba bean (*Vicia faba*). Food Chem. 43, 383–385.

Wheeler, E.I., Ferrel, R.E., 1971. A method for phytic acid determination in wheat and wheat fractions. Cereal Chem. 48, 312–316.

Youssef, M.M., Abd El-Aal, M.H., Shekib, L.A.E., Ziena, H.M., 1987. Effects of dehulling, soaking, on chemical composition, mineral elements and protein patterns of faba beans (*Vicia faba* L.). Food Chem. 23, 129–138.

Zdunczyk, Z., Frejnagel, S., Wróblewska, M., Juśkiewicz, J., Oszmiański, J., Estrella, I., 2002. Biological activity of polyphenol extracts from different plant sources. Food Res. Int. 35, 183–186.

CHAPTER

41

Processing of Millet Grains and Effects on Non-Nutrient Antioxidant Compounds

Fereidoon Shahidi[*], *Anoma Chandrasekara*[†]

[*]Department of Biochemistry, Memorial University of Newfoundland, St. John's, Canada, [†]Department of Applied Nutrition, Wayamba University of Sri Lanka, Gonawila, Sri Lanka

CHAPTER POINTS

- Millets are rich in bioactive phytochemicals such as phenolics, polyphenolics, and phytic acids which are potent antioxidants in biological systems.
- The content of non-nutrient antioxidative compounds of millet grains as well as their activities were affected by different processing conditions experienced during food preparations.
- Mechanical operations, hydrothermal processing, soaking, fermentation, and germination/malting affect the content of phenolic compounds and phytates.
- Limited data are available on the antioxidant activities of phytochemicals present in millet grains after submitting to different processing operations.
- It is necessary to investigate the effect of different processing conditions on the bioactive content of millet grains in order to optimize their antioxidant properties.

INTRODUCTION

Millets are the oldest domesticated cereals known to humans from the beginning of civilization. Some archeobotanical evidences have shown that common millet was domesticated as a staple food 10,000 years ago in Northern China. Millets include a group of small seeded cereal grains belonging to the family *Poaceae*. They serve as a major source of calories and proteins for millions of economically disadvantaged groups of people in Africa and Asia.

Millets have additional benefits as subsistence crops due to their unique ability to grow under harsh environmental conditions where other crops yield poorly. Millets are subsistence crops tolerable under drought and low fertile soil conditions. In addition, there are early maturing millet cultivars of 60–90 days and resistance to disease and pest attacks in the field. Furthermore, millet grains can be stored with minimum pest attacks thus constituting a candidate crop for improving food security of vulnerable populations, especially during famines when grain is in short supply.

Oxygen, the indispensable element for life, seems to be a paradox due to both its beneficial and toxic effects. A small proportion (1–3%) of oxygen we breathe is freed from the energy production system of oxidative phosphorylation in mitochondria and forms what we collectively refer to as reactive oxygen species (ROS). ROS include free radicals and can be defined as species capable of independent existence that contain one or more unpaired electrons (Halliwell, 2007). Free radicals are generally unstable, but very reactive, thus are capable of initiating a series of damaging chemical changes in biological and food systems. In living systems, non-radical molecules generate radicals when they are brought into contact with radicals and transition metal ions such as those of iron and copper which can move electrons (Halliwell, 2007). At low or moderate levels, ROS exert beneficial effects in cellular signaling systems and the body's immune function (Halliwell, 2007). At high concentrations, however, ROS generate oxidative stress in biological systems and play a key role in the development of degenerative disorders such as cancer, arthritis, autoimmune disorders, type 2 diabetes, cardiovascular

Processing and Impact on Active Components in Food
http://dx.doi.org/10.1016/B978-0-12-404699-3.00041-X

and neurodegenerative diseases, and aging. Hence, maintenance of the delicate balance of ROS *in vivo* is most important for optimum health.

The antioxidants function in biological systems by scavenging free radicals and singlet oxygen, or by stabilizing transition metal ions, hence preventing, delaying or inhibiting deteriorative reactions induced by free radicals (Shahidi and Nazck, 2004). Antioxidants may act as chain breaking or oxidation preventing compounds. The body has a number of mechanisms to counteract the oxidative stress caused by internal as well as external factors and when they are overwhelmed under pathophysiological conditions, exogenous antioxidants supplied through foods or supplements are necessary. In foods, processing and storage may lead to the destruction of endogenous antioxidative compounds, thus they may be added to products in order to maintain food quality.

Antioxidants include nutritive compounds such as minerals, vitamins, proteins, carbohydrates, and non-nutritive phytochemicals such as phenolics that are capable of donating hydrogen atoms or electrons to a pro-oxidant. This chapter focuses on the effects of different processing and food preparation practices on the content as well as the activities of non-nutritive antioxidants of millet grains.

TYPES OF MILLET

Millet is a generic name used to denote a set of highly variable small seeded cereals. They do not belong to a single species or a single genus. Foremost millet genera are *Pennisetum, Elusine, Setaria, Panicum, Paspalum, Eragrostis, Echinochola,* and *Digitaria.* Based on global production, the major millet type is pearl millet (*Pennisetum glaucum*) which accounts for approximately 46% followed by foxtail (*Setaria italica)*, proso (*Panicum milliaceum),* and finger (*Elusine coracana)* millets (FAOSTAT, 2011). Other minor millets include kodo (*Paspalum scrobiculatum)*, little (*Panicum sumatrense)*, Japanese barnyard (*Echinochola crus-galli)*, fonio (*Digitaria exilis)*, and teff (*Eragrostis tef*) millets.

In general, the importance of individual millet type varies with respect to agro-ecological systems, where they grow, and the food culture of the people. Evidences in Northern China demonstrated that some 4000 years ago noodles were made from proso and foxtail millets. At present, millets occupy the sixth place as the most important cereal in the world. However, to the western world, millets are grains for feed and forage for livestock. In the USA, pearl millet grows as summer grazing and hay crops while proso millet is cultivated for grains in the Great Plains mainly in Dakotas, Colorado, and Nebraska. The significant millet crop in Canada is proso millet cultivated in Canadian prairies (Alberta, Saskatchewan, and Manitoba) and in Southwestern Ontario. Their popularity as alternating crops in tobacco farms is emerging in Canada. Furthermore, use of millets in multigrain products and in niche markets for gluten-free products and organic cereals is growing in the markets of North America and Europe.

Millets have similar nutritional composition to other economically important cereals, namely maize, rice, wheat, barley, sorghum, oat, and rye (FAO, 1995). The protein content of different millet grains ranges from 7–11% which is comparable to that of other cereals such as maize, rice, and wheat (Klopfenstein and Hoseney, 1995). Pearl millets contain the highest fat content of 5%, making them more vulnerable for off-flavour and off-odour development due to oxidative deterioration upon grinding into flour when compared to other cereals. Millets are rich in minerals such as iron and phosphorus as well as B complex vitamins, namely niacin, thiamin, and riboflavin. Furthermore, finger millet has a high calcium content of 350 mg/100 g which is the highest amount reported among cereals. The crude fibre content of millets is 2– 7% which is generally higher than that in other cereals (Klopfenstein and Hoseney, 1995).

In addition, millets are gluten-free grains which can be tolerated by coeliac patients. Coeliac disease is a condition characterized by damage to the mucosa of the small intestine caused by ingestion of certain proteins in wheat, rye and barley. The gliadins and glutenins of wheat gluten have been shown to cause the problem which can be prevented only by consuming gluten-free cereals. Thus, millets are potential candidate cereals for coeliacs and add variety in their diet.

PRODUCTION AND DISTRIBUTION OF MILLETS

According to the 2009 statistics, global millet production is about 27 million tonnes; approximately 90% of which is utilized in developing countries for food (FAOSTAT, 2011). The major millet producing continents in the world are Africa and Asia as they produce 56 and 41% of the total world production, respectively (Figure 41.1). World millet production has been stable during recent decades due to the subsistence nature of the crop in mainly cultivated areas. It is noted that about two thirds of millets produced are consumed as food while the rest is used for planting seeds, animal feed, beer production, and as bird seeds (FAO, 1995).

PROCESSING OF MILLETS

In Africa, East-Asia, and the Indian subcontinent, where millets are traditionally grown, they are prepared for consumption in different ways using the grains as flour/meal and malt. Therefore, millet grains are usually

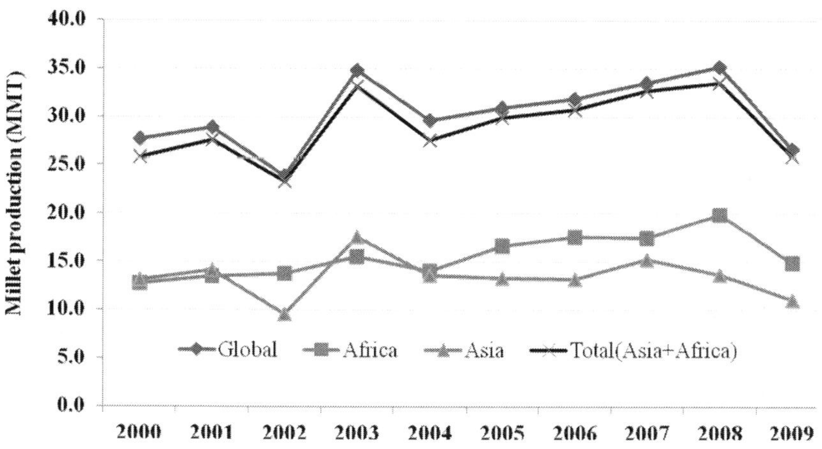

FIGURE 41.1 Global millet production in million metric tonnes (MMT). *Data adopted from FAOSTAT, 2011.*

subjected to a wide range of processing conditions such as soaking, decortication, flaking, grinding, malting, fermentation, and heat treatment using traditional or advanced technologies. Thus, millet grains undergo substantial changes in composition during initial post-harvest practices such as threshing (separation of grains from stalks), and milling as well as different food manufacturing conditions. In addition, foods prepared from millets vary across continents, countries, and regions within the same country by the nature of preparation as well as other ingredients added.

Traditionally, millet is processed by malting or by fermentation and the resultant flour or extracts derived are used in the preparation of weaning foods, infant foods, supplementary food formulations, and beverages such as beer. In general, the different foods prepared from millets include porridges, steam-cooked products, fermented and unfermented breads, boiled rice-like products, alcoholic and non-alcoholic beverages, and snacks (Murty and Kumar, 1995). In African countries and the Indian subcontinent, millets play an integral role as a major component of many meals. Millets are consumed as 'Couscous' (steam-cooked product), 'To' (stiff porridges) and 'Ogi' (thin porridges) which can also be used as a complementary food for infants and young children. In addition, regional specific snack foods such as 'Halepe' in Sri Lanka are prepared with finger millet flour.

COMPOUNDS IN MILLET GRAINS WITH ANTIOXIDATIVE PROPERTIES

Phenolic Compounds

Cereal phenolics are known to possess *in vitro* antioxidative activities (Shahidi and Naczk, 2004; Chandrasekara and Shahidi, 2010, 2011a). The regular consumption of whole grains is emphasized to contribute to the risk reduction of non-communicable disease load and overall well-being in populations. Millet grains contain phenolic acids which are located in the pericarp, testa, aleurone layer, and endosperm. Free and conjugated forms of phenolic acids, which include derivatives of hydroxybenzoic and hydroxycinnamic acids, have been identified in millets. Hydroxybenzoic acids reported in different millet grains are gallic, protocatechuic, gentisic, vanillic, syringic, and *p*-hydroxybenzoic acids (Chandrasekara and Shahidi, 2011b). Hydroxycinnamic acids and their derivatives that are found in millet grains include ferulic, *p*-coumaric, sinapic, caffeic, and cinnamic acids. In addition, dimers and trimers of ferulic acid that have demonstrated higher antioxidant activities than those of monomers were also identified in several millet grains (Chandrasekara and Shahidi, 2011b).

Flavonoids found in millets belong to the anthocynidins, flavanols, flavones, flavanones, chalcones, and aminophenolic group of compounds. Phenolic acids and flavonoids are found in different parts of the grain and their content and composition vary depending on the type of millet grain (Chandrasekara and Shahidi, 2011b). It is noted that flavonoids are abundantly present in the free and esterified forms in finger millet and kodo millet grains, among others (Chandrasekara and Shahidi, 2011b). However, composition of flavonoids showed remarkable differences between the two millet types. Catechin, gallocatechin, epicatechin, and epigallocatechin were reported as major flavonoids in finger millets whereas apigenin, vitexin, isovitexin , and luteolin were reported in kodo millet grains (Chandrasekara and Shahidi, 2011b).

Condensed tannins, also known as proanthocyanidins are also present in millets. These oligomeric or polymeric flavonoids consist of flavan 3-ol units. Proanthocyanidins are biologically active and are potent antioxidants. Chandrasekara and Shahidi (2010) showed that proanthocyanidin content of millet grains as determined by the vanillin assay ranged from 311– 3 μmol of catechin equivalents (CE)/g of defatted meal. Finger millet had the highest content of proanthocyanidins among the examined millet types, namely proso, foxtail, pearl, finger and little millets. Procyanidin B1 and B2 were reported as major dimers in finger millets.

Phytic Acid

Phytic acid, also known as myo-inositol hexaphosphoric acid, is an ubiquitous phytochemical found in foods including cereals as their primary storage form of both phosphate and inositol in seeds. In general, phytates are found in the outer layers of the grain. However, depending on the type of millet, distribution within the grain may vary. Phytate content of pearl millet grain was 752 mg/100 g in germ against 86 and 278 mg/100 g for endosperm and bran, respectively (Simwemba *et al.*, 1984).

Phytic acid is generally considered an antinutrient due to its complex formation with essential dietary minerals such as calcium, magnesium, iron, and zinc, to decrease their bioavailability in humans and animals. However, due to its chelating ability of free iron, phytic acid is a potent inhibitor of iron-catalyzed hydroxyl radical formation (Shahidi, 1997). Phytic acid is known to reduce the incidence of colonic cancer by suppressing the oxidative damage caused to the epithelium in the colon (Graf and Eaton, 1990). Thus, phytic acid is a potent antioxidant in certain food sources, including cereals.

EFFECT OF PROCESSING ON PHENOLICS AND THEIR ANTIOXIDANT ACTIVITIES OF MILLET GRAINS

Generally, antioxidative non-nutrient compounds and nutrients are not distributed evenly in the grain. Thus, processing of millet grains may exert a noted effect on their bioactivities. The outermost layers of the grains possess a high phenolic content and antioxidant activity (Chandrasekara and Shahidi, 2011c; Chandrasekara *et al.*, 2012). In addition, a variety of processing methods applied during food preparation may exert varying effects on the antioxidant activities of millet grains.

Cooking is a common hydrothermal treatment employed during cereal grain preparation as foods. N'Dri *et al.* (2012) recently demonstrated that cooking of millets may increase the total content of phenolic compounds, whereas some specific phenolic acids may be lost. In pearl millet, the cooking process significantly increased the soluble phenolic acid content as determined by high performance liquid chromatography (HPLC). The soluble phenolic acid contents of raw and cooked pearl millets were 64.8 and 115.2 mg/kg of flour, respectively. However, TPC as determined by Folin–Ciocalteu colorimetric method showed a significant decrease in cooked millets compared to that of their raw counterparts.

TABLE 41.1 Antioxidant Activity of Phenolics from Whole (WG), Dehulled (DG), and Cooked (CG) Millet Grains

	Finger Millet			Pearl Millet		
Antioxidant Activity	**WG**	**DG**	**CG**	**WG**	**DG**	**CG**
DPPH* radical scavenging activity[a]	35.7	31.6	30.9	13.8	13.8	11.2
Hydroxyl radical scavenging activity[a]	26.8	9.69	3.0	56.0	2.1	1.5
Hydrogen peroxide scavenging activity[a]	38.9	37.2	31.1	52.9	57.2	46.2
Oxygen radical absorbance capacity[b]	101	129	116	60.3	95.9	65.5
Superoxide radical scavenging activity[c]	1.0	0.7	0.6	1.41	1.42	1.09

**2,2-diphenyl-1-picryllhydrazyl;*
[a]determined as μmol ferulic acid equivalents/g defatted meal;
[b]determined as μmol trolox equivalents/g defatted meal;
[c]determined as mmol ferulic acid equivalents/g defatted meal.
Data adopted from Chandrasekara et al., *2012.*

Furthermore, fonio millets showed a slight, but significant decrease in almost all soluble phenolic acids along with an increase in bound phenolic compounds (N'Dri *et al.*, 2012). In addition, Chandrasekara *et al.* (2012) demonstrated that TPC of cooked millet grains did not differ significantly from their uncooked counterparts except for finger millets. Cooked finger millets showed 11–36% reduction of TPC and the authors suggested that the reduced TPC in finger millet could be due to the degradation of phenolics upon cooking or leaching of phenolics into the endosperm, thus making them less extractable (Chandrasekara *et al.*, 2012). Furthermore, antioxidant activities presented as scavenging of hydroxyl radical, superoxide radical and hydrogen peroxide, and oxygen radical absorbance capacity of dehulled, cooked grains varied depending on the type of millet (Table 41.1).

Pradeep and Guha (2011) showed that germination, steaming and roasting of little millet increased the TPC compared to the unprocessed grains. The TPC of native, germinated, steamed, and roasted were 430, 453, 486, and 521 mg of gallic acid equivalents (GAE)/100 g of sample (dry basis), respectively. Furthermore, there was an increase in the content of total flavonoids as well as tannins. The increase in TPC during roasting of little millet grains may be due to the increase in the release of insoluble bound phenolics by the thermal degradation of cellular constituents.

Dehulling, a process for removing the outer layers of the grain, also reduced the total phenolic content by 22% from its original value (Hag *et al.*, 2002). A recent study showed that dehulling and removal of outer layers of the grain decreased the TPC of whole grain millets (Chandrasekara *et al.*, 2012). Hulls had higher TPC than

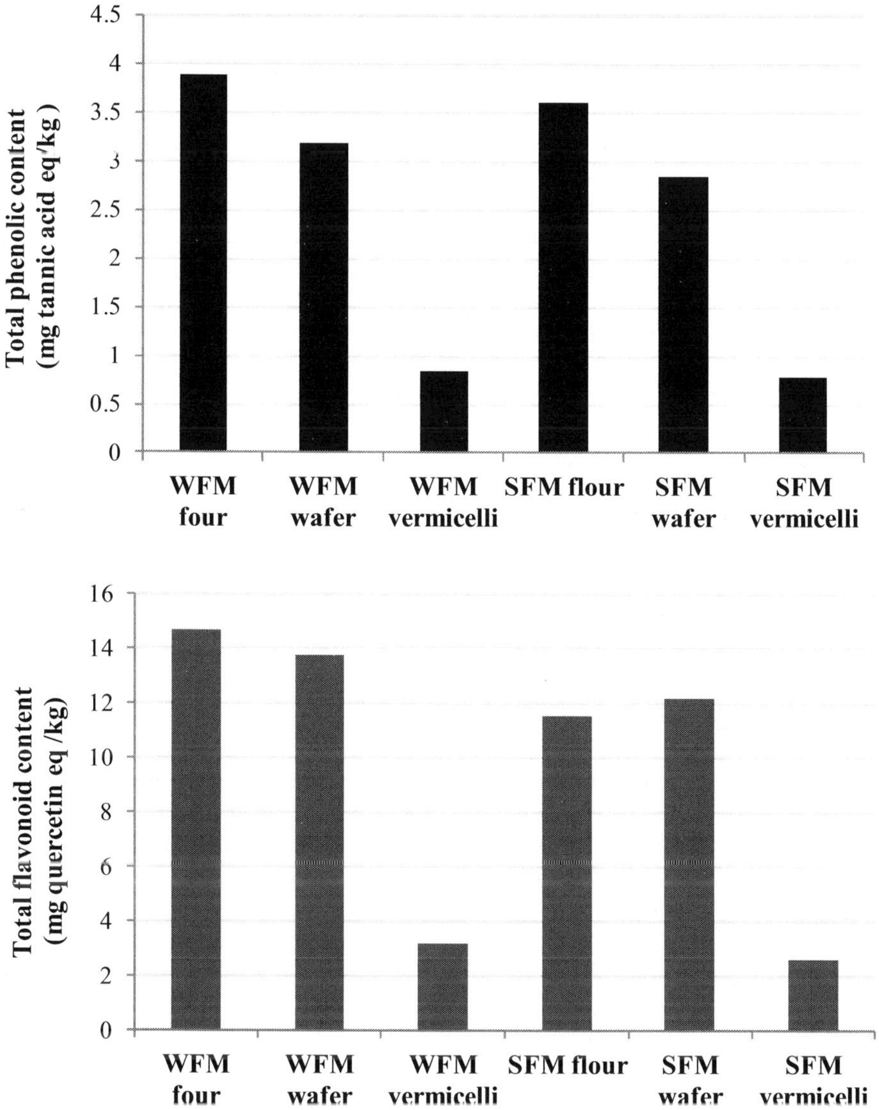

FIGURE 41.2 Effect of sieving and food preparation on total phenolic and flavonoid contents of whole finger millet (WFM), sieved finger millet (SFM) and their products. *Data adopted from Oghbaei and Prakash, 2012.*

those of dehulled grains in kodo, foxtail, proso, pearl, little, and finger millets. Furthermore, TPC of dehulled and cooked grains of millet varieties did not differ significantly except for finger millets (Chandrasekara *et al.*, 2012). Phenolic acids were mainly concentrated in the hulls of millet grains (Chandrasekara and Shahidi, 2011c). Kodo and pearl millet hulls had 34- and 4-fold more ferulic acid, respectively, compared to their dehulled counterparts (Figure 41.2). In addition, *p*-coumaric acid content in hulls of kodo and pearl millet grains was seven times higher than their corresponding dehulled grains (Chandrasekara and Shahidi, 2011c). Furthermore, dehulling affects the antioxidant activity exerted by millet grains. Dehulled grains showed less inhibition of low density lipoprotein (LDL) cholesterol, DNA strand scission induced by peroxyl radicals and liposome oxidation (Table 41.2).

Malting changed the content of free and bound phenolic acids in finger millet. The content of bound phenolic acid continuously decreased whereas free phenolic acid content increased upon malting of finger millet grains for 96 h (Rao and Muralikrishna, 2002). Ferulic, caffeic and coumaric acids were found as major bound phenolic acids in native finger millet and upon malting for 96 h, a one- to two-fold decrease in their contents was reported (Rao and Muralikrishna, 2002). Furthermore, malting of finger millets for 96 h decreased the protocatechuic acid content from 45 to 16 mg/100 g. However, the content of other free phenolic acids, namely *p*-coumaric, gallic and ferulic acids was increased upon 96 h of malting by 2-, 4-, and 10-fold, respectively (Rao and Muralikrishna, 2002).

Hag *et al.* (2002) demonstrated that fermentation and dehulling of pearl millet reduced the TPC as determined by Folin-Denis method. The TPC of standard cultivar of

TABLE 41.2 Antioxidant Activity of Millet Phenolics as Affected by Dehulling

Antioxidant Activity	Kodo Millet			Pearl Millet		
	WG[a]	DG[b]	Hull	WG	DG	Hull
Inhibition of LDL cholesterol oxidation (%)	36	28	58	5	1.5	9
Inhibition of DNA strand scission (%)	95	87	85	43	31	75
Inhibition of liposome oxidation (%)	74	63	65	45	44	69

[a]*WG, whole grain*
[b]*DG, dehulled grain.*
Data adopted from Chandrasekara and Shahidi, 2011c.

pearl millet decreased from 304 to 122 mg/100 g upon fermentation for 14 h. The combination of different processing methods used in the preparation for consumption of millet grains may change their phenolic contents. Gupta and Nagar (2010) showed that cooking, fermentation, dehulling and type of utensils used in the preparation of pearl millet flour *rabadi* had an effect on flavonoid (quercitin and pelargonidin) content. Quercitin content increased when *rabadi* was prepared using two stages of fermentation (4 h before cooking and 16 h after cooking) and fermented in clay pots.

Progressive decortication from 0–50% and cooking reduced the concentration of *C*-glycosylflavones (vitexin and orientin) of pearl millets (Akingbala, 1991). In addition, conversion of pearl millet (*Pennisetum americanum*) into Ogi, a thin porridge which is a traditional Nigerian fermented food, reduced the flavanol content of the native grain by 38% (Akingbala *et al.*, 2002); its flavanol content was the least and was 643 mg glycosylvitexin equivalents/kg dry matter. The bran fraction contains a higher amount (37%) of *C*-glycosylflavanol (2985 mg glycosylvitexin equivalents/kg dry matter) than the endosperm. Thus, the removal of pericarp and the germ as the bran which contained the highest concentration of flavanol was responsible for this observed reduction (Akingbala *et al.*, 2002).

According to Matuschek *et al.* (2001), cooking, soaking and germination reduced the total phenolic content (TPC) of sorghum udo and finger millet. Fermentation and germination of finger millets decreased their 2,2-diphenyl-1-pcryllhydrazyl (DPPH) radical scavenging ability compared to their raw counterparts (Siripriya *et al.*, 1996). According to Shobana and Malleshi (2004) hydrothermal treatment and decortication of finger millets reduced polyphenolic content by 14 and 74%, respectively. In addition, Towo *et al.* (2003) showed that hydrothermal treatment of finger millet reduced the TPC by 1.7 times compared to that of the raw grain. Processing may increase the bioavailability of bioactive compounds in the grains; but may decrease their levels. Several studies have shown that the outermost layers of grains possess a high phenolic content and antioxidant activity. In a recent study, Oghbaei and Prakash (2012) demonstrated that sieving decreased the content of polyphenols and flavonoids in whole finger millet flour. However, sieved finger millet flour showed increased *in vitro* bioaccessibility. Sieving of whole finger millet flour separates the outer coating of the grain which is resistant to fine grinding. However, the outer coating is the fraction of grain rich in bioactive phytochemicals including phenolic compounds and tannin. Wafers and vermicelli are basically prepared by mixing the flour with water, heating the slurry (for gelatinization), extruding or pressing the hot paste, and finally sun drying (Oghbaei and Prakash, 2012). The sieving of finger millet flour and preparation of different food products significantly reduced the TPC (Figure 41.3).

Opoku *et al.* (1981) showed that germination of pearl millet (*Pennisetum typhoides*) decreased the level of tannins from 1.6 to 0.83%. Conversely, in a recent study Pradeep and Guha (2011) showed that germination of little millet (*Pannicum sumatrense*) grains increased the tannin content by 17% compared to the native grain, the reason for which remains unclear. Tannins are capable of binding to proteins, carbohydrates, and minerals, thereby reducing digestibility of these nutrients.

Pushparaj and Urooj (2011) showed that different food processing methods such as boiling, pressure cooking, roasting, and germination could affect the available tannin content of foods prepared with pearl millets. Semi-refining of pearl millet flour lowered the tannin levels. In addition, bran-rich millet flour had high tannin content attributed to the high concentration of tannins in the seed coat of the grain. Tannin contents were reduced upon wet and dry heat treatments due to their possible loss upon leaching into cooking water, as tannins readily dissolve in water. In addition, during heat processing, tannins may bind to proteins, carbohydrates, or minerals, thus making their extraction difficult (Pushparaj and Urooj, 2011).

EFFECT OF PROCESSING ON PHYTIC ACID CONTENT

The distribution of phytates in the millet grain affects their reduction after submitting to different processing conditions. According to Lestienne *et al.* (2007), 4 and 8% of the phytates of pearl millet grains were removed from Gampela and IKMP-5 cultivars, respectively, at 12% decortication. The initial phytate content was 0.72 and 0.80 g/100 g of dry matter for Gampela and IKMP-5 cultivars, respectively. Thus, findings of this study further revealed that phytates are mainly distributed in the starchy endosperm and the germ of pearl millet grains (Lestienne *et al.*, 2007).

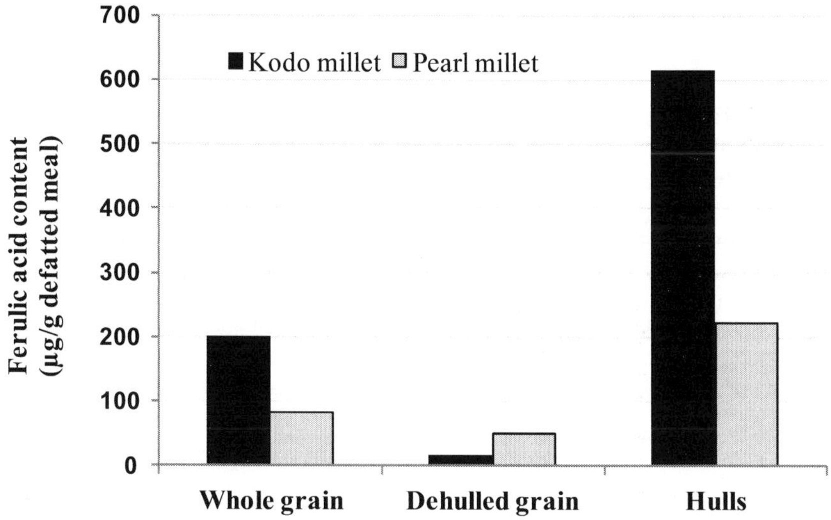

FIGURE 41.3 Effect of dehulling on the content of ferulic acid of millet grains. *Data adopted from Chandrasekara and Shahidi, 2011c.*

Eyzaguirre *et al.* (2006) demonstrated that cooking alone did not reduce phytate content of wholemeal flour of pearl millet (variety IKMP-5). Cooking in kanwa slightly reduced the level of phytates, possibly due to the formation of complexes with the minerals during cooking. Kanwa is a naturally occurring alkaline rock salt, mainly composed of sesquicarbonates (Na_2CO_3, $NaHCO_3.xH_2O$) containing various elements such as Ca, Fe, S, Cl, Si, P, K, and Al (Eyzaguirre *et al.*, 2006). In West and Central Africa, kanwa is used as a tenderizer and for reducing the cooking time of beans and other tough foods.

Soaking of pearl millet (*Pennisetum glaucum*) grains followed by wet grinding and cooking reduced phytic acid to IP5 (inositol pentaphosphate) and IP4 (inositol tetraphosphate) configurations (Eyzaguirre *et al.*, 2006). Soaking grains in water can result in passive diffusion of water-soluble phytate, which will be removed by decanting the water.

Germination and lactic acid fermentation can partially decompose phytic acid. However, phytate reduction by the action of endogenous phytases is inhibited at low pH during fermentation due to the production of low-molecular-weight organic acids such as citric, and malic, lactic acids. The phytate content of whole grain of pearl millet was reduced by soaking for 24 h in water from 419 to 250 mg/100 g on a dry weight basis (Eyzaguirre *et al.*, 2006). Furthermore, soaking for 24 h and germination for 4 days decreased the phytate content by 39% of the original value of the whole grains of pearl millets (Eyzaguirre *et al.*, 2006). In the same study authors showed that phytase treatment of whole meal flour of pearl millet for 4 h reduced phytate content by 18% (Figure 41.4). Germination/malting processes increase the activity of endogenous phytase through *de novo* synthesis, activation of intrinsic phytase, or by both actions.

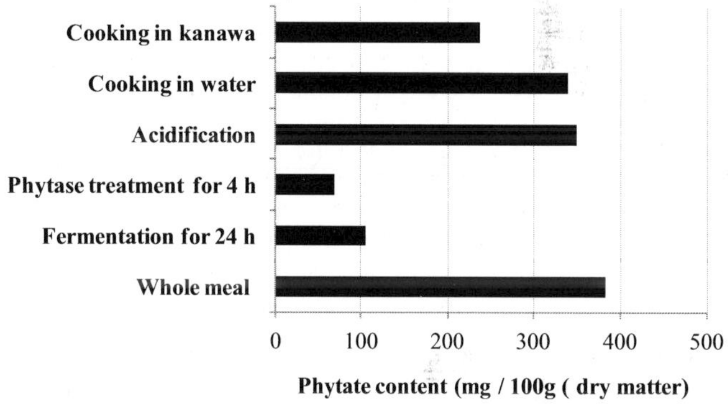

FIGURE 41.4 Effect of different processing conditions on phytate content of pearl millet grains. *Data adopted from Eyzaguirre* et al., *2006.*

SUMMARY

Millets are the first cereals domesticated by the human civilization, yet they remain underutilized even in countries where they are produced. In developed countries, millets are utilized for feeding animals whereas they are the main staple for millions of economically disadvantaged people in parts of Africa and Asia. Millets are rich in bioactive phytochemicals such as phenolics, polyphenolics, and phytic acids which are potent antioxidants in biological systems. Furthermore, millets may serve as staple food substitutes for coeliac patients as they are gluten free. A wide variety of traditional foods may be prepared using millet grains and this is dictated by the culture of the people in the region.

This contribution showed that the content of non-nutrient antioxidative compounds of millet grains as well as their activities were affected by different processing conditions experienced during food preparations. These include mechanical operations (dehulling, decortication, grinding, sieving), hydrothermal processing (boiling, pressure cooking, roasting, steaming), soaking, fermentation, and germination/malting. Depending on the type of millet, processing operations affect the content of phenolic compounds and phytates. Limited data are available on the antioxidant activities of phytochemicals present in millet grains after submitting to different processing operations to evaluate their significance as functional food ingredients in the reduction of ailments such as hypertension, cardiovascular disease, and type 2 diabetes. Thus, it would be interesting to investigate the effect of different processing conditions on the bioactive content of millet grains in order to optimize their antioxidant properties.

References

Akingbala, J.O., 1991. Effect of processing on flavonoids in millet (*Pennisetum americanum*) flour. Cereal Chem. 68, 180–183.

Akingbala, J.O., Uzo-Peters, P.I., Jaiyeoba, N.C., Baccus-Taylor, G.S.H., 2002. Changes in the physical and biochemical properties of pearl millet (*Pennisetum americanum*) on conversion to Ogi. J. Sci. Food Agric. 82, 1458–1464.

Chandrasekara, A., Shahidi, F., 2010. The content of insoluble bound phenolics in millets and their contribution to antioxidant capacity. J. Agric. Food Chem. 58, 6706–6714.

Chandrasekara, A., Shahidi, F., 2011a. Inhibitory activities of soluble and bound millet seed phenolics on free radicals and reactive oxygen species. J. Agric. Food Chem. 59, 428–436.

Chandrasekara, A., Shahidi, F., 2011b. Determination of antioxidant activity in free and hydrolyzed fractions of millet grains and characterization of their phenolic profiles by HPLC-DAD-ESI-Ma^n. J. Functional Foods 3, 144–158.

Chandrasekara, A., Shahidi, F., 2011c. Bioactivities and antiradical properties of millet grains and hulls. J. Agric. Food Chem. 59, 563–9571.

Chandrasekara, A., Naczk, M., Shahidi, F., 2012. Effect of processing on the antioxidant activity of millet grains. Food Chem. 133, 1–9.

Eyzaguirre, R.Z., Nienaltowska, K., Jong, L., de., E.Q., Hasenack, B.B.E., Nout, M.J.R., 2006. Effect of food processing of pearl millet (*Pennisetum glaucum*) IKMP-5 on the level of phenolics, phytate, iron and zinc. J. Sci. Food Agric. 86, 1391–1398.

FAO, 1995. Sorghum and millets in human nutrition. FAO, Rome.

FAOSTAT, 2011. http:// www. faostat.fao.org. (accessed 12. 02. 11.).

Graf, E., Eaton, J.W., 1990. Antioxidant functions of phytic acid. Free Radic. Biol. Med. 8, 61–69.

Gupta, N., Nagar, R., 2010. Effect of cooking, fermentation, dehulling and utensils on antioxidants present in pearl millet *rabadi* — a traditional fermented food. J. Food Sci. Technol. 47, 73–76.

Hag, M.E.E., Tinay, A.H.E., Yousif, N.E., 2002. Effect of fermentation and dehulling on starch, total polyphenols, phytic acid content and in vitro protein digestibility of pearl millet. Food Chem. 77, 193–196.

Halliwell, B., 2007. Biochemistry of oxidative stress. Biochem. Soc. Trans. 35, 1147–1150.

Klopfenstein, C.F., Hoseney, R.C., 1995. Nutritional properties of sorghum and the millets. In: Dandy, D.A.V. (Ed.), Sorghum and Millets: Chemistry and Technology. American Association of Cereal Chemists, Inc., St. Paul, MN, USA, pp. 125–166.

Lestienne, I., Buisson, M., Lullien-Pellerin, V., Picq, C., Treche, S., 2007. Losses of nutrients and anti-nutritional factors during abrasive decortication of two pearl millet cultivars (*Pennisetum glaucum*). Food Chem. 100, 1316–1323.

Matuschek, E., Towo, E., Savanberg, U., 2001. Oxidation of polyphenols in phytate-reduced high-tannin cereals: Effect on different phenolic groups and on *in vitro* accessible iron. J. Agric. Food Chem. 49, 5630–5638.

Murty, D.S., Kumar, K.A., 1995. Traditional uses of sorghum and millets. In: Dandy, D.A.V. (Ed.), Sorghum and Millets: Chemistry and Technology. American Association of Cereal Chemists, Inc., St. Paul, MN, USA, pp. 185–221.

N'Dri, D., Mazzeo, T., Zaupa, M., Ferracane, R., Fogliano, V., Pellegrini, N., 2012. Effect of cooking on the total antioxidant capacity and phenolic profile of some whole-meal African cereals. J. Sci. Food Agric. DOI:10.1002/jsfa.5837.

Oghbaei, M., Jamuna Prakash, J., 2012. Bioaccessible nutrients and bioactive components from fortified products prepared using finger millet (*Eleusine coracana*). J. Sci. Food Agric. 92, 2281–2290.

Opoku, A.R., Ohenhen, S.O., Ejiofor, N., 1981. Nutrient composition of millet (*Pennisetum typhoides*) grains and malts. J. Agric. Food Chem. 29, 1247–1248.

Pradeep, S.R., Guha, M., 2011. Effect of processing methods on the nutraceutical and antioxidant properties of little millet (*Panicum sumatrense*) extracts. Food Chem. 126, 1643–1647.

Pushparaj, F.S., Urooj, A., 2011. Influence of processing on dietary fibre, tannin and in vitro protein digestibility of pearl millet. Food Nutr. Sci. 2, 895–900.

Rao, M.V.S.S.T.S., Muralikrishna, G., 2002. Evaluation of the antioxidant properties of free and bound phenolic acids from native and malted finger millet (Ragi, *Elucine coracana* Indaf-15). J. Sci. Food Agric. 50, 889–892.

Shahidi, F., 1997. Natural antioxidants: An overview. In: Shahidi, F. (Ed.), Natural Antioxidants: Chemistry, Health Effects, and Applications. American Oil Chemists Society Press, Washington, DC, USA, pp. 1–11.

Shahidi, F., Nazck, M., 2004. Phenolics in Food and Nutraceuticals. pp 1–82. CRC press, Boca Ratón, FL, USA.

Shobana, S., Malleshi, N.G., 2004. Preparation and functional properties of decorticated finger millet (*Eleusine coracana*). J. Food Eng. 79, 529–538.

Simwemba, C.G., Hoseney, R.C., Varriano-Marston, E., Zeleznak, K., 1984. Certain B vitamin and phytic acid contents of pearl millet [*Pennisetum americanum* (L.) Leeke]. J. Agric. Food Chem. 32, 31–34.

Sripriya, G., Chandrasekaran, K., Murty, V.S., Chandra, T.S., 1996. ESR spectroscopic studies on free radical quenching action of finger millet (*Elusine coracana*). Food Chem. 57, 537–540.

Towo, E.E., Svanberg, U., Ndossi, G.D., 2003. Effect of grain pretreatment on different extractable phenolic groups in cereals and legumes commonly consumed in Tanzania. J. Sci. Food Agric. 83, 980–986.

CHAPTER

42

Influence of Processing on Nutraceuticals of Little Millet (*Panicum sumatrense*)

Manisha Guha, Yadahally N. Sreerama, N.G. Malleshi

Department of Grain Science and Technology, Central Food Technological Research Institute, Council of Scientific and Industrial Research (CSIR), Mysore, India

CHAPTER POINTS

- The physical separation of different morphological parts by milling process has a significant impact on the content of nutrients in little millet.
- Debranning process removed substantial amounts of micronutrients from brown grain.
- Milled grain and defatted bran are good sources of soluble and total dietary fiber, respectively.
- Processing the millet significantly improved the nutraceutical properties by increasing its content in phenolic compounds and also antioxidant activities.
- Carotenoids and tocopherols present in some varieties of little millet may play an important role in human health in reducing the risk of various degenerative diseases.
- Soaking and germination processes were effective in increasing the content of GABA, a non-protein amino acid which has health promoting properties.
- Judicious processing of little millet significantly improves its nutraceutical characteristics and qualifies the utilization of millet in health food formulations.
- Further elucidating the full potential of the health promoting capabilities of little millet bioactive compounds may continue to enhance and advance the discipline of functional foods and nutraceutical research.

INTRODUCTION

Cereals including millets contribute a significant amount of energy, protein, selected micronutrients, and non-nutrients in the diet of populations around the globe in both developed and developing countries. Millets are warm-season minor cereals that are used as food, feed, and fodder in various parts of the world. Millets are generally consumed as whole grain products and provide many health-promoting components. Nowadays, consumers are increasingly interested in deriving health benefits through foods and have begun to look beyond the basic nutrition of food. Epidemiological evidence suggests that regular consumption of whole grains and whole-grain products is associated with the reduced risk of several non-communicable chronic diseases (Jones and Engleson, 2010).

Little millet (*Panicum sumatrense*) is an important minor millet in low- and moderate-rainfall areas. In India, this millet is grown primarily in the drought-prone arid and semiarid zones and used mostly for food purposes especially by people of economically weaker sectors. Besides India, it is cultivated in Nepal, Western Burma, and also in some African countries. It has a good nutritional profile and is comparable to staple cereals, namely, rice and wheat in terms of protein, minerals, and vitamins (Gopalan *et al.*, 1987). Little millet is also a good source of nutraceuticals such as phenolics, gama-amino butyric acid (GABA), lignans, resistant starch, sterols, and phytates. The additive and synergistic effects of these bioactive nutraceuticals in millet-based foods may offer several health benefits.

Nutraceuticals are a class of food-based bio-molecules and exhibit hypglycemic, hypocholesteromic, hypolipidemic and anti-glycation properties. Furthermore, they

Processing and Impact on Active Components in Food
http://dx.doi.org/10.1016/B978-0-12-404699-3.00042-1

also possess free radical scavenging, enzyme inhibitory activities and also metal chelating characteristics (Prathapan *et al.*, 2011). There are a wide range of oxygen-free radicals and other reactive oxygen species (ROS), including free radicals such as superoxide anion radicals (O_2^-), hydroxyl radicals ($HO^•$), and non-free-radical species, namely, hydrogen peroxide (H_2O_2) and singlet oxygen ($^1O_2$). These oxygen-free radicals and other reactive oxygen species induce lipid peroxidation that causes deterioration of foods. Furthermore, these free radicals also attack biological molecules such as lipids, protein, and DNA, and these consequently initiate several chronic diseases. Antioxidant agents can act against free radicals either by retarding their formation or by inactivating radicals in the reaction medium (Shahidi and Chandrashekara, 2013a). Currently, the natural antioxidant α-tocopherol and some synthetic antioxidants, such as butylated hydroxytoluene, butylated hydroxyanisole, and propyl gallate, are commonly used to act against free radicals in food and biological systems. However, the use of these synthetic antioxidants in food products is under strict regulation owing to their potential health hazards.

Phenolic compounds in cereals and millets with nutraceutical properties are gaining a lot of importance because of their health promoting abilities. Previous studies have shown that the millet phenolic compounds exhibit a wide range of pharmacological and medicinal properties, including anti-inflammatory, anti-carcinogenic, and vasodilatory actions, which have been mostly attributed to their free radical scavenging, metal chelating, and antioxidant activities (Arpornsuwan *et al.*, 2010). They may also play an important role in human health, reducing the risk of various degenerative diseases, such as cardiovascular diseases, osteoporosis, and cancer.

GABA, an important inhibitory neurotransmitter in mammalian central nervous system provides several health benefits. A lot of reports are available on the GABA content of germinated rice. But such information is scant on little millet. On the other hand, considerable information is available on millet polyphenols. Processing such as milling, hydrothermal treatments, soaking, fermentation, malting, and other indigenous processing conditions also affect the content of nutraceuticals and their bioactivities in millets (Shahidi and Chandrasekara, 2013b). However, there is a dearth of information on the specific health beneficial components in little millet with special reference to phenolic compounds and other nutraceuticals. Furthermore, millets remain underutilized cereals even in regions where they are grown due to their limited inclusion in processed foods, and lack of research and innovative product development. Therefore, the aim of this review is to examine the influence of processing on the stability and activity of bioactive nutraceuticals in little millet and the *in vitro* evidences about their antioxidant properties.

HOW COMPOSITION IS ALTERED

Milling and Distribution of Nutrients in Milling Fractions

Millets in general and little millet in particular are commonly processed by dehulling or milling which involves detachment of the seed coat (husk) and separation of the bran. The milled millet exhibits improved cooking, sensory attributes, digestion, assimilation, and absorption. There are no exclusive mills for milling the little millets. Normally, dehusking and debranning of millets is carried out in existing cereal milling machines with only a little modification in the process. Dehusking and debranning in abrasive mills like rice hullers damages and breaks the endosperm. Whereas, dehusking in centrifugal shellers followed by debranning in hullers, yields milled grain of satisfactory quality. Studies on milling of little millet in a Satake rubber roll sheller yielded 83.1% of brown grain and subsequent polishing of brown grains provided 77.0% of polished grains. Nutrients are unevenly distributed in the different morphological parts of little millet (Table 42.1). The brown grains and the milled grain contained 11.4% and 9.5% protein, respectively. Brown grain contained higher fat (5.4%) compared to milled grain (3.1%). Total carbohydrate content increased in milled grain as a result of the debranning process. The calcium and phosphorus contents of brown grains were 13.1 and 412.4 mg/100 g,

TABLE 42.1 Milling Yields and Distribution of Nutrients in Little Millet Fractions

Milled Fraction	Milling Yield (%)	Moisture (%)	Protein (%)	Carbohydrate (%)	Fat (%)	Ash (%)	Calcium (mg/100 g)	Phosphorus (mg/100 g)
Brown grain	83.1	6.8	11.4	74.8	5.4	1.6	13.1	412.4
Milled grain	77.0	7.6	9.5	78.9	3.1	0.9	4.4	262.7

Carbohydrate content was shown as difference of 100 − (moisture + protein + ash + fat). Results are average of four determinations.
Data adapted from Hadimani and Malleshi (1993).

respectively. Milling removes nearly 66% calcium and about 36% of phosphorus of brown grain (Hadimani and Malleshi, 1993)

The total dietary fiber content of debranned millet was 15.9%, indicating that little millet, even after removal of husk and major portion of bran, contained appreciable amounts of dietary fiber (Hadimani and Malleshi, 1993). The soluble dietary fiber content of milled grain was 36% of the total dietary fiber, which represents a fairly higher proportion as compared to the rice and wheat (Wisker *et al.*, 1981). This indicates that little millet forms an important source of soluble dietary fiber. It may be mentioned here that accordingly to Ranhotra *et al.*, 1990, the soluble fraction of the dietary fiber has a beneficial effect in rats with respect to hypocholesterolemic activity. The millet bran appears to be a good source of dietary fiber. The defatted bran contained 48.7% total dietary fiber, out of which about 8% was soluble fraction (Hadimani and Malleshi, 1993). Investigation with regard to hypocholesteremic effect of millets may reveal useful information for diversified uses of their bran.

Effect of Indigenous Processing Methods on the Nutraceutical Content of Little Millet

Primary processing technologies such as soaking, cooking, roasting, and germination are widely used for improving the cooking properties, palatability and also reduction of antinutritional compounds. These processing methods also affect the content and activities of nutraceuticals in millets (Shahidi and Chandrasekara, 2013b). Table 42.2 shows the total phenolic contents (TPC) of native and processed little millet extracts. The TPC increased during processing up to 521.0 mg GAE/100 g. The maximum TPC was observed in roasted millet followed by steamed and germinated millet (Pradeep and Guha, 2011). The increase in TPC during roasting may be due to better retention of polyphenols during processing and subsequent polishing compared to native polished millets. The increase in TPC during processing has also been reported by Duenas *et al.*, (2009) on lupin seeds, and on barley by Gallegos-Infante *et al.*, (2010).

There was also a similar increase in the flavonoid contents on processing (Table 42.2). The germinated millet showed least increase (4.6%), whereas roasted millet produced a considerable increase (25.5%) in flavonoid content compared to native millet (334.9 mgCE/100 g). The lower level of flavonoids on germination may be attributed to the biochemical changes of millet during germination, leading to the conversion of these plant secondary metabolites. Steaming the millet also increased the extractability of flavonoids considerably.

The tannin contents of the little millet extracts also increased during processing (Table 42.2). Among the different processing methods, the roasted millet extract showed a maximum increase in tannin contents (18.8%) compared to the unprocessed polished millet. It is reported that the outermost layers of grain possess high tannin content (Chandrasekara and Shahidi 2011a). The various processing conditions used for little millet may have facilitated the migration of soluble tannins towards endosperm, which renders better retention of tannins during subsequent polishing, compared to unprocessed polished millet. The increase in the tannin contents may also be due to the degradation of a high molecular weight insoluble polymer into a low molecular weight soluble polymer, during processing.

Processing Induced Compositional Variation of Phenolic Compounds

The reverse phase HPLC profiles of phenolic compounds isolated and identified in native and processed little millet are presented in Figure 42.1. Processing conditions induced variations in the content and composition of phenolic compounds. The phenolic acid content of native and processed little millet extracts is shown in Table 42.3. Processing of the millet showed significant effect on the extractability of the ferulic acid as the extract from steamed (1.5 kscg) millet content was

TABLE 42.2 Nutraceutical Properties of Native and Processed Little Millet Extracts

Samples	Total Phenolics (mg/100 g)A	Total Flavonoids (mg/100 g)B	Tannins (mg/100 g)C
Native	429.9 ± 0.40^{a}	334.9 ± 0.89^{a}	283.4 ± 0.41^{a}
Germinated	453.3 ± 1.76^{b}	350.2 ± 0.75^{b}	332.1 ± 1.12^{b}
Steamed (0 kscg)	462.9 ± 4.20^{c}	398.7 ± 1.02^{c}	284.9 ± 2.34^{a}
Steamed (1.5 kscg)	485.7 ± 2.54^{d}	410.1 ± 2.10^{d}	308.6 ± 0.65^{c}
Roasted	521.0 ± 0.15^{e}	420.2 ± 4.73^{e}	336.8 ± 1.14^{db}

Results are mean of three determinations ± SD. Values with same letters (a, b, c, d, e, within columns) are not significantly different at $p < 0.05$. (A) mg gallic acid equivalents (GAE)/100 g of sample, dry basis. (B) mg catechin equivalents (CE)/100 g of sample, dry basis. (C) mg catechin equivalents (CE)/100 g of sample, dry basis.

Table adapted from Pradeep and Guha (2011).

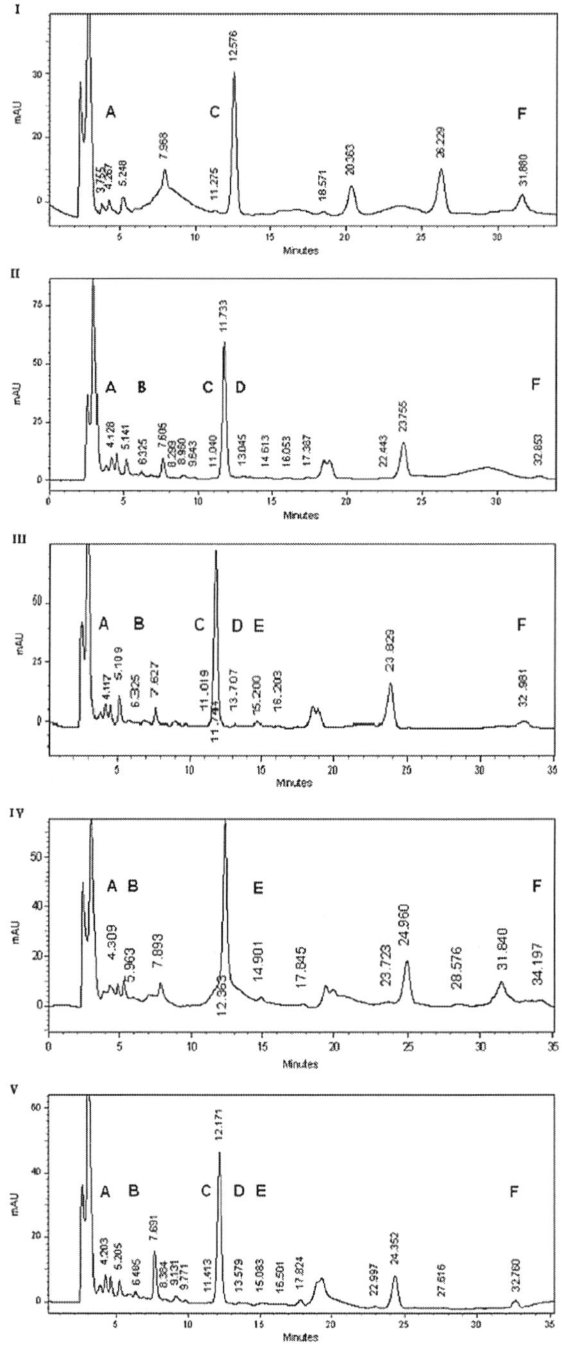

FIGURE 42.1 Fractionation of phenolics from native and processed little millet extracts. I, native; II, germinated; III, 0 kscg steamed; IV, 1.5 kscg steamed; V, roasted—A, gallic acid; B, proto-catechic acid; C, gentisic acid; D, vanillic acid; E, syringic acid; F, ferulic acid. *Figure from Pradeep and Guha (2011).*

18.29 mg/100 g, whereas the germinated millet content was only 1.27 mg/100 g. Gallic acid was the other major phenolic acid detected in native and processed millet extracts and its concentration increased during processing compared to the native millet extract. Moreover, p-hydroxybenzoic acid, kaempferol, and *trans*-cinnamic acids were not detected in both the native and processed millet extracts (Pradeep and Guha., 2011).

Carotenoid and Tocopherol Content in Little Millet

Some of the little millet varieties are slightly yellowish colored, indicating the possibility of presence of carotenoids. It was reported that the total carotenoid content in little millet was in the range of 51–104 μg/100 g, however, the presence of β-carotene has not been reported. The total

TABLE 42.3 Phenolic Acid Composition of Native and Processed Little Millet Extracts

Phenolic acid	Native	Germinated	Steamed (0 kscg)	Steamed (1.5 kscg)	Roasted
Gallic acid	1.74 (4.2)	2.65 (4.1)	2.50 (4.1)	1.90 (4.3)	5.81 (4.2)
Proto-catechuic acid	ND	0.14 (6.3)	0.27 (6.3)	0.24 (5.9)	0.30 (6.4)
p-Hydroxy benzoic acid	ND	ND	ND	ND	ND
Gentisic acid	0.65 (11.2)	0.21 (11.0)	0.55 (11.0)	ND	0.18 (11.4)
Vanillic acid	ND	0.57 (13.0)	0.20 (13.7)	ND	0.95 (13.5)
Syringic acid	ND	ND	0.05 (15.2)	0.50 (14 .9)	0.19 (15.0)
Kaempferol	ND	ND	ND	ND	ND
Ferulic acid	13.54 (31.8)	1.27 (32.8)	2.22 (32.9)	18.29 (31.8)	3.88 (32.7)
trans-cinnamic acid	ND	ND	ND	ND	ND

The results are expressed as mg/100g extract. Values given in parenthesis indicate retention time (min). ND, not detected.
Table taken and adapted from Pradeep and Guha (2011).

tocopherol content in little millet has been reported to be about 1.3 mg/100 g and it is mainly δ-tocopherol, although the α and γ isomers of tocopherols have been detected as minor constituents. Total antioxidant capacity in a few little millet varieties was in the range of 3.1–6.3 mM tocopherol equivalent/g (Asharani *et al.*, 2010).

Gamma-Aminobutyric Acid (GABA) Content in Little Millet

GABA is a four-carbon non-protein amino acid, produced primarily by the decarboxylation of L-glutamic acid, catalyzed by α-glutamate decarboxylase during the germination process (Bown *et al.*, 1999). GABA is a neurotransmitter in the brain and spinal cord of mammals and induces hypotensive, diuretic, and tranquillizing effects (Watanabe *et al.*, 2002, Akama *et al.*, 2009). It is generally believed that in plants, GABA content increases by stress, such as hypoxia, cytosolic acidification, cold shock, mechanical stimulation, water stress, and darkness. Besides these, GABA content also increases during soaking and germination of food grains. Karladee and Suriyong, (2012) observed maximum level of GABA content in 24 h germinated brown rice in all the 21 rice varieties they studied. The inhibitory action of GABA extracts on leukemia cell proliferation, stimulatory action on the cancer cell apoptosis has been reported by Oh and Oh, (2004). Hence the food formulations containing GABA enriched cereals can serve as a source of nutraceuticals and may reduce the complications of sleeplessness, depression, and autonomic disorder, and stimulate immune cells (Okada *et al.*, 2000).

The GABA content of various commonly consumed millets and legumes isolated and characterized followed by (TLC) and HPTLC respectively are presented in Figure 42.2. GABA was identified in all the tested millets and legumes. The GABA content of native, soaked and germinated little millet in comparison to other millets and legumes is shown in Table 42.4, which indicates the increase in GABA content both during soaking and germination. In general, millets contained more GABA compared to legumes in their native form. GABA content was maximum 84.6 nm/100 mg in the little millet at 24 h of germination and it decreased subsequently on further germination (Table 42.4). These results indicate that the soaking and germination process influences the GABA production significantly in millets and legumes in general, and little millet in particular. Increase in GABA content during soaking and germination may be due to the enzymatic conversion of glutamate to GABA by α glutamate decaroboxylase. Synthesis of GABA during germination may have triggered due to the limited availability of oxygen during soaking of the seeds (Roberts *et al.*, 1984).

Antioxidant Properties of Processed Little Millet

Processed millet extracts exhibited higher 2,2′-diphenyl-1-picrylhydrazyl (DPPH) radical scavenging activity compared to the native millet extracts. The roasted millet extract showed the higher radical scavenging activity followed by steamed and geminated millet extracts (Pradeep and Guha, 2011). The higher phenolic content in roasted millets may have contributed to higher radical scavenging activity. In addition, Chandrasekara and Shahid (2011b) have shown strong and positive association between phenolic content and scavenging of DPPH radical and reactive oxygen

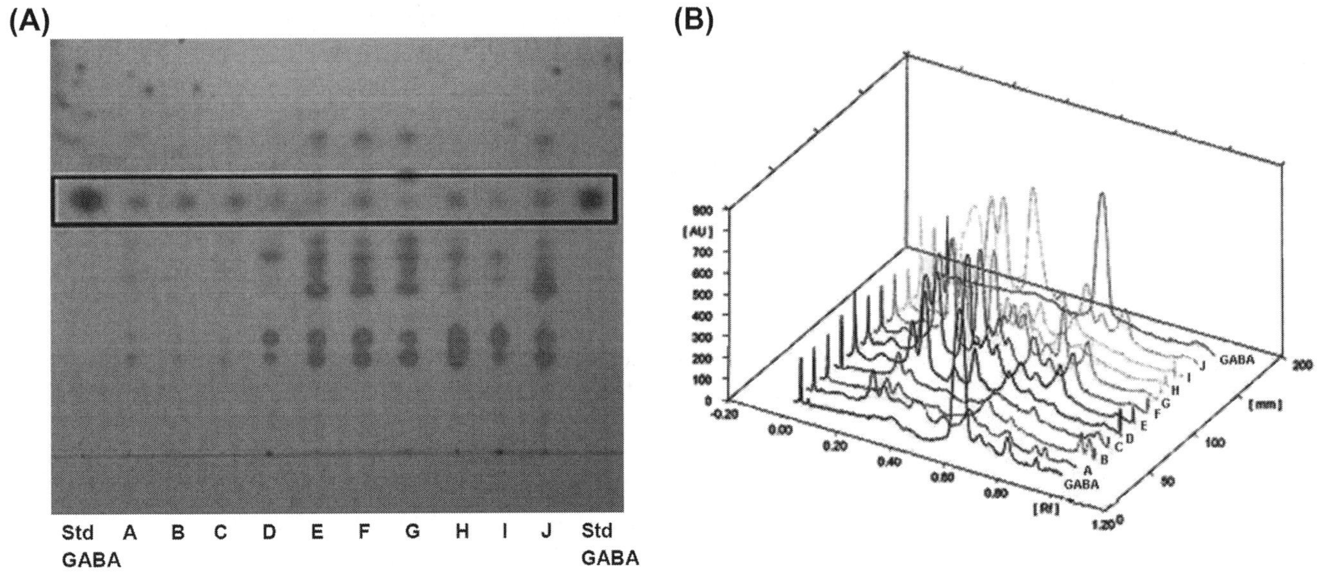

FIGURE 42.2 (A) Thin-layer chromatogram of standard GABA and extracts of millets and legumes. A, Sorghum; B, Finger millet; C, Little millet; D, Bengal gram; E, Green gram; F, Black gram. (B) High performance thin layer chromatogram of standard GABA and extracts of millets and legumes. A, Sorghum; B, Finger millet; C, Little millet; D, Bengal gram; E, Green gram, F, Black gram; G, Horse gram; H, Lentil; I, Moth bean; J, Green pea. *Data taken and adapted from Pradeep* et al. *(2011).*

TABLE 42.4 Gamma-Aminobutyric Acid Content in Native, Soaked, and Germinated Millets and Legumes

		Germination period (h)				
Samples	Native	0	24	48	72	96
Sorghum	$24.9^{a} \pm 1.5$	$88.7^{b} \pm 2.0$	$68.9^{c} \pm 4.1$	$59.3^{dc} \pm 4.5$	$51.6^{ed} \pm 5.1$	$49.0^{fde} \pm 4.3$
Finger millet	$21.4^{a} \pm 1.1$	$83.8^{b} \pm 2.5$	$256.5^{c} \pm 3.2$	$311.2^{d} \pm 6.3$	$345.0^{e} \pm 3.9$	$361.8^{f} \pm 6.7$
Little millet	$38.2^{a} \pm 0.8$	$77.2^{b} \pm 3.2$	$84.6^{cb} \pm 5.7$	$65.6^{d} \pm 5.9$	$59.0^{ed} \pm 1.2$	$55.4^{fde} \pm 3.0$
Bengal gram	$10.9^{a} \pm 2.7$	$312.4^{b} \pm 5.5$	$238.1^{c} \pm 2.9$	$290.0^{d} \pm 3.0$	$323.1^{e} \pm 3.8$	$367.7^{f} \pm 4.4$
Green gram	$20.6^{a} \pm 2.8$	$385.0^{b} \pm 3.4$	$610.4^{c} \pm 6.9$	$644.5^{d} \pm 4.7$	$699.6^{e} \pm 3.7$	$781.0^{f} \pm 5.6$
Black gram	$11.1^{a} \pm 1.6$	$349.4^{b} \pm 5.7$	$208.9^{c} \pm 1.1$	$312.2^{d} \pm 3.1$	$352.8^{eb} \pm 3.9$	$385.5^{f} \pm 3.3$
Horse gram	$16.9^{a} \pm 2.1$	$63.5^{b} \pm 3.6$	$104.2^{c} \pm 1.6$	$162.7^{d} \pm 1.9$	$197.7^{e} \pm 5.1$	$241.0^{f} \pm 5.7$
Lentil	$12.6^{a} \pm 3.2$	$139.5^{b} \pm 3.3$	$207.9^{c} \pm 4.7$	$301.4^{d} \pm 5.6$	$347.4^{e} \pm 1.1$	$384.6^{f} \pm 6.8$
Moth bean	$20.7^{a} \pm 2.8$	$531.6^{b} \pm 3.0$	$947.6^{c} \pm 4.9$	$1242.5^{d} \pm 5.3$	$1313.4^{e} \pm 5.8$	$1401.7^{f} \pm 7.7$
Green pea	$14.2^{a} \pm 0.9$	$89.0^{b} \pm 1.8$	$305.6^{c} \pm 2.3$	$447.6^{d} \pm 3.5$	$500.0^{e} \pm 3.9$	$573.1^{f} \pm 5.8$

Results are mean of three determinations ± SD. Mean values in the same row followed by different superscript are significantly different ($p < 0.05$). Values expressed as nm/100 mg, dry basis.
Data adapted from Pradeep et al. *(2011).*

species in little millet. It can be speculated that phenolic compounds present in the extracts may exert their antioxidant activity individually as well as synergistically, Furthermore, Millard reaction products formed during the high temperature short time roasting conditions may have played a role in scavenging the free radicals. Similar findings have been reported for coffee brews during roasting by Nicoli *et al.* (1997) and also for common bean during pressure cooking by Rocha-Guzman *et al.* (2007). Similar to DPPH free radical scavenging activity, ferrous reducing antioxidant power of processed millet varied depending on the nature of processing; activity was higher in roasted followed by steamed, germinated, and native millets (Pradeep and Guha, 2011). Presence of reductones and high reducing power has also reported for the soluble and bound phenolics of little millet (Chandrasekara and Shahidi, 2010). The effects of different processing conditions on the contents of non-nutrient antioxidative compounds of various millet grains as well as their activities have been described in detail in Chapter 41 of this book (Shahidi and Chandrasekara, 2013b). Since the processed little millet extracts exhibited higher antioxidant properties in terms of free radical scavenging activity and also iron reducing powerassay, these processed millets could be used as ingredients in health food formulations.

OTHER WAYS IN WHICH COMPOSITION IS ALTERED

Although native millets could be stored for long periods without substantial quality changes, their quality rapidly deteriorates after they are ground into meal. Poor storage quality has been attributed largely to hydrolytic changes associated with the action of lipolytic enzymes. Milled millet storage stability is improved by dry-milling processes that remove the major lipid-containing portions of the grains (Abdelraham *et al.*, 1983). Little millet grains can be partially or fully cooked in extruder or roll cookers to inactivate hydrolytic and oxidative enzymes to improve storage stability. This thermal inactivation of enzymes resulted in substantial improvement in yield and nutritional quality of cereals and oils seeds (Bookwalter, 1983). Furthermore, heat-processing of whole proso millet at 97°C in a steam-jacketed conveyer inactivated lipid enzymes and extended storage life. Germ fractions could be retained in the final milled product, which resulted in much higher protein, lipid, thiamine, riboflavin, niacin, iron. zinc, available lysine, and protein efficiency ratios in the product (Bookwalter *et al.*, 1987). These results suggest that improvement in yield and nutritional quality can be achieved by judicious thermal-processing and milling of little millet.

ANALYTICAL TECHNIQUES

Milling and Nutrient Composition of Milled Fractions

Little millet grains were dehulled in a rubber roll sheller, and husk was aspirated off. The brown grains were then polished (debranned) in an abrasive polisher and the bran was separated by sieving. Brown grains and debranned grains were powdered in an Udy cyclone mill and their defatted flours were analysed for their nutrient composition by standard AACC procedures (1983).

Extraction and Quantification of Total Phenolic, Total Flavonoid, and Tannin Contents

Total phenolics from defatted native and processed millet flours were extracted by refluxing with methanol: 1% HCl at 60°C and their total phenolic content was determined as per the method of Singleton and Rossi (1965). The result was expressed as mg of gallic acid equivalents (GAE) per 100 gram of sample. The total flavonoid content (TFC) of flours was measured by the aluminum chloride colorimetric assay method of Kim *et al.* (2003) and was expressed as mg of catechin equivalent (CE) per 100 g of sample. The tannin contents of millet flours were determined according to the vanillin-HCl procedure described by Price *et al.* (1978) and results were expressed as mg of CE per 100 g of sample.

Fractionation of Phenolics

The phenolic components in native and processed little millet extracts were fractionated and quantified by reverse phase HPLC using an isocratic solvent system, water : acetic acid : methanol (80 : 5 : 15, v/v) as described by Pradeep and Guha (2011).

Spectrophotometric Quantification and Identification of GABA by TLC and High Performance Thin Layer Chromatography

Gamma-aminobutyric acid content of the millet and legumes samples was determined using the method of Zhang and Bown, (1997) and was identified by TLC followed by HPTLC (on activated high performance silica gel plates using n-butanol : acetic acid : water (5 : 2 : 2, v/v) as solvent system, by ascending mode) according to Saravana Babu *et al.* (2007).

Determination of Antioxidant Properties

DPPH Free Radical Scavenging Activity

Free radical scavenging capacities of the millet was determined by reaction with the 2,2′-diphenyl-1-picrylhydrazyl (DPPH) radical, according to the method adapted from Brand-Williams *et al.* (1995). Decreasing the absorbance of the DPPH solution indicates an increase in DPPH radical scavenging activity. The absorbance was measured at 515 nm and the result was expressed as a percentage of inhibition of the DPPH radical.

Ferrous reducing antioxidant power (FRAP)

The ferrous reducing power of millet sample was determined by mixing the extract with potassium ferricyanide and measuring the absorbance at 700 nm of the color developed by adding ferric chloride as per the procedure of Oyaizu (1986). Increase in absorbance of the reaction mixture indicates increase of reducing power.

References

AACC, 1983. Approved Methods, eighth ed. American Association of Cereal Chemists, St. Paul, Minnesota Asp NG, Johansson.

Abdelrahman, A., Hoseney, R.C., Varriano-Marston, E., 1983. Milling process to produce low-fat grits from pearl millet. Cereal Chern. 60, 189.

Akama, K., Kanetou, J., Shimusaki, S., Kawakami, K., Tsuchikura, S., Takaiwa, F., 2009. Seed- specific expression of truncated osGAD2 produces GABA- enriched rice grains that influence a decrease in blood pressure in spontaneously hypertensive rats. Transgenic Res. 18, 865–876.

Arpornsuwan, T., Changsri, K., Roytraku, S., Punjanon, T., 2010. The effects of the extracts from *Carthamus tinctorius* L. on gene expression related to cholesterol metabolism in rats. Songklanakarin J. Sci. Tech. 32, 129–136.

Asharani, V.T., Jayadeep, A., Malleshi, N.G., 2010. Natural antioxidants in edible flours of selected small millets. Int. J. Fd. Properties. 13, 41–50.

Bookwalter, G.N., 1983. World feeding strategies utilizing cereals and other commodities. Cereal Foods World 28, 507.

Bookwalter, G.N., Slyle, S.A., Warner, K., 1987. Millet processing for improved stability and nutritional quality without functionality changes. J. Food Sci. 52, 399–402.

Bown, A.W., McLean, M.D., Shelp, B.J., 1999. Metabolism and functions of gamma-aminobutyric acid. Trends Plant Sci. 4, 446–452.

Brand-Williams, W., Cuvelier, M.E., Berset, C., 1995. Use of free radical method to evaluate antioxidant activity. LWT Food Sci. Tech. 28, 25–30.

Chandrasekara, A., Shahidi, F., 2010. The content of insoluble bound phenolics in millets and their contribution to antioxidant capacity. J. Agric. Food Chem. 58, 6706–6714.

Chandrasekara, A., Shahidi, F., 2011a. Bioactivities and antiradical properties of millet grains and hulls. J. Agric. Food Chem. 59, 9563–9571.

Chandrasekara, A., Shahidi, F., 2011b. Inhibitory activities of soluble and bound millet seed phenolics on free radicals and reactive oxygen species. J. Agric. Food Chem. 59, 428–436.

Duenas, M., Hernandez, T., Estrella, I., Fernandez, D., 2009. Germination as a process to increase the polyphenol content and antioxidant activity of lupin seeds *(Lupinus angustifolius* L.). Food Chem. 117, 599–607.

Gallegos-Infante, J.A., Rocha-Guzman, N.E., Gonzalez-Laredo, R.F., Pulido-Alonso, J., 2010. Effect of processing on the antioxidant properties of extracts from Mexican barley (*Hordeum vulgare*) cultivar. Food Chem. 119, 903–906.

Gopalan, C., Ramasashtri, B.V., Balasubramanian, S.C.N., 1987. Nutritive Value of Indian Foods. National Institute of Nutrition, Hyderabad, India, pp. 60.

Hadimani, N.A., Malleshi, N.G., 1993. Studies on milling, physico-chemical properties, nutrient composition and dietary fibre content of millets. J. Food Sci. Technol. 30, 17–20.

Jones, J.M., Engleson, J., 2010. Whole grain: benefits and challenges. Annu. Rev. Food Sci. Technol. 1, 19–40.

Karladee, D., Suriyong, S., 2012. γ-amino butyric acid (GABA) content in different varieties of brown rice during germination. Sci. Asia 38, 13–17.

Kim, D., Jeong, S., Lee, C.Y., 2003. Antioxidant capacity of phenolic phytochemicals from various cultivars of plums. Food Chem. 81, 321–326.

Nicoli, M.C., Anese, M., Manzocco, L., Lerici, C.R., 1997. Antioxidant properties ofcoffee brews in relation to the roasting degree. Lebensmittel Wiss. Technol. 30, 292–297.

Oh, C.H., Oh, S.H., 2004. Effects of germinated brown rice extracts with enhanced levels of GABA on cancer cell proliferation and apoptosis. J. Med. Food 7, 19–23.

Okada, T., Sugishita, T., Murakami, T., Murai, H., Saikusa, T., Horino, T., Onoda, A., Kajimoto, O., Takahashi, R., Takahashi, T., 2000. Effect of the defatted rice germ enriched with GABA for sleeplessness, depression, autonomic disorder by oral administration. Nippon shokuhin kagaku kogaku kaishi 47, 596–603.

Oyaizu, M., 1986. Studies on product of browning reaction prepared from glucose amine. Jpn. J. Nutri. 44, 307–315.

Pradeep, S.R., Guha, M., 2011. Effect of processing methods on the nutraceutical and antioxidant properties of little millet (*Panicum sumatrense*) extracts. Food Chem. 126, 1643–1647.

Pradeep, S.R., Malleshi, N.G., Guha, M., 2011. Germinated millets and legumes as a source of gamma-aminobutyric acid. World Appl. Sci. J. 14, 108–113.

Prathapan, A., Singh, M.K., Anusree, S.S., Soban Kumar, D.R., Sundaresan, A., Raghu, K.G., 2011. Antiperoxidative, free radical scavenging and metal chelating activities of boerhaavia diffusa l. J. Food Bio-Chem. 35, 1548–1554.

Price, M.L., Van Scoyoc, S., Butler, L.G., 1978. A critical evaluation of the vanillin reaction as an assay for tannin in sorghum grain. J. Agric. Food Chem. 26, 1214–1218.

Ranhotra, G.S., Gelroth, J.A., Astroth, K., Rao, C.S., 1990. Relative lipidemic responses in rats fed with oat bran or oat bran concentrate. Cereal Chem. 76, 509–511.

Roberts, J.K., Callis, J., Wemmer, D., Walbot, V., Jardetzky, O., 1984. Mechanism of cytoplasmic pH regulation in hypoxic maize root tips and its role in survival under hypoxia. P. Nat. Acad. Sci. USA 81, 3379–3383.

Rocha-Guzman, N.E., Gonzalez-Laredo, R.F., Ibarra-Perez, F.J., Nava-Berumen, C.A., Gallegos-Infante, J.A., 2007. Effect of pressure cooking on the antioxidant activity of extracts from three common bean (*Phaseolus vulgaris L.*) cultivars. Food Chem. 100, 31–35.

Saravana Babu, C., Sunil, A.G., Vasanthi, H.R., Muthswamy, V.S., Ramanathan, M., 2007. Development and validation of an HPTLC method for simultaneous estimation of excitatory neurotransmitters in rat brain. J. Liq. Chromatogr. 30, 2891–2902.

Shahidi, F., Chandrasekara, A., 2013a. Millet grain phenolics and their role in disease risk reduction and health promotion: A review. J. Functional Foods 5, 570–581.

Shahidi, F., Chandrasekara, A., 2013b. Processing of millet grains and effects on non-nutrient antioxidant compounds. In: Preedy, V.R. (Ed.), Processing and impact on active components in food. Elsevier B.V., Netherlands.

Singleton, V.L., Rossi, J.A., 1965. Colorimetry of total phenolics with phosphomolybdic-phosphotungstic acid reagents. Ame. J. Enol. Viticul. 16, 144–158.

Watanabe, M., Maemura, K., Kanbara, K., Tamayama, T., Hayasaki, H., 2002. GABA and GABA receptors in the central nervous system and other organs. Int. Rev. Cytol. 213, 1–47.

Wisker, E., Feldheim, W., Pomeranz, Y., Meuser, 1981. Dietary fibre in cereals. In: Pomeranz, Y. (Ed.), Advances in Cereal Science and Technology, Vol.VII, American Association of Cereal Chemists. St. Paul, Minnesota, USA, pp. 183–185.

Zhang, G., Bown, A.W., 1997. The rapid determination of γ-amino butyric acid. Phytochem 44, 1007–1009.

CHAPTER

43

Processing Oats and Bioactive Components

Apollinaire Tsopmo

Food Science and Nutrition Program, Carleton University, Ottawa, Canada

CHAPTER POINTS

- Commercial processing like steaming and flaking effects on the content of phenolic acids is dependent of chemical structures. There is therefore an increase in concentrations of some phenols and a decrease in concentration of others.
- In germinated oats, the content of most avenanthramides increased, while the amount of lysine, a limiting amino acid in cereals, increased and therefore increasing the amino acid balance of geminated oats.
- Germination resulted in a greater decreased of total fat in hull-less than hulled oats.
- In wet processing, water-alcohol mixtures are the preferred solvents for free and soluble phenols. The yields of phenolic acids from alkaline hydrolysis are higher compared to acid hydrolysis.
- The wet processing procedures are used to obtain beta-glucans of up to 95% purity.

INTRODUCTION

Oats are notable amongst other cereals because they are often consumed as whole grain and therefore provide the human body with nutrients and bioactive compounds. Oats have been grown for thousands of years, mainly as an animal feed crop, but were not widely accepted as part of the human diet until the 19th century (Webster, 1996). Livestock feed is still the primary use of oat crops, accounting for about 74% of the world's total oat usage (Strychar, 2011). For a long time, oats were generally consumed as oatmeal, but today oats are present in breakfast cereals, beverages, bread, and infant foods (Yao *et al.*, 2006; Zhang *et al.*, 2007). Oat world production currently ranks sixth in the grain-production statistics, following corn, wheat, barley, sorghum, and millet in decreasing order. Russia, Canada, the United States, the European Union (EU), and Australia's production account for nearly 77% of the world's supply of whole grains, seeds, and industrial-grade oats. Canada is the largest global commercial producer and exporter of oats, accounting for 15% of total global production and roughly 60% of global exports (Strychar, 2011). *Avena sativa* L. also referred to as covered or hulled oat is the most cultivated oat species in the world (White, 1995). The other cultivar *A. sativa* var. *nuda* referred to as naked or hull-less variety has a loosely attached hull attached to the groat, good grain quality, but is more prone to mechanical damage (Lasztity, 1998).

The increased human consumption of oats has led to many research investigations aimed at identification of its bioactive compounds. Molecules of interest are vitamins, sterols, polyphenols, dietary fibers, proteins, and peptides. Biological activities of most of these compounds such as antioxidants, cholesterol lowering, and anti-inflammatory properties have been reported. The aim of this chapter is to provide information on how different processing methods affect the availability and biological activity of oat compounds.

OAT PROCESSING

Processing techniques are used in the food industry to improve the nutritional, texture, and sensory characteristics that will attract consumers. Oats processing involves: (i) cleaning and heat treatment; (ii) dehulling and cutting; and (iii) flaking or milling. There may be some variations depending on the sample (e.g., covered, naked) or on the desired final oat product. Oat cereal is therefore available in many forms, such as whole groats, steel-cut, rolled, quick oats, flour, and bran.

Processing and Impact on Active Components in Food
http://dx.doi.org/10.1016/B978-0-12-404699-3.00043-3

The cleaning of oats is to remove coarse field trash, other grains, chaff, or dust that may disrupt further processing. Oats lipids contain mostly polyunsaturated fatty acids that may be hydrolyzed by lipases present in the grain, or oxidized by lipoxygenases, therefore causing rancidity. Oats intended for food purposes are then heat treated to inactivate these enzymes (Girardet and Webster, 2011). Heat treatment contributes to the destruction of microorganisms and development of oat aroma, however, it may reduce protein solubility and increase lipid oxidation (Lehtinen *et al.*, 2003).

The cleaned raw oat grains are forced by centrifuge against an impact ring or rubber line that causes the separation of the hull from the groat (or kernel). Air aspiration then separates the dehulled groats from the hulls (Girardet and Webster, 2011). Groats broken during dehulling are sorted on sizing systems. Groats are also steel cut and sorted into small and large pieces using sifters.

Steam (99–104°C) is added to oat groats to increase their moisture content and to soften the kernel so that they can form flakes with minimum breakage. It also completes the inactivation of undesired enzymes and gives oats their characteristic flavours. Flaking of intact oat groats produces rolled oats (Girardet and Webster, 2011). Oat flour can be produced by milling oat flakes or groats using an impact-type hammer mill where exhaust air is drawn through the system to prevent the relatively high fat oat flour sticking to the sides of the mill. Milling is also performed to separate the anatomical parts of the grain into oat bran and flour without bran (Wang *et al.*, 2007). Drum drying of oats is a process where rolled oats are milled and mixed with water and applied to steam-heated rotating rollers. The thin film produced is milled and used, for example, in instant baby-foods (Bryngelsson *et al.*, 2002).

EFFECT OF COMMERCIAL PROCESSING ON OAT ACTIVE COMPOUNDS

Rolled oats, wholemeal, sifted flour, and bran are the major commercially manufactured oats food products. They are commonly used as ingredients in breakfast cereals, infant foods, bread, cookies, and snacks. Commercial manufacturing of oat products also provides low-value fractions such as hulls, always removed prior to food production and polish waste which is obtained when groats are polished in order to prevent condensation of water within the trichomes and thus inhibit fungal growth during storage.

Effect of Commercial Processing on Phenolic Compounds

Oat contains several phenolic molecules like ferulic acid and its derivatives, flavonoids, anthocyanins, and avenanthramides, a group of phenolic molecules present in oats but not in other cereals (Collins, 1989). The structures of the main phenolic compounds are shown in Figure 43.1. Bryngelsson *et al.* (2002) investigated the effect of three commercial processing methods—steaming, autoclaving (Figure 43.2), and drum drying—on levels of phenolic compounds in oats. They showed that concentrations of *p*-coumaric acid and vanillin were unaffected during the steam-polishing step. Meanwhile, there was an increase in ferulic acid and a decrease in caffeic acid. During

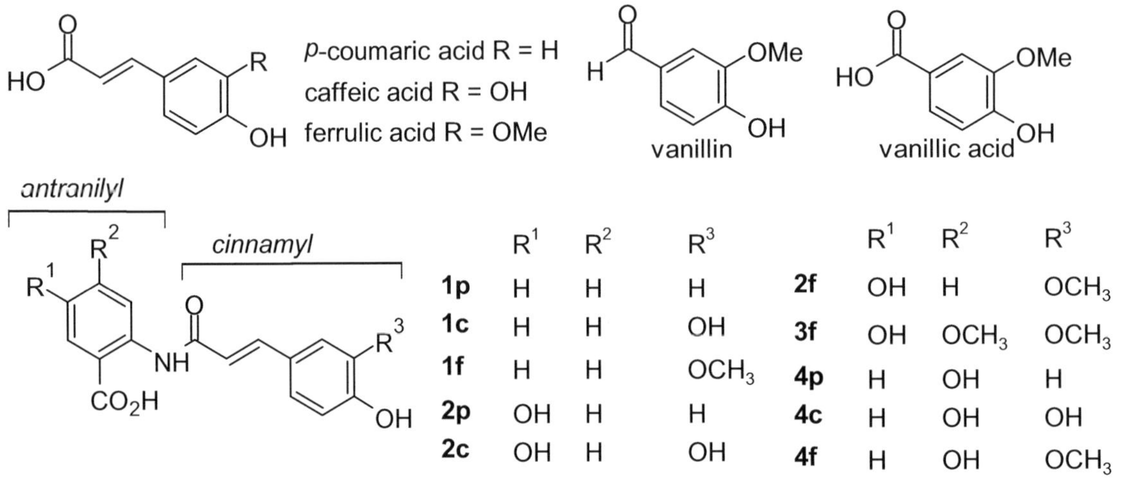

FIGURE 43.1 Chemical structures of common oat phenolic acids and avenanthramides (AVs). AVs are commonly referred as 1p, 1c, 1f, 2p, 2c, 2f, 3f, 4p, 4c, and 4f. Numbers correspond to antranilic acid (1), 5-hydroxyantranillic acid (2), 5-hydroxy-4-methoxyantranillic acid (3), and 4-hydroxyantranillic (4) while letters correspond to cinnamic acid moieties *p*-coumaric acid (p), caffeic acid (c) and ferulic acid (f).

the steam-flaking step, there was a further increase in ferulic acid and vanillin. The avenanthramides 2f and 2c were not significantly affected during preparation of rolled oats, while there was a reduction in level of 2p by 45% in steam-rolled oats compared to raw groats. The steam-polishing of oat provided a waste fraction that was demonstrated to contain a lower level of avenanthramides but a very high level of *p*-coumaric acid and vanillin compared to raw hulls from steamed oats. The effect of autoclaving on levels of *p*-coumaric, ferulic, and caffeic acids, and vanillin in groats was much larger compared to steaming. Caffeic acid decreased to non-detectable levels, whereas the levels of ferulic acid, *p*-coumaric acid, and vanillin increased by 222%, 1137%, and 1044%, respectively. Drying after autoclave caused a significant decrease in vanillin but not in ferulic and *p*-coumaric acids. The avenanthramides 2c and 2p decreased both during autoclaving and drying, whereas 2f was unaffected by autoclaving but decreased during drying. The autoclave was used to produce wholemeal that had higher concentrations of *p*-coumaric acid and vanillin than rolled oats from steaming. The autoclaved hulls had the higher levels *p*-coumaric acid, and vanillin than raw hulls and steam-polish waste while levels of avenanthramides did not differ. Drum drying on milled rolled oats and wholemeal caused a decrease in concentrations of ferulic acid, *p*-coumaric acid, and vanillin. Concentrations of avenanthramides 2f, 2c, and 2p decreased after drum drying of milled rolled oats but not after drum drying of wholemeal oats. However, final levels of avenanthramides in the two drum-dried products were approximately the same.

Effect of Commercial Processing on Starch and Fibers

Four different processing methods—normal pressure steaming (NPS), autoclaved steaming (AS), hot-air roasting (HAR), and infrared roasting (IR)—were used to evaluate changes that may occur to oat starch and fibers (e.g., beta-glucans) in the kernel (Hu *et al.*, 2010). It was found that the NPS treatment decomposed large starch granules into small pieces, which adhered to each other to form irregular shapes. Only minor transformation of starch granules was observed in AS-treated kernels; while in HAR treatment, large starch granules were very rare, and protein networks very loose. The thermo-effect of infrared radiation during IR-processing resulted in large starch ball disintegration and formation of small multi-angular starch pieces. It was concluded that deactivation treatments changed the shape of starch granules, crumbled large starch granules, and reduced the connection between protein network and starch granules. These changes had an effect on gelatinization as the breakdown, final viscosity and setback levels in any of the processed samples were significantly lower than in the control group. It was believed that heat in processing treatments broke hydrogen bonds connecting starch molecules that favored starch gelatinization. No difference in beta-glucan content was observed in control and samples processed using NPS, AS, HAR, and IR procedures, an indication that glucanases were equality inactivated. It has been demonstrated that peeling treatment in excess of 20s may lead to damage of beta-glucan-rich aleurone layer and beta-glucan loss (Hu *et al.*, 2009). Three mechanisms of oat milling were tested

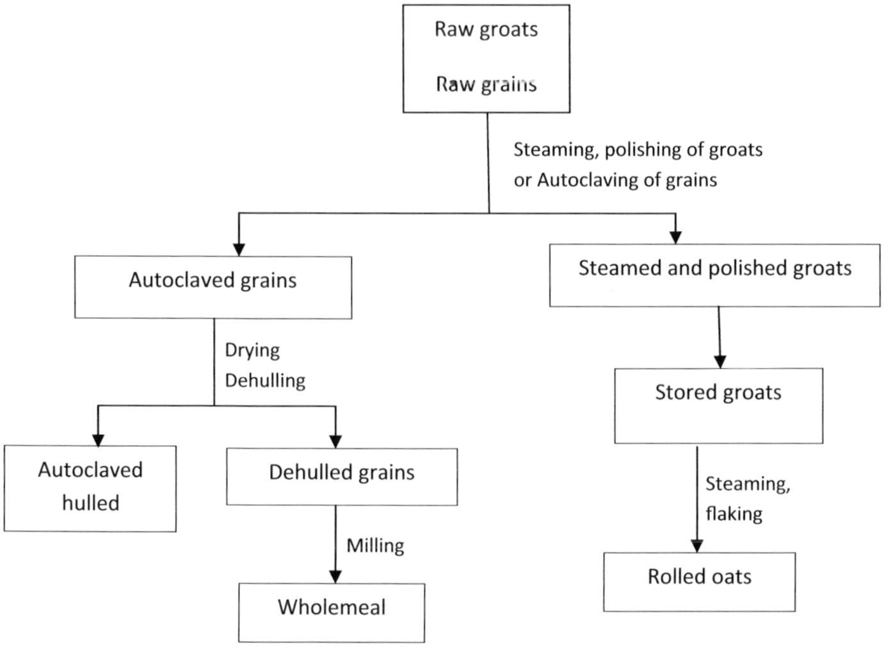

FIGURE 43.2 Processing methods used to compared the levels of phenolic acids and avenanthramides in steamed and flaked, and autoclaved oat samples. *Adapted from Bryngelsson* et al., *2002.*

for laboratory-scale oat bran production (Doehlert and Moore, 1997). Bran obtained with either roller-milling or impact-milling of groats, followed by sieving to retain larger particles was 1.7-fold enriched in beta-glucan compared to bran from pearling mill. Pin ball milling at different speeds, air-classified, and sieving of fractions from defatted oat bran showed that the highest content of beta-glucan was obtained in fractions with >30 μm particle size. As particle size decreased, yield of >30 μm fraction and starch content decreased, however this was associated with an increase in beta-glucan content (Wu and Doehlert, 2002).

Effect of Commercial Processing on Free Fatty Acids

Lipids content in oat ranges from 5–9% which is higher than the amount present in other cereals. Most oat lipids are unsaturated making them susceptible to hydrolysis by lipase to free fatty and glycerol but also to oxidation to hydroperoxides by lipoxygenase and lipoperoxidase (Ekstrand *et al.*, 1992). Steaming followed by drying of oat groats at 88–98°C for 100 min and superheat steaming (110°C, 10 and 14 min) did not cause any change in levels of free fatty acids even after storage for 12 and 14 weeks. This was believed to be due to complete inactivation of lipolytic enzymes (Head *et al.*, 2011). Also, no change in lipids of oat kernel was observed when naked oats were processed using normal pressure steaming (NPS), autoclaved steaming (AS), hot-air roasting (HAR), and infrared roasting (IR) because of complete destruction of lipase and peroxidases specifically in NPS, AS, and IR processing (Hu *et al.*, 2010). The effect of IR treatment on lipase activity however, appeared to vary depending on the cultivar.

EFFECT OF GERMINATION ON OAT ACTIVE COMPOUNDS

Cereal seeds have been germinated for centuries to soften the kernel structure, to increase nutrient content, and to decrease anti-nutritional compounds (Kaukovirta-Norja *et al.*, 2004). Germination is initiated by steeping (i.e., soaking) where dry grains absorb water to a moisture content of 43–45%, and metabolic activity resumes. During the subsequent germination process, enzyme synthesis and kernel modification take place (Skoglund *et al.*, 2008). The use of oat malt is currently still very minor however, chemical changes occurring during the drying of germinated grains enhanced sensory attributes of oat products and increased availability of some important micronutrients (Heinio *et al.*, 2002).

Influence of Germination on Phenolic Compounds

The effect of germination has mainly been studied for avenanthramides. Bryngelsson *et al.* (2003a) reported an increase of avenanthramide concentrations in steeped intact raw groats that was time and temperature dependent with maximum levels obtained after 10 h at 20°C. The increase was due to *de novo* synthesis as there was a concurrent increase in hydroxycinnamoyl-CoA:hydroxyanthranilate *N*-hydroxycinnamoyl transferase, an enzyme involved in avenanthramide biosynthesis and also to the release of the bound form of these oat phenolics. HPLC analysis demonstrated a large increase in concentration of many avenanthramides and other compounds in steeped oats, particularly for 2p, 2f, and 3f (Figure 43.3) (Skoglund *et al.*, 2008). Some variations were observed between cultivars. Contrary to oat

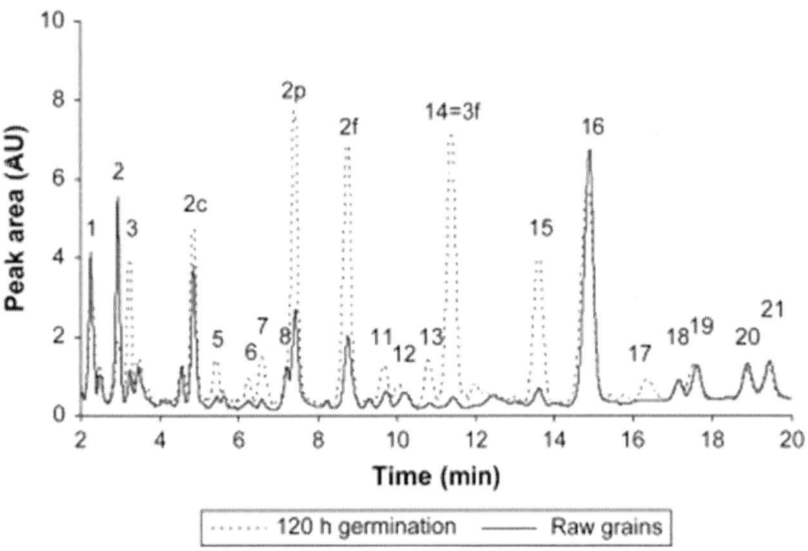

FIGURE 43.3 Changes in avenanthramides and other phenols during germination of the oat cultivar Dane at 16°C. Raw grains (straight line) and sample after 120 h of germination (dotted line). Peaks for the avenanthramides 2c, 2p, 2f and 3f are identified. Peaks 1–21 were not identified. *Reprinted with permission from Skoglund* et al., *2008.*

seeds, steeping of oat flour resulted in a decrease of avenanthramide levels and this was suggested to be due to the action of the polyphenoloxidase that oxidizes phenols to polymeric derivatives. The decrease was more pronounced for avenanthramides containing caffeic and *p*-coumaric acids than for those comprising synapic and ferulic acids (Bryngelsson *et al.*, 2003b).

Apart from avenanthramides, effects of germination on total phenolic content (TPC) of oats have been investigated. Tian *et al.* (2010) reported that TPC of oats increased from 0.20 mg GAE/g in raw seed to 0.91 mg GAE/g in germinated seeds. This change of more than 4-fold was attributed to better extractability of phenolic compounds from the kernel structures after germination and to the increase in activity of alpha-amylase that may have released bound phenolic compounds. Similar increases of three- to four-fold in total phenolic content was reported in other studies (Kaukovirta-Norja *et al.*, 2004).

Effect of Germination on Starch and Polysaccharides

Starch is the main chemical compound in oat seeds, accounting for up to 60% of dry weight. Oat starch content decreased significantly from 60% to nearly 20% after 48 h of germination. The decrease was attributed to greater total amylase activity, up to 6-fold in germinated oats relative to raw oat seeds (Tian *et al.*, 2010). Other studies have reported increased carbohydrase enzyme activities during germination but have concluded that starch degradation was very limited. The main cell component of oats is beta-glucans. The concentration in raw seeds ranges from 2–7% on a dry weight basis. Germination generally causes the breakdown of beta-glucans, which is undesirable with respect to their functional properties. Wilhelmson *et al.* (2001) demonstrated that germination for 72 h at 15°C followed by oven drying produced germinated oats with reduced beta-glucan loss. Under these conditions, oat retained 55–60% of the beta-glucans compared with the native oat. In another study, it was observed that beta-glucan concentrations quickly decreased from 4.2% in raw oats to 2.6% after 72 h germination. When oat samples were germinated for 96 h only 0.2% of beta-glucan was detected (Huebner *et al.*, 2010). The decrease in beta-glucan concentrations was attributed to an increase in activity of beta-glucanase. Germination temperatures have an effect on glucanase activity (Wilhelmson *et al.*, 2001).

Effect of Germination on Phytate

Phytate or phytic acid is the principal storage form of phosphorus in many plant tissues, especially bran and seeds. It can form complexes with metals or proteins and therefore reduce their bioavailability in the gastrointestinal tract. Phytate is thus considered an anti-nutrient compound in oats. Germination has the potential to increase phytase activity and subsequent degradation of phytates. Phytate content of oat seeds decreased from 0.35% in native seeds to 0.11% during germination and this was believed to be due to an increase in phytase activity as germination progressed (Tian *et al.*, 2010; Larsson and Sandberg, 1992). Environmental factors and cultivars did not affect the level of phytic acid in oat flours (Miller *et al.*, 1980). Larsson and Sandberg (1992) found that germination conditions for maximum reduction of phytates in oat seeds were 17 h, pH 4–4·5, temperature 37–40°C as they successfully reduced the phytate content in seeds by almost 79%. In humans, the use of germinated (or malted) oat in the preparation of breakfast porridge significantly increased iron and zinc uptake confirming the anti-nutritional properties attributed of phytates.

Effect of Germination on Amino Acids and Proteins

Oats proteins (15–20%) have better nutritional value compared to other cereals because of their high lysine content and lower level of prolamin proteins (Kaukovirta-Norja *et al.*, 2004). Germination has been demonstrated to influence both amino acids and proteins encountered in oats. In general, there is a slight increase in proteins during germination because of degradation of other kernel compounds for respiration. The protein concentration of oat seeds slowly increased from 19.0% to 22.0% during germination (Tian *et al.*, 2010). It was reported that the content of albumin, rich in essential amino acids lysine and tryptophan increased while there was a decrease in content of globulin and prolamin, two proteins with little lysine. Germinated oats therefore have a better amino acid balance relative to native seeds. Lower amounts of globulin in germinated seeds are believed to be due to the action of endoproteinases which hydrolyze globulin into intermediate peptides that do not accumulate in seeds during the germination process. The degradation of prolamin and avenin-type proteins also occurred during germination as a result of the action of cysteine proteinases. Klose *et al.* (2009) used a LabChip to monitor the proteolytic degradation of prolamin, globulin, and glutenin proteins in unmalted, germinating, and malted oat grains and found that all three proteins were degraded by almost 47%.

Free amino acid contents of oats increased continuously to 0.37% during germination, a value that was almost 10-fold higher compared with raw oats (Tian *et al.*, 2010). The threonine content of germinated oat seeds increased 30%, lysine 35% after 144 h, while valine content increased 22% during the same time period. Various degrees of increase were also observed for other amino acids.

Effect of Germination on Lipid

The degradation of lipids in 39 hulled and 10 hull-less malted oat varieties was investigated. Before germination, higher total lipids content ranging from 5.7–10.2% was detected in hull-less samples compared to 4.4–9.6% in hulled samples (Kaukovirta-Norja *et al.*, 2004). This translated into a higher proportion of individual free fatty acids in hull-less samples relative to hulled samples. During germination, total lipids decreased in hull-less by 15% while there was only a 5% decrease in hulled samples (Peterson, 1998). Meanwhile, there was an increase in free fatty acids during germination as a resulted of increase enzymatic degradation of lipids and this was more pronounced in hulled samples.

WET PROCESSING

Wet processing technologies often use meal, flour, or bran as raw-material, and the separation and concentration of active compounds are performed after extraction with solvents or mixtures of solvents (Kahkonen *et al.*, 1999). Cereals including oats contain a wide range of phenolic acids that mainly exist in bound forms and are typical components of complex structures such as lignins, hydrolysable tannins, and organic acids. Phenolic acids found in the cell wall are linked to hemicelluloses such as 2-O-5′-O-(E)-feruluoyl-β-D-xylopyranosyl)-(1→4)-D-xylopyranose. Ferulic acid, one of the major phenolic compounds in oat, is attached to the cell wall arabinoxylan and can also form dimers through oxidative cross-linking that may serve as an attachment to cell-wall polymers. Different procedures have been used to obtain these molecules (fibers, lipids, phenols, proteins) from oats.

Phenols

Extraction of soluble phenols from oat is commonly performed using a combination of methanol/water, methanol/acetone/water, or ethanol/water. A comparison of eight different solvent systems including petroleum ether, diethyl ether, chloroform-diethyl ether (2:1), chloroform-dichloroethane-ether (1:1:2), methanol, chloroform- methanol-dichloroethane (1:1:2) for extraction of antioxidant molecules from groats and hulls of several oat lines found that methanol was the most efficient solvent (Duve and White, 1991). Methanol 80% extracted substantially more phenolic compounds (17.6 μg catechin/mg extract) and exhibited higher antioxidant capacity than water extracts (8.2 μg catechin/mg) of whole oats (Zielinski and Kozlowska, 2000). Methanol extraction of milled oat groats had total phenolics 1.5-fold higher than isopropanol (Auerbach and Gray, 1999). To obtain free phenolic acids, alcohol or aqueous alcohol extracts are dried, dissolved in water at pH 2 followed by partition in ethyl acetate or ether. In this process, free phenolic acids are extracted with the organic solvent. A comparative analysis of four different studies showed that there were some variations in concentrations of free phenolic acids like caffeic acid, ferulic acid, *p*-coumaric acid, vanillic acid, and vanillin due to solvent, duration of the extraction, or samples (whole groats, debranned four) (Peterson, 2001). A comparison of 25% water in methanol, ethanol, and acetone for extraction of free phenolic compounds in cultivar and commercial oat samples found that acetone yielded the highest amount of *p*-hydroxbenzaldehyde, caffeic acid, and vanillin, while methanol extract had the highest amount of other phenols (e.g., ferulic, *p*-coumaric acids, avenanthramides 2c, 2p, and 2f) (Verardo *et al.*, 2011). It has been demonstrated that up to 80% of phenols in oat are bound to other molecules. To release them, free or soluble phenols are first extracted and then the residue is treated with a base (usually 4 or 2 M NaOH) or with hydrochloric acid (usually 2–3 mol/l). The release phenols are recovered at pH2 by partition with ethyl acetate. In general base hydrolysis is more efficient than acid hydrolysis. A comparison of base and acid hydrolysis procedures on the release of 15 phenols found that the concentrations of 13 of them were higher under alkaline hydrolysis conditions (Verardo *et al.*, 2011).

Beta-Glucans

Wet separation techniques for beta-glucan concentration have been divided into four: (i) aqueous or aqueous alkali extraction; (ii) aqueous alcohol enzymatic treatment; (iii) aqueous enzymatic treatment; and (iv) aqueous thermo-mechanical treatment. The main advantage of wet separation techniques is that the beta-glucan content of the fiber concentrates obtained through these processes can reach 95% (w/w). These processes have been described by Vasanthan and Temelli (2008). Fiber concentrates produced through aqueous technologies are high purity beta-glucans, however their viscosity in solution is low. This has been mainly attributed to the degradation of beta-glucan molecules by endogenous enzymes such as cellulase and beta-glucan ase that are naturally present in the flour. The aqueous alcohol based enzymatic process can be used to obtain fiber concentrates containing up to 65% beta-glucan with high molecular weight and greater viscosity compared to those obtained through other aqueous technologies. The aqueous enzymatic and aqueous thermo-mechanical approaches are used to produce fiber concentrates with up to 25% beta-glucan content. Fibre concentrated from the aqueous enzymatic process is used as a fat replacer in various food products. Analysis of beta-glucan content of 93 Swedish and 41 American oat samples using hot water containing $CaCl_2$ and thermostable alpha-amylase showed a large variation in both the content of extractable beta-glucan (0.76–3.68%) and the average molecular weight [$(1.25–1.78) \times 10^6$ g/mol] (Ajithkumar *et al.*, 2005). The authors showed that the extractable beta-glucan content was a heritable trait, while the molecular weight depended more on environmental factors.

ANALYTICAL METHODS

Several analytical methods are used for separation or quantification of phenols, fibers, proteins, peptides, and amino acids present in oats.

The most commonly used technique for quantification of total phenol compounds in oats and other foods is a colorimetric procedure using the Folin-Ciocalteu reagent. The reaction involves the reduction of phosphomolybdic-phosphotungstic acid by phenols under alkaline conditions to a blue colored complex easily detected at 760 nm (Robbins, 2003). However, quantification is problematic, in that other components in the food extracts behaved as reducing agents. For characterization of individual phenolic molecules, HPLC with reverse phase column technology is the preferred analytical method for phenolic acids (Robbins, 2003). There is a great diversity of columns available, however for the determination of phenolic acids, packing materials are almost exclusively composed of C18. Detection of oat phenols by HPLC has mostly been performed in the UV-Vis range with photodiode array detection. The most commonly used wavelength is 254 nm, however some phenols are also detected at 270, 280, and 360 nm. Peak assignments are made with both the retention times as well as by comparing the UV-Vis spectra of analytes to available standards. Other methods of detection include electrochemical, fluorescence, and mass spectrometry (Robbins, 2003).

Three analytical methods that are used for determination of dietary fibers have been summarized by Elleuch *et al.* (2011). Non-enzymatic gravimetric procedure was the earliest and can measure crude, acid detergent, and neutral detergent fibers, but not water-soluble fiber, and therefore underestimate total dietary fibers content. The second method, enzymatic-gravimetric, uses enzymes such as amylase to remove starch and proteins followed by precipitation of fibers in aqueous ethanol. The dietary fiber is then weighted and corrected for proteins and ash. The third method, enzymatic-chemical, measures fiber as non-polysaccharides. Starch and proteins are removed enzymatically, then precipitation with aqueous ethanol or dialysis separates the soluble dietary fiber from low-molecular sugars and starch hydrolysis products. Dialysis is preferred to avoid fiber loss (Elleuch *et al.*, 2011). Soluble beta-glucans isolated from bran and whole-grain oats were hydrolyzed with lichenase, an enzyme specific for (1→3),(1→4)-β-D-β-glucans and characterized by chromatography (GC, HPLC), capillary electrophoresis, as well as by nuclear magnetic resonance spectroscopy (Johansson *et al.*, 2000).

Total nitrogen contents of oat samples have been determined using combustion methods. To evaluate protein changes during the malting process, a novel analysis technique, the Lab-on-a-Chip capillary electrophoresis, was used together with two-dimensional gel electrophoresis of extracted protein fractions. The free amino acid profile of the milled germinated oat samples was measured using high performance cation exchange column (Klose *et al.*, 2009). When there is a need to determine the sequences of proteins or peptides, tandem mass spectrometry or MS/MS is the technique of choice. High resolution MS/MS has sensitivity that does not require an external standard for amino acid or amino acid variant to find a protein or peptide sequence. Difficulties however exist with isobaric or near isobaric masses (e.g., telling lysine from glutamine) on low resolution or low mass accuracy mass spectrometers. Peptide or protein sequence is derived from spacing (in Dalton) between adjacent product ions of the same series. Interpreted spectra are searched against theoretical product ion spectra via a number of database algorithms like MASCOT search engine and matched to the published sequences of proteins from the National Center for Biotechnology Information non-redundant database (NCBInr) (Jodayree *et al.*, 2012).

References

Ajithkumar, A., Andersson, R., Aman, P., 2005. Content and molecular weight of extractable beta-glucan in American and Swedish oat samples. J. Agric. Food Chem. 53, 1205–1209.

Auerbach, R.H., Gray, D.A., 1999. Oat antioxidant extraction and measurement—towards a commercial process. J. Sci. Food Agric. 79, 385–389.

Bryngelsson, S., Ishihara, A., Dimberg, L.H., 2003a. Levels of avenanthramides and activity of hydroxycinnamoyl- CoA: Hydroxyanthranilate N-hydroxycinnamoyl transferase (HHT) in steeped or germinated oat samples. Cereal Chem. 80, 356–360.

Bryngelsson, S., Dimberg, L.H., Kamal-Eldin, A., 2002. Effects of commercial processing on levels of antioxidants in oats (Avena sativa L.). J. Agric. Food Chem. 50, 1890–1896.

Bryngelsson, S., Sunnerheim, K., Holm, C., Dimberg, L.H., 2003b. Tentative avenanthramide-modifying enzyme in oats. Cereal Chem. 80, 361–364.

Collins, F., 1989. Oat phenolics: avenanthramides, novel Substituted N-cinnamoylanthranilate alkaloids from oat groats and hulls. J. Agric. Food Chem. 37, 60–66.

Doehlert, D.C., Moore, W., 1997. Composition of oat bran and flour prepared by three different mechanisms of dry milling. Cereal Chem. 74, 403–406.

Duve, K.J., White, P.J., 1991. Extraction and identification of antioxidants in oats. J. Am. Oil Chem. Soc. 68, 365–370.

Ekstrand, B., Gangby, I., Akesson, G., 1992. Lipase activity in oats — distribution, pH-dependence, and heat inactivation. Cereal Chem. 69, 379–381.

Elleuch, M., Bedigian, D., Roiseux, O., Besbes, S., Blecker, C., Attia, H., 2011. Dietary fibre and fibre-rich by-products of food processing: Characterisation, technological functionality and commercial applications: A review. Food Chem. 124, 411–421.

Girardet, N., Webster, F.H., 2011. Oat milling: specifications, storage, and processing. In: Webster, F.H., Wood, P.J. (Eds.), Oats: chemistry and technology. AACC International, Inc., pp. 301–319.

Head, D., Cenkowski, S., Arntfield, S., Henderson, K., 2011. Storage stability of oat groats processed commercially and with superheated steam. Lwt-Food Sci. Technol. 44, 261–268.

Heinio, R.L., Lehtinen, P., Oksman-Caldentey, K.M., Poutanen, K., 2002. Differences between sensory profiles and development of rancidity during long-term storage of native and processed oat. Cereal Chem. 79, 367–375.

Hu, X., Xing, X., Ren, C., 2010. The effects of steaming and roasting treatments on beta-glucan, lipid and starch in the kernels of naked oat (Avena nuda). J. Sci. Food Agric. 90, 690–695.

Hu, X., Wei, Y., Ren, C., Zhao, J., 2009. Relationship between kernel size and shape and lipase activity of naked oat before and after pearling treatment. J. Sci. Food Agric. 89, 1424–1427.

Huebner, F., O'Neil, T., Cashman, K.D., Arendt, E.K., 2010. The influence of germination conditions on beta-glucan, dietary fibre and phytate during the germination of oats and barley. Euro. Food Res. Technol. 231, 27–35.

Jodayree, S., Smith, J.C., Tsopmo, A., 2012. Use of carbohydrase to enhance protein extraction efficiency and antioxidative properties of oat bran protein hydrolysates. Food Res. Int. 46, 69–75.

Johansson, L., Virkki, L., Maunu, S., Lehto, M., Ekholm, P., Varo, P., 2000. Structural characterization of water soluble β-glucan of oat bran. Carbohydr. Polym. 42, 143–148.

Kahkonen, M.P., Hopia, A.I., Vuorela, H.J., Rauha, J.P., Pihlaja, K., Kujala, T.S., Heinonen, M., 1999. Antioxidant activity of plant extracts containing phenolic compounds. J. Agric. Food Chem. 47, 3954–3962.

Kaukovirta-Norja, A., Wilhelmson, A., Poutanen, K., 2004. Germination: A means to improve the functionality of oat. Agric. Food Sci. 13, 100–112.

Klose, C., Schehl, B.D., Arendt, E.K., 2009. Fundamental study on protein changes taking place during malting of oats. J. Cereal Sci. 49, 83–91.

Larsson, M., Sandberg, A.S., 1992. Phytate reduction in oats during malting. J. Food Sci. 57, 994–997.

Lasztity, R., 1998. Oat grain—A wonderful reservoir of natural nutrients and biologically active substances. Food Rev. Int. 14, 99–119.

Lehtinen, P., Kiiliainen, K., Lehtomaki, K., Laakso, S., 2003. Effect of heat treatment on lipid stability in processed oats. J. Cereal Sci. 37, 215–221.

Miller, G.A., Youngs, V.L., Oplinger, E.S., 1980. Environmental and cultivar effects on oat phytic acid concentration. Cereal Chem. 57, 189–191.

Peterson, D., 2001. Oat antioxidants. J. Cereal Sci. 33, 115–129.

Peterson, D., 1998. Malting oats: Effects on chemical composition of hull-less and hulled genotypes. Cereal Chem. 75, 230–234.

Robbins, R.J., 2003. Phenolic acids in foods: An overview of analytical methodology. J. Agric. Food Chem. 51, 2866–2887.

Skoglund, M., Peterson, D.M., Andersson, R., Nilsson, J., Dimberg, L.H., 2008. Avenanthramide content and related enzyme activities in oats as affected by steeping and germination. J. Cereal Sci. 48, 294–303.

Strychar, R., 2011. World oat production, trade, and usage. In: Webster, F.H., Wood, P.J. (Eds.), Oats: chemistry and technology. AACC International., Inc., pp. 1–10.

Tian, B., Xie, B., Shi, J., Wu, J., Cai, Y., Xu, T., Xue, S., Deng, Q., 2010. Physicochemical changes of oat seeds during germination. Food Chem. 119, 1195–1200.

Vasanthan, T., Temelli, F., 2008. Grain fractionation technologies for cereal beta-glucan concentration. Food Res. Int. 41, 876–881.

Verardo, V., Serea, C., Segal, R., Caboni, M.F., 2011. Free and bound minor polar compounds in oats: Different extraction methods and analytical determinations. J. Cereal Sci. 54, 211–217.

Wang, R., Koutinas, A.A., Campbell, G.M., 2007. Dry processing of oats — Application of dry milling. J. Food Eng. 82, 559–567.

Webster, F.H., 1996. Oats. In: Henry, R.J., Kettlewell, P.S. (Eds.), Cereal grain chemistry. Chapman and Hall, London, UK, pp. 179–203.

White, E.M., 1995. Structure and development of oats. In: Welch, R.W. (Ed.), The oat crop: Production and utilization. Chapman and Hall, London, UK, pp. 88–119.

Wilhelmson, A., Oksman-Caldentey, K., Laitila, A., Suortti, T., Kaukovirta-Norja, A., Poutanen, K., 2001. Development of a germination process for producing high beta-glucan, whole grain food ingredients from oat. Cereal Chem. 78, 715–720.

Wu, Y.V., Doehlert, D.C., 2002. Enrichment of beta-glucan in oat bran by fine grinding and air classification. LWT-Food Sci.Technol. 35, 30–33.

Yao, N., Jannink, J., Alavi, S., White, P.J., 2006. Physical and sensory characteristics of extruded products made from two oat lines with different beta-glucan concentrations. Cereal Chem. 83, 692–699.

Zhang, H., Onning, G., Oste, R., Gramatkovski, E., Hulthen, L., 2007. Improved iron bioavailability in an oat-based beverage: the combined effect of citric acid addition, dephytinization and iron supplementation. Eur. J. Nutr. 46, 95–102.

Zielinski, H., Kozlowska, H., 2000. Antioxidant activity and total phenolics in selected cereal grains and their different morphological fractions. J. Agric. Food Chem. 48, 2008–2016.

CHAPTER

44

Soybean Mineral Composition and Glyphosate Use

Krishna N. Reddy, Stephen O. Duke†*

*USDA-ARS, Crop Production Systems Research Unit, Jackson, TN, USA, †USDA-ARS, Natural Products Utilization Research Unit, MS, USA

CHAPTER POINTS

- Considering the global importance of GR crops and glyphosate, the small amount of published data on effects of glyphosate on mineral nutrition of GR soybeans is surprising.
- Any herbicide could influence almost all physiological parameters, including mineral nutrition, at a high enough dose in a susceptible species.
- Although glyphosate will chelate mineral cations, it is a weak chelator compared to endogenous, natural chelators found in plants (e.g., review by Duke *et al.*, 2012b).
- The critical question is whether glyphosate significantly affects mineral content of the seed of GR soybean when used at recommended rates in the field. Unfortunately, there are only two studies that address this question directly (Table 44.2), and these studies found no effects on N and several minerals.
- Zobiole *et al.* (2010a–d, 2011, 2012) consistently find that glyphosate is harmful to GR soybeans in many other ways than reducing mineral content, e.g., strong inhibition of photosynthesis and growth. This is inconsistent with the findings of many researchers that glyphosate application to GR soybeans at recommended rates does not influence growth or yield (e.g., Duke *et al.*, 2012a; Bellaloui *et al.*, 2008; Henry *et al.*, 2011; Reddy *et al.*, 2000; Reddy and Whiting, 2000) and that GR soybean is approximately 50-fold less sensitive to glyphosate than non-GR soybeans (Nandula *et al.*, 2007).
- Furthermore, the averaged linear increases in USA soybean yields since GR soybeans were introduced in 1996, argues against harmful effects to GR soybeans by glyphosate.
- Nevertheless, rigorous field studies of the effects of glyphosate on the mineral content of GR soybeans in different soil types and climates are needed to establish with certainty whether glyphosate can significantly alter the mineral nutritional value of GR soybeans under any conditions.

INTRODUCTION

Glyphosate [*N*-(phosphonomethyl) glycine] has become the most widely used herbicide in the world, in large part because of it use in transgenic, glyphosate-resistant (GR) crops (Duke, 2012; Duke and Powles, 2008, 2009). This herbicide is considered by some as the most perfect herbicide ever devised (Duke and Powles, 2008). It is non-selective; that is, it is effective on almost all plant species. It is inactivated by soil, so it can only be used as a postemergence, foliar spray. Once taken up by the plant, it is readily translocated via the phloem, and to a lesser extent the xylem, to metabolically active tissues, such as meristems and any other growing parts of the plant (reviewed by Duke, 1988). Compared to herbicides introduced before glyphosate, it is slower acting, giving time for the herbicide to translocate after being taken up. This allows all growing meristems to be killed, so that regrowth of the weed does not occur.

Processing and Impact on Active Components in Food
http://dx.doi.org/10.1016/B978-0-12-404699-3.00044-5

Glyphosate is a non-halogenated glycine analogue that kills by inhibiting the enzyme 5-enolpyruvylshikimate-3-phosphate synthase (EPSPS) of the shikimate pathway (reviewed by Duke *et al.*, 2003). The shikimate pathway produces essential aromatic amino acids (phenylalanine, tyrosine, and tryptophan). In addition to causing shortages of these amino acids, glyphosate deregulates the shikimate pathway, drawing in critical carbon backbones from carbon fixation to go into shikimic acid and its derivatives, resulting in inhibited photosynthesis and other metabolic disruption. Furthermore, many of the plant defenses to pathogens are dependent on products of the shikimate pathway, so glyphosate-treated plants are highly susceptible to plant diseases to which they are exposed (reviewed by Duke *et al.*, 2007). Thus, glyphosate predisposes weeds to "biocontrol" by plant pathogens. Glyphosate is less effective in sterile soil and much more effective in the presence of plant pathogens (Schafer *et al.*, 2012).

Animals do not have EPSPS, and glyphosate is considered a very safe herbicide in terms of mammalian toxicity (Williams *et al.*, 2000). In fact, its LD_{50} for rats is higher than that of aspirin or table salt, and the vast preponderance of long-term studies show no chronic effects at doses higher than one might expect to obtain from the highest allowable residues in food. The ecotoxicological risk of glyphosate is also considered to be quite low (Geisy *et al.*, 2000).

Several recent papers have claimed that GR soybeans, including both the plant vegetation and the harvested seed, are deficient in certain minerals, particularly Mn (reviewed by Duke *et al.*, 2012a,b). Some of these minerals are important for proper mineral nutrition of the animals (livestock and humans) that consume them. This chapter discusses these claims in the context of the entire body of relevant published research.

SOYBEAN NUTRITIONAL VALUE (IN GENERAL)

Soybean is one of the most widely grown oilseed crops in the world and in particular the USA. In 2010, about 265 million tons of soybean were produced across the world, of which about 91 million tons—nearly a third—were produced in the USA (USDA, 2012a). Soybean is primarily used for the oil and meal products as well as whole bean products. A limited amount of soybean is grown for use of the entire above ground plant as forage for cattle and other ruminants. Soybeans form a very important part of human and animal diet in many parts of the world. Soybean seed consist of about 40% protein, 22% oil, 33% carbohydrates, and 5% ash of the dry weight of soybean seed (USDA, 2011). In addition soybean also contains minerals, vitamins, and isoflavones. Among the major mineral components in soybean, K (17970 ppm) is found in the highest concentration followed by P (7040 ppm), Mg (2800 ppm), Ca (2770 ppm), and S (1690 ppm). The minor minerals present in soybeans are Fe (157 ppm), Zn (49 ppm), Mo (30 ppm), B (26 ppm), Mn (25 ppm), Na (20 ppm), Cu (16 ppm), and Se (0.2 ppm) (Samarah *et al.*, 2004; USDA, 2011).

Soybean seed composition can be affected by genetic diversity, environmental factors, and agronomic production practices as well as interaction between these factors. Pesticides are routinely used for weed, insect, and plant pathogen control in soybean production. Since the introduction of GR soybean, glyphosate has become the most widely used herbicide for weed control in soybean. There has been speculation and some published data to suggest that glyphosate affects accumulation of minerals in GR soybean by affecting plant growth and nutrient uptake and mobility in the plant (summarized by Duke *et al.*, 2012b). The effect of glyphosate on soybean mineral composition is summarized below.

COMMERCIALIZATION OF GLYPHOSATE-RESISTANT SOYBEAN

Since its commercialization in 1974, glyphosate has been used extensively throughout the world as a non-selective, systemic, broad-spectrum, postemergence herbicide. Because of its lack of selectivity, its use in soybean crops was initially limited to preplant, post-directed, and post-harvest applications for weed control. There was limited use of glyphosate as a harvest aid to kill and desiccate the soybean crop before harvest. GR soybean was created by stable integration of a transgene (*cp4 epsps*) from an *Agrobacterium* species that encodes glyphosate-insensitive EPSPS. Expression of insensitive EPSPS enzyme in GR soybean allows maintenance of normal aromatic amino acid levels in GR soybean treated with glyphosate.

In the USA, GR soybean cultivars were first commercialized for planting in 1996. GR soybean, commercially known as 'Roundup Ready®' soybean, remains unaffected when treated with recommended rates of glyphosate for weed management. Transgenic soybean varieties made resistant to the herbicide glufosinate [(±)-2-amino-4-(hydroxymethylphosphinyl) butanoic acid] are also available in the USA, but their adoption rate has been only a small fraction of that of GR soybeans. The most sensitive indicator of an effect of glyphosate on the shikimate pathway is elevated shikimate levels (Singh and Shaner, 1998), and glyphosate-treated soybeans do not have this symptom (e.g., Reddy *et al.*, 2004). GR soybeans are approximately 50-fold less sensitive to glyphosate than non-GR soybeans (Nandula *et al.*, 2007).

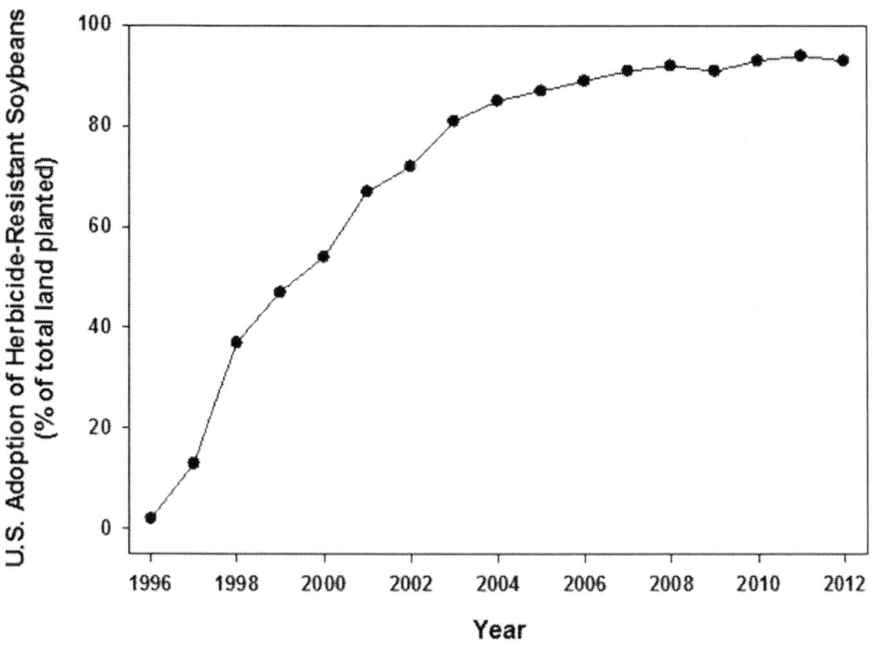

FIGURE 44.1 Percent of herbicide-resistant soybeans planted since GR soybeans were introduced in 1996. *Data are from the U.S. Economic Research Service of the United States Dept. of Agriculture (http://www.ers.usda.gov/data-products/adoption-of-genetically-engineered-crops-in-the-us.aspx).*

GR soybean has been widely adopted by farmers in North America and in several other countries, particularly Brazil and Argentina. The soybean area planted with GR cultivars has increased from 2% in 1996 to 93% in 2012 (USDA, 2012a) (Figure 44.1). Globally, 75.4 million ha or 75% of soybean area was planted to GR soybean cultivars in 2011 (James, 2011). GR soybeans cover more area than any other transgenic crop.

Where available, this technology was adopted more rapidly than any other agricultural technology in history. There were multiple reasons for this, including economic gains, superior weed control, and simplicity and flexibility of the glyphosate/GR soybean technology (Gianessi, 2008; Duke and Powles, 2009; Hurley *et al.*, 2009; Fernandez-Cornejo *et al.*, 2012). Furthermore, in addition to other environmental benefits (Brookes and Barfoot, 2010) adoption of this postemergence herbicide technology allowed farmers to reduce or eliminate tillage, leading to the environmental benefits of reduced soil erosion by wind and water (Cerdiera and Duke, 2006; Givens *et al.*, 2009) and reduced expenditure of fossil fuels (Brookes and Barfoot, 2012). Soybean yields in the USA have continued to increase since GR soybeans were introduced in 1996 (Duke *et al.*, 2012b).

GLYPHOSATE USE ON GR SOYBEAN

The rapid adoption of GR soybean by farmers has shifted herbicide use patterns in favor of glyphosate. Glyphosate is used either alone or with other herbicides for managing weeds in GR soybean. As glyphosate use in soybeans increased, the use of other herbicides decreased concomitantly in the USA (Reddy, 2001). In the USA, the total active ingredient of glyphosate use has increased in soybean from 2.9 million kg/year in 1995 (the year before GR soybean was commercialized) to 41.7 million kg/year in 2006 (USDA, 2012b). This represents a 14-fold increased use of glyphosate in GR soybean since commercialization of GR technology. Also, the average amount of glyphosate use has increased from 0.68 kg ai/ha/year in 1995 to 1.49 kg ai/ha/year in 2006, and the average number of glyphosate applications has increased from 1 in 1995 to 1.7 in 2006 in soybean production (USDA, 2012b). Increased use (both frequency and amount) of glyphosate in GR soybean has been of concern to some scientists because glyphosate is a metal ion chelator (reviewed by Duke, 1988; Duke *et al.*, 2012b), and, being a systemic herbicide, it could potentially affect mineral uptake and mobility in plants. Several researchers have examined the effects of glyphosate on mineral composition in soybean and their results are summarized below.

GLYPHOSATE DRIFT

Glyphosate is usually applied multiple times in a year, both as a preplant treatment to kill existing vegetation before planting spring-seeded crops and as one or more postemergence treatments in GR crops using ground or aerial equipment. As with all pesticide applications, a small fraction of glyphosate can drift downwind and can be deposited on non-target plants. Glyphosate drift is of particular concern because it is a non-selective herbicide and can cause toxic effects on sensitive plant species at low doses. In 2012, 93% of soybean hectareage was planted to GR soybean in the USA. Seventeen years after introduction of GR soybean, about 7% of soybean area is still planted to non-GR cultivars. Glyphosate drift complaints are common in the Mississippi delta; 55 cases of glyphosate drift onto non-target crops were reported in Mississippi in 2011 (Vollor, L. T. Mississippi Department of Agriculture and Commerce, personal communication, 2012). In a simulated drift study, 12.5% of the common use rate of 0.84 kg ae/ha glyphosate caused chlorosis and decreased nitrogen fixation and assimilation, but had no affect on yield of non-GR soybean (Bellaloui *et al.*, 2006). This study did not examine effects on mineral nutrition of the crop.

There has always been a problem with drift of herbicides from sites of intended use to sensitive crops. The mode of action of no herbicide is directly associated with plant mineral nutrition (Fedtke and Duke, 2005), but any herbicide to which a crop is sensitive will secondarily influence almost all physiological functions of the crop, including mineral nutrition, at a high enough dose. The next section examines the potential effect of glyphosate at drift concentrations on non-GR soybean.

MINERAL COMPOSITION IN NON-GR SOYBEAN

We should emphasize that glyphosate is not applied directly to non-GR soybean for weed management. Thus, non-GR soybean exposure to glyphosate is limited only to off-target drift from ground and aerial applications. Several researchers have examined the direct application of glyphosate at potential drift rates on soybean and observed physiological and metabolic disturbances, however, soybean recovered over time without reduction in yields (Bellaloui *et al.*, 2006; Cakmak *et al.*, 2009; Reddy *et al.*, 2000). Effects of foliar-applied glyphosate on mineral content of non-GR soybean are summarized in Table 44.1. Glyphosate applied at a drift rate of 105 g ae/ha had no effect on leaf N content in non-GR soybean under field conditions (Bellaloui *et al.*, 2006). In a greenhouse study, glyphosate applied at 189 μM had no effect on P, K, Ca, Zn, and Cu content, but decreased Mg and Mn in non-GR soybean (Cakmak *et al.*, 2009). Glyphosate effect on leaf Fe was mixed, with no effect in a greenhouse study (Cakmak *et al.*, 2009), but with decreases in a field study (Bellaloui *et al.*, 2009).

Glyphosate increased concentrations of K, Zn, and Cu, reduced concentrations of Ca, Mg, Fe, and Mn, and had no effect on P in non-GR soybean seed (Bellaloui *et al.*, 2009; Cakmak *et al.*, 2009) (Table 44.1). Glyphosate effects on non-GR soybean seed N were mixed, with no effect in a field study (Bellaloui *et al.*, 2006) and increases in a greenhouse study (Cakmak *et al.*, 2009). The decreased concentrations of Fe, Mg, and Mn in glyphosate treated non-GR soybean leaves and seeds were attributed to reduced root uptake and reduced translocation, possibly due to formation of weak glyphosate-metal complexes, although this mechanism is not supported by what we know of the relatively weak chelating properties of glyphosate, compared to endogenous, natural phytochemical chelators found in plants (summarized by Duke *et al.*, 2012b). Foliarly-applied glyphosate translocates rapidly to roots, where it inhibits growth and other processes. Since young, growing roots are responsible for mineral uptake from soil (e.g., Harrison-Murray and Clarkson, 1973), it is not surprising to find that glyphosate can have significant effects on mineral nutrition in sensitive plants. Because glyphosate is not directly used in non-GR soybean, literature is sparse on glyphosate effects on mineral content in non-GR soybean.

MINERAL COMPOSITION IN GR SOYBEAN

The published literature on this topic is conflicting and inadequate. The first information available to the public on this topic was that generated as part of the required compositional equivalence studies for GR crops by a company that sells GR soybean seed. Such studies, using good laboratory practices and at multiple locations, are required by USA regulatory agencies to assess potential effects of the glyphosate resistance transgene on the crop. None of these studies found any effect of the transgene conferring glyphosate resistance on mineral nutrition of GR crops (reviewed by Duke *et al.*, 2012b). An university study by Loecker *et al.* (2010) found no effect of the transgene on Mn uptake or any response to Mn in the absence of glyphosate. Unfortunately, the two such published studies on GR soybean treated with glyphosate do not provide mineral data (Harrigan *et al.*, 2007; Lundry *et al.*, 2008), making it impossible to determine whether any exposure of GR crops to glyphosate affects mineral composition of the crop. In studies with other crops (e.g., maize; Ridley *et al.*, 2011) treated with glyphosate, mineral levels (Ca, Cu, Fe, Mg, Mn, P, K, and Zn) were within the 99% tolerance level of conventional, non-GR hybrids.

There are several studies that only dealt with the effects of glyphosate on potted GR soybeans. Such studies cannot be extrapolated to field conditions because the root system of potted plants is limited by the pot, resulting in mineral uptake that is unlike what occurs in the field. Results of such studies are variable. For example, Bott *et al.* (2008) found that glyphosate reduced Mn and Zn in vegetative GR soybean tissues when applied at 0.9 and 1.8 kg ae/ha. Similarly, Zobiole *et al.* (2010c, d) found that application of 0.6 kg ae/ha applied twice or 1.2 kg ae/ha applied once reduced Zn, Mn, Fe, B, Ni, and even Ca in leaves of greenhouse-grown GR soybeans. Zobiole *et al.* (2010b)

TABLE 44.1 Glyphosate Effect on Components in Non-Glyphosate-Resistant (Non-GR) Soybean Leaves and Seed

	Non-GR Soybean Leaves			Non-GR Soybean Seeds			
Components	**Non-treated**	**Glyphosate treated**	**Glyphosate effect on component**	**Non-treated**	**Glyphosate treated**	**Glyphosate effect on component**	**Reference**
N (%)	5.48	5.56	No effect	5.97	6.02	No effect	Bellaloui *et al.* (2006)
N (%)	NA	NA		4.6	6.2	Increased	Cakmak *et al.* (2009)
P (g/kg)	2.3	2.3	No effect	5.9	6.1	No effect	
K (g/kg)	26.3	25.2	No effect	14.1	14.7	Increased	
Ca (g/kg)	22.3	15.6	No effect	3.9	2.9	Decreased	
Mg (g/kg)	8.8	5.8	Decreased	2.4	2.1	Decreased	
Fe (mg/kg)	49	40	No effect	71	36	Decreased	
Mn (mg/kg)	232	121	Decreased	56	31	Decreased	
Zn (mg/kg)	93	65	No effect	44	58	Increased	
Cu (ppm)	5.5	4.1	No effect	11	13	Increased	
Fe (mg/kg)	150	57	Decreased	85	54	Decreased	Bellaloui *et al.* (2009)

TABLE 44.2 Reported Effects of Glyphosate Treatment on Mineral Content of Glyphosate-Resistant (GR) Soybean Seed in Field Studies

Glyphosate Treatment	Effects	Reference
1.12 or 3.36 kg ae/ha at 4 and 6 weeks after planting	No effect on seed N content	Bellaloui *et al.* (2008)
0.86 kg ae/ha at both 3 and 6 weeks after planting	No effects on Mg, Ca, K, Sr, Ba, Mn, Fe, Ni, Cu, Zn, Cd, Cr, Co, or Se content in soybean seed	Duke *et al.* (2012a)

found 1.2 kg ae/ha of glyphosate applied 30 days after sowing GR soybean in the greenhouse to reduce P (–16%), Ca (–18%), Mn (–22%), Zn (–4%), and Cu (–13%) in immature seed. There were no effects on N, K, Mg, S, Fe, or B. In an almost identical study Zobiole *et al.* (2010a) found linear decreases in Fe, Zn, Mn, Fe, Cu, Mo, Co, B, Cu, Mg, S, Ca, P, N, and K in foliage of greenhouse-grown, GR soybeans treated with linear increases in glyphosate dose (0.6 to 2.4 kg ae/ha). Serra *et al.* (2011) found no effect on the efficiency of uptake or translocation of N, Mn, Cu, Zn, and Fe in vegetation of soybean plants treated with up to 2.6 kg ae/ha glyphosate, but growth reductions resulted in reductions in the amount of mineral per plant. N, P, K, Ca, Mg, S, Zn, Mn, Fe, and Cu were reported to be reduced in foliage of second generation GR soybeans with doses of 0.8 to 2.4 kg ae/ha in the greenhouse (Zobiole *et al.*, 2011). In an extensive greenhouse study, six different formulations of glyphosate, each applied at 0.96 kg ae/ha to two different GR cultivars, gave mixed results on N, P, K, S, B, Ca, Mn, Fe, Mg, Zn, and Cu content of vegetative tissues, with inconsistent increases and decreases in mineral concentrations (Cavalieri *et al.*, 2012). In another greenhouse study, there were no effects of glyphosate on any of the minerals measured, including Ca, Mg, Zn, Fe, Cu, Sr, Ba, Al, Cd, Cr, Co, Ni, Mn, and Zn, in vegetative or harvested, mature seeds of plants treated twice with 0.86 kg ae/ha (Duke *et al.*, 2012a). Likewise, in studies done in Brazil (Rosolem *et al.*, 2010; Andrade and Rosolem, 2011) found no effect of glyphosate at 0.75–0.96 kg ae/ha on Mn in foliage of greenhouse-grown plants.

Unfortunately, there are only two field studies of glyphosate effects on mineral content of mature GR soybean seeds, and only one of these deals with minerals that form divalent cations. Results of these studies are summarized in Table 44.2. The details of the only study on metallic mineral content are provided in Figure 44.2. There were no effects of glyphosate in either study. Other field studies have only examined effects of glyphosate on mineral content of GR soybean vegetative tissue. In one of these studies, Henry *et al.* (2011) found 0.84 kg ae/ha to cause no consistent effects on N, P, K, S, Mg, Ca, B, Zn, Mn, Fe, Cu, and Al. They concluded that there was no evidence of mineral deficiencies. Similarly, Stefanello *et al.* (2011) found no effect of three applications of glyphosate totaling 2.4 kg ae/ha on foliar content of Mn, Fe, Zn, or Cu. Ebelhar *et al.* (2006) found in a two-year field study that applications of glyphosate up to four times (3.46 kg ae/ha) the recommended rate had no effect on foliar Mn content. In the only field study reporting an

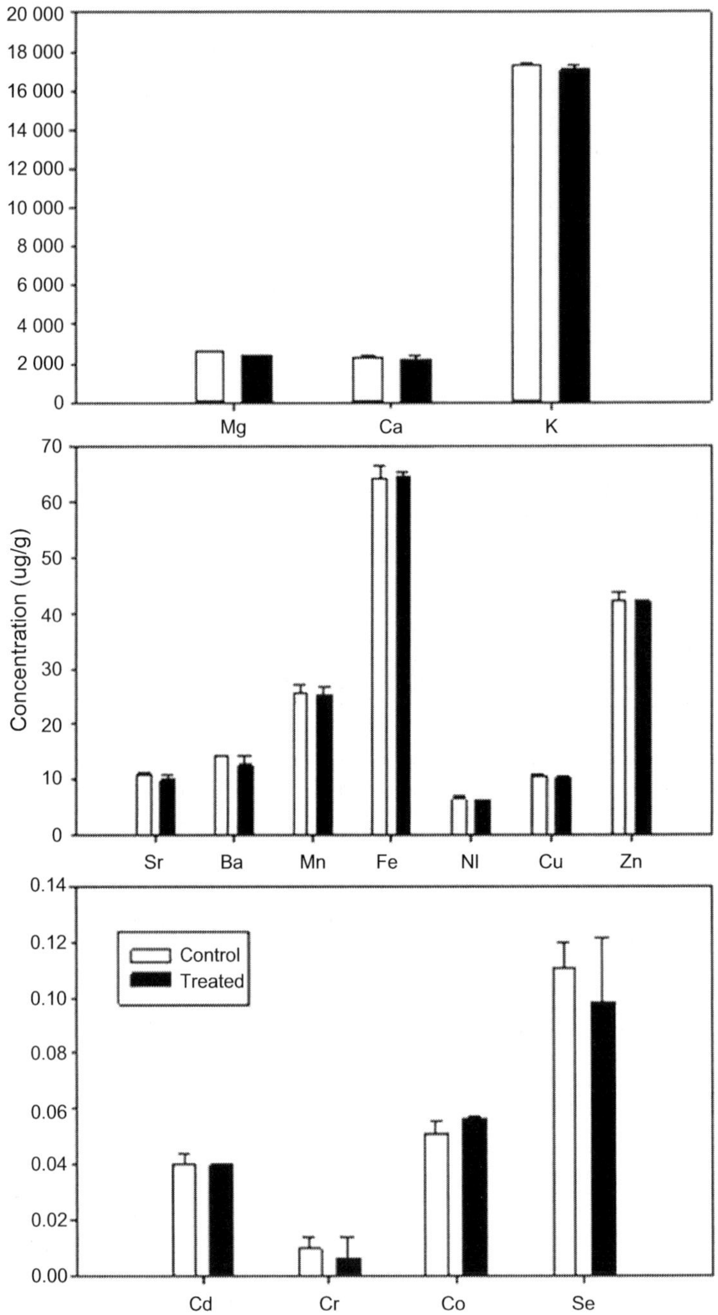

FIGURE 44.2 Effects of different two glyphosate treatments (0.86 kg ai/ha at both 3 and 6 weeks after planting) on metal content of mature seed of field-grown GR soybean plants. Error bars are ±1 SE of the mean. *Reprinted with permission from Duke* et al. *(2012a), Copyright 2012, American Chemical Society.*

effect, Zobiole *et al.* (2012) reported that 0.8 to 2.4 kg ae/ha of glyphosate applied at 10, 20, and 34 days after sowing reduced N, P, K, Ca, Mg, S, Zn, Mn, Fe, Cu, and B in vegetative tissue.

References

Andrade, G.J.M., Rosolem, C.A., 2011. Uptake of manganese in RR soybean under glifosate application. Rev. Bras. Cienc. Solo. 35, 961–968.

Bellaloui, N., Reddy, K.N., Zablotowicz, R.M., Abbas, H.K., Abel, C.A., 2009. Effects of glyphosate application on seed iron and root ferric (III) reductase in soybean cultivars. J. Agric. Food Chem. 57, 9569–9574.

Bellaloui, N., Reddy, K.N., Zablotowicz, R.M., Mengistu, A., 2006. Simulated glyphosate drift influences nitrate assimilation and nitrogen fixation in non-glyphosate-resistant soybean. J. Agric. Food Chem. 54, 3357–3364.

Bellaloui, N., Zablotowicz, R.M., Reddy, K.N., Abel, C.A., 2008. Nitrogen metabolism and seed composition as influenced by glyphosate application in glyphosate-resistant soybean. J. Agric. Food Chem. 56, 2765–2772.

Bott, S., Tesfamariam, T., Candan, H., Cakmak, I., Römheld, V., Neumann, G., 2008. Glyphosate-induced impairment of plant growth and micronutrient status in glyphosate-resistant soybean (*Glycine max* L.). Plant Soil 312, 185–194.

Brookes, G., Barfoot, P., 2010. Global impact of biotech crops: Environmental effects, 1996–2008. AgBioForum 13, 76–94.

Brookes, G., Barfoot, P., 2012. Global impact of biotech crops: Environmental effects, 1996–2010. GM Crops & Food 3, 129–137.

Cakmak, I., Yazici, A., Tutus, Y., Ozturk, L., 2009. Glyphosate reduced seed and leaf concentrations of calcium, manganese, magnesium, and iron in non-glyphosate-resistant soybean. Europ. J. Agronomy 31, 114–119.

Cavalieri, S.D., Velini, E.D., Silva, F.M.L., São José, A.R., Andrade, G.J.M., 2012. Nutrient and shoot dry matter accumulation of two GR soybean cultivars under the effect of glyphosate formulations. Planta Daninha 30, 349–358.

Cerdeira, A.L., Duke, S.O., 2006. The current status and environmental impacts of glyphosate-resistant crops: A review. J. Environ. Qual. 35, 1633–1658.

Duke, S.O., 1988. Glyphosate. In: Kearney, P.C., Kaufman, D.D. (Eds.), Herbicides-Chemistry, Degradation and Mode of Action Vol. III. Marcel Dekker, Inc., New York, pp. 1–70.

Duke, S.O., 2012. Why have no new herbicide modes of action appeared in recent years? Pest Manag. Sci. 68, 505–512.

Duke, S.O., Powles, S.B., 2008. Glyphosate: A once in a century herbicide. Pest Manag. Sci. 64, 319–325.

Duke, S.O., Powles, S.B., 2009. Glyphosate-resistant crops and weeds: Now and in the future. AgBioForum 12, 346–357.

Duke, S.O., Baerson, S.R., Rimando, A.M., 2003. Herbicides: Glyphosate. In: Plimmer, J.R., Gammon, D.W., Ragsdale, N.N. (Eds.), Encyclopedia of Agrochemicals. John Wiley & Sons, New York. http://www.mrw.interscience.wiley.com/eoa/articles/agr119/frame.html.

Duke, S.O., Wedge, D.E., Cerdeira, A.L., Matallo, M.B., 2007. Interactions of synthetic herbicides with plant disease and microbial herbicides. In: Vurro, M., Gressel, J. (Eds.), Novel Biotechnologies for Biocontrol Agent Enhancement and Management. Springer, Dordrecht, The Netherlands, pp. 277–296.

Duke, S.O., Reddy, K.N., Bu, K., Cizdziel, J.V., 2012a. Effects of glyphosate on mineral content of glyphosate-resistant soybeans (*Glycine max*). J. Agric. Food Chem. 60, 6764–6771.

Duke, S.O., Lydon, J., Koskinen, W.C., Moorman, T.B., Chaney, R.L., Hammerschmidt, R., 2012b. Glyphosate effects on plant mineral nutrition, crop rhizophere microbiota, and plant disease in glyphosate-resistant crops. J. Agric. Food Chem. 60, 10375–10397.

Ebelhar, S.A., Varsa, E.C., Hart, C.D., 2006. Soil pH and manganese effects on yield of Roundup Ready® soybeans. Illinois Fert. Conf. Proc. 2006, 54–65.

Fedtke, C., Duke, S.O., 2005. Herbicides. In: Hock, B., Elstner, E.F. (Eds.), Plant Toxicology. Marcel Dekker, New York, pp. 247–330.

Fernandez-Cornejo, J., Hallahan, C., Nehring, R., Wechsler, S., 2012. Conservation tillage, herbicide use, and genetically engineered crops in the United States: The case of soybeans. AgBioForum 15, 231–241.

Geisy, J.P., Dobson, S., Solomon, K.R., 2000. Ecotoxicological risk assessment for Roundup® herbicide. Rev. Environ. Toxicol. 167, 35–120.

Gianessi, L.P., 2008. Economic impacts of glyphosate-resistant crops. Pest Manag. Sci. 64, 346–352.

Givens, W.A., Shaw, D.R., Kruger, G.R., Johnson, W.G., Weller, S.C., Young, B.G., Wilson, R.G., Owen, M.D.K., Jordan, D., 2009. Survey of tillage trends following the adoption of glyphosate-resistant crops. Weed Technol. 23, 150–155.

Harrison-Murray, R.S., Clarkson, D.T., 1973. Relationships between structural development and the absorption of ions by the root system of. *Cucurbita pepo*. Planta 114, 1–16.

Harrigan, G.G., Ridley, W.P., Riordan, S.G., Nemeth, M.A., Sorbet, R., Trujillo, W.A., Breeze, M.L., Schneider, R.W., 2007. Chemical composition of glyphosate-tolerant soybean 40–3-2 grown in Europe remains equivalent with that of conventional soybean (*Glycine max* L.). J. Agric. Food Chem. 55, 6160–6168.

Henry, R.S., Wise, K.A., Johnson, W.G., 2011. Glyphosate's effect upon mineral accumulation in soybean. Crop Manag. Network Crop Manag. DOI: 10.1994/CM-2011-1024-01-RS.

Hurley, T.M., Mitchell, P.D., Frisvold, G.B., 2009. Effects of weed-resistance concerns and resistance-management practices on the value of Roundup Ready® crops. AgBioForum 12, 291–302.

James, C., 2011. Global Status of Commercialized Biotech/GM Crops: 2011. ISAAA Brief No. 43–2011: Executive Summary. ISAAA, Ithaca, NY. http://www.isaaa.org/resources/publications/briefs/43/executivesummary/default.asp (accessed 27.08.12.).

Loecker, J.L., Nelson, N.O., Gordon, W.B., Maddux, L.D., Janssen, K.A., Schapaugh, W.T., 2010. Manganese response in conventional and glyphosate resistant soybean. Agron. J. 102, 606–611.

Lundry, D.R., Ridley, W.P., Meyer, K.J., Riordan, S.G., Nemeth, T.A., Trujillo, W.A., Breeze, M.L., Sorbet, R., 2008. Composition of grain, forage, and processed fractions from second-generation glyphosate-tolerant soybean, MON 89788, is equivalent to that of conventional soybean (*Glycine max* L.). J. Agric. Food Chem. 56, 4611–4622.

Nandula, V.K., Reddy, K.N., Rimando, A.M., Duke, S.O., Poston, D.H., 2007. Glyphosate-resistant and -susceptible soybean (*Glycine max*) and canola (*Brassica napus*) dose response and metabolism relationships with glyphosate. J. Agric. Food Chem. 55, 3540–3545.

Reddy, K.N., 2001. Glyphosate-resistant soybean as weed management tool: opportunities and challenges. Weed Biol. Manag. 1, 193–202.

Reddy, K.N., Hoagland, R.E., Zablotowicz, R.M., 2000. Effect of glyphosate on growth, chlorophyll, and nodulation in glyphosate-resistant and susceptible soybean (*Glycine max*) varieties. J. New Seeds. 2, 37–52.

Reddy, K.N., Whiting, K., 2000. weed control and economic comparisons of glyphosate-resistant, sulfonylurea-tolerant, and conventional soybean (*Glycine max*) systems. Weed Technol. 14, 204–211.

Reddy, K.N., Rimando, A.M., Duke, S.O., 2004. Aminomethylphosphonic acid, a metabolite of glyphosate causes injury in glyphosate-treated, glyphosate-resistant soybean. J. Agric. Food Chem. 52, 5139–5143.

Ridley, W.P., Harrigan, G.G., Breeze, M.L., Nemeth, M.A., Sidhu, R.S., Glenn, K.C., 2011. Evaluation of compositional equivalence for multitrait biotechnology crops. J. Agric. Food Chem. 59, 5865–5876.

Rosolem, C.A., de Andrade, G.J.M., Lisboa, I.P., Zoca, S.M., 2010. Manganese uptake and redistribution in soybean as affected by glyphosate. Rev. Bras. Cienc. Solo 34, 915–1922.

Samarah, N., Mullen, R., Cianzio, S., 2004. Size distribution and mineral nutrients of soybean seeds in response to drought stress. J. Plant Nutr. 27, 815–835.

Schafer, J.R., Hallett, S.G., Johnson, W.G., 2012. Response of giant ragweed (*Ambrosia trifida*), horseweed (*Conyza canadensis*), and common lambsquarters (*Chenopodium album*) biotypes to glyphosate in the presence and absences of soil microorganisms. Weed Sci. 60, 641–649.

Serra, A.P., Marchetti, M.E., da Silva Candido, A.C., Ribiero Dias, A.C., Christoffoleti, P.J., 2011. Glyphosate influence on nitrogen, manganese, iron, copper and zinc nutritional efficiency in glyphosate resistant soybean. Ciênc. Rural, Santa Maria 41, 77–84.

Singh, B.K., Shaner, D.L., 1998. Rapid determination of glyphosate injury to plants and identification of glyphosate-resistant plants. Weed Technol. 12, 527–530.

Stefanello, F.F., Marchetti, M.E., da Silva, E.F., Stefanello, J., Doreto, R.B.S., Novelino, J.O., 2011. Effect of glyphosate and manganese on nutrition and yield of transgenic glyphosate-resistant soybean (in Portuguese). Ciênc. Agrár. 32, 1007–1014.

U.S. Department of Agriculture, Agricultural Research Service, 2011. USDA National Nutrient database for standard reference, Release 24. Nutrient Data Laboratory Home Page. http://www.ars.usda.gov/ba/bhnrc/ndl (accessed 20.08.12.).

U.S. Department of Agriculture, 2012a. National Agricultural Statistics Service. Acreage [Online]. Available at http://usda.mannlib.cornell.edu/MannUsda/viewDocumentInfo.do?documentID=1000 (accessed 08.17.12).

U.S. Department of Agriculture, 2012b. National Agricultural Statistics Service. Chemical use data in Quick Stats 2.0, advanced database search [Online]. Available at http://www.nass.usda.gov/Statistics_by_Subject/Environmental/index.asp (accessed 02.10.12.).

Williams, G.M., Kroes, R., Munro, I.C., 2000. Safety evaluation and risk assessment of the herbicide Roundup and its active ingredient, glyphosate, for humans. Regul. Toxicol. Pharmacol. 31, 117–165.

Zobiole, L.H.S., Oliveira Jr., R.S., Kremer, R.J., Muniz, A.S., Oliviera Jr., A., 2010a. Nutrient accumulation and photosynthesis in glyphosate-resistant soybeans is reduced under glyphosate use. J. Plant Nutr. 33, 1860–1873.

Zobiole, L.H.S., Oliviera Jr., R.S., Visentainer, J.V., Kremer, R.J., Bellaloui, N., Yamada, T., 2010b. Glyphosate affects seed composition in glyphosate-resistant soybean. J. Agric. Food Chem. 58, 4517–4522.

Zobiole, L.H.S., Oliveira Jr., R.S., Kremer, R.J., Constantin, J., Yamada, T., Castro, C., Oliveira, F.A., Oliveira Jr., A., 2010c. Effect of glyphosate on symbiotic N_2 fixation and nickel concentration in glyphosate-resistant soybeans. Appl. Soil Ecol. 44, 176–180.

Zobiole, L.H.S., Oliveira Jr., R.S., Huber, D.M., Constantin, J., Castro, C., Oliveira, F.A., Oliveira Jr., A., 2010d. Glyphosate reduces shoot concentrations of mineral nutrients in glyphosate-resistant soybeans. Plant Soil 328, 57–69.

Zobiole, L.H.S., Kremer, R.J., Oliveira Jr., R.S., Constantin, J., 2011. Glyphosate affects chlorophyll, nodulation and nutrient accumulation of "second generation" glyphosate-resistant soybean (*Glycine max* L.). Pestic. Biochem. Phyisol. 99, 53–60.

Zobiole, L.H.S., Kremer, R.J., Oliveira Jr., R.S., Constantin, J., 2012. Glyphosate effects on photosynthesis, nutrient accumulation, and nodulation in glyphosate-resistant soybean. J. Plant Nutr. Soil Sci. 175, 319–330.

CHAPTER

45

Drought and Heat Stress Effects on Soybean Fatty Acid Composition and Oil Stability

Nacer Bellaloui, Krishna N. Reddy†, Alemu Mengistu**

*USDA-ARS, Crop Genetics Research Unit, Jackson, TN, USA, †USDA-ARS, Crop Production Systems Research Unit, Jackson, TN, USA

CHAPTER POINTS

- Higher oleic acid and lower linoleic and lower linolenic fatty acids are desirable for oil stability, long shelf storage, and processing.
- Fatty acid profiling is significantly affected by genotype and environmental factors, especially drought and high heat.
- To ensure the stability of a desired fatty acid trait across environments, further research on genotype and environment interactions is needed.
- Breeders and biotechnology researchers continue to develop soybean with high levels of healthy oils such as oil with *trans*-fat free, reduced saturated fat soybean oil, and increased stability for food processing.
- Advances in genetic modifications continue, and soybeans with designed fatty acid profile in seed oil may dominate the market in the future.

INTRODUCTION

Soybean is a major source of vegetable oil in the world, and soybean oil represents 56% of world production (Soystat, 2008). Oil concentration in soybean seeds ranges from 8.3 to 28%, with a mean of 19.5% (Wilson, 2004). Soybean oil contains five major fatty acids, two saturated fatty acids (palmitic, C16:0; stearic, C18:0) and three unsaturated fatty acids (oleic, C18:1; linoleic, C18:2; linolenic C18:3). The concentration of saturated fatty acids in soybean oil ranges from 10–12% for palmitic acid and 3–5% for stearic acid. The concentration of unsaturated fatty acids in oil is about 24% for oleic acid, 54% for linoleic acid, and 8.0% for linolenic acid. Lower saturated fatty acids and higher unsaturated fatty acid contents are desirable for human health, although higher oleic acid and lower linolenic acids are desirable for oil stability, long shelf-storage, and processing because a higher concentration of oleic acid contributes to oxidative stability. An oxidation reaction with polyunsaturated fatty acids such as linoleic and linolenic acids occurs when they are exposed to oxygen, forming oxygenated species such as hydroperoxide and subsequent degraded products. The level of oil degradation depends on the fatty acid composition of the parent oil.

Soybean oil with higher linoleic and linolenic acid contents and lower oleic acid requires an additional industrial process, hydrogenation, to increase saturation and improve oil stability and utility properties. It is now known that the process of hydrogenation produces *trans*-fatty acids, as opposed to the naturally occurring *cis* fatty acids. *Trans*-fatty acids have negative effects on human health, increasing the risk of coronary heart disease and are associated with higher LDL-cholesterol and lower HDL-cholesterol (Federal Register, 2003). The partial hydrogenation converts polyunsaturated fatty acids such as linolenic acid to oleic and stearic acids, reducing polyunsaturated fatty acids to about 18% and linolenic acid to below 2% (Clemente and Cahoon, 2009). This has increased demand for low-linolenic soybean oil that allows food processors to reduce the need for hydrogenation, thus reducing *trans*-fatty acids from oil in foods.

Processing and Impact on Active Components in Food
http://dx.doi.org/10.1016/B978-0-12-404699-3.00045-7

DROUGHT EFFECTS ON SEED OIL, FATTY ACID COMPOSITION, AND OIL STABILITY

Drought stress during seed-fill (R5-R6) stage can alter chemical composition of soybean seed, and severe drought can result in up to12.4% lower total oil (Dornbos and Mullen, 1992). These authors also indicated that drought stress increased stearic acid, but decreased oleic acid (Dornbos and Mullen, 1992), and the decrease or increase in oil may depend on the level of drought (Specht *et al.*, 2001). Lee *et al.* (2008) studied the effects of irrigation on elevated oleic acid, and/or reduced linolenic acid genotypes under field conditions. They found that irrigation did not significantly affect unsaturated fatty acid concentration in soybean genotypes with modified higher oleic acid and/or modified low linolenic acid. When comparing irrigated with non-irrigated soybean genotypes, Lee *et al.* (2008) indicated that oleic acid tended to increase and linolenic acid tended to decrease in genotypes with high oleic acid and low linolenic acid. They concluded that near optimum irrigation during the growing season could result in the highest levels of oleic acid and lowest levels of linolenic acid. Boydak *et al.* (2002) reported that cultivar Asgrow 3935 had higher linoleic acid and lower oleic acid when soybeans were irrigated every 12th day after emergence. Table 45.1 shows that seed of soybean varieties (maturity group MG IV) grown in drought stressed (−90 to −100 kPa soil water potential) conditions (Bellaloui *et al.*, 2011) had higher oil and oleic acid concentrations and lower linoleic and linolenic concentrations compared with irrigated soybeans (−15 to −20 kPa soil water potential). The alteration of fatty acids depended on genotype (Table 45.1) and MG (Figure 45.1). List *et al.* (1980) reported that poor quality oil, extracted from soybean seed damaged in the field or exposed to frost, heat, and moisture, resulted in low refining oil quality, poor refined-bleached color, lower flavor, and oxidative stability. In addition, drought caused shrinking and cracking, making dehulling much more difficult and less efficient for processing. For example, Brumm and Hurburgh (1990) reported that in the Midwest USA drought in 1988 resulted in soybeans with shriveled and wrinkled seed coats, and processors expressed concerns about the processing of the seed due to size and shape. Carrera *et al.* (2009) investigated the effect of water deficit on soybean seed protein, oil, and fatty acids during the reproductive period, and found that when precipitation minus potential evapotranspiration was less than 70 mm oil increased linearly with the increase of the average daily mean temperature during seed-fill and with increase in water deficit. They concluded that water stress during seed-fill stage is important due to its effects on the inverse relationship between oil and protein with temperature. Scherder *et al.* (2008) studied the stability of soybean lines with modified elevated oleic acid, combined with modified low saturated fatty acids, 1% linolenic acid, or both low saturates and 1% linolenic acid. They found, when these lines were grown in 11 environments (one in Missouri and four in Iowa, USA) in a 2-year experiment, that the mean oleic concentration was 590 g/kg in elevated oleic acid and low saturated fatty acid lines, 521 g/kg in elevated oleic acid and 1% linolenic acid lines, and 557 g/kg in elevated oleic acid and low saturated fatty acids and low linolenic acid lines, and 269 g/kg for conventional/normal cultivars. They found that the lines with the highest concentration of oleic acid generally had the most variation across environments, but the lines with the highest mean of oleic acid were most probably meeting or exceeding 500 g/kg across environments. Gao *et al.* (2009) evaluated the effect of precipitation on soybean oil and fatty acid composition, and found that oleic acid and linoleic acid contents were significantly affected by seasonal precipitation, but palmitic, stearic, and linolenic contents were relatively stable. They also found that oil quality ratio of oleic/(linoleic + linolenic) had

TABLE 45.1 Effect of Drought Stress Conditions (−90 to −100 kPa Soil Water Potential) on Soybean Seed Total Oil and Fatty Acids Composition (g/kg) in Genotypes of Maturity Group IV Under Greenhouse Conditions[a]

	Watered Soybean						Drought Stressed Soybean					
Variety	Oil	C16:0	C18:0	C18: 1	C18: 2	C18: 3	Oil	C16: 0	C18: 0	C18: 1	C18: 2	C18: 3
Hutcheson	209 c	124 a	43 a	244 b	556 b	69 b	238 ab	111 b	46 a	266 b	488 a	54 bc
DT97-4290	211 c	117 c	41 b	243 b	565 a	68 b	234 b	111 b	45 ab	267 b	468 b	52 c
AG 4403	218 b	118 c	43 a	247 ab	560 ab	67 b	239 a	120 a	43 b	280 a	482 a	57 a
AG 4903	225 a	121 b	44 a	251 a	536 c	74 a	239 a	118 a	38 c	278 a	457 b	57 a

Drought stress soybeans were grown at −90 to −100 kPa soil water stress and compared with watered soybean grown at field capacity at −15 to −20 kPa, according to Bellaloui *et al.*, (2011). Soybeans were grown under greenhouse conditions similar to those in Bellaloui *et al.* (2011). Values are means of four replicates. Values within columns sharing a letter are not significantly different (P > 0.05) using Fishers' test.

[a] *Seed composition analysis was conducted at harvest maturity (R8 stage) according to Bellaloui* et al. *(2009).*

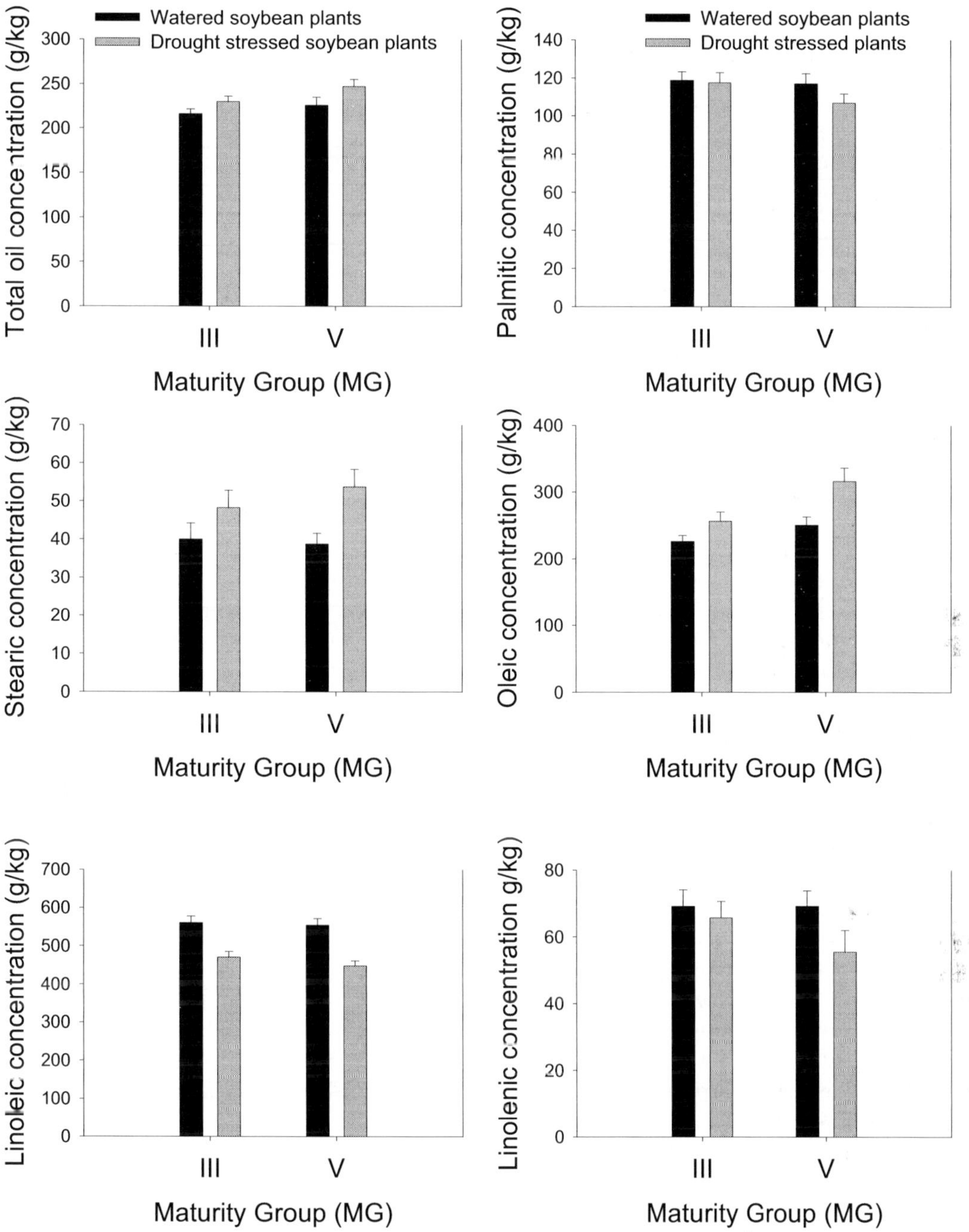

FIGURE 45.1 Effects of high temperature on oil composition in maturity group (MG) III and V soybean genotypes. The graph shows the alteration of oil composition by temperature "Unpublished." Values are mean of four replicates ±SE. Seed composition analyses were conducted on harvest maturity seed (R8 stage) according to Bellaloui *et al.* (2009).

a quadratic relationship with precipitation. These studies showed that drought decreases total oil (Dornbos and Mullen, 1992), alters fatty acid composition (Gao *et al.*, 2009), and affects oil stability and oil processing (Brumm and Hurburgh, 1990). Therefore, development of drought tolerant soybeans with stable high oleic and low linolenic acid genes across geographical locations is critical to maintain the stability of oil production and desirable fatty acid composition.

EFFECT OF HEAT STRESS ON TOTAL OIL, FATTY ACID COMPOSITION, AND OIL STABILITY

The wide range of oil concentration in soybean seed is due to genotype, the environmental factors under which soybeans grow, and geographic location. It was found that genetics and temperature during seed-fill stage (Oliva *et al.*, 2006; Bellaloui *et al.*, 2009) and drought

TABLE 45.2 Effect of Normal (25/20°C) and Warmer (26/28°C) Temperature on Soybean Seed Total Oil and Fatty Acids Composition (g/kg) in Genotypes of Maturity Group IV Under Growth Chamber Conditions[a]

	25/20°C						36/28°C					
Variety	**Oil**	**C16:0**	**C18:0**	**C18:1**	**C18:2**	**C18:3**	**Oil**	**C16:0**	**C18:0**	**C18:1**	**C18:2**	**C18:3**
Hutcheson	195 c	125 b	39 bc	213 a	561 a	80 b	233 b	114 b	44 a	268 b	461 b	50 c
DT97-4290	205 a	126 a	43 a	218 a	482 c	76 c	245 a	113 b	44 a	306 a	482 a	45 d
AG 4403	198 bc	114 c	40 b	213 a	545 b	75 c	242 a	119 a	35 c	276 b	476 ab	55 b
AG 4903	201 ab	114 c	38 c	201 b	560 a	87 a	242 a	117 a	39 b	278 b	472 ab	61 a

In heat stress experiments, soybean genotypes were grown under growth chamber conditions with photon flux density of about 1000 μmol•m^{-2}•s^{-1}, supplied with a combination of 10 high pressure sodium and metal halide lights, each of 400 W. Values are means of four replicates. Values within columns sharing a letter are not significantly different (P > 0.05) using Fishers' test.

[a]*Seed composition analysis was conducted at harvest maturity (R8 stage) according to Bellaloui* et al. *(2009).*

(Bellaloui *et al.*, 2011) play a major role in oil deposition and oil composition. Previous research has shown that concentrations of linoleic and linolenic acids in soybean seed developed under higher temperature decreased, but oleic acid concentrations increased (Dornbos and Mullen, 1992). Oliva *et al.* (2006) studied the effects of average temperature over the final 30 days of the reproductive period on oleic and linolenic acids in 17 normal and modified fatty acid genotypes across 10 environments. They found that the concentration of oleic acid in modified mid-oleic genotypes were less stable than in those of reduced oleic acid, but found mid-oleic (50–60% oleic acid) lines, N98-4445A and N97-3363-4 with stability coefficients of 3.28 and 2.53, respectively, were the most unstable genotypes, and the elevated oleic acid line M23 (up to 80% oleic acid) was the most stable (stability coefficient of 0.13). Conversely, studying the stability of linolenic acid concentration in reduced linolenic acid genotype IA 3017 (a line with 1% linolenic acid), Oliva *et al.* (2006) found that linolenic concentration was the most stable across environments, and the higher linolenic acid genotypes were less stable. They concluded that soybean lines with stable oleic acid and linolenic acid across environments could be used as a source of crops with desirable fatty acids composition. In an experiment, where normal (Century cultivar) and low-linolenic soybean genotypes (C1640 and 9509: genotypes differ in linolenic acid due to alleles at the fan locus), Wilcox and Cavins (1992) found that palmitic acid concentration tended to decrease and stearic acid tended to increase with later planting, and linolenic acid was more sensitive to this change than other fatty acids due to temperature changes coinciding with late planting.

Howell and Collins (1957) found that linolenic acid concentration decreased from 105 to 66 g/kg as daytime temperatures in the greenhouse increased from 21 to 29°C. Linolenic acid decreased from 164 to 50 g/kg, linoleic acid decreased from 558 to 403 g/kg, and oleic acid increased from 131 to 387 g/kg as day/night temperature increased from 18/13 to 33/28°C (Wolf *et al.*, 1982). There were no changes in palmitic and stearic acids with temperature changes (Wolf *et al.*, 1982). Our research showed that seed of similar and different MGs, with different genotypes grown at normal (25/20°C, day/night) and at warmer temperature (36/28°C) had higher total oil and oleic acid concentrations and lower linoleic and linolenic concentrations at warmer temperature compared with those grown at normal temperatures (Table 45.2; Figure 45.2). Therefore, alteration of fatty acids depended on both MG and genotype (Table 45.2; Figure 45.2). Rennie and Tanner (1989) evaluated fatty acid concentration in five low-linolenic acid lines, and found at day/night temperatures ranging from 15/12 to 40/30°C, a decrease in linoleic and linolenic acids, increase in oleic acid, and relatively stable palmitic and stearic acid concentrations in the five lines. It was also reported that growing soybean cultivars near the northern range of their adaptation resulted in 10 to 20 g/kg higher linolenic acid concentration and 30 to 60 g/kg higher linoleic acid than those grown at the southern range of their adaptation (Collins and Sedgewick, 1959), reflecting influences of temperature and geographical location.

The instability of conventional cultivars and mid-oleic acid germplasm across environment created a challenge, mainly due to temperature changes that influence the enzymes controlling biosynthesis of soybean seed fatty acids, especially at seed-fill stage (Wilcox and Cavins, 1992; Bachlava and Cardinal, 2009). The explanations of the effects of temperature on oleic and linolenic acid were that temperature may affect oleate and linoleate desaturases (Burton, 1991), decrease oleyl and linoleyl desaturase activities at 35°C (Cheesbrough, 1989), and ω-6 desaturase enzyme, encoded by the FAD2-1A gene, was degraded at high growth temperatures of 30°C (Tang *et al.*, 2005). Studying the transcript level of the functional GmFAD2 isoforms, FAD2-1A, FAD2-1B, FAD2-2B, and FAD2-2C, Schlueter *et al.* (2007) found that only significant increase in expression was observed in FAD2-2C when soybean was grown at 18/12°C day/night during the pod development stage. Upchurch and Ramirez (2010)

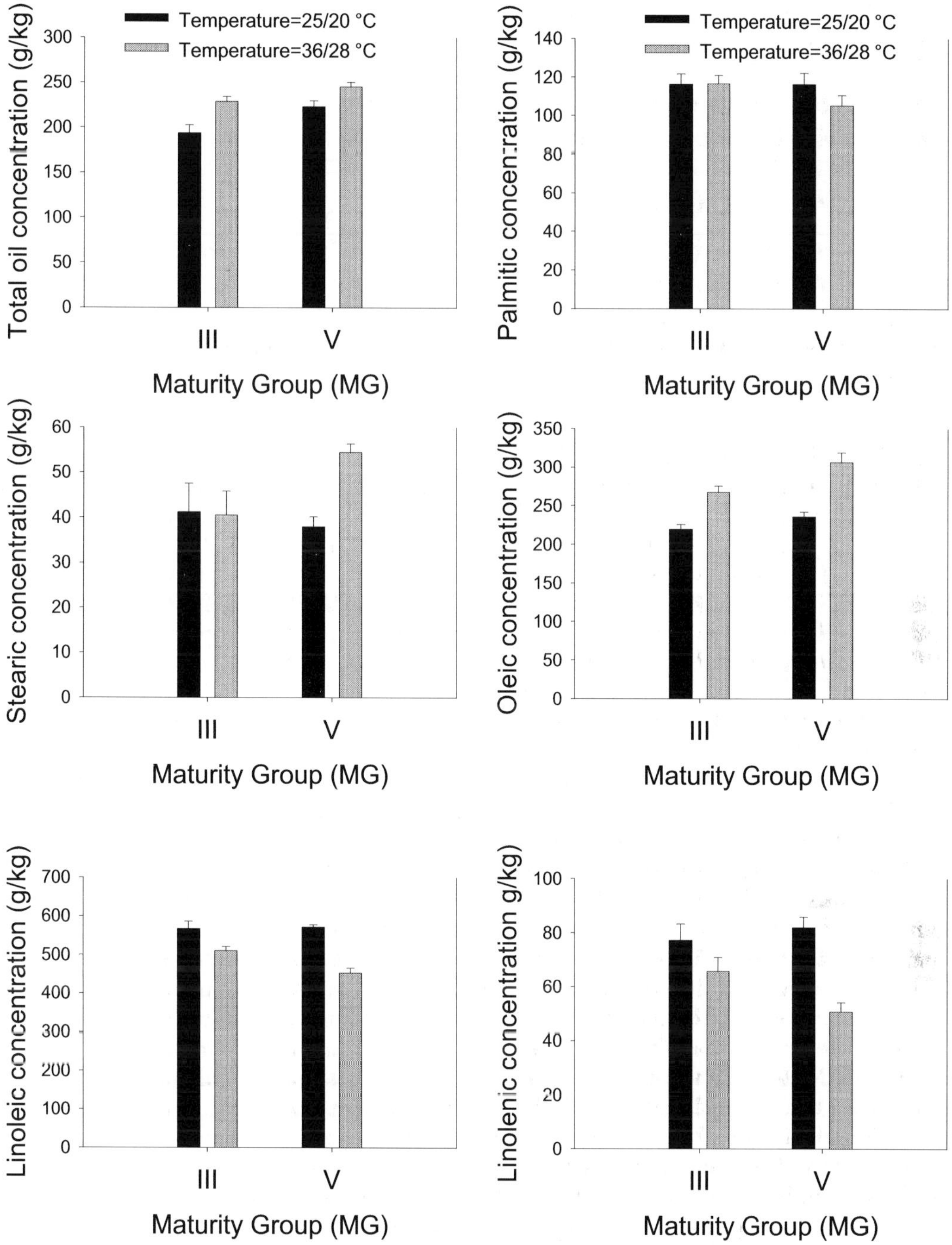

FIGURE 45.2 Effects of drought stress on oil composition in maturity group (MG) III and V soybean genotypes. The graph shows the drought stress (−90 to −100 kPa soil water potential) effects on total oil and fatty acids composition "Unpublished." Watered soybeans were grown under field capacity conditions (−15 to −20 kPa soil water potential). Values are mean of four replicates ± SE. Soybeans were grown according to Bellaloui *et al.* (2011). Seed composition analyses were conducted on harvest maturity seed stage (R8) according to Bellaloui *et al.* (2009).

studied the expression levels of ω-6 desaturase GmFAD2 genes and genes that determine seed oleic acid concentration in modified mid-oleic acid soybean mutant M23 (a deletion of GmFAD2-1A, an allele of the microsomal ω-6 oleate desaturase gene) with 50–60% oleate. They found that there were lower expressions of GmFATB1a, GmFAD2-1A, GmFAD2-1B, GmFAD2-2, and GmFAD2-3 and higher expression of GmSACPD-C. They suggested that several soybean genomic loci known or suspected could be involved in oleic acid phenotype, and some may encode genes involved in regulation of oleate biosynthetic genes.

Pham *et al.* (2010), through their breeding program, were able to create and identify high oleic acid trait of

80% oleic acid concentration in soybeans seed by identifying and combining mutations in two delta-twelve fatty acid desaturase genes, FAD2-1A and FAD2-1B. They found three polymorphisms in the FAD2-1B alleles of two soybean lines. The mutant FAD2-1B alleles were associated with an increase in oleic acid concentration, but the FAD2-1B mutant alleles alone could not produce high oleic unless the existing FAD2-1A mutations were combined with the novel mutant FAD2-1B alleles. This combined mutation resulted in high (80%) oleic acid that was recovered only for lines which were homozygous for both of the mutant alleles. The developed conventional soybean lines with 80% oleic acid were stable in two production environments (Lee *et al.*, 2012). Therefore, heat stress is a major environmental factor that affects total oil and alters fatty acid composition. Further research is needed to establish that non-GMO or GMO soybeans with desirable fatty acid composition do not compromise production and other seed quality traits across environments and geographic locations.

GENOTYPE × DROUGHT × HEAT STRESS INTERACTIONS ON OIL QUALITY AND STABILITY

The effects of genotype × environment interactions on total oil and fatty acids are major concerns because of their influence on oil level and fatty acid composition. Although seed oil level and composition are genetically controlled, environment plays a major role in altering the level and composition of oils. Since drought and temperature are major environmental factors altering seed oil level and composition, we will discuss the environment effects in the context of drought and temperature and how drought and temperature interact with genotype or maturity group (MG). The effect of environment on seed composition has been extensively studied (Piper and Boote, 1999; Dardanelli *et al.*, 2006). However, the reasons for the variability of oil under different environments are still not completely understood. Hou *et al.* (2006) evaluated fatty acids composition in a recombinant inbred line (RILs) population, developed from a cross between a modified low palmitic acid line and a high stearic acid parent. They conducted this experiment for two years and in three environments in Canada, and found that linolenic acid was more sensitive to the environment, and year effects were greater than the effects of location in palmitic, stearic, and oleic acid concentration. However, location was more important for the relative concentrations of linoleic, and linolenic acids. Dardanelli *et al.* (2006) evaluated oil level in six MG (II–III, IV, V, VI, VII, VIII–IX) in 14 to 24 environments from 2000–2003 in Argentina, using water deficit and temperature data from 1970–2003 for each location. They found that the oil variations among environments depended mainly on MG, and showed that oil concentrations were higher in MG II, III, and IV in every environment than in other groups in every environment, suggesting broad adaptations of early MG, and that high temperature during seed-fill could explain the consistent pattern of higher oil concentration across seasons and environments in early MGs. Another example of the environmental and geographical influences on oil concentration in soybean seed is that oil concentration in the northern and western growing area in the USA is greater than in those of southern USA (Hurburgh *et al.*, 1990), which is the opposite trend of seed protein due to the inverse relationship between oil and protein (Dornbos and Mullen, 1992; Bellaloui *et al.*, 2009). This suggests that oil level and its composition can be altered by protein also.

Serretti (1993) reported that oil concentration increased 6.7 g/kg/°C with the increase of mean temperature, and Kane *et al.* (1997) found that oil concentration increased 5.2–6.6 g/kg/°C among six cultivars. In growth chamber and greenhouse experiments, Gibson and Mullen (1996) and Dornbos and Mullen (1992) found that oil concentration increased with increasing temperature with an optimum at 25–28°C, above which the oil concentration declined. It was concluded that oil concentration increased linearly with the increase of mean temperature (Gibson and Mullen, 1996; Wolf *et al.*, 1982). Recently, Bellaloui *et al.* (2009) investigated the effect of genotypic background and environment and their interactions on seed oil in two sets of near isogenic lines (Clark and Harosoy isolines), where each set has the same genotypic background, but differ in maturity genes. Clark (e1 E2 E3 e5) and Harosoy (e1 e2 E3 e5) are different only in the E2/e2 locus for the maturity genes. Bellaloui *et al.* (2009) found that year, maturity, and year × maturity interactions were significant ($P < 0.0001$) for seed oil in the Clark isoline set, but in the Harosoy isoline set there was no year × maturity effect for oil, suggesting that the response of the two genotypic backgrounds to the environmental factors in each growing season was different. They also found that E-gene (maturity genes), E-gene × genotypic background, and E-gene × year were the major contributors to oil variability. The physiological and biochemical mechanisms of how these variables interact with each other and their effects on seed composition are still not understood (Bellaloui *et al.*, 2009).

ALTERED SOYBEAN FATTY ACIDS AND OIL ON THE HORIZON

Private and public sectors have been conducting intensive research to develop soybean with desirable fatty acids profile. Pioneer, a DuPont company, is developing soybean using accelerated yield technology under the brand Plenish™ with high oleic acid, low linolenic acid, less saturated fat, and high monounsaturated fat compared to commodity soybean oil. The Plenish™ high oleic soybean oil has zero *trans*-fat with high stability for extended high temperature use with nutritional benefits. The Plenish™ high oleic soybean is currently in the field and oil testing phase and was projected for a limited launch in 2012 (Pioneer, 2012). Monsanto has developed through traditional breeding techniques and launched Vistive® low linolenic soybean with low levels of linolenic acid and low *trans*-fat in processed oil which increase shelf life and flavor stability in processed foods. Monsanto is currently developing Vistive® Gold soybeans with *trans*-fat free and reduced saturated fat soybean oil with increased stability for food processing. The ultra-low linolenic acid (1%) non-GMO varieties IA2096, IA2097, IA3042, and Roundup Ready variety A2098RR, and low non-GMO saturated fatty acids varieties IA1024 and IA2095, were also developed (Warner and Fehr, 2008; Healthier Soybean Oils, 2008). Monsanto is also developing stearidonic acid (SDA) ω-3 soybeans rich with SDA, which the body converts to heart-healthy 1 eicosapentaenoic acid (EPA), one of the three ω-3 fatty acids used by the body. Both Vistive Gold and SDA ω-3 soybeans are developed through biotechnology and are currently in phase IV of the research and development pipeline, and are not commercially available yet (Monsanto, 2012). As advances in genetic modifications continue, it is more than likely that soybeans with designed fatty acid profile in seed oil could dominate the market in the future.

Acknowledgments

The authors are thankful to Dr David Pilbeam and Dr Linghe Zeng for reviewing the manuscript, and to Sandra Mosley for managing the reference list. Mention of trade names or commercial products in this publication is solely for the purpose of providing specific information and does not imply recommendation or endorsement by the US Department of Agriculture. The US Department of Agriculture (USDA) prohibits discrimination in all its programs and activities on the basis of race, color, national origin, age, disability, and where applicable, sex, marital status, familial status, parental status, religion, sexual orientation, genetic information, political beliefs, reprisal, or because all or part of an individual's income is derived from any public assistance program. (Not all prohibited bases apply to all programs.) Persons with disabilities who require alternative means for communication of program information (Braille, large print, audiotape, etc.) should contact USDA's TARGET Center at (202) 720-2600 (voice and TDD). To file a complaint of discrimination, write to USDA, Director, Office of Civil Rights, 1400 Independence Avenue, S.W., Washington, D.C. 20250-9410, or call (800) 795-3272 (voice) or (202) 720-6382 (TDD). USDA is an equal opportunity provider and employer.

References

Bachlava, E., Cardinal, A.J., 2009. Correlation between temperature and oleic acid seed content in three segregating soybean populations. Crop Sci. 49, 1328–1335.

Bellaloui, N., 2011. Effect of water stress and foliar boron application on seed protein oil fatty acids and nitrogen metabolism in soybean. Amer. J. Plant Sci. 2, 692–701.

Bellaloui, N., Smith, J.R., Ray, J.D., Gillen, A.M., 2009. Effect of maturity on seed composition in the early soybean production system as measured on near-isogenic soybean lines. Crop Sci. 49, 608–620.

Boydak, E., Alpaslan, M., Hayta, M., Gercek, S., Simsek, M., 2002. Seed composition of soybeans grown in the Harran region of Turkey as affected by row spacing and irrigation. J. Agric. Food. Chem. 50, 4718–4720.

Brumm, T.J., Hurburgh, C.R., Johnson, L.A., 1990. Cracking and dehulling shriveled and wrinkled soybeans. J. Am. Oil Chem. Soc. 67, 750–756.

Burton, J.W., 1991. Recent developments in breeding soybeans for improved oil quality. Fat Sci. Technol 93, 121–128.

Carrera, C., Martínez, M.J., Dardanelli, J., Balzarini, M., 2009. Water deficit effect on the relationship between temperatures during the seed fill period and soybean seed oil and protein concentrations. Crop Sci. 49, 990–998.

Cheesbrough, T.M., 1989. Changes in the enzymes for fatty acid synthesis and desaturation during acclimation of developing soybean seeds to altered growth temperature. Plant Physiol. 90, 760–764.

Clemente, T.E., Cahoon, E.B., 2009. Soybean oil: genetic approaches for modification of functionality and total content. Plant Physiol. 151, 1030–1040.

Collins, F.I., Sedgwick, V.E., 1959. Fatty acid composition of several varieties of soybeans. J. Am. Oil Chem. Soc. 36, 641–644.

Dardanelli, J.L., Balzarini, M., Martinez, M.J., Cuniberti, M., Resnik, S., Ramunda, S.F., Herrero, R., Baigorri, H., 2006. Soybean maturity groups, environments, and their interaction define mega-environments for seed composition in Argentina. Crop Sci. 46, 1939–1947.

Dornbos Jr., D.L., Mullen, R.E., 1992. Soybean seed protein and oil contents and fatty-acid composition adjustments by drought and temperature. J. Am. Oil Chem. Soc. 69, 228–231.

Federal Register-68 FR 41433 July 11, 2003. Food labeling; trans fatty acids in nutrition labeling, consumer research to consider nutrient content and health claims and possible footnote or disclosure statements; final rule and proposed rule. Verified on October 7, 2012. http://www.fda.gov/food/labelingnutrition/labelclaims/nutrientcontentclaims/ucm110179.htm.

Gao, J., Hao, X., Thelen, K.D., Robertson, G.P., 2009. Agronomic management system and precipitation effects on soybean oil and fatty acid profiles. Crop Sci. 49, 1049–1057.

Gibson, L.R., Mullen, R.E., 1996. Soybean seed composition under high day and night growth temperatures. Ibid 73, 733–737.

Healthier Soybean Oils, Iwoa State University, 2008. Verified on October 12, 2012, http://www.notrans.iastate.edu/.

Hou, G., Ablett, G.R., Pauls, K.P., Rajcan, I., 2006. Environmental effects on fatty acid levels in soybean seed oil. J. Am. Oil Chem. Soc 83, 759–763.

Howell, R.W., Collins, F.I., 1957. Factors affecting linolenic and linoleic acid content of soybean oil. Agron. J. 49, 593–597.

Hurburgh, C.R., Brumm, T.J., Guin, J.M., Hartwig, R.A., 1990. Protein and oil patterns in U.S. and world soybean markets. J. Am. Oil Chem. Soc. 67, 966–973.

Kane, M.V., Steele, C.C., Grabau, L.J., MacKown, C.T., Hildebrand, D.F., 1997. Early-maturing soybean cropping system: III. protein and oil contents and oil composition. Agron. J. 89, 464–469.

Lee, J.D., Bilyeu, K.D., Pantalone, V.R., Gillen, M.G., So, Y.S., Shannon, J. G, 2012. Environmental stability of oleic acid concentration in seed oil for soybean lines with *FAD2-1A* and *FAD2-1B* mutant genes. Crop Sci. 52, 1290–1297.

Lee, J.D., Oliva2, M.L., Sleper, D.A., Shannon, J.G., 2008. Irrigation has little effect on unsaturated fatty acid content in soya bean seed oil within genotypes differing in fatty acid profile. J. Agron. Crop Sci. 194, 320–324.

List, G.R., Erickson, D.R., Pryde, E.H., Brekke, O.L., Mounts, T.L., Falb, R.A., 1980. Handbook of soy oil processing and utilization. AOCS Press, Champaign, Illinois 267–354.

Company, Monsanto, 2012. Soybean seeds. Available. Verified on October 18, 2012. http://www.monsanto.com/products/Pages/soybean-seeds.aspx.

Oliva, M.L., Shannon, J.G., Sleper, D.A., Ellersieck, M.R., Cardinal, A.J., Paris, R.L., Lee, J.D., 2006. Stability of fatty acid profile in soybean genotypes with modified seed oil composition. Crop Sci. 46, 2069–2075.

Pham, A.T., Lee, J.D., Shannon, J.G., Bilyeu, K.D., 2010. Mutant alleles of FAD2-1A and FAD2-1B combine to produce soybeans with the high oleic acid seed oil trait. BMC Plant Biol. 10, 195.

Pioneer, 2012. Plenish™ high oleic soybeans. Available. Verified on October 18, 2012. http://www.pioneer.com/home/site/about/products/product-traits-technology/plenish/.

Piper, E.L., Boote, K.J., 1999. Temperature and cultivar effects on soybean seed oil and protein concentrations. J. Am. Oil Chem. Soc. 76, 1233–1242.

Rennie, B.D., Tanner, I.W., 1989. Fatty acid composition of oil from soybean seeds grown at extreme temperatures. J. Am. Oil Chem. Soc. 66, 1622–1624.

Scherder, C.W., Fehr, W.R., Shannon, J.G., 2008. Stability of oleate content in soybean lines derived from M23. Crop Sci. 48, 1749–1754.

Schlueter, J.A., Vasylenko-Sanders, I.F., Deshpande, S., Yi, J., Siegfried, M., Roe, B.A., Schlueter, S.D., Scheffler, B.E., Shoemaker, R.C., 2007. The FAD2 gene family of soybean: Insights into the structural and functional divergence of a paleopolyploid genome. Crop Sci. 47, S14–S26.

Seed Today, 2012. Pioneer is positive about Plenish™ high oleic soybeans. Verified on 10/11/2012. http://www.seedtoday.com/articles/Pioneer__Is_Positive_About_Plenish_trade__High_Oleic_Soybeans-119615.html.

Serretti, C., 1993. Influence of high protein, genotype, and environment on protein quality of soybean. Master's Thesis. Kansas State University, Manhattan, Kansas.

Soystat, 2008. World soybean production 2007. United Soybean Board. Available at. Verified on 10/8/2012. http://www.soystats.com/2008/Default-frames.htm.

Specht, J.E., Chase, K., Macrander, M., Graef, G.L., Chung, J., Markwell, J.P., Orf, H.H., Lark, K.G., 2001. Soybean response to water: a QTL analysis of drought tolerance. Crop Sci. 41, 493–509.

Tang, G.Q., Novitzky, W.P., Griffin, H.C., Huber, S.C., Dewey, R.E., 2005. Oleate desaturase enzymes of soybean: Evidence of regulation through differential stability and phosphorylation. Plant J. 44, 433–446.

Upchurch, R.G., Ramirez, M.E., 2010. Gene expression profiles of soybeans with mid-oleic acid seed phenotype. J. Am. Oil Chem. Soc. 87, 857–864.

Warner, K., Fehr, W., 2008. Mid-oleic/ultra low linolenic acid soybean oil: A healthful new alternative to hydrogenated oil for frying. J. Am. Oil. Chem. Soc. 85, 945–951.

Wilcox, J.R., Cavins, J.F., 1992. Normal and low linolenic acid soybean strains: Response to planting date. Crop Sci. 32, 1248–1251.

Wilson, R.F., 2004. Seed composition. In: Boerma, H.R., Specht, J.E. (Eds.), Soybeans: Improvement, Production, and Uses. ASA Monogr. 16. ASA, Madison, WI, pp. 621–669.

Wolf, R.B., Cavins, J.F., Kleiman, R., Black, L.T., 1982. Effect of temperature on soybean seed constituents: Oil, protein, moisture, fatty acids, amino acids and sugars. J. Am. Oil Chem. Soc. 59, 230–232.

CHAPTER

46

Processing Sesame Seeds and Bioactive Fractions

Ranjana Das, Chiranjib Bhattacharjee
Jadavpur University, Kolkata, India

CHAPTER POINTS

- Sesame is an important oil seed crop among generally consumed vegetable oil seeds.
- Various parts of sesame seed can be utilized as a source of bioactive components.
- Sesame hull can be utilized as a potential source of antioxidants.
- Sesame cake can be utilized as a source of bioactive peptides.
- Sesame cake is a source of hydrophilic antioxidants like sesame glucosides.
- Sesame unsaponifiables are a source of non polar glucosides like sesame lignans.
- Sesame bioactive modifications are a novel approach of product development.

INTRODUCTION

Sesame (*Sesamum indicum* L.) is a traditional health food, generally used to improve nutritional status and in the prevention of various diseases in Asian countries for thousands of years. Sesame (*Sesamum indicum* L.) comes from the family of Pedalliaceae and the genus Sesamum. It ranks ninth among the top thirteen oilseed crops which make up 90% of the world production of edible oil (Adeola *et al.*, 2010). It is called 'sesame' internationally, while it is called 'benni seed' in West Africa; 'simsim' in East Africa and 'Till' in India. It is generally called 'ridi' in the Northern States. Sesame seed with about 50% oil content is a good source of protein (about 20%) and lignans (e.g., sesamin and sesamolin) (up to 1.5%). The residue after oil extraction from dehulled oil seeds (seed meal/cake), is considered a cheaper source of protein, reported as less cholesterolemic and less atherogenic than animal protein (Kritchevsky, 1979), along with fiber and important phytochemicals, precisely phenolic antioxidants. Sesame meal based products are also found suitable for diabetics as they contain less carbohydrate and high protein (Bigoniya *et al.*, 2012). Sesame protein is very important as a protein source for human consumption due to the presence of balanced amino acid composition, especially sulfur containing amino acid methionine, which is very rare in other plant proteins. Thus, a diet enriched with this protein was found beneficial to patients suffering from kwashiorkor (Evans *et al.*, 1967). In sesame globulin fraction (67.3%), the ratio of the acids, lysine, and arginine is 0.67, which reported as more appropriate for cholesterol metabolism (Vioque *et al.*, 2000) as compared to casein with higher ratio of 2.

In the last two decades fundamental studies have opened a new field of research related to bioactive food components that not only help to ensure adequate nutrition, but may exert regulative activities in the human organism beyond basic nutrition, which are claimed to be health-enhancing components used to reduce the risk of disease or to enhance certain physiological functions like ACE inhibitor, opioid, immunomodulatory, mineral binding, antimutagenic, antimicrobial, and cytomodulatory effects (Nakano *et al.*, 2006). Proteolytic enzyme modifications of sesame protein isolate to protein hydrolysates have emerged as potential sources of bioactive peptides. Some nutraceutical characteristics of sesame seeds have been identified; including antioxidant (Mohamed *et al.*, 1998), hypocholesterolemic (Visavadiya *et al.*, 2008) hepatoprotective (Cengiz *et al.*, 2013) effects, as well as prevention of hypertension (Nakano *et al.*, 2006). Studies have revealed the potential of sesame meal as a source of diverse types of hydrophilic phyto-nutrients with well defined

Processing and Impact on Active Components in Food
http://dx.doi.org/10.1016/B978-0-12-404699-3.00046-9

functional properties, mostly lignan glucosides which are recognized as the natural antioxidant (Katsuzaki *et al.*, 1992). Sesame unsaponifiables are also known as a good source of non polar antioxidants basically sesame lignans (Mohamed *et al.*, 1998). Elluech *et al.* (2007) made detailed studies on the quality characteristics of sesame seed and some byproducts which evidenced the immense potential of sesame seed. The present review objective is to highlight some of the bio-actives present in sesame seed, their utilization pattern and processing routes for better applications and sustainable product developments.

HOW COMPOSITION IS ALTERED

Processing of Sesame Hull to Obtain Bioactives

Agricultural products are now receiving attention for sustainable product development. Various parts of sesame seeds can be consumed for nutraceuticals applications. Adequate studies have been conducted to evaluate the nutritional aspects of the sesame hull (Elluche *et al.*, 2007; Shahidi *et al.*, 2006). Prospects of sesame coat as a source of dietary fiber, polyphenols, and essential minerals (potassium, magnesium, phosphorus, and iron) have been evaluated by Elluche *et al.*, (2007). They have performed a detailed study on the chemical composition and importance of each component in the seed coat (Table 46.1). They studied the composition of the raw seed coat and the roasted seed coat. Dietary fiber and polyphenol content was reported to be reduced to a considerable extent after roasting, whereas mineral content of the roasted hull increased. Oil from "raw seed coat" and "roasted seed coat" showed higher content of chlorophylls, free fatty acids, and sesamol than raw seed oil. Shahidi *et al.* (2006), studied the antioxidant potential of the seed coat extract by "Trolox equivalent antioxidant capacity," "2, 2-diphenyl-1-picrylhydrazyl (DPPH) radical scavenging activity," and "chelation capacity." According to the authors, the total polyphenol content (TPC) of whole black sesame seeds and hull extracts were 29.9 ± 0.6 and 146.6 ± 0.6 mg catechin equivalents/g crude ethanolic extract, respectively whereas, those for white sesame were 10.6 ± 1.6 and 29.7 ± 0.9 mg catechin equivalents/g crude ethanolic extract. They reported the highest radical scavenging capacity of 14.4 ± 0.9, 2.5 ± 0.4, 94.9 ± 0.8 and 25.1 ± 0.4, for white sesame hulls and white sesame seeds, black sesame hulls, black sesame seeds, respectively at dosing level of 40 mg/mL. LDL oxidation inhibition at 100 ppm level was reported highest for black sesame hulls (96.7%) followed by those for white sesame hulls (84.6%), black sesame (78.4%), and white sesame seeds (57.3%). Sesame hull extract displayed good ferrous ion chelating capacities, in the range of 12%–46% and 17%–62% at 50 and 100 ppm levels, respectively. Dietary fiber characteristics and the antioxidant activity of the seed coat have also been analyzed by Elluche *et al.*, (2012). A considerable amount of the neutral sugar (15.11 g/100 g seed coat dry matter) was obtained in the dietary fiber of sesame seed coat with some insoluble uronic acid (10.52 g/100 g seed coat dry matter) and lignan (5.42 g/100 g seed coat dry matter). They also reported high polyphenol content of 9.9 mg/g of seed coat dry matter. Aqueous methanol, ethanol, and acetone have been attempted as extractants of the polyphenolic antioxidant from the seed coat.

Sesame coat was also utilized as a source of bioactive nano-crystalline cellulose with an objective to value addition to agricultural waste (Zhang *et al.*, 2007; Purkait *et al.*, 2011). As stated by associated authors, cellulose-based polymer products are favorable, being "ecosafe", with very high "tensile strength," and a high "Young's modulus." Attempts have been made by several authors to develop spherical cellulose nano-particles for various applications such as regenerative medicine (Fleming *et al.*, 2001) and optical applications (Revol *et al.*, 1998). Though the authors efficiently prepared the nano-cellulose particle (NCP), they have pointed out some of the limitations of its practical applications. The major problem was reported to involve drying of the nano-cellulose, as the fibers have a strong tendency to re-aggregate, which may cause partial loss of the nano-size effect. Another problem is associated with the dispersion of hydrophilic nano-cellulose particles into hydrophobic polymer matrixes. They therefore suggested

TABLE 46.1 Physiochemical Characteristics of Raw Sesame Seed and Byproducts (edited and adopted from Elleuch *et al.*, 2007)

		Byproducts	
Components	**RS**	**SC1**	**SC2**
Dry matter %	95.29 ± 0.19	83.79 ± 0.04	97.02 ± .09
Oil[A]	52.24 ± 0.34	12.21 ± 0.02	32.84 ± 2.32
Protein[A]	25.77 ± 1.02	10.23 ± 0.32	18.35 ± 0.16
Total fibers[A]	19.33 ± 1.97	42.03 ± 0.60	32.34 ± 1.36
Ash[A]	4.68 ± 0.20	23.90 ± 1.04	13.7 ± 0.844
Soluble sugars[A]	2.48 ± 0.09	0.97 ± 0.09	1.34 ± 0.05
Starch[A]	0.88 ± 0.01	1.33 ± 0.01	0.84 ± 0.01
Polyphenols[B]	87.77 ± 3.15	598.2 ± 4.47	260.6 ± 3.99

[A]*In %dry matter*
[B]*in mg/100 g dry matter.*
All given values are means of three determinations ± standard deviation. RS, raw sesame seed; SC1, Sesame seed coat; SC2, roasted sesame seed coat.

surface modification of nano-particles to increase their suitability as reinforcing fillers in polymer matrixes.

Processing of Sesame Oil for Bioactive Isolation

Except for direct consumption as edible oil, it can also be utilized as a source of sesame lignans, the major components in sesame oil unsaponifiables. Unsaponifiable lipid constituents of seed oils naturally contain hydrocarbon terpene alcohols, sterol, tocopherol, and other phenolic constituents which may act as an oxidation inhibitor under a range of conditions. The effectiveness of the lipid unsaponifiable matter in retarding oil deterioration has been demonstrated by many investigators. A combination of the number of minor constituents such as tocopherol, sterol, sesamol, squaline, and anti-polymerisable sterol was reported to have a synergistic role in increasing the oxidation stability of oils (Mohamed *et al.*, 1998). Beneficial health effects of sesame are primarily attributed to lignans and their glucosides. The main lignans in sesame oil are sesamin and sesamolin, which are reported to be in the concentration range of 1.0–2.0%. Sesame lignans such as episesamin, sesamolin, and sesamin have been implicated as having estrogenic and/or anti-estrogenic, anti-tumorigenic, and antioxidant properties. Sesame lignans also affect lipid metabolism and reduce cholesterol biosynthesis (Reshma *et al.*, 2010). They reported in detail the mode of extraction, separation, and characteristics of the sesame oil lignan for nutraceuticals application. Sesame oil was subjected to sequential extraction with methanol under selected conditions of temperature (70°C), time (100 min) and solvent : oil ratio (1 : 1). On HPLC analysis, the methanol concentrate showed a total lignan content of 9.32 ± 0.19% (6.54 ± 0.12% sesamin and 2.78 ± 0.31% sesamolin). The concentrate on low temperature crystallization at 4°C for the separation of lignan crystals, yields 51% of the lignans in the oil with 94.4% purity. Saponification of the removed crystals followed by desolventization of methanolic concentrate yielded unsaponifiable matter with 64% lignan. Chattopadhyaya *et al.* (2010) studied the effect of high and low dosage of sesame lignans (natural antioxidant) on nicotine-induced oxidative damage and genotoxicity on the liver tissue of rats. The authors designed the work to assess the natural antioxidants having a strong protective role against nicotine-induced toxicity. Experiments were conducted on male albino rats by injecting nicotine tartrate (3.5 mg/kg body wt/day for 15 days) subcutaneously and thereby supplementing sesame lignans (0.1 g/100 g diet and 0.2 g/100 g diet) orally to them. According to their study, a significant ($P < 0.01$) inhibiting effect to increase of triglyceride, LDL-cholesterol, total cholesterol, and VLDL-cholesterol, decrease of HDL-cholesterol and lipid peroxidative product was observed in plasma. Thus, the isolation of the sesame lignan from sesame oil unsaponifiable may be a potential processing route to obtain valuable bioactives.

Processing of Sesame Meal/Cake for Isolation of Bioactives

Sesame Cake as a Source of Flavonoids and Polyphenolic Antioxidant

As well as providing a highly stable oil, sesame seed also provides meal/cake that are normally used in sweetmeats and confectionery foods, and has a variety of medicinal applications (Kapoor, 2001). The seeds are reported to be used as demulcent in respiratory affections, infantile cholera, diarrhea, dysentery, and other bowel affections and bladder diseases. The seed powder is useful in amenorrhea, dysmenorrhea, ulcers, and bleeding piles (Kapoor, 2001). Additionally, the effect of whole sesame seed powder as a dietary supplement was assessed on 'normocholesterolaemic' animals with a view to investigate and compare its influence on lipid metabolism in both normal and hyperlipidaemic animals (Visavadiya *et al.* 2008). Sesame seed supplementation was reported to show significant beneficial effects on hyperlipidaemic/hypercholesterolaemic animals under oxidative stress conditions (Visavadiya *et al.*, 2008). This hypocholesterolaemic and antioxidant efficacy of the sesame seed could be attributed to the valuable dietary components (i.e. fibers, sterols, polyphenols, and flavonoids) and other phytoconstituents in seed cake. The major bioactive components found in defatted sesame meal are lignan glycosides, including sesaminol glucosides, pinoresinol glucosides, and sesamolinol glucosides (Katsuzaki *et al.* 1992). Sesame products containing these hydrophilic compounds have been reported to possess anti-oxidative, anti-carcinogenic, and neuroprotective activities. Flavonoids from sesame have hypolipidaemic and hypoglycemic activity which profoundly effects hepatic fatty acid oxidation and serum triacylglycerol level altering lipid metastasis in a potentially beneficial manner. Bigoniya *et al.* (2012) observed the preventive effect of sesame cake on hyperglycemia and obesity. In the oral glucose tolerance test, rats fed with high fructose diet supplemented with 2% and 4% sesame seed cake significantly reduced the plasma glucose after 120 min of glucose dosing, indicating an enhanced glucose tolerance with a probability that the intake of sesame seed cake supplementation can be taken as a therapeutic agent for the prevention of obesity and hyperglycemia. According to their observations, the animals with high fructose diet showed a significant ($p < 0.05$) reduced weight of liver ($p < 0.001$, $p < 0.05$) and kidney ($p < 0.001$) with sesame seed cake dosing of 2–4 g/kg. They reported the sesame seed coat treatment to significantly reduce the levels of total cholesterol ($p < 0.001$), TG ($p < 0.05$), LDL ($p < 0.001$) and serum-insulin ($p < 0.01$) but non-significant increasing effect on HDL cholesterol level. They concluded the anti-diabetic activity was a consequence of the antioxidant potential and to some extent due to the presence of dietary fibre and phytosterol. Dietary sesame powder has been shown to possess hypocholesterolaemic and enhanced antioxidant capacity (Visavadiya *et al.*, 2008). A literature survey reveals anti-diabetic activity of sesame on genetically diabetic rat (Bigoniya *et al.*, 2012). The hypoglycemic agent in defatted sesame was reported as pinoresinol by Wikul *et al.* (2012). Pinoresinol is exclusively encountered in sesame seeds, considered as a principal lignan associated with reduced risk of particular cancers and cardiovascular diseases. According to their study, the intake of defatted sesame seeds not only enhances glycemic control in patients but also suppresses the incidence of complications associated with diabetes mellitus. (+)-Pinoresinol showed inhibitory action against rat intestinal maltase with an IC_{50} value of 34.3 μM.

Attempts have been made by several groups of authors to explore the antioxidant capacity of sesame cake as a source of natural antioxidant (Suja *et al.*, 2004; Mohdaly *et al.*, 2011). They tried to assess the protective effect of sesame cake extracted antioxidants on storage stability of vegetable oils. They also reported the potential of the sesame cake extract as a source of natural antioxidants on different models such as β-carotene bleaching method, linoleic acid emulsion system by thiocyanate method, in DPPH free radical system. As reported, the sesame cake extract (SCE), at concentrations of 5, 10, 50, and100 ppm in vegetable oils, could significantly lower the 'diene value', 'p-anisidine value,' and 'peroxide value' of oil during storage at 60°C. Results revealed that crude extract was effective at 100 and 200 ppm levels and equivalent to butylated hydroxy toluene (BHT) at 200 ppm, whereas, purified extract showed comparable or better activities at 5, 10, 50, 100, and 200 ppm levels. They presented the HPLC profile of the cake extract which evidenced the presence of sesamol, sesamin, sesamolin, sesaminol diglucoside, and sesaminol triglucoside in the extract. Several biological effects of these lignan glycosides have been studied. Defatted sesame flour containing 1% sesaminol glucoside has been reported to decrease susceptibility to oxidative stress in hypercholesterolaemic rabbits and has potential as an anticancer agent (Liao *et al.*, 2010). Mohdaly *et al.* (2011) have evaluated the antioxidant effectiveness of sesame cake extract during oxidation of sunflower oil and soybean oil under accelerated oxidation conditions and compared the antioxidant activity with commercially available antioxidants (BHA, BHT, and TBHQ). They obtain the total phenolic, flavonoid and flavonol contents in sesame coat extract as 1.94 (mg gallic acid equivalent (GAE)/g dry weight (dry weight), 0.88 (mg quercetain equivalent/g dry weight), and 0.40 (mg quercetain/g dry weight), respectively. 2,2-azinobis (3-ethylbenzthiazoline sulfonate) (ABTS) radical scavenging activity, 1,1-diphenyl-2-picrylhydrazyl (DPPH) radical scavenging capacity, and β-carotene/linoleic acid test model were used to assess the antioxidant efficacy of sesame coat extract. They studied the effect of the sesame cake extract on the "peroxide value" (measure of the concentration of peroxides and hydroperoxides formed in the initial stages of lipid oxidation), the "p-anisidine value" (a measure of secondary lipid oxidation), and stabilized "conjugated diene and triene" content of the oil with results comparable to commercially available antioxidants at all concentration ranges.

Sesame Cake as Source of Protein Based Bioactives

Sesame cake is recognized as being a good source of food grade protein (40–50%) with well balanced amino acid composition and excellent physiological importance (Visavadiya and Narasimhacharya, 2008). A study was conducted to examine the dose dependant effects of sesame seed powder as a dietary supplement on hypercholesterolaemic and oxidative stress conditions in male albino rats at 5% and 10% dose levels along with normal and hypercholesteraolaemic diet. Authors reported that, feeding of sesame seed powder to hypercholesterolaemic rats causes a significant decline in plasma lipid, hepatic total lipid, cholesterol levels, and plasma LDL-cholesterol levels and an increase in plasma HDL-cholesterol levels with increased fecal excretion of neutral sterol, cholesterol, and bile acid, along with increases in hepatic HMG-CoA reductase activity and bile acid content. Sesame seed application improved the hepatic antioxidant status (catalase and SOD enzyme activities) with a reduction in lipid peroxidation (Visavadiya and Narasimhacharya, 2008). Studies have revealed the nutraceuticals potential of sesame protein hydrolysate (Nakano *et al.*, 2006). Sesame peptides powder exhibited angiotensin I-converting enzyme (ACE) inhibition activity, and temporarily significant decrease in systolic blood pressure in spontaneously hypertensive rats suggesting that, sesame peptides powder can be a beneficial ingredient for preventing and providing therapy against hypertension and its related diseases (Nakano *et al.*, 2006). A recent study has revealed the antioxidant activity and antihyperlipidemic effect of sesame protein isolate in albino rats (Biswas *et al.*, 2010). As per their studies, 49% and 64% lowering of plasma lipid peroxidation as well as 36% and 56% lowering of "lipoprotein oxidation susceptibility" was observed in sesame protein isolate and isolate containing cholesterol group, compared to control (i.e. casein and casein containing cholesterol) groups. A significant lowering of erythrocyte membrane lipid peroxidation (68% and 63% lowering in sesame protein isolate and isolate containing cholesterol groups) and liver lipid peroxidation (61% and 76% lowering in the sesame protein isolate and isolate containing cholesterol group than the corresponding control casein groups) was also reported which evidenced the nutraceutical relevance of the sesame protein. Enzymatic modification of sesame protein to enhance functional properties has already been reported (Bandyopadhyay and Ghosh, 2002) and study also conducted on antioxidant potential and "fibrinolytic activity" of sesame protein hydrolysate (Liu and Chiang, 2008). Antibacterial activity of sesame peptide has also been assessed by Das *et al.* (2012a), on pathogenic bacterial growth. They emphasized on the synthesis of sesame protein hydrolysate and fractionation of protein hydrolysate to get the most potent peptide fraction (molecular weight <1 g/mole), and identification of the potent peptide with amino acid composition and pharmacological evaluation on pathogenic bacteria. Inhibition to the growth of both gram negative bacterium (*P. aeruginosa*) and gram positive bacterium (*B. subtilis*), was indicative of its bacteriostatic effect and the probability to incorporate in the therapeutic formulation as bacteriostat (Table 46.2). In the study of Wang *et al.* (2012), the metal chelating activities of sesame peptides has been evaluated. They developed a method for separating and identifying the Zn chelating peptide from trypsin derived sesame protein hydrolysate. Some potential metal chelating peptides (Ser-Met, Leu-Ala-Asn and Asn-Cys-Ser) were synthesized with a peptide 'Asn-Cys-Ser', which exhibited higher zinc and iron chelating abilities than glutathione. All the studies made so far suggest the probability of using sesame meal based protein products as bioactive ingredients of the future.

TABLE 46.2 Antibacterial Effect of Sesame Peptide with Varying Concentrations

Sample Concentration (μg)	*Pseudomonous aeruginosa* Inhibition zone length(mm)			*Bacillus subtilis* Inhibition zone length(mm)		
	$^{a}P_1$	$^{b}P_2$	$^{c}P_3$	$^{a}P_1$	$^{b}P_2$	$^{c}P_3$
10	tr	1	3	tr	<1	2
20	1	3	4	<1	1	3
40	1	4	6	1	>1	3
50	>1	4	>6	1	>1	>3

[a]P1, Peptides in permeate of 5-kDa ultrafiltration step; [b]P2, Peptides in permeate of 5-kDa ultrafiltration step; [c]P3, Peptides in permeate of 5-kDa ultrafiltration step.
tr, trace.
Adapted from Das et al., *2012a.*

OTHER WAYS IN WHICH COMPOSITION IS ALTERED

Sesame seed can also be utilised for some speciality applications, though studies related to unique applications are scanty but hope to be explored in the near future by researchers. Utilisation of sesame cake in bioprocesses for the production of industrial bioproducts can add value to the oil cakes utilisation pattern. Their application in bioprocesses also assists in bioremediation and biological detoxification of hazardous compounds. Their application in fermentation technology has resulted in the production of bulk-chemicals and value-added products such as enzymes, amino acid, organic acids, single-cell protein, mushrooms, and biologically active secondary metabolites. These can be used in the production of enzymes, biopesticides, antibiotics, biochemicals, growth supplement for nematodes, and as a source of protein hydrolysate (Ramachandran *et al.*, 2007). Preparation and utilisation of the nano/ submicron range sesame glucoside also add value to the sesame meal. In pharmaceutical industries, the application of nanotechnology is well established in enhancing the bioavailability of drug components. Some attempts have also been made on the preparation of nano-size lignan-glycosides from sesame meal with enhanced transport and absorption efficiency in Caco-2 cell monolayer (Liao *et al.*, 2010). Nano-/sub-microsized lignan glycosides from sesame meal (LGSM) were produced using a "wet-milling" process. N-LGSM showed 30.0–64.1% higher transport and absorption efficiency than LGSM at each time point in Caco-2 cell monolayer model. Authors assumed the nanonization to be an useful way to enhance the efficiency of intestinal absorption and bioavailability of sesaminol triglucoside. Hung *et al.* (2013) assessed the effectiveness of the nano-particles in practical application. They aimed to evaluate the mutagenicity and oral toxicity of nano-/ submicro-sized ingredient, lignan glycosides from sesame meal (N-LGSM). They made conclusive statement that neither LGSM nor N-LGSM caused mutagenicities nor oral toxicities following sub-acute administration at the levels up to 1000 and 200 mg/ kg of body weight for 28 consecutive days indicating a safe and prospective processing method for sesame lignan glucosides. There are several aspects of the sesame seed that are yet to be explored. Nano size protein also emerges as a new route of sesame protein utilization in drug delivery and food industries. Though sufficient data is not available at present, nano-size sesame protein or peptide or nano-encapsulation of the peptide hope to add value to the sesame meal industries and may be a new approach to sustainable product development.

ANALYTICAL TECHNIQUES FOR RECOVERY OF BIOACTIVES FROM SESAME SEED

Recovery of Antioxidants from Hull

Isolation of the polyphenol is primarily based on the solvent extraction principle using varying proportions of the methanol and water from defatted sesame coat. As reported by Elluche *et al.* (2007), polyphenols were isolated by extraction with methanol:water (60:40, v/v) three times, from an oil-in-hexane (5g:10ml) solution. Sesamol was isolated by extraction with potassium hydroxide solution (potassium hydroxide:ethanol:water, 10:20:80, w/v/v) three times, from an oil-in-isooctane solution. Shahidi *et al.* (2006) used a similar approach for isolation of the polyphenolic antioxidant from defatted seed coat using 80% aqueous methanol.

Recovery of Lignan from Sesame Oil and Sesame Cake

As the literature reviewed, the isolation processes of lignans are largely based on solvent extraction route and supercritical extraction route (Hu *et al.* 2004). As reported in the study of Reshma *et al.* (2010), an immiscible organic solvent for extracting the lignans from sesame obtained a lignan content of only 32.23% in the extract, based on total lignans in oil. In supercritical extraction, the reported percent yield of extracts ranged from 21% to 24%. Similarly, other techniques have also been tried, wherein the oil was subjected to liquid extraction, followed by saponification and precipitation at 4°C; with very poor yield of extraction. The process adopted by Reshma *et al.* (2010) described a simple and less cumbersome procedure of extraction and crystallization, with saponification only as an additional step, making the process most economical and effective for producing lignans with high yield and purity lignans from the extract. Figure 46.1 represents the steps of the recovery of sesame lignans from sesame oil and lignan glucosides from sesame cake (Shyu *et al.*, 2002).

Recovery of Nano-Cellulose from Sesame Husk

Recovery of nano-cellulose particle (NCP) from sesame husk has been presented by Purkait *et al.* (2012) and the steps are shown in Figure 46.2. After chemical delignification (0.7% $NaClO_2$ treatment) followed by proper washing with sodium bisulfate and 17.5% NaOH the light straw yellow residue was acid digested and the

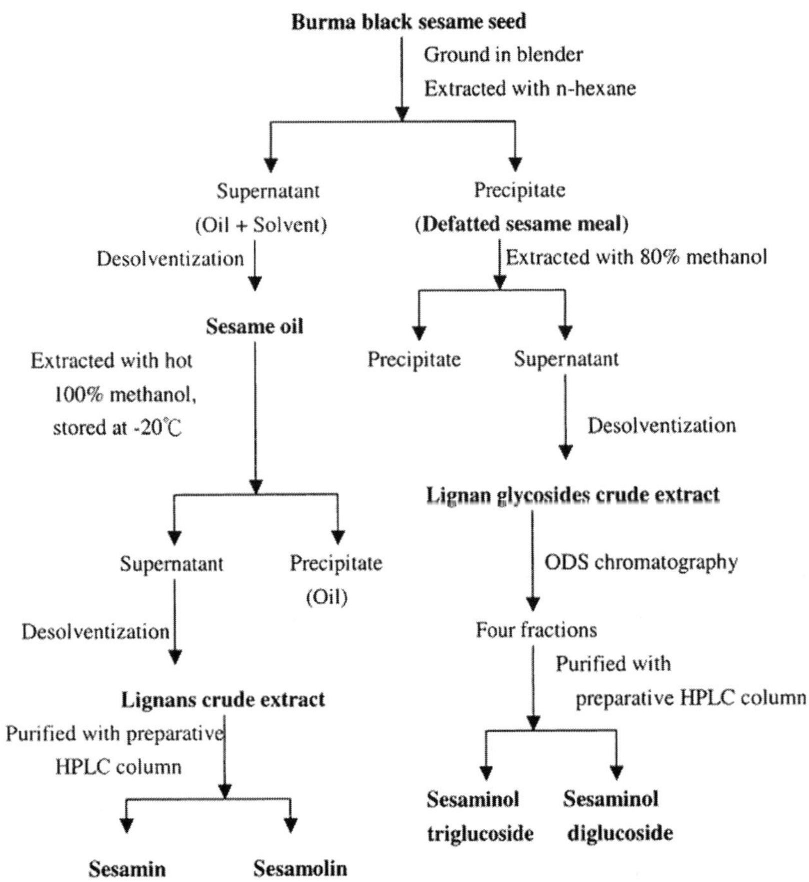

FIGURE 46.1 Schematic diagram for the isolation of the sesame lignin and the lignin glucosides from sesame seed. *Adapted from Shyu* et al., *2002.*

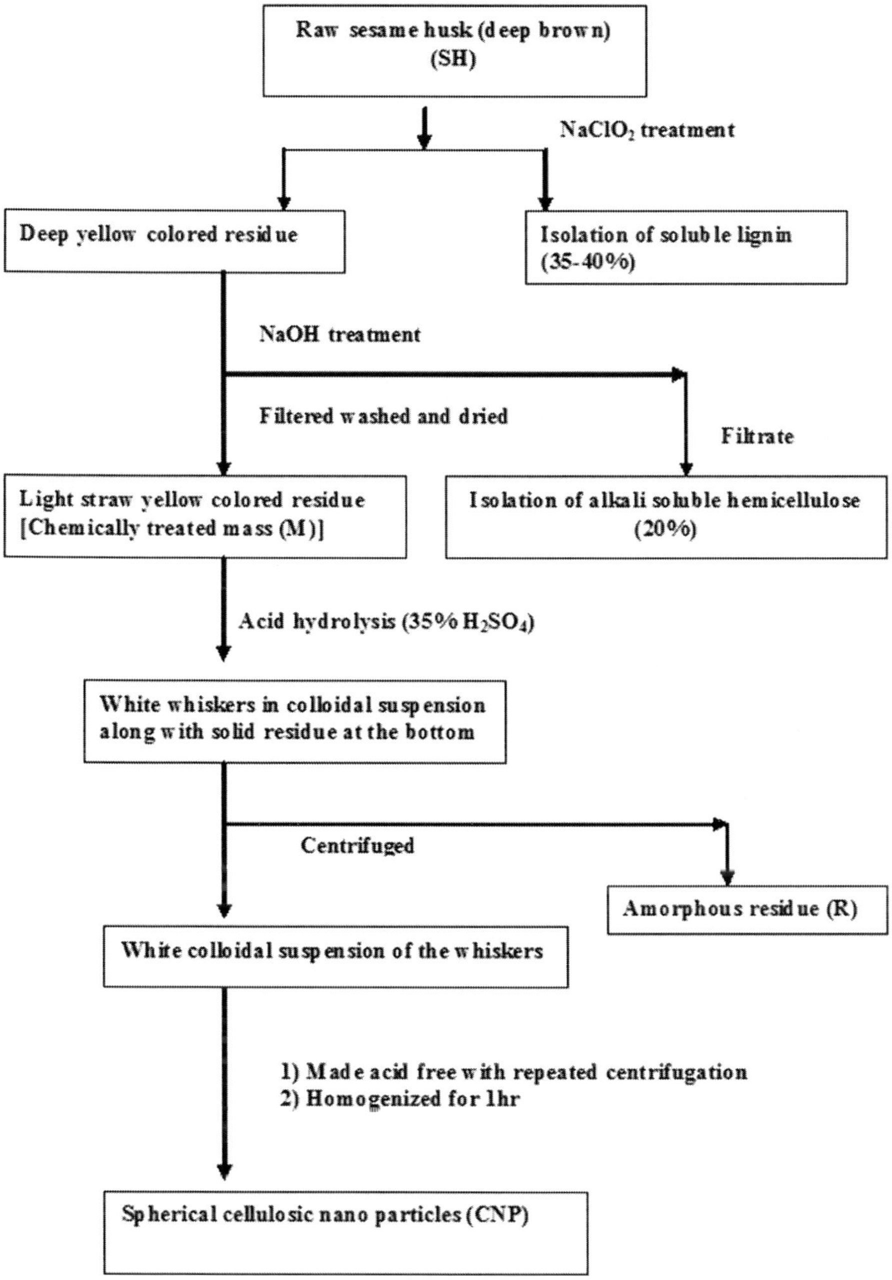

FIGURE 46.2 Flow-sheet diagram showing the preparation of cellulose nanoparticles from sesame husk. *Adapted from Purkait* et al., *2011.*

white colloidal solution made acid free with multiple washing and finally homogenized to get the spherical nano-cellulose particle.

Recovery of Hydrophilic Antioxidants from Sesame Meal

Das *et al.* (2011) have reported an ultrafiltration- nanofiltration based technique to recover the polar hydrophilic antioxidants from sesame meal. This study describes a novel approach for recovery and concentration of sesame glucosides from defatted sesame cake using membrane separation route. Aqueous extraction of the seed meal followed by two step membrane fractionation (ultrafiltration-nanofiltration) was performed to obtain sesame glucoside concentrate. This method described the recovery of sesame glucosides from sesame meal by aqueous extraction method without using any organic solvents. A comparative study was made regarding the efficiency of the extracted antioxidants with conventional solvent extracted antioxidants. In this process, about 0.7%±0.29 recovery of total meal glucoside was achieved along with co-product sesame protein isolate at yield value of 36.50%. Figure 46.3 shows the steps involved in the aqueous extraction of the sesame glucosides from sesame meal with the potential to recover seed protein isolate as a

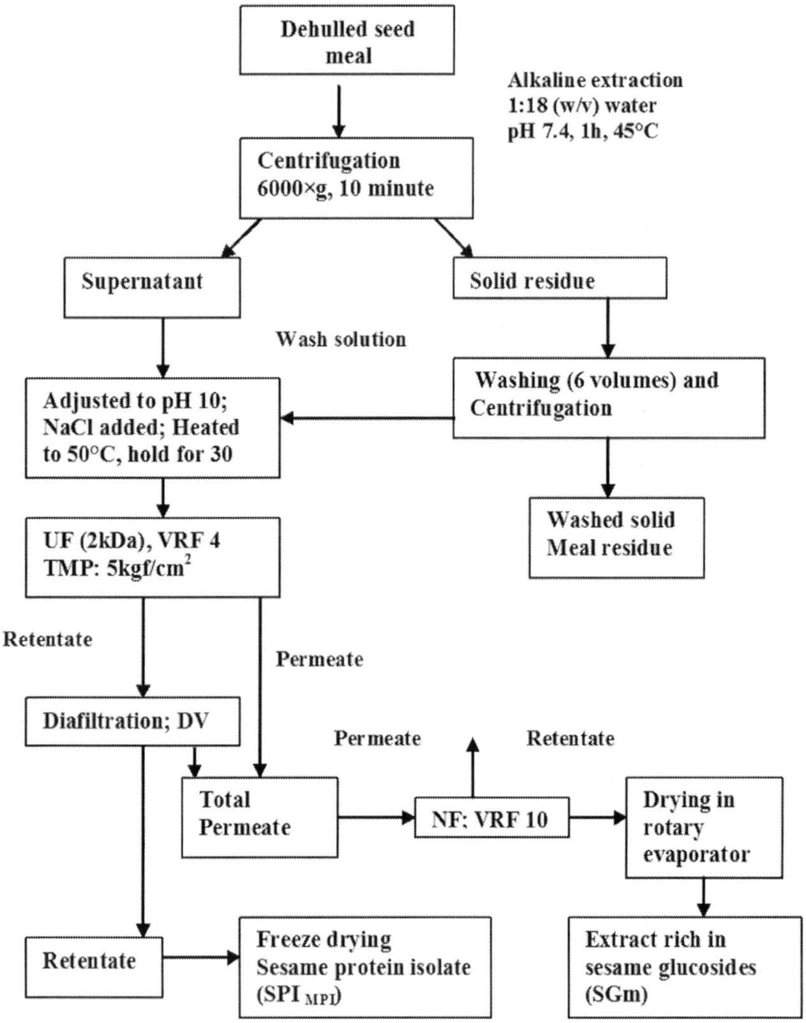

FIGURE 46.3 Scheme for recovery of sesame glucosides from seed meal using ultrafiltration-nanofiltration based technique. *Adopted from Das* et al., *2011.*

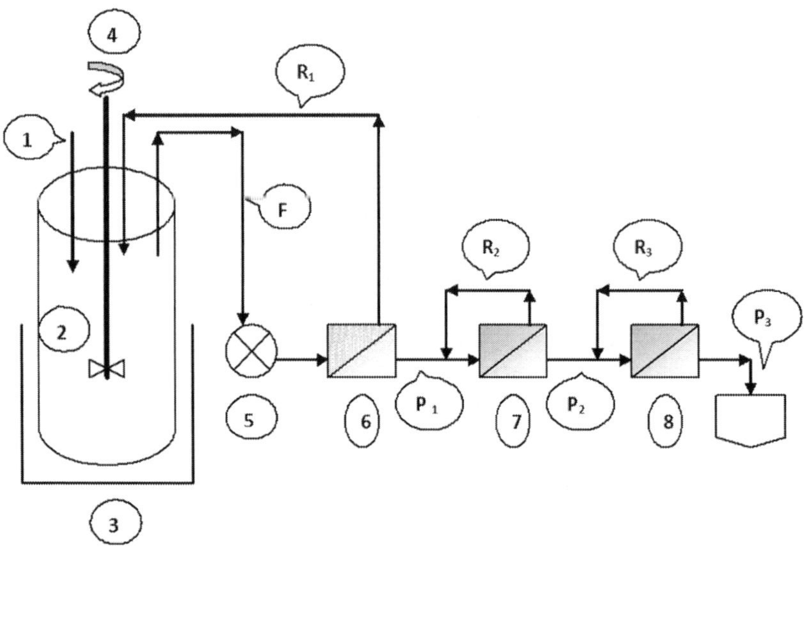

FIGURE 46.4 Operational set up for hydrolysis of sesame protein isolate in EMR (side stream recycle membrane bioreactor) and subsequent UF fractionation. (1) Feeding line for protein isolate; (2) reaction vessel; (3) constant temperature bath; (4) stirrer; (5) peristaltic pump; (6) cross flow UF with 5 kDa membrane; (7) UF with 2 kDa membrane; (8) UF with 1 kDa membrane; F-feed; P_1, P_2, P_3 permeates; R_1, R_2, R_3 retentate. *Adapted from Das* et al., *2012b.*

process byproduct. However, there are several technical hurdles that will need to be circumvented before this technique becomes reality and it requires scaling up to assess the feasibility.

Recovery of Sesame Peptides with Antioxidant and Antibacterial Activity from Sesame Cake

Das *et al.* (2012b) reported an enzyme membrane reactor (EMR) based continuous method (Figure 46.4) for selective isolation of the antioxidative peptides and the peptides with bacteriostatic activity. In this approach, after enzyme dosing, hydrolysis continued for 10 min in recycle mode (with both permeate and retentate recycled back to reaction vessel) at 50 ± 1°C, followed by permeate collection (product removal) through 5 kDa membrane for rest of reaction period (20 min). Retentate was recycled continuously in reaction vessel (without any bleed) for better hydrolysis and permeate was collected for successive fractionation steps. Permeate from 5 kDa UF fractionation was further fractionated by 2 kDa and 1 kDa membranes successively for more selective fractionation of peptide mixtures. Each isolated fraction was analyzed in high performance liquid chromatography to assess the molecular weight pattern and amino acid profiles.

Recovery of Zn Binding Peptide from Sesame

To prepare sesame protein hydrolysate, enzymatic hydrolysis was performed by using Papain (0.5–21/mg, pH 7.0, 50°C), alcalase (2.4 Au/g, pH 8.5, 60C) and trypsin (2501/mg, pH 8.0, 55°C) at their optimal conditions (Wang *et al.*, 2012). Separation of the protein hydrolysate to four fractions was performed by elution of buffer (0.1 M of Na_2HPO_4) at pH 6.0, containing NH_4Cl. Fraction 1 was reported having the strongest antioxidant activity, while Fraction 2 and 3 reported higher metal chelating activities. Fraction 4 was reported to have less activity than Fraction 2 and 3 due to the fewer amounts.

References

Adeola, Y.B., Augusta, C.O., Oladejo, T.A., 2010. Proximate and mineral composition of whole and dehulled Nigerian sesame seed. African J. Food Sci. Technol. 1 (3), 071–075.

Bandyopadhyay, K., Ghosh, S., 2002. Preparation and characterization of papain modified sesame (*Sesamum indicum* L.) protein isolates. J. Agric. Food Chem. 50, 6854–6857.

Bigoniya, P., Nishad, R., Singh, Shekhar, 2012. Preventive effect of sesame seed cake on hyperglycemia and obesity against high fructose-diet induced Type 2 diabetes in rats. Food Chem. 133, 1355–1361.

Biswas, A., Dhar, P., Ghosh, S., 2010. Antihyperlipidemic effect of sesame (*Sesamum indicum* L.) protein isolate in rats fed a normal and high cholesterol diet. J. Food Sci. 75, H274–H279.

Cengiz, N., kavak, S., Guzel, A., Ozbek, H., Bektas, H., Him, A., Erdogan, E., Balahoroglu, R., 2013. Investigation of the hepatoprotective effects of sesame (*Sesamum indicum* L.) in carbon tetrachloride-induced liver toxicity. J. Membrane Biol. 246, 1–6.

Chattopadhyay, K., Mondal, S., Chattopadhyay, B., Ghosh, S., 2010. Ameliorative effect of sesame lignans on nicotine toxicity in rats. Food Chem. Toxicol. 48, 3215–3220.

Das, R., Bhattacharjee, C., Ghosh, S., 2011. Novel Approach to Recover Natural Antioxidants from Oil Seed Meal in Ultrafiltration-Nanofiltration-Based Technique. Indust. Eng. Chem. Res. 50, 12124–12133.

Das, R., Dutta, A., Bhattacharjee, C., 2012a. Preparation of sesame peptide and evaluation of antibacterial activity on typical pathogens. Food Chem. 131, 1504–1509.

Das, R., Ghosh, S., Bhattacharjee, C., 2012b. Enzyme membrane reactor in isolation of antioxidative peptides from oil industry waste: A comparison with non-peptidic antioxidants. LWT- food Sci. Technol. 47, 238–245.

Elleuch, M., Bedigian, D., Besbe, S., Blecker, C., Attia, H., 2012. Diatary fiber characteristics and antioxidant activity of sesame seed coats (Testae). Int. J. Food Properties 15, 25–37.

Elleuch, M., Besbes, S., Roiseux, O., Blecker, C., Attia, H., 2007. Quality characteristics of sesame seeds and by-products. Food Chem. 103, 641–650.

Evans, R.J., Bandemer, S.L., 1967. Nutritive value of some oil seed protein. Cereal Chem. 44, 417–426.

Fleming, K., Gray, D.G., Matthews, S., 2001. Cellulose crystallites. Chem. Euro. J. 7, 1831–1835.

Hu, Q., Xu, J., Chen, S., Yang, F., 2004. Antioxidant Activity of Extracts of Black Sesame Seed (*Sesamum indicum* L.) by Supercritical Carbon Dioxide. J. Agric. Food Chem. 52, 943–947.

Hung, W.L., Lu, C.H., Liao, C.D., Hwang, L.S., 2013. Safety evaluation of nano/sub-microsized lignan glycosides from sesame meal. Food Control 30, 129–136.

Kapoor, L.D., 2001. Hand book of Ayurvedic Medicinal Plants. Herbal Reference Library Edition. CRC Press, New York.

Katsuzaki, H., Kawasumi, M., Kawasaki, S., Osawa, T., 1992. Structure of Novel Antioxidative Lignan Glucosides Isolated from Sesame Seed. Biosci. Biotechnol. Biochem. 56, 2087–2088.

Kritchevsky, D., 1979. Vegetable protein and atherosclerosis. J. Am. Oil Chem. Soc. 56, 135–140.

Liao, C.-D., Hung, W.-L., Jan, K.-C., Yeh, An-I., Ho, C.-T., Hwang, L.S., 2010. Nano/sub-microsized lignan glycosides from sesame meal exhibit higher transport and absorption efficiency in Caco-2 cell monolayer. Food Chem. 119, 896–902.

Liu, B.L., Chiang, P.S., 2008. Production of hydrolysate with antioxidative activity and functional properties by enzymatic hydrolysis of defatted sesame (*Sesamum indicum* L.). Int. J. Appl. Sci. Eng. 6, 73–83.

Mohdaly, Adel A.A., Smetanska, I., Ramadan, M.F., Sarhan, M.A., Mahmoud, A., 2011. Antioxidant potential of sesame (*Sesamum indicum*) cake extract in stabilization of sunflower and soybean oils. Indust. Crops Products 34, 952–959.

Mohamed, H.M.A., Awatiff, I.I., 1998. The use of Sesame unsaponifiable matter as a natural antioxidant. Food Chem. 62, 269–276.

Nakano, D., Ogura, K., Miakoshi, M., Ishii, F., Kawanishi, Kurumazuka, H.D., Kwak, C., Ikemura, K., Takaoka, M., Moriguchi, S., Iino, T., Kusumoto, A., Asami, S., Shibata, H., Kiso, Y., Matsumura, Y., 2006. Antihypertensive effect of angiotensin I-converting enzyme inhibitory peptides from a sesame protein hydrolysate in spontaneously hypertensive rats. Biosci. Biotechnol. Biochem. 70, 1118–1126.

Purkait, B.S., Ray, D., Sengupta, S., Kar, T., Mohanty, A., Misra, M., 2011. Isolation of Cellulose Nanoparticles from Sesame Husk. Indust. Eng. Chem. Res. 50, 871–876.

Ramachandran, S., Kumar Singh, S., Larroche, C., Soccol, C.R., Pandey, A., 2007. Oil cakes and their biotechnological applications—A review. Bioresource. Technol. 98, 2000–2009.

Reshma, M.V., Balachandran, C., Arumughan, C., Sunderasan, A., Sukumaran, D., Thomas, S., Saritha, S.S., 2010. Extraction, separation and characterisation of sesame oil lignan. Food Chem. 120, 1041–1046.

Revol, J.F., Godbout, L., Gray, D.G., 1998. Solid self-assembled films of cellulose with chiral nematic order and optically variable properties. J. Pulp and Paper Sci. 24, 146–149.

Shahidi, F., Liyana-Pathirana, C.M., Wall, Dana S., 2006. Antioxidant activity of white and black sesame seeds and their hull fractions. Food Chem. 99, 478–483.

Shyu, Y.-S., Hwang, L.S., 2002. Antioxidative activity of the crude extract of lignan glycosides from unroasted Burma black sesame meal. Food Res. Int. 35, 357–365.

Suja, K.P., Abraham, John T., Thamizh, Selvam N., Jayalekshmy, A., Arumughan, C., 2004. Antioxidant efficacy of sesame cake extract in vegetable oil protection. Food Chem. 84, 393–400.

Vioque, J., Sanchez-Vioque, R., Clemente, A., Pedrocheand, J., Millan, F., 2000. Partially hydrolysed rapeseed protein isolate with improved functional properties. J. Am. Oil Chem. Soc. 77, 447–450.

Visavadiya, N.P., Narasimhacharya, A.V., 2008. Sesame as a hypocholesteraemic and antioxidant dietary component. Food Chem. Toxicol. 46, 1889–1895.

Wang, C., Bo, L., Jing, A., 2012. Separation and identification of zinc-chelating peptides from sesame protein hydrolysate using IMAC-Zn^{2+} and LC-MS/MS. Food Chem. 134, 1231–1238.

Wikul, A., Damsud, T., Kataoka, K., Phuwapraisirisan, P., 2012. (+)-Pinoresinol is a putative hypoglycemic agent in defatted sesame (*Sesamum indicum*) seeds though inhibiting a-glucosidase. Bioorganic & Med. Chem. Letters 22, 5215–5217.

Zhang, J., Elder, T.J., Pu, Y., Ragauskas, A.J., 2007. Facile synthesis of spherical cellulose nanoparticles. Carbohydrate Polymers 69, 607–611.

CHAPTER

47

Processing and Utilization of Jackfruit Seeds

Charu Lata Mahanta, Dipankar Kalita

Tezpur University, Assam, India

CHAPTER POINTS

- Jackfruit seed flour has potential in food formulation because of its high starch, dietary fiber, protein, and mineral content. It is an abundant and cheap source of these compounds.
- Seeds have an appreciable amount of phenolic content and antioxidant activity
- Jackfruit seed starch can withstand thermal and mechanical shear and therefore is suitable in food industries.
- The seed starch has high swelling power and solubility and can be utilized as a thickener and stabilizer in food products.
- Differently modified starches have altered properties indicating their specific end use.
- Gel of native starch has good freeze-thaw stability which is lost on acid alcohol modification.

INTRODUCTION

Jackfruit (*Artocarpus heterophyllus* Lam.) is a tropical fruit native to India, Thailand, and Indonesia and is now found in many parts of Asia, Africa, South America, and northern Australia. Jackfruit belongs to the Moraceae family and grows in warm and moist regions. Ripe jackfruits are large sized fruits measuring between 22–90cm in length and 13–50cm in diameter with weight ranging from 2–36kg. The fruit may contain between 100 and 500 seeds which represents 18–25% (db) of the fruit weight while the pulp represents 30% of the fruit weight (Madrigal-Aldrina *et al.*, 2011). The raw fruit is eaten as a vegetable. Ripened jackfruit pulp is eaten both in raw and processed forms like canned juice (Seow and Sanmugam, 1992), and fruit leather. The pulp is cream in color due to the presence of carotenoids. The shape of the seeds of the ripe fruit varies from oval to oblong ellipsoidal to round. The seeds are light brown in color and are generally 2–3cm in length and 1–1.5cm in diameter. Seeds of the ripe fruit have around 55% moisture and therefore do not keep well for long. Seeds are eaten after boiling, steaming, and roasting. Boiled seeds are also canned in brine and in tomato sauce.

In recent years there has been a growing search for alternative sources of starch to meet industry demands, because of which jackfruit seed is gaining the attention of researchers. The seeds have considerable amounts of starch and protein which can be industrially exploited. Though jackfruit seed is not used by the industry much, recent studies, which are few in number, have highlighted the properties of starch and its utilization. This chapter reviews the effect of processing on starch that can provide indicators for its utilization.

CHEMICAL COMPOSITION OF JACKFRUIT SEED

The reported values for the nutrients in jackfruit seed seem to vary. These variations can be attributed to the differences in source, variety, environment, and ripeness of the fruit. Nonetheless, the seed is a fairly good source of starch and protein. Table 47.1 presents the chemical composition reported by several workers. It has been demonstrated by ICP-OES (Gupta *et al.*, 2011) that jackfruit seed flour has high levels of potassium (786.6ppm, db) followed by calcium (29.5ppm, db), and sodium (28.4ppm, db).

The phenolic compounds are preferentially located in the seed. The phenolic compounds are estimated at 2770mg/100g and carotenoid content at 1910µg/100g (Ayala-Zavala *et al.*, 2011). The phenolic compounds are responsible for the antioxidant properties of the seeds as evident from the appreciable DPPH and ABTS radical

Processing and Impact on Active Components in Food
http://dx.doi.org/10.1016/B978-0-12-404699-3.00047-0

TABLE 47.1 Chemical Composition of Jackfruit Seed Reported by Different Researchers

Moisture	Protein	Fat	Ash	Fiber	Carbohydrate	Energy	References
54.06 ± 0.85 (% as is basis)	21.99 ± 0.51 (% as is basis)	0.70 ± 0.01 (% as is basis)	3.14 ± 0.22 (% as is basis)	7.32 ± 0.14 (% as is basis)	ND	ND	Madrigal-Aldana *et al.* (2011) (seed with spermoderm)
6.09 ± 0.01 (%db)	13.50 ± 0.06 (%db)	1.27 ± 0.01 (%db)	2.70 ± 0.02 (%db)	3.19 ± 0.01 (%db)	79.34 ± 0.06 (%db)	382.79 ± 1.20 (Kcal/100 g)	Ocloo *et al.* (2010) (seed without spermoderm)
5.10 (%db)	17.20 (%db)	2.20 (%db)	3.60 (%db)	3.06 (%db)	74.00 (%db)	ND	Singh *et al.* (1991) (seed with spermoderm)

[a]ND = not determined

scavenging effects and metal ion chelating activity (Gupta *et al*, 2011). On freeze drying, the AEAC and FRAP values tested for antioxidant activity were 25.4 μmol/g and 2.8 μmol/g, respectively, and the phenolic content was 88.2 mg/g expressed as gallic acid equivalent (Soong and Barlow, 2004). The dietary fiber content in jackfruit seeds is around 28% (db) with around 2% soluble dietary fiber and 26% insoluble dietary fiber. This good amount of dietary fiber will have positive physiological effects.

The seeds contain antinutritional factors such as lectins, trypsin inhibitor, saponin, tannin, phytin, and oxalates (Ocloo *et al.*, 2010). More than 50% of the protein in jackfruit seed is comprised of jacalin, a lectin having molecular weight of 65 kDa. Jacalin is a tetramer with two chains, the α-chain with 133 amino acids and the β-chain with 20–21 amino acids (Kabir, 1998). Lectins are considered to be antinutritional as they bind with carbohydrates and escape digestion. The undigested lectins bind with the glycosyl groups of the cells lining the digestive tract that affects the turnover and loss of gut epithelial cells, damaging the luminal membranes of the epithelium, interfering with nutrient digestion and absorption, changing the gut microflora, and lowering the immune state of the digestive tract (Vasconcelos and Oliveira, 2004). The practice of processing that involves heat treatment prior to consumption reduces/destroys the antinutritional factors in the seed.

JACKFRUIT SEED STARCH

Native jackfruit seed starch is off-white in color, odorless and fine with poor to medium flow property. The starch content of the seed is found to increase with maturity (Rahman *et al.*, 1999) and as the fruit ripens, some starch is hydrolyzed into sugars. Sucrose is the predominant sugar besides glucose and fructose. Starch yield varies between 12–28% (Oates and Powell, 1996; Kittipongpatana and Kittipongpatana, 2011; Tondang, 2008).

Extraction of Starch

The seeds after removal from the bulbs are washed with tap water to remove the cohesive material from the surface and air-dried for 2 days. The arils are manually separated from the seeds and the spermoderms are removed on lye peeling in 0.6 M potassium hydroxide solution for 3 h at room temperature. The pale-white cotyledons that are obtained are washed several times with distilled water to remove any traces of alkali on the seed surface. The seeds are ground with distilled water (1 : 3, seed to water) for 2 min and filtered through 200 μm standard sieve. The filtrate is centrifuged at 2000 rpm for 4 min and the sediment is resuspended in 0.5 M sodium thiosulfate solution (1 : 1, residue to solution) for 36 h. The suspension is intermittently stirred to remove any protein fractions adhering to the starch granules and centrifuged at 3000 rpm for 5 min. The brown layer comprising proteins and cell wall material is scraped off. The residue is neutralized with 0.1 M hydrochloric acid and washed twice with distilled water followed by 50% ethanol to remove any ions. The residue is dried, ground to pass through a 100-μm sieve to obtain starch powder (Dutta *et al.*, 2011). The starch slurry during the extraction process becomes pink in color and some workers use 0.5% sodium bisulfite at 0.05% concentration to decolorize.

Physicochemical Properties of Native Starch

Amylose Content

Amylose content in the starch appears to vary widely as reported by researchers. Amylose content ranges from 22–32% (Kittipongpatana and Kittipongpatana, 2011; Dutta *et al.*, 2011; Rengsutthi and Charoenrein, 2011), which is similar to that in potato (Singh *et al.*, 2003). The wide variation may be due to the effect of growing environment, maturity, and variety.

TABLE 47.2 Pasting Properties of Native and Acid-Modified Jackfruit Seed Starch

Properties	Native	ST15	ST22	ST30	LT1	LT5	LT10	LT15
PV (cP)[a]	3483[g]	46[c]	27[b]	17[a]	1747[f]	217[e]	98[d]	19[a]
HPV(cP)[a]	2658[g]	35[c]	10[b]	3[a]	1307[f]	68[e]	46[d]	4[a]
BV (cP)[a]	824[g]	10[a]	17[c]	14[b]	440[f]	151[e]	50[d]	15[bc]
CPV(cP)[a]	5073[f]	74[b]	38[a]	27[a]	2432[e]	295[d]	203[c]	32[a]
SV (cP)[a]	2440[g]	39[c]	28[b]	23[a]	1124[f]	227[e]	155[d]	26[ab]

ST15, ST22, and ST30 mean alcohol-acid modified starch for 15, 22, and 30 min, respectively. LT1, LT5, LT10, and LT15 mean alcohol-acid modified starch for 1, 5, 10, and 15 min, respectively.
[a]*Means with the same superscript letter do not differ significantly from one another* ($P < 0.05$).
Source: Dutta et al., *2011.*

GT

The GT, however, is different—being higher than most starches. The starch paste forms a rigid gel that is desirable in pie fillings, canned soups, and confectionery fillings and moulds. The gel has extremely high apparent rigidity value. Jackfruit seed starch has elevated GT range from 75–80°C which is higher than the GT of other starch sources such as rice, cassava, corn, wheat, and potato (Santos *et al.*, 2009).

Pasting Properties

Pasting properties are influenced by the extent of granule swelling, friction between the swollen granules, the amount of leached amylose, starch crystallinity, chain length of amylose and amylopectin molecules, and their breakdown and extent of gelatinization. HPV, i.e., the minimum viscosity at 95°C, is influenced by the extent of amylose leaching, granule swelling and competition for water by the leached amylose and swollen granules. CPV, i.e., the final viscosity at 50°C, is mainly the effect of retrogradation of the leached amylose that occurs during cooling. Breakdown is the reduction of viscosity during the cooking process that is affected by temperature and mixing stress. Setback is defined as the degree of reassociation between the starch molecules involving amylose. The pasting properties of jackfruit seed starch slurry at 8% concentration studied in a Brabender Viscoamylograph shows that the pasting temperature is 74.5°C and the paste shows almost negligible breakdown. Jackfruit seed starch is therefore very resistant to mechanical and thermal shear even after a high degree of swelling. In recent years, Brabender Viscoamylograph has been replaced by the RVA for measuring the pasting properties because of the small quantity of sample and short timeframe required for analysis. The RVA profile of native jackfruit seed starch is presented in Table 47.2 as reported by Dutta *et al.* (2011). The very low breakdown is indicative of the resistance of the starch granules to shear thinning. The low setback means either there is less leaching out of the amylose from the granule, or there may be differences in the molecular weight of the amylose and its ability to retrograde. Seed starch viscosity does not significantly change with pH change indicating the acid resistance of granules (Santos *et al.*, 2009).

Light Microscopy

It is observed from light microscopy (Figure 47.1a) that the granule size ranges from 5 to 12 μm (Tondang, 2008; Dutta *et al.*, 2011; Kittipongpatana and Kittipongpatana, 2011). The maltese cross is at the center dividing the granule into four equal segments (Figure 47.1b). The granule size is important because the size influences its functional properties such as swelling, solubility, and viscosity, that in turn influence its industrial application. The amylose content and percent light transmittance are strongly correlated with the granule size. The variation in granule size may be due to varietal difference, climatic conditions, and starch extraction methods. Extraction process may also cause granule damage.

SEM

SEM is used to study the transverse section of jackfruit seed and morphology of starch granules. The starch granules in the seeds are present in clusters in the cells separated by well-defined membranes made up of protein (Figure 47.1c). The membranes stain bluish purple with bromocresol blue. The SEM picture of the isolated starch granules reveals basically four types of structures, namely round, dome-shaped, trigonal, and tetragonal (Figure 47.1d) (Dutta *et al.*, 2011). The surface of the granules is smooth. The largest granules are dome-shaped, while the smallest ones are observed to be the trigonal types. All four granule shapes are almost equally represented.

XRD

Jackfruit seed starch exhibits A-type crystalline pattern with diffraction peaks at 2θ value of 15°, 17°, 18°, and 23°. XRD shows that the starch crystallinity varies from 28–35% (Kittipongpatana and Kittipongpatana, 2011; Madrigal-Aldana *et al.*, 2011). Being A-type starch

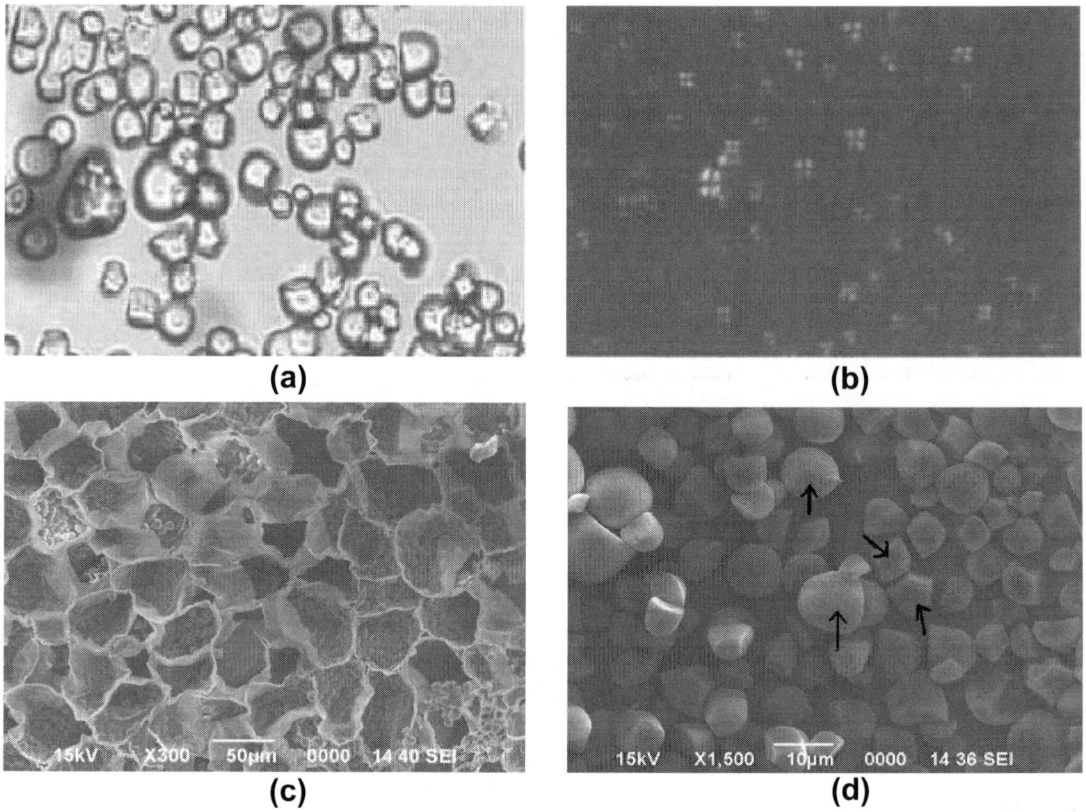

FIGURE 47.1 Micrographs of starch granules isolated from jackfruit seeds. (a) Light microscopy and (b) polarized-light microscopy (Source: Madrigal-Aldana *et al.*, Starch/Starke 2011, 63, 364–372); SEM pictures of (c) transverse section of jackfruit seed and (d) native. Arrow marks (↓) indicate shapes of starch granules in (d) (Source: Dutta *et al.*, Food Chemistry 2011, 128, 284–291).

similar to cereal starch, jackfruit starch is also expected to have a high amount of slowly digestible starch that has a beneficial effect in the body (Zhang *et al.*, 2006).

Thermal Properties

The thermal properties of jackfruit seed starch were assessed in a DSC. DSC is a technique that measures both the loss of molecular order during the gelatinization process when starch is heated in the presence of water and molecular reorganization which occurs when the gelatinized starch undergoes retrogradation during the cooling and aging period. The temperature of the characteristic transitions, namely onset temperature is 81°C, peak temperature is 84°C, and end temperature is 87°C. The enthalpy of transition (ΔH) is 17 J/g (Tongdang, 2008). The enthalpy for order–disorder for any native starch is between 10–20 J/g, which for jackfruit seed starch is on the higher side similar to waxy and normal corn. This means that more energy is required to gelatinize jackfruit seed starch which confirms the strong molecular organization within the granule. However, the lower thermal temperatures and enthalpy value for jackfruit seed starch than the above-mentioned values have been reported indicating the effect of variety, environment, maturity, and method of extraction of starch.

Functional Properties of Starch and Flour

Swelling Power and Solubility

Swelling power and solubility of jackfruit seed starch, like all native starches, increase with an increase in temperature when heated in water (Madrigal-Aldana *et al.*, 2011) and the increase is rapid after GT is reached (Tongdang, 2008). Granules swell and absorb water at higher temperatures due to hydrogen bonding between the hydroxyl groups of the glucose in starch and water. Increase in solubility at higher temperatures is associated with the leaching out of the amylose chains from the swollen granules.

Water Absorption Capacity and Oil Absorption Capacity

Jackfruit seed flour possesses good ability to bind with both water and oil. The water absorption capacity of 200% and oil absorption capacity of 90% provides functional properties that are desirable in the food industry.

Freeze Stability

Jackfruit starch gel is observed to possess a very high degree of freeze stability (Dutta *et al.*, 2011). There was no syneresis of the gel stored at 4°C for 5 days, which is better than the gels made from rice, potato, tapioca, and arrowroot starches.

PROCESSING AND UTILIZATION OF JACKFRUIT SEED FLOUR AND STARCH

The food industry requires starches to be able to withstand mechanical shear, temperature, and pH changes, as well as to possess good water-holding and gel-forming abilities. Starches are subjected to physical and chemical modification for improved physicochemical and functional properties to those of native starches. Hydroxypropyl starches are widely used in food products where they provide viscosity and freeze thaw stability. Oxidized starches are used as binding agents in food, such as batter applied to meat before frying. Carboxymethyl starches, also called starch ethers, are less prone to damage by heat and bacteria and their gel remains soft as the starch becomes more hydrophilic. Jackfruit seed starch has been physically and chemically modified. Hydroxypropylated jackfruit seed starch has a significant decrease in gelatinization temperature and cross-linked starch has a slight decrease in GT. Conversely, carboxymethylated starch does not show any transition peak, which means that the seed starch on carboxymethylation undergoes gelatinization at room temperature. Carboxymethyl starch exhibits high water uptake while cross-linked and hydroxypropylated starches are similar to native starch in water uptake at room temperature. Cross-linked starch has limited swelling and solubility in water. Hydroxypropylated starch has high swelling and solubility at temperatures above 70°C. Based on the effect of shear and temperature, modified jackfruit seed starch can accordingly be utilized in food industries.

Jackfruit seed starch granules can withstand thermal and mechanical shear. They also can be modified by acid alcohol treatment. Both short-term and long-term acid modification causes drastic loss in viscosity (Table 47.2) and the very low HPV and CPV indicates the possibility that acid thinned jackfruit starch can be utilized in confectionery fillings and gum candies.

Jackfruit seed starch maintains the pH, titratable acidity, and total soluble solids in a high acid product like chili sauce when incorporated in the chili sauce at 1% level as a thickener and stabilizer. The incorporated chili sauce has higher consistency and no serum separation on storage. Serum separation is less because the jackfruit seed starch has low breakdown and setback (Rengsutthi *et al.*, 2011). The quality attributes of the chili sauce incorporated with jackfruit seed starch are better than those incorporated with corn starch (Figure 47.2). The seed starch has been demonstrated to suitably replace up to 16% of wheat flour in the production of reduced calorie chocolate cake

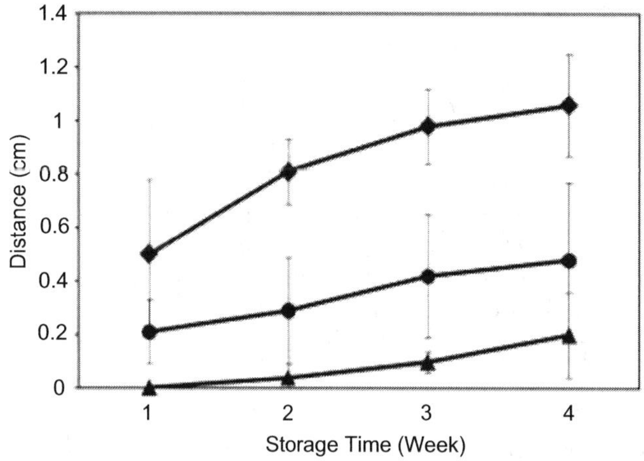

FIGURE 47.2 Serum separation data of chilli sauce at 37°C for 4 weeks. Chilli sauces with —◆— no starch, —▲— Jackfruit seed starch and —●— corn starch. *Source: Rengsutthi and Charoenrein, 2011.*

along with the substitution of sucrose with polydextrose at 11% level (Siti Faridah and Noor Aziah, 2012).

Valorization of fresh seeds into flour can effectively enhance its utilization in food products because of the high carbohydrate, dietary fiber, protein, and minerals present. The seed flour blends well with wheat flour with good water and oil absorption capacities that have possibilities of its incorporation in food product development. Jackfruit seed flour at 11% level with 27% jackfruit dehydrated pulp is successfully incorporated in cereal bars having good sensory and nutritional properties (Santos *et al.*, 2011). Jackfruit seed powder has been successfully utilized as a carbon source for the production of stable *Monascus* pigments through solid state fermentation using a fungal culture of *Monascus purpureus*. The pigment is red in color and has the potential to be used as a food colorant. The color is stabilized by the buffering action of jackfruit seed powder over a wide range of initial pH of the substrate (Babitha *et al.*, 2006). Such utilization will replace synthetic coloring agents.

Jackfruit seed has a considerable amount of non-reducing sugars which can be utilized as prebiotics. Laboratory scale production of prebiotics was optimized with Response Surface Methodology. An extraction time of 15 min at 60°C and solvent to raw material ratio of 10:1 (v/w) gives the optimum level of non-reducing sugar content (491.70 mg/g) in the extract. An ethanol concentration of 50% is used as the extraction solvent (Bhornsmithikun *et al.*, 2010).

Jackfruit flour at 2% level is used as a cheap source of carbon for the extracellular production of pullulan by *Aureobasidium pullulans* MTCC2195. Pullulan has industrial application in food coating, packaging, and starch hydrolysis (Govindasamy *et al.*, 2012). Jackfruit seed starch is a cheaper source of carbon for the bacteria *B. sphaericus* NCIM 5149 for the production of polyhydroxybutyrate

(PHB). Jackfruit seed starch is hydrolyzed and liquefied by alpha amylase and saccharified with glucoamylase and the hydrolysate obtained is subjected to submerged fermentation for the production of PHB, a bioplastic that degrades easily in the environment (Ramdas *et al.*, 2010).

Extracts of the seed flour extracted in different solvents demonstrate antibacterial activity against multidrug resistant Methicillin Resistant *S. aureus* in the laboratory (Karthy *et al.*, 2009). Study results remain to be extended to animal models to validate its use.

References

Ayala-Zavala, J.F., Vega-Vega, V., Rosas-Dominguez, C., Palafox-Carlos, H., Villa-Rodriguez, J.A., Wasim-Siddiqui, Md., Davila-Avina, J.E., Gonzalez-Aguilar, G.A., 2011. Agro-industrial potential of exotic fruit byproducts as a source of food additives. Food Res. Int. 44, 1866–1874.

Babitha, S., Soccol, C.R., Pandey, A., 2006. Jackfruit seed—A novel substrate for the production of *Monascus* pigments through solid state fermentation. Food Technol. Biotechnol. 44, 465–471.

Bhornsmithikun, V., Chetpattananondh, P., Yamsaengsung, R., Prasertsit, K., 2010. Continuous extraction of prebiotics from jackfruit seeds. Songklanakarin J. Sci. Technol. 32, 635–642.

Dutta, H., Paul, S.K., Kalita, D., Mahanta, C.L., 2011. Effect of acid concentration and treatment time on acid-alcohol modified jackfruit seed starch properties. Food Chem. 128, 284–291.

Govindasamy, S., Chandrasekaran, M., Godhaviya, N., 2013. Extracellular biopolymer production by *Aureobasidium pullulans* MTCC 2195 using jackfruit seed powder. J. Polymers Environ. 21, 487–494.

Gupta, D., Mann, S., Sood, A., Gupta, R.K., 2011. Phytochemical, nutritional and antioxidant activity evaluation of seeds of jackfruit (*Artocarpus heterophyllus* Lam.). Int. J. Pharma Bio. Sci. 2, 336–345.

Kabir, S., 1998. Jacalin: a jackfruit (*Artocarpus heterophyllus*) seed-derived lectin of versatile applications in immunobiological research. J. Immunol. Methods 212, 193–211.

Karthy, E.S., Ranjitha, P., Mohankumar, A., 2009. Antimicrobial potential of plant seed extracts against multi drug resistant Methicillin Resistant Staphylococcus aureus (MDR-MRSA). Int. J. Biol. 1, 34–40.

Kittipongpatana, O.S., Kittipongpatana, N., 2011. Preparation and physicochemical properties of modified jackfruit starches. LWT-Food Sci. Technol. 44, 1766–1773.

Madrigal-Aldana, D.L., Tovar-Gomez, B., Mata-Montes de Oca, M., Sayago-Ayerdi, S.G., Gutierrez-Meraz, F., Bello-Perez, L.A., 2011. Isolation and characterization of Mexican jackfruit (*Artocarpus heterophyllus* L) seeds atrch in two mature stages. Starch/Starke 63, 364–372.

Oates, C.G., Powell, A.D., 1996. Bioavailability of carbohydrate material stored in tropical fruit seeds. Food Chem. 56, 405–414.

Ocloo, F.C.K., Bansa, D., Boatin, R., Adom, T., Agbemavor, W.S., 2010. Physico-chemical, functional and pasting characteristics of flour produced from Jackfruits (*Artocarpus heterophyllus*) seeds. Agri. Biology J. North Am. 1, 903–908.

Rahman, M.A., Nahar, N., Mian, A.J., Mosihuzzaman, M., 1999. Variation of carbohydrate composition of two forms of fruit from jack tree (*Artocarpus heterophyllus* L.) with maturity and climatic conditions. Food Chem. 65, 91–97.

Ramdas, N.V., Soccol, C.R., Pandey, A., 2010. A statistical approach for optimization of polyhydroxybutyrate production by Bacillus Spaericus NCIM 5149 under submerged fermentation using Central Composite Design. Appl. Biochem. Biotechnol. 162, 996–1007.

Rengsutthi, K., Charoenrein, S., 2011. Physico-chemical properties jackfruit seed starch (*Artocarpus heterophyllus*) and its application as a thickener and stabilizer in chilli sauce. LWT- Food Sci. Technol. 44, 1309–1313.

Santos, C.T., Bonomo, R.F., Fontan, R.C.I., Bonomo, P., Veloso, C.M., Fontan, G.C.R., 2011. Characterization and sensorial evaluation of cereal bars with jackfruit. Acta Scientiarum. Technol. 33, 81–85.

Santos, L.S., Bonomo, R.C.F., Fontan, R.C.I., Santos, W.O., Silva, A.A.L., 2009. Gelatinisation temperature and acid resistance of jackfruit seed starch. CyTA J. Food 7, 1–5.

Seow, C.C., Shanmugan, G., 1992. Storage stability of canned jackfruit (*Artocarpus heterophyllus*) juice at tropical temperature. J. Food Sci. Technol. 29, 371–374.

Singh, A., Kumar, S., Singh, I.S., 1991. Functional properties of jackfruit seed flour. LWT- Food Sci. Technol. 24, 373–374.

Singh, N., Singh, J., Kaur, L., Sodhi, N.S., Gill, B.S., 2003. Morphological, thermal and rheological properties of starches from different botanical sources. Food Chem. 81, 219–231.

Siti Faridah, M.A., Noor Aziah, A.A., 2012. Development of reduced calorie chocolate cake with jackfruit seed (*Artocarpus heterophyllus* Lam.) flour and polydextrose using response surface methodology (RSM). Int. Food Res. J. 19, 515–519.

Soong, Y.Y., Barlow, P.J., 2004. Antioxidant activity and phenolic content of selected fruit seeds. Food Chem. 88, 411–417.

Tondang, T., 2008. Some properties of starch extracted from three aromatic food seeds. Starch/Starke 60, 199–207.

Vasconcelos, I.M., Oliveira, J.T.A., 2004. Antinutritional properties of plant lectins. Toxicon 44, 385–403.

Zhang, G., Venkatachalam, M., Hamaker, B.R., 2006. Structural basis for the slow digestion of native cereal starches. Biomacromolecules 7, 3259–3266.

CHAPTER

48

The Stability of DNA in Genetically Modified Rice During Food Processing

Hongmei Xiao, Shangxin Song

Key Laboratory of Meat Processing and Quality Control, Nanjing Agricultural University, Nanjing, China

CHAPTER POINTS

- DNA in GM rice food products is subjected to degradation in the course of food processing steps containing boiling, drying, frying, fermentation, etc.
- The degradation effect of heating process for GM rice food varies, and may be contributed to by different temperature, processing time, or heat transfer media.
- As the special microorganism effect exists in fermentation rice foods, it has caught the attention of some researchers to investigate the DNA degradation of transgenic components.
- It has been proved that DNA damage would happen in the DNA extraction process which often contains a number of complex chemical and physical steps.
- DNA-based PCR technology has been the analytical method of priority to detect DNA fragmentation in GM foods, due to its high efficiency and specificity, ease of use and applicability both as a qualitative and quantitative method.

INTRODUCTION

Rice (*Oryza sativa* L.) is one of the most important crops in the world and is the staple food of more than half of the world's population (Chen *et al.*, 2009). Since the first transgenic rice was produced in 1988, the genetic transformation of rice has taken rapid strides (Toriyama *et al.*, 1988). However, only a few transgenic rice varieties have been approved for commercial cultivation at present, including LL 62, LL 06, and LL 601 in the USA (Babekova *et al.*, 2009), and LL 62 in the European Union (GMO Compass, 2007). With the continuously increasing area of cultivated transgenic plants, the risk of contamination with unauthorized genetically modified (GM) material is increasing too. According to the investigations of the GM Contamination Register, from 1997 to 2010, there have been 318 cases of GM contamination occurring in some 57 countries involving almost all parts of the food chain, and among these, 30% of the total involved GM rice (Gene Watch UK and Greenpeace International, 2011).

There is now a widespread fear of consuming transgenic foods containing exogenous genes which may pose a threat to human health. For example, it was feared that these foreign genes may spread to intestinal and food-associated bacteria in humans. However, for such DNA transfer, it requires a minimal length of homology between the donor and recipient DNA (Kharazmi *et al.*, 2003). Therefore, an intact DNA molecule of a certain size is generally a prerequisite for the transfer event, and any degradation of DNA during food processing should reduce the number of DNA molecules that are capable of transferring to other cells.

DNA is subjected to degradation in the course of food processing which often contains a number of complex steps such as boiling, drying, frying, and fermentation. Throughout the world it has become an important part of safety evolution of transgenic food to investigate their DNA degradation during food processing. Regarding the analytic method, PCR has become the main analytic tool of DNA degradation of GM food. Previously, most

Processing and Impact on Active Components in Food
http://dx.doi.org/10.1016/B978-0-12-404699-3.00048-2

researchers focused on the analysis of transgenic corn or soybean, but now some scientists have begun to focus on the safety evolution of transgenic rice.

HOW COMPOSITION IS ALTERED

The basic studies of DNA stability demonstrated that factors including temperature, chemical agents, and enzymes change the primary structure of DNA and cause hydrolysis, oxidation, and deamination of DNA (Lindahl, 1993). These factors are also effective in food production. Although food processing may lead to increased homogeneity, it will result in significant degradation of DNA or removal of DNA from the sample. Until now, there have been several papers studying the DNA degradation of transgenic rice during food processing.

High Temperature

The effect of high temperatures on DNA degradation is most obvious and has, therefore, attracted most attention by researchers. Temperature processing results in physical degradation of DNA (denaturation). The mechanism of DNA destruction by heat is based on depurination or deamination (Gryson, 2010), and results in a decrease in the average genomic DNA fragment size due to random cleavage of these macro-molecules.

There are various heating processes including boiling, drying, baking, microwaving, and frying. The intensity of degradation effect of processing varies, which may be contributed to by different temperature, processing time, or heat transfer media. The effect of food processing on rice DNA degradation is illustrated in Table 48.1. The temperature for drying was 80°C for 2h, baking was at 230°C for 3min, frying was at 180°C for 30s and microwaving was at 800W for 3min. According to results from transgenic rice crackers (Song *et al.*, 2011), the treatments of boiling, drying, and microwaving each contributed to the initial degradation reactions of the genes, while the process of baking resulted in further degradations, and frying led to the most severe changes. It was suggested that temperature played a greater role than processing time in DNA degradation. It also suggested that oil was a better media for heat transfer than air and water; therefore, DNA degradation under the frying processing was always more acute than boiling in water, which was also supported by the results of Shen *et al.* (2006) in Table 48.1.

However, according to the study of Ren *et al.*, (2006) whose findings seem quite different from the results presented above, no significant differences were observed in terms of degradation either among the infant rice cereal, or between the infant rice cereal and the milled rice (Table 48.1). This might be due to differences in processing of these products, because the process of infant rice cereal products involves more gentle and physical steps, i.e., grinding.

The degradation of both transgenic and endogenous reference gene sequences during rice processing was detected by researchers. It was generally considered that DNA regions containing high GC contents are to be more stable when exposed to high temperatures. However, according to a study by Song *et al.* (2011), the differences between the GC content of the amplicons did not completely correspond with the stability of the target genes. The GC content of the amplicons of *NOS* (*nopaline synthase*) terminator was less than 35%, whereas that of *Bar* (*bialaphos resistance*) gene was more than 68% (Table 48.2). Whilst, according to the relative length of DNA fragments preserved after food processing, *Bar* gene was least stable during the heat treatment, *NOS* terminator was more stable than *Bar* gene (Table 48.2). There may be other factors contributing to the different levels of DNA degradation, such as ingredients having different properties, and the different susceptibilities of DNA to various processing conditions (Gryson, 2010). Therefore, predicting the stability and degradation of foreign DNA in food processing is quite difficult or almost impossible. Thus it is essential to analyse by using scientific and effective methods for GM food safety control.

Fermentation

In China, there are a wide rang of traditional fermented foods such as soy sauce, sufu, vinegar, distilled spirits, rice wine, fermented vegetable and meat products. These fermented foods have improved the quality of the diet of Chinese people and also enriched dishes all over the world (Han *et al.*, 2001). Fermented alcoholic beverages, such as rice wine, beer, wine, etc., have a long history and have been enjoyed by people all over the world. To make sweet rice wine, rice is first boiled (for about 30min), cooled, and then mixed with fermentation starter (microorganism) and incubated in a constant temperature incubator at 37°C for several days (Song *et al.*, 2011). During the fermentation, the starch in rice is broken down into sugars by the glucoamylase from *Rhizopus*. Soluble solids content and free water produced by saccharification of *Rhizopus* forms the juice of rice wine (Liu *et al.*, 2007).

As the special microorganism effect exists in fermentation foods, it has caught considerable attention from researchers investigating the DNA degradation of transgenic components in fermented foods, especially for the fermented transgenic soybean. The first study on fermented transgenic rice food was carried out by Song *et al.* in 2011. It revealed that *CAMV35S* (*cauliflower mosaic virus 35S*) promoter and *NOS* terminator were degraded

TABLE 48.1 The Effect of Food Processing on Rice DNA Degradation

Process or Product	Details	Target and Length[1]	Reference
Boiling of rice	100°C, 15–30 min	*SPS*: 916, 456, 277, 168 bp; *CAMV35S*: 735, 446, 195 bp; *Bar*: 568, 370, 175 bp; *NOS*: 217, 180 bp	Song *et al.* (2011)
Drying of cooked rice	80°C, 2.5 h	*SPS*: 916, 456, 277, 168 bp; *CAMV35S*: 735, 446, 195 bp; *Bar*: 568, 370, 175 bp; *NOS*: 217, 180 bp	
Microwaving of dried cooked rice, microwaved cracker	800 W, 3 min	*SPS*: 916, 456, 277, 168 bp; *CAMV35S*: 735, 446, 195 bp; *Bar*: 370, 175 bp; *NOS*: 180 bp	
Baking of dried cooked rice, baked cracker	230°C, 3 min	*SPS*: 277, 168 bp; *CAMV35S*: 446, 195 bp; *Bar*: 175 bp; *NOS*: 180 bp	
Frying of dried cooked rice, fried cracker	180°C, 30 s	*SPS*: 168 bp; *CAMV35S*: 446, 195 bp; *Bar*: 175 bp; *NOS*: nd	
Cooked rice	Normal condition	Hpt: 418, 310, 236 bp Rbcl: 433 bp	Shen *et al.* (2006)
Congee	Normal condition	*Hpt*: 310, 236bp *Rbcl*: 433 bp	
Crispy rice	Frying of cooked rice	*Hpt*: nd[2] *Rbcl*: nd	
Popcorn-like rice	Normal condition	*Hpt*: nd *Rbcl*: nd	
Infant rice cereal-plain	Heating and grinding	MIPS: 850 bp SSR markers:120–400 bp	Ren *et al.* (2006)
Infant rice cereal-with egg yolk			
Infant multi cereal-with oat, wheat, beef and vegetables			
Infant rice cereal-with fish and vegetables			
Milled rice	Milling	*MIPS:* 850 bp *SSR markers:*120–400 bp	

Target and length[1]: the detectable fragments of genes under food processing conditions. For example, the detectable fragments lengths of *SPS* gene under the processes of boiling, drying, and microwaving were 916, 456, 277, and 168 bp. While they were 277 and 168 bp after baking and only 168 bp after frying. It indicated that the treatments of boiling, drying, and microwaving contributed to the initial degradation of the *SPS* gene. The process of baking resulted in further degradations, and frying led to the most severe changes; nd[2], not detected.

by the third day of fermentation, whereas the exogenous gene *Bar* remained unaffected at the end of 4 days of fermentation. At the end of fermentation, the sweet rice wine had the largest detectable fragments of *SPS* gene, *CAMV35S* promoter, *Bar* gene, and *NOS* terminator to be 916, 446, 568, and 180 bp, respectively (Table 48.3).

OTHER WAYS IN WHICH COMPOSITION IS ALTERED

Polymerase chain reaction (PCR) is the main DNA-based analytical method to detect genetically modified organisms (GMO) and to determine their content in food

TABLE 48.2 The Effect of Fermentation on DNA Degradation of Sweet Rice Wine (*Song et al., 2011*)

Fermenting Day	Target and Length[1]
Day 1	*SPS*: 916, 456, 277, 168 bp; *CAMV35S*: 735, 446, 195 bp; *Bar*: 568, 370, 175 bp; *NOS*: 217, 180 bp
Day 2	*SPS*: 916, 456, 277, 168 bp; *CAMV35S*: 735, 446, 195 bp; *Bar*: 568, 370, 175 bp; *NOS*: 217, 180 bp
Day 3	*SPS*: 916, 456, 277, 168 bp; *CAMV35S*: 735, 446, 195 bp; *Bar*: 568, 370, 175 bp; *NOS*: 217, 180 bp
Day 4	*SPS*: 916, 456, 277, 168 bp; *CAMV35S*: 446, 195 bp; *Bar*: 568, 370, 175 bp; *NOS*: 180 bp

Target and length[1]: the detectable fragments of genes during the fermentation of sweet rice wine. For example, the detectable fragments lengths of *CAMV35S* during the first three-day's fermentation were 735, 446, and 195 bp. However, at the end of 4 days fermentation, the detectable fragments of *CAMV35S* were 446 and 195 bp. It indicated that *CAMV35S* had been degraded to a larger extent on the fourth day of fermentation.

TABLE 48.3 Length and GC% of Amplicons of *SPS*, CAMV35S Promoter, Bar and NOS Terminator (Song et al., 2011)

Primer Name	Length of Amplicons (bp)	Relative Length[2]	GC% of Amplicons
SPS (7150 bp)[1]	3148	0.44	44.3
	1952	0.27	45.9
	916	0.13	43.6
	456	0.06	45.6
	277	0.04	62.5
	168	0.02	64.3
CAMV35S (835 bp)	735	0.88	44.2
	446	0.53	45.7
	195	0.23	50.3
Bar (615 bp)	568	0.92	68.2
	370	0.60	69.4
	175	0.28	69.1
NOS (253 bp)	217	0.86	34.1
	180	0.71	32.8

SPS (7150 bp)[1]: the number in bracket means the full length of the gene; relative length[2]: the length of amplicons was divided by the full length of genes.

or feed; therefore, DNA extraction is an essential step for PCR detection. However, there are a number of complex chemical and physical treatments in the extraction process for DNA isolation and purification. Thus it is undoubted that DNA would be damaged during extraction. Table 48.4 summarizes different studies in which DNA degradation of rice food products during DNA extraction are analyzed using agarose gel electrophoresis. It shows that when using the CTAB (Cetyltrimethylammonium Bromide) method, there was no amplification of rice endogenous DNA sequence *SPS* (*Sucrose Phosphate Synthase*) gene (277 bp) for microwaved and baked rice crackers. In contrast, there were 277 bp bands of *SPS* gene with the alkaline lysis method. While for the fried rice crackers, 277 bp fragment of *SPS* gene were degraded in the process of both CTAB and alkaline lysis method. The research conducted by Ren *et al.* (2006) indicated that no significant differences were observed in terms of degradation among the three conventional extraction methods [CTAB, SDS (sodium dodecyl sulfate), and a commercial kit].

TABLE 48.4 The Effect of DNA Extraction on Rice DNA Degradation

Rice Products	DNA Extraction Method	Target and Length[1]	Reference
Microwaved cracker	CTAB method Alkaline lysis method	*SPS*, nd[2] *SPS*, 277 bp	Song *et al.* (2011)
Baked cracker	CTAB method Alkaline lysis method	*SPS*, nd *SPS*, 277 bp	
Fried cracker	CTAB method Alkaline lysis method	*SPS*, nd *SPS*, nd	
Infant rice cereal-plain Infant rice cereal-with egg yolk Infant multi cereal-with oat, wheat, beef and vegetables Infant rice cereal-with fish and vegetables	CTAB method SDS method SAAS method	*MIPS:* 850 bp *SSR markers:* 120–400 bp	Ren *et al.* (2006)

Target and length[1]: the detectable fragments of genes under different DNA extracting methods. For example, for the microwaved cracker, there was no amplification of the *SPS* gene when using the CTAB method. In contrast, there was 277 bp of *SPS* gene with the alkaline lysis method; nd[2], not detected.

ANALYTIC TECHNIQUES

Due to its high specificity, ease of use and applicability both as a qualitative and quantitative method, PCR has been a widely used technique considered as the method of first choice in GMO detection. Since PCR is a DNA-based technology, the selection of a suitable

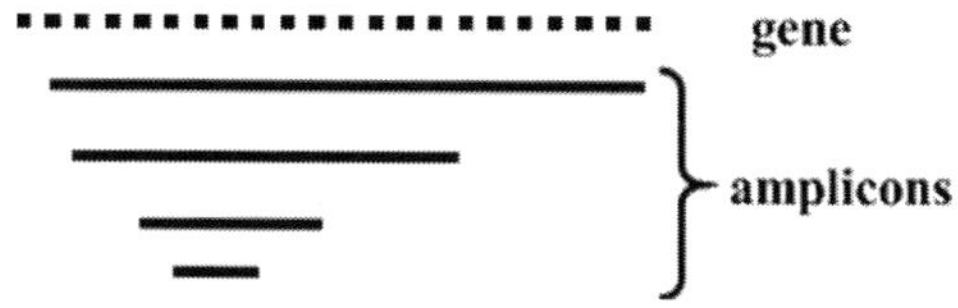

FIGURE 48.1 The design strategies of primers for the qualitative PCR detection of the GM food. The dash line represents complete length of some genes; solid lines represent the designed amplicons for different fragment size PCR method.

DNA extraction method is of great importance, especially for the PCR detection of transgenic food products. Normally, DNA quality is assessed by a PCR test for an endogenous DNA sequence which is present in the plant genome, whether GM or not. This assay verifies the PCR quality status of the extracted DNA and guards against false negatives. Nowadays, different fragment size PCR method has been commonly used as a qualitative assay (the target gene is absent or present) to analyze the food processing steps which result in the loss of target DNA molecules (Murray *et al.*, 2007). Moreover, the degree of DNA degradation of transgenic food adversely affects the limit of detection by qualitative PCR, which is dependent on the effective concentrations of amplifiable DNA sequences. For the primer designation, the most popular strategy is using nested primer pairs amplifying different fragment-sizes of DNA (see Figure 48.1).

In addition, it is important to develop quantitative methods for GMO analysis in foods, because in a number of countries, the need for labeling has been linked to certain thresholds. Real-time PCR is the preferred approach to perform GMO quantitative testing (Engel *et al.*, 2006). Generally, the relative content of GMO can be estimated from the ratio of copy numbers of transgene DNA and endogenous reference DNA using real-time PCR (Kakihara *et al.*, 2007). However, different degradation levels of the reference DNA and recombinant DNA during food processing will reduce the sensitivity of the analysis and affect limits of detection and quantification (Murray *et al.*, 2009). Generally, real-time PCR assays are directed at target molecules in the range of 80–200 bp. Moreover, both targets for the endogenous reference DNA and recombinant DNA should be of similar length in order to obtain similar PCR efficiencies for each target (Gryson, 2010).

References

Babekova, R., Funk, T., Pecoraro, S., Engel, K.-H., Busch, U., 2009. Development of an event-specific Real-time PCR detection method for the transgenic Bt rice line KMD1. Eur. Food Res. Technol. 228, 707–716.

Chen, H., Lin, Y.J., Zhang, Q.F., 2009. Review and prospect of transgenic rice research. Chin. Sci. Bull. 54, 4049–4068.

Shen, L., Wu, Y., Zhang, J., Zhou, P., Wei, X., Zhu, Z., 2006. Stability of hpt marker gene in transgenic rice in different food matrices and under varying food processing conditions. J. Hyg. res. 35, 431–434.

Engel, K.H., Moreano, F., Ehlert, A., Busch, U., 2006. Quantification of DNA from genetically modified organisms in composite and processed foods. Trends. Food Sci. Technol. 17, 490–497.

Gene Watch, U.K., Greenpeace International, 2011. Illegal rice Bt63 from China contaminates food products. GM Contamination Register. http://www.gmcontamination register.org/index.php?content=default.

GMO Compass, 2007. GMO Database. European Union. http://www.gmo- compass.org/eng/gmo/db/.

Gryson, N., 2010. Effect of food processing on plant DNA degradation and PCR-based GMO analysis: A review. Anal. Bioanal. Chem. 396, 2003–2022.

Han, B.Z., Rombouts, F.M., Nout, M.J.R., 2001. A Chinese fermented soybean food. Int. J. Food Microbiol. 65, 1–10.

Kakihara, Y., Matsufuji, H., Chino, M., Yamagata, K., 2007. Detection of recombinant DNA of genetically modified (GM) soybeans in heat-treated GM soybeans and commercial natto. Food Control 18, 1289–1294.

Kharazmi, M., Bauer, T., Hammes, W.P., Hertel, C., 2003. Effect of food processing on the fate of DNA with regard to degradation and transformation capability in Bacillus subtilis. Syst. Appl. Microbiol. 26, 495–501.

Lindahl, T., 1993. Instability and decay of the primary structure of DNA. Nature 362, 709–715.

Liu, F., He, Y., Wang, L., Pan, H., 2007. Feasibility of the use of visible and near infrared spectroscopy to assess soluble solids content and pH of rice wines. J. Food Eng. 83, 430–435.

Murray, S.R., Butler, R.C., Hardacre, A.K., Timmerman-Vaughan, G.M., 2007. Use of quantitative real-time PCR to estimate maize endogenous DNA degradation after cooking and extrusion or in food products. J. Agri. Food Chem. 55, 2231–2239.

Murray, S.R., Butler, R.C., Timmerman-Vaughan, G.M., 2009. Quantitative real-time PCR assays to detect DNA degradation in soy-based food products. J. Sci. Food Agri. 89, 1137–1144.

Ren, X., Zhu, X., Warndorff, M., Bucheli, P., Shu, Q., 2006. DNA extraction and fingerprinting of commercial rice cereal products. Food Res. Int. 39, 433–439.

Song, S., Zhou, G., Gao, F., Zhang, W., Qiu, L., Dai, S., Xu, X., Xiao, H., 2011. Degradation of transgene DNA in genetically modified herbicide-tolerant rice during food processing. Food Chem. Toxicol. 49, 3174–3182.

Toriyama, K., Arimoto, Y., Uchimiya, H., Hinata, K., 1988. Transgenic rice plants after direct gene transfer into protoplasts. Nature Biotechnol. 6, 1072–1074.

CHAPTER

49

Winter Melon (*Benincasa hispida*) Seeds and Impact of Extraction on Composition

Mandana Bimakr[¶,*], *Russly A. Rahman*[†,‡,*], *Ali Ganjloo*[¶]

[¶]Department of Food Science and Technology, Faculty of Agriculture, University of Zanjan, Zanjan, Iran, *Department of Food Technology, Faculty of Food Science and Technology, Universiti Putra Malaysia, Selangor, Malaysia, [†]Halal Product Research Institute, Universiti Putra Malaysia, Selangor, Malaysia, [‡]Department of Process and Food Engineering, Faculty of Engineering, Universiti Putra Malaysia, Selangor, Malaysia

CHAPTER POINTS

- From the results, it is clear that composition of the *B. hispida* seeds extracts obtained using different extraction methods is different.
- Extractions employing supercritical carbon dioxide (SC-CO_2) technology offer new opportunities for resolving the Soxhlet extraction (SE) drawbacks. It may be possible to concentrate bioactive flavonoids in *B. hispida* seeds extracts by manipulating SC-CO_2 extraction condition.
- SC-CO_2 extraction reveals advantages over the SE namely a shorter extraction time, better quality of extracts and avoiding consumption of polluting solvents.
- SC-CO_2 extracts obtained from *B. hispida* seeds are a rich source of bioactive flavonoid compounds. However, some novel flavonoid compounds of *B. hispida* seeds may be extracted by using the other process parameters.
- Essential fatty acids contributed to a high percentage (around 68%) of total fatty acids which confirmed the value of the *B. hispida* seeds extracts.
- It is observed that the fatty acid compositions of extracts were hardly affected by the application of different extraction methods.
- Application of SC-CO_2 extraction for separation of natural compounds has a promising future as a green and clean technique. However, for exploiting industrial applications of the novel extraction techniques, more research is needed.
- A better understanding of the extraction mechanisms can be achieved by modeling of novel extraction processes, and it is useful to optimize the extraction conditions and scale-up any designs of the processes.

INTRODUCTION

Extraction of valuable compounds from plant materials is one of the most sustainable approaches that may be employed. Effective separation of valuable compounds (high crude extraction yield and concentration of bioactive compounds) from a plant matrix is a difficult procedure due to the degradation of valuable compounds and co-extraction of unwanted compounds. For extraction of valuable compounds, Soxhlet extraction (SE) is considered as a standard procedure and main reference to assess the performance of different extraction techniques (Wang and Weller, 2006). The application of this method is limited due to the high amount of organic solvents usage which has adverse

Processing and Impact on Active Components in Food
http://dx.doi.org/10.1016/B978-0-12-404699-3.00049-4

effects on human health and the environment. Furthermore, natural bioactive valuable compounds are thermosensitive and may decompose at higher temperatures for long extraction times, which is common in SE (Zhang *et al.*, 2008). In past decades, scientists have been interested in developing environmentally friendly techniques in line with the 'Green Chemistry' concept (Kamran Khan *et al.*, 2010). Therefore, an improved or better extraction technique is necessary and favored. Supercritical carbon dioxide (SC-CO_2) extraction (SCE) is a green extraction technique which has recently been addressed extensively in medical and food industries. Carbon dioxide (CO_2) is commonly used as a solvent due to its non-explosive, non-flammable, non-toxic, and cost-efficient properties, and the ease with which it is removed from the solutes (Chiu *et al.*, 2002). It has been stated that different compounds from different natural sources have been separated favorably by application of supercritical fluid extraction (SFE) (Sanchez-Vicente *et al.*, 2009).

Benincasa hispida, which belongs to the family *Cucurbitaceae*, is called winter melon, white gourd, ash pumpkin, tallow gourd, white pumpkin, ash gourd, wax gourd, gourd melon and Chinese watermelon, or Chinese preserving melon in English (Marr *et al.*, 2007; Mohd Zaini *et al.*, 2011; Morton, 1971). The name 'wax gourd' comes from the white wax covering on mature fruit which enhances long-term post-harvest life (Grover *et al.*, 2001). The winter melon which has been cultivated for at least 2000 years originated from south-east Asia. A young fruit is characterized by its fleshy, succulent, and hairy attributes, while the mature one has thickly deposited hairs with an easily removable waxy bloom (Mohd Zaini *et al.*, 2011). This fruit is very large and seedy with white, spongy flesh. The Index of Nutritional Quality (INQ) data shows that *B. hispida* has been valued as a high-quality vegetable (Mingyu *et al.*, 1995; Mohd Zaini *et al.*, 2011). In this chapter, some theoretical and practical aspects of the SCE of winter melon seed oil have been presented and compared with those published for SE (Bimakr *et al.*, 2012).

HOW COMPOSITION IS ALTERED

The extraction efficiency and the quality of products are affected by different parameters, such as moisture content and particle size of the seed sample, extraction temperature and pressure, extraction time, solvent flow rate, and the use of a modifier. Therefore, the following results and discussion emphasize the effect of these processing variables on the crude extraction yield of extracts obtained from *B. hispida* seeds.

Effect of Sample Preparation on Extraction of Valuable Compounds from *B. hispida* Seeds

Particle Size

The extraction of valuable and bioactive compounds from *B. hispida* seeds can be facilitated by grinding samples to small particles to increase the surface area and shorten the path lengths over which the solutes have to travel to reach the bulk fluid medium. In general, during the fast extraction period, the fresh solvent comes in contact with the feed and the released solutes on the surface of the particles are solubilized and extracted. The extraction rate is decreased during a slow extraction period. During this period the SC-CO_2 has to diffuse inside the particles and solubilize the solutes and solvent + solutes have to release out which slows the process. The amount of crude extract obtained at the end of the fast extraction period depends on the particle size (Ozkal, 2005). *B. hispida* seeds were ground prior to the extraction processes, so that the surface area was increased or a greater number of cell walls were ruptured, resulting in good recovery of extracts. It was reported that good extraction rates were obtained from ground or thinly flaked soybeans (<0.10 mm) (Snyder *et al.*, 1984). In the current study, some rigid inert materials such as glass beads were packed with the ground seed to maintain a desired permissibility of the particle bed. For *B. hispida* seed oil extraction, a particle size of 0.80–1.20 mm resulted in the highest crude extraction yield (CEY).

Moisture Content

The moisture content of feed also affects the extraction efficiency by conditioning the surface structure (Stahl *et al.*, 1998). A more permeable cell membrane is obtained due to the cell expansion with moisture absorption, resulting in solutes and solvents being able to pass more quickly (Martinez, 2008). However, high moisture content has a negative effect on extraction efficiency. High moisture content reduces SC-CO_2-sample contact due to the pasty consistency of samples, and moisture acts as a barrier to the diffusion of SC-CO_2 in the sample as well as the diffusion of solutes outside the sample (Sahena *et al.*, 2009). Although water is only about 0.3% soluble in SC-CO_2, highly water-soluble solutes would prefer to partition into the aqueous phase, resulting in low efficiency of SCE (Lang and Wai, 2001). Therefore, it was necessary to dry the seeds before extraction. When the moisture content of *B. hispida* seeds was increased from 5 to 20%, the CEY was decreased by up to 38%. Further dehydration of samples below 4% was not necessary, because oven drying is very slow and shorter drying times would decrease the risk of bioactive valuable compound loss.

Effects of the Most Influential Operational Variables on Extraction of Valuable Compounds from *B. hispida* Seeds

Pressure and Temperature

Figure 49.1(a–c) presents the effect of temperature at different pressure levels (150, 300, and 450 bar) on the CEY of *B. hispida* seeds during 180 min of dynamic extraction time. The CEY obtained under the given conditions varied from 112.40 ± 0.56 to 162.30 ± 1.03 mg/g suggesting the potentiality of SCE process to extract valuable compounds from *B. hispida* seeds. It was clear that at lower pressure (150 bar) using temperatures of 40 and 50°C, the CEY increased until 180 min and reached 128.94 ± 1.32 and 144.6 ± 0.66 mg/g. Application of higher pressure (300 and 450 bar) at 40 and 50°C led to an increase of CEY with increasing dynamic extraction time until it reached 120 min, after that non-significant ($p > 0.05$) changes were detected. These results revealed the effective role of higher pressure to reduce the required extraction time. It has been documented that at higher pressure the CO_2 density increases, leading to higher solvent power and mass transfer (Rajaei *et al.*, 2005; Wang and Weller, 2006; Liza *et al.*, 2010). It can be concluded that at 150 bar pressure, due to the lower CO_2 density, the solvent power of SC-CO_2 density is reduced and maximum CEY was obtained at 180 min. Referring to scientific documents, at higher pressures solvent power of CO_2 increases due to its higher density, leading to higher extraction rates (Reverchon and De Marco, 2006). Therefore, in the current study, higher CEY values were obtained at higher pressures (300 and 450 bar) during 120 min of dynamic extraction time. The results were also consistent with those reported by Topal *et al.* (2006). Moreover, the same behavior was reported by Liu *et al.* (2009a) in SCE of pomegranate (*Punica granatum* L.) seeds. They found that an elevation of pressure (up to around 320 bar) caused significant ($p < 0.05$) increase of crude yield extract. They mentioned that this result was most likely due to the improvement of solute solubility resulting from the increased solvent density (Abbasi *et al.*, 2008; Liu *et al.*, 2009a). Referring to the Figure 49.1(a–c), it is clear that at a given pressure (for example 300 bar) the CEY values obtained at 50°C (161.94 ± 0.97 mg/g) were higher than those obtained at 60°C (140.42 ± 1.08 mg/g) during constant dynamic extraction time (e.g., 120 min). The CEY under different temperatures depended on a complex balance between the varying of the SC-CO_2 density and the solubility of compounds. The effect of temperature on solute solubility is different at pressures in the critical range. Near the system critical pressure, the fluid density is very sensitive to temperature. This might be the reason why the yield changed significantly when temperature was changed over the range of 40–60°C. A moderate increase in temperature can lead to a large decrease in fluid density, with a consequent reduction in solute solubility (Roop *et al.*, 1989, Liza *et al.*, 2010, Wang *et al.*, 2008). However, the increase in temperature will also accelerate mass transfer and improve the extraction yield. The increase of temperature can increase the vapor pressure of the extractable compounds. Thus, the tendency of the compounds to be extracted is increased to pass in the supercritical fluid phase (Reverchon and De Marco, 2006). For a volatile solute, there is competition between its solubility in SC-CO_2 and its volatility (Pourmortazavi

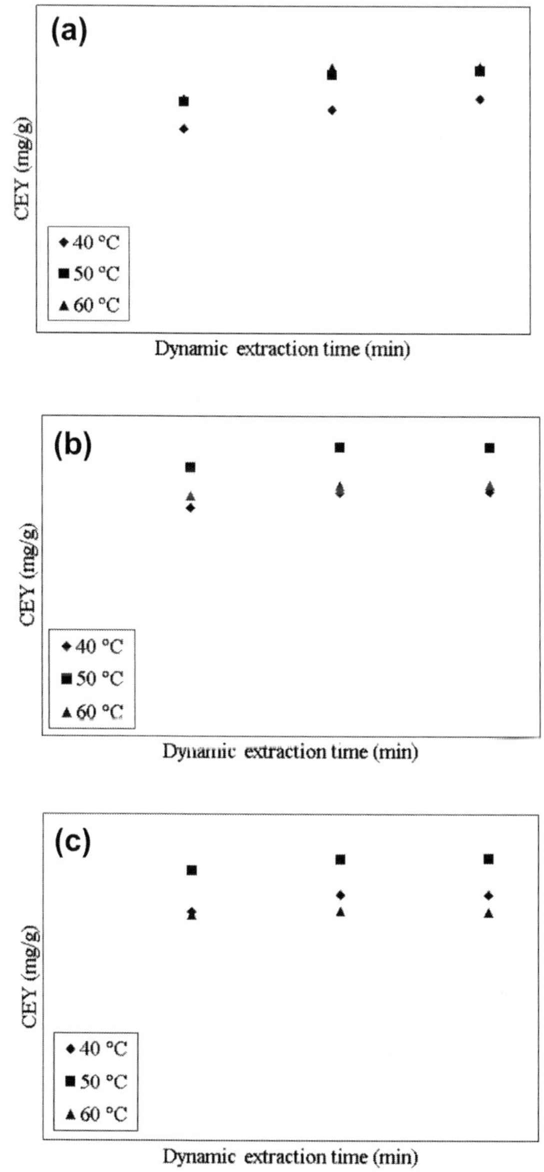

FIGURE 49.1 Effect of temperature at (a) 150 bar pressure; (b) 300 bar pressure; and (c) 450 bar pressure on the CEY at different dynamic extraction times (standard deviation bars are smaller than the symbol size).

and Hajimirsadeghi, 2007). Therefore, it is difficult to predict the effect of temperature. In another study conducted by Yi *et al.* (2009), the same results on the effect of temperature were obtained. In the current study, the CEY obtained using 450 bar pressure at the constant temperature is lower than that obtained using 300 bar. One possible explanation for the low efficiency of the SCE process at high pressure is due to the compression of seed coats at higher pressure resulting in reduced ability for the solvent used to diffuse inside the seed particles (Luengthanaphol *et al.*, 2004). Mitra *et al.* (2009) reported that the highest extraction yield of pumpkin seeds was obtained using 350 bar pressure and above this range no further increase was observed. It is interesting to know that by using a pressure of 450 bar and high temperature (60°C) the CEY of *B. hispida* reduced with increasing dynamic time from 120 (126.72 ± 0.86 mg/g) to 180 min (125.83 ± 0.24 mg/g), however, it was not significant ($p > 0.05$). This unexpected result, as explained by other researchers, is due to the volatility and sensitivity of extracted compounds to high pressure or more likely due to thermo sensitivity of valuable compounds presented in extracts which are degraded using high temperature over a long time (Rezaei and Temelli, 2000; Kanatt et al., 2007). Finally, the best results in terms of high value of CEY were obtained at 300 bar, 50°C and 120 min (161.94 ± 0.97 mg/g).

Flow Rate

The solvent-to-feed ratio could increase by increasing the solvent flow rate which results in higher solutes mass transfer (Martinez, 2008). The CEY of *B. hispida* seed oil increased from 6.3% to 17.8% with an increase in solvent flow rate from 5 g/min to 10 g/min. However, with a further increase to 13 g/min, the CEY changed only from 17.8% to 18.0%. The effect of flow rate was also studied for the extraction of different natural compounds from plants (Ozkal *et al.*, 2005).

Extraction Time

The extraction rate and duration of extraction are affected by temperature, pressure, flow rate, and co-solvent consumption. For the *B. hispida* seed oil extraction it was observed that the CEY increased with dynamic time until 180 min applying 150 bar. However, at higher pressures (300 and 450 bar) the CEY increased with dynamic extraction time until 120 min. This could be due to the fact that at higher pressures the extraction rate is higher and as a consequence the duration of extraction process is shortened to 120 min (Reverchon and De Marco, 2006). Therefore, as shown in Figure 49.1, the highest CEY was achieved at 120 min dynamic extraction time. Furthermore, the effect of extraction time on recovery of different valuable compounds from plants was studied (Ozkal *et al.*, 2005; Liza *et al.*, 2010).

Co-Solvent Addition

The limiting step of application of SC-CO_2 is extraction of polar compounds. This problem could be overcome by adding polar solvents as a co-solvent or modifier. Methanol, ethanol, acetonitrile, acetone, water, and ethyl ether were studied as supercritical solvents for extraction of polar compounds. Among all of them, methanol is an effective polar modifier but in supercritical fluid extraction of nutraceuticals ethanol may be a better choice due to its lower toxicity (Lang and Wai, 2001). Therefore, ethanol (EtOH, 99.5%) was selected as the supercritical fluid extraction co-solvent to extract bioactive flavonoid compounds from *B. hispida* seeds. An increasing co-solvent amount resulted in a higher extraction yield, while the CEY reached a maximum when co-solvent amount was up to a certain value, with no further significant improvement ($p > 0.05$). In the present study, addition of ethanol (1 g/min) as a cosolvent significantly ($p < 0.05$) enhanced the CEY. Applying 300 bar pressure and 50°C temperature, the CEY increased from 73.34 mg/g to 161.92 mg/g by adding cosolvent (1 g/min). This result suggested the improved solvent power of SC-CO_2 by adding co-solvent for extraction of polar compounds. In another study which was conducted by Wang *et al.* (2008), they found a similar result regarding the effect of co-solvent flow rate on the flavonoid yield from *Pueraria lobata*. This could be explained by the fact that the polar constituents in the plant would be easier to extract with a more polar solvent.

Identification and Quantification of the Major Bioactive Flavonoids

From the results, the best SC-CO_2 extraction conditions to obtain the highest value of CEY from *B. hispida* seeds were 50°C, 300 bar, and 120 min. The product obtained under this condition was analyzed by high-performance liquid chromatography (HPLC) in order to determine the contents of the main flavonoid compounds in the *B. hispida* seeds extracts. Furthermore, the product of another two SC-CO_2 extraction conditions which included minimum (150 bar, 40°C, and 60 min) and maximum (450 bar, 60°C, and 180 min) level of each studied parameters were analyzed as control runs. The results of flavonoid content of different SC-CO_2 extracts are shown in Table 49.1 and compared with Soxhlet extracts. Referring to the results the variations in the concentration of flavonoid compounds is obvious. The SC-CO_2 extraction using 300 bar, 50°C and 120 min was more efficient than minimum and maximum levels of studied parameters for obtaining bioactive flavonoid compounds from *B. hispida* seeds. Furthermore, the composition of the SC-CO_2 products and Soxhlet flavonoids was significantly different. The bioactive flavonoids are thermosensitive compounds which decompose at high

TABLE 49.1 Identification and Quantification of the Flavonoids from *B. hispida* Seed Extracts using Different Conditions

Extraction Mode	CEY (mg/g)[a]	Flavonoid Content (mg/g)				
		Gallic Acid	Catechin	Naringenin	Myrecitin	Quercetin
SC-CO_2 EXTRACTION CONDITIONS						
100 bar, 40°C and 60 min	112.43 ± 0.42	0.210	—	0.120	—	0.110
300 bar, 50°C and 120 min	161.91 ± 0.20	0.610	0.250	0.260	0.140	0.230
450 bar, 60°C and 180 min	125.80 ± 0.33	0.540	0.220	0.270	0.110	0.160
SOXHLET EXTRACTION CONDITIONS						
EtOH (99.5%), 6 h	250 ± 1.30	—	—	—	—	—

[a] *Values are mean ± SD of duplicate runs.*

temperature for long extraction time which is common in the Soxhlet method. The most important limiting steps of Soxhlet extraction are long extraction time, losses of volatile compounds, and degradation of bioactive compounds, resulting in unfavorable off-flavor compounds due to heat (Grigonis *et al.*, 2005).

Identification and Quantification of the Fatty Acid Compositions

The fatty acid composition of a kind of extract is its most useful chemical feature. Many of the chemical tests for extract identity or purity can be related to their fatty acid content (Pritchard, 1991, Liu *et al.*, 2009b). The fatty acid composition analysis of extracts obtained using SE and the best extraction condition of SC-CO_2 are analyzed by GC and the results are presented in Table 49.2. It is observed that both of the extracts analyzed are very rich in unsaturated fatty acids (making up to 83% for SCE and 75% for SE). The major fatty acids detected in extracts obtained using Soxhlet and SC-CO_2 are oleic acid (14.10 and 13.92% of total fatty acids, respectively) and linoleic acid (60.60 and 67.11% of total fatty acids, respectively), and relatively low content in the saturated fatty acids. Essential fatty acids (including 18:2 and 18:3) were detected in extracts obtained using the SC-CO_2 technique. Omega-3 fatty acids were introduced as effective dietary supplements for the management of different diseases by enhancing the mechanisms of antioxidant defence (Hu *et al.*, 2007). The presence of essential fatty acids in extracts of *B. hispida* seeds demonstrated its importance in the human body since the essential fatty acids, such as linoleic acid, are not easily synthesized in the human system and must be supplied externally through the diet (Mitra *et al.*, 2009). The saturated fatty acids (SFAs) extracted by the Soxhlet method (24.35% of total fatty acid) were higher than those extracted by SC-CO_2 technique (16.66% of total fatty acids). Rajaei *et al.* (2005) obtained the same results for the SC-CO_2 extraction of tea seeds. They pointed out that higher temperatures appear to be unfavorable for the extraction of UFAs. The results indicate that *B. hispida* seeds have a high degree of unsaturated fatty acids, so the extracts obtained using SC-CO_2 extraction must be stored at lower temperatures, avoiding exposure to air.

TABLE 49.2 Fatty Acid Compositions (% of total fatty acids) of *Benincasa hispida* Seeds

Fatty Acid	SC-CO_2 Extract	SE Extract
Myristic acid (C14:0)	1.33 ± 0.04	1.62 ± 0.22
Palmitic acid (C16:0)	9.55 ± 0.07	15.33 ± 0.12
Stearic acid (C18:0)	5.78 ± 0.12	7.40 ± 0.13
ΣSFA	16.66	24.35
Palmitoleic acid (C16:1)	1.14 ± 0.05	0.68 ± 0.09
Oleic acid (C18:1)	13.92 ± 0.16	14.10 ± 0.10
ΣMUFA	15.06	14.78
Linoleic acid (C18:2)	67.11 ± 0.11	60.60 ± 0.17
Linolenic acid (C18:3)	0.98 ± 0.03	—
ΣPUFA	68.09	60.60

Values are means ± standard error (n = 3). SFA, saturated fatty acid; PUFA, polyunsaturated fatty acid; and MUFA, monounsaturated fatty acids.

SC-CO_2 *versus* Soxhlet Extracts

Data concerning the value of CEY, bioactive flavonoids content, and fatty acid composition obtained using SC-CO_2 and Soxhlet extractions are summarized in Tables 49.1 and 49.2. The content of the main bioactive flavonoids in the *B. hispida* seeds extracts by SE is lower, although its CEY is significantly ($p < 0.05$) higher compared with the SC-CO_2 extraction technique. Flavonoid compounds have usually been extracted by other conventional extraction methods such as solvent extraction, or steam distillation (Grigonis *et al.*, 2005). Some of the main disadvantages of these methods include long extraction times, losses of volatile compounds, and degradation of thermosensitive compounds, resulting in unfavorable off-flavor compounds due to heat (Lang and Wai, 2001). Higher selectivity of SC-CO_2 extraction results in extracts with higher concentrations of bioactive flavonoid compounds. Furthermore, short processing times and lower consumption of hazardous solvents are the other important advantages of SC-CO_2 extraction. Superiority of SC-CO_2 extraction compared with conventional methods has been reported for black pepper oleoresin (Tipsrisukond *et al.*, 1998), eucalyptus leaves (Fadel *et al.*, 1999) and *Lippia alba* leaves and stems (Stanshenko *et al.*, 2004), in terms of higher concentrations of active compounds. It should be noted that the fatty acid contents of the extracts from different extraction methods (Soxhlet and SC-CO_2) are different; however, the differences are minor. It has been shown that in all *B. hispida* seed extracts, the linoleic acid was dominant (60–67% of total fatty acids).

ANALYTICAL TECHNIQUES

Supercritical Carbon Dioxide Extraction of *B. hispida* Seeds Oil

Whole winter melon (*Benincasa hispida*) fruits were purchased from a local market in Serdang, Selangor, Malaysia. The fruits were cut, and then seeds were separated manually and washed under tap water. Seeds were dried at 40°C in a ventilated drying oven (1350FX, USA) for 24 h and then stored at ambient temperature in the dark. The samples were ground in a grinder mill (MX-335, Panasonic, Malaysia) for 10 s to produce a powder with an approximate size of 1.5–2.5 mm before extraction. SCEs were carried out in a batch system using a supercritical fluid apparatus (ABRP200, Pittsburgh, PA, USA). Figure 49.2 is a schematic diagram of the SCE system (500 ml sample capacity). The CO_2 flow rate, the extraction temperature, and the pressure were adjusted by the equipment software, and the extraction time was measured by stopwatch. Samples of ground seeds (40 g) and glass beads (90 g, 2.00 mm in diameter) were weighed accurately, mixed, and then put into the extraction vessel. The liquefied CO_2 was transferred to the extraction vessel by a high-pressure pump (P-50, Thar designs, Inc. Pittsburg, PA, USA). The supercritical CO_2 flow rate was maintained at 10 g/min and the duration of static extraction time was fixed to 30 min. Ethanol (EtOH, 99.5%) was applied as a co-solvent due to its good miscibility with carbon dioxide, its non-toxicity, and because it is allowed for use in the food and pharmaceutical industries (Sanchez-Vicente *et al.*, 2009). After extraction, the extract was collected and the co-solvent (ethanol) was removed with vacuum rotary evaporator (Eyela, A-1000 S, Japan) under reduced

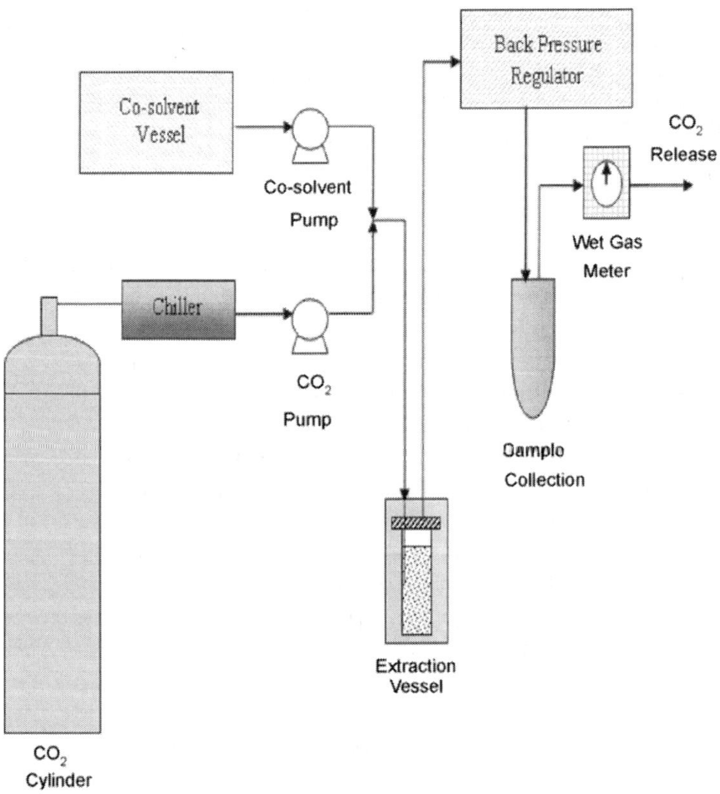

FIGURE 49.2 Schematic diagram of supercritical fluid extractor. *From Liza* et al., *2010.*

pressure at 40°C (water-bath temperature). The extract was collected in glassware and yield was determined gravimetrically (Eq. 1) and reported in mg/g. The extractions were performed in duplicate.

$$CEY = \frac{m_e}{m_s} \times 1000 \quad (1)$$

where m_e is the crude extract mass (g) and m_s is the extracted sample mass (g).

HPLC Analysis

HPLC was composed of a Water 600 pump Controller, 9486 tuneable absorbance UV detector and equipped with an Eclipes XDR-C18 reversed-phase column (25 cm × 4.6 mm × 5 μm, Supelco, USA). The volume of the injection loop was 20 μl. The mobile phase used for analysis was solvent A: TFA (triflouroacetic acid) pH2.5 in deionized water; and solvent B: pure methanol (HPLC grade). The flow rate was set at 1.0 ml/min. The flavonoid compounds were detected at 280 nm. All of the main flavonoid compounds were identified by matching their retention time against those of standard compounds.

Preparation of Fatty Acid Methyl Esters

The extracts were brought to a temperature of 50–60°C and homogenized thoroughly before taking a test sample. A 100-μl aliquot of the test sample was mixed with 1 ml *n*-hexane in a 2-ml vial. A 1-μl aliquot of sodium methoxide was added to the vial which was mixed vigorously with a vortex mixer. The clear upper layer of methyl ester was injected into the GC.

Fatty Acid Composition Analysis

Fatty acid composition analysis was performed in a Hewlett-Packard 6890 gas chromatograph (GC) (Wilmington, DE) equipped with a flame ionization detector (FID) and a capillary column BPX70 (30 m × 0.25 mm × 0.25 μm, Victoria, Australia). Oven temperature was programed isothermally to 115°C during 2 min, then was raised at 4°C/min to 163°C and then at 1°C/min to 170°C. Finally, temperature increased to 200°C at 10°C/min and held at this temperature for 2 min. Helium was used as a carrier gas and flowed at a rate of 1 ml/min. The injection volume was 1 μl. Standard methyl esters of fatty acids were used as authentic samples. The peak areas were obtained from the computer and the percentages of the fatty acids were calculated as the ratio of partial area to total area.

References

Abbasi, H., Rezaei, K., Rashidi, L., 2008. Extraction of essential oils from the seeds of pomegranate using organic solvents and supercritical CO_2. J. Am. Oil Chem. Soc. 85, 83–89.

Bimakr, M., Russly, A.R., Farah, S.T., Noranizan, M.A., Zaidul, I.S., Ganjloo, A., 2012. Antioxidant activity of winter melon (*Benincasa hispida*) seeds using conventional soxhlet extraction technique. Int. Food Res. J. 19, 229–234.

Chiu, K.L., Cheng, Y.C., Chen, J.H., Chang, C.J., Yang, P.W., 2002. Supercritical fluids extraction of *Ginkgo ginkgolides* and flavonoids. J. Supercrit. Fluids 24, 77–87.

Fadel, H., Marx, F., El-Sawy, A., El-Ghorab, A., 1999. Effect of extraction techniques on the chemical composition and antioxidant activity of *Eucalyptus camaldulensis* var. *brevirostris* leaf oils. Eur. Food Res. Technol. 208, 212–216.

Grigonis, D., Venskutonis, P.R., Sivik, B., Sandahl, M., Eskilsson, C.S., 2005. Comparison of different extraction techniques for isolation of antioxidants from sweet grass (*Hierochloe odorata*). J. Supercrit. Fluids 33, 223–233.

Grover, J.K., Adiga, G., Vats, V., Rathi, S.S., 2001. Extracts of *Benincasa hispida* prevent development of experimental ulcers. J. Ethnopharmacol. 78, 159–164.

Hu, Q., Pan, B., Xu, J., Sheng, J., Shi, Y., 2007. Effects of supercritical carbon dioxide extraction conditions on yields and antioxidant activity of *Chlorella pyrenoidosa* extracts. J. Food Eng. 80, 997–1001.

Kamran Khan, M., Abert-Vian, M., Fabiano-Tixier, A.S., Dangles, O., Chemat, F., 2010. Ultrasound-assisted extraction of polyphenols (flavanone glycosides) from orange (*Citrus sinensis* L.) peel. Food Chem. 119, 851–858.

Kanatt, S.R., Chander, R., Sharma, A., 2007. Antioxidant potential of mint (*Mentha spicata* L.) in radiation-processed lamb meat. Food Chem. 100, 451–458.

Lang, Q., Wai, C.M., 2001. Supercritical fluid extraction in herbal and natural product studies — A practical review. Talanta 53, 771–782.

Liu, G., Xu, X., Hao, Q., Gao, Y., 2009a. Supercritical CO_2 extraction optimization of pomegranate (*Punica granatum* L.) seed oil using response surface methodology. LWT 42, 1491–1495.

Liu, W., Fu, Y.J., Zu, Y.G., Tong, M.H., Wu, N., Liu, X.L., 2009b. Supercritical carbon dioxide extraction of seed oil from *Opuntia dillenii* Haw and its antioxidant activity. Food Chem. 114, 334–339.

Liza, M.S., Abdul Rahman, R., Mandana, B., Jinap, S., Rahmat, A., Zaidul, I.S.M., 2010. Supercritical carbon dioxide extraction of bioactive flavonoid from *Strobilanthes crispus* (Pecah Kaca). Food Bioprod. Proc. 88, 319–326.

Luengthanaphol, S., Mongkholkhajornsilp, D., Douglas, S., Douglas, P.L., Pengsopa, L., Pongamphai, S., 2004. Extraction of antioxidants from sweet Thai tamarind seed coat, preliminary experiments. J. Food Eng. 63, 247–252.

Marr, K.L., Xia, Y.M., Bhattarai, N.K., 2007. Allozymic, morphological, phenological, linguistic, plant use, and nutritional data of *Benincasa hispida* (Cucurbitaceae). Eco. Botany 61, 44–59.

Martinez, J.L., 2008. Supercritical fluid extraction of nutraceuticals and bioactive compounds. CRC Press.

Mingyu, D., Mingzhang, L., Qinghong, Y., Weiming, F., Jianxiang, X., Weiming, X., 1995. A study on *Benincasa hispida* contents effective for protection of kidney. J. Agr. Sci. 11, 46–52.

Mitra, P., Ramaswamy, H.S., Chang, K.S., 2009. Pumpkin (*Cucurbita maxima*) seed oil extraction using supercritical carbon dioxide and physicochemical properties of the oil. J. Food Eng. 208, 208–213.

Mohd Zaini, N.A., Anwar, F., Abdul Hamid, A., Saari, N., 2011. Kundur [*Benincasa hispida* (Thunb.) Cogn.]: A potential source for valuable nutrients and functional foods. Food Res. Int. 14, 2368–2376.

Morton, J.F., 1971. The wax gourd, a year-round Florida vegetable with unusual keeping quality. Proc. Fla. State Horticultural Soc. 84, 104–109.

Ozkal, S.G., Yener, M.E., Bayindirli, L., 2005. Response surface of apricot kernel oil yield in supercritical carbon dioxide. LWT 38, 611–616.

Pourmortazavi, S.M., Hajimirsadeghi, S.S., 2007. Supercritical fluid extraction in plant essential and volatile oil analysis. J. Chromatogr. A 1162, 2–24.

Pritchard, J.L.R., 1991. Analysis and properties of oil seeds. In: Rossel, J.B., Pritchard, J.I.R. (Eds.), Analysis of Oil Seeds, Fats and Fatty Foods. Elsevier Science, Oxford, pp. 305–308.

Rajaei, A., Barzegar, M., Yamini, Y., 2005. Supercritical fluid extraction of tea seed oil and its comparison with solvent extraction. Eur. Food Res. Technol. 220, 401–405.

Reverchon, E., De Marco, I., 2006. Review: Supercritical fluid extraction and fractionation of natural matter. J. Supercrit. Fluids 38, 146–166.

Rezaei, K., Temelli, F., 2000. Using supercritical fluid chromatography to determine the binary diffusion coefficient of lipids in supercritical CO_2. J. Supercrit. Fluids 17, 35–44.

Roop, R.K., Akgerman, A., Dexter, B.J., Irvin, T.R., 1989. Extraction of phenol from water with supercritical carbon dioxide. J. Supercrit. Fluids 2, 51–56.

Sahena, F., Zaidul, I.S.M., Jinap, S., Karim, A.A., Abbas, K.A., Norulaini, N.A.N., 2009. Application of supercritical CO_2 in lipid extraction—A review. J. Food Eng. 95, 240–253.

Sanchez-Vicente, Y., Cabanas, A., Renuncio, J.A.R., Pando, C., 2009. Supercritical fluid extraction of peach (*Prunus persica*) seed oil using carbon dioxide and ethanol. J. Supercrit. Fluids 49, 167–173.

Snyder, J.M., Friedrich, J.P., Christianson, D.D., 1984. Effect of moisture and particle size on the extractability of oils from seeds with supercritical CO_2. J. Am. Oil Chem. Soc. 61, 1851–1855.

Stahl, E., Quirin, K.W., Gerard, D., 1998. Dense gases for extraction and refining. Springer-Verlag, New York, p. 176.

Stashenko, E.E., Beatriz, E.J., Jairo, R.M., 2004. Comparison of different extraction methods for the analysis of volatile secondary metabolites of *Lippia alba* (Mill.) N.E. Brown, grown in Colombia, and evaluation of its in vitro antioxidant activity. J. Chromatogr. A 1025, 93–103.

Tipsrisukond, N., Fernando, L.N., Clarke, A.D., 1998. Antioxidant effects of essential oil and oleoresin of black papper from supercritical carbon dioxide extractions in ground pork. J. Agr. Food Chem. 46, 4329–4333.

Topal, U., Sasaki, M., Goto, M., Hayakawa, K., 2006. Extraction of lycopene from tomato skin with supercritical carbon dioxide: Effect of operating conditions and solubility analysis. J. Agr. Food Chem. 54, 5604–5610.

Wang, L., Li, D., Bao, C.L., You, J.Y., Wang, Z.M., Shi, Y.H., 2008. Ultrasonic extraction and separation of anthraquinones from *Rheum palmatum* L. Ultrason. Sonochem. 15, 738–746.

Wang, L., Weller, C.L., 2006. Recent advances in extraction of nutraceuticals from plants. Trends Food Sci. Technol. 17, 300–312.

Yi, C., Shi, J., Xue, S.J., Jiang, Y., Li, D., 2009. Effects of supercritical fluid extraction parameters on lycopene yield and antioxidant activity. Food Chem. 113, 1088–1094.

Zhang, Z.S., Wang, L.J., Li, D., Jiao, S.S., Chen, X.D., Mao, Z.H., 2008. Ultrasound-assisted extraction of oil from flaxseed. Sep. Purif. Technol. 62, 192–198.

SECTION 7

MARINE FOODS

CHAPTER

50

Isoelectric Processing of Seafood Products: Amino Acid and Fatty Acid Profiles

Jacek Jaczynski[*], *Reza Tahergorabi*[†]

[*]West Virginia University, Division of Animal and Nutritional Sciences, Morgantown, WV, USA, [†]North Carolina A&T State University, Greensboro, NC, USA

CHAPTER POINTS

- Isoelectric solubilization/precipitation (ISP) allows efficient recovery of lipids and proteins from seafood by-products.
- ISP does not alter the fatty acid profile of recovered lipids.
- Amino acid profile of recovered proteins shows high nutritional value.
- ISP-recovered proteins and lipids appear to be nutritious for human and animal consumption.

INTRODUCTION

Processing of raw foods into food products inevitably entails generation of some types of by-products, and seafood processing is no exception in this regard. In fact, utilization of the seafood by-products is more important for economical viability of this sector than most other sectors (Gildberg, 2002). Although it varies significantly, mechanical filleting of 100 kg of fish typically yields 40 kg of fillets and 60 kg of by-products. The 60 kg of by-products contain approximately 20 kg of fish meat and some fish oil (Torres *et al.*, 2007). Fish processing by-products are usually reduced to compost or animal feed, or are land filled. While it is not uncommon to just 'grind-and-discard' the 60–70% of fish processing by-products, this practice should be considered an irresponsible utilization of natural resources. Instead, the nutrients available in the by-products should be extracted and used for human consumption (Torres *et al.*, 2007). Fish by-catch and discards fall into the by-product category. The by-catch is particularly high in shrimp, reaching 90% of the total catch. However, on average, the by-catch and discards account for 30% of the total world capture fisheries. This translates into approximately 30 million metric tons of available resource that is not utilized (Torres *et al.*, 2007). Significant value could be added if proteins and lipids were recovered from the fish processing by-products for subsequent use in human food products. Therefore, the development of an efficient protein and lipid recovery technology is highly desirable.

Surimi technology could be a good alternative for recovery of functional muscle proteins (mainly myofibrillar proteins). However, surimi processing can not be applied to recover proteins from fish processing by-products. In addition, surimi processing requires large volumes of water. Relatively recently, an isoelectric solubilization/precipitation (ISP) has been developed to recover protein from fish muscle, resulting in improved recovery yield when compared to surimi processing (Hultin and Kelleher, 1999). The ISP also results in concurrent separation of lipids from proteins. The proteins recovered by this novel approach also retain functional properties such as gel-forming ability (Kristinsson and Hultin, 2003). Although proteins and lipids can be efficiently recovered from seafood processing by-products by ISP, it is crucial to investigate the effects of this technique on amino acid and fatty acid composition of the recovered proteins and lipids, respectively. This information is critical to initiate the assessment of the nutritional quality of potential fish protein and lipid for human and animal consumption.

Fish meat is mainly composed of water, muscle proteins, and intramuscular lipids (fats and oils). Therefore, basic understanding of these components will be

Processing and Impact on Active Components in Food
http://dx.doi.org/10.1016/B978-0-12-404699-3.00050-0

necessary in order to develop new recovery technologies or to maximize current recovery yields.

Proteins in Seafood

Seafood proteins can be divided into three major groups according to their water solubility characteristics: myofibrillar proteins, sarcoplasmic proteins, and stroma proteins.

Myofibrillar proteins are soluble in saline solutions (ionic strength≥0.6) as well as having extremely low ionic strength. Water solubility of myofibrillar proteins varies depending on the temperature, pH, and ionic strength. Myofibrillar proteins are composed of myosin, actin, and regulatory proteins such as tropomyosin, troponin, and actinin. Generally, seafood myofibrillar proteins are less thermally stable than the proteins isolated from terrestrial animals. The pH and ionic strength affect thermal stability of seafood myofibrillar proteins, and hence, heat-induced denaturation and gelation. Sarcoplasmic proteins contain several individual types of water-soluble proteins called myogen. Since sarcoplasmic proteins are water soluble, they are isolated from fish muscle by simply pressing the fish muscle tissue or by extraction with low-ionic-strength saline solution. Stroma proteins form connective tissue in the muscle structure. These proteins are completely water insoluble. The components of stroma proteins are collagen and elastin (Tahergorabi *et al.*, 2011).

Functional Properties of Seafood Proteins

There are 20 different amino acids (AAs) which are building blocks of all proteins. As per their name, all AAs contain amino and acid (carboxyl) groups that are bonded to a central carbon atom (Figure 50.1). The condensation of α-amino and α-carboxyl groups on adjacent AAs forms a strong covalent bond, known as a peptide bond. A polypeptide consists of several AAs joined together by peptide bonds. Peptide bonds are much stronger than the bonds that hold water dipoles together or result in protein–water interactions. Amino and carboxyl groups that are not involved in peptide bonds can be electrostatically charged in solutions; and therefore, participate in protein–water interactions via weak hydrogen bonds (i.e., water solubility) (Figure 50.1). The bonding energy of a hydrogen bond is low (2–40 kJ/mol); however, when present in high numbers, these bonds are primarily responsible for stabilization of the complex secondary, tertiary, and quaternary structures of proteins. AAs share a common amino and carboxyl group, but differ in their functional groups, commonly referred to as side groups (or chains) (Figure 50.1). The different side groups, unique to each AA, give rise to the vastly diverse characteristics of different food proteins.

Lipids and Fatty Acids found in Seafood

The intramuscular lipids in fish of importance to food technologists are mainly composed of a glycerol backbone covalently bonded via ester bonds to three fatty acids (FAs) chains of various lengths and saturation levels. Because of the number of FAs attached to glycerol, the intramuscular lipids are often referred to as triglycerides. The functional properties of fish oils depend on the composition of FAs attached to glycerol. Overall, lipids are hydrophobic and lighter (i.e., lower specific density) than water; and therefore, when fish oils are in water, they tend to join together (i.e., coalesce) and float to the top if enough time is allowed or sufficiently high *g*-force is applied. Unlike fish muscle proteins, fish oil (i.e., triglycerides) does not have charges present on its surface and can not bond with water. For this reason, lipids are frequently called apolar (or hydrophobic) compounds. Due to hydrophobic characteristics of lipids, they can interact with hydrophobic side groups (R) of AAs in fish muscle proteins; and therefore, create weak lipid–protein hydrophobic bonds. Another type of lipid that is probably more important to fish technologists than triglyceride is phospholipid. Phospholipids are an integral component of cell membranes. Phospholipids are also made of the glycerol backbone; however, instead of three FAs only two are covalently attached via ester bonds to the glycerol backbone. The third position on glycerol is occupied by a nitrogen-containing compound

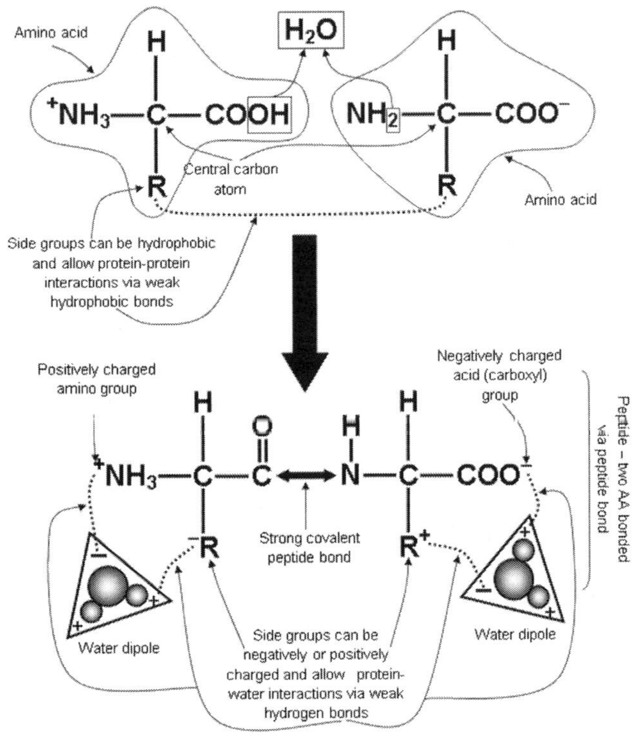

FIGURE 50.1 Protein structure and its interaction with water. *Source: Adapted from Torres et al., 2007.*

such as choline connected to the glycerol backbone via a phosphate group; and hence, the name phospholipids. Fish oils compared to the fats of terrestrial origins have higher levels of unsaturation (Gehring *at al.*, 2011).

The omega-3 (ω-3) polyunsaturated FAs (PUFAs) are considered 'essential' (EFAs) from the human nutrition stand point, because similarly to EAAs, the EFAs can not be synthesized by the human body. Therefore, the EFAs must be provided in the diet for proper development and functioning.

COMPOSITIONAL CHANGES OF PROTEINS AND LIPIDS FROM SEAFOOD DURING ISOELECTRIC SOLUBILIZATION/PRECIPITATION

Isoelectric Solubilization/Precipitation Processing

The ISP of fish muscle proteins with concurrent separation of lipids was proposed by Hultin and Kelleher (1999, 2001). Following these pioneering developments along with earlier works by Meinke, Rahman, and Mattil (1972) as well as Meinke and Mattil (1973), several food science laboratories began active research on ISP. In fish muscle homogenates, myofibrillar proteins are present as aggregates that are held by weak protein–protein hydrophobic interactions (Undeland *et al.*, 2003). However, the protein side-groups can assume different electrostatic charges depending on the conditions that the fish muscle proteins are subjected to (Figure 50.2). This means that the solubility of fish muscle proteins can be 'turned on' or 'off' by providing conditions which either favor or disfavor protein solubility, respectively.

In general, there are five steps in recovering proteins and lipids from fish processing by-products with ISP (Figure 50.3). The first step is to homogenize a 1:6 (wt/wt) dilution of by-products in water in order to provide reaction medium and increase available surface area for the subsequent protein solubilization reaction. In muscle homogenates, myofibrillar proteins are present as aggregates that are induced by hydrophobic forces (Undeland *et al.*, 2003). In the second step, fish muscle proteins are dissolved at either acidic or alkaline conditions. As the pH moves further away from the isoelectric point (pI), fish muscle proteins assume a more negative or positive surface charge for alkaline or acidic conditions, respectively (Figure 50.2). Charge shift results in protein–protein electrostatic repulsion, which weakens protein–protein hydrophobic interactions (i.e., protein aggregates) with a simultaneous increase of protein–water electrostatic interactions. Ultimately, these changes lead to protein solubility in water. Solubilization of fish

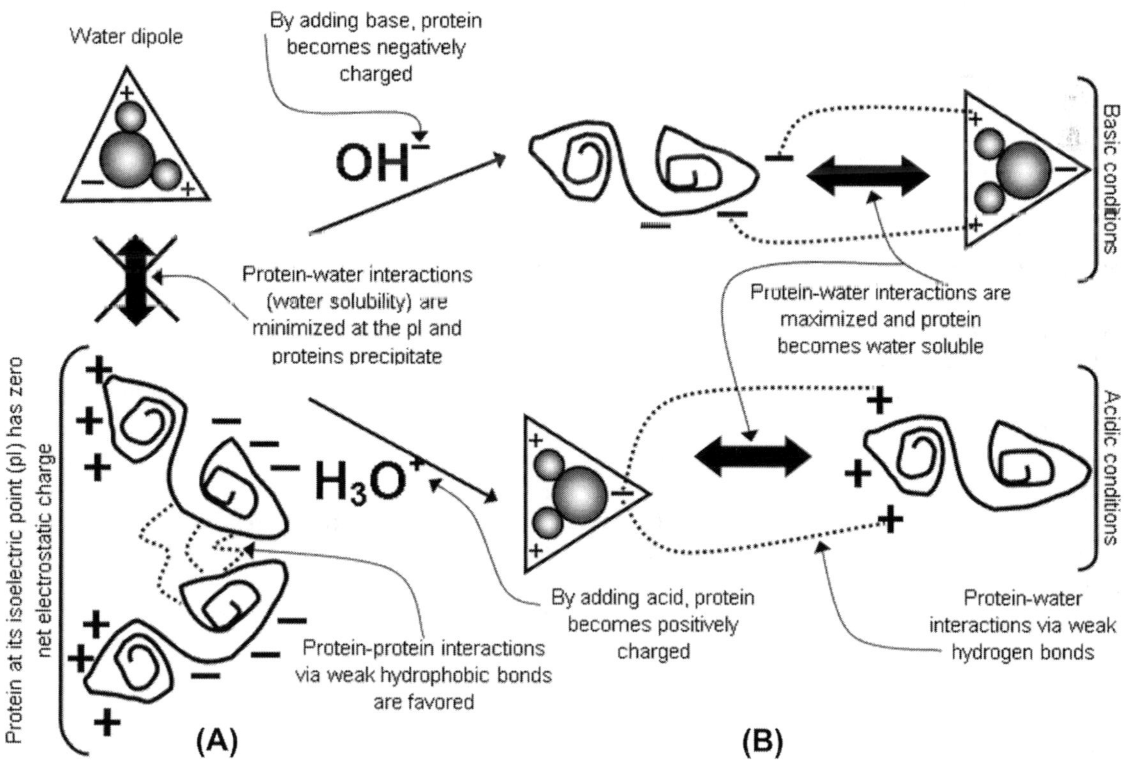

FIGURE 50.2 A protein at its isolectric point (pI) has a zero net electrostatic charge. A. At its pI, protein-water interactions are at its minimum, while protein-protein interactions via weak hydrophonic bonds are at its maximum, causing protein precipitation. B. Protein-water interactions prevail under acidic or basic conditions far from the pI, resulting in protein solubility in water. *Source: Adapted from Gehring et al., 2011.*

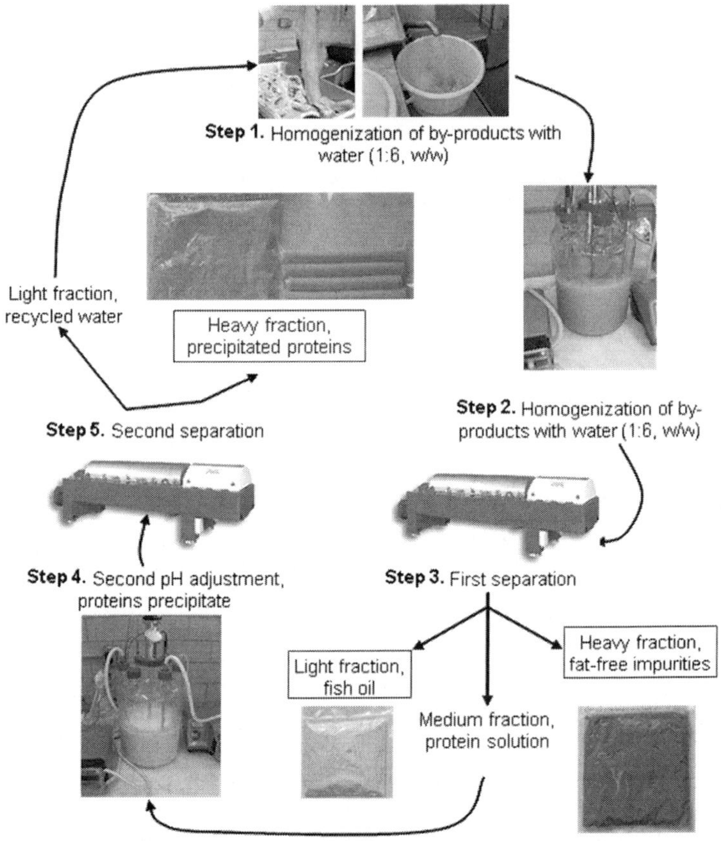

FIGURE 50.3 Diagram of the isolectric solubilization/precipitation technology with concurrent oil separation proposed for fish processing by-products. *Source: adapted from Tahergorabi et al., 2011.*

muscle proteins is ascribed to protonation of aspartyl and glutamyl (p*K*a = 3.8 and 4.2, respectively) residues at acidic pH and deprotonation of lysyl, tyrosyl, and cysteinyl (p*K*a = 9.5–10.5, 9.1–10.8, and 9.1–10.8, respectively) residues at alkaline pH. The accumulation of a net positive or negative charge causes electrostatic repulsion and an increased hydrodynamic volume due to expansion and swelling (Undeland *et al.*, 2003). While the muscle proteins are dissolved in water, the third step, separation, is applied. Typically, centrifugation separates the solution for light, medium, and heavy fractions containing fish oil, dissolved muscle proteins, and insoluble impurities (bones, scale, skin, insoluble proteins, and etc.), respectively. The medium fraction, containing dissolved fish muscle proteins is recovered and subjected to the second pH adjustment in step 4. The pH is adjusted to the average pI of the fish muscle proteins (pH 5.5). At pH 5.5, fish muscle proteins precipitate due to increased protein–protein hydrophobic interactions and decreased protein–water interactions as well as decreased protein–protein electrostatic repulsion. The precipitated fish muscle proteins are separated from the process water typically by centrifugation.

The muscle proteins retain gel-forming-ability; and therefore, can be used as a functional and major ingredient in human food products such as surimi seafood (commonly referred to as crab-flavored seafood). Functional muscle proteins from fish have thus far been recovered at the laboratory scale using a batch mode (Undeland *et al.*, 2003) and pilot plant scale (Mireles DeWitt *et al.*, 2007). The ISP processing in a continuous mode has also been applied to fish processing by-products (Chen *et al.*, 2007), krill (Chen and Jaczynski, 2007), and whole fish (Taskaya *et al.*, 2009a,b).

Amino Acid Profile of Seafood Proteins Recovered with Isoelectric Solubilization/Precipitation

While all 20 AAs are indispensable for normal growth and sustaining metabolic processes, nine AAs can not be synthesized by adult humans. Thus, the nine AAs are referred to as 'essential' (EAAs). The biological value (BV) of a protein measures its efficiency in supporting the human body needs. Egg proteins are regarded as a reference protein and have a BV of 100, meaning that 100% of the protein is digested/absorbed and retained in the human body. Milk, beef, fish, corn, and rice proteins have BVs of 93, 75, 75, 72, and 59, respectively. The BV of krill proteins has been reported as slightly higher than the BV of milk proteins (Suzuki and Shibata, 1990).

To assess protein quality, it is important to determine the amount of EAAs in proteins recovered with ISP. Table 50.1 shows the content of EAAs in the proteins

TABLE 50.1 EAA Content (mg/g) in Fish Processing By-Products and in the Protein Recovered from Fish Processing By-Products by Solubilization at Different pH and Precipitation at pH 5.5 as compared to the Reference Protein of Whole Egg and the Recommended Values for Adults and Infants

	Protein Recovered from Fish Processing By-Products at Different Treatments (pHvalue) (mg/g protein)						Fish Processing By-Products	Whole egg	FAO/WHO/ UNU (1985) Adults (Infants)
EAAs	2.0	2.5	3.0	12.0	12.5	13.0	(mg/g Protein)	(mg/g Protein)	(mg/g Protein)
Isoleucine	38.91 ± 0.50b	36.31 ± 0.20c	39.52 ± 0.75b	42.11 ± 0.36a	41.49 ± 0.43a	42.81 ± 0.28a	18.27 ± 0.64d	63	13 (46)
Leucine	65.51 ± 0.08c	60.47 ± 0.20d	65.89 ± 0.31c	69.5 ± 0.00b	69.08 ± 0.32b	71.02 ± 0.37a	30.57 ± 0.86e	88	19 (93)
Lysine	73.61 ± 0.40c	66.96 ± 0.23d	73.38 ± 0.35c	76.34 ± 0.24b	76.28 ± 0.39b	78.35 ± 0.42a	35.13 ± 0.95e	70	16 (66)
Methionine	26.30 ± 0.50a	22.40 ± 0.15b	25.53 ± 0.04a	26.21 ± 0.54a	26.14 ± 0.67a	26.21 ± 0.37a	14.27 ± 0.23c	56*	17 (42)*
Phenylalanine	33.94 ± 0.35d	30.96 ± 0.48e	34.17 ± 0.13cd	35.48 ± 0.43b	35.06 ± 0.12bc	36.79 ± 0.07a	16.27 ± 0.44f	98**	19 (72)**
Threonine	36.85 ± 0.78c	34.25 ± 0.33d	36.77 ± 0.60c	38.38 ± 0.84c	39.00 ± 0.36ab	40.59 ± 0.18a	18.33 ± 0.26e	49	9 (43)
Tryptophan	10.24 ± 0.20c	9.25 ± 0.38d	9.02 ± 0.15d	11.00 ± 0.21ab	10.65 ± 0.28bc	11.50 ± 0.14a	5.27 ± 0.22e	16	5 (17)
Valine	46.48 ± 0.84c	42.88 ± 0.26d	47.01 ± 0.83c	49.93 ± 0.35ab	48.87 ± 0.45b	50.83 ± 0.24a	22.23 ± 0.68e	72	13 (55)
Histidine	21.25 ± 0.20c	19.03 ± 0.13d	21.02 ± 0.15c	22.61 ± 0.12b	22.47 ± 0.18b	23.44 ± 0.12a	11.73 ± 0.23e	24	16 (26)

**Methinine + Cyesteine.*
***Phenylalanine + Tyrosine.*
Data are given as mean ± SEM (n = 3). Mean values in rows with different letters indicate significant differences ($P < 0.05$).
Adapted from Chen et al. *(2007).*

recovered from fish processing by-products using ISP at various pH values as compared to the starting material (i.e., fish processing by-products), whole egg, and the recommended values by FAO/WHO/UNU for adults and infants (FAO/WHO/UNU, 1985). The ISP at basic pH generally yielded a higher amount of EAAs than ISP at acidic pH (Chen *et al.*, 2007, Taskaya *et al.*, 2009c). This was likely caused by the least pH-induced proteolysis at basic pH. The proteolysis may lead to partially hydrolyzed proteins and eventually liberation of free AAs. Partially hydrolyzed fish muscle proteins and free AAs are typically water-soluble. Polar AAs (i.e., hydrophilic), lysine and threonine exhibit high water solubility and, therefore, are likely to be lost if proteolysis occurs. More ($P < 0.05$) lysine and threonine were retained with proteins recovered using ISP at basic pH (Table 50.1). Therefore, some of the hydrophilic EAAs may be recovered to a lesser degree by ISP at acidic pH.

Proteins recovered from fish processing by-products using ISP showed higher BV as compared to soybean protein concentrate and similar BV to the milk protein, casein. Lysine is often considered a limiting EAA. Therefore, it needs to be emphasized that proteins recovered from fish processing by-products with ISP had a similar ($P > 0.05$) concentration of lysine to whole egg; while the concentration of lysine was greater ($P < 0.05$) in proteins recovered from whole carp and whole krill (Taskaya *et al.*, 2009c). The nutritional quality of proteins recovered with ISP, although lower than that of egg protein, is comparable to or higher than that of soy protein and milk protein. The nutritional quality of proteins recovered from whole krill using ISP has been determined (Gigliotti *et al.*, 2008; Tou *et al.*, 2007). Table 50.2 shows protein quality measurements for krill protein recovered with ISP as compared to casein. Currently, the protein information on nutritional food labels is assessed using

TABLE 50.2 Nutritional Quality of Protein Recovered from Whole Krill using ISP at Alkaline pH

Protein quality Measurement	Casein	KPC
Digestibility (%)	93.3 ± 2.0	93.2 ± 1.0
PDCAAS	1	1
PER (g body weight/g protein)	1.57 ± 0.05	1.44 ± 0.15
Final body weight (g)	235.4 ± 5.8	241.8 ± 3.4
Total body weight gain (g/4 weeks)	73.3 ± 3.0	80.0 ± 3.5
Total food intake (g)	451.1 ± 11.5	462.7 ± 11.6
Feed efficiency (g body weight/g food intake)	0.16 ± 0.005	0.17 ± 0.009

Adapted from Gigliotti et al. *(2008).*
PDCAAS, Protein Digestion Corrected Amino Acid Score; PER, Protein Efficiency Ratio.

the Protein Digestion Corrected for Amino Acid Score (PDCAAS) as the preferred measurement for protein quality. The FDA specifies the use of casein as the reference protein for the Protein Efficiency Ratio (PER). Therefore, the ability of krill protein to support growth equal to that of rats fed with casein indicates that krill protein recovered with ISP is a complete protein (Gigliotti *et al.*, 2008; Tou *et al.*, 2007).

Proving that a novel protein source has a high nutritional value is only half of the needed consideration. Safety associated with consumption of the novel protein also needs to be addressed. It has been demonstrated that different protein sources affect mineral deposition in the kidneys (i.e., nephrocalcinosis), an initial phase in renal disease. Interestingly, experimental rats fed with krill protein recovered with ISP displayed much lower ($P < 0.05$) Ca and P content in the kidney as compared to rats fed with casein (Gigliotti *et al.*, 2011a). Histological evaluation revealed a decrease in renal tissue mineral deposits in experimental rats fed with krill protein as compared to casein. Rats fed with krill protein did not develop microtubular calcification in kidneys as opposed to rats fed with casein. Therefore, krill protein recovered by ISP appears safe to consume and may protect against nephrocalcinosis which increases the risk of kidney stones. However, more comprehensive and long-term studies are needed to more fully assess the safety of seafood proteins recovered with ISP.

Fatty Acid Profile of Seafood Lipids Recovered with Isoelectric Solubilization/Precipitation

Alpha-linolenic (ALA, 18:3 ω3), eicosapentaenoic (EPA, 20:5 ω3) and docosahexaenoic acids (DHA, 22:6 ω3) are the main ω-3 PUFAs, while linoleic (LA, 18:2 ω6) and arachidonic acids (AA, 20:4 ω6) are the main ω-6 PUFAs in aquatic animals (Chen *et al.*, 2006). Although seafood lipids have shown numerous health benefits and are used in various functional foods, they have not received as much attention as proteins in the ISP field. Lipid recovery by ISP should be investigated more thoroughly because of their health benefits; and hence, higher potential value than the recovered proteins. It has been demonstrated that ISP allows efficient recovery of fish lipids. ISP likely enhances the recovery of apolar (i.e., neutral) triglycerides due to increased polarity when pH is altered during protein solubilization. Some published works have shown that ISP at basic pH allows more efficient lipid recovery than acidic pH. The pH during solubilization of fish muscle proteins generally does not affect the composition of ω-3 or ω-6 PUFAs; and consequently, the ω-3/ω-6 ratio remains unaltered (Taskaya *et al.*, 2009c). The FA profile of lipids recovered with ISP at acidic or basic pH is generally similar to the FA profile in the starting material (Figure 50.4). Similar

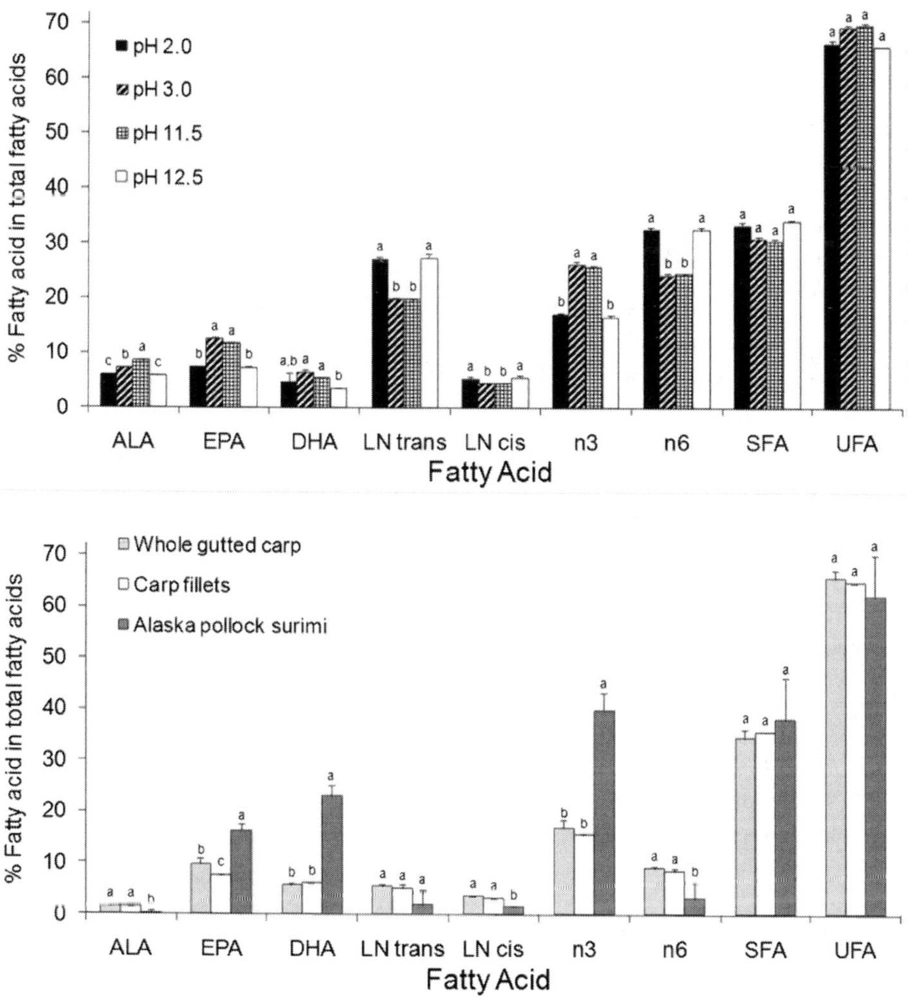

FIGURE 50.4 Major fatty acids (FAs) (% FA in total FAs) in the lipids recovered from whole gutted carp using solubilization at different pH and precipitation at pH = 5.5 (top graph) as compared to whole gutted carp fillets, and Alaska Pollock surimi (bottom graph). Data are given as mean values ± SD (n=3). Different letters on data bars indicate significant differences (P < 0.05) between mean values. ALA – linolenic acid (18:3 ω3), EPA – eicosapentaenoic acid (20:5ω3), DHA – docosahexaenoic acid(22:6ω3), LA – linoleic acid (18:2ω6), n3 – total omega-3 fatty acids (ω-3 FA), n6 – total (ω-6 FA, SFA – total saturated FA, and UFA - total unsaturated FA. *Source: Adapted from Taskaya et al., 2009c.*

results were obtained by Okada and Morrissey (2007), demonstrating that pH-induced protein solubilization does not affect FA composition of the recovered lipids. Likely, the low temperature and short time during ISP processing prevent a possible pH-induced degradation of PUFAs.

Although ISP allows efficient recovery of fish lipids, it is unlikely that phospholipids are recovered to a great extent (Torres *et al.*, 2007). Due to amphiphilic characteristics of fish phospholipids, particularly from krill, centrifugation at 20,000 × *g* did not result in their efficient recovery (Chen and Jaczynski, 2007). Therefore, the ISP-recovered krill protein concentrate used by Gigliotti *et al.* (2008) and Bridges *et al.* (2010) contained 8% of krill lipids (dry weight basis). Krill lipids are a good source of DHA and EPA that are bound mainly in phospholipids. The ω-3 PUFAs (mainly EPA and DHA) were 27% of total lipids. Total blood cholesterol was lower (P < 0.05) in rats fed the ISP-recovered krill protein concentrate (with 8% lipids) compared to casein-fed rats. However, lower total blood cholesterol was accompanied by decreased HDL (Gigliotti *et al.*, 2008). In a different study it was demonstrated that hyperlipidemic subjects consuming krill oil had increased HDL and decreased LDL compared to subjects consuming fish oil (Bunea *et al.*, 2004). These authors suggested that the effects of krill oil could be due to the ω-3 PUFAs associated with phospholipids; whereas, the ω-PUFAs in fish are associated with triglycerides. Consumption of ω-PUFAs in the form of phospholipids or triglycerides may lead to different health and physiological effects. The ω-3 PUFAs in the ISP-recovered krill protein concentrate increased (P < 0.05) the DHA concentration in the brain of young rats (Bridges *et al.*, 2010). Feeding the ISP-recovered krill protein concentrate to young rats also decreased (P < 0.05) pro-inflammatory metabolites, while there was no increase (P > 0.05) in lipid oxidation or total antioxidant capacity in the tissues of experimental rats. Therefore, even the low amount of ω-3 PUFAs in the ISP-recovered krill protein concentrate provided beneficial effects of increasing tissue EPA and DHA deposition and reduced pro-inflammatory metabolites without increasing lipid oxidation.

Fatty Acid Profile of Omega-3-Fortified Food Products Developed from Fish Protein Isolate

Western diet is characterized by the increased dietary intake of saturated fat, ω-6 FAs, and trans-FAs, as well as inadequate intake of ω-3 FAs. As a result, the ratio of ω-6 FAs to ω-3 FAs is at 15–20 : 1, instead of the suggested 1 : 1 (Simopoulos, 1999). Therefore, there is increasing interest in the fortification of food products with ω-3 PUFAs because of their health benefits, especially the reduction of cardiovascular disease (CVD).

Since the seafood products developed from the ISP-recovered fish protein isolate would be formulated products associated with aquatic sources, they are a logical vehicle for increasing the consumption of ω-3 PUFAs; and therefore, addressing the diet-driven CVD without the need for dietary supplements in a pill or capsule form. Thus, Tahergorabi *et al.* (2012) studied the FA profile of functional seafood products developed from the ISP-recovered fish protein isolate fortified with ω-3 PUFAs. In this study, different sources of ω-3-rich oils such as flaxseed, fish, algae, krill, and a blend (flaxseed : fish : algae, 8 : 1 : 1) oils were added at 10% to the protein isolates. They found that the ω-3-fortified functional seafood products had higher content ($P < 0.05$) of total ω-3 PUFAs (34–51% of total FAs) than the control samples (20%). However, the control samples were made of the ISP-recovered fish protein isolate without added ω-3 PUFAs-rich oil. The ISP isolates normally contain very low amounts of fat, usually below 2% ('as-is' basis) (Taskaya *et al.*, 2009a). Thus, the control samples in that study had correspondingly lower amounts of fat than the ω-3 fortified products. Although the control gels had 20% of ω-3 PUFAs in total FAs, they had very low absolute amount of these health-beneficial FAs due to the low amount of fat in the ISP isolate. In contrast, the ω-3-fortified products had 10% of ω-3-rich oil that was added to the ISP isolate. The highest ($P < 0.05$) content of ω-3 PUFAs was in samples fortified with flaxseed oil (51%), followed by blend (49%), krill (46%), algae (43%), and fish oil (34%). Flaxseed oil is a rich source of ALA (52–55%) (Ansorena and Astiasaran, 2004). Thus, the samples fortified with flaxseed oil and the blend (8% flaxseed oil) had the highest ($P < 0.05$) content of ALA (49 and 39%, respectively) compared to other ω-3-fortified samples (fish, 2.5%; krill, 1.2%; and algae, 0.5%). Pietrowski *et al.* (2011) reported similar results for surimi gels fortified with comparable levels of ω-3 rich oils. Surimi gels fortified with krill and fish oil had the greatest ($P < 0.05$) content of EPA (24 and 16%, respectively), whereas all other gels contained less than 3% of EPA. Krill and fish oils are very good sources of EPA and DHA; while algal oil contains DHA as their main ω-3 PUFA (Gigliotti *et al.*, 2011b; Tou *et al.*, 2007). The DHA content was the highest ($P < 0.05$) in surimi gels fortified with algal oil (41%), followed by krill (19%), fish (11%), blend (7%), and flaxseed oil (2%).

CONCLUSIONS

ISP allows efficient recovery of functional and nutritious muscle proteins and lipids from sources difficult to process through conventional means. These resources include seafood processing by-products as well as low-value and under-utilized aquatic species. ISP with alkaline pH results in the recovery of muscle proteins with higher concentrations of EAAs than acidic pH. Based on the EAA profile, the nutritional quality of muscle proteins recovered from seafood processing by-products with ISP appears to be better than that of soy protein, comparable to milk protein, but slightly below the egg protein. The FA profile of the lipids recovered with ISP at acidic or alkaline pH is similar to the FA profile in the starting material. Therefore, ISP does not alter the FA profile or the ω-3/ω-6 FA ratio in the recovered seafood lipids. Based on the studies discussed in this chapter, proteins and lipids obtained with ISP appear to be nutritious and safe sources for human and animal consumption. Further studies are needed to determine the health benefits associated with consuming seafood protein concentrates/isolates obtained with ISP.

References

Ansorena, D., Astiasaran, I., 2004. The use of linseed improves nutritional quality of lipid fraction of dry fermented sausages. Food Chem. 87, 69–74.

Bridges, K.M., Gigliotti, J.C., Altman, S., Jaczynski, J., Tou, J.C., 2010. Determination of digestibility, tissue deposition, and metabolism of ω-3 fatty acid content of krill protein concentrate in growing rats. J. Agri. Food Chem. 58 (5), 2830–2837.

Bunea, R., El Farrah, K., Deutsch, L., 2004. Evaluation of the effects of Neptune krill oil on the clinical course of hyperlipidemia. Altern. Med. Rev. 9 (4), 420–428.

Chen, Y.C., Nguyen, J., Semmens, K., Beamer, S., Jaczynski, J., 2006. Enhancement of omega-3 fatty acid content in rainbow trout (*Oncorhynchus mykiss*) fillets. J. Food Sci. 71, C383–389.

Chen, Y.C., Tou, J.C., Jaczynski, J., 2007. Amino acid, fatty acid, and mineral profiles of materials recovered from rainbow trout (*Oncorhynchus mykiss*) processing by-products using isoelectric solubilization/precipitation. J. Food Sci. 72 (9), C527–C535.

Chen, Y.C., Jaczynski, J., 2007. Protein recovery from rainbow trout (*Oncorhynchus mykiss*) processing by-products via isoelectric solubilization/precipitation and its gelation properties as affected by functional additives. J. Agri. Food Chem. 55 (22), 9079–9088.

FAO/WHO/UNU expert consultation, 1985. World Health Organization Technical Report Series 724. WHO, Geneva, Switzerland.

Gehring, C.K., Gigliotti, J.C., Moritz, J.S., Tou, J.C., Jaczynski, J., 2011. Functional and nutritional characteristics of proteins and lipids recovered by isoelectric processing of fish by-products and low-value fish: A review. Food Chem. 124 (2), 422–431.

Gildberg, A., 2002. Enhancing returns from greater utilization. In: Bremner, H.A. (Ed.), Safety and Quality Issues in Fish Processing. Woodhead Publishing Limited, Cambridge (U.K.).

Gigliotti, J.C., Jaczynski, J., Tou, J.C., 2008. Determination of the nutritional value, protein quality and safety of krill protein concentrate isolated using an isoelectric solubilization/precipitation technique. Food Chem. 111 (1), 209–214.

Gigliotti, J.C., Smith, A.L., Jaczynski, J., Tou, J.C., 2011a. Consumption of krill protein concentrate prevents early renal injury and nephrocalcinosis in female Sprague–Dawley rats. Urological Res. 39 (1), 59–67.
Gigliotti, J.C., Davenport, M.P., Beamer, S.K., Tou, J.C., Jaczynski, J., 2011b. Extraction and characterization of lipids from Antarctic krill (*Euphausia superba*). Food Chem. 125 (3), 1028–1036.
Hultin, H.O., Kelleher, S.D., 1999. Process for isolating a protein composition from a muscle source and protein composition. U.S. Patent No. 6,005,073.
Hultin, H.O., Kelleher, S.D., 2001. Process for isolating a protein composition from a muscle source and protein composition. U.S. Patent No. 6,288,216.
Kristinsson, H.G., Hultin, H.O., 2003. Changes in conformation and subunit assembly of cod myosin at low and high pH and after subsequent refolding. J. Agri. Food Chem. 51 (24), 7187–7196.
Meinke, W.W., Rahman, M.A., Mattil, K.F., 1972. Some factors influencing the production of protein isolates from whole fish. J. Food Sci. 37, 195–198.
Meinke, W.W., Mattil, K.F., 1973. Autolysis as a factor in the production of protein isolates from whole fish. J. Food Sci. 38, 864–866.
Mireles DeWitt, C.A., Nabors, R.L., Kleinholz, C.W., 2007. Pilot plant scale production of protein from catfish treated by acid solubilization/isoelectric precipitation. J. Food Sci. 72, E351–E355.
Okada, T., Morrissey, M.T., 2007. Recovery and characterization of sardine oil extract by pH adjustment. J. Agri. Food Chem. 55, 1808–1813.
Pietrowski, B., Tahergorabi, R., Beamer, S., Matak, K., Tou, J., Jaczynski, J., 2011. Chemical properties of surimi seafood nutrified with ω-3 rich oils. Food Chem. 129, 912–919.
Simopoulos, A.P., 1999. Genetic variation and evolutionary aspects of diet. In: Papas, A. (Ed.), Antioxidants in nutrition and health. CRC Press, Boca Raton, pp. 65–88.
Suzuki, T., Shibata, N., 1990. The utilization of Antarctic krill for human food. Food Rev. Int. 6, 119–147.
Tahergorabi, R., Hosseini, S.V., Jaczynski, J., 2011. Seafood Proteins. In: Philips., G.O., Williams, P.A. (Eds.), Handbook of Food Proteins. Woodhead Publishing Limited, pp. 116–143.
Tahergorabi, R., Beamer, S.K., Matak, K.E., Jaczynski, J., 2012. Isoelectric solubilization/precipitation as a means to recover protein isolate from striped bass (*Morone saxatilis*) and its physicochemical properties in a nutraceutical seafood product. J. Agri. Food Chem. 60 (23), 5979–5987.
Taskaya, L., Chen, Y.C., Jaczynski, J., 2009a. Functional properties of proteins recovered from whole gutted silver carp (*Hypophthalmichthys molitrix*) by isoelectric solubilization/precipitation. LWT—Food Sci. Technol. 42 (6), 1082–1089.
Taskaya, L., Chen, Y.C., Beamer, S., Jaczynski, J., 2009b. Texture and color properties of proteins recovered from whole gutted silver carp (*Hypophthalmichthys molitrix*) using isoelectric solubilisation/precipitation. J. Sci. Food Agri. 89 (2), 349–358.
Taskaya, L., Chen, Y.C., Beamer, S., Tou, J.C., Jaczynski, J., 2009c. Compositional characteristics of materials recovered from whole gutted silver carp (*Hypophthalmichthys molitrix*) using isoelectric solubilization/precipitation. J. Agri. Food Chem. 57 (10), 4259–4266.
Tou, J.C., Jaczynski, J., Chen, Y.C., 2007. Krill for human consumption: nutritional value and potential health benefits. Nutrition Rev. 65 (2), 63–77.
Torres, J.A., Chen, Y.C., Rodrigo-Garcia, J., Jaczynski, J., 2007. Recovery of byproducts from seafood processing streams. In: Shahidi, F. (Ed.), Maximising the value of marine by-products. CRC Press., Boca Raton, USA, pp. 65–90.
Undeland, I., Kelleher, S.D., Hultin, H.O., McClements, J., Thongraung, C., 2003. Consistency and solubility changes in herring (*Clupea harengus*) light muscle homogenates as a function of pH. J. Agri. Food Chem. 51, 3992–3998.

CHAPTER

51

The Effect of Carbon Monoxide on Slaughter and Processing of Fish

*Anna Concollato**, *Gry Aletta Bjørlykke*†, *Bjørn Olav Kvamme*†, *Oddvin Sørheim*‡, *Erik Slinde*†, §, *Rolf Erik Olsen**

*Institute of Marine Research, Matre Research Station, Matredal, Norway, †Institute of Marine Research, Bergen, Norway, ‡Nofima, Ås, Norway, §Department of Chemistry, Biotechnology and Food Science, Norwegian University of Life Science, Ås, Norway

CHAPTER POINTS

- CO is an efficient sedative and anesthetic agent in fish.
- CO affects the brain directly by binding to heme proteins in *Saccus vasculosis* and neuroglobin.
- CO can be detected together with gases like nitrogen, oxygen, and carbon dioxide using gas chromatography
- Visible and near infrared spectroscopy can be used to study color and protein, fat and water content online.
- CO stabilizes the color of red fish muscle.
- CO increases product stability by inhibition of microbial growth and lipid oxidation.

INTRODUCTION

Fish are regarded as a highly perishable food, since they are very susceptible to microbial and chemical decay. The type and rate of decay varies with fish species and is significantly influenced by the immediate handling before and after slaughter, and by processing and packaging systems. Optimizing these processes will lower the two main challenges, bacterial growth and lipid oxidation, the latter being particularly challenging in fatty fish species.

In the aquaculture industry, there is a growing awareness of maintaining animal welfare all the way through the slaughter process. The ethics involved in fish husbandry require that the process proceed with a minimum amount of strain. Normally, the fish should be anesthetized before being slaughtered and bled. This has proven to be a challenge for the industry. The use of chemicals like clove oil had been suggested as a nontoxic anesthetic (Iversen *et al.*, 2003). However, the use of chemicals is troublesome as traces may remain in the flesh at consumption and will cause concern to some consumer groups. Further, the use of chemicals is likely to be banned in some countries while allowed in others. This creates a challenge when fish are sold on a global market. Consequently, in most countries, the aquaculture industry relies on non-chemical anesthetic methods. In Norway, only percussion and electrical stunning are allowed by the authorities, while other countries also use liquid ice and carbon dioxide (CO_2). Many of these methods have potential welfare issues that may limit their use in the future. For example, electrical stunning may cause muscular contraction promoting rapid onset of *rigor mortis*. However, when properly applied on sedated fish, the method is very useful (Roth *et al.*, 2003). Immersion in liquid ice is considered stressful by many authors and is a questionable approach (Robb 2001; Kestin *et al.*, 1991). The use of CO_2 is basically to asphyxiate the fish, and the exposure generally elicits a flight response, causing the fish to swim erratically, trying to escape; this method is therefore regarded as unacceptable (EFSA, 2009).

Carbon monoxide has been used in animal euthanasia for a long time (Smith, 2001) but is not widely used in fish. However, recent data in Atlantic salmon (Bjørlykke *et al.*, 2011, 2012), tilapia (Mantilla *et al.*, 2008), pollack

Processing and Impact on Active Components in Food
http://dx.doi.org/10.1016/B978-0-12-404699-3.00051-2

(*Pollachius pollachius*), herring (*Clupea harengus*), and mackerel (*Scomber scombrus*) (Slinde *et al.*, unpublished data) suggest that CO is an excellent fish sedative agent that does not appear to cause any visible stress response.

The mode of action of CO as sedative is not fully understood. However, it is well known that CO will bind to the heme group of hemoglobin and myoglobin displacing oxygen and producing carboxy-myoglobin (COMb) and carboxy-hemoglobin (COHb) that are incapable of oxygen transport. Both COMb and COHb are stable compounds, and it has been assumed that the animal will die due to oxygen shortage without sensing the deficiency. Recent data also suggest that CO binds to the oxygen-storage proteins in *Saccus vasculosus*, and neuroglobin (Ngb) (Figure 51.1) of the brain. It is believed that *Saccus vasculosis* is an oxygen depot with functions during hypoxia and stress (Burmester and Hankeln 2009), while Ngb is an oxygen transporter mainly located in neurons of the central and peripheral nervous systems (Figure 51.1A, B) and in some endocrine tissues (Reuss *et al.*, 2002). Blocking these with CO may induce immediate sedation and unconciousness in fish.

HOW FISH COMPOSITION IS ALTERED

Processing Methods

The dominant commercial method for applying CO to fish during processing is by pretreatment with filtered smoke (Kowalski 2006). Fish rich in red muscle containing heme proteins, like tuna (various *Thunnus*) and mahi mahi (*Coryphaena hippurus*), are suitable for this technology. Filtered smoke is generated from natural sawdust by removal of some taste and odor components, carcinogen compounds, and gases. Usually filtered smoke contains 15–40% CO, and the fish are treated in chambers for 2–48 h, depending on the size and thickness of fillets. Thereafter, the treated fillets are vacuum packaged, frozen, and transported to the markets.

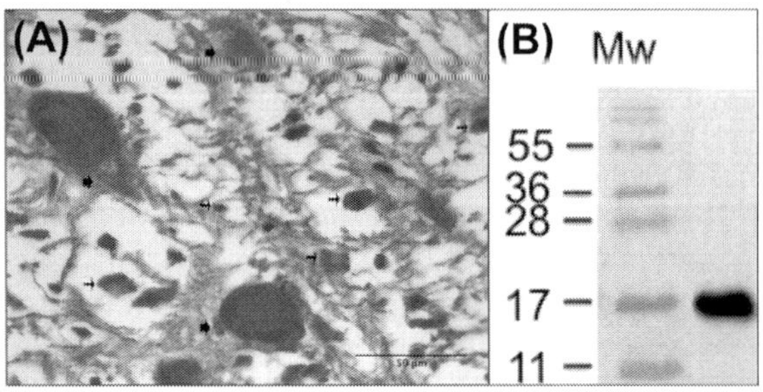

FIGURE 51.1 (A) Immunostaining using anti salmon neuroglobin in 1:70 dilution. Neuroglobin is found in perikardion (thick marker), and axon (thin marker) in thalamus of brain of Atlantic salmon. The arrows mark some of the positive staining. (B) Western blot analysis using anti-salmon neuroglobin to detect recombinant salmon neuroglobin (Bjørlykke *et al.*, Unpublished data) at the expected MW of 17,000.

In addition to filtered smoke, fish may be pretreated, packaged or stored in high concentrations of CO, close to 100%. Packaging of fresh meat with low levels of CO, up to 0.4%, combined with high levels CO_2 and no oxygen is well established, in particular in the USA (Cornforth and Hunt, 2008). Based on the beneficial experiences obtained with low CO packaging of meat over the last two to three decades, there is a potential for implementing this technology in the fish processing industry, yielding better color, longer shelf-life and inhibition of lipid oxidation. The application of CO already to the live fish is in this connection regarded as beneficial.

Flesh Color

In normal tissue, most color is caused by myoglobin. But some hemoglobin will also be present, particularly if fish have not been bled. Fresh tissue contains only oxymyoglobin/oxyhemoglobin (OMb/OHb) which has a bright-red color. Shortly after death, oxygen is lost producing deoxyglobins (DMb/DHb) that have a dark-red color. With further decay, iron is oxidized to its ferric state, producing the meth-form of the globins (MMb/MHb) which has a brown color. This color change is an unattractive feature to consumers, and producers therefore aim to maintain the bright-red color as long as possible.

When CO is added it binds directly to oxy- or deoxyglobins displacing oxygen, producing COMb/COHb that has a cherry red color. They are stable compounds and the degradation to MMb/MHb takes a long time (Chow *et al.*, 1998) and will thus prevent discoloration. The attribute has been used by some producers to maintain color in products like tuna for a long period of time. The process (gas or filtered wood smoke) may also be used to stabilize globins of white flesh fish, improving the color appearance over time (Mantilla *et al.*, 2008; Kowalski, 2006). Table 51.1 shows redness, a* values, of salmon, herring, and mackerel treated with CO compared to control groups. In herring and mackerel the COMb in the red muscle show a persistent cherry-red color due to binding of CO. CO-treated fish did not develop the typical rancid smell even after 6 days of storage as was the case for the controls. Storage of Atlantic salmon in 100% CO and consequent binding to heme demonstrated that

TABLE 51.1 Color (a*, i.e., change in redness) of cold-room-stored (4°C) Atlantic salmon (1.30 kg), herring (0.16 kg), and mackerel (0.56 kg) after being anesthetized with CO

	Salmon		Herring		Mackerel	
Day	Control	CO	Control	CO	Control	CO
1	22.3 ± 0.7	23.0 ± 2.0	2.7 ± 1.1	10.0 ± 2.0	9.5 ± 1.0	8.9 ± 2.0
6	19.4 ± 1.0	21.9 ± 2.0	3.7 ± 0.3	10.3 ± 2.0	5.1 ± 0.9	10.0 ± 0.7

The L* and b* values were very similar for treated and untreated fish.

this pigment contributes slightly to the color in addition to the dominant astaxanthin pigment (Bjørlykke *et al.*, 2011; Ottestad *et al.*, 2011).

Lipid Oxidation

After slaughter, oxidative processes will start in meat. Oxymyoglobin is a relatively unstable compound and has the potential to contribute significantly to oxidation through several pathways. For example, OMb (Fe^{2+}) is easily oxidized to MMb (Fe^{3+}) producing superoxide anion that can dismutate to hydrogen peroxide and thus initiate lipid peroxidation. Next peroxides (lipid peroxides and hydrogen peroxide) are strong oxidizers that can oxidize OMb $Fe^{2+} > Fe^{3+}$ (Tajima and Shikama, 1987) and thus propagate peroxidation. Finally, many peroxidation products like aldehydes, e.g., 4-hydroxy-2-noenal, may themselves attack sites on myoglobin facilitating its oxidation (Faustman *et al.*, 2010). COMb does not contain oxygen and will thus not facilitate the production of superoxide radicals. Furthermore, introducing a ligand like CO to the 6th coordination orbital increases the stability of the Fe^{2+} in the heme moiety. This will increase the shelf-life, and reduce lipid oxidation and browning of the product (Cornforth and Hunt, 2008; Hsieh *et al.*, 1998). This is especially important in fatty fish where the high level of unsaturated fatty acids makes them more susceptible to lipid oxidation.

Microbial Growth

CO is known to have an inhibitory effect on microbial growth at levels above 5% (Gee and Brown, 1980). Fish treated with filtered smoke benefits from this by having extended microbiological shelf-lives (Kowalski, 2006). In a study of aerobic bacteria in stored yellowfin tuna (*Thunnus albucares*), filtered smoke efficiently reduced bacteria caused by high levels of CO, carbon dioxide, and smoke components (Kristinsson *et al.*, 2007). The storage of tuna under 100% CO reduced bacterial growth, but to a smaller extent than filtered smoke. The mechanism of CO induced inhibition on bacterial growth is still relatively unclear, however. CO will affect cell respiration through inhibition of many enzymes (e.g., cytochromes) with heme groups similar to Hb and Mb. With cytochromes, CO inhibits oxidative phosporylation and thus aerobic bacteria respiration and survival (Prescott *et al.*, 1996). During prolonged storage of marine fish with red muscles, bacteria may penetrate the flesh and convert free histidine to histamine. Although histamine is toxic at very low concentrations, it does not cause appreciable visual or organoleptic changes. This increases the risk of intoxication by the consumer. Treating (directly or indirectly) fish with CO can reduce aerobic bacterial growth and histamine formation and increase shelf-life (Garner and Kristinsson, 2004).

ANALYTICAL TECHNIQUES

Visible (VIS)/Near Infrared Spectroscopy (NIRs) Spectra of CO in Fish and SRI Dissolved Gas Analyzer–Gas Chromatography (DGA-GC) System

Ottestad *et al.* (2011) used spectroscopic measurements on mackerel muscle to study how spectral changes correspond to color variations under three different storage conditions; air, vacuum and CO (Figure 51.2). The spectral color properties were dominated by myoglobin (and hemoglobin) at different oxidation states and bound to different ligands. The formation of COMb was positively correlated to the a* value on the L*a*b* scale (lightness, redness and yellowness). This implies the presence of different myoglobin species in fish, as reported by other authors (Mantilla *et al.*, 2008; Smulevich *et al.*, 2007). It will be interesting for future studies to use spectroscopy as a non-destructive way for online measurements of water, lipid, and protein (Folkestad *et al.*, 2008) together with visible color.

Detecting the soluble gasses like CO in the water (Figure 51.3) has been challenging both for scientific trials and as a dosage-control system in the industry. Using gas–liquid chromatography may offer long-awaited solutions to these challenges. For example, using a dissolved gas analyzer–gas chromatography (DGA-GC) it is now possible to quantify the amount of CO in the water, in addition to oxygen and waste gaseous products.

PRACTICAL CONSIDERATIONS

Human Toxicity

It is important to be aware of the possible toxic effect of CO on humans. It is a colorless, odorless, tasteless, and non-irritant gas. Inhalation of CO decreases the amount of O_2 delivered to the tissues. The affinity of hemoglobin for CO is over 200-times higher than its affinity for O_2. However, a low concentration of CO is not considered a hazard. The uptake of CO to hemoglobin is reversible and the half-life of COHb is 4–6 h. The rate of absorption and excretion of CO from the body is relatively slow. When working with CO a security alarm should be worn at all times. The administrative Norwegian working norm of CO is 25 ppm.

Legal Issues

Presently, CO is not permitted for foods in the EU and Norway. The regulations in the use of CO in treatment and processing of muscle foods however do differ between countries and regions. United States Food and Drug Administration stated 'tasteless smoke' or filtered smoke as GRAS (Generally Recognized as Safe) in 2000

FIGURE 51.2 Absorption spectra from mackerel fillet stored in vacuum, air, and CO. The Soret maximum in Visible Spectra for mackerel packed in air was 421 nm; in CO 423 nm; and in vacuum 431 nm. This shows the presence of the various myoglobin species depending on storage conditions. Fillet stored in vacuum (black solid line), air (dotted line), and CO (hyphened line). *Courtesy of Jens Petter Wold, Nofima.*

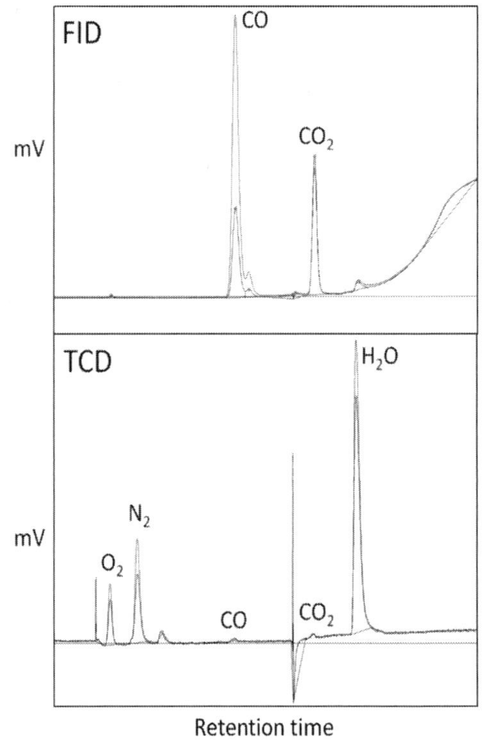

FIGURE 51.3 Detection of soluble gasses in water using an SRI DGA-GC system equipped with a Thermal Conductivity Detector (TCD) and a Flame Ionisation Detector (FID). Two different standards (i.e., water samples) with different amounts of CO were used. For each detector the individual traces are overlaid, with the upper trace representing the sample having twice the CO content compared to the lower trace sample. This detection method shows excellent reproducibility and allows detailed analysis of the water gas atmosphere. A detailed explanation of the system can be found at the SRI homepage (http://www.srigc.com).

(USFDA, 2000). Later, packaging of meat with up to 0.4% CO has been permitted in the USA to master-bags and case-ready meat. Until 2004, low CO concentrations were widley used for packaging of meat in Norway, but at that time CO packaging was prohibited due to trade agreements with the EU. The positive effects of CO might cause a change in legislation in the EU in the future. However, there are no regulations in the EU on the use of CO as a sedative or anesthetic component for fish.

References

Bjørlykke, G.A., Kvamme, B.O., Raae, A.J., Roth, B., 2013. Slaughter of Atlantic salmon (*Salmo salar*, L.) in the presence of carbon monoxide. Fish Physiol. Biochem. 39, 871–879 .

Bjørlykke, G.A., Roth, B., Sorheim, O., Kvamme, B.O., Slinde, E., 2011. The effects of carbon monoxide on Atlantic salmon (*Salmo salar* L.). Food Chem. 127, 1706–1711.

Burmester, T., Hankeln, T., 2009. What is the function of neuroglobin? J. Exp. Biol. 212, 1423–1428.

Chow, C.J., Hsieh, P.P., Tsai, M.L., Chu, Y.J., 1998. Quality changes during iced and frozen storage of tuna flesh treated with carbon monoxide gas. J. Food Drug Anal. 6, 615–623.

Cornforth, D., Hunt, M., 2008. Low oxygen packaging of fresh meat with carbon monoxide: Meat quality, microbiology and safety. White papers no. 2. American Meat Science Association, Champaign, IL, USA, pp. 1–10.

EFSA, 2009. Species-specific welfare aspects of the main systems of stunning and killing of farmed Atlantic salmon. Scietific Opinoin of the Panel of Animal Health and Welfare. In: The EFSA Journal. http://www.efsa.europa.eu/(accessed 29.11.11.).

Faustman, C., Sun, Q., Mancini, R., Suman, S.P., 2010. Myoglobin and lipid oxidation interactions: Mechanistic bases and control. Meat Sci. 86, 86–94.

Folkestad, A., Wold, J.P., Rørvik, K.A., Tschudi, J., Haugholt, K.H., Kolstad, K., Mørkøre, T., 2008. Rapid and non-invasive measurements of fat and pigment concentrations in live and slaughtered Atlantic salmon (*Salmo salar* L.). Aquaculture 280, 129–135.

Garner, K., Kristinsson, H.G., 2004. Quality of Spanish mackerel (*Scomberomorous maculatus*) muscle as affected by carbon monoxide and filtered smoke gas treatment. In: IFT annual meeting, Las Vegas Abstact 49B–12.

Gee, D.L., Brown, W.D., 1980. The effect of carbon monoxide on bacte rial growth. Meat Sci. 5, 215–222.

Hsieh, P.P., Chow, C.J., Chu, Y.J., Chen, W.L., 1998. Change in color and quality of tuna during treatment with carbon monoxide gas. J. Food Drug Anal. 6, 605–613.

Iversen, M., Finstad, B., McKinley, R.S., Eliassen, R.A., 2003. The efficacy of metomidate, clove oil, Aqui-S (TM) and Benzoak (R) as anaesthetics in Atlantic salmon (*Salmo salar* L.) smolts, and their potential stress-reducing capacity. Aquaculture 221, 549–566.

Kestin, S.C., Wotton, S.B., Gregory, N.G., 1991. Effect of slaughter by removal from water on visual evoked activity in the brain and reflex movement of rainbow trout (*Oncorhynchus mykiss*). Vet. Rec. 128, 443–446.

Kowalski, B., 2006. Tasteless smoke sources, specifications, and controls. In: Otwell, E.S., Kristinsson, H.G., Balaban, M.O. (Eds.), Modified atmosphere processing and packaging of fish—filtered smokes, carbon monoxide and reduced oxygen packaging. Blackwell Publishers, Gainesville, USA, pp. 117–126.

Kristinsson, H.G., Crynen, S., Yagiz, Y., 2007. Effect of a filtered wood smoke treatment compared to various gas treatments on aerobic bacteria in yellowfin tuna steaks. Lebensm. —Wiss. Technol. 41, 746–750.

Mantilla, D., Kristinsson, H.G., Balaban, M.O., Otwell, W.S., Chapman, F.A., Raghavan, S., 2008. Carbon monoxide treatments to impart and retain muscle color in tilapia fillets. J. Food Sci. 73, C390–C399.

Ottestad, S., Sørheim, O., Heia, K., Wold, J.P., 2011. Effects of storage atmosphere and heme state on the color of visible reflectance spectra of salmon (*Salmo salar*) fillets. J. Agric. Food Chem. 59, 7825–7831.

Prescott, L.M., Harley, J.P., Klein, D.A., 1996. Microbiology. In: Wm. C. Brown Publishers, Dubuque.

Reuss, S., Saaler-Reinhardt, S., Weich, B., Wystub, S., Reuss, M., Burmester, T., Hankeln, T., 2002. Expression analysis of neuroglobin mRNA in rodent tissues. Neuroscience 115, 645–656.

Robb, D.H.F., 2001. Measurement of fish flesh colour. In: Kestin, S.C., Wariss, P., (Eds.), Farmed fish quality. Blackwell Science, Oxford, U.K., pp. 298–306.

Roth, B., Imsland, A., Moeller, D., Slinde, E., 2003. Effect of electric field strength and current duration on stunning and injuries in market-sized Atlantic salmon held in seawater. N. Am. J. Aquacult. 65, 8–13.

Smith, A.S., 2001. Laboratory animal science. In: Hem, A., Eide, D.M., Engh, E., Smith, A.S. (Eds.), Oslo, Norway.

Smulevich, G., Droghetti, E., Focardi, C., Coletta, M., Ciaccio, C., Nocentini, M., 2007. A rapid spectroscopic method to detect the fraudulent treatment of tuna fish with carbon monoxide. Food Chem. 101, 1071–1077.

Tajima, G.-I., Shikama, K., 1987. Autoxidation of oxymyoglobin. J. Biol. Chem. 262, 12603–12606.

USFDA, 2000. Agency response letter GRAS notice no. GRN 000015. United States Food and Drug Administration. http://seafood.ucdavis. edu/guidelines/grn000015. htm (accessed March 2000.).

CHAPTER

52

Sodium Nitrite, Salt-Curing and Effects on Carotenoid and N-Nitrosoamines in Marine Foods

Jørgen Lerfall

Department of Food Technology, Sør-Trøndelag University College, Trondheim, Norway

CHAPTER POINTS

- Salt-curing of salmonid fillets affect physical ablution and decomposition of carotenoids through processing.
- There is no evidence that addition of nitrite salt affects the retention of carotenoids in salmonids. However, addition of nitrite increases the content of 9-*cis* and 13-*cis* astaxanthin.
- Improved redness of nitrite treated fish is caused by formation of a nitroso-heme complex, either nitrosomyoglobin or nitrosohemoglobin, or a combination of both.
- Fresh fish contain only small traces of naturally occurring nitrite. However, nitrite can be formed during processing, which allows small amounts of N-nitrosoamines to be formed.
- Use of nitrite as a food additive increases the content of residual nitrite, which further can induce formation of N-nitrosoamines in the product or endogenously.
- No studies have yet reported increased levels of N-nitrosoamines in salmonid fillets treated with nitrite as compared to fillets treated without. However, one should be more vigilant not to add nitrite to marine species with significant levels of naturally occurring secondary amines.
- Chromatographic analyses of carotenoids are complex, and the choice of method relevant to a particular study depends on the information at hand.
- Several chromatographic methods have been used to analyse N-nitrosoamines. They all include a high definition detector such as a thermal energy analyser (TEA) or a mass spectroscopic system (MS).

INTRODUCTION

Salt-curing is an excellent method to lower water-activity and extend the shelf-life of marine foods. Nowadays, marine food often appears as lightly preserved products which might not be fully protected against pathogens such as *Listeria monocytogenes* and *Clostridium botulinum*. These pathogens thrive in lightly salted products (e.g., smoked salmon), low temperatures (>3.3°C), and in vacuum packages, and are therefore a potential threat against food safety (Craig *et al.*, 1968; Heinitz and Johnson, 1998). An effective food additive to prevent pathogens is nitrite salt. This is commonly used in meat curing and is also allowed in salt curing of salmonid fish in the USA (CFR, 2009). Nitrite gives the products a stable red color (nitroso-heme), and it acts as an antioxidant. In addition, it increases lactic acid bacteria growth (Lyhs *et al.*, 1998) which again prevents growth of the spoilage- and pathogenic bacterial flora (Honikel, 2008; Pelroy *et al.*, 1994). Nitrite ($NO^-{}_2$) is unstable at acidic pH and can disproportionate into nitrous acid (HNO_2), which again decomposes to nitric oxide (NO). Furthermore, NO might react with secondary amines to form volatile N-nitrosoamines (Yurchenko and Mölder, 2006). Use of nitrite salt in salt preservation of marine species

Processing and Impact on Active Components in Food
http://dx.doi.org/10.1016/B978-0-12-404699-3.00052-4

is controversial because of a generally high natural content of secondary and tertiary amines. However, there are differences between various species. Freshwater and salmonid fish have a low content of amines (Bellatti and Parolari, 1982) and are therefore not as vulnerable when using nitrite in curing as saltwater fish.

Several studies have shown that salt-curing affects the red color in salmonid fish (Birkeland *et al.*, 2004; Lerfall *et al.*, 2011a). The characteristic red flesh color of Atlantic salmon is mainly due to the presence of the carotenoid astaxanthin (3,3′-dihydroxy-β,β-carotene-4, 4′-dione), which is the most commonly used carotenoid for muscle pigmentation of farmed salmonid fish. Astaxanthin is present in the muscle in its unesterified form (Henmi *et al.*, 1987) where the geometrical *trans* isomer predominates (>90% of total astaxanthin) (Bjerkeng *et al.*, 1997). Astaxanthin is believed to bind weakly to hydrophobic sites of actomyosin in salmon flesh (Henmi *et al.*, 1987). However, recent research has indicated astaxanthin to be associated to other muscle proteins as well as actomyosin (Birkeland and Bjerkeng, 2004; Saha *et al.*, 2005) and it has been concluded that the major astaxanthin binding protein in salmon muscle is α-actinin (Matthews *et al.*, 2006).

HOW COMPOSITION IS ALTERED

Effects of Salt-Curing on Carotenoids in Salmonids

The contents of carotenoids in the fillet surface of cold-smoked salmon flesh are significantly altered by the salting method applied (Lerfall *et al.*, 2011a). However, a study by Birkeland *et al.* (2004) did not find any significant differences in astaxanthin retention between dry-salted and injection-salted fillets. Furthermore, only minor effects of sodium nitrite on astaxanthin stability in cold smoked Atlantic salmon are reported (Lerfall and Østerlie, 2011b). Immediately after salting and during smoking, dry salted fillets of Atlantic salmon have NaCl concentrations up to 8–10% in the fillet surface. This affects retention of astaxanthin and idoxanthin (a metabolite of astaxanthin) negatively as compared to salmon fillets where NaCl were injected (Lerfall *et al.*, 2011a). However, lutein seems to be more stable throughout processing and storage as compared to astaxanthin and idoxanthin (Lerfall *et al.*, 2011a). Carotenoids are labile compounds where the stability in salmonid muscle is related to the strength of the protein–carotenoid binding (Latscha, 1990), which may differ between carotenoids due to their chemical properties and be affected by high NaCl concentrations in the muscle.

Birkeland *et al.* (2004) concluded that coloration of dry salted salmon fillets may be positively affected if large amounts of water are used and prolonged exposure of diluted brine occurs. Since carotenoids are bounded to myofibrillar- and sarcoplasmatic proteins in salmonid flesh (Birkeland *et al.*, 2004; Matthews *et al.*, 2006; Saha *et al.*, 2005), it is likely that carotenoids are being extracted with brines with low salt content. Therefore, high salt concentrations (>20%) used during injection and brine salting may prevent astaxanthin–protein extraction explaining higher carotenoid retention in the fillet surface of injection-salted salmon fillets as compared to dry-salted fillets (Lerfall *et al.*, 2011a). However, physical ablution is not the only factor that affects the retention of carotenoids during processing. Loss of astaxanthin by chemical decomposition is also likely, where the first step may be a release of astaxanthin from the protein-complex influenced by a high NaCl content (Lerfall *et al.*, 2011a) and thereafter decomposition of carotenoids caused by light, oxygen, and probably chemical components absorbed from the smoke (Schiedt and Liaaen-Jensen, 1995).

In addition to physical ablution and decomposition of carotenoids during salt-curing, isomerization of carotenoids may occur. Moreover, nitrite salt is found to have an isomerising effect on astaxanthin during processing (Table 52.1). Isomerization of astaxanthin by nitrite occurs because of the reaction power of nitrite itself or of nitrite-related compounds like nitrous acid, nitric oxide, nitrogen dioxide radical, and peroxynitrite formed under conditions obtained in the smoking process. Antioxidant properties of astaxanthin are mainly due to physical quenching of singlet oxygen ($^1O_2$), where energy absorbed from $^1O_2$ to produce triplet oxygen ($^3O_2$) is converted to rotary and vibratory energy by the

TABLE 52.1 Astaxanthin Isomers in the Fillet Surface and in the Rest of the Muscle of Cold-Smoked Atlantic Salmon Treated with Regular Sodium Chloride and Nitrite Salt, Respectively

	Fillet Surface			Rest of the Muscle		
Treatment	***trans*-astaxantin**	***9-cis*-astaxanthin**	**13*Z*-astaxanthin**	***AllE*-astaxantin**	**9*Z*-astaxanthin**	**13*Z*-astaxanthin**
Sodium chloride	90.5%	2.8%	6.7%	92.8%	1.9%	5.2%
Nitrite salt*	84.4%	4.7%	10.9%	90.0%	3.2%	6.8%

* *Nitrite salt [nutrient quality added 0.6% sodium nitrite ($NaNO_2$) in sodium chloride (NaCl)].*

chromophore system of the carotenoid (Stahl and Sies, 1993). Furthermore, astaxanthin and other carotenoids can protect the muscle matrix against superoxidants like nitrogen dioxide radical and peroxynitrite (Yoshioka *et al.*, 2006).

Due to color appearance *cis*-isomers of carotenoids show a hypsochromic shift (displacement of λ_{max} to shorter wavelength) of 2–6 nm for mono-*cis* isomers which indicate a slightly more yellowish appearance as compared to all-*trans* carotenoids (Britton, 1995).

Contents of Nitrite and Nitrate in Marine Products

Levels of naturally occurring nitrite and nitrates determined in marine species are low, and data reported is limited. However, there are indications that nitrite may be formed during smoking. A preliminary study by Matforsk (2006) reported traces of residual nitrite up to 2.3 mg/kg in dry salted and smoked Atlantic salmon. After 45 days of refrigerated storage, the content ranged from 2.1 to 2.6 mg/kg. Levels reported for raw unprocessed Atlantic salmon are <1.3 mg/kg (Lerfall *et al.*, 2011b; Matforsk, 2006). Moreover, Karimzadeh *et al.* (2010) have reported nitrite and nitrate levels in smoked Caspian kutum (*Rutilus frisii kutum*) ranging from 2.3 to 9.0 and 15.5 to 17.5 mg/kg, respectively, and in smoked mullet (*liza Auratus*) ranging from 0.9 to 8.0 and 14.0 to 19.3 mg/kg, respectively.

Use of sodium nitrite as a food additive result in relative high amounts of residual nitrite in the product (Table 52.2). However, measured levels range between 3 and 80% of theoretical levels which is due to chemical decomposition and formation of other nitro compounds during processing and storage (Lerfall *et al.*, 2011b; Matforsk, 2006).

No sensory evaluation of nitrite-treated cold-smoked salmon has yet been done. A study by Cuppet (1988) did not find any differences in taste of smoked Great Lakes Whitefish treated with and without nitrite after 14 days of storage. Moreover, a study by Froehlich *et al.* (1983) on ham supported the assertion that sensory properties are probably not affected by low levels of nitrite.

Residual nitrite in marine products may give several unwanted reactions. Formation of different N-nitrosoamines are likely for saltwater species with high naturally contents of amines (Sen *et al.*, 1970). In salmonids, however, contents of amines are low, and the most serious unwanted effect of eating nitrite-treated salmonids is probably related to a high intake of nitrite. A high intake of nitrite is often related to health problems such as methemoglobinemia, cancer, cardiac and vascular diseases, circulatory shock, and inflammation (Pacher *et al.*, 2007).

Formation of N-Nitrosoamines

Several studies have detected traces of N-nitrosoamines in marine species where naturally occurring nitrites and nitrates in the raw material or nitro compounds formed during processing are nitroso agents (Iyengar *et al.*, 1976; Sen *et al.*, 1985; Yurchenko and Mölder, 2006). Processing of marine foods with impure salt, high temperatures, and low pH enhance the formation of N-nitrosoamines, whereas pure NaCl inhibits their formation (Bulushi *et al.*, 2009). Moreover, high temperatures obtained during hot-smoking and frying increase the contents of N-nitrosoamines in the product. In literature, only a few studies deal with the contents of N-nitrosoamines in various fish products (Iyengar *et al.*, 1976; Yurchenko and Mölder, 2006). A survey of the results obtained by Yurchenko and Mölder, 2006 are presented in Table 52.3.

Data on effects of nitrite-curing of marine species is limited, and only a few studies describe formation of N-nitrosoamines after such treatments. A study reported by Lerfall *et al.* (2011b) describe the effects of nitrite-curing and cold-smoking of Atlantic salmon fillets. This study did not report increased levels of N-nitrosoamines in nitrite-cured fillets as compared to fillets treated with pure NaCl. Another recent study by the same authors (Lerfall and Østerlie, 2013) did

TABLE 52.2 Residual Nitrite in Atlantic Salmon Treated with Nitrite Salt [Nutrient Quality Added 0.6% Sodium Nitrite ($NaNO_2$) in Sodium Chloride (NaCl)]

	Day 1		Day 24		Day 44**/45*	
Sample	**NaCl (g/100 g)**	**$NaNO_2$ (mg/kg)**	**NaCl (g/100 g)**	**$NaNO_2$ (mg/kg)**	**NaCl (g/100 g)**	**$NaNO_2$ (mg/kg)**
Injected fillets*		75.0 (108)				3.6 (108)
Injected fillets*		49.2 (72)				8.9 (72)
Dry salted fillets (fillet surface)**	7.0	342.7 (420)	2.5	101.1 (150)	2.3	74.9 (138)
Dry salted fillets (rest of muscle)**	1.4	61.7 (84)	2.4	76.3 (144)	2.4	45.7 (144)

Theoretical values are presented in parentheses.
* *Matforsk (2006).*
** *Lerfall* et al., *2011b.*

TABLE 52.3 Effects of Different Processing on Formation of Volatile N-Nitrosoamines in Different Fish Products (Yurchenko and Mölder, 2006)

Treatments	Fish Species	N-Nitrosoamines* (μg/kg)	Comments
Untreated fresh fish	Herring, Silver hake, Mackerel, Sea perch, Sardine	Not detected	N-nitrosoamines was not reported in fresh untreated fish
Cold-smoked fish	Mackerel, Salmon, Herring	Range 1.92–2.15	Hot-smoking in general generate higher amounts of N-nitrosoamines as compared to cold-smoking due to higher temperatures during processing
Hot-smoked fish	Mackerel, Herring, Trout, Bream, Cod, Sprat, Salmon, Burbot, Silver hake, Sea perch, Pike	Range 3.57–4.85	
Fried fish	Herring in tomato sauce, Sprats in tomato sauce	Range 6.96–9.62	High temperatures induce formation of N-nitrosoamines
Pickled fish	Herring	5.37	
Salted/dried fish	Roach	3.81	

* *Sum of five N-nitrosoamines: N-nitrosodimethylamine; N-nitrosodiethylamine; N-nitrosopyrrolidine; N-nitrosopiperidine; N-nitrosodibuthylamine.*

not find increased levels of N-nitrosoamines during chilled storage (44 days). However, formation of N-nitrosoamines occurs in other species, and especially saltwater species which contain higher levels of naturally occurring amines (Sen *et al.*, 1970). A study by Fiddler *et al.* (1993) deals with the opportunity of using Atlantic Menhaden (*Brevortia tyrannus*) mince and surimi as partial meat substitutes in nitrite-treated frankfurters. Atlantic Menhaden is a member of the herring family which may form N-nitrosoamines when treated with nitrite salt. After 15 and 50% substitution of the meat by washed and unwashed mince and surimi, only small amounts of N-nitrosodimethylamine (NDMA) were found (range 0.43–3.07 μg/kg). The significantly highest content of NDMA was found in frankfurters with added 50% unwashed Menhaden mince (3.07 μg/kg). Moreover, another study by the same authors found a significant effect of the heating method applied on the formation of NDMA in nitrite-treated Alaskan Pollock (*Theragra chalcogramma*) fish-meat frankfurters. Contents of NDMA ranged between 0.3 μg/kg (15% added surimi, cocked by microwave) and 18.1 μg/kg (50% added unwashed mince, fried) (Fiddler *et al.*, 1992).

OTHER WAYS IN WHICH COMPOSITION IS ALTERED

Colorimetric Alteration

Altered composition in the fillet surface of nitrite-cured salmon fillets affects color appearance due to formation of a nitroso–heme complex. This red complex potentially covers a low intensity of astaxanthin (VIS, $\lambda_{max}=470$ nm), resulting in increased color evenness and intensity of redness on the fillet surface. Formation of this nitroso–heme complex was confirmed in an experiment with Cod (*Gadus murhua*) and herring (*Clupea harengus*) fillets treated with nitrite salt.

Nitrite curing results in significantly lower reduction of a^* and H°_{ab} from raw to smoked fillets, i.e., better color stability in nitrite-treated fillets as compared to respective fillets treated with pure NaCl (Lerfall *et al.*, 2011b). The fish muscle has low levels of myoglobin and other heme compounds (Olsen *et al.*, 2006), but clearly sufficient levels to produce nitrosomyoglobin and/or nitrosohemoglobin (pink color) and increase the color quality of cold-smoked salmon fillets. The improved color of nitrite-treated fillets faded during cold storage, and after 24 days differences in a^*- and H^0_{ab}-values were no longer significant. However, the translucence was still significantly lower at days 24 and 44, indicating a darker fillet (lower L^*-values). Relative colorimetric values also indicated that the improved color observed right after smoking still has some significance after 44 days cold storage (Figure 52.1).

ANALYTICAL TECHNIQUES

Measurements of Carotenoids

In nature, more than 600 carotenoids exist (Britton, 1995), where five to six of them are relevant today in carotenoid analyses of salmonid flesh. In addition to the wide variety of carotenoids in nature, carotenoids occur in different stereo isomers (*cis*- and *trans*-) and optical isomers (*R* and *S*) (Figure 52.2). Extraction of carotenoids from muscle tissue is quite challenging. These components are rather labile and can easily decompose or be isomerized when exposed to high temperatures, light, and chemicals, and be oxidized when exposed to oxygen or acids (Schiedt *et al.*, 1995). It is possible to increase the stability of carotenoids by adding an antioxidant or a piece of solid CO_2 into the sample (Schiedt *et al.*, 1995). Another critical point is the lipids, which are extracted together with the carotenoids. Stabilization of samples prior to analysis is possible by adding butylated hydroxytoluene (BHT) and/or rinsing on a solid-phase

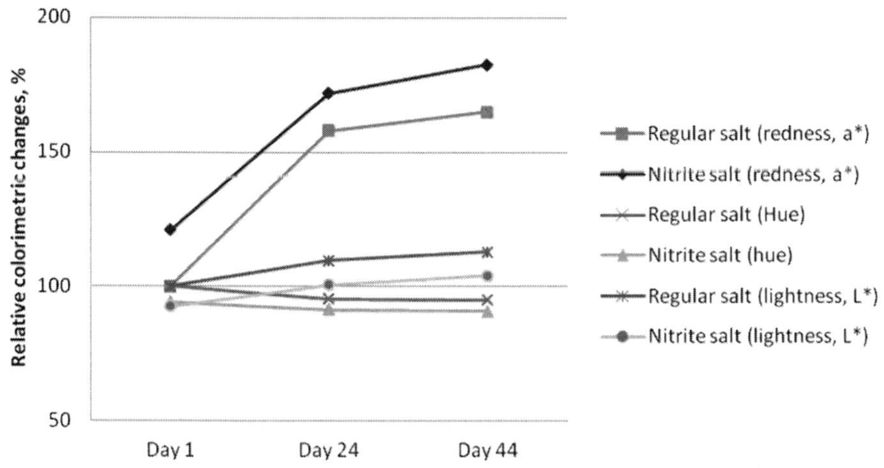

FIGURE 52.1 Relative colorimetric changes (% of initial colorimetric value) in the fillet surface of cold-smoked Atlantic salmon treated with regular salt (99.9% NaCl) and nitrite salt (0.6% $NaNO_2$ in NaCl) and stored in vacuum packages for 44 days (4°C).

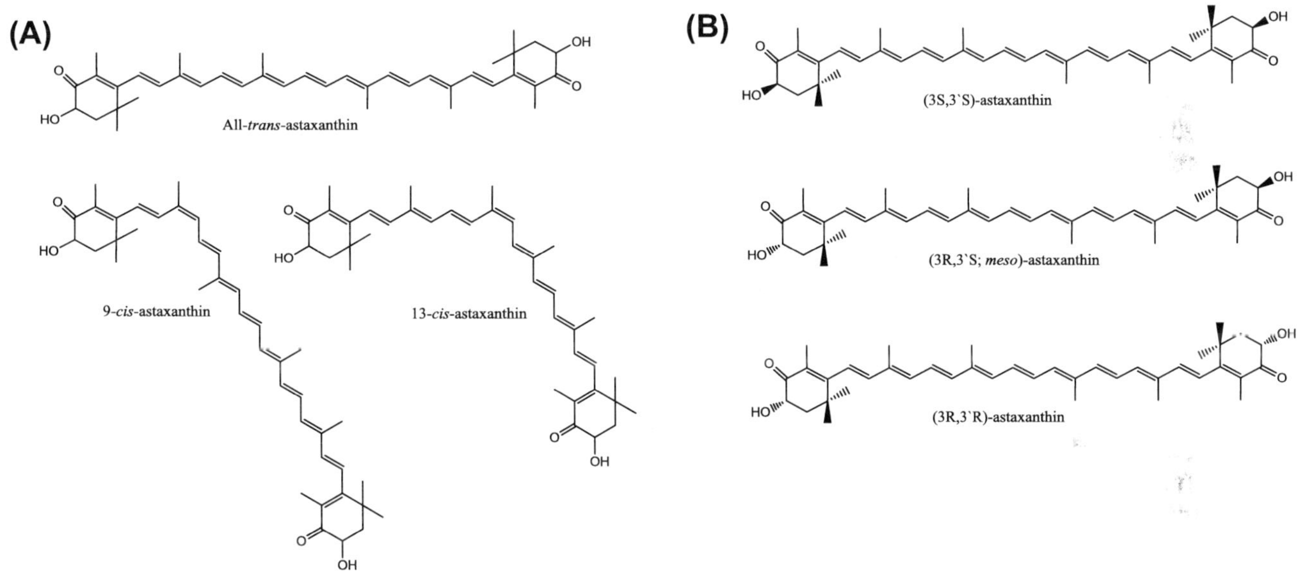

FIGURE 52.2 (A) Geometrical isomers of astaxanthin. (B) Optical isomers of astaxanthin.

extraction (SPE) column as described by Schierle and Frey (2005). To obtain good results, it is also important to protect samples from heat and light during the homogenization and extraction procedures.

Chromatographic analyses of carotenoids are complex, and the choice of method relevant to a particular study depends on the information at hand. Astaxanthin is often analyzed by a method after Vecchi *et al.* (1987) on a $H_3PO_4^-$-modified silica gel HPLC column and detected at 470 nm. This system separates *cis-* and *trans*-isomers of astaxanthin and other actual carotenoids like canthaxanthin, zeaxanthin, and lutein. One of the limitations of this system is that it is difficult to detect and impossible to separate *cis-* and *trans*-isomers of idoxanthin. *Cis-* and *trans*-isomers of idoxanthin can be analyzed by a method after Aas *et al.* (1997). In addition, total amounts of other carotenoids like astaxanthin and lutein can be quantified. A new method for homogenizing, extracting and quantifying astaxanthin, canthaxanthin, lutein, and zeaxanthin from fish flesh was developed in 2010 by a scientific group put together by the Norwegian Seafood Federation, FHL. This method is based on Schierle *et al.* (2005) and Vecchi *et al.* (1987) and may present itself as valid in future analyses. For specific information on stereoisomerism (except for astaxanthin) and optical isomerism, methods adapted to the specific topic of exploration must be chosen.

Measurements of N-Nitrosoamines

Contents of N-nitrosoamines are usually analyzed by gas chromatography (GC) capillary separation combined with either a thermal energy analyser (TEA) or a mass spectroscopic system (MS). Other methods such as gas–liquid chromatography (GLC) with TEA or Coulson electrolytic conductivity detector have also been used

(Yurchenko and Mölder, 2006). In marine species several N-nitrosoamines might occur, dependent on the spectra of amines present in the raw material. Specific care has to be taken due to the potential carcinogenic effects of the components.

References

Aas, G.H., Bjerkeng, B., Hatlen, B., Storebakken, T., 1997. Idoxanthin, a major carotenoid in the flesh of Arctic charr (*Salvelinus alpinus*) fed diets containing astaxanthin. Aquaculture 150, 135–142.

Bellatti, M., Parolari, G., 1982. Aliphatic secondary amines in meat and fish products and their analysis by high pressure liquid chromatography. Meat Sci. 7, 59–65.

Birkeland, S., Bjerkeng, B., 2004. Extractabilities of astaxanthin and protein from muscle tissue of Atlantic salmon (*Salmo salar*) as affected by brine concentration and pH. Food Chem. 85, 559–568.

Birkeland, S., Haarstad, I., Bjerkeng, B., 2004. Effects of Salt-curing Procedure and Smoking Temperature on Astaxanthin Stability in Smoked Salmon. J. Food Sci. 69, FEP198–FEP203.

Bjerkeng, B., Følling, M., Lagocki, S., Storebakken, T., Olli, J.J., Alsted, N., 1997. Bioavailability of all-E-astaxanthin and Z-isomers of astaxanthin in rainbow trout (*Oncorhynchus mykiss*). Aquaculture 157, 63–82.

Britton, G., 1995. UV/Visible spectroscopy. In: Britton, G., Liaaen-Jensen, S., Pfander, H. (Eds.), Carotenoids, Isolation and Analysis, vol. 1A. Birkhäuser, Basel, pp. 13–62.

Bulushi, I.A., Poole, S., Deeth, H.C., Dykes, G.A., 2009. Biogenic Amines in Fish: Roles in Intoxication, Spoilage, and Nitrosamine Formation—A Review. Crit. Rev. Food Sci. Nutr. 49, 369–377.

CFR., 2009. Food and Drugs – Part 172: Food additives permitted for direct addition to food for human consumption, Subpart B: Food Preservatives, 21(3), Sec 172.175 Sodium nitrite. Code of Federal Regulations.

Craig, J.M., Hayes, S., Pilcher, K.S., 1968. Incidence of Clostridium botulinum type E in salmon and other marine fish in the Pacific Northwest. Appl. Microbiol. 16, 553–557.

Cuppet, S.L., 1988. The effect of processing variables on the safety and acceptability of smoked Great Lakes Whitefish, Dissertation Abstracts International. 207.

Fiddler, W., Pensabene, J.W., Gates, R.A., Hale, M., Jahncke, M., 1992. N-Nitrosodimethylamine Formation in Cooked Frankfurters Containing Alaska Pollock (*Theragra chalcogramma*) Mince and Surimi. J. Food Sci. 57, 569–571.

Fiddler, W., Pensabene, J.W., Gates, R.A., Jahncke, M.L., Hale, M.B., 1993. Atlantic menhaden (*Brevoortia tyrannus*) mince and surimi as partial meat substitutes in frankfurters: Effect on N-nitrosamine formation. J. Agri. Food Chem. 41, 2238–2241.

Froehlich, D.A., Gullett, E.A., Usborne, W.R., 1983. Effect of Nitrite and Salt on the Color, Flavor and Overall Acceptability of Ham. J. Food Sci. 48, 152–154.

Heinitz, M.L., Johnson, J.M., 1998. The Incidence of Listeria spp., Salmonella spp., and Clostridium botulinum in Smoked Fish and Shellfish. J. Food Prot. 61, 318–323.

Henmi, H., Iwata, T., Hata, M., Hata, M., 1987. Studies on the carotenoids in the muscle of salmons. I. Intracellular distribution of carotenoids in the muscle. Tohoku J. Agri. Res. 37, 101–111.

Honikel, K.O., 2008. The use and control of nitrate and nitrite for the processing of meat products. Meat Sci. 78, 68–76.

Iyengar, J.R., Panalaks, T., Miles, W.F., Sen, N.P., 1976. A survey of fish products for volatile N-nitrosamines. J. Sci. Food Agri. 27, 527–530.

Karimzadeh, L., Koohdani, F., Mahmodi, M., Safari, F., Babaee, Z., 2010. Determination of Nitrate and Nitrite Residues in Smoked Caspian Kutum, *Rutilus frisii kutum* and Mullet, *Liza auratus* in the North of Iran. World J. Fish Mar. Sci. 2, 62–65.

Latscha, T., 1990. Carotenoids—their nature and significance in animal feeds Department of animal nutrition and health. F. Hoffman-La Roche Ltd., Basel.

Lerfall, J., Akse, L., Østerlie, M., Birkeland, S., 2011a. Salting method affects the retention of carotenoids in the fillet surface of cold-smoked Atlantic salmon (*Salmo salar* L.). Int. J. Food Sci. Technol. 46, 2218–2223.

Lerfall, J., Østerlie, M., 2011b. Use of sodium nitrite in salt-curing of Atlantic salmon (*Salmo salar* L)—Impact of product quality. Food Chem. 124, 759–766.

Lerfall, J., Østerlie, M., 2013. The use of sodium nitrite in cold-smoke processing of farmed Atlantic salmon - effect on storage stability. Int. J. Food Sci. Technol. 48, 1985–1990.

Lyhs, U., Björkroth, J., Hyytiä, E., Korkeala, H., 1998. The spoilage flora of vacuum-packaged, sodium nitrite or potassium nitrate treated, cold-smoked rainbow trout stored at 4°C or 8°C. Int. J. Food Microbiol. 45, 135–142.

Matforsk, 2006. Level of residual nitrite by smoking of smoked salmon.

Matthews, S.J., Ross, N.W., Lall, S.P., Gill, T.A., 2006. Astaxanthin binding protein in Atlantic salmon. Comp. Biochem. Physiol. Part B: Biochem. Mol. Biol. 144, 206–214.

Olsen, S.H., Sorensen, N.K., Stormo, S.K., Elvevoll, E.O., 2006. Effect of slaughter methods on blood spotting and residual blood in fillets of Atlantic salmon (*Salmo salar*). Aquaculture 258, 462–469.

Pacher, P., Beckman, J.S., Liaudet, L., 2007. Nitric Oxide and Peroxynitrite in Health and Disease. Physiol. Rev. 87, 315–424.

Pelroy, G., Peterson, M., Paranjpye, R., Almond, J., Eklund, M., 1994. Inhibition of Listeria-Monocytogenes in Cold-Process (Smoked) Salmon by Sodium-Nitrite and Packaging Method. J. Food Prot. 57, 114–119.

Saha, M.R., Ross, N.W., Gill, T.A., Olsen, R.E., Lall, S.P., 2005. Development of a method to assess binding of astaxanthin to Atlantic salmon *Salmo salar* L. muscle proteins. Aquaculture Res. 36, 336–343.

Schiedt, K., Liaaen-Jensen, S., 1995. Isolation and analysis. In: Britton, G., Liaaen-Jensen, S., Pfander, H. (Eds.), Carotenoids, Isolation and Analysis, vol. 1A. Birkhäuser, Basel, pp. 81–108.

Schierle, J., Frey, C., 2005. Determination of astaxanthin in fish flesh. Version 1.0. DSM Nutritional Products Ltd.

Sen, N.P., Smith, D.C., Schwinghamer, L., Howsam, B., 1970. Formation of nitrosamines in nitrite-treated fish. Can. Inst. Food Technol. J. 3, 66–69.

Sen, N.P., Tessier, L., Seaman, S.W., Baddoo, P.A., 1985. Volatile and nonvolatile nitrosamines in fish and the effect of deliberate nitrosation under simulated gastric conditions. J. Agri. Food Chem. 33, 264–268.

Stahl, W., Sies, H., 1993. Physical Quenching of Singlet Oxygen and cis-trans Isomerization of Carotenoidsa. Ann. N. Y. Acad. Sci. 691, 10–19.

Vecchi, M., Glinz, E., Meduna, V., Schiedt, K., 1987. HPLC separation and determination of astacene, semiastacene, astaxanthin, and other keto-carotenoids. J. High Resolution Chromatogr. 10, 348–351.

Yoshioka, R., Hayakawa, T., Ishizuka, K., Kulkarni, A., Terada, Y., Maoka, T., Etoh, H., 2006. Nitration reactions of astaxanthin and β-carotene by peroxynitrite. Tetrahedron Letters 47, 3637–3640.

Yurchenko, S., Mölder, U., 2006. Volatile N-Nitrosamines in various fish products. Food Chem. 96, 325–333.

CHAPTER

53

Marinating and Salting of Herring, Nitrogen Compounds' Changes in Flesh and Brine

Mariusz Szymczak, Grzegorz Tokarczyk, Katarzyna Felisiak

Department of Food Science and Technology, Faculty of Food Science and Fisheries, West Pomeranian University of Technology, Poland

CHAPTER POINTS

- Protein content of ripened herring is similar to that of raw material. In turn, non-protein nitrogen (NPN) content increases to 350–600 mg in marinades and 320–390 mg in salted fish, which represents 10–15% of total nitrogen of raw material.
- The content of NPN depends more on freezing than fish dressing, and more on salt concentration than acid concentration in the brine.
- An important factor influencing nitrogen content of marinated and salted herring is the process of diffusion to brine. Losses of total nitrogen account for 5–25% and 7–14% of raw material nitrogen during marinating and salting processes, respectively. The total content of NPN in brine exceeds the total content of these fractions in flesh.
- Tissue of frozen/thawed fish, compared to the fresh tissue, responds differently to increasing concentrations of salt or acid in the brine.
- Marinades and salted fish from fresh and frozen herring are characterized by different quantitative and qualitative composition of nitrogen fractions, due to various dynamics of nitrogen fractions formation and different extents of their losses.
- NPN fractions content in salted herring flesh and brine depends on the fishing season.

INTRODUCTION

Herring contribution to the global consumption of fish products is significant due to its abundance in marine waters and its high nutritive value. It is usually consumed salted or marinated.

Typical sensory traits of salted (salting brine) and marinated (cold marinades) fish are developed upon enzymatic ripening of tissue in a solution of vinegar, salt, and various seasonings. Differences in ripening and composition of nitrogen fractions between salted and marinated fish are mostly due to various compositions of brine and specificity of proteases. Brine salting of carcasses/fillets involves the addition of organic acids that reduce the pH value to 5.4–6.0 and activate muscle tissue enzymes or enzymatic preparations, whilst in the marinating process, the penetrating acid reduces the pH value to 4.0–4.5, strongly activates tissue enzymes (cathepsins), and dissolves and denaturates other proteins.

Mild enzymatic hydrolysis results in the formation of protein hydrolysis products (polypeptides, peptides and free amino acids) with high nutritive and functional values. Nitrogen compounds with a lower molecular mass (to 15 kDa) impart typical taste and odor attributes of ripe flesh to marinades (Shenderyuk and Bykowski, 1990). The higher nutritive value of hydrolyzed than native protein is also due to the presence of biologically active peptides directly absorbable by cells of the gastrointestinal tract.

The quantitative and qualitative composition of nitrogen fractions in a food product depends on many factors linked with raw material and technology (Sikorski, 2004;

Processing and Impact on Active Components in Food
http://dx.doi.org/10.1016/B978-0-12-404699-3.00053-6

Kołakowski *et al.*, 1993). Resources of fish used for marinating and salting are observed to diminish, hence technological factors are contemporarily becoming of key significance. Apart from the sensory quality and stability of product, an increasing problem is posed by the weight yield. For this reason, producers of salted and marinated fish express an increasing interest in the diffusion of nitrogen fractions from flesh to brine. Nitrogen losses affect its lower content in the tissue as well as the sensory, technological, nutritive, and functional quality of the finished product, and also greater contamination of the environment.

HOW COMPOSITION IS ALTERED

Nitrogen in Flesh

Marinating

Herring flesh contains 2.6–3.2g of nitrogen, depending on fatty tissue content and dressing degree. Protein content of marinated herring is similar to that in the raw material. Fresh herring contains 250–300mg of non-protein nitrogen (NPN), whereas NPN content after frozen storage is higher by ca. 50mg (Szymczak, 2011). It is due to proteolysis before freezing and particularly during thawing. After 2 days of marinating, as a result of diffusion to brine, the NPN content may decrease by 50–100mg, remain unchanged, or increase by 50mg/100g flesh. Ripened marinades contain 350–450mg, whereas very ripened marinades contain as much as 600mg. NPN constitutes 10–15% of protein and does not depend on total nitrogen (TN) content in raw material. Marinated carcasses, unlike fillets, contain more NPN (by 30–50mg), which may affect significant higher sensory scores. Carcasses of Baltic herring and fillets from Atlantic herring are characterized by higher contents and faster NPN increase in tissue, compared to fillets from Baltic herring (Szymczak *et al.*, 2009). It results from a lower ratio of raw material mass to its surface that facilitates higher dynamics of brine components diffusion to flesh and greater nitrogen losses from flesh to brine. The higher NPN concentration in the frozen material is reflected in its higher content in marinated herring, however after 10–14 marinating days, a higher NPN quantity was determined in marinades from fresh herring (Szymczak, 2011).

Higher concentrations of salt and acid inhibit the ripening rate. It is due to high susceptibility of acidic proteases, cathepsin D in particular, to salt concentration above 4–5% (Klomklao *et al.*, 2008). An increase in salt concentration from 5 to 15% causes NPN content to decrease by as much as 110mg/100g fresh herring flesh (Figure 53.1a). In turn, an increase in acetic acid concentration in the range of 3–8% affects NPN content by only 30–50mg (Figure 53.1b). The highest NPN content was reported when using brine containing 7% salt and 5% acetic acid. This corresponds to salt concentration of 2.5–3% and acid concentration of 1.6–1.8% in flesh and to pH4.2–4.3. Tissue of frozen-thawed herring is more susceptible to the negative impact of salt and acetic acid (Szymczak *et al.*, 2012).

In NPN, the greatest contribution is noted for α-amine nitrogen, followed by peptides, and the lowest is for free amino acids. The content of nitrogen of these fractions in fresh herring reaches 20–30mg, 2–7mg, and 0.2–0.4mg,

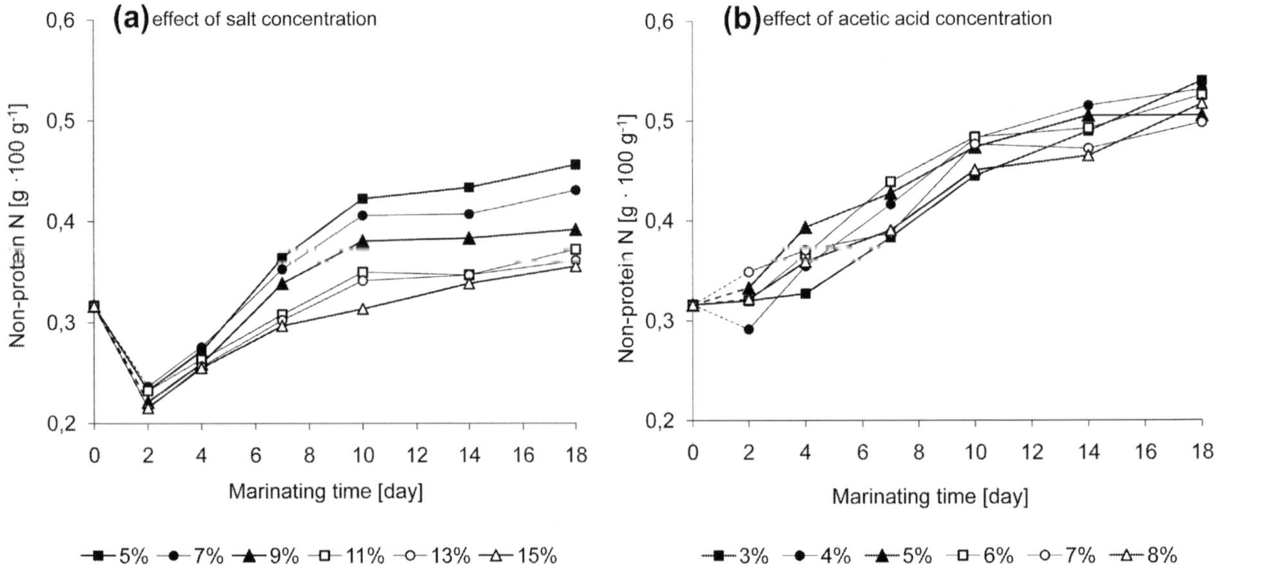

FIGURE 53.1 Concentration of non-protein nitrogen in herring flesh during marinating in brine with different concentrations of (a) salt 5–15% and (b) acetic acid 3–8%. Constant concentration of (a) acetic acid was 5% and of (b) salt was 6% in brine.

whereas in marinated herring it reaches 70–130, 60–100, and 0.5–1.0 mg N/100 g, respectively. The content of the peptide fraction during marinating increases in the logarithmic function, contrary to NPN, tyrosine, and α-amine N whose contents increase in the linear function. The process of frozen storage causes a two- to four-fold increase in peptides content, whilst only a two-fold increase in α-amine N and tyrosine contents. Despite that, the marinades from frozen herring contain more α-amine N, a similar content of tyrosine, and less peptides than these from fresh herring. It stems from the effect of freezing on proteases whose activity diminish by 20–80% depending on cathepsin type (Melendo *et al.*, 2001). It indicates that marinade flavor is also significantly dependent on the qualitative and quantitative composition of NPN.

After increasing salt content in brine from 5 to 15%, the dynamics of peptides increase in flesh is six-fold lower than that of tyrosine. The effect of salt on the products of protein hydrolysis (PHB) is two or more times greater after 4 days. In contrast, the impact of the increasing content of acetic acid on individual nitrogen fractions is significantly smaller.

Brine Salting

TN content of flesh of Baltic herring carcasses, salted with the brine method, ranges from 2.3 to 3.1 g/100 g flesh within a year. The most rapid decrease in TN content was noted in flesh of salted herring from the pre-spawning period (Table 53.1), which was linked with its highest content in the raw material (Tokarczyk, 2003).

Acetic acid increase in brine from 0 to 7% causes an increase in the TN content of salted Baltic herring flesh that results from diminished water absorbability of flesh (Kołakowski and Bednarczyk, 2003). The greatest increase in the TN content (by 26%) after one week of ripening was noted upon the use of 2% of acetic acid in brine. The extension of salting time to two weeks causes a reduction in TN content.

During ripening of salted Baltic herring, a steady quantitative increase is observed in NPN, α-amine nitrogen, peptides of PHB [PHB(R)] and tyrosine of PHB [PHB(A)]. The highest dynamics of NPN and PHB(R) increase is reported in the spawning season, and that of PHB(A) in the pre-spawning period. The dynamics of α-amine nitrogen synthesis reaches its maximum in the pre-spawning and spawning period. At the final stage of herring ripening, retardation is observed in an increase in nitrogen fractions, which results from stronger diffusion of nitrogen substances from flesh to brine. The mean NPN content in flesh reaches 320 mg/100 g (12% TN) once industrial maturity is reached (8–10 days of salting) and 390 mg (15% TN) once the consumption ripeness is reached. Hence, the daily increase of NPN is 7 mg/100 g flesh (Tokarczyk, 2003).

Contents of amine nitrogen and peptides are increasing faster than that of polypeptides. A lower salt concentration contributes to higher contents of NPN, amino acids, and peptides. In contrast, salt concentration increase in brine from 8 to 16% reduces the content of amine nitrogen by 70%, that of peptides by 40%, and of polypeptides by 16%. In turn, an increase in acetic acid

TABLE 53.1 The Amount of Total Nitrogen in Flesh of Headed and Gutted Salted Baltic Herring and its Dynamic Changes Indicated by Linear Equation Coefficients Values

	Total Nitrogen Content in Flesh (g/100 g)		Linear Equations Coefficients ($y=a+bx$)		
Month of Catching	**The Beginning of Salting**	**The End of Salting**	**a**	**b**	**R^2**
January	2.87 ± 0.06	2.38 ± 0.02	2.93	−0.029	0.77
February	2.75 ± 0.02	2.54 ± 0.04	2.75	−0.015	0.78
March	2.98 ± 0.04	2.33 ± 0.01	2.99	−0.034	0.90
April	2.94 ± 0.15	2.30 ± 0.06	2.81	−0.036	0.73
May	3.09 ± 0.02	2.29 ± 0.02	3.03	−0.038	0.87
June	2.73 ± 0.01	2.41 ± 0.04	2.68	−0.026	0.89
July	2.80 ± 0.18	2.60 ± 0.08	2.91	−0.028	0.83
August	2.34 ± 0.05	2.11 ± 0.01	2.54	−0.010	0.76
September	2.73 ± 0.03	2.55 ± 0.04	2.67	−0.014	0.71
October	2.78 ± 0.01	2.53 ± 0.08	2.71	−0.020	0.74
November	2.79 ± 0.09	2.62 ± 0.05	2.91	−0.028	0.83
December	2.65 ± 0.04	2.40 ± 0.02	2.64	−0.006	0.70

concentration in brine from 0 to 7% causes an increase in NPN content of flesh by 40% in the second week of salted herring ripening (especially at 1% addition of acetic acid). After two weeks, the highest NPN content was reported in the sample with 2% addition of acetic acid. The percentage contribution of NPN in TN decreases along with an increasing acetic acid addition from 13.5% to 10% after 1 week of salting, and from 18% to 11% after two weeks.

The freezing of raw material inhibits the increase of nitrogen fractions. The rate of PHB(R) increase decreases by 90%, whereas that of PHB(A) decreases by 10%. After the addition of acetic acid to brine, the effect of freezing on PHB(A) is greater. In thawed salted Baltic herring, a high content of PHB is noted at the initial stage of ripening, followed by a rapid decrease of these fractions in the consecutive period of fish ripening in a 12% solution of NaCl. In herring salted in acidified brine, the content of PHB(A) increases significantly after 4 days of ripening compared to the herring salted in brine without acetic acid addition.

A longer period of frozen storage of raw material affects a decrease in PHB(R) content in NPN from 15% to 10%. This results in an increased content of amine nitrogen from 16% to 20%. In addition, the increase in the relative NPN content in the TN points to the transformation of soluble fractions with a higher molecular mass (peptides) into fractions with a lower molecular mass (amino acids), and to degradation of muscle proteins.

Nitrogen in Brine

Marinating

Acetic acid penetrating tissue causes considerable swelling of structural proteins and dissolution of some collagen fractions in the connective tissue and muscle membranes. Moreover, salt in the first stage of penetration dissolves some of the proteins, denaturing them later on. It causes reduced flesh water-holding capacity and release of water from fish to marinating brine. Together with water, proteins, and other nitrogen fractions, diffuse into the brine.

After 2 days of marinating, the loss ranges from 5% in fresh carcasses to 17.6% in frozen-thawed fillets of Baltic herring (Szymczak and Kołakowski, 2012). Higher losses are noted for frozen herring than for fresh ones. Freezing of carcasses increases the losses by 2–3 percentage points, whereas in fillets the difference is twice as big. Extension of marinating time to 9–10 days contributes to the growth of TN loss, up to 11–24%. Furthermore, a higher nitrogen content in raw fish has not been proved to influence higher nitrogen losses in a brine.

Losses of NPN from herring marinated for 4–5 days constitute *ca.* 56% and 90% of TN losses in fresh fillets and carcasses of Baltic herring, respectively. Freezing of herring before marinating reduces this amount. After 9–10 days of marinating, NPN amounts to 53% and 96% of TN losses in fresh fillets and carcasses, respectively, in comparison to 50% and 74% in frozen ones. The decrease in NPN is caused by the increase in protein content. A smaller NPN amount may also result from a lower activity of enzymes in the frozen material.

In respect of the TN of raw material, α-amine nitrogen constitutes 30% of NPN, whereas PHB(R) constitutes only 1.2%, and PHB(A) only 0.12% (Figure 53.2a). Alpha-amine nitrogen losses, such as free amino acids and oligopeptides, increase with marinating time, by 0.5–1 in fillets (fresh-frozen), by 1.2–1.8 in carcasses of Baltic herring, and by 2.5 percentage points in fillets of Atlantic herring. Another method used to determine the structure of losses is to compare contents of particular nitrogen fractions in the whole brine adequate to the

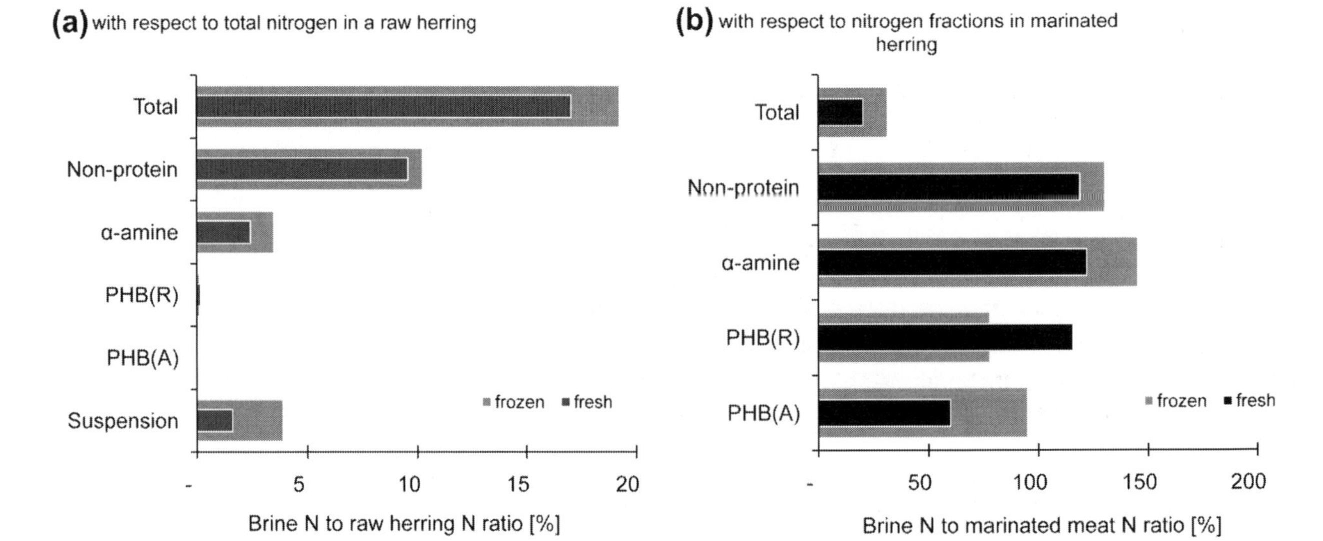

FIGURE 53.2 Losses of nitrogen fractions from fresh and frozen herring to brine after 5 days marinating in brine containing 6% salt and 5% acetic acid with respect to (a) total nitrogen in a total raw herring and (b) nitrogen fractions in total marinated flesh (%).

fractions in the whole flesh (Figure 53.2b). Losses of NPN and α-amine N exceed (120–150%) nitrogen content of these fractions in marinated herring. In turn, PHB(R) and PHB(A) nitrogen represents only 60 and 110%. The freezing process significantly increases losses and changes their structure. Therefore, the poorer taste of marinades from frozen herring (Szymczak, 2011) may be due to higher losses of amine nitrogen.

The nitrogen content of suspension floating in the brine increased very little during fresh herring marinating. However, the content of suspension nitrogen from frozen fillets increased rapidly until it reached industrial maturity, and then decreased after reaching commercial maturity. This may be due to secondary hydrolysis of the protein suspension, and/or increase in the solubility of the protein by decreasing the concentrations of salt and acid in the brine with increasing the marinating time. Loss of this fraction constituted 1.5–2.5% of fresh fish nitrogen and 4–5% of frozen fish nitrogen (Szymczak, 2011).

Increasing the salt concentration in a brine solution from 5% to 15% lowers the TN loss by 1–2 percentage points and increases it by 0.2–4 percentage points from fresh to frozen herring, respectively. The effect of salt on reducing the loss from fresh fish results mainly from a limited loss of non-protein nitrogen. The greater loss from frozen fish is, to a large extent, connected to a higher protein content in the brine. The extension of marinating time intensifies the above-mentioned influence of salt. In case of α-amine, PHB(R) and PHB(A), the amount of these fractions' nitrogen in brine decreases with an increasing salt concentration.

The increase in acetic acid concentration in brine from 3% to 8% after 4–5 days of marinating results in higher losses of TN from frozen fillets by 4–6 percentage points and decreases losses from fresh fillets by 4 percentage points. In case of herring ripening for twice as long, the lowest nitrogen losses are observed for the highest addition of acid in Baltic herring and for the lowest addition of acid in Atlantic herring.

Despite the differences between herring, it can be noted that qualitative changes of losses with an increasing acid concentration are, in the case of industrially mature fish, mainly caused by protein nitrogen loss, and in the case of commercially mature fish—by both protein and NPN loss. NPN fractions more often show higher nitrogen losses when 5% of acetic acid is added than when 3% or 8% is used. As far as losses of nitrogen present in suspension are concerned, it is observed that a higher acid concentration reduces the loss from fresh fillets and increases the loss from frozen ones.

Brine salting

As a result of nitrogen diffusion from flesh, a steady increase of its content was observed in brine. The dynamics of TN increase in the spawning period was higher by 28% than before spawning. TN content after day 2 of salting reaches 100 mg/100 ml (3.7% of flesh TN). At the final state of ripening, nitrogen loss from herring to brine accounts for 5–17 mg/100 ml/day, thus reaching from 200 to 400 mg of nitrogen per 100 ml, which represents 7–14% of flesh TN (Tokarczyk, 2003). Studies by other authors demonstrate nitrogen losses at a level of 8–10% of raw material nitrogen (Sikorski, 2004; Chakrabarti, 2006). Considering nitrogen losses during salting, it can be concluded that TN losses during marinating are two- to three-times higher.

NPN losses in brine are the most rapid in the first week of ripening, whereas in the second week nitrogen diffusion declines. The most rapid increase of NPN was also observed during spawning, and the rate of diffusion was higher by 40% than before spawning. In the first days of salting, NPN content constituted 80% of brine TN and 3% of flesh TN. After the first week of ripening, NPN content increased two to three times reaching 180–230 mg/100 ml (90 and 8–9% of brine and flesh TN) (Tokarczyk, 2003).

Changes in the content of α-amine nitrogen in brine during ripening of salted Baltic herring reflect changes of this fraction in flesh. The rate of nitrogen increase is the highest in the pre-spawning (50%) and spawning (30%) period. Losses of α-amine nitrogen from flesh to brine at the initial fish salting reach 14–50%, whereas at the end of ripening increase to 60–80%. The content of α-amine nitrogen in brine after one-week of fish ripening amounts to 50 mg/100 ml (23% NPN of brine and 1.8% of flesh TN) and increases to 80 mg/100 ml in a completely ripened product (27% NPN of brine and 3.1% of flesh TN).

In the spawning period, the rate of peptide increase in brine is observed to decrease by 20%. In contrast, in the pre-spawning period, a 14–26% increase is reported in this respect. During ripening of salted herring, the content of peptide nitrogen in the brine is higher than or similar to its content in flesh (v/m). In the first days of salting, PHB(R) content reaches 50 mg, which constitutes 8% of brine TN and 0.5% of flesh TN. At the end of the salting process, the content of peptides increases significantly to 13% of total brine nitrogen and 3% of total flesh nitrogen.

The increase in acetic acid concentration in brine does not evoke significant differences in TN content of the brine, but contributes to changes in the ratio of non-protein to protein nitrogen. An increasing addition of acetic acid is accompanied by a decrease in the content of protein nitrogen and by an increase in non-protein nitrogen. These processes intensify along with salting period progression. NPN losses are increasing with an increasing concentration of acetic acid and salting time—with the greatest losses observed in the sample containing 3% of acid. Protein diffusing to brine without acetic acid addition is mainly composed of protein nitrogen, whereas nitrogen of acidified brine contains mainly polypeptides

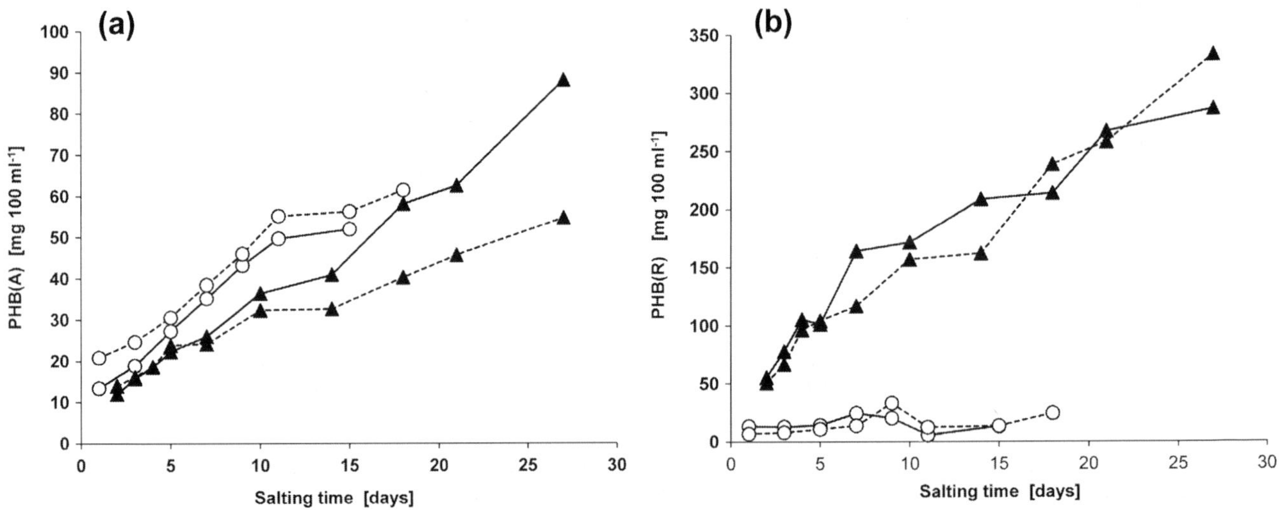

FIGURE 53.3 Effect of raw material freezing and brine composition on content of (a) PHB(A) and (b) PHB(R) in brine. (–○–), unfrozen (12% NaCl); (---○---), frozen-thawed (12% NaCl); (–▲–), unfrozen (12% NaCl + 0.7% CH3COOH); (---▲---), frozen–thawed (12% NaCl + 0.7% CH3COOH).

soluble in trichloroacetic acid (TCA). The brine of herring salted in 12% NaCl was characterized by a high content of PHB(A) and a low content of PHB(R) (Figure 53.3), whereas brine of herring salted in 12% NaCl with the addition of 0.7% CH_3COOH contained very high quantities of PHB(R) and PHB(A) (Tokarczyk *et al.*, 2010).

OTHER WAYS IN WHICH COMPOSITION IS ALTERED

Applying brine circulation during marinating increases salt and acid concentration in tissue. The mixing also facilitates diffusion of nitrogen fractions from flesh to brine, thus causing the losses to increase.

Differences with marinades in various cover brines are due, most of all, to penetration of brine to flesh and diffusion of nitrogen substances from flesh to cover brine. Losses of TN to vinegar brine range from 8% in Rollmops to 15% in Bismarck (Szymczak *et al.*, 2013).

ANALYTICAL TECHNIQUES

Proteins and NPN compounds have a different and complex structures, therefore they are analyzed with a number of various methods. The Lowry method combines the burette reagent with the Folin–Ciocalteau phenol reagent, which reacts with tyrosine and tryptophan residues in proteins. Despite high sensitivity and simplicity, this method is sensitive to interfering substances present in the samples. The Lowry method modified by Kołakowski (2005) enables avoiding simultaneous determination of peptides with other interfering compounds. It consists in the read out of absorbance at 750 nm with and without copper ion addition. The difference in values corresponds to absorbance of [PHB(R)] peptides, whereas the absorbance without copper ions is attributable to tyrosine and other nitrogen compounds with a low molecular mass [PHB(A)], that yield a positive reaction with the Folin–Ciocalteu reagent. In turn, the content of PHB nitrogen from free amino acids to oligopeptides up to 14 kDa from well-soluble complexes may be assayed with the Pope and Stevens method (Tokarczyk and Kołakowski, 2003).

A separate group of protein analyses includes chromatographic methods. Tokarczyk (2003) was assaying protein hydrolysis products in TCA extracts using a Hi-Pore RP 318 column (4.6 × 250 mm), absorbance at 195–300 nm and a solution of acetonitrile, water and trifluoroacetic acid as a mobile phase. The course of protein proteolysis during marinades and salted fish ripening is also monitored with electrophoretic methods, in particular SDS–PAGE and CZE. The change in the profile of compounds soluble in 5% TCA or ethanol (mainly amino acids and peptides) was found to be useful in determining the degree of salted and marinated fish ripeness (Felisiak, 2007).

References

Chakrabarti, R., 2006. Enzymatic bioprocessing of tropical seafood wastes. Losses during salting. In: Shetty, K., Paliyath, G., Pometto, A., Levin, R.E. (Eds.), Food biotechnology, second ed. CRC, Taylor & Francis, Boca Raton, p. 1612.

Felisiak, K., 2007. Optimisation of conditions for capillary zone electrophoresis (CZE) separation of trichloroacetate extracts of salted fish meat. Pol. J. Food Nutr. Sci. 57, 177–184.

Klomklao, S., Kishimura, H., Benjakul, S., 2008. Endogenous proteinases in true sardine (*Sardinops melanostictus*). Food Chem. 107, 213–220.

Kołakowski, E., 2005. Analysis of proteins, peptides, and amino acids in foods. In: Ötles, S. (Ed.), Methods of analysis of food components and additives. CRC Taylor & Francis, Boca Raton, pp. 59–96.

Kołakowski, E., Bednarczyk, B., 2003. Changes in headed and gutted Baltic herring during immersed salting in brine with the addition of acetic acid. Part 2. Intensity of proteolysis. Electron. J. Pol. Agric. Univ. 6, 1.

Kołakowski, E., Bortnowska, G., Lachowicz, K., 1993. Effect of fishing season on the rate of maturation of Baltic herring marinades. XXIV Sesja Naukowa Komitetu Technologii i Chemii Żywności PAN, Wrocław, 36, 84–89.

Melendo, J., Beltrán, J.A., Roncalés, P., 2001. Preservation of the proteolytic activity of a bovine spleen lysosomal-enriched extract using various freezing conditions. Enzyme Microb. Technol. 28, 453–459.

Shenderyuk, V.I., Bykowski, P.J., 1990. Salting and marinating of fish. In: Sikorski, Z. (Ed.), Seafood: resources, nutritional composition, and preservation. CRC Press. Inc., Boca Raton, pp. 147–162.

Sikorski, Z., 2004. Salting and marinating of fish. In: Fish and shelfish. Obtaining, properties and processing. WNT, Warszawa, pp. 311–340.

Szymczak, M., 2011. Comparison of physicochemical and sensory changes in fresh and frozen herring (*Clupea harrengus* L.) during marinating. J. Sci. Food Agric. 91, 68–74.

Szymczak, M., Kołakowski, E., 2012. Losses of nitrogen fractions from herring to brine during marinating. Food Chem. 132, 237–243.

Szymczak, M., Kołakowski, E., Tokarczyk, G., 2009. Marinating process dynamics of meat of thawed Baltic and Atlantic herring. Folia Pomeranae Univ. Technol. Stetin. 272, 83–94.

Szymczak, M., Kołakowski, E., Felisiak, K., 2012. Influence of salt concentration on properties of marinated meat from fresh and frozen herring (*Clupea harengus* L.). Int. J. Food Sci. Technol 47, 282–289.

Szymczak, M., Szymczak, B., Koronkiewicz, A., Felisiak, K., Bednarek, M., 2013. Effect of cover brine type on the quality of meat from herring marinades. J. Food Sci. 78, 619–625.

Tokarczyk, G., Kołakowski, E., Bednarczyk, B., Szymczak, M., Przybylska, S., Waliłko, A., Lament, K., Krzywiński, T., Bienkiewicz, G., 2010. Effect of freezing on ripening process of headed and gutted Baltic herring (*Clupea harengus* L.) salted in brine. Folia Pomeranae Univ. Technol. Stetin. 278, 81–106.

Tokarczyk, G., 2003. Effect of fishing season on the dynamics of nitrogen substances changes during ripening of Baltic herring salted in brine, PhD Thesis. Agricultural University, Szczecin.

Tokarczyk, G., Kołakowski, E., 2003. Specificity of Pope Stevens method of α-amine nitrogen determination. In: Quality of Polish food prior to accession to EU. XXXIV Sesja Naukowa Komitetu Nauk o Żywności PAN, 10–11 September, Wrocław, 359.

CHAPTER

54

Shellfish (Mussel) Processing and Components

Sergio Almonacid[*,†], *Joselyn Bustamante*[*], *Ricardo Simpson*[*,†], *Marlene Pinto*[*]

[*]Department of Chemical and Environmental Engineering, Universidad Técnica Federico Santa María, Valparaíso, Chile,
[†]Centro Regional para el Estudio de Alimentos Saludables (CREAS), Valparaíso, Chile

CHAPTER POINTS

- Integrated average harvest of mussels in China, Spain, Chile, and Thailand, was 1,272,131 t, 65% of the total world production.
- Raw mussels are a good source of proteins, vitamin B_{12}, Fe, P, and omega-3 fatty acids (DHA + EPA).
- Mussels are mainly processed through pre-cooking, freezing, or canning.
- The main effect of processing on mussel components is concentration due to water loss. Leaching and reaction have a minor influence.
- The consumption of 100 g of canned or frozen mussel meat can satisfy more than 20% of the recommended daily allowance of proteins, vitamin B_{12}, Fe, P, and omega-3 fatty acids (DHA + EPA).

INTRODUCTION

Crustaceans and mollusks have increased substantially in the last 50 years, from 166,975 t in 1950 to approximately 2 million t in 2010. This growth is almost entirely due to the increase in aquaculture production, which was particularly rapid in the 1990s. World production of mussel aquaculture or farming increased from approximately 1 million t in 1990 to nearly 1.8 million t in 2010, which represents a value of US$ 1.5 billion, with an average growth rate of 3.5% annually during this period. It is important to highlight that current world production of mussels relies on farming, representing 95% of the total production in 2010 (FAO, 2007). The mussel farming industry reached a significant production level in the Netherlands with the aquaculture cultivation of the common mussel, also known as the blue shell mussel (*Mytilus edulis*). In the 1950s, while countries such as Spain, China, and Italy showed a production of less than 3000 t, the Netherlands harvested annually more than 50,000 t of mussels. The remarkable development of mussel harvesting in the Netherlands achieved a maximum of 120,000 t/year in the 1970s, which has since declined to 40,000–100,000 t/year (FAO, 2012).

China is the main producer of mussels, at 700,000 t/year in 2010, mainly of the species commonly known as sea mussel nei (*Mytilus galloprovincialis*). The production of mussels in China has grown rapidly in the past 30 years, especially in the 1980s, with an average of 22.1% annual growth rate. In the period between 1980 and 2010, the average annual growth rate was 9.7% (FAO, 2012). Mussel aquaculture is highly concentrated in China, Spain, Chile, and Thailand, whose total average harvest, between 2005 and 2010, accounted for 65% of the total world production, 1,272,131 t/year in 2010.

The diversity of cultivated species under the denomination of mussels includes 13 species, and according to the world aquaculture production of the Mytilidae, the species *Mytilus galloprovincialis* is among the most important considering that China and Spain produced 796,698 t of it, on average, between 2005 and 2010. However, in terms of economical implications, the main species is the Chilean mussel (*Mytilus chilensis*), with a total annual production of US$ 416,561, but the blue mussel (*Mytilus edulis*) may be the most relevant species given its ratio of US$ per ton of mussel, evaluated at 1.75. This species of mussel reaches its commercial size of two inches after about three years growth. The shells of fresh mussels are hermetically closed or they close when touched (FAO, 2012). Common

Processing and Impact on Active Components in Food
http://dx.doi.org/10.1016/B978-0-12-404699-3.00054-8

mussels are widely distributed in European waters, extending from the White Sea off of Russia as far south as the Atlantic coast of south France. *Mytilus edulis* has a very wide distribution, mainly due to its ability to withstand wide fluctuations in salinity, temperature, and oxygen tension. The species is well acclimated to a temperature range of 5–20°C, with an upper limit of thermal tolerance approximately 29°C for adults (FAO, 2004).

HOW COMPOSITION IS ALTERED

Nutritional Composition of Mussels

Marine-derived foods are a good source of omega-3 fatty acids and have numerous nutrition and health benefits. Their proteins possess a well balanced amino acid composition, which may be used directly as a source of amino acids or in preparations such as dispersions or hydrolyzed solutions (Shahidi, 2004). From a nutritional perspective, mussels represent an excellent seafood choice. They are low in calories and fat and high in protein, as well as rich in many micronutrients, such as vitamins and minerals. Table 54.1 shows the proximal composition of raw mussels, including micronutrients.

As with most seafood, mussels are rich in protein, 12% wet base (Table 54.1). They are similar to beef protein in quantity and quality, but with less saturated fat, 0.4 as compared with 1.2g lipid/100g of product (USDA, 2012). The recommended daily protein intake is 0.75g/kg body weight (García-Gabarra, 2006). For example, for a person of 70kg, the daily protein intake should be 52.5g, which means that with 100g of mussels, this person would satisfy approximately 23% of his/her requirements.

Although mussels are low in lipids, these are of high quality and provide a healthy contribution to the diet, based on their significant concentration of polyunsaturated and monounsaturated fatty acids (Table 54.1). Today, the amount of fat consumed is considered important, as is its quality. It is desirable that foods include an adequate amount and proportion of essential fatty acids (EFA), omega-6/omega-3, an appropriate amount of monounsaturated fatty acids, a low amount of saturated fatty acids and, ideally, no fatty acids with trans-isomers. Mussels are low in total and saturated fat, and more important, this type of mollusk is one of the richest sources of omega-3 fatty acids.

Some important traits of omega-3 fatty acids are (Byrd-Bredbenner *et al.*, 2009):

- Essential for normal development and function of the retina.
- Help to regulate nerve transmission.
- Regulate blood pressure, blood clotting, body temperature, inflammation or hypersensitivity reactions, and immune and allergic responses.
- Maintain normal kidney function and fluid balance.

Compared with other sources of eicosapentaenoic acid (EPA)+docosahexaenoic acid (DHA), blue mussels are a good source of these essential fatty acids, 0.44g/100g of fresh produce, contrasted with 0.4 and 0.5g/100g for Crustaceans and general Mollusks, respectively (USDA, 2012). In 2009, the European Food Safety Authority (EFSA) published recommendations with regard to PUFAs, advising a dose of 2g/day of fatty omega-3 alpha-linolenic acid (ALA) and 250mg/day of the long-chain omega-3 fatty acids EPA and DHA (Bresson *et al.*, 2009). The Japan Society for Lipid Nutrition has recommended a reduction of energy intake of 3–4% for Japanese people, whose average consumption of omega-3 fatty acids of 2.6g/day, including approximately 1g/day of EPA+DHA (Hamazaki and Okuyama, 2003). Considering that the recommended daily allowance (RDA) of EPA+DHA is 1g/day, 100g of fresh mussel provides 44% of the recommended dose.

Mussels incorporate important amounts of trace elements and minerals necessary to the diet (Do Amaral *et al.*, 1998). Table 54.1 shows information regarding the content of vitamins and minerals provided by 100g of fresh mussels, compared with the RDA and adequate intake (AI). The per cent contributions to the RDA of mussels' micronutrients are shown in the same table.

The contribution of vitamin B_{12} is the most significant in proportion to the RDA, whose contribution by 100g of fresh mussels is 12µg, or 500% of the RDA (Table 54.1). Vitamin B_{12} helps the body maintain sheaths around nerve fibers, helps activate another B-vitamin called folic acid, and participates in many cellular processes (Wardlaw *et al.*, 2009). García-Gabarra (2006) reported that the upper-level (UL) or maximum intake of vitamin B_{12} in adults aged between 19 and 65 years is 1000µg, a quantity 100-times greater than the content of vitamin B_{12} in 100g of fresh Mussel. No toxic or adverse effects have been associated with large intake of vitamin B_{12}, either from food or supplements, in healthy people (Linus Pauling Institute, 2012).

Processing

Despite the above considerations, mussels are not consumed raw but are processed in different ways, and their nutritional characteristics may be affected by processing. Mussel processing, regardless of the final product, consists of a first, common stage of pre-cooking, followed by a second stage of processing,

TABLE 54.1 Proximate and RDA of Nutrients Found in Mussel

Nutrient	Unit	100 g raw Mussel	RDA man	RDA Woman	100 g raw Mussel	% Contribution Man/Woman
PROXIMATE						
Water	g	80.58				
Energy	kcal	86				
Protein	g	11.9				
Total lipid (fat)	g	2.24				
Carbohydrate	g	3.69				
Fiber, total dietary	g	0				
Sugar, total	g	0				
MINERALS						
Calcium, Ca	mg		1000*	1200*	26	2.6/2.2
Iron, Fe	mg		8	18	3,95	49/22
Magnesium, Mg	mg		400 (420[a])	310 (320[a])	34	8/1
Phosphorus, P	mg		700	700	197	28/28
Potassium, K	g		4,7*	4,7*	0.32	7/7
Sodium, Na	g		1,5 (1,3/1,2)*	1,5 (1,3/1,2)*	0.29	24/24
Zinc, Zn	mg		11	8	1,6	15/20
VITAMINS						
Vitamin C, total ascorbic acid	mg		90	75	8	9/11
Thiamin	mg		1,2	1,1	0,16	13/15
Riboflavin	mg		1,3	1,1	0,21	16/19
Niacin	mg		16	14	1,6	10/11
Vitamin B_6	mg		1,3/1,7	1,3/1,5	0,05	3/3
Folate, DFE	μg		400	400	42	11/11
Vitamin B_{12}	μg		2,4	2,4	12	500/500
Vitamin A, RAE	μg		900	700	48	5/7
Vitamin A, IU	IU				160	
Vitamin E (alpha-tocopherol)	mg		15	15	0,55	4/4
Vitamin D (D_2 + D_3)	μg		5 (10/15)	6 (10/15)	0	—
Vitamin D	IU				0	—
Vitamin K (phylloquinone)	μg		120	90	0,1	—
LIPIDS						
Fatty acids, total saturated	g	0.425				
Fatty acids, monounsaturated	g	0.507				
Fatty acids, polyunsaturated	g	0.606				
Cholesterol	g	0.028				

Data from USDA, 2012. AI, Adequate Intake.

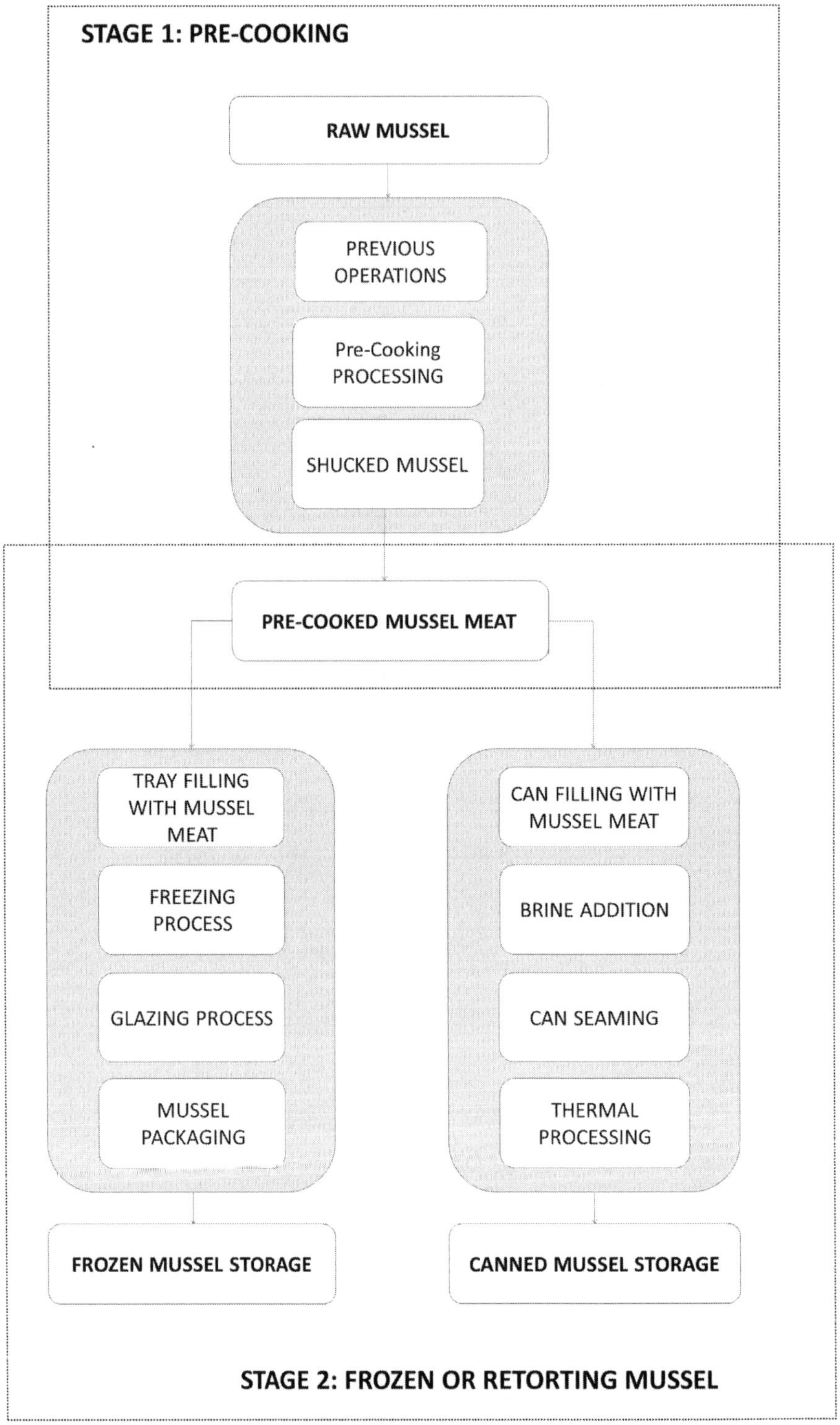

FIGURE 54.1 Schematic representation of mussel processing.

which is usually freezing or sterilization (canning) (Figure 54.1). Bello *et al.* (2012) described these processes in mussels in some detail. They involve the above steps plus some auxiliary operations that result in the production of cooked mussel meat from fresh mussels. Raw, fresh material is collected from hatcheries and transported to the processing plant. At the plant, they are inspected for quantity and quality, and the selected units continue to the following steps of storage, washing, and conditioning. In the final stage, mussels are either frozen or commercially sterilized (Figure 54.1). When sterilized, mussel meat that has been cooked and shucked is ready for canning. According to Almonacid *et al.* (2012), this stage is divided into five operations: filling cans with mussels, adding coverage liquid (salted brine), seaming the containers and, finally, heat-processing to commercial sterility (retorting). Once mussel meats are packed into the open cans, hot liquid brine, consisting of an aqueous solution of 1–3% salt and 0.1% citric acid is added to each can at a temperature of 75°C (Almonacid *et al.*, 2012). Thermal sterilization of filled and sealed cans is carried out in saturated steam at temperatures in the range of 110–130°C for the specified time to achieve a final process lethality of F_0=6min (or the time specified by the process authority). After heating, a final cooling step is applied with water at ambient temperature (19–21°C). Finally, cans are stored at room temperature and have a shelf-life of at most 4 years under these conditions (Holdsworth and Simpson, 2007).

The mussel-freezing process starts with cooked and shucked meat. It is grouped into blocks of about one inch, which are laid out on trays. The freezing process is usually carried out in blast freezers at −40°C for a long enough time to obtain a temperature of −20°C in the geometrical center of the mussel meat block. The process normally takes approximately 50min. After that, the unwrapped meat is glazed to mitigate the water loss by evaporation.

Finally, glazed and not necessarily packed meats are stored in a cold room at a temperature of −20°C. Their shelf-life under these conditions should be 8–9 months (Shafiur, 2012).

The processing effects on mussel composition are mainly caused by two mechanisms: material exchange between mussel meat and processing media (water, steam, brine, air), and thermal inactivation by high temperature (molecular lysis or reaction) when mussels are sterilized. Most water loss takes place during the first, common stage of pre-cooking, and the main effect on the proximal composition of mussel meat is by concentration. Slabyj (1977) reported a dry weight increase from 18.8% to 25.4% due to water lost during pre-cooking of blue mussels, which means a decrease in water content from 81.2 to 74.7%. Krzynowek and Wigginn (1979) used boiling water for 2min and observed a change in moisture from 78.7 to 76.8%. The difference in water loss was due to the pre-cooking method, steaming for 6min in the first case and cooking in boiling water for 2min in the second case. Water loss is significantly affected by the pre-cooking method; Bustamante (2009) found that water loss due to pre-cooking ranged from approximately 50–60g/100g of raw mussel meat, depending on the cooking time (1, 4, or 7min). The working conditions depend on the characteristics of each specific load (especially on mussel size) and on the desired final product (Bello *et al.*, 2012). As described by Almonacid *et al.* (2012), pre-cooking can be performed under different conditions. Casales *et al.* (1988) used cooking times of 7–10min in atmospheric steam to open the shells and facilitate removal of mussel meats. Turan *et al.* (2007) pre-cooked cleaned mussels in boiling water at 100°C for approximately 6min to open the shells and free the mussel meat, but they used cooking times of 1, 4, and 7min at 100°C, and the equipment used for this pre-cooking step was a hot water bath that heated the water up to 100°C by injection of atmospheric steam. At the end of the prescribed pre-cooking time, the mussels are cooled immediately under a spray of potable water for 3min (Almonacid *et al.*, 2012). In addition to being cooked, mussels also must be opened. Cooked mussels can be discharged in a cooling pool containing treated seawater, where they remain until their temperature decreases enough to enable manipulation. This water is at room temperature, but its temperature can increase by 1 or 2°C after receiving a discharge of cooked mussels. The cooked and cooled mussels are transported to two consecutive pools with previously prepared brine. In these pools the viands are separated from the shells (the salty water favors their separation), and because of a density difference, the meat flows to a different depth than shells. The brine flowing between the pools favors the separation of the viands and the shells. The remaining shells are manually removed while the mussels are transported to the next stage (Bello *et al.*, 2012). The amount of water loss is important because it determines the concentrations of mussel meat compounds and their possible leaching into processing media. Table 54.2 shows a general summary of mussel proximate compositions at all stages of processing, comparing many reference sources. Moisture data from Slabyj (1977) are transformed to wet basis for comparison purposes. In general, the moisture content falls from approximately 81% (raw mussel) to 75% (pre-cooked), and then, regardless of the subsequent processing, the moisture content does not change significantly.

As shown in Table 54.2, the protein concentration increases from approximately 11% in raw to 18% in pre-cooked meat (wet basis). However, the original data from Slabyj (1977) show an apparent increase of protein on a dry basis of 63–70% due to decreases in dry-basis carbohydrate and ash concentrations. The author concluded that there is no significant difference in protein content after pre-cooking, that there is no leaching of protein and that the main effect is that of concentration by water loss. Protein and fat concentrations in the final product, frozen or canned, are depicted in the same table, where no significant differences from pre-cooked meat can be observed in either frozen or canned meat. For canned mussel, this means that proteins can be denatured by high temperature, but the amino acid concentration is not affected (Casales, 1988). Casales (1988) found no effect on protein or amino acid concentration by different time–temperature conditions (110.5 and 118°C) considering F_0 values of 3, 7, and 10min. No effect on nutritional value or amino acid composition

TABLE 54.2 Proximal Composition of Mussel Meat from Raw to Processed

	Raw Meat[1]	Raw Meat[2]	Raw Meat[3]	Raw Meat[5]	Pre-Cooked[3]	Pre-Cooked[5]	Frozen[3]	Canned[3]	Canned[4]
Moisture	80,6	81,5	81,2	78.8	74,6	76.8	76,8	75,7	75,1
Protein	11,9	10	11,8	12.4	17,7	15	16,2	17,2	17
Fat	2,3	2,1	2,9	1.7	4,0	2.8	3,7	4,3	3
Ash		3,4	1,5	2.1	1,6	1.7	1,6	1,6	3
CH	3,7		2,7		2,2		1,7	1,2	1,9
Sum	98,5*	97**	100,0		100		100	100	100

[1]*USDA (2012)*
[2]*Fuentes* et al. *(2009)*
[3]*Slabyj (1977)*
[4]*DTU (2009)*
[5]*Krznowek and Wiggin (1979).*
**Difference should be Ash*
***Difference should be CH.*

of canned mussels could be found in the literature for canned mussels. It should be noted that no significant differences in amino acid composition have been reported for other seafoods, such as canned urchin (*Paracentrotus lividus*), albacore tuna, and scallop adductor muscle (de la Cruz-García *et al.*, 2000; Pérez-Martín *et al.*, 1988; Kawashima and Yamanaka, 1996). Frozen storage can have a large effect on the protein and fat concentrations, especially if mussel meat is not properly glazed and/or packed, allowing for moisture lost and affecting organoleptic factors rather than nutritional ones (Gokoglu *et al.*, 2000).

Table 54.3 shows that mussels' fatty acid composition, specifically EPA + DHA, makes an important contribution to the human diet. The intake of 100 g of mussel meat can fill 44% of the RDA for these compounds. Otles and Sengor (2005) reported on the fatty acid composition of mussel meat (*Mytilus galloprovincialis* L.) after various technological processes, including the pre-processing step of steaming for 10 min, polyunsaturated fatty acids represent approximately 38% of fat material in raw mussel, which increases to 57% in pre-cooked mussels, mainly as a consequence of the decrease in monounsaturated and saturated fatty acids. Half of EPA + DHA is retained after pre-cooking, falling from 26.4 to 13.2%. As shown in the proximal analysis of mussel after pre-cooking (Table 54.2), the total fat concentration after pre-cooking changes mainly because of water loss, meaning that the relative change in fatty acid composition is a result of chemical or biochemical reactions due to high-temperature processing, oxidative reactions or long-chain fatty acid hydrolysis by lipolytic enzymes. Senturk *et al.* (2000) showed that fatty acids present in fish shatter during thermal processing, such as cooking and canning, resulting in relative compositional changes. Importantly, DTU (2009) reported that the concentration of EPA + DHA in canned mussel shows the excellent value of 1 g/100 g of mussel meat, which would satisfy 100% of the RDA. It should be noted that a mussel can has approximately 180 g of drained weight. Another mollusk, the sea urchin, has been analyzed for fatty acid composition after canning (de la Cruz-García *et al.*, 2000). Canning in water did not significantly affect the amount of each fatty acid, and the retention of EPD + DHA after canning was 83.1% in a process without pre-cooking, which means that the retention of this important bioactive compound differs between shellfish.

TABLE 54.3 Most Significant Contributions of Mussel Meat

Type of Nutrient	Compound	% RDA
Macronutrients	Protein	23
Micronutrients	Fe	49/22*
	P	28
	B_{12}	500
Bioactive	DHA + EPA	44

**Man/Woman.*

The effects of processing on micronutrients, minerals, and vitamins have been studied by Slabyj (1977) for the pre-cooking, freezing, and canning of mussel meat. Table 54.3 shows the most significant micronutrients provided by mussels, vitamin B_{12}, Fe, and P. In general, some elements leach out during pre-cooking, such as I, K, and Na, and the rest of elements are not significantly affected by pre-cooking. Further processing results in leaching of Mg and K, mainly during canning, but also as a consequence of freezing. The elements Fe and P show no significant losses through all the processing steps: Fe changes from 29 to 35 and 42 ppm dry weight in frozen and canned processing, respectively. The increase is due to concentration by water loss. Similarly, P increases from 10,500 (raw mussel) to 12,000 and 11,300 ppm for frozen and canned, respectively. With regard to vitamin B_{12}, the only available data are for canned mussel, which shows a value of 10 μm/100 g of drained meat, compared with 12 μm/100 g in raw mussel (DTU, 2009). From these data, it is possible to conclude that processed mussels are still a good source of the components shown in Table 54.3 and most of the components of the raw mussel as well.

OTHER WAYS IN WHICH COMPOSITION IS ALTERED

Due to the biological characteristics of mussels (they have a high capacity to carry out bioconcentration phenomena from the marine environment in which they grow), health problems may occur if toxic compounds are present in the environment. Under these conditions, suitable checks of the water quality and implementation of purification processes for mollusks intended for human consumption must be considered (Parejo,1989). One of the most abundant types of pollutants in marine environments, which are linked to human activity, are metals. These compounds can generate toxicological problems with significant health impacts (Pérez-López, 2003). In this sense, it is important to make a clear distinction between the elements that have essential functions for life, as it is the case of copper and zinc (Copa-Rodriguez and Basadre-Pampín, 1994), and those whose intake at low doses triggers to toxicity, such as lead and cadmium (Voegborlo *et al.*, 1999).

The tolerable daily intakes of zinc and copper are 40 and 10 mg/adult, respectively (García-Gabarra, 2006). Perez-Lopez *et al.* (2003) carried out a study to determine the contents of heavy metals in canned mussels and found the following average concentrations (expressed in wet weight):

Zinc: 311.8 ppm
Copper: 6.335 ppm
Cadmium: 0.333 ppm
Lead: 0.714 ppm.

The recommended dietary intake of zinc is 11 mg for an adult (Table 54.1). Considering a mussel can of 100 g drained weight, it is possible to estimate the metal intake from the consumption of one can. The estimated Zn intake is 31.3 mg so one would cover the RDA and would not exceed the upper limit of 40 mg. The estimated copper intake is 0.634 mg, clearly under the limit of tolerance of 10 mg/person per day (or 15 cans/day). On the other hand, if the needs of copper are estimated at 3.2 mg/person per day (Copa-Rodriguez and Basadre-Pampín, 1994), one would need to eat five mussel cans per day to meet this requirement.

With regard to the other two metals, a level of tolerable intake of 400–500 μg per week, for Cd has been reported by Voegborlo *et al.* (1999) and 1.5 mg per week for Pb by WHO/FAO (1999). Perez-Lopez (2003) concluded that one would need to consume 20 cans of this mollusk per week to reach the upper level for Cd and 30 cans in the same period for Pb. From these data, it can be concluded that mussels grown in properly controlled areas do not constitute a danger to human health from heavy metals.

ANALYTICAL TECHNIQUES

Proximate analyses are determined according to standard methods from AOAC, and other important components such as fatty acid composition, amino acid profile and micronutrients utilize gas chromatography, HPLC, and Atomic Absorption Spectrophotometry, respectively, as shown by Fuentes *et al.* (2009).

References

Almonacid, S., Bustamante, J., Simpson, R., Urtubia, A., Pinto, M., Teixeira, A., 2012. Commercially Sterilized Mussel Meats (*Mytilus chilensis*): A Study on Process Yield. J. Food Sci. 77, 127–135 (Nr. 6).

Bello, P., Stupak, A., Cristóbal, L., Torres, R., 2012. Material Flow Analysis in a cooked mussel processing industry. J. Food Eng. 113, 100–117.

Bresson, J., Flynn, A., Heinonen, M., Hulshof, K., Korhonen, H., Lagiou, P., Løvik, M., Marchelli, R., Martin, A., Moseley, B., Przyrembel, H., Salminen, S., Strain, S., Strobel, S., Tetens, I., Van den Berg, H., Van Loveren, H., Verhagen, H., 2009. European Food Safety Authority. Scientific Opinion: Labelling reference intake values for n-3 and n-6 polyunsaturated fatty acids. EFSA J. 1176, 1–11.

Bustamante, 2009. Estudio del proceso de elaboración de conservas de choritos (*Mytilus Chilensis*): Efecto en el peso drenado. Master Thesis, Chemical Engineering, Universidad Técnica Federico Santa María.

Byrd-Bredbenner, C., Moe, G., Beshgetoor, D., Berning, J., 2009. Perspectives in Nutrition, eighth ed. McGraw-Hill, New York, p. 686.

Casales, M.R., Del Valle, C.E., Soulé, C.L., 1988. Changes in composition of mussels due to thermal processing. J. Food Sci. 53 (1), 282–283.

Copa-Rodríguez, F.J., Basadre-Pampín, M.I., 1994. Determination of iron, copper and zinc in tinned mussels by inductively coupled plasma atomic emission spectrometry (ICP-EAS). Fresenius J. Anal. Chem. 348, 390–395.

De la Cruz-García, C., López-Hernández, J., González-Castro, M., Rodríguez-Bernaldo De Quirós, A., Simal-Lozano, J., 2000. Protein, amino acid and fatty acid contents in raw and canned sea urchin (*Paracentrotus lividus*) harvested in Galicia (NW Spain). J. Sci. Food Agric. 80, 1189–1192.

Do Amaral, M., fields, b., C., De Morais, C., Ostini, a., S., 1998. Proximate composition and caloric value of mussel *Perna perna*, cultivated in Utatuba, São Paulo State, Brazil. Food Chem. 62 (4), 473–475.

DTU, 2009. Technical University of Denmark. National Food Institute. Available from http://www.foodcomp.dk/v7/fcdb_details.asp?FoodId=0188 (accessed September 2012.).

FAO, 2004. Programa de información de especies acuáticas. *Mytilus edulis*. Programa de información de especies acuáticas. Actualizado 1 January 2004. Available at: http://www.fao.org/fishery/culturedspecies/Mytilus_edulis/es.

FAO, 2007. Estado actual del cultivo y manejo de moluscos bivalvos y su proyección futura. Factores que afectan su sustentabilidad en América Latina. ISSN 2071–1026.

FAO, 2012. FishStatJ version 2.0.0. Software for fishery statistical time series, between 1950 and 1980. Available at: www.fao.org/fishery/statistics/software/fishstatj/en.

Fuentes, A., Fernández-Segovia, I., Escriche, J.A., Serra, J.A., 2009. Comparison of physico-chemical parameters and composition of mussels (*Mytilus galloprovincialis* Lmk.) from different Spanish origins. Food Chem. 112, 295–302.

García-Gabarra, A., 2006. Ingesta de Nutrientes: Conceptos y Recomendaciones Internacionales (2ª Parte). Nutrición Hospitalaria 21, 437–447 (N°4).

Gokoglu, N., Erkan, N., Ozden, O., 2000. The effect of frozen storage on proximate composition and quality of mussels (*Mitylus galloprovincialis*). J. Aquatic Food Product Technol. 9 (2), 83–88.

Hamazaki., T., Okuyama, H., 2003. The Japan Society for Lipid Nutrition recommends to reduce the intake of linoleic acid. A review and critique of the scientific evidence. World Rev. Nutr. Diet. 92, 109–132.

Holdsworth, D., Simpson, R., 2007. Canning Operations. In: Thermal Processing of Packaged Foods, second ed. Springer, New York.

Kawashima, K., Yamanaka, H., 1996. Free amino acids responsible for browning of cooked scallop adductor muscle. Fisheries Sci 62, 293–296.

Krzynowek, J., Wiggin, K., 1979. Seasonal variation and frozen storage stability of blue mussels (*Myths edulis*). J. Food Sci. 44, 1644–1645.

Linus Pauling Institute, 1012. Available from: http://lpi.oregonstate.edu/infocenter/vitamins/vitaminB12/ (accessed November 2012.).

Otles, S., Sengor, G., 2005. Effect of Various Technological Processes on the Fatty Acid Composition of Mussel (*Mytilus galloprovincialis, L.*). Int. J. Food Eng. 1 (3), ISSN (Online), 1556–3758.

Parejo, C.B., 1989. In: Mundi-Prensa, (Ed.), Moluscos: Tecnología de Cultivo: Madrid, p. 167–173.

Pérez-López, M., Bellón, E., Alonso, J., Melgar, M.J., 2003. Contenido en metales pesados (Zn, Cd, Pb y Cu) en conservas de mejillón (*Mytilus* sp). ALIMENTARIA (349), 73–80.

Perez-Martin, R.I., Franco, J.M., Aubourg, S., Gallardo, J.M., 1988. Changes in free amino acid content in albacore (*Thunnus alalunga*) muscle during thermal processing. Z Lebensm Untersuch Forsch 187, 432–435.

Senturk, A., Baygar, T., Gokoglu, N., Kaplan, M., Sentruk, I., 2000. Determination of the changes with some chemical and microbiological quality of Yellowfin Tuna (*Thunnus albacores, L. 1758*) during canning process. Ministry of agriculture and rural affairs of Turkey. The control of directorship of Dardanlles Province. Project number: TAGEM/GY/98/07/01/015.

Slabyj, B.M., 1977. Processing effect on proximate composition and mineral content of meats of blue mussel (*Mytilus edulis*). J. Food Sci. 42 (5), 1153–1155.

Shafiur, M., 2012. Chilling, Freezing and Thawing Process Design. In: Ahmed, J., Shafiur, M. (Eds.), Handbook of food Process Design, vol. 1. Wiley–Blackwell, Oxford, pp. 430–455.

Shahidi, F., 2004. Functional seafood products. ACS Symposium Series 702, 29–49.

Turan, H., Sönmez, G., Yesim, C., elikM, Yalc¸in Kaya, Y., 2007. Effects of different salting process on the storage quality of Mediterranean mussel (*Mytilus galloprovincialis* L. 1819). J. Muscle Foods 18, 380–390.

USDA, 2012. The U.S. Department of Agriculture. National Nutrient Database for Standard Reference, Release 25. Home Page, http://www.ars.usda.gov/ba/bhnrc/ndl. Food list available at: http://ndb.nal.usda.gov/ndb/search/list.

Voegborlo, R.B., El-Methnani, A.M., Abedin, M.Z., 1999. Mercury, cadmium and lead content of canned tuna fish. Food Chem. 67, 341–345.

WHO/FAO, 1999. Summary of the 53rd meeting of the joint FAO/WHO expert committee on food additives (JECFA), agenda item 4ª.

CHAPTER

55

Influence of Ozone Depuration on the Physical Properties of Fresh American oysters (*Crassostrea virginica*)

Violeta Pardío-Sedas

Facultad de Medicina Veterinaria y Zootecnia, Universidad Veracruzana, Colonia Unidad Veracruzana, Veracruz, México

CHAPTER POINTS

- Ozonization represents an alternative technique for enhancing the value of raw oysters. From a public health perspective, oysters depurated with 0.4 and 0.6 mg O_3/l during 6 h may be safe for raw consumption. These findings are of importance as *V. cholerae* non-O1 populations were decreased to non-culturable levels.
- At 0.4 and 0.6 mg O_3/l, ozone may influence the functional properties of protein as a result of structural changes by protein hydrolysis, which depended on the amount of ozone. This denaturation was detected by panelists as a decrease in firmness of the meat oyster.
- Correlation results suggest that denaturation of muscle proteins during ozone-depuration have effects on free amino acid (FAA), shear force, and texture. Thus, ozone should be carefully dosed to avoid undesired oxidation of proteins.
- According to these findings, FAA content could be considered as indicator of freshness and quality of bivalve mollusks.
- Further investigations are needed in order to determine the changes in structural and functional properties of oyster proteins.

INTRODUCTION

Globally, seafood provides more than 3 billion people with at least 15% of their average per capita intake of animal protein (FAO, 2011). Nevertheless, bivalve mollusks (oysters, clams, mussels, cockles, and scallop) are highly perishable and risky, and may cause human illnesses because: (i) their immobility and their habitat in coastal estuaries render them susceptible to fecal contamination and environmental pollution in the water; (ii) their filter-feeding behavior leads to the accumulation of pathogens naturally present in growing areas contaminated with polluted waters; (iii) consumers often eat the shell fish whole and raw, including the gastrointestinal tract where pathogens are concentrated. Raw fresh oysters are sold either as whole in-shell unpackaged or shucked packaged in polyethylene pouches along with tap water, having a shelf-life of about 6 days under refrigeration.

To improve and ensure their microbiological safety, raw shellfish are depurated by removing primary human pathogens and the load of commensal bacteria that may also be pathogenic to humans such as *Vibrios*, reducing the likelihood of transmitting infectious agents to consumers, and to prevent recontamination with food-borne pathogens and spoilage organisms. Depuration is one of the major treatment processes applied to moderated sewage-contaminated shellfish bivalves to increase the availability and supply of safe and nutritious oysters, mussels, and clams, eaten raw. The depuration process exploits the natural physiological functions of the gastrointestinal tract of live animals and the success of the process is dependent on the animal wellbeing, the health status, the physiological activity, and pumping rate and behavioral responses (Schneider *et al.*, 2009).

Although markets have been growing for the treated seafood, a substantial number of consumers still prefer live, whole, raw, or slightly cooked mollusks. There is a strong growing consumer demand for minimally processed, high-quality foods that are additive-free, with

Processing and Impact on Active Components in Food
http://dx.doi.org/10.1016/B978-0-12-404699-3.00055-X

their natural freshness and sensorial characteristics, and extended shelf-life, and safe products without quality degradation; this has led to the development of several novel and innovative food-processing methods of microbial control known as non-thermal technologies introduced to replace or complement the well-established thermal processes, such as the emerging technologies of ohmic heating, ultraviolet light, and ozonization (Novak and Yuan, 2007).

HOW COMPOSITION IS ALTERED

The main purpose of depuration is to maintain the shelf-life and to reduce or minimize the loss of flavor and texture during storage in highly perishable fresh shellfish, such as bivalve mollusks (Cheney, 2010). To increase the efficacy of depuration, it is combined with disinfectants such as ozone (O_3). Ozone is very unstable both in the gaseous phase and in solution, decomposing into highly reactive free radicals such as hydroxy ($^{\bullet}OH$), hydroperoxy ($^{\bullet}HO_2$), and superoxide ($^{\bullet}O_2$). The powerful and effective antimicrobial activity comes from the molecule's inherent instability, attributed to the great oxidizing power of these free radicals, killing microorganisms through oxidation of their cell membranes (Kim *et al.*, 1999). Considered as generally recognized as safe (GRAS) as an antimicrobial agent for use as a disinfectant or sanitizer for foods (USDA, 1997), ozone has a high germicidal effectiveness against a wide range of pathogenic organisms including both Gram-positive and Gram-negative bacteria, viruses, fungi, and protozoa (Skall and Olesen, 2011).

Microbiological, biochemical, and sensory changes are associated with the deterioration of mollusk quality that depends on many factors, such as handling and storage, intrinsic factors of the animals (species, age and size, fat composition, feeding and physiological status), and the qualitative and quantitative compositions of the initial microbial flora (Gram and Hus, 2000). Ozone depuration may promote a faster physiological deterioration, biochemical changes, and microbial degradation of oyster even when only minor processing operations are used.

Limited data is available in the literature on the application of ozone to molluskan shellfish preservation, and different process conditions and concentrations of ozone have been employed to depurate several molluskan shellfish. In order to study the influence of ozone on the physical characteristics of American oyster (*Crassostrea virginica*), 750 organisms (5–6 g, 9 cm long) collected, by diving, from banks of harvesting areas in a coastal lagoon system, were immediately transported to the laboratory in coolers at 4°C (NOM-109-SSA1-1994) (SSA, 2012). Oysters were scrubbed and rapidly washed by high-pressure water spray to remove fouling organisms, mud and surface debris, and attached algae adhering to the shell, damaged gaping animals were discarded, and immediately placed in the recirculating seawater system (9 l/min, $27^0/_{00}$ salinity, 90% dissolved oxygen, 25.0°C, and pH 8.0) within 2 h of harvesting, and depurated with clean artificial seawater for 2–4 h to promote oyster activity. Afterwards, seawater was disinfected with ozone generated by corona discharge at 0.2, 0.4, and 0.6 mg/l. Each tank was loaded with 250 oysters previously estimated and placed into plastic mesh baskets as a single overlapping layer. Duplicate samples of 30 oysters from aleatory locations in each tank were collected at 0, 2, 4, and 6 h for microbial analyses within 2 h of collection. According to Table 55.1, *E. coli* levels declined in the first 2 h of depuration, and after 4 h the MPN decreased to 88, 50, and 64% of its original concentration, and 88, 50 and 82% after 6 h of depuration when 0.2, 0.4, and 0.6 mg O_3/l were applied, respectively.

E. coli in oyster control samples decreased by 33.3% during 4 h of depuration, and increased to 650 MPN/100 g at 6 h. Meanwhile, *Vibrio cholerae* non-O1 was not detected after 6 h of depuration with 0.2 mg/l and after 4 h of depuration with 0.4 and 0.6 mg/l, reaching the legally bacteriological limit of no detection of *V. cholerae* non-O1 in 50 g of oyster flesh. As can be observed, oysters released *E. coli* rapidly after 2 h of depuration and reached the legally bacteriological limits of *E. coli* < 230 MPN/100 g (NOM-242-SSA1-2005), indicating that *E. coli* was more sensitive than *V. cholera* non-O1. Schneider *et al.* (1991) reported that 3 mg/l ozonated

TABLE 55.1 Effect of Ozone Depuration on American Oyster Naturally Contaminated with *Escherichia coli* and *Vibrio cholera*

	Ozone Concentration (mg O_3/l)			
Time (hour)	**Control**	**0.2**	**0.4**	**0.6**
E. coli (MPN/100 g)				
Fresh	600[a]	340[b]	85[c]	135[c]
0	600[a]	340[b]	40[c]	110[c]
2	200[a]	60[b]	20[c]	40[c]
4	200[a]	40[b]	20[c]	70[b]
6	650[a]	40[b]	20[c]	20[c]
V. cholerae non-O1/non-O139 (% ISOLATION)				
Fresh	100	100	100	100
0	100	100	100	100
2	100	100	100	100
4	100	100	0	0
6	100	0	0	0

Means with different letter are significantly different (P < 0.05) among columns.

recirculated artificial seawater applied to *Mercenaria campechiensis*, reduced *V. vulnificus* in the shellfish meats by an average of 2 log units. *E. coli* in *Mytilus galloprovincialis* experimentally contaminated was reduced to 42% of its original concentration with 50 mg O_3/h after 5 h of depuration and *V. cholerae* O1 was reduced by 1 log after 24 h of depuration, a level that remained for 44 h (Croci *et al.*, 2002); according to Meloni *et al.* (2008) *E. coli* counts in *Mytilus galloprovincialis* decreased after 8 h depuration, meanwhile Vibrios declined at a slower rate. Maffei *et al.* (2009) working in ozone open-circuit seawater-disinfection system at the Bivalve Depuration Center (Cattolica, Rimini, Italy) reported a mean *E. coli* reduction (62%) during depuration of *Chamelea gallina*; however *V. parahaemolyticus* and *V. alginolyticus* were not eliminated. Cozzi *et al.* (2009) depurated blue mussels (*Mytilus galloprovincialis*) using a combined system of UV light (55 W) and ozone (50 mg/h). After 24 h of treatment, both *E. coli* and *V. parahaemolyticus* concentrations decreased by approximately 2 log while *V. vulnificus* decreased slower only by 1 log. After 72 h the depuration process reduced *E. coli* and *Vibrio* contamination to levels close or below the detection limit of the methods (10 CFU/g). When Borazjani *et al.* (2003) depurated whole in-shell oysters (*Crassostrea virginica*) for 30 min with 2–4 mg O_3/l, *V. vulnificus* counts in oyster's meats were not reduced. Rong *et al.* (2010) treated oysters with 5 µg/l of ozonated water for 2 min and this resulted in a less than 1-log decrease of indigenous microbiota.

This finding confirms that *E. coli* is an inadequate index of microbiological safety of molluskan shellfish, not only because its presence does not correlate with the occurring Vibrios, but also as a means of assessing the efficacy of the depuration process. In consequence, shorter depuration treatment times of mollusks with high initial loads of Vibrios could lead to an insufficient reduction in the pathogen loads, failing to guarantee the safety of consumers and representing a potential health hazard.

While the destruction of bacteria by ozone has been studied extensively, relatively little information is available on the effect of ozone on various food components. Oyster is one of the most important and abundant harvested bivalves and highly valued worldwide for its rich nutrition and desirable taste. The popularity of oysters in the diet continues to increase because they constitute a healthy low-fat contribution of proteins, healthy polyunsaturated fatty acids, and phytosterols into a balanced diet. Depending on sex, maturity, water temperatures, food supply, stress, and other environmental parameters, 100 g of raw oyster contains 77.4–90.2 g moisture, 5.6–10.9 g protein, 1.9–4.7 g carbohydrate, 0.7–2.4 g total fat, 0.7– 2.9 g ash, 37–58 mg cholesterol and 75 calories. It provides 100% of vitamin B12 and zinc and about 35% iron (Sáenz, 2007). As fresh seafood, oysters are more perishable than other high-protein foods and have a short shelf-life. Fresh oysters are popular consumed raw as they have more a delicate flavor and texture than cooked oysters and retain more nutrients. For a consumer, 'eating quality' of mollusks, e.g., oysters, clams, and scallops, is associated with freshness that indicates typical aquatic fresh odour and flavor, meat with brilliant appearance and colour, and texture characteristics of the species in good condition; thus, fresh shellfish products should be marketed as fast as possible since their shelf-life is very short (Gram and Huss, 2000).

The possible pro-oxidant effect of ozone on shellfish constituents has not been extensively studied up to now, although previous reports have shown a potential negative effect on phospholipids, polyunsaturated fatty acids, and membrane proteins (Cataldo, 2003). Given the high reactivity of ozone, it is difficult to predict its reaction as it may oxidize or ionize a substrate or spontaneously decompose to oxygen and free radicals. Ozone is not universally beneficial, and in some cases may promote surface oxidation, discoloration, or even development of undesirable odors, resulting in degradation of the color, texture, and flavor of the food (Khadre *et al.*, 2001); moreover, ozone treatments may negatively affect the nutritional quality of the product. Unfortunately, little is known about the changes in physical, biochemical, or sensory properties of bivalve mollusks.

Aquatic multicellular animals contain collagen in the edible tissues such as muscle, bone, and skin. A close relationship has recently been reported between texture and collagen content for the muscles of many aquatic species, suggesting the importance of this protein in association with physical properties of marine foods (Mizuta *et al.*, 2005). Information on the biochemical properties and components of muscle collagens of bivalve mollusks is still limited. The mantle of a bivalve is a thin, sheet-like membranous organ that covers the inside of the animal's shell and biomineralization of shell is considered as its primary function. Collagen fiber is mainly distributed in the muscle connective tissues (epimysium, perimysium, and endomysium) of both mantles and adductors, and in the inner connective tissue matrix of the mantles. The collagen contents of the edible tissues varied considerably; the mantles have much higher collagen content than the adductors (Mizuta *et al.*, 2004). Mizuta *et al.* (2005) reported that the major collagen in the mantle of the oyster *Crassostrea gigas* has a heterotrimer structure $(\alpha 1)_2\alpha 2$, with a content of 1.1% of protein, and the major collagen showed relatively low content of alanine and high content of hydroxylysine.

Several studies have been carried out to investigate the basic interaction of ozone with different proteins and the interactions of specific amino acids in proteins. Although all amino acids are potential targets for oxidation by reactive oxygen, the major aromatic amino acids tyrosine, tryptophan, phenylalanine, the sulfur-containing amino acids cysteine, methionine as well as the aliphatic amino acids arginine, lysine, proline, and histidine, appear especially sensitive to oxidation (Cataldo, 2003). Ozone reacts with proteins and causes the oxidation of the polypeptide backbone of the protein, peptide bond cleavage, protein–protein cross-linking and a range of amino acid side-chain modifications. Helices, sheets, coils, and folded branches which define the secondary and tertiary structures of proteins seem to be modified by the reaction of ozone with proteins (Hicran *et al.*, 2012). The sensitivity of a given kind of amino acid residue to oxidation by ozone varies from one protein to another and this variability is due to differences in the structures of the proteins (Sharma and Graham, 2010). The oxidation of amino acids may induce an increase in local flexibility or rigidity to the protein chain causing an alteration of its secondary and tertiary structure. As a result of ozone attack the protein molecules undergo changes in their visual folding and binding ability which influence their functionality (Cataldo, 2003).

Curran *et al.* (1984) reported that oxygen-derived free radicals from ozone (i) can directly degrade soluble collagen, and (ii) at low levels can potentiate the destructive role of proteinases by increasing the susceptibility of collagen to degradation by proteinases. These authors established that oxidants can directly degrade soluble collagen, and at low levels can modify collagen, making it susceptible to proteolytic degradation. Fujimori (1985) showed that upon prolonged exposure even at low levels, the initial modification of the most ozone-sensitive collagen regions may be followed by further extensive oxidation.

OTHER WAYS IN WHICH COMPOSITION IS ALTERED

Although scientific information is available in relation to the action of ozone on proteins, little is known about the change in physical properties of proteins in food systems after an ozonization process that may affect the textural properties of the food product. To understand the effect of ozone in oyster texture when used as an alternative processing technology to improve the quality of fresh oysters, the effect of ozone on proteins was detected by the proteolysis, Warner–Bratzler shear force, and texture analysis of oyster meats (Table 55.2).

Ozone depuration with 0.4 mg/l concentration did not produce significant changes ($P > 0.05$) in lysine (LYS) levels between control fresh oysters before (113.38 mg/g) and after 6 h of depuration (112.48 mg/g); however LYS levels increased significatively ($P < 0.05$) from 113.38 to 133.42 mg/g after 6 h of depuration, being ozonated samples significatively ($P < 0.05$) higher than control samples (133.42 and 112.48 mg/g, respectively), indicating an increase in free amino acids (FAA) as ozone reacted with proteins causing proteolysis. In the case of 0.6 mg/l depuration, LYS levels between control oyster samples before and after 6 h depuration (119.10 and 117.86 mg/g) were not statistically different ($P > 0.05$); nevertheless, LYS levels of ozonated oyster samples (96.95 mg/g) were significantly higher than fresh control oysters (119.10 mg/g) and control oysters (117.87 mg/g) after 6 h of depuration. It can therefore be surmised that the protein quality of oysters was affected by the ozone-depuration processes.

The Warner–Bratzler shear force values for oyster muscle and mantle of fresh oyster samples (0.430 kg_f) were significantly higher ($P < 0.05$) than shear values for ozonated samples (0.402 kg_f) after 6 h of depuration with 0.4 mg/l. A similar trend was observed when oysters were depurated with 0.6 mg O_3/l, being the shear force values for fresh oyster meat (0.460 kg_f) significantly higher ($P < 0.05$) than ozonated oyster shear values (0.422 kg_f), and both ozonated oyster meats with 0.4 and 0.6 mg O_3/l were significantly lower than control oyster meats after 6 h of depuration. Depuration with 0.4 mg/l produced a 6.5% decrease in the shear force values and 0.6 mg/l treatment produced a decrease of 8.3%. According to Curran *et al.* (1984), at low levels of ozone, collagen and elastin can be modified, making them susceptible to proteolytic degradation, generating peptides and amino acids. The intermolecular crosslinks in collagen, the main constituent of fish and shellfish connective tissue, are thought to be responsible for the stability, physical strength, and mechanical properties of the connective tissue and other components of the extracellular matrix (Hultmann and Rustad, 2004). Thus, the breakdown of the connective tissue by ozone may lead to undesirable textural changes in the oyster and the decreased shear strength may have been due to the denaturation of the protein fraction as FAA content increased, as proteins are broken down to peptides and FAA by ozone. Moreover, seafoods have less extensive crosslink formations in collagenous tissue in comparison with their mammalian counterparts. Additionally, the nutritional quality of the protein may be changed as FAA content increased, and the quantity of the essential amino acids in food is connected to the nutritional quality of protein.

When oysters were depurated with 0.4 and 0.6 mg O_3/l of ozone, the sensory scores for texture of control and ozonated samples with 0.4 mg/l were significantly different ($P < 0.05$) after 6 h of depuration. Texture scores of ozonated oysters were significantly higher than fresh and depurated-control oysters, indicating that panelists detected a softer or slightly firm consistency. When oysters were depurated with 0.6 mg/l, a significantly higher ($P < 0.05$) increase in texture scores was observed. The panelists detected a softer texture (3.45 less firm and

TABLE 55.2 Effect of Ozone Depuration on Proteolysis, WB Shear Force (kg_f), and Texture of American Oyster Samples

Time (hour)	Control	Ozonated 0.4 mg/l	Control	Ozonated 0.6 mg/l
PROTEOLYSIS (mg lysine/g d.w.)				
Fresh	$113.38 \pm 0.63^{a,y}$	$113.38 \pm 0.63^{a,y}$	$119.10 \pm 2.45^{a,y}$	$119.10 \pm 2.45^{a,y}$
6	$112.48 \pm 0.63^{a,y}$	$133.42 \pm 1.84^{b,z}$	$117.86 \pm 2.59^{a,y}$	$127.00 \pm 2.00^{c,z}$
WARNER–BRATZLER SHEAR FORCE (kg_f)				
Fresh	$0.430 \pm 0.96^{a,y}$	$0.430 \pm 0.96^{a,y}$	$0.460 \pm 0.51^{b,y}$	$0.460 \pm 0.51^{b,y}$
6	$0.428 \pm 0.11^{a,y}$	$0.402 \pm 0.22^{b,z}$	$0.459 \pm 0.35^{a,y}$	$0.422 \pm 0.36^{c,z}$
TEXTURE SCORES				
Fresh	$1.75 \pm 0.21^{a,y}$	$1.75 \pm 0.21^{a,y}$	$1.60 \pm 0.11^{b,y}$	$1.60 \pm 0.11^{b,y}$
6	$1.60 \pm 0.14^{a,y}$	$2.60 \pm 0.28^{b,z}$	$1.59 \pm 0.15^{a,y}$	$3.45 \pm 0.21^{c,z}$

[a,b]*Means with different letter are significantly different (P < 0.05) among columns.*
[x,y]*Means with different letter are significantly different (P < 0.05) among rows.*

elastic) in ozonated oysters with 0.6 mg/l than fresh, control, and ozonated oyster samples with 0.4 mg/l (2.60), although the loosening pattern of firmness was observed in both ozonated oyster treatments. This texture degradation phenomenon may be due to protein denaturation as a result of ozone treatment and noted by the panelists.

Although the Eastern oyster or *Crassostrea virginica* is one of the most economically and ecologically important shellfish species along the eastern seaboard, very few scientific studies have been conducted to evaluate its sensory characteristics. Besides aroma, texture is considered to be one of the important quality attributes of seafoods which determines consumer acceptance and, hence, the marketability of these products. Collagen, because of its occurrence and characteristics, plays a factor in the tenderness of the oyster meat. The content of collagen is important to consumers because when oyster has more collagen it has a crunchy, fleshy, chewier texture, and is wonderfully plump and meaty as shown in Figure 55.1.

Correlations were obtained to examine possible relationships between the texture attribute, proteolysis, and the instrumental texture measurements. As can be observed in Table 55.3, texture attributes for oyster meat (mantle and adductor muscle) show a pattern of positive correlation ($r = 0.550$) with proteolysis, indicating that panelists detected a reduction in the chewiness of the meat of the oyster due to the protein denaturation as a result of ozone treatment. Negative correlations were obtained between texture and shear force ($r = -0.724$), indicating that instrumental values of shear force decreased as the panelists detected a softer meat ever since samples were scored by panelists using a 10-cm line scale starting from 0 (firm) to 10 (friable) to evaluate the changes in texture. Meanwhile, WB shear force was negatively correlated ($r = -0.469$) to proteolysis as FAA content increased after ozonation treatment and WB shear force diminished.

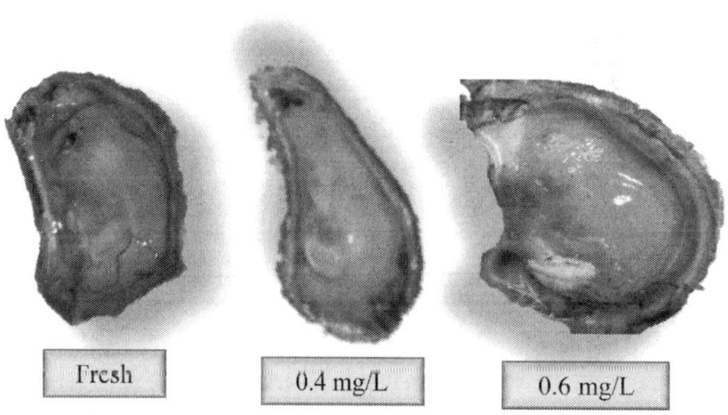

FIGURE 55.1 The Meat of an Oyster Sample Fresh and Depurated 6 h with 0.4 and 0.6 mg/l of Ozone.

TABLE 55.3 Correlation Matrix of the Proteolysis, Shear Force, and Texture for Oyster Meat Depurated with 0.4 and 0.6 mg/l of Ozone

Variables	Proteolysis (mg lysine/g d.w.)	WB Shear Force (kg_f)	Texture Attribute
Proteolysis (mg lysine/g d.w.)	1.00	−0.469	0.550
WB Shear force (kg_f)	−0.469	1.00	−0.724
Texture attribute	0.550	−0.724	1.00

Values in bold are different from 0 with a significance level $\alpha = 0.05$.

According to these findings, the Warner–Bratzler shear force measurement may be useful for quality-control purposes for determining firmness in oystesr, and FAA content could be considered as an indicator of freshness and quality of bivalve mollusks, as objective quality indicators are useful when they can predict the shelf-life before a fresh shellfish becomes spoiled.

ANALYTICAL TECHNIQUES

A variety of biochemical, physical, and microbiological methods have been used to assess freshness, although the sensory evaluation is still the most satisfactory method to achieve such a goal.

The requirement that purified shellfish must meet (MPN *E. coli* < 230/100 g of shellfish) and the presence of *V. cholera*, *V. parahaemolyticus*, and *V. vulnificus* detection are generally performed according to US Food and Drug Administration methodology and expressed as per cent of isolation (FDA, 2004). Presumptive *V. cholerae* isolates could be confirmed with polyvalent antisera Serobac (Difco Laboratories Ltd, Franklin Lakes, NJ, USA) to identify the O1 serogroup, and the serotype is identified using Inaba and Ogawa specific antisera. If there is no agglutination with the polyvalent antisera, the presence of *V. cholerae* non-O1 is reported.

Unlike other muscle foods, seafoods have less connective tissue content and less extensive crosslink formation in collagenous tissue in comparison with their mammalian counterparts. These intermolecular crosslinks in collagen, the main constituent of shellfish connective tissue, are thought to be responsible for the stability, physical strength, and mechanical properties of the connective tissue and other components of the extracellular matrix (Hultmann and Rustad, 2004). The effect of ozone on proteins could be carried out by the proteolysis of oyster meats measured by using the trinitro-benzene-sulfonic acid (TNBS) method based on the reaction of primary amino groups with TNBS reagent (Adler-Nissen 1979). The absorbance is read against water at 340 nm and results are reported as mg lysine/g dry weight. Another physical characteristic is texture, measured by cutting strength of the ventral part of the shucked oyster meat with a texture analyzer (TA-XT2i, Stable Micro Systems, Goldaming, UK) equipped with a Warner–Bratzler blade operating at a speed of 4.00 mm/s, a cutting distance of 35 mm, and a force penetration of 5.0 g. Cutting strength is taken as the mean of three measurements on each oyster per treatment and per trial and expressed as the force (kg_f) required for cutting 1 mm into oyster tissue. The effect of ozone on texture could be determined by descriptive sensory analysis as well. Candidates for this type of panel are selected on the basis of both availability and liking of seafood. To assess abilities of candidates, candidates are subjected to sequential analysis ($p_0 = 0.30$, $p_1 = 0.60$, α and $\beta < 0.05$). Selected panelists are trained on randomly selected days in 30-min training sessions. They are oriented with the fresh, ozonated, and rotten samples. During 1-h training sessions, panelists are asked to generate a list of texture descriptors. To evaluate the changes in texture, samples of treatments and control are compared in terms of degree of deviation of intensity from the fresh sample attributes using a 10-cm line scale with three-anchor points: 0 for firm (elastic and chewy), 5 for soft (mild, slightly firm), and 10 for friable (mushy, lax). The analyses are performed once a week in individual booths for sensory panel and samples are scored under red light. The samples must be shucked in aseptic conditions and kept over crushed ice until served to the panelists in individually plastic capped cups, coded with three-digit numbers, and presented to panelists in one row in random order. Each judge receives three dishes in each session trial: the raw fresh sample (reference), the depurated raw control samples, and ozonated oyster samples in order to evaluate the degree of firmness. Distilled water and soda crackers are used as palate cleansers and the panelists are instructed to discard the samples after every evaluation (Pardío *et al.*, 2011).

References

Adler-Nissen, J., 1979. Determination of the degree of hydrolysis of food protein hydrolysates by trinitrobensenesulfonic acid. J. Agric. Food Chem. 27, 1256–1262.

Borazjani, A., Andrews, L., Veal, Ch, D., 2003. Novel nonthermal methods to reduce *Vibrio vulnificus* in raw oysters. J. Food Safety, 23,179–187.

Cataldo, F., 2003. On the action of ozone on proteins. Polymer Degradation and Stability 82, 105–114.

Cheney, D.P., 2010. Bivalve shellfish quality in the USA: from the hatchery to the consumer. J. World Aquaculture Soc. 41, 192–206.

Cozzi, L., Suffredini, E., Ciccaglioni, G., Croci, L., 2009. Depuration treatment of mussels experimentally contaminated with *V. parahaemolyticus* and *V. vulnificus*. www.symposcience.org [September 2012].

Croci, L., Suffredini, E., Cozzi, L., Toti, L., 2002. Effects of depuration of molluscs experimentally contaminated with *Escherichia coli, Vibrio cholerae* O1 and *Vibrio parahaemolyticus*. J. Appl. Microbiol. 92, 460–465.

Curran, S.F., Amoruso, M.A., Goldstein, B.D., Berg, R.A., 1984. Degradation of soluble collagen by ozone or hydroxyl radicals. FEBS 176, 155–160.

Food Agriculture Organization UN, 2011. FAOSTAT. http://faostat.fao.org/ [September 2012].

Food and Drug Administration (FDA), 2004. Bacteriological Analytical Manual. Vibrio. C.A. Kaysner and A. DePaola. Chapter 9. U.S. Department of Health & Human Services. http://www.fda.gov/Food/ScienceResearch/LaboratoryMethods/BacteriologicalAnalyticalManualBAM/default.htm [September 2012].

Fujimori, E., 1985. Changes induced by ozone and ultraviolet light in type I collagen Bovine Achilles tendon collagen versus rat tail tendon collagen. Fur. J. Ihochem. 152, 299–306.

Gram, L., Huss, H.H., 2000. Fresh and processed fish and shellfish. In: Lund, B.M., Baird-Parker, T.C., Gould, G.W. (Eds.), The Microbiological Safety and Quality of Food, vol. 1. Aspen Publishers Inc, Gaithersburg, Maryland, pp. 472–506.

Hicran, U., Ibanoglu, E., Catal, H., Ibanoglu, S., 2012. Effects of ozone on functional properties of proteins. Food Chem. 134, 647–654.

Hultmann, L., Rustad, T., 2004. Iced storage of Atlantic salmon (*Salmo salar*) — effects on endogenous enzymes and their impact on muscle proteins and texture. Food Chem. 87, 31–41.

Khadre, M.A., Yousef, A.E., Kim, J.G., 2001. Microbiological aspects of ozone applications in food: a review. J. Food Sci. 6, 1242–1252.

Kim, J.-G., Yousef, A.E., Dave, S., 1999. Application of ozone for enhancing the microbiological safety and quality of foods: A review. J. Food Protect. 62, 1071–1087.

Maffei, M., Vernocchi, P., Lanciotti, R., Guerzoni, M.E., Belleti, N., Gardini, F., 2009. Depuration of striped Venus clam (*Chamelea gallina* L.): Effects on microorganisms, sand content, and mortality. J. Food Sci. 74, M1–M7.

Meloni, D., Mureddu, A., Pisanu, M., Serra, S., Piras, A., Virgilio, S., Mazzette, R., 2008. Efficacia della depurazione sulla sicurezza di mitili (*Mytilus galloprovincialis*) allevati nel Golfo di Olbia. A.I.V.I. 1, 53–56.

Mizuta, S., Miyagi, T., Nishimiya, T., Yoshinaka, R., 2004. Partial characterization of collagen in several bivalve species. Food Chem. 87, 83–88.

Mizuta, S., Miyagi, T., Nishimiya, T., Yoshinaka, R., 2005. Characterization of the quantitatively major collagen in the mantle of oyster *Crassotrea gigas*. Fisheries Sci. 71, 229–235.

Novak, J.S., Yuan, J.T., 2007. The ozonation concept: adventages of ozone treatment and commercial developments. In: Tewari, G., Juneja, V. (Eds.), Advances in Thermal and Non-Thermal Food Preservation. Blackwell Publishing Professional, Ames, Iowa, USA.

Pardío, V.T., Waliszewski, K.N., Zuñiga, P., 2011. Biochemical, microbiological and sensory changes in shrimp (*Panaeus aztecus*) dipped in different solutions using face-centred central composite design. Int. J. Food Sci. Technol. 46, 305–314.

Rong, C., Qi, L., Bang-Zhong, Y., Lan-Ian, Z., 2010. Combined effect of ozonated water and chitosan on the shelf-life of Pacific oyster (*Crassostrea gigas*). Innov. Food Sci. Emerg. Technol. 11, 108–112.

Sáenz, M.P., 2007. Catálogo de pescados y mariscos del litoral veracruzano: el ostión. Rev. La Ciencia y el Hombre 20, 17–18. http://www.uv.mx/cienciahombre/revistae/vol20num2/articulos/mariscos/index.html. [September 2012].

Schneider, K.R., Steslow, F.S., Sierra, F.S., Rodrick1, G.E., Noss, C.I., 1991. Ozone depuration of *Vibrio vulnificus* from the southern quahog clam, *Mercenaria campechiensis*. J. Invert. Pathol. 57 (2), 184–190.

Schneider, K.R., Cevallos, J., Rodrick, G.E., 2009. Molluscan shellfish depuration. In: Shumway, S.E., Rodrick, G.E. (Eds.), Shellfish safety and quality. Woodhead Publishing, pp. 508–540.

Secretaría de Salud (SSA), 2012. www.salud.gob.mx [September 2012].

Sharma, V.K., Graham, N.J.D., 2010. Oxidation of amino acids, peptides and proteins by ozone: A Review. Ozone: Sci. Eng. 32, 81–90.

Skall, H.F., Olesen, N.J., 2011. Treatment of wastewater from fish slaughterhouses. National Veterinary Institute, Technical University of Denmark, pp. 89–96.

USDA, 1997. Code of Federal Regulations, Title 9, Part 381.66—poultry products; temperatures and chilling and freezing procedures. Office of the Federal Register National Archives and Records Administration, Washington, DC.

CHAPTER

56

Fish Roe Lipids: Composition and Changes During Processing and Storage

*P.G. Prabhakara Rao**, *K. Balaswamy**, *T. Jyothirmayi**, *M.S.L. Karuna*†, *R.B.N. Prasad*†

*CSIR-Central Food Technological Research Institute (CFTRI), Resource Centre, Habshiguda, Hyderabad, India, †Centre for Lipid Research, CSIR-Indian Institute of Chemical Technology (IICT), Tarnaka, Hyderabad, India

CHAPTER POINTS

- Fatty acid composition of roe lipids varies with species, season, and domestication.
- Neutral lipid is a major lipid class in roes accounting for more than 50% and phospholipids account for 25–40%.
- The major fatty acid present in roe lipids is palmitic acid while docosahexaenoic acid (DHA), eicosapentaenoic acid (EPA), and arachidonic acid account for 30–50% in the phospholipid fraction.
- Saturated fatty acids increase and polyunsaturated fatty acids (PUFA) decrease during spawning months.
- Thermal inhibition of enzymatic hydrolysis, oxidation of roe lipids and salting can successfully enhance storage stability of roe products.
- Cholesterol content of roe lipids varies between 2 and 15%, and it tends to reduce during processing.
- Smoking, salting, and freezing are the major methods of processing of roes.
- Reduction in EPA and DHA contents is observed during processing.
- Lipid hydrolysis is observed due to the presence of inherent lipases.
- Frozen storage at −20°C is the best option for retention of nutritionally important n-3 PUFA in roe products.

INTRODUCTION

Fish is considered to be the best source of lean meat. The discovery of n-3 fatty acids has boosted the importance of fish because they prevent the formation of atherosclerotic plaques. Useful fish by-products such as protein, collagen, gelatin, oil from different parts and calcium from bones have been recovered (Kim and Mendis, 2006).

Fish roes are fully ripe ovaries. Caviar is the term used for processed roes. Products from roes are relished worldwide and are mainly derived from cooking, frying or grilling (Bledsoe *et al.*, 2003). Steamed eggs in China are popular. Hilsa fish roes are considered a delicacy in India and Bangladesh. In Iran, ashpal roe is consumed in grilled, cured, and salted forms. Ikura, Sujiko from salmon roe, Kazunoko from herring and Mentaiko, Tarako from Alaska pollock roe and Tobiko from flying fish in Japan; smoked and salted cod roe paste (kaviar) in Sweden are popular. Herring roe is sold in British supermarkets. Deep-fried roe is served in Peru seafood restaurants. Roes, after recovery of lipid, form a rich protein mass and have several applications for development of protein isolates, hydrolysates, and protein-rich foods. Stability of roes during processing and storage forms an important aspect in retaining the nutrients for promotion of human health. The important components of fish roe lipids are saturated fatty acids (SFA), polyunsaturated fatty acids (PUFA) and cholesterol. This chapter deals with lipid composition of roes, and how the lipid composition varies with season, diet, maturity, culturing, and processing.

Processing and Impact on Active Components in Food
http://dx.doi.org/10.1016/B978-0-12-404699-3.00056-1

COMPOSITION OF ROE LIPID

The roe lipid consists of waxes, triacylglycerols (TAG), phospholipids (PL), and cholesterol. PUFA was found in substantial quantities in fish roes. Southwest Atlantic Hake (*Merluccius hubbsi*) roe lipids (6.6%) contain waxes (27.6%), TAG (42%), PL (14%), and cholesterol (5.7%) (Mendez *et al.*, 1992). The presence of PUFA (45% of the total lipid) makes this marine lipid an attractive source in a normal diet. Similar results were observed in the roes of *Cyprinus carpio* and *Carassius auratusa* (Lili *et al.*, 2007), where total lipid (TL) was 6.4 and 2.8%, respectively. These lipids consisted of neutral lipids (NL) 69.5 and 76%; polar lipids 30.5 and 24%, respectively. The ratio of n-3/n-6 in these lipids was higher and possessed arachidonic acid (AA) and docosahexaenoic acid (DHA) in abundance. The lipid content in the roes was higher than in muscle. Similar observations were made on the fatty acid profile of flesh and roe of white fish (*Coregonus albula*) (Kaitaranta, 1980). The fresh roe contained 9.8% lipid with a high degree of unsaturation compared to flesh, with 3% lipid. Fatty acid composition was similar in both tissues with PUFA accounting for 50%, while the remaining 50% was made up of monounsaturated fatty acid (MUFA) and SFA. A study on NL and polar lipids of mature marine roes of herring, cod, haddock, whiting siathe, sand eel, and capelin (Tocher and Sargent, 1984) has shown that TL ranged between 10 and 26% with NL accounting for 77 and 49% in sand eel and capelin, respectively. PUFA in these species accounted for 50% of total lipids. Roe lipids of immature pollock were found to contain palmitic acid (21.2%), DHA (21.1%), eicosaenoic acid (19%), and oleic acid (11.6%) as the main fatty acids (Bechtel *et al.*, 2007).

Some of the roe lipids were found to contain wax esters in major quantities. For example, mullet roe lipids were found to contain wax esters with even-numbered chain-lengths, predominantly C32 and C34. Roe lipids were found to contain higher quantities of wax esters (63%) and lower TAG (9.9%) compared to muscle (66%) and trace amounts of wax esters (Joh *et al.*, 1988). Roe wax esters were significantly rich in MUFA such as palmitoleic acid, heptadecenoic acid, oleic acid, and EPA. Acid moiety in the wax ester showed 60% of palmitoleic acid and oleic acid as the main fatty acids; alcohol moiety was predominantly palmitic acid (Mitsu *et al.*, 1974). Similarly, roe of parshey (*Mugil parshey*), was found to contain a major portion of wax ester (Parimal *et al.*, 1997). The roes possessed lower TAG and PLs in all lipid classes.

Roes were found to be good sources of PLs. Roe lipids of *Wallagu attu* and *Ompok pabda* were found to contain PL to an extent of 28% and 56%, respectively (Mukhopadhyay and Ghosh, 2007). Phosphatidylcholine (PC) and phosphatidylethanolamine (PE) accounted for more than 80% in the PL of roe of *Cyprinus carpio* and *Carassius aurantusa* (Lili *et al.*, 2007). The PL fraction of Atlantic halibut (*Hippoglossus hippoglossus*) roe was 70.8%, which contained PC (62.2%) and PE (6.7%) with a high concentration of n-3 PUFA namely, EPA, and DHA (Peterson *et al.*, 1986).

A study on mature salmon roe lipids (Kazuo *et al.*, 1988) showed the presence of TAG (63%), PLs (30%), sterols (4.2%), and steryl esters (SE) (0.7%). Presence of furan fatty acids was observed in the total lipids of salmon roes, i.e., 12,15-epoxy-13,14-dimethyleicosa-12,14-dienoic acid to an extent of 3.8 and 0.6% in the fatty acids of the PL and SE fractions, while they were absent in TAGs. Larger amounts of furan acids were observed in PC than in PE.

Some freshwater fishes namely *Labeo rohita* and *Channa striatus* have shown the presence of NLs (43.8 and 72.9%); GLs (12.7 and 9.4%) and PLs (43.5 and 17.7%), respectively, in their roe lipids (Prabhakara Rao *et al.*, 2010). Roe lipids consisted of PUFA namely EPA and DHA at 1.5, 0.6% and 11.8, 6.1%, respectively, apart from the palmitic acid at 30.2 and 30.4%. Composition of roe lipids with a special emphasis on n-3 fatty acids such as linolenic acid, EPA and DHA of some common fishes is given in Table 56.1.

ANALYTICAL TEHCNIQUES

A variety of chromatographic techniques have been employed for determining the composition of fish roe lipids. Lipid composition of roes, muscle, and viscera of *Liza carinata* were studied using column chromatography, thin-layer chromatography (TLC) and gas–liquid chromatography (GLC) (Joh *et al.*, 1988). Bell (1989) studied the molecular species of PLs namely PC, PE, and phosphatidylinositol (PI) of ripe roes of cod fish using reverse-phase HPLC coupled with a UV detector using 3,5-dinitrobenzoyl derivatives. Molecular species of palmitic acid/EPA, oleic acid/EPA, palmitic acid/DHA, oleic acid/DHA contributed to 67 and 61.8% of PC and PE, whereas oleic acid/AA were found to account for 23% of PI and 36.7% PC *versus* 0.7% of PE. TLC separation of PL fraction of *Calichrous pabda* revealed the presence of myristic acid, palmitic acid, palmitoleic acid, oleic acid, linoleic acid, AA, EPA, and DHA (Ghosh, 1997). Salmon roe contained 27.2% PL with PC as the major component (80%) as determined by a solvent-partitioning technique (Ando *et al.*, 2003). DHA and EPA were reported to be major fatty acids in PC and PE, whereas AA was abundant in PI. Orange roughy (*Hoplostethus atlanticus*) roe oil subjected to chromatographic techniques revealed that TAGs and PLs were predominant with palmitic acid, oleic acid, eicosenoic acid, EPA, and DHA, although wax esters constituted their major component (Body, 2006). Cod roe lipid was characterized by chromatography on solid aluminum oxide and

TABLE 56.1 Important PUFA Composition (wt. %) of some Fish Roe Lipids

	PUFA Composition (wt. %)			
Fish roe	18:3	20:5	22:6	Reference
Orange roughy	—	4.90	15.0	Body (1985)
Salmon, mature roe	0.6	6.86	24.32	Bekhit *et al.* (2009)
Pollock, immature roe	0.21	18.98	21.17	Bechtel *et al.* (2007)
Common snook	0.28	2.38	13.73	Roca *et al.* (2009)
White sturgeon	1.1	3.10	9.2	Czesny *et al.* (2000)
Lake sturgeon	0.5	1.90	4.8	
Striped bass	0.59	3.74	11.56	Harrell and Woods (1995)
Herring	0.52	13.72	21.65	Huynh *et al.* (2007)
Skipjack	0.53	3.80	23.46	Intarasirisawat *et al.* (2011)
Tongol	0.56	4.62	26.19	
Bonito	90.64	4.62	20.53	
Sea bass	0.9	1.40	9.0	Navas *et al.* (2001)
Rohu	5.5	0.8	12.7	Prabhakara Rao *et al.* (2010)
Murrel	7.2	0.8	5.3	
Catla	1.8	2.0	6.9	Prabhakara Rao *et al.* (2013)
Mrigal	9.4	—	3.6	
Mullet roe 1	0.47	12.75	12.93	Rosa *et al.* (2009)
Mullet roe 2	0.43	4.33	8.29	
Salted *Rutilus frisii* roe product	0.29	0.94	0.61	Taghiof *et al.* (2012)

18:3, Linolenic acid; 20:5, eicsapentaenoic acid; 22:6, docosahexaenoic acid.

found to contain TAG (3.4%), SE (0.2%), two major PLs, PC (76.1%); PE (10.3%), and cholesterol (7.1%) (Israelsen, 2008).

Mahmoud *et al.* (2008) reported higher yields of roe lipids (9.7%) when chemical methods were employed compared to proteolytic enzyme hydrolysis method (3.7%) using alkalase, nutrase, or protamex in rainbow trout (*Oncorhynchus mykiss*). PUFA content of the oils were 41.8% with DHA and EPA accounting for 25.6 and 10.9% of TL, respectively.

HOW COMPOSITION ALTERED

Effect of Diet, Season, and Spawning

Lipid composition of fish diets has been shown to have significant effects on the composition of fish roe lipids. Fatty acid composition and lipid content varied in the diets, and it was observed that variation of lipid content had no effect on fatty acid composition (Navas *et al.*, 2001). Spawning seasons were shown to affect the fatty acid composition. Body lipids decreased and lipids of gonads or milt increased during spawning. It was estimated that 40% migration of lipids occur during spawning seasons. Lipid content of flesh (10.78%) and liver (10.54%) of non-spawning herring decreased during spawning to 2.41 and 1.92%, respectively (Huynh *et al.*, 2007). SFA (34.5%), MUFA (31.5%), and PUFA (34.9%) were reported in Common snook (*Centropomous undecimalis*) roe. SFA increased, while n-3 fatty acids and PUFA decreased during spawning months of May to September (Roca *et al.*, 2009). Results suggested that all fish species follow similar trends of fatty acid profiles during spawning seasons. Table 56.2 depicts the major changes observed in the lipid profiles, highlighting the components namely, SFA, MUFA, EPA, and DHA.

TABLE 56.2 Comparative Fatty Acid Composition (wt %) of Flesh and Roe During Spawning and Non-Spawning Seasons

	Huynh *et al.* (2007)			Roca *et al.* (2009)	
	Flesh		Roes	Roes (spawning Months)	
Fatty acid	Spawning	Non-spawning	Spawning	May	September
SFA	28.7	26.21	27.45	31.0	40.0
MUFA	46.54	48.12	26.82	—	—
n-3 PUFA	19.73	20.59	38.76	27.5	19.0
n-6 PUFA	2.99	2.93	3.62	—	—
EPA	7.52	8.74	13.72	—	—
DHA	9.98	8.58	21.65	17.5	10.0

Effect of Domestication

Significant variations were observed in fish roe lipid composition in eggs of species of the same fish from marine, river, and domesticated samples. Harrell and Woods (1995) reported higher total lipid, PUFA, EPA, and DHA in the wild striped bass (*Morone saxatilis*). Fatty acid profile and n-3/n-6 ratio of domesticated fish eggs were in concurrence with freshwater fish species. A similar study was carried out on fatty acid compositions of NL and PL fractions of domesticated, wild sturgeon (*Acipenser transmontanus*), and lake sturgeon (*Acipenser fulvescens*) roes (Czesny *et al.*, 2000). In NL, palmitoleic acid was found to be higher in freshwater wild fishes in comparison to white sturgeon, which possessed higher DHA. Variation in total lipid content, EPA, and DHA for domesticated and wild species of some sea fish roes are presented in Table 56.3.

TABLE 56.3 Fatty Acid Composition (wt. %) of Domesticated and Wild Fish Roes

Fish Species	Total Lipid (wt. %) on a Dry Weight Basis	Fatty Acid Composition (wt. %)			References
		EPA	DHA	Total n-3 PUFA	
Striped bass domesticated	5.67	2.27	4.24	7.25	Harrell and Woods (1995)
Striped bass wild	6.45	3.73	5.99	11.56	
White sturgeon domesticated in sea farm	13.8	5.50	10.7	21.40	Czesny *et al.* (2000)
White sturgeon domesticated in university of California	13.8	4.6	9.0	18.6	
White sturgeon wild from Sacramento river	13.8	3.1	9.2	16.2	
Lake sturgeon, wild	10.8	1.9	4.8	11.5	

Armstrong *et al.* (1994) studied the influence of geographical location and seasonal variation on fatty acid profiles of fish catches of Australian origin. They selected seasons of autumn and spring and fish catches from three different latitudes. n-3 PUFA increased as they moved southwards and to colder months and the n-3/n-6 ratio was higher than the fish caught from northern waters. Maximum variation (3.7%) coincided with the period from August to November. Significant differences between large- and medium-sized fish were observed and DHA content was lower (26% *versus* 33%) in larger fishes. Liver oil in female fish possessed higher quantities of eicosenoic acid (4.5–14.9%) and erucic acid (1.8–12.3%) in late summer and fall. However, this trend was not observed in the male fish. Fatty acid composition of the milt was similar to the roe with few exceptions of higher oleic and lower palmitoleic acid contents.

OTHER WAYS IN WHICH COMPOSITION ALTERED

Effect of Processing on Roe Lipid Composition

Processing plays an important role in lipid composition. Bledsoe *et al.* (2003) described composition of various products based on caviar available in the market. Among popular products, salted and frozen roe retained longer shelf-life. Mullet roe prepared by soaking in brine, pressing, and air drying was found to be rich in protein (35.5%) and lipids (25.7%). Dried roe from *species* of *Mugil, Thunnus, Molva;* salted roes of *Acipenser cepalin, Cyclopterus lumpus, Gadus cepelanus* and *Salmo salar* were found to be nutritious due to the presence of n-3 PUFA and optimum ratio of SFA:MUFA (Cengarle *et al.*, 2000). Analysis of fresh and smoked products of cod (*Gadus morhua*) showed that smoking increased total lipid, but did not affect the lipid class or fatty acid composition (Peter *et al.*, 1967). However, they observed decreases in EPA and DHA contents.

The effects of processing parameters and storage conditions for rainbow trout (*Salmo gairdneri*) and whitefish (*Coregonus albula*) roe products was determined by acid value for 12 months (Kaitaranta, 1982) and hydrolysis of 75% in NLs of trout fish and 60% in whitefish was observed. Lipid hydrolysis among the phospholipids, PC, PE and sphingomyelin showed that PE got hydrolyzed to the extent of 35–40% in trout and whitefish during cold storage, while hydrolysis in PC was only 15–20%. Pasteurization at 80°C for 40 min controlled hydrolysis whereas salt (14%) could slow down the rate of hydrolysis at refrigerated conditions. Short-time cooking of hake (*Merluccius hubbsi*) roe resulted in minor changes in TAGs, diacylglycerols, PLs, and waxes, which might be due to lower availability of oxygen (Mendez *et al.*, 1992).

Roes and salted roe products (bottarga) from *Mugil cephalus* were assessed for oxidative stability during storage at −20°C for 60 days and at room temperature exposed to air and light (Rosa *et al.*, 2009). Oxidative stability of the products was measured by analyzing EPA, DHA, and conjugated diene hydroperoxides. Oxidative stability was highest for raw roes followed by whole bottarga and grated bottarga. Salting and drying of roes did not influence the oxidation of n-3 PUFA when stored at −20°C. The grated products stored at ambient temperature showed significant increases in hydroperoxides after 32 days. They concluded that raw mullet (*Mugil cephalus*) roes are more stable compared to cured products stored at −20°C towards oxidation.

Popular fish roe products namely Ikura (salmon), Tarakon (pollock), Tobiko (flying fish), and Kazunoko (herring) were found to be rich in palmitic acid, oleic acid, EPA, and DHA (Shirai *et al.*, 2005). TL, SE, TAG, PL, PC, and PE fractions contained DHA in good quantities in all the roe products. PC and PE fractions of Tobiko were richer in DHA than Ikura, Tarako, and Kazunoko.

An increase in FFA and a reduction in TAG and PL was observed during ripening of cod roes at RT (17°C), chilling

(3°C), using benzoate as preservative (0, 20 mg/100 g) and citric acid for reduction of pH over a period of 4 months (Guimaraes *et al.*, 2011). Cholesterol content decreased in ripened cod roes when compared to fresh roe lipids. Lower temperatures and acidic pH protected PL to a greater extent, though interactions between citric acid and benzoate affected PL. Vacuum-packed and refrigerated dry-salted Caspian kutum (*Rutilus frisii*) roe during 5, 60, and 120 days of storage retained EPA, DHA, whereas maximum losses were observed in unpacked samples stored at room temperature (Taghiof *et al.*, 2012).

Fresh roes are highly susceptible to microbial spoilage due to high moisture, protein, and lipid contents. Treatments such as pasteurization, blanching, and freezing are effective against spoilage. Lactic acid bacteria, *Enterobactereaceae* and *Vibrio* species were responsible for odd flavors and spoilage in lightly salted *lump* fish roe stored at 5°C. Salt has been shown to effectively control microbial spoilage. Salted mullet roe products from Taiwan were analyzed for the presence of histamine and histamine-forming bacteria. Aerobic plate count (APC) in all samples were in permissible limits (1.0–7.1 log CFU/g), and contained no coli forms (Kung *et al.*, 2008). Salt blanched roes of *Labeo rohita* were found to be suitable for preparation of shelf-stable fish roe pickle (Balaswamy *et al.*, 2010).

Herring roe lipids were more stable towards oxidation compared to commercial sardine and tuna fish oils due to the presence of major quantities of PL (Moriya *et al.*, 2007). Storage stability of salted and dried mullet roe containing 16% moisture, 40% fat and 43% crude protein was assessed during a 9-week period (Joh *et al.*, 1988). SFA increased from 40.8 to 45.3 mg/100 g, MUFA decreased from 25.3 to 14.1 mg/100 g and PUFA content decreased from 25.6 to 1.3 mg/100 g during storage. Joh *et al.* (1988) reported that palmitic, oleic and linoleic acids were abundant, which accounted for three-quarters of fatty acids bound to GLs during processing and storage of salted and dried mullet roes. A decrease in essential fatty acid content during storage from 4.32 mg/100 g to 1.46 mg/100 g over 9 weeks was noticed. Changes in lipid components such as FFA, TAG, PL, SE, and cholesterol in a few fresh roes and products under refrigerated storage are presented in Table 56.4. The effects of processing parameters such as salting and drying on essential fatty acids was studied and compared with fresh roe of *Mullet japonica* (Sang and Jung, 1993). The ratio of unsaturated to saturated fatty acids decreased from 1.53 to 0.34 after storage of 9 weeks at 0–10°C and essential fatty acids were completely lost during processing. Heat treatment at 80°C for 1.5–2 min resulted in effective inactivation of lipoxygenase as assayed by thiobarbituric acid method (Wang *et al.*, 1991). Oxidative stability of lipids could be enhanced by heat treatment under oxygen-free atmosphere.

TABLE 56.4 Effect of Processing/Storage on Lipid Class Components

Class of Lipids	Composition (%)		Reference
	Fresh cod roe	*Processed products (ripened cod roe)*	
Free fatty acids	15.3	40.9	Guimares *et al.* (2011)
Triacylglycerols	32.2	21.3	
Phospholipids	28.2	11.2	
Sterol esters	3.3	5.6	
Cholesterol	15.2	10.0	
	Fresh roe	*Commercial salted roe product 'Bottarga'*	
Fatty alcohols	70	72	Rosa *et al.* (2009)
Saturated fatty acids	15	17	
Monounsaturated fatty alcohols	18	20	
Monounsaturated fatty acids	30	22	
Polyunsaturated fatty acids	44	44	
Cholesterol	3.2	3.2	
	Rainbow trout fish	*After storage for 52 weeks at 2°C*	
Free fatty acids	—	1.6	Kaitaranta (1982)
Triacylglycerols	46.4	42.6	
Phospholipids	51.0	52.6	
Sterol esters	0.2	0.3	
Cholesterol	2.4	2.2	
	Whitefish	*After pasteurization and storage for 52 weeks at 2°C*	
Triacylglycerols	65.6	63.8	Kaitaranta (1982)
Phospholipids	31.3	32.5	
Sterol esters	1.5	2.1	
Cholesterol	1.3	1.6	

References

Ando, Y., Samoto, H., Sasaki, S., 2003. Preparation of salmon roe polar lipid concentrate as an HUFA enricher of *Artemia nauplii*. J. Oleo Sci. 52 (8), 443–447.

Armstrong, S.G., Wyllie, S.G., Leach, D.N., 1994. Effects of season and location of catch on the fatty acid composition of some Australian fish species. Food Chem. 51 (3), 295–305.

Balaswamy, K., Prabhakara Rao, P.G., Jyothirmayi, T., Rao, D.G., 2010. Effects of pretreatments and salt concentration on rohu (*Labeo rohita*) roes for preparation of roe pickle. J. Food Sci. Technol. 47 (2), 219–223.

Bechtel, P.J., Chantarachoti, A.C.M., Sathivel, S., 2007. Characterisation of protein fractions from immature Alaska Walleye Pollock (*Theragra chalcogramma*) roe. J. Food Sci. 72 (5), S338–S343.

Bekhit, A.E.A., Morton, J.D., Dawson, C.O., Zhao, J.H., Lee, H.H.Y., 2009. Impact of maturity on the physico-chemical and biochemical properties of Chinook salmon roe. Food Chem. 117, 318–325.

Bell, M.V., 1989. Molecular species analysis of phosphoglycerides from the ripe roes of cod (*Gadus morhua*). Lipid 24, 585–588.

Bledsoe, G.E., Bledsoe, C.D., Rasco, B., 2003. Caviars and fish roe products. Crit. Rev. Food Sci. Nutri. 43 (3), 317–356.

Body, D.R., 2006. Food, the composition of orange roughy (*Hoplostethus atlanticus*) roe lipids. J. Sci. Food Agric. 36 (8), 679–684.

Cengarle, L., Carta, A., Pinna, L., Marceddu, M.F., Tilloca, G., 2000. Lipid fraction of fish eggs I. Semipreserves. Riv. Ital – delle Sostanze Grasse 77 (2), 77–83.

Czesny,S.,Dabrowsky,K.,Christenson,J.E.,Eenennaam,J.V.,Doroshov,S., 2000. Discrimination of wild and domestic origin of sturgeon ova based on lipids and fatty acid analysis. Aquaculture 189, 145–153.

Ghosh, M., 1997. Principle fatty acids of phospholipid classes of an Indian fresh water fish (*C. pabda*). J. Food Lipid 4 (3), 189–197.

Guimaraes, J.L., Trattner, S., Pickova, J., 2011. Effect of processing and amine formation and the lipid profile of cod (*Gadus morhua*) roe. Food Chem. 129, 716–723.

Harrell, R.M., Woods, L.C., 1995. Comparative fatty acid composition of eggs from domesticated and wild striped bass (*Morone saxatiliis*). Aqualculture 133, 225–233.

Huynh, M.D., Kitts, D.D., Hu, C., Trites, A.W., 2007. Comparative Biochem. Physiol. Part B 146, 504–511.

Israelsen, D.A., Eva, F., Turid, R., 2008. Solid phase extraction as a tool to separate lipid classes and study deterioration of marine lipids. J. Aquatic Food Prod.Technol. 17 (1), 39–59.

Intarasirisawat, R., Benjakul, S., Visessanguan, W., 2011. Chemical compositions of the roes from skipjack, tongol and bonito. Food Chem. 124, 1328–1334.

Joh, Y.G., Lee, K.H., Cho, Y.J., 1988. Lipid composition of roe, muscle, and viscus of *Liza carinata*, a species of the Mugilidae family. Korean J. Food Sci. Technol. 20 (5), 674–682.

Kaitaranta, J.K., 1980. Lipids and fatty acids of a whitefish (*Coregonus albula*) flesh and roe. J. Sci. Food Agric. 31 (12), 1303–1308.

Kaitaranta, J.K., 1982. Hydrolytic changes in the lipids of fish roe products during storage. Int. J. Food Sci. Technol. 17 (1), 87–98.

Kazuo, I., Haruo, O., Youji, O., Hiroshi, W., 1988. Studies on furan fatty acids of salmon roe phospholipids. J. Biochem. 103 (5), 836–839.

Kim, S.K., Mendis, E., 2006. Bioactive compounds from marine processing byproducts – A review. Food Res. Int. 39, 383–393.

Kung, H.F., Chien, L.T., Liao, H.J., Lin, C.S., Liaw, E.T., Chen, W.C., Tsai, Y.H., 2008. Chemical characterisation and histamine-forming bacteria in salted mullet roe products. Food Chem. 110 (2), 480–485.

Lili, L., Shunsheng, C., Yinghong, Q., Zhuoping, D., Wenhui, W., 2007. Lipids and fatty acids composition of roe from *Cyprinus carpio* (carp) and *Carassius auratus* (crucian carp). Yingyang Xuebao 29 (4), 415–416.

Mahmoud, A.S.K., Linder, M., Fanni, J., Parmentier, M., 2008. Characterisation of the lipid fractions obtained by proteolytic and chemical extractions from rainbow trout *(Oncorhynchus mykiss)* roe. Process Biochem. 43 (4), 376–383.

Mendez, E., Fernandez, M., Pazo, G., Grompone, M.A., 1992. Hake roe lipid: Composition and changes following cooking. Food Chem. 45 (3), 179–181.

Mitsu, K., Ikuo, H., Yasuyuki, I., 1974. Fish roe lipids, especially on mullet roe wax esters. Yukagaku 23 (5), 290–295.

Moriya, H., Hosokawa, M., Miyashita, K., 2007. Combination effect of herring roe lipids and proteins on plasma lipids and abdominal fat weight of mouse. J. Food Sci. 72 (5), 231–234.

Mukhopadhyay, T., Ghosh, S., 2007. Lipid profile and fatty acid composition of two silurid fish eggs. J. Oleo Sci. 56 (8), 399–403.

Navas, J.M., Thrush, M., Januy, S., Ramos, J., Bromage, N., Carrillo, M., 2001. Total lipid in the brood stock diet did not affect fatty acid composition and quality of eggs of sea bass (Dicentrarchus labrax L.). Sci. Marina 65 (1), 11–19.

Parimal, S.C., Dutta, J., Amitabha, G., 1997. Lipid composition of parshey roe (*Mugil parsia*). J. Indian Chem. Soc. 74 (6), 470–473.

Peter, J.M., Ackman, R.G., Sipos, J.C., 1967. Seasonal changes in fatty acid composition of cod liver, flesh, roe and milt lipids. J. Fisheries Res. Board Canada 24 (3), 613–627.

Peterson, F.S., Peterson, F.I.B., Sargent, J.R., Haug, T., 1986. Lipid class and fatty acid composition of eggs from the Atlantic halibut (*Hippoglossus hippoglossus*). Aquaculture 52 (3), 207–211.

Prabhakara Rao, P.G., Balaswamy, K., Jyothirmayi, T., Karuna, M.S.L., Prasad, R.B.N., 2013. Lipid classes, fatty acid and phospholipid composition of fish roe lipids from *Catla catla and Cirrhinus mrigala*. Int. Food Res. J 20 (1), 275–279.

Prabhakara Rao, P.G., Jyothirmayi, T., Karuna, M.S.L., Prasad, R.B.N., 2010. Studies on lipid profiles and fatty acid composition of roe from *rohu (Labeo rohita)* and *murrel (Channa striatus)*. J. Oleo Sci. 59 (10), 515–519.

Roca, C.Y., Rhody, N., Nystrom, M., Main, K.L., 2009. Effects of fatty acid composition and spawning season patterns on egg quality and larval survival in common snook (*Centropomous undecimalis*). Aqualculture 287, 335–340.

Rosa, A., Scano, P., Melis, M.P., Deiana, M., Atzeri, A., Dessi, M.A., 2009. Oxidative stability of lipid components of mullet (*Mugil cephalus*) roe and its product "bottarga". Food Chem. 115, 891–896.

Sang, J.J., Jung, H.J., 1993. Changes in the fatty acid composition of phospholipids in dried and salted mullet roe during processing and storage. J. Korean Soc. Food Nutr. 22 (3), 286–290.

Shirai, N., Higuchi, T., Suzuki, H., 2005. Analysis of lipid classes and the fatty acid composition of the salted fish roe food products, Ikura, Tarako, Tobiko and Kazunoko. Food Chem. 94 (1), 61–67.

Taghiof, M., Nikoo, M., Pourshamsian, K., 2012. Proximate and fatty acid composition of salted Caspian Kutum (*Rutilus frisii*) roes influenced by storage temperature and vacuum packaging. World J. Marine Sci. 4 (5), 525–529.

Tocher, D.R., Sargent, J.R., 1984. Analyses of lipids and fatty acids in ripe roes of some northwest European marine fish. Lipid 19 (7), 492–499.

Wang, Y.J., Miller, L.A., Addis, P.B., 1991. Effect of heat inactivation of lipoxygenase in lipid oxidation in lake herring (*Coregonus artedii*). J. American Oil Chem. Soc. 68 (10), 752–757.

SECTION 8

BEVERAGES

CHAPTER

57

Use of Oak Wood to Enrich Wine with Volatile Compounds

*Pedro M. Pérez-Juan**, *María Dolores Luque de Castro*[†]

*Bodega y Viñedos Vallebravo, Puente de Génave, Pol. Ind. La Vicaría, Jaén, Spain, [†]Department of Analytical Chemistry, Annex Marie Curie Building, Campus of Rabanales, University of Córdoba, Córdoba, Spain

CHAPTER POINTS

- Placement in contact with oak wood is the most widely used wine-aging method.
- The aging process causes the transfer of wood components to the wine and the transformation of some wine components.
- Aging in the presence of wood results in typical changes in wine flavor and aroma.
- There is a current trend to conducting in-depth studies of wine changes during aging by using cutting-edge analytical equipment.
- The use of alternative wood types and materials such as vine shoots to age wine is another current trend.

List of Abbreviations:

CE Capillary Electrophoresis
DAD Diode Array Detector
ESI Electro-Spray Ionization
FID Flame Ionization Detector
GC Gas Chromatography
HPLC High Performance Liquid Chromatography
LC Liquid Chromatography
LPME Liquid-Phase Micro-Extraction
MALDI–TOF Matrix-Assisted Laser Desorption–Ionization–Time-Of-Flight
MS Mass Spectrometry
SBSE Stir Bar Sorptive Extraction
SPE Solid Phase Extraction
SPME Solid Phase Micro-Extraction
UV–VIS Ultra-Violet–Visible

INTRODUCTION

The Greco–Roman civilization used clay vessels to produce, store, and transport wine. Subsequently, wood, used by the Celtic culture to obtain beer, gradually replaced clay for storing and transporting wine by virtue of its being lighter, less fragile, more resistant to compression, and easier to handle and stack, and also because of its having a favorable influence on the physicochemical properties of wine.

High-capacity wooden containers known as 'fudres' were, together with earthenware jars, the most widely used wine containers for fermentation and storage in wineries before the advent of cement, iron, plastic, and stainless-steel containers. With time, steel has superseded all previous materials on account of its low cost, easy maintenance, and cleanliness, and because it facilitates simple control of wine fermentation.

In former times, wood containers capable of holding 250–5001 were used to transport wine from wineries to outlets. During transportation, some wine properties changed due to the effect of contact with the wood, the environmental conditions (temperature and humidity, mainly), and the properties of the wood. The specific changes and their extent were a function of climate changes and the distance traveled during transportation.

Consumers' appreciation of wine characteristics due to contact with wood during transportation or storage in cellars promoted the extensive use of barrels. Initially, the type of wood used to make containers was a matter of availability; with time, oak wood was favored because it conferred optimal properties on the wine. At present, however, wood barrels are only used in

Processing and Impact on Active Components in Food
http://dx.doi.org/10.1016/B978-0-12-404699-3.00057-3

some winemaking areas. One case in point is La Palma (Canary Islands, Spain), which produces 'tea wine' with special sensory properties due to the effect of aging in *Pinus canariensis* wood barrels (European Commission, 2010).

According to Vivas (2005), the widespread use of oak wood to hold aging wine has resulted from: (a) its easy deformation by heating, which allows wood staves to be bent into the characteristic shape of a barrel; (b) its high gas porosity and limited permeability to liquids; (c) its low density, which facilitates construction of low-weight/high-capacity barrels; (d) its great hardness and resistance to microbial attack; (e) its low contents in bitter compounds, which can be easily removed by drying and natural leaching of the barrels prior to use; (f) the presence of astringent substances in the wood that are leached by wine during aging; and (g) that of odorant substances coming from the wood or produced during construction of the barrel which diffuse into the wine and alter its olfactory profile as a result.

Oak trees belong to the genus *Quercus*, which comprises more than 250 species of trees and bushes distributed mainly in the temperate zone of the northern hemisphere and differing widely in their properties and suitability for wine aging. Such properties depend not only on the particular species, but also on the pedoclimatic conditions during tree life.

The major species of oak have traditionally been classified as a function of geographical origin into French oaks (e.g., *Quercus robur* and *Q. petraea*), which, however, are more often distinguished by growing region than by species, and American oaks (*Q. alba*, mainly). A third, more recent group named 'Eastern Europe oaks' comprising *Q. robur* and *Q. petraea* has been established to include trees of which the properties as materials for holding aging wine depend on the particular pedoclimatic conditions (Vivas, 2005).

One crucial characteristic of wine barrels is wood grain (Vivas, 2005). This depends on how fast trees grow, which in turn is a function of the specific pedoclimatic zone – the primary factor affecting annual ring growth. Rings are the combination of the growth vessels of spring and summer–autumn, and their thickness determines grain.

The wooden staves used to make barrels require a previous treatment involving natural or artificial drying which causes not only dehydration but also, potentially, simultaneous leaching, oxidation, and fungal degradation of natural wood components. The specific compounds that are transferred to wine during aging depend to a great extent on this drying treatment (Vivas, 1995b). During barrel fabrication, staves are heated in order to make the wood malleable enough to adopt a barrel shape. Finally, the wood is roasted at a variable temperature for a variable time depending on the maker.

Storage in barrels has various effects on wine the most significant of which are as follows (Hidalgo, 2003; Flanzy, 2000; Vivas, 2005; Zamora, 2003; Ribéreau-Gayon *et al.*, 1999):

(a) A fast, efficient increase in limpidity by effect of the removal of residual CO_2 from the fermentation medium, and flocculation and precipitation of colloids (proteins and phenols, mainly).

(b) Precipitation of salts (particularly tartaric, but also ferric salts).

(c) Redox reactions involving oxygen slowly but continuously diffusing through barrel walls; wine components (particularly ethanol and the phenolic fraction) and hydrolyzable tannins and ellagitannis from the wood.
Acetaldehyde, a product from the oxidation of ethanol, plays a key role in the evolution and stabilization of color and flavor in red wines. This aldehyde acts as a link between anthocyans and variably polymerized flavonols, and also between flavonols themselves. In addition, acetaldehyde is an essential ingredient in the production of pyranoanthocyanins, which contribute to color in aged red wine. Also, the antioxidant character of tannins from the wood modulates oxidation, protects anthocyans, and facilitates their combination with acetaldehyde.
Free anthocyans are partly responsible for wine color as they can undergo redox reactions yielding colorless products.

(d) Reactions between anthocyans and flavonols in monomeric or polymeric forms without acetaldehyde mediation that also contribute to wine color and flavor.

(e) Degradation of lignin in the acid, ethanolic medium provided by wine, to simple phenols from cinnamic compounds such as syringaldehyde, sinapaldehyde, vanillin, and coniferaldehyde, and their corresponding acids (syringic, sinapic, vanillic, and ferulic).

(f) Potential microbiological alterations by yeasts of the genus *Brettanomyces* or bacteria of the genus *Acetobacter* yielding products with significant effects on wine aroma and flavor.

(g) Diffusion into the wine of substances from the wood that can alter the flavor and aroma of the former. Some such substances (e.g., ellagitannins) additionally influence the evolution of colored compounds in the wine.

Wine–wood interactions are not equally strong throughout a barrel's useful life; in fact, they peak in new barrels and fall to very low levels after 5 years of use (Vivas, 1999). This requires periodic renewal of barrels. In some aging processes, however, wood components diffuse less markedly into the wine and the containers can be used for more than 50 years in some instances. Such is the case with *Porto* wines, which undergo oxidative reactions, such as those produced in Western Andalusia (e.g., *oloroso, amontillado, palo cortado, Pedro Ximénez*), and also with so-called *fino* wines, which are aged for some time under a flor film of yeasts also in the same Spanish region.

Singleton and Draper, cited by Ribéreau-Gayon *et al.* (1976), proposed direct addition of oak chips in doses from 1 to 10 g/l to wine in order to confer on it 'wood character' in as short a time as possible while avoiding losses of wine, the shortcomings associated with its handing and the cost of barrels. This winemaking practice, which has only recently been accepted by the European Union (European Commission, 2009), is commonly used in other regions with less stringent legislation. In order to ensure that consumers are accurately informed, the European Union (European Commission 2010) has banned the use of words typically associated with wood barrel-aged wines (e.g., *crianza, reserva, gran reserva, roble, barrica*) on the labels of wines aged in the presence of wood chips or held in barrels during the process.

The wood chips added to the wine are wood residues from staves used to construct barrels and their quality must be on a par with those of the staves. Chips are dried similarly to staves, milled to the desired particle size and roasted in a rotary drum similarly to coffee or cocoa (Chatonnet, 2008). A wide variety of oak chips differing in botanical and geographic origin, tree species, degree of roasting, and particle size is currently available. Also, winemakers can use commercial oak wood extracts in powder form or in solution for direct addition to the wine in order to obtain the benefits of contact with barrel wood in a fast, economical way. Powder wood extracts are obtained by recirculating water–ethanol mixtures through a bed of chips at ambient temperature and pressure for several days, and then subjecting the extract evaporation with an extractant, atomization, or lyophilization to obtain the solid powder. The characteristics of the end-product are dictated by the type of wood, extractant and technique used to obtain the powder (Álvarez, 2007; OIV, 2012). González-Rodríguez *et al.* (2003) proposed using superheated extractants to accelerate the conventional method without altering the characteristics of the final powder.

One alternative to wood chips currently under study is vine shoots. Application of a superheated extractant to this vine pruning residue provides an extract containing most of the components of oak extracts in relative concentrations depending on the particular cultivar (Luque-Rodríguez *et al.*, 2006; Delgado-Torre *et al.*, 2012a, b).

HOW WINE COMPOSITION IS ALTERED BY AGING IN OAK BARRELS

The specific compounds extracted into wine from oak barrel walls or chips depend on the particular wood but not on its physical state. Table 57.1 lists the most salient compounds that can be transferred from wood to wine (Zamora, 2003; Moreno-Vigara and Peinado-Amores, 2010).

Heartwood timber is the oak wood typically used to fabricate barrels and obtain oak powder from chips. This material contains both structural components of cell walls forming woody fibers and extractables (a group of organic and inorganic substances bound to woody fibers). These substances can be extracted by solvents of variable polarity (e.g., ethanol, acetone, ether, hexane, water) or by steam stripping. Although the chemical composition of wood differs between *Quercus* species, geographic origins, pedoclimatic conditions, and tree parts, it is widely accepted that the structural components cellulose, hemicellulose, and lignin account for 40, 20, and 25%, respectively, of the wood dry weight, extractable substances constituting approximately 15% of the remaining material. Quantitatively, the most important extractable compounds are ellagitannins, which play a key role in flavor and color stability in wine. Minor compounds of diverse nature such as simple phenols (aldehydes and benzoic acids, allylphenols, and cinnamic aldehydes), coumarins, flavanols, lignans, condensed tannins, galotannins, lactones, norisoprenoids, terpenes, fatty acids, hydrocarbons, and sterols also have a decisive influence on wine flavor and aroma, either because they diffuse from wood to wine or because they are the substrates of transformations occurring during fabrication of barrels or obtainment of chip powder (Browning, 1967; Fengel and Wegener, 1984; Zamora, 2003, Vivas, 2005; Puech *et al.*, 2000). Wood treatments cause drastic changes in the chemical composition of the original wood and facilitate the transfer of compounds significantly modifying the sensory profile of the wine. In fact, the typical 'woody odor' is due to compounds formed during the drying and roasting of the wood.

Although cellulose and hemicellulose polysaccharides are dissimilarly soluble in wine – the former are virtually insoluble and the latter are found as such in wine – both undergo significant thermal degradation during wood roasting due to Maillard reactions that yield aroma compounds of which the olfactory descriptors include 'toasted almond', 'roasted', 'candy', 'toast', 'cocoa', 'nuts', and 'coffee', and have the chemical structure of a furan, pyrone or dimethylpyrazine (Table 57.1). These reactions develop to an extent dependent on the temperature and duration of the process, which additionally causes hemicellulose hydrolysis and the release of acetyl groups as a result. All these compounds diffuse

TABLE 57.1 Main Compounds Transferred from Wood to Wine and Sensory Effects on the Wine

Original Wood Component		Process	Compound		Odor Descriptor	Flavor
Type	Name		Chemical Family	Name		
Structural	Cellulose and hemicellulose	Roasting	5-Atom oxygenated heterocycles	Furfural	Almonds	Undetectable
				Methyl-5-furfural	Roasted almonds	
				Hydroxymethyl-5-furfural		
				Furfuryl alcohol	Wet hay	
			6-Atom oxygenated heterocycles	Maltol	Caramel, roasted nuisance	
				Dihydromaltol		
				3-Methyl-3-cyclopenten-2-one		
				3-Methyl-3-cyclohexen 2-one		
				3-Methyl-2-furanone		
			6-Atom nitrogenated heterocycles	2,3-Dimethylpyrazine	Cocoa, hazelnut, toast, coffee	
				2,5-Dimethylpyrazine		
				2,5-Dimethylpyrazine		
	Hemicellulose		Acids	Acetic acid	Vinegar	
	Lignin		Simple phenols	Phenol	Ink	
				Guaiacol	Smoke	
				Syringol		
				Methy-4-syringol		
			Alkyl and vinyl phenols	Methy-4-guaiacol	Roasted wood	
				Ethyl-4-guaiacol		
				4-Vinylguaiacol	Carnation	
				4-Vinylphenol	Pharmacy	
				4-Ethylphenol	Undetectable	
			Benzoic aldehydes	Vanillin	Vanilla	
				Syringaldehyde	Undetectable	
			Phenyl ketones	Acetophenone	Vanilla	
				Acetovanillone		
				Propiovanillone		
				Butyrovanillone		
			Allylphenols	Eugenol	Clove spices	
				Allyl-4-syringol	Wood smoke	
			Cinnamic aldehydes	Sinapaldehyde	Undetectable	
				Coniferaldehyde		

TABLE 57.1 Main Compounds Transferred from Wood to Wine and Sensory Effects on the Wine—Con'd

Original Wood Component		Process	Compound		Odor Descriptor	Flavor
Type	**Name**		**Chemical Family**	**Name**		
Extractable	Fresh wood (Lipids) Fresh wood		Lactones	*cis*-β-Methyl-γ-octolactone *trans*-β-Methyl-γ-octolactone	Coconut, woody fresh	
			Benzoic aldehydes	Vanillin	Vanilla	
			Allylphenols	Eugenol	Clove spices	
			Coumarins (glycosides)	Scopoline		Bitter
				Esculin		
			Coumarins (aglycones)	Scopoletin		Acid and harsh
				Esculetin		
			Hydrolyzable tannins	Ellagitannins		Astringent and bitter
				Gallotannins		Acid, astringent and bitter
			Phenol acids	Gallic acid		Acid and astringent
				Ellagic acid		Acid
			Flavanols, procyanidins and condensed tannins	(+)Catechin, B3 dimers		Astringent and bitter

Modified from Ribéreau-Gayon et al. *(1999); Zamora (2003); Moreno-Vigara and Peinado-Amores (2010).*

into wine in higher proportions in new barrels, where they increasing the levels of acetic acid – a characteristic marker of wine alteration at high enough concentrations (Boidron *et al.*, 1988; Chatonnet *et al.*, 1989; Vivas, 1995a; Zamora, 2003; Vivas, 2005).

Lignin is a three-dimensional phenolic polymer strongly influencing the mechanical properties of wood. Similarly to structural polysaccharides, the roasting step promotes destruction of the polymeric structure of lignin and leads to the formation of various products such as simple phenols, ethyl and vinyl phenols, benzoic aldehydes (vanillin, mainly), acetophenones, allylphenols, and cinnamic aldehydes. Table 57.1 shows some of the odor compounds formed and their associated descriptors ('vanilla', 'burnt wood', 'smoke', 'spice', 'ink' and 'pharmacy', among others) (Monties, 1980; Boidron *et al.*, 1988; Zamora, 2003; Vivas, 2005). The concentration of simple phenols with descriptors associated with 'burnt wood' and 'smoke' increases with increasing roasting (Moreno-Vigara & Peinado-Amores, 2010).

Ellagic tannins or ellagitannins are polymers of ellagic acid and glucose that are released into wine products by the effect of their acid or basic hydrolysis; therefore, they belong to hydrolyzable tannins. The ellagitannins contained in *Quercus* are mainly monomers such as castalagin, vescalagin, grandinin, and roburin E, but also dimers of the last (*viz.* roburins A to D). Vescalagin and castalagin are the two main ellagitannin isomers in oak used for cooperage; their partial hydrolysis, which causes the loss of hexahydroxydiphenic acid, produces vescalin and castalin, respectively (Ribéreau-Gayon *et al.*, 1999; Vivas, 2005, Fernández-Simón *et al.*, 2009). Sensorily, ellagitannins endow wine with astringency and bitterness.

Natural drying often exposes wood to rain, which has an efficient washing effect on ellagitannins (Viriot *et al.*, 1994; Vivas, 2005) not obtained with artificial drying – the latter, however, is faster and more economical (Vivas *et al.*, 1997). As a result, the bitterness and astringency of wines subjected to the influence of naturally dried wood – and then used to construct barrels or obtain chips or extracts – are lower than those of wines aged with artificially dried wood provided all other variables are identical (Fernández-Simón and Cadahía, 2007). Ellagitannins dissolve rapidly in dilute alcohol media such wines and brandies (Moutounet *et al.*, 1989), and play a prominent role in the aging of red and white wines in oak barrels by virtue of their easy oxidation (Vivas and Glories, 1993, 1996) and flavor-endowing properties (Pocock *et al.*, 1994).

Coumarins are lactones from *o*-hydroxy-*Z*-cinnamic acids with a characteristic benzopyran ring. The most abundant coumarins in oak wood are scopoletin and esculetin, which are not found in free form but as their glycosylated derivatives, scopoline and esculin, respectively, both with a strongly bitter taste. Natural wood drying not only causes hydrolysis of some compounds, but also favors development of fungi such as *Trichoderma harzianum* and *Aerobasidium pullulans*, which hydrolyze combined forms and release aglycons, thereby reducing the concentrations of bitter compounds transferable to wine (Vivas, 1995b; Zamora 2003; Fernández-Simón and Cadahía, 2007).

Other groups of hydrolyzable tannins include gallotannins, flavanols, procyanidins, and condensed tannins, all present at significant concentrations in tree bark and leaves, and also, at low concentrations, in heartwood timber; therefore, their contribution to sensory characteristics is very limited (Vivas, 2005). No information about the effect of wood treatments on these compounds has been reported. Vivas, cited by Zamora (2003), discussed their effect on the sensory profile of barrel-aged wines (Table 57.1).

Fresh wood contains different volatiles, the concentration of which is increased by roasting, which have a significant impact on the olfactory profile of wood-aged wine (Puech *et al.*, 2000; Zamora, 2003; Moreno-Vigara and Peinado-Amores, 2010). The most salient compounds among such volatiles are as follows:

(a) The *cis*- and *trans*- isomers of β-methyl-γ-octolactone or whisky-lactone, where the *cis*-isomer is the stronger odorant. These compounds are the greatest contributors to fresh wood aroma, with 'coconut' and 'wood' notes.

(b) Eugenol, a phenylpropanoid possessing 'clove' and 'spice' notes that is the second most abundant component of this fraction.

(c) Vanillin, a benzoic aldehyde with 'vanilla' as its main odor descriptor but also possessing other notes such as 'coffee', 'chocolate', and 'smoked'.

(d) Various volatiles contributing to wine aroma, particularly in American oak, including terpenes such as linalool, β-terpineol and β-ionone norisoprenoids, and β-damascenone.

The particular plant species influences the volatiles content of fresh wood. Thus, *Q. alba* (American oak) is significantly richer in whisky-lactone, eugenol, and vanillin than the French oak *Q. petraea* and, especially, *Q. robur* (Vivas, 2005). In barrels fabricated under similar conditions from these three oak species, *Q. alba* releases the greatest amounts of aroma compounds to wine (particularly *cis*-β-methyl-γ-octolactone and vanillin, two substances with strong notes of 'coconut' and 'vanilla', and small amounts of ellagic tannins). Conversely, *Q. robur* releases small amounts of aroma compounds but the greatest amount of tannins. Finally, *Q. petraea* releases intermediate amounts of tannins and aroma compounds, and leads to a more complex, balanced, smart profile with substantial notes of 'coconut', 'vanilla', and 'spice' (Masson *et al.*, 1997; Chatonnet and Dubourdieu, 1998).

As stated above, grain represents the width of growth rings in the wood. This parameter, which depends on the particular oak species and pedoclimatic conditions, dictates the chemical composition of wood and hence what it can transfer to wine. Irrespective of botanical species, fine-grained oaks release more aroma compounds and less tannins than coarse-grained oaks; the former are thus favored for winemaking purposes (Chatonnet, 1995; Vivas, 1995c).

With regard to the influence of natural and artificial wood drying on the olfactory profile of wine, the former increases the concentration of vanillin, eugenol, and the more odorant isomer of β-methyl-γ-octolactone – the *cis*-isomer (Ribéreau-Gayon *et al.*, 1999; Zamora, 2003). Increased roasting decreases the levels of β-methyl-γ-octolactones and ellagic tannins, and increases those of furans and phenols (Ribéreau-Gayon *et al.*, 1999; Zamora, 2005). Specifically, wood roasting influences the sensory characteristics of wine as follows: light roasting has the strongest impact on wine aroma, with marked 'coconut' notes and large amounts of ellagic tannins, which leads to excessive astringency and dryness in the mouth. Conversely, heavy roasting dramatically reduces aroma and introduces strong 'toast' and 'smoked' notes, in addition to excessively heavy 'roasted' notes and a weak tannic component. Finally, medium roasting provides the best balance, with 'coconut', 'vanilla', 'roasted', 'smoked', and 'spices' notes, and large enough amounts of ellagic tannins to ensure taste and color balance in the aged wine.

Experiments using oak chips to endow wines with woody character revealed the presence of substantial amounts of aroma compounds from the barrels ('vanilla', 'coconut') that increased with increasing roasting time and decreasing particle size. However, wood chips and tannin extracts should be used with caution since they release not only aroma compounds, but also other wood components such as hydrolyzable tannins, which increase astringency and bitterness to levels deteriorating wine quality (Vivas and Glories, 1996; Oliva *et al.*, 2001; Zamora, 2003).

Bertrand *et al.* (1997) examined the effects of various forms of oak (toasted splints, untoasted splints, liquid wood flavoring, chips, toasted granules, untoasted granules, commercial tannins, and shavings) on red wine and found the treated wines to have a strong woody character with little finesse or complexity, as well as 'burnt' overtones when the wood fragments had been toasted. Also, extraction of wood components was generally fast and strong because of the large contact surface.

There are two main current trends with regard to studies on wine aging in barrels. One involves thoroughly characterizing the chemical composition of the wine and the influence of aging on its sensory properties; and the other involves assessing the impact of new wood types with regard to botanical variety or geographic origin as alternatives to the traditional choices.

OTHER WAYS IN WHICH WINE COMPOSITION IS ALTERED BY AGING IN NO OAK BARRELS

The almost complete replacement of other wood types – those with a substantial weight in the Mediterranean basin included – by oak wood in the past has promoted assessment of potential alternative species such as acacia, chestnut, cherry, and mulberry to obtain wines with a personality of their own or even pharmacological use (De Rosso *et al.*, 2009; Sanz *et al.*, 2012a, b; Alañón *et al.*, 2012a). Also, other European oaks (Hungarian, Russian, German and Polish) have been used to expand available technological information with a view to improving and diversifying the production of aged wines (Alañón *et al.*, 2011).

One other present trend is the optimization of the use of chips, wooden staves, tannin powders, and liquid extracts obtained from plants of diverse botanical and geographic provenance, and roasted to a variable degree. Some authors have also studied the addition of oxygen and/or lees to the wine in wood barrels or stainless steel containers. Clearly, the aim is to obtain wines with a competitive wood aroma aimed at a market segment of modest purchasing power (García *et al.*, 2012); Del Barrio-Galán *et al.*, 2012; Gallego *et al.*, 2012; Harbertson *et al.*, 2012; Alañón *et al.*, 2012b; Schumacher *et al.*, 2013).

There is another trend for using vinegrowing residues such as prunings to obtain enologically useful extracts as proposed by Luque-Rodríguez *et al.* (2006). These authors have used vine shoots from a number of cultivars and compared the resulting extracts with those from commercial oak chips. The results with regard to phenol compounds of low molecular weight and antioxidant power were very similar. Therefore, vine shoots can be an effective alternative to commercial chips, powdered tannins and liquids extracts from oak wood (Delgado-Torre *et al.*, 2012a, b).

ANALYTICAL METHODS FOR DETERMINING THE VOLATILE FRACTION OF WINE

A number of methods currently exist for determining wine composition in general and antioxidant capacity, total antioxidant content, and the volatile fraction or individual components thereof. These methods are either (a) officially endorsed or well-established methods used in combination with chemometric approaches, or (b) methods using leading-edge analytical equipment to extract supplementary information. The two differ in their sample preparation steps.

Common wine parameters such as ethanol content, pH, total dry extract, titratable acidity, volatile acidity, total and free SO_2, optical density (OD), total phenolics, and lactic and malic acids, are usually determined with officially endorsed tools such as the European Official Methods (ECC Commission Regulation, 1990) and their results processed chemometrically. Such was the case with a study by Castellari *et al.* (2001) on the effect of the aging conditions on the previous parameters in Sangiovese red wine as determined with official analytical methods and their sensory analysis after 320 days of storage at two different temperatures in various containers and barrels from several types of wood. The results were subjected to different chemometric treatments including MANOVA to establish the influence of the aging conditions on the sensory properties of the wine (Figure 57.1A) and PCA to assess the weight of chemical parameters such as color, polymeric phenolics, gallic acid, total phenolics, tannic phenolics, total anthocyanins and (+)-catechin (Figure 57.1B).

The antioxidant capacity of wine and oak wood used to age wine can be assessed with various methods for subsequent correlation with the phenol content, namely: by measuring the scavenging capacity against a given radical such as ABTS or DPPH; the oxygen radical absorbance capacity (ORAC) or the ferric reducing antioxidant power (FRAP). Another very common way of determining total phenols is with the Folin–Ciocalteu method. These classical methods are all based on spectrophotometric measurements. Although their lack of selectivity can result in overestimated phenol contents, they afford fast, inexpensive determinations that are especially suitable for high throughput analyses. Vibrational spectroscopy is being increasingly used for quantitation of wine compounds at all stages of winemaking from grape harvesting to bottling of the wine (Villiers *et al.*, 2012).

Reflectance and color tristimulus measurements are commonly used to monitor wine aging, and so is HPLC in combination with a photometric detector (usually a diode array detector, DAD) to obtain individual information on simple and polymeric phenols, and simple and condensed tannins and anthocyanins (Mazauric and Salmon, 2006), among other compounds. Capillary electrophoresis (EC) has also been used, albeit less often, in combination with different detectors to determine common parameters such as ethanol, sulfite, inorganic and organic anions, amino acids and amines, sugars,

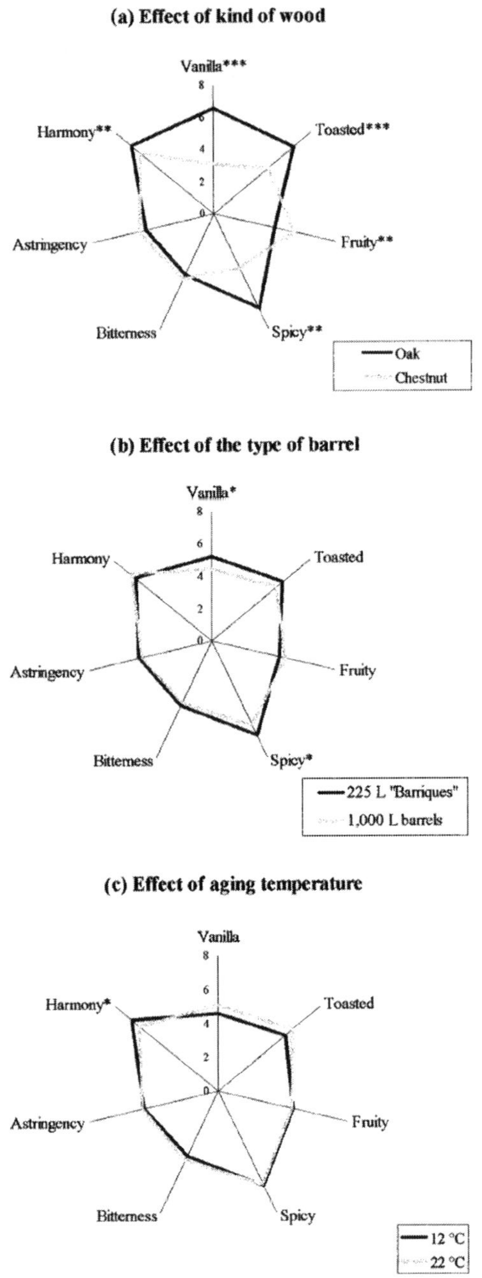

FIGURE 57.1 (A) MANOVA results for the descriptive sensory analysis of Sangiovese red wines after 320 days of aging in wooden barrels. Spider webs represent the mean values ($n=20$ for each level of the variable) for (a) wood type, (b) barrel type and (c) aging temperature. Asterisks indicate the separation of means at $p<0.05$ (*), $p<0.01$ (**) or $p<0.001$ (***). (B) PCA of the wines after 320 days of storage. Factorial weights higher than 0.70 are represented as vectors. A, color hue; B, polymeric phenolics; C, gallic acid; D, total phenolics; E, tannic phenolics; F, total anthocyanins; G, (+)-catechin. *Reproduced with permission of the American Chemical Society.*

phenols, proteins, or yeasts (Sádecká and Polonský, 2000). Volatiles in wine are most often determined by gas chromatography coupled to flame ionization detection (GC–FID) (Palomero *et al.*, 2009).

As an alternative to the classical methods described above, new, more powerful analytical equipment based mainly on mass spectrometry (MS) coupled to LC or GC is being increasingly used to analyze the aroma fraction of wine. For example, LC–MS is currently the best analytical approach to studying polyphenols in wine and also the most effective tool for structural studies of anthocyanins or the characterization of the complex structures of grape polyphenols such as procyanidins, proanthocyanidins, prodelphinidins, and tannins; in addition, LC–MS has provided experimental evidence for previously hypothesized structures. The MS/MS combination enables the characterization of anthocyanin aglycone and sugar moieties. Also, matrix-assisted laser-desorption ionization–time-of-flight (MALDI-TOF) MS affords the detection of molecules of a higher molecular weight with increased accuracy; this has been used to determine procyanidin oligomers consisting of up to seven monomers in the reflectron mode and up to nine in the linear mode. LC-MS has also been used to determine resveratrol in wine following sample preparation by solid phase microextraction (SPME) (Flamini, 2003).

In some instances, the optimum choice is using LC-DAD and LC-MS in combination. Such was the case with quantitation of anthocyanins in wine aged over lees of *Schizosaccharomyces* or *Saccharomycodes* yeasts in addition to the monitoring of wine color by UV-VIS spectrophotometry and the identification of anomalies of potential sensory importance by GC-FID (Palomero *et al.*, 2009). The need for three different techniques testifies to the high complexity of wine and its aging process. One other typical case involving the joint use of several analytical techniques in combination for improved results is the study of the changes in the composition of polyphenols and volatile compounds in red wine caused by aging in acacia, chestnut, cherry, mulberry, and oak barrels conducted by De Rosso *et al.* (2009). LC-DAD was used to monitor changes in anthocyanins (Figure 57.2) and LC/ESI-MS^n revealed the presence of substantial amounts of *cis-* and *trans-*piceatannol (a phytoalexin with antileukemia and antimelanoma activity) in mulberry-aged wine.

Volatiles in wine have traditionally been determined by GC-FID (Palomero *et al.*, 2009) but are increasingly being quantified by multidimensional gas chromatography–mass spectrometry (Ochiai and Sasamoto, 2011), occasionally further coupled to head space (HS) and TOF (HS-SPME–GC × GC–TOF-MS) (Weldegiorgis *et al.*, 2011). This has enabled the identification or more than 800 volatiles (particularly alcohols, aldehydes, ketones, acids, esters, geosmin, haloanisoles, heterocyclic compounds, phenols, norisoprenoids, terpenes, furanones and furans, lactones, norisoprenoids, isoprenoids, and thiols) in wine.

Sample preparation for determining volatiles in wine is lately being carried out by using various microtechniques but especially solid-phase microextraction (SPME)

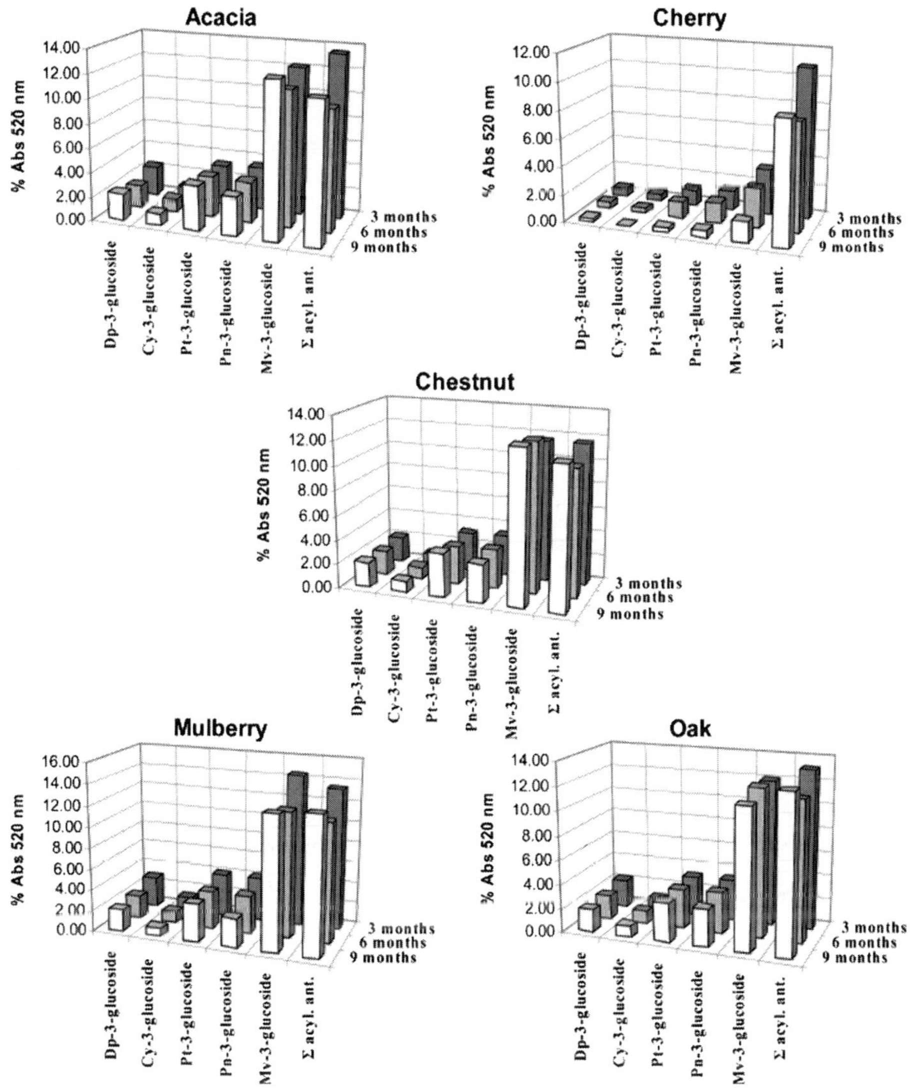

FIGURE 57.2 Anthocyanins in Raboso Piave wine aged in acacia, cherry, chestnut, mulberry and oak wood barrels. Each compound is expressed as a percentage of the combined peak areas in the LC/UV chromatograms at 520 nm. Σacyl.ant is the combination of acetate and *p*-coumarate anthocyanins. *Reproduced with permission of the American Chemical Society.*

and, to a much lesser extent, SPE, stir bar sorptive extraction (SBSE) or liquid-phase microextraction (LPME) (Jelen *et al.*, 2012).

The unstoppable advances in analytical equipment are providing new platforms for extracting more precise information about the wood components that are transferred to wine (Wamhoff and Gribble, 2012) and the effect on such components on the composition of wine prior to aging (Rodríguez-Rodríguez and Gómez-Plaza, 2012); developing sensitive, selective methods for better monitoring of wine changes during aging (Prieto *et al.*, 2012); and correlating the concentrations of wood components transferred to wine and the sensory characteristics of aged wines (Parker *et al.*, 2012).

Acknowledgments

Funding of this work by Spain's Ministry of Economy and Competitiveness and the FEDER Program through project CTQ2012-37428 is gratefully acknowledged.

References

Alanon, M., Castro-Vázquez, L., Díaz-Maroto, M., Gordon, M.H., Pérez-Coello, M., 2011. A study of the antioxidant capacity of oak wood used in wine ageing and the correlation with polyphenol composition. Food Chem. 128, 997–1002.

Alañón, M., Diaz-Maroto, M., Pérez-Coello, M., 2012a. Analysis of volatile composition of toasted and non-toasted commercial chips by GC-MS after an accelerated solvent extraction method. Internat. J. Food Sci. Technol. 47 (4), 816–826.

Alañón, M., Castro-Vázquez, L., Díaz-Maroto, M., Pérez-Coello, M., 2012b. Aromatic potential of Castanea saitva Mill. compared to Quercus species to be used in cooperage. Food Chem. 130, 875–881.

Álvarez, J.M., 2007. Tanino: la revolución enológica. Mito o realidad. Revista Enología 2, 1–15.

Bertrand, A., Barbe, J., Gazeau, O., 1997. Aromatisation de vins par différents produits elaborés à partir de bois de chêne. J. Sci. Tech. Tonn. 3, 37–47.

Boidron, J., Chatonnet, P., Pons, M., 1988. Influence du bois sur certaines substances odorantes des vins. Conn. Vigne Vin 22, 275–294.

Browning, L. (Ed.), 1967. Methods of Wood Chemistry. Intersci. Public. N.Y, London.

Castellari, M., Piermattei, B., Arfelli, G., Amati, A., 2001. Influence of aging conditions on the quality of red Sangiovese wine. J. Agric. Food Chem. 49, 3672–3676.

Chatonnet, P., Boidron, J., Pons, M., 1989. Incidence de traitement thermique du bois de chêne sur sa composition chimique. 2e partie: Evolution de certains composés volatils en fonction d l'intesité du brûlage. Conn. Vigne Vin 23, 223–250.

Chatonnet, P., 1995. Principales origines et caracteristiques des chênes destinés à l'élevage des vins. Revue Oenologues 75, 15–18.

Chatonnet, P., 2008. Productos alternativos a la crianza en barrica de los vinos. Influencia de los parámetros de fabricación y de uso. Revista Enología n° 3, 2–24.

Chatonnet, P., Dubourdieu, D., 1998. Comparative study of the characteristics of American white oak (*Quercus alba*) and European oak (*Quercus petraea* and *Q. robur*) for production of barrels used in barrel ageing of wines. Am. J. Enol. Vitic. 49, 79–85.

De Rosso, M., Panighel, A., Dalla Vedova, A., Stella, L., Flamini, R., 2009. Changes in chemical composition or a red wine aged in acacia, cherry, chesnut, mulberry and oak wood barrels. J. Agric. Food. Chem. 57, 1915–1920.

Del Barrio-Galán, R., Pérez-Mararino, S., Ortega-Heras, M., 2012. Effect of the aging on lees and other alternative techniques on the low molecular weight phenols of Tempranillo red wine aged in oak barrels. Anal. Chim. Acta 732, 53–63.

Delgado-Torre, M.P., Priego-Capote, F., Luque de Castro, M., 2012a. Evaluation of the composition of vine shoots and oak chips for oenological purposes by superheated liquid extraction and high-resolution liquid chromatography-time-of-flight/mass spectrometry analysis. J. Agric. Food. Chem. 60, 3409–3417.

Delgado-Torre, M.P., Ferreiro-Vera, C., Priego-Capote, F., Pérez-Juan, P., Luque de Castro, M., 2012b. Comparison of accelerated methods for the extraction of phenolic compounds from different vine-shoot cultivars. J. Agric. Food. Chem. 60, 3051–3060.

ECC Commission Regulation VO. 2676/90, 1990. Concerning the establishment of common analytical methods in the sector of wine. Off. J. Eur. Communities Legis. October 3, No. L 272, pp. 1–192.

European Commission, 2009. Reglamento (CE) N° 606/2009 de la Comisión de 10 de julio de 2009. DOCE L193, 1.

European Commission, 2010. Art. 1, 1) Reglamento (UE) N° 401/2010 de la Comisión de 7 de mayo que modifica y corrige el Reglamento (CE) n° 607/2009. DOCE L117, 13.

Fengel, D., Wegener, G. (Eds.), 1984. Wood Chemistry, Ultraestructure Reaction. Walter de Gruytier, Berlin.

Fernández-Simón, B., Cadahía, E., 2007. Tratamiento de la madera de roble para tonelería. Revista Enología n° 4. 21–18.

Fernández-Simón, B., Cadahía, E., Sanz, M., Poveda, P., 2009. Comportamiento de Quercus pyrenaica durante los procesos de tonelería. *5° Congreso forestal español*. Ávila: Sociedad Española de Ciencias Forestales y Junta de Castilla y León.

Flamini, R., 2003. Mass spectrometry in grape and wine chemistry. Part I: Polyphenols. Mass Spectrom. Rev. 22, 218–250.

Flanzy, C., 2000. En *Enología: Fundamentos Científicos y Tecnológicos*. AMV and Mundi Prensa, Madrid, Spain.

Gallego, L., Del Álamo, M., Nevres, I., Fernández, J., De Simón, B., Cadahia, E., 2012. Phenolic compounds and sensorial characterization of wines aged with alternative to barrel products made of Spanish oak wood (*Quercus pyrenaica* Willd.). Food Sci.Technol. Internat. 18 (2), 151–165.

García, R., Soares, B., Dias, C., Freitas, A., Cabrita, M., 2012. Phenolic and furanic compounds of Portuguese chesnut and French, American and Portuguese oak wood chips. Eur. Food Res.Technol. 235 (3), 457–467.

González-Rodríguez, J., Pérez-Juan, P., Luque de Castro, M.D., 2003. Extraction of wood compounds by use of subcritical fluids. Chromatographia 57, 363–368.

Harbertson, J., Parpinello, G., Heymann, H., Downey, M., 2012. Impact of exogenous tannin additions on wine chemistry and wine sensory character. Food Chem. 131 (3), 999–1008.

Hidalgo, L. (Ed.), 2003. Tratado de Enología, Vol II. Mundi Prensa, Madrid, Spain.

Jelen, H.H., Majcher, M., Dziadas, M., 2012. Microextraction techniques in the analysis of food flavor compounds: A review. Anal. Chim. Acta 738, 13–26.

Luque-Rodríguez, J.M., Pérez-Juan, P., Luque de Castro, M.D., 2006. Extraction of polyphenols from vine shoots of *Vitis vinifera* by superheated ethanol–water mixtures. J. Agric. Food Chem. 54, 8775–8781.

Masson, G., Guichard, E., Fournier, N., Puech, J., 1997. Teneurs en stéréo-isomeres de la beta-metil-gamma-octolactone des bois de chêne européens et américains. Application aux vins et aux eaux-de-vie. J. Sci. Tech. Tonn. 3, 1–8.

Mazauric, J.P., Salmon, J.M., 2006. Interactions between Yeast Lees and Wine Polyphenols during Simulation of Wine Aging: II. Analysis of Desorbed Polyphenol Compounds from Yeast Lees. J. Agric. Food Chem. 54, 381–387.

Monties, B., 1980. Les lignines. In: Monties, B. (Ed.), Les Polymères Végétaux. Polymères Pariétaux et Alimentaires non Azotés. Bordas, Paris, pp. 122–155.

Moreno-Vigara, J.J., Peinado-Amores, R.A., 2010. Crianza Química (pp. 463–477. AMV Ediciones and Mundi Prensa, Madrid, Spain.

Moutounet, M., Rabier, P., Puech, J., Verette, E., Barillère, J., 1989. Analysis by HPLC of extractable substances in oak wood. Application to a Chardonnay wine. Sci. Aliments 9, 35–51.

Ochiai, N., Sasamoto, K., 2011. Selectable one-dimensional or two-dimensional gas chromatography–olfactometry/mass spectrometry with preparative fraction collection for analysis of ultra-trace amounts of odor compounds. J. Chromatogr. A 1218, 3180–3185.

OIV, 2012. Codex Oenologique International. OIV, Paris.

Oliva, J., Azorín, P., Cámara, M., Barba, A., Pardo, F., 2001. Incidencia de la adición de distintos tipos de taninos enológicos en el color de vinos tintos de Monastrell. Alimentación, equipos y tecnología 20, 87–92.

Palomero, F., Morata, A., Benito, S., Calderón, F., Suárez-Lepe, J.A., 2009. New genera of yeasts for over-lees aging of red wine. Food Chem. 112, 432–441.

Parker, M., Osidacz, P., Baldock, G., Hayasaka, Y., Black, C., Pardon, K., et al., 2012. Contribution of several volatile phenols and their glycoconjugates to smoke-related sensory properties of red wine. J. Agric. Food Chem. 60 (10), 2629–2637.

Pocock, K., Sefton, M., Williams, P., 1994. Taste thresholds of phenolic extrac of french and amarican oackwood: The influence of oak phenols on wine flavor. Am. J. Enol. Viticult. 45 (4), 429–434.

Prieto, N., Rodríguez-Méndez, M., Leardi, R., Oliveri, P., Hernando-Esquisabel, D., Iniquez-Crespo, M., 2012. Application of multi-way analysis to UV-visible spectroscopy, gas chromatography and electronic nose data for wine ageing evaluation. Anal. Chim. Acta 719, 43–51.

Puech, J.L., Feuillat, M., Boulet, J.C., Feuillat, F., Keller, R., Masson, G., Moutounet, M., Naudin, R., Peyron, D., 2000. Crianza de los vinos. In: Flanzy, C.E. (Ed.), Enología: Fundamentos Científicos y Tecnológicos. AMV and Mundi Prensa, Madrid, Spain, pp. 608–638.

Ribéreau-Gayon, J., Peynaud, E., Ribéreau-Gayon, P., Sudraud, P., 1976. Vieillissement des Vins. Vinifications. Transformations du vin. Sciences et Techniques du Vin, vol. 3., France. Dunod, Paris.

Ribéreau-Gayon, P., Glories, Y., Maujean, A., Dubourdieu, D., 1999. Aging Ted Wines in Vat and Barrel. Phenomena Occurring During Aging. In: Ribéreau-Gayon, P., Glories, Y., Maujean, A., Dubourdieu, D. (Eds.), Handbook of Enology. The Chemistry of Wine Stabilization and Treatments, vol. 2.. U.K. John Wiley & Sons, Ltd, Chichester, pp. 352–391.

Rodríguez-Rodríguez, P., Gómez-Plaza, E., 2012. Dependence of oak-related volatile compounds on the physicochemical characteristics of barrel-aged wines. Food Technol. Biotechnol. 50 (1), 59–65.

Sádecká, J., Polonský, J., 2000. Electrophoretic methods in the analysis of beverages. J. Chromatogr. A 880, 243–279.

Sanz, M., Fernández de Simón, B., Esteruelas, E., Muñoz, A., Cadahia, E., Hernandez, M., et al., 2012a. Polyphenols in red wine aged in acacia (*Robinia pseudoacacia*) and oak (*Quercus petraea*) wood barrels. Anal. Chim. Acta 732, 83–90.

Sanz, M., Simón, B., Cadahia, E., Esteruelas, E., Muñoz, A., Hernández, M., et al., 2012b. Polyphenolic profile as a useful tool to identify the wood used in wine aging. Anal. Chim. Acta 732, 33–45.

Schumacher, R., Alañón, M., Castro-Vázquez, L., Pérez-Coello, M., Díaz-Maroto, C., 2013. Evaluation of oak chips treatment on volatile composition and sensory characteristics of Merlot wine. J. Food Quality ISSN, 1745–4557.

Villiers, A., Albertsa, P., Tredouxb, A.G.J., Nieuwoudt, H.H., 2012. Analytical techniques for wine analysis: An African perspective; a review. Anal. Chim. Acta 730, 2–23.

Viriot, C., Scalbert, A., Hervé Du Penhoat, C.L.M., Moutounet, M., 1994. Ellagitannins in woods of Sessile oak and sweet chestnut. Dimerization and hydrolysis during wood ageing. Phytochem. 36, 1253–1260.

Vivas, N., 1995a. La qualité du bois de chêne et son utilisation pour la vinification et l'elevage des vins. J. Sci. Tech. Tonn. 1, 1–8.

Vivas, N., 1995b. Fenómenos ligados a la maduración de la madera de roble durante su secado. X Cursos Rioja, 97–109. Haro, Spain. Ayuntamiento de Haro.

Vivas, N., 1995c. Sur la notion de grain en tonnellerie. J. Sci. Tech. Tonn. 1, 17–32.

Vivas, N., 1999. Influencia de la duración de la utilización de las barricas sobre sus aportaciones a los vinos. Tecnología de la elaboración de vinos tintos e innovaciones en la estabilización de los vinos, 6–24. Firavi. Vilafranca del Penedés, Spain.

Vivas, N., 2005. Manual de tonelería. Mundi Prensa, Madrid, Spain.

Vivas, N., Glories, Y., 1993. Les phénomènes d'oxydoréduction liés à l'élevage en barrique des vins rouges: aspects technologiques. Rev. Fr. Oenol. 142, 33–38.

Vivas, N., Glories, Y., 1996. Role of oak wood ellagitannins in the oxidation process of red wines during ageing. Am. J. Enol. Vitic. 47, 103–107.

Vivas, N., Glories, Y., Doneche, B., Gueho, E., 1991. Observation sur la flore fongique du bois de chêne au cours de son sechage naturel. Ann. Sc. Nat. Bot. 13, 149–153.

Vivas, N., Saint-Cricq, N., Doneche, B., Glories, Y., 1997. The duration effect of natural seasoning of *Quercus petraea* Liebl. and *Quercus robur* L. on the diversity of existing fungiflora and some aspect of its ecology. J. Sci. Tech. Tonnell 3, 27–35.

Wamhoff, H., Gribble, G., 2012. Wine Heterocycles. Advances Heterocyclic Chem. 106, 185–225.

Weldegiorgis, B.T., Villiers, A., McNeish, C., Seethapathy, S., Mostafa, A., Górecki, T., Crouch, A.M., 2011. Characterisation of volatile components of Pinotage wines using comprehensive two-dimensional gas chromatography coupled to time-of-flight mass spectrometry (GC x GC-TOFMS). Food Chem. 129, 188–195.

Zamora, F., 2003. Influencia de la crianza sobre las características organolépticas del vino. In: F. Zamora, *Elaboración y Crianza del Vino Tinto: Aspectos Científicos y Prácticos*. AMV Ediciones and Mundi Prensa, Madrid, Spain.

Zamora, F., 2005. Los aromas que el roble aporta el vino. Influencia del grado de tostado de las duelas. Enólogos 35, 42–45.

CHAPTER

58

Influence of Diammonium Phosphate Addition to Fermentation on Wine Biologicals

Mar Vilanova, Isak S. Pretorius*†*, Paul A. Henschke*‡*

*Misión Biológica de Galicia, CSIC, Pontevedra, Spain, †Macquarie University, Sydney, Australia, ‡The Australian Wine Research Institute, Glen Osmond, Australia

CHAPTER POINTS

- Nitrogen status of grape vines impacts on must composition and therefore contributes to wine quality.
- Supplementing grape must with inorganic nitrogen (diammonium phosphate, DAP) has a number of consequences for resultant wine quality. The principal volatile compounds affected are derived from yeast sugar metabolism, and include higher alcohols, ethyl and acetate esters, and volatile sulfur compounds. The non-volatile compounds glycerol, acidity, and phenolics are also modified by yeast metabolism, and together with volatiles affect overall wine quality.
- In general, musts with low available nitrogen result in lower production of yeast aroma compounds, with the exception of the higher alcohols, to give a low aromatic intensity of more complex but less fruity attributes. Such musts are also at greater risk of slow or stuck fermentation and wines acquiring 'reductive' attributes.
- When nitrogen is added, it increases the balance of fruity and floral esters with respect to higher alcohols. Large additions of nitrogen should only be made with great caution; when inorganic nitrogen (DAP) is used there is an increased risk of high residual phosphate and a large risk of excessive acetate production.

INTRODUCTION

Nitrogen is the most abundant soil-derived macronutrient in grapevine, and plays a major role in many of the biological functions of grapevine and fermentative microorganisms, including wine yeast and malolactic bacteria (Bell and Henschke, 2005). Nevertheless, in many viticultural regions across the world, limited soil nitrogen availability constrains grapevine development, which also restricts accumulation of nitrogen in the grape berry. Consequently, the variable abundance of this nutrient is associated with a variety of complex biological processes in the vineyard and in the winery, which ultimately affects wine flavor and quality.

Grape must is the principle source of nutrients, including nitrogen, required to support the growth of yeast needed to carry out the fermentation of grape sugars to produce alcohol and carbon dioxide, as well as flavor-active metabolites. Nitrogen is present in grapes and must in a wide variety of chemical forms but only inorganic (ammonia-N) and organic (alpha-amino) nitrogen provide yeast with assimilable forms of nitrogen, commonly referred to as yeast-assimilable nitrogen (YAN) (Figure 58.1).

Nitrogen-deficient musts restrict the growth of fermentative microorganisms, which can negatively impact on wine production and quality. Low yeast biomass leads to low catabolic rate, which is evident as a slow or incomplete ('stuck') fermentation. Insufficient must YAN induces metabolic stress, which is also associated with undesirable hydrogen sulfide (H_2S) production. An appropriate YAN concentration promotes good alcoholic fermentation, which is associated with good quality wine.

Processing and Impact on Active Components in Food
http://dx.doi.org/10.1016/B978-0-12-404699-3.00058-5

$YAN = FAN + NH_3\text{-}N$

YAN: Yeast assimilable nitrogen
FAN: Free amino nitrogen
NH_3-N: Ammonia nitrogen

FIGURE 58.1 Yeast assimilable nitrogen.

Two major options for correcting nitrogen deficiency are available: application of nitrogen in the vineyard or in the winery (Bell and Henschke, 2005). This chapter will only consider the more practical, immediate, and flexible strategy, which entails supplementing grape musts with ammonium salts, most commonly diammonium phosphate (DAP). Wine law in most regions permits DAP addition to musts but usually restricts maximum addition. DAP is a powerful fermentation stimulant, which provides a widely adopted strategy for ameliorating the negative effects of nitrogen deficiency. Nevertheless, despite its wide application, the impact that DAP supplementation has on wine flavor composition has only received limited attention, especially perceived sensory effects (Bell and Henschke, 2005). Several studies, including a number associated with our laboratories, have shown that addition of inorganic nitrogen to fermentation affects a wide range of volatile and non-volatile compounds in wine (Bell and Henschke, 2005; Carrau *et al.*, 2008; Hernández-Orte *et al.*, 2006; Mendes-Ferreira *et al.*, 2010; Rapp and Versini, 1996; Torrea *et al.*, 2011; Ugliano *et al.*, 2007, 2010; Vilanova *et al.*, 2007, 2012).

General aspects of the various roles of yeast in wine flavor development have been recently reviewed by Bell and Henschke (2005), Bartowsky and Pretorius (2008), Ugliano and Henschke (2009), and Pretorius *et al.* (2012). This chapter describes the main changes that are associated with must nitrogen and the effect that DAP addition to white and red must has on wine composition.

HOW COMPOSITION IS ALTERED

Nitrogen in the Must: Implications for Fermentation and Wine Composition

Freshly prepared grape must typically contains all of the essential nutrients required for yeast growth and complete fermentation of sugars. The YAN component, which is distributed across the skin, pulp, and seeds, is, after crushing, largely released in soluble form to contribute to the nutrient content of grape must (Bell and Henschke, 2005). A variety of grape harvest and must-processing practices can lead to YAN loss, which can impact on fermentation, as summarized by Bell and Henschke (2005).

Although indigenous yeasts can be encouraged to carry out fermentation, inoculation with a selected strain of *Saccharomyces cerevisiae* is widely used to promote the fermentation of sugar to ethanol, carbon dioxide, and flavor-active secondary compounds (Pretorius *et al.*, 2012). Because the amount of yeast, either indigenous or inoculated, is too low for active fermentation, a period of rapid growth is needed to build up sufficient biomass to produce a high rate of fermentation and it is this step that depends on nutrient content. Furthermore, metabolite formation depends on the extent of growth.

When all other nutrients are present in adequate amounts, yeast growth and fermentation rate responds primarily to YAN availability (Salmon, 1989; Bely *et al.*, 1990). YAN is accumulated early in the growth phase, a proportion being incorporated directly into protein, some being degraded to synthesize amino acids, nucleic acids, and other nitrogenous metabolites as required, and any surplus is stored in the vacuole. Yeast can accumulate approximately 400 mg N/L in nitrogen-rich musts whereas the threshold concentration for low-risk fermentation has been estimated to be around 140 mg N/L, as established by a variety of experimental approaches (Bell and Henschke, 2005; Jiranek *et al.*, 1995a). Below this threshold value, yeast biomass is low and results in a slow rate of fermentation, which may not proceed to completion.

For several decades, it has been observed that wines produced from grapes having higher YAN, or from musts supplemented with nitrogen to increase YAN, have higher concentrations of fruity and floral esters and lower concentrations of less desirable compounds like higher alcohols and hydrogen sulfide. The higher ratio of desirable to undesirable compounds has provided, at least in part, an explanation for these observations (Henschke and Jiranek, 1993; Rapp and Versini, 1996).

Grape must contains a variety of nitrogen sources of which the most important for yeast growth are ammonium, primary amino acids, and small peptides. Nitrogen status of the vine and grape must directly or indirectly affect numerous primary and secondary grape and yeast metabolites, such as alcohols, acids, esters, carbonyls, volatile sulfur compounds, terpenes, phenols, and other compounds (Rapp and Versini, 1996; Bell and Henschke, 2005). Moreover, YAN also indirectly affects the metabolism of grape compounds, which form the basic composition of wine and its varietal character. Therefore, in order to optimize fermentation for producing the desired flavor profile it is necessary to measure YAN. Figure 58.2 illustrates the influence of the nitrogen status of the must and effects on wine quality.

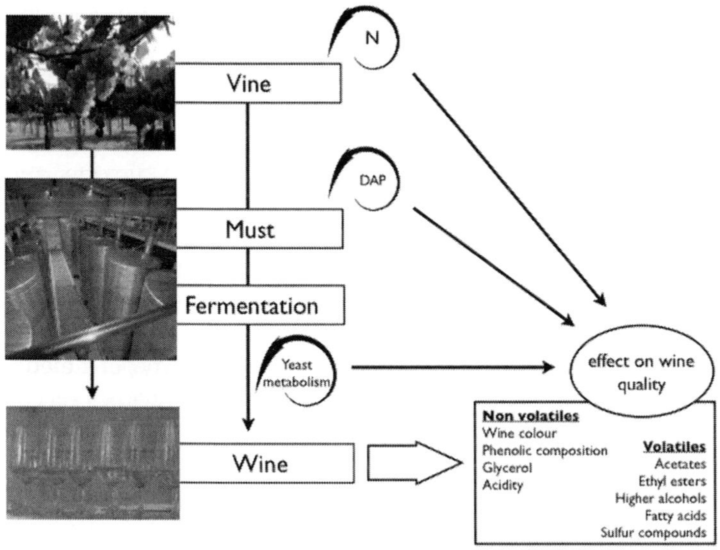

FIGURE 58.2 Scheme of principal sources of nitrogen and impact on wine quality.

Influence of DAP Addition to Must on Wine Quality

Since the earliest writings on the science of wine fermentation, yeast type and fermentation conditions have been regarded as playing an important role in wine quality (Pretorius *et al.*, 2012). However, it has proven problematic to associate must nutrients with wine flavor constituents until the last decade or so because no facile assay methods were available (Henschke and Jiranek, 1993). Nevertheless, through vineyard and must supplementation studies an association between the nitrogen status of must and the properties of wine emerged. The yeast volatile metabolites including esters, higher alcohols, acids, carbonyls, and sulfur compounds, which were found to vary according to must nitrogen status, were considered to play a role in the nitrogen associated changes in wine flavor (Henschke and Jiranek, 1993; Rapp and Versini, 1996). In particular, several compelling associations between, for example, certain amino acids and higher alcohols and nitrogen availability and hydrogen sulfide production have served to underpin this hypothesis (Rapp and Versini, 1996; Vos and Gray, 1979). Therefore, it is surprising that although DAP has been extensively used in winemaking for some four decades, only relatively recently have systematic investigations into sensory effects and their chemical bases been undertaken (Bell and Henschke, 2005).

Grape must typically contains a lower ratio of ammonium-N:amino acid-N, ranging from 1:1–3, which depends on various viticultural factors, especially cultivar and soil nitrogen (Bell and Henschke, 2005). Addition of DAP therefore increases the ratio and, for a low-nitrogen must, this ratio can exceed 3:1. This large variation in type of nitrogen source, independent of concentration, has important consequences for yeast metabolism. In essence, ammonium-rich media requires the synthesis of organic nitrogenous metabolites (especially amino acids and nucleic acid bases) from carbon precursors, derived from glycolysis and tricarboxylic acid pathways, and inorganic nitrogen via transamination reactions. Growth and fermentation rates also tend to be lower on ammonium compared to mixed amino acids of equivalent nitrogen content, with single amino acids giving intermediate to low rates depending on the amino acid supplied (Albers *et al.*, 1996; Beltran *et al.*, 2004; Gutiérrez *et al.*, 2012; Henschke and Jiranek, 1993; Mendes-Ferreira *et al.*, 2010; Torrea *et al.*, 2011). Moreover, a different profile of yeast metabolites are produced during growth on ammonium-N, which involves amino acids biosynthetic reactions when compared to growth on a mixture of amino acids, such as contained in grape must, which involves a greater proportion of amino acid degradation reactions.

There is limited systematic information on the relative impacts of inorganic (e.g., DAP) *versus* organic nitrogen (mixtures of amino acids) on volatile and non-volatile yeast and grape-derived metabolites under winemaking conditions. Most investigations have focused on nitrogen concentration, generally achieved by supplementing low-nitrogen, clarified grape must, or synthetic medium with an ammonium salt (Carrau *et al.*, 2008; Torrea *et al.*, 2011, Vilanova *et al.*, 2007, 2012) or using high solids, red must supplemented with DAP (Ugliano *et al.*, 2007; Ugliano and Henschke, 2009).

Most studies have investigated the impact of initial YAN concentration on fermentation metabolite formation, achieved by adding different amounts of DAP-N before inoculation. This experimental approach results in the control or reference medium, which is composed of predominantly amino acids, therefore having low ammonium-N:amino acid-N, becoming progressively high in ammonium-N : amino acid-N ratio. This experimental approach nevertheless best models practical winemaking in which DAP is typically added to a grape must before inoculation.

In general, YAN is found to modulate the concentration of many grape and yeast metabolites, which are likely to contribute to the wine properties. Of the major wine compounds, ethanol, glycerol, organic acids, sulfur dioxide, and phenolics are affected, as are the yeast volatile compounds higher alcohols, fatty acids, acetates and ethyl esters, carbonyls, and volatile sulfur compounds (Figures 58.3 and 58.4). The grape-derived aromatic metabolites, terpenes, norisoprenoids, and polyfunctional thiols, and the flavor metabolites, phenolics, are also affected but generally to a lesser extent. The trends in metabolites variability in response to DAP addition are complex, which reflects the multitude of metabolic and regulatory pathways that are involved. Responses of the more important metabolites to DAP supplementation follow.

Ethanol and Glycerol

Reduction in ethanol yield in response to DAP supplementation of model and authentic grape juice is variable and relatively small suggesting that DAP supplementation is not a reliable or effective strategy for lowering ethanol content of wine (Ugliano and Henschke, 2009). Ethanol yield decreased by up to 10% in synthetic media in which ammonium-N rather than amino acid-N served as sole nitrogen source (Albers *et al.*, 1996). This reduction is attributed to increased glycerol production, which

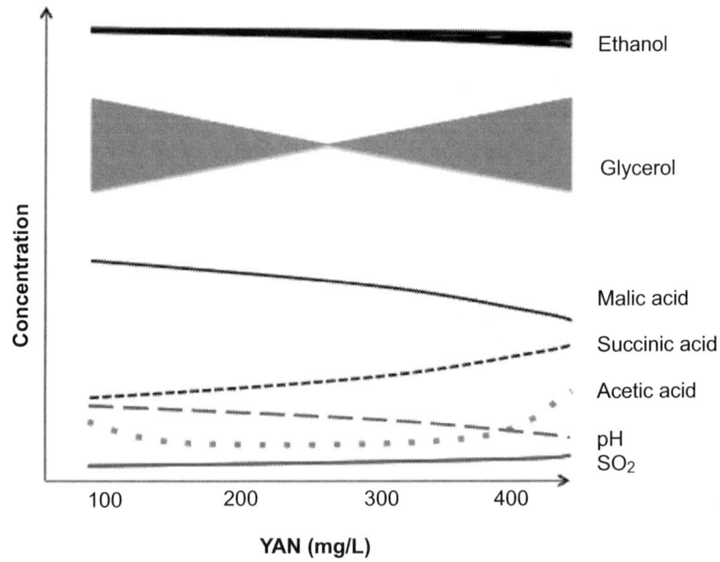

FIGURE 58.3 Effect of must YAN (DAP supplementation) on major products of yeast sugar fermentation and sulfur assimilation. *Data from Ugliano* et al. *(2007).*

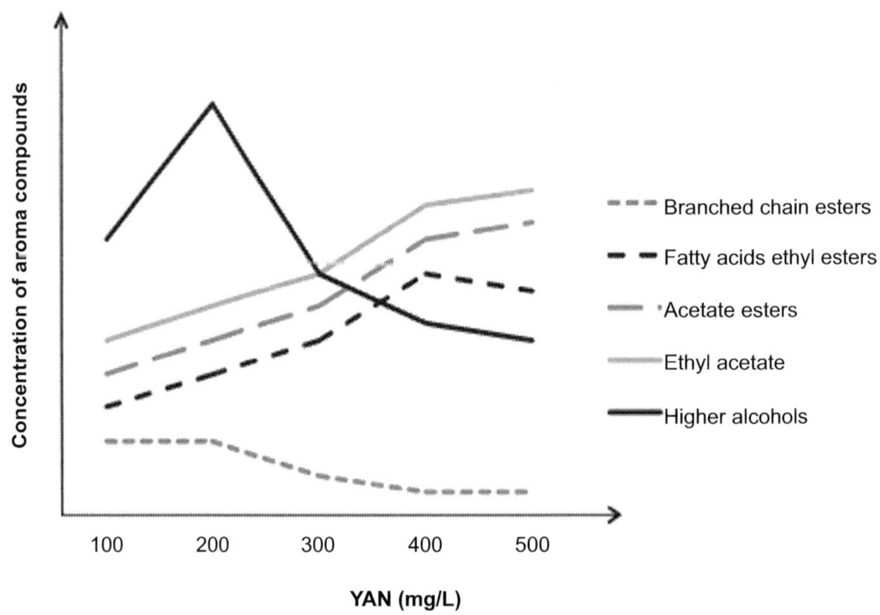

FIGURE 58.4 Effect of must YAN on production of yeast aroma compounds. *Data from Ugliano* et al. *(2007).*

results from the oxidation of NAD(P)H arising from the need to synthesize each amino acid from simple sugars and inorganic nitrogen. However, in real and model grape juices ethanol yields are highly variable, ranging from no change to 0.7% (v/v) reduction (Arias-Gil *et al.*, 2007; Hernández-Orte *et al.*, 2006; Torrea *et al.*, 2011; Vilanova *et al.*, 2007, 2012).

Compared to ethanol, glycerol production varies widely in response to DAP supplementation (Figure 58.2) (Ugliano *et al.*, 2007; Torrea *et al.*, 2011; Vilanova *et al.*, 2007, 2012). Glycerol production varies to the greatest extent at low and high YAN concentrations but is also affected by other factors, including yeast strain and must composition, such as variable osmotic stress imposed by grape maturity or grape solids. These findings suggest that DAP supplementation can not be used reliably to modulate wine glycerol content without knowledge of yeast characteristics and must composition.

Organic Acids

Wine contains a relatively high concentration of organic acids resulting from the high tartaric and malic acid content of grapes. Fermentation yeasts are unable to metabolize tartaric acid but degrade a small proportion of L-malate (Torija *et al.*, 2003). Some yeast can form malic acid in low YAN musts whereas degradation is enhanced by higher YAN (Bell and Henschke, 2005) (Figure 58.2). The principal non-volatile acid, succinic acid, is produced by yeast in variable concentrations in response to DAP addition but can be produced in abnormal amounts from certain amino acids, especially γ-amino butryric acid (Albers *et al.*, 1996; Beltran *et al.*, 2004; Torrea *et al.*, 2011; Vilanova *et al.*, 2012).

The complex effects that DAP supplementation has on organic acid metabolism is reflected in variability of wine pH and titratable acidity (Figure 58.3), but in general these parameters tend to decrease in response to DAP but not to amino acid supplementation (Arias-Gil *et al.*, 2007; Torija *et al.*, 2003; Ugliano *et al.*, 2010; Vilanova *et al.*, 2007, 2012). The extent of changes depends on the amount of DAP added, the acids present and their buffer capacity, but also on the strain of yeast, must composition and fermentation conditions. One disadvantage of DAP as a supplement is the acidification that can result, leading to lower than expected wine pH, due to differences in proton flux associated with ammonium cation compared to amino acid uptake and phosphate anion accumulation.

Grape Phenolics

Phenolic compounds are extracted and modified from grapes during must processing and fermentation. Yeast interactions with phenolics compounds during fermentation are complex and include promoting the formation of stable pyroanthocyanin pigments by reaction of anthocyanins with yeast carbonyl compounds and formation of ethyl-bridge linked polymeric pigments by condensation reactions between anthocyanins and flavanol-3-ols or proanthocyanidins with acetaldehyde (Ugliano and Henschke, 2009). These reactions can, however, be limited by either added or yeast-produced sulfite, which forms adducts with carbonyl compounds. Yeast can also decarboxylate hydroxycinnamic acids to form vinyl derivatives, which can promote formation of pyroanthocyanin pigments.

There is limited information on how nitrogen supplementation affects yeast-phenolic interactions in red wine fermentation carried out by maceration on grape solids. In one study DAP supplementation resulted in wines with higher concentrations of malvidin-3-glucoside, higher color intensity, and altered color tonality (hue) but no change in polymeric pigments or tannins. The mechanism favoring higher anthocyanins content and color intensity by DAP addition was not studied but might involve earlier extraction of pigments and their protection against degradation by greater fermentation vigor (Ugliano *et al.*, 2007).

Higher Alcohols

Higher alcohols or fusel alcohols are quantitatively major volatile by-products of fermentation and are thought to contribute to the aromatic complexity of wine; higher concentrations negatively impact on wine aroma by contributing harsh aroma and taste. They also appear to play a role in varietal character (Bell and Henschke 2005; Torrea *et al.*, 2011).

Higher alcohols are derived primarily from sugar-carbon during the biosynthesis of branched-chain amino acids, with a smaller proportion produced during the degradation of corresponding amino acids. In red wines higher alcohol concentration is largely determined by grape solids whereas in clarified musts nitrogen is an important regulator of their production. An inverse relationship between nitrogen and higher alcohols production is observed except in very low YAN juice, in which there is a direct relationship (Figure 58.4). The relationship between higher alcohols and DAP-N depends on various factors, including yeast strain and fermentation conditions (Bell and Henschke 2005; Carrau *et al.*, 2008; Vilanova *et al.*, 2007). No clear relationship has been shown between the composition of amino acids in a juice and the concentration of corresponding higher alcohols; for example, the concentration of leucine does not determine the formation of 3-methylbutanol, the corresponding alcohol.

Volatile Fatty Acids

Apart from acetic acid, the role of short to medium chain and the branched chain fatty acids in wine aroma is not well understood (Ugliano and Henschke, 2009). Acetic acid, being the major contributor to volatile acidity, is generally not considered desirable. Under nitrogen stress conditions, such as those imposed by very low or very high YAN acetic acid increases sharply (Figure 58.3). Excessive DAP addition, compared to that of equivalent amino acid nitrogen, strongly induces acetic acid formation (Arias-Gil *et al.*, 2007; Torrea *et al.*, 2011; Vilanova *et al.*, 2007).

The medium-chain fatty acids, which are only produced in relatively low concentrations, are associated with complex flavor characteristics. Like acetic acid their production is lowest at moderate YAN. Conversely, the branched-chain acids are probably sensorially negative, contributing 'yeasty', 'cheesy' and 'sweaty' aromas (Torrea *et al.*, 2011; Ugliano *et al.*, 2010). These acids are formed by the oxidation of corresponding ketoacids derived from branched chain amino acid biosynthesis. Their formation is inversely related to DAP nitrogen (Figure 58.4) (Carrau *et al.*, 2008; Ugliano *et al.*, 2008; Vilanova *et al.*, 2007)

Acetate and Fatty Acids Ethyl Esters

Esters represent the largest and most important group of aroma compounds produced by yeast during fermentation contributing 'fruity' characteristics (Ugliano and Henschke, 2009). Two major groups of esters are formed, these are the acetate esters and the ethyl esters of fatty acids. The acetate esters tend to produce estery and fruity aromas whereas the fatty acids ethyl esters, which give more fruity and floral aromas, are considered relatively more desirable.

DAP supplementation is an extremely powerful tool for modulating the production of fermentation esters (Torrea *et al.*, 2011; Ugliano *et al.*, 2010; Vilanova *et al.*, 2007). In general, the fatty acids ethyl esters, as well as the acetate esters, including ethyl acetate, increase, though at different rates, when DAP is added to must prior to alcoholic fermentation (Figure 58.4). This can have positive implications for wine flavor because both the fatty acids ethyl esters and acetates are largely responsible for the fruity character of wine. Conversely, the ethyl esters of the branched-chain fatty acids tend to decrease with DAP supplementation (Hernández-Orte *et al.*, 2006; Vilanova *et al.*, 2007). However, ethyl acetate, when present at very high concentrations, decreases the wine quality. Compared to amino acid-N equivalent DAP-N more strongly induces ethyl acetate production (Torrea *et al.*, 2011). However, while DAP nitrogen can be used to strongly modulate ester production there is considerable variability amongst yeast strains as to their ester response to nitrogen (Miller *et al.*, 2007; Ugliano *et al.*, 2010).

Sulfur Compounds

The volatile sulfur compounds produced by yeast are generally considered negative to wine quality (Mestres *et al.*, 2000; Rauhut, 2008). Must nitrogen in part regulates their formation. Of the sulfur off-flavors produced in fermentation, several sulfides, such as hydrogen sulfide (H_2S) and dimethyl sulfide (DMS) are commonly found in wine as are the thiols methyl mercaptan, which has a negative odor (Mestres *et al.*, 2000; Ugliano and Henschke, 2009).

Hydrogen sulfide is formed as a metabolic intermediate, in the reduction of the inorganic sulfur compounds, sulfate, and sulfite, in the biosynthesis of the sulfur amino acids cysteine and methionine. Physiological regulation of H_2S production in wine fermentation is not entirely understood; however, nitrogen limitation is a commonly observed factor, whereby insufficient nitrogen prevents the sequestration of H_2S, which is formed during active growth when amino acid biosynthetic demand is high (Vos and Gray, 1979; Jiranek *et al.*, 1995b). DAP is widely used to reduce the risk of H_2S formation but is not always effective since many other factors can be involved, including other nutrient deficiencies and strain genetic differences (Mendes-Ferrera *et al.*, 2010; Ugliano and Henschke, 2009). Conversely, DAP can lead to increased SO_2 production (Figure 58.3), which can affect induction of malolactic fermentation.

Whereas DAP is added to lower the risk of residual H_2S in wine, it actually increases the formation of thiols and sulfides, such as methyl mercaptan and dimethyl sulfide (Figure 58.5). H_2S, thiols, and sulfides can lead to undesirable 'reductive' odors in wine whereas dimethyl sulfide is considered desirable in some wine types (Ugliano and Henschke, 2009).

DAP Addition to White Musts: Chardonnay and Albariño

Albariño is an aromatic white grape variety, which requires fermentation to reveal its varietal potential (Vilanova *et al.*, 2012). Supplementation of *Albariño* musts with inorganic nitrogen significantly affected wine composition (Vilanova *et al.*, 2012). In general, glycoside precursors of volatile varietal compounds, ethyl esters, volatile fatty acids and higher alcohols responded to DAP supplementation, whereas total free varietal compounds and acetate esters were relatively unresponsive to YAN concentration (Figure 58.6). However, several individual free varietal compounds were responsive to DAP-N, revealing their more complex metabolism. Wines made with moderate nitrogen supplementation presented potentially greater ethyl esters and C_{13}-norisoprenoids (Vilanova *et al.*, 2012).

The aroma profile of non-floral varieties, such as Chardonnay, depends largely on grape C_{13}-norisoprenoids and yeast-derived aroma compounds. Chardonnay is a good example in which to study yeast contribution to wine aroma (Francis and Newton, 2005). Supplementation of clarified, low-nitrogen Chardonnay musts with DAP, were characterized by increased total ester and reduced higher alcohol content compared to wines made from unsupplemented juice (Figure 58.6) (Torrea *et al.*, 2011). Wine made

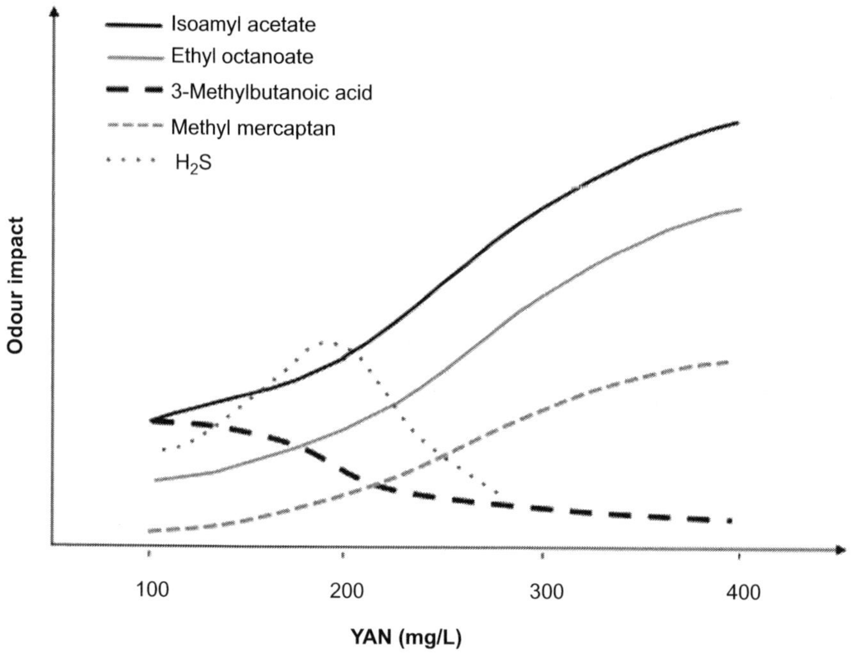

FIGURE 58.5 Odor impact of YAN on some volatile compounds produced by yeast. *Adapted from Ugliano* et al. *(2007)*

from unsupplemented juice had more complex aromas, with relatively higher concentrations of higher alcohols and volatile fatty acids. Excessive addition of DAP, but not of amino acids, produced wines with intense estery, solvent and volatile acidity aromas. This suggests that DAP supplementation of low YAN musts should be used cautiously whereas musts naturally high in YAN produce wines with intense but balanced aroma content.

DAP Addition to Red Must: Shiraz

It is generally believed that the conditions normally adopted for the production of red wine render fermentations less susceptible to slow or stuck fermentations. The grape berry skin contributes a significant proportion of the total amino acids. When skins are included in the must for fermentation in red grapes they can make a significant contribution to the yeast assimilable amino acid nitrogen concentration of the must (Stines *et al.*, 2000).

DAP supplementation of Shiraz strongly influences formation of various yeast volatile metabolites (Figure 58.6), especially higher production of fatty acids ethyl esters and acetate esters, whereas higher alcohols are scarcely affected (Ugliano *et al.*, 2010). Nitrogen-related variations in the concentration of other yeast metabolites, such as ethanol and 2- and 3-methylbutanoic acids, affect perceived fruitiness.

Analytical parameters related to color intensity and hue vary with DAP supplementation (Ugliano *et al.*, 2010). This effect might be ascribable to various aspects of yeast metabolism that are known to modulate wine color and phenolics composition. These factors include variations in the rate of ethanol production, absorption of anthocyanins on yeast cell walls, or reactions with yeast-derived metabolites such as pyruvic acid and acetaldehyde to form pigmented polymers (Romero and Bakker, 1999).

OTHER WAYS IN WHICH COMPOSITION IS ALTERED

Wine composition is the result of complex interactions between must chemical and nutrient composition, yeast strain and choice of fermentation conditions (Ugliano and Henschke, 2009). In order to best manage wine quality, it is necessary to determine the must-assimilable nitrogen content (YAN) and use DAP according to the style of wine required, but also taking into account the genetically determined responses that yeast show to nutrient supplementation. Development of new yeast strains with new or highly modified metabolic capabilities holds greatest promise for developing wines with new and exciting aroma profiles (Pretorius *et al.*, 2012).

ANALYTICAL TECHNIQUES

Nitrogen Supplemented Fermentations

Musts (synthetic, Albariño, Chardonnay and Shiraz) of low YAN content were supplemented with inorganic nitrogen to produce juices or musts with a range of initial nitrogen contents (Vilanova *et al.*, 2007; Torrea *et al.*, 2011; Vilanova *et al.*, 2012; Ugliano *et al.*, 2010). Fermentations were initiated with *Saccharomyces cerevisiae* AWRI 796 or MO5.

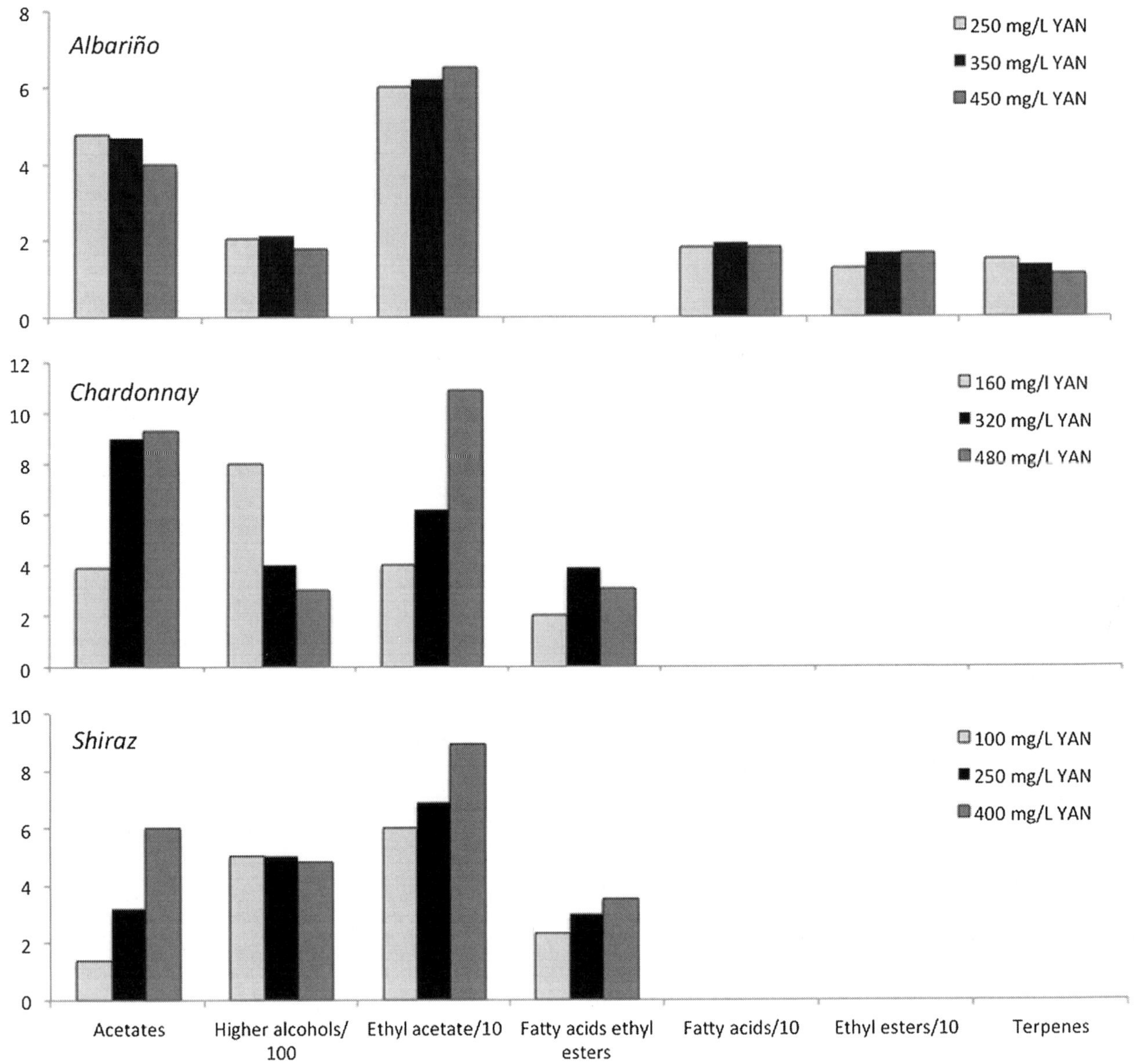

FIGURE 58.6 Effect of juice YAN (DAP supplementation) on volatile composition of *Albariño* (Vilanova *et al.*, 2012) *Chardonnay* (Torrea *et al.*, 2011) and *Shiraz* wine (Ugliano *et al.*, 2010)

Analysis of Basic, Non-Volatiles, and Volatiles Wine Composition

Fermentations were monitored and wines were analyzed as described by Torrea *et al.* (2011) for Chardonnay wines, Vilanova *et al.* (2007) for synthetic wines, Vilanova *et al.* (2012) for Albariño wines, and Ugliano *et al.* (2009) for Shiraz wines. Volatile fermentation products were analyzed with a HP 6890 gas chromatograph coupled to a HP 5973N mass selective detector which was used for gas chromatography–mass spectrometry. Instrument control and data analysis were carried out with the HPG1701CA ChemStation software.

References

Albers, E., Larsson, C., Liden, G., Niklasson, C., Gustafsson, L., 1996. Influence of the nitrogen source on *Saccharomyces cerevisiae* anaerobic growth and product formation. Appl. Environ. Microbiol. 62, 3187–3195.

Arias-Gil, M., Garde-Cerdan, T., Ancin-Azpilicueta, C., 2007. Influence of addition of ammonium and different amino acid concentrations on nitrogen metabolism in spontaneous must fermentation. Food Chem. 103, 1312–1318.

Bartowsky, E.J., Pretorius, I.S., 2008. Microbial formation and modification of flavour and off-flavour compounds in wine. In: König, H., Unden, G., Frölich, J. (Eds.), Biology of microorganisms on grapes, in must and wine. Springer, Heidelberg, Germany, pp. 211–233. Chapter 11.

Bell, S.J., Henschke, P.A., 2005. Implications of nitrogen nutrition for grapes, fermentation and wine. Aust. J. Grape Wine Res. 11, 242–295.

Beltran, G., Novo, M., Rozès, N., Mas, A., Guillamón, J.M., 2004. Nitrogen catabolite repression in *Saccharomyces cerevisiae* during Wine fermentations. FEMS Yeast Res. 4, 625–632.

Bely, M., Sablayrolles, J.M., Barre, P., 1990. Automatic detection of assimilable nitrogen deficiencies during alcoholic fermentation in enological conditions. J. Ferment. Bioeng. 70, 246–252.

Carrau, F.M., Medina, K., Farina, L., Boido, E., Henschke, P.A., Dellacassa, E., 2008. Production of fermentation aroma compounds by *Saccharomyces cerevisiae* wine yeasts: Effects of yeast assimilable nitrogen on two model strains. FEMS Yeast Res. 8, 1196–1207.

Francis, I.L., Newton, J.L., 2005. Determining wine aroma from compositional data. Aust. J. Grape Wine Res. 11, 114–126.

Gutiérrez, A., Chiva, R., Sancho, M., Beltran, G., Arroyo-López, F.N., Guillamon, J.M., 2012. Nitrogen requirements of commercial wine yeast strains during fermentation of a synthetic grape must. Food Microbiol. 31, 25–32.

Henschke, P.A., Jiranek, V., 1993. Yeasts metabolism of nitrogen compounds. In: Fleet, G.H. (Ed.), Wine Microbiology and Biotechnology. Harwood Academic Publishers, Chur, Switzerland, pp. 77–164.

Hernández-Orte, P., Bely, M., Cacho, J., Ferreira, V., 2006. Impact of ammonium additions on volatile acidity, ethanol and aromatic compound production by different *Saccharomyces cerevisiae* strains during fermentation in controlled synthetic media. Aust. J. Grape Wine Res. 12, 150–169.

Jiranek, V., Langridge, P., Henschke, P.A., 1995a. Amino acid and ammonium utilization by *Saccharomyces cerevisiae* wine yeasts from a chemically defined medium. Am. J. Enol. Vitic 46, 75–83.

Jiranek, V., Langridge, P., Henschke, P.A., 1995b. Regulation of hydrogen sulfide liberation in wine-producing *Saccharomyces cerevisiae* strains by assimilable nitrogen. Appl. Environ. Microbiol. 61, 461–467.

Mendes-Ferreira, A., Barbosa, C., Inês, A., Mendes-Faia, A., 2010. The timing of diammonium phosphate supplementation of wine must affects subsequent H_2S release during fermentation. J. Appl. Microbiol. 108, 540–549.

Mestres, M., Busto, O., Guasch, J., 2000. Analysis of organic sulfur compounds in wine aroma. J. Chromatogr. A 881, 569–581.

Miller, A.C., Wolff, S.R., Bisson, L., Ebeler, S.E., 2007. Yeast strain and nitrogen supplementation: dynamics of volatile ester production in *Chardonnay* Juice fermentations. Am. J. Enol. Vitic. 58, 470–483.

Pretorius, I.S., Curtin, C.D., Chambers, P.J., 2012. The winemaker's bug: From ancient wisdom to opening new vistas with frontier yeast science. Bioengineered Bugs 3, 1 10.

Rapp, A., Versini, G., 1996. Influence of nitrogen on compounds in grapes on aroma compounds in wines. J. Int. Sci. Vigne Vin 51, 193–203.

Rauhut, D., 2008. Usage and formation of sulphur compounds. Chapter 10, In: König, H., Unden, G., Frölich, J. (Eds.), Biology of microorganisms on grapes, in must and wine. Springer, Heidelberg, Germany, pp. 181–207.

Romero, C., Bakker, J., 1999. Effect of storage temperature and pyruvate on kinetics of anthocyanin degradation, vitisin A derivative formation, and color characteristics of model solutions. J. Agric. Food Chem. 48, 2135–2141.

Salmon, J.M., 1989. Effect of sugar transport inactivation in Saccharomyces cerevisiae on sluggish and stuck enological fermentations. Appl. Environ. Microbiol. 55, 953–958.

Stines, A.P., Grubb, J., Gockowiak, H., Henschke, P.A., Høj, P.B., van Heeswijck, R., 2000. Proline and arginine accumulation in developing berries of *Vitis vinifera* L. in Australian vineyards: Influence of vine cultivar, berry maturity and tissue type. Aust. J. Grape Wine Res 6, 150–158.

Torija, M.J., Beltran, G., Novo, M., Poblet, M., Rozès, N., Mas, A., Guillamón, J.M., 2003. Effect of organic acids and nitrogen source on alcoholic fermentation: study of their buffering capacity. J. Agric. Food Chem. 51, 916–922.

Torrea, D., Varela, C., Ugliano, M., Ancin-Azpilicueta, C., Francis, I.L., Henschke, P.A., 2011. Comparison of inorganic and organic nitrogen supplementation of grape juice—effect on volatile composition and aroma profile of a Chardonnay wine fermented with *Saccharomyces cerevisiae* yeast. Food Chem. 127, 1072–1083.

Ugliano, M., Henschke, P., Herdrich, M.J., Pretorius, I.S., 2007. Nitrogen management is critical for wine flavour and style. Aus. N.Z. Wine Industry J. 22 (6), 24–30.

Ugliano, M., Henschke, P.A., 2009. Yeast and wine flavour. Chapter 8D, In: Moreno-Arribas, V., Polo, M.C. (Eds.), Wine Chemistry and Biochemistry. Springer, New York, pp. 313–392.

Ugliano, M., Travis, B., Francis, I.L., Henschke, P.A., 2010. Volatile composition and sensory properties of *Shiraz* wines as affected by nitrogen supplementation and yeast species: Rationalizing nitrogen modulation of wine aroma. J. Agric. Food Chem. 58, 12417–12425.

Vilanova, M., Siebert, T.E., Varela, C., Pretorius, I.S., Henschke, P.A., 2012. Effect of ammonium nitrogen supplementation of grape juice on wine volatiles and non-volatiles composition of the aromatic grape variety Albariño. Food Chem. 133, 124–131.

Vilanova, M., Ugliano, M., Varela, C., Siebert, T., Pretorius, I.S., Henschke, P.A., 2007. Assimilable nitrogen utilisation and production of volatile and on volatile compounds in chemically defined medium by *Saccharomyces cerevisiae* wine yeasts. Appl. Microbiol. Biotechnol. 77, 145–157.

Vos, P.J.A., Gray, R.S., 1979. The origin and control of hydrogen sulfide during fermentation of grape must. Am. J. Enol. Vitic. 30, 187–197.

CHAPTER

59

The Application of Foliar Urea on the Concentration of Amines in Wine

*Carmen Ancín-Azpilicueta**, *Rodrigo Nieto-Rojo**, *Julio Gómez-Cordón*†

*Department of Applied Chemistry, Universidad Pública de Navarra, Pamplona, Spain, †Avanzare Innovación Tecnológica, Logroño, Spain

CHAPTER POINTS

- Treatment with foliar urea (2 kg N/ha and 4 kg N/ha) increased significantly ($p < 0.05$) the concentrations of histamine and spermine in Tempranillo wine.
- The concentrations of histamine in the wines coming from the grape treated with urea were above the threshold of concentration where this amine could provoke toxic effects in the consumer.
- Treatment with foliar urea did not increase the concentrations of tyramine, nor putrescine nor phenylethylamine + spermidine.
- Between the volatile amines with secondary amino groups, the concentrations of pirrolidine and diethylamine increased to a significant degree after treatment with foliar urea.
- Further studies are needed in order to relate wine amine concentration with nitrogen treatments applied to the vineyards, both at foliar level as well as at vineyard soil level.

INTRODUCTION

Nitrogen is an essential macronutrient and is required in relatively high quantities for growth in plants as it is a necessary component for chlorophyll as well as for protein synthesis, nucleic acids, and growth hormones. Foliar fertilization does not substitute the traditional fertilization of grape production, but it does serve as a help to complete and guarantee the nutritional requirements of a cultivar which is often difficult to supply only through soil fertilization. In addition, this practice permits the application of relatively small amounts of the product in relation to the quantities which need to be supplied to the soil.

Amines are formed in wine during alcoholic fermentation due to the metabolism of yeast strains (Torrea and Ancín, 2002), and during malolactic fermentation due to the proliferation of lactic acid bacteria, and amine concentration is also modified during wine aging in barrels (Jiménez-Moreno and Ancín-Azpilicueta, 2004). Biogenic amines in wines are formed through the action of decarboxylase enzymes produced by microorganisms from the corresponding amino acid precursor (ten Brink *et al.*, 1990). There are also volatile amines in wine that, with the exception of phenylethylamine, appear to come from the amination of non-nitrogenous compounds such as aldehydes and ketones. Furthermore, a great variability in the concentration of amines exists between different wines. This is due to the many factors which have an influence, to a greater or lesser extent, in the formation of these substances. Among these factors, the content of nitrogenous compounds in the grape stands out and depends, among other factors, on the level of maturation and on the nitrogenous fertilization.

Some amines can be toxic for people who are sensitive to these substances. The main biogenic amines implicated in food poisoning are histamine, tyramine, and phenylethylamine. Symptoms of intoxication by histamine are vomiting, diarrhoea, headaches, facial flushes, itching, and throat inflammation (Silla-Santos, 1996). Tyramine and phenylethylamine form part of the group of vasoconstrictor amines and so they can provoke hypertensive crisis and they are also responsible for alimentation migraine. Putrescine and cadaverine, although they

Processing and Impact on Active Components in Food
http://dx.doi.org/10.1016/B978-0-12-404699-3.00059-7

do not have any toxic activity, can inhibit the action of the monoamine oxidase (MAO) and diamine oxidase (DAO), enzymes which are responsible for the degradation of histamine, tyramine, and phenylethylamine and so, they can reinforce the toxic activity of these amines. Conversely, those amines with secondary amino groups (spermine, spermidine, dimethylamine, pirrolidine…) can react with nitrous acid and its salts to form nitrosamines, compounds with carcinogenic, mutagenic, and teratogenic activity.

The impact of different factors related to vinification on the concentration of amines in wine has been studied. However, there is little literature about the influence of the foliar fertilization of the vineyard on the concentration of these compounds in wines. In this study the scant literature available on the influence of fertilization with foliar urea on the amine concentration in wine is provided as well as some other factors which have an influence on the concentration of these substances in the product.

THE INFLUENCE OF FERTILIZATION WITH UREA FOLIAR ON THE CONCENTRATION OF BIOGENIC AMINES IN WINES

The main nutrients required for the yeasts during must fermentation are the carbonate compounds and the nitrogen compounds. A limitation of these compounds and of the membrane lipids as well as an accumulation of toxic compounds for the yeasts during fermentation are the principal reasons for stuck fermentations. The nitrogen compounds assimilable by the yeasts are principally ammonium, amino acids and, to a lesser degree, small peptides (Ancín-Azpilicueta *et al.*, 2008). Among the nitrogen compounds present in must, the concentration of amino acids is that which has the greatest importance in the formation of amines. Some amino acids are direct precursors of biogenic amines (histidine is a precursor of histamine, tyrosine of tyramine, phenylalanine of phenylethylamine, lysine of cadaverine, ornithine of putrescine, spermine and spermidine) and, in addition, they constitute a source of nutrients for the development of bacteria which can carry out secondary fermentations (Soufleros *et al.*, 1998). Those treatments which increase the concentration of nitrogen compounds in must could alter the concentration of biogenic amines in wine, although in some studies carried out, no direct relationship was found between the formation of biogenic amines on wine and the consumption of their precursor amino acids by yeasts during alcoholic fermentation. In this respect, Ancín-Azpilicueta *et al.* (2004) using Garnacha must found no relationship between the content of biogenic amines in wine and the utilization of their precursor amino acids during fermentation. In a later work, González-Marco *et al.* (2006) did not find any increase in the concentration of amines in wine during alcoholic fermentation, after adding amino acids to the must from autolysate yeast. In this same study, no correlation was found between the consumption of amino nitrogen or of precursor amino acids of amines with the formation of biogenic amines in wine after alcoholic fermentation. Herbert *et al.* (2005) studied the relationship between the quantity of free amino acids in musts of different varieties from the Alentejo region in Portugal with the total quantity of biogenic amines in red wine that had undergone both alcoholic and malolactic fermentation. Unlike the results of these previous studies, they found that the grape variety with the highest quantity of amino acids also showed the highest concentration of amines in the wine. Therefore, it is difficult to establish a correlation between the concentration of biogenic amines and the consumption of their precursor amino acids during alcoholic fermentation. This could be due to the fact that during alcoholic fermentation, yeasts mainly use amino acids directly in biosynthesis or as a nitrogen source and, to a lesser extent, for decarboxylation reactions where biogenic amines are formed. It is also necessary to take into account that the formation of amines is highly influenced by the aminogenic capacity both of yeasts as well as of bacteria that carry out the alcoholic and malolactic fermentations. This would suggest that, for improved knowledge of the nitrogenous metabolism, including the formation of biogenic amines, it would be important to carry out studies in synthetic media and with selected strains in order to eliminate variables from the natural medium.

In order to study the influence of fertilization with foliar urea on the content of amines in wines, *Vitis vinífera* var. Tempranillo grapes were used (2008 vintage) from three adjacent plots within the same vineyard, located in The Rioja region, in the North of Spain. Nitrogen fertilizer of 21 kg N/ha was added to the soil of the whole vineyard. Two of the plots were treated with foliar application of urea, in one of the plots 2 kg N/ha was applied while in the other plot 4 kg N/ha was applied. The third plot was not treated with urea (control plot). Nitrotecnia-20® was used for foliar urea application with 20% p/p of total N from the urea. The urea treatment was started near to veraison as the aim was for the urea nitrogen to pass directly to the grape and not to increase the foliage of the plant. The grapes were collected from the three plots. After the harvest, the grapes were destemmed and crushed, and afterwards they underwent pressing. Potassium metabisulphite (Aldrich, Gillingham, UK) was added to the must up to a concentration of 50 mg/l of total SO_2. A study of amines was made in the must at the beginning of the alcoholic fermentation (M), must at 6 days after the start

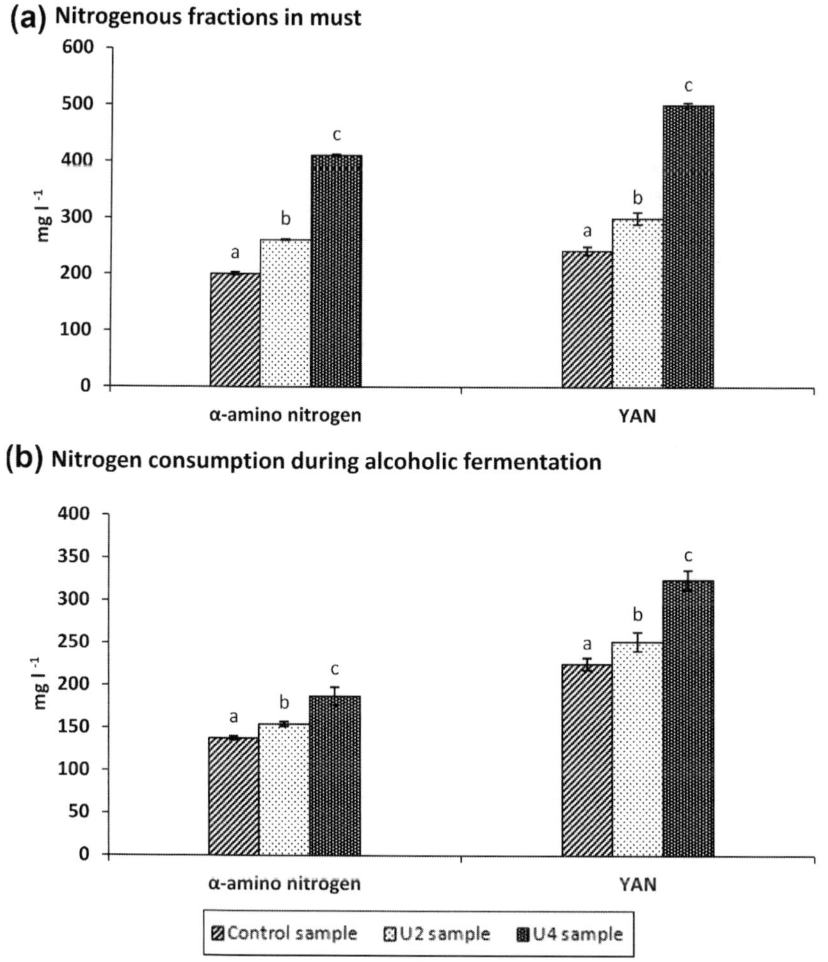

FIGURE 59.1 Nitrogenous fractions (α-amino nitrogen and yeast assimilable nitrogen, YAN) in must and nitrogen consumption during alcoholic fermentation. The graphics show the mean and the SD of four sets of analyses of three samples with and without foliar urea fertilization. Bars with different letters are significantly different ($p<0.05$). *With permission from Ancín-Azpilicueta* et al. *(2011).*

of fermentation (M6), must at 8 days after the start of fermentation (M8), and wine at the end of alcoholic fermentation (W). Fermentation lasted a total of 11 days. Those samples coming from the grapevines treated with 2 kg N/ha of foliar urea were given the name U2; while the samples coming from the grapevines treated with 4 kg N/ha of foliar urea were given the name U4 and the samples from the grapevines not treated with foliar urea were the control samples.

When foliar urea was applied, the concentration of amino nitrogen and yeast-assimilable nitrogen (YAN) significantly increased ($p<0.05$) in the must respect to control Tempranillo must (Figure 59.1a). In this sample, the concentration of amino nitrogen (201 mg/l) was similar to the average value (199 mg/l) found by Henschke and Jiranek (1993) for seven Australian varieties of grapes. It was observed that foliar urea application increased the concentration of amino nitrogen in must by 29% when 2 kg N/ha was applied and by 104% when 4 kg N/ha was applied (Figure 59.1a). YAN was also much higher in the must from the grape treated with 4 kg N/ha of foliar urea (500 mg/l) and in the must from the grape treated with 2 kg N/ha of foliar urea (300 mg/l) than in the control must (242 mg/l). These results indicate that must treated with urea accumulated a greater quantity of arginine in relation to proline than in the control must. Furthermore, the consumption both of amino nitrogen as well as of YAN during fermentation was significantly higher in the musts treated with foliar urea than in the control must (Figure 59.1b). The concentration of total biogenic amines did not show any significant differences ($p< 0.05$) between the three wine samples (control, 33.5 mg/l; U2, 34.7 mg/l; U4, 30.9 mg/l). Consequently, no correlation was found between the concentration of YAN in the must and its consumption, with the concentration of total biogenic amines in the wine.

With regard to the individual quantities of each amine under study from the Tempranillo must used in this work, small quantities of histamine, tyramine, putrescine, and spermine were found. The rest of the biogenic amines were formed in the second half of alcoholic fermentation, although the concentration of cadaverine decreased at the end of fermentation (Figure 59.2). Some amines at low levels such as putrescine, spermidine, and histamine are normal constituents of the raw materials

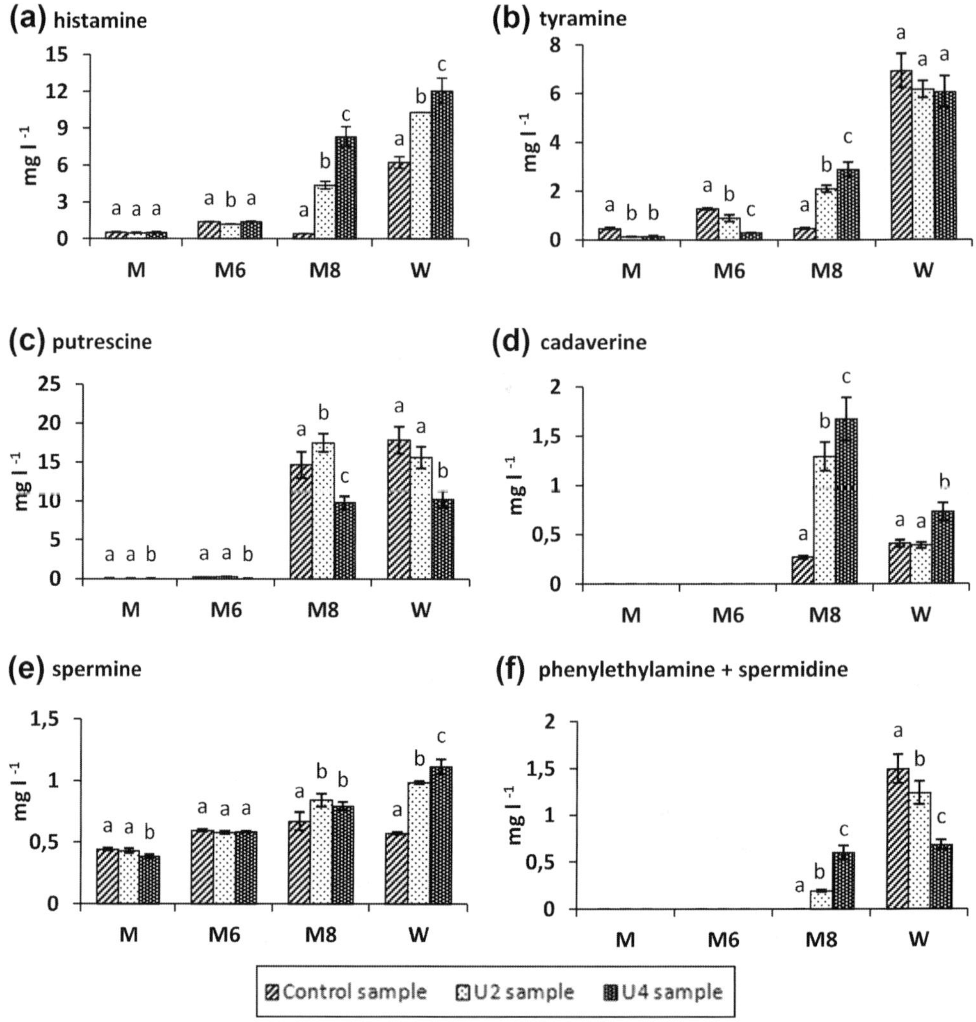

FIGURE 59.2 Evolution of the concentration of biogenic amines during alcoholic fermentation (M: must, M6: must at 6 days after the start of fermentation, M8: must at 8 days after the start of fermentation) and in wine (W). The graphics show the mean and the SD of four sets of analyses of three samples with and without foliar urea fertilization. Bars with different letters are significantly different ($p<0.05$). *With permission from Ancín-Azpilicueta* et al. *(2011).*

because they are indispensable components of all living cells. These compounds may also be a key protective factor for stressed cells.

The concentrations of histamine, cadaverine and spermine (Figure 59.2a, d, e) were significantly higher ($p<0.05$) in the wine coming from the must treated with the highest concentration of foliar urea (U4 wine) than in the control wine as well as in the U2 wine. The concentrations of histamine and spermine also increased with the application of 2 kg N/ha of urea in comparison with the control sample. Bertrand *et al.* (1991) found that histamine, putrescine, and cadaverine concentrations in wines from nitrogen-treated vines were 1.3- to 3.8-times higher than those of wines from untreated vines. The histamine concentration (Figure 59.2a) in both wines from musts that had been treated with urea is within the range of concentrations (8–20 mg/l) where this amine could provoke toxic effects in the consumer (Soufleros *et al.*, 1998). The effects produced by histamine are the most well-known as they are implicated in most cases of food poisoning. This amine is freed in certain tissues as a result of allergic hypersensitivity or of inflammations and has a potent vasodilatory action which could cause important drops in blood pressure. Histamine has traditionally been the amine most related to the so-called wine intolerance that some sensitive people have to this product. The toxicity of histamine on the general population is very difficult to establish since, in some cases, toxic effects have been associated with even small amounts of this amine (Wantke *et al.*, 1996), as there are some individuals who are especially sensitive to the action of histamine. Generally, these individuals show a deficiency of DAO or are being treated with monoamine oxidase inhibitors (MAOI). There also exist strengtheners of the toxicity of histamine, such as other biogenic amines present in food (tyramine, phenylethylamine, tryptamine, putrescine or cadaverine), which compete for the same endogen metabolic pathways and strengthen the absorption of histamine in the intestine (Veciana-Nogués and Vidal-Carou, 2000). Likewise, it has been observed that some compounds present in food and drinks such as alcoholic ones, or the acetaldehyde of wine, or the polyphenols from tea

TABLE 59.1 Concentration in Must of Amino Acid Precursors of Biogenic Amines

	Control	U2 (2 kg N/ha)	U4 (4 kg N/ha)
Histidine	192.6[a]	119.2[b]	266.8[c]
Arginine	242.9[a]	507.6[b]	735.0[c]
Tyrosine	17.0[a]	16.2[a]	15.6[a]
Phenylalanine	5.8[a]	0.0[a]	0.0[b]
Lysine	2.5[a]	36.0[b]	36.1[b]

This table shows the mean of four analyses of each sample. Means within the same row followed by different letters are significantly different ($p < 0.05$).
With permission from Ancín-Azpilicueta et al. *(2011).*

all strengthen the toxic action of histamine. Concentration of putrescine (Figure 59.2c) was significantly inferior ($p < 0.05$) in the wine from the must treated with the highest concentration of foliar urea (U4 wine) than in the control and in the U2 wines. In our case, the concentration of arginine was very high (Table 59.1) in the U4 must. In this sample, arginine would have been preferably used as a nitrogen source and, to a lesser extent than in the other samples, its decarboxylation would have been produced to form putrescine. Thus, the concentration of putrescine in the U4 wine was significantly lower ($p < 0.05$) than in the control wine and in the U2 wine. Conversely, the U4 must, where the yeasts had higher amounts of a good nitrogen source such as arginine, lysine, and histidine, would not have followed the biosynthesis pathways and they would have decarboxylated in greater measure to produce their corresponding biogenic amines (cadaverine and histamine) than in the other samples. The yeasts have a mechanism known as general amino acid control (Hinnesbuch, 1988). The precursor amino acids of histamine, cadaverine, and putrescine (histidine, lysine, and arginine, respectively) are all under this general amino acid control mechanism, so that their metabolic pathways are correlated and depend on the concentration of each of them. High concentration of arginine stimulates arginase activity, this regulatory strategy allows a rapid switch from arginine biosynthesis to arginine catabolism. However, *S. cerevisiae* uses the nitrogen compounds at the start of fermentation without modification. This yeast, once it has covered its specific demands for each amino acid, assimilates those that are in excess and uses them as nitrogen sources (Bisson, 1991).

The concentration of tyramine (Figure 59.2b) did not show any significant differences ($p < 0.05$) between the control sample and the wine samples from grapes with foliar urea fertilization. The concentration of the precursor amino acid of this amine, tyrosine, did not show any significant differences ($p < 0.05$) either between any of the three must samples (Table 59.1). In all the samples, the concentrations of tyramine were found to be below the range at which wines could have toxic effects for the consumer (10–80 mg/l), so that these wines do not present any toxic risk due to this amine. The concentration of phenylethylamine + spermidine (Figure 59.2f) also decreased in the wine on applying foliar urea. Phenylethylamine + spermidine were formed during alcoholic fermentation and its concentration was significantly lower ($p < 0.05$) in the wines coming from the must treated with urea than in the control sample. The precursor amino acid of phenylethylamine (phenylalanine) disappeared in the must coming from the grapes treated with foliar urea (Table 59.1). In all the wine samples, the concentration of phenylethylamine + spermidine was so low that it would be necessary to drink more than 21 of these wines to ingest a dose of 3 mg, above which phenylethylamine causes migraines (Sandler *et al.*, 1974).

The Influence of the Fertilization with Foliar Urea on the Concentration of Volatile Amines

Volatile amines have been found both in grape and in must. Accordingly, Ough *et al.* (1981) found that volatile amines were present in the grapes and unlike biogenic amines, most of the volatile amines were found in California musts before alcoholic fermentation. González-Marco *et al.* (2006), just like the above-mentioned authors, also found volatile amines (ethylamine, diethylamine, dimethylamine, pyrrolidine, and isobutylamine) in Chardonnay Spanish must. The influence of different vinification processes on the concentration of wine volatile amines has not been investigated as much as the influence that they have on the concentration of biogenic amines. Ough and Daudt (1981) found that most of the volatile amines in Pinot Noir and White Riesling wines decreased during alcoholic fermentation. Kiss and Sass-Kiss (2005) found that some amines increased during the wine-making process in botrytized wines and Herbert *et al.* (2005) concluded that the vintage had a great effect on total volatile amines.

In the case of the Tempranillo musts coming from grapes with foliar fertilization, pyrrolidine, ethylamine, and diethylamine were found in all the musts (Figure 59.3). The concentration of these three amines in the must was significantly higher ($p < 0.05$) in the U4 must than in the other two musts. The rest of the volatile amines were formed during alcoholic fermentation (Figure 59.3). Pyrrolidine and diethylamine (Figures 59.3d and f) were found in greater concentrations in the U4 wine. In the case of dimethylamine (Figure 3a), its concentration decreased in the wines coming from grape treated with urea. Dimethylamine, pyrrolidine, and diethylamine contain secondary amino groups in their molecular structure and so they can form nitrosamines.

The concentrations of isobutylamine and amylamine (Figures 59.3c and g) hardly showed any significant differences ($p < 0.05$) after the treatment with foliar urea.

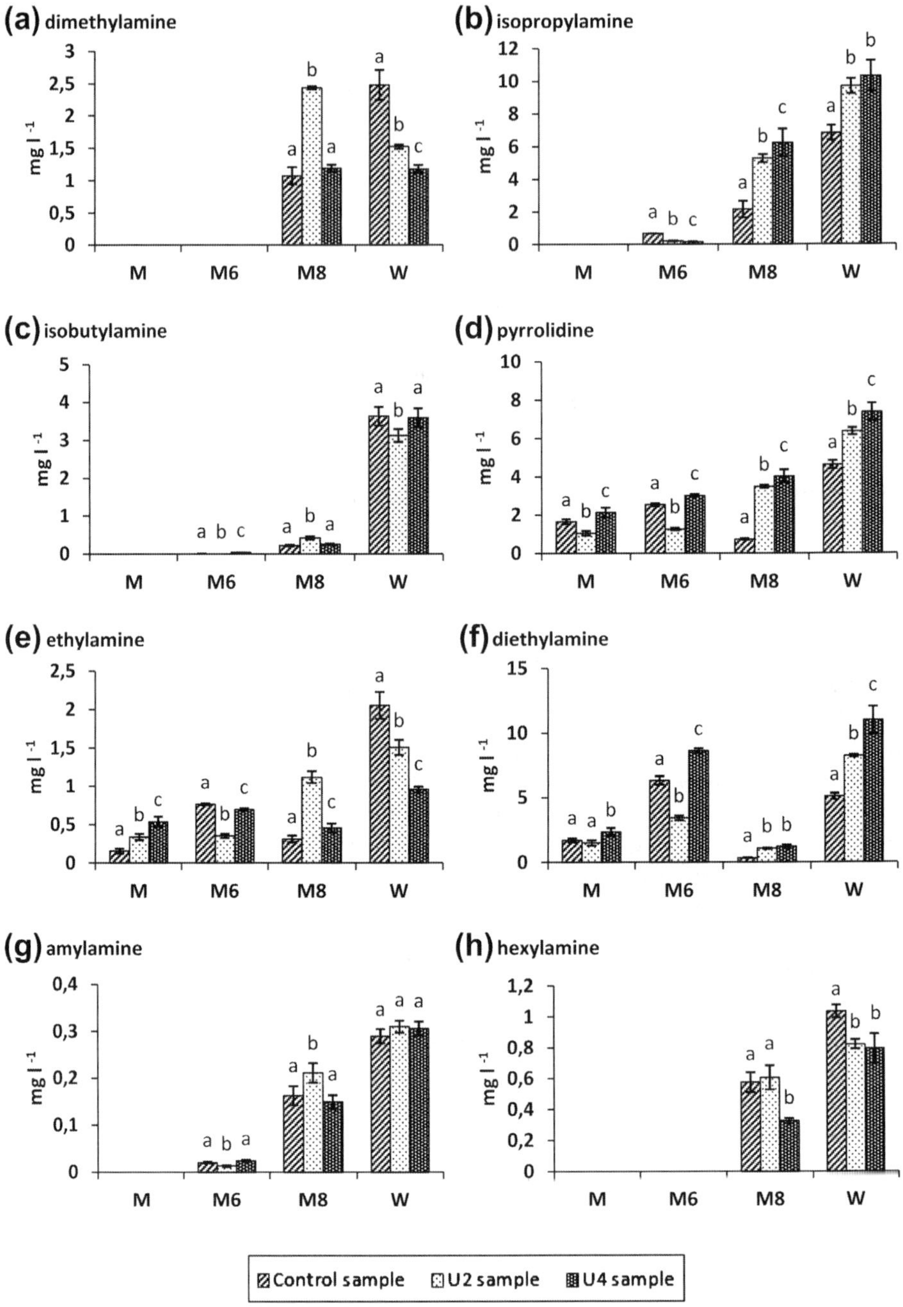

FIGURE 59.3 Evolution of the concentration of volatile amines during the alcoholic fermentation (M: must, M6: must at 6 days after the start of fermentation, M8: must at 8 days after the start of fermentation) and in wine (W). The graphics show the mean and the SD of four sets of analyses of three samples with and without foliar urea fertilization. Bars with different letters are significantly different ($p<0.05$). *With permission from Ancín-Azpilicueta* et al. *(2011).*

Only the concentration of isobutylamine was significantly lower ($p<0.05$) in the U2 wine than in the control wine. The concentrations of ethylamine and hexylamine (Figure 59.3e and h), both decreased after foliar fertilization. Conversely, the concentration of isopropylamine (Figure 59.3b) increased with foliar fertilization, while no significant differences ($p<0.05$) were observed between the different doses of treatment with foliar urea. These amines are less important from the point of view of toxicology as they do not show secondary amine groups and consequently they do not form nitrosamines. However, they could have an undesirable effect on the quality of the wine as they could have a negative effect on the aroma of the product.

Influence of Nitrogen Application in the Vineyard and the Other Factors in the Concentration of Amines in Wine

Few studies have been made on changes in the amine concentration in wine when different quantities of nitrogen are applied to the vineyard soil. Bertrand *et al.* (1991), Spayd *et al.* (1994), and Conradie (2001) found that nitrogen application in the vineyard increases the concentration of histidine (precursor of histamine) in juice and wine. Bertrand *et al.* (1991) studied the composition of must and wines from ungrafted Merlot and from Merlot on three different rootstocks (SO4, 3309, and Riparia). Their data show that excessive nitrogen fertilization caused an increase in the content of the two major amines (histamine and putrescine) in the wines coming from different harvests. The concentration of histamine in Merlot wines from nitrogen-treated vines was approximately double (4.38–8.18 mg/l) that of the concentration of this amine in wines which did not receive any nitrogen treatment. They also observed that phenylethylamine and cadaverine were practically non-existent in wines from unfertilized vines, and appeared significantly in the other samples. Nitrogen treatment in the soil of the vineyard also increased the concentration of methylamine, ethylamine, and tyramine, although their concentrations were lower than that of histamine.

The greater concentration of amines that are generally found in red wine more than rosé or white wine, are due to the influence of different factors. This is due to the fact that in red wine vinification there are high fermentation temperatures, maceration of the solid parts of the grape, excessively non-acid pHs, high yeast biomass, development of malolactic fermentation and low levels of free SO_2. These factors could favor the growth of microorganisms with high aminogenic activity and, conversely, they could increase the total content of amino acids in must and wines. It should also be pointed out that, once the amines have been formed, their elimination would be difficult without changing the wine composition. Accordingly, the most advisable thing would be to control, as far as it is possible, the factors that have an influence on the formation of these substances.

ANALYTICAL TECHNIQUES

The toxicology and sensorial importance of the amines has resulted in the proliferation of analysis methods in use to quantify these compounds in must and in wine. The quantitative determination of these substances in the products, either individually or simultaneously, may be calculated through diverse analytical methods which, in most cases, require previous treatment of the sample given their complexity and, moreover, due to the low concentration in which they are found. At first, the analysis of amines was made through spectrofluorimetry techniques; although, at present, chromatographic methods are now more frequently used. In addition, there are alternative techniques which can be used to determine amines, such as the technique of capillary electrophoresis or the enzymatic methods.

Determination of Amines by HPLC

Chromographic liquid is by far the most important technique to determine amines in wine. More precisely, the most-used technique is the-reverse phase high-pressure liquid chromatography (RP-HPLC), due to its high resolution, sensitivity, and great versatility (Lehtonen, 1996). In order to determine amines through HPLC, the derivatization of these compounds is required, since these substances do not possess chromophore groups in their structure, and it is necessary to form derivates which absorb in UV or which are fluorescent so that they can be detected and quantified with a fluorescence detector. The principal derivatizating agents used for the quantification of amines are o-phtaldialdehyde (OPA) (Bover-Cid *et al.*, 2006), dansyl chloride (Busto *et al.*, 1994), and 6-aminoquinolyl-N-hydroxysuccinimidyl carbamate (AQC) (Torrea and Ancín, 2001). While in the case of the dansyl chloride and with the AQC, the derivatization is made pre-column, in the case of OPA, the reaction may be carried out equally before or after the column.

Given that the concentration of amines in wine is very low, a very common method used is that of detection through flouresence. OPA is an equally common derivatization method. It reacts quickly in the presence of a mercaptan (for example, 2-mercaptoethanol) to form a derived fluorescent, but it only does this with primary amines, although the simultaneous addition of NaOCl allows for the derivatization of secondary amines. However, the problem is that the OPA derivates obtained are highly unstable and they degrade quickly to species which do not display fluorescence. This problem may be solved by automating the injection of the sample. A method used for amine determination was carried out by Torrea and Ancín (2001). Samples are ultrafiltered before the precolumn derivatization of the amines is carried out through the addition of AQC. The detection was made by means of a fluorescent detector, using 250 and 395 nm as the excitation and emission wavelengths, respectively. A reverse-phase column with a stationary phase of dimethyloctadecylsilyl bonded to amorphous silica is used. This method allows the simultaneous quantification of fourteen biogenic and volatile amines.

Determination of Amines by Capillary Electrophoresis

Capillary electrophoresis (CE) is a separating technique based on the different mobility of the analytes (normally charged) within an electrophoretic medium established inside a capillary. This technique is a viable alternative for the determination of amines due to its numerous advantages, as it achieves greater efficiency and requires less volume from the sample and it consumes lower quantities of dissolvent than the chromatographical methods (Ma *et al.*, 2004). In the bibliography, examples of amine analysis have been found in food, including wine through CE coupled to different methods of detection (Önal, 2007). Among the most used detection methods in the capillary electrophoresis there is that of indirect chemiluminescence, indirect UV detection, electrochemiluminescence, pulsed amperometric detection and the coupling of CE with mass spectrometry (MS) (Al-Hadithi and Saad, 2011).

Determination of Amines by Enzymatic Methods

The enzymatic determination of amines is carried out through biosensors. Biological recognition is used by means of enzymatic specificity. These constitute a cheap and efficient alternative to the traditional methods. The electrodes are constituted by amino oxidase enzymes (AO) which catalyze the oxidative deamination primary and secondary amines in order to produce aldehydes, ammonia, and hydrogen peroxide (Carelli *et al.*, 2007). The quantification of amines is made through the amount of oxygen consumed or the amount of hydrogen peroxide liberated. The enzymatic biosensors reduce the time needed for the analysis and offer a fast method of diagnosis to carry out routine industrial tests on food quality.

References

Al-Hadithi, N.N., Saad, B., 2011. Determination of underivatized polyamines: A review of analytical methods and applications. Anal. Lett. 44, 2245–2264.

Ancín-Azpilicueta, C., Torrea, D., Fraile, P., Garde, T., 2004. Amino acids and volatiles compounds in the fermentation of inoculated musts: biogenic amines in the wines. In: Shahidi, F., Weerasinghe, D.K. (Eds.), Nutraceutical Beverages. Chemical, Nutrition and Health effects. ACS Symposium Series 871, Washington, pp. 302–313.

Ancín-Azpilicueta, C., González-Marco, A., Jiménez-Moreno, N., 2008. Current knowledge about the presence of amines in wine. Crit. Rev. Food Sci. Nutr. 48, 257–275.

Ancín-Azpilicueta, C., Nieto-Rojo, R., Gómez-Cordón, J., 2011. Influence of fertilisation with foliar urea on the content of amines in wine. Food Addit. Contam.: Part A. 28, 877–884.

Bertrand, A., Ingargiola, M.C., Delas, J., 1991. Effects of nitrogen fertilisation and grafting on the composition of must and wine from Merlot grapes, particularly on the presence of ethyl carbamate. In: Rantz, J.M. (Ed.), International Symposium on Nitrogen in Grapes and Wine. American Society for Enology and Viticulture, Davis, pp. 215–220.

Bisson, L.F., 1991. Influence of nitrogen on yeast and fermentation of grapes. In: Rantz, J.M. (Ed.), International Symposium on Nitrogen in Grapes and Wine. American Society for Enology and Viticulture, Davis, pp. 78–89.

Bover-Cid, S., Miguélez-Arrizado, M.J., Latorre-Moratalla, M.L., Vidal-Carou, M.C., 2006. Freezing of meat raw materials affects tyramine and diamine accumulation in spontaneously fermented sausages. Meat Sci. 72, 62–68.

Busto, O., Valero, Y., Guasch, J., Borrull, F., 1994. Solid phase extraction applied to the determination of biogenic amines in wines by HPLC. Chromatographia 38, 571–578.

Carelli, D., Centonze, D., Palermo, C., Quinto, M., Rotunno, T., 2007. An interference free amperometric biosensor for the detection of biogenic amines in food products. Biosens. Bioelectron. 23, 640–647.

Conradie, W.J., 2001. Timing of nitrogen fertilisation and the effect of poultry manure on the performance of grapevines on sandy soil. Leaf analysis, juice analysis and wine quality. S. Afr. J. Enol. Vitic. 22, 60–68.

González-Marco, A., Jiménez-Moreno, N., Ancín-Azpilicueta, C., 2006. Influence of addition of yeast autolysate on the formation of amines in wine. J. Sci. Food Agric. 86, 2221–2227.

Henschke, P.A., Jiranek, V., 1993. Yeasts-metabolism of nitrogen compounds. In: Fleet, G.H. (Ed.), Wine Microbiology and Biotechnology. Harwood Academic Publishers, Chur, pp. 77–163.

Herbert, P., Cabrita, M.J., Ratola, N., Laureano, O., Alves, A., 2005. Free amino acids and biogenic amines in wines and musts from the Alentejo region. Evolution of amines during alcoholic fermentation and relationship with variety, sub-region and vintage. J. Food Eng. 66, 315–322.

Hinnesbuch, A.G., 1988. Mechanisms of gene regulation in the general control of amino acid biosynthesis in Saccharomyces cerevisiae. Microbiol. Rev. 52, 248–273.

Jiménez-Moreno, N., Ancín-Azpilicueta, C., 2004. Influence of wine turbidity on the accumulation of biogenic amines during aging. J. Sci. Food Agric. 84, 1571–1576.

Kiss, J., Sass-Kiss, A., 2005. Protection of originality of Tokaji Aszú: amines and organic acids in botrytized wines by high-performance liquid chromatography. J. Agric. Food Chem. 53, 10042–10050.

Lehtonen, P., 1996. Determination of amines and amino acids in wine—A review. Am. J. Enol. Vitic. 47, 127–133.

Ma, Y., Liu, G., Du, M., Stayton, I., 2004. Recent developments in the determination of urinary cancer biomarkers by capillary electrophoresis. Electrophoresis 25, 1473–1484.

Önal, A., 2007. A review: Current analytical methods for the determination of biogenic amines in foods. Food Chem. 103, 1475–1486.

Ough, C.S., Daudt, C.E., 1981. Quantitative determination of volatile amines in grapes and wines. I. Effect of fermentation and storage temperature on amine concentrations. Am. J. Enol. Vitic. 32, 185–188.

Ough, C.S., Daudt, C.E., Crowell, E.A., 1981. Identification of new volatile amines in grapes and wines. J. Agric. Food Chem. 29, 938–941.

Sandler, M., Youdin, M.B.H., Hanington, E., 1974. A phenylethylamine oxidizing defect in migraine. Nature 250, 335–336.

Silla Santos, M.H., 1996. Biogenic amines: their importance in foods. Int. J. Food Microbiol. 29, 213–231.

Soufleros, E., Barrios, M.L., Bertrand, A., 1998. Correlation between the content of biogenic amines and other wine compounds. Am. J. Enol. Vitic. 49, 266–278.

Spayd, S.E.., Wample, R.G., Evans, R., Stevens, R., Seymour, B.J., Nagel, C.W., 1994. Nitrogen fertilisation of white Riesling grapes in Washington. Must and wine composition. Am. J. Enol. Vitic. 45, 34–41.

ten Brink, B., Damink, C., Joosten, H.M.L.J., Huis in't Veld, J.H.J., 1990. Occurrence and formation of biologically active amines in foods. Int. J. Food Microbiol. 11, 73–84.

Torrea, D., Ancín, C., 2001. Influence of yeast strain on biogenic amines content in wines: relationship with the utilization of amino acids during fermentation. Am. J. Enol. Vitic. 52, 185–190.

Torrea, D., Ancín, C., 2002. Content of biogenic amines in a Chardonnay wine obtained through spontaneous and inoculated fermentations. J. Agric. Food Chem. 50, 4895–4899.

Veciana-Nogués, M.T., Vidal-Carou, M.C., 2000. Dieta restrictiva en histamina. In: Salas-Salvador, J., Bonada-Santjaume, A., Trallero-Casañes, R., Saló-Solà, E. (Eds.), Nutrición y dietética clínica. Elsevier and Masson, Barcelona, pp. 405–408.

Wantke, F., Hemmer, W., Haglmüller, T., Götz, M., Jarisch, R., 1996. Histamine in wine. Bronchoconstriction after a double-blind placebo-controlled red wine provocation test. Int. Arch. Allergy Imm. 110, 397–400.

CHAPTER

60

Effect of Wine Production Techniques on Wine Resveratrol and Total Phenolics

Jelena Cvejić, Milica Atanacković

Department of Pharmacy, Faculty of Medicine, University of Novi Sad, Novi Sad, Serbia

CHAPTER POINTS

- Resveratrol and phenolics contribute to the beneficial effects of moderate wine consumption on human health.
- Various factors influence the content of resveratrol and phenolics in wine.
- Winemaking procedures have diverse effects on the content of biologically active compounds.
- Factors increasing resveratrol and total phenolic content in wine include fermentation temperature, thermovinification, extended maceration, must freezing, and pectolytic enzymes addition.
- Processes influencing phenolic content include yeast selection, runoff process, carbonic maceration, cold-soak treatment, and sulfur dioxide level.
- Understanding of factors influencing preservation of resveratrol and total phenolic content is essential for increasing health benefits and nutritional properties of wine.

INTRODUCTION

The beneficial effects of wine on human health, mainly resulting from phenolic compounds, are well known. Among them, resveratrol has been studied most extensively (*trans*-3,5,4′-trihydroxystilbene). These compounds (phenolic acids, stilbenes, flavanols, flavonols, anthocyanins, etc.) have potential protective effects on diseases mediated by oxidation including coronary diseases, inflammation, and cancer (Cimino *et al.*, 2007; Soleas *et al.*, 1997). Phenolics present in wine contribute also to their sensory characteristics, such as color, flavor, and astringency. Total phenolic content of red wines usually range from 900 to 2500 mg/l gallic acid equivalents (GAE) and can reach up to 6000 mg/l GAE (Cimino *et al.*, 2007; Woraratphoka *et al.*, 2007).

Resveratrol is a stilbene produced by grapevines, primarily located in grape skin and usually present in red wine. Numerous biological activities of this compound have been demonstrated (Frémont, 2000). The naturally occurring form of resveratrol in grape is the *trans*-isomer. Also, it exists as the *cis*-isomer as well as resveratrol-glucoside (Figure 60.1). It is considered that red wine on average contains 1.9 mg/l of *trans*-resveratrol ranging from non-detectable up to 14.3 mg/l. Concentrations of *cis*-resveratrol are in general lower and can reach 5 mg/l (Sterbvo *et al.*, 2007).

During the winemaking process, phenolic compounds are partially transferred from grapes to must and to wine. The final composition of phenolic compounds in wine depends on grape variety and maturity, environmental factors, extraction parameters, winemaking technology, as well as chemical reactions that occur during wine aging (Paixao *et al.*, 2007). Wine phenolics, mostly resveratrol, have been widely studied in many different publications. However, factors influencing their content are not investigated to any great extent.

HOW COMPOSITION IS ALTERED

Effects of Winemaking Technologies on Resveratrol Content

Being a compound with a numerous potential biological activities, resveratrol could contribute to the beneficial and health-promoting properties of wine. Therefore, elucidation of specific factors influencing the amount of

Processing and Impact on Active Components in Food
http://dx.doi.org/10.1016/B978-0-12-404699-3.00060-3

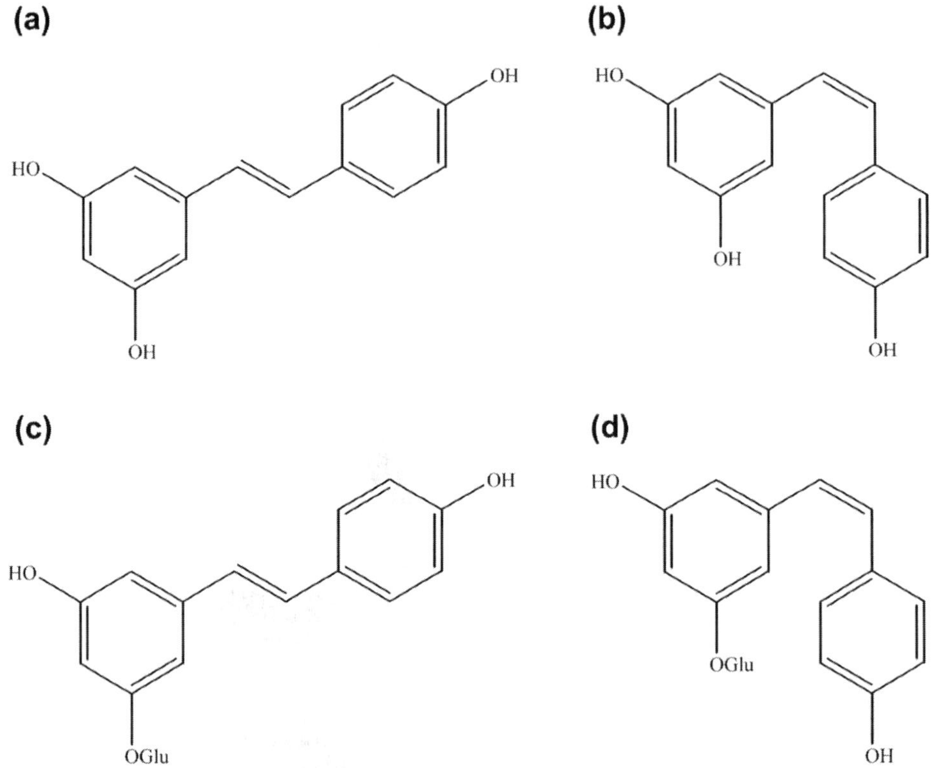

FIGURE 60.1 Structures of resveratrol and its forms: (a) *trans*-resveratrol; (b) *cis*-resveratrol; (c) *trans*-resveratrol glucoside (*trans*-piceid); (d) *cis*-resveratrol glucoside (*cis*-piceid).

this compound in wine is important. Resveratrol content in wine depends on different factors (variety, harvest year, climatic conditions, UV light, etc.), and winemaking technology is one among them that can be controlled.

The most important results concerning this subject are summarized in Table 60.1.

Maceration Time

Generally, it is well established that prolonged maceration time and higher ethanol content positively affect the extraction of resveratrol and piceid, which is due to their better solubility. A previously published study analyzed the application of different maceration times (3, 6, and 10 days) on resveratrol and piceid content. The highest concentrations of resveratrol and piceid were observed in wines produced with 10 days of maceration. Conversely, wines with relatively low contents of these compounds reached maximal concentrations after shorter periods of time. This could be due to enzymatic cleavage of the glycosidic bond of piceid as well as the possibility of metabolism of resveratrol by yeasts (Kostadinović *et al.*, 2012). The same authors observed that longer maceration time increased the antioxidant activity of wines, which could be due to higher content of other phenolics. Short contact of skin and seeds, which involves lower ethanol concentration, results in an incomplete extraction of phenolic compounds. In another study, prolonged maceration increased resveratrol content in wines made from Cabernet Sauvignon, while the opposite effect (decrease of resveratrol content) was observed in varieties Cynthiana and Noble (Threllfall *et al.*, 1999). Moreover, the study conducted by Sun *et al.* (2003) showed that prolonged maceration did not affect concentration of resveratrol in wines made from the Castelao variety.

These results suggest that the influence of maceration time could vary depending on the grape cultivar used. Resveratrol concentration can increase during fermentation on the skins, but it is important to emphasize that the final concentration of this compound in wine depends also on the grape variety as well as on enological conditions.

Thermovinification

Thermovinification is a process that requires intact or crushed grape (usually for the red-wine-making process) to be heated (50°C to 87°C) for a short time (Clarke and Bakker, 2004). Atanacković *et al.* (2012) analyzed the influence of two different thermovinification methods (60°C for 30 min and 80°C for 3 min) on resveratrol content in four different grape varieties (Merlot, Cabernet Sauvignon, Pinot Noir, and Prokupac). The results demonstrated that resveratrol content depends mainly on the grape variety. However, compared with control samples, applied processes increased resveratrol content only in the Pinot Noir variety. In another study, Netzel *et al.* (2003) showed that this procedure increases total resveratrol content in wines made from Pinot Noir, Lemberger, and Cabernet Franc, probably due to better extraction.

TABLE 60.1 Effects of Winemaking Technologies on Resveratrol Content in Wine—An Overview

Reference	Experiment	Analyzed Parameter	Effect	Comment
Netzel *et al.* (2003)	Fermentation on skin Mash heating (65°C) Combination of both	*Trans*- and *cis*-resveratrol *Trans*/*cis* piceid	Combination of heating with skin fermentation increased total resveratrol up to 400% times	Applied procedure also increased antioxidant activity and some other phenolics content (anthocyanins, flavan-3-ols and flavonols)
Sun *et al.* (2003)	Carbonic maceration Skin fermentation with/without stems	*Trans*- and *cis*-resveratrol	Augmentation of resveratrol content after skin fermentation comparing to carbonic maceration; no influence of stems	Prolonged maceration has no influence on resveratrol content
Atanacković *et al.* (2012)	Thermovinification 60°C for 30 min, 80°C for 3 min	*Trans*- and *cis*-resveratrol	Increase of *trans*-resveratrol up to 40% for Pinot Noir variety	Significant influence of grape variety on resveratrol content
Kostadinović *et al.* (2012)	Duration of maceration (3, 6, 10 days), Merlot wine	*Trans*-resveratrol *Trans*-piceid	Maximal concentrations detected after 10 days in comparison with 3-day maceration piceid increased100%, resveratrol nearly 600% increased	Duration of maceration time had strong influence on antioxidative activity
	French and Macedonian yeast, Merlot and Vranec wines	*Trans*-resveratrol *Trans*-piceid	Piceid concentrations 100% higher with French yeast Resveratrol concentrations up to 400%	Wines produced with Macedonian yeast had higher antioxidative activity than those with French yeast

The table shows effects produced by variation of different parameters (e.g., temperature and duration of fermentation) during wine production on resveratrol concentration.

These findings suggest that thermovinification could be responsible for the enhancement of concentration of resveratrol in wine, but this effect, as shown in some studies, may also differ according to grape variety.

Yeast selection

Another important factor, which can significantly influence resveratrol content in wine, is yeast selection. However, only a few published reports address this topic. Vacca *et al.* (1997) showed that yeasts used for fermentation of must are among the factors responsible for the decrease of resveratrol content in wines. This was explained by the absorption of the compound on cell walls or absorption/metabolism by the yeast cells. The fact that different yeast types could influence resveratrol content was also confirmed by Kostadinović *et al.* (2012). The authors noticed an increase of resveratrol (up to four times) and piceid (up to two times) in Merlot wines produced using French yeast when compared with Macedonian yeast.

It was observed (authors' unpublished data) that when compared with spontaneous fermentation, addition of selected yeasts could result in an augmentation as well as a decrease of resveratrol content, depending on yeast variety and properties.

In addition, it has been established that yeasts' β-glucosidase activity depends on maceration time (Vrhovsek *et al.*, 1997), thus the final content of resveratrol and piceid is a result of yeast selection as well as maceration period. Addition of β-glucosidase from *Aspergillus niger* in Sicilian wines increased the *trans*-resveratrol content up to 75% (Todaro *et al.*, 2008).

Effects of Winemaking Technologies on Total Phenolic Content

It is well known that phenolic compounds present in wine can show a beneficial effect on human health. Therefore, clarification of the influence of different enological practices on phenolic level in wine is essential and this topic has been addressed for many years in scientific publications. The main results from selected publications focused on this topic are summarized in Table 60.2.

Fermentation Temperature

There are several published reports that confirm that higher fermentation temperatures increase phenolic extraction in wine. As far as the authors are aware, Ough and Amerine (1961) published the first observation

TABLE 60.2 Effects of Winemaking Technologies on Total Phenolic Content in Wine—An Overview

Reference	Experiment	Analyzed Parameter	Effect	Comment
Timberlake and Bridle (1976)	Fermentation on the skins Thermovinification (60°C, 30 min, then 45°C, 30 min) Carbonic maceration	Total phenolics (TP)	Carbonic maceration wine almost 20% higher TP than thermovinificated wine	Thermovinificated wine was much more colored
Girard *et al.* (2001)	Cold fermentation 15°C Ambient fermentation 20°C High temperature 30°C Modified cold fermentation 15°C	Total phenolics (TP)	High temperature of fermentation lead to almost 50% increase in total phenolics compared with modified cold fermentation	Higher fermentation temperature; increase permeability of the hypodermal cell; releasing anthocyanin; increased solubility of other phenolics in the wine solution
	Different vinification yeasts *Saccharomyces bayanus* *Saccharomyces cerevisiae*	Total phenolics (TP)	For modified cold fermentation *S. cerevisiae* increased TP by 50%	
Spranger *et al.* (2004)	Skin fermentation with stems (7, 21 days) 25°C Skin fermentation without stems (7 days) 25°C Carbonic maceration 35°C (21 days)	Total phenolic index (TP) Total anthocyanins (TA)	Duration had no significant influence Carbonic maceration wines significantly lower TA and TP than skin contact (average 20% lower for TP and 50 % lower for TA) Stem contact slightly increased TP	Although higher temperature may induce degradation-condensation of anthocyanins, it favors diffusion of other phenols during carbonic maceration
Atanacković *et al.* (2012)	Thermovinification 60°C for 30 min, 80°C for 3 min	Total phenolics (TP)	Increased total phenolic content in all thermovinificated samples (up to 100%)	Only for Pinot Noir, higher temperature (80°C) produced more phenolic compounds in wine Increased total phenolic lead to increased antioxidative activity of wine samples

The table shows effects produced by variation of different parameters (e.g., thermovinification, temperature, and yeast selection) during wine production on total phenolic content.

of fermentation temperature influence on phenolic content. It was noticed that the color of wines produced from Pinot Noir and Cabernet Sauvignon was more intense as the fermentation temperature increased. Some later reports for Pinot Noir confirmed that increasing temperature from 15 to 30°C significantly increases total phenolic content in wine (Girard *et al.*, 2001). In another report by Spranger *et al.* (2004), it was observed that increased temperature (35°C) during carbonic maceration induces degradation or condensation of anthocyanins, but it can favor improved diffusion of other polyphenols from grape to wine and increase total phenolic content.

Thermovinification

After heating of intact or crushed grapes during thermovinification process, the grapes are pressed and must is inoculated with yeast strain for fermentation. Before fermentation, grape skins and seeds could be removed from must or left in during this process. Atanacković *et al.* (2012) examined total phenolic content in wines made by two different thermovinification methods (Table 60.2) from four grape varieties (Merlot, Cabernet Sauvignon, Pinot Noir, and Prokupac). It was observed that increased temperature lead to higher phenolic content in all cultivars and accordingly increased antioxidative capacity of wine samples. Netzel *et al.* (2003) observed that heating of must (skins with liquid) results in higher concentration of anthocyanins, flavan-3-ols and flavonols in Pinot Noir, Lemberger, and Cabernet Franc varieties. Conversely, some other previously published results reported that thermovinification leads to improved color extraction (anthocyanins), but much lower total phenolic extraction (Auw *et al.*, 1996). A similar trend was also observed for some Italian grape varieties (Stella *et al.*, 1991).

It is generally recognized that heat treatment can be effective in increased phenolic extraction if it is applied on must (containing grape skins and seeds) (Sacchi *et al.*, 2005).

Carbonic Maceration

During carbonic maceration, whole berries and clusters are held under a carbon dioxide atmosphere and partial fermentation occurs because of the activity of glycolytic enzymes present in grapes (Flanzy, 1935). In order to complete fermentation, after 1–2 weeks, the grapes are pressed and the juice is inoculated.

Timberlake and Bridle (1976) found that wines produced by carbonic maceration contained the highest amount of phenolic compounds and less anthocyanin and polymeric pigments when compared with wines made by thermovinification and fermentation on the skin. Similar results were obtained by Sun *et al.* (2001). Conversely, the study by Spranger *et al.* (2004) showed that carbonic maceration of Castelao wines resulted in the lowest phenolic as well as anthocyanin content. A study by Pellegrini *et al.* (2000) confirmed that the influence of carbonic maceration on total phenolics depends on grape variety.

Maceration Time

Extended maceration prolongs skin contact and the hypothesis that this process could increase extraction of phenolic compounds is supported by current literature. It has been shown that longer pomace contact increased the concentration of anthocyanins and tannins (10 *versus* 4 or 5 days). Also, aging in the bottle (1 year) of wine produced by longer pomace contact results in higher content of polymeric pigments (Gomez-Plaza *et al.*, 2001). Prolonged maceration time for Merlot wines lead to an increase of total phenols up to 36 days, while pigments increased up to day 4, and then slowly decreased (Yokotsuka *et al.*, 2000). On the contrary, another study did not reveal any significant difference between total phenols within 7 and 21 days' skin fermented wine. Moreover, a decrease in anthocyanins was observed, probably because of their absorption on yeast or/and condensation and degradation (Spranger *et al.* 2004).

Yeast

There are a few published studies pertaining to the effect of yeast on phenolic content in wine. Obtained results showed that selected yeast could produce variable effects depending mainly on the grape variety studied. It has been observed that, applied to the same variety, differences could be obtained according to yeast strain (Cuinier, 1997; authors' unpublished data). However, in some studies no major differences were observed in phenolic content as regards the use of different yeast. Zini *et al.* (2003) showed that application of three yeast strains on one variety gave similar phenolic content. Likewise, significant differences were not observed when comparing phenolic content in wines obtained from four grape varieties using five yeast strains (Nicolini *et al.*, 2003).

OTHER WAYS IN WHICH COMPOSITION IS ALTERED

Other factors affecting phenolic content during the wine-making process were also investigated. Besides those previously mentioned, the factors that could have a positive effect on the increase of biologically active compounds in wine (when applied during wine making) are summarized below.

Freezing the must before fermentation could influence the content of phenolics in wine. This process breaks cell membranes and thus can increase the extraction of anthocyanins and tannins. This was confirmed by Couasnon (1999), who observed a 50% increase in anthocyanins and tannins in three grape varieties (Merlot, Cabernet Sauvignon, and Cabernet Franc). Conversely, freezing the pomace separated from the juice resulted in a slight increase of anthocyanins, while the impact on total phenolic compounds varied.

Low-temperature extraction in the absence of alcohol (cold soak) produced little difference in phenolic content. It has been shown that this process generally has negative or no effects (decrease of anthocyanins and flavonols) on phenolics. However, when combined with addition of sulfur dioxide, the increase of total phenolics and anthocyanins is pronounced (Gerbaux, 1993).

Prefermentation juice run-off is a process aimed at increasing skin to juice ratio by removing juice before fermentation. The procedure is expected to increase concentration of phenolics from skins and seeds in wine. This was supported by the results obtained by Zamora *et al.* (1994) where increases in flavonoid and anthocyanin concentrations was observed when 10% juice was removed. An increase in phenolic content was also observed after applying the run-off technique by Gerbaux (1993). However, some studies showed different results. It has been noticed that 30% and 50% juice run-off (30% and 50% removal) results in variable effects on phenolic content in wines from different grape varieties (Atanacković *et al.*, 2012). Inconsistent results after 10% and 20% juice removal were observed also by Gawel *et al.* (2001). It could be partially explained by solubility limitation, since there is less liquid for dissolution of targeted compounds.

Addition of pectolytic enzymes can increase total phenols as well as wine color intensity. It has been demonstrated that pectinases improve extraction of phenolics (Daudt and Polenta 1999). In most cases, applied enzymes do not increase anthocyanis, but can increase the amount of other phenolic compounds (e.g., tannins).

Additionally, it has been reported that higher sulfur dioxide levels can produce enlargement of pigment extraction during early fermentation, but final concentration of pigments as well as total phenolics in wines were not significantly higher (Bakker *et al.*, 1998).

Application of increased fermentation temperature, thermovinification, must freezing, and extended maceration during the wine-making process have been reported to increase phenolic content in wine. Conversely, other investigated factors, such as sulfur dioxide levels and cold-soak treatments have frequently been shown to have little or no lasting effects, or to lead to a decrease in phenolic levels.

ANALYTICAL TECHNIQUES

Siemann and Creasy (1992) conducted the first analysis of resveratrol content in wine. The described HPLC analytical method included a complex extraction procedure and large sample volumes. Nowadays, numerous studies are focused on this topic. In the last 20 years, interest in resveratrol and quantification of this compound in different samples has rapidly increased, and a number of publications covering this topic have been published. The most commonly used method for adequate determination of all resveratrol forms is HPLC analysis. Some of the techniques applied are listed in Table 60.3. Successful identification and quantification of both resveratrol isomers (*trans-*, *cis-*) has been reported (Atanacković *et al.*, 2012) as well as other resveratrol forms (piceid, resveratrol-glucuronide, etc.) (Soleas *et al.*, 1997; Kostadinović *et al.*, 2012). Reference standards for *trans*-resveratrol and *trans*-piceid are commercially available, while corresponding *cis*-forms are usually obtained after *in situ* UV light exposure of *trans*-isomer solution (Cvejić *et al.*, 2010; Netzel *et al.*, 2003; Sun *et al.*, 2003).

The methods used could require sample preparation, as frequently used solid-phase extraction or classic liquid/liquid extraction (Sun *et al.*, 2003). Conversely, the increasingly used direct injection of wine samples has also been shown to provide satisfactory results (Netzel *et al.*, 2003).

There is a variety of detection methods for resveratrol determination using liquid chromatography. Most commonly used method is UV/DAD detection performed with or without sample preparation. Detection wavelengths are usually near the absorption maximum of investigated compounds (around 305 nm and 280 nm for *trans-* and *cis*-resveratrol, respectively). In addition, due to the physico-chemical properties of resveratrol, use of the fluorescent detector (FLD) is possible. This type of highly sensitive detection could allow measurement of small quantities of resveratrol (Atanacković *et al.*, 2012). Also, a more specific mode of detection comparing to UV/DAD, FLD could be easily used for resveratrol quantification from a complex matrix such as wine, without any sample preparation. Mass spectrometry (MS) detection has been coupled with HPLC for reliable confirmation of resveratrol isomers and metabolites in different samples (Kostadinović *et al.*, 2012).

The most commonly used stationary phase for HPLC analysis of resveratrol is easily available octadecil reverse phase. Mobile phases applied for successful separation

TABLE 60.3 Analytical Procedures for Identification and Quantification of Resveratrol

Reference	Analyte	Sample Preparation	Method	Conditions	Detection
Threlfall *et al.* (1999)	*Trans*-resveratrol	Quantification by standard addition method	HPLC Phenomenex bondclone phenyl column	Gradient elution Methanol/water pH 2.5	UV 310 nm
Netzel *et al.* (2003)	*Trans*-resveratrol *Trans*-piceid	Direct injection	HPLC RP C18 column	Gradient elution Acetonitrile/water/formic acid	UV/DAD 310 nm
Sun *et al.* (2003)	*Trans-* *Cis-* resveratrol	SPE Liquid/liquid extraction	HPLC RP C18 column	Gradient elution Acetonitrile/water	UV 307 nm *trans* 285 nm *cis*
Todaro *et al.* (2008)	*Trans-* *Cis-* resveratrol	Liquid/liquid extraction	HPLC RP C18 column	Gradient elution Acetonitrile/phosphoric acid	UV/DAD Wavelength not specified
Atanacković *et al.* (2012)	Trans- *Cis-* resveratrol	Direct injection	HPLC RP C18 column	Gradient elution Methanol/water/formic acid	FLD Ex330 Em374 *trans* UV/DAD 280 nm *cis*
Kostadinović *et al.* (2012)	*Trans*-resveratrol *Trans*-piceid	Direct injection	HPLC RP C18 column	Gradient elution Acetonitrile/acetic acid/water	MS

The table shows the most commonly used analytical procedures for determination of resveratrol isomers in wine.
HPLC, high-performance liquid chromatography; RP, reverse phase; C18, octadecyl bonded silica; UV/DAD, ultraviolet diode array detection; FLD, fluorescence detection.

could vary, but consist mainly of a combination of frequently used solvents for HPLC analysis such as methanol, ethanol, acetonitrile, formic acid, and water.

Another approach for resveratrol determination, which is not as frequently used as HPLC, is application of gas chromatography coupled with mass spectroscopy. Sample preparation includes solid-phase extraction or liquid extraction after which trimethylsylil resveratrol derivates are formed.

The total phenolic amount is commonly determined with UV/Vis spectroscopy by applying Folin–Ciocalteu's reaction. The results are expressed either as total phenolic index (Sun *et al.*, 2001) or more frequently, as gallic acid equivalents in 1 l of wine (mg/l GAE) (Atanacković *et al.*, 2012; Woraratphoka *et al.*, 2007). Since antioxidative capacity is closely related to total phenolic content of wine, it is very often included in studies concerning phenolic content. It is usually estimated by DPPH (Atanacković *et al.*, 2012) or TEAC tests (Netzel *et al.*, 2003; Kostadinović *et al.*, 2012).

The individual phenolic compounds most abundant in wine (e.g., catehin, gallic acid, quercetin, etc.) are usually successfully identified and quantified with HPLC analysis using DAD. Nevertheless, analysis of different factors, which influence individual phenolic compounds in wine, is not extensively described in the literature. With respect to the importance of these natural compounds and their assumed health benefits, this subject is currently expanding.

References

Atanacković, M., Petrović, A., Jović, S., Gojković-Bukarica, Lj., Bursać, M., Cvejić, J., 2012. Influence of winemaking technique on the resveratrol content, total phenolic content and antioxidant potential of red wines. Food Chem. 131 (2), 513–518.

Auw, J.M., Blanco, V., O'Keefe, S.F., Sims, C.A., 1996. Effect of processing on the phenolics and color of Cabernet Sauvignon, Chambourcin, and Noble wines and juices. Am. J. Enol. Vitic. 47, 279–286.

Bakker, J., Bridle, P., Bellworthy, S.J., Garcia-Viguera, C., Reader, H.P., Watkins, S.J., 1998. Effect of sulphur dioxide and must extraction on colour, phenolic composition and sensory quality of red table wine. J. Sci. Food Agric. 78, 297–307.

Cimino, F., Sulfaro, V., Trombetta, D., Saija, A., Tomaino, A., 2007. Radical-scavenging capacity of several Italian red wines. Food Chem. 103, 75–81.

Clarke, R.J., Bakker, J., 2004. Wine-Flavour Chemistry. Blackwell Publishing Ltd., Oxford.

Couasnon, M.B., 1999. Une nouvelle technique: La maceration préfermentaire à froid-extraction à la neige carbonique. Premièr partie: Résultats oenologiques. Rev. Oenol. 92, 26–30.

Cuinier, C., 1997. Ceppi di lievito e composizione fenolica dei vini rossi. Vignevini 7/8, 39–42.

Cvejic, J.M., Djekic, S.V., Petrovic, A.V., Atanackovic, M.T., Jovic, S.M., Brceski, I.D., Gojkovic–Bukarica, Lj.C., 2010. Determination of *trans*- and *cis*-resveratrol in Serbian commercial wines. J. Chromatogr. Sci. 48, 229–234.

Daudt, C.E., Polenta, G.A., 1999. Les phénols dans les moûts de Cabernet Sauvignon et d'Isabel soumis à divers traitements. J. Sci. Tech. Tonnellerie 5, 47–55.

Flanzy, M., 1935. Nouvelle méthode de vinification. C.R. Acad. Agric. Fr. 21, 935–938.

Frémont, L., 2000. Biological effects of resveratrol. Life Sci. 66 (8), 663–673.

Gawel, R., Iland, P.G., Leske, P.A., Dunn, C.G., 2001. Compositional and sensory differences in Syrah wines following juice run off prior to fermentation. J. Wine Res. 12, 5–18.

Gerbaux, V., 1993. Etude de quelques conditions de cuvaison susceptibles d'augmenter la composition polyphénolique des vins de Pinot noir. Rev. Oenol. 69, 15–18.

Girard, B., Yuksel, D., Cliff, M.A., Delaquis, P., Reynolds, A.G., 2001. Vinification effects on the sensory, colour, and GC profiles of Pinot noir wines from British Colombia. Food Res. Int. 34, 483–499.

Gómez-Plaza, E., Gil-Muñoz, R., López-Roca, J.M., Martínez-Cutillas, A., Fernández-Fernández, J.I., 2001. Phenolic compounds and color stability of red wines: Effect of skin maceration time. Am. J. Enol. Vitic. 52, 266–270.

Kostadinović, S., Wilkens, A., Stefova, M., Ivanova, V., Vojnovski, B., Mirhosseini, H., Winterhalter, P., 2012. Stilbene levels and antioxidant activity of Vranec and Merlot wines from Macedonia: Effect of variety and enological practices. Food Chem. 135, 3003–3009.

Lopez-Velez, M., Martinez-Martinez, F., Del Valle-Ribes, C., 2003. The study of phenolic compounds as natural antioxidants in wine. Crit. Rev. Food Sci. Nutr. 43 (3), 233–244.

Nicolini, G., Mattivi, F., Larcher, R., Volpini, A., 2003. Vinificazioni in rosato ed in rosso con lieviti selezionati: observazioni circa il colore. Enologo 39, 79–83.

Ough, C.S., Amerine, M.A., 1961. Studies on controlled fermentation.V. Effects on color, composition, and quality of red wines. Am. J. Enol. Vitic. 12, 9–19.

Paixao, N., Perestrelo, R., Marques, J.C., Camara, J.S., 2007. Relationship between antioxidant capacity and total phenolic content of red, rose´ and white wines. Food Chem. 105, 204–214.

Pellegrini, N., Simonetti, P., Gardana, C., Brenna, O., Brighenti, F., Pietta, P., 2000. Polyphenol content and total antioxidant activity of vini novelli (young red wines). J. Agric. Food Chem. 48, 732–735.

Sacchi, K.L., Bisson, L.F., Adams, D.O., 2005. A Review of the Effect of Winemaking Techniques on Phenolic Extraction in Red Wines. Am. J. Enol. Vitic. 56 (3), 197–206.

Siemann, E.H., Creasy, L.L., 1992. Concentration of the Phytoalexin Resveratrol in Wine. Am. J. Enol. Vitic. 43 (1), 49–52.

Soleas, G., Diamandis, E.P., Goldberg, D.M., 1997. Resveratrol: A molecule whose time has come? And gone? Clin. Biochem. 30, 91 113.

Spranger, M.I., Climaco, M.C., Sun, B., Eiriz, N., Fortunato, C., Adelina, N., Leandro, M.C., Avelar, L., Belchior, A.P., 2004. Differentiation of red winemaking technologies by phenolic and volatile composition. Anal. Chim. Acta 513, 151–161.

Stella, C., Testa, F., Viviani, C., Sabatelli, M.P., 1991. Trattamento a caldo delle uve: influenza su vini da vitigni diversi. Vignevini 12, 47–50.

Stervbo, U., Vang, O., Bonnesen, C., 2007. A review of the content of the putative chemopreventive phytoalexin resveratrol in red wine. Food Chem. 101, 449–457.

Sun, B., Ferrao, C., Spranger, M.I., 2003. Effect of wine style and winemaking technology on resveratrol levels in wines. Cienc. Tec. Vitivinic. 18 (2), 77–91.

Sun, B., Spranger, I., Roque-do Vale, F., Leandro, C., Belchior, P., 2001. Effect of Different Winemaking Technologies on Phenolic Composition in Tinta Miúda Red Wines. J. Agric. Food Chem. 49, 5809–5816.

Threlfall, R.T., Morris, J.R., Mauromoustakos, A., 1999. Effect of Variety, Ultraviolet Light Exposure, and Enological Methods on the *trans*-Resveratrol Level of Wine. Am. J. Enol. Vitic. 50 (1), 57–64.

Timberlake, C.F., Bridle, P., 1976. The effect of processing and other factors on the colour characteristics of some red wines. Vitis 15, 37–49.

Todaro, A., Palmeri, R., Barbagallo, R.N., Pifferi, P.G., Spagna, G., 2008. Increase of *trans*-resveratrol in typical Sicilian wine using b-Glucosidase from various sources. Food Chem. 107, 1570–1575.

Vacca, V., Leccis, L., Fenu, P., Pretti, L., Farris, G.A., 1997. Wine yeasts and resveratrol content. Biotechnol. Lett. 19 (6), 497–498.

Vrhovsek, U., Wendelin, S., Eder, R., 1997. Effects of various vinification techniques on the concentration of *cis* and *trans* resveratrol and resveratrol glucoside isomers in wine. Am. J. Enol. Vitic. 48, 214–219.

Woraratphoka, J., Intarapichet, K.O., Indrapichate, K., 2007. Phenolic compounds and antioxidative properties of selected wines from the northeast of Thailand. Food Chem. 104, 1485–1490.

Yokotsuka, K., Sato, M., Ueno, N., Singleton, V.L., 2000. Colour and sensory characteristics of Merlot red wines caused by prolonged pomace contact. J. Wine Res. 11, 7–18.

Zamora, F., Luengo, G., Margalef, P., Magriña, M., Arola, L., 1994. Nota. Efecto del sangrado sobre el color y la composición en compuestos fenólicos del vino tinto. Rev. Esp. Cienc. Tecnol. Aliment. 34, 663–671.

Zini, S., Canuti, V., Siliani, A., Bertuccioli, M., 2003. Criomacerazione in Toscana: Esperienze sul Sangiovese. Ind. Bevande 32, 16–23.

CHAPTER

61

Arsenic in Wines and Beers from European Markets: Alert of Arsenic Species in Response to Processing

J.-H. Huang, Kan-Nian Hu†, Jörg Ilgen‡, Gunter Ilgen§, Christine Alewell¶*

***Institute of Environmental Geosciences, University of Basel, Basel, Switzerland, †Presentation Road, Brighton, MA, USA, ‡Wein-Bastion Ulm, Ulm, Germany, §Central Analytic, Bayreuth Center for Ecology and Environmental Research (BayCEER), University of Bayreuth, Bayreuth, Germany, ¶Institute of Environmental Geosciences, University of Basel, Basel, Switzerland**

CHAPTER POINTS

- Arsenic concentrations are low in wines and beers nowadays.
- Inorganic arsenic predominates in wines and beers.
- Late-harvested grape berries result in high arsenic concentrations in wines.
- Particle removal decreases arsenic concentration.
- Fermentation changes not only total arsenic concentrations but also arsenic speciation.
- Arsenic in different chemical forms is stable due to the acidity in wines and beers.

INTRODUCTION

Arsenic in different food materials has drawn significant public attention due to its high toxicity (Uneyama *et al.*, 2007). Therefore, many efforts are made to reduce arsenic concentrations during food and beverage production. In particular, arsenic in wines and beers is one of the major public concerns due to the incidence of hundreds of deaths worldwide due to arsenic poisoning from beer and wine consumption in the first half of the 20th century and for the high daily consumption of wine and beer consumption, especially in Europe (Coelho *et al.*, 2003; Fiket *et al.*, 2011; Huang *et al.*, 2012). The toxicity of arsenic depends strongly on its chemical form aside from its quantity (Figure 61.1) (Brown *et al.*, 1997; Dopp *et al.*, 2004). In general, inorganic arsenic is more toxic than organic arsenic. Trivalent arsenic exhibits higher toxic effects than pentavalent arsenic. Thus, arsenic speciation is essential for assessing arsenic toxicity in food materials. This article addressed the occurrence of different arsenic species in beers and wines acquired from European markets. One of the major focuses is the influence of wine- and beer-making procedures on the arsenic concentration and speciation.

RESULTS AND DISCUSSION

Arsenic Concentrations in Beers and Wines

The arsenic concentrations in beers and wines from European markets are generally low according to the literature published in the past decade or so (Figure 61.2) (Barbaste *et al.*, 2003; Coelho *et al.*, 2003; Donadini *et al.*, 2008; Fiket *et al.*, 2011; Huang *et al.*, 2012; Karadjova *et al.*, 2005; Martinez *et al.*, 2001; Moreira *et al.*, 2011; Moreno *et al.*, 2000; Segura *et al.*, 1999; Tasev *et al.*, 2005), which shows only 27 samples among 299 wines contain arsenic with concentrations higher than 10 ppb (WHO and EU drinking water limit). In comparison, the arsenic concentration in beers seems to be averagely higher than in wines, with 13 of 49 samples showing arsenic concentrations higher than 10 ppb. The arsenic concentration in

Processing and Impact on Active Components in Food
http://dx.doi.org/10.1016/B978-0-12-404699-3.00061-5

FIGURE 61.1 Chemical structure of inorganic and organic arsenic compounds found in wines and beers and their pKa values (Aguilar *et al.*, 1987b) and oral LD_{50} for rats (Brown *et al.*, 1997).

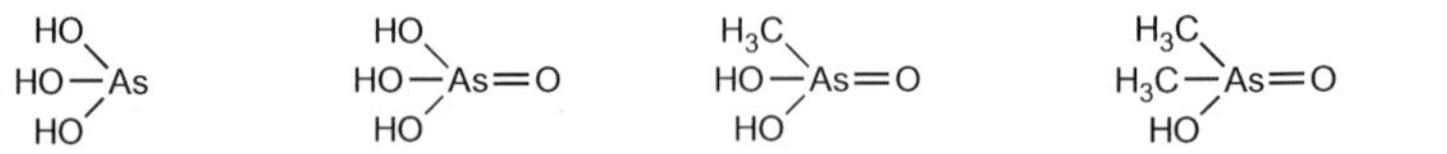

Arsenous acid/ Arsenite	Arsenic acid/ Arsenate	Monomethylarsonic acid/ Monomethylarsonate	Dimethylarsinic acid/ Dimethylarsinate
pKa_1=9.23	pKa_1=2.25	pKa_1=4.26	pKa=6.25
pKa_2=12.1	pKa_2=7.25	pKa_2=8.25	
pKa_3=13.4	pKa_3=12.30		
Oral LD_{50} for rats 41 mg kg^{-1}	Oral LD_{50} for rats 100 mg kg^{-1}	Oral LD_{50} for rats 961 mg kg^{-1}	Oral LD_{50} for rats 644 mg kg^{-1}

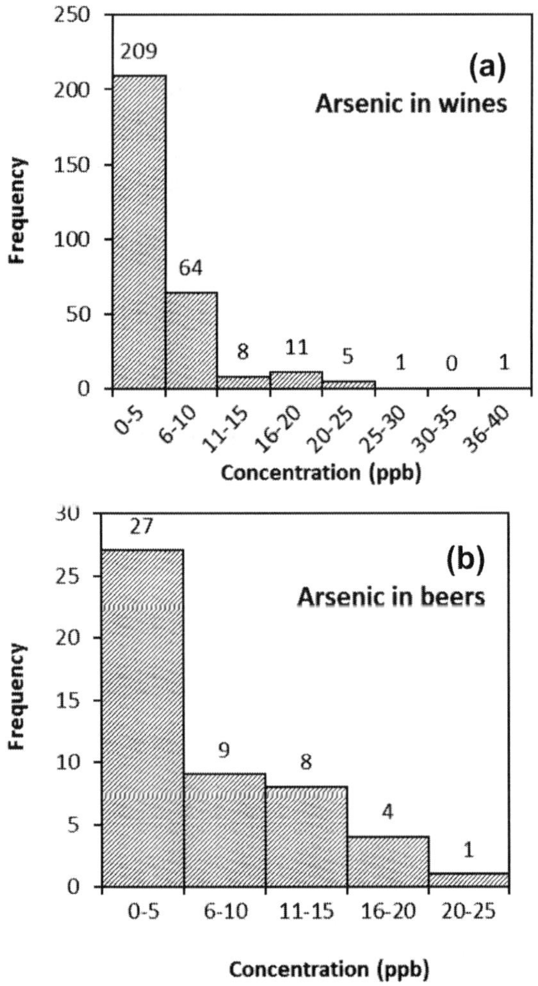

FIGURE 61.2 Histogram of total arsenic concentrations in (a) wines and (b) beers from European markets inclusive of Switzerland, Germany, Austria, France, Italy, Spain, Bulgaria, and Macedonia (Barbaste *et al.*, 2003; Coelho *et al.*, 2003; Donadini *et al.*, 2008; Fiket *et al.*, 2011; Huang *et al.*, 2012; Karadjova *et al.*, 2005; Martinez *et al.*, 2001; Moreira *et al.*, 2011; Moreno *et al.*, 2000; Segura *et al.*, 1999; Tasev *et al.*, 2005). The number of samples is 299 for wines and 49 for beers.

45 wines and 21 beers from Spain are also low with a median of around 8 ppb (Herce-Pagliai *et al.*, 1999, 2002). The averagely low arsenic concentrations reflect the successful control of arsenic in beers and wines nowadays, despite some fluctuation of the data acquired from wines and beers of different origins and sorts.

It is plausible that arsenic concentrations in wines may reflect the bioavailable soil arsenic concentrations. This might be true if the other factors affecting arsenic concentration in wines are similar. Burow *et al.* (1988) investigated total arsenic concentrations in wines from different regions in Germany. The wines from Rheinhessen and Rheinpfalz had apparently higher arsenic concentrations (7.2 and 8.7 ppb) on average than those from the other regions (averagely 6.0 ppb), which coincides well with the concentrations of bioavailable arsenic in soils (Burow *et al.*, 1988).

Summarizing the total arsenic concentrations in European wines from 1750 to 2008 may allow us to understand the historic development of arsenic in wines (Figure 61.3a). In general, the total arsenic concentrations in wines produced before 1960 were mostly above 10 ppb and slightly higher than those after 1960. The highest total arsenic concentrations were found in wines from 1937 and 1947 produced in Frankonia, Germany (392 and 289 ppb). Such extremely high values resulted from the use of arsenic-containing insecticides, which were prohibited in 1942 in Germany and generally worldwide today (Burow *et al.*, 1988). Apparently, the use of arsenic-containing insecticides has the strongest impact on arsenic concentrations in the history of wine. Counterintuitively, without the two extreme values, there is significant negative correlation between the concentration of total arsenic and the year of wine production ($r=-0.73$, $p=0$) (Figure 61.3b). The industrialization of the past 200 years has accelerated the release of geological arsenic into the surface environment with activities including burning fossil fuels, mining, and agricultural use of arsenic compounds. The

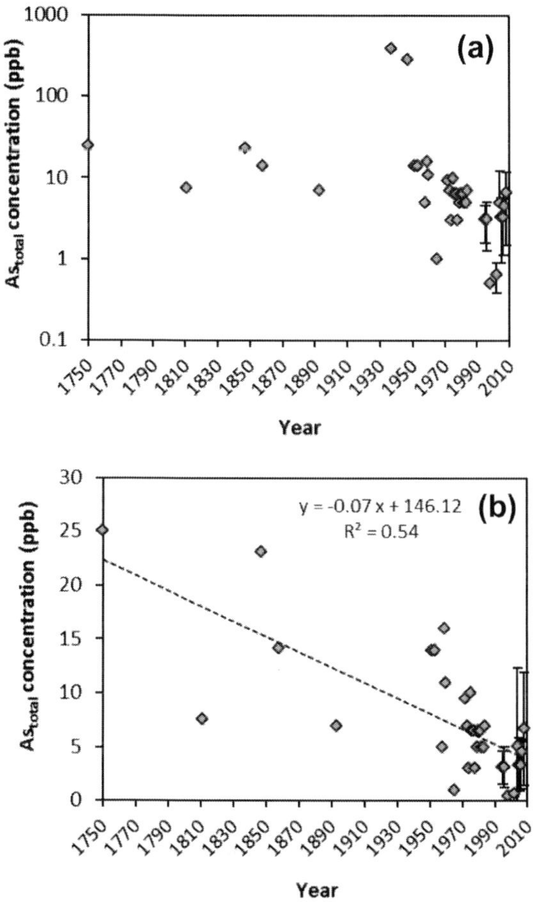

FIGURE 61.3 Historic arsenic concentrations in wines. (a) Overview of average total arsenic concentrations in European wines from different years. The arsenic concentrations are shown in logarithmic scale. (b) Correlation between total arsenic concentration and the year of wine production excluding the two extreme outliers. *Data were summarized from Barbaste et al. (2003), Burow et al. (1988), and Huang et al. (2012).*

soil is an open system and can thus accumulate arsenic from atmospheric deposition and anthropogenic activity, resulting in elevated soil arsenic concentrations worldwide (Shotyk *et al.*, 1996). The arsenic concentration in wine is supposed to increase with the production year due to increasing available arsenic in soils with time. However, decreasing total arsenic in wines was observed with the production year, probably due to the advanced wine production techniques. We will discuss the influence of different wine-making procedures on the arsenic concentration in the later sections.

Arsenic Speciation in Beers and Wines

The primary chemical form of arsenic in beers and wines is still under debate. Herce-Pagliai *et al.* (1999, 2002) pointed out the predominance of either monomethylarsonate or dimethylarsinate among different arsenic species in beers and wines from Spain. The occurrence of methylated arsenic might result from fermentation as detoxification induced by yeasts. This hypothesis was based on the observation of higher proportions of methylated arsenic in alcoholic beers than alcohol-free beers (Herce-Pagliai *et al.*, 1999). Moreover, a positive proportion of alcoholic content with methylated arsenic concentrations was found in wine samples (Herce-Pagliai *et al.*, 2002). Such an explanation seems to be plausible; however, it is not supported by the findings of most of the other studies concerning arsenic speciation in beers and wines, which all showed the predominance of inorganic arsenic (Coelho *et al.*, 2003; Huang *et al.*, 2012; Karadjova *et al.*, 2005; Martinez *et al.*, 2001; Moreira *et al.*, 2011; Moreno *et al.*, 2000; Tasev *et al.*, 2005). Monomethylarsonate and dimethylarsinate were either in low proportion to total arsenic or not found.

A recent study found trace amounts of unknown arsenic compounds in wines, especially in late harvest and ice wines, whose chemical structures were not able to be identified with current analytical techniques (Huang *et al.*, 2012). By cationic exchange chromatography in comparison to the standards (Zorbax 300 SCX with a mobile phase with 20 mM pyridine, pH 2.6), the unknown arsenic compounds should not be trimethylarsine oxide, tetramethylarsonium ion, arsenobetaine, or arsenocholine.

Influence of the Wine-Making Procedure on the Arsenic Concentration and Speciation in Wine

Aguilar *et al.* (1987b) investigated the arsenic concentration and speciation in must, rosé, and red wine samples from Spain. The average total arsenic concentrations decreased from 6.6 ppb in must samples to 3.5 ppb in rosé samples and 4.6 ppb in red wines, indicating that different wine-making procedures are able to decrease arsenic concentrations. In must samples, arsenate was the predominant species by concentration, occupying averagely 35% of total arsenic. Apparently, the exposure to oxygen during must production has resulted in the oxidation of arsenite, which is usually the prevalent arsenic species in plants (Zhao *et al.*, 2009). Through wine production, the proportion of arsenite increased from 20% of total arsenic in must samples to 26% and 41% of total arsenic in rosé and red wines, respectively. Arsenite became especially dominant in red wines. In addition to redox transformation of arsenic during wine making, the concentrations of dimethylarsinate decreased pronouncedly. The proportion of dimethylarsinate to total arsenic decreased from 40% in must to 26% in red wines. There seems to be net arsenic reduction and demethylation during wine making. However, arsenic speciation in must and wine liquids was also influenced by the partition between liquid and solid precipitate, which will be removed during later wine production procedures. In summary, the wine-making procedures influence not only the concentrations of arsenic in wines but also its speciation (Aguilar *et al.*, 1987a).

Step 1. Harvesting (Grape Ripeness)

Generally, it is not trivial to evaluate the factors influencing arsenic concentrations in wines of different types and origins using statistics of even a large data set due to the underlying complexity. Nevertheless, analytical data from wines with different harvesting conditions showed clearly that the longer grape berries stayed on the vine, the more arsenic was concentrated along with the sugar as water escaped. This trend was implied by the arsenic concentration level found in the order: ice wine > late harvest wine > red and white wine (Huang *et al.*, 2012). Late harvest wine is made from grapes that are left longer on the vine in order to concentrate aromas and sugar in the berry. For example, a grape in Germany needs > 76°Oe (degree Oechsle, an indication of grape ripeness and sugar content) to call itself a late harvest (Cluster–Ernährung). For an ice wine, the berries are left even longer on the vine until > 110°Oe is achieved. In the case of ice wine, the berries are left longer on the vine and then harvested and pressed in a frozen state. Thus, while generating more concentrated juice in grapes, the frozen processes also elevate arsenic concentrations because both sugar and dissolved solutes are expelled from the ice.

Step 2. Crushing and Pressing

Crushing and pressing slightly decreased arsenic concentrations by removing stalk and stem by desteming (Aguilar *et al.*, 1987a).

Step 3. Fermentation

Fermentation is one of the most effective procedures of decreasing arsenic concentrations. Removal of ~34% and ~36% of arsenic in red and rosé wines after fermentation was evidenced (Aguilar *et al.*, 1987a). Arsenic volatilization was speculated as the major cause of arsenic loss during fermentation. This may be a plausible explanation since arsenic volatilization is one of the most important processes in arsenic biogeochemical cycling (Chilvers and Peterson, 1987). Microbial arsenic volatilization has been widely identified in both oxic and anoxic conditions, although anoxic conditions are more favorable. Many microorganisms in nature have the ability to volatilize arsenic. To date, more than 125 bacterial and 16 archaeal ArsM homologs have been identified (Qin *et al.*, 2006). Sedimentation of arsenic-containing colloids formed during fermentation may be another pathway for arsenic loss during fermentation, which subsequently stick to the yeast present in the skin.

Aguilar *et al.* (1987a) suggested that the extent of arsenic concentration decrease depended on the intensity of fermentation. To clarify the net impact of fermentation on the total arsenic concentration and arsenic speciation, a model experiment was performed by spiking 0.1 μg arsenite, arsenate, monomethylarsonate, and dimethylarsinate to separate aliquots in wine must in the presence of wine yeast (*Saccharomyces cerevisiae* vars. elliposideus) (Aguilar *et al.*, 1987b). The arsenic speciation was checked after the fermentation stopped after around 2 weeks. Monomethylarsonate was the species with most decreased concentrations. Aguilar *et al.* (1987b) postulated that the loss of monomethylarsonate was due to lipoic acid in yeast cell walls. Monomethylarsonate is able to react with lipoic acid at pH ranging from 3 to 9 (Aguilar *et al.*, 1987b). In comparison, arsenite, arsenate, and dimethylarsinate were known to react with lipoic acid under slight basic condition (pH 8.5). Since wines are usually acidic (pH 2.8–3.8) (Hawesko Weinlexikon), the reactivity of arsenite, arsenate, and dimethylarsinate to lipoic acid has not been probed.

There is both arsenic reduction and oxidation during wine fermentation as a significant amount of arsenite was found in the batch spiked with arsenate and *vice versa* (Aguilar *et al.*, 1987b). In the arsenite spiked experiment, only 63% arsenite of the total arsenic was measured while 19% was arsenate after two weeks' incubation. In comparison, the experiment spiked with arsenate showed only 39% arsenate of the total arsenic while 40% was determined to be arsenite. The above experiments showed greater tendency of arsenate reduction than that of arsenite oxidation. In addition, fermentation seemed to promote arsenate reduction when one study showed that the fermentation experiment spiked with either arsenite or arsenate resulted in an equilibrium proportion between arsenite and arsenate of 3:2 in wines (Burow *et al.*, 1988). This finding corroborates with the fact that arsenite is the most dominant species in wines (Aguilar *et al.*, 1987b; Huang *et al.*, 2012; Karadjova *et al.*, 2005; Moreno *et al.*, 2000).

Arsenic has a high affinity for thiol functional groups. Several thiols and alkyl sulfides, which are responsible for off-aroma in wine were capable of reducing arsenate to arsenite as shown in the literature (Messens and Silver, 2006). Thus, the thiols, sulfides, and alkylsulfides (e.g., lipoic aicd) in must may serve as the reductant in arsenate reduction. In the case of living plants, the thiol-containing phytochelatins are responsible for *in vivo* arsenate reduction for subsequent complexation as a detoxicifation mechanism (Zhao *et al.*, 2009). This explains the predominance of arsenite in freshly pressed grape juice (Huang *et al.*, 2012).

Arsenic methylation and demethylation seem to occur simultaneously during wine fermentation. Around 3 and 10% of measurable arsenic were methylated arsenic after 2 weeks' incubation in the experiment spiked with arsenite and arsenate, respectively (Aguilar *et al.*, 1987b). Six percent of detectable arsenic appeared in the incubation with monomethylarsonate. Conversely, in the batch spiked with monomethylarsonate and dimethylarsinate, 1–2% of measureable arsenic was presented as inorganic arsenic. In summary, the rate of arsenic methylation seemed to be higher than that of demethylation during wine fermentation. In those observations, arsenic methylation is more facilitated under anoxic conditions because organic arsenic may be used as a carbon source for microbial respiration under oxic conditions. Moreover, the tendency of methylation for different arsenic forms during wine fermentation is in the order: arsenate > monomethylarsonate > arsenite (Aguilar *et al.*, 1987b). Despite the presence of arsenic methylation, the amount of methylated arsenic production from fermentation is small and deemed insignificant.

Hydrogen sulfides are produced by yeast during the reduction of sulfate, sulfite, and organosulfur compounds during fermentation. Although copper is usually used in the process to precipitate sulfide, arsenate reduction and formation of arsenic–sulfur complexes and arsenic sulfide precipitates (Burow *et al.*, 1988) still can not be excluded.

Step 4. Purification

The clarity of wine is among the essential qualities required by consumers. Residual particles in wine may disturb the presentation and affect the flavor. Several procedures in wine production, such as crushing, pressing, and fermentation, may generate solid particles of different sizes. The clarity of wine can be restored by racking, protein fining, bentonite, centrifugation and filtration.

Clarification after fermentation can remove notable amounts of arsenic associated with colloids, bacterial remains, as well as amorphous particles (Aguilar *et al.*, 1987a). Colloids are charged particles with a repulsion force that prevents them from becoming larger aggregates for flocculation and sedimentation. The size of colloids ranges from 1 nm to 1 μm thus provides high surface areas for the adsorption of inorganic and organic substances. Proteins, cellulose fibers, yeast cells and bacteria, colloidal coloring matter, ferric phosphate, copper sulfide, ferric ferrocyanide, and bentonide are potential colloidal particles found during wine making. Positively charged colloids such as protein and cellulose fibers are of special interest for arsenate adsorption. Based on pKa (Figure 61.1), arsenate is negatively charged under acidic conditions in wine. In comparison, monomethylarsonate may be partially negatively charged. Dimethylarsinate and arsenite are uncharged. Therefore, they have in principle much lower affinity to colloidal particles. This may also partly explain the species-dependent removal efficiency as monomethylarsonate (31% removed) > dimethylarsinate (16%) > arsenite (15%) (Aguilar *et al.*, 1987b) and the predominance of arsenite in wines. Moreover, bacterial cells and copper sulfide are known as good sorbents for arsenic although the adsorption is decreased by the acidic condition. In summary, removal of solid particles of different sizes contributes to most of the reduction of arsenic concentrations during wine production (Aguilar *et al.*, 1987a).

Some of the materials used for clarification are even able to remove additional arsenic from wine due to their strong adsorption affinity to arsenic, e.g., bentonite (a hydrated aluminum silicate clay), diatomaceous earth (siliceous sedimentary rock), and perlite (spherical, pearlshaped, aluminum silicate particles made by processing volcanic rock). Racking also provides suitable conditions for oxygen to dissolve in the wine to eliminate some unpleasant hydrogen sulfide odors and iron-containing species. In addition, the exposure of wine with oxygen may contribute to arsenite oxidation, which explains the significant amounts of arsenate found in wine.

Influence of the Brewing Procedure on the Arsenic Concentration and Speciation in Beer

The brewing process is usually consistent with mashing, lautering, boiling, fermenting, conditioning, and filtering. It is expected that lautering, fermentation, and filtering will decrease total arsenic concentrations in beer. In contrast to arsenite as the major arsenic species in wine, arsenate is usually the dominant species in beer (Coelho *et al.*, 2003; Huang *et al.*, 2012). Although the stability of arsenite in hot water has been confirmed (Narukawa and Chiba, 2010), its oxidation caused by longer exposure to atmospheric oxygen and heating (boiling) during brewing can not be completely excluded. Incomplete understanding of arsenic in beer requires future research on the evolution of the arsenic concentration and speciation during different brewing procedures.

Arsenic Stability in Wines and Beers

The acidic condition in wines and beers helps to stabilize both inorganic and organic arsenic species because microbial activities are suppressed and arsenite is thermodynamically stable under acidic conditions. A stability test by spiking arsenite, arsenate, and monomethylarsonate and dimethylarsinate in opened wines and beers demonstrated that monomethylarsonate and dimethylarsinate were stable at least for 2 months at 4°C (Huang *et al.*, 2012). There was slight oxidation of arsenite in wines over 6 months' storage at 4°C.

ANALYTICAL TECHNIQUES

Speciation of inorganic and organic arsenic in wines and beers is frequently performed with hyphenation techniques consistent with chromatographic separation of arsenic species and subsequent detection using inductively coupled plasma–mass spectrometry (Huang *et al.*, 2012; Moreira *et al.*, 2011), atomic fluorescence spectrometry (Coelho *et al.*, 2003) or atomic absorption spectroscopy (Herce-Pagliai *et al.*, 1999, 2002). Different strategies were developed to minimize the potential impact of wine and beer matrix on the analytical accuracy including dilution (Moreira *et al.*, 2011), hydride generation (Martinez *et al.*, 2001), and standard addition (Herce-Pagliai *et al.*, 1999, 2002). Hydride generation in combination with atomic fluorescence spectrometry or atomic absorption spectroscopy provides a simpler alternative to measure inorganic arsenic, arsenite, and arsenate, without chromatographic separation (Martinez *et al.*, 2001; Tasev *et al.*, 2005).

CONCLUSION

The processing for wine production has been confirmed to influence the arsenic concentrations and speciation in wine strongly. Such a trend of influence may be applicable in beer products. The wine production procedures of interest here include: (1) the harvest timing that significantly alters the arsenic concentration in grape berries; (2) the crushing, pressing, and purification that decreases arsenic concentrations by removing particle-associated arsenic; (3) the fermentation that decreases not only the arsenic concentrations through volatilization but also changes arsenic speciation through redox transformation, methylation, and demethylation in reaction with the wine matrix. In addition, the geological arsenic background of the viticultural area may influence arsenic concentration in wine. However, it seems to be less important than the aforementioned factors. Finally, the use of arsenic-containing insecticides elevated arsenic concentrations to the maximum level in wine history, but fortunately it is generally banned today.

References

Aguilar, M.V., Martinez, M.C., Masoud, T.A., 1987a. Arsenic content in some Spanish wines—influence of the wine-making technique on arsenic content in musts and wines. Z. Lebensm. Unters. Forsch. 185, 185–187.

Aguilar, M.V., Para, M.C.M., Masoud, T.A., 1987b. Effect of alcoholic fermentation on arsenic species content. Am. J. Enol. Vitic. 38, 282–286.

Barbaste, M., Medina, B., Perez-Trujillo, J.P., 2003. Analysis of arsenic, lead and cadmium in wines from the Canary Islands, Spain, by ICP/MS. Food Addit. Contam. 20, 141–148.

Brown, J.L., Kitchin, K.T., George, M., 1997. Dimethylarsinic acid treatment alters six different rat biochemical parameters: Relevance to arsenic carcinogenesis. Teratogenesis Carcinog. Mutagen. 17, 71–84.

Burow, M., Stoeppler, M., Eschnauer, H., 1988. Arsenic in German and foreign wines. Deut. Lebensm. -Rundsch. 84, 16–19.

Chilvers, D.C., Peterson, P.J., 1987. Global cycling of arsenic. In: Hutchinson, T.C., Meema, K.M. (Eds.), Lead, mercury, cadmium and arsenic in the environment, vol. Scope 31. John Wiley & Sons, New York, pp. 279–301.

Cluster-Ernährung. Wein—Qualitätsstufen, Bayerisches Staatsministerium für Ernährung, Landwirtschaft und Forsten. Available at http://www.cluster-bayern-ernaehrung.de/cms/front_content.php?idart=231.

Coelho, N.M.M., Parrilla, C.N., Cervera, M.L., Pastor, A., de la Guardia, A., 2003. High performance liquid chromatography—atomic fluorescence spectrometric determination of arsenic species in beer samples. Anal. Chim. Acta 482, 73–80.

Donadini, G., Spalla, S., Beone, G.M., 2008. Arsenic, cadmium and lead in beers from the Italian market. J. Inst. Brew. 114, 283–288.

Dopp, E., Hartmann, L.M., Florea, A.M., von Recklinghausen, U., Pieper, R., Shokouhi, B., Rettenmeier, A.W., Hirner, A.V., Obe, G., 2004. Uptake of inorganic and organic derivatives of arsenic associated with induced cytotoxic and genotoxic effects in Chinese hamster ovary (CHO) cells. Toxicol. Appl. Pharmacol. 201, 156–165.

Fiket, Z., Mikac, N., Kniewald, G., 2011. Arsenic and other trace elements in wines of eastern Croatia. Food Chem. 126, 941–947.

Hawesko Weinlexikon. Available at http://www.hawesko.de/wein-abc.html.

Herce-Pagliai, C., Gonzalez, G., Camean, A.M., Repetto, M., 1999. Presence and distribution of arsenical species in beers. Food Addit. Contam. 16, 267–271.

Herce-Pagliai, C., Moreno, I., Gonzalez, G., Repetto, M., Camean, A.M., 2002. Determination of total arsenic, inorganic and organic arsenic species in wine. Food Addit. Contam. 19, 542–546.

Huang, J.-H., Hu, K.N., Ilgen, J., Ilgen, G., 2012. Occurrence and stability of inorganic and organic arsenic species in wines, rice wines and beers from Central European market. Food Addit. Contam. Part A-Chem. 29, 85–93.

Karadjova, I.B., Lampugnani, L., Onor, M., D'Ulivo, A., Tsalev, D.L., 2005. Continuous flow hydride generation-atomic fluorescence spectrometric determination and speciation of arsenic in wine. Spectroc. Acta Pt. B-Atom. Spectr. 60, 816–823.

Martinez, A., Morales-Rubio, A., Cervera, M.L., de la Guardia, M., 2001. Atomic fluorescence determination of total and inorganic arsenic species in beer. J. Anal. At. Spectrom. 16, 762–766.

Messens, J., Silver, S., 2006. Arsenate reduction: Thiol cascade chemistry with convergent evolution. J. Mol. Biol. 362, 1–17.

Moreira, C.M., Duarte, F.A., Lebherz, J., Pozebon, D., Flores, E.M.M., Dressler, V.L., 2011. Arsenic speciation in white wine by LC-ICP-MS. Food Chem. 126, 1406–1411.

Moreno, E., Camara, C., Corns, W.T., Bryce, D.W., Stockwell, P.B., 2000. Arsenic speciation in beverages by direct injection-ion chromatography—hydride generation atomic fluorescence spectrometry. J. Autom. Method Manag. 22, 33–39.

Narukawa, T., Chiba, K., 2010. Heat-assisted aqueous extraction of rice flour for arsenic speciation analysis. J. Agric. Food Chem. 58, 8183–8188.

Qin, J., Rosen, B.P., Zhang, Y., Wang, G.J., Franke, S., Rensing, C., 2006. Arsenic detoxification and evolution of trimethylarsine gas by a microbial arsenite S-adenosylmethionine methyltransferase. Proc. Natl. Acad. Sci. USA. 103, 2075–2080.

Segura, M., Madrid, Y., Camara, C., 1999. Evaluation of atomic fluorescence and atomic absorption spectrometric techniques for the determination of arsenic in wine and beer by direct hydride generation sample introduction. J. Anal. At. Spectrom. 14, 131–135.

Shotyk, W., Cheburkin, A.K., Appleby, P.G., Fankhauser, A., Kramers, J.D., 1996. Two thousand years of atmospheric arsenic, antimony, and lead deposition recorded in an ombrotrophic peat bog profile, Jura Mountains, Switzerland. Earth Planet. Sci. Lett. 145, E1–E7.

Tasev, K., Karadjova, I., Stafilov, T., 2005. Determination of inorganic and total arsenic in wines by hydride generation atomic absorption spectrometry. Microchim. Acta 149, 55–60.

Uneyama, C., Toda, M., Yamamoto, M., Morikawa, K., 2007. Arsenic in various foods: Cumulative data. Food Addit. Contam. 24, 447–534.

Zhao, F.J., Ma, J.F., Meharg, A.A., McGrath, S.P., 2009. Arsenic uptake and metabolism in plants. New Phytol. 181, 777–794.

CHAPTER

62

Aflatoxin B1 in Beer at Different Stages of Production

Antonio Ruiz-Medina, Mª Luisa Fernández-de Córdova

Department of Physical and Analytical Chemistry, University of Jaén, Jaén, Spain

CHAPTER POINTS

- The foods most susceptible to aflatoxin B1 contamination are corn, rice, cereals in general, nuts, pistachios, peanuts, dried fruits, and spices.
- The appearance of aflatoxin B1 in beer is the result of contamination from malted grain or food additives.
- The European Union has not set a maximum allowable limit for aflatoxin B1 in beer, at present, although there is a limit in barley and malt.
- The analytical determination of aflatoxin B1 in beer is complicated by both the complexity of the sample matrix and the low levels present in the sample.
- The two most important groups of methods applied to aflatoxin B1 determination are immunologic ones and chromatographic ones.
- Aflatoxin B1 is not completely removed from the raw materials in brewing and it might be converted to some unknown products.

INTRODUCTION

Mycotoxins are secondary metabolites produced by several species of fungi in the genera *Aspergillus*, *Penicillium*, and *Fusarium*, and are found in several kinds of food, especially cereals and cereal products. Mycotoxins are known to cause severe effects in both humans and animals through their carcinogenic, teratogenic, nephrotoxic,and hepatotoxic properties, and pose serious problems for food safety worldwide. The maximum levels and codes of practice for prevention and reduction have been instituted by the Codex Alimentarius Commission in order to minimize mycotoxin levels in various foods. These regulations have been adopted for foods in many countries; the European Union (EU), especially, places strict regulations on many different kinds of mycotoxins in a wide range of foods.

The best-known and most intensively researched mycotoxins in the world are probably aflatoxins (AFs) which are produced by different species of *Aspergillus* (mainly *A. flavus*, *A. parasitous*, and *A. nomius*), a fungus which is especially found in areas with hot and humid climates. The factors implicated in the growth of this fungus in foods are as much those relating to the environment in which they develop (pH, composition of the food or water activity) as to extrinsic factors: ambient humidity, storage temperature, and microbial competition. They can occur in foods, such as groundnuts, tree nuts, maize, corn, rice, figs, and other dried foods, spices, crude vegetable oils, and cocoa beans, as a result of fungal contamination before and after harvest, during storage if water is allowed to exceed critical values for mold growth. AFs, like mycotoxins, are potentially toxic, carcinogenic, mutagenic, and immunosuppressive agents. They were discovered at the end of the 1950s and beginning of the 1960s as a result of an investigation conducted to evaluate the high mortality rate in poultry and other food-producing animals as a consequence of ingesting feed containing peanuts originating from South America. The ability of AFs to cause cancer and related diseases in humans given their seemingly unavoidable occurrence in food and feed makes the prevention and detoxification of these mycotoxins one of the most challenging toxicology issues of today. The Joint FAO/WHO Expert Committee on Food Additives (JECFA) and the International Agency for Research on Cancer have evaluated AFs on several occasions since

Processing and Impact on Active Components in Food
http://dx.doi.org/10.1016/B978-0-12-404699-3.00062-7

1987 and have recommended that, due to their carcinogenic potential, dietary exposure to AFs should be minimized as much as possible. In this way, the 2007 report by the Panel on Contaminants of the European Food Safety Authority (EFSA) indicated that exposure to AFs from any food source should be kept as low as is reasonably possible, due to their genotoxic and carcinogenic properties.

Among 18 different types of AFs identified, major members are AFs B1 (AFB1), B2, G1, G2, M1, and M2. The EU introduced measures to minimize the presence of all these AFs in different foodstuffs (cereals, oilseeds, spices, foods for infants and young children, dietary foods, etc.). Maximum levels of AFs are laid down in European Commission (EC) Regulation No 1881/2006. As a result, the products specified in the regulations which contain AFs in levels exceeding the established maximum content must not be marketed as such, even after being combined with other food products, and must not be used as ingredients in other foods. AFB1 is the most common in food and amongst the most potent genotoxic and carcinogenic AFs. It has a condensed coumarinic structure with one bifurane and one pentanone, as shown in Figure 62.1.

Since cereals are raw materials of beer and beer-based drinks, it is necessary to consider that the risk of contamination also exists in beer and beer-based drinks, produced in this case by *Aspergillus carbonarius*. Studies with different mycotoxins added at various stages of the brewing process show that AFB1 may be transmitted from contaminated grains into beer. They originate from the malted grain or from food additives. The natural occurrence in hops of AFB1, maize for example, has been reported in various studies; so, as a consequence of the use of these adjuncts in the brewing process, AFB1 may potentially occur in beer. Reducing the fungal contamination of malt and barley is the most promising strategy for reducing AFB1 in beer. Maximum residue limits (MRL) have been established for AFB1 and the sum of B1, B2, G1, and G2 in different foods; nevertheless, at present, the European Union has not set a maximum allowable limit for AFB1 in beer, although there is a limit in barley and malt [EC no 2174/2003 of 12 December 2003].

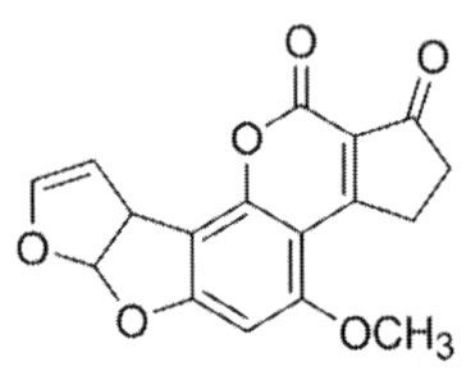

FIGURE 62.1 Structure of aflatoxin B1.

EVOLUTION OF AFB1 ALONG THE BREWING PROCESS

Beer's preparation process and transfer of AFB1 in this matrix will be evaluated.

Brewing Process

All beer is brewed from malted barley, hops, yeast and water, although other ingredients such as fruit, wheat and spices are sometimes used. The yeast turns sugars in the malt into alcohol and the hops provide the bitter flavors in beer and the flowery aroma. The flavor of the beer depends on many things, including the types of malt and hops used, other ingredients, and the yeast variety. Getting the yeast right is essential as each variety has its own distinctive effect on the beer. The basic process may be simple but the execution is highly sophisticated. The complete stages, shown in Figure 62.2, are as follows.

Malting

Malting is the process of readying barley to be used in brewing. Firstly, the grain is steeped (soaked) in a vat of water for about 40 h, and then germination of the barley grain is produced. At the end of this process, about three days, the starch has become soft and the enzymes have not started converting the starch into sugar yet. The barley grain is called green malt. The germination is halted by drying the green malt on metal racks in the kiln house at 50°C. The temperature is then raised to around 85°C for a light malt, or higher for a dark malt. Although malted barley is the primary ingredient, unmalted corn, rice, or wheat are sometimes added, to produce different beer flavors.

Milling

Milling is the cracking of the grain which the brewer chooses for the particular batch of beer. Milling the grain allows it to absorb the water it will eventually be mixed with, in order for the water to extract sugars from the malt.

Mashing

Mashing is the process of turning the finely ground malt, the grist, into a sweetened liquid. Mashing converts the starches, which were released during the malting stage, to sugars that can be fermented. The milled grain is dropped into warm water then gradually heated to around 75°C in a large cooking vessel called the mash tun. In this mash tun, the grain and heated water mix creating a cereal mash to dissolve the starch into the water, transforming it into sugar—mainly maltose. Because water is such a vital part of the brewing process, the water itself

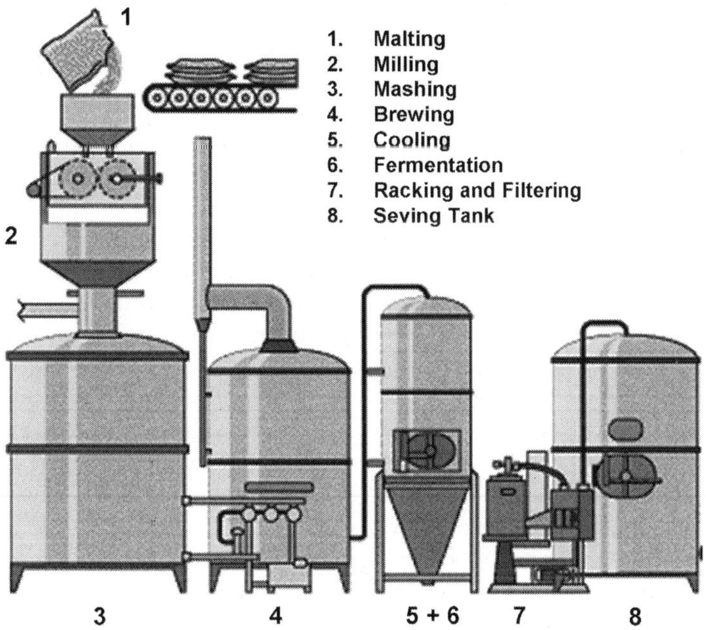

FIGURE 62.2 Stages in the brewing process.

is a key ingredient. This sugar-rich water is then strained through the bottom of the mash and is now called wort.

Brewing

The spent grains are filtered out and the wort is ready for boiling, which involves many technical and chemical reactions. During this stage, important decisions will be made affecting the flavor, color, and aroma of the beer. Certain types of hops are added at different times during the boil for either bitterness or aroma and to help preserve it. Normally, maize grit is used. The wort is boiled for 1–2h to sterilize and concentrate it, and extract the necessary essence from the hops. The pH is adjusted to 5.5 ± 0.5 by means of phosphoric acid.

Cooling

The wort is transferred quickly from the brew kettle through a device to filter out the hops, and then on to a heat exchanger to be cooled. The heat exchanger basically consists of tubing inside of a tub of cold water. It is important to quickly cool the wort to a point where yeast can safely be added, because yeast does not grow in high temperature. The hopped wort is saturated with air, essential for the growth of the yeast in the next stage.

Fermentation

After passing through the heat exchanger, the cooled wort goes to the fermentation tank. The brewer now selects a type of yeast and adds it to the fermentation tank. The yeast ferments the wort sugars into alcohol and flavor, and also produces carbon dioxide. This process of fermentation takes 10 days, approximately. The wort finally becomes beer. Each brewery has its own strains of yeast, and it is these that largely determine the character of the beer. In some yeast varieties, the cells rise to the top at the end of fermentation, and are then skimmed off. This is called top fermentation, and ales are brewed in this way. When at the end of fermentation the yeast cells sink to the bottom, the process is known as bottom fermentation, used for lager or pils.

Racking (Maturation) and Filtering

The beer has now been brewed, but it can still be improved through *maturation*. During this phase, the brewer moves, or racks, the beer into a new tank called the conditioning tank. The brewer then waits for the beer to complete its aging process. The taste ripens. The liquid clarifies as yeast and other particles settle. Finally, the beer is filtered and carbonated. Further *filtering* gives the beer a sparkling clarity (diatomeaus earth and polyvinyl polypyrrolidine can be used as the main filter aids). Subsequently, dilution to the final beer alcohol content is performed by means of de-aerated water.

Serving Tank

The beer is moved to a holding tank where it remains until it is bottled, pasteurized (kept at ~65°C for 20min and then cooled to room temperature), canned, or put into kegs. Filling techniques ensure air does not come into contact with the beer, and can not be trapped within the container.

Transfer of AFB1

Since beer significantly contributes to the diet of the population, control of the presence of mycotoxins, including AFB1, in this commodity is very important. The first experiments to trace the AFB1 fate in the brewing process were carried out in the 1970s when the authors (Chu *et al.*, 1975) added purified standard of AFB1 at concentrations of 1 and 10 mg/g of malt before the malting process. A great portion of AFB1 was partially lost during the brewing. It was relatively stable in the cooker mash step but was more sensitive to later treatments, especially the wort boiling and final fermentation. Since AFB1 is relatively stable to heat, it is not surprising that more than 90% remained in the sample after cooker mash treatment. Removal of toxin from the product was very significant in the malt mash (with a loss of 12% for a 10 mg/g sample and 24–27% for a 1-mg/g sample), in boiled wort (20–30% loss of material added to the malt mash), and in the final fermentation (another 20–30% loss from the boiled wort step) steps. The mechanism of AFBl removal, however, is not known. The significant loss of AFB1 between the combined mash and boiled wort steps may be due to the non-specific interaction or adsorption of AFB1 by the solid particles removed by the filtration process.

In another interesting study, the transfer of AFB1 from naturally contaminated raw materials to beer was evaluated (Pietri *et al.*, 2010). AFB1 ranged from 0.31 to 14.85 μg/kg in maize grit; it was also found in malted barley, at levels ranging from 0.20 to 4.07 μg/kg. Malted barley and maize grit used for brewing showed concentrations higher than 2 μg/kg, which is the maximum level fixed by EC Regulation for all cereals and all products derived from cereals, including processed cereal products. For each stage of the brewing, the AFB1 percentage in the product, referring to the total amount in the raw materials (TARM), was calculated. A relatively small amount of the toxin was transferred to the liquid phase during mashing; only 14% of the AFB1 TARM was found in the liquor after mashing: this is probably due to the low solubility of the toxin in water. In the sweet wort the AFB1 level was reduced both because of the dilution of the product, and mainly because of the filtration step: only an average of 8% of the AFB1 TARM was found in the sweet wort. In the following steps, the reductions were more moderate (boiling −28%, cooling −3%, fermentation 14%), while a further strong decrease (−60%) was observed passing from green to bottled beer. The percentages of AFB1 TARM found in the bottled beer varied from 0.6 to 2.2% (mean 1.5%) and the concentrations, because of the high dilution, were at the low parts-per-trillion level.

Before 1991 AFs, including AFB1, were not detected in European beers; later, in the 1990s trace amounts were very rarely found in several samples of beers imported from south Europe. They occurred in beers coming from South America or equatorial Asian countries, at levels lower than 0.09 mg/l (Scott, 1996). In a subsequent survey on 304 samples of Canadian and imported beer, the toxin was found only in some samples from India (highest value 0.23 mg/l), Spain, Portugal, and Mexico (all lower than 0.02 mg/l) (Mably *et al.*, 2005). The presence of AFB1 in raw materials and beers from tropical countries is common; increased aflatoxin concentration is frequently reported in sorghum, a traditional African cereal, and beers brewed from it.

In 2008–2011 a total set of 333 samples of brewing raw materials and beer were analyzed (Beneŝová *et al.*, 2012) for the presence of AFB1 and AFs B2, G1, and G2. In seven of 216 samples of brewing raw materials (3.2%), AFs were found at trace concentrations to 1.2 mg/kg. Conversely, six positive beer samples (5.1%) contained AFs in the values exceeding the quantitation limit (LOQ). In five of them, only AFB1 at maximum concentration 10.6 ng/l was detected. The sixth positive beer sample contained 31.0 ng/l of the other three AFs. Values in barley and malt did not exceed the maximum allowable limit set by the EU.

The occurrence of AFB1 and other mycotoxins in a sample of 106 beers produced in several European countries was also investigated (Bertuzzi *et al.*, 2011). AFB1 was not detected in any sample, whereas ochratoxin A, deoxynivalenol, and fumonisins were found in a relatively high number of samples. Finally, a study for the selective determination of AFB1 in different types of beer was recently developed (Molina-García *et al.*, 2012). Of the 17 beers analyzed, including normal (n = 12) and non-alcoholic (n = 5) beers, only three samples contained traces of the target compound (0.86, 0.28, and 0.17 mg/l). AFB1 was not detected in the remaining beer samples investigated, including non-alcoholic beers.

ANALYTICAL TECHNIQUES

The analytical determination of AFB1 is complicated by two main factors: (1) the complexity of the sample matrix in which it normally appears; and (2) the low levels present in the sample. Analytical methodologies must be designed to address these requirements. Firstly, to avoid matrix interferences, a typical analysis of AFB1 requires a pre-treatment of the beer. Secondly, with respect to its detection, a reliable and sensitive method must be selected for the screening and determination of this compound.

Pre-treatment

This involves a liquid–liquid extraction (using solvents such as methanol, acetonitrile, and/or their combinations), followed by a clean-up step, e.g., multifunctional or immunoaffinity columns. Solid-phase extraction (SPE) with immunoaffinity material is very popular in mycotoxin analysis because it is a very selective and time-saving sample cleanup tool (Visconti *et al.*, 2000) for removing matrix compounds. This approach has been mainly used for the analysis of one mycotoxin or family of mycotoxins. However, one of the problems of the immunoaffinity materials is the high cost, and other alternatives have been checked with conventional sorbents such as C_{18}, hydrophilic–lipophilic balanced copolymers, or ion exchangers (Romero-González *et al.*, 2009). The clean-up of this extraction procedure prevents the instrumental system from being contaminated, and the cartridges are less expensive than immunoaffinity columns.

Another alternative is to use the QuEChERS (Quick, Easy, Cheap, Effective, Rugged, and Safe) method (Anastassiades *et al.*, 2003). This methodology, which has recently attracted attention as a simple clean-up method for multiple pesticide residues analysis, was examined in beer-based drinks (Tamura *et al.*, 2011) (beer, low-malt beer, new genre, and nonalcoholic). The characteristic features of the QuEChERS method are as follows: (1) extraction with acetonitrile in a disposable tube, followed by the salting out and removal of water from acetonitrile using sodium chloride and anhydrous magnesium sulfate; and (2) purification with dispersive solid-phase extraction, in which the extract is processed by shaking with either a primary–secondary amine (PSA) silica gel alone, or PSA plus C_{18} or graphite carbon black. The greatest advantage of using this method is that the time required to perform the assay is reduced, since there are only two steps involved.

Finally, other alternative procedures such as solid-phase microextracion (SPME) and liquid-phase microextraction (LPME) have also been proposed for the determination of mycotoxins in alcoholic beverages. These procedures have been used for the determination of mycotoxins such as ochratoxin A but no application to AFB1 has been developed up to date.

Determination of AFB1

As indicated above, AFB1 gets into beer either from contaminated input materials or adjuncts added in the course of brewing. For this purpose, reliable analytical methods for fast and effective monitoring of this compound in the beer production chain are needed. Different methods have been established for the determination of AFB1 (Turner *et al.*, 2009), including capillary electrophoresis (Peña *et al.*, 2002) and thin-layer chromatography (Braicu *et al.*, 2008).

Nevertheless, the two most important groups of methods applied to AFB1 determination are immunologic ones and chromatographic ones (gas or liquid chromatography). Kits to perform the enzyme-linked immunosorbent assays (ELISA) (Ghali *et al.*, 2008) are available from different suppliers and these methods are often used for screening purposes and in the field. Different versions exist that vary in the rapidity of the procedure and the quantification of the results, although all of them require enzymatic reactions, washing, and separation of bound and free label. A very important step of validation is the detection of matrix effects, and since false positives can occur, positive results could require confirmation by more expensive chromatographic methods.

The chromatographic methods measure the compound specifically after sample extraction and extract clean-up followed by a separation by gas or liquid chromatography (LC) (Gilbert and Vargas, 2003; Zollner and Mayer-Helm, 2006) and a specific detection. Nowadays, LC coupled to mass spectrometry (MS) or tandem mass spectrometry (MS/MS) is mainly used for the simultaneous determination of several classes of mycotoxins because it provides more accurate identification, better selectivity, and higher sensitivity than other detection techniques although it requires expensive equipment. The number of publications in this field has increased in recent years (Sforza *et al.*, 2006), including applications to beer samples (Romero González *et al.*, 2009; Rudrabhatla *et al.*, 2008; Ventura *et al.*, 2006; Zachariasova *et al.*, 2010). Characteristics of the system and figures of merit of different published methods are included in Table 62.1.

Despite the advantages of MS detection, the AOAC official method for the determination of AFB1 in foods and agriculture commodities is based on high-performance liquid chromatography (HPLC) with fluorescence detection (native fluorescence of AFB1) (AOAC Official method, 2000). Firstly, an extraction step with acetonitrile and deionized water followed by a clean-up procedure through a packing material are realized. The purified extract is transferred quantitatively from top of column to screw cap vial (derivatization vial) and is evaporated under nitrogen. Secondly, after adding n-hexane to redissolve AFB1, trifluoroacetic acid and acetonitrile are added to produce the aflatoxin derivatization. Finally, fluorescence is employed as detection in HPLC. The excitation and emission wavelengths of 365 and 435 nm, respectively, are set during analysis.

Sensors may be also a good choice for the analysis of AFB1 due to their fast, simple, and low-cost detection capabilities. Electrochemical sensors (Owino *et al.*, 2008; Tan *et al.*, 2009) can be found in literature for the determination of this aflatoxin but they have not been applied to alcoholic beverages. The use of spectrofluorimetric detection would be a possibility to develop an optical sensor but it is difficult due to the complexity of the matrix, which shows a great variety of natural fluorescent compounds whose spectra often overlap the analyte signal. As an alternative to measure native fluorescence of AFB1, a method based on the on-line generation of a fluorescent photoproduct from this one by

TABLE 62.1 Chromatographic Methods with Fluorescence or Mass Detection

Method	Instrumentation	Sample Preparation	LOD[a] (μg/l)	LOQ[b] (μg/l)	Sample	Reference
HPLC–fluorescence detection	Column: Superspher RP-18 (Merk)	Immunoaffinity clean-up	0.001	0.003	Beer	Bertuzzi *et al.* (2011)
HPLC–MS	Column: Synergi Hidro RP 80A	Immunoaffinity clean-up	0.0015	0.005	Brewing materials and beer	Benešová *et al.* (2012)
U-HPLC–MS/MS	Column: Acquity UPL (Waters)	SPE clean-up	0.04	0.13	Beer	Romero González *et al.* (2009)
U-HPLC–MS/MS	Column: Acquity UPL (Waters)	SPE clean-up	0.1	—	Beer	Ventura *et al.* (2006)
LC–Fluorescence detection	—	Immunoaffinity clean-up	0.019	—	Beer	Scott *et al.* (1997)
LC–Fluorescence detection	Column: HP SB-c18 (Chromatog. Specialities)	Immunoaffinity clean-up	0.004	0.0007	Wine and beer	Mably *et al.* (2005)
LC–MS/MS	Column: Pro Star 210 (Varian)	Immunoaffinity clean-up	—	—	Beer and sake	Rudrabhatla *et al.* (2008)
U-HPLC–orbitrapMS	Column: Acquity UPLC HSS T3 (Waters)	Liquid extraction (acetonitrile)	—	0.5	Beer	Zachariasova, *et al.* (2010)
U-HPLC–MS/MS	Column: Acquity UPLC (Waters)	QuEChERS	—	1.0	Beer-based drinks	Tamura *et al.* (2011)

[a]*LOD: Detection limit.*
[b]*LOQ: Quantitation limit.*

UV-irradiation and its monitoring when adsorbed onto C_{18} silica gel has been proposed (Molina-García *et al.*, 2012). The measurement of the photoproduct retained on this solid support, packed in the flow-cell placed in the detection area, was adopted for improving both detection limit and selectivity. The use of multicommutation as an alternative to conventional flow injection analysis, is introduced due to its favorable intrinsic advantages such as low-cost equipment, high sample throughput, and simplicity, as long as automation is complete.

CONCLUSIONS

The findings and the changes that are occurring in the field of nutrition make it necessary to rethink nutritional policies. In the case in question, both the introduction into the diet of foods originating from different countries, which were not previously a habitual part of the diet, and the changes in environmental conditions, can lead to changes in fungal development or the production of mycotoxins. On this point, the European Food Safety Authority has recently initiated the process for studying the potential increase in AFB1 presence in foods as a consequence of climate change, using predictive models to indicate an emerging potential food contamination. The most important tools for the prevention and limiting of the appearance of AFB1 in foods are good practice in agricultural production and food storage. Foods should be stored at relative temperatures and humidities that discourage both fungal growth and aflatoxin production by potentially aflatoxin-producing fungi. It is also extremely important to avoid cross-contamination between different batches of foods.

In the case of beer, AFB1 is not completely removed from the raw materials in brewing and it might be converted to some unknown products which might also be toxic. In general, in order that barley germinates and grows (malt) satisfactorily, and therefore meets the requirements of the malting industry, it must be stored under conditions which would not promote the growth of storage molds. Moreover, detoxication methods such as heat and fermentation processes can go along way in reducing the levels of AFB1.

We can conclude that beers brewed from quality, purified, and well stored raw materials do not represent any health risk. If raw materials comply with the limits fixed by the EC Regulation, the contribution of a moderate daily consumption of beer to AFB1 intake does not contribute significantly to the exposure of the consumer.

References

Anastassiades, M., Lehotay, S.J., Stajnbaher, D., Schenck, F.J., 2003. Fast and easy multiresidue method employing acetonitrile extraction/partitioning and "dispersive solid-phase extraction" for the determination of pesticide residues in produce. J. AOAC Int. 86, 412–431.

AOAC Official Method 994.08, 2000. Aflatoxins in corn, almonds, Brazil nuts, peanuts, and pistachio nuts, multifunctional column (Mycosep) method. Nat. toxins-chapter 49, 26–27. Official Methods of Analysis of AOAC International, seventeenth ed. Volume II, AOAC International, Gaithersburg, Maryland, USA.

Benešová, K., Běláková, S., Mikulíková, R., Svoboda, Z., 2012. Monitoring of selected aflatoxins in brewing materials and beer by liquid chromatography/mass spectrometry. Food Control 25, 626–630.

Bertuzzi, T., Rastelli, S., Mulazzi, A., Amedeo, G.D., 2011. Mycotoxin occurrence in beer produced in several European countries. Food Control 22, 2059–2064.

Braicu, C., Puia, C., Bodoki, E., Socaciu, C., 2008. Screening and quantification of aflatoxins and ochratoxin A in different cereals cultivated in Romania using thin layer chromatography–densitometry. J. Food Qual. 31, 108–120.

Chu, F.S., Chang, C.C., Ashoor, S.H., Prentice, N., 1975. Stability of aflatoxin B1 and ochratoxin A in brewing. Appl. Microbiol. 29, 313–316.

Ghali, R., Hmaissia-khlifa, K., Ghorbel, H., Maaroufi, K., Hedili, A., 2008. Incidence of aflatoxins, ochratoxin A and zearalenone in Tunisian foods. Food Control 19, 921–924.

Gilbert, J., Vargas, E.A., 2003. Advances in sampling and analysis for aflatoxins. J Toxicol. 22, 381–422.

Mably, M., Mankotia, M., Cavlovic, P., Tam, J., Wong, L., Pantazopoulos, P., Calway, P., Scott, P.M., 2005. Survey of aflatoxins in beer sold in Canada. Food Addit. Contam. 22, 1252–1257.

Molina-García, L., Fernández-de Córdova, M.L., Ruiz-Medina, A., 2012. Indirect determination of aflatoxin B1 in beer via a multi-commuted optical sensor. Food Addit. Contam. 29, 392–402.

Owino, J.H.O., Arotiba, O.A., Hendricks, N., Songa, E.A., Jahed, N., Waryo, T.T., Ngece, R.F., Baker, P.G.L., Iwuoha, E.I., 2008. Electrochemical immunosensor based on polythionine/gold nanoparticles for the determination of aflatoxin B1. Sensors 8, 8262–8274.

Peña, R., Alcaraz, M.C., Arce, L., Rios, A., Valcarcel, M., 2002. Screening of aflatoxins in feed samples using a flow system coupled to capillary electrophoresis. J. Chromatogr A. 967, 303–314.

Pietri, A., Bertuzzi, T., Agosti, B., Donadini, G., 2010. Transfer of aflatoxin B1 and fumonisin B1 from naturally contaminated raw materials to beer during an industrial brewing process. Food Addit. Contam. 27, 1431–1439.

Romero-González, R., Martínez-Vidal, J.L., Aguilera-Luiz, M.M., Garrido-Frenich, A., 2009. Application of conventional solid-phase extraction for multimycotoxin analysis in beers by ultrahigh-performance liquid chromatography–tandem mass spectrometry. J. Agric. Food Chem. 57, 9385–9392.

Rudrabhatla, M., George, J.E., Hill, N.R., Siantar, D.P., 2008. Detection of relevant mycotoxins in wheat beer and sake by LC-MS/MS using prototype immunoaffinity column clean-up: a preliminary study. Food Contaminants: Mycotoxins Food Allergens 1001, 241–254.

Sforza, S., Dall'Asta, C., Marchelli, R., 2006. Recent advances in mycotoxin determination in food and feed by hyphenated chromatographic techniques/mass spectrometry. Mass Spectrom. Rev. 25, 54–76.

Scott, P.M., 1996. Mycotoxins transmitted into beer from contaminated grains during brewing. J. AOAC Int. 79, 875–882.

Scott, P.M., Lawrence, G.A., 1997. Determination of aflatoxins in beer. J. AOAC Int. 80, 1229–1234.

Tamura, M., Uyama, A., Mochizuki, N., 2011. Development of a multi-mycotoxin analysis in beer-based drinks by a modified QuEChERS method and ultra-high-performance liquid chromatography coupled with tandem mass spectrometry. Anal. Sci. 27, 629–635.

Tan, Y., Chu, X., Shen, G.L., Yu, R.Q., 2009. A signal-amplified electrochemical immunosensor for aflatoxin B1 determination in rice. Anal. Biochem. 387, 82–86.

Turner, N.V., Subrahmanyam, S., Piletsky, S.A., 2009. Analytical methods for determination of mycotoxins: a review. Analytica Chimica Acta 632, 168–180.

Ventura, M., Guillén, D., Anaya, I., Broto-Puig, F., Lliberia, J.P., Agut, M., Comellas, L., 2006. Ultra-performance liquid chromatography/tandem mass spectrometry for the simultaneous analysis of aflatoxins B1, B2, G1, G2 and ochratoxin A in beer. Rapid Commun. Mass Spectrom. 20, 3199–3204.

Visconti, A., Pascale, M., Centonze, G., 2000. Determination of ochratoxin A in domestic and imported beers in Italy by immunoaffinity clean-up and liquid chromatography. J. Chromatogr. A 888, 321–326.

Zachariasova, M., Cajka, T., Godula, M., Malachova, A., Veprikova, Z., Hajslova, J., 2010. Analysis of multiple mycotoxins in beer employing (ultra)-high-resolution mass spectrometry. Rapid Commun. Mass Spectrom. 24, 3357–3367.

Zollner, P., Mayer-Helm, B., 2006. Trace mycotoxin analysis in complex biological and food matrices by liquid chromatography–atmospheric pressure ionisation mass spectrometry. J. Chromatogr. A 1136, 123–169.

CHAPTER

63

Behavior of Triazole Fungicide Residues from Barley to Beer

Simón Navarro[*], *Gabriel Pérez-Lucas*[†], *Nuria Vela*[‡], *Ginés Navarro*[*]

*Departamento de Química Agrícola, Geología y Edafología, Facultad de Química, Universidad de Murcia, Campus Universitario de Espinardo, Murcia, Spain, †Departamento de Química Agrícola, Geología y Edafología, Facultad de Ciencias Sociosanitarias, Avenida de las Fuerzas Armadas, Lorca, Murcia, Spain, ‡Facultad de Enfermería, Universidad Católica San Antonio de Murcia, Campus de Los Jerónimos, Guadalupe, Murcia, Spain

CHAPTER POINTS

- Several pests and diseases can attack cultivated barley and pesticides are widely used in different combinations at many stages of barley cultivation and also during post-harvest storage.
- Azole compounds (triazole- and imidazole-derivatives) play a key role as antifungal agents in agriculture because of their broad therapeutic window, wide spectrum of activity, and low toxicity.
- Azole fungicides inhibit targets within the fungal sterol biosynthesis. They are by far the most important class of sterol biosynthesis inhibitor (SBI) fungicides since the 1980s.
- Residues of azole compounds on barley may pass to malt, wort and beer during the common beer brewing operations (malting, mashing, boiling, and fermentation).
- Nowadays, analysis of triazole and imidazole residues in malt and beer compounds is mostly carried out by liquid chromatography in tandem with mass spectrometry.
- Fermentation rate and quality of the beer may be affected by the presence of triazole residues on the sweet wort.
- The amounts of triazole residues found in beer and spent grains must be safe for consumers and animals, respectively, and must be as low as possible to avoid toxicological risks.

INTRODUCTION

As a consequence of infestation by pests and diseases on crops, plant protection products are used before and after harvest. The amounts of residues in food must be safe for consumers and must be as low as possible. Control of pesticide residues in crops is generally based on Maximum Residue Limits (MRLs) which are set using field trials for a particular pesticide to arrive at the highest residue levels expected under use according to Good Agricultural Practice (GAP). To evaluate the behavior of pesticide residues, not only the results of residue tests carried out on plant and plant products are necessary, but also the results of residue tests carried out on processed plant products (processing studies) are important and necessary. Depending on the type of process involved and on the chemical nature of the residue in the raw agricultural commodities (RACs), differences in the nature of the residue in the processed commodities may have to be determined. The major factors influencing the stability of residues in typical processes are temperature, pH, water content, and chemical nature of the residue. According to Timme and Walz-Tylla (2004), hydrolysis is most likely to affect the nature of the residue during most processing operations, because processes such as heating would generally inactivate enzymes present in the substrate, leaving simple hydrolysis as a degradation mechanism.

Because of their high content of starch and storage proteins, barley grains especially represent an attractive source of nutrients for insects and microbial pathogens. As a consequence, several pests and diseases can attack cultivated barley and pesticides are widely used

Processing and Impact on Active Components in Food
http://dx.doi.org/10.1016/B978-0-12-404699-3.00063-9

in different combinations at many stages of barley cultivation and also during post-harvest storage. As reported by some authors (Navarro and Vela, 2009), if barley contains pesticide residues they may remain in the malt (germinated barley), wort, and beer obtained after common brewing operations (malting, mashing, boiling, and fermentation). Their transfer will depend on the process used, although it should be noted that the removal of material in the form of trub and spent grains tends to reduce the level of pesticides, which are often relatively insoluble in water (Miyake *et al.*, 1999; Hengel and Shibamoto, 2002; Navarro *et al.*, 2005; 2011a). Depending on the stage involved, the physical location of residues and their physical–chemical properties (mainly K_{OW} value, water solubility, hydrolytic rate constant, vapor pressure and Henry's Law constant), differences in their final fate are observed (Holland *et al.*, 1994). Also, if pesticide residues are present in the brewer wort they can alter the normal fermentative process, being able to cause in certain cases sluggish and even stuck fermentation resulting in the alteration of some organoleptic properties of the beer (Jones *et al.*, 1988; Navarro 2007b, 2011b). Therefore, the knowledge of the fate of pesticide residues during brewing is an essential feature of the modern brewing industry.

Especially, azole compounds (triazole- and imidazole-derivatives) play a key role as antifungal agents in agriculture because of their broad therapeutic window, wide spectrum of activity, and low toxicity. Triazole derivatives used in agriculture are successfully used against mildews and rust of cereal grains, fruits, vegetables, and ornamentals. They are sterol biosynthesis-inhibiting (SBI) fungicides and their antifungal activity is based on their capability to interfere with steroid biosynthesis and thereby with the formation of fungal walls (Koller, 1988).

STEROL BIOSYNTHESIS INHIBITOR (SBI) FUNGICIDES IN AGRICULTURE

For more than the past three decades, fungicides that inhibit targets within the fungal sterol biosynthesis have been the most important group of specific fungicides worldwide (Kuck *et al.*, 2012). Fungi have specific sterols that differ from those found in plants and animals. The sterols in nature come from a common source, cyclopentane perhydro phenanthrene, also known as sterane, precursor of diverse steroids like cholestane, stigmastane, or ergostane that are those responsible for the formation of cholesterol (in animals) and stigmasterol and ergosterol (in vegetables), respectively (Figure 63.1). As a consequence, selective inhibitors that cover a broad spectrum of plant pathogens have been developed. In most pathogens, fungal cell membranes are characterized belonging to the Ascomycetes and Basidiomycetes by a common dominant sterol component, ergosterol, a crystalline sterol synthesized by yeast from sugars or derived from ergot and converted to vitamin D_2 when exposed to ultraviolet radiation.

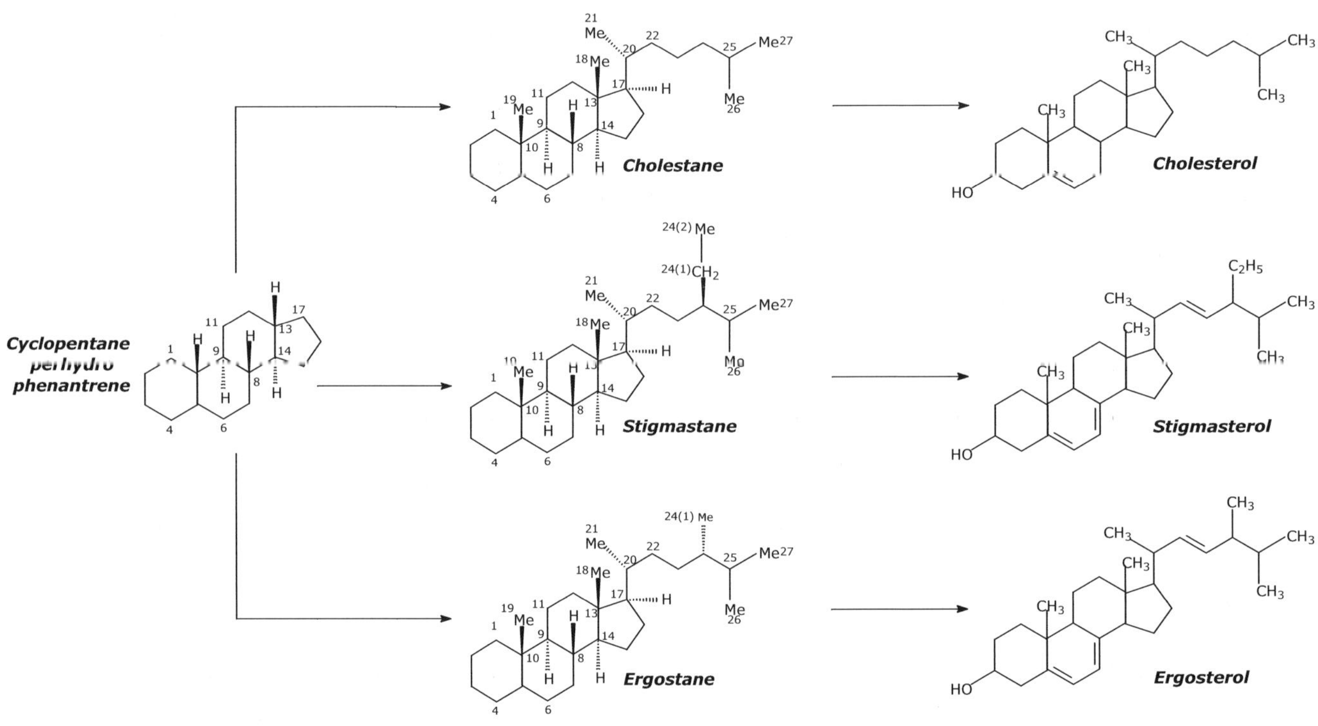

FIGURE 63.1 Precursors of sterols in nature.

Since 1980, the SBI fungicides have become the most important class overall. The fact that more than 40 SBI fungicides have reached market stage is proof of the interesting properties of the inhibitors of this biosynthetic pathway (Kuck *et al.*, 2012). The intensive wheat and barley cultivation in Europe consumes a great fraction of the worldwide SBI production. Latin America, especially the Brazilian market, has gained a considerable importance for triazole fungicides (TZFs) during the last few years. The success of SBI fungicides is essentially the success of the TZFs. All other SBI fungicides play only a limited role in terms of sales.

CLASSIFICATION OF SBI FUNGICIDES

There are four classes of fungicides that comprise the SBI. Only three classes (G1 to G3) are used as agricultural fungicides. All classes inhibit targets within fungal sterol biosynthesis but differ with regard to the precise target sites they inhibit. Table 63.1 shows the different classes of SBIs (FRAC, 2012).

The DMI fungicides are by far the most important mode of action class. Moreover, within the DMIs, one chemical class, azole (imidazole and triazole fungicides), dominates not only in the market but also in the number of compounds that have reached market level.

PROPERTIES AND ANALYSIS OF AZOLE FUNGICIDES

The azoles constitute an important class of systemic fungicides. They are used for control and treatment at low rates of a very wide range of fungal diseases, including rust, mildews and smut. Azoles in agricultural use have diverse structures in which the presence of a triazole or imidazole ring is the common structure. Their synthesis is usually based on the introduction of the azole group into the molecule by way of an activated form such as trimethylsilylimidazole or acylated azoles. Metabolic pathways generally involve the reactions of the groups attached to the triazole moiety (Roberts and Hutson, 1999). Azoles are thermally semi-labile compounds, non-ionized, and are stable to hydrolysis. They are slightly soluble in water and more soluble in common organic solvents with vapor pressures generally quite low (AERU, 2013).

Gas chromatography (GC) with nitrogen-phosphorous (NPD), electron capture (ECD) or mass spectrometric (MS) detection has been widely used to develop multiresidue methods where some azolic compounds are included. However, taking into account the low volatility and thermal stability of the target compounds, liquid chromatography (LC) with diode array (DAD) and mass spectrometric (MS) detectors seems to be the best choice. Although liquid–liquid extraction (LLE) is frequently recommended, nowadays QuEChERS, solid-phase extraction (SPE), and solid-phase microextraction (SPME) are preferred. Recently, several methods for the control of about 300 pesticides including TZFs in malt and beer using LC-MS have been published (Omote *et al.*, 2006; Inoue *et al.*, 2011).

MODE OF ACTION OF AZOLE COMPOUNDS

Azole compounds are believed to act by inhibiting the cytochrome P450 (CYP51, lanosterol C14 α-demethylase) -mediated conversion of lanosterol to ergosterol, a crystalline sterol synthesized by yeast from sugars, resulting in accumulation of sterols still bearing α-C14 methyl groups, changing the exact shape and physical properties of the fungal membrane, and causing permeability changes and malfunction of membrane-imbedded proteins (Buchenauer, 1987). Figure 63.2 shows the schematic inhibition of ergosterol biosynthetic pathway.

TABLE 63.1 Classes of SBI Fungicides

	SBI Group			
Cluster	**G1**	**G2**	**G3**	**G4**
Group name	DeMethylation Inhibitors (DMIs)	Amines (formerly morpholines)	Hydroxyanilides	Squalene epoxidase inhibitors
Target in SB	Sterol C14 demethylase	Δ14 reductase and Δ7→Δ8 isomerase	3-Keto reductase	Squalene epoxidase
Chemistry	Piperazines, pyridines, pyrimidines, imidazoles, triazoles	Morpholines, piperidines, spiroketalamines	Hydroxyanilides	Thiocarbamates, alkylamines
Uses	Agriculture (fungicide), Medicine (antimycotic)	Agriculture (fungicide)	Agriculture (fungicide)	Agriculture (herbicide), Medicine (antimycotic)

Adapted from Fungicide Resistance Action Committee (FRAC, 2012).

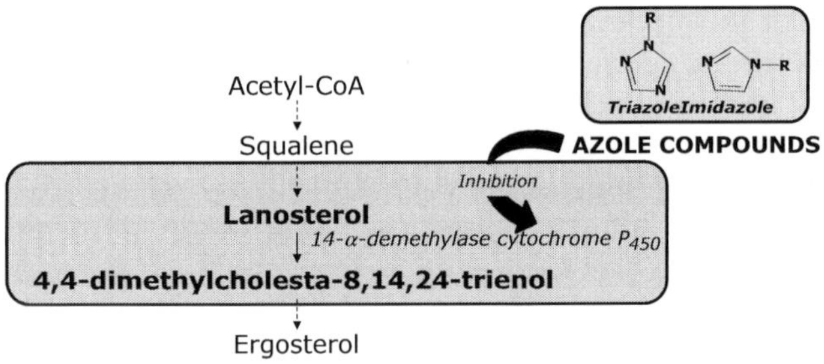

FIGURE 63.2 Schematic inhibition of the ergosterol biosynthesis by azole compounds.

FATE OF TRIAZOLE FUNGICIDE RESIDUES DURING MALTING, MASHING AND BOILING STAGES OF BEERMAKING

Residues of water-soluble pesticides on barley tend to be eliminated in the greatest amount after steeping. Conversely, hydrophobic residues remain in steeped grain. Table 63.2 shows the percentage remaining of some triazole fungicides after each stage of the malting process. Some TZFs, such as myclobutanil and propiconazole, are removed from the barley grains (after steeping) in proportions of about 41%, which is expected, bearing in mind their respective coefficients between *n*-octanol and water (K_{OW}) (Navarro *et al.*, 2007a). Miyake *et al.* (2002) found higher percentages of elimination for some azole fungicides, specifically 50, 62 and 76% for propiconazole, triflumizole, and triadimefon, respectively. Miyake *et al.* (1999) suggest that brewers should pay attention to the residues of hydrophilic pesticides on malt with log $K_{OW}<4$ because they can be carried over into beer being of special interest to the steeping stage of the malting process. Another study carried out with five TZFs (Navarro *et al.*, 2011a) shows that the percentage removed after steeping ranged from 26 to 55% for diniconazole (log $K_{OW}=4.3$) and flutriafol (log $K_{OW}=2.3$), respectively. Although in smaller proportion than steeping, germination and kilning are significant stages on the fungicide dissipation as can be seen in Table 63.2. Finally, malt storage does not significantly reduce fungicide residues (Navarro *et al.*, 2007a).

The carryovers of residues for several TZFs after mashing are shown in Table 63.2. During mashing, the soluble substances (sugars, amino acids and peptides) produced in malting and mashing are extracted into the liquid fraction (sweet wort), which is then separated from the residual solid particles (spent grains). According to Navarro *et al.* (2005), at the end of the mashing phase, the remaining percentages of two fungicides (myclobutanil, and propiconazole) were below 10% of the amount recorded in malt, with propiconazole showing the greatest decrease (to 4%). On the contrary, the retained amounts on spent grain were relatively high (38 and 42% for myclobutanil and propiconazole, respectively, both compounds having $K_{OW}>2$). Some TZFs (cyproconazole, diniconazole, epoxiconazole, flutriafol, myclobutanil, propiconazole, tebuconazole, and triadimenol) were recovered from the sweet wort in proportions ranging from 3 to 36% for diniconazole and triadimenol, respectively, while about 40–50% of the initial mass on malt was found in spent grains (Miyake and Tajima, 2003; Navarro *et al.*, 2005, 2011a). The calculated transfer factors (TFs) after mashing indicate the strong dissipation of triazole fungicide residues from malt to sweet wort (TFs$\leq$0.02). As previously indicated, a greater amount of residues is retained on the spent grains. As a consequence, TFs for spent grains are markedly high (0.66 to 0.89 for flutriafol and diniconazole, respectively) (Navarro *et al.*, 2011a). It is important to bear in mind that spent grain, the main by product of the brewing industry, is widely used as a buffer or as forage or concentrate replacer.

As can be seen in Table 63.2, a very small decrease (<10%) has been observed in the residual content after the wort had been boiled for myclobutanil, and propiconazole residues, which points to the stability of both compounds at temperatures higher than 100°C (Navarro *et al.*, 2005). Similar behavior was observed for other TZFs (Navarro *et al.*, 2011a).

EFFECT OF TZF RESIDUES ON THE FERMENTATION RATE AND QUALITY OF THE BEER

With regard to the influence of the fermentative process on the elimination of pesticide residues (Table 63.2), a significant reduction has been observed for propiconazole residues (48% of the content recorded in brewer wort) but much less for other fungicides like myclobutanil (around 20%) (Navarro *et al.*, 2007b). Also, some TZFs had relatively high residue recoveries (53–82%) at the end of fermentation with top-fermenting yeasts (Navarro *et al.*, 2011b). Flutriafol (log $K_{OW}=2.3$) and cyproconazole (log $K_{OW}=3.1$) remained in the beer after fermentation, in proportions of about 80%. On the contrary, tebuconazole, epoxiconazole, and diniconazole (log $K_{OW}=3.4$–4.3) were removed from the beer in higher proportions, mainly with the trub. These results would suggest that there is a relationship between log K_{OW} and the amount of pesticides found in the young beer and the trub. The losses during fermentation may be attributed to biotic metabolism of the yeast and abiotic degradation from the anaerobic environment created by fermentation. Also, according to the Henry's Law constants (the tendency of a compound to volatilize from aqueous solution to air) those pesticides with high vapor pressure and low solubility in water may escape to the atmosphere. This process is also favored by the constant evolution of CO_2 during the first days of the fermentation.

TABLE 63.2 Percentage Remaining of Some Triazole and Imidazole Fungicide Residues after Each Stage of Beermaking

	Beermaking stages				
	Malting				
Fungicides	**Steeping**	**Germination**	**Kilning**		**References**
Cyproconazole[1]	47	38	31		Navarro *et al.* (2011a)
Diniconazole[2]	70	61	39		Navarro *et al.* (2011a)
Epoxiconazole[3]	62	53	38		Navarro *et al.* (2011a)
Flutriafol[4]	43	35	30		Navarro *et al.* (2011a)
Myclobutanil[5]	59	42	36		Navarro *et al.* (2007a)
Propiconazole[6]	50	10	55		Miyake *et al.* (2002)
	55	43	30		Navarro *et al.* (2007a)
Tebuconazole[7]	56	45	37		Navarro *et al.* (2011a)
Triadimefon[8]	24	5	30		Miyake *et al.* (2002)
Triadimenol[9]	36	13	47		Miyake *et al.* (2002)
Triflumizole[10]	38	11	9		Miyake *et al.* (2002)
	Mashing and Boiling				
	Sweet Wort	**Spent Grains**	**Brewer Wort**	**Spent Hops**	
Cyproconazole[1]	10	40	9	ND[c]	Navarro *et al.* (2011a)
Diniconazole[2]	4	49	3	ND	Navarro *et al.* (2011a)
Epoxiconazole[3]	8	44	7	ND	Navarro *et al.* (2011a)
Flutriafol[4]	13	36	10	ND	Navarro *et al.* (2011a)
Myclobutanil[5]	9	38	8	ND	Navarro *et al.* (2005)
Propiconazole[6]	4	42	4	ND	Navarro *et al.* (2005)
Tebuconazole[7]	8	44	7	ND	Navarro *et al.* (2011a)
Triadimenol[9]	36	ND	ND	ND	Miyake and Tajima (2003)
	Fermentation				
	Beer	**Spent yeast**			
Cyproconazole[1]	79[b]	ND			Navarro *et al.* (2011b)
Diniconazole[2]	53[b]	ND			Navarro *et al.* (2011b)
Epoxiconazole[3]	59[b]	ND			Navarro *et al.* (2011b)
Flutriafol[4]	82[b]	ND			Navarro *et al.* (2011b)
Myclobutanil[5]	78[a]	ND			Navarro *et al.* (2005)
Propiconazole[6]	52[a]	ND			Navarro *et al.* (2005)
Tebuconazole[7]	55[a]	58			Hengel and Shibamoto (2002)
	67[b]				Navarro *et al.* (2011b)
Triadimenol[9]	43[a]	ND			Miyake and Tajima (2003)

Log K_{OW} (MacBean, 2012): 1 = 3.1. 2 = 4.3. 3 = 3.4. 4 = 2.3. 5 = 2.9. 6 = 3.6. 7 = 3.7. 8 = 3.1. 9 = 3.1. 10 = 4.4.
[a]*Values obtained using bottom-yeast;*
[b]*Values obtained using top-yeast;*
[c]*Not Determined.*

Some pollutants, such as pesticides, can alter the normal fermentative process being able to cause in certain cases sluggish and even stuck fermentation. In consequence the organoleptic properties of the beer should be modified as occur in other fermented beverages as wine (Navarro, 2000). According to Navarro *et al.* (2007b), a marked influence of some TZF residues in the fermentation rate (lager and ale fermentation) were observed as can be seen in Figure 63.3. From the fourth day onwards, the lager fermentation prematurely ceases (stuck fermentation; i.e., the premature termination of fermentation before all fermentable sugars have been metabolized) in samples with propiconazole residues compared with the blank. Minimum differences in the evolution of specific gravity were found for samples fermented with myclobutanil residues (Navarro *et al.*, 2007b). Other results show that at the end of ale fermentation the mean values of specific gravity for the blank samples were significantly different ($p<0.05$) from those determined in the samples containing residues of TZFs (Navarro *et al.*, 2011b).

The sugars in wort are not all fermented in the same proportions. No significant differences were observed between the blank sample and those treated with myclobutanil, while in the case of propiconazole assay, a delay in the glucose consumption was observed after four days. Sucrose was easily fermented by yeast in all cases because the enzyme which decomposes it, invertase, is located in the cell wall and sucrose is therefore treated as a start of fermentation sugar by the yeast. No significant differences were observed between the blank and the other samples although assimilation of this sugar was

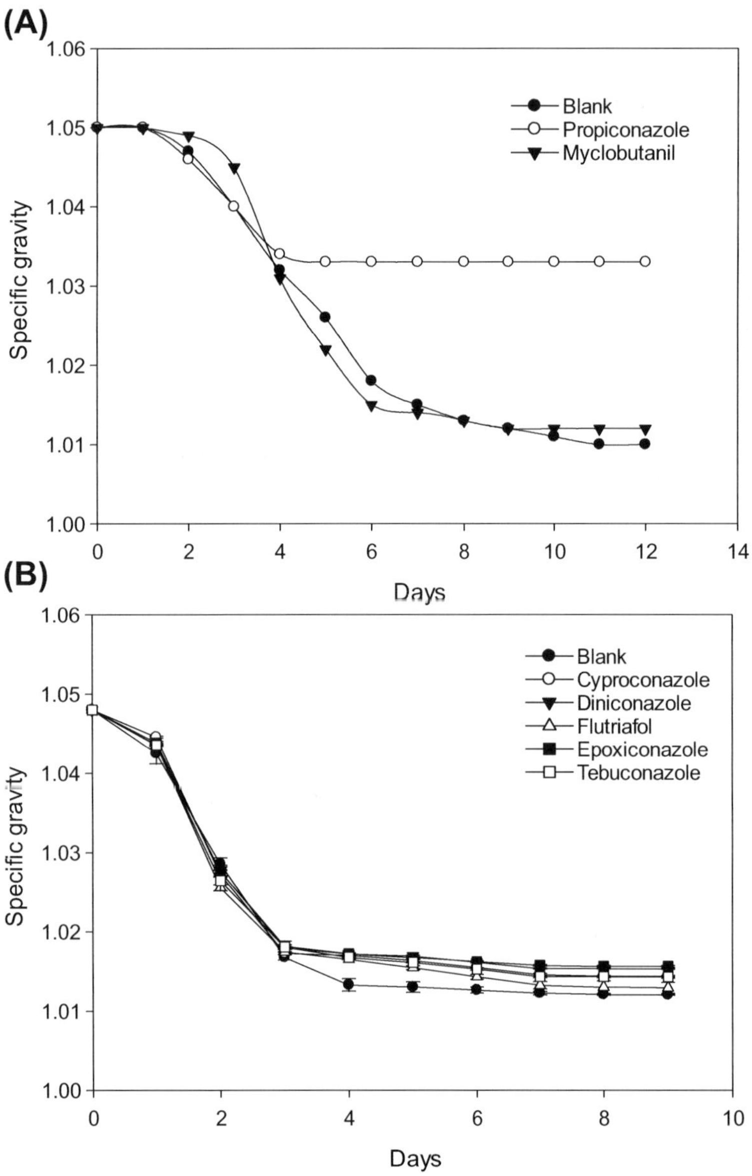

FIGURE 63.3 Progress of specific gravity *versus* time during fermentation: (A) bottom-fermenting yeasts, *Saccharomyces carlbergensis*; (B) top-fermenting yeasts, *Saccharomyces cerevisiae*. *Source: Navarro* et al. *(2007b, 2011b).*

slower in the blank sample during the first 48h. Fructose assimilation follows a different behavior to those of glucose and sucrose. Samples with nuarimol and fenarimol (pyrimidine fungicides) residues consume this sugar faster than those with propiconazole and myclobutanil (TZFs) residues. The slowest assimilation corresponds to the blank sample. In all cases, the higher consumption occurs from 24 to 216h. Similar behavior exhibits maltose although the biggest consumption takes place between 96 and 216h, during the main fermentation. It is important to remark that after the fourth day it decreases the consumption of this sugar drastically by the yeasts in the sample with propiconazole residues, which is logical bearing in mind that fermentation was stopped at this time. Finally, maltotriose is the last sugar assimilated for the yeasts. TZFs, especially propiconazole, have a marked influence on the assimilation of this sugar by the yeasts. Similar findings were found during the fermentation of young ale beer in the presence of other TZF residues (mainly diniconazole and epoxiconazole), where a higher amount of residual sugars (mainly maltose and maltotriose) was recovered in the beer (Navarro *et al.*, 2011b).

The effect of some pesticide residues on the pH and color of beer has also been observed (Navarro *et al.*, 2007b). Thus, the pH values at the end of the fermentation were 4.1 for the blank sample and 3.0 and 3.7, for those containing residues of propiconazole and myclobutanil, respectively. Also in this case, the presence of propiconazole alters the final quality of the beer. Minimal differences were observed in the case of epoxiconazole (Navarro *et al.*, 2011b).

Navarro *et al.* (2007b) have pointed out that the color of the beer falls by about 1–1.5 EBC units during lager fermentation. This is possibly due to the decoloration of some substances caused by the drop in pH, and absorption of highly colored compounds in the yeast cells or precipitation in the vessel bottom (Kunze, 2010). For ale fermentation in the presence of TZF residues, the color intensity was lower and tint higher than the values in blank samples (Navarro *et al.*, 2011b).

Regarding total polyphenol and flavonoid contents found after fermentation, significant differences have been observed between the samples containing residues of some TZFs, especially in the case of propiconazole due to the stuck fermentation after four days from the beginning. However, no significant differences were found for the pH and the polyphenol content after fermentation of young ale beer (top-fermenting yeasts) among the blank and the treated samples with five TZFs (Navarro *et al.*, 2011b).

In some cases pesticide metabolites produced during the brewing phases have the same or more toxicity than their parent compounds and they can persist during fermentation. Agrochemical metabolites are generally water soluble because most of them have hydroxyl or amine groups. This is the case of triadimenol and TF-6-1, metabolites of triadimefon (triazole) and triflumizole (imidazole), respectively. Both have been found in beer (Miyake and Tajima, 2003). In particular, 1H-1,2,4-triazole (TA), a metabolite of TZFs, is a compound with high solubility (700g/l) and stable in water (pH5–9, 25°C) for more than 30 days. Additionally, TA is known to cause a problem in reproduction and development (AERU, 2011).

References

Agriculture and Environment Research Unit (AERU) at the University of Hertfordshire, 2013. The Pesticide Properties DataBase, PPDB. http://www.herts.ac.uk/aeru/footprint.

Buchenauer, H., 1987. Mechanism of action of triazolyl fungicides and related compounds. In: Lyr, H. (Ed.), Modern Selective Fungicides: Properties, Applications and Mechanism of Action. Longman Scientific & Technical, UK, pp. 205–231.

FRAC (Fungicide Resistance Action Committee), 2012. FRAC Code List©: Fungicides sorted by mode of action (including FRAC Code numbering). Available at: http://www.frac.info/frac/publication/anhang/FRAC%20Code%20List%202011-final.pdf.

Hengel, M.J., Shibamoto, T., 2002. Method development and fate determination of pesticide-treated hops and their subsequent usage in the production of beer. J Agri. Food Chem. 50, 3412–3418.

Holland, P.T., Hamilton, D., Ohlin, B., Skidmore, M.W., 1994. Effects of storage and processing on pesticide residues in plant products. Pure Appl. Chem. 66, 335–356.

Inoue, T., Nagatomi, Y., Suga, K., Uyama, A., Mochizuki, N., 2011. Fate of pesticides during beer brewing. J. Agric. Food Chem. 59, 3857–3868.

Jones, R.D., Kavanagh, T.E., Clarke, B.J., 1988. Determination of carbaryl residues in malt and beer and their impact on beer quality. J. Am. Soc. Brew. Chem. 46, 43–50.

Koller, W., 1988. Sterol demethylation inhibitors: Mechanism of action and resistance. In: Delp, C.J. (Ed.), Fungicide Resistance in North America. APS Press, St. Paul, MN, pp. 79–88.

Kuck, K.H., Stenzel, K., Vors, J.P., 2012. Sterol Biosynthesis Inhibitors. In: Krämer, W., Schirmer, V., Jeschke, P., Witschel, M. (Eds.), Modern Crop Protection Compounds. Second, Revised and Enlarged Edition, Vol. II. Wiley-VCH, Hoboken, pp. 761–830.

Kunze, W., 2010. Technology of Brewing and Malting. (4th International English Edition). The Research and Teaching Institute for Brewing in Berlin (VLB). VLB's Publishing Department, Berlin.

MacBean, C.D.S., 2012. The Pesticide Manual. A World Compendium, sixteen ed. British Crop Protection Council, Hampshire, UK.

Miyake, Y., Koji, K., Matsuki, H., Tajima, R., Ono, M., Mine, T., 1999. Fate of agrochemical residues, associated with malt and hops, during brewing. J. Am. Soc. Brew. Chem. 57, 46–54.

Miyake, Y., Hashimoto, K., Matsuki, H., Ono, M., Tajima, R., 2002. Fate of insecticide and fungicide residues on barley during storage and malting. J. Am. Soc. Brew. Chem. 69, 110–115.

Miyake, Y., Tajima, R., 2003. Fate of pesticide metabolites on malt during brewing. J. Am. Soc. Brew. Chem. 61, 33–36.

Navarro, S., 2000. Pesticide Residues in Enology. In: Mohan, R.M. (Ed.), Research Advances In Agricultural and Food Chemistry, vol. 1. Global Research Network, Kerala, India, pp. 101–112.

Navarro, S., Pérez, G., Vela, N., Mena, L., Navarro, G., 2005. Behaviour of myclobutanil, propiconazole, and nuarimol residues during lager beer brewing. J. Agric. Food Chem. 53, 8572–8579.

Navarro, S., Pérez, G., Navarro, G., Vela, N., 2007a. Decline of pesticide residues from barley to malt. Food Add. Contam. 24, 851–859.

Navarro, S., Pérez, G., Navarro, G., Mena, L., Vela, N., 2007b. Influence of fungicide residues on the primary fermentation of young lager beer. J. Agric. Food Chem. 55, 1295–1300.

Navarro, S., Vela, N., 2009. Fate of pesticide residues during brewing. In: Preedy, V.R. (Ed.), Beer in Health and Disease Prevention. Elsevier Inc., San Diego CA, pp. 415–428.

Navarro, S., Vela, N., Navarro, G., 2011a. Fate of triazole fungicide residues during malting, mashing and boiling stages of beermaking. Food Chem. 124, 278–284.

Navarro, S., Vela, N., Pérez, G., Navarro, G., 2011b. Effect of sterol-inhibiting (SBI) fungicides on the fermentation rate and quality of young ale beer. Food Chem. 126, 623–629.

Omote, M., Harayama, K., Sasaki, T., Mochizuki, N., Yamashita, H., 2006. Analysis of simultaneous screening for 277 pesticides in malt and beer by liquid chromatography with tandem mass spectrometry. J. Am. Soc. Brew. Chem. 64, 139–150.

Roberts, T., Hutson, D. (Eds.), 1999. Metabolic Pathways of Agrochemicals. Part two: Insecticides and Fungicides. The Royal Society of Chemistry. Cornwall, UK.

Timme, G., Waltz-Tylla, B., 2004. Effects of food preparation and processing on pesticide residues in commodities of plant origin. In: Hamilton, D., Crossley, S. (Eds.), Pesticide Residues in Food and Drinking Water: Human Exposure and Risks. John Wiley & Sons, Ltd, West Sussex, UK, pp. 121–148.

CHAPTER

64

Effects of Processing Stages on the Profile of Phenolic Compounds in Beer

Haifeng Zhao

College of Light Industry and Food Sciences, South China University of Technology, Guangzhou, China

CHAPTER POINTS

- This review is the first comprehensive summary of the effects of processing stages on the profile of phenolic compounds during beer brewing.
- Phenolic compounds in barley malt undergo a serial of brewing processes, thus high levels of phenolic compounds can easily survive into the final beer, which are extremely important to the quality and nutritional property of beer.
- The levels of phenolic compounds in beer were influenced by the brewing process clearly, and their contents generally increased significantly during malting and mashing, but decreased markedly during the subsequent fermentation and storage.
- Optimization of the brewing process is one of the most efficient measures to increase the levels of phenolic compounds in beer, but it requires a clear insight into the phenolic compounds.
- HPLC with reverse-phase column coupled to fluorimetry, DAD, ECD, or MS detectors is the most widespread chromatographic technique in the investigations of phenolic compounds.
- Knowledge of the phenolic compounds and their bioactivity in beer can be used to develop appropriate technological innovation processes to control the flavor stability or to reinforce the nutritional property of beer.

INTRODUCTION

Beer is one of the most universally nutritious and less harmful alcoholic beverages in the world, which motivates brewers to continuously improve beer quality and organoleptic stability. The shelf-life of beer is largely determined by its microbiological, colloidal, foam, and flavour stability. The progress of brewing science and technology in the 20th century has resolved the microbiological and colloidal stability of beer, which makes flavor stability one of the important limiting factors in shelf-life of beer. Over the past decades, preventing oxygen pick-up during the brewing process and avoiding oxidation of the beer after fermentation has been adopted as efficient measures to resolve the beer aging issue. However, flavor stability remains hard to control despite extensive science and engineering research for years. The flavor stability of beer is increasingly found to be determined to a large extent by the endogenous antioxidant activity of beer itself (Zhao *et al.*, 2010). Actually, beer itself with high endogenous antioxidant activity can prevent the generation of free radicals and these free radicals are formed only after the antioxidants in beer are deactivated. As the major endogenous antioxidants of beer, phenolic compounds are of particular interest to brewers because they play a key role in the brewing process by delaying, retarding, or preventing the oxidation processes.

Phenolic compounds, products of secondary metabolism of plants, are generally considered as one group of important compounds responsible for both colloidal and flavor stability of beer. About 80% of phenolic compounds in beer originate from barley malt, and the remaining 20% come from hop. Phenolic compounds in barley exist in both free and bound forms and can be classified into simple phenols, phenolic acids, hydroxycinnamic acid derivatives, and flavonoids, depending on their structure. Apart from the characteristics of malting barley (Figure 64.1), the brewing process (Figure 64.2) and the conditions of storage have profound impacts on the phenolic compounds of final beer. Thus, this overview on the effects of processing stages on the phenolic compounds in beer aims to illustrate their formation and evolution mechanisms during beer brewing.

Processing and Impact on Active Components in Food
http://dx.doi.org/10.1016/B978-0-12-404699-3.00064-0

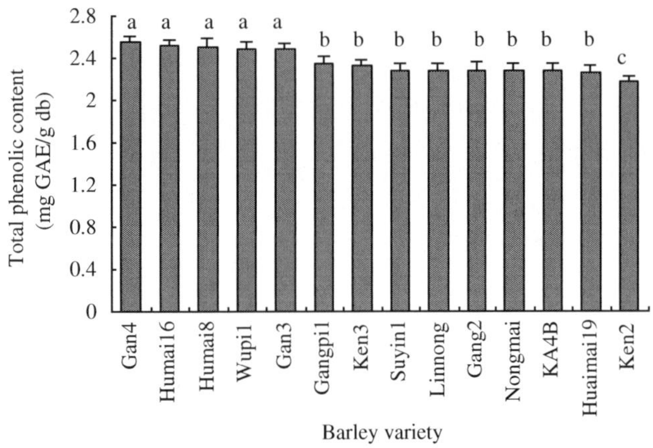

FIGURE 64.1 Total phenolic contents (mg GAE/g) of different malting barley varieties. Vertical bars represent the standard deviation of each data point ($n=3$). Locations for each barley variety marked by the different letter are significantly different ($P<0.05$). *Adapted from Zhao* et al. *(2008).*

EFFECTS OF THE VARIETY AND CHARACTERISTICS OF BARLEY ON PHENOLIC COMPOUNDS

The content of phenolic compounds in barley depends to a large extent on the genetic and environmental factors but appears to be less variable in phenolic profile. Depending on the variety of barley, place of growth, and altitude, significant differences in phenolics content were observed, and the values of the total phenolic contents (TPC) and the total ferulic acid contents in different barley varieties ranged from 2.17 to 2.56 mg GAE/g db (dry barley) (Figure 64.1; Zhao *et al.*, 2008), and from 343.2 µg/g to 579.7 µg/g d.w. (dry weight) (Zupfer *et al.*, 1998), respectively. Barley contains hundreds of phenolic compounds including benzoic and cinnamic acid derivatives, flavonoids, chalcones, quinines, and amino phenolic compounds. Flavonoids are the major free phenolic compounds in barley, while the bound phenolic compounds are ester-linked to cell wall polymers and mainly are phenolic acids (Dvoráková *et al.*, 2008). As far as the profile of phenolic compounds in barley is concerned, no great difference is found among varieties. Indeed, Maillard *et al.* (1996) found that although the variety of barley had significant impacts on the total contents of phenolic compounds, the proportion of individual phenolic groups was not influenced. A study from Holtekjølen *et al.* (2006) showed that there was no significant difference in the portion of dimers to total flavanol contents in barley type, while the percentages of *p*-coumaric acid and ferulic acid in total phenolic acids were significantly different among varieties.

The types of barley, such as six- and two-row, hulled and hull-less, and normal and atypical starch, also had some impact on the content of phenolic compounds. There were no evident differences in phenolic profile and content between six- and two-row barley, whereas the higher contents of catechin and procyanidin were observed in the waxy hull-less variety, which indicated that the presence of the waxy gene in barley does affect the content of phenolics (Holtekjølen *et al.*, 2006). Moreover, presence or lack of hull was considered as an important factor affecting the phenolic compounds in barley. Holtekjølen *et al.* (2006) found that the content of phenolic acids in the hulled varieties was significantly higher than those in hull-less barley, whereas a comparative study carried out by Dvorakova *et al.* (2008) indicated that catechin and proanthocyanidin content as well as TPC was higher in hull-less barley varieties than those in hulled barley genotypes. Dabina-Bicka *et al.* (2011) found that the differences of TPC between hulled barley and hull-less barley lines were not significant. Furthermore, phenolic compounds are selectively stored in some morphological structures on the basis of their function or physiologic state, thus different morphological fractions of barley would influence phenolic compounds greatly. Sharma and Gujral (2010) reported that the content of phenolic compounds in whole barley and its milling fraction decreased in the following order: bran > whole barley > refined barley flour. In another study, two barley varieties were separated by pearling into seven fractions and results showed that the outer fractions contained higher amounts of phenolic acids, while the inner fractions contained substantially low amounts (Madhujith *et al.*, 2006). Moreover, smaller grains contain higher levels of hydroxycinnamic acid due to the higher surface-to-volume ratio resulting in a higher percentage of outer layers (Vanbeneden *et al.*, 2007). However, there was no correlation between the 1000 grain weight of different barley cultivars and their ferulic acid concentration (Zupfer *et al.*, 1998).

EFFECTS OF MALTING ON PHENOLIC COMPOUNDS

Malting is a three-stage process involving steeping, germination, and kilning, to convert barley to malt suitable for brewing. Lu *et al.* (2007) studied the evolution of phenolic compounds during malting, and found that the levels of phenolic compounds decreased significantly during steeping and the early stages of germination and then increased remarkably during the later stages of germination and subsequent kilning (Table 64.1). Generally, malt shows higher phenolic content than its corresponding barley. Dvoráková *et al.* (2008) investigated the contents of phenolic compounds in different barley varieties and corresponding malts, and found that the TPC are in all cases higher in malt than in barley. Dabina-Bicka *et al.* (2011) further proved that phenolic content in malt

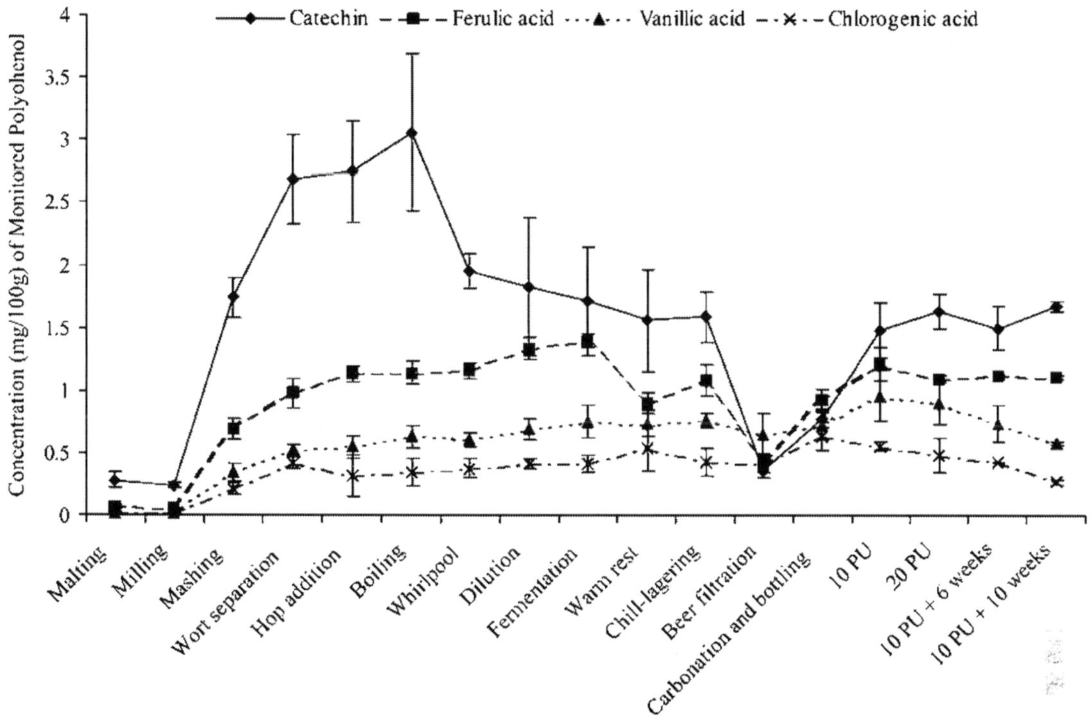

FIGURE 64.2 The effect of brewing stages on concentrations of monitored phenolic compounds (mg/100 g). Error bars represent standard deviation. PU, pasteurization units (as defined by the European Brewing Convention). *Adapted from Pascoe* et al. *(2003).*

was higher than in unprocessed barley, and the increase ranged from 28% to 65%. However, a study performed by Leitao *et al.* (2012) indicated that the malting resulted in a sharp decrease in the contents of catechin, prodelphinidin B3, and the procyanidin.

Steeping generally caused a significant decrease in the content of phenolic compounds (Lu *et al.*, 2007; Szwajgier *et al.*, 2005a), which is due to the leaching of phenolic compounds or the formation of insoluble complexes with proteins. However, there were some discrepancies in the effects of germination on the phenolic compounds of barley. The studies carried out by Lu *et al.* (2007) and Sharma and Gujral (2010), showed that the level of phenolics decreased significantly during the early stage of germination, and then increased remarkably during the later stage of germination (Table 64.1). Contrary to these observations, Gallegos-Infante *et al.* (2010) examined the effect of germination on the antioxidant capacity of a Mexican barley cultivar, and found that the germination reduced the TPC of barley. Moreover, the application of 'Activated Germination Malting' technology with an elevated temperature and a lower pH can increase significantly the free ferulic and coumaric acid contents in malts (Szwajgier *et al.*, 2005a).

Several studies have shown that the kilning process and regime influence the phenolic compounds contents of malts significantly. Maillard and Berset (1995) reported that the levels of individual phenolics increased by about 130% until 80°C and then decreased by 20% until 90°C during kilning, but the proportions of the different phenolics are not significantly changed. Inns *et al.* (2011) investigated the effect of the kilning regime on levels of free ferulic acid in malt, and suggested that the content of free ferulic acid in finished malt was affected by the kilning regime, and the coolest kilning regime (45°C for 2 h) increased the ferulic acid content in the finished malt. Gallegos-Infante *et al.* (2010) compared the effect of traditional cooking, roasting, and germination on the TPC of a Mexican barley cultivar, and results showed that cooking and roasting barley exhibited higher TPC than that of unprocessed barley.

EFFECTS OF MASHING ON PHENOLIC COMPOUNDS

Mashing is a procedure that involves a series of rests at various temperatures, and the types of mashing are divided into two methods: infusion mashing and decoction mashing. Both the process and the type of mashing had been found to significantly affect the content of phenolics.

Generally, the mashing process results in significant increases in phenolic content, and since mashing takes place under conditions suitable for enzyme action, it is likely to lead to the release of bound phenolics. Indeed, all monitored phenolic compounds including catechin,

TABLE 64.1 Evolution of Individual Phenolic Compounds during Malting

MALTING PROCESS	μg/g of Dry Weight									
	(+)-Catechin	Ferulic Acid	Syringic Acid	(−)-Epicatechin	Caffeic Acid	Vanillic Acid	*p*-Coumaric Acid	Gallic Acid	Protocatechuic Acid	SPC
RB	59.71 ± 0.85b	16.64 ± 0.33d	11.58 ± 0.11b	15.91 ± 0.13b	7.25 ± 0.03d	4.51 ± 0.11d	1.85 ± 0.10a	2.53 ± 0.06d	1.69 ± 0.05b	121.67
ES	53.05 ± 0.16c	15.90 ± 0.14e	11.12 ± 0.15bc	15.37 ± 0.16c	7.02 ± 0.02e	4.00 ± 0.15e	1.70 ± 0.03b	2.38 ± 0.03e	1.56 ± 0.03c	112.10
G1	50.88 ± 0.55d	15.46 ± 0.05f	10.88 ± 0.05c	15.12 ± 0.37c	7.08 ± 0.02e	4.25 ± 0.06e	1.61 ± 0.08c	2.57 ± 0.09d	1.38 ± 0.08d	109.23
G2	49.23 ± 0.32d	15.07 ± 0.19g	10.25 ± 0.06e	14.78 ± 0.10d	7.15 ± 0.03d	4.52 ± 0.13d	1.55 ± 0.07c	2.65 ± 0.06cd	1.34 ± 0.02d	106.54
G3	45.97 ± 1.28e	14.97 ± 0.10g	10.15 ± 0.08e	14.68 ± 0.16d	7.30 ± 0.05cd	4.60 ± 0.08d	1.50 ± 0.04c	2.78 ± 0.05c	1.45 ± 0.06d	103.40
G4	45.05 ± 0.90e	14.69 ± 0.21g	10.45 ± 0.05d	15.05 ± 0.24c	7.54 ± 0.09c	4.82 ± 0.05c	1.50 ± 0.08c	2.89 ± 0.04c	1.40 ± 0.04d	103.39
G5	48.91 ± 0.47d	15.17 ± 0.23fg	10.50 ± 0.09d	15.25 ± 0.08c	7.68 ± 0.06c	4.95 ± 0.07c	1.58 ± 0.07c	3.25 ± 0.09ab	1.58 ± 0.03c	108.87
K1	52.07 ± 0.87c	16.65 ± 0.13d	10.73 ± 0.12c	15.83 ± 0.06b	7.77 ± 0.07bc	5.17 ± 0.09bc	1.64 ± 0.03bc	3.30 ± 0.06a	1.60 ± 0.05bc	114.76
K2	58.46 ± 0.26b	19.92 ± 0.34a	11.47 ± 0.10b	16.08 ± 0.09ab	7.98 ± 0.10b	5.32 ± 0.07b	1.66 ± 0.05b	3.10 ± 0.05b	1.72 ± 0.02b	125.71
K3	64.31 ± 0.50a	17.34 ± 0.25c	11.84 ± 0.07a	16.43 ± 0.14a	7.84 ± 0.02b	5.84 ± 0.10a	1.79 ± 0.07ab	3.15 ± 0.03b	1.85 ± 0.04a	130.39
K4	63.28 ± 0.82a	18.26 ± 0.12b	10.98 ± 0.03c	16.23 ± 0.25a	8.90 ± 0.08a	5.88 ± 0.14a	1.82 ± 0.08a	3.40 ± 0.08a	1.69 ± 0.05b	130.44

Each value is the mean ± standard deviation of triplicate determinations. Means with different letters in the same column for each barley variety are significantly different ($p < 0.05$). RB, raw barley; ES, end of steeping; G1–G5, 1st–5th day of germination; K1, 1–8 h of kilning (45°C); K2, 9–16 h of kilning (55°C); K3, 17–19 h of kilning (65°C); K4, 20–22 h of kilning (85°C).
Adapted from Lu et al. *(2007).*

ferulic, vanillic, chlorogenic, and *p*-coumaric acids (Figure 64.2) were significantly increased during mashing (Pascoe *et al.*, 2003). Zhao and Zhao (2012) examined the evolution of phenolic content during mashing and presented that the TPC increased during the early stage of mashing and then decreased during the mashing at higher temperature, and the whole mashing process resulted in a 6.0% of increase in TPC (Figure 64.3). Other studies regarding the evolution of free and bound ferulic acid during the wort production process performed by Vanbeneden *et al.* (2008) and Szwajgier *et al.* (2005b) also indicated that the ferulic acid concentration increased continuously from the start of mashing-in until the heating of the wort at 62°C or 75°C. And the highest amount of ferulic acid was released during the first 15 min of mashing and no additional ferulic acid was released during the subsequent operations. However, Szwajgier (2009) observed a decrease in the phenolic acid content from the very beginning of the mashing process. There was a controversy regarding the effects of boiling on the phenolic contents. For example, Vanbeneden *et al.* (2008) found that boiling the wort with hops caused a slight decrease in bound ferulic acid concentration, while no significant influence in free ferulic acid content was observed; whereas Szwajgier *et al.* (2005b) reported a net increase in the ferulic acid content during wort boiling.

Recently, Jurková *et al.* (2012) found that the type of mashing had a strong impact on the content of phenolics, and worts produced with decoction mashing showed higher phenolic content than those produced with infusion mashing. The effects of mashing variables including mashing-in temperature, time and pH, mash thickness, grist coarseness and composition, and stirring regime on the release of ferulic acid during mashing had been examined by Vanbeneden *et al.* (2008), and they found that mash thickness, the grist coarseness and composition, and the stirring regime had a pronounced effect on the release of ferulic acid during mashing, and

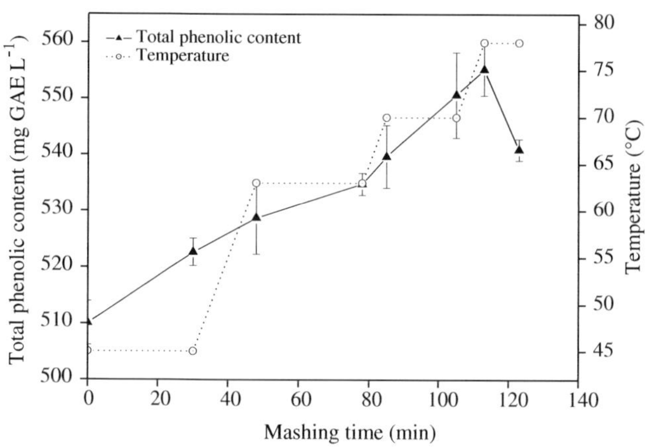

FIGURE 64.3 Evolution of total phenolic content (mg GAE/l) during mashing. *Adapted from Zhao and Zhao (2012).*

an optimal mashing condition of temperature of 40°C and pH of 5.8 was observed.

Some measures to control the content of phenolics in wort have been reported. A number of works have examined the effects of stabilizing agents such as polyvinylpolypyrrolidone (PVPP), formaldehyde, hexamethylenetetramine, and sulfite on the phenolic content of wort. Macey *et al.* (1966) demonstrated that addition of formaldehyde during mashing can remove anthocyanogens from wort. Andersen and Skibsted (2001) found that hexamethylenetetramine reduced the contents of catechin, prodelphinidin B3, and procyanidin B3 in wort, whereas the content of ferulic acid was unaffected and sulfite had only a minor impact on the contents of phenolics in wort. On the contrary, the application of ferulic acid esterase is considered as an efficient method for improvement of the phenolics content in wort. It was found that use of ferulic acid esterase caused more effective release of free ferulic, vanillic, *p*-OH-benzoic, syringic, protocatechuic, caffeic, and chlorogenic acids into the wort in comparison to the wort without any preparations; moreover, the release of these phenolic acids is temperature dependent (Szwajgier *et al.*, 2005b, 2011). Also, according to Vanbeneden *et al.* (2007), the release of ferulic acid during mashing did not only depend on the malt esterase activity, but also on the amount of ester-bound ferulic acid initially present in the wort and on its endoxylanase activity.

EFFECTS OF FERMENTATION ON PHENOLIC COMPOUNDS

Fermentation is the process by which fermentable carbohydrates are converted by yeast into alcohol, carbon dioxide, and numerous byproducts. The fermentation process had no significant effects on the levels of catechin, epicatechin, prodelphinidin, proanthocyanidin B3, and phenolics acids (Figure 64.2). However, Szwajgier (2009), Gromus and Lustig (2001), and Szwajgier *et al.* (2005b) found a substantial decrease of phenolics content in beers in comparison to the corresponding worts. They speculated that this might be due to the precipitation of tannins and non-tannin phenolics from worts and beers and the absorption by the yeast during fermentation.

EFFECTS OF FILTRATION ON PHENOLIC COMPOUNDS

Beer filtration was found to cause a significant decrease in the amount of phenolic compounds (Pascoe *et al.*, 2003; Szwajgier *et al.*, 2005b). During beer filtration, yeast cells, permanent and chill-haze are removed, leading to a loss of phenolics. Moreover, the decrease of phenolic content during beer filtration appeared to be filtration or stabilization material dependent. Fumi *et al.* (2011) found that the reduction in total phenols was 5%, 12%, and 14% in beer after Kieselguhr filter (flat-sheet filter), PVPP filter (coated screen filter), and Trap filter (polypropylene cartridge filter), respectively. The reduction in phenolic compounds was found mainly on tannins, and PVPP filter and Trap filter resulted in 37% and 66% of decreases in tannins, respectively.

EFFECTS OF STORAGE ON PHENOLIC COMPOUNDS

Numerous studies have mentioned a significant decrease in phenolics content during storage due to the fact that the phenolic compounds participate in the formation of colloidal hazes, and are oxidatively polymerized to higher-molecular-weight species. Szwajgier *et al.* (2005b) noted 2–10% decreases in both free and bound ferulic acid contents after 1 month of beer storage, and the level of free ferulic acid also decreased more rapidly than that of the bound ferulic acid in most of beers. In another study, Siqueira *et al.* (2011) presented a similar evolution trend of phenolics, finding that the TPC in both beers decreased significantly after 16 days of storage, and then remained constant until the end of the 2 months of storage. Moreover, the levels of chlorogenic acid and vanillic acid significantly decreased by 37% and 35%, respectively, after 10 weeks of storage (Figure 64.2), but no change in the amounts of other monitored phenolic compounds was observed during storage (Pascoe *et al.*, 2003).

ANALYTICAL TECHNIQUES

Methods for identification and quantification of the main phenolic compounds in beer and its raw materials have been generally based on high-performance liquid chromatography (HPLC) coupled to fluorimetry, diode array detection (DAD), electrochemical detection (ECD), or mass spectrometry (MS). Reversed phase columns, composed of a C18 stationary phase with the most common internal diameter of 4.6 mm and the particle size of 3 or 5 μm, can separate nearly all phenolic compounds. Gradient elution systems are more frequently employed than their isocratic counterparts. Binary elution systems are used including an aqueous acidified polar medium (aqueous acetic acid, phosphoric acid, or formic acid) and a less polar organic system (predominantly methanol and acetonitrile, but propanol, butanol, and ethyl acetate are employed as well). Photodiode array allowing for real-time UV-Vis spectra at wavelengths from 190 to 380 nm are the most prevalent detection method. Considering the multitude of components and complexity of beer matrices, identification of peaks is a difficult issue. Other detection methods have been investigated, namely electrochemical and fluorometric systems with the aim of increasing specificity. ECD is a highly sensitive and selective tool for the characterization of phenolic compounds in complicated matrices and provided the additional information for the identification of phenolic acids not previously identified with LC/UV. Moreover, HPLC separation techniques are more and more frequently hyphenated to MS, and LC/MS has become a very powerful for the analyses of complex matrixes. The mass analyzers used can be quadrupoles, ion traps, or time-of-flight systems, and various analysis modes exist such as Full-MS, SIM, MS/MS, and multiple reaction monitoring (MRM). For very selective quantitative analyses, SIM or MRM is recommended, but MS/MS is more informative for structure determination (Callemien and Collin, 2010).

Acknowledgments

I gratefully acknowledge the National Natural Science Foundation of China (No. 31000810) and the Key Technology R&D Program of Guangdong Province (Nos. 2012A080107005 and 2011A020102001) for their financial support.

References

Andersen, M.L., Skibsted, L.H., 2001. Modification of the levels of polyphenols in Wort and Beer by Addition of Hexamethylenetetramine or Sulfite during Mashing. J. Agric. Food Chem. 49, 5232–5237.

Callemien, D., Collin, S., 2010. Structure, organoleptic properties, quantification methods, and stability of phenolic compounds in beer — a review. Food Rev. Int. 26, 1–84.

Dabina-Bicka, I., Karklina, D., Kruma, Z., 2011. Polyphenols and vitamin E as potential antioxidants in barley and malt. Foodbalt, 121–126.

Dvoráková, M., Guido, L.F., Dostálek, P., Skulilová, Z., Moreira, M.M., Barros, A.A., 2008. Antioxidant properties of free, soluble ester and insoluble-bound phenolic compounds in different barley varieties and corresponding malts. J. Inst. Brew. 114, 27–33.

Dvorakova, M., Moreira, M.M., Dostalek, P., Skulilova, Z., Guido, L.F., Barros, A.A., 2008. Characterization of monomeric and oligomeric flavan-3-ols from barley and malt by liquid chromatography–ultraviolet detection–electrospray ionization mass spectrometry. J. Chromatogr. A 1189, 398–405.

Fumi, M.D., Galli, R., Lambri, M., Donadini, G., Faveri, D.M.D., 2011. Effect of full-scale brewing process on polyphenols in Italian all-malt and maize adjunct lager beers. J. Food Compos. Anal. 24, 568–573.

Gallegos-Infante, J.A., Rocha-Guzman, N.E., Gonzalez-Laredo, R.F., Pulido-Alonso, J., 2010. Effect of processing on the antioxidant properties of extracts from Mexican barley (*Hordeum vulgare*) cultivar. Food Chem. 119, 903–906.

Gromus, J., Lustig, S., 2001. Influence of polyphenols and reducing substances on beer quality and their fate during beer production. Brauwelt Int. 3, 208–213.

Holtekjølen, A.K., Kinitz, C., Knutsen, S.H., 2006. Flavanol and bound phenolic acid contents in different barley varieties. J. Agric. Food Chem. 54, 2253–2260.

Inns, E.L., Buggey, L.A., Booer, C., Nursten, H.E., Ames, J.M., 2011. Effect of modification of the kilning regimen on levels of free ferulic acid and antioxidant activity in malt. J. Agric. Food Chem. 59, 9335–9343.

Jurková, M., Horák, T., Hašková, D., Culík, J., Cejka, P., Kellner, V., 2012. Control of antioxidant beer activity by the mashing process. J. Inst. Brew. 118, 230–235.

Leitao, C., Marchioni, E., Bergaentzlé, M., Zhao, M., Didierjean, L., Miesch, L., Holder, E., Miesch, M., Ennahar, S., 2012. Fate of polyphenols and antioxidant activity of barley throughout malting and brewing. J. Cereal Sci. 55, 318–322.

Lu, J., Zhao, H., Chen, J., Fan, W., Dong, J., Kong, W., Sun, J., Cao, Y., Cai, G., 2007. Evolution of phenolic compounds and antioxidant activity during malting. J. Agric. Food Chem. 55, 10994–11001.

Macey, A., Stowell, K.C., White, H.B., 1966. Use of formaldehyde for the reduction of the anthocyanogen content of wort. J. Inst. Brew. 72, 29–35.

Maillard, M.N., Berset, C., 1995. Evolution of antioxidant activity during kilning: role of insoluble bound phenolic acids of barley and malt. J. Agric. Food Chem. 43, 1789–1793.

Maillard, M.N., Soum, M.H., Boivin, P., Berset, C., 1996. Antioxidant activity of barley and malt: relationship with phenolic content. Lebensm. Wiss. Technol. 29, 238–244.

Madhujith, T., Izydorczyk, M., Shahidi, F., 2006. Antioxidant properties of pearled barley fractions. J. Agric. Food Chem. 54, 3283–3289.

Pascoe, H.M., Ames, J.M., Chandra, S., 2003. Critical stages of the brewing process for changes in antioxidant activity and levels of phenolic compounds in ale. J. Am. Soc. Brew. Chem. 61, 203–209.

Sharma, P., Gujral, H.S., 2010. Antioxidant and polyphenol oxidase activity of germinated barley and its milling fractions. Food Chem. 120, 673–678.

Siqueira, P.B., Bolini, H.M.A., Macedo, G.A., 2011. Polyphenols and antioxidant properties in forced and naturally aged Brazilian beer. J. Brew. Distill. 2, 45–50.

Szwajgier, D., 2009. Content of individual phenolic acids in worts and beers and their possible contribution to the antiradical activity of beer. J. Inst. Brew. 115, 243–252.

Szwajgier, D., 2011. The use of an extracellular ferulic acid esterase from Lactobacillus acidophilus K1 for the release of phenolic acids during mashing. J. Inst. Brew. 117, 427–434.

Szwajgier, D., Pielecki, J., Targoński, Z., 2005a. Changes of free ferulic and coumaric acid contents during malting of barley grain. Pol. J. Food Nutr. Sci. 14/55, 423–429.

Szwajgier, D., Pielecki, J., Targoński, Z., 2005b. The release of ferulic acid and feruloylated oligosaccharides during wort and beer production. J. Inst. Brew. 111, 372–379.

Vanbeneden, N., Roey, T.V., Willems, F., Delvaux, F., Delvaux, F.R., 2008. Release of phenolic flavor precursors during wort production: Influence of process parameters and grist composition on ferulic acid release during brewing. Food Chem. 111, 83–91.

Vanbeneden, N., Gils, F., Delvaux, F., Delvaux, F.R., 2007. Variability in the release of free and bound hydroxycinnamic acids from diverse malted barley (*Hordeum vulgare* L.) cultivars during wort production. J. Agric. Food Chem. 55, 11002–11010.

Zhao, H., Chen, W., Lu, J., Zhao, M., 2010. Phenolic profiles and antioxidant activities of commercial beers. Food Chem. 119, 1150–1158.

Zhao, H., Fan, W., Dong, J., Lu, J., Chen, J., Shan, L., Lin, Y., Kong, W., 2008. Evaluation of antioxidant activities and total phenolic contents of typical malting barley varieties. Food Chem. 107, 296–304.

Zhao, H., Zhao, M., 2012. Effects of mashing on total phenolic contents and antioxidant activities of malts and worts. Int. J. Food Sci. Technol. 47, 240–247.

Zupfer, J.M., Churchill, K.E., Rasmusson, D.C., Fulcher, R.G., 1998. Variation in ferulic acid concentration among diverse barley cultivars measured by HPLC and microspectrophotometry. J. Agric. Food Chem. 46, 1350–1354.

CHAPTER

65

Furan and Other Furanic Compounds in Coffee: Occurrence, Mitigation Strategies, and Importance of Processing

Monica Anese

Dipartimento di Scienze degli Alimenti, University of Udine, Udine, Italy

CHAPTER POINTS

- Furan and 5-hydroxymethylfurfural (HMF) are volatile compounds which may form in heated foods.
- Furan and HMF have been reported to be toxic and probably carcinogenic to humans.
- Coffee is the major source of furan and HMF for regular coffee consumers.
- Furan and HMF intake depend on many variables, such as degree of coffee roast, beverage preparation method, and consumer handling.
- Reduction of furan and HMF levels in coffee is a challenging issue, because it is very difficult to mitigate their levels without affecting coffee acceptability.
- Strategies potentially exploitable for furan and HMF mitigation in coffee are those based on the reduction of the thermal input and vacuum removal.

INTRODUCTION

The term 'coffee' refers to the beverage prepared by brewing roasted and ground coffee beans (Buffo and Cardelli-Freire, 2004). Roasted coffee is obtained by subjecting the green beans to a roasting process which is generally carried out at temperatures higher than 200°C (Clarke, 1987; Eggers and Pietsch, 2001). During roasting, complex chemical reactions, i.e., the Maillard reaction, caramelization, and pyrolysis, take place leading to deep chemical, physical and sensory changes of coffee beans (Clarke, 1987). In particular, these can be regarded as visible and invisible changes, as shown in Figure 65.1. Among the visible changes, color and flavor development are the ones mostly affecting coffee acceptability. More than 800 volatile compounds with a wide variety of functional groups have been identified in coffee (Clarke, 1990; Grosch, 2001; Flament and Bessiere-Thomas, 2002). However, less than 20 of them have been found to be relevant to coffee aroma (Buffo and Cardelli-Freire, 2004). The main components of coffee aroma are furans, followed by pyrazines, pyrroles, and ketones (Flament and Chevallier, 1988). Furan compounds are a wide class of heterocyclic molecules containing a five membered aromatic ring with four carbon atoms and one oxygen atom, with one or more substitutes. The latter include aldehydes, ketones, esters, alcohols, acids, thiols, sulphides and in combination with pyrazines and pyrroles (Buffo and Cardelli-Freire, 2004). The number and nature of the functional groups markedly affect the chemical and physical properties of furans, such as their solubility and volatility. This would influence their absolute amount in the headspace and thus perception threshold. Important furan compounds occurring in the aroma of roasted coffee are 2-furfurylthiol, 2-methyl-3-furanthiol; 5-hydroxymethylfurfural (HMF), furfural and furfuryl alchol (Grosh, 2001; Buffo and Cardelli-Freire, 2004; Moon and Shibamoto, 2009). In particular, 2-furfurylthiol is considered as a key aroma compound of coffee, presenting roasty, coffee-like notes (Grosh, 2001). The levels of furan compounds in the roasted coffee correlate well with the roast color, as they increase with roasting of coffee from light to dark degrees (Guenther *et al.*, 2010). However, potential furans formation seems not to significantly

Processing and Impact on Active Components in Food
http://dx.doi.org/10.1016/B978-0-12-404699-3.00065-2

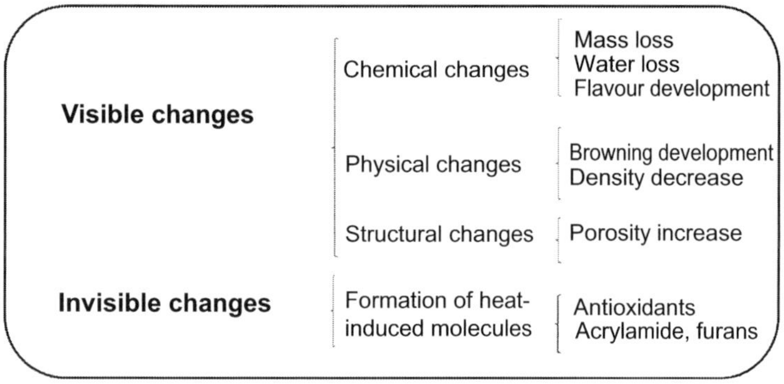

FIGURE 65.1 Consequences of coffee roasting process

vary for different green coffee types (La Pera *et al.*, 2009; Guenther *et al.*, 2010).

Among furans, furan and HMF have been reported to be toxic and probably carcinogenic to humans (IARC, 1995; Glatt and Sommer 2006). Several studies showed that HMF may induce genotoxic and mutagenic effects in bacterial and human cells and promote colon and liver cancer in rats and mice (Zhang *et al.*, 1993; Glatt and Sommer 2006; Monien *et al.*, 2012). However, *in vitro* studies on HMF genotoxicity and mutagenicity have given controversial results and the toxicological relevance of its exposure has not yet been clarified since the mechanisms by which HMF exerts its genotoxicity remain unclear (Husøy *et al.*, 2008). Furan was found to induce hepatocellular adenomas and carcinomas in rats and mice, and high incidence of cholangicarcinomas in rats at doses higher than 2 mg/kg body weight (Moro *et al.*, 2012). Up until today, a final conclusion concerning genotoxic or mutagenic activity of these molecules has not been drawn yet (EFSA, 2010). Nevertheless, furan and HMF have generated great concern due to their widespread presence in foods. Indeed, as coffee has been reported as the major source of furan and HMF in the adult diet (USFDA, 2006; Husøy *et al.*, 2008; Morehouse *et al.*, 2008; Rufian-Henares and de la Cueva, 2008; ESFA, 2011), the scientific community is aware that efforts should be taken to keep furan and HMF levels as low as is reasonably achievable.

HOW COMPOSITION IS ALTERED

In-depth descriptions of precursors and mechanisms of furan and HMF accumulation in food have been published by Maga (1979), Yaylayan (2006), Crews and Castle (2007), Morales (2009), and Capuano and Fogliano (2011). Furan may originate from multiple sources according to different pathways. These include (i) thermal degradation of reducing sugars, (ii) thermal degradation of amino acids, (iii) Maillard reaction involving reducing sugars and amino acids, (iv) oxidation of polyunsaturated fatty acids and carotenoids, (v) oxidation of ascorbic acid and ascorbates. HMF can be formed as an intermediate in the Maillard reaction, which occurs when reducing sugars are heated in the presence of amino acids or proteins, or, alternatively, by thermal dehydration of a sugar. The latter does not require the presence of an amino group and is strongly enhanced under acid conditions.

Furan and HMF levels in coffee roast and beverage are reported to vary into a wide range of concentrations depending on the degree of roasting as well as grinding and brewing procedures (Guenther *et al.*, 2010).

According to EFSA (2011), among the heat-treated commercial foods monitored between 2004 and 2009 by 18 EU countries, the highest furan levels were found in solid coffee. Furan concentration in roasted beans has been reported to be in the range from 1000 to 6900 μg/kg (Reinhard et al., 2004; Kuballa *et al.*, 2005; Hasnip *et al.*, 2006; ESFA 2011). Lower furan contents were found in ground coffee (Kuballa *et al.*, 2005; Hasnip *et al.*, 2006; Guenther *et al.*, 2010; EFSA, 2011), while instant coffee showed the lowest levels (Kuballa *et al.*, 2005; Hasnip *et al.*, 2006; Zoller *et al.*, 2007; EFSA, 2011). Furan concentration in coffee brews has been found to be much lower than in roasted beans (Hasnip *et al.*, 2006; Zoller *et al.*, 2007; Kim *et al.*, 2009). Processing steps subsequent to roasting, e.g., grinding, packaging, brew preparation, and consumer handling, may affect furan content in the final product and consequently consumer intake (Guenther *et al.*, 2010). In fact, furan level in the coffee cup may vary with degassing degree, particle size, percolation time and temperature, and coffee/water ratio. Guenther *et al.* (2010) estimated that only approximately 10% of the initially generated furan during roasting can be found in the final cup of coffee. This is probably due to the high volatility of furan (boiling point, 31°C at 760 mmHg), that may evaporate during coffee preparation (Kim *et al.*, 2009; Guenther *et al.*, 2010). Coffee degassing and grinding have been found to account for 20 and 40% furan losses, respectively (Guenther *et al.*, 2010). Additionally, coffee powder size may affect furan content in the brew. Beverages made with fine grained coffee generally contain more furan than those made from normal or coarse-grained coffee (Crews *et al.*, 2009). Furan levels in the cup were also found to be affected by the brewing procedure (Table 65.1). In general terms, the use of closed extraction systems (e.g., fully automatic and espresso machines, machines based on coffee pads, disks or capsules) combined with high brewing temperature, pressure, and time, as well as high coffee powder/water ratio are responsible for higher levels of the toxic molecule in the coffee beverage (Zoller *et al.*, 2007; La Pera *et al.*, 2009; Guenther *et al.*, 2010). By contrast, brewing operations favoring furan evaporation, such as infusion in hot water, allow beverages with reduced furan content to be obtained. The lowest concentrations were found in beverages prepared by using instant coffee, due to the much lower furan concentration in instant coffee powder compared with ground coffee beans (Hasnip *et al.*, 2006).

HMF concentration was found to vary from 26 to 1900 mg/kg in roasted coffee to about 2500 mg/kg in soluble coffee and 500–2300 mg/kg in roasted torrefacto (i.e., sucrose added before roasting) (Kanjahn *et al.*, 1996; Murkovich and Pichler, 2006; Arribas-Lorenzo and Morales, 2010; Teixido *et al.*, 2011). Due to its chemical instability, HMF content during roasting increases exponentially, reaches a maximum, and then decreases. The degradation is rather quick and commercially roasted coffee beans contain less HMF than expected (Murkovic and Bornik, 2007).

OTHER WAYS IN WHICH COMPOSITION IS ALTERED

Despite the publication of a great number of papers dealing with furans in coffee, especially in the last 10 years (Figure 65.2), up until today scarce information is available in the literature about the possible routes to reduce furan and HMF levels in coffee, and thus consumer intake.

Processing strategies for reducing furan and HMF contents in coffee and coffee brews can be regarded as preventive and removal interventions (Anese and Suman, 2013). The preventive strategies are aimed at keeping furan and HMF formation as low as possible during roasting by acting on precursors and mechanistic pathways, and thus creating less favorable reaction conditions.

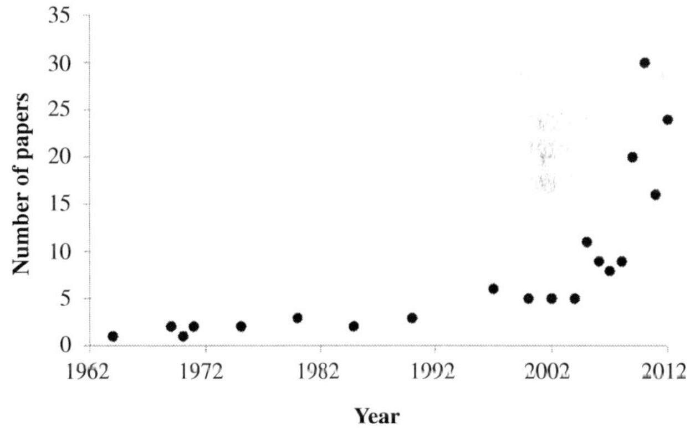

FIGURE 65.2 Number of publications on furans in coffee since 1962 (source, Isi Web of Knowledge)

TABLE 65.1 Furan Concentration in Coffee Brews Prepared According to Different Methods

Furan Concentration (μg/kg) ± sd						
Infusion in Hot Water	**Drip Filter**	**Espresso Machine**	**Fully Automated Machine**	**Pad-based Machine**	**Instant**	**Reference**
	50 ± 12 [6]	44 ± 11 [7]	88 ± 30 [5]		11 ± 7 [7]	Kuballa *et al.* (2005)
115 ± 75 [12]		45 ± 14 [2]				La Pera *et al.* (2009)
	13 [1]	102 ± 33 [4]	76 ± 42 [2]			Zoller *et al.* (2007)
	62 ± 41(light)[a] [6] 102 ± 41(medium)[a] [6] 133 ± 65(dark)[a] [8]					Pavesi-Arisseto *et al.* (2011)
	41 ± 16 [15] 22 ± 2(light)[a] 53 ± 3(medium)[a] 138 ± 8(dark)[a]	80 ± 29 [15]		166 ± 45 [9]	25 ± 9 [5]	Altaki *et al.* (2011)

Squared brackets: number of samples analyzed.
[a] *Light, medium, dark: coffee degree of roast.*

The removal strategies are aimed to move away the already formed molecules from the finished product. As the removal interventions are thought of as post-process interventions, in principle they are not affected by the previous process steps. Relevant technological strategies of mitigation are summarized in Table 65.2. They can be applied in one or more steps of coffee production, from the roasting process, to the brew preparation and consumption. Actually, some of these strategies are hardly applicable. For instance, although brewing of roast and ground coffee in drip-filter machines (i.e., associated with lower furan and HMF levels in the coffee cup) is still typical for the majority of European coffee consumption, the use of fully automatic machines, espresso machines and machines based on coffee pads, disks, or capsules (i.e., closed systems limiting furan and HMF losses from the brew) has greatly increased in the last few years. Also, coffee holding time and temperature as well as stirring before consumption may vary. Typically, coffee prepared by means of the closed machines are intended for immediate consumption. Therefore, lower losses of furans as well as flavor compounds would occur. Paradoxically, these coffees which retain excellent aroma and flavor may have higher amounts of the heat-induced toxic molecules. Reduction of furan and HMF in coffee is actually a challenging issue. Indeed, a limiting factor for the exploitation of the mitigations strategies is the existence of a close relationship between the formation of the undesired furan and HMF and the development of the desired sensory properties of coffee. As the formation of these heat-induced molecules is concomitant with that of color, flavor, and texture, it is very difficult to minimize their formation without affecting the coffee's sensory acceptability.

In this respect, potentially applicable mitigation strategies are discussed here, i.e., low thermal input-based roasting and vacuum treatments. The former can be regarded as a preventive strategy, the latter as a removal intervention.

Preventive Strategies

As has already been pointed out, the levels of furan and, to some extent, HMF in coffee increase with the increase of the roasting degree (Guenther *et al.*, 2010; Pavesi-Arisseto *et al.*, 2011). Therefore, in principle, decreasing the thermal

TABLE 65.2 Preventive and Removal Strategies of Furan and 5-Hydroxymethylfurfural in Coffee, Mechanisms of Action, and Relevant References

Mitigation Strategy	Process Step	Type of Intervention	Mechanism of Action	Coffee Product	Reference
Prevention	Roasting	Roasting to light or medium degree	Decrease of formation rate	Beans	Guenther *et al.* (2010); Pavesi-Arisseto *et al.* (2011)
		Low temperature–long time	Decrease of formation rate	Beans	Altaki *et al.* (2011)
		Microwave–hot air	Decrease of formation rate	Beans	Akkarachaneeyakorn *et al.* (2010)
	Formulation	Use of instant coffee	Low amount in the powder	Brew	Hasnip *et al.* (2006)
		Low amount of coffee for high water volume	Dilution	Brew	Zoller *et al.* (2007); Guenther *et al.* (2010)
		Use of normal or coarse ground coffee	Low extraction degree	Brew	Crews *et al.* (2009)
Removal	Degassing	Small coffee particle size Prolonged exposure to air at relatively high temperature	Evaporation	Beans	Guenther *et al.* (2010)
	Brewing	High extraction temperature (>85°C) Use of open systems	Evaporation	Brew	Hasnip *et al.* (2006); Zoller *et al.* (2007); Kim *et al.* (2009); La Pera *et al.* (2009); Guenther *et al.* (2010)
	Post-processing	Vacuum removal	Evaporation	Beans	Quarta and Anese (2012)
		Keeping coffee hot/stirring	Evaporation	Brew	Morehouse *et al.* (2008); La Pera *et al.* (2009); Guenther *et al.* (2010)

input, i.e., the heat amount provided to the coffee during roasting, could represent an effective way of furan and HMF minimization during roasting. Thermal input reduction can be obtained by applying prolonged heating at lower temperatures. Altaki *et al.* (2011) compared furan levels in coffee brews obtained from beans roasted to the same degree by using different temperature–time conditions, i.e., high temperature–short time (200°C, 6 min), intermediate temperature–intermediate time (170°C, 12 min) and low temperature–long time (140°C, 20 min). They found that the low temperature–long time roasting process allowed the lowest furan concentration to be obtained. However, the authors did not evaluate whether the coffee flavor profile was affected by the application of these roasting conditions.

Alternatively, a reduction of the thermal input can be obtained by applying dielectric heating. Akkarachaneeyakorn *et al.* (2010) used combined microwave–hot air roasting to produce roasted malts, which are used in the brewery industry to give particular colors and flavors to beers. This technology proved to be effective in minimizing furan and related compounds in black malts as compared to the conventional process. As is well known, heating of foods by microwave energy sources is instantaneous and uniform, in contrast to conventional heating that transfers thermal energy from the product surface toward its center. Heat is generated within the product due to the frictional interactions of polar dielectric molecules (i.e., water) trying to align to the electric field associated with electromagnetic radiation oscillating at very high frequencies (Mudgett, 1989). As the microwaves enter the product, they interact with regions of positive and negative charges on water molecules, and heat is generated at any site of the food where water is present. Therefore, foods processed by microwave heating undergo lower thermal input than the conventionally processed ones. Although research is needed to identify proper process parameters, this technology could represent an innovative tool for roasting coffee.

Roasting to lower degrees could represent a further option to decrease furan levels. However, this strategy should take into account the differences in taste and preference among people in different countries. Light and medium degrees of roast are well accepted in American and North European countries, whereas they do not meet the preference of South European countries. It is noteworthy that roasting to a light degree to prevent furan formation may increase at the same time the level of acrylamide, which is also a probable heat-induced carcinogenic molecule (Guenther *et al.*, 2010).

Removal Strategies

Reduction of furan and HMF levels in coffee could be accomplished by exploiting their high volatility. A number of studies actually deal with furan loss due to evaporation during different steps of coffee production and preparation (Kuballa *et al.*, 2005; Zoller *et al.*, 2008; Guenther *et al.*, 2010). Based on these observations, one could exploit the high volatility of furan compounds to remove them by applying proper temperature, time, and pressure conditions. The results of a recent study showed that HMF can be effectively removed from coffee powder by means of vacuum treatments (Quarta and Anese, 2012). The removal of HMF turned out not to be possible from dry coffee due to the hindering effect of the glassy matrix. On the contrary, the vacuum treatment turned out to be effective on coffee powder hydrated to high water activity (around 0.7) prior to the vacuum treatment. Indeed, an increase in water content may induce transitions from the glassy to a rubbery state, responsible for volatile release. By contrast, the vacuum treatment allowed only furan to be removed from a high moisture food matrix, probably due to its higher volatility in comparison with HMF (Anese and Suman, 2013). Unfortunately, the vacuum treatments may cause a significant decrease in the coffee headspace total volatiles, which in turn may also be responsible for lower coffee odor intensity (Quarta and Anese, 2012). Nevertheless, the vacuum technology represents a promising technology for removal of furan and HMF from coffee. In order to make it exploitable at the industrial level, further studies are needed to identify process conditions able to minimize the loss of sensory properties. These could be proper pressure, temperature, and time combinations, or, alternatively, a flavor enrichment step to be performed on the vacuum-treated coffee just before packaging.

ANALYTICAL TECHNIQUES

Different analytical techniques are currently used to determine furan and HMF in foods. Furan levels are generally quantified by means of headspace (HS) or headspace solid-phase microextraction (HS-SPME) coupled to gas chromatography–mass spectrometry (GC-MS) (Bianchi *et al.*, 2006; FDA, 2006; Zoller *et al.*, 2007; Altaki *et al.*, 2007; Morehouse *et al.*, 2008). In headspace analysis a portion of the headspace gas is taken and injected into the GC-MS. In contrast, in SPME a fiber of polymeric material is exposed to the headspace of sample vial vapor and then desorbed by exposing the fiber in the GC injector.

Headspace analysis requires partitioning of furan from the food sample into the vial headspace. It is affected by many variables, such as time and temperature of partitioning as well as food composition. Van Lancker *et al.* (2009) found that lipophilic compounds, such as oil in coffee, are responsible for significant furan retention. Therefore, headspace procedure has to be carried out at a sufficiently high temperature to obtain complete furan volatilization in the gas phase of the vial headspace. Quantification of furan is carried out by means of standard addition or external calibration graph incorporating a deuterium-labeled internal standard (i.e., furan-d_4). For MS detection in selected ion monitoring mode, furan and furan-d_4 quantification is carried out by using *m/z* 68 and *m/z* 72, respectively. Analytical recovery for coffee has been reported to be higher than 90% with limits of detection of less than 1 μg/kg (Crews and Castle, 2007).

Critical issues may affect reliable furan quantification. Indeed, because of its volatility, furan can be released by the matrix or generated during analysis. To avoid losses, sample preparation is generally carried out at chilling temperature, while the headspace incubation temperature is, in most cases, settled at 50°C or below to prevent extra furan formation (Crews and Castle, 2007).

A wide number of methods for HMF analysis in coffee are described in the literature, including colorimetric, spectrophotometric, and chromatographic approaches (Morales, 2009). Among these, the chromatographic ones are mostly preferred as they provide accurate and reliable quantification of HMF. Before chromatographic detection coffee has to be subjected to a cleanup step, consisting of water extraction and subsequent filtration. Clarification of coffee extract is recommended. It is generally carried out by using solid-phase extraction cartridges (Murkovic and Pichler, 2006). HMF detection is usually performed by reversed-phase liquid chromatography coupled with UV or photodiode array detector set at 280 nm.

References

Akkarachaneeyakorn, S., Laguerre, J.C., Tattiyakul, J., Neugnot, B., Boivin, P., Morales, F.J., Birlouez-Aragon, I., 2010. Optimization of combined microwave-hot air roasting of malt based on energy consumption and neo-formed contaminants content. J. Food Sci. 74, E201–E207.

Altaki, M.S., Santos, F.J., Galceran, M.T., 2007. Analysis of furan in foods by headspace solid-phase microextraction–gas chromatography–ion trap mass spectrometry. J. Chrom. A. 1146, 103–109.

Altaki, M.S., Santos, F.J., Galceran, M.T., 2011. Occurrence of furan in coffee from Spanish market: contribution of brewing and roasting. Food Chem. 126, 1527–1532.

Anese, M., Suman, M., 2013. Mitigation strategies of furan and 5-hydroxymethylfurfural in food. Food Res. Int. 51, 257–264.

Arribas-Lorenzo, G., Morales, F.J., 2010. Estimation of dietary intake of 5-hydroxymethylfurfural and related substances from coffee to Spanish population. Food. Chem. Toxicol. 48, 644–649.

Bianchi, F., Careri, M., Mangia, A., Musci, M., 2006. Development and validation of a solid phase micro-extraction–gas chromatography–mass spectrometry method for the determination of furan in baby-food. J. Chrom. A. 1102, 268–272.

Buffo, C., Cardelli-Freire, C., 2004. Coffee flavor: an overview. Flavour Fragr. J. 19, 99–104.

Capuano, E., Fogliano, V., 2011. Acrylamide and 5-hydroxymethylfurfural: A review on metabolism, toxicity, occurrence in food and mitigation strategies. LWT Food Sci. Technol. 44, 793–810.

Clarke, R.J., 1987. Roasting and grinding. In: Clarke, R.J., Macrae, R. (Eds.), Coffee. Volume 2: Technology. Elsevier Applied Science Publ. Ltd, London, pp. 73–107.

Clarke, R.J., 1990. The volatile compounds of roasted coffee. It. J. Food Sci. 2, 79–88.

Crews, C., Castle, L., 2007. A review of the occurrence, formation and analysis of furan in heat-processed foods. Trends Food Sci. Technol. 18, 365–372.

Crews, C., Roberts, D., Lauryssen, S., Kramer, G., 2009. Survey of furan in foods and coffees from five European Union countries. Food Addit. Contam.: Part B 2, 95–98.

EFSA, European Food Safety Authority, 2010. Update of results on the monitoring of furan levels in food. EFSA J. 8 (7), 1702–1720.

EFSA, European Food Safety Authority, 2011. Update on furan levels in food from monitoring years 2004–2010 and exposure assessment. EFSA J. 9 (9), 2347–2380.

Eggers, R., Pietsch, A., 2001. Technology I: Roasting. In: Clarke, R.J., Vizthum, O.G. (Eds.), Coffee: Recent Developments. Blackwell Science Ltd, London,, pp. 90–107. Chapter 4.

Flament, I., Chevallier, C., 1998. Analysis of volatile constituents of coffee aroma. Coffee Ind. 18, 592–596.

Flament, I., Bessiere-Thomas, Y., 2002. Coffee Flavour Chemistry. Wiley, New York.

Glatt, H., Sommer, Y., 2006. Health risk of 5-hydroxymethylfurfural (HMF) and related compounds. In: Skog, K., Alexander, J. (Eds.), Acrylamide and other hazardous compounds in heat-treated foods. Woodhead Publ. Ltd, Boca Raton, FL, pp. 328–357.

Grosh, W., 2001. Chemistry III: Volatile compounds. In: Clarke, R.J., Vitzthum, O.G. (Eds.), Coffee: Recent Developments. Blackwell Science Ltd, London, pp. 68–89. Chapter 3.

Guenther, H., Hoenicke, K., Biesterveld, S., Gerhard-Rieben, E., Lantz, I., 2010. Furan in coffee: pilot studies on formation during roasting and losses during production steps and consumer handling. Food Addit. Contam. Part A 27, 283–290.

Hasnip, S., Crews, C., Castle, L., 2006. Some factors affecting the formation of furan in heated foods. Food Addit. Contam. 23, 219–227.

Husøy, T., Haugen, M., Murkovic, M., Jöbstl, Stølen, L.H., Bjellaas, T., Rønningborg, C., Glatt, H., Alexander, J., 2008. Dietary exposure to 5-hydroxymethylfurfural from Norvegian food abd correlations with urine metabolites of short-term exposure. Food Clin. Toxicol. 46, 3697–3702.

IARC, 1995. Dry cleaning, some chlorinated solvents and other industrial chemicals. In: Monographs on the Evaluation of Carcinogenic Risks to Humans. International Agency for Reasearch on Cancer, Lyon, France, pp. 394–407. n.63.

Kanjahn, D., Jarms, U., Maier, H.G., 1996. Hydroxymethylfurfural and furfural in coffee and related beverages. 1. Evaluation of the method. Deutsche Lebesm. Runds 92, 328–331.

Kim, T.K., Lee, Y.K., Park, Y.S., Lee, K.G., 2009. Effects of cooking or handling conditions on the furan levels of processed foods. Food Addit. Contam. 26, 767–775.

Kuballa, T., Stier, S., Strichow, N., 2005. Furan in Kaffee und Kaffeegetränken. Deutsche Lebensmittel-Rundschau 101, 229–235.

La Pera, L., Liberatore, A., Avellone, G., Fanara, S., Dugo, G., Agozzino, P., 2009. Analysis of furan in coffee of different provenance by headspace solid phase microextraction gas chromatography–mass spectrometry: effect of brewing procedures. Food Addit. Contam. 26, 786–792.

Maga, J.A., 1979. Furans in food. Crit. Rev. Food Sci. Nutr. 11, 355–400.

Monien, B.H., Engst, W., Barknowitz, G., Seidel, A., Glatt, H., 2012. Mutagenicity of 5-hydroxymethylfurfural in V79 cells expressing human SULT1A1: identification and mass spectrometric quantification of DNA adducts formed. Chem. Res. Toxicol. 25, 1484–1492.

Moon, J.-K., Shibamoto, T., 2009. Role of roasting conditions in the profile of volatile flavour chemicals formed from coffee. J. Agric. Food Chem. 57, 5823–5831.

Morales, F.J., 2009. Hydroxymethylfurfural (HMF) and related compounds. In: Stadler, R.H., Lineback, D.R. (Eds.), Process-induced food toxicants. John Wiley & Sons, Inc, New York, pp. 135–174.
Morehouse, K.M., Nyman, P.J., McNeal, T.P., Di Novi, M.J., Perfetti, G.A., 2008. Survey of furan in heat processed foods by headspace gas chromatography/mass spectrometry and estimated adult exposure. Food Addit. Contam. Part A 25, 259–264.
Moro, S., Chipman, J.K., Wegener, J.-W., Hamberger, C., Dekant, W., Mally, A., 2012. Furan in heat-treated foods: formation, exposure, toxicity, and aspects of risk assessment. Mol. Nutr. Food Res. 56, 1197–1211.
Mudgett, R.E., 1989. Microwave food processing. Food Technol. 1, 117–126.
Murkovic, M., Pichler, N., 2006. Analysis of 5-hydroxymethylfurfural in coffee, dried fruits and urine. Mol. Nutr. Food Res. 50, 842–846.
Murkovic, M., Bornik, M.-A., 2007. Formation of 5-hydroxymethyl-2-furfural (HMF) and 5-hydroxymethyl-2-furoic acid during roasting of coffee. Mol. Nutr. Food Res. 51, 390–394.
Pavesi-Arisseto, A., Vicente, E., Soares Ueno, M., Verdiani Tfouni, S.A., De Figueiredo Toledo, M.C, 2011. Furan levels in coffee as influenced by species, roast degree, and brewing procedures. J. Agric. Food Chem. 59, 3118–3124.
Quarta, B., Anese, M., 2012. Furfurals removal from roasted coffee powder by vacuum treatment. Food Chem. 130, 610–614.
Reinhard, H., Sager, F., Zimmermann, H., Zoller, O., 2004. Furan in foods on the Swiss market. Methods and results. Mitt. Lebensm. Hyg 95, 532–535.
Rufian-Henares, J.A., de la Cueva, S.P., 2008. Assessment of hydroxymethylfurfural intake in the Spanish diet. Food Addit. Contam. Part A 25, 1306–1312.
Teixido, E., Nunez, O., Santos, J.F., Galceran, T.M., 2011. 5-Hydroxymethylfurfural content in foodstuffs determined by micellar electrokinetic chromatography. Food Chem. 126, 1902–1908.
US FDA–Food and Drug Administration, 2006. Determination of furan in food. Available from http://www.fda.gov/Food/FoodSafety/FoodContaminantsAdulteration/ChemicalContaminants/Furan/ucm078400.htm (accessed in January 2013).
Van Lancker, F., Adams, A., Owczarek, A., De Meulenaer, B., De Kimpe, N., 2009. Impact of various food ingredients on the retention of furan in foods. Mol. Nutr. Food Res. 53, 1505–1511.
Yaylayan, V.A., 2006. Precursors, formation and determination of furan in food. J. Cons. Prot. Food Saf. 1, 5–9.
Zhang, X.M., Chan, C.C., Stamp, D., Minkin, S., Archer, M.C., Bruce, W.R., 1993. Initiation and promotion of colonic aberrant crypt foci in rats by 5-hydroxymethyl-2-furaldehyde in thermolyzed sucrose. Carcinogen 14, 773–775.
Zoller, O., Sager, F., Reinhard, H., 2007. Furan in food: Headspace method and product survey. Food Addit. Contam. 24 (S1), 91–107.

CHAPTER

66

Mineral Composition Variability of Coffees: A Result of Processing and Production

Rebeca Cruz[*], *Simone Morais*[†], *Susana Casal*[*]

[*]REQUIMTE, Faculdade de Farmácia, Universidade do Porto, Porto, Portugal, [†]REQUIMTE, Instituto Superior de Engenharia do Porto, Instituto Politécnico do Porto, Portugal

CHAPTER POINTS

- Distinct elemental composition in coffee beans is observed when the plant is cultivated in different geographical areas and in organic or conventional fertilized soils.
- Elemental pattern and isotope analysis of coffee are discriminators to its geographical origin and species.
- Mild coffees usually present lower total soluble solids (TSS) content than natural ones, by diffusion from the bean to the water during fermentation and washing steps.
- Each beverage preparation method involves a coffee/water ratio, percolation temperature and time, grinding and roasting degree, which, alongside each element extractability yield, results in a distinct elemental composition.
- Coffee by-products are still rich in several minerals leading to interesting applications; furthermore, they present great adsorption potential for heavy metals decontamination.
- Atomic absorption spectrometry is one of the most widely applied techniques due to its high sensitivity, selectivity, robustness, wide linear range, low operational costs, and simplicity of operation. Inductively coupled plasma with atomic emission spectrometry (ICP-AES) applications are increasing due its truly multi-element character and faster analysis.

INTRODUCTION

Coffee is a widespread commodity, being the result of a complex sequence of technological processes, starting in the field and finishing in the cup. All steps involved, such as cultivation, processing, transport, and marketing, provide employment to millions of people, thus playing a key role in economies and politics of both producing and consuming countries.

Two coffee species attain higher commercial importance—*Coffea arabica* and *Coffea canephora*, usually known as Arabica and Robusta, respectively. Despite being indigenous to Ethiopia, coffee production has spread around the world where adequate temperature, humidity and altitude are available for its cultivation (Clarke, 1989). The actual locations of the foremost producing and exporting countries are shown in Figure 66.1.

Elemental food profiling has been performed mostly due to minerals' nutritional relevance, with significant physiological activities acknowledged for 23 elements. Since these micronutrients have proven bioactivity, it is vital to assess their course during production and processing because the final elemental composition of a food is determined, not only by its physiological and genetic characteristics, but also by the elements made available during its development and by the technological processes applied.

Green coffee beans usually contain around 4% (dry weight) of total minerals with K as its major constituent, representing 40% of the oxide ash (Clarke, 1989). Coffee mineral composition variability will be detailed in the present chapter, highlighting not only the production and technological factors involved, but also the applicability of mineral profiling and the analytical methodologies used.

Processing and Impact on Active Components in Food
http://dx.doi.org/10.1016/B978-0-12-404699-3.00066-4

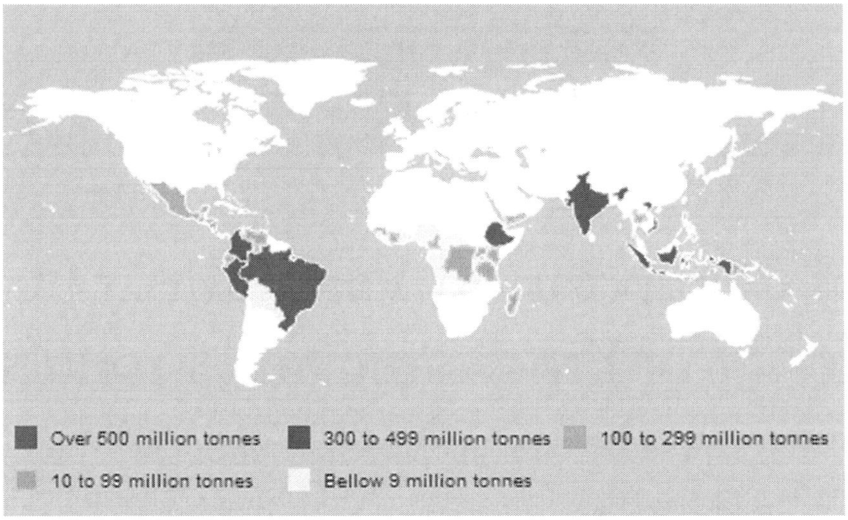

FIGURE 66.1 Main coffee producing countries. *Adapted from ICO (2012).*

AGRICULTURAL PRACTICES

The mineral composition of the coffee plant, and consequently its seeds, the coffee beans, is influenced by both environmental and agronomical conditions. An adequate element balance and acidity (pH5–6) are determinant to produce high-quality coffee. For instance, the endogenous level of K is vital for the beans, development, also influencing total sugars and citric acid content, while suitable Ca amounts provide increased resistance to pathogen attack (Illy and Viani, 2005).

These elements are usually absorbed from the soil and complemented by fertilizers. Despite being cultivated in equatorial regions, soil characteristics are naturally diversified, varying from volcanic soils with high K amounts (e.g., Indonesia or Central America), to those rich in basal and granite with a low K status (e.g., Brazil, West Africa or India) (Illy and Viani, 2005). These soil elements will influence coffee beans' mineral content, providing a basis for the development of authenticity tools for geographical discrimination, as will be detailed below.

However, increasing fertilizer application is a major source of soil contamination with foreign elements, including potentially harmful ones that may induce health risks for the consumers. Santos *et al.* (2009) evaluated the metal content in Brazilian coffees cultivated by conventional and organic agriculture, suggesting that Cd, Cu, Zn, and other toxic elements present in some inorganic fertilizers used in conventional coffee farms, might increase the amounts in the crop soil and pass into the food chain. A conversion period is therefore recommended for genuine organic coffee and mineral profiling can be used as an authenticity and safety analytical tool.

PROCESSING METHODS

Wet and Dry Processing

The coffee plant fruit, known as the coffee cherry due to its reddish taint when mature, usually encloses two seeds, the raw coffee beans, beneath a mucilage layer and a hard shell. For separating these beans, different post-harvest processing methods are applied, usually chosen according to water availability and tradition of the producing region. The method applied, however, will have a marked influence on the quality of the coffee cup, and particularly on the beans' mineral content.

Post-harvest handling of the cherries is carried out by the dry (natural), wet (washed), or semi-wet (pulped-natural) methods, as represented in Figure 66.2. In the dry process, the whole cherries (beans, mucilage, pulp, and shell) are dried under the sun or in mechanical-dryers, and finally the dry shell (husk) is removed. In the wet process, the cherries are mechanically pulped and the mucilage remains (pectin and sugars) are removed by microbial fermentation, being thereafter washed and dried under the sun or in mechanical dryers, as for the dry method. In the pulped-natural method, an intermediate process between natural and washed processes, the cherries are pulped and dried, without the fermentation step.

The major difference in the chemical composition of washed and natural coffee is their total soluble solids (TSS) content, which is higher in the latter. Being dried within the cherry, a concentration gradient is established from the mucilage into the bean, minerals included. Instead, significant amounts of TSS diffuse from the bean to the water in the wet process, both during fermentation and washing steps. Leached compounds include those of major water-solubility, such as sugars and minerals. Consequently, dry-processed coffee's soluble solids are

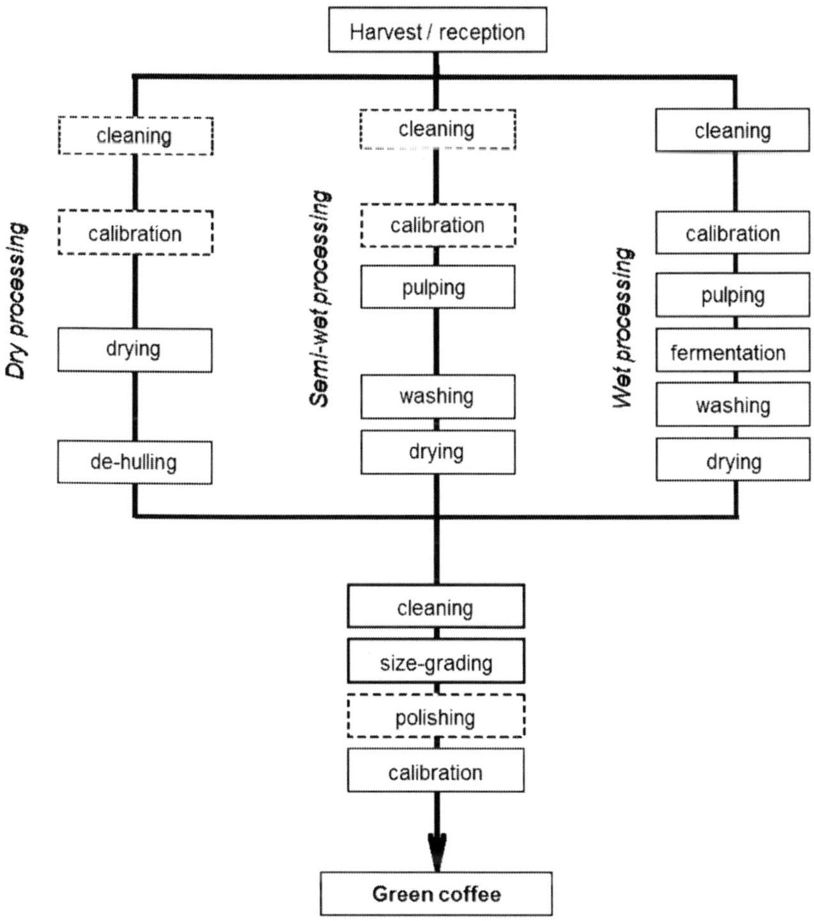

FIGURE 66.2 Green coffee processing methods.

0.4–0.9% higher than the wet-processed ones (Illy and Viani, 2005), thus presenting higher mineral contents. Apart from nutritional implications, which are reduced, the increased TSS content is determinant for the body of the beverage, an important feature in some coffee brews, particularly espresso.

A summary of the mineral profile of green, roasted, and instant coffee is presented in Table 66.1. Potassium is the major element, independently of the coffee species. The differences between Arabica and Robusta green coffee are not significant, though total mineral amounts are slightly lower in Arabica beans (3.5–4.5 g/100 g), when compared to the Robusta ones (3.9–4.5 g/100 g) (Clarke, 1989). Still, these amounts are affected by the processing method applied. When Arabica and Robusta coffees are both dry-processed, the differences are reduced, with total mineral content of 4.11 to 4.27 g/100 g for Arabica, and from 4.14 to 4.39 g/100 g for Robusta (Clarke, 1989). Only when dry- and wet-processed Arabica are directly compared are the differences highlighted, with wet-processed Arabica mineral amounts reduced to 3.58–3.95 mg/100 g (Clarke, 1989), providing supporting evidence for the fact that technological process is a determinant factor in the green coffee mineral content. Except for K, which follows the same pattern described for total mineral amounts (Clarke, 1989), no details on individual elements are found in the literature, with an overall reference for green coffee, without specifying the processing method.

Decaffeination

Some coffees are industrially decaffeinated in order to attend to some consumer's requirements. In order to minimize flavor and aroma loss, this process is usually performed before roasting. In simplified terms, the beans are initially swelled in water to solubilize the caffeine–potassium–chlorogenate complex and to make caffeine more available for extraction. A variety of solvents and conditions can be used, ranging from chemical solvents to hot water extraction (followed, in this case, by caffeine removal with adsorbents) or, more recently, to supercritical fluid extraction. The solvent residues are removed and the beans are dried. Being a non-selective solvent-based extraction, numerous soluble compounds might leach out. However, mineral solubility in the chemical solvents used is reduced and most recent water techniques use the principle of extractive solutions saturated in coffee-soluble components, reducing losses, which is crucial to ensuring the coffee characteristics.

TABLE 66.1 Multi-Elemental Composition (mg/100 g) of Green and Roasted Coffee Beans, and Instant Coffees

Samples	Species	K	Mg	P	Ca	Na	Fe	Mn	Cu	Zn	References
GREEN COFFEE											
n = 4	Arabica	1630–1750	190–220	166–174	100–140	9–12	3.7–6.1	3.2–6.0	1.3–2.4	0.6–1.4	Ferreira *et al.* (1971)
n = 28	Arabica	1211–1828	172–206	141–170	95–137	3–12	2.5–5.5	1.6–5.0	1.4–7.7	0.4–6.1	Martin *et al.* (1998)
n = 14	Robusta	1690–1810	160–210	135–194	100–140	7–9	3.2–8.6	1.3–3.3	0.5–1.8	0.5–1.3	Ferreira *et al.* (1971)
n = 13	Robusta	1545–1896	160–197	176–220	93–162	2–10	2.9–9.3	1.4–2.0	1.5–2.6	0.5–2.0	Martin *et al.* (1998)
	Variability	**1211–1896**	**153–220**	**117–220**	**92–162**	**2–12**	**2.3–9.3**	**1.3–6.0**	**0.5–37**	**0.4–6.1**	
ROASTED COFFEE											
n = 9	Arabica	1273–1493	173–194	124–144	88–109	1.1–3.3	4.1–5.3	1.3–4.5	1.2–1.8	0.5–3.7	Martin *et al.* (1999)
n = 60	Arabica	1137–2275	75–93	215–332	67–149	3.1–11.3	1.8–5.4	1.7–4.7	0.9–3.1	0.6–1.4	Grembecka *et al.* (2007)
n = 9	Robusta	1389–1541	146–182	158–196	94–135	0.7–3.4	5.1–7.3	1.2–2.0	1.3–1.7	0.5–3.2	Martin *et al.*,1999
n = 5	n.i.	1360–2910	200–310	—	49–97	—	1.6–9.2	0.7–1.3	0.4–1.6	0.3–0.9	Suseela *et al.* (2001)
n = 160	n.i.	1750–1960	206–241	171–211	93–123	1.0–147	1.2–3.1	1.9–3.9	1.2–1.8	0.6–0.8	Anderson and Smith (2002)
n = 75	n.i.	1175–1585	80–284	176–403	51–162	3.6–170	2.0–6.8	1.6–4.1	1.2–2.0	0.3–1.6	Grembecka *et al.* (2007)
n = 9	n.i., decaffeinated	1342–1534	201–252	220–250	87–102	39.2–93.9	3.8–8.3	1.7–2.3	1.6–2.0	0.3–0.7	Grembecka *et al.* (2007)
	Variability	**1137–3710**	**75–310**	**158–403**	**51–220**	**0.7–170**	**1.2–9.2**	**1.2–5.0**	**0.9–3.1**	**0.1–3.7**	
INSTANT COFFEE											
n = 21	n.i.	3250–5170	212–415	223–410	106–189	27–666	1.4–45.1	0.4–3.9	0.05–0.2	0.3–1.5	Santos and Oliveira (2001)
n = 5	n.i.	1360–2910	200–310	—	49–97	—	1.5–9.2	0.7–1.3	0.04–1.6	0.2–0.9	Suseela *et al.* (2001)
n = 27	n.i.	1871–3324	85–451	348–598	65–265	2.8–347	1.9–5.9	0.9–3.5	0.03–0.1	0.2–1.1	Grembecka *et al.* (2007)
n = 8	n.i.	3717–6150	307–547	2618–4290	65–96	26–289	1.8–10.8	1.0–2.4	—	—	Oliveira *et al.* (2012)
	Variability	**1360–6150**	**85–546**	**223–598**	**65–265**	**3–666**	**1.4–45.1**	**0.1–3.9**	**0.03–1.6**	**0.2–1.5**	

n.i., not identified.

No significant losses in mineral content are expected during this pre-roasting step, but the original mineral profile might be slightly altered from diffusion from previous water-decaffeinated batches. However, literature data are scarce, with no references to the modifications induced by water or solvent decaffeination processes. Only Grembecka and coworkers (2007) detail the mineral composition of decaffeinated coffees, but only after roasting (Table 66.1). Still, the values reported are within those presented by the remaining roasted coffee samples, indicating that, if present, mineral leaching should be reduced.

Roasting

Roasting conveys significant physical and chemical changes in the raw coffee beans. During this step, a portion of their mass is lost, including water vapor, carbon dioxide, and other volatile compounds as a result of pyrolysis. With temperatures usually up to 240°C, no losses are expected to occur in the minerals. However, as the global beans, mass is greatly reduced during the process, the mineral content is expected to increase proportionally on a mass basis. When green and roasted coffees are compared (Table 66.1), an increase in most elements is observed, being particularly evident for the main ones, potassium and magnesium. The roasting degree will also have a marked influence on the extractable soluble solids, with implications in the beverage preparation and coffee soluble industry yield. Values presented in Table 66.1 correspond to total elements for roasted beans, and not necessarily to the amounts that will be extracted into the brew, as will be detailed below.

Instant Coffee

The instant coffee market has been increasing over recent years, both due to economical restrains and convenience. Simultaneously, mixtures of instant coffee with soluble cereal extracts are popular, again due to economic reasons or to reduce caffeine ingestion. The soluble coffee industrial preparation follows the same principles as beverage, with roasting, grinding and extraction, followed, in this case, by a concentration into a dry soluble powder. However, this extractive process requires maximum extractability for economical profit, using battery column percolators and water usually above 100°C. Robusta beans are usually preferred due to their higher TTS and lower market price.

In the production of soluble coffee extracts most minerals are extracted into the concentrated 'brew'. Therefore, the elements amount in the final product will depend mostly on the green coffee initial contents, conditioned by the extraction yield achieved. Still, some minerals are naturally more water-soluble than others, while some form complexes with the matrix, reducing their extractability. When diverse instant coffee samples are compared on a direct mass basis with roasted coffees (Table 66.1), a clear increase is visible for the major elements, particularly K and Mg. Other elements such as Mn, Cu, or Zn seem to be lower, a probable indication of poorer extractability under the processing conditions used in the manufacture of the instant coffees. Water characteristics will also contribute to the soluble coffee mineral content, particularly in what concerns to the Ca and Na amounts (Gillies and Birkbeck, 1983).

BEVERAGE PREPARATION

Coffee brews are usually prepared with an Arabica coffee or Arabica/Robusta blend, presenting themselves to consumers as roasted beans, ground roasted coffee or instant coffee. Beverage preparation results from the direct contact of roasted beans with hot water, the soluble substances being extracted, while the oils are partially emulsified. Hence, it is worth mentioning that, water being the second-most important ingredient of coffee brews, its pH, hardness, and absence of unpleasant foreign flavors (such as chlorination) should be controlled (Illy and Viani, 2005), as it will interfere with beverage quality and, in this particular case, with its mineralization pattern. Additionally, the beverage chemical composition is dependent on the extractive efficiency, which, in turn, relies on several variables, including the roasting degree, grinding grade, coffee/water ratio, and percolation temperature (Illy and Viani, 2005).

Together with all these variables, beverage preparation type plays a preponderant role in brewing yield and concentration. Three major groups of extraction methods can be assembled: (i) decoction methods, embracing boiled coffee, Turkish coffee, percolator coffee, and vacuum coffee; (ii) infusion methods, such as filter coffee and Napoletana coffee; and (iii) pressure methods, which include plunger, moka, and espresso coffee. The former method has a limitative extraction yield, since its rate decreases as the brew concentration increases, though higher temperatures favor extraction yields and rates. Regarding infusion methods, a much more effective extraction of water-soluble substances, such as minerals, is observed, as a short contact time to every elementary volume of clean liquid is practiced. Finally, the latter methods, by being subjected to higher pressures during extraction, give rise to a brew of distinct chemical, and thus, sensorial properties (Petracco, 2001).

Total soluble solids assay is a common parameter evaluated in roasted coffee, which estimates the amount of substances potentially extracted during beverage preparation, and consequently contributing to its sensorial properties, such as body, flavor, and aroma (Illy and Viani, 2005). Knowing that TTS in coffee represent an average value of 30%, usually higher in Robusta, and that total mineral content corresponds to 15% of the former (Illy and Viani, 2005), a high mineral extractability in the brew is predictable. Nevertheless, not all elements present the same extractive behavior. For instance, K is promptly transferred into the brew, yet only about one-half of Na and Mn do so under the same brewing conditions (Clarke, 1989), which is in accordance with the aforementioned statements for soluble coffee.

Also, with finer grinding, the transfer of emulsifiable and soluble substances into the brew is increased, including minerals (Illy and Viani, 2005). Brewing temperature is also directly related to yield variability, and darker roasting allows higher yields. Also, high percolation time and low coffee/water ratio will produce an over-extracted brew (Illy and Viani, 2005), with, therefore, a greater mineral content.

An example of this technological influence can be seen in Table 66.2, for the concentrations achieved with the same coffee samples processed by direct infusion or filtered coffee, attaining higher amounts of K, Mg, Ca, and Mn in the former (Varo *et al.*, 1980). Still, lower amounts of Fe and Cu were observed, indicating that these compounds are not favorably leached out. The mineral content of espresso coffee brew varies between 7000 to 7200 mg/l, and K amounts range from 2900 to 3200 mg/l, higher than those reported for other coffee beverages prepared by different methods due to its concentrated character, being therefore a shorter beverage (Illy and Viani, 2005).

When the extractive efficiencies (%) during beverage preparation are compared (Table 66.2), high extractabilities can be achieved for K, Mg, Ca, and Na while only up to 8% Cu is extracted or a maximum of 21% for Fe. The extractability achieved with instant coffee preparation is, as expected, almost complete (Table 66.2).

COFFEE REMAINS

As mentioned above, the coffee processing chain includes several technological steps, where huge amounts of residues are generated. In the producing countries, coffee husks and pulp are obtained after dehulling of the coffee cherries during dry or wet processing, respectively. The former usually presents 3–7 g/100 g of total minerals, on a dry basis, while the latter exhibits 6–10 g/100 g (Franca and Oliveira, 2009). Both residues are particularly rich in K (~4 g/100 g), which has encouraged the study of the application as organic fertilizers. Coffee husks performance as adsorbents for the removal of heavy metal ions from aqueous solutions has been also evaluated, with positive outcomes (Franca and Oliveira, 2009).

Spent coffee grounds are generated by the soluble coffee industry or after beverage preparation in cafeterias, restaurants, or at home (Cruz *et al.*, 2012). The industrial residues are particularly exhausted of soluble compounds, with a mineral content of 0.3–1 g/100 g, on a dry basis, with K as its major constituent (~350 mg/100 g), followed by P, Mg, Ca, Al, Fe, Mn, Cu, and Zn (Mussatto *et al.*, 2011). Conversely, the composition of the spent coffee after coffee brewing depends on the extraction yield inherent to the brewing procedure, but retains a higher mineral content. A detailed mineral profiling of espresso spent coffee was performed by the authors (Cruz *et al.*, 2012) attaining higher total mineral content (~2 g/100 g), K, P, and Mg than the instant coffee residues, making them potentially interesting as soil amendments, among other possibilities.

ORIGINS AND SPECIES

The geographical origin, species, and variety, may also lead to differences between products, which are particularly important from the economical point of view.

The assurance of geographic origin is an active research area, mainly due to a growing demand from consumers for authentic food products. A linkage between elemental compositions of green coffee to its country of origin was primarily performed by Krivan *et al.* (1993). From 20 elements analyzed, Mn was found to be the best discriminator between the eight production areas studied (Columbia, Costa Rica, Cuba, El Salvador, Mexico, Nicaragua, Panama, and Papua New Guinea). Afterward, Anderson and Smith (2002) conducted a study on 18 elements of 160 roasted coffee samples from the three major coffee-growing regions: Indonesia, East Africa, and Central/South America. The global results allowed a differentiation between origins, with a particular contribution of Cu, Na, Mn, and Fe. A more recent study proved that a combination of the isotopic systems of O and Sr was a good approach to discriminate coffees according to their geographical origin in smaller competitive regions, such as South America or Africa (Rodrigues *et al.*, 2011).

Coffee species discrimination constitutes an authenticity specification in coffee trading. Arabica coffees are generally considered to be of better quality, and thus of higher market prices. Early studies reported that Arabica coffee usually reveals higher amounts of Mn (2.5–6.0 mg/100 g) when compared with Robusta (1.0–3.3 mg/100 g) (Clarke, 1989). Later, Martín *et al.* (1998, 1999) evaluated the metal content (11 elements) of green and roasted coffee samples concluding that the content

TABLE 66.2 Multi-Elemental Composition (mg/l) and Extractability Yield (%) in Coffee Infusions Beverages Prepared by Different Brewing Methods

Brewing Method	Coffee/ Water Ratio	K	Mg	P	Ca	Na	Fe	Mn	Cu	Zn	References
Filter	60 g/l	740–856	73–80	—	25–29	—	0.54–0.60	0.5–0.7	0.03–0.06	0.10–0.5	Varo *et al.* (1980)
Infusion	60 g/l	900–1100	110–120	—	35–39	—	0.35–0.39	0.9–1.2	0.01–0.03	0.13–0.16	Varo *et al.* (1980)
Filter	21 g/l	740–856	43–52	—	27.8–28.6	8–13	0.11–0.33	0.02–0.16	0.06–0.58	0.17–0.18	Gillies and Birkbeck (1983)
Boiled	30 g/l	350–390	28–30	—	13–20	5–7	0.10–0.20	0.2–0.3	0.02–0.04	0.2–0.3	Ashu and Chandravanshi (2011)
	Variability	**195–1100**	**26–120**	**—**	**13–39**	**5–13**	**0.10–0.39**	**0.02–0.8**	**0.02–0.58**	**0.1–0.3**	
Instant	8.5 g/l	—	405–406	8.7–12.9	44–45	21	0.10–17	0.24–0.3	0.14–0.15	0.16–0.48	Gillies and Birkbeck (1983)
		Extractability yield (%)									
n.i.	n.i.	57–96	34–80	33–70	2–71	6–100	2–21	9–52	2–7	6–90	Grembecka *et al.* (2007)
Boiled	30 g/l	83–88	48–48	—	50–65	38–44	8–13	28–34	6–7	53–58	Ashu and Chandravanshi (2011)
	Variability	**57–96**	**34–80**	**33–70**	**2–71**	**6–100**	**2–21**	**9–39**	**2–8**	**6–90**	
Instant	8.5 g/l	90–109	88–100	87–103	86–103	90–103	96	89–119	74–110	69–106	Gillies and Birkbeck (1983)

n.i., not identified.

of P, Mn, and Cu can be good species discriminators. Moreover, higher amounts of Cu, Sr, and Ba were also detected in Robusta than in Arabica (Illy and Viani, 2005).

Recently, Santato and coworkers (2012) focused on characterizing the composition of 62 green Arabica and Robusta coffee beans grown in South and Central America, Africa and Asia. By using the δ^2H, Mg, Fe, Co, and Ni contents, they were able to assign correctly 95% of green coffee beans to the appropriate variety. Simultaneously, canonical discriminant analysis was performed using several elements and correctly traced the origin of 98% of coffee beans.

As for roasted or ground coffee, almost all commercialized soluble coffee products result from a blend of coffee species from different regions. None the less, the elemental pattern of coffee surrogates, mostly roasted cereals, has been found to be significantly different from plain coffee, being important to the authenticity of these products. In particular, Mg might provide a potential estimator for the presence of coffee in substitute blends (Oliveira *et al.*, 2012).

In summary, despite the changes resulting from processing, it is still possible to track species, geographical origin, and coffee surrogates on the basis of the elemental composition.

ANALYTICAL APPROACHES

Numerous approaches have been described in the literature for mineral composition analysis of coffees (Table 66.3).

Regarding sample preparation for coffee bean analysis, the conventional methods include dry ashing (Suseela *et al.*, 2001; Grembecka *et al.*, 2007; Bertrand *et al.*, 2008), wet digestion (Martín *et al.*, 1998), and microwave-assisted acid digestion (Krivan *et al.*, 1993; Asfaw and Wibetoe, 2005; Oleszczuk *et al.*, 2007; Rodrigues *et al.*, 2011; Cruz *et al.*, 2012), or extraction and derivatization for metalloid quantification (Asfaw and Wibetoe, 2005), all being preceded by grinding. Previous studies showed that graphite furnace atomic absorption spectrometry (GFAAS) also allows, besides neutron activation analysis (NAA) and X-ray fluorescence spectrometry (XRF), direct solid sampling analysis of green coffee beans with minimum sample preparation, which compares favorably

TABLE 66.3 Most Usual Techniques Applied for Minerals Composition Analysis of Coffees

Technique	Sample	Analyzed Elements	References
Atomic absorption spectrometry (electrothermal, GFAAS, or flame, FAAS)	GC	Ca,Mg,Zn,Cr,Cu,Fe,Mn,Sr	Krivan *et al.* (1993)
	GC	Mn,Co,Cu	Oleszczuk *et al.* (2007)
	RC, DRC, IC, FC	Ca,Mg,K,Na,Co,Mn,Fe,Cr,Ni, Zn,Cu,Cd,Pb	Grembecka *et al.* (2007)
	IC	Ca,Cr,Fe,K,Mg,Mn,Ni,Sr,Zn	Suseela *et al.* (2001)
	IC	Ca,Mg,K,Na,Fe,Mn,Cr,Ni	Oliveira *et al.* (2012)
Inductively coupled plasma with atomic emission spectrometry (ICP-AES)	GC	Zn,P,Mn,Fe,Mg,Ca,Na,K, Cu,Sr,Ba	Martín *et al.* (1998)
	GC	P,K,Ca,Mg,Fe,Cu,Zn,B	Bertrand *et al.* (2008)
	RC	K,Mg,Ca,Na,Al,V,Cr,Mn,Fe, Co,Ni,Cu,Zn,Mo,S,Cd,Pb,P	Anderson and Smith (2002)
	RC,IC	Al,Ca,Cu,Fe,K,Mg,Mn,Na,Pb, S,Se,Si,Sn,Sr,Zn	Fernandes *et al.* (2005)
	IC	Se,Ca,Mg,K,P,S,Zn	Asfaw and Wibetoe (2005)
Inductively coupled plasma with mass spectrometry (ICP-MS)	GC	Sr	Rodrigues et al. (2011)
X-ray fluorescence spectrometry (XRF)	IC,DIC	Br,Ca,Cr,Cu,Fe,K,Mn,Ni,Rb, Sr,Zn	Haswell and Walmsley (1998)
Neutron activation analysis (NAA)	GC	K,Mn,Na,Ba,Br,Ca,Co,Cr,Cs,Fe, La,Rb,Sc,Zn	Krivan *et al.* (1993)
Electrochemical method	IC	Pb,Cd,Cu	Suseela *et al.* (2001)

DIC, decaffeinated instant coffee; DRC, decaffeinated roasted coffee; FC, flavored coffee; GC, green coffee; IC, instant coffee; RC, roasted coffee.
Adapted from Draghici et al. *(2010).*

with the inductively coupled plasma instrumentation (Oleszczuk *et al.*, 2007). Soluble instant coffee samples are usually dissolved in hot water and analyzed directly after acidification and sample dilution (Oliveira *et al.*, 2012).

Atomic absorption spectrometry (flame, FAAS, and graphite furnace, GAAFS) and inductively coupled plasma with atomic emission spectrometry (ICP-AES) are the most widely applied techniques (Table 66.3). Although being significantly more expensive than the second, ICP-AES operates in a wider working range and also has the advantage of allowing the analysis of P and S (Anderson and Smith, 2002; Fernandes *et al.*, 2005). They are applicable for the determination of up to 60–70 elements. Macronutrients (Ca, K, Mg, Na, S, and P), micronutrients (Co, Cr, Cu, Fe, Mn, Mo, Ni, Se, and Zn), nonessential (B, Ba, Sr, Si, Al, and V), and toxic elements (Sn, Cd, and Pb) were determined in green, roasted coffee beans, decaffeinated, soluble and flavored coffees (Table 66.3).

Inductively coupled plasma—mass spectrometry (ICP-MS), a more recent technology involving an ICP ion source coupled with a mass detector, was applied by Rodrigues *et al.* (2011) to Sr isotope fingerprinting of green coffee beans. Radiochemical methods such as XRF (Haswell and Walmsley, 1998) and NAA (Krivan *et al.*, 1993) were also applied in some pioneer studies regarding discrimination of green coffee beans and instant coffees based on elemental composition (Haswell and Walmsley, 1998; K, Mn, and Na: Krivan *et al.*, 1993). The main advantages of these quantitative techniques are that they are non-destructive and do not require preparation of solid samples.

Stripping voltammetry was also applied in soluble instant coffee samples (Suseela *et al.*, 2001). Associated with its high sensitivity (10^{-10} mol/l in some cases), low cost, relative broad scope, and ability to measure several elements simultaneously, it also offers the possibility of studying metal speciation. The main limitations of stripping analysis are its restriction to about 30 elements and the susceptibility of the electrode surface to fouling by sample surface active compounds.

Acknowledgments

This work received financial support from the European Union (FEDER funds through COMPETE) and National Funds (FCT, Fundação para a Ciência e Tecnologia) through project Pest-C/EQB/LA0006/2013.

References

Anderson, K.A., Smith, B.W., 2002. Chemical profiling to differentiate geographic growing origins of coffee. J. Agric. Food Chem. 50, 2068–2075.

Asfaw, A., Wibetoe, G., 2005. Simultaneous determination of hydride (Se) and non-hydride-forming (Ca, Mg, K, P, S and Zn) elements in various beverages (beer, coffee, and milk), with minimum sample preparation, by ICP–AES and use of a dual-mode sample-introduction system. Anal. Bioanal. Chem. 382, 173–179.

Ashu, R., Chandravanshi, B.S., 2011. Concentration levels of metals in commercially available Ethiopian roasted coffee powders and their infusions. Bull. Chem. Soc. Ethiop. 25, 11–24.

Bertrand, B., Villarreal, D., Laffargue, A., Posada, H., Lashermes, P., Dussert, S., 2008. Comparison of the Effectiveness of Fatty Acids, Chlorogenic Acids, and Elements for the Chemometric Discrimination of Coffee (*Coffea arabica* L.) Varieties and Growing Origins. J. Agric. Food Chem. 56, 2273–2280.

Clarke, R.J., 1989. Water and mineral contents. In: Clarke, R.J., Macrae, R. (Eds.), Coffee chem., vol. 1. Elsevier Science Publishers, Ltd, pp. 42–82.

Cruz, R., Cardoso, M.M., Fernandes, L., Oliveira, M., Mendes, E., Baptista, P., Morais, S., Casal, S., 2012. Espresso Coffee Residues: A Valuable Source of Unextracted Compounds. J. Agric. Food Chem. 60, 7777–7784.

Draghici, C., Coman, G., Jelescu, C., Dima, C., Chirila, E., 2010. Heavy metals determination in environmental and biological samples. In: Environmental Heavy Metal Pollution and Effects on Child Mental Development– Risk Assessment and Prevention Strategies. NATO Advanced Research Workshop.

Fernandes, A.P., Santos, M.C., Lemos, S.G., Ferreira, M.M.C., Nogueira, A.R.A., Nobrega, J.A., 2005. Pattern recognition applied to mineral characterization of Brazilian coffees and sugar-cane spirits. Spectrochim. Acta. Part B 60, 717–724.

Ferreira, L.A.B., Fragoso, M.A., Peratta, M.F., Silvo, M.C.C., Rebeto, M.C., 1971. Mineral constituents in coffees of Angola. 5th Colloq Sci Int Café. . 51–62.

Franca, A.S., Oliveira, L.S., 2009. In: Ashworth, G.S., Azevedo, P. (Eds.), Agricultural Wastes. Nova Science Publishers, Inc, pp. 155–189.

Gillies, M.E., Birkbeck, J.A., 1983. Tea and coffee as sources of some minerals in the New Zealand diet. Am. J. Clin. Nutr. 38, 936–942.

Grembecka, M., Malinowska, E., Szefer, P., 2007. Differentiation of market coffee and its infusions in view of their mineral composition. Sci. Total Environ. 383, 59–69.

Haswell, S.J., Walmsley, A.D., 1998. Multivariate data visualisation methods based on multi-elemental analysis of wines and coffees using total reflection X-ray fluorescence analysis. J. Anal. At. Spectrom. 13, 131–13438.

ICO, 2012. International Coffee Organization. http://www.ico.org/ [online accessed in 11.19.12].

Illy, A., Viani, R., 2005. Espresso Coffee: The Science of Quality. Elsevier Academic Press.

Krivan, V., Barth, P., Morales, A.F., 1993. Multielement Analysis of Green Coffee and Its Possible Use for the Determination of Origin. Microchim. Acta. 110, 217–236.

Martín, M.J., Pablos, F., González, A.G., 1998. Characterization of green coffee varieties according to their metal content. Anal. Chim. Acta., 358,177–183.

Martín, M.J., Pablos, F., González, A.G., 1999. Characterization of arabica and robusta roasted coffee varieties and mixture resolution according to their metal content. Food Chem. 66, 365–370.

Mussatto, S.I., Carneiro, L.M., Silva, J.P., Roberto, I.C., Teixeira, J.A., 2011. A study on chemical constituents and sugars extraction from spent coffee grounds. Carbohydr. Polym. 83, 368–374.

Oleszczuk, N., Castro, J.T., Da Silva, M.,M., Korn, M.G.A., et al., 2007. Method development for the determination of manganese, cobalt and copper in green coffee comparing direct solid sampling electrothermal atomic absorption spectrometry and inductively coupled plasma optical emission spectrometry. Talanta 73, 862–869.

Oliveira, M., Casal, S., Morais, S., Alves, C., Dias, F., Ramos, S., Mendes, E., Delerue-Matos, C., Oliveira, M.B.P.P., 2012. Intra- and interspecific mineral composition variability of commercial instant coffees and coffee substitutes: Contribution to mineral intake. Food Chem. 130, 702–709.

Petracco, M., 2001. Technology IV: Beverage Preparation: Brewing Trends for the New Millennium. In: Clarke, R.J., Vitzthum, O.G. (Eds.), Coffee Recent Developments. Blackwell Science, Ltd, pp. 140–164.

Rodrigues, C., Máguas, C., Prohaska, T., 2011. Strontium and oxygen isotope fingerprinting of green coffee beans and its potential to proof authenticity of coffee. Eur. Food Res. Technol. 232, 361–373.

Santato, A., Bertoldi, D., Perini, M., Camin, F., Larcher, R., 2012. Using elemental profiles and stable isotopes to trace the origin of green coffee beans on the global market. J. Mass Spectrom. 47, 1132–1140.

Santos, E.J., Oliveira, E., 2001. Determination of mineral nutrientes and toxic elements in Brazilian soluble coffee by ICP-AES. J. Food Comp. Anal. 14, 523–531.

Santos, J.S., Santos, M.L.P., Conti, M.M., Santos, S.N., Oliveira, E., 2009. Evaluation of some metals in Brazilian coffees cultivated during the process of conversion from conventional to organic agriculture. Food Chem. 115, 1405–1410.

Suseela, B., Bhalke, S., Kumar, A.V., Tripathi, R.M., Sastry, V.N., 2001. Daily intake of trace metals through coffee consumption in India. Food Addit. Contam., 18,115–120.

Varo, P., Nuurtamo, M., Saari, E., Koivistoinen, P., 1980. Mineral element composition of Finnish foods. IX. Beverages, confectionaries, sugar and condiments. Acta Agric. Scand. 22, 127–139.

CHAPTER

67

Free Radical Processes in Coffee I—Solid Samples

Bernard A. Goodman, Chahan Yeretzian*[†]

*State Key Laboratory for Conservation and Utilization of Subtropical Agro-Bioresources, Guangxi University, Nanning, Guangxi, China, [†]Zurich University of Applied Sciences, Institute of Chemistry and Biological Chemistry, Wädenswil, Switzerland

CHAPTER POINTS

- Free radicals are produced during coffee roasting.
- EPR spectroscopy shows that radical production occurs in distinct stages.
- Free radical reactions occur during coffee storage and are influenced by storage conditions.
- Lipid peroxidation and 8-oxocaffeine are associated with free radical reactions during storage.

INTRODUCTION

Coffee is rich in antioxidant molecules, and for many people it contributes a major proportion of their dietary antioxidant intake (e.g., Pulido *et al.*, 2003; Svilaas *et al.*, 2004). Antioxidants and free radical processes are closely linked, since $^3O_2$, the stable form of oxygen, is a free radical. For this reason, antioxidants are often considered to be free radical scavengers, although not all free radicals are oxidizing agents, and many oxidizing agents are not free radicals. Furthermore, $^3O_2$ is actually a diradical, and the products of its one-electron reduction are also free radicals. Thus, the role of antioxidants in controlling oxidizing free radical reactions is an essential one in the life and food sciences. However, despite its abundance of antioxidants, free radical reactions still occur in coffee, and furthermore, such free radical reactions may be beneficial (e.g., for flavor development during the roasting process), or detrimental (e.g., loss of sensory properties during storage in air) to the perceived quality of the final product.

In this chapter, we review our current knowledge of the formation of free radicals in the various stages of the coffee production process from the green bean to the ground product after roasting, and their impact on coffee quality criteria during storage; the following chapter (Chapter 68) describes free radical processes in liquid coffees and the roles they play in the production of molecules that are related to quality characteristics of the beverage.

ANTIOXIDANTS, FREE RADICALS, AND REACTIVE OXYGEN SPECIES—DEFINITIONS AND FUNCTIONS

The role of an antioxidant molecule is to inhibit oxidation, and in foods this can be achieved in a number of ways:

(i) prevention/inhibition of the formation of an oxidizing agent;
(ii) direct reaction with the primary oxidizing agent (scavenging) to prevent/inhibit its action;
(iii) reaction with a reactive product produced by action of the primary oxidizing agent, thus preventing/inhibiting any further reaction.

The broadest definition of a free radical is that it is a molecule with one or more unpaired electrons, although most people exclude paramagnetic transition metal ions. Because of the tendency for electrons to form pairs, free radicals are often, but not always, highly reactive, but there are also many examples of stable paramagnetic materials. Many of these are high-molecular-mass polymers, such as the melanoidins which make a major contribution to the sensory properties of coffee, but some can be small, $^3O_2$ and NO being perhaps the best examples of stable low-molecular-mass free radicals.

Processing and Impact on Active Components in Food
http://dx.doi.org/10.1016/B978-0-12-404699-3.00067-6

Reactive oxygen species (ROS) is a general term used for molecules related to oxygen, but which behave as more powerful oxidizing agents than O_2 itself. Some ROS, such as the hydroxyl radical $HO^{\bullet}$, the superoxide radical anion, $O_2^{\bullet -}$, and its protonated form $HOO^{\bullet}$, are free radicals, but others, such as $^1O_2$ and H_2O_2, are diamagnetic. An important reaction of H_2O_2 is its breakdown under acidic conditions to produce hydroxyl radicals in the presence of low-oxidation-state transition metal ions, such as Fe(II) [the Fenton reaction (Prousek, 2007)]:

$$Fe(II) + H_2O_2 \rightarrow Fe(III) + HO^- + HO^{\bullet}$$

Since antioxidants are also reducing agents, they can recycle the metal back to its lower oxidation state and thus help catalyze further hydroxyl radical production, thereby exhibiting a pro-oxidant property. Consequently, the type of reaction in which an antioxidant molecule participates is strongly dependent on its chemical environment.

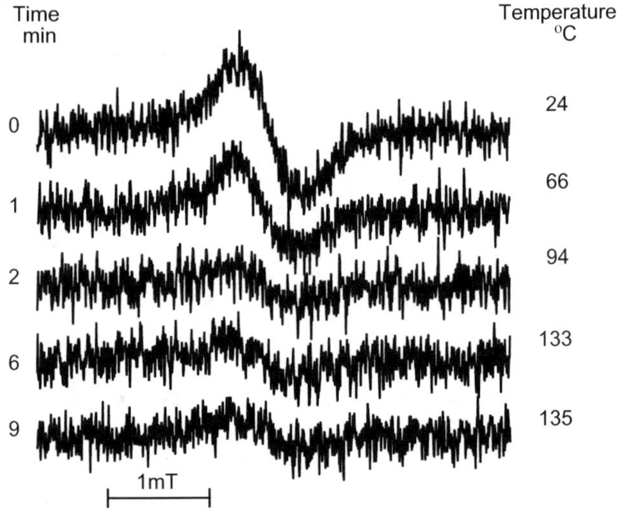

FIGURE 67.1 EPR spectra from a Robusta coffee bean whilst being heated prior to roasting. *Adapted from Goodman* et al. *(2011).*

DIRECT OBSERVATION OF FREE RADICALS USING ELECTRON PARAMAGNETIC RESONANCE (EPR) SPECTROSCOPY

EPR spectroscopy is based on the fact that an electron can exist in two spin states whose energies are unequal in the presence of a magnetic field. Transitions between these states can be induced by the absorption of electromagnetic radiation with energy equal to their energy difference, which is in the microwave region of the electromagnetic spectrum; an EPR spectrum is usually recorded by varying the magnetic field whilst holding the microwave frequency constant, and it is commonly displayed as the first derivative of the microwave absorption. The position of this absorption produces a constant known as the g-value, which is a characteristic of the molecule under investigation, and is determined by the equation:

$$h\nu = g\mu_B B$$

where ν is the microwave frequency, and B is the magnetic field at resonance, h is Planck's constant (6.626069×10^{-34} Js), and μ_B is the Bohr magneton (9.274009×10^{-28} J/T).

FREE RADICAL PROCESSES IN GREEN COFFEE BEANS

Although most of the literature concerning free radicals in coffee has concentrated on roasted beans, there are free radicals in green coffee, and in common with other seeds they are influenced by age and storage conditions. There has been some dispute in the literature concerning the classification of coffee seeds, but recent research indicates that they do not tolerate desiccation (da Rosa *et al.*, 2011), and consequently stored green coffee beans are likely to experience free radical processes that are similar to those in other recalcitrant seeds. Since Hendry *et al.* (1992) have reported that increases in lipid peroxidation and free radical contents accompany desiccation of mature seeds of *Quercus robur* L., it might be expected that extended storage of dried green coffee beans would be accompanied by some peroxidation of their linoleic acid and polyunsaturated fatty acid components. However, we are unaware of any evidence that supports this is supposition. Nevertheless, there are differences between the composition of mature and immature beans (Clifford and Kazi, 1987) and between damaged and undamaged beans that affect the quality of the roasted product (Franca *et al.*, 2005).

The initial stage of the coffee roasting process involves complete desiccation of the beans. As shown in Figure 67.1, this also results in a considerable decrease in the EPR free radical signal intensity, but with no significant change in either the linewidth or g-value. The g-value for the green coffee bean in Figure 67.1 was 2.0053, which is in the range commonly seen for O-centered radicals (Springer Materials), although this assignment is not definitive. Furthermore, this result suggests that the indigenous radicals participate in chemical reactions during this 'drying' stage of the roasting process, and that it does not simply involve the loss of water.

FREE RADICAL FORMATION DURING COFFEE BEAN ROASTING

Free radical formation is a common feature of roasted foods, and the presence of a stable free radical signal in roasted coffee has been known for more than half a century (O'Meara *et al.*, 1957). Although most studies of

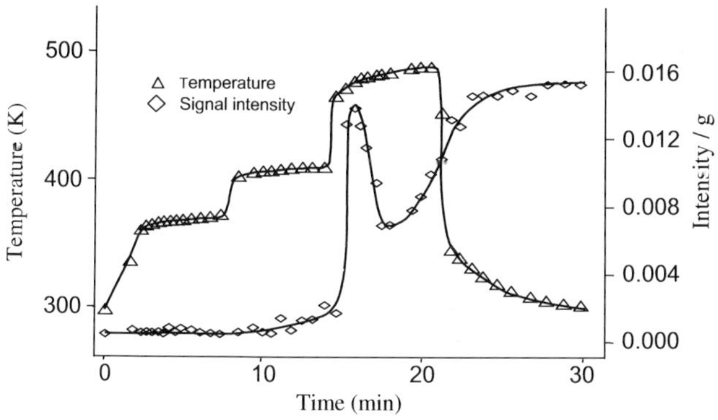

FIGURE 67.2 Variation with time and temperature of the free radical signal intensity during roasting a Robusta coffee bean in N_2. *Adapted from Goodman* et al. *(2011).*

coffee roasting have focused on the product produced by roasting, Goodman *et al.* (2011) have recently studied the free radical dynamics during the roasting process, and Figure 67.2 illustrates the variation of the EPR free radical signal intensity during the roasting phase of the same sample as that shown in Figure 67.1. Apart from the large increase in intensity, there was also a shift in g-value to 2.0039. This lower g-value suggests that it could be either a N- or C-centered radical, although the absence of any ^{14}N hyperfine structure would tend to discount the former. The EPR spectra in Figure 67.2 were recorded in a N_2 atmosphere in order to preserve the anaerobic conditions that exist in commercial roasters as a result of CO_2 release from carbohydrate breakdown. When a single coffee bean was roasted in air in the EPR spectrometer, the anaerobic conditions were lost because of the high gas flow rate in the variable temperature system. This resulted in massive free radical generation during the cooling phase (see Goodman *et al.*, 2011), a result which suggests that there is a practical limit to the gas flow rate that can be used in rapid roasting without appreciably altering the chemical characteristics of the final product.

CHARACTERIZATION OF THE FREE RADICALS IN ROASTED COFFEE

The EPR free radical signal in roasted coffee beans is associated with colored products, known generally as melanoidins, which are formed as a result of the Maillard and caramelization reactions. The Maillard reaction involves condensation between amino acids or proteins and reducing sugars or carbohydrates (Nursten, 2005), whereas the caramelization reaction is just based on reactions of sugars or carbohydrates (Golon and Kuhnert, 2012). The free radical intensity trace in Figure 67.2 shows an initial rapid burst of free radical activity at about 200°C. This free radical is not stable and its intensity declines rapidly before the formation of the more stable radical which is typically seen in roasted coffee. Since the Maillard reaction is energetically favored over caramelization (e.g., Pałasiński *et al.*, 1985), the first radical burst is almost certainly due to the Maillard reaction. It is not possible to conclude from these EPR measurements whether the second burst is due to caramelization, a secondary stage of Maillard reaction chemistry, or a combination of the two. However, real time measurements of the emission of volatile organic compounds by on-line proton-transfer-reaction mass spectrometry strongly suggest that both type of reaction take place up to the end of the roasting process (Dorfner *et al.*, 2003, 2004; Wieland *et al.*, 2012).

The production of free radicals during melanoidin formation by the Maillard reaction has been studied extensively (e.g., Namiki and Hayashi, 1981, 1983; Hofmann *et al.*, 1999, 2002), and although this is an extremely complex set of reactions, an early stage involves cleavage of the carbohydrate skeleton and liberation of a glycoaldehyde imine. After dimerization to the 1,4-dialkylpyrazinium diquaternary salt (diquat) and oxidation, this yields 1,4-dialkylpyrazinium radical cations (Namiki and Hayashi, 1983). Free radical generation has been shown to precede color formation, and (Hofmann *et al.*, 2002) have proposed that the protein-bound 1,4-bis(5-amino-5-carboxy-1-pentyl) pyrazinium radical cation (CROSSPY) has a fundamental role in melanoidin formation. Therefore, it seems feasible that the initial free radical burst in Figure 67.2 could correspond to CROSSPY and/or related radicals, but at the present time this is simply speculation.

Coffee melanoidins show strong antioxidant activity (Bekedam *et al.*, 2008); this increases with increasing degree of roast and occurs predominantly in the high-molecular-mass fraction. Thus the melanoidins in dark roasted coffee may be associated more with caramelization than Maillard reaction products, though we are unaware of any specific chemical data to indicate that this is indeed the case. Sugar degradation in the caramelization reaction is also extremely complex, and a large number of compounds are formed as a result of a small number of unselective and chemoselective reactions (Golon and Kuhnert, 2012). Caramelization products include oligomers with up to six carbohydrate units formed through unselective glycosidic bond formation, dehydration products of oligomers losing up to a maximum of eight water molecules, hydration products of sugar oligomers, disproportionation products, and colored aromatic products. Furthermore, Ginz *et al.* (2000) have shown that the heat-induced breakdown of sucrose also leads to the formation of formic, acetic, glycolic, and lactic acids, which may make an appreciable contribution to the acidity of aqueous extracts of roasted coffee, and presumably offset at least in part the pH increase that would occur from the loss of chlorogenic acid during roasting.

FACTORS AFFECTING THE EPR FREE RADICAL SIGNAL INTENSITY IN ROASTED COFFEE BEANS

In coffee beans roasted under commercial conditions, Baesso *et al.* (1990) reported a direct relationship between the intensity of the free radical EPR signal and the darkness of the roast, and a similar result was reported recently by Yeretzian *et al.* (2012). Since Bekedam *et al.* (2008) observed an increase in antioxidant activity with increasing darkness of roast, this is further evidence that coffee free radicals are not scavenged by its antioxidants. Furthermore, Pascual *et al.* (2005) found no correlation between free radical concentration and antioxidant activity as measured by the FRAP assay (Benzie and Strain, 1996) in a range of commercial instant coffees (Figure 67.3). Thus it seems that antioxidant and stable free radical formation are the result of independent processes during coffee roasting.

Yeretzian *et al.* (2012) observed little influence of roasting atmosphere (in a small-scale convectional hot air roaster, which had a low gas flow rate) on the EPR signal intensity for samples (from the same batch of coffee beans) roasted for the same length of time, a result which is consistent with anaerobic conditions being maintained during roasting in air as a result of CO_2 release from the breakdown of carbohydrate components. This latter result contrasts with the observations of Goodman *et al.* (2011) for single bean samples prepared with high gas flow rates, where a large effect of the atmosphere in the cooling phase was observed.

By combining the results from the on-line EPR roasting experiments on single beans and the off-line experiments on batch-roasted coffee, the following integrated interpretation can be proposed. During roasting, chemical reactions inside the beans lead to the formation and release of large amounts of inorganic gases (mainly CO_2 and CO), which create a protective layer and prevent oxidation. Once roasting has ceased, this gas emission process rapidly decreases, and O_2 can reach the bean (thus initiating oxidation processes). We have observed differences in free radical behavior depending on whether beans were roasted under a very high flow of hot air (higher than that currently implemented in industrial fluidized bed roasters) or with a more moderate flow. If beans are roasted with a very high flow of hot air, a very strong increase in the EPR signal is observed during the cooling phase, whereas a batch of coffee roasted with a moderate flow of air showed only a small increase in EPR signal after roasting. We speculate that the roasting with a very high flow of hot air results in the lipids being removed and leaving the beans vulnerable to oxidation at the end of the roasting, whereas roasting at lower gas flows leaves an intact surface lipid layer on the beans. This then protects the beans from oxidation once roasting is completed. In support of this interpretation, we observed a thin layer of lipid deposited on the surface of the roasting tube, in which the single beans experiments were conducted with high hot air flows.

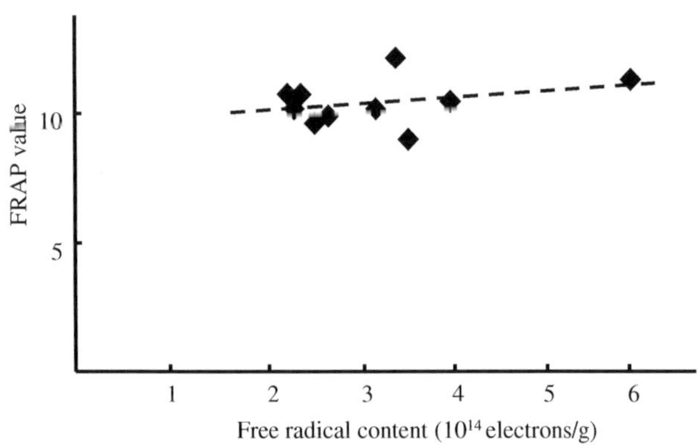

FIGURE 67.3 Relationship between free radical concentration and antioxidant content (as determined by the FRAP assay) for a collection of commercial soluble coffees. *Adapted from Pascual* et al. *(2005).*

In an examination of factors that influence free radical generation during coffee bean roasting, Nebesny and Budryn (2006) reported that there was little difference between the effects from convective and microwave heating, whereas the moisture content of the original beans prior to roasting had an appreciable influence. This was in the form of an inverse relationship between the moisture contents of coffee beans prior to roasting, and the generation of free radicals during roasting and subsequent storage. The measurements of Goodman *et al.* (2011) showed much higher free radical concentrations in roasted Arabica beans than with Robusta beans roasted under similar conditions, although there was reasonable agreement between individual beans from the same lot. Since measurements were only performed on beans from a single batch of each type, it could not be concluded whether these differences were due to the coffee genetics or to other factors in their production history. However, large differences in the free radical concentrations in individual beans were observed after roasting a blend, which consisted of a mixture of six varieties of Arabica and Robusta beans, thus eliminating the possibility of variation in the roasting conditions being responsible for the observed differences, and adding support for the conclusions of Nebesny and Budryn (2006). In contrast to these results obtained with a blend of coffee beans, measurements made on individual Arabica beans from a single lot of Columbian origin showed relatively minor variation and little effect of roasting atmosphere (Figure 67.4).

RADICAL REACTIONS DURING STORAGE OF ROASTED COFFEE BEANS

The operation of oxidation processes during storage of roasted coffee has been studied extensively. For example, Nicoli *et al.* (1993, 2009) and Anese *et al.* (2006) reported that coffee staling is accompanied by a loss of CO_2 and headspace volatiles, and these authors and Vila *et al.* (2005) showed that lipid peroxidation increased during the storage in air of roasted coffee samples. Changes in free radical contents also occur during storage of roasted coffee beans as first reported by Baesso *et al.* (1990), who found that the EPR signal intensity decreased during storage in polyethylene bags. This result clearly indicates that roasting-induced free radicals are not stable indefinitely, and that free radical concentrations change during the coffee staling process, though the two are not necessarily related. In contrast to the result of Baesso *et al.* (1990), Nebesny and Budryn (2006) reported that free radical contents increased during the first four months of storage of roasted and ground (R&G) coffee beans in polyethylene bags before reaching a maximum and then decreasing. This effect was interpreted as being due to the presence of some O_2 entrapped in the porous coffee material, even though the samples were vacuum packed. It must also be borne in mind that the relationship between free radical concentrations and storage time can be quite complex in food samples of biological origin. Contributions can come from free radicals present in the original material and radicals generated during the grinding process, in addition to reactions in the ground material, which may be dependent on or independent of the atmosphere. Furthermore, reaction with O_2 (i.e., during storage in air) can result in either the generation or loss of free radicals. However, support for the concept of O_2 being responsible for an increase in the free radical contents

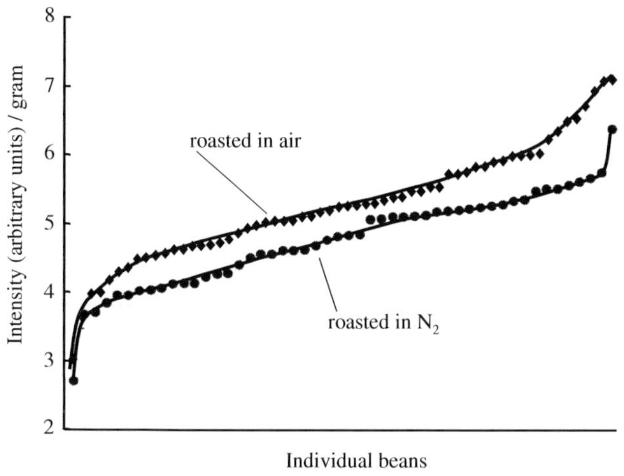

FIGURE 67.4 Free radical contents of individual coffee beans from the same lot and roasting. *Adapted from Goodman* et al. *(2011).*

during storage of roasted coffee beans was obtained recently by Yeretzian *et al.* (2012). In an investigation of changes in the EPR free radical signal intensity during storage in air of whole, half, and ground roasted coffee beans (Figure 67.5), large differences were observed between whole and half beans, but there was little difference between half and ground beans; thus exposure to O_2 and not the grinding process is the major factor responsible for the free radical increase in ground beans.

Collectively, these various studies suggest that there are (at least) two independent free radical processes operating during storage of roasted coffee beans—a slow O_2-independent decay of the original EPR signal (Baesso *et al.*, 1990), and an O_2-dependent production of a second signal with a similar EPR g-value (Yeretzian *et al.*, 2012). These observations are further complicated by the fact that the free radical signal is dependent on the degree of roasting (Baesso *et al.*, 1990; Yeretzian *et al.*, 2012), on the water content of the green beans (Nebesny and Budryn, 2006), and is strongly dependent on the type and/or history of the coffee beans (Goodman *et al.*, 2011). Thus, the absolute value of the EPR signal intensity is influenced by many factors, which limits the value of single measurements, but the conduct of a series of measurements over extended periods of time can shed light on the influences of processing and storage conditions, and changes in the EPR free radical signal intensity may complement existing aroma indexes based on, e.g., ratios of volatiles in the headspace of coffee (Marin *et al.*, 2008) for indicating storage-related aging processes.

FREE RADICAL REACTION PRODUCTS IN STORED COFFEE

Because of the instability of many free radicals that are important for oxidative processes, the existence of free radical reactions is often inferred by determining their diamagnetic end-products. Some examples are discussed in the following paragraphs.

Oxidized Lipids

Although coffee is well-known for its ability to inhibit lipid oxidation, coffee beans also have appreciable lipid contents made up of primarily triglycerides, diterpenes, and tocopherols (Folstar, 1985). The main fatty acids in coffee are the saturated palmitic (C16:0) and stearic (C18:0) acids, along with the monounsaturated oleic (C18:1) acid and the polyunsaturated linoleic (C18:2) acid, although the relative amounts differ appreciably between different types of bean (Martín *et al.*, 2001). As mentioned above, the staling of coffee beans is accompanied by lipid peroxidation (e.g., Nicoli *et al.*, 1993, 2009; Vila *et al.*, 2005; Anese *et al.*, 2006), and as with other

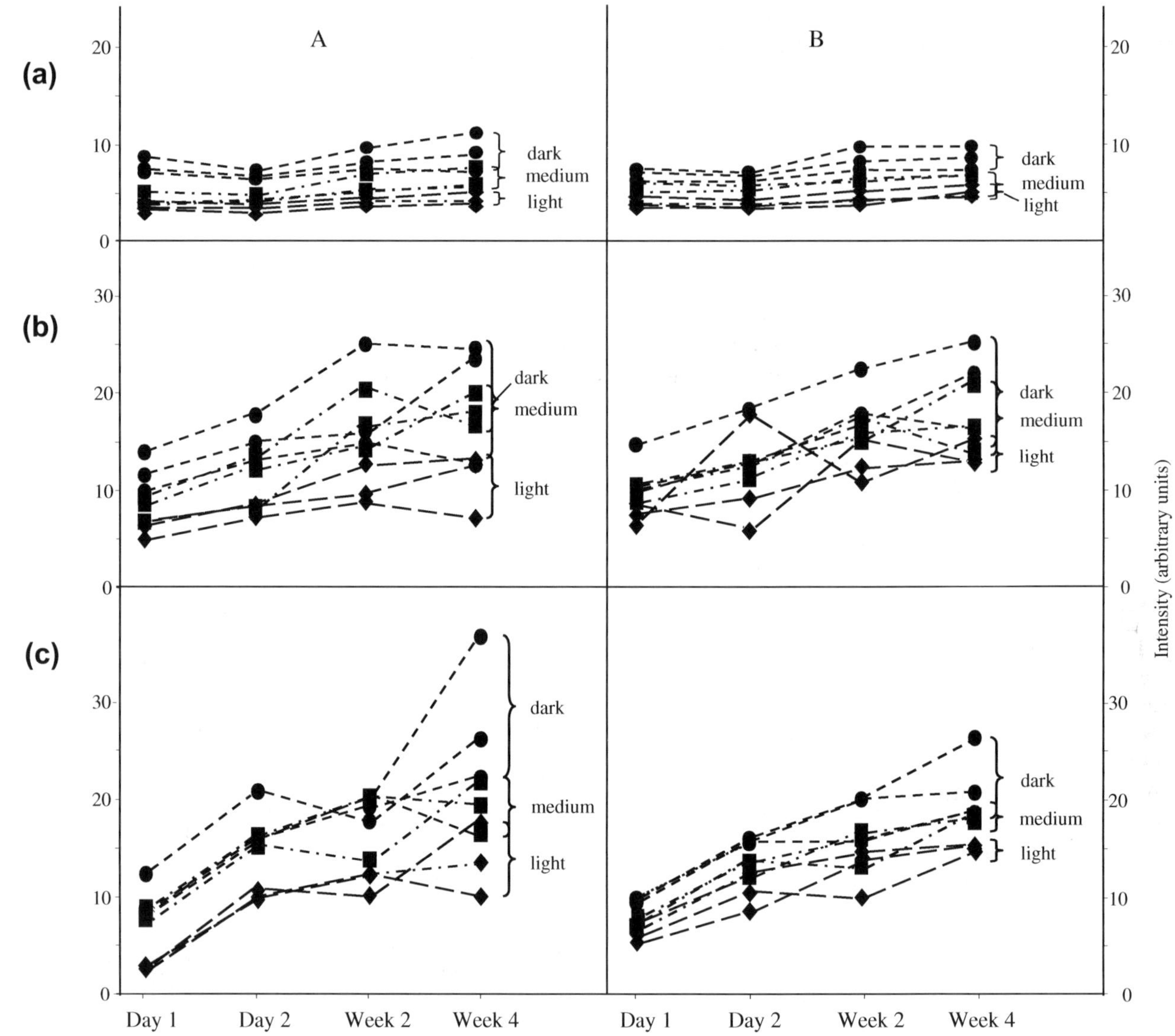

FIGURE 67.5 Variation with time and physical condition of the free radical contents of individual coffee bean samples roasted in (A) air or (B) N_2, then stored in air as (a) whole beans, (b) half beans, and (c) ground beans. *Adapted from Yeretzian* et al. *(2012).*

food products, the degree of oxidation of unsaturated fatty acids can be used to assess the oxidative challenge to which a coffee sample has been subjected. Furthermore, Toci *et al.* (2013) have recently reported changes in the proportions of triacylglycerols (TAG) and free fatty acids (FFA) during storage of roasted *Coffea arabica* beans. These were found to be affected directly by the roasting process and subsequent changes during storage were influenced by temperature and atmosphere. These changes involved large increases in FFA during the first month of storage, followed by subsequent progressive decreases; these decreases were lower for lower storage temperatures and for N_2 compared to air atmosphere, suggesting that oxidative changes are responsible for the loss of FFA during storage.

8-Oxocaffeine

8-Oxocaffeine (1,3,7-trimethyluric acid) is produced by the reaction of caffeine with hydroxyl radicals; its formation in coffee subjected to oxidative stress has been reported by Stadler and Fay (1995), and shown to be dependent on O_2 tension, hydrogen peroxide and transition metal availability. This observation is consistent with Fenton reaction chemistry resulting from an initial enzymatic conversion of O_2 to H_2O_2 (Obinger *et al.*, 1997), which then undergoes transition metal catalyzed reduction to yield hydroxyl radicals (Prousek, 2007). The 8-oxocaffeine contents of unoxidized coffee beans are negligible (Stadler *et al.*, 1996), an observation which suggests that the ratio of 8-oxocaffeine:caffeine is proportional to the hydroxyl radical induced oxidation in coffee, since hydroxyl radicals react virtually indiscriminately with organic molecules.

DISCUSSION AND CONCLUSIONS

Free radical processes play important roles at all stages of coffee production from the storage of green beans through the roasting process, to storage of the roasted product. These are influenced by many factors including the plant genetics and exposure to O_2, and various factors related to the roasting conditions. None of these has as yet been investigated thoroughly, although evidence is being generated on several factors related to coffee production. It is now clear that a number of coffee quality characteristics are influenced by oxidative free radical processes, but this is a complex relationship, and it is inappropriate with the present state of knowledge to try to equate simple free radical intensity measurements to a quality parameter. In fact it appears that the intensity of the 'stable' free radical EPR signal that is observed in solid coffee samples is irrelevant for the perceived quality of beverages made from them. Nevertheless, free radical processes play fundamental roles in the formation of melanoidins via the Maillard and caramelization reactions, which are also instrumental in the formation of many sensory active molecules, especially those associated with coffee aroma. In addition, free radical processes lead to a deterioration in quality (staling) of coffee stored in air, which is invariably the atmosphere used by domestic consumers. During aerobic storage, however, whole roasted beans are considerably more stable with respect to free radical processes than the corresponding cut or ground beans, and it seems likely that beverage quality will be maintained for considerably longer periods if roasted beans are stored whole, and only ground immediately prior to the preparation of the brew.

Recent roasting experiments on single bean and batch roasting under various gas flow conditions have provided insight into the dynamics of oxidation and as well as the protection from oxidation during and after roasting. These experiments indicate that EPR could be a powerful tool for studying the impact of processing parameters, and it is proposed that future work explores in more detail the link between changes in the EPR signal and the stability of coffee during storage.

References

Anese, M., Manzocco, L., Nicoli, M.C., 2006. Modeling the secondary shelf life of ground roasted coffee. J. Agric. Food Chem. 54, 5571–5576.

Baesso, M.L., Da Silva, E.C., Vargas, H., Cortez, J.G., Pelzl, J., 1990. J. Use of electron spin resonance for the determination of staling of roast coffee in polyethylene bag packs. Z. Lebensm. Unters. F. A 191, 24–27.

Bekedam, E.K., Schols, H.A., Cämmerer, B., Kroh, L.W., van Boekel, M.A., Smit, G., 2008. Electron spin resonance (ESR) studies on the formation of roasting-induced antioxidative structures in coffee brews at different degrees of roast. J. Agric. Food Chem. 56, 4597–4604.

Benzie, I.F.F., Strain, J.J., 1996. The ferric reducing ability of plasma (FRAP) as a measure of "antioxidant power": the FRAP assay. Anal. Biochem. 239, 70–76.

Clifford, M.N., Kazi, T., 1987. The influence of coffee bean maturity on the content of chlorogenic acids, caffeine and trigonelline. Food Chem. 26, 59–69.

da Rosa, S.D.V.F., Carvalho, A.M., McDonald, M.B., von Pinho, E.R.V., Silva, A.P., Veiga, A.D., 2011. The effect of storage conditions on coffee seed and seedling quality. Seed Sci. Technol. 39, 151–164.

Dorfner, R., Ferge, T., Kettrup, A., Zimmermann, R., Yeretzian, C., 2003. Real-time monitoring of 4-vinylguaiacol, guaiacol and phenol during roasting by resonant laser ionisation time-of-flight mass-spectrometry. J. Agric. Food Chem. 51, 5768–5773.

Dorfner, R., Ferge, T., Yeretzian, C., Kettrup, A., Zimmermann, R., 2004. Laser Mass Spectrometry as On-Line Sensor for Industrial Process Analysis: Process Control of Coffee Roasting. Anal. Chem. 76, 1386–1402.

Folstar, P., 1985. Lipids. In: Clarke, R.J., Macrae, R. (Eds.), Coffee Volume 1 Chemistry. Elsevier, London, pp. 203–222.

Franca, A.S., Oliveira, L.S., Mendonça, J.C.F., Silva, X.A., 2005. Physical and chemical attributes of defective crude and roasted coffee beans. Food Chem. 90, 89–94.

Ginz, M., Balzer, H.H., Bradbury, A.G.W., Maier, H.G., 2000. Formation of aliphatic acids by carbohydrate degradation during roasting of coffee. Eur. Food Res. Technol. 211, 404–410.

Golon, A., Kuhnert, N., 2012. Unraveling the chemical composition of caramel. J. Agric. Food Chem. 60, 3266–3274.

Goodman, B.A., Pascual, E.C., Yeretzian, C., 2011. Real-time monitoring of free radical processes during the roasting of coffee beans using electron paramagnetic resonance spectroscopy. Food Chem. 125, 248–254.

Goodman, B.A., Yeretzian, C., 2014. Free radical processes in coffee II—Liquids. In: Preedy, V.R. (Ed.), Processing and Impact on Active Components in Food. Chapter 68.

Hendry, G.A.F., Finch-Savage, W.E., Thorpe, P.C., Atherton, N.M., Buckland, S.M., Nilsson, K.A., Seel, W.E., 1992. Free radical processes and loss of seed viability during desiccation in the recalcitrant species *Quercus robur* L. New Phytol. 122, 273–279.

Hofmann, T., Bors, W., Stettmaier, K., 1999. Radical assisted melanoidin formation during thermal processing of foods as well as under physiological conditions. J. Agric. Food Chem. 47, 391–396.

Hofmann, T., Bors, W., Stettmaier, K., 2002. CROSSPY: a radical intermediate of melanoidin formation in roasted coffee. In: Morello, M.J., Shahidi, F., Ho, C.T. (Eds.), Free Radicals in Food: Chemistry, Nutrition and Health Effects. ACS Symposium Series, vol. 807. American Chemical Society, Washington, DC, pp. 49–68.

Marin, K., Požrl, T., Zlatić, E., Plestenjak, A., 2008. A new aroma index to determine the aroma quality of roasted and ground coffee during storage. Food Technol. Biotech. 46, 442–447.

Martín, M.J., Pablos, F., González, A.G., Valdenebro, M.S., León-Camacho, M., 2001. Fatty acid profiles as discriminant parameters for coffee varieties differentiation. Talanta 54, 291–297.

Namiki, M., Hayashi, T., 1981. Formation of novel free radical products in an early stage of Maillard reaction. Prog. Food Nutr. Sci. 5, 81–91.

Namiki, M., Hayashi, T., 1983. A new mechanism of the Maillard reaction involving sugar fragmentation and free radical formation. In: Waller, G.R., Feather, M.S. (Eds.), The Maillard Reaction in Foods and Nutrition. ACS Symposium Series, vol. 215. American Chemical Society, Washington, DC, pp. 21–46.

Nebesny, E., Budryn, G., 2006. Changes in free radicals content in coffee beans throughout convective roasting and microwaving and during storage. Deut. Lebensm-Rundsch. 102, 526–530.

Nicoli, M.C., Innocente, N., Pittia, P., Lerici, C.R., 1993. Staling of roasted coffee: volatile release and oxidation reactions during storage. ASIC 15th Int. Sci. Colloquium Coffee, 557–566.

Nicoli, M., Calligaris, S., Manzocco, L., 2009. Shelf-life testing of coffee and related products: uncertainties, pitfalls, and perspectives. Food Eng. Rev. 1, 159–168.

Nursten, H.E., 2005. The Maillard Reaction: Chemistry, Biochemistry and Implications. Royal Society of Chemistry, Cambridge.

Öbinger, C., Burner, U., Ebermann, R., 1997. Generation of hydrogen peroxide by plant peroxidases mediated thiol oxidation. Phyton 37, 219–226.

O'Meara, J.P., Truby, F.K., Shaw, T.M., 1957. Free radicals in roasted coffee. Food Res. 22, 96–101.

Pałasíński, M., Tomasik, P., Wiejak, S., 1985. Thermolysis of carbohydrates in oxygen-free atmosphere. Part I. Caramelization of mono- and disaccharides. Starch 37, 308–313.

Pascual, E.C., Yeretzian, C., Pirker, K.F., Goodman, B.A., 2005. Antioxidant, pro-oxidant and free radical processes in coffee. In: Zhao, B., Liu, G., Packer, L. (Eds.), Natural Antioxidants and Micronutrients. Medimond, Bologna, Italy, pp. 119–125.

Prousek, J., 2007. Fenton chemistry in biology and medicine. Pure Appl. Chem. 79, 2325–2338.

Pulido, R., Hernández-García, M., Saura-Calixto, F., 2003. Contribution of beverages to the intake of lipophilic and hydrophilic antioxidants in the Spanish diet. Eur. J. Clin. Nutr. 57, 1275–1282.

Springer Materials The Landolt–Börnstein Database. Magnetic properties of free radicals. www.springermaterials.com.

Stadler, R.H., Fay, L.B., 1995. Antioxidative reactions of caffeine: formation of 8-oxocaffeine (1,3,7-trimethyluric acid) in coffee subjected to oxidative stress. J. Agric. Food Chem. 43, 1332–1338.

Stadler, R.H., Richoz, J., Turesky, R.J., Welti, D.H., Fay, L.B., 1996. Oxidation of caffeine and related methylxanthines in ascorbate and polyphenol-driven Fenton-type oxidations. Free Radical Res. 24, 225–240.

Svilaas, A., Kaur Sakhi, A., Frost Andersen, L., Svilaas, T., Ström, E.C., Jacobs Jr., D.R., Ose, L., Blomhoff, R., 2004. Intakes of antioxidants in coffee, wine, and vegetables are correlated with plasma carotenoids in humans. J. Nutr. 134, 562–567.

Toci, A.T., Neto, V.J.M.F., Torres, A.G., Farah, A., 2013. Changes in triacylglycerols and free fatty acids composition during storage of roasted coffee. LWT Food Sci. Technol. 50, 581–590.

Vila, M.A., Andueza, S., Paz de Peña, M., Cid, C., 2005. Fatty acid evolution during the storage of ground, roasted coffees. J. Am. Oil Chem. Soc. 82, 639–646.

Wieland, F., Gloess, A.M., Keller, M., Wetzel, A., Schenker, S., Yeretzian, C., 2012. On-line monitoring of coffee roasting by proton-transfer-reaction time-of-flight mass-spectrometry (PTR-ToF-MS): towards a real-time process control for a consistent roast profile. Analyt. Bioanalyt. Chem. 402, 2531–2543.

Yeretzian, C., Pascual, E.C., Goodman, B.A., 2012. Effect of roasting conditions and grinding on free radical contents of coffee beans stored in air. Food Chem. 131, 811–816.

CHAPTER

68

Free Radical Processes in Coffee II—Liquids

Bernard A. Goodman, Chahan Yeretzian*†

*State Key Laboratory for Conservation and Utilization of Subtropical Agro-Bioresources, Guangxi University, Nanning, Guangxi, China, †Zurich University of Applied Sciences, Institute of Chemistry and Biological Chemistry, Wädenswil, Switzerland

CHAPTER POINTS

- Coffee extracts contain free radicals with EPR spectra similar to those seen in solid roasted coffee.
- EPR spectroscopy can be used to investigate the free radical dynamics of coffee solutions.
- Unstable radicals in coffee solutions are detected by EPR spectroscopy in combination with spin trapping.
- Hot coffee solutions have both pro- and antioxidant properties.
- Sensory properties are degraded during the storage of coffee solutions.

INTRODUCTION

As discussed in the Chapter 67, antioxidant molecules and free radical processes are closely linked, since $^3O_2$, the stable form of oxygen, is a free radical, as are several products of oxidation reactions. Although not all free radicals are oxidizing agents, and not all oxidizing agents are free radicals, antioxidants play important roles in controlling oxidizing free radical reactions. For this reason they are considered to be essential for life, and are regarded as important components of a healthy diet (e.g., Scalbert and Williamson, 2000; Scalbert *et al.*, 2005). Coffees are rich in antioxidant molecules, and their contents are higher than in most other food products. Indeed, for many people coffee contributes a major proportion of their dietary antioxidant intake (e.g., Pulido *et al.*, 2003; Svilaas *et al.*, 2004). Yet despite this abundance of antioxidants, free radical reactions still occur in coffee beverages, and in aqueous solutions, coffee is essentially unstable.

In this chapter, we review our current knowledge of the roles played by free radicals in the chemistry of liquid coffees, their detection, and their influence on stability, especially with respect to sensory properties and molecules that have potential health implications. However, it should also be recognized that this science is currently in a state of rapid development, and our knowledge of the relevant chemical processes is far from complete. Thus we also indicate topics which we believe need greater attention if we are to achieve a fuller understanding of the factors which ultimately contribute to the quality characteristics of the coffee beverage.

OBSERVATION OF FREE RADICALS USING EPR SPECTROSCOPY

In Chapter 67, we described the detection and measurement of variations in free radical contents of various solid coffee products by the use of EPR spectroscopy. A similar signal is obtained in the EPR spectra of coffee solutions, and there is little, if any, change in its spectral characteristics between the fluid and frozen states. In addition to this (relatively) stable free radical, unstable free radicals can be generated in coffee solutions, but with lifetimes that are too short and steady state concentrations too small for direct observation of their EPR spectra. However, they can be studied with the aid of a technique known as spin trapping; this involves addition to solutions of a diamagnetic molecule (spin trap), which reacts with unstable free radicals to produce a (more) stable radical (often a radical adduct of the spin trap). Many spin traps are nitrones, which produce nitroxides after reaction with free radicals (e.g., Figure 68.1), and the spectral parameters of the resulting nitroxides provide evidence for the type of radical 'trapped' (NIEHS, 2002). The spin trapping approach has proved to be very useful for detecting the production of unstable free radicals in the life sciences, and also in

Processing and Impact on Active Components in Food
http://dx.doi.org/10.1016/B978-0-12-404699-3.00068-8

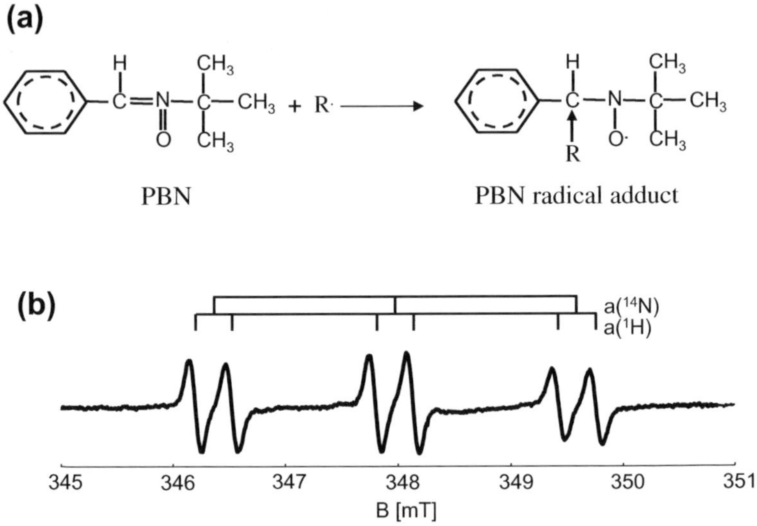

FIGURE 68.1 (a) Formation of an adduct by the reaction of a spin trap (PBN) with a radical (R·), and (b) its EPR spectrum.

various types of foodstuff, and spin traps are especially useful for the detection of hydroxyl and superoxide radicals (e.g., Buettner and Mason, 2003). Various nitrone spin traps have advantages and disadvantages over one another; cyclic traps, such as DMPO and its related compounds have high reactivity with superoxide and hydroxyl radicals and are less readily hydrolyzed than aryl nitrones, such as 4-POBN or PBN, although the latter have better reactivity towards carbon-centered radicals and also have longer shelf lives. For further information on spin trapping the reader should consult one of the many texts available on the subject. However, it should also be recognized that nitrones are reactive compounds, and can form nitroxides by mechanisms other than radical trapping, so caution must always be exercised when interpreting results using spin trapping technology.

RADICAL REACTIONS IN COFFEE SOLUTIONS INCLUDING SOLUBLE COFFEE

As mentioned in the Chapter 67, free radicals are formed in substantial quantities during the roasting of green coffee beans, and are primarily associated with the melanoidins, which are formed as a result of Maillard and caramelization reactions. These are extremely complicated reactions, and both produce a wide range of products. However, one of the early stages of the Maillard reaction is the formation of 1,4-dialkylpyrazinium diquaternary salts (diquat), and Hofmann *et al.* (1999) showed that free radical formation was instantaneous when diquat was dissolved in water. These authors found that free radical concentrations subsequently decreased in heated solutions by ~50% over ~10 min, after which they stabilized.

Results illustrating the variations in intensity of the free radical signal, in soluble coffee solutions as a function of time and temperature are illustrated in Figure 68.2. These show that, although the signal remained relatively constant at 65°C (Figure 68.2a), at 21°C it dropped to about 50% of its original intensity in about 20 min, and then remained virtually constant for the following 2 h (Figure 68.2b). A closer examination of the data in Figure 68.2a shows that there was also a small decline in the spectral intensity during the first 20 min at 65°C before it stabilized.

These results can be explained in terms of an initial temperature-independent decline in spectral intensity after dissolution of the coffee followed by a temperature-dependent increase in spectral intensity. A plot of the equilibrium intensity (after 1 h) for intermediate temperatures shows constant intensity for temperatures ≤30°C and a linear increase with increasing temperature for values ≥40°C (Figure 68.3). Thus it appears that it is only at temperatures above normal ambient storage that there is appreciable generation of the secondary radical.

PRO-AND ANTIOXIDANT PROPERTIES OF COFFEE SOLUTIONS

Coffee solutions show both pro- and antioxidant properties, although it is the latter which have received the most attention in the scientific literature, because of their dietary importance (e.g., Manach *et al.*, 2004; Williamson and Holst, 2008). However, pro-oxidant activity of coffee solutions was reported almost 20 years ago, when it was observed that coffee solutions destabilized ascorbic acid relative to solutions in water for temperatures >40°C (Goodman *et al.*, 1994). The oxidation of ascorbic acid to the ascorbate radical follows a similar curve to that reported for the generation of the

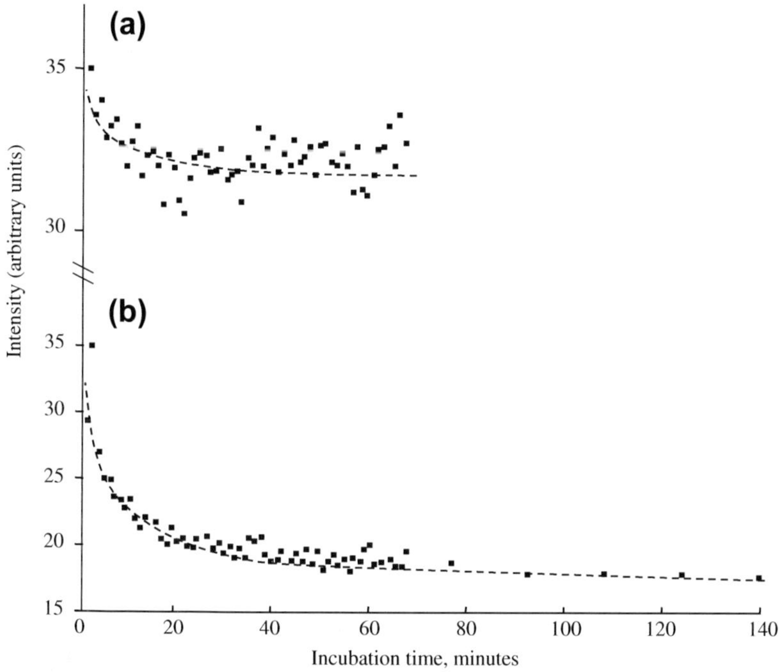

FIGURE 68.2 EPR free radical intensity in 5% soluble coffee solutions as a function of time at (a) 65°C, and (b) 21°C. *Adapted from Pascual* et al. *(2002).*

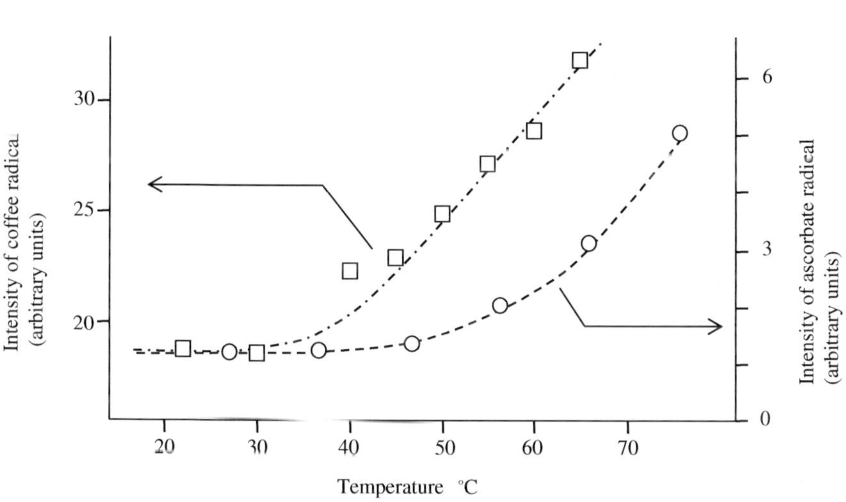

FIGURE 68.3 Comparison of pro-oxidant activity of coffee and its free radical concentration as a function of temperature. *Based on data from Goodman* et al. *(1994) and Pascual* et al. *(2002).*

temperature-dependent free radical in coffee solutions (Figure 68.3), albeit shifted to slightly higher temperatures, a result which suggests that this free radical may be the oxidizing agent.

In contrast to the small number of studies on the pro-oxidant activity of coffee, there have been many investigations of antioxidant activity of coffee solutions (e.g., Preedy, 2014), thus suggesting that experimental designs might be chosen to test for properties that might appeal to funding bodies rather than being completely objective. Nevertheless, there is overwhelming evidence that coffee solutions have high antioxidant activity, and there is growing evidence that coffee consumption is associated with health benefits (Dorea and da Costa, 2005; Goodman, 2012).

ROLE OF O_2 IN THE FREE RADICAL CHEMISTRY OF COFFEE SOLUTIONS

The role of O_2 in the formation of unstable free radicals in coffee solutions is illustrated by the spin trapping data of Pascual *et al.* (2002) for the heating of 5% instant coffee solutions in air in the presence of the spin trap PBN (Figure 68.4). The rate of adduct formation for fresh aerated solutions (Figure 68.4a) was ~2.5-times faster than for solutions that were initially deoxygenated (Figure 68.4b). The results from aged solutions were essentially identical (Figure 68.4c, d) and very similar to Figure 68.4b, a result which indicates that coffee reacts to remove O_2 from solution. These results can be interpreted in terms of an initial conversion of O_2 to H_2O_2

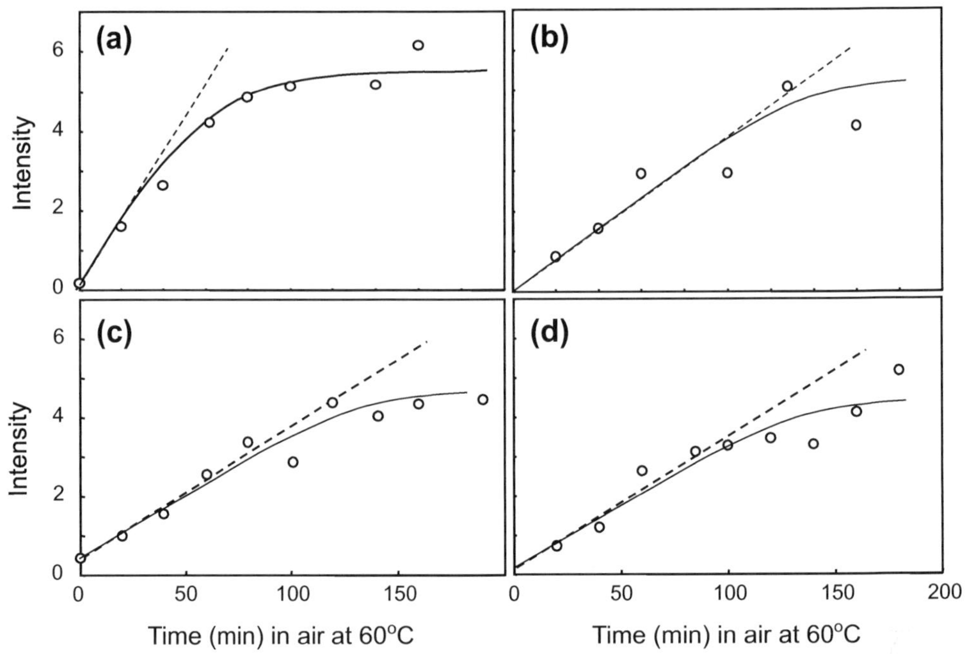

FIGURE 68.4 Spin trap free radical adduct formation during forced oxidation of coffee solutions with different storage histories (a) fresh solution natural O_2 level, (b) fresh solution deoxygenated, (c) aged solution natural O_2 level, and (d) aged solution deoxygenated. *Adapted from Pascual* et al. *(2002).*

(Obinger *et al.*, 1997), followed by the transition metal catalyzed breakdown of H_2O_2 to yield $^{\bullet}OH$, which then reacts indiscriminately with organic molecules in the coffee solution to produce C-centered radicals which are trapped and whose EPR spectra are recorded.

The free radicals trapped in the measurements shown in Figure 68.4 are not those responsible for the temperature-dependent single peak signal in Figure 68.3. This was demonstrated by Pascual *et al.* (2002), who showed that the single peak resonance is present along with that from the adduct, and as shown in Figure 68.5, its intensity is virtually identical to that seen in the absence of the spin trap.

FACTORS AFFECTING THE STABILITY OF COFFEE CONCENTRATES

The development of the ready-to-drink and concentrates segments of the coffee business has been hampered by problems of flavor and physical instability, which result in unacceptably short shelf-lives for the beverage, and consequently represent a major obstacle in the development of these products. Currently, these stability problems are addressed by the addition of buffers, stabilizers, antioxidants, and commercial flavors, but there is a desire to develop high-quality products without additives that have acceptable shelf-lives for storage at ambient temperatures.

As mentioned above, one of the main reasons for the inherent instability of liquid coffee products is the existence of Fenton reaction chemistry, which results in the production of hydroxyl radicals from hydrogen peroxide. Furthermore, this reaction can be stimulated by antioxidant molecules such as polyphenols, because of their ability to function as reducing agents, and thus decrease the oxidation state of transition metals which catalyze the Fenton reaction. For example:

$$\text{Fe (III)} + \text{polyphenol} \rightarrow \text{Fe (II)} + \text{Semiquinone}$$

$$\text{Fe (II)} + H_2O_2 \rightarrow \text{Fe (III)} + HO^{\bullet} + HO^{-}$$

The presence of antioxidants in anaerobically stored coffee will thus ensure that this reaction continues until all of the hydrogen peroxide is used up. However, since O_2 can be reduced enzymatically to H_2O_2 (e.g., Obinger *et al.*, 1997), the production of the hydroxyl radical may continue in air *whilst there are antioxidants available to ensure the recycling of the metal oxidation states*, provided the metal remains in soluble chemical forms.

On the basis of the hypothesis described above, it might be expected that chlorogenic acid-rich products would have lower stability than coffee products with lower contents of the polyphenol (i.e., darker roasts), but to the best of our knowledge this hypothesis has not been tested. However, Yeretzian *et al.* (2013) reported that variations occur in the concentrations of free radicals and paramagnetic metal ions in concentrated aqueous extracts of roast and ground (R&G) coffee as a function of the presence of O_2 during the cooling/grinding, extraction, and retorting/processing stages. In that paper, free radical concentrations were observed to decrease by 10–50% during storage at 4°C for one month, but the smallest changes were observed with samples that had been extracted with deoxygenated

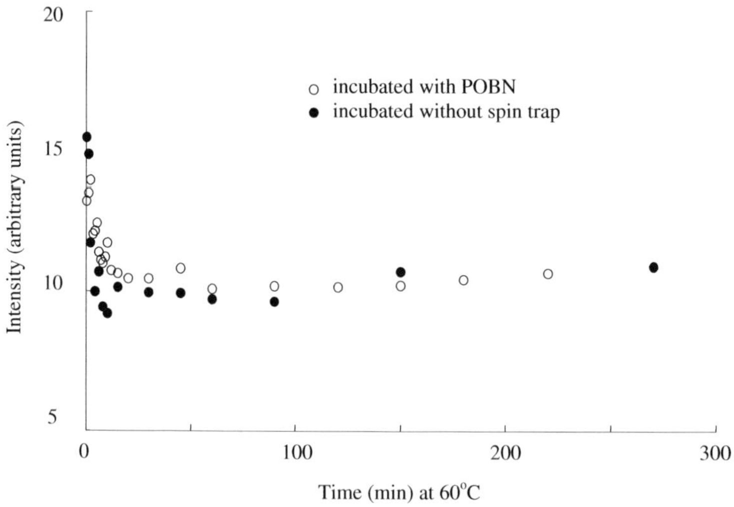

FIGURE 68.5 Intensity of single peak free radical signal in a 5% soluble coffee solutions during incubation at 60°C either with or without added spin trap 4-POBN. *Adapted from Pascual* et al. *(2002).*

water. The presence of O_2 during cooling/grinding the roasted beans also had a significant effect, but the free radical signal was insensitive to the atmosphere used for the retorting process. The intensity of the Fe(III) (g = 4.3) EPR signal and the percentage change in the free radical EPR signal during storage were directly related, and this observation could form the basis for the development of an assay for early prediction of shelf-life. However, it should be noted that this role for O_2 on the concentrations of stable free radicals in the coffee solutions is the opposite to its effect on the storage of R&G coffee (Yeretzian *et al.*, 2012), where it is responsible for increased free radical concentrations during storage. At the present time, we do not have an understanding of the detailed chemical processes responsible for these observations, and more work is needed before such scientific information can be applied to the development of new coffee products.

INFLUENCE OF O_2 ON THE STABILITY OF WHOLE ROASTED COFFEE BEAN EXTRACTS

As mentioned in the previous paragraph, development of liquid coffee products of good quality with extended shelf-lives is hampered by rapid quality degradation as a result of both physical and flavor instability. One approach that is being considered for extending the shelf lives of liquid coffees is that of supplementing the beverage with a very slow continuous extraction from intact roasted beans that are held within the aluminium can used to store the beverage, since this could lead to a continual replenishment of desirable sensory molecules that become depleted during storage. Recently, Pascual *et al.* (2013) have reported using EPR spectroscopy to examine the effects of key parameters that affect the stability of liquid coffee prepared from aqueous extracts from whole roasted coffee beans; these were the O_2 content of the extraction water and headspace during extraction and storage, and the storage temperature. It was found that the magnitude of the free radical signal increased during storage, as with storage of roasted coffee beans (Yeretzian *et al.*, 2012), but in contrast to observations with liquid concentrates prepared from R&G coffee where decreases were observed (see previous paragraph). The free radical signal intensity was sensitive to the O_2 content of the extraction water and storage temperature, but not the storage atmosphere (headspace), whereas the intensity of the Fe(III) (g = 4.3) signal in frozen coffee solutions was only affected by the O_2 content of the extraction water, and the Mn signal was insensitive to all of the parameters investigated. Thus, as with the work involving coffee concentrates, the O_2 content of the extraction water seems to be the most critical factor affecting the stability of the coffee products, even though it results in the opposite trends in free radical intensity in the two types of product.

FORMATION OF FREE RADICAL REACTION PRODUCTS

Lipid Peroxidation

The production of oxidized lipids during the storage of roasted coffee products has been described in the previous chapter (Goodman and Yeretzian, 2014), Such reactions would also be expected to occur in liquid coffee products, and indeed might be expected to be more rapid in the liquid compared to the solid phase. Evidence to support this hypothesis is obtained from the fact that the aroma of solutions of soluble coffee evolves and degrades rapidly (see below), whereas in the powder form it is highly stable.

Oxocaffeine and Other Caffeine Degradation Products

Oxocaffeine (1-methyl uric acid) is produced by the reaction of caffeine with hydroxyl radicals (Zbaida *et al.*, 1987), and thus can be used as a marker for Fenton reaction chemistry, and the oxidative challenge to which a liquid coffee product has been subjected. However, hydroxyl radicals react virtually indiscriminately, so all major components of coffee are affected by Fenton reaction chemistry, and as stated above this reaction is probably responsible for flavor and physical instability of coffee during storage in the liquid form. However, such changes are not necessarily detrimental to the quality of the product and Stadler *et al.* (1996) have reported that hydroxyl radicals are responsible for demethylation reactions of caffeine to produce theophylline, paraxanthine, and theobromine, all of which are bioactive molecules.

DEGRADATION OF COFFEE FLAVOR AND AROMA DURING STORAGE IN THE LIQUID FORM

As indicated in the standard text on food chemistry by Belitz *et al.* (2009), the preservation of good flavor and aroma during the storage of food products is a challenging problem. Although many studies have investigated changes in sensory properties as a result of storage of ground coffee powder, only fragmentary information is available on the reasons for their degradation in coffee beverages (Grosch, 2001). However, investigations combining instrumental analyses with human olfactory perception (e.g., GC/olfactometry) have revealed a rapid decrease in the concentrations of odor-active thiols at elevated temperature. Examples include the manufacturing of instant coffee (Semmelroch and Grosch, 1995), heat sterilization of coffee beverages (Kumazawa *et al.*, 1998), or keeping a freshly prepared coffee brew warm in a thermos flask (Hofmann *et al.*, 2001). Concentrations of 2-furfurylthiol, a key coffee odorant in R&G coffee, are significantly reduced, and a strong decrease of the sulfury-roasty odor quality in the overall aroma of coffee beverages is observed (Hofmann *et al.*, 2001). Furthermore, reactions involving the melanoidin components have been implicated in these changes (Hofmann *et al.*, 2001; Hofmann and Schieberle, 2002).

The chemical processes involved in aroma and flavor instability are exceedingly complex (Flament, 2002). This complexity has acted as a deterrent to much research using conventional methods, but one process that is frequently invoked to explain flavor degradation is free radical formation and reactions (e.g., Tsuji *et al.*, 1991; Blank *et al.*, 2000, 2002; Hofmann and Schieberle, 2002). Once radical species are formed, they may affect the flavor in various ways, for example, by decreasing the concentration of specific flavor impact molecules, by revealing an intrinsic off-flavor characteristic, and by generating new off-flavor compounds via reactions with a variety of compounds including other flavor-active radicals and non-volatiles in the coffee matrix, as well as by initiating a chain of radical reactions. Sediment formation is another major problem in the development of high-quality liquid coffee products with extended shelf-lives, and in addition a progressive increase in acidity is frequently observed during storage of liquid coffee beverages. A first step towards understanding this complex series of events is the identification of the radicals themselves and of their possible roles (Pascual *et al.*, 2000; Blank *et al.*, 2002; Munro *et al.*, 2003).

DISCUSSION AND CONCLUSIONS

We are now starting to arrive at an understanding of the various chemical processes in liquid coffee extracts that are controlled by free radical reactions, and to gain an appreciation of their significance for quality aspects of the beverage. First of all, it is important to appreciate that there are vast differences in the behavior of different types of free radical, and that the common approach of using a single name to describe them is inadequate for understanding their biological effects and their roles in food quality. Often when people in the life sciences talk about free radicals, they actually mean those that are ROS (and sometimes RNS), but when free radicals are measured by EPR spectroscopy and related techniques, it is all molecules with unpaired electrons that are considered. The latter represents a much broader group of chemical compound, and in routine measurements the EPR results are dominated by the most stable species.

The history of the investigations of free radicals in coffee represents a good model that could be used for the development of our knowledge in other food systems, especially those that involve appreciable heating. Thus the 'stable' free radical that is generated during roasting has been known for more than half a century, yet its impact on coffee quality is still essentially unknown, although as a component of the melanoidins it probably is involved in the development of flavor. Conversely, knowledge of the generation of unstable free radicals in coffee beverages is relatively recent.

The mechanisms controlling the concentrations of the various types of free radical in coffee are complex, and the underlying chemical processes are currently largely unelucidated. Flavor formation during roasting and flavor stability/degradation of roasted products have been discussed in the previous chapter (Goodman and Yeretzian, 2014), but free radical reactions in the liquid phase also have an appreciable impact on the quality of the beverage. Amongst these, antioxidant activity has received the greatest attention (e.g., del Castillo *et al.*, 2002; Moreira *et al.*, 2005; Cämmerer and Kroh, 2006; Lopez-Galilea *et al.*, 2007; Brezová *et al.*, 2009), but coffee is also an important free radical scavenger (Morrice *et al.*, 1993; Goodman *et al.*, 1994; del Castillo *et al.*, 2005), and coffee solutions react readily with O_2 (Pascual *et al.*, 2002).

Thus, O_2 has a critical role in the free radical chemistry of coffee, and in the liquid beverage it is responsible for initiating many of the changes that are observed after brewing. With solutions of soluble coffee, the EPR signal from the initial free radical was observed to decrease progressively over a period of ~20 min in air, but then stabilize (Pascual *et al.*, 2002). Decreases in EPR free radical signal intensities have also been observed during the storage of coffee concentrates (Yeretzian *et al.*, 2013), and this decrease was faster if O_2 was present in the extracting water, than if it was absent. In contrast to these observations, the results of Pascual *et al.* (2002) indicate free radical formation in coffee solutions heated in air, and also that this increases with increasing temperature. The generation of this latter radical seems to correlate roughly with the pro-oxidant activity of coffee, as measured by its ability to destabilize ascorbic acid to produce the ascorbate radical. Conversely, there appears to be no relationship between the intensity of the EPR free radical signal intensity and antioxidant activity as measured by the FRAP assay (Pascual *et al.*, 2005).

Thus, although coffee quality characteristics are influenced by oxidative processes this is a complex relationship, and it is inappropriate with the present state of knowledge to try to equate simple free radical intensity measurements to a quality parameter. Furthermore, exposure to O_2 results in increases in the free radical concentrations in whole coffee bean extracts, but decreases in coffee concentrates prepared from extracts of R&G coffee. However, in both systems the rates of change were faster in the presence of O_2 than in its absence. Thus it appears that careful control of O_2 exposure, especially at the extraction stage, can optimize the resistance to free-radical-induced changes during subsequent storage of liquid coffee products, although our knowledge of the chemical mechanisms responsible for this behavior is far from complete.

Finally, although the free radical chemistry of coffee beverages is exceedingly complex, there is real business potential to be gained from understanding how to stabilize such products, because of their huge convenience. Indeed the development of high-quality, shelf-stable liquid coffee beverages and coffee concentrates at room and elevated temperatures is very much dependent on a better understanding of the chemical processes underlying flavor degradation in the brew.

References

Belitz, H.-D., Grosch, W., Schieberle, P., 2009. Food Chemistry, fourth ed. Springer-Verlag, Berlin, Germany.

Blank, I., Pascual, E.C., Fay, L.B., Stadler, R.H., Goodman, B.A., Yeretzian, C., 2000. Degradation of the coffee flavor compound furfuryl mercaptan in Fenton-type model systems. In: Parliment, T.H., Ho, C.-T., Schieberle, P. (Eds.), Caffeinated Beverages, Health Benefits, Physiological Effects, and Chemistry. American Chemical Society, Washington, D.C, pp. 230–240.

Blank, I., Pascual, E.C., Devaud, S., Fay, L.B., Stadler, R.H., Yeretzian, C., Goodman, B.A., 2002. Degradation of the coffee flavor compound furfuryl mercaptan in model *Fenton*-type reaction systems. J. Agric. Food Chem. 50, 2356–2364.

Brezová, V., Slebodová, A., Stasko, A., 2009. Coffee as a source of antioxidants: an EPR study. Food Chem. 114, 859–868.

Buettner, G.R., Mason, R.P., 2003. Spin-trapping methods for detecting superoxide and hydroxyl free radicals in virto and in vivo. In: Cutler, R.G., Rodriguez, H. (Eds.), Critical Reviews of Oxidative Stress and Aging: Advances in Diagnostics and Intervention, vol. 1. World Scientific, pp. 27–38.

Cämmerer, B., Kroh, L.W., 2006. Antioxidant activity of coffee brews. Eur. Food Res. Technol. 223, 469–474.

del Castillo, M.D., Ames, J.M., Gordon, M.H., 2002. Effect of roasting on the antioxidant activity of coffee brews. J. Agric. Food Chem. 50, 3698–3703.

del Castillo, M.D., Gordon, M.H., Ames, J.M., 2005. Peroxyl radical-scavenging activity of coffee brews. Eur. Food Res. Technol. 221, 471–477.

Dorea, J.G., da Costa, T.H., 2005. Is coffee a functional food? Br. J. Nutr. 93, 773–782.

Flament, I., 2002. Coffee Flavor Chemistry. John Wiley & Sons Ltd, Chichester, England.

Goodman, B.A. (Ed.), 2012. Coffee Consumption and Health. Nova Science Publishers, Inc, Hauppauge, New York.

Goodman, B.A., Glidewell, S.M., Deighton, N., Morrice, A.E., 1994. Free radical reactions involving coffee. Food Chem. 51, 399–403.

Goodman, B.A., Yeretzian, C., 2014. Free radical processes in coffee I—Solid samples. In: Preedy, V.R. (Ed.), Processing and Impact on Active Components in Food. Chapter 67.

Grosch, W., 2001. Chapter 3: Volatile compounds. In: Clark, R.J., Vitzthum, O.G. (Eds.), Coffee: Recent Developments. Blackwell Science, Oxford, U.K, pp. 68–89.

Hofmann, T., Bors, W., Stettmaier, K., 1999. Studies on radical intermediates in the early stage of the nonenzymatic browning reaction of carbohydrates and amino acids. J. Agric. Food Chem. 47, 379–390.

Hofmann, T., Czerny, M., Calligaris, S., Schieberle, P., 2001. Model studies on the influence of coffee melanoidins on flavor volatiles of coffee beverages. J. Agric. Food Chem. 49, 2382–2386.

Hofmann, T., Schieberle, P., 2002. Chemical interactions between odor-active thiols and melanoidins involved in the aroma staling of coffee beverages. J. Agric. Food Chem. 50, 319–326.

Kumazawa, K., Masuda, H., Nishimura, O., Hiraishi, S., 1998. Change in flavor of coffee drink during heating. Nippon Shokuhin Kagaku Kogaku Kaishi 45, 108–113.

Lopez-Galilea, I., De Pena, M.P., Cid, C., 2007. Correlation of selected constituents with the total antioxidant capacity of coffee beverages: influence of the brewing procedure. J. Agric. Food Chem. 55, 6110–6117.

Manach, C., Scalbert, A., Morand, C., Rémésy, C., Jiménez, L., 2004. Polyphenols: food sources and bioavailability. Am. J. Clin. Nutr. 79, 727–747.

Moreira, D.P., Monteiro, M.C., Ribeiro-Alves, M., Donangelo, C.M., Trugo, L.C., 2005. Contribution of chlorogenic acids to the iron-reducing activity of coffee beverages. J. Agric. Food Chem. 53, 1399–1402.

Morrice, A.E., Deighton, N., Glidewell, S.M., Goodman, B.A., 1993. Free radical scavenging reactions involving coffee. ASIC 15th Int. Sci. Colloquium Coffee, 644–649.

Munro, L.J., Curioni, A., Andreoni, W., Yeretzian, C., Watzke, H.J., 2003. The elusiveness of coffee aroma: new insights from a non-empirical approach. J. Agric. Food Chem. 51, 3092–3096.

National Institute of Environmental Health Services (NIEHS), 2002. Spin trap database. http://epr.niehs.nih.gov last updated: September 23 2002.

Öbinger, C., Burner, U., Ebermann, R., 1997. Generation of hydrogen peroxide by plant peroxidases mediated thiol oxidation. Phyton 37, 219–226.

Pascual, E.C., Blank, I., Goodman, B.A., Yeretzian, C., 2000. The detection and characterization of free radicals generated during the decomposition of solutions of coffee flavour compound furfuryl mercaptan. ASIC 18th Int. Sci. Colloquium Coffee, 50–57.

Pascual, E.C., Goodman, B.A., Yeretzian, C., 2002. Characterisation of free radicals in soluble coffee by electron paramagnetic resonance spectroscopy. J. Agric. Food Chem. 50, 6114–6122.

Pascual, E.C., Yeretzian, C., Goodman, B.A., 2013. Free radical processes during the storage of aqueous extracts from whole roasted coffee beans. J. Agric. Food Chem. submitted for publication.

Preedy, V.R. (Ed.), 2014. Processing and Impact on Antioxidants in Beverages.

Pulido, R., Hernández-García, M., Saura-Calixto, F., 2003. Contribution of beverages to the intake of lipophilic and hydrophilic antioxidants in the Spanish diet. Eur. J. Clin. Nutr. 57, 1275–1282.

Scalbert, A., Williamson, G., 2000. Dietary intake and bioavailability of polyphenols. J. Nutr. 130, 2073S–2085S.

Scalbert, A., Johnson, I.T., Saltmarsh, M., 2005. Polyphenols: antioxidants and beyond. Am. J. Clin. Nutr. 81, 215S–217S.

Semmelroch, P., Grosch, W., 1995. Analysis of roasted coffee powders and brews by gas chromatography–olfactometry of headspace samples. LWT Food Sci. Technol. 28, 310–313.

Stadler, R.H., Richoz, J., Turesky, R.J., Welti, D.H., Fay, L.B., 1996. Oxidation of caffeine and related methylxanthines in ascorbate and polyphenol-driven Fenton-type oxidations. Free Radical Res. 24, 225–240.

Svilaas, A., Kaur Sakhi, A., Frost Andersen, L., Svilaas, T., Ström, E.C., Jacobs Jr., D.R., Ose, L., Blomhoff, R., 2004. Intakes of antioxidants in coffee, wine, and vegetables are correlated with plasma carotenoids in humans. J. Nutr. 134, 562–567.

Tsuji, S., Shibata, T., Ohara, K., Okada, N., Ito, Y., 1991. Studies on the factors affecting the formation of hydrogen peroxide in coffee. J. Food Hyg. Soc. Jpn. 32, 504–512.

Williamson, G., Holst, B., 2008. Dietary reference intake (DRI) value for dietary polyphenols: are we heading in the right direction? Brit. J. Nutr. 99 (Suppl. S3), S55–S58.

Yeretzian, C., Pascual, E.C., Goodman, B.A., 2012. Effect of roasting conditions and grinding on free radical contents of coffee beans stored in air. Food Chem. 131, 811–816.

Yeretzian, C., Pascual, E.C., Goodman, B.A., 2013. Effects of O_2 during various processing steps on free radical concentrations in hot aqueous extracts of roast & ground coffee and their changes during storage. Pro. PACCON 2013, Bangsaen, Thailand in press.

Zbaida, S., Kariv, R., Fischer, P., Gilhar, G., 1987. Reactions of theophylline, theobromine and caffeine with Fenton's reagent-simulation of hepatic metabolism. Xenobiotica 17, 617–621.

CHAPTER

69

Acrylamide in Coffee: Influence of Processing

*Cristina M.D. Soares**, *Rita C. Alves*[†, ‡], *M.Beatriz P.P. Oliveira*[†]

*REQUIMTE, Departamento de Química e Bioquímica, Faculdade de Ciências da Universidade do Porto, Portugal, [†]REQUIMTE, Departamento de Ciência Químicas, Faculdade de Farmácia da Universidade do Porto, Porto, Portugal, [‡]REQUIMTE, Instituto Superior de Engenharia do Porto, Instituto Politécnico do Porto, Porto, Portugal

CHAPTER POINTS

- Acrylamide is a heat-generated food toxicant.
- Acrylamide is formed via the Maillard reaction between asparagine and reducing sugars during coffee roasting.
- Acrylamide levels are high in the early stages of coffee beans roast and then tend to decrease.
- For a similar roast degree, Robusta coffee contain higher acrylamide levels than Arabica species.
- Immature beans contain a high amount of asparagine, an acrylamide precursor.
- The depulping of green coffee fruits reduces asparagine levels.
- The way the coffee beverage is prepared influences its acrylamide content.

INTRODUCTION

Thermal processing of foods (domestic or industrial) improves their organoleptic properties, preservation, and microbiological safety. Heat induces chemical changes and the formation of new molecules in foods, some of which impart positive health effects, while others are claimed to have carcinogenic and, in some cases, mutagenic properties (Studer *et al.*, 2004). Concern regarding acrylamide was raised when Swedish researchers reported the presence of this compound in foodstuffs (Tareke *et al.*, 2002). Its formation occurs at temperatures above 120°C and it is dependent on pH, water activity, and foodstuffs composition. French fries, potato chips, bread, and crispbread presented the highest levels of acrylamide but, subsequently, it was also detected in coffee and coffee surrogates. After the initial finding of acrylamide in foods, several independent studies have estimated overall exposure levels of the average consumer. They realized that there are considerable differences between many of the European countries, due to the differences in food consumption patterns and cooking traditions (Dybing *et al.*, 2005). Depending on the population's age, eating habits, and processing preferences, the intake of foodstuffs rich in this substance varies greatly between regions. Potato products (French fries, potato chips, roasted and baked potatoes) and bread are extensively consumed worldwide, but in some European countries, for instance, the contribution of coffee to acrylamide exposure reaches close to 40% (Dybing *et al.*, 2005). Acrylamide has been classified as "probably carcinogenic to humans" (class 2A) by the International Agency for Research on Cancer and a high level exposure causes damage to the nervous system. Although there is conflicting evidence about its potential harmfulness, the concern is due to its presence in such highly consumed foodstuffs. Nowadays, the main efforts are directed to mitigation and reduction strategies of the compound during food processing.

HOW COMPOSITION IS ALTERED

The chemical pathway leading to the formation of acrylamide in foodstuffs derives from the Maillard reaction that occurs between reducing sugars and amino acids. One of the main contributing compounds seems to be asparagine, an amino acid frequently present in foods (Yaylayan *et al.*, 2003). The cooking process, in particular frying or roasting at high temperatures, then induces the formation of acrylamide (Figure 69.1). More and more studies reported the details in many aspects of acrylamide formation. As the studies concerning acrylamide formation pathways proceed, 3-aminopropionamide (3-APA) was recently suggested to be an intermediate in the

Processing and Impact on Active Components in Food
http://dx.doi.org/10.1016/B978-0-12-404699-3.00069-X

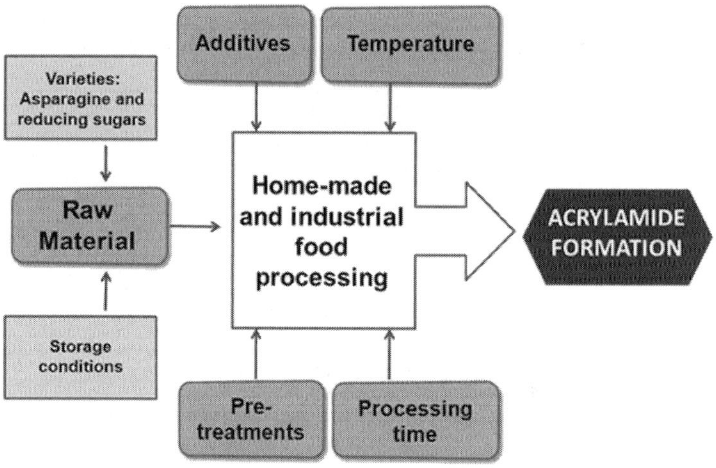

FIGURE 69.1 Factors affecting acrylamide formation during food processing.

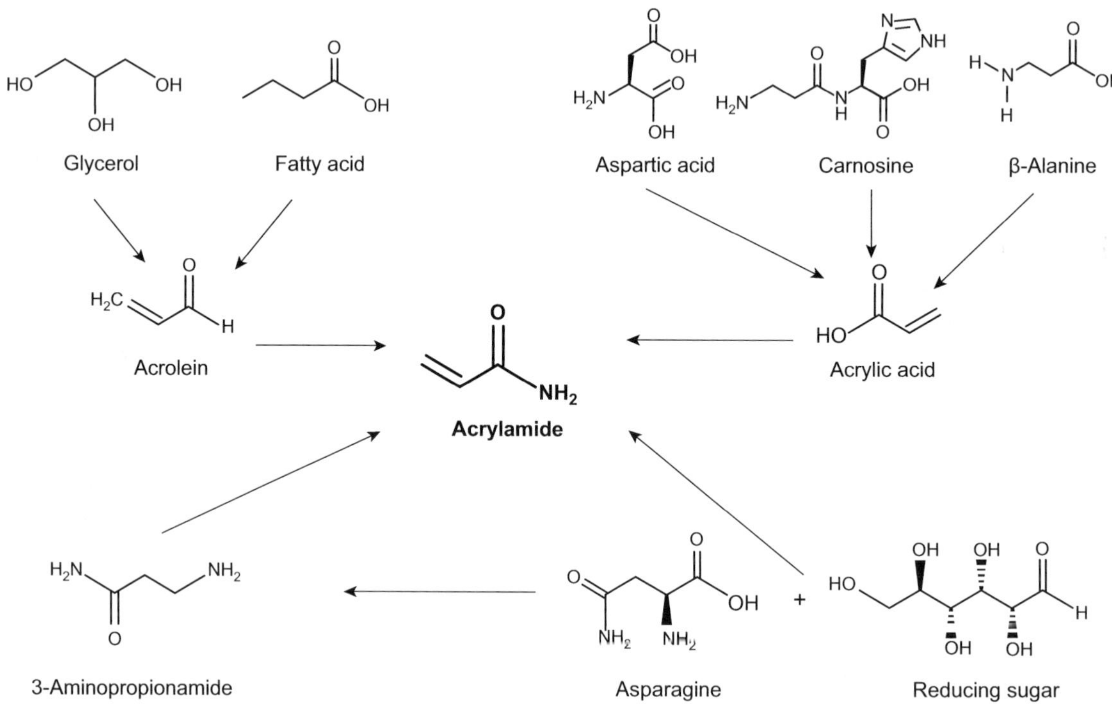

FIGURE 69.2 Acrylamide formation pathways. Nowadays the most important pathway seems to be through reaction between asparagine and a reducing sugar.

formation of acrylamide in the Maillard reaction of asparagine with reducing carbohydrates (Granvogl *et al.*, 2004). 3-APA can also be formed in raw foodstuffs by enzymatic decarboxylation of asparagine. It was described that in the presence of 3-APA in the food matrix acrylamide can be formed easily, even under aqueous conditions during heating. In this pathway, the Maillard reaction is not taking place; acrylamide can be formed even if there are no reducing sugars in the system. This can explain why acrylamide is found in raw materials containing low amounts of asparagine (Granvogl *et al.*, 2004). Figure 69.2 resumes the proposed acrylamide formation pathways. Nowadays, the most important one seems to be the previously described reaction between asparagine and a reducing sugar. Over the last few years, scientists have made efforts to find chemical or physical ways to reduce the amount of acrylamide formed in foodstuffs.

Acrylamide in Coffee

Coffee is generally roasted at temperatures in the range of 220–250°C, and the time and speed of roasting play an important role on the sensorial properties, such as color, aroma, and taste. These are carefully fine-tuned

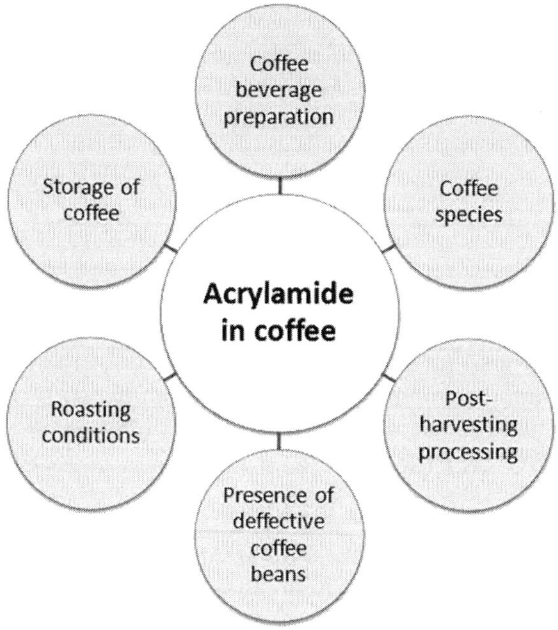

FIGURE 69.3 Factors affecting acrylamide formation in coffee.

to a characteristic profile leading to the clear identity of the coffee product. Coffee beans are thus subjected to relatively higher temperatures than other foods, and one can expect more than one pathway leading to acrylamide besides the Maillard reaction (Stadler and Scholz, 2004). The free asparagine concentration in green coffee beans lies within a very narrow range, typically from 30 to 90 mg/100 g. Conversely, experiments have shown that acrylamide is degraded or eliminated during roasting, and the profile of acrylamide formation during the roasting of coffee reflects this effect very clearly (Taeymans *et al.*, 2004). In coffee, acrylamide is formed at the very beginning of the roasting step, reaching more than 7 mg/kg, and then declines sharply toward the end of the roasting cycle due to higher rates of elimination. These and other factors affecting acrylamide formation in coffee (Figure 69.3) are detailed below.

Influence of Coffee Species

Coffea arabica and *Coffea canephora* var. *Robusta*, commonly known as Arabica and Robusta coffees, respectively, are the two coffee species of highest agronomical importance. They differ from each other in several aspects, including bean morphology, organoleptic properties, and chemical composition. The quality of the coffee beverage is usually dependent on the proportion of both varieties in the blend, Arabica being a higher-value product (Alves *et al.*, 2011).

The main precursors of acrylamide in coffee seem to be sucrose and asparagine. The concentration of sucrose in green coffee is lower in Robusta and higher in Arabica varieties, while the concentration of free asparagine in Robusta coffees is normally higher than in Arabica ones (Murkovic and Derler, 2006). Nevertheless, this may vary significantly due to cultivation practices and processing.

Asparagine amount in the raw material seems to be a limiting factor for the acrylamide formation in coffee. In fact, the quantity of acrylamide in roasted coffee is usually higher in Robusta than in Arabica coffees related to the higher amount of this precursor. Approximate amounts of 230 and 500 μg of acrylamide/kg have been described for medium-roasted Arabica and Robusta coffees, respectively (Alves *et al.*, 2010).

Influence of Post-Harvest Processing

The post-harvest processing of green beans embraces a succession of steps by which the coffee seeds are removed from the cherries and dried to provide a stable product for exportation. Three main procedures are known: the dry method ('natural'), the wet process ('washed'), and the 'semi-washed' process. In the former method, the cherries are dried, usually in the sun, and afterwards they are mechanically dehusked. This process is used for most Robusta coffees and some Arabica varieties. Washed coffees are usually of better quality attaining higher prices in international markets. The fresh mature cherries are depulped, fermented (decomposition of mucilage remains), washed, and dried. This method is essentially used for Arabica coffees and for some high-quality Robusta cultivars. In the third procedure, also known as 'natural depulped', the fruits are depulped and directly dried without being subjected to any fermentation step (Alves *et al.*, 2011).

It is well known that the type of post-harvest coffee processing determines the quality of the corresponding green beans and, thus, the final coffee quality, establishing the characteristic flavor differences. However, the main distinction between the chemical composition of washed and natural coffee is the soluble solids content, which is higher in the case of natural coffee. Natural depulped beans present an intermediate position (Illy and Vianni, 2005). Furthermore, specific and highly low-molecular flavor precursors, i.e., carbohydrates (Knopp *et al.*, 2006) and free amino acids (Bytof *et al.*, 2005), are dissimilar in differentially processed green coffees. Indeed, the chemical, biochemical, and physiological alterations that occur during bean processing can result in variations in the quality and quantity of flavor and aroma precursors. These changes may be related, among other factors, to the presence or absence of the coffee fruit husk.

While amino acids, in general, contribute to the development of the typical coffee aroma during roasting, which is considered the major attribute related to coffee quality, asparagine, in particular, will produce acrylamide. Recently, it was shown that the depulping of green coffee fruits enables a reduction in the levels of asparagine, which will consequently reduce the acrylamide amount formed during roasting (Dias, 2010).

Influence of Defective Coffee Beans

A factor that could contribute to a relatively higher acrylamide level is the number of defective beans used in the batch production. This parameter is a central component of quality control and, in most cases, the green beans are classified accordingly. Defective coffee beans, in particular immature ones, are characterized by significantly higher amounts of free asparagine (Mazzafera, 1999). Work recently carried out showed that the peeling of immature fruits allows a reduction in the levels of asparagine (Dias, 2010) which can represent a way to reduce the formation of acrylamide in coffee.

Influence of Roasting Conditions

Although coffee beans are roasted at quite high temperatures, the amounts of acrylamide found in the roasted beans and ground coffee are reported to be low. In coffee, acrylamide is formed in high concentrations during the first minutes of roasting, resulting in >7 mg/kg. The increase in roasting time leads to the degradation of acrylamide but also to the formation of undesirable off-flavor compounds. Therefore, the optimization of the roasting conditions with respect to a reduction of the acrylamide formation and maintaining the product quality has not been realized yet (Guenther *et al.*, 2007). Roasted coffee beans contain acrylamide in the range of 40–400 μg/kg with a mean value of about 200 μg/kg (Murkovic, 2004) while dry instant coffee may contain more than 500 μg/kg (Andrzejewski *et al.*, 2004). Granby and Fagt (2004) found no significant differences between preparation techniques but determined higher acrylamide contents in coffee from medium-roasted beans (about 10 μg/l) compared to dark-roasted beans (about 5 μg/l). Therefore, light-roasted coffees may contain relatively higher amounts of acrylamide than very-dark-roasted beans. The temperature *per se*, however, did not show a significant effect in the formation of acrylamide. According to the EU Commission database, roast coffee finished products have a median acrylamide level of 265–290 mg/kg. Because coffee is a product with very stringent quality attributes, it seems that there are not many options currently available for mitigation of acrylamide during roasting. As aroma is a result of the roasting process, which is related to the chemical composition of the raw material, every change in the raw material or the roasting process leads to totally different product characteristics (Bagdonaite *et al.*, 2008). The importance of the roasting process for flavor and color and the relatively narrow range for commercial products as well as the impact of elimination make mitigation for coffee particularly complicated. For coffee, acrylamide formation does not appear to be affected by the reducing sugars content and seems to be only correlated with asparagine concentration in green beans.

Acrylamide does not accumulate during coffee roasting. Rather, formation and degradation occur simultaneously. Only 20% of initially formed acrylamide survives to the roasting process (Lantz *et al.*, 2006).

Influence of Storage

Two research groups have found that acrylamide is not stable in commercial coffee stored in its original container (Andrzejewski *et al.*, 2004; Delatour *et al.*, 2004). Losses of 40 to 60% have been recorded in ground coffees stored at room temperature predominantly through binding of acrylamide to constituents of the ground and roasted coffee matrix (Lantz *et al.*, 2006).

Influence of Beverage Preparation

Acrylamide is highly soluble in water (Andrzejewski *et al.*, 2004), and is, therefore, easily extracted to coffee beverage during its preparation. Nevertheless, the type of percolation (including preparation method and coffee/water ratio), as well as the blend composition and its roast degree, will significantly influence the acrylamide amount present in the final beverage (Alves *et al.*, 2007). Many ways to make coffee brews are known, usually based on three different processes: decoction (boiled, Turkish, vacuum, and percolation), infusion (filter and napoletana), and pressure (press-pot or French press, mocha, and espresso coffee).

In general, acrylamide levels in common coffee beverages (such as plunger pot and filtered coffee) range between 2 and 25 μg/l (Andrzejewski *et al.*, 2004; Dybing *et al.*, 2005; Granby and Fagt, 2004). Compared with other coffee beverages, the brewing procedure of espresso coffee extracts acrylamide incompletely due to the short contact time between coffee and water (Lantz *et al.*, 2006). This brew is prepared by a special brewing technique in which a limited amount (20–50 ml) of hot water under high pressure (9 ± 2 atm, 90 ± 5°C) is percolated in a very short time (30 ± 5 s) through a ground coffee cake (6.5 ± 1.5 g). The result is a concentrated and intensely flavored brew covered by a dense foam layer, which should be tasted at the exact moment of extraction (Alves *et al.*, 2007). This particular type of beverage is highly appreciated in some European countries like Italy, Portugal, or France, but its consumption is increasing worldwide.

Although acrylamide is incompletely extracted during espresso coffee preparation, the final concentration of this brew is usually higher than other coffee beverages due to the higher ground coffee/water ratio used to prepare it (from 6.5 g/20 ml to 6.5 g/70 ml), a feature that varies with consumers' preferences and geographical habits: 20 g/l in USA (Andrzejewski *et al.*, 2004) and 40 g/l in Northern Europe (Granby and Fagt, 2004).

Indeed, acrylamide levels of 0.32–1.76 μg/cup (30 ml), corresponding to concentrations of 10.7–58.5 μg/l, are described in literature for standard espresso coffees (Soares *et al.*, 2006, Alves *et al.*, 2010). Recently, coffee capsules have emerged as a simple and easy way to produce espresso coffee in adapted machines. The convenience of preparation and the consistency of brew quality are much appreciated by consumers. With regard to acrylamide levels, Alves *et al.* (2010) found no significant differences between these types of espresso coffees [1.00–1.66 μg/ cup (30 ml) or 33.4–55.3 μg/l] and the standard ones. Also, similar acrylamide contents have been described for decaffeinated espresso coffees [0.74–1.49 μg/cup (30 ml) or 24.8–49.5 μg/l] suggesting that the decaffeination process does not influence acrylamide precursors (reducing sugars and asparagine) in green coffee beans.

With regard to the composition of the coffee blend that gives rise to the beverage, the amount of Arabica *versus* Robusta species highly influences the acrylamide content, as expected. For the same roast degree, brews prepared only with Robusta coffee contain approximately double the amount of this compound per cup [1.71–2.92 μg/cup (30 ml)] than those prepared only with Arabica species [0.87 and 1.52 μg/cup (30 ml)]. In turn, no significant differences were found between espresso coffees prepared with a dry-processed sample from Brazil and other wet-processed Arabicas from other origins (Alves *et al.*, 2010).

As acrylamide is formed during the early stages of coffee roasting and subsequently its content tends to decrease (Taeymans *et al.*, 2004; Alves *et al.*, 2010), the roast degree of the coffee batch from which the brews are prepared will consequently affect the acrylamide amount ingested by consumers. Accordingly, an average decrease of 25% was found when medium-roasted beverages were compared with their dark-roasted counterparts (Alves *et al.*, 2010). Figure 69.4 shows the acrylamide contents of espresso coffees prepared with Arabica and Robusta coffee beans roasted to different degrees. Although coffee species and roasting degree affect the acrylamide content of the brew, they do not seem to influence acrylamide extractability [(~80%, for standard espresso coffees (30 ml), and 100% for other beverages)](Lantz *et al.*, 2006; Alves *et al.*, 2010). As mentioned, contrary to other percolation methods the espresso process incompletely extracts acrylamide from ground coffee. Nevertheless, extending the time of extraction [and consequently increasing the volume of the brew from 20 ml (short) to 70 ml (long)], it was possible to increase acrylamide extractability from ~60 to ~99%, respectively. In turn, although the final content of acrylamide increases with volume, the brew concentration simultaneously decreases due to a reduction in the coffee/water ratio: from 108.2 to 50.1 μg/l (Robusta espressos) and from 56.3 to 24.2 μg/l (Arabica espressos).

Referring to the acrylamide intake, it will obviously depend on the type and amount of beverage consumed. While espresso coffee is, generally, a very short beverage, higher volumes per cup of other coffee beverages are usually consumed. For example, a cup of filter coffee may achieve 200 ml, since it is considered a light brew (Alves *et al.*, 2007). Therefore, acrylamide intake through coffee brews will essentially depend on the consumption habits (type, strength, and volume of beverage, together with intake frequency), factors that are influenced by cultural and personal preferences of consumers.

OTHER WAYS IN WHICH COMPOSITION IS ALTERED

Within the commercial roasting range, the acrylamide amount was reduced by a factor of approximately 10 compared with the deepest level recorded during the complete roasting cycle (Stadler and Scholz, 2004). An option to reduce the amount of acrylamide ingested through coffee beverages is to select commercial blends

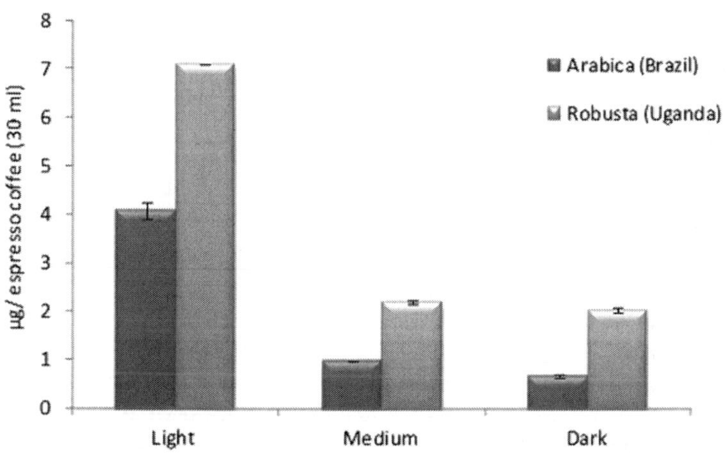

FIGURE 69.4 Acrylamide contents of espresso coffees prepared with Arabica and Robusta coffee beans (from Brazil and Uganda, respectively) roasted to different degrees (light, medium, and dark).

TABLE 69.1 Brief Overview of Published Methodologies based on LC-MS to Quantify Acrylamide in Coffee

Sample Preparation	Chromatographic Method	Acrylamide Levels (μg/kg)	Reference
Liquid chromatography methods			
Extraction: H_2O SPE clean-up: Oasis HLB and Bond Elut–Accucat cartridges	LC-MS/MS	Instant coffee (*n* = 12): 169–539 Ground coffee (*n* = 31): 45–374 Brewed coffee (*n* = 40): 6–11	Andrzejewski *et al.* (2004)
Extraction: H_2O Clean-up: Carrez I and II, DCM salting out with NaCl and extraction with EtAc; SPE clean-up: ISOLUTE Multimode cartridge	LC-MS/MS	Ground coffee (*n* = 2): 203–312 Soluble coffee powder (n = 1): 771	Delatour *et al.* (2004)
Extraction: H_2O SPE clean-up: ISOLUTE Multimode cartridge	LC-MS/MS	Brewed coffee (n = 30): 2–16	Granby and Fagt (2004)
Extraction: MeOH Clean-up: Carrez I and Carrez II; SPE clean-up: Oasis HLB cartridge	LC- MS	Ground coffee (n = 11): 14 –29 Turkish coffee (n = 5): 29–75 Filter coffee (n = 1): 50 Instant coffee (n = 3): 42–338	Senyuva and Gökmen (2005)
Extraction: H_2O Clean-up: DCM; SPE clean-up: aminopropyl cartridge	LC-MS/MS	Ground coffee (n = 3): 43–465	Águas *et al.* (2006)
Extraction: H_2O Clean-up: Carrez I and Carrez II, EtAc; SPE clean-up: ISOLUTE Multimode cartridge	LC-MS/MS	Instant coffe (n = 1): 715 Ground coffee (n = 1): 206	Arisseto *et al.* (2008)
Extraction: H_2O after deffating with hexane SPE clean-up: Bond Elut–Accucat cartridge Derivatization: 2-Mercapto benzoic acid (pH 8, 3 h)	LC-MS/MS	Ground coffee (n = 12): 299–762	Bagdonaite *et al.* (2008)
Extraction: H_2O Clean-up: hexane, MeCN, $MgSO_4$ and NaCl Dispersive-SPE clean up: $MgSO_4$ and PSA	LC-MS/MS	Ground coffee (n = 1): 145	Mastovska and Lehotay (2006)
Extraction: Pressurized fluid extraction (PFE) with MeCN Clean-up: Florisil inside the PFE extraction cell	LC-MS/MS	Ground coffee (n = 20): 62–287	Pardo *et al.* (2007)

DCM, dicholomethane; EtAc, ethyl acetate; EtOH, ethanol; MeCN, acetonitrile; MeOH, methanol; MSPD, matrix solid phase dispersion; Oasis HLB, hydrophilic–lipophilic-balanced reversed-phase sorbent; PFE, pressurized fluid extraction; PSA, primary secondary amine; SPE: solid-phase extraction

with higher Arabica percentages and darker degrees of roasting and, simultaneously, to perform shorter brews instead of long ones, but this will obviously depend on the consumers' preferences. Nevertheless, deeper roasting as a potential choice to reduce acrylamide could generate other undesirable compounds and negatively impact the sensory properties of the product. Consequently, no applicable solutions are today in practice that would reduce acrylamide levels in beans and concomitantly retain the quality characteristics of coffee, because the roasting step can not be completely changed. There is little that can be done from a technological point of view to reduce the high levels of acrylamide in coffee beans. One potential strategy for cutting acrylamide levels is to reduce either the sugar or the asparagine content. This can be done in dough products by introducing an enzyme that reduces the level of asparagine but it is not feasible in the case of coffee because the reactants are located largely in the coffee bean and, therefore, are not accessible from the outside. Another option would be to change the roasting conditions, as has been done to some extent in potato products, but again this is not a suitable strategy for coffee. Changing roasting conditions creates products that would not be accepted by the consumer. However, scientists have been suggesting other solutions based in agronomy, and coffee plants with low asparagine levels could eventually be bred.

ANALYTICAL TECHNIQUES

Over recent years, several methods for acrylamide's quantification in ground coffee and coffee beverages have been reported using mainly LC-MS/MS and, less frequently, GC-MS. Tables 69.1 and 69.2 show a bibliographic compilation of the referred methods (these are not exhaustive). A brief description of the sample preparation, including clean-up, chromatographic

TABLE 69.2 Brief Overview of Published Methodologies based on GC/MS to Quantify Acrylamide in Coffee

Sample Preparation	Chromatographic Method	Acrylamide Levels (μg/kg)	Reference
Gas chromatography methods			
Extraction: H_2O Clean-up: EtOH, Carrez I and II SPE clean-up: ISOLUTE Multimode/C18 cartridges Derivatization: bromination (1 h)	GC-MS	Espresso (n = 18): 11–36 Soluble coffee (n = 5): 47–95	Soares *et al.* (2006)
Extraction: MSPD and water Clean-up: MSPD combined with SPE (Isolute MultiMode and C_{18}) Derivatization: bromination (1 h)	GC-MS	Ground coffee (n = 3): 220–305 Coffee brew (n = 5): 47–66	Soares *et al.* (2010)

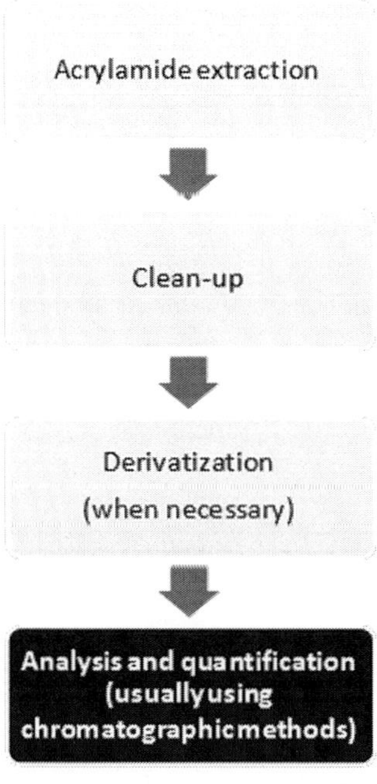

FIGURE 69.5 Rough sketch of the analytical method for acrylamide quantification.

methodology, and range of concentrations obtained is presented. The great majority of these techniques mainly present differences in sample extraction and clean-up procedures. An efficient clean-up is required to obtain an extract that provides final chromatograms almost free of interferences. Acrylamide is usually extracted with water and clean-up is performed with Carrez solutions (deproteination) and solid-phase extraction (SPE) cartridges with different adsorbents. Figure 69.5 depicts a simplified analytical schematic for acrylamide quantification.

Acknowledgments

This work has been supported by Fundação para a Ciência e a Tecnologia (FCT) (PEst-C/EQB/LA0006/2013) and by QREN (NORTE-07-0124-FEDER-000069-CIÊNCIA DO ALIMENTO). C. Soares and R. C. Alves are grateful to FCT for a PhD grant (SFRH/BD/39360/2007) and a post-doc (SFRH/BPD/68883/2010) grant, respectively, financed by POPH-QREN and subsidized by FSE and MCTES.

References

Aguas, P.C., Fitzhenry, M.F., Giannikopoulos, G., Varelis, P., 2006. Analysis of acrylamide in coffee and cocoa by isotope dilution liquid chromatography–tandem mass spectrometry. Anal. Bioanal. Chem. 385, 1526–1531.

Alves, R.C., Oliveira, M.B.P.P., Casal, S., 2011. Coffee authenticity. In: Oliveira, M.B.P.P., Mafra, I., Amaral, J.S. (Eds.), Current Topics on Food Authentication. Transworld Research Network, Kerala, India, pp. 57–72.

Alves, R.C., Soares, C., Casal, S., Fernandes, J., Oliveira, M.B.P.P., 2010. Acrylamide in espresso coffee: Influence of species, roast degree and brew length. Food Chem. 119, 929–934.

Alves, R.C., Casal, S., Oliveira, M.B.P.P., 2007. Factors influencing the norharman and harman contents in espresso coffee. J. Agric. Food Chem. 55, 1832–1838.

Andrzejewski, D., Roach, J.A., Gay, M.L., Musser, S.M., 2004. Analysis of coffee for the presence of acrylamide by LC–MS/MS. J. Agric. Food Chem. 52, 1996–2002.

Arisseto, A.P., Toledo, M.C.F., Govaert, Y., van Loco, J., Fraselle, S., Degroodt, J.-M., 2008. A modified sample preparation for acrylamide determination in cocoa and coffee products. Food Anal. Met. 1, 49–55.

Bagdonaite, K., Derler, K., Murkovic, M., 2008. Determination of acrylamide during roasting of coffee. J. Agric. Food Chem. 56, 6081–6086.

Bytof, G., Knopp, S.E., Schieberle, P., Teutsch, I., Selmar, D., 2005. Influence of processing on the generation of g-aminobutyric acid in green coffee beans. Eur. Food Res. Technol. 220, 245–250.

Delatour, T., Périsset, A., Goldmann, T., Riediker, S., Stadler, R.H., 2004. Improved sample preparation to determine acrylamide in difficult matrixes such as chocolate powder, cocoa, and coffee by liquid chromatography tandem mass spectroscopy. J. Agric. Food Chem. 52, 4625–4631.

Dias, E.C., 2010. Efeito do processamento de frutos imaturos e da torração na ocorrência e na formação de compostos relevantes para a qualidade e segurança em café arábica. Lavras, Universidade Federal de Lavras. PhD Thesis (in pPortuguese), Brazil.

Dybing, E., Farmer, P., Andersen, M., Fennell, T., Lalljie, S., Müller, D., Olin, S., Petersen, B., Schlatter, J., Scholz, G., Scimeca, J., Slimani, N., Törnqvist, M., Tuijtelaars, S., Verger, P., 2005. Human exposure and internal dose assessments of acrylamide in food. Food Chem. Toxicol. 43, 365–410.

Granby, K., Fagt, S., 2004. Analysis of acrylamide in coffee and dietary exposure to acrylamide from coffee. Anal. Chim. Acta 520, 177–182.

Granvogl, M., Jezussek, M., Koehler, P., Schieberle, P., 2004. Quantitation of 3-Aminopropionamide in Potatoes—A Minor but Potent Precursor in Acrylamide Formation. J. Agr. Food Chem. 52, 4751–4757.

Guenther, H., Anklam, E., Wenzl, T., Stadler, R.H., 2007. Acrylamide in coffee: review of progress in analysis, formation and level reduction. Food Addit. Contam. 24, 60–70.

Illy, A., Viani, R., 2005. Espresso coffee: the science of quality. Academic Press, London, UK.

Knopp, S.E., Bytof, G., Selmar, D., 2006. Influence of processing on the content of sugars in green Arabica coffee beans. Eur. Food Res. Technol. 223, 195–201.

Lantz, I., Ternite, R., Wilkens, J., Hoenicke, K., Guenther, H., van der Stegen, G., 2006. Studies on acrylamide levels in roastings, storage and brewing of coffee. Mol. Nutr. Food Res. 50, 1039–1046.

Mastovska, K., Lehotay, S.J., 2006. Rapid sample preparation method for LC-MS/MS or GC-MS analysis of acrylamide in various food matrices. J. Agric. Food Chem. 54, 7001–7008.

Mazzafera, P., 1999. Chemical composition of defective coffee beans. Food Chem. 64, 547–554.

Murkovic, M., 2004. Acrylamide in Austrian foods. J. Biochem. Bioph. Meth. 61, 161–167.

Murkovic, M., Derler, K., 2006. Analysis of amino acids and carbohydrates in green coffee. J. Biochem. Biophys. Met. 69, 25–32.

Pardo, O., Yusà, V., Coscollà, C., León, N., Pastor, A., 2007. Determination of acrylamide in coffee and chocolate by pressurised fluid extraction and liquid chromatography–tandem mass spectrometry. Food Addit. Contam. 24, 663–672.

Senyuva, H.Z., Gökmen, V., 2005. Study of acrylamide in coffee using an improved liquid chromatography mass spectrometry method: Investigation of colour changes and acrylamide formation in coffee during roasting. Food Add. Contam. 22, 214–220.

Soares, C.M., Alves, R.C., Casal, S., Oliveira, M.B.P.P., Fernandes, J.O., 2010. Development and validation of a matrix solid phase dispersion method to determine acrylamide in coffee and coffee substitutes. J. Food Sci. 75, 57–63.

Soares, C., Cunha, S., Fernandes, J., 2006. Determination of acrylamide in coffee and coffee products by GC–MS using an improved SPE clean-up. Food Add. Contam. 23, 1276–1282.

Stadler, R.H., Scholz, G., 2004. Acrylamide: An Update on Current Knowledge in Analysis, Levels in Food, Mechanisms of Formation, and Potential Strategies of Control. Nutr. Rev. 62 , 449–467 (n°12).

Studer, A., Blank, I., Stadler, R.H., 2004. Thermal processing in foodstuffs and potential strategies of control. Czech J. Food Sci. 22, 1–10.

Taeymans, D., Wood, J., Ashby, P., Blank, I., Studer, A., Stadler, R.H., Gondé, P., Van Eijck, P., Lalljie, S., Lingnert, H., Lindblom, M., Matissek, R., Müller, D., Tallmadge, D., O'Brien, J., Thompson, S., Silvani, D., Whitmore, T., 2004. A review of acrylamide: An industry perspective on research, analysis, formation, and control. Crit. Rev. Food Sci. Nutr. 44, 323–347.

Tareke, E., Rydberg, P., Karlsson, P., Eriksson, S., Törnqvist, M., 2002. Analysis of acrylamide, a carcinogen formed in heated foodstuffs. J. Agric. Food Chem. 50, 4998–5006.

Yaylayan, V., Wnorowski, A., Locas, C.P., 2003. Why Asparagine Needs Carbohydrates to Generate Acrylamide. J. Agr. Food Chem. 51, 1753–1757.

CHAPTER

70

Use of Microwaves to Extract Chlorogenic Acids from Green Coffee Beans

Lingamallu Jagan Mohan Rao, Kulathooran Ramalakshmi

Plantation Products, Spices and Flavour Technology Department, Central Food Technological Research Institute, Mysore, India

CHAPTER POINTS

- Microwave-assisted extraction (MAE) showed obvious advantages in terms of short duration and high efficiency in the recovery of chlorogenic acid from raw plant materials in comparison with conventional heat-reflux extraction.
- The mechanism of the enhanced extraction by microwave assistance was studied by observing cell destruction of plant material after MAE treatment by scanning electron microscopy. The enhanced extraction was related partly to a greater extent of cell rupture of the plant materials, and this was observed by scanning electron microscopy. The plant materials were significantly destroyed due to the cell rupture during MAE treatment.
- In general, extraction is the first step in the preparation of pharmaceutical formulations from raw plant material and significantly affects the cost of the whole manufacture process.
- MAE has been considered as a potential alternative to conventional solvent extraction for the isolation of phenolic compounds from plants.
- Under optimum conditions of time (5 min), temperature (50°C), and wattage (800 W), the maximum quantities of chlorogenic acids and caffeine could be extracted with water as solvent.
- The extracts contained chlorogenic acids and caffeine in the ranges of 31–62% and 22–40%, respectively.
- The yields of MAE under optimum conditions were higher than those from the conventional solvent extraction at 5 min and 50°C and the extracts showed radical-scavenging activity of >75%, even at a concentration of 25 ppm.
- The MAE process can thus be predicted and controlled for industrial application.

INTRODUCTION

Electromagnetic Radiation

Electromagnetic radiation is a form of energy that is transmitted through a space at an enormous velocity. It is said to have a dual nature, exhibiting both wave and particle characteristics. As the name implies, electromagnetic radiation is an alternating electrical and associated magnetic force field in space. Thus, an electromagnetic wave has an electric component and a magnetic component. Both the components oscillate in planes perpendicular to the direction of the propagation of the radiation. The entire range over which electromagnetic radiation exists is known as electromagnetic spectrum. The electromagnetic spectrum consists of gamma, X-ray, ultraviolet, visible, infrared, microwave, radiofrequency, and ultrasound radiations (Ali and Hardiman, 2005; Susanne, 2012).

Infrared Radiation

Infrared radiation (IR) is a part of electromagnetic spectrum with a wavelength in the range of 0.78 μm to 1000 μm. Application of infrared heating has recently been employed in certain food-processing

Processing and Impact on Active Components in Food
http://dx.doi.org/10.1016/B978-0-12-404699-3.00070-6

applications, because of its superiority in terms of costs and the product quality (color, organoleptic and nutritional value) as compared with conventional heating. Infrared heating offers many advantages over conventional heating under similar processing temperatures. Both the situations involve simultaneous heat and mass transfer, and in the case of conventional heating both are in opposite directions. These are in the same (from within to outside) direction in case of IR, which results in a higher rate of heat transfer to the material (Robyn *et al.*, 2000).

Microwave Radiation

Microwave radiation (MW; wavelength in the range of 1 mm to 1 m) lies between IR and radiofrequency (RF) and is known as dielectric radiation. Material exposed to MW is heated due to dipolar rotation and ionic polarization. Hence, material containing a polar substance such as water is heated faster with MW. The typical domestic MW oven has a frequency of 2450 MHz. The applications of MW in food processing are blanching, drying, pasteurization, baking, etc. Compared to standard microwave technology the need to use higher frequencies such as 24,150 MHz for industrial applications has to be carefully verified with respect to special physical/engineering advantages.

MICROWAVE-ASSISTED EXTRACTION

MAE is a process that uses microwave energy along with solvent to extract target compounds from various matrices. The volumetric heating, localized temperature, and pressure can cause selective migration of target compounds from the material to the surroundings at a more rapid rate and with similar or better recoveries compared with conventional extractions. MAE was used for extraction of selected components from a wide variety of sample matrices and has been used as a promising alternative sample preparation technique for a number of applications (Zhou and Liu, 2006; Gao *et al.*, 2006; Zhang and Xu, 2007; Rostagno *et al.*, 2007). MAE can considerably reduce both extraction time and solvent consumption compared to conventional methods.

Radiofrequency Waves

RF waves are also being used for thermal processing food materials. In a radio frequency heating system, the RF generator creates an alternating electric field between two electrodes. The material is placed between the electrodes where the alternating energy causes polar molecules in the material to continuously reorient them to face opposite poles in much the same way that bar magnets behave in an alternating magnetic field. The friction resulting from molecular movement causes the material to rapidly heat throughout its entire mass. For those applications that require uniform heating and precise temperature control, RF offers many advantages over conventional drying methods, namely uniform heating through the entire thickness, 2- to 20-times faster than conventional drying methods, energy efficient, uniform moisture profiling, low maintenance, and eco-friendly process. The permitted frequencies for industrial applications are 13, 27, and 40 MHz. A few reports on the existence of a non-thermal effect of RF energy on microorganisms are available. Freshness and nutrients are retained in foods when RF energy inactivates microorganisms non-thermally (Ali and Hardiman, 2005).

Generally foodstuffs are complex mixtures of different large biochemical molecules (simple sugars, amino acids, vitamins etc), biochemical polymers (complex sugars, proteins, lipids etc.), inorganic salts and water. The strongest absorption is often localizable within the so-called group frequency, which is generated by the vibrations of these molecular aggregates (or molecular structural groups). The wavelength region of interest in discussing the interaction with foodstuffs can be restricted to the infrared range 0.78–15 μm even for high-temperature radiators, the spectral flux, [(W/m^2)/μm] ascribed to wavelengths larger than 15 μm represents a small fraction of total radiation power.

HOW THE COMPOSITION ALTERED

Chlorogenic Acids

Chlorogenic acid (5-*O*-caffeoyl-quinic acid), an ester of caffeic acid with quinic acid, has received considerable attention for its wide distribution and potential biological effects (Clifford, 1999). It is one of the most abundant polyphenols in the human diet with coffee, fruits, and vegetables as its major sources. It is also an important bioactive compound and rich in selected raw materials used in traditional Chinese medicine, such as flowers and buds of *Lonicera japonica* Thumb, and the leaves of *Eucommia ulmodies*, both which have been used for the treatment of exopathogenic wind-heat or epidemic febrile disease at the early stage, carbuncles, furuncles, and swellings for centuries.

The chlorogenic acids (CGA) are an important group of compounds in green coffee beans. Although 71 different species of CGA have now been identified (Rohit and Jaganmohanrao, 2013) the vast majority of the compounds belong to three classes: caffeoylquinic acids (CQA; 3CQA, 4CQA, and 5CQA), di-caffeoylquinic acids (diCQA; 3,4diCQA, 3,5diCQA, and 4,5diCQA) and feruloylquinic acids (FQA) as shown in Figure 70.1. Other quinic acids are CiQA, cinnamoyl quinic acids; GQA, galloyl quinic acids; iFQA, isoferuloyl quinic acids; oCoQA, o-coumaroyl quinic acids; SQA, sinapoyl quinic

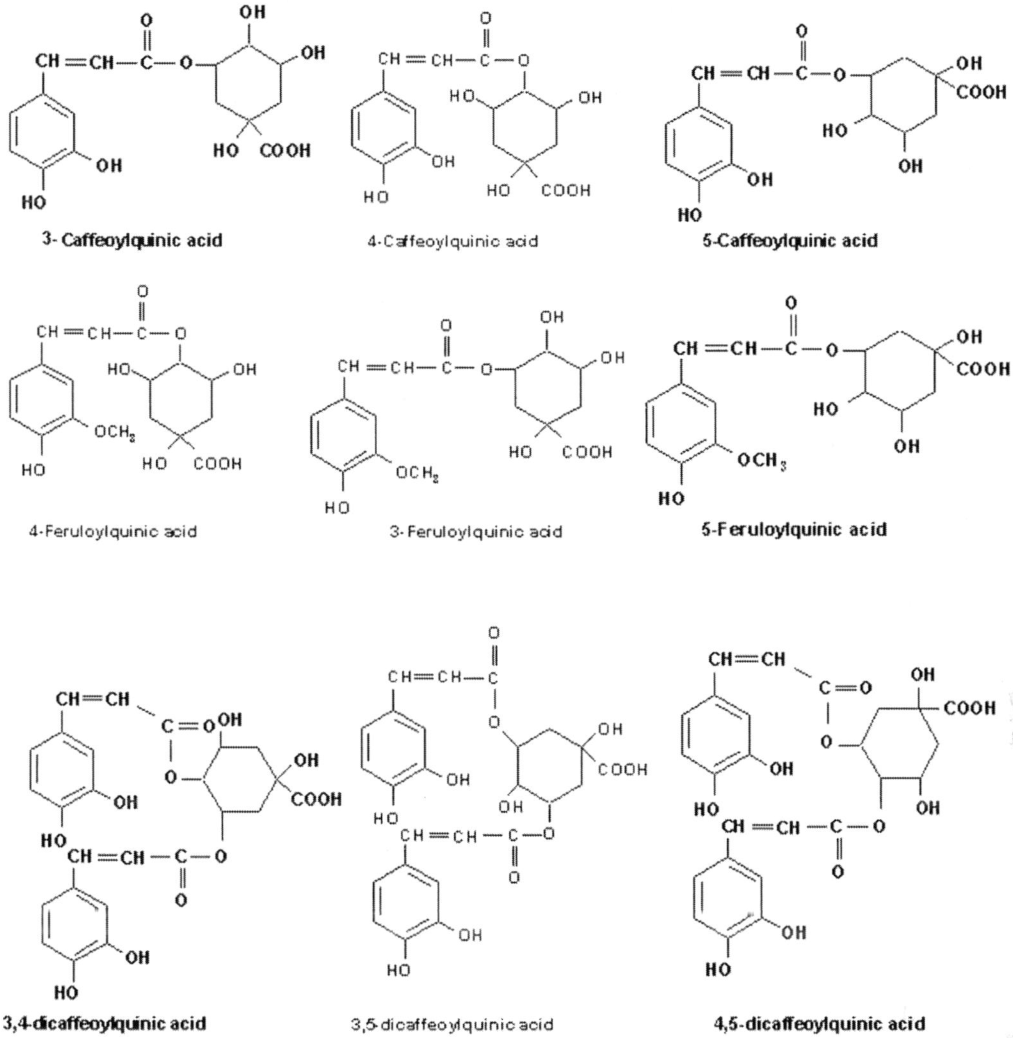

FIGURE 70.1 Chlorogenic acid isomers.

acids, which are known either naturally or synthetically. Although CGA (primarily as 5-caffeoylquinic acid) are widely distributed in plant materials, their content in green coffee is among the highest found in plants, ranging from 4 to 14% (Trugo and Macrae, 1984; Farah *et al.*, 2008). The levels of CGA in green coffee beans have been reported to vary from approximately 7.88 to 14.4% dry matter for *Coffea canephora* (*robusta*) and approximately 3.4 to 4.8% for *Coffea arabica* (Ky *et al.*, 2001).

The analytical methods for CGA estimation include chromatographic and spectroscopic techniques. Chromatographic techniques reported are thin-layer chromatography, gas chromatography, and liquid chromatography (Bicchi *et al.*, 1995; Rehwald *et al.*, 1994). Spectral techniques include spectrophotometry, IR spectrometry, ^{1}H NMR spectroscopy and chemiluminescence.

Microwave Extraction

In general, CGA are extracted from dried and powdered plant materials using solvent/aqueous extraction. However, this technique is very time consuming as well as requiring high energy. This drawback prevents the application of the reported method. Development of a new technique for separation and purification of CGA is essential. Growing advancements in the isolation/separation techniques in recent years have made the purification and quantification of CGA easy, sensitive, and specific, with marked improvements in yield in much less time compared to traditional methods. Generally, absorption of the microwave energy increases with the dielectric constant of the molecule, resulting in power dissipation inside the solvent and plant materials and then generating more effective molecular movement and heating. As a polar solvent, water can efficiently absorb microwave energy and leads to efficient heating. MAE was used for extraction of interested components from a wide variety of sample matrices and has been used as a promising alternative sample preparation technique for a number of applications. Compared to conventional methods, MAE can considerably reduce both extraction time and solvent consumption.

Chlorogenic Acids from Flower Buds of Lonicera japonica *Thunb*

An efficient MAE technique has been developed to recover CGA from flower buds of *Lonicera japonica*, using a mixture of water and ethanol (Gao *et al.*, 2003; Dan *et al.*, 2003). Ethanol (50%) as MAE solvent gave the best yield of CGA in 5 min at a temperature of 60°C. For MAE, the choice of extraction solvent takes into account not only its solubility for target component but also its ability to absorb microwave energy. The yield of CGA increased with the increase of material: solvent ratio (amount of plant material: volume of extraction solvent) up to 1:10 (w/v) and further increase has no effect on the yield (Criado *et al.*, 2004).

The yield of CGA increased with time up to 30 min at 40°C whereas the same yield of CGA is obtained within 5 min at 60°C under microwave extraction and further increase in temperature has no effect on the yield. The yield of CGA reached 6.14% at 60°C within 5 min, which was significantly higher than that at 40°C. There was no obvious difference on yield of CGA between 60°C and 80°C; therefore, it was reported that the MAE temperature of 60°C was suitable for CGA extraction from flower buds.

A conventional heat-reflux extraction of CGA from raw plant materials was carried out at 60°C. The yield of CGA reached 5.19% at an optimal ethanol concentration of 50% as extraction solvent within 5 min; further increase in the yield is slow and reached a maximum of 6% within 30 min. In comparison to the heat-flux extraction, the MAE showed obvious advantages in terms of short duration and high efficiency to extract CGA. This is mainly due to the fact that microwave energy is delivered efficiently to materials through molecular interaction with the electromagnetic field and offers a rapid transfer of energy to the extraction solvent and raw plant materials.

Control and extracted samples of raw materials were subjected to Scanning Electron Microscopy. The changes in the raw material during heat-reflux extraction were not considerably different from the untreated sample, and only few slight ruptures happened on its surface. The surface of the MAE sample was greatly destroyed because that microwave irradiation accelerated cell rupture by sudden temperature rise and internal pressure increase inside the cells of plant sample. During the rupture process, a rapid exudation of the chemical substance within the cells into the surrounding solvents took place.

Chlorogenic Acids from Coffee

Coffee, an aromatic, non-alcoholic brew is known for its stimulating and refreshing taste. Two species are of significant economic importance namely, Arabica (*Coffea arabica*) and Robusta (*Coffea canephora*) (Varnam and Sutherland, 1994). In recent years, due to the increasing interest in finding physiologically functional foodstuffs, the relationship between coffee and health has been extensively investigated (George *et al.*, 2008). Coffee is reported to exhibit a number of bioactivities, such as antioxidant (Ramalakshmi *et al.*, 2008, 2009, 2011), anticarcinogenic, and antimutagenic activity (Giovannucci, 1998).

Coffee is known to show a protective effect on cancer and other cardiovascular diseases due to its antioxidant activity. It also protects low-density-lipoproteins from oxidation. This protective effect is the result of the action of several polyphenolic constituents. The physical and chemical properties of individual phenolics, strongly affect their antioxidant activities. In addition, these molecules can have a synergistic or antagonistic effect when present in complex mixtures. Indeed, the polyphenol composition of the beverage varies in species of coffee and tea (Richelle, 2001). In the case of coffee, robusta exhibits a higher antioxidant activity than arabica, which could be due to the higher amount of CGA. Following light roasting, the antioxidant activities of both coffees decreased markedly but further roasting produced the heterocyclic compounds having the antioxidant activity, and as these polyphenolic rich beverages are often consumed with milk, the effect of milk on different beverages (coffee, cocoa, black tea) were evaluated and coffee was found to have high antioxidant activity. An understanding of the protective role of dietary antioxidants *in vivo* requires a better characterization of the polyphenol composition of the antioxidant matrix as well as quantitative data on their absorption, their tissue distribution, their metabolism, and their biological actions. Indeed, after consumption, polyphenols have to cross the intestinal wall but must also resist further catabolism. The metabolism of polyphenols involves two important organs—the liver, where biotransformation enzymes convert them or their metabolites into conjugated forms such as glucuronides or sulphates, and the colon where microorganisms degrade unabsorbed ones. At present, only a little information is available on the absorption of the vast diversity of polyphenols present in these beverages (Richelle, 2001). So the antioxidant activity of coffee has to be considered while estimating the daily ingested dose of polyphenols. Thus, the beneficial effects of coffee may be attributed in part to polyphenols and caffeine serving as antioxidants. These positive properties of coffee on human health are attributed to the presence of phenolic compounds such as CGA with strong antioxidant and radical-scavenging activities present in green coffee.

Earlier reports (Clifford and Wilson, 1985; Debabrata *et al.*, 2011) suggest that the content of CGA in *C. canephora* varies from 7 to 10% (mfb), whereas *C. arabica* varies from 5 to 7.5% (mfb). In addition to having different total CGA levels, the relative amounts of CQA, diCQA, and FQA have also been shown to vary in mature *C. canephora* and *C. arabica* grain. For example Ky *et al.* (2001) found that CQA, diCQA and FQA represent roughly 67, 20, and 13%, respectively, of the total CGA content of *C. canephora* variety *versus* 80, 15, and 5%, respectively, of the total CGA content of *C. arabica* variety. In addition to being found in coffee, these compounds are also found at significant levels in plant foods such as apples, pears, tomato, potato, and eggplant (Niggeweg *et al.*, 2004). Total CGA content of green and roasted beans of different varieties is provided in Table 70.1.

Extraction of Chlorogenic Acids using Microwaves

MAE showed obvious advantages in terms of short duration and high efficiency to recover CGA from raw plant materials in comparison with conventional heat-reflux extraction. The mechanism of the enhanced extraction by microwave assistance was studied by observing cell destruction of plant material after MAE treatment by scanning electron microscopy. The enhanced extraction was related partly to a greater extent of cell rupture of the plant materials, and this was observed by scanning electron microscopy. The plant materials were significantly destroyed due to the cell rupture during MAE treatment. In this context, conditions for microwave-assisted extraction of green coffee beans were optimized and the quality of the extracts was monitored through determination of the contents of CGA, caffeine, and total polyphenols as well as evaluating the radical scavenging activity. A comparative analysis between MAE and conventional heat reflux methods was also made.

TABLE 70.1 Chlorogenic Acid Content in Green and Roasted Coffee Varieties

Coffee Variety	Bean Variety	Chlorogenic Acid (mg/g)
Coffea arabica	Dry green coffee beans	68–70
	Medium roasted	26–30
	Dark roasted	22–25
	Very dark roasted	7–9
Coffea canephora	Dry green coffee beans	85–88
	Medium roasted	35–37
	Dark roasted	20–22
	Very dark roasted	6–7

EXTRACTION METHODOLOGY

Robusta cherry green coffee beans (Figure 70.2) were ground and sieved (<720 μm) using Hammer mill and then defatted. Defatted coffee powder was extracted with different solvents such as water, methanol, and ethanol using Microwave lab station [(Model: STARTS configuration with control terminal 260, Milestone, Italy); built-in focused IR sensor and magnetic stirrer; Magnetron: SN: 133613; Frequency 50 Hz] in a closed system under different sets of conditions of temperature and wattage for specific periods of time. The slurry obtained was filtered to get a clear extract that was used for quantitative analysis. The quantitative analyses of extracts for percentage of yield, caffeine, CGA, and total polyphenols (TPP) contents, and radical scavenging activity (RSA) of MAE-treated defatted ground robusta cherry coffee were presented. The caffeine, CGA, RSA, and TPP content were analyzed spectrophotometrically at λ_{max}275, 325, 517, and 760 nm, respectively. The percentage yield of analyte under different conditions/parameters of MAE, i.e., time, microwave wattage, and temperature, are presented in the Table 70.2.

YIELD OF MAE EXTRACTS

The yields are in the range of 15–20% with water as solvent of extraction and found to be higher than alcohol

FIGURE 70.2 Robusta coffee berries (*Coffea robusta*).

TABLE 70.2 Microwave-Assisted Extraction of Green Coffee using Aqueous Medium

Variable Parameter	Constant Parameter	Variables	Caffeine (%)	Chlorogenic acids (%)	Yield (%)
Time	Temperature (50°C) Wattage (800 W)	2 min	3.75 ± 0.07	6.05 ± 0.07	17.15 ± 0.07
		5 min	4.60 ± 0.14	6.75 ± 0.07	17.50 ± 0.14
		10 min	4.40 ± 0.14	6.35 ± 0.07	17.05 ± 0.07
Wattage	Time (5 min) Temperature (50°C)	400 W	3.70 ± 0.14	4.70 ± 0.14	15.40 ± 0.14
		600 W	4.25 ± 0.07	5.80 ± 0.14	17.75 ± 0.07
		800 W	7.25 ± 0.07	8.40 ± 0.28	18.10 ± 0.14
Temperature	Time (5 min) Wattage (800 W)	30°C	6.68 ± 0.07	6.65 ± 0.07	17.60 ± 0.07
		50°C	7.25 ± 0.07	8.10 ± 0.04	18.05 ± 0.07
		70°C	4.75 ± 0.07	6.75 ± 0.07	17.60 ± 0.14
		90°C	4.65 ± 0.07	6.35 ± 0.07	16.75 ± 0.07

Rohit et al. *(2012).*

extracts (8.5–9.5%), indicating better extraction efficiency in water than in organic solvents. The reason could be that the dielectric constant of water is higher than hat of the alcohol, which helps in absorption of microwaves. It was found that the yield was highest (18%) with water as solvent for extraction under MAE conditions of 5 min, 800 W, and 50°C compared to other conditions. Chen and Spiro (1995) reported that the yield of extracts from rosemary leaves improved with an increase in microwave power and the induction stage appeared. However, Chemat *et al.* (2005) confirmed that there is not much difference in the extraction yield when the microwave power was increased from 50 to 150 W during the extraction of caraway seeds. The difference may be due to the nature of the plant material and the components to be extracted.

CHLOROGENIC ACIDS IN EXTRACT

The percentage yields of CGA in the aqueous extracts obtained at different MAE parameters/conditions are presented in Table 70.2. The yield of CGA is in the range of 4.7–8.1% in aqueous extractions, whereas it is in the range of 4.9–5.6% in alcohols. The yield of CGA increases up to 5 min. The yield of CGA increases with increased wattage up to 800 W; experiments could not be carried out beyond 800 W, due to limitations with the equipment. Extraction yield of CGA is increased along with temperature up to 50°C, and decreased at higher temperatures, which could be due to degradation at higher temperatures. The highest yield of CGA (8.4 ± 0.28%) was obtained at MAE conditions of 5 min, 800 W and 50°C and water as solvent of extraction.

CAFFEINE IN EXTRACT

The caffeine content in the aqueous extracts obtained at MAE times of 2, 5, and 10 min were 3.75 ± 0.07%, 4.6 ± 0.14%, and 4.4 ± 0.14%, respectively. The yield of caffeine did not increase after 5 min. With MAE wattage of 400, 600, and 800 W, the yields in aqueous extracts were 3.7 ± 0.14%, 4.25 ± 0.07%, and 7.25 ± 0.07%, respectively. This indicates that the extraction of caffeine increases with increase in wattage. Experiments could not be carried out beyond 800 W, due to limitations with the equipment. At MAE temperature of 30, 50, 70, and 90°C, the yields obtained were 6.35 ± 0.07%, 7.25 ± 0.07%, 4.75 ± 0.07, and 4.65 ± 0.07%, respectively. Extraction yield of caffeine is increased along with temperature up to 50°C, and decreased at higher temperatures; this may be due to the degradation of caffeine at higher temperatures. When ethanol and methanol were used as extraction solvents, the yield obtained in the extracts were 3.05 ± 0.07% and 3.75 ± 0.07% respectively at MAE conditions of 5 min, 800 W and 50°C and lower than water as solvent. The highest yield (7.25%) of caffeine obtained MAE conditions of 5 min, 800 W, and 50°C in aqueous extraction. The comparative analysis of percentage extraction yields of caffeine at different MAE parameters is shown in Table 70.2. Pan *et al.* (2003) conducted experiments of MAE for polyphenols and caffeine at the microwave power of 700 W from green tea leaves. It is reported that MAE for 4 min provided maximum extraction of caffeine and polyphenols from tea leaves.

TPP IN EXTRACT

The TPP content in coffee extract is measured in terms of gallic acid equivalents. During the estimation, Folin–Ciocalteu Reagent (FCR) was added to the extract. The FCR reacts with polyphenol compounds in the alkaline medium (sodium carbonate) and forms complexes. The various polyphenols that are present in the coffee are caffeic acid, protocatecuic acid, gallic acid, quinic acid, pyrogallol, etc., apart from CGA.

The total amounts of polyphenol content in extracts under different MAE conditions using different solvents are presented in Table 70.3. TPP content was observed in the range of 2.6–5.5 mg/g gallic acid equivalents in aqueous extractions, where as it is in the range of 0.95–1.5 mg/g gallic acid equivalents in alcohol extracts. TPP increased with time up to 5 min. TPP content is decreased along with increase in wattage as well as temperature. This may be due to degradation of other phenolic compounds at higher wattages as well as at higher temperatures. Also, the TPP content was found to be higher in methanol extracts compared to ethanol extracts indicating more solubility of polyphenols in methanol than ethanol. The greater polyphenols content will help to increase the antioxidant potential of coffee extracts.

TABLE 70.3 Microwave-Assisted Extraction of Green Coffee using Different Solvents

Solvent	Extraction Conditions	Caffeine (%)	Chlorogenic Acids (%)	Yield (%)
Ethanol	Time (5 min)	3.05 ± 0.07	4.95 ± 0.07	8.95 ± 0.07
Methanol	Temperature (50°C)	3.44 ± 0.07	5.60 ± 0.14	9.05 ± 0.07
Water	Wattage (800 W)	7.25 ± 0.07	8.40 ± 0.28	18.10 ± 0.14

Rohit et al. *(2012).*

RADICAL SCAVENGING ACTIVITY OF COFFEE EXTRACTS

RSA in coffee extract was measured in terms of percentage inhibition of DPPH free radicals. RSA of different extracts under different MAE conditions using different solvents are presented in the Table 70.4. RSA in extracts were found to be in the range of 75–81% even at 25 ppm concentration. The RSA in water extracts was higher compared to ethanol (38%) and methanol (45%) at 25 ppm concentration. It was also found to be higher in methanol extracts compared to ethanol extracts indicating better antioxidant activity in methanol extracts. The greater polyphenols content will help to increase the antioxidant potential of coffee extracts.

On the basis of results obtained from different sets of experiments conducted using MAE, it was observed that the following sets of conditions give the optimum extraction of CGA and caffeine from defatted grounded robusta cherry coffee samples. MAE conditions, to provide higher quantities of CGA and caffeine were as follows: Time: 5 min; microwave wattage: 800 W;

TABLE 70.4 TPP and Radical Scavenging Activity (RSA) of Extracts using Microwave-Assisted Extraction

Exp set	Variable Parameter	Variables	TPP (mg/g)	RSA (%) 25 ppm	RSA (%) 50 ppm
A	FOR AQUEOUS EXTRACT				
I	Time*	2 min	2.95 ± 0.07	77.15 ± 0.07	79.50 ± 0.14
		5 min	3.45 ± 0.07	78.55 ± 0.21	82.10 ± 0.14
		10 min	3.10 ± 0.14	77.80 ± 0.14	82.20 ± 0.14
II	Wattage**	400 W	3.50 ± 0.14	78.35 ± 0.21	81.55 ± 0.21
		600 W	2.60 ± 0.14	78.70 ± 0.14	81.95 ± 0.07
		800 W	2.20 ± 0.14	81.35 ± 0.21	83.10 ± 0
III	Temperature***	30°C	4.20 ± 0.14	78.50 ± 0	81.20 ± 0.14
		50°C	3.85 ± 0.07	79.60 ± 0.14	82.35 ± 0.21
		70°C	3.60 ± 0.14	77.60 ± 0.14	80.10 ± 0.40
		90°C	2.85 ± 0.07	75.10 ± 0.14	78.75 ± 0.07
B	FOR SOLVENT EXTRACT+				
IV	Ethanol		0.95 ± 0.07	38.5 ± 0.14	56.4 ± 0.14
V	Methanol		1.50 ± 0.14	45.65 ± 0.07	61.7 ± 0.14

#As gallic acid equivalents.
**Temperature, 50°C; wattage, 800 W.*
***Temperature, 50°C; time, 5 min*
****Wattage, 800 W; time, 5 min;*
+Temperature, 50°C; wattage, 800 W; time, 5 min.
Rohit et al. (2012).

TABLE 70.5 Comparative Analyses between Conventional Heat-Reflux Extraction and MAE

Method of Extraction	Parameters	%Yield Caffeine	Chlorogenic acids
Conventional heat reflux method of extraction	Time (5 min) Temperature (50°C) Sample : solvent (1 : 4)	3.05 ± 0.21	3.95 ± 0.21
Microwave-Assisted Extraction (MAE)	Time (5 min) Temperature (50°C) Wattage (800 W) Sample : solvent (1 : 4)	7.25 ± 0.07	8.4 ± 0.28

Rohit et al. *(2012).*

and Temperature: 50°C. The extraction efficiency in MAE was shown to be higher in water than in organic solvents, namely ethanol and methanol. This may due to the different dielectric constants of the respective solvents.

COMPARATIVE ANALYSIS OF MAE AND CONVENTIONAL EXTRACTS

The comparative analysis of extraction efficiency between conventional heat-reflux and MAE at optimal extraction conditions is presented in Table 70.5. It is evident that MAE has shown significantly higher and better extraction yield than conventional heat reflux for both CGA and caffeine at optimal extraction conditions.

References

Ali, Z., Hardiman, C., 2005. Electromagnetic radiation and human health: A review of sources and effects. High Freq. Electron. 16–26 July.

Bicchi, C.P., Binello, A.E., Pellegrino, G.M., Vanni, A.C., 1995. Characterization of Green and Roasted Coffees through the Chlorogenic Acid Fraction by HPLC-UV and Principal Component Analysis. J. Agri. Food Chem. 14, 1549.

Chemat, S., Amar, H.A., Lagha, A., Esveld, D.C., 2005. Microwave-assisted extraction kinetics of terpenes from caraway seeds. Chem. Eng. Proc. 44, 1320–1326.

Chen, S.S., Spiro, M., 1995. Kinetics of microwave extraction of rosemary leaves in hexane, ethanol and a hexane+ethanol mixture. Flavour Fragrance J. 10, 101–112.

Clifford, M.N., 1999. Chlorogenic acids and other cinnamates—nature, occurrence and dietary burden. J. Sci. Food Agri. 79, 362–372.

Clifford, M.N., Wilson, K.C., 1985. Coffee—Botany, Biochemistry and Production of Bean and Beverages. Croom Helm, London.

Criado, M.R., Torre, S.P., Pereiro, I.R., Torrijos, R.C., 2004. Optimization of a microwave-assisted derivatization-extraction procedure for the determination of chloro phenols in ash samples. J. Chromatogr. A 1024, 155–163.

Dan, L., Zhao, G.L., Liu, J.J., 2003. Study on extraction process of chlorogenic acid from *Flos lonicerae*. Nat. Prod. Res. Drug Dev. 15, 124–126.

Debabrata Patra, K., Ramalakshmi, L., Jagan Mohan Rao. Recent trends in coffee in Recent trends in soft beverages, 2011. Woodhead Publishing Limited, New Delhi, http://www.woodheadpublishing.com/en/book.

Farah, A., Perrone, D., Donangelo, C.M., de Paulis, T., Martin, P.R., 2008. Comprehensive analysis of major and minor chlorogenic acids and lactones in economically relevant Brazilian coffee cultivars. Food Chem. 106, 859–867.

Gao, J.R., Hu, Sun, J.S., 2003. Study on extraction process of chlorogenic acid from *Flos lonicerae*. J. China Agri. Univ. 8, 5–8.

Gao, M., Song, B.Z., Liu, C.Z., 2006. Dynamic microwave-assisted extraction of flavonoids from *Saussurea medusa* Maxim cultured cells. Biochem. Eng. J. 32, 79–83.

George, S.E., Ramalakshmi, K., Jagan Mohan Rao, L., 2008. A perception on health benefits of coffee. Crit. Rev. Food Sci. Nutr. 48, 264–286.

Giovannucci, E., 1998. Meta-analysis of coffee-consumption and risk of colorectal cancer. Am. J. Epidemiol. 147, 1043–1052.

Ky, C.L., Louarn, J., Dussert, S., Guyot, B., Hamon, S., Noirot, M., 2001. Caffeine, trigonelline, chlorogenic acids and sucrose diversity in wild Coffea arabica L. and C-canephora P. accessions. Food Chem. 75, 223–230.

Niggeweg, R., Michael, A.J., Martin, C., 2004. Engineering plants with increased levels of the antioxidant chlorogenic acid. Nat. Biotechnol. 22, 746–754.

Pan, X., Niu, G., Liu, H., 2003. Microwave-assisted extraction of tea polyphenols and tea caffeine from green tea leaves. Chem. Eng. Process. 42, 129–133.

Ramalakshmi, K., Jagan Mohan Rao, L., Yuko Takano-Ishikawa, Masao Goto, 2009. Bioactivities of low-grade green coffee and spent coffee in different *in vitro* model systems. Food Chem. 115, 79–85.

Ramalakshmi, K., Rahath Kubra, Jagan Mohan Rao, L., 2008. Antioxidant potential of low grade coffee beans. Food Res. Int. 41, 96–103.

Ramalakshmi, K., Hithamani, G., Asha, K.R., Jagan Mohan Rao, L., 2011. Separation and characterization of chlorogenic acid rich conserves from low grade coffee beans and their radical scavenging potential. Int. J. Food Sci. Technol. 46, 109–115.

Rehwald, A., Meier, M., Stiche, O., 1994. Qualitative and quantitative reversed-phase high-performance liquid chromatography of flavonoids in Passiflora incarnate L. Pharma. Acta Helvetiae 69, 153–158.

Richelle, M., 2001. Effect of various antioxidants in vegetable oils. J. Agri. Food Chem. 49 3438–3344.

Robyn, A.L., David, R.M., Jocelyn, A.L., Peter, W.F., 2000. Biological effects of electromagnetic fields—Mechanisms for the effects of pulsed microwave radiation on protein conformation. J. Theortical Biol. 206, 291–298.

Rohit, U., Jaganmohanrao, L., 2013. An outlook on chlorogenic acids—occurrence, chemistry, technology, and biological activities. Crit. Rev. Food Sci. Nutr. DOI: 10.1080/10408398.2011.576319.

Rohit, U., Ramalakshmi, K., Jaganmohanrao, L., 2012. Microwave-assisted extraction of chlorogenic acids from green coffee beans, Food Chem. 130, 184–188.

Rostagno, M.A., Palma, M., Barroso, C.G., 2007. Microwave assisted extraction of soyisoflavones. Anal. Chim. Acta 588, 274–282.

Susanne, N.M.H., 2012. Students' Conceptions about 'Radiation': Results from an Explorative Interview Study of 9th Grade Students. J. Sci. Educ. Technol. 21, 826–834.

Trugo, L.C., Macrae, R., 1984. A study of the effect of roasting on the chlorogenic acid composition of coffee using HPLC. Food Chem. 15, 219–227.

Varnam, A.H., Sutherland, J.P., 1994. Beverages: Technology, chemistry and microbiology. Boundary Row, London: Chapman and Hall. pp. 191–255.

Zhang, W.B., Xu, S.Y., 2007. Microwave-assisted extraction of secoisolariciresinol diglucoside from flaxseed hull. J. Sci. Food Agri. 87, 1455–1462.

Zhou, H.Y., Liu, C.Z., 2006. Rapid determination of solanesol in tobacco by high performance liquid chromatography with evaporative light scattering detection following microwave-assisted extraction. J. Chromatogr. B 835, 119–122.

CHAPTER

71

Tea Leaf Age, Shade and Characteristic Levels of L-Theanine, Caffeine, (−)-Epigallocatechin Gallate (EGCG), (−)-Epigallocatechin (EGC), (−)-Epicatechin (EC), and (−)-Epicatechin Gallate (ECG)

Kehau A. Hagiwara, Anthony D. Wright

Daniel K. Inouye College of Pharmacy, Department of Pharmaceutical Sciences, University of Hawaii at Hilo, Hawaii, USA

CHAPTER POINTS

- It is evident that growing conditions, including shade levels, can and do affect the chemical composition of tea leaves quite dramatically.
- In some cases change in composition of an individual component can be of the order of 6% or more of dry-leaf weight depending on the part of the plant examined.
- For fine teas, where utmost care is taken to hand select only the bud and first two leaves for production of the final product, this finding is undoubtedly significant as such increases can clearly affect the flavor of the resultant infused tea.
- So far there are no studies into the flavor of tea grown in such shaded conditions, so it is still undetermined as to whether shade-grown tea is of a better flavor quality than tea grown in full light, but there is clearly a difference.
- Probably the most significant findings to date from studies relating to shade, especially for growers establishing new crops, is the fact that productive plants grow successfully in shade levels of up to 60% and that the quality of the final product appears to be uncompromised in any way, and might actually be enhanced.
- Tea is unquestionably one of the most complex drinks of natural origin.
- The complexity of its flavor is infinitely variable, now potentially more so with the introduction of product grown in shade, a variable previously seldom considered important in tea production.

INTRODUCTION

One of the world's oldest non-alcoholic beverages, if not the oldest, is tea. After water, tea is the most popular drink in the world today. Tea drinking almost certainly has its roots in Chinese culture and is said to have been one of the seven basic daily necessities, the others being fuel, rice, oil, salt, soy sauce, and vinegar. Legend has it that the Chinese Emperor Shen Nong accidentally invented tea, the drink, in 2737 B.C. The emperor was a scholar and herbalist, as well as a creative scientist and patron of the arts. Among his beliefs, he held that drinking boiled water contributed to good health. By his decree, his subjects and servants had to boil their water before drinking it as a precaution against illness. On a summer day while visiting a distant region, he and his entourage stopped to rest. The servants began to boil water for the Emperor and his subjects to drink. Coincidentally, dried leaves from a nearby Camellia bush fell

Processing and Impact on Active Components in Food
http://dx.doi.org/10.1016/B978-0-12-404699-3.00071-8

into the boiling water. The emperor was interested in the new liquid because of its pleasing aroma. He drank the infusion and discovered that it was very refreshing and had a delightful flavor. He declared that tea gives vigor to the body; that was probably the moment tea, the drink, was invented. However, it seems likely that up until approximately 300 AD tea was considered only as a medicinal beverage rather than the refreshing drink we know today.

Production of tea begins with generation of the raw material, the tea leaves. The best tea leaves should be grown and harvested to produce high quality and, in consequence, high-value tea. For this reason, many tea farmers and producers have sought to optimize the quality of their teas by selecting prime growth parameters and treatments for the final product. The optimal environmental conditions for *Camellia sinensis* cultivation are areas with tropical to subtropical climates, high rainfall, slightly acidic soil and good drainage (Chan *et al.*, 2007). For more than a thousand years, tea has been grown with the notion that the best tea grows in forests, on sunbathed hills with canopy cover (Song *et al.*, 2012). However, tea is almost exclusively grown on farms at high elevation without any shade. Regardless of the environmental conditions involved in the cultivation of the tea-plants, the most prized parts of the plant for tea production are the buds and the first two leaves.

In order for collected leaves to become tea, they must undergo a certain level of processing. Almost all true teas are derived from the same plant species, whether it is white, green, oolong, or black tea. The differences between the types of tea are a result of the type and level of processing that the leaves undergo (Figure 71.1). White teas are composed of tea leaves that have been withered or steamed and dried. Green teas are generally withered, then steamed or fried, then rolled or dried. Oolong teas undergo a bruising step after the leaves have been withered. The leaves are then fermented for a short while and are subsequently pan-fried and dried. Black teas are processed via the same method as oolong teas, the only difference is that black teas are allowed to fully ferment or oxidize, whereas the oolong teas are limited in this step. There are variations in the length of time for each step, additional steps, and, or other details that contribute to the processing of tea that modify the chemical constituents of the leaves and consequently, the aroma and taste of the final product. The flavor of tea, or more correctly an aqueous extract of leaves thereof, is as variable and complex as that of wine, and derives from the complex mixture of primary and secondary metabolites that all play a role in the unique and varied flavor of this highly diverse drink.

HOW COMPOSITION IS ALTERED

When considering the areas of the world that grow and produce tea, it is evident that there is little or no consideration given to the role that shade may play in assisting or hindering the production of key chemicals produced by the leaves harvested and used in production of the final product, dried and/or fermented tea. This is quite surprising when the amount (millions of tons) and commercial value (hundreds of millions of dollars) of tea produced each year is considered. In general, tea is grown at relatively high altitudes in plantations that offer no shade to the plants, like the one shown in Figure 71.2. The reasons for this mode of production are probably historical and until recently no apparent thought was given to the production of the crop in anything other than this fully exposed format. With the advent of tea production in Hawaii, questions were

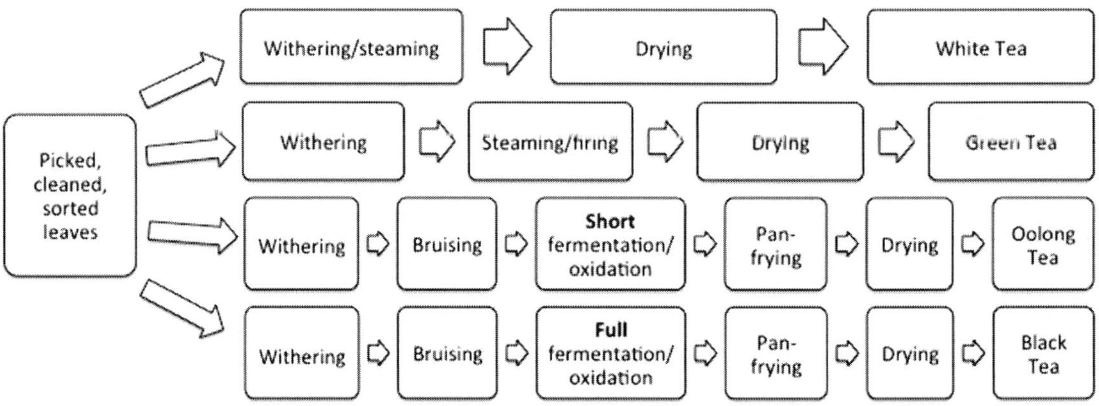

FIGURE 71.1 A schematic of general processes for production of white, green, oolong and black teas. Specific tea processes can include different time periods for each step and, or incorporate additional steps for differentiating the final product.

asked by many new producers of tea relating to whether or not it was necessary to clear all of the vegetation, particularly trees, from the areas where they wanted to establish their new crop. Such questions were posed for two main reasons, one economic, the cost of clearing land prior to establishment of a plantation, and the other environmental, complete removal of all potentially interfering plants may/will dramatically affect the microbial make up of the soil and in so doing reduce the value of the land for future production for many years to come, particularly for growers who wished to use little or no extraneous chemicals in production of their product.

FIGURE 71.2 A fully exposed tea plantation on the Big Island of Hawaii. Approximately 3 km outside of Kurtistown on Volcano Road towards the town of Volcano, elevation approximately 300 m.

Some growers actually believed there was no need to clear-fell the areas where they wanted to plant their crop and so planned their plantations accordingly.

With these issues in mind, a study was undertaken to determine if shade did in any way influence the production of six well-known natural products produced by teas (Song *et al.*, 2012). The constituents investigated were the four major catechins; (−)-epigallocatechin gallate (EGCG, **1**), (−)-epigallocatechin (EGC, **2**), (−)-epicatechin (EC, **3**), and (−)-epicatechin gallate (ECG, **4**), caffeine (**5**), and L-theanine (**6**) (Figure 71.3; note that bold numbers throughout the text from this point refer to the structures shown in Figure 71.3), all of which are reported to affect the flavor and health giving properties of tea (Morley *et al.*, 2005; Chan *et al.*, 2007; Gradisar *et al.*, 2007). One of the catechins, EGCG, is often reported as the most abundant compound of this class found in tea leaves (Dalluge *et al.*, 1998; Lu *et al.*, 2009). Caffeine (**5**) and L-theanine (**6**), one of the biosynthetic precursors of the catechins such as EGCG (Kito *et al.*, 1968), are two other naturally occurring substances also reported to have positive health effects. Caffeine is probably best known for its ability to stimulate the central nervous system (Lovett, 2005), while L-theanine is an amino acid that has been found to reduce mental and physical stress (Kimura *et al.*, 2007), to have potential as an immune system stimulant by being able to boost production of anti-bacterial proteins (Kamath *et al.*, 2003), and also be one of the components responsible for the characteristic taste of green tea (Lee *et al.*, 2010). Recent studies of tea leaves have shown there to be a relationship between leaf age and concentrations of natural constituents, based on young leaves being

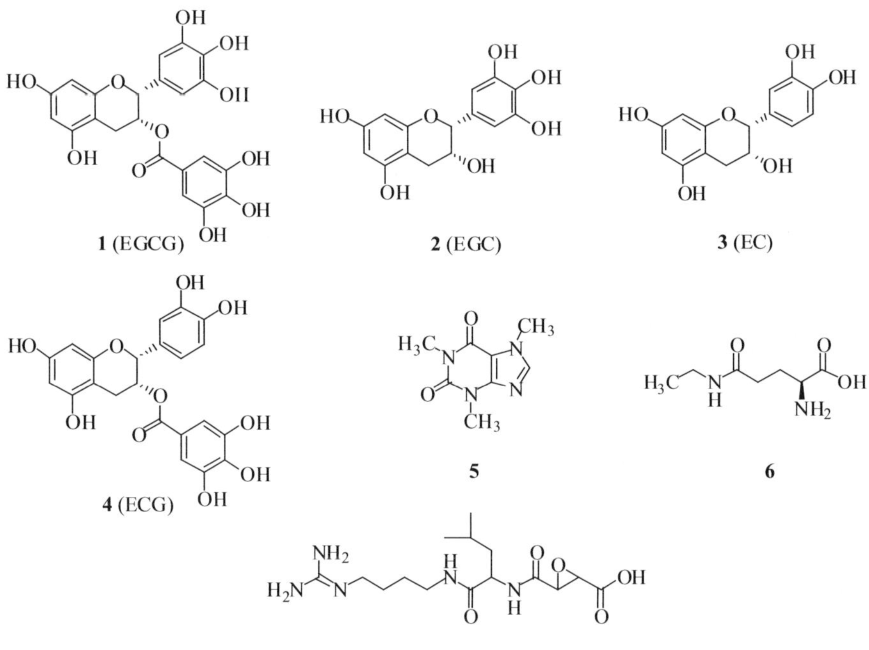

FIGURE 71.3 Chemical structures of (−)-epigallocatechin gallate (EGCG), (−)-epigallocatechin (EGC, **2**), (−)-epicatechin (EC, **3**), and (−)-epicatechin gallate (ECG, **4**), and caffeine (**5**), L-theanine (**6**) and *trans*-epoxysuccinyl-L-leucylamido-(4-guanidino)-butane (E-64, **7**).

defined as the bud and the two leaves underneath the bud (Figure 71.4), as compared to other leaves that are considered to be old leaves (Lin *et al.*, 1996; Santana-Rios *et al.*, 2001; Chiu and Lin, 2005; Lu *et al.*, 2009). Three of these studies showed that young leaves contain more antioxidants than older leaves (Lin *et al.*, 1996; Santana-Rios *et al.*, 2001; Chiu and Lin, 2005).

Numerous studies have been conducted to determine how levels of the catechin polyphenols in tea vary according to position from where leaves are picked, variety, agricultural conditions, and processing (Lee *et al.*, 2011; Khokhar and Magnusdottir, 2002; Lee and Ong, 2000; Singh *et al.*, 1999). However, no similar studies that evaluate changes in the levels of polyphenols, caffeine, and L-theanine using a single analytical method have been undertaken until recently (Song *et al.*, 2012). In this study, employing a high performance liquid chromatography (HPLC) method to investigate young, freshly picked, minimally processed tea leaves, the concentrations of the most abundant polyphenols, EGCG, EGC, EC, and ECG (**1–4**), as well as caffeine (**5**) and L-theanine (**6**) in extracts of the bud and the first and second leaves of selected tea-plants (Figure 71.4) were determined and correlated to shade levels to determine if shade did indeed influence levels of the natural products. The antioxidant potential of leaf extracts was also determined employing the ferric reducing antioxidant power (FRAP) assay (Benzie and Strain, 1999) to see if this variable, often associated with health related properties of food, could also be correlated with leaf age.

For the actual study, fresh leaves were collected from "Mauna Kea Tea", a tea plantation on the Big Island of Hawaii. The plantation has an average elevation of about 590 m, with an atmospheric temperature typically between 10 and 25°C, and an annual rainfall of 196 to 245 mm, the majority falling between December and April. The selected study area had an approximate area of 300 m^2, and an apparent northeast-facing slope. The plantation was mostly open to the sky, except on the northwest and southwest sides where Ohia lehua (*Metrosideros polymorpha*) trees provided varying degrees of shade. Depending on the season, the northwestern tree canopy could block direct afternoon sunlight.

The results of HPLC analyses of extracts of the buds and first two leaves (Table 71.1 and Figure 71.5A), clearly showed the levels of each of the leaf components examined, **1–6**, changed in a statistically significant way on moving from the bud to the first and second leaf and also with respect to the level of shade (not all results shown). In particular, the levels of the four catechins (**1–4**) increased in going from bud to the first and second leaf while caffeine (**5**) and L-theanine (**6**) levels decreased. The change in the level of EGC was most dramatic; increasing by at least 6% of dry leaf weight in most cases. Changes for all other metabolites were all less than 2% in comparing the bud to the second leaf.

Determination of the FRAP activity of extracts of the bud to the first and second leaves (Figure 7.5B), showed the antioxidant power to increase in going from the bud to the first and second leaf. This finding correlated well with the trend observed for the four catechin compounds (**1–4**), all well known as antioxidants (Xu *et al.*, 2004).

Aside from the fact that shade was shown to correlate statistically with all of the discussed variables, its true effect is likely to be more complex and needs further, more detailed, investigation. However, it is apparent that shade levels up to 60% do not negatively impact plant growth or secondary metabolite production. Therefore, plantations can be established without the need for clear felling of vegetation prior to

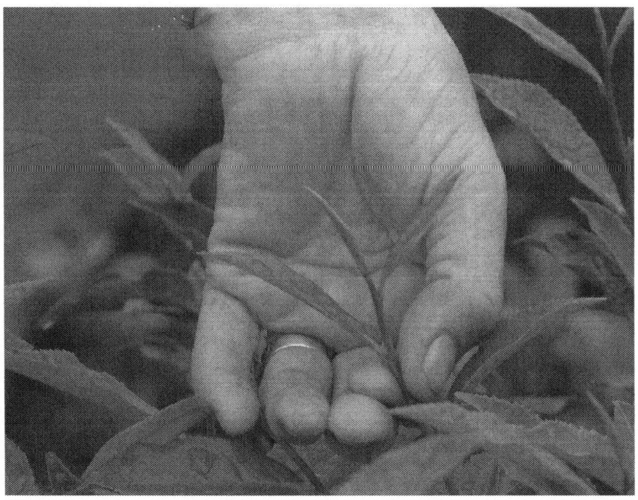

FIGURE 71.4 Picking of tea, bud and first two leaves. *Picture compliments of Cam Muir and Eliah Halpenny of "Big Island Tea."*

TABLE 71.1 Dry Tea Leaf (bud, first and second leaf) Levels of EGCG (**1**)[a], EGC (**2**)[b], EC (**3**)[c], ECG (**4**)[d], Caffeine (**5**) and L-theanine (**6**), as Mean % of Dry Weight (±SD[e]) of Methanol : Water (70 : 30) Extracts

Tea Compound	Bud	First Leaf	Second Leaf
EGCG	6.33 ± 1.34	7.97 ± 1.57	7.68 ± 1.59
EGC	0.78 ± 0.24	4.02 ± 1.10	7.02 ± 1.02
EC	0.45 ± 0.22	0.59 ± 0.13	0.83 ± 0.12
ECG	0.2 ± 0.09	0.34 ± 0.06	0.41 ± 0.08
Caffeine	2.94 ± 0.51	2.80 ± 0.41	1.98 ± 0.37
L-Theanine	0.74 ± 0.18	0.52 ± 0.15	0.38 ± 0.11

[a]*EGCG, Epigallocatechin gallate;*
[b]*EGC, Epigallocatechin;*
[c]*EC, Epicatechin;*
[d]*ECG, Epicatechin gallate;*
[e]*SD, Standard deviation.*
All data extracted from Song et al. *(2012).*

planting. The effects on flavor of teas produced in areas of shade still remains to be assessed. However, if sales of Hawaii-grown teas are any indicator it seems as if shade may have a positive influence.

During the antioxidant (FRAP) assays part of the investigation described previously, it became evident that one of the reference standards (EGC, **2**) (Figure 71.3) did not have some chemical properties consistent with it being the suggested material, as it had zero antioxidant activity. Spectroscopic examination of this material employing a variety of techniques, primarily ^{1}H-NMR spectroscopy (Figures 71.6A and B), showed the reference standard to be a completely other class of material altogether (E64, **7**, Figure 71.3). This material, although being quite different in chemical structure from EGC happened to have the same HPLC behavior in the system used for the analyses. This finding was a disconcerting but important finding as it led to adoption of strict quality and identity control of all externally acquired chemicals prior to their use (Kelman and Wright, 2012).

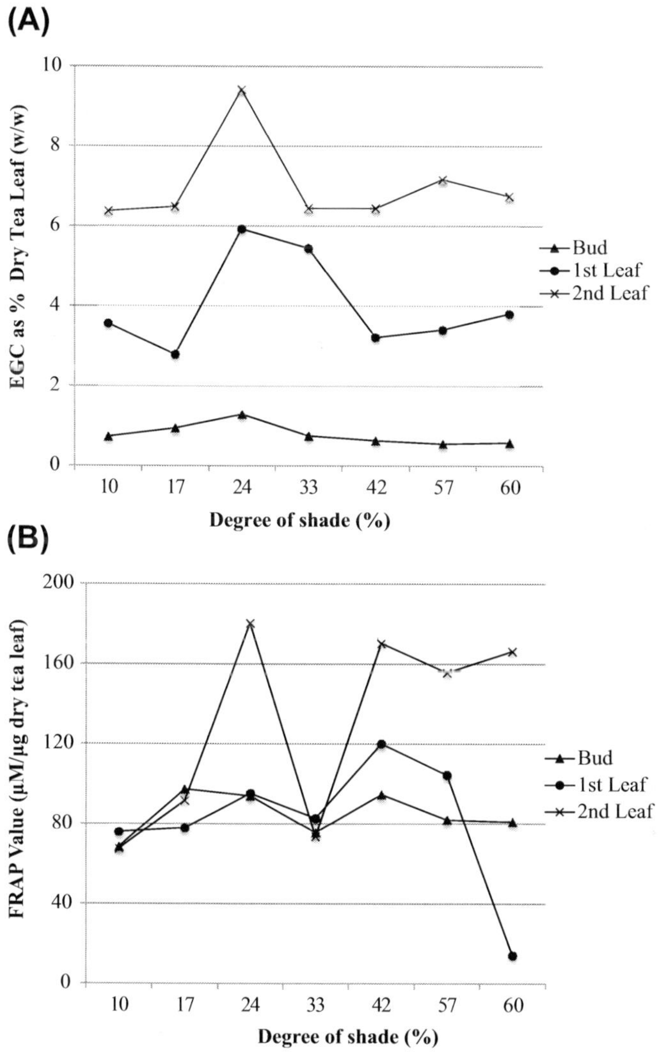

FIGURE 71.5 (A) Epigallocatechin (EGC, **2**) levels as % of dry weight of methanol : water (70 : 30) extracts from tea plant parts (leaves) of various ages (bud, first and second leaf) and degree of shade. (B) Total antioxidant capacity of methanol : water (70 : 30) extracts from tea plant parts (leaves) of various ages (bud, first and second leaf) and degree of shade, presented as FRAP value in µM/µg dry tea leaf. *Data extracted from Song* et al., *2012.*

OTHER WAYS IN WHICH COMPOSITION IS ALTERED

Tea is such a complex material that post-harvest activities are critical to the production of what could ultimately be a fine or merely an ordinary product. The best starting materials, although the ideal starting point, are no guarantee of a market-accepted fine product. Most tea in the world is fully oxidized (withered, rolled, fully fermented, dried/fried) or black tea. This is probably because it is the simplest and least expensive to produce, the whole process typically being fully mechanized.

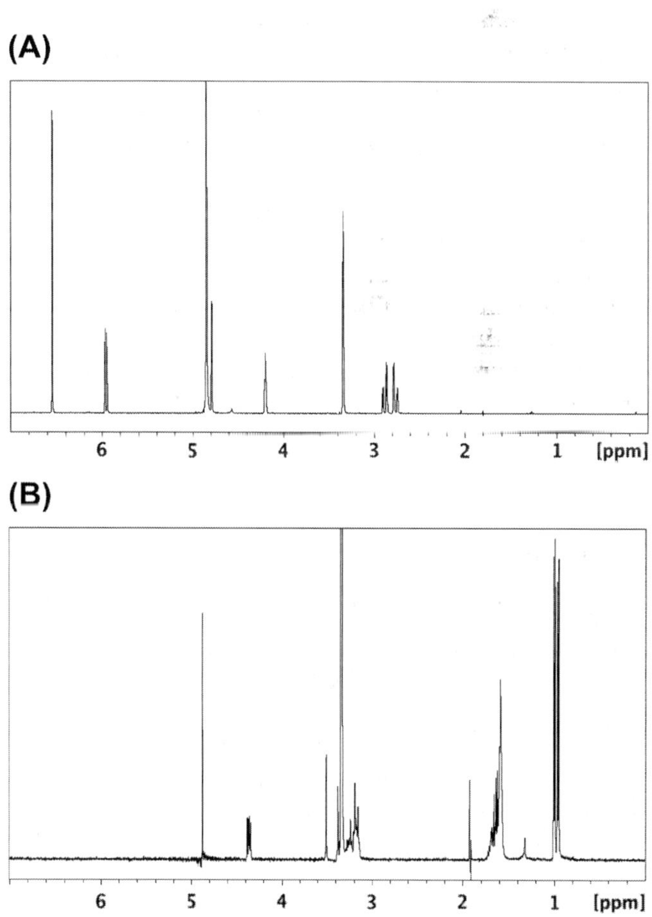

FIGURE 71.6 (A) ^{1}H-NMR (Proton nuclear magnetic resonance) spectrum of epigallocatechin (EGC, **2**), recorded at 400 MHz in CD_3OD. (B) ^{1}H-NMR (Proton nuclear magnetic resonance) spectrum of supposed epigallocatechin (EGC, **2**) but actually *trans*-epoxysuccinyl-L-leucylamido-(4-guanidino)-butane (E-64, **7**), recorded at 400 MHz in CD_3OD.

The more highly prized teas tend to be those where leaf selection has been more careful than for black tea and that vary in their degree of processing from minimally processed (white tea—steamed then dried), onwards to green (withered, steamed, fried, rolled, fried, dried), and oolong (withered, rolled, briefly fermented, fried, dried). For each of these types of tea, black included, the chemical content of the starting material is critical in determining final flavor, and each of the steps in the process modifies the chemistry within the processed leaf and hence the final flavor.

ANALYTICAL TECHNIQUES

Tea leaf samples discussed in this review were made in triplicate and represent buds, first, and second leaves. Shade level (canopy closure) was determined with a concave densitometer. Samples were collected from plants ranging in height from 0.8–1.0 m. Samples were inspected and only ones undamaged by pests, and composed of a complete bud and the first two fully expanded leaves underneath were used for further investigation (Figure 71.4). After microwave treatment to halt all enzymatic processes within the leaves they were dried at 85°C in a laboratory oven. Tea produced in this way is considered as minimally processed or white tea. After drying samples were powdered, weighed and extracted with a methanol water mixture (70:30 v/v). Resultant samples were filtered and analyzed by HPLC according to a published method (Song *et al.*, 2012). FRAP assays were also undertaken according to a published method (Benzie and Strain 1999). All samples were strictly validated prior to use employing a variety of analytical techniques including LC, LC-MS, IR, and most importantly ^{1}H-NMR (Kelman and Wright, 2012).

References

Benzie, I.F.F., Strain, J.J., 1999. Ferric reducing antioxidant power assay: direct measure of total antioxidant activity of biological fluids and modified version for simultaneous measurement of total antioxidant power and ascorbic acid concentration. Methods Enzymol. 299, 15–27.

Chan, E.W.C., Lim, Y.Y., Chew, Y.L., 2007. Antioxidant activity of *Camellia sinensis* leaves and tea from a lowland plantation in Malaysia. Food Chem. 102, 1214–1222.

Chiu, F.L., Lin, J.K., 2005. HPLC analysis of naturally occurring methylated catechins, 3″- and 4″-methyl-epigallocatechin gallate, in various fresh tea leaves and commercial teas and their potent inhibitory effects on inducible nitric oxide synthase in macrophages. J. Agric. Food Chem. 53, 7035–7042.

Dalluge, J.J., Nelson, B.C., Thomas, J.B., Sander, L.C., 1998. Selection of column and gradient elution system for the separation of catechins in green tea using high-performance liquid chromatography. J. Chromatogr. A 793, 265–274.

Gradisar, H., Pristovsek, P., Plaper, A., Jerala, R., 2007. Green tea catechins inhibit bacterial DNA gyrase by interaction with its ATP binding site. J. Med. Chem. 50, 264–271.

Kamath, A.B., Wang, L., Das, H., Li, L., Reinhold, V.N., Bukowski, J.F., 2003. Antigens in tea-beverage prime human Vγ2Vδ2 T cells *in vitro* and *in vivo* for memory and nonmemory antibacterial cytokine responses. Proc. Nat. Acad. Sci. USA 100, 6009–6014.

Kelman, D., Wright, A.D., 2012. The importance of ^{1}H-nuclear magnetic resonance spectroscopy for reference standard validation in analytical sciences. PLoS ONE 7 (7), e42061.

Khokhar, S., Magnusdottir, S.G.M., 2002. Total Phenol, Catechin, and Caffeine Contents of Teas Commonly Consumed in the United Kingdom. J. Agric. Food Chem. 50, 565–570.

Kimura, K., Ozeki, M., Juneja, L., Ohira, H., 2007. L-Theanine reduces psychological and physiological stress responses. Biol. Psychol. 74, 39–45.

Kito, M., Kokura, H., Izaki, J., Sasaoka, K., 1968. Theanine, a precursor of the phloroglucinol nucleus of catechins in tea plants. Phytochem. 7, 599–603.

Lee, J.-E., Lee, B.-J., Hwang, J.-A., Ko, K.-S., Chung, J.-O., Kim, E.-H., Lee, S.-J., Hong, Y.-S., 2011. Metabolic Dependence of green tea on plucking positions revisited: a metabolomic study. J. Agric. Food Chem. 59, 10579–10585.

Lee, B.-L., Ong, C.N., 2000. Comparative analysis of tea catechins and theaflavins by high-performance liquid chromatography and capillary electrophoresis. J. Chromatogr. A. 881, 439–447.

Lee, J.-E., Lee, B.-J., Chung, J.-O., Hwang, J.-A., Lee, S.-J., Lee, C.-H., Hongj, Y.-S., 2010. Geographical and climatic dependencies of green tea (*Camellia sinensis*) metabolites: A ^{1}H NMR-based metabolomics study. J. Agric. Food Chem. 58, 10582–10589.

Lin, Y.L., Juan, I.M., Chen, Y.L., Liang, Y.C., Lin, J.K., 1996. Composition of polyphenols in fresh tea leaves and associations of their oxygen-radical-absorbing capacity with antiproliferative actions in fibroblast cells. J. Agric. Food Chem. 44, 1387–1394.

Lovett, R., 2005. Coffee: The demon drink? New Sci. 2518, 38–41.

Lu, J., Wang, D., Shi, X., Yang, D., Zheng, X., Ye, C., 2009. Determination of purine alkaloids and catechins in different parts of *Camellia assamica* var. *kucha* by HPLC-DAD/ESI-MS/MS. J. Sci. Food Agric. 89, 2024–2029.

Morley, N., Clifford, T., Salter, L., Campbell, S., Gould, D., Curnow, A., 2005. The green tea polyphenol (−)-epigallocatechin gallate and green tea can protect human cellular DNA from ultraviolet and visible radiation-induced damage. Photodermat. Photoimmuno. Photomed. 21, 15–22.

Santana-Rios, G., Orner, G.A., Amantana, A., Provost, C., Wu, S.Y., Dashwood, R.H., 2001. Potent antimutagenic activity of white tea in comparison with green tea in the Salmonella assay. Mut. Res. 495, 61–74.

Singh, H.P., Ravindranath, S.D., Singh, C., 1999. Analysis of tea shoot catechins: spectrophotometric quantitation and selective visualization on two-dimensional paper chromatograms using diazotized sulfanilamide. J. Agric. Food Chem. 47, 1041 1045.

Song, R., Kelman, D., Johns, K.L., Wright, A.D., 2012. Correlation between leaf age, shade levels, and characteristic beneficial natural constituents of tea (*Camellia sinensis*) grown in Hawaii. Food Chem. 133, 707–714.

Xu, J.Z., Yeung, S.Y.V., Chang, Q., Huang, Y., Chen, Z.-Y., 2004. Comparison of antioxidant activity and bioavailability of tea epicatechins with their epimers. Br. J. Nutr. 91, 873–881.

CHAPTER

72

The Impact of Variety, Environment and Agricultural Practices on Catechins and Caffeine in Plucked Tea Leaves

Hao Cheng, Kang Wei, Liyuan Wang

Tea Research Institute Chinese Academy of Agricultural Sciences (TRICAAS), Zhejiang, China

CHAPTER POINTS

- Recent studies demonstrated that there were significant effects of variety, environment, and agricultural practices on catechins and caffeine in plucked tea leaves.
- Assam varieties generally have higher catechins and caffeine contents than China and Cambod varieties.
- Climate factors are closely correlated with the variation of catechins and caffeine over locations and seasons.
- Suitable nitrogen and high amount of phosphorus and potassium facilitate the accumulation of catechins.
- Young tea leaves generally accumulate more catechins and caffeine.
- Other factors, including altitude, plucking frequency, shade treatment, and nutrient elements such as sulfur, magnesium, and boron should also be taken into account in tea production.

INTRODUCTION

Tea is one of the most widely consumed beverages in the world. Generally there are three categories of tea on the basis of processing procedures, namely green tea (non-fermented), oolong tea (semi-fermented), and black tea (fully fermented). Consumer preference for tea is not only attributed to its attractive flavor and pleasant taste, but also for many health benefits, such as lowering cholesterol levels, anti-obesity and anti-carcinogenic functions (Chen *et al.*, 2008). These beneficial effects are mainly associated with its chemical composition, such as catechins and purine alkaloids (mainly caffeine). Catechins, which account for 6–16% of dry tea leaves, are the most researched chemicals for their health potential. The major catechins in tea include (−)-Epigallocatechin gallate (EGCG), (−)-epigallocatechin (EGC), (−)-epicatechin gallate (ECG), and (−)-epicatechin (EC) (Liang *et al.*, 2001). Among them, EGCG is generally the major constituent, while the content of EC is relatively low. Most tea catechins are oxidized and dimerized during fermentation process to form theaflavins, which contribute to the bright orange red color, astringency, and briskness in oolong and black tea; while green tea catechins remain in their original forms due to an initial heating process. Therefore, the content and composition of catechins in fresh tea leaves are of quality importance for all types of tea. Meanwhile, purine alkaloids, mainly caffeine, are also important quality components (Wang *et al.*, 2011). Caffeine can stimulate the central nervous system and enhance mental and physical processes in the human body. It also has a bitter taste and contributes significantly to the briskness of tea. It was found that tea plants produce a high amount of caffeine, with an average level of 3% of dry tea leaves (Graham, 1992). Other purine alkaloids include methylxanthines, theobromine, and theophylline in very small amounts.

There are many factors influencing catechin and caffeine content in plucked tea leaves. These include tea variety, climate, field condition, growing season, leaf age, and agricultural practices. To help meet the

Processing and Impact on Active Components in Food
http://dx.doi.org/10.1016/B978-0-12-404699-3.00072-X

TABLE 72.1 Summary of Recently Reported Variations of Catechins and Purine Alkaloids in Different Tea Genotypes

EGC	EC	EGCG	ECG	Total Catechins	Caffeine	Theobromine	Number of Genotypes	Reference
(mg/g)								
—	—	—	—	93.7–268.2	16.4–54.6	—	About 1500	Takeda (1994)
12.7–27.3	12.1–21.7	95.1–138.6	8.8–20.9	138.3–203.9	—	—	27	Saravanan *et al.* (2005)
—	—	—	—	81.9–262.7	12–59	—	596	Chen and Zhou (2005)
9.4–52.5	1.9–10.4	67.7–219.8	5.9–36.3	124.6–296.6	18.4–31.6	0–1.6	22	Wang *et al.* (2008)
15.8–17.5	5.7–6.6	45.8–55.6	16.6–20.7	103.2–108.9	—	—	3	Wei *et al.* (2011)
16.5–63.1	5.7–21.2	76.7–163.6	20.2–62.6	141.7–274.0	—	—	30	Sabhapondit *et al.* (2012)

The table shows the variations of catechin components, total catechins, caffeine and theobromine among different genetic resources (including varieties) in previous reports. EC, (–)-epicatechin; ECG, (–)-epicatechin gallate; EGC, (–)-epigallocatechin; EGCG, (–)-Epigallocatechin gallate.

requirements of consumers and producers of dietary tea products, understanding their impacts on catechins and purine alkaloids is a major goal in recent tea study.

VARIETAL EFFECT

Tea variety has a large effect on catechins and purine alkaloids. The variations of catechin components and caffeine among different genetic resources (including varieties) in previous reports are summarized in Table 72.1. Taxonomically, tea plants could be classified into Assam, China, and Cambod varieties according to their morphological characteristics. Sabhapondit *et al.* (2012) found Assam varieties had the highest level of catechins (on average 231±7 mg/g) as compared with Cambod (202±5 mg/g) and China varieties (157±4 mg/g). Moreover, Assam teas generally had higher caffeine content than China teas. As high catechin and caffeine contents in green leaves are considered to be correlated with high quality potential of black tea, Assam varieties are widely used to produce black tea (Obanda *et al.*, 1997). In terms of individual catechins, it is well established that the formation of dihydroxylated (EC and ECG) to trihydroxylated catechins (EGC and EGCG) is under genetic control (Marles *et al.*, 2003). Saravanan *et al.* (2005) reported the ratio of dihydroxylated to trihydroxylated catechins in Assam teas was 1:4, while that in China teas was 1:5. Furthermore, high EGC content was found to be a characteristic of Assam teas, suggesting tea variety also affects catechin composition (Sabhapondit *et al.*, 2012).

On the other hand, chlorophylls seem to be important internal factors affecting catechin content. Wei *et al.* (2012) reported albino tea cultivar 'Anji baicha' had lower catechin and caffeine contents than normal green tea cultivar 'Lonjing 43,' irrespective of growth region. Moreover, correlation analysis showed chlorophyll contents were significantly and negatively correlated with (+)-catechin (C), EGC, and EGCG, indicating chlorophylls might be involved in the regulation of catechin biosynthesis.

CLIMATE

Wei *et al.* (2011) reported the variation of catechins among different growth regions were larger than those among cultivars, which indicates the importance of environmental effects on catechins. The effects of growth region on catechins and purine alkaloids might be mainly associated with climate and field condition.

The climatic effect on catechins and caffeines in tea has long been known. However, until now, a direct relationship between catechins, caffeine, and climate factors has not been well studied. Light is the most well studied environmental parameter influencing catechin biosynthesis (Mariya *et al.*, 2003). Kito *et al.* (1968) found the conversion of theanine to catechins in tea shoots was controlled by light. All catechin fractions in tea leaves increase when exposed to sunlight but are retarded under shade treatment, suggesting the positive effects of light irradiation on catechin biosynthesis (Table 72.2, Wang *et al.*, 2012). Caffeine content seems not be significantly affected by light (Suzuki and Waller, 1985; Wang *et al.*, 2011).

On the other hand, low temperature would facilitate the accumulation of catechins, particularly for the levels of EC, EGC, and EGCG (Chen *et al.*, 2010; Lee *et al.*, 2010; Wei *et al.*, 2012). This was considered to be attributed to the growth of tea leaves, as high temperature accelerates leaf growth which dilutes the contents of secondary metabolites such as catechins (Gulati and Ravindranath, 1996; Yao *et al.*, 2005). Furthermore, tea plucked during slow growth conditions might contain a lower proportion of gallate ester type catechins to free catechins (Hilton *et al.*, 1973).

TABLE 72.2 Concentrations (mg/g d.w.) of Catechins and Chlorophyll in Leaves of Tea Plants Grown in Shade or Directly Exposed to Sunlight

	Sunlight			Shade		
Compound	**Bud and First Leaf**	**Second Leaf**	**Third Leaf**	**Bud and First Leaf**	**Second Leaf**	**Third Leaf**
Chlorophyll ($p=0.002$)*	2.33±0.26	3.54±0.76	4.05±0.72	2.95±0.09	5.79±0.19	7.93±0.65
Catechins ($p<0.001$)*						
Gallocatechin	3.36±0.29	3.76±0.63	4.33±0.41	1.37±0.68	0.90±0.63	1.26±1.17
EGC	30.41±6.37	42.01±12.66	50.53±2.13	19.62±2.64	31.50±0.89	30.20±4.28
C	3.72±3.43	1.20±1.00	0.41±0.70	3.15±2.75	0.93±0.61	0.67±0.16
EC	15.17±3.19	20.45±4.85	19.01±2.75	7.68±1.30	10.55±1.34	9.46±0.67
EGCG	82.97±4.91	75.89±8.17	61.17±7.39	79.86±8.98	75.83±10.62	62.78±4.11
ECG	40.74±3.30	30.15±1.38	24.53±3.61	30.25±5.49	23.83±5.18	17.96±2.82
Total catechins	176.38±13.70	173.46±7.05	159.97±13.61	151.37±19.35	150.72±18.93	122.34±5.28

Tea cv. 'Shuchazao' plants were grown under a shade net or full sunlight for 21 days to make comparison. This table shows the chlorophyll and catechins contents in leaves of different treatments and positions. Data are means ±SD of duplicate assays.

**significant correlation ($p<0.05$). P-values are derived from the paired sample comparison between light-exposed leaves versus shaded.*

Source: Wang et al., *2012, with permission from Elsevier.*

TABLE 72.3 Comparative Analyses of Contents of Catechins of 'Huang Zhi Xiang' Oolong Tea (mg/g d.w.).

	Spring Oolong Tea		Autumn Oolong Tea		
	Altitude:500 m	**Altitude:350 m**	**Altitude:500 m**	**Altitude:350 m**	**LSD($P<0.05$)**
GC	8.2±0.5b	10.1±0.9a	8.0±0.1b	8.0±0.7b	1.2
EGC	16.8±2.0b	20.3±2.3a	17.6±1.9ab	16.0±0.1b	3.4
C	5.9±0.1bc	7.9±1.2a	7.0±0.1ab	5.6±0.9c	1.4
EC	4.2±0.6a	5.4±1.2a	5.5±0.6a	4.7±0.6a	1.5
EGCG	102.8±0.7a	76.6±5.6c	103.4±3.1a	90.6±5.5b	8.0
ECG	20.8±3.1b	20.9±3.1b	25.2±1.3ab	25.6±1.0a	4.4
TC	158.3±2.1ab	142.0±11.4c	167.2±2.5a	151.5±5.5bc	12.3

This table exhibits variation in catechin contents of 'Huang Zhi Xiang' oolong tea at various growing altitudes and seasons. The experimental results were analyzed in one-way analysis of variance with Duncan's Multiple Range Test (DMRT) ranking—means ±SD within a row followed by the same letter are not significantly different at $P<0.05$ according to DMRT.

Chen et al. *(2010), with permission from Elsevier.*

Precipitation is considered to be negatively correlated with the accumulation of all catechin fractions except (+)-C (Lee *et al.*, 2010; Wang *et al.*, 2012). The effect of precipitation might be partly due to the reduced sunlight exposure. However, to what extent the effect of precipitation has on catechins and purine alkaloids still remains elusive.

Altitude is another factor markedly affecting catechins and purine alkaloids in teas. The quality of teas cultivated at high altitude is generally better than those grown at low altitude, which is mainly correlated with the accumulation of secondary metabolites (Owuor *et al.*, 1990). As shown in Table 72.3, the EGCG and total catechin (TC) contents of oolong teas grown at a high altitude were significantly higher than those grown at a low altitude, irrespective of growing seasons (Chen *et al.*, 2010). However, for spring oolong tea, the contents of (+)-gallocatechin (GC) and EGC from a low altitude were significantly higher than those from a high altitude. The C content of spring oolong tea cultivated at a low altitude was significantly higher than that at a high altitude. In contrast, an opposite tendency was observed in the autumn oolong tea, suggesting C is easily affected by environment. In addition, there were no significantly changes in EC and ECG content in the oolong teas grown at different altitudes. These results indicate the variation in simple catechin content in the oolong tea grown at different altitudes was different from that of catechin gallates. These differences were also considered to be mainly associated with relative climates (Chen *et al.*, 2010).

TABLE 72.4 Catechins and Caffeine Concentrations (mg/g) in Young Shoots of Tea Plants as Affected by Different Concentrations of Nitrogen (N) in the Nutrient Solution

	Nitrogen Supply			
Levels of nitrogen	**0.30 mmol/l**	**0.75 mmol/l**	**1.50 mmol/l**	**4.50 mmol/l**
(−)-EGCG	72.33 ± 5.29ab	76.83 ± 8.80ab	83.54 ± 3.60b	73.87 ± 1.74a
(−)-EGC	27.51 ± 3.48b	27.51 ± 2.45b	31.08 ± 1.86b	22.84 ± 1.17a
(−)-ECG	23.35 ± 1.53b	22.90 ± 1.65b	22.13 ± 0.95b	16.25 ± 0.58a
(−)-EC	13.11 ± 0.76b	12.69 ± 0.72b	12.23 ± 0.94b	8.17 ± 0.46a
(+)-C	1.81 ± 0.23b	1.92 ± 0.10b	1.82 ± 0.16b	1.45 ± 0.09a
Total catechins	139.49 ± 10.83b	143.20 ± 12.35b	151.98 ± 6.99b	124.18 ± 2.46a
Caffeine	23.19 ± 2.42a	29.02 ± 0.89b	33.22 ± 1.48c	37.52 ± 1.12d

Catechins and caffeine concentrations of young shoots were investigated in tea samplings cultivated hydroponically with 0.3, 0.75, 1.5 or 4.5 mmol/l nitrogen. Data are means ± SD of four replicates. Different letters following data within a line indicate a significant ($P < 0.05$) difference by LSD.
Source: Ruan et al., *2010, with permission from John Wiley and Sons.*

AGRICULTURAL PRACTICE

Fertilizers are widely used in tea cultivation to raise yield and quality of tea. The effects of nutrient elements, mainly nitrogen (N), phosphorus (P), and potassium (K) on catechins and caffeine have been frequently reported (Geng and Wang, 2005; Matsunaga *et al.*, 2009; Lin *et al.*, 2012). Large doses of N fertilizer are usually applied for high tea production. It was found that high rates of nitrogen application resulted in reduction of all catechins in tea leaves (Table 72.4, Ruan *et al.*, 2010). However, when N status was relative low, EGCG, EGC, and TC contents would slightly increase under N treatment; whereas caffeine content increases with increasing N supply (Table 72.4, Owuor and Odhiambo, 1994; Matsunaga *et al.*, 2009; Ruan *et al.*, 2010). It was thought decreased quality of tea under excessive N application might be partly correlated with the decrease of catechins.

P is essential for both tea plant growth and development. P-deficiency not only limits plant growth but also influences catechin composition. Lin *et al.* (2012) reported P-deficient tea decreased levels of GC, GCG, ECG, and EC, but increased C content. The concentrations of EGCG, EGC, CG, and TC did not change significantly in response to P supply (Figure 72.1).

The promoting effect of K on catechins and caffeine has previously been demonstrated (Ruan *et al.*, 1999; Venkatesan *et al.*, 2004). K fertilizer could improve overall quality of tea and partly offset the deleterious effect of excessive N application. Mudau *et al.* (2007) found a combined application of 300 kg/ha N, 300 kg/ha P, and 200 kg/ha K maximized total polyphenols (catechins) in tea, indicating the combined effects of different fertilizers on catechins and caffeine should not be ignored.

Furthermore, nutrient elements such as sulfur (Barbora, 1995), magnesium (Ruan *et al.*, 1999) and boron (Ali *et al.*, 1997) were also reported to be able to affect catechin and caffeine contents in tea.

There are also other agricultural practices affecting catechins and caffeine contents, such as plucking frequency and shade treatments. Long plucking intervals reduced the caffeine and theaflavin content in black tea (Owuor and Odhiambo, 1994). In terms of shade treatment, Wang *et al.* (2012) found all catechin fractions in leaves exposed to full sunlight were higher than those in shaded leaves (Table 72.2). Similar results were reported by Saijo (1980) and Ku *et al.* (2010). These results suggest shade treatment would seriously inhibit the accumulation of catechins. However, shade treatment seems not to influence caffeine content (Suzuki and Waller, 1985).

PLUCKING SEASON AND LEAF AGE

The seasonal effect on quality of tea has long been known. There are many reports about seasonal changes of catechins and caffeine in tea. Yao *et al.* (2006) found the levels of EGCG, ECG, and CG were higher, but EGC and total catechins were lower in samples collected in the warm months. While Wang *et al.* (2011) reported EGC, EC, EGCG, and total catechin contents were significantly and positively correlated to temperature. Other climatic factors including precipitation, sunshine hours, and relative humidity would also affect catechins and caffeine content. Chen *et al.* (2010) reported the seasonal changes of catechins in oolong tea grown at low altitude were different from those at high altitude. These contrasting results indicating the effect of plucking season on catechins and caffeine contents might be correlated with many factors, including climate, nutrient status, and plantation management. Thus, more detailed studies in this field are worth trying.

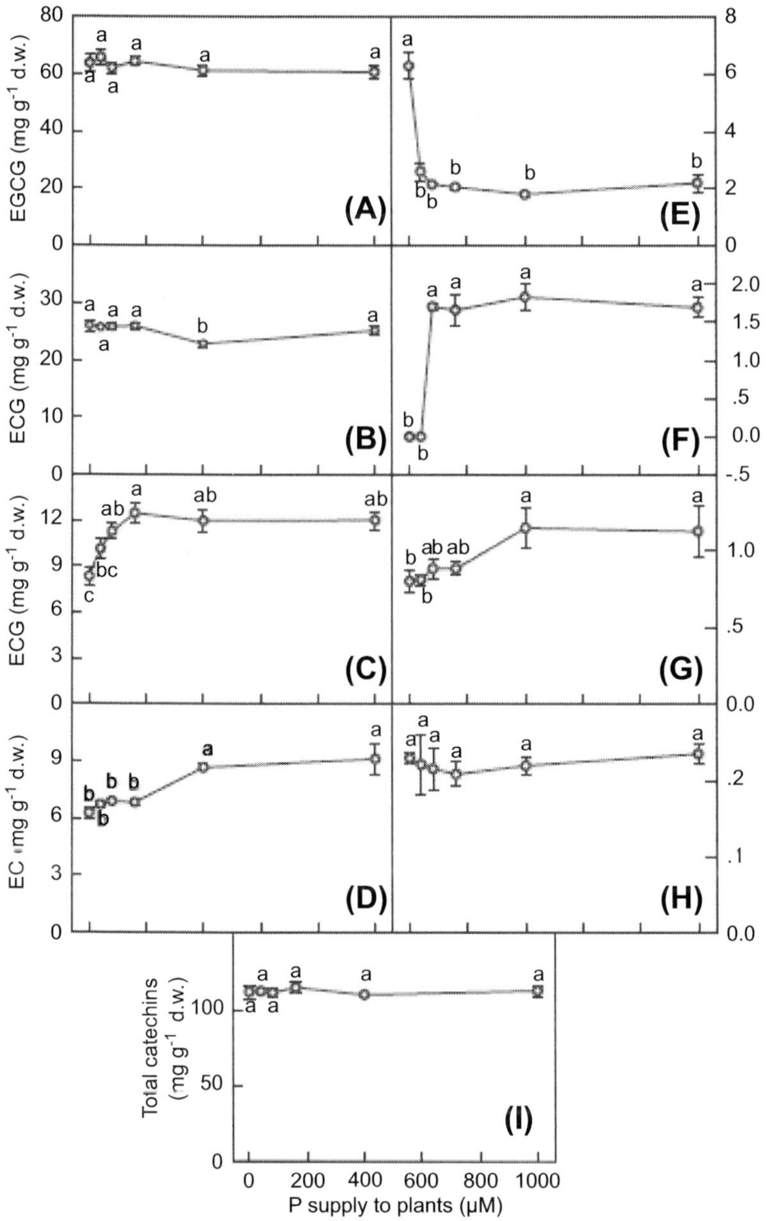

FIGURE 72.1 Effects of phosphorus (P) supply on the concentrations of epigallocatechin gallate (EGCG, A), epigallocatechin (EGC), epicatechin gallate (ECG, C), epicatechin (EC, D), catechin (C, E), gallocatechin (GC, F), gallocatechin gallate (GCG, G), catechin gallate (CG, H), and total catechins (I) in green tea. Self-rooted, 10-month-old, uniform tea [*Camellia sinensis* (L.) O. Kuntze cv. Huangguanyin] plants were supplied with 0, 40, 80, 160, 400, or 1000 μM phosphorus (P) for 17 weeks to determine how P-deficiency affects the quality of green tea. Each point is the mean ±SD of three replicates. Different letters above or below the standard error bars indicate significant differences at $P<0.05$. *Lin* et al. *(2012), with permission from Elsevier.*

Leaf stage is another important parameter affecting tea quality. One leaf and a bud to two leaves and a bud stage are preferred for making green tea. While older tea leaves are required for oolong and black tea. The dynamic changes of catechins in two green tea cultivars 'Wuniuzao' and 'Longjing 43' harvested from different leaf stages are shown in Figure 72.2. A significant decline of C and a general increase of EGC and EC was found during young leaf development for both cultivars (Figure 72.2). In addition, the opposite dynamic changes of the chlorophyll and C contents during different leaf stages for both cultivars are consistent with the trend of changes in different leaf positions (Table 72.2, Wang *et al.*, 2012).

Caffeine content is also seriously affected by the position plucked. We previously found caffeine content dropped dramatically from the first to fifth leaf (Figure 72.3). Meanwhile, caffeine level in leaves was always higher than that in stems at the same position, indicating caffeine is mainly biosynthesized in young tea leaves.

ANALYTICAL TECHNIQUES

There is a wide range of analytical techniques used to determine catechins and caffeine contents, including liquid chromatography-photodiode array-electrospray ionization-mass spectrometry (LC-PDA-ESI/MS) (Ku *et al.*, 2010), high-performance liquid chromatography (HPLC) (Chen *et al.*, 2010), and ^{1}H nuclear magnetic resonance spectroscopy (^{1}H NMR) (Lee *et al.*, 2010). The

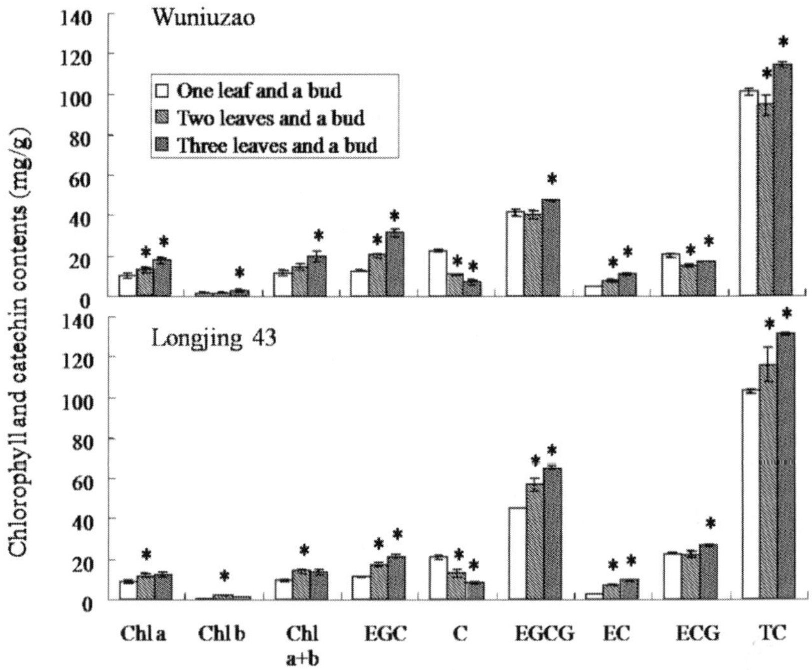

FIGURE 72.2 The dynamic changes of chlorophyll and catechin contents (harvested on 24th March, 9th April, and 17th April, respectively) in Wuniuzao and Longjing 43. Tea samples were plucked from three different leaf stages (one leaf and a bud, two leaves and a bud, and three leaves and a bud) in two tea cultivars (cv. Wuniuzao and Longjing 43) in Hangzhou to study the dynamic changes of chlorophylls and catechins contents. Data are means ±SD of three replicates. *Significant difference at 95% probability level. Chl a, Chl b, Chl a+b represent chlorophyll a, chlorophyll b, and total chlorophylls respectively. *Wei* et al. *(2011), with permission from Elsevier.*

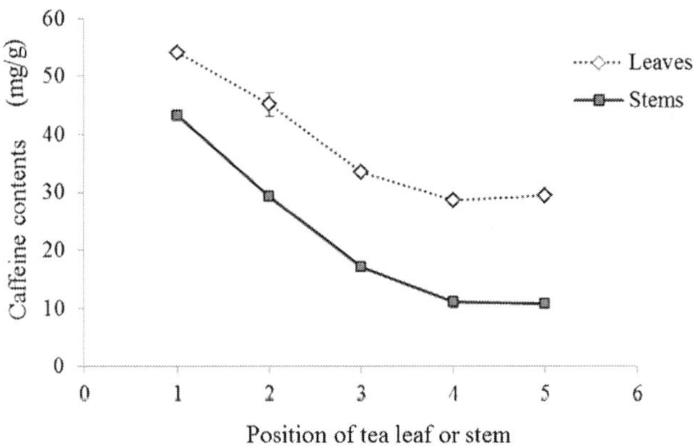

FIGURE 72.3 Caffeine level decreased from the first to fifth leaf and stem in cultivar Zhongcha108. Tea samples (cv. Zhongcha108) of five leaves and a bud stage were plucked in Spring, 2010. Leaves and stems in different positions were separated to determine caffeine contents. Data are means ±SD of three replicates (unpublished).

large variation in data previously reported should be partly attributed to the different analytical techniques used, which makes a fairer comparison very hard. We considered HPLC method according to ISO 14 505–2 as the state of the art recently (Anonymous, 2005). Meanwhile, Kelman and Wright (2012) reported it was important to use ^{1}H NMR spectroscopy to verify HPLC results, indicating further improvement in standard technique is required in this field.

LIST OF ABBREVIATIONS

C (+)-catechin;
Chl a chlorophyll a;
Chl a+b the total chlorophylls;
Chl b chlorophyll b;
EC (−)-epicatechin;
ECG (−)-epicatechin gallate;
EGC (−)-epigallocatechin;
EGCG (−)-Epigallocatechin gallate;
GC (+)-gallocatechin;
HPLC high-performance liquid chromatography;
K potassium;
LC-PDA-ESI/MS liquid chromatography-photodiode array-electrospray ionization-mass spectrometry;
N nitrogen;
NMR nuclear magnetic resonance;
P phosphorus;
TC total catechins.

References

Ali, T.S., Deka, A., Barbora, A.C., 1997. Influence of nitrogen, sulphur and boron on certain quality parameters of black tea. J. Interacademicia 1, 233–239.

Anonymous, 2005. Determination of substances characteristic of green and black tea. Part 2: Content of catechins in green tea. Method using high-performance liquid chromatography. International standard ISO 2005, 14502–2. International Organization for Standardization, Geneva.

Barbora, A.C., 1995. Sulphur management for tea in North Eastern India. Sulphur Agri. 19, 9–15.

Chen, D., Milacic, V., Chen, M.S., Wan, S.B., Lam, W.H., Huo, C., Landis-Piwowar, K.R., Cui, Q.C., Wali, A., Chan, T.H., Dou, Q.P., 2008. Tea polyphenols, their biological effects and potential molecular targets. Histol Histopathol. 23, 487–496.

Chen, L., Zhou, Z., 2005. Variations of main quality components of tea genetic resources (*Camellia sinensis* (L.) O. Kuntze) preserved in the China national germplasm tea repository. Plant Food Hum. Nutr. 60, 31–35.

Chen, Y., Jiang, Y., Duan, J., Shi, J., Xue, S., Kakuda, Y., 2010. Variation in catechin contents in relation to quality of "Huang Zhi Xiang" Oolong tea (*Camellia sinensis*) at various growing altitudes and seasons. Food Chem. 119, 648–652.

Geng, J., Wang, W., 2005. Research progress of potassium in tea tree. Chinese Agri. Sci. Bull. 21, 175–177.

Graham, H.N., 1992. Green tea composition, consumption, and polyphenol chemistry. Prev. Med. 21, 334–350.

Gulati, A., Ravindranath, S.D., 1996. Seasonal Variations in Quality of Kangra Tea (*Camellia sinensis* (L) O Kuntze) in Himachal Pradesh. J. Sci. Food Agri. 71, 231–236.

Hilton, P.J., Palmer-Jones, R., Ellis, R.T., 1973. Effects of season and nitrogen fertilizer upon the flavanol composition and tea making quality of fresh shoots of tea (*Camellia sinensis* L,) in central Africa. J. Sci. Food Agri. 24, 819–826.

Kelman, D., Wright, A.D., 2012. The importance of ^{1}H-nuclear magnetic resonance spectroscopy for reference standard validation in analytical sciences. PloS One 7, e42061.

Kito, M., Kokura, H., Izaki, J., Sasaoka, K., 1968. Theanine, a precursor of the phloroglucinol nucleus of catechins in tea plants. Phytochem. 7, 599–603.

Ku, K.M., Choi, J.N., Kim, J., Kim, J.K., Yoo, L.G., Lee, S.J., Hong, Y.S., Lee, C.H., 2010. Metabolomics Analysis Reveals the Compositional Differences of Shade Grown Tea (*Camellia sinensis* L.). J. Agri. Food Chem. 58, 418–426.

Lee, J.E., Lee, B.J., Chung, J.O., Hwang, J.A., Lee, S.J., Lee, C.H., Hong, Y.S., 2010. Geographical and climatic dependencies of green tea (*Camellia sinensis*) metabolites: A ^{1}H NMR-based metabolomics study. J. Agri. Food Chem. 58, 10582–10589.

Liang, Y.R., Ma, W.Y., Lu, J.L., Wu, Y., 2001. Comparison of chemical composition of *Ilex latifolia* Thumb and *Camellia sinensis* L. Food Chem. 75, 339–343.

Lin, Z.H., Qi, Y.P., Chen, R.B., Zhang, F.Z., Chen, L.S., 2012. Effects of phosphorus supply on the quality of green tea. Food Chem. 130, 908–914.

Mariya, J.K.M., Sasikumar, R., Balasubramanian, M., Saravanan, M., RajKumar, R., 2003. Influence of light on catechin biosynthesis in tea. Tea 24, 80–86.

Marles, M.A.S., Ray, H., Gruber, M.Y., 2003. New perspectives on proanthocyanidin biochemistry and molecular regulation. Phytochem. 64, 367–383.

Matsunaga, A., Saba, T., Nesumi, A., 2009. Effect of nitrogen fertilizers on composition of catechins in shoots of tea cultivars. Tea Res. J. 108, 19–27.

Mudau, F.N., Soundy, P., du Toit, E.S., 2007. Effects of nitrogen, phosphorus and potassium nutrition on total polyphenol content of bush tea (*Athrixia phylicoides* L.) leaves in shaded nursery environment. Hortscience 42, 334–338.

Obanda, M., Owuor, P.O., Taylor, S.J., 1997. Flavanol composition and caffeine content of green leaf as quality potential indicators of Kenyan black teas. J. Sci. Food Agri. 74, 209–215.

Owuor, P.O., Odhiambo, H.O., 1994. Response of some black tea quality parameters to nitrogen fertilizer rates and plucking frequencies. J. Sci. Food Agri. 66, 555–561.

Owuor, P.O., Obaga, S.O., Othieno, C.O., 1990. The effects of altitude on the chemical composition of black tea. J. Sci. Food Agri. 50, 9–17.

Ruan, J., Wu, X., Härdter, R., 1999. Effects of potassium and magnesium nutrition on the quality components of different types of tea. J. Sci. Food Agri. 79, 47–52.

Ruan, J., Haerdter, R., Gerendas, J., 2010. Impact of nitrogen supply on carbon/nitrogen allocation: a case study on amino acids and catechins in green tea [*Camellia sinensis* (L.) O. Kuntze] plants. Plant Biol. 12, 724–734.

Sabhapondit, S., Karak, T., Bhuyan, L.P., Goswam,i, B.C., Hazarika, M., 2012. Diversity of catechin in northeast Indian tea cultivars. Sci. World J. 2012, 485193.

Saijo, R., 1980. Effect of shade treatment on biosynthesis of catechins in tea plants. Plant Cell Physiol. 21, 989–998.

Saravanan, M., Maria John, K.M., Raj Kumar, R., Pius, P.K., Sasikumar, R., 2005. Genetic diversity of UPASI tea clones (*Camellia sinensis* (L.) O. Kuntze) on the basis of total catechins and their fractions. Phytochem. 66, 561–565.

Suzuki, T., Waller, G.R., 1985. Effects of light on the production and degradation of caffeine in *Camellia sinensis* L. seedlings. Plant Cell Physiol. 26, 765–768.

Takeda, Y., 1994. Differences in caffeine and tannin contents between tea cultivars, and application to tea breeding. Jpn. Agri. Res. Quarterly. 28, 117–123.

Venkatesan, S., Murugesan, S., Ganapathy, M.N., Verma, D.P., 2004. Long-term impact of nitrogen and potassium fertilizers on yield, soil nutrients and biochemical parameters of tea. J. Sci. Food Agri. 84, 1939–1944.

Wang, D., Lu, J., Miao, A., Xie, Z., Yang, D., 2008. HPLC-DAD-ESI-MS/MS analysis of polyphenols and purine alkaloids in leaves of 22 tea cultivars in China. J. Food Composition Anal. 21, 361–369.

Wang, L.Y., Wei, K., Jiang, Y.W., Cheng, H., Zhou, J., He, W., Zhang, C.C., 2011. Seasonal climate effects on flavanols and purine alkaloids of tea (*Camellia sinensis* L.). Eur. Food Res. Technol. 233, 1049–1055.

Wang, Y.S., Gao, L.P., Shan, Y., Liu, Y.J., Tian, Y.W., Xia, T., 2012. Influence of shade on flavonoid biosynthesis in tea (*Camellia sinensis* (L.) O. Kuntze). Sci. Horticulturae 141, 7–16.

Wei, K., Wang, L., Zhou, J., He, W., Zeng, J., Jiang, Y., Cheng, H., 2011. Catechin contents in tea (*Camellia sinensis*) as affected by cultivar and environment and their relation to chlorophyll contents. Food Chem. 125, 44–48.

Wei, K., Wang, L., Zhou, J., He, W., Zeng, J., Jiang, Y., Cheng, H., 2012. Comparison of catechins and purine alkaloids in albino and normal green tea cultivars (*Camellia sinensis* L.) by HPLC. Food Chem. 130, 720–724.

Yao, L., Caffin, N., D'arcy, B., Jiang, Y., Shi, J., Singanusong, R., Liu, X., Datta, N., Kakuda, Y., Xu, Y., 2005. Seasonal Variations of Phenolic Compounds in Australia-Grown Tea (*Camellia sinensis*). J. Agri. Food Chem. 53, 6477–6483.

Yao, L., Liu, X., Jiang, Y., Caffin, N., D'Arcy, B., Singanusog, R., Datta, N., Xu, Y., 2006. Compositional analysis of teas from Australian supermarkets. Food Chem., 94,115–122.

CHAPTER

73

Cocoa Processing and Impact on Composition

Jasminka Giacometti, Slavica Mazor Jolić†, Djuro Josić*, ‡*

*University of Rijeka, Rijeka, Croatia, †Kraš,d.d., Zagreb, Croatia, ‡Brown University, Rhode Island, USA

CHAPTER POINTS

- Fermentation, drying, roasting, and alkalization are the main processes in the cocoa chain where bioactive substances can be lost.
- Standardization of farming practices is important for quality and high bioactive compounds content of cocoa products.
- Appropriate starter cultures are useful for controlled cocoa bean fermentation processes.
- Presence of fungi in cocoa is undesirable and often related to the formation of off-flavors, spoilage, and accumulation of mycotoxins.
- 'Omics' methods are a new challenge in the development of traceability and authenticity, quality, and safety methods.

INTRODUCTION

Cocoa is an important crop around the world: it is an important cash crop for growing countries, as well as an important branch of the food industry for processing and consuming countries (Table 73.1) (Cocoa Market Update, 2012).

Most of the cocoa beans produced worldwide are produced by small-sized farms. Fragmentation of farms often causes the heterogeneity of batches of cocoa beans (Saltini *et al.*, 2013). The expectations of chocolate manufacturers directed on the qualitative parameters by country origin aim at having uniform and constant raw materials to produce chocolate.

Quality and flavor of cocoa products strongly depend on the various stages of cocoa processing (Saltini *et al.*, 2013). These processes begin very early with cocoa farming, storage, fermentation, drying, and packing the cocoa beans and continue with the manufacturing of chocolate (Figure 73.1). In recent years, interest in the implementation of traceability and authenticity of foods has increased as a result of a growing concern with food quality and food safety (van der Vorst, 2005), leading to the introduction of legislations in some parts of the world, such as the European Union. However, cocoa beans are produced in non-European countries and EU regulations do not apply to the participants at the beginning of the cocoa supply chain.

Cocoa and its derived products (cocoa powder, cocoa liquor, and chocolate) are a very rich source of bioactive components such as polyphenols (Wollgast and Anklam, 2000; Figure 73.2). The main groups of cocoa polyphenols are catechins (37%), anthocyanins (4%), and proantocyanidins (58%) (Hii *et al.*, 2009). The amounts of polyphenols vary due to the type of cultivar and the country of origin, as well as production processes of cocoa bean and chocolate manufacturing processes (Cooper *et al.*, 2007; Table 73.2).

Flavor compounds and flavor precursors like free amino acids and reducing sugars are formed during fermentation. The amount of these compounds is a confirmation of the well-fermented process and is responsible for the flavor profile of the finished chocolate (Afoakwa *et al.*, 2008; Lefeber *et al.*, 2012). During the roasting process, due to the Maillard reaction, flavor precursors are transformed into flavor compounds (Jumnongpon *et al.*, 2012).

The impact of cocoa processing on polyphenol composition and suitable flavors and/or flavor precursors of cocoa beans and chocolate is discussed in this chapter. Furthermore, our review also highlights the direction of current and future investigations related to the impact of processing cocoa bioactive compounds.

BIOACTIVE COMPONENTS IN COCOA CULTIVARS

In recent years numerous studies have focused on cocoa polyphenols, especially flavonoids, as potent antioxidants for human health. Therefore, the important

Processing and Impact on Active Components in Food
http://dx.doi.org/10.1016/B978-0-12-404699-3.00073-1

TABLE 73.1 Worldwide Cocoa Market

	Cocoa Bean Producers		
Production of cocoa beans (10^3 tonnes)	**2009–2010**	**2010–2011**	**2011–2012**
Africa	2428 (68.%)	3076 (73.3%)	2801 (70.3%)
Cameroon	205	230	220
Cote d'Ivoire	1184	1668	1400
Ghana	740	860	870
Nigeria	230	240	230
Other Africa	69	78	81
Asia and Oceania	642 (18%)	563 (13.4%)	623 (15.6%)
Indonesia	530	450	500
Malaysia	20	18	18
Other Asia	92	95	105
Latin America	499 (14%)	558 (13.3%)	563 (14.1%)
Brazil	159	197	185
Ecuador	150	160	170
Other	189	201	208
World Total	3569	4197	3987
	Cocoa Bean Processors		
Grindings of cocoa beans (10^3 tonnes)	**2009–2010**	**2010–2011**	**2011–2012**
Europe	1492 (41.2%)	1595 (41.7%)	1554 (39.1%)
Germany	361	439	421
Netherlands	500	525	490
Other	631	631	643
Americas	801 (22.1%)	839 (22.0%)	865 (21.8%)
Brazil	223	236	243
United States	380	390	400
Other	198	213	222
Asia and Oceania	689 (19%)	770 (20.1%)	897 (22.6%)
Indonesia	120	170	270
Malaysia	298	305	312
Other	271	295	315
Africa	642 (17.7%)	618 (16.2%)	657 (16.5%)
Cote d'Ivoire	390	340	380
Ghana	200	220	222
Others	52	58	55
World Total	3624	3822	3973

Source: ICCO; Reuters; LMC Report, February 2012.

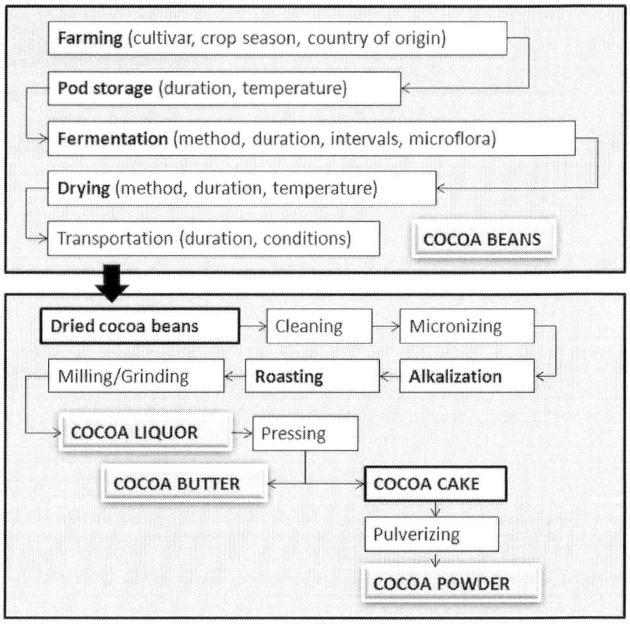

FIGURE 73.1 Flowchart of cocoa bean production and manufacturing processes.

topics in cocoa research are directed at the processing of cocoa beans, manufacturing processes of cacao products, and analytical technologies for the determination of cocoa polyphenols. Some studies on cocoa polyphenols were also related to biochemical changes during the processing and obtaining of the characteristic cocoa flavor (Rodriguez-Campos *et al.*, 2011; Zahouli *et al.*, 2010).

Cultivars

The cocoa tree belongs to the species *Theobroma cacao*. Different subspecies can be classified within four cultivars: Criollo, Forastero, Trinitario, and Nacional (Counet *et al.*, 2004). The Criollo cultivar has a high level of procyanidins (Counet *et al.*, 2004; Elwers *et al.*, 2009). Criollo beans also have a high level of flavor precursors, reducing sugars and amino acids that can result in a very high level of pyrazines. Other aromatics are formed during the roasting process in reactions between reducing sugars and amino acids (Misnawi *et al.*, 2004a; Elwers *et al.*, 2009). Compared to Criollo, the chocolate produced with Forastero and Trinitario beans showed lower procyanidins levels (Counet *et al.*, 2004) and similar amount of pyrazines, while Trinitario showed strong basic chocolate characters as well as some typical winery type of aroma that distinguishes this cultivar from the other varieties (Counet *et al.*, 2004). Unlike the others, the Nacional cultivar is known to have a very low concentration of reducing sugars after fermentation (Elwers *et al.*, 2009) and also, low levels

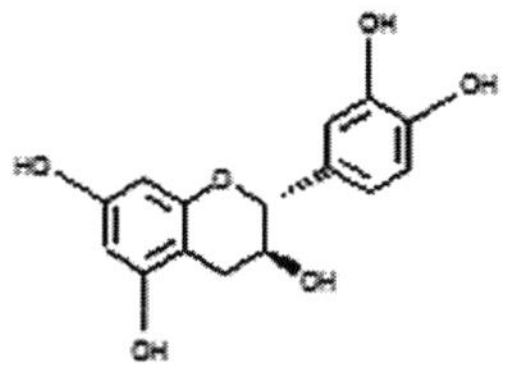

(+)-Catechin (2R, 3S)

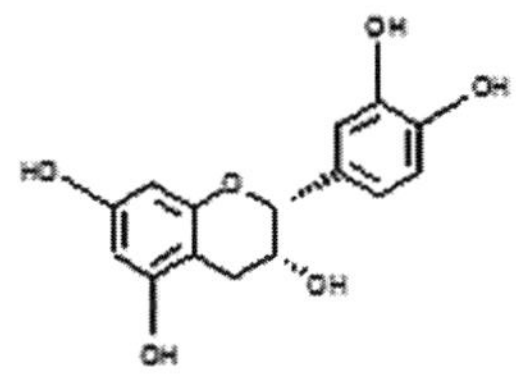

(-)-Epicatechin (2R, 3R)

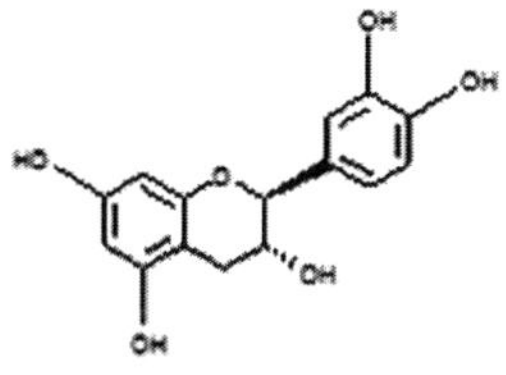

(-)-Catechin (2S, 3R)

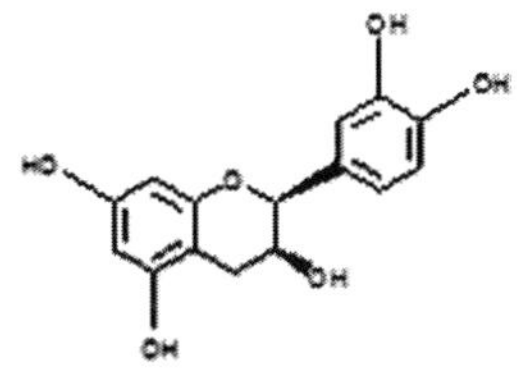

(+)-Epicatechin (2S, 3S)

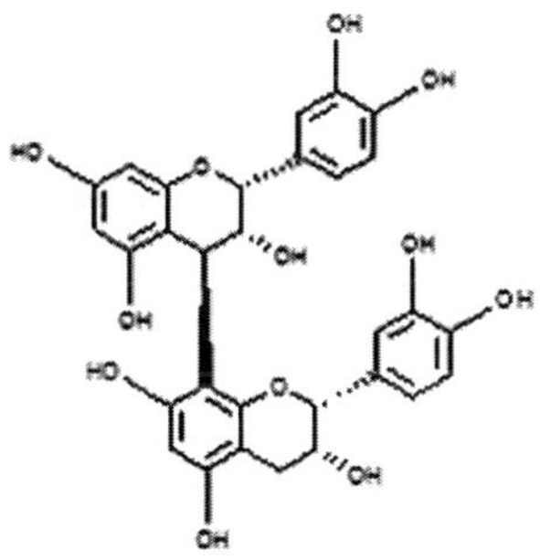

Procyanidin B2 (dimer B2),
epicatechin-(4β - 8)-epicatechin

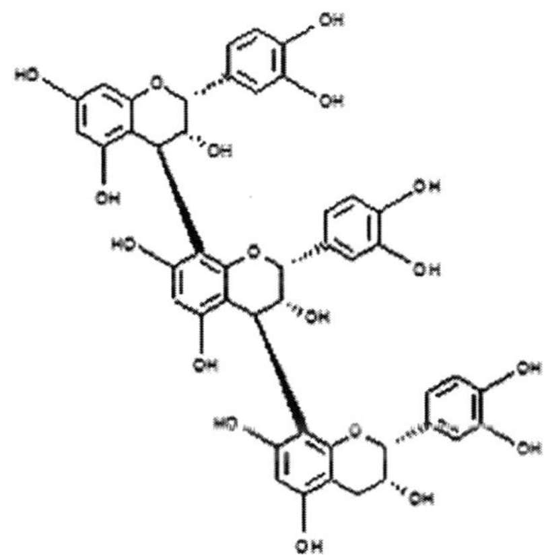

Procyanidin C1 (trimer C1),
epicatechin-(4β → 8)$_2$-epicatechin

FIGURE 73.2 Structures of the catechin/epicatechin enantiomers, procyanidin B2 and procyanidins C1.

of pyrazines found after roasting (Jinap *et al.*, 2004; Misnawi *et al.*, 2004).

Factors Affecting the Quantity and Quality of Polyphenols in Cocoa Beans

A number of spontaneous enzymatic reactions occurred during fermentation of cocoa beans. Fermentation and drying are two very important steps involved in the production of cocoa beans. The amounts and composition of amino acids change during the fermentation process, as well as reducing sugars such as glucose and fructose, polyphenols, and the pH profile. The presence of yeast, lactic acid bacteria (LAB), and acetic acid bacteria (AAB) during the fermentation process positively affect the concentration of polyphenols and development of flavor precursors, while aerobic spores and moulds have a negative effect.

Inadequate procedures during fermentation and drying processes can significantly reduce polyphenols and allow the development of undesirable off-flavors.

Fermentation

FERMENTATION METHOD

Different fermentation methods are used for fermenting cocoa beans depending on farmers, (Guehi *et al.*, 2010) areas, and countries (Camu *et al.*, 2008; Schwan, 1998). Platforms, heaps, baskets, and boxes are the most used fermentation methods. The platform method has a quite low fermentation rate, adequate for Criollo beans which require short fermentation, but inappropriate for Forastero which requires longer fermentation. This longer fermentation often induces the growth of unwanted molds and consequently off-flavors (Guehi *et al.*, 2010; Schwan and Wheals, 2004).

TABLE 73.2 Polyphenols and Total Antioxidant Capacities of Cocoa and Chocolate Products

Product	Catechins (mg/g)	Procyanidins (mg/g)	Total Flavanols and Flavonols (mg/g)	Total Polyphenols* (mg/g)	ORAC (μmol of TE/g)#
Milk chocolate	0.23–0.32 (Gu *et al.*, 2006)	2.16–3.14 (Gu *et al.*, 2006) 0.43–0.90 (Miller *et al.*, 2008)		3.25–5.38 (Miller *et al.*, 2008)	68–86 (Gu *et al.*, 2006)
Dark chocolate	0.77–1.58 (Gu *et al.*, 2006)	8.52–19.85 (Gu *et al.*, 2006) 2.78–4.10 (Miller *et al.*, 2008)		11.73–14.88 (Miller *et al.*, 2008)	211–349 (Gu *et al.*, 2006)
Natural powder	2.90–3.48 (Gu *et al.*, 2006)	32.19–48.70 (Gu *et al.*, 2006)	2.11–3.06 (Andres–Lacueva *et al.*, 2008)		871–899 (Gu *et al.*, 2006)
Alkalized cocoa powder (Dutched powder)	0.41–0.73 (Gu *et al.*, 2006)	7.02–10.82 (Gu *et al.*, 2006)	0.85–1.15 (Andres-Lacueva *et al.*, 2008)		397–406 (Gu *et al.*, 2006)
Cocoa powder		19.28–23.71 (Miller *et al.*, 2008)		45.30–60.20 (Miller *et al.*, 2008)	
Chocolate syrup		0.37–0.91 (Miller *et al.*, 2008)		3.66–4.79 (Miller *et al.*, 2008)	

**Results expressed as gallic acid equivalents (GAE) on a dry weight basis.*
#Data expressed as micromoles of Trolox equivalents per gram (μmol of TE/g).

The fermentation in boxes shows a relatively low concentration of sugars, ethanol, and acetic acid, as well as a high pH. Given the heterogeneity in the size of the cocoa beans, as well as the construction material of the boxes, this method of fermentation significantly affects the pH value, tannins and sugar content, and the presence of purple beans (Wallace and Giusti, 2011). In contrast to the box method of fermentation, the temperature increases faster at the beginning of the process in the heap method (Guehi *et al.*, 2010). It is considered to be the reason for a smaller number of matching purple beans (Guehi *et al.*, 2010).

MICROBIAL ACTIVITY

The role of microbiology in the cocoa bean fermentation is described by Schwan and Wheals (2004). The changes in pH values during fermentation are very important for microbial activity. Briefly, the initial low pH of the pulp (3.6), presence of citric acid, and low oxygen levels favor yeast colonization which leads to ethanol production and secretion of pectolytic enzymes within the first 24h, after which the process slowly decreases. The resulting conditions are favorable to the growth of LAB, which reach their peak after around 36h from the beginning of fermentation, when glucose is degraded to lactic acid. LAB population decreases and AAB increases in the next period of fermentation. In these conditions, the temperature increases up to about 50°C. It is assumed that the condition induced by AAB causes the diffusion and hydrolysis of proteins, and AAB probably plays a key role in the formation of flavor precursors. After decreasing AAB (5–6 days), aerobic spores and fungi occur. Then, the period that adversely affects the quality and content of polyphenols and flavors begins (Afoakwa *et al.*, 2012; Camu *et al.*, 2008; Counet *et al.*, 2004).

ENZYMES

The enzymes endoproteinase and also other enzymes such as carboxypeptidase, aminopeptidase, invertase, polyphenol oxidase (PPO), or glycosidases break down the cell wall proteins into peptides of different chain length and free amino acids (de Brito *et al*, 2004; Misnawi *et al.*, 2003).

Unfermented cocoa beans contain a very low amount of free amino acids (Misnawi *et al.*, 2002). Low pH at the beginning of fermentation favors proteolysis and the liberation of free amino acids, while too low pH reduces the flavor precursors and leads to over-acidification of the final product (Camu *et al.*, 2008). To know the content and composition of the released amino acids is extremely important for predicting the synthesis of aromatic compounds in the later stages of the production of chocolate. Low amounts of D-amino acids were detected in fermented cocoa beans (Brückner *et al.*, 1993). Taking into account that the bacteria are important sources of D-amino acids in fermented foodstuffs (Brückner *et al.*, 1993), it seems that bacteria do not penetrate the cocoa beans during the microbial fermentation of the pulp. The hydrophobic amino acids such as leucine (Leu), alanine (Ala), phenylalanine (Phe), valine (Val), isoleucine (Ile), and tyrosine (Tyr) were found in the highest amount in free amino acids profile during fermentation (Rohsius *et al.*, 2006). Finally, knowledge of the content and composition of total and free amino acids can give useful information of the flavor potential of cocoa beans and thus on the final cocoa products.

The optimal concentration of reducing sugars in cocoa beans as well as the optimal concentration of free amino acids is found after 4–5 days of the fermentation process (Saltini *et al.*, 2013). Rodriguez-Campos *et al.* (2012) found that the optimal conditions for the fermentation and drying of cocoa beans were 6 days of the beginning of fermentation, followed by drying at 70°C.

During fermentation and drying, polyphenols diffuse from their storage cells and oxidize to become condensed high molecular compounds, mostly insoluble tannins. PPO is involved in the catalysis of these processes (Rohsius *et al.*, 2006). Its activity is strongly reduced during the first day of fermentation. The epicatechin as the major cocoa phenolic compound is then reduced 10–20%, and at the same time the enzymatic activity of PPO is strongly decreased from 50% to 6% (Wollgast and Anklam, 2000). The decreased activity in PPO is related to the significantly low level of polyphenols. It is believed that the non-enzymatic oxidation of polyphenols causes inactivation of PPO (Wollgast and Anklam, 2000). The consequence of the oxidation of polyphenols is also the production of black, brown and red pigments of cocoa beans (de Brito *et al.*, 2002). These results lead to a better understanding of the fermentation process that is aimed at the alteration of biologically active components such as polyphenols and flavor compounds.

Polyphenols are involved in the flavor profile of cocoa products, giving astringency and bitterness, and increasing antioxidative capacity. Raw cocoa beans contain a higher content of phenolics than those processed. In unfermented cocoa, 60% of total phenolics were flavanol monomers and procyanidin oligomers (Tomas-Barberan *et al.*, 2007). The epicatechin content was observed to decrease between days two and three during fermentation, which could indicate a loss of fluids or formation of large tannins (Afoakwa *et al.*, 2012). The brown color of cocoa beans is related to the production of tannins during the optimal fermentation.

Drying

Drying of cocoa beans is the final farming process, which uses heating and reduces the moisture content of the beans to less than 7.5% (w/w) (Zahouli *et al.*, 2010). Many investigations have been carried out to find the optimal drying methods. Therefore, the drying conditions, temperature and duration of drying, drying rate, and grade were studied.

Drying continues during the process of oxidation that follows after the fermentation process and therefore plays an important role in the composition of polyphenols (Payne *et al.*, 2010; Wollgast and Anklam, 2000).

PPO is also sensitive to sun drying, reducing its enzymatic activity about 2% (Wollgast and Anklam, 2000). The increased temperature during the drying process is also an important factor in the loss of polyphenols. Epicatechin decreased approximately 50% in unfermented beans after 2 days of sun drying (Wollgast and Anklam, 2000).

Too fast drying affects the excessive production of acids, including acetic acid, which is deleterious to the flavor, while too slow drying results in lower pH, the absence of the optimal color of cocoa beans and increased growth of molds (Zahouli *et al.*, 2010; Rodriguez-Campos *et al.*, 2012; Hurst *et al.*, 2011). The quality of cocoa beans drying depends on the use of conventional hot air dryers and sun drying (Hii *et al*, 2011). Natural sun drying is still widely used, especially in West Africa. Step-up-dried samples showed the best flavor profile with a high level of cocoa flavor, low in sourness and not excessive in bitterness and astringency (Hii *et al.*, 2011).

Factors that Affect the Quantity and Quality of Polyphenols in Chocolate Production

Chocolate characteristics originate not only in flavor precursors present in cocoa beans, but are generated during post-harvest treatments and transformed into desirable odor notes in the manufacturing process as well. Complex biochemical modifications of bean constituents are further altered by thermal reactions in roasting and conching, and in alkalization (Afoakwa *et al.*, 2008). Alkalization and roasting are two processes that contribute to the flavor and color of the semi-finished products (excluding cocoa butter).

Roasting

Roasting of the fermented cocoa beans is carried out due to mainly two purposes: the removal of undesired compounds with low boiling points, such as acetic acid, and the formation of the typical roasty, sweet odorants of cocoa (Oliviero *et al.*, 2009). The cocoa roasting process is carried out at high temperatures (120–150°C). However, the synthesis of flavor precursors, such as free amino acids and reducing sugars, occur before the roasting process by enzymatic reaction (proteolysis and hydrolysis) during the fermentation process. Non-enzymatic browning, known as Maillard reaction, is involved in numerous reactions including the formation of brown pigments and synthesis of pyrazines. The concentration of pyrazines increases rapidly during the roasting process until reaching a maximum value, after which these constant values are maintained.

Misnawi *et al.* (2004a) studied sensory properties of cocoa liquor as affected by the polyphenol concentration and duration of roasting. They found that cocoa polyphenols had a negative effect on flavor properties. A lower concentration of amino acids and reducing sugars was related to a higher polyphenol concentration and lower pH value (Noor-Soffalina *et al.*, 2009).

The Criollo cultivar has higher pyrazines than those of Forastero (Elwers *et al.*, 2009). On the other hand, Forastero roasted cocoa beans have three times higher amount of pyrazines than Criollo (Counet *et al.*, 2004). During roasting, amino acids and sugars are reduced (amino acids between 24–72%, sugars between 47–60%) and vary depending on the cultivar, type of fermentation method, temperature, and dynamics of roasting (Misnawi *et al.*, 2004b). Unlike sugars, where the concentration continuously decreases until all the sugar is decomposed, the content of amino acids decreased in the first 30 minutes after the beginning of the process. This decrease in the concentration of amino acids depends on the roasting temperature, where higher temperature favors pyrazine synthesis (Misnawi *et al.*, 2004b).

By the action of proteases and PPO during fermentation, additional brown products are formed from quinones and free amino acids. Activity of these enzymes is strongly correlated with amount of water and temperature during roasting, thus reducing bitterness and astringency of cocoa products (Misnawi *et al.*, 2004a).

Cocoa polyphenols are thermally labile to some degree; however, the exact stability of epicatechin in chocolate will likely depend on stabilizing interactions with other components of the cocoa nib. Generally, high processing temperatures and /or longer processing times reduce the amount of polyphenols available in cocoa beans (Hurst *et al.*, 2011). Those degradations are known to be related to enzymatic and nonenzymatic oxidations that occur during these manufacturing processes (Tomas-Barberán *et al.*, 2007). Recently, Payne *et al.* (2010) found that when cacao beans were roasted to 120°C, the catechin level in beans increased 696% in unfermented beans, 650% in Ivory Coast beans, and 640% in Papua New Guinea fermented beans compared to the same unroasted beans. These results suggest that roasting in excess of 70°C generates significant amounts of (−)-catechin (Payne *et al.*, 2010). (−)-Catechin is generally formed during the manufacturing process by epimerization which converts (−)-epicatechin into its epimer (−)-catechin (Hurst *et al.*, 2011). The manufacturing process could equally reduce the amount of (−)-epicatechin as compared to procyanidins B2, B5, C1, and D. Epicatechin concentrations can be used to predict the content of other polyphenols, especially B2 and C1, and total polyphenols content. Finally, the (−)-catechin content is not predictable from the epicatechin content, and it is the main form of polyphenol that varies according to manufacturing conditions and cocoa origin (Cooper *et al.*, 2007).

Conching is typically applied with gentle grinding to improve the flavor characteristics and reduce the concentration of free acids and other volatile by-products from the cocoa bean (Sulistyowati and Misnawi, 2008).

Alkalization

Cocoa nibs, cocoa liquor, and cocoa powder can be modified by treatment with alkali, also known as alkalization. Although originally performed to make the powder not agglomerate or sink to the bottom, when it was added to milk or water-based drinks, the alkalization process is nowadays mainly applied to modify the flavor and color of cocoa powders (Andres-Lacueva *et al.*, 2008). It was established that alkalization caused a progressive reduction of polyphenols as well as their antioxidant activity (Miller *et al.*, 2008). Reduction of the polyphenol antioxidant activity was suggested to be triggered by heat and alkali synergistically (Sulistyowati and Misnawi, 2008). Recently, a study by Payne *et al.* (2010) found that compared to natural cocoa powders, alkalization caused a loss in both epicatechin (up to 98%) and catechin (up to 80%). Andres-Lacueva *et al.* (2008) described a decrease in epicatechin and catechin of 67% and 35%, respectively, as a result of the alkalization of cocoa powders. In the case of flavonols, quercetin presented the highest loss (86%), whereas quercetin-3-glucuronide, quercetin-3-arabinoside, and isoquercetin showed a similar decrease (58, 62, and 61%, respectively) (Andres-Lacueva *et al.*, 2008). In general, changes occurring as the result of alkalization treatment could be attributed to the oxidation of phenolic compounds under basic pH conditions, leading to the brown pigments that are polymerized to different degrees. In particular, secondary reactions involving *o*-quinones previously formed during the fermentation stage by polyphenol oxidase are probably involved in further reactions responsible for the browning developed during alkalization (Miller *et al.*, 2008).

ANALYTICAL TECHNIQUES

High performance liquid chromatography (HPLC) is usually the method of choice for characterisation and quantification of polyphenols in chocolate. This method is often used coupled with a variety of detectors such as UV/Vis, photodiode array (PDA), fluorescence, electrochemical (ECD), and mass spectrometry (MS) (Wollgast and Anklam, 2000). Cooper *et al.* (2007) improved this method by using ultra-performance liquid chromatography (UPLC).

Antioxidant activity (ORAC), total polyphenols (TP), and flavanols of natural and processed cocoa powders including the monomeric, oligomeric, polymeric, and total flavanols of cocoa powders were studied by Miller *et al.* (2008). Oxygen radical absorbance capacity (ORAC) is the fluorescent method for assessing antioxidant capacity (Miller *et al.*, 2008). Total polyphenols are determined by the colorimetric method. Flavanols were measured in cocoa and chocolate products using a normal phase HPLC-ESI/MS based on the separation of the flavanol monomers and polymers (procyanidins) by Gu *et al.* (2006).

PERSPECTIVES AND CHALLENGES

As described previously, fermentation, drying, roasting, and alkalization are the main processes in the cocoa chain where bioactive substances can be lost.

Optimization of these processes is one of the tasks. Due to standardization of farming practices, some of the world's chocolate manufacturers now have their own plantations in the countries of origin.

The restriction of yeast, LAB, and AAB species diversity is crucial for successful cocoa bean fermentation process and is responsible for formation of the necessary color and flavor precursors of well-fermented dry cocoa beans (Schwan and Wheals, 2004). This knowledge concerning microbial species diversity and their activities during spontaneous cocoa bean fermentation processes can be implemented using appropriate starter cultures for controlled cocoa bean fermentation processes (Lefeber *et al.*, 2012).

In general, the presence of fungi in cocoa is undesirable and often related to the formation of off-flavors, spoilage, and accumulation of mycotoxins (Copetti *et al.*, 2011).

The use of proteomics for process development and validation in food technology and food biotechnology as well as corresponding quality control of starting materials and final products provide new opportunities (Gašo-Sokač *et al.*, 2010). This methodology has not yet been used in the evaluation of cocoa processing.

Verifying the quality and safety of cocoa products in terms of their composition, processing or origin is a new challenge and an important one in protecting consumers.

These practical challenges relate to developing traceability and authenticity methods, where the most successful are 'omics' methods. Transcriptomics, proteomics, metabolomics, and glycomics termed as 'omics' technologies can be used in monitoring cocoa processing in order to manufacture cocoa products with a high content of bioactive substances and appropriate flavor properties.

References

Afoakwa, E.O., Paterson, A., Fowler, M., Ryan, A., 2008. Flavor formation and character in cocoa and chocolate: a critical review. Crit. Rev. Food Sci. Nutr. 48 (9), 840–857.

Afoakwa, E.O., Quao, J., Takrama, F.S., Budu, A.S., Saalia, F.K., 2012. Changes in total polyphenols, o-diphenols and anthocyanin concentrations during fermentation of pulp pre-conditioned cocoa (*Theobroma cacao*) beans. Int. Food Res. J. 19 (3), 1071–1077.

Andres-Lacueva, C., Monagas, M., Khan, N., Izquierdo-Pulido, M., Urpi-Sarda, M., Permanyer, J., Lamuela-Raventós, R.M., 2008. Flavanol and flavonol contents of cocoa powder products: influence of the manufacturing process. J. Agr. Food Chem. 56 (9), 3111–3117.

Brückner, H., Becker, D., Lüpke, M., 1993. Chirality of amino acids of microorganisms used in food technology. Chirality 5, 385–392.

Camu, N., González, A., De Winter, T., Van Schoor, A., De Bruyne, K., Vandamme, P., Takrama, J.S., Addo, S.K., de Vuyst, L., 2008. Influence of turning and environmental contamination on the dynamics of populations of lactic acid and acetic acid bacteria involved in spontaneous cocoa bean heap fermentation in Ghana. Appl. Environ. Microbiol. 74 (1), 86–98.

Cocoa Market Update, 2012. (pp. 1-7). Retrieved from www.worldcocoa.org.

Cooper, K.A., Campos-Gimánez, E., Jiménez Alvarez, D., Nagy, K., Donovan, J.L., Williamson, G., 2007. Rapid reversed phase ultra-performance liquid chromatography analysis of the major cocoa polyphenols and inter-relationships of their concentrations in chocolate. J. Agr. Food Chem. 55 (8), 2841–2847.

Copetti, M.V., Iamanaka, B.T., Frisvad, J.C., Pereira, J.L., Taniwaki, M.H., 2011. Mycobiota of cocoa: from farm to chocolate. Food Microbiol. 28 (8), 1499–1504.

Counet, C., Ouwerx, C., Rosoux, D., Collin, S., 2004. Relationship between procyanidin and flavor contents of cocoa liquors from different origins. J. Agr. Food Chem. 52 (20), 6243–6249.

de Brito, E.S.D., Horacio, N., García, P., Amancio, A.C., 2004. Biology and Technology Use of a Proteolytic Enzyme in Cocoa (*Theobroma cacao* L.). Proc. Braz. Arch. Biol. Technol. 47 (August), 553–558.

de Brito, E.S., Garcia, N.H.P., Amancio, A.C., 2002. Effect of polyphenol oxidase (PPO) and air treatments on total phenol and tannin content of cocoa nibs. Ciênc. Tecnol. Aliment. Campinas. 22 (1), 45–48.

Elwers, S., Zambrano, A., Rohsius, C., Lieberei, R., 2009. Differences between the content of phenolic compounds in Criollo, Forastero and Trinitario cocoa seed (*Theobroma cacao* L.). Eur. Food Res. Technol. 229 (6), 937–948.

Gašo-Sokač, D., Kovač, S., Josić, D., 2010. Application of Proteomics in Food Technology and Food Biotechnology: Process Development, Quality Control and Product Safety. Food Technol. Biotechnol. 48 (3), 284–295.

Gu, L., House, S.E., Wu, X., Ou, B., Prior, R.L., 2006. Procyanidin and catechin contents and antioxidant capacity of cocoa and chocolate products. J. Agric. Food Chem. 54, 4057–4061.

Guehi, T.S., Dadie, A.T., Koffi, K.P.B., Dabonne, S., Ban-Koffi, L., Kedjebo, K.D., Nemlin, G.J., 2010. Performance of different fermentation methods and the effect of their duration on the quality of raw cocoa beans. Int. J. Food Sci. Technol. 45 (12), 2508–2514.

Hii, C.L., Law, C.L., Suzannah, S., Misnawi, Cloke, M., 2009. Polyphenols in cocoa (*Theobroma cacao* L.). J. Food Ag-Ind. 2 (04), 702–722.

Hii, Ching L., Law, C.L., Cloke, M., Sharif, S., 2011. Improving Malaysian cocoa quality through the use of dehumidified air under mild drying conditions. J. Sci. Food Agr. 91 (2), 239–246.

Hurst, W.J., Krake, S.H., Bergmeier, S.C., Payne, M.J., Miller, K.B., Stuart, D.A., 2011. Impact of fermentation, drying, roasting and Dutch processing on flavan-3-ol stereochemistry in cacao beans and cocoa ingredients. Chem. Cent. J. 5 (1), 53.

Jumnongpon, R., Chaiseri, S., Hongsprabhas, P., Healy, J.P., Meade, S.J., Gerrard, J.A., 2012. Cocoa protein crosslinking using Maillard chemistry. Food Chem. 134 (1), 375–380.

Lefeber, T., Papalexandratou, Z., Gobert, W., Camu, N., De Vuyst, L., 2012. On-farm implementation of a starter culture for improved cocoa bean fermentation and its influence on the flavor of chocolates produced thereof. Food Microbiol. 30 (2), 379–392.

Miller, K.B., Hurst, W.J., Payne, M.J., Stuart, D.A., Apgar, J., Sweigart, D.S., Ou, B., 2008. Impact of alkalization on the antioxidant and flavanol content of commercial cocoa powders. J. Agr. Food Chem. 56 (18), 8527–8533.

Misnawi, Jinap, S., Nazamid, S., Jamilah, B., 2002. Activation of remaining key enzymes in dried under-fermented cocoa beans and its effect on aroma precursor formation. Food Chem. 78 (4), 407–417.
Misnawi, Jinap, S., Jamilah, B., Nazamid, S., 2003. Effects of incubation and polyphenol oxidase enrichment on colour, fermentation index, procyanidins and astringency of unfermented and partly fermented cocoa beans. Int. J. Food Sci. Technol. 38, 285–295.
Misnawi, Jinap, S., Jamilah, B., Nazamid, S., 2004a. Sensory properties of cocoa liquor as affected by polyphenol concentration and duration of roasting. Food Qual. Prefer. 15 (5), 403–409.
Misnawi, Jinap, S., Jamilah, B., Nazamid, S., 2004b. Effect of polyphenol concentration on pyrazine formation during cocoa liquor roasting. Food Chem. 85 (1), 73–80.
Noor-Soffalina, S.S., Jinap, S., Nazamid, S., Nazimah, S.A.H., 2009. Effect of polyphenol and pH on cocoa Maillard-related flavor precursors in a lipidic model system. Int. J. Food Sci. Technol. 44 (1), 168–180.
Oliviero, T., Capuano, E., Cämmerer, B., Fogliano, V., 2009. Influence of roasting on the antioxidant activity and HMF formation of a cocoa bean model systems. J. Agr. Food Chem. 57 (1), 147–152.
Payne, M.J., Hurst, W.J., Miller, K.B., Rank, C., Stuart, D.A., 2010. Impact of fermentation, drying, roasting, and Dutch processing on epicatechin and catechin content of cacao beans and cocoa ingredients. J. Agr. Food Chem. 58 (19), 10518–10527.
Rodriguez-Campos, J., Escalona-Buendía, H.B., Orozco-Avila, I., Lugo-Cervantes, E., Jaramillo-Flores, M.E., 2011. Dynamics of volatile and non-volatile compounds in cocoa (*Theobroma cacao* L.) during fermentation and drying processes using principal components analysis. Food Res. Int. 44 (1), 250–258.
Rodriguez-Campos, J., Escalona-Buendía, H.B., Contreras-Ramos, S.M., Orozco-Avila, I., 2012. Effect of fermentation time and drying temperature on volatile compounds in cocoa. Food Chem. 132 (1), 277–288.
Rohsius, C., Matissek, R., Lieberei, R., 2006. Free amino acid amounts in raw cocoas from different origins. Eur. Food Res. Technol. 222 (3–4), 432–438.
Saltini, R., Akkerman, R., Frosch, S., 2013. Optimizing chocolate production through traceability: A review of the influence of farming practices on cocoa bean quality. Food Control 29 (1), 167–187.
Schwan, R.F., 1998. Cocoa fermentations conducted with a defined microbial cocktail inoculum. Appl. Environ. Microbiol. 64 (4), 1477–1483.
Schwan, R.F., Wheals, A.E., 2004. The microbiology of cocoa fermentation and its role in chocolate quality. Crit. Rev. Food Sci. Nutr. 44 (4), 205–221.
Sulistyowati and Misnawi, 2008. Effects of alkali concentration and conching temperature on antioxidant activity and physical property of chocolate. Int. Food Res. J. 15 (3), 297–304.
Tomas-Barberan, F.A., Cienfuegos-Jovellanos, E., Marín, A., Muguerza, B., Gil-Izquierdo, A., Cerda, B., Zafrilla, P., Morillas, J., Mulero, J., Ibarra, A., Pasamer, M.A., Ramon, D., Espin, J.C., 2007. A new process to develop a cocoa powder with higher flavonoid monomer content and enhanced bioavailability in healthy humans. J. Agr. Food Chem. 55 (10), 3926–3935.
van der Vorst, J.G.A.J., 2005. Product traceability in food-supply chains. Accred. Qual. Assur. 11 (1–2), 33–37.
Wallace, T.C., Giusti, M.M., 2011. Selective removal of the violet color produced by anthocyanins in procyanidin-rich unfermented cocoa extracts. J. Food Sci. 76 (7), C1010–C1017.
Wollgast, J., Anklam, E., 2000. Review on polyphenols in *Theobroma cacao*: changes in composition during the manufacture of chocolate and methodology for identification and quantification. Food Res. Int. 33 (6), 423–447.
Zahouli, G.I.B., Guehi, S.T., Fae, A.M., Nemlin, J.G., 2010. Effect of Drying Methods on the Chemical Quality Traits of Cocoa Raw Material. Adv. J. Food Sci. Technol. 2 (4), 184–190.

CHAPTER

74

Tagatose Stability in Beverages as Impacted by Composition and Thermal Processing

Leonard N. Bell

Auburn University, Auburn, Alabama, USA

CHAPTER POINTS

- Tagatose, a prebiotic monosaccharide, can undergo chemical degradation.
- Tagatose degradation rates are influenced by pH, beverage composition, and temperature.
- Greater extents of tagatose loss occur at higher pH levels.
- Phosphate buffer catalyzes tagatose loss more so than citrate buffer.
- Higher buffer concentrations enhance tagatose degradation.
- Tagatose can react with amino acids through the Maillard reaction.
- Pasteurization results in tagatose losses of less than 1%.
- Tagatose remains stable in acidic beverages for 8 months.
- Refrigerated storage would optimize tagatose retention in higher pH beverages.

INTRODUCTION

Tagatose Background

Tagatose is a monosaccharide naturally found in small amounts in some fruits and dairy products (Skytte, 2006). Chemically, tagatose is a ketose (Figure 74.1). The only structural difference between tagatose and fructose is the reversed orientation of the hydroxyl group on the fourth carbon (Levin, 2002). As a food ingredient, tagatose must be synthesized due to economic constraints associated with extracting its low quantities from natural sources. Tagatose can be produced through either chemical synthesis or enzymatic methods.

Interest in tagatose results from studies suggesting this monosaccharide has prebiotic properties. A prebiotic substance has been defined as a "nondigestible food ingredient that beneficially affects the host by selectively stimulating the growth and/or activity of one or a limited number of bacteria in the colon, and thus improves host health" (Gibson and Roberfroid, 1995). In the upper gastrointestinal tract, the absorption of tagatose is low (Livesey and Brown, 1996; Bertelsen *et al.*, 1999; Laerke and Jensen, 1999). The microflora of the large intestines ferment tagatose into short chain fatty acids, such as butyrate and propionate (Bertelsen *et al.*, 1999; Laerke and Jensen 1999; Laerke *et al.*, 2000; Bertelsen *et al.*, 2001; Venema *et al.*, 2005) In humans, tagatose was shown to reduce the number of coliform bacteria while promoting an increase in lactic acid bacteria (Bertelsen *et al.*, 1999).

The allowance of tagatose in foods has been approved by the European Union, Australia, New Zealand, South Africa, Korea, and Brazil (Skytte, 2006). In the United States, tagatose incorporation must not exceed specified limits, for example 33% in ready-to-eat breakfast cereals, 3% in non-diet soft drinks, and 10% in cookies (Cheeseman, 2011).

Monosaccharide Degradation

Monosaccharides, such as tagatose, can proceed through different reaction pathways, leading to their structural degradation. The composition of the monosaccharide solution, its pH, and temperature all impact the type and rate of the degradation reactions.

In the absence of amine-containing compounds, monosaccharides can undergo degradation initiated by ionization and mutarotation (Wong, 1989; Robyt, 1998).

Processing and Impact on Active Components in Food
http://dx.doi.org/10.1016/B978-0-12-404699-3.00074-3

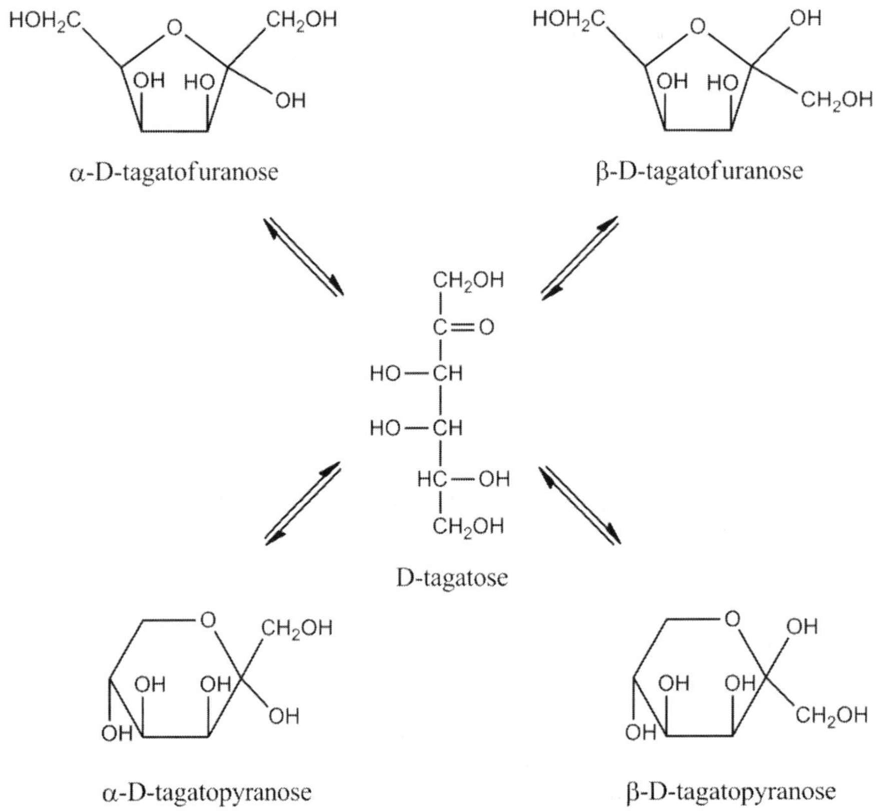

FIGURE 74.1 Chemical structures of tagatose.

Structural rearrangement yields an enediol anion that continues through other degradation pathways, including cleavage, to produce small carboxylic acids (De Bruijn *et al.*, 1986). This degradation reaction is enhanced by elevated pH (Robyt, 1998) and temperature (Wolfenden and Yuan, 2008).

In the presence of amine-containing compounds, such as amino acids, monosaccharides can react through the Maillard reaction (Hodge, 1953; Wong, 1989; Friedman, 1996). At elevated pH levels, the acyclic monosaccharide formed via mutarotation undergoes nucleophilic attack by the unprotonated amine (Wong, 1989). This non-enzymatic reaction culminates with the production of brown pigmentation (Hodge, 1953; Friedman, 1996). As the reaction progresses, the monosaccharide is structurally degraded.

Detailed mechanistic studies specifically involving tagatose degradation have not been reported. However, observations made during tagatose stability studies are consistent with the above discussion. Tagatose degradation has therefore been hypothesized to follow mechanisms similar to those of other monosaccharides (Dobbs and Bell, 2010; Luecke and Bell, 2010).

Monosaccharide degradation has been modeled using first-order kinetics (Bamford *et al.*, 1950; Wolfenden and Yuan, 2008). The first-order model was also applied to tagatose degradation in solution (Dobbs and Bell, 2010; Luecke and Bell, 2010). A graphical example of this model is shown in Figure 74.2 for tagatose degradation in 0.1 M phosphate buffer. By regressing the degradation data to the following first-order kinetic equation:

$$[\mathrm{T}] = [\mathrm{T}]_o e^{-k_{obs}t} \tag{74.1}$$

the first-order rate constant, k_{obs}, can be determined. In equation (74.1), [T] is the tagatose concentration at time, t, and $[\mathrm{T}]_o$ is the initial tagatose concentration. Rate constants give an indication of reactivity and are used to predict the extent of tagatose loss. Bell (2007) provides more details on kinetic modeling, stability testing, and shelf-life prediction.

HOW COMPOSITION IS ALTERED

When tagatose is added to beverages, its stability to thermal processing and long-term storage depends upon the characteristics of the beverage. The pH and interactions with other dissolved substances influence tagatose retention in model solutions and beverages. These aspects affecting tagatose stability are discussed subsequently.

pH

Tagatose in buffer solutions degraded faster at pH7 than pH3. In citrate and phosphate buffer, degradation occurred an average of 14 and 37 times faster,

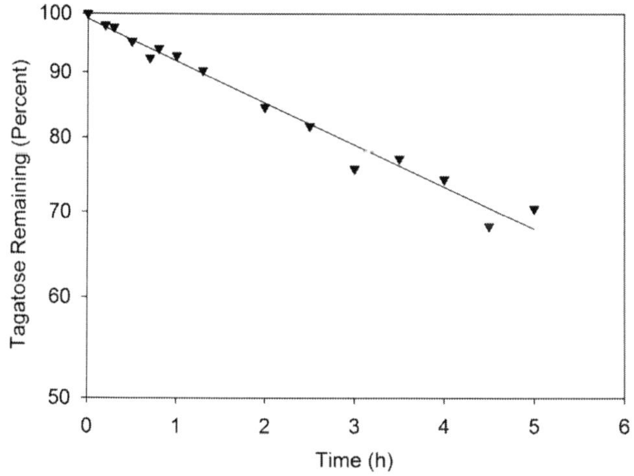

FIGURE 74.2 First-order plot for tagatose degradation in 0.1 M phosphate buffer at pH7 and 80 °C. *Adapted from Luecke and Bell (2010).*

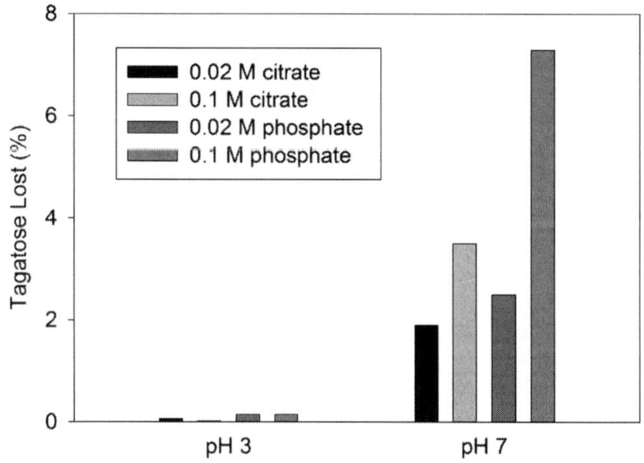

FIGURE 74.3 Extent of tagatose degradation in buffer solutions held 1 h at 80°C. *Percentages of tagatose lost were predicted using the kinetic data of Luecke and Bell (2010).*

respectively, at the higher pH (Dobbs and Bell, 2010; Luecke and Bell, 2010). The initial ionization step leading to monosaccharide degradation requires removing a proton (Wong, 1989). This proton removal occurs more efficiently at pH7 than pH3, leading to faster tagatose loss in higher pH solutions.

Bell and Luecke (2012) examined tagatose stability in two beverages: diet lemonade and reduced-fat milk. Degradation was faster in milk, at pH6.6, than in lemonade, at pH2.8. For example, the time required for a 10% tagatose loss at 81°C was 13 h in milk compared to 95 h in lemonade. It must be recognized that, in addition to differences in pH, the milk also contained proteins which could cause tagatose degradation via the Maillard reaction. Differences in both pH and beverage composition would be expected to impact the degradation rates. A discussion regarding tagatose and the Maillard reaction will be presented later.

Results regarding the effect of pH on tagatose stability are consistent between the three studies (Dobbs and Bell, 2010; Luecke and Bell, 2010; Bell and Luecke, 2012). Tagatose loss occurs more rapidly in solutions having a higher pH. However, other ingredients such as buffer salts and proteins must also be considered.

Buffer Type and Concentration

The effect of buffer type and concentration on tagatose stability was evaluated in 0.02 and 0.1 M phosphate and citrate buffers (Dobbs and Bell, 2010; Luecke and Bell, 2010). Tagatose degradation was faster in phosphate buffer than citrate buffer at pH7 (Dobbs and Bell, 2010; Luecke and Bell, 2010). In 0.1 M buffer, degradation proceeded approximately six times faster in phosphate buffer than citrate buffer at 20–40°C (Dobbs and Bell, 2010). Tagatose degradation was 2–3 times faster in phosphate buffer than citrate buffer at 60–80°C (Luecke and Bell, 2010). Only minimal differences existed between tagatose loss in phosphate and citrate buffers at pH3. These studies suggested a component present in the pH7 phosphate buffer was significantly more catalytic than citrate buffer.

In unbuffered solutions at pH3–5, no tagatose loss was observed after heating for 5 h at 100°C (Ryu *et al.*, 2003). In citrate and phosphate buffer solutions at pH7, tagatose degradation was faster at higher buffer concentrations than lower concentrations (Dobbs and Bell, 2010; Luecke and Bell, 2010). Citrate and phosphate buffer salts catalyzed the degradation reaction at pH7. Trends were less clear in pH3 buffer solutions.

The dibasic phosphate anion (HPO_4^{-2}) has been implicated in previous kinetic studies as being highly catalytic (Bell and Wetzel, 1995; Bell, 1997; Pachapurkar and Bell, 2005). Monosaccharide degradation involves proton transfers. The small dibasic phosphate anion can facilitate these proton transfers more effectively than the larger citrate anion. As buffer concentration and/or pH increase, the concentration of HPO_4^{-2} also increases, causing faster rates of tagatose degradation. At pH 3, HPO_4^{-2} is virtually non-existent so the effect of buffer concentration on tagatose degradation is more obscure. Thus, the stability of tagatose depends upon a complex interaction between pH and buffer composition.

An overview of the combined effects of pH, buffer type, and buffer concentration on tagatose degradation is visualized in Figure 74.3, using data collected at 80°C. Degradation is more extensive at pH7 than 3, in phosphate buffer than citrate buffer, and at higher buffer concentrations.

Amine-containing Compounds

In addition to degradation via intramolecular rearrangement and cleavage, tagatose may also be lost by its participation in the Maillard reaction. Ryu *et al.* (2003) described browning produced in various

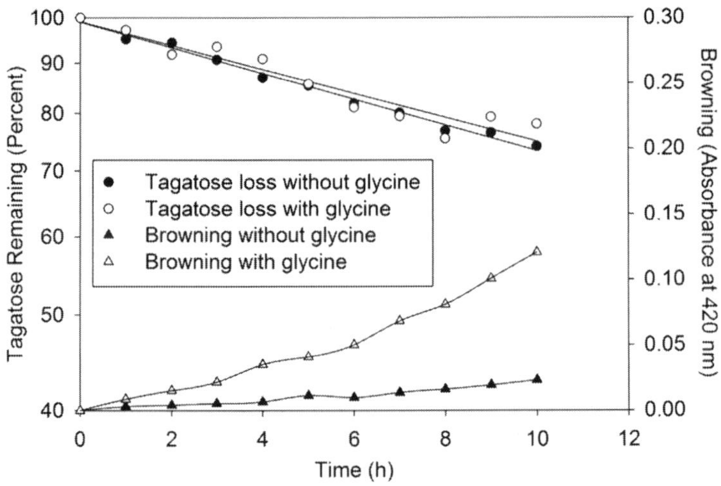

FIGURE 74.4 Tagatose degradation and accompanied browning in 0.1M phosphate buffer with and without glycine at pH7 and 70°C. Tagatose losses were similar in the two solutions, but browning was greater in the presence of glycine. *Data are from Luecke (2009).*

tagatose-glycine model systems heated at 70–100°C. As either temperature or pH increased, tagatose-glycine solutions became increasingly dark. In the absence of glycine at pH3–5, no tagatose loss was observed after heating at 100°C for 5h. Thus, Ryu *et al.* (2003) concluded that the browning came from glycine reacting with tagatose through the Maillard reaction. However, because tagatose concentrations were not specifically measured in the glycine-containing samples, the extent of its loss is unknown.

Dobbs (2008) evaluated tagatose stability and browning in buffer solutions containing equimolar tagatose and glycine at 20–40°C. Although the glycine-containing solutions showed much more browning, the rates of tagatose degradation were similar between the solutions with and without glycine. Luecke (2009), evaluating tagatose thermal stability at 60–80°C, also noted enhanced browning when glycine was added to the solution; like Dobbs (2008), glycine was found to have minimal effects on the degradation rates of tagatose. Figure 74.4 shows an example of tagatose loss and browning in the presence and absence of glycine.

The Maillard reaction between tagatose and glycine would generally be expected to enhance the loss of tagatose in comparison to similar systems without glycine. Neither the studies by Dobbs (2008) or Luecke (2009) showed enhanced tagatose degradation. Similarly, Ajandouz *et al.* (2001) reported fructose degradation in pH7 phosphate buffer was not enhanced by the presence of lysine. The complex mechanisms associated with the intramolecular rearrangement of tagatose and the Maillard reaction both begin with the acyclic sugar derived through mutarotation. Low concentrations of tagatose equilibrate into its acyclic structure (Figure 74.1). Acyclic tagatose then rearranges into an enediol, which may (a) proceed through a series of reactions leading to small organic acids, (b) isomerize into an aldose, which then reacts reversibly with glycine through the Maillard reaction, or (c) revert back to the ketose, which may react reversibly with glycine. The current hypothesis is that an undetermined rate-limiting step exists in the early transformation pathways of acyclic tagatose such that its degradation rate does not depend upon subsequent faster reactions. Consistent with data in Figure 74.4, rates of tagatose loss would be similar with and without the amino acid, but browning would be more pronounced via the Maillard reaction pathway.

As discussed earlier, tagatose degradation was faster in reduced-fat milk than diet lemonade at 20–80°C (Bell and Luecke, 2012). The milk became dark brown during the experiments whereas the lemonade did not show noticeable darkening (Bell, unpublished observation). Based on the study by Ryu *et al.* (2003), it was speculated that the Maillard reaction was occurring within the milk samples. Tagatose degradation in milk was undoubtedly enhanced by its higher pH; however, whether the rate of tagatose loss was also affected by the dairy proteins cannot be stated with certainty.

OTHER WAYS IN WHICH COMPOSITION IS ALTERED

In addition to the characteristics of the tagatose solution, processing and storage temperatures affect tagatose retention as well. The effect of temperature on tagatose degradation is governed by the activation energy, which is also used for shelf-life predictions (Bell, 2007). In order to predict either the product shelf-life or extent of tagatose loss, degradation data must be collected at several temperatures.

Three studies evaluated tagatose degradation in solutions as a function of temperature (Dobbs and Bell, 2010; Luecke and Bell, 2010; Bell and Luecke, 2012). Because the extents of degradation and reaction rates were extremely low in acidic solutions, activation energies could not be

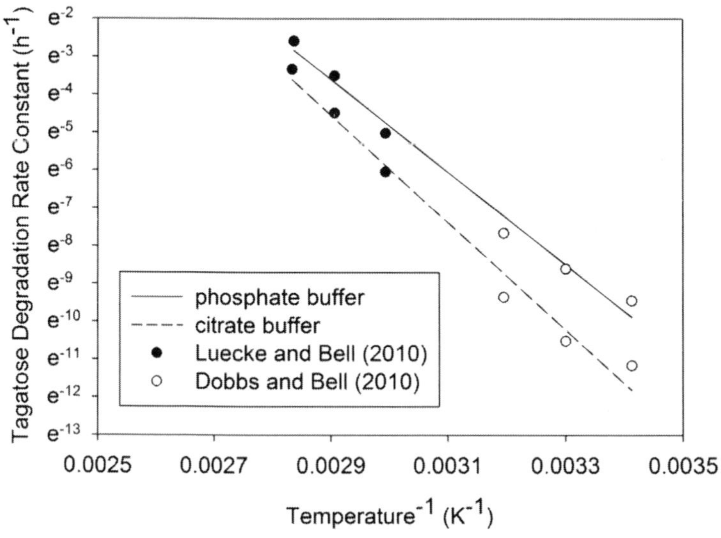

FIGURE 74.5 Effect of temperature on tagatose degradation in 0.1 M buffer at pH 7. *Data from Dobbs and Bell (2010) and Luecke and Bell (2010) were combined to construct these Arrhenius plots.*

reliably determined. Dobbs and Bell (2010) examined the storage stability of tagatose at 20, 30, and 40°C; activation energies for tagatose degradation in buffer solutions at pH 7 ranged from 12.6–16.4 kcal/mol. Luecke and Bell (2010) examined the thermal stability of tagatose at 60, 70, and 80°C; activation energies for its degradation in pH 7 buffer were 31.4–38.7 kcal/mol. To improve the reliability of the regression, the two data sets were combined. Figure 74.5 shows the collective data for tagatose degradation in 0.1 M phosphate buffer and citrate buffer at pH 7 in the form of Arrhenius plots. Using Arrhenius kinetics (Bell, 2007), activation energies for tagatose degradation in phosphate and citrate buffer over the 20–80°C temperature range were calculated. In phosphate buffer, activation energies were 23.3–24.4 kcal/mol while in citrate buffer they were 26.2–28.3 kcal/mol. The lower activation energies for tagatose degradation in phosphate buffer are due to the catalytic ability of the dibasic phosphate anion, as discussed earlier. These activation energies are similar to the value of 24.6 kcal/mol for tagatose degradation in reduced-fat milk at pH 6.6 (Bell and Luecke, 2012).

Using these activation energies, the extent of tagatose degradation under various scenarios can be determined. Table 74.1 lists the amount of tagatose lost in four buffer solutions and milk during various types of processing and storage conditions. Pasteurization (basic, HTST, UHT) of tagatose-containing solutions at pH 7 would cause losses of less than 1%. Because tagatose degradation rate constants were much lower at pH 3 (Dobbs and Bell, 2010; Luecke and Bell, 2010), less tagatose would be lost during the pasteurization of acidic beverages. The thermal processing required for sterilization would cause greater amounts of tagatose loss. During refrigerated storage, very little tagatose would be lost after 6 months. However, at 25–30°C, large amounts of tagatose would degrade in pH 7 solutions during 6 months of storage.

TABLE 74.1 Tagatose Loss (%) from Various Solutions after Specified Treatments

	Phosphate Buffer, pH 7		Citrate Buffer, pH 7		Reduced-Fat Milk
Treatment	**(0.02 M)**	**(0.1 M)**	**(0.02 M)**	**(0.1 M)**	**(pH 6.6)**
Processing					
63 °C/30 min[a]	0.13	0.54	<0.1	0.18	<0.1
72 °C/15 sec[b]	<0.1	<0.1	<0.1	<0.1	<0.1
135 °C/5 sec[c]	0.16	0.94	0.19	0.86	<0.1
118 °C/12 min[d]	6.5	31	6.4	24	3.2
Storage					
4 °C/180 d	0.67	2.0	0.15	0.20	0.16
25 °C/180 d	12	36	4.0	7.0	3.6
30 °C/180 d	22	59	8.1	15	7.1

[a]*Basic pasteurization.*
[b]*High temperature short time (HTST) pasteurization.*
[c]*Ultrahigh temperature (UHT) pasteurization.*
[d]*Sterilization.*
Values are calculated using data from Dobbs and Bell (2010), Luecke and Bell (2010), and Bell and Luecke (2012).

Table 74.1 also shows the improvement in tagatose retention by lowering buffer concentration and using citrate buffer instead of phosphate buffer. Again, for acidic beverages, much less loss would occur during storage. For example, virtually no tagatose was lost from diet lemonade held at 40°C for 250 days (Bell and Luecke, 2012).

ANALYTICAL TECHNIQUES

To study the stability of a biologically active substance in beverages, a reliable analytical method is required. BeMiller (2003) described methods appropriate for sugar analysis. For quantification of sugars, chromatographic methods are frequently employed. High performance liquid chromatography (HPLC) can analyze sugars directly while gas chromatography requires derivatization into volatile compounds. Therefore, HPLC has become the method of choice. Separation and analysis can occur using anion-exchange, cation exchange, normal phase, or reversed-phase chromatography. Electrochemical detection can be used with anion-exchange chromatography, however refractive index detection is most commonly used for analyzing sugars via HPLC.

The analysis of tagatose in solutions has recently involved reversed-phase HPLC with refractive index detection (Luecke and Bell, 2010; Bell and Luecke, 2012). The stationary phase consisted of a 250 × 4.6 mm LUNA 5 μ amino column (Phenomenex, Torrance, California, USA). The mobile phase of 89% (w/w) acetonitrile and 11% (w/w) water flowed isocratically at 3 mL/min. The column and detector were both maintained at a temperature of 40°C. Tagatose eluted around 7 min. For analyzing tagatose in buffer solutions, the above procedure produced average recoveries of 99.5–102.6% with coefficients of variation less than 2% (Luecke and Bell, 2010). Recovery of tagatose from actual beverages was 93.6% in milk and 104.1% in lemonade, with coefficients of variation less than 5% (Bell and Luecke, 2012). The primary limitation observed with this HPLC method is its inability to separate tagatose from fructose. Thus, analysis of tagatose in beverages also containing fructose would require a different methodology.

References

Ajandouz, E.H., Tchiakpe, L.S., Ore, F.D., Benajiba, A., Puigserver, A., 2001. Effects of pH on caramelization and Maillard reaction kinetics in fructose-lysine model systems. J. Food Sci. 66, 926–931.

Bamford, C.H., Bamford, D., Collins, J.R., 1950. Kinetic studies on carbohydrates in alkaline conditions. II. The kinetics of the rearrangements of glucose and fructose in alkaline solution. Proc. Royal Soc. London. Ser. A. Math. Phys. Sci. 204 (1076), 85–98.

Bell, L.N., 1997. Maillard reaction as influenced by buffer type and concentration. Food Chem. 59, 143–147.

Bell, L.N., 2007. Nutraceutical stability concerns and shelf life testing. In: Wildman, R.E.C. (Ed.), Handbook of Nutraceuticals and Functional Foods, second ed. CRC Press, Boca Raton, FL, pp. 467–483.

Bell, L.N., Luecke, K.J., 2012. Tagatose stability in milk and diet lemonade. J. Food Sci. 71 (1), H36–H39.

Bell, L.N., Wetzel, C.R., 1995. Aspartame degradation in solution as impacted by buffer type and concentration. J. Agric. Food Chem. 43, 2608–2612.

BeMiller, J.N., 2003. Carbohydrate analysis. In: S.S. Nielsen (Ed.), Food Analysis, third ed. Kluwer Academic, New York, pp. 143–174.

Bertelsen, H., Andersen, H., Tvede, M., 2001. Fermentation of D-tagatose by human intestinal bacteria and dairy lactic acid bacteria. Microb. Ecol. Health. Dis. 13, 87–95.

Bertelsen, H., Jensen, B.B., Buemann, B., 1999. D-tagatose—a novel low-calorie bulk sweetener with prebiotic properties. World Rev. Nutr. Diet. 85, 98–109.

Cheeseman, M.A., 2011. Agency Response Letter GRAS Notice No. GRN 000352. Available at: http://www.fda.gov/Food/FoodIngredientsPackaging/GenerallyRecognizedasSafeGRAS/GRASListings/ucm245241.htm (accessed 20.11.12.).

De Bruijn, J.M., Kieboom, A.P.G., van Bekkum, H., van der Poel, P.W., 1986. Reactions of monosaccharides in aqueous alkaline solutions. Sugar Technol. Rev. 13, 21–52.

Dobbs, C.M., 2008. Storage stability of tagatose in buffer solutions of various composition. M.S. Thesis. Auburn University, Auburn, Alabama.

Dobbs, C.M., Bell, L.N., 2010. Storage stability of tagatose in buffer solutions of various composition. Food Res. Int. 43, 382–386.

Friedman, M., 1996. Food browning and its prevention: an overview. J. Agric. Food Chem. 44, 631–653.

Gibson, G.F., Roberfroid, M.B., 1995. Dietary modulation of the human colonic microbiota: introducing the concept of prebiotics. J. Nutr. 125, 1401–1412.

Hodge, J.E., 1953. Chemistry of browning reactions in model systems. J. Agric. Food Chem. 1, 928–943.

Laerke, H.N., Jensen, B.B., 1999. D-tagatose has low small intestinal digestibility but high large intestinal fermentability in pigs. J. Nutr. 129, 1002–1009.

Laerke, H.N., Jensen, B.B., Hojsgaard, S., 2000. In vitro fermentation pattern of D-tagatose is affected by adaptation of the microbiota from the gastrointestinal tract of pigs. J. Nutr. 130, 1772–1779.

Levin, G.V., 2002. Tagatose, the new GRAS sweetener and health product. J. Med. Food 5, 23–36.

Livesey, G., Brown, J.C., 1996. D-tagatose is a bulk sweetener with zero energy determined in rats. J. Nutr. 126, 1601–1609.

Luecke, K.J., 2009. Thermal stability of tagatose in solution. M.S. thesis. Auburn University, Auburn, Alabama.

Luecke, K.J., Bell, L.N., 2010. Thermal stability of tagatose in solution. J. Food Sci. 75 (4), C346–351.

Pachapurkar, D., Bell, L.N., 2005. Kinetics of thiamin degradation in solutions under ambient storage conditions. J. Food Sci. 70, C423–426.

Robyt, J.F., 1998. Essentials of Carbohydrate Chemistry. Springer-Verlag, New York.

Ryu, S.Y., Roh, H.J., Noh, B.S., Kim, S.Y., Oh, D.K., Lee, W.J., Yoon, J.R., Kim, S.S., 2003. Effects of temperature and pH on the non-enzymatic browning reaction of tagatose-glycine model system. Food Sci. Biotechnol. 12, 675–679.

Skytte, U.P., 2006. Tagatose. In: Mitchell, H. (Ed.), Sweeteners and Sugar Alternatives in Food Technology. Blackwell Publishing, Ames, Iowa, pp. 262–294.

Venema, K., Vermunt, S.H.F., Brink, E.J., 2005. D-tagatose increases butyrate production by the colonic microbiota in healthy men and women. Microb. Ecol. Health Dis. 17, 47–57.

Wolfenden, R., Yuan, Y., 2008. Rates of spontaneous cleavage of glucose, fructose, sucrose, and trehalose in water, and the catalytic proficiencies of invertase and trehalas. J. Am. Chem. Soc. 130, 7548–7549. Supplemental information available at: http://pubs.acs.org/doi/suppl/10.1021/ja802206s. (accessed 27.07.11.).

Wong, D.W.S., 1989. Mechanism and Theory in Food Chemistry. Van Nostrand Reinhold, New York.

CHAPTER

75

High-Pressure Processing, Strawberry Beverages, and Composition of 'Bioactives'

Ravi Kiran Tadapaneni, Indika Edirisinghe, Britt Burton-Freeman

Institute for Food Safety and Health, Illinois Institute of Technology, Illinois, USA

INTRODUCTION

Part of the secret of success in life is to eat what you like and let the food fight it out inside. ***Mark Twain***

Provocative indeed; however, within only 100 years of Mark Twain's death, modern lifestyle and dietary habits have changed remarkably. Habits have changed so much so that careers are built on the science, policy, and practice of food and nutrition due to its importance in health and disease risk. Even among the lay public, food, nutrition, and health are a central focus. Paradoxically, current statistics indicate an alarming number of deaths (approximately 1.7 million in 2008) due to poor dietary habits; specifically highlighting that fruits and vegetables are consumed well below recommendations for health (Global Health Observatory, 2008;USDA and USDHHS, 2010). Several reasons have been suggested to explain the low rates of fruit and vegetable intake in America and globally. Convenience, cost, availability (of quality produce) are among the cited reasons; however, in some parts of the world, under-developed farming practices to meet the needs of citizens year-round is another barrier. Preservation technology has a role to play in all of these circumstances; however, care in its application to maintain quality is of paramount importance.

The art of processing food has been in practice since the existence of ancient civilizations when sun-drying was used to extend the shelf-life of fresh fruits and vegetables. With technological advances, processing techniques have evolved; delivering a wide variety of safe, extended shelf-life and shelf-stable food products to the marketplace, including offering greater availability of perishable food commodities in communities with limited access. Accordingly, modern food processing techniques have improved the ease of access to certain foods as well as increased availability of non-seasonal food products year-round.

In as much as food processing adds value to the safety of our food supply and access to a wider variety of foods throughout the year, a major issue with processing is that it can negatively impact nutrient content; and hence, nutritional quality of products. This is a well-known issue and has been the impetus behind enrichment laws and practices globally. Enrichment is synonymous with fortification except it refers to replacing nutrients that are lost during processing (FAO/WHO, 1994). In the US, the addition of thiamin, niacin, folic acid, riboflavin, and iron to white flour is required by law (US Food and Drug Administration, 1941) due to the processing losses and public health concerns of nutrient deficiency, since flour is a staple in the American diet. Other examples of processing-induced losses may be less apparent on a public health scale, but nonetheless, may be important when cumulative intake across product categories is considered. Historically, replacement of processing-induced nutrient losses has focused on essential micronutrients. However, current knowledge suggests that we should also be concerned with losses of non-essential nutrient-like components of plants foods. Dietary fiber and co-existing essential and non-essential bioactive whole-grain components are examples where losses have been reconsidered in the refining process. The potential loss of polyphenols and other non-essential biologically active plant components should also be considered when processing foods. This chapter provides insight into recent developments and supporting research in food processing from the perspective of delivering on nutritional quality as well as food safety demands.

BIOACTIVE COMPOUNDS AND HEALTH

'Bioactive' compounds have been defined as "the extra-nutritional constituents that typically occur in small quantities of foods" (Kris-Etherton *et al.*, 2002).

Processing and Impact on Active Components in Food
http://dx.doi.org/10.1016/B978-0-12-404699-3.00075-5

The essential nutrients are also biologically active, but in the context of this chapter, bioactives will be used in reference to biologically active phytochemcials, unless otherwise indicated. Chemical characterization of bioactive compounds in foods has been a major area of research emphasis in recent years. Fruits and vegetables contain a varied composition of 'bioactive' compounds, many of which fall into the large family of compounds known as polyphenols. Early reports suggested that polyphenols are anti-nutritional components of plants (Bravo, 1998). Nowadays, there are sufficient data to support the supposition that polyphenolic compounds have a strong potential to alter disease risk profiles in humans through their ability to modulate various biological pathways (Vauzour *et al.*, 2010; González-Gallego *et al.*, 2010; Joseph *et al.*, in press).

POLYPHENOLS IN STRAWBERRY

Strawberries are one of the most well-liked fruits globally. Strawberries contain appreciable amounts of essential nutrients as well as non-essential bioactive polyphenols, including ellagitannins, flavanols, and anthocyanins (Wu and Prior 2005; Edirisinghe *et al.*, 2011; Seeram *et al.*, 2006) (Figure 75.1). Anthocyanins are highly pigmented polyphenolic compounds found in many colored (red, purple, and blue) fruits and vegetables, such as strawberries and other berries, red grapes, red cabbage, purple carrots. Anthocyanins are glycoside and acyl-glycoside derivatives of six common anthocyanidins: pelargonidin, cyanidin, delphinidin, peonidin, petunidin, and malvidin. Strawberries are naturally high in anthocyanins (Wu and Prior 2005). Strawberries are particularly rich in pelargonidin-based anthocyanins (Table 75.1 and Figure 75.2), making them one of the most significant contributors of pelargonidin anthocyanins in the diet. However, their concentrations can differ from the raw fruit when included in a formulated matrix, such as a dairy based beverage (Table 75.1). Studies with anthocyanins have suggested that these brightly colored compounds are associated with a wide variety of health benefits, likely due to their effects on processes underlying inflammation and oxidative stress (Edirisinghe *et al.*, 2011; Halvorsen *et al.*, 2006; Burton-Freeman *et al.*, 2010). Anti-inflammatory and anti-thrombotic effects have also been reported after 6 weeks supplementation (Ellis *et al.*, 2011). These data provide the impetus for innovation in processing to deliver the rich bioactive profile of strawberries in a convenient, available form so that consumers have more opportunities to consume strawberries. Similar considerations for other fruits may be appropriate where science is emerging showing health benefits, and where fresh fruit may not be practical to deliver in all markets.

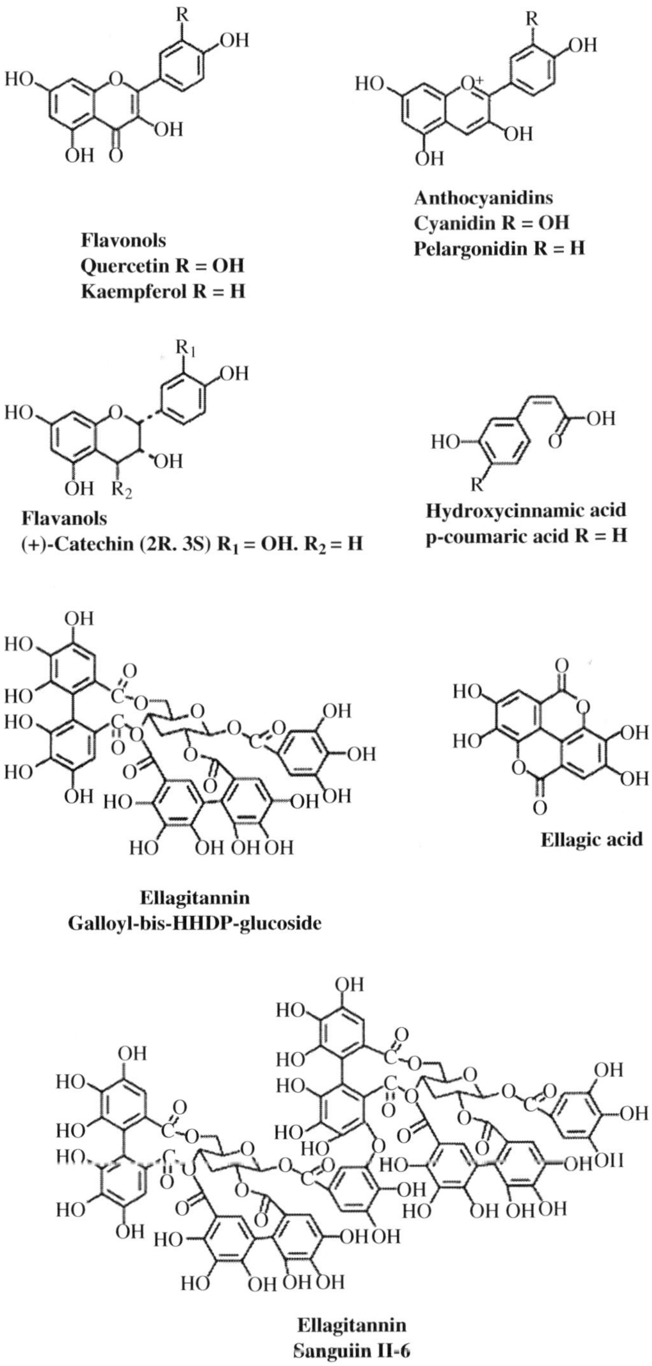

FIGURE 75.1 Isolated phenolic compounds from strawberry. *Reproduced with permission from Seeram* et al. *(Seeram, N. P., Lee, R., Samuel Scheuller, H., and David, H.* Food Chem. *97, 1–11). Copyright (2006) Elsevier.*

PROCESSING EFFECTS ON FOOD/ BEVERAGE QUALITY

Traditionally, the food industries have been using thermal technologies to pasteurize and sterilize foods. Thermal processes are the established industry standard and involve subjecting foods/beverages to elevated temperatures for short periods of time with the goal of

TABLE 75.1 Bioactive Composition of Freeze Dried Strawberry Powder and its Beverages as Reported by Edirisinghe *et al.* (2011)

Compound	Ion (+/–)	MS m/z	MS2 m/z	Strawberry Powder	Strawberry Beverage	Placebo Beverage
1. Pelargonidin-3-diglucoside	+	595	433, 271	0.37 ± 0.41[a]	0.55 ± 0.01[a]	ND
2. Ellagitannin	–	789	775, 753, 633	0.40 ± 0.41[a]	0.00 ± 0.00[b]	ND
3. Galloyl-HHDP-glucose	–	633	481, 301	6.52 ± 0.19[a]	11.44 ± 0.23[b]	0.15 ± 0.04[c]
4. bis-HHDP-glucose	–	783	481, 301, 257, 229	2.64 ± 0.23[a]	2.21 ± 0.13[a]	0.022 ± 0.06[b]
5. Proanthocyanidin B1	–	577	425, 407	0.71 ± 0.06[a]	0.61 ± 0.01[a]	ND
6. Quercetin	–	301	151, 179	1.88 ± 0.06[a]	1.50 ± 0.05[a]	0.019 ± 0.06[c]
7. Cyanidin-3-*O*-glucoside	+	449	287	0.80 ± 0.02[a]	1.61 ± 0.04[b]	0.066 ± 0.04[c]
8. (+)-Catechin	–	289	245	1.97 ± 0.4[a]	1.08 ± 0.01[a]	ND
9. p-Coumaroylhexose-4-*O*-hexoside	–	487	325, 163, 119	0.08 ± 0.0[a]	0.11 ± 0.01[a]	ND
10. Pelargonidin-3-*O*-glucoside	+	433	271	67.99 ± 5.1[a]	28.11 ± 0.72[b]	1.246 ± 0.02[c]
11. Proanthocyanidin trimer	–	865	739, 695	0.49 ± 0.01[a]	0.12 ± 0.00[a]	ND
12. p-Coumaroylhexose	–	325	163, 119	6.65 ± 0.38[a]	18.34 ± 0.26[b]	ND
13. Pelargonidin-3-*O*-rutinoside	+	579	433, 271	2.35 ± 0.13[a]	5.43 ± 0.12[b]	0.312 ± 0.18[c]
14. Quercetin-3-*O*-glucoside	–	463	301, 151, 179	0.57 ± 0.04[a]	0.61 ± 0.06[a]	ND
15. Kaempferol	–	285	285	0.64 ± 0.04[a]	0.24 ± 0.01[b]	ND
16. 5-Pyranopelargonidin-3-glucoside	+	501	339, 295	0.42 ± 0.03[a]	0.54 ± 0.01[b]	0.003 ± 0.20[c]
17. Pelagonidin-3-malonylglucoside	+	519	433, 271	9.48 ± 0.58[a]	2.48 ± 0.06[b]	0.056 ± 0.05[c]
18. Ellagic acid pentoside	–	433	301, 229, 257	2.88 ± 0.50[a]	0.20 ± 0.00[b]	ND
19. Ellagic acid (EA)	–	301	257, 229, 185	1.01 ± 0.04[a]	1.13 ± 0.09[a]	ND
20. Quercetin-3-*O*-rutinoside	–	609	301, 151, 179	0.26 ± 0.04[a]	0.78 ± 0.06[b]	ND
21. Galloyl-bis-HHDP-glucoside	–	935	633, 301	6.40 ± 0.50[a]	4.54 ± 0.10[a]	0.012 ± 0.09[b]
22. Quercetin-3-*O*-glucuronide	–	477	301, 151, 179	10.46 ± 1.48[a]	10.99 ± 0.17[a]	0.412 ± 0.27[b]
23. Pelagonidin-3-malonylrhamnoside	+	503	271	0.24 ± 0.00[a]	0.32 ± 0.00[a]	0.002 ± 0.05[b]
24. Quercetin-3-malonylglucoside	–	549	301, 151, 179	0.13 ± 0.01[a]	0.23 ± 0.01[a]	0.056 ± 0.17[b]
25. Kaempferol-3-*O*-glucuronide	–	461	285	0.72 ± 0.08[a]	0.66 ± 0.00[a]	0.012 ± 0.07[b]
26. Kaempferol-3-malonylglucoside	–	533	285	0.15 ± 0.02[a]	0.13 ± 0.00[a]	ND
27. Kaempferol-3-coumaroylglucoside	–	593	447, 285	0.22 ± 0.04[a]	0.64 ± 0.01[b]	ND

(Continued)

TABLE 75.1 Bioactive Composition of Freeze Dried Strawberry Powder and its Beverages as Reported by Edirisinghe *et al.*—cont'd

Compound	Ion (+/−)	MS m/z	MS2 m/z	Strawberry Powder	Strawberry Beverage	Placebo Beverage
Total detected polyphenols				126.45 ± 9.2[a]	94.66 ± 2.17[b]	2.37 ± 0.00[c]
Total detected Anthocyanins				81.65 ± 5.90[a]	39.04 ± 0.96[b]	1.68 ± 0.00[c]
ORAC Value (μM Trolox equivalents)				7700 ± 20[a]	5163 ± 176[b]	1006 ± 79.3[c]

ND, not detected, Values in the same row with different superscript are significantly different $P < 0.05$, n = 3.
Reproduced with permission from Edirisinghe et al. *(Edirisinghe, I., Banaszewski, K., Cappozzo, J., Krishnankutty, S., Ellis, C. L., Tadapaneni, R. K., Kappagoda, C. T., and Burton-Freeman, B. M.).* Br. J. Nutr. *106, 913–922). Copyright (2011). Cambridge University Press.*

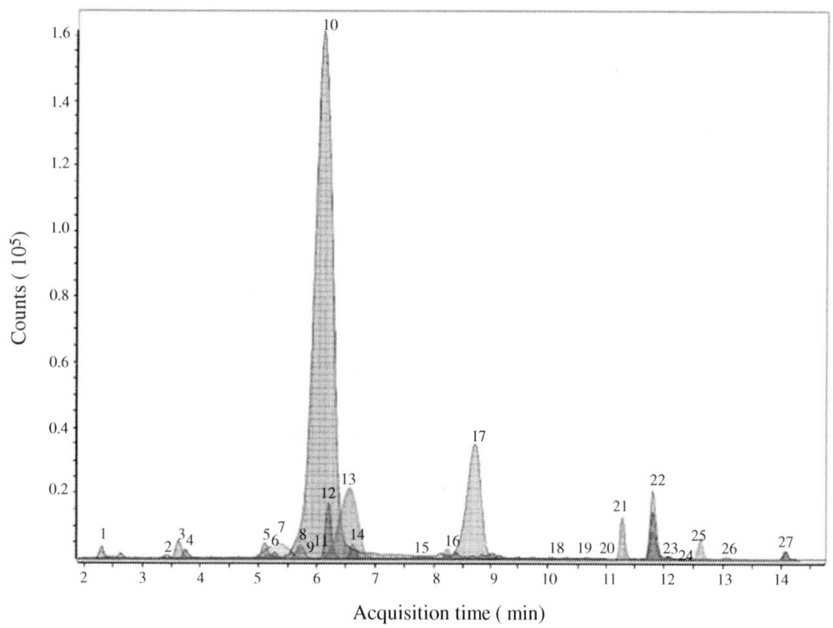

FIGURE 75.2 Phenolic composition of strawberry-dairy beverage containing 10 g freeze dried strawberry powder, as determined by LC-MS/MS; Phenolic compounds were identified by retention times relative to external standards and MS² fragmentation patterns. Peaks 1–27 denote different compounds detected in freeze dried strawberry powder; available reference standards were used to quantify identified peaks. *Reproduced with permission from Edirisinghe* et al. *(Edirisinghe, I., Banaszewski, K., Cappozzo, J., Krishnankutty, S., Ellis, C. L., Tadapaneni, R. K., Kappagoda, C. T., and Burton-Freeman, B. M.* Br. J. Nutr. *106, 913–922). Copyright (2011) Cambridge University Press.*

preventing growth of pathogenic microorganisms and spoilage-causing yeast and mold, extending protection over a period of time. With the successful application of thermal processing, more people have access to a greater variety of safe foods and beverages. However, the advantages of thermal treatment have a trail of some serious detrimental alterations in food. While effective in killing detrimental microbiological intruders in foods and beverages, thermal processing can also destroy the functionality and flavors of many foods/beverages. Heating techniques can also significantly destroy heat-labile nutrients such as certain vitamins or bioactive phytochemicals. Ascorbic acid, for example, is heat liable and therefore canned fruits have a lower content of ascorbic acid than fresh fruits (Rickman *et al.*, 2007). Thermal applications can also induce unintended chemical interactions resulting in new compounds that may have unfavorable health effects or render the parent forms inactive (van Boekel *et al.*, 2010). Quality aspects related to flavor and color can also be compromised when thermally processing fruit into juice concentrate and storing at room temperature (Lundhal *et al.*, 1989). Patras *et al.* (2009) reported reduced anti-radical power of thermally processed strawberry purée compared to unprocessed purée (1.55 to 1.16 g/l) as measured using 2,2-diphenyl-1-picrylhydrazyl radical (DPPH•) assay. Likewise, work by Aaby *et al.* (2007) reported that exposure to higher temperatures (75–80°C) reduces the anthocyanin and total ascorbic acid content when strawberry is processed into purée.

At a time when consumers are more informed and have a greater appreciation for nutrition and the impact of diet on health and quality of life, they are demanding foods and beverages that are fresher, have higher nutrient content, are more natural, additive-free, and minimally processed. Alternative food processing strategies and novel technologies are emerging and are under broad development to meet consumer demands.

THERMAL PROCESSING ALTERNATIVES

In order to reduce the negative effects of heat treatments in foods, alternative processing technologies capable of inactivating microorganisms at temperatures below those used during thermal processing are being sought (Farkas, 1998; Gervilla *et al.*, 1997). Non-thermal technologies have emerged as attractive mild preservation alternatives to thermal processing techniques. The objective of non-thermal processes is to produce more fresh-like, flavorful, colorful and nutrient rich food products. High-pressure processing (HPP), pulsed electric fields treatment (PEF), light technologies (UV/Pulsed UV), cold plasma, and bio-preservatives are examples of mild preservation, non-thermal technologies. They are usually applied at (or below) room temperature and have a trivial impact on the quality and freshness of food products (Hugas *et al.*, 2002) and may also preserve nutrients. Klopotek *et al.* (2005) suggested that the decreases they observed in total anthocyanins and hydrophilic antioxidant activity in strawberry beverages, particularly the nectar and juice beverage after thermal treatments may be minimized if milder processing techniques could be used. The temperatures to which foods are exposed in these newer advanced processes are relatively low and may be below pasteurization temperatures (Butz and Tauscher, 2002; Gerlach *et al.*, 2008; Oey *et al.*, 2008). For this reason, there is tremendous potential for production of superior quality products, especially as it relates to higher nutrient content.

Research on processing technologies that retain fresh-like appearance and high nutrient values has made tremendous progress in the last decade and will continue to develop in the future. In addition to nutritional and food safety quality, other benefits of these novel technologies may include reduction in material, energy and time costs (Farkas, 2011). Taste retention, shelf-life extension, and fewer or no chemical additives are also among the benefits touted, particularly among HPP users; making this a technology to watch as trends develop in minimally processed foods.

HIGH-PRESSURE PROCESSING (HPP) AS AN EMERGING TECHNOLOGY

Among the non-thermal approaches, HPP is considered to be a promising technology for the food industry; as it allows for developing new shelf stable foods and extending the shelf-life of current foods while preserving high nutritional value and excellent organoleptic characteristics (Préstamo *et al.*, 1999). The introduction of HPP as a food processing technique dates back to the late nineteenth century when B. Hite from West Virginia University had treated milk with HPP. He claimed that pressure can be used as an alternative preservation method to thermal processing (Hite, 1899). In the twentieth century, more in-depth research was conducted using HPP on a broader range of food products and commodities. With the concept that 'pressure' could be used as a mode to process food, products treated with HPP have increasingly been introduced to the consumer market (Hayashi, 1996). Formation of research consortiums across the globe and technological support from HPP manufacturing companies such as Avure Technologies Inc, Elmhurst Research Inc, NC Hyperbaric, and many other established firms, has helped aid the acceptance and implementation of HPP as a viable commercial alternative (to thermal) processing method. In HPP, foods are subjected to high pressures by a hydraulic system in a pressure sustainable cylinder, generally in the range of 400–800 MPa with low holding time at a lower temperature than thermal treatments (Patazca *et al.*, 2007) (Figure 75.3). At these elevated pressures micro-organisms and enzymes are inactivated without degradation of flavor and presumably nutrients, which is normally observed after conventional processing with high temperature (Señorans *et al.*, 2003). Current commercial applications of HPP include low acid food products such as fruit juices in addition to other products like lunch meats and ready to eat meals, seafood and dairy products.

IMPACT OF HPP ON BIOACTIVES

The research studies investigating the effects of HPP on nutritive quality of foods, especially fruits and vegetables, suggest that HPP could be an effective processing method for retaining health promoting compounds such as anthocyanins and other polyphenolic compounds. In contrast to losses of bioactive composition in fruit juices due to thermal processing (Wang and Xu, 2007), HPP offers the benefit of retaining the nutritive value in juices without compromising food safety. Matser *et al.* (2004) claimed to preserve heat-sensitive bioactive phytochemicals by sterilizing herbs and essential oils using HPP treatment in comparison to conventional retort process. In a study investigating total antioxidant capacity of reconstituted orange juice after processing, HPP compared to conventional thermal pasteurization processing resulted in higher total antioxidant retention under varied storage conditions (Polydera *et al.*, 2004). Others have indicated that the bioactive composition of strawberries

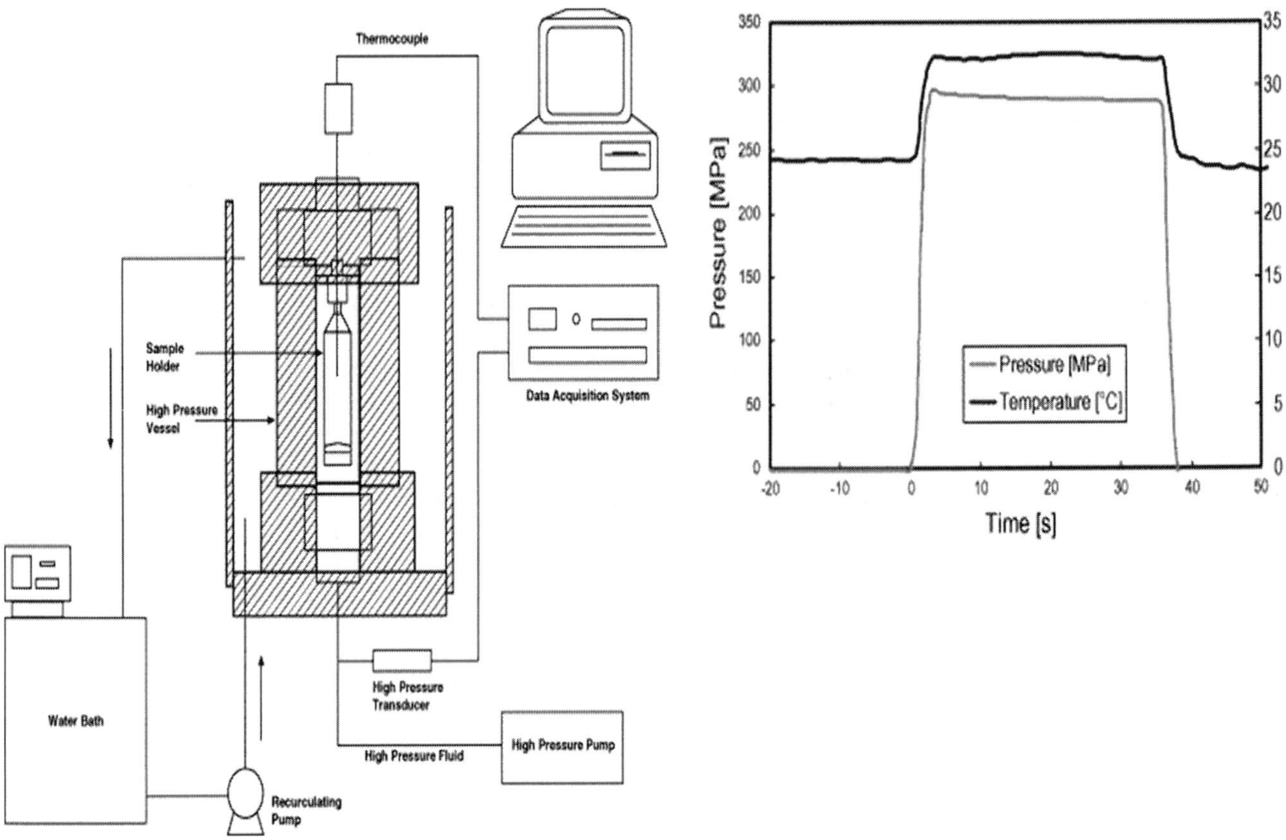

FIGURE 75.3 Schematic presentation of HPP assembly unit with time-temperature and pressure profiles of hydraulic medium in pressurized chamber. *Reproduced with permission from Patazca* et al. *(Patazca, E., Koutchma, T., and Balasubramaniam, V. M.* J. Food Eng. *80, 199–205). Copyright (2007) Elsevier.*

is stable under pressures of 400–600 MPa; although slight reductions in the anthocyanin content, reasoning to the interactions with flavanoid compounds, are detectable under higher pressures (Tiwari *et al.*, 2009). Patras *et al.* (2009) demonstrated that processing strawberry purée with HPP resulted in minor color (red) reductions in the fruit product in contrast to thermal processed purées. Furthermore, Patras *et al.* reported that HPP treatment of strawberry purées resulted in more than 90% ascorbic acid retention, whereas marked losses of ascorbic acid were evident after heat processing. At a pressure of 600 MPa, total phenolics concentrations in strawberry purée were increased by 9.8% and at 400–600 MPa the anthocyanin content was unaffected compared to thermal processing (Patras *et al.*, 2009) (Figure 75.4).

The effect of matrix may also play a role in HPP processing outcomes. We studied the effects of thermal versus HPP of strawberry-based beverages formulated using milk *versus* water on antioxidant capacity as measured by the oxygen radical absorbance capacity (ORAC) and the ferric reducing antioxidant power (FRAP) methods (Tadapaneni *et al.*, 2012). Further, we also assessed anthocyanins and quercetin content. The matrix alone significantly altered antioxidant capacity and flavonoid content of the beverages. Strawberry beverages containing dairy had significantly lower antioxidant capacity as measured by both ORAC and FRAP, and lower flavonoid content compared to the strawberry beverage prepared with water (non-dairy). After processing with high temperatures for pasteurization, antioxidant capacity and flavonoid content decreased significantly compared to the unprocessed beverages regardless of matrix. However, differential responses to HPP were evident depending on matrix. Strawberry beverages without milk showed no significant difference in ORAC values after HPP with pressures up to 600 MPa compared to unprocessed beverages, whereas applying higher pressures (700–800 MPa) resulted in significantly lower ORAC compared to unprocessed beverages. In contrast, beverages formulated with milk exhibited significantly higher ORAC values when treated with HPP up to 600 MPa compared to unprocessed beverages (Tadapaneni *et al.*, 2012) (Figure 75.5, ORAC). Measurements in antioxidant capacity by FRAP showed that the strawberry beverage without milk had significantly lower FRAP values after HPP at all pressures tested (200–800 MPa). However, formulating the strawberry beverage with

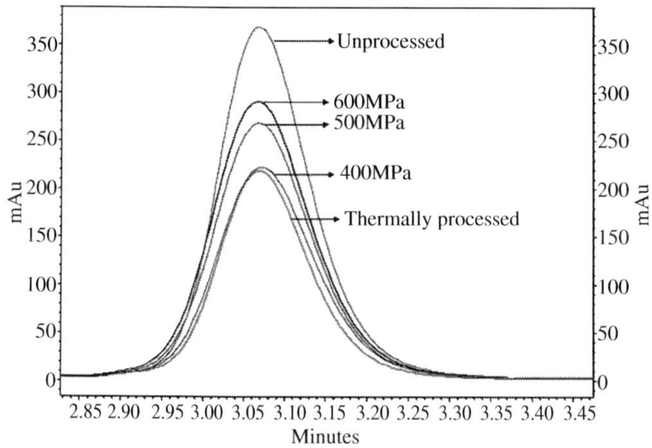

FIGURE 75.4 Chromatogram peaks of major anthocyanin in strawberry, pelargonidin-3-glucoside for freeze dried strawberry purée under different process conditions; thermally processed: 70°C for 2 min/HPP: 400–600 MPa at 20°C. *Reproduced with permission from Patras* et al. *(Patras A., Brunton N. P., Da Pieve S., and Butler F.* Innovat. Food Sci. Emerg. Tech. *10, 308–13). Copyright (2009) Elsevier.*

milk resulted in no change in FRAP after HPP compared to the unprocessed beverage (Tadapaneni *et al.*, 2012) (Figure 75.5, FRAP). Effects of HPP on flavonoid composition tended to decrease after applying HPP (compared to unprocessed beverages); however, reductions were modest (compared to high temperature processing) and relatively stable until pressures of ~700–800 MPa were applied, at which point the reduced concentrations of flavonoids were not different to those imposed by the high temperature processing (Tadapaneni *et al.*, 2012). Interestingly, the sensitivity of flavonoids to different pressures appeared dependent on the chemistry of the bioactive, e.g., pelargonidin anthocyanins and flavonols linked to glucoside or malonylglucoside sugars were more stable at lower pressures than pelargonidin and flavonol compounds linked to rutinoside or coumaroylglucoside. The biological importance of any of the differences observed requires further investigation.

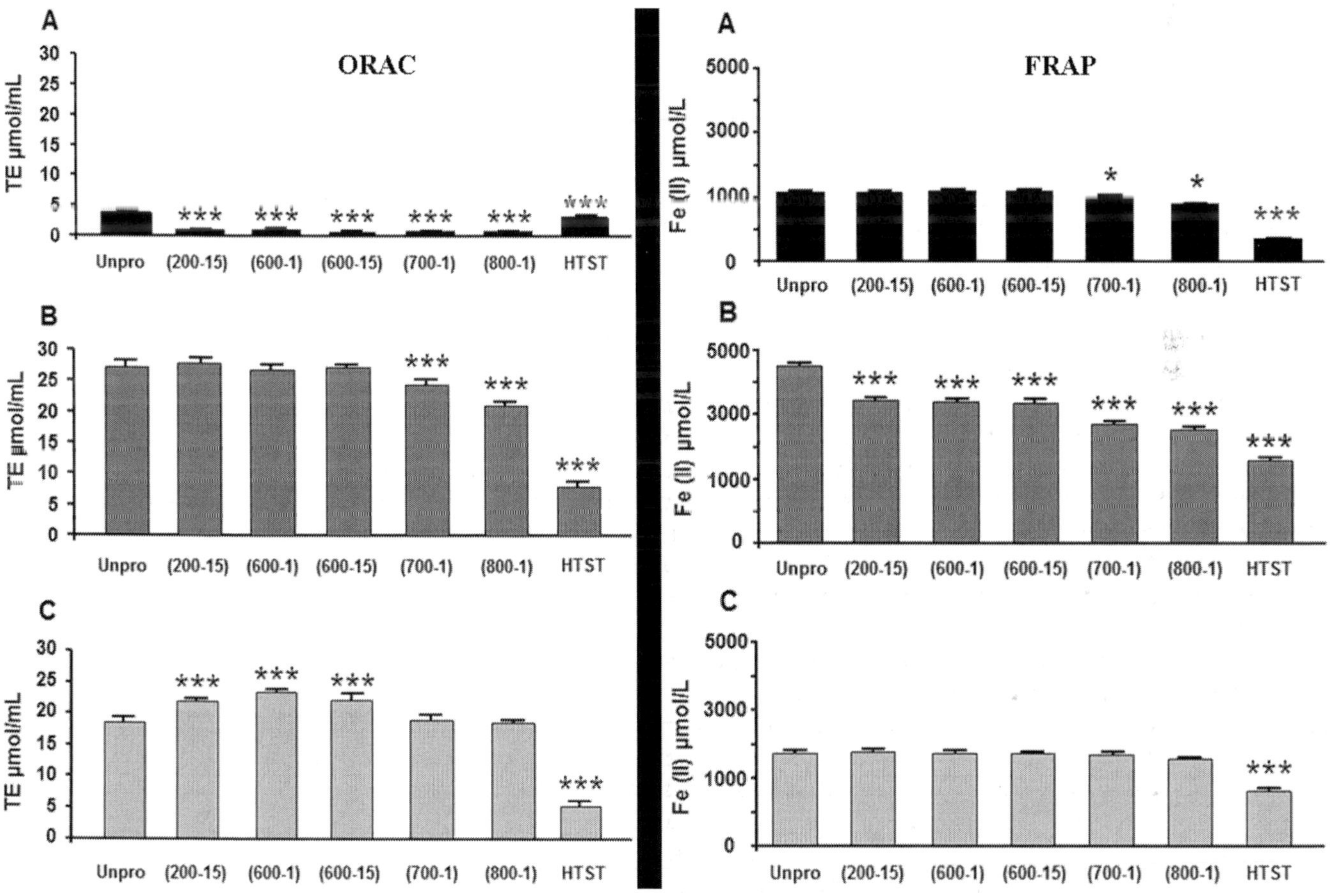

FIGURE 75.5 Antioxidant activity analyzed [with references to Oxygen Radical Absorbance Capacity (ORAC) and Ferric Reducing Antioxidant Power (FRAP) assays] for strawberry beverages with different formulations when treated with HPP and thermal processing conditions. Where A denotes beverage with dairy ingredient but no strawberry powder; B denotes beverage with water and strawberry powder; and C denotes beverage with strawberry powder and dairy ingredient. Unpro, unprocessed, 200–15 and 600–15 = 200 and 600 MPa for 15 min holding respectively, 600–1, 700–1 and 800–1 = 600, 700, and 800 MPa for 1 min holding respectively; HTST, high temperature–short time (72°C/20 s); values represent mean ± SD; n = 4; $^{***}P < 0.001$ compared to Unpro. *Reproduced with permission from Tadapaneni* et al. *(Tadapaneni R. K., Banaszewski K., Patazca E., Edirisinghe I., Cappozzo J., Jackson L., Burton-Freeman B. M.). Copyright (2012) American Chemical Society.*

SUMMARY AND CONCLUSIONS

Food processing strategies continue to evolve and improve our access to safe, convenient foods. Among the newest technologies, HPP may offer the greatest promise in the early twenty-first century for delivering on consumer demands of health and safety. Improvements in fruit and vegetable intake are a major public health objective; and processing to extend shelf-life, increase availability, affordability, and provide more options for people to choose fruits and vegetables and/or their derived products with the expected nutrient quality will be central to this cause. Strawberries are a model fruit for testing options and opportunities in processing to ensure bioactive compounds are preserved and sensory qualities are maintained. Work with strawberries in a processing environment has lead to practical considerations relative to bioactives and delivery of health promoting processed products which can be summarized:

1. Traditional thermal processing methods have been reported to exhibit detrimental effects on the water-soluble essential nutrient and bioactive composition of foods/beverages;
2. Alternative non-thermal technologies to process food/beverages that minimize nutrient and bioactive losses and enhance sensory attributes have been the subject of research and development;
3. HPP, among the emerging novel technologies, may present niche processing opportunities to commercial processors;
4. Application of HPP to process food/beverage products has been reported by several researchers to retain health promoting bioactives and other quality attributes compared to thermal processing; however, higher pressures can impact nutritive quality and organoleptic properties (Gamlath and Wakeling, 2011; Lambert *et al.*, 1999).
5. The biological importance of nutrients and bioactives retention imparted by new technologies remains to be shown.

Overall, there is a critical need to deliver more opportunities to consumers to eat fruits and vegetables. Food processing industries with application of novel technologies can deliver more conveniences to consumers taking care to ensure the nutrients and bioactives content of fresh food. In some cases, processing may enhance bioavailability of plant components and therefore greater benefit may be afforded by the processed product or small losses or changes in composition may be irrelevant biologically. Given the importance of diet, and particularly fruits and vegetables, to health, there is no doubt that the field will continue to evolve. The recognized needs to support focused, efficient progress forward to deliver the next generation of foods/beverages with health opportunities include sufficient funding for research and development, the interaction of process engineers, food scientists, food safety specialists, chemists, nutrition scientists, and cooperation between government and the food industry to bring these new opportunities to reality sooner than later.

References

Aaby, K., Wrolstad, R., Ekeberg, D., Skrede, G., 2007. Polyphenol composition and antioxidant activity in strawberry purees; impact of achene level and storage. J. Agric. Food Chem. 55, 5156–5166.

Bravo, L., 1998. Polyphenols: chemistry, dietary sources, metabolism, and nutritional significance. Nutr. Rev. 56, 317–333.

Burton-Freeman, B., Linares, A., Hyson, D., Kappagoda, C.T., 2010. Strawberry Modulates LDL Oxidation and Postprandial Lipemia in Response to High-fat Meal in Overweight Hyperlipidemic Men and Women. J. Am. Coll. Nutr. 29, 46–54.

Butz, P., Tauscher, B., 2002. Emerging Technologies: Chemical Aspects. Food Res. Int. 35, 279–284.

Edirisinghe, I., Banaszewski, K., Cappozzo, J., Krishnankutty, S., Ellis, C.L., Tadapaneni, R.K., Kappagoda, C.T., Burton-Freeman, B.M., 2011. Strawberry anthocyanin and its association with postprandial inflammation and insulin. Br. J. Nutr. 106, 913–922.

Ellis, C.L., Edirisinghe, I., Kappagoda, C.T., Burton-Freeman, B.M., 2011. Attenuation of meal induced inflammatory and thrombotic responses in overweight men and women after 6-week daily strawberry (*Fragaria*) intake. J. Atheroscler. Thromb, 18, 318–327.

FAO/WHO, 1994. Codex Alimentarius, second ed, vol. 4.

Farkas, J., 1998. Irradiation as a Method for Decontaminating Food: A Review. Int. J. Food Micro. 44, 189–204.

Farkas, D.F., 2011. High pressure processing pathways to commercialization. In: Zhang, H.Q., Barbosa-Canovas, G.V., Balasubramaniam, V.M., Dunne, C.P., Farkas, D.F., Yuan, J.T. (Eds.), Nonthermal processing technologies for foods. Wiley-Blackwell & Institute of Food Technologists, pp. 28–35.

Gamlath, S., Wakeling, L., 2011. Trends in high pressure processing of foods: food quality and bioactive components. In: New topics in food engineering. Nova Science Publishers, Hauppauge, N.Y., pp. 121–149.

Gerlach, D., Alleborn, N., Baars, A., Delgado, A., Moritz, J., Knorr, D., 2008. Numerical Simulations of Pulsed Electric Fields for Food Preservation: A Review. Innovat. Food Sci. Emerg. Tech. 9, 408–417.

Gervilla, R., Felipe, X., Ferragut, V., Guamis, B., 1997. Effect of High Hydrostatic Pressure on *Escherichia coli and Pseudomonas* Fluorescens Strains in Ovine Milk. J. Dairy Sci. 80, 2297–2303.

Global Health Observatory (GHO), 2008. Unhealthy Diet. Available at: http://www.who.int/gho/ncd/risk_factors/unhealthy_diet_text/en/index.html.

González-Gallego, J., García-Mediavilla, M.V., Sánchez-Campos, S., Tuñón, M.J., 2010. Fruit polyphenols, immunity and inflammation. Br. J. Nutr. 104, S15–S27.

Halvorsen, B.L., Carlsen, M.H., Philips, K.M., Holte, S.K., Jacobs, D.R., Blomhoff, R., 2006. Content of Redox-active Compounds (ie, Antioxidants) in Foods Consumed in the United States. Am. J. Clin. Nutr. 84, 95–135.

Hayashi, R., 1996. An overview of the use of high pressure in bioscience and biotechnology. In: Hayashi, R., Balny, C. (Eds.), High Pressure Bioscience and Biotechnology, vol. 13. Elsevier, Amsterdam, pp. 1–6.

Hite, B.H., 1899. The Effect of Pressure in the Preservation of Milk. Bul. West Virginia Univ. Exper. Station 58, 15–35.

Hugas, M., Garriga, M., Monfort, J.M., 2002. New Mild Technologies in Meat Processing: High Pressure as a Model Technology. Meat Sci. 62, 359–371.

Joseph, S., Edirisinghe, I., and Burton-Freeman, B. (in press). Fruit polyphenols: a review of anti-inflammatory effects in humans. Crit. Rev. Food Sci. Nutr.

Klopotek, F., Otto, K., Böhm, V., 2005. Processing strawberries to different products alters contents of vitamin C, total phenolics, total anthocyanins, and antioxidant capacity. J. Agric. Food Chem. 53, 5640–5646.

Kris-Etherton, P.M., Hecker, K.D., Bonanome, A., Coval, S.M., Binkoski, A.E., Hilpert, K.F., Griel, A.E., Etherton, T.D., 2002. Bioactive compounds in foods: their role in the prevention of cardiovascular disease and cancer. Am. J. Med. 13, 71–88.

Lambert, Y., Demazeau, G., Largeteau, A., Bouvier, J.M., 1999. Changes in Aromatic Volatile Composition of Strawberry after High Pressure Treatment. Food Chem. 67, 7–16.

Lundahl, D.S., McDaniel, M.R., Wrolstad, R.E., 1989. Flavor, Aroma, and Compositional Changes in Strawberry Juice Concentrate Stored at 20°C. J. Food Sci. 54, 1255–1258.

Matser, A.M., Krebbers, B., Van Den Berg, R.W., Bartels, P.V., 2004. Advantages of High Pressure Sterilisation on Quality of Food Products. Trends Food Sci. Tech. 15, 79–85.

Oey, I., Lille, M., Loey, A.V., Hendrickx, M., 2008. Effect of High-pressure Processing on Color, Texture and Flavor of Fruit- and Vegetable-based Food Products: a Review. Trends Food Sci. Tech. 19, 320–328.

Patazca, E., Koutchma, T., Balasubramaniam, V.M., 2007. Quasi-adiabatic temperature increase during high pressure processing of selected foods. J. Food Eng. 80, 199–205.

Patras, A., Brunton, N.P., Da Pieve, S., Butler, F., 2009. Impact of High Pressure Processing on Total Antioxidant Activity, Phenolic, Ascorbic Acid, Anthocyanin Content and Color of Strawberry and Blackberry Purées. Innovat. Food Sci. Emerg. Tech. 10, 308–313.

Polydera, A.C., Stoforos, N.G., Taoukis, P.S., 2004. The Effect of Storage on the Antioxidant Activity of Reconstituted Orange Juice Which Had Been Pasteurized By High Pressure or Heat. Int. J. Food Sci. Tech. 39, 783–791.

Préstamo, G., Sanz, P.D., Fonberg-Broczek, M., Arroyo, G., 1999. High Pressure Response of Fruit Jams Contaminated with *Listeria monocytogenes*. Applied Micro. 28, 313–316.

Rickman, J.C., Bruhn, C.M., Barrett, D.M., 2007. Review: Nutritional comparison of fresh, frozen, and canned fruits and vegetables. Part I. Vitamins C and B and phenolic compounds. J. Sci. Food Agric. 87, 930–944.

Seeram, N.P., Lee, R., Samuel Scheuller, H., David, H., 2006. Identification of Phenolic Compounds in Strawberries by Liquid Chromatography Electrospray Ionization Mass Spectroscopy. Food Chem. 97, 1–11.

Señorans, J., Ibáñz, E., Cifuentes, A., 2003. New Trends in Food Processing. Crit. Rev. Food Sci. Nutr. 43, 507–526.

Tadapaneni, R.K., Banaszewski, K., Patazca, E., Edirisinghe, I., Cappozzo, J., Jackson, L., Burton-Freeman, B., 2012. Effect of high-pressure processing and milk on the anthocyanin composition and antioxidant capacity of strawberry-based beverages. J. Agric. Food Chem. 60, 5795–5802.

Tiwari, B.K., O'Donnell, C.P., Cullen, P.J., 2009. Effect of non-thermal processing technologies on the anthocyanin content of fruit juices. Trends Food Sci. Technol. 20, 137–145.

US Department of Agriculture and US Department of Health and Human Services, 2010. Dietary Guidelines for Americans, 2010, seventh ed. US Government Printing Office, Washington, DC.

US Food and Drug Administration, 1941. Wheat flour and related products: definitions and standards of identity. Fed. Regist. 6, 2574–2582.

van Boekel, M., Fogliano, V., Pellegrini, N., Stanton, C., Scholz, G., Lalljie, S., Somoza, V., Knorr, D., Jasti, P.R., Eisenbrand, G., 2010. A review on the beneficial aspects of food processing. Mol. Nutr. Food Res. 54, 1215–1247.

Vauzour, D., Rodriguez-Mateos, A., Corona, G., Oruna-Concha, M.J., Spencer, J.P., 2010. Polyphenols and human health: prevention of disease and mechanisms of action. Nutrients 2, 1106–1131.

Wang, W., Xu, S., 2007. Degradation kinetics of anthocyanins in blackberry juice and concentrate. J. Food Eng. 82, 271–275.

Wu, X., Prior, R.L., 2005. Identification and Characterization of Anthocyanins by High-Performance Liquid Chromatography–Electrospray Ionization–Tandem Mass Spectrometry in Common Foods in the United States: Vegetables, Nuts, and Grains. J. Agric. Food Chem. 53, 3101–3113.

CHAPTER

76

Processing Pomegranates for Juice and Impact on Bioactive Components

Nallely Nuncio-Jáuregui*, Ángel Calín-Sánchez*, Laura Vázquez-Araújo†, Antonio José Pérez-López‡, María J. Frutos-Fernández*, Ángel A. Carbonell-Barrachina*

*Miguel Hernández University, Alicante, Spain, †University of Vigo, Ourense, Spain, ‡Catholic University of San Antonio, Murcia, Spain

CHAPTER POINTS

- Fresh pomegranates are a good source of valuable substances with high antioxidant capacity.
- Punicalagins, hydrolysable tannins, anthocyanins, and ellagic acid are the main compounds responsible for the antioxidant capacity of pomegranate juice.
- Manufacturing of pomegranate juice causes relevant changes on the bioactive compounds.
- Pomegranate juice has higher contents of antioxidant compounds than the fresh fruit.
- During manufacturing, punicalagins increase while ellagic acid and anthocyanins decrease.
- Terpenes are the predominant group in the pomegranate fresh juice aroma while furans are only present in processed juices.
- The flavor of pomegranate juice is characterized by esters, alcohols, and terpenes.

INTRODUCTION

Pomegranate belongs to the Punicaceae family and is one of the oldest known edible fruits. It is an interesting and promising species for different world areas, because it adapts quite well to arid and semi-arid soils and dry weather conditions (Melgarejo and Salazar, 2003). Nowadays it is mainly cultivated in Iran, Afghanistan, India, USA, Mediterranean countries (Tunisia, Turkey, Egypt, Spain, and Morocco) and to some extent China, Japan, and Russia. Spain is the main European pomegranate producer and its production is mainly located in the provinces of Alicante and Murcia (Melgarejo and Salazar, 2003).

Pomegranates are a well-known source of many valuable substances, such as hydrolyzable tannins (e.g., punicalagins and punicalins) (Gil *et al.*, 2000), condensed tannins (e.g., proanthocyanidins) (Poyrazoglu *et al.*, 2002), anthocyanins (Hernández *et al.*, 1999), phenolic acids (e.g., gallic acid, ellagic acid) (Mousavinejad *et al.*, 2009), and organic acids (e.g., malic acid) (Poyrazoglu *et al.*, 2002) (Table 76.1); all these compounds show high antioxidant activity (García-Alonso *et al.*, 2004) and have been correlated with health benefits against cancer, cardiovascular, and other diseases (Sun *et al.*, 2004).

Pomegranate arils (seeds) can be consumed as a fresh product; however, there is a significant percentage of fruits whose quality is not good enough to be marketed as fresh items. The edible portion (arils) are of optimal quality, but the appearance of the fruits is not appropriate for their commercialization, mainly due to defects caused by farming issues (e.g., low development of husk color), ripening (fruits maturation is not homogeneous) and physiopathies (e.g., cracking) (Melgarejo and Salazar, 2003). For this reason, it is necessary to find other commercial options within the agro-food industry for those fruits not marketable as fresh items. Without any doubt, pomegranate juice is the most popular and studied pomegranate-based product (Carbonell-Barrachina *et al.*, 2012; Calín-Sánchez *et al.*, 2011). Other products that can be prepared using pomegranate, and that can

Processing and Impact on Active Components in Food
http://dx.doi.org/10.1016/B978-0-12-404699-3.00076-7

TABLE 76.1 Bioactive Compounds Found in Pomegranate Fruits and Juice

Compound	Chemical Group	Reference
Punicalagins	Hydrolysable tannins	Gil *et al.* (2000)
Punicalins	Hydrolysable tannins	Gil *et al.* (2000)
Proanthocyanidins	Condensed tannins	Poyrazoglu *et al.* (2002)
Cyanidin 3-glucoside	Anthocyanins	Gil *et al.* (2000)
Delphinidin 3-glucoside	Anthocyanins	Gil *et al.* (2000)
Cyanidin 3,5-diglucoside	Anthocyanins	Gil *et al.* (2000)
Delphinidin 3,5-diglucoside	Anthocyanins	Gil *et al.* (2000)
Gallic acid	Phenolic acids	Mousavinejad *et al.* (2009)
Ellagic acid	Phenolic acids	Mousavinejad *et al.* (2009)
Malic acid	Organic acids	Poyrazoglu *et al.* (2002)
Citric acid	Organic acids	Poyrazoglu *et al.* (2002)

List of bioactive compounds and scientific publications in which authors have described their presence in pomegranate fruits and juice.

be found in international markets include, among others, jams, wines, and dried arils. Pomegranate husk or peel is a richer source of antioxidants than the edible arils (Li *et al.*, 2006) and could be used as a nutraceutical supplement due to its elevated content in ellagitannins and ellagic acid (Calín-Sánchez *et al.*, 2012). Dry and ground pomegranate husk or even its extract can also be used as an antioxidant ingredient for fruit juices and at the same time increase its vitamin C content, without significant changes in the juice sensory attributes (Navarro *et al.*, 2011).

From now on, we will concentrate on the effects of processing on the bioactive compounds on juice because this is the most popular and demanded pomegranate product.

HOW COMPOSITION IS ALTERED DUE TO POMEGRANATE JUICE MANUFACTURING

Changes in Antioxidant Capacity, Punicalagins, and Ellagic Acid during Pomegranate Juice Processing

Antioxidants in fruit and vegetables are of interest for many reasons; they can protect components of the food itself against oxidative damage and later they can be absorbed into the human body and could have beneficial health effects. Most of the antioxidants from vegetal sources can be classified as phenolic compounds, which act as chain-breaking antioxidants because their –OH group scavenges reactive radicals, such as peroxyl radicals ($RO_2^{\bullet}$) (Halliwel, 2002). There are several methods and analytical techniques to evaluate the antioxidant capacity of foods, including fruit and vegetables. The most widely used methods in pomegranate juices are (i) the *DPPH method* which uses 2,2-diphenyl-1-picrylhydrazyl (DPPH$^{\bullet}$), a free radical that measures the ability of a compound to donate an electron (Mena *et al.*, 2012; Calín-Sánchez *et al.*, 2011); (ii) the *FRAP method* developed to measure the ability to reduce ferric complex with the molecule tripyridyl s-triazine (TPTZ) to its ferrous form at low pH (Zaouay *et al.*, 2012); (iii) the *ORAC assay*, which is based on the inhibition of the peroxyl-radical-induced oxidation initiated by thermal decomposition of azo-compounds (Navarro *et al.*, 2011); and (iv) finally by the *ABTS method*, antioxidants are added previously to the generation of the ABTS radical and the inhibition in the radical formation is evaluated (Mena *et al.*, 2012; Carbonell-Barrachina *et al.*, 2012). Using all these methods, Seeram (2008) concluded that the antioxidant capacity of polyphenol-rich beverages followed the order: pomegranate juice > red wine > grape juice > blueberry juice > blackberry juice = cranberry juice > orange juice = iced tea beverages = apple juice. The exact values of the antioxidant potency depend on many factors, including cultivar, maturity index, geographical source, irrigation regime, etc. However, general values for the antioxidant capacity of pomegranate could be estimated at 6–15 mM Trolox (DPPH), 10–40 mM Trolox (FRAP), 6–25 mM Trolox (ORAC) and 10–40 mM Trolox (ABTS) (Seeram, 2008; Mena *et al.*, 2011). Very recently Carbonell-Barrachina *et al.* (2012) demonstrated that much of the pomegranate antioxidant capacity has a hydrophilic nature (~80%) (Carbonell-Barrachina *et al.*, 2012) (Figure 76.1). The main compounds responsible for the antioxidant capacity of pomegranate are punicalagins, anthocyanins, and ellagic acid (Gil *et al.*, 2000). In general, those compounds are detected and quantified by high-performance liquid chromatography (HPLC) equipped with a diode-array detector. These compounds are well-known for their ability to eliminate free radicals and inhibit lipid oxidation. During the manufacturing of fresh and commercial pomegranate juices, the whole fruit is pressed (arils and rind) and it is expected that a large amount of phenolic compounds present in the rind or husk migrate to the juice (Tezcan *et al.*, 2009). In this way, it is normal to find higher contents of antioxidant compounds in commercial juices than in the pomegranate arils themselves (consumed as fresh products).

The correlation between the total content of phenolic compounds and the antioxidant capacity has been widely evaluated (Wu *et al.*, 2004). Tezcan *et al.* (2009) reported that both hydrolysable tanning and anthocyanins from the husk increased the antioxidant capacity of

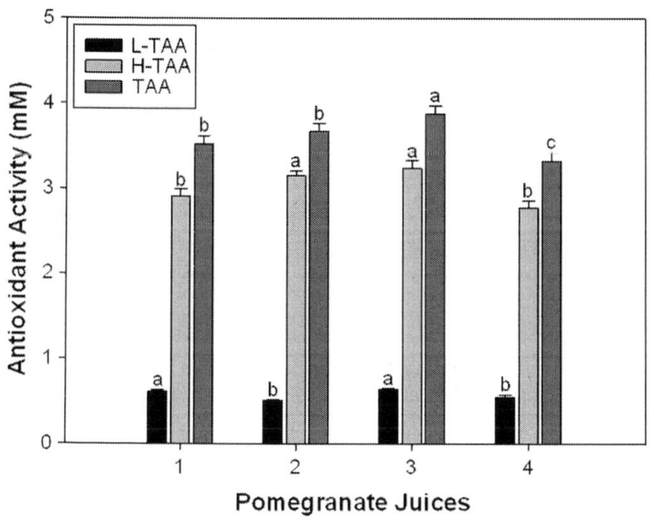

FIGURE 76.1 Total antioxidant activity of pomegranate juices. The histogram shows the mean ± standard error of three replicates of pomegranate juices. Total (TAA), hydrophilic (H-TAA), and lipophilic (L-TAA) antioxidant capacity determined in fresh and commercial pomegranate juices using the ABTS method. *Data to prepare this figure has been obtained from Carbonell-Barrachina* et al. *(2012).*

commercial pomegranate juices. However, results from Gil *et al.* (2000) and Tzulker *et al.* (2007) concluded that while punicalagins played an important role in the antioxidant capacity, anthocyanins only contributed played a minor role. Tzulker *et al.* (2007) reported that during manufacturing of pomegranate juice, the total content of polyphenols increased about 6.5 times, while the antioxidant capacity increased much more, about 20 times.

Different technological approaches have been developed and used to preserve pomegranate juice (Yildiz *et al.*, 2009); thermal pasteurization is still the most widely used because it is inexpensive, efficient and environmentally friendly. Punicalagins isomers (α and β) are the main ellagitannins, non-colored phenolic compounds found in pomegranate juices (Figure 76.2), and are responsible for a high percentage of the antioxidant capacity of this product (Gil *et al.*, 2000). Gil *et al.* (2000) suggested that the punicalagin content is generally higher in commercial juices than in fresh ones due to processing; this increase is not only due to migration from the husk but to thermal treatment. More recently, Mena *et al.* (2012) reported that punicalagins were present in low concentrations in fresh juices before pasteurization; however, they significantly increased for all studied juices after the thermal treatment (high temperature–short time and low temperature–long time). These same authors hypothesized that the punicalagin formation could be related to the degradation of other ellagitannins with higher molecular weight. Conversely, pasteurization resulted in a significant reduction (~60%) of ellagic acid (Mena *et al.*, 2012). A similar conclusion was reached by Yildiz *et al.* (2009).

Long storage at 5 and 25°C of pomegranate juices negatively affected the contents of punicalagins (reduction of ~70%), ellagic acid (~70%), anthocyanins (~15%), and vitamin C (~30%), but a significant increase was observed in punicalins (~55%) (Mena *et al.*, 2012); punicalins came from the degradation or breakdown of ellagitannins.

Changes in Color and Anthocyanin Content during Pomegranate Juice Processing

Several parameters may influence food color and its changes during processing; these parameters include enzymatic and non-enzymatic browning, heat treatment (short time–high temperature or long tim –low temperature), storage conditions (time, temperature, and humidity), pH, and packaging material (Ahmed and Shivhare, 2001). Anthocyanins are potent antioxidant flavonoids and provide pomegranate juice with its characteristic bright and intense dark red color. Individual anthocyanin identification and quantification is generally performed by HPLC analysis (Mena *et al.*, 2012; Gil *et al.*, 2000; Hernández *et al.*, 1999). The pomegranate color intensity increases during ripening (Hernández *et al.*, 1999), and drastically decreases after squeezing. The intensity of the dark red color depends on total concentration of anthocyanins and on the anthocyanin profile (Miguel *et al.*, 2004). However, anthocyanins are not only important because of their color but also for their antioxidant properties (Gil *et al.*, 2000). Hernández *et al.* (1999) studied the changes in the contents of the individual anthocyanins during ripening and found that basically six anthocyanin pigments were responsible for the dark red color of the pomegranate juice. They were identified as delphinidin 3-glucoside, 3,5-diglucoside, cyanidin 3-glucoside, 3,5-diglucoside, pelargonidin 3-glucoside, and 3,5 diglucoside (Figure 76.2). Generally, there was an increase in juice pigmentation during fruit ripening. In the early fruit ripening stages, delphinidin 3,5-diglucoside was the main pigment, followed by cyanidin 3,5-diglucoside; while in the later stages, the monoglucoside derivatives cyanidin 3-glucoside (30–35%) and delphinidin 3-glucoside (>20%) significantly increased and become predominant. The pelargonidin derivatives were always present in small amounts (<5%).

The content of total anthocyanins generally decreases during juice manufacturing; this has a direct effect on the dark red color intensity. The effect of the extraction method of pomegranate juice on its quality and anthocyanin stability was evaluated by Miguel *et al.* (2004); they stated that the separation of arils (deseeding), manually juicing and further centrifugation drastically decreased the content of cyanidin 3,5-diglucoside in the juice compared to the squeezing of fruit halves with an electric squeezer.

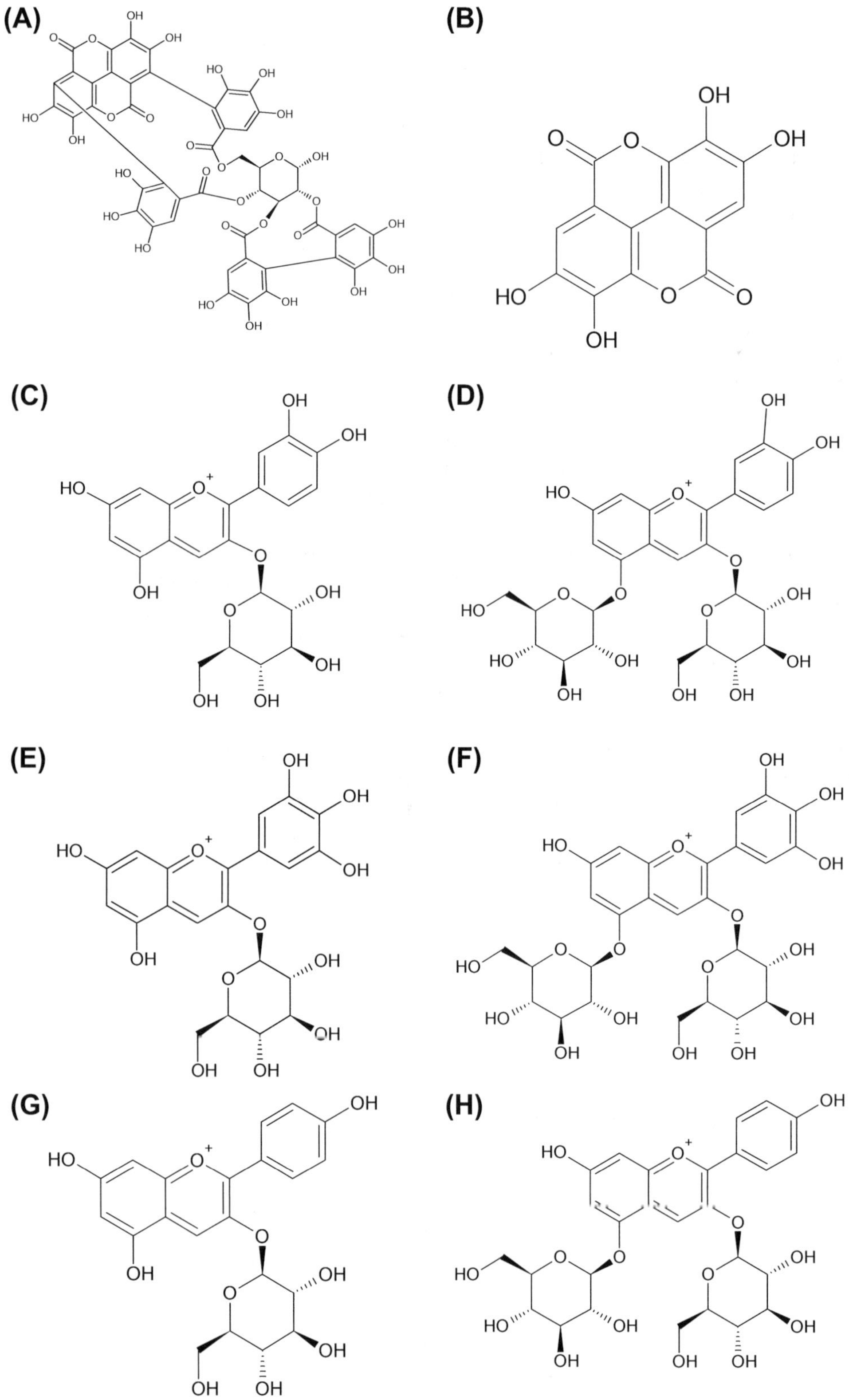

FIGURE 76.2 Principal phenolic compounds present in pomegranate juices. Molecular structure of the main phenolic compounds present in pomegranate fruit and juice: (A) punicalagin; (B) ellagic acid; (C) cyanidin 3-glucoside; (D) cyaniding 3,5-diglucoside; (E) delphinidin 3-glucoside; (F) delphinidin 3,5-diglucoside; (G) pelargonidin 3-glucoside; and (H) pelargonidin 3,5-diglucoside.

Mena *et al.* (2012) compared two thermal treatments [high temperature–short time (90°C, 5 s) and low temperature–long time (65°C, 30 s)] and the addition of lemon juice as acidifier in blended pomegranate juices (cultivars *Mollar de Elche* and *Wonderful*). These authors concluded that pasteurization, independently of the treatment, did not affect the total content of anthocyanins. The addition of lemon juice improved the preservation of the bioactive compounds responsible for the pomegranate juice color. However, the concentration of these chemicals significantly diminished during storage, although the decrease was much more intense at room temperature (15°C) than under refrigeration (5°C). Turfan *et al.* (2011) evaluated the effects of the clarification and pasteurization steps on anthocyanins and color in pomegranate juice prepared from arils and whole fruits. Their results proved that clarification caused losses of 4% and 19% of the anthocyanins content in juice from arils and whole fruits, respectively. Again, the effects of pasteurization on the anthocyanins were minimal.

Even though pasteurization is the most popular preservation treatment of juices, the effects of non-thermal treatments on the contents of bioactive compounds has also been investigated in pomegranate juices. For instance, Pala and Toklucu (2011) suggested that pomegranate juice treated with UV-C radiation, a non-thermal technology, significantly reduced the content of anthocyanins and consequently affected its polymeric color, but preserved quite well other quality parameters, such as bioactive compounds, nutritional, microbiology, and sensory quality. The effects of the UV-C treatment resulted in anthocyanin decreases ranging between 8.1 and 16.3%. However, anthocyanin content of pomegranate juice was significantly affected by heat treatment (90°C, 2 min) and decreases were higher, 15.4 to 28.3%. A similar behaviour (higher differences found in heat treated juices than in UV-C treated ones compared to the untreated controls) was found for polymeric color. The different behaviour reported in the studies by Mena *et al.* (2012) and Pala and Toklucu (2011) may be related to the longer time used in the last one (2 min) compared to the relatively shorter one of the first one (5 s) for the pasteurization at 90°C. However, in a previous study, Yildiz (2009) reported a better efficacy of thermal treatment (ohmic heating) compared to non-thermal treatments in pomegranate juice.

OTHER WAYS IN WHICH COMPOSITION OF POMEGRANATE JUICE IS ALTERED DUE TO THE MANUFACTURING

Effect of Pomegranate Juice Processing on its Volatile Composition and Sensory Properties

Volatile compounds are collected in a concentrated hydrophobic liquid commonly known as essential oil. Essential oils, and consequently volatile compounds, have shown antimicrobial and antifungal activities, and antioxidant capacity (Bozin *et al.*, 2006). The functionality of essential oils is basically related to their content of terpenes, monoterpenes, sesquiterpenes, and sesquiterpenoids.

It is expected that during juice manufacturing, the volatile composition and therefore, the functionality associated with terpenes and related chemical groups, changes as well. These changes will be mainly related to oxidation and enzymatic reactions, activated by cell rupture (Belitz *et al.*, 2009). For example, esters are significant aroma constituents of many fruits and plants and are synthesized only by intact cells, but during the processing of the plant material, esters are rapidly hydrolyzed by enzymes and the fruity aroma flattens (Belitz *et al.*, 2009). This is the main reason why the flavor of fruit juices is different from those of the fresh fruits; additionally, differences are more pronounced after the application of thermal treatments, such as pasteurization.

Different studies studied the volatile composition of fresh pomegranate and pomegranate juices (Melgarejo *el at.*, 2011; Calín-Sánchez *et al.*, 2011; Vázquez-Araújo *et al.*, 2011b). Volatile compounds and essential oils of fruits and vegetables are identified by gas chromatography with a mass spectrometry detector (GC-MS), while the quantification is performed by gas chromatograph with a flame ionization detector (GC-FID). Studies reported above used different techniques for the extraction of the volatile compounds, such as hydrodistillation or head space-solid phase microextraction (HS-SPME). Table 76.2 provides information about some of the volatile compounds that have been reported in fresh pomegranates and which remain in commercial pomegranate juices. Although some volatiles found in commercial juices come from the pomegranate fruits, Vázquez-Araújo *et al.* (2011b) reported an important presence of benzene derivatives in these commercial items (e.g., benzaldehyde). These compounds come from the degradation of lignin and phenolic acids, and their aromatic notes depend on the specific compound being formed (usually vanilla, woody, smoky, or spicy). Vázquez-Araújo *et al.* (2011b) studied the volatile composition of 13 different commercial pomegranate juices, comparing them with a freshly squeezed control juice, and found that two families better describing the differences among those juices were terpenes and furans (e.g., furfural, 5-methyl-furfural). Terpenes were the predominant group in the fresh juice, while furans were only present in processed juices. These findings agree quite well with results from Calín-Sánchez *et al.* (2011) and Melgarejo *et al.* (2011). Some terpenes with hydroxy groups (terpenoids) are naturally present in fruit juice, at least in part, as glycosides. These terpene glycosides hydrolyze, either enzymatically (β-glucosidase) or because of the low pH of the juices. The latter process is strongly accelerated by heat treatments (Belitz *et al.*, 2009), changing the terpene profile in the juices during processing, especially if thermal stages (pasteurization) are part of the processing of pomegranate. Furans are produced from carbohydrate degradation as secondary products of the Maillard reaction; therefore, these compounds were only found in

TABLE 76.2 Volatile Compounds Simultaneously Found in Fresh Fruits and Pomegranate Juices

Compound	Sensory Descriptor	References
Hexanal	Fatty, green, grassy, powerful	Melgarejo *et al.* (2011), Calín-Sánchez *et al.* (2011), Vázquez-Araújo *et al.* (2011a), Carbonell-Barrachina *et al.* (2012)
cis-3-Hexenal	Apple, grape, floral, green, vegetable	Melgarejo *et al.* (2011), Calín-Sánchez *et al.* (2011), Carbonell-Barrachina *et al.* (2012)
cis-3-Hexenol	Fresh, green grass	Melgarejo *et al.* (2011), Calín-Sánchez *et al.* (2011), Vázquez-Araújo *et al.* (2011a), Vázquez-Araújo *et al.* (2011b), Carbonell-Barrachina *et al.* (2012)
1-Hexanol	Mint, grass	Melgarejo *et al.* (2011), Calín-Sánchez *et al.* (2011), Vázquez-Araújo *et al.* (2011a), Vázquez-Araújo *et al.* (2011b), Carbonell-Barrachina *et al.* (2012)
α-Pinene	Sharp, pine	Melgarejo *et al.* (2011), Calín-Sánchez *et al.* (2011), Vázquez-Araújo *et al.* (2011a), Vázquez-Araújo *et al.* (2011b), Carbonell-Barrachina *et al.* (2012)
β-Pinene	Woody, pine	Melgarejo *et al.* (2011), Calín-Sánchez *et al.* (2011), Vázquez-Araújo *et al.* (2011a), Vázquez-Araújo *et al.* (2011b), Carbonell-Barrachina *et al.* (2012)
Limonene	Mild, citrus, sweet, orange, lemon	Melgarejo *et al.* (2011), Calín-Sánchez *et al.* (2011), Vázquez-Araújo *et al.* (2011a), Vázquez-Araújo *et al.* (2011b), Carbonell-Barrachina *et al.* (2012)
γ-Terpinene	Herbaceous, citrus	Melgarejo *et al.* (2011), Calín-Sánchez *et al.* (2011), Vázquez-Araújo *et al.* (2011a), Vázquez-Araújo *et al.* (2011b), Carbonell-Barrachina *et al.* (2012)
α-Terpineol	Fragrant, floral, lilac	Melgarejo *et al.* (2011), Calín-Sánchez *et al.* (2011), Vázquez-Araújo *et al.* (2011a), Vázquez-Araújo *et al.* (2011b), Carbonell-Barrachina *et al.* (2012)
4-Terpineol	Grapefruit, lemon, lime, pepper, herbaceous	Calín-Sánchez *et al.* (2011), Vázquez-Araújo *et al.* (2011a), Vázquez-Araújo *et al.* (2011b), Carbonell-Barrachina *et al.* (2012)
β-Myrcene	Sweet, balsamic	Melgarejo *et al.* (2011), Calín-Sánchez *et al.* (2011), Carbonell-Barrachina *et al.* (2012)

List of volatile compounds and scientific publications in which authors have described their presence in pomegranate fruits and juice.

processed juices and could be used as indicators of the intensity of the heat treatment. Besides, furans will also be found in other pomegranate products such as jellies or jams. Carbonell-Barrachina *et al.* (2012) also studied some commercial juices and did not detect furans in the volatile compositions, which indicated that the intensity of the thermal treatment (98°C for 20s) of those juices was low.

Calín-Sánchez *et al.* (2011) studied the relationship among instrumental parameters of pomegranate fresh juices and overall liking of consumers. Overall liking of the juices seemed to be related to the attributes "fresh flavor" and "fresh odor", which in turn seemed to be related to the presence of some volatile compounds, mainly terpenes (α-pinene, β-pinene, β-myrcene, limonene, and γ-terpinene). During storage of pomegranate juices, the amounts of ethanol, ethyl acetate (from the esterification of ethanol), and sesquiterpenes (e.g., β-caryophyllene, α-bergamotene, and β-farnesene) significantly increases and simultaneously the consumer acceptance decreases. In this way, the "flavor life" of pomegranate juices is often shorter than their "storage life" as describe by physico-chemical and microbiological quality parameters.

Vázquez-Araújo *et al.* (2011a) reported that after macerating pomegranate juices with pomegranate albedo homogenate, the total phenolic content of the juice increased significantly, but the sensory characteristics of the obtained product were similar to those of the original juice. Thus, the total phenolic content is not always reflected in the sensory attributes of the juice or its acceptance by consumers. Previously, Borochov-Neori *et al.* (2009) had studied 11 different cultivars of pomegranate and reported that the antioxidant activity and a sensory attribute called "sensory quality" were not correlated, and that both parameters were dependent on cultivar and climatic conditions during fruit maturation and ripening.

To finish this chapter, it is important to demonstrate that all the technological approaches described in previous sections about manufacturing of pomegranate juice are also reflected in changes of sensory attributes. Sensory quality will have an important effect on the success of the product in the market (product acceptability) and it is therefore of the utmost importance for farmers and food companies. In this way, it is expected that the processing of fruits, including pomegranate, should maintain the maximum levels of functional ingredients, but at the same time should also maintain their sensory quality.

Koppel and Chambers (2010) studied 33 commercial pomegranate juices and developed a sensory lexicon to describe the main sensory attributes of these products, finding large variations among the different juices and classifying the samples in five different clusters characterized by: (i) berry, dark-fruity flavors, and a toothetch mouth feel; (ii) grape, cranberry, and wine-like flavor; (iii) fermented flavor and a toothetch mouthfeel; (iv)

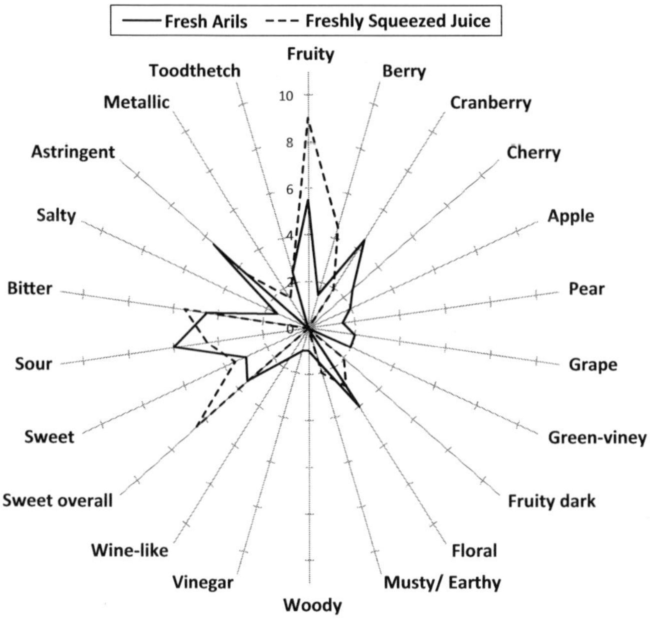

FIGURE 76.3 Flavor profiles of pomegranate arils and freshly squeezed juice (cv. Wonderful). Spider plots of the flavor profiles of pomegranate products obtained using 15 points numeric scale with 0.5 increments. *Sensory attributes were selected from Vázquez-Araújo* et al. *(2011b).*

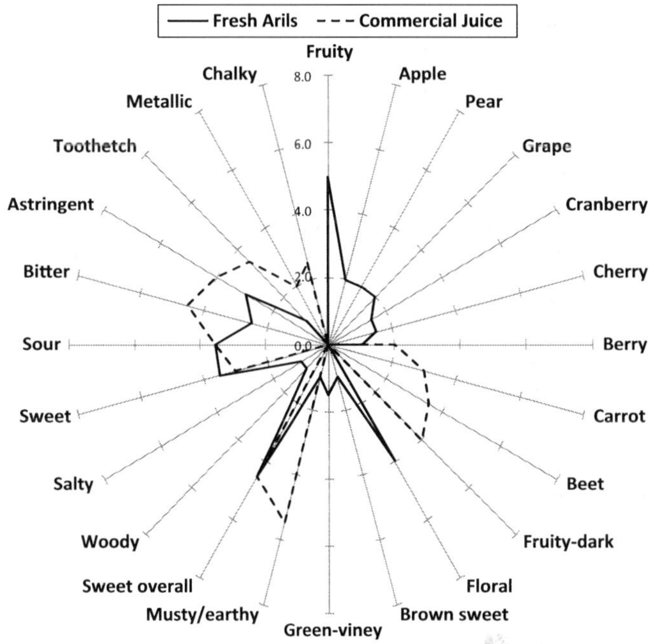

FIGURE 76.4 Flavor profiles of pomegranate arils and commercial juice (cv. *Mollar de Elche*). Spider plots of the flavor profiles of pomegranate products obtained using 15 points numeric scale with 0.5 increments. *Sensory attributes were selected from Vázquez-Araújo* et al. *(2011b).*

brown color and musty/earthy flavor; and (v) candy-like and sweet overall flavors. Some of the sensory differences, such as astringency, bitterness, or toothetch might be caused by processing steps (use of clarification, concentration, pasteurization, etc.), presence/absence of some preservatives, or presence/absence of added flavorings in the juices (Vázquez-Araújo *et al.* 2011a). The detailed lexicon developed by Koppel and Chambers, (2010) had more than 30 descriptors, including terms related to fruity notes (e.g., cranberry, cherry, grape, etc.) and other terms related to flavor notes which may come from processing (e.g., molasses or candy-like). Vázquez-Araújo *et al.* (2011a,b) used this lexicon as a base to study sensory characteristics of commercial and freshly squeezed pomegranate juices, and also to relate those sensory attributes with the volatile composition of the juices. Figures 76.3 and 76.4 show the sensory profile of pomegranate juices at different stages of the manufacturing process: fresh arils (cv. *Mollar de Elche* and *Wonderful*), freshly squeezed juice (cv. *Wonderful*) prepared exclusively from arils, and commercial juice (cv. *Mollar de Elche*). As shown in Figure 76.3, sensory properties of pomegranate arils, cv. *Wonderful*, changed immediately even at the first stage of processing (squeezing); flavor notes present in the arils, such as cherry, apple, pear, etc., disappeared completely while the fruity note significantly increased. These changes could be related to some of the volatile compounds being lost in the solid residue generated during the squeezing step and the oxidation reactions started during squeezing. Logically, sensory changes were more notable when comparing arils with the final processed commercial juice of pomegranates cv. *Mollar de Elche* (Figure 76.4). In the commercial juice it was observed that some notes were also lost (cherry, apple, etc.), but at the same time other attributes (carrot, beet, fruity-dark, cooked) appeared and reached significant scores. In accordance with these results, Vázquez-Araújo *et al.* (2011b) reported that juice from freshly squeezed arils was characterized by having more floral, fruity, and sweet overall notes than commercial juices (not from concentrate juice), which had higher intensities of cranberry, wine-like, and grape notes. Juices from concentrate were highlighted by fruity dark notes and the presence, in some samples, of other notes typical in processed foods (molasses and candy-like).

More studies which correlate bioactive compounds with specific sensory attributes and consumer liking would be really useful for companies, orienting the manufacture to products in which the bioactive and nutritive compounds are present in high concentrations, but not affecting in a negative way the sensory characteristics of the product and/or its acceptance by consumers. At the same time, improving pomegranate juice extraction and preservation, or even developing artificial pomegranate flavorings/aromas, would help in the progress of improving the quality of the products that already can be found in the market, increasing the consumption of pomegranate juice (Vázquez-Araújo *et al.* 2011b).

References

Ahmed, J., Shivhare, U.S., 2001. Thermal kinetics of color change, rheology, and storage characteristics of garlic puree/paste. J. Food Sci. 66 (5), 754–757.

Belitz, H.D., Grosch, W., Schieberle, P., 2009. Food Chemistry, fourth ed. Springer, Berlin (Germany).

Borochov-Neori, H., Judeinstein, S., Tripler, E., Harari, M., Greenberg, A., 2009. Seasonal and cultivar variations in antioxidant and sensory quality of pomegranate (*Punica granatum* L.) fruit. J. Food Comp. Anal. 22, 189–195.

Bozin, B., Mimica-Dukic, N., Simin, N., Anackov, G., 2006. Characterization of the volatile composition of essential oils of some *Lamiaceae* species on the antimicrobial and antioxidant activities of the entire oils. J. Agric. Food Chem. 54, 1822–1828.

Calín-Sánchez, A., Figiel, A., Hernández, F., Melgarejo, P., Lech, K., Carbonell-Barrachina, A.A., 2012. Chemical composition, antioxidant capacity, and sensory quality of pomegranate (*Punica granatum* L.) arils and rind as affected by drying method. Food Bioprocess. Tech. (in press).

Calín-Sánchez, A., Martínez, J.J., Vázquez-Araújo, L., Burló, F., Melgarejo, P., Carbonell-Barrachina, A.A., 2011. Volatile composition and sensory quality of Spanish pomegranates (*Punica granatum* L.). J. Sci. Food Agric. 91, 586–592.

Carbonell-Barrachina, A.A., Calín-Sánchez, A., Bagatar, B., Hernandez, F., Legua, P., Martínez-Font, R., Melgarejo, P., 2012. Potential of Spanish sour-sweet pomegranates (cultivar C25) for juice industry. Food Sci. Technol. Int. 18 (2), 129–138.

García-Alonso, M., De Pascual-Teresa, S., Santos-Buelga, C., Rivas-Gonzalo, J.C., 2004. Evaluation of the antioxidant properties of fruits. Food Chem. 84, 13–18.

Gil, M.I., Tomas-Barberan, F.A., Hess-Pierce, B., Holcroft, D.M., Kader, A.A., 2000. Antioxidant activity of pomegranate juice and its relationship with phenolic composition and processing. J. Agric. Food Chem. 48, 4581–4589.

Halliwell, B., 2002. Food-derived antioxidants: how to evaluate their importance. In: Cadenas, E., Packer, L. (Eds.), Food and in Vivo in Handbook of Antioxidants, second ed. Mercel Dekker, New York (USA), pp. 18–63.

Hernández, F., Melgarejo, P., Tomás-Barberán, F.A., Artés, F., 1999. Evolution of juice anthocyanins during ripening of new selected pomegranate (*Punica granatum*) clones. Eur. Food Res. Technol. 210, 39–42.

Koppel, K., Chambers, E.IV., 2010. Lexicon to describe appearance and flavor of pomegranate juice. J. Sens. Stud. 25 (6), 819–837.

Li, Y., Guo, C., Yang, J., Wei, J., Xu, J., Cheng, S., 2006. Evaluation of antioxidant properties of pomegranate peel extract in comparison with pomegranate pulp extract. Food Chem. 96, 254–260.

Melgarejo, P., Salazar, D.M., 2003. Tratado de Fruticultura para Zonas Áridas y Semiáridas. In: Mundi-Prensa, (Ed.), Madrid (Spain).

Melgarejo, P., Calín-Sánchez, A., Vázquez-Araújo, L., Hernández, F., Martínez, J.J., Legua, P., Carbonell-Barrachina, A.A., 2011. Volatile composition of pomegranates from 9 Spanish cultivars using head space-solid phase microextraction. J. Food Sci. 76 (1) 144–120.

Mena, P., Martí, N., Saura, D., Valero, M., García-Viguera, C., 2012. Combinatory effect of thermal treatments and blending on the quality of pomegranate juices. Food Bioproc. Tech. (in press).

Miguel, G., Dandlen, S., Antunes, S., Neves, A., Martins, D., 2004. The effect of two methods of pomegranate (*Punica granatum* L) juice extraction on quality during storage at 4°C. J. Biomed. Biotechnol. 5, 332–337.

Mousavinejad, G., Emam-Djomeh, Z., Rezaei, K., Khodaparast, M.H.H., 2009. Identification and quantification of phenolic compounds and their effects on antioxidant activity in pomegranate juices of eight Iranian cultivars. Food Chem. 115, 1274–1278.

Navarro, P., Nicolas, T.S., Gabaldon, J.A., Mercader-Ros, M.T., Calín-Sánchez, A., Carbonell-Barrachina, A.A., Pérez-López, A.J., 2011. Effects of cyclodextrin type on vitamin C, antioxidant activity, and sensory attributes of a mandarin juice enriched with pomegranate and goji berries. J. Food Sci. 76, S319–S324.

Pala, Ç.U., Toklucu, A.K., 2011. Effect of UV-C on anthocyanin content and other quality parameters of pomegranate juice. J. Food Comp. Anal. 24 (6), 790–795.

Poyrazoglu, E., Gökmen, V., Nevzat, A., 2002. Organic Acids and Phenolic Compounds in Pomegranates (*Punica granatum* L.) Grown in Turkey. J. Food Comp. Anal. 15, 567–575.

Seeram, N.P., 2008. Berry fruits: compositional elements, biochemical activities, and the impact of their intake on human health, performance, and disease. J. Agric. Food Chem. 56, 627–629.

Sun, J., Chu, Y.F., Wu, X., Liu, R.H., 2004. Antioxidant and antiproliferative activities of common fruits. J. Agric. Food Chem. 50, 7449–7454.

Tezcan, F., Gültekin-Özgüven, M., Diken, T., Özçelik, B., Erim, F.B., 2009. Antioxidant activity and total phenolic, organic acid and sugar content in commercial pomegranate juices. Food Chem. 115, 873–877.

Turfan, O., Türkyılmaz, M., Yemis, O., Mehmet Özkan, M., 2011. Anthocyanin and color changes during processing of pomegranate (*Punica granatum* L., cv. Hicaznar) juice from sacs and whole fruit. Food Chem. 129 (4), 1644–1651.

Tzulker, R., Glazer, I., Bar-Ilan, I., Holland, D., Aviram, M., Amir, R., 2007. Antioxidant activity, polyphenol content, and related compounds in different fruit juices and homogenates prepared from 29 different pomegranate accessions. J.f Agric. Food Chem. 55, 9559–9570.

Vázquez-Araújo, L., Chambers, E.-IV., Adhikari, K., Carbonell-Barrachina, A.A., 2011a. Physico-chemical and sensory properties of pomegranate juices with pomegranate albedo and carpellar membranes homogenate. LWT-Food Sci. Technol. 44, 2119–2125.

Vázquez-Araújo, L., Koppel, K., Chambers, E.I.V., Adhikari, K., Carbonell-Barrachina, A.A., 2011b. Instrumental and sensory aroma profile of pomegranate juices from the USA: differences between fresh and commercial juice. Flavor Frag. J. 26, 129–138.

Wu, X., Beecher, G.R., Holden, J.M., Haytowitz, D.B., Gebhardt, S.E., Prior, R.L., 2004. Lipophilic and hydrophilic antioxidant capacities of common foods in the United States. J. Agric. Food Chem. 53, 4026–4037.

Yildiz, H., Bozkurt, H., Icier, F., 2009. Ohmic and conventional heating of pomegranate juice: effects on rheology, color, and total phenolics. Food Sci. Technol. Int. 15 (5), 503–512.

Zaouay, F., Mena, P., Garcia-Viguera, C., Mars, M., 2012. Antioxidant activity and physico-chemical properties of Tunisian grown pomegranate (*Punica granatum* L.) cultivars. Ind. Crop Prod. 40, 81–89.

CHAPTER

77

Carotenoids in Nonthermally Treated Fruit Juices

I. Odriozola-Serrano, G. Oms-Oliu, R. Soliva-Fortuny, O. Martín-Belloso

Department of Food Technology, University of Lleida, Lleida, Spain

CHAPTER POINTS

- Nonthermal technologies allow obtaining fruit juices with minor changes or increased content in carotenoids.
- Temperature during processing and storage period is a determinant factor affecting the carotenoid content of fruit juices.
- Compared to thermally-pasteurized tomato juices, nonthermally-treated fruit juices kept higher carotenoids content during storage.
- Different results may be obtained for different fruit juices because the presence of antagonistic agents in the matrix may substantially compromise carotenoids stability.
- Future research should focus on identifying the role of biochemical transformations of carotenoid precursors as influenced by processing.

INTRODUCTION

Consumption of plant-based minimally processed foods such as fruit juices is increasing because of the evidence about the many benefits for human health that have been attributed to the dietary intake of fresh-like fruits and vegetables (Kaur and Kapoor, 2001). It is widely known that these products are outstanding sources of bioactive compounds, such as carotenoids. Carotenoids (carotens and xanthophylls) are yellow, orange, and red pigments present in many commonly eaten fruits and vegetables. Their importance to human health is related to their role as provitamin A, antioxidants, cell differentiation and proliferation regulators, cell-cell communication stimulators, immune function and carcinogen metabolism modulators, and blue light filters. There are many studies showing a strong correlation between carotenoids intake with both the improvement of the immune system and the reduction of the incidence of some degenerative diseases such as cancer, cardiovascular diseases, cataract and macular degeneration (Cooper *et al.*, 2004). However, carotenoids are highly unsaturated compounds with an extensive conjugated double-bonds system and are therefore susceptible to oxidation, isomerization, and other chemical changes during processing and storage (Shi and Le Maguer, 2000). Despite carotenoids being affected by abusive temperatures, thermal processing remains the most commonly used technology for inactivating microorganisms and enzymes in processed food. In this context, nonthermal technologies such as high intensity pulsed electric fields (HIPEF) and high-pressure (HP) are being introduced by food processors as alternative or complementary to conventional thermal treatments. Both technologies have potential to improve the quality and freshness character of processed food while assuring its microbiological safety and stability (Odriozola-Serrano *et al.*, 2009; Plaza *et al.*, 2011). In recent years, the impact of nonthermal technologies on health-related compounds has been extensively evaluated. Based on these results, the effect of HIPEF and HP on the stability of carotenoids in fruit juices will be reviewed.

HOW COMPOSITION IS ALTERED

Pulsed Electric Field Treatment

Processing by pulsed electric fields may allow keeping the stability of carotenoids. Some studies have demonstrated that carotenoid content is significantly increased

Processing and Impact on Active Components in Food
http://dx.doi.org/10.1016/B978-0-12-404699-3.00077-9

after HIPEF processing compared to the untreated juice (Table 77.1). Odriozola-Serrano *et al.* (2007) reported an enhancement of up to 46.2% in the lycopene concentration of tomato juices after applying different HIPEF treatments (35 kV/cm for 1000 μs). Consistently, slight increases in lycopene content of HIPEF-treated watermelon juices had been previously reported by Oms-Oliu *et al.* (2009). It has been hypothesized that thermal treatments may imply an increase in some individual carotenoids, owing to greater stability, inactivation of oxidative and hydrolytic enzymes, and unaccounted losses of moisture, which concentrate the sample (Rodriguez-Amaya, 1997). Nguyen and Schwartz (1999) suggested that homogenization and heat treatment disrupt cell membranes and protein-carotenoids complex, making carotenoids more accessible for extraction. Results obtained by Odriozola-Serrano *et al.* (2009) offer some controversy, provided that changes in the relative amounts of carotenoids are not consistent for similar compounds. Although the reason for these results is not well understood, it was speculated that carotenoid conversions could be triggered by PEF treatments. This would explain the increase in lycopene at the expense of its precursors. HIPEF processing may accelerate lycopene synthesis in tomatoes, involving the conversion of geranyl-geranyl diphosphate (GGPP) to phytoene by phytoene synthase and the conversion of phytoene to phytofluene, *β*-carotene, and lycopene by phytoene desaturase (Fraser, *et al.*, 2004). Cortés *et al.* (2006a) stated that the effects of treatments on carotenoid composition mainly depended on processing conditions. Process parameters such as electric field strength and treatment time are the most relevant variables to be controlled in HIPEF treatments for attaining microbial reduction and have also been shown to play a key role in carotenoids retention. In this regard, the greater the electric field strength and treatment time, the higher the carotenoid content (Odriozola-Serrano *et al.*, 2008). Consistently, Cortés *et al.* (2006b) observed that the carotenoids content in HIPEF-treated orange juice increased significantly when the most intensive treatments (35 or 40 kV/cm) were conducted. Carotenoids concentration in orange-carrot juice rose as after prolonged HIPEF treatments at 25 or 30 kV/cm (Torregrosa *et al.*, 2005). Furthermore, other treatment variables such as pulse frequency, pulse width, and polarity may also matter. Higher frequency and pulse width resulted in a greater carotenoid content in tomato juices compared to untreated samples, whereas the use of bipolar pulses led to a greater rise in the carotenoids content of tomato juices with respect to juices subjected to monopolar treatments (Odriozola-Serrano *et al.*, 2007).

TABLE 77.1 Effect of High Intensity Pulsed Electric Field Treatment on Carotenoid Compounds of Fruit Juices

Product	Treatment Conditions	Major Finding	Reference
Orange juice	35–40 kV/cm/30–340 μs	Slight increase in the concentration of some carotenoids (13-*cis*-violaxanthin and 7,8,7′,8′-tetrahydrolycopene)	Cortés *et al.* (2006b)
	40 kV/cm/30–60 μs	Increase in Vitamin A content (4.5–20%)	Torregrosa *et al.* (2005)
Tomato juice	35 kV/cm/1000 μs (bipolar 7-μs pulses at 250 Hz)	Enhancement of lycopene concentration (46.2%)	Odriozola-Serrano *et al.* (2007)
	35 kV/cm/1000 μs (bipolar 4-μs pulses at 100 Hz)	Increase in several individual carotenoids such as lycopene (10%), *β*-carotene (38%) and phytofluene (5%)	Odriozola-Serrano *et al.* (2009)
Carrot juice	35 kV/cm/1500 μs (bipolar 6-μs pulses at 200 Hz)	Substantial increase in *β*-carotene concentration (23%)	Quitão-Teixeira *et al.* (2009)
Watermelon juice	35 kV/cm/50 μs (bipolar 7-μs pulses at 200 Hz)	Slight increase in lycopene content (113%)	Oms-Oliu *et al.* (2009)

Other studies have been focused on determining the influence of HIPEF on vitamin A precursors such as *β*-carotene, α-carotene, γ-carotene, and α and *β*-cryptoxanthin. HIPEF-treated carrot juice exhibited higher *β*-carotene concentrations than untreated juice, which is related to greater vitamin A contents in the samples (Quitão-Teixeira *et al.*, 2009). Odriozola-Serrano *et al.* (2009) observed that *β*-carotene in treated tomato juice underwent a significant increase (31%–38%), whereas γ-carotene content depleted (3%–6%) when HIPEF treatments (35 kV/cm for 1000 μs) were applied. It was suggested that γ-carotene may undergo cyclization to form six membered rings at one end of the molecule, giving *β*-carotene as a product after HIPEF processing. In contrast, other studies have not reported major effects of HIPEF treatments on the carotenoid content of orange juices (Sánchez-Moreno *et al.*, 2005).

Several authors have studied the evolution of some carotenoids in HIPEF-treated juices during storage, reporting higher stability of these health-related compounds in comparison to thermally-pasteurized tomato juices (Min *et al.*, 2003; Odriozola-Serrano *et al.*, 2007), orange juices (Cortés *et al.*, 2006a), carrot juices (Quitão-Teixieira *et al.*, 2009), and fruit juice-milk beverages (Morales-de la Peña *et al.*, 2011). Most differences between HIPEF and heat treatments can be explained in terms of the temperature/time binomial. Hence, mild processing temperatures during HIPEF-processing might explain the higher retention of carotenoids in fruit juice samples. According to Kidmose *et al.* (2002) the major cause of carotenoid losses in vegetable products is the oxidation of the highly unsaturated carotenoid structure. Oxidation may occur by autooxidation, which is a spontaneous free-radical chain reaction in the presence of oxygen, or by photooxidation produced by oxygen in the presence of light. These oxidative reactions may result in carotene bleaching, which is the cause of formation of colourless end-products (Gross, 1991). During autooxidation of carotenoids, alkylperoxyl radicals are formed and these radicals attack the double bonds resulting in formation of epoxides. The severity of oxidation depends on the structure of carotenoids and the environmental conditions, and the compounds being formed depend on the oxidation process and the carotenoids structure (Ramakrishnan and Francis, 1980). Lutein has been shown to be the most susceptible carotenoid to isomerisation or oxidation processes, thus exhibiting the greatest losses in HIPEF and thermally treated orange juice after 40 days of storage at 4°C (Plaza *et al.*, 2011). Zulueta *et al.* (2007) mentioned that lutein and zeaxanthin are highly susceptible to degradation during thermal treatments due to the presence of oxygen in their chemical structures. Regarding carotenes, Odriozola-Serrano *et al.* (2009) pointed out that lycopene content in HIPEF and heat processed tomato juices decreased much more considerably than other carotenes (neurosporene, γ-carotene, ξ-carotene, and *β*-carotene) through storage at 4°C (Figure 77.1).

High Pressure Processing

Several research efforts have assessed the suitability of HP technology for fruit juice processing (Sánchez-Moreno *et al.*, 2009). It seems that high pressure influences the extraction yield of carotenoids in fruit juices (Table 77.2). HP-processed orange juice exhibited a significant increase in total carotenoids (49.19%) and vitamin A content (30.89%) compared to an untreated juice (Plaza *et al.*, 2011). According to De Ancos *et al.* (2002) orange juices treated at 350 MPa/30°C/5 min exhibited a higher carotenoid content (α-carotene, 60%; *β*-carotene, 50%; α-cryptoxanthin, 63%; *β*-cryptoxanthin, 42%) than freshly squeezed juices, which was attributed to the pressure-induced denaturation of carotenoid-binding proteins. HP processing (500–700 MPa, 100°C) increased the extractability of lycopene from tomato juice, whereas thermal treatment had a negligible effect on its concentration (Gupta *et al.*, 2011). These results are in agreement with those found by Hsu (2008), who reported a significant increase in lycopene and total carotenoids (up to 60 and 62%, respectively) as a result of HP treatment of tomato juice (300–500 MPa/25°C/10 min). Carotenoids are localized in subcellular organelles (plastids), in chloroplasts and chromoplasts, mainly associated with macromolecules such as proteins and membrane lipids (Gärtner *et al.*, 1997). It has been reported that high pressure treatments above 300 MPa cause irreversible protein denaturation at room temperature leading to irreversible damage in vegetable cell membranes (Hendrickx *et al.*, 1998). Therefore, HP treatment could improve the extractability of carotenoids owing to a better release of these bioactive compounds from the fruit matrix, leading to different release rates for specific carotenoid types.

Storage may cause the instability of the polyene chain of carotenoids. As a consequence, these compounds may undergo geometric isomerisation, which is promoted by heat, light, and acids, as well as oxidation, which is stimulated by light, heat, metals, enzymes, and peroxides, and inhibited by antioxidants (Rodríguez-Amaya, 1997). Non significant changes in carotenoids concentration have been reported for at least 10 days of refrigerated storage in an orange juice treated at 100 MPa/60°C/5 min, whereas substantial losses have been found at the

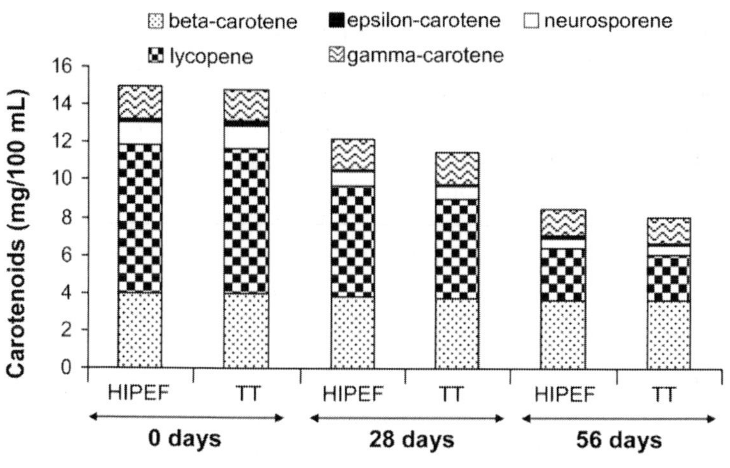

FIGURE 77.1 Contents of the main carotenoids in HIPEF-treated and thermally-treated tomato juice through storage at 4 °C. HIPEF: 35 kV/cm for 1500 μs. TT: 90 °C for 1 min. Data shown are mean 4 set of analysis of tomato juices. (Unpublished).

TABLE 77.2 Effect of High Pressure Processing on Carotenoids Compounds of Fruit Juices

Product	Treatment Conditions	Major Finding	Reference
Tomato juice	300 MPa/4°C and 25°C	62 and 56% increase in total carotenoids and lycopene contents, respectively	Hsu (2008)
Orange juice	50–350 MPa/30–60°C/2.5–15 min	Up to 23% increase in total carotenoids after 100- to 350-MPa treatments	De Ancos *et al.* (2002)
	100–400 MPa/30–60°C/1–5 min	10–32% enhancement of total carotenoids content Zeaxanthin content increased regardless the treatment applied	Sánchez-Moreno *et al.* (2003)
	600 MPa/20°C/60 s	No modifications in *β*-carotene levels	Bull *et al.* (2004)
	400 MPa/40°C/1 min	Increase in the extractability of individual carotenoids	Plaza *et al.* (2011)

end of the storage period of samples processed at 350 MPa/30°C/2.5 min or 400 MPa/40°C/1 min (20.56% and 9.16%, respectively) (Sánchez-Moreno *et al.*, 2003). According to De Ancos *et al.* (2002), the vitamin A content of HP-treated orange juice significantly decreased throughout storage (42% and 24% reduction after 30 days storage of samples treated at 50 MPa/30°C/5 min and 100 MPa/30°C/5 min, respectively). This decrease significantly surpassed the 9% reduction observed in untreated juice samples. In contrast, Bull *et al.* (2004) reported that the *β*-carotene content of HP-treated (600 MPa/20°C/60 s) and heat-treated (85°C, 25 s) orange juices did not significantly decrease during storage at 4°C. Thus, the extent of the oxidation depends on the carotenoid involved. HP treatments have been reported to better maintain carotenoids than traditional thermal treatments in many juices. Gupta *et al.* (2010) reported a higher retention of carotenoids in HP treated tomato juice compared to thermally processed samples. HP-treated orange juice exhibited higher carotenoids concentration than heat-pasteurized juice during refrigerated storage at 4°C. In consequence, vitamin A values increased above 40% with respect to those in untreated samples (Plaza *et al.*, 2011) (Figure 77.2). The inactivation of enzymes involved in carotenoid losses during storage

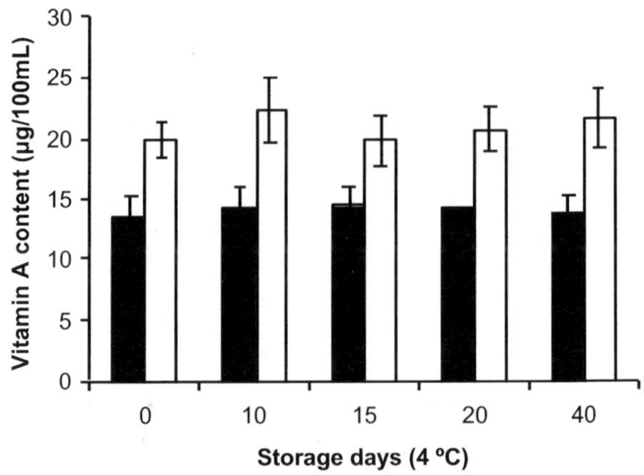

FIGURE 77.2 Vitamin A value of thermally processed and HP-treated orange juices. Vitamin A content: retinol activity equivalents per 100 mL. Data shown are mean ± standard deviation of 4 set of analysis of orange juices. Thermally processed: 70°C/30s (black square). HP-treated: 400 MPa/4°C/1 min (white square). (Unpublished)

and the facilitation of extraction due to HP treatments are the reasons that some authors have put forward to explain those results (De Ancos *et al.*, 2002).

The *all-trans* isomers of carotenoids are the predominant form in nature (94–96%) in fruits and vegetables. However, the high concentrations of *in vivo cis* isomers suggest that they may be more bioavailable and/or more bioactive than the *all-trans* form, possibly due to the higher solubility and/or preferential incorporation of the *cis* isomers into chylomicrons. Varma *et al.*, (2010) reported that HP processing causes conformational changes from the *all-trans* to *cis* isomer form, indicating that high pressure application can induce isomerization, increasing the availability of the carotenoids in the sample.

ANALYTICAL TECHNIQUES

Carotenoids have a long chain of unsaturated double bonds which confers instability and facilitates their reaction, namely with acids, oxygen, and active radicals in the presence of light. To avoid artifacts formation and quantitative losses, the analysis should be completed within the shortest time possible, at relatively low temperature, in exclusion of oxygen, using protecting atmosphere, under protection from light, excluding the light wavelengths absorbed by carotenoids, avoiding contact with acids, using high-purity solvents free from harmful impurities such as chlorinated compounds and peroxides, and using material that does not adsorb carotenoids (Britton, 1991). Applicability of an analytical method depends on the matrix, the analytes present, and their levels. Various methods have been reported for the determination of carotenoids in nonthermally processed fruit juices including chromatographic and spectroscopic procedures.

Extraction Techniques

For extracting carotenoids from fruit juices, there is not an universally accepted method, mainly due to the great number of possible carotene/matrix combinations. For food samples that contain large amounts of water such as fruit juices, it is advisable to use organic solvents that are miscible with water, to allow a better release of carotenoids from the matrix. Diluted solutions of ethanol, hexane, ethyl acetate, acetone, and tetrahydrofuran, alone or in combination, are generally used for the extraction of carotenoids from fruit juices. The conjugated double bounds system found in carotenoids is the main source of problems associated with carotenoids manipulation. To avoid this issue, butylhydroxytoluene (BHT) is used to avoid oxidation at concentrations ranging between 0.05 and 0.1% (Cortés *et al.*, 2006a,b; Odriozola-Serrano *et al.*, 2007; Oms-Oliu *et al.*, 2009).

Saponification

Saponification (alkaline hydrolysis) has been widely used to facilitate carotenoid analysis because it is an effective method to remove chlorophylls and unwanted lipids that may interfere in the chromatographic separation, or to hydrolyze the carotenoid esters, therefore simplifying the chromatographic profiles. In several studies related to nonthermally treated juices, saponification is conducted by mixing carotenoids with methanolic potassium hydroxide (5–10%) under nitrogen at room temperature for a few hours in darkness. After this, the carotenoid solution is transferred to a separatory funnel, diethyl ether is added, the solution is then washed several times with water to remove the alkali and eventually evaporated (De Ancos *et al.*, 2002; Sánchez-Moreno *et al.*, 2005; Cortés *et al.*, 2006a,b; Morales-De la Peña *et al.*, 2011)

Separation, Identification and Quantification by HPLC

High-performance liquid chromatography (HPLC) methods are advantageous compared to direct measurement techniques, especially in what regards to specificity, sensitivity, or easiness of operation. Reversed-phase C_{18} columns have been widely used for carotenoids separation because of the weak hydrophobic interaction with the analytes, their compatibility with most carotenoid solvents, and the polarity range of carotenoids. Mobile-phases used in HPLC methods for the determination of carotenoids in different fruit juices are summarized in Table 77.3. Carotenoids are easily identified at wavelengths between 290 nm and 500 nm. Quantification can be carried out by comparison with internal or external standards.

Spectrophotometric Methods

Carotenoids may be easily identified since the majority exhibit a maximum in the long UV range. Absorbance spectroscopy is the simplest method to identify carotenoids because the absorbance spectrum is the fingerprint of each carotenoid molecule. Total lycopene content has been measured spectrophotometrically in nonthermally treated tomato juices

TABLE 77.3 Mobile Phase Used for Carotenoids Analysis of Nonthermally Treated Fruit Juices

Product	Carotenoids	Mobile Phase	Reference
Tomato juice	Lycopene	MeOH:methyl tert-butyl ether	Min *et al.* (2003) Gupta *et al.* (2010, 2011)
	Total carotenoids and lycopene	1-butanol:ACN: dichloromethane	Hsu (2008)
	Lycopene, neurosporene, γ-, ζ-, β-, phytofluene and phytoene	ACN:MeOH:hexane Dichloromethane	Odriozola-Serrano et al. (2009)
Carrot juice	β-Carotene	ACN:MeOH:hexane dichloromethane	Quitão-Teixeira et al., (2009)
Orange juice	Lutein, zeaxanthin, α-and β-criptoxanthin, α- and β-carotene	ACN:MeOH:water: dichloromethane	De Ancos *et al.* (2002) Sánchez-Moreno *et al.* (2003) Plaza *et al.* (2011)
	β-Carotene	Acetonitrile:MeOH:1, 2-dichloroethane	Bull *et al.* (2004)
	Lutein, zeaxanthin, isolutein, β-criptoxanthin, phytoene + phytofluene, α-β and ζ-carotene	MeOH: methyl tert-butyl ether:water	Cortés *et al.* (2006ab)
Orange–carrot juice	Antheraxanthin, mutatoxanthin, violaxanthin, lutein, zeaxanthin, isolutein, β-criptoxanthin, phytoene + phytofluene, α-β and ζ-carotene	MeOH: methyl tert-butyl ether:water	Torregrosa *et al.* (2005)

(Odriozola-Serrano *et al.*, 2007) and watermelon juices (Oms-Oliu *et al.*, 2009). Extraction of lycopene from juices was carried out with a mixture of acetone, ethanol, and hexane. The lycopene content of each sample was estimated using the absorbance at 503 nm and the sample weight.

References

Britton, G., 1991. Carotenoids. Meth. Plant. Biochem. 7, 473–518.

Bull, M., Zerdin, K., Howe, E., Goicoechea, D., Paramanandhan, P., Stockman, R., Sellahewa, J., Szabo, E.A., Johnson, R., Stewart, C.M., 2004. The effect of high pressure processing on the microbial, physical and chemical properties of Valencia and Navel orange juice. Innov. Food. Sci. Emerg. Technol. 5, 135–149.

Cooper, D.A., 2004. Carotenoids in health and disease: recent scientific evaluations, research recommendations and the consumer. J. Food Nutr. 134, 221S–224S.

Cortés, C., Torregrosa, F., Esteve, M.J., Frígola, A., 2006a. Carotenoid profile modification during refrigerated storage in untreated and pasteurized orange juice and orange juice treated with high-intensity pulsed electric fields. J. Agric. Food Chem. 54, 6247–6254.

Cortés, C., Esteve, M.J., Rodrigo, D., Torregrosa, F., Frígola, A., 2006b. Changes of colour and carotenoids contents during high intensity pulsed electric field treatment in orange juices. Food Chem. Toxicol. 44, 1932–1939.

De Ancos, B., Sgroppo, S., Plaza, L., Cano, M.P., 2002. Possible nutricional and health-related value promotion in orange juice preserved by high pressure treatment. J. Sci. Food Agric. 82, 790796.

Fraser, P.D., Bramley, P.M., 2004. The biosynthesis and nutritional use of carotenoids. Prog. Lipid Res. 43 (3), 228–265.

Gupta, R., Balasubramaniam, V.M., Schwartz, S.J., Francis, D.M., 2010. Storage stability of lycopene in tomato subjected to combined pressure-heat treatments. J. Agric. Food Chem. 58, 8305–8313.

Gupta, R., Kopec, R.E., Schwartz, S.J., Balasubramaniam, V.M., 2011. Combined pressure-temperature effects on carotenoid retention and bioaccessibility in tomato juice. J. Agric. Food Chem. 59, 7808–7817.

Gross, J., 1991. Pigments in Vegetables. Chlorophylls and carotenoids. Van Nostrand Reinold, New York.

Hendrickx, M., Ludikhuyze, L., Van den Broeck, I., Weemaes, C., 1998. Effects of high pressure on enzymes related to food quality. Trends Food Sci. Technol. 9, 197–203.

Hsu, K.-C., 2008. Evaluation of processing qualities of tomato juice induced by thermal and pressure processing. LWT-Food Sci. Technol. 41, 450–459.

Kaur, C., Kapoor, H.C., 2001. Antioxidants in fruits and vegetables-the millennium's health. Intern. J. Food Sci. Technol. 36, 703–725.

Kidmose, U., Edelenbos, M., Nøbæk, R., Christensen, L.P., 2002. Colour stability in vegetables. In: MacDougall, D.B. (Ed.), Colour in Food: Improving quality. CRC Press, Boca Raton, Florida, pp. 179–232.

Min, S., Jin, Z.T., Zhang, Q.H., 2003. Commercial scale pulsed electric field processing of tomato juice. J. Agric. Food Chem. 51, 3338–3344.

Morales-de la Peña, M., Salvia-Trujillo, L., Rojas-Graü, M.A., Martín-Belloso, O., 2011. Changes on phenolic and carotenoid composition of high intensity pulsed electric field and thermally treated fruit juice–soymilk beverages during refrigerated storage. Food Chem. 129, 982–990.

Nguyen, M.L., Schwartz, S.J., 1999. Lycopene: chemical and biological properties. Food Technol. 53 (2), 38–45.

Odriozola-Serrano, I., Aguiló-Aguayo, I., Soliva-Fortuny, R., Gimeno-Añó, V., Martín-Belloso, O., 2007. Lycopene, vitamin C, and antioxidant capacity of tomato juice as affected by high-intensity pulsed electric fields critical parameters. J. Agric. Food Chem. 55, 9036–9042.

Odriozola-Serrano, I., Soliva-Fortuny, R., Gimeno- Añó, V., Martín-Belloso, O., 2008. Modeling changes in health-related compounds of tomato juice treated by high-intensity pulsed electric fields. J. Food Eng. 89, 210–216.

Odriozola-Serrano, I., Soliva-Fortuny, R., Hernández-Jover, T., Martín-Belloso, O., 2009. Carotenoid and phenolic profile of tomato juice processed by high intensity pulsed electric fields compared to conventional thermal treatments. Food Chem. 112, 258–266.

Oms-Oliu, G., Odriozola-Serrano, I., Soliva-Fortuny, R.C., Martín-Belloso, O., 2009. Effects of high-intensity pulsed electric field processing conditions on lycopene, vitamin C and antioxidant capacity of watermelon juice. Food Chem. 115, 1312–1319.

Plaza, L., Sánchez-Moreno, C., De Ancos, B., Elez-Martínez, P., Martín-Belloso, O., Cano, M.P., 2011. Carotenoid and flavanone content during refrigerated storage of orange juice processed by high-pressure, pulsed electric fields and low pasteurization. LWT-Food Sci. Technol. 44, 834–839.

Quitão-Teixeira, L.J., Odriozola-Serrano, I., Soliva-Fortuny, R., Martín-Belloso, O., 2009. Comparative study on antioxidant properties of carrot juice stabilised by high-intensity pulsed electric fields or heat treatments. J. Sci. Food Agric. 89, 2636–2642.

Ramakrishnan, T.V., Francis, F.J., 1980. Autooxidation of carotenoids and their relative polarity. J. Food Qual. 3, 25–34.

Rodriguez-Amaya, D.B., 1997. Carotenoids and food preservation: the retention of provitamin A carotenoids in prepared, processed and stored food. USAID, OMNI Project.

Sánchez-Moreno, C., Plaza, L., De Ancos, B., Cano, M.P., 2003. Effect of high-pressure processing on health-promoting attributes of freshly squeezed orange juice (*Citrus sinensis* L.) during chilled storage. Eur. Food Res. Technol. 216, 18–22.

Sánchez-Moreno, C., Plaza, L., Elez-Martínez, P., De Ancos, B., Martín-Belloso, O., Cano, M.P., 2005. Impact of high pressure and pulsed electric fields on bioactive compounds and antioxidant activity of orange juice in comparison with traditional thermal processing. J. Agric. Food Chem. 53, 4403–4409.

Sánchez-Moreno, C., De Ancos, B., Plaza, L., Elez-Martínez, P., Cano, M.P., 2009. Nutritional approaches and health-related properties of plant foods processed by high pressure and pulsed electric fields. Critical Rev. Food Sci. Nutr. 49, 552–576.

Shi, J., Le Maguer, M., 2000. Lycopene in tomatoes: chemical and physical properties affected by food processing. Critical Rev. Food Sci. Nutr. 40 (1), 1–42.

Torregrosa, F., Cortés, C., Esteve, M.J., Frígola, A., 2005. Effect of high-intensity pulsed electric fields processing and conventional heat treatment on orange-carrot juice carotenoids. J. Agric. Food Chem. 53, 9519–9525.

Varma, S., Karwe, M.V., Lee, T.C., 2010. Effect of high hydrostatic pressure processing on lycopene isomers. Int. J. Food Eng. 6, 1–20.

Zulueta, A., Esteve, M.J., Frígola, A., 2007. Carotenoids and color of fruit juice and milk beverage mixtures. J. Food Sci. 72, C457–C463.

SECTION 9

HERBS AND SPICES AND MISCELLANEOUS VEGETATION

CHAPTER

78

Dehydration of Basil Leaves and Impact of Processing Composition

Raffaella Boggia, Paola Zunin, Vilma Hysenaj, Aldo Bottino, Antonio Comite

University of Genoa, Genoa, Italy

CHAPTER POINTS

- Sweet basil (*Ocimum basilicum* L.) is a very important culinary herb and a functional ingredient mainly marketed dried to ensure microbiological safety and to extend shelf-life.
- Decreased flavor, shrinkage, unappetizing texture, color damage, and browning effects are the most evident drawbacks of dried basil leaves.
- As far as flavor is concerned, there is a decrease of the intensities of floral and herbaceous notes while there is an increase of spicy, earthy, and woody notes.
- Adverse changes can be minimized by choosing among the mildest drying technologies and by appropriately designing the process conditions using design of experiments (DOE) in a comprehensive way, also considering pre-drying and post-drying processes.
- Convective drying (CD) usually implies a significant reduction in the content of most of the volatiles together with a significant variation in their relative proportions according to changes in air temperature.
- Convective pre-drying and vacuum-microwave finishing drying lead to high-quality dried products in terms of total volatiles in a relatively short time (about 250 min).
- Osmotically dehydrated basil leaves show a higher quality than those treated with traditional methods.
- Vacuum–microwave drying (VMD) often minimizes the leaf shrinkage and improves their texture.
- Usually osmotic dehydration (OD) and freeze drying of herbs minimize the thermal damage to color by preventing enzymatic browning.
- Nowadays, there are only a few studies on the effects of drying processes on the mineral content of basil leaves.

INTRODUCTION

The genus *Ocimum*, belonging to the *Lamiaceae* family, comprises annual and perennial aromatic herbs and is characterized by a great genetic diversity, with 65–150 species reported in the literature, based on differences in growth habit, leaf color, size, shape, and aromatic composition (Makri and Kintzios, 2008).

Sweet basil (*Ocimum basilicum* L.), the Mediterranean type and the most popular species, is assumed to originate from India, Africa, and/or the Middle East (Figure 78.1). It is the most common basil type in the Western hemisphere and has the greatest economic importance. It is utilized as a food ingredient, remedy, cosmetic ingredient, and for ornamental purpose. It is considered one of the most versatile and highly used herbs in the culinary world and is popular in the cooking of a large variety of foods in many types of cuisines, from oriental to Italian, particularly in the famous pesto sauce from Genoa.

Recently, sweet basil essential oils and/or herb have attracted a great deal of scientific interest due to their potential use as functional ingredients, since a high concentration of natural antioxidants and biologically active compounds have been documented (Tepe *et al.*, 2006).

The aromatic leaves of basil are used fresh, stored in a cool place or in the refrigerator for a very short time. Since

Processing and Impact on Active Components in Food
http://dx.doi.org/10.1016/B978-0-12-404699-3.00078-0

FIGURE 78.1 Fresh and dried sweet basil (*Ocimum basilicum* L.) leaves.

the high percentage of water (about 92%) makes basil quite perishable, it is usually marketed dried or frozen.

Dehydration is the most common and cheapest way of preserving basil leaves ensuring microbial safety and extending their shelf-life. However, even if it prevents several biochemical reactions, it may decrease the nutritional value of basil, may introduce undesirable changes in its appearance and may cause modifications of texture, flavor, and color.

As far as aroma attributes are concerned, fresh basil is usually described as having a fresh, herbaceous, and floral odor, while the dried samples have a mentholated, spicy, earthy, and hay-like odor. This fact mainly depends on the modification of the volatile fraction, since volatile compounds are particularly perishable. Moreover, during dehydration, loss of green color is commonly detected (Di Cesare *et al.* 2003).

HOW COMPOSITION IS ALTERED

In the literature the botanical identity of basil together with the exact number of species in the genus is a controversy (Makri and Kintzios, 2008). Depending upon different authors, there are 65–150 *Ocimum* species and at least 60 cultivars of *Ocimum basilicum*.

Leaves vary in their size, shape, and texture by species. In particular, leaf textures range from smooth and shiny to curled and hairy; leaf color can vary from green to blue/purple, depending on the species.

Basil has low calorific value and high nutritional values. It contains carotene, vitamins A, B_6, C, K, as well as potassium, calcium, magnesium, phosphorus, sodium, and iron.

Table 78.1 reports an average chemical composition in nutrients of both generic fresh basil leaves and dried ones, as published in the USDA National Nutrient Database for Standard Reference.

The basil herb also contains flavonoids and it is considered an antioxidant (Kwee and Niemeyer, 2011). The essential oil (less than 1%) is of complex and variable composition. There are several types of basil oil in international commerce, each principally derived from different cultivars or chemotypes of sweet basil.

In particular, sweet basil essential oil contains mainly monoterpenes, sesquiterpenes, phenylpropanoids, alcohols, aldehydes, ketones, esters, and miscellaneous compounds (Lee *et al.*, 2005), whose contents affect sensory qualities (Table 78.2).

Not only the type of cultivar, but also the agronomical practices, environmental conditions, seasonal changes, time of harvest, and even different growth stages affect the composition of sensory important compounds (Miele *et al.*, 2001; Jirovetz *et al.*, 2003; Hussain *et al.*, 2008).

Regardless of these factors, the major essential oil components, i.e., those which comprise 20% or more of the total in at least one genotype, are especially variable in occurrence and concentration among the different accessions, ranging from absent in some genotypes to more than 90% of the total essential oil composition in others. Those major constituents (whose content is determined by genotype and influenced by geographical and agronomic conditions) are: linalool, methyl chavicol (estragole), eugenol, methyl eugenol, and eucalyptol (1,8-cineole).

Sweet basil is also known to have strong antioxidant activity (Makri and Kintzios, 2008), since its leaves contain phenolic compounds, especially rosmarinic acid, chicoric acid, caftaric acid (all derivatives of caffeic acid in order from the most abundant to the least), and cinnamic acid, caffeic acid, sinapic acid, and ferulic acid (Lee and Scagel, 2009; Surveswaran *et al.*, 2007; Javanmardi *et al.*, 2003).

In 2012 Calín-Sánchez and co-authors (Calín-Sánchez *et al.* 2012) reported that the most important compounds are affected by the geographical origin. In particular, in

TABLE 78.1 Average Chemical Composition of Fresh and Dried Basil Leaves

		Fresh Basil Leaves	Dried Basil Leaves
Nutrients	**Unit**	**Value per 100.0 g**	**Value per 100.0 g**
Water	g	92.06	10.35
Energy	kcal	23	233
Protein	g	3.15	22.98
Total lipid (fat)	g	0.64	4.07
Carbohydrate, by difference	g	2.65	47.75
Fiber, total dietary	g	1.6	37.7
Sugars, total	g	0.3	1.71
Minerals			
Calcium, Ca	mg	177	2240
Iron, Fe	mg	3.17	89.80
Magnesium, Mg	mg	64	711
Phosphorus, P	mg	56	274
Potassium, K	mg	295	2630
Sodium, Na	mg	4	76
Zinc, Zn	mg	0.81	7.10
Vitamins			
Vitamin C, total ascorbic acid	mg	18	0.8
Thiamin	mg	0.034	0.080
Riboflavin	mg	0.076	1.200
Niacin	mg	0.902	4.900
Vitamin B-6	mg	0.155	1.340
Folate, DFE	μg	68	310
Vitamin B-12	μg	0	0.00
Vitamin A, RAE	μg	264	37
Vitamin A, IU	IU	5275	744
Vitamin E (alpha-tocopherol)	mg	0.8	10.70
Vitamin D (D2 + D3)	μg	0	0.0
Vitamin D	IU	0	0
Vitamin K (phylloquinone)	μg	414.8	1714.5
Lipids			
Fatty acids, total saturated	g	0.041	2.157
Fatty acids, total monounsaturated	g	0.088	1.238
Fatty acids, total polyunsaturated	g	0.389	0.498
Cholesterol	mg	0	0

U.S. Department of Agriculture, Agricultural Research Service in the National Nutrient Database for Standard Reference, Release 25 (2012). Nutrient Data Laboratory Home Page, http://www.ars.usda.gov/ba/bhnrc/ndl.

TABLE 78.2 Volatile Compounds Identified in Fresh and Dried Sweet Basil Samples

Monoterpenes
α-Pinene
Camphene
Sabinene
β-Myrcene
β-Pinene
α-Phellandrene
α-Terpinene
Limonene
trans- β-ocimene
γ-Terpinene
Terpinolene
Monoterpenoids
1,8-Cineole (Eucalyptol)
Linalool
Camphor
1-Terpineol
Borneol
4-Terpineol
α-Terpineol
Geraniol
Thymol
Bornyl acetate
Phenylpropanoids
Estragole (methyl chavicol)
Eugenol
Methyleugenol
Sesquiterpenes
α-Cubebene
Copaene
β-Cubebene
α-Bergamotene
Caryophyllene
α-*trans*-Bergamotene
α-Farnesene
α-Humulene
β-Farnesene
Germacrene
α-Selinene
γ-Cadinene
Sesquiterpenoids
γ-Eudesmol
trans-Cadinol
Others
cis-3-Hexen-1-ol
1-Octen-3-ol
Octanal
1-Octen-3-yl-acetate

Egyptian sweet basil (Karawya *et al.*, 1974), linalool and estragole are dominant; in Israeli sweet basil (Fleisher, 1981), linalool, estragole, and eugenol are the major compounds; in Italian sweet basil, eugenol together with methyleugenol, eucalyptol, and linalool (Di Cesare *et al.*, 2003) are the most important volatiles; linalool and eugenol are commonly dominant in Spanish sweet basil (Díaz-Maroto *et al.*, 2004).

In summary, it is evident that the cultivar has a significant effect on the aromatic profile and, in particular, several compounds (i.e., linalool, methyl chavicol, eugenol, methyl eugenol and methyl cinnamate, bergamotene) can be regarded as the dominant volatile components, whose contents can be considered as cultivar markers.

The increasing rejection of chemicals is making the problem of preserving foodstuff more complex, since the new products require an ever longer shelf-life and a higher degree of protection against pathogenic microorganisms.

In the literature, innovative processes have been reported with their advantages and drawbacks (Antonelli *et al.*, 1998; Yousif *et al.*, 1999; Di Cesare *et al.*, 2003; Barbieri *et al.*, 2004; Diaz-Maroto *et al.*, 2004; Hussain *et al.*, 2008; Calín-Sánchez *et al.*, 2012; Boggia *et al.*, 2012). All the proposed methods aim to both preserve basil leaves by decreasing their water activity and to produce high quality products at competitive costs and energy consumption.

Drying processes applied to basil leaves can be classified as follows (Tortoe, 2010)

Natural drying (sun or natural dryers). Natural drying has many disadvantages due to the inability to achieve consistent quality standards.

Atmospheric drying (convective drying by hot and dry gas, usually air; it could be either continuous or batch). The main drying parameters to monitor for product quality are the following: air temperature; air relative humidity; air velocity; drying time. Hot-air-drying, however, can cause heat damage.

Subatmospheric drying (includes vacuum shelf dryer, continuous vacuum dryer, and freeze dryer). Since the boiling point of water is reduced as the pressure is reduced, the temperatures used in the drying processes are decreased and, in this way, it is possible to obtain better quality products with respect to atmospheric drying. The main drying parameters to monitor in order to guarantee product quality are: pressure; temperature; drying time. In particular, among the sub-atmospheric methods, freeze drying is one of the most sophisticated processes even if it is quite expensive.

Novel drying technologies (e.g., microwave drying, superheated steam drying, osmotic dehydration etc.)

It has been clearly demonstrated that the high temperature used during the drying process involves color alteration, aroma degradation, and a decrease of nutritional quality. All these negative effects are further intensified by the relatively long drying time. This has led to the development and/or the application of less invasive technologies.

In particular, microwave drying (MD) is a technique that allows rapid dehydration and can be applied to basil leaves (Zhang *et al.*, 2006). The high capacity of penetration of microwaves allows heating not only the leaves' surface but also inside them. Thus dehydration time can be drastically reduced improving the quality of dried products by replacing hot air with microwave energy, at least in some steps of the drying process. In fact, according to many authors, the combined use of microwaves and hot air drying improves the quality of the final product (Contreras *et al.*, 2005; Raghavan and Silveira, 2001).

Besides, the use of low pressure can further shorten the drying time. This technique is usually called vacuum–microwave drying (VMD). VMD combines the advantages of both microwave heating and vacuum drying with less heat and oxygen damage than air-drying (Zhang *et al.* 2006). The resulting products may be comparable in terms of tissue structure to freeze-dried foods.

Another interesting novel technological alternative to the use of hot air employs low pressure superheated steam (LPSS) to dry basil leaves. Many authors assess that LPSS produces less damage to dried food as it works at lower temperatures and oxygen is virtually absent. The retention of aroma compounds in basil dried with LPSS was reported (Barbieri *et al.*, 2004). In the meantime, according to literature, LPSS seems to be attractive from an economical point of view, since it usually requires less energy than hot air-drying.

Recently osmotic dehydration (OD) process, as another alternative preservation technique, received more attention due to the consumer demand for minimally processed products (Singh *et al.* 2009). In osmotic dehydration, foods are immersed or soaked in hypertonic solutions, the brines, whose composition (i.e., sugar, salt, sorbitol or glycerol) is compatible with food, at a specific temperature and for a time long enough to achieve samples dehydration. The main drying parameters that are usually monitored are the qualitative and quantitative composition of the brine, the process temperature, the sample size, the speed of agitation, and the time of immersion (dipping time). Several authors report that OD has several advantages, such as quality improvement in terms of color, flavor, texture, better energy efficiency, significant packaging and distribution cost reduction, no chemical pre-treatment, higher product stability, and retention of nutrients during storage.

Sometimes some pre-treatment such as blanching, sulphiting, starching, freezing, sucrose or sodium chloride dipping (osmotic dehydration), etc., can be applied to foodstuffs before the drying process with the aim of improving mass and heat transfer, as well as product appearance and nutritional value.

As far as basil is concerned, blanching is often applied. The objective of blanching is to produce several favorable effects on processed materials, such as the inactivation of enzymes present, the expulsion of air trapped between cells, the reduction of possible initial infection, and the removal of undesirable flavors and aromas.

Finally, regardless of the type of technology used, it is important to emphasize that the process of dehydration should always be pre-optimized, possibly by statistical experimental design, also called design of experiments (DOE). In particular, DOE can be used in problems in which the influence of several process factors such as temperature, pressure, time, etc., on product quality is to be monitored and optimized (Leardi, 2009).

Heat and mass transport phenomena may deeply impact on the quality of dried basil leaves. Texture, appearance, color, flavor, taste, and nutritional value are all modified by the drying process and have to be controlled.

Volatile Fraction Impact

The changes in concentrations of the volatile compounds during drying depend on several factors, such as the drying method and its parameters. In particular, drying of sweet basil leaves decreases the intensities of floral and herbaceous notes while increasing spicy, hay-like, sweet, earthy, woody, and infusion notes.

According to Diaz-Maroto and co-authors (Diaz-Maroto *et al.*, 2004), the total quantity of volatiles of fresh basil samples decreases considerably during oven drying (at 45°C) and freeze-drying. The basil leaves used by these authors were characterized by a high content of linalool, absence of methylchavicol, and a relatively large quantity of eugenol in their volatile fraction, Conversely, air drying of the same basil samples at room temperature brought about only small losses of volatile components and even an increase was observed in the content of some sesquiterpenes. This increase in sesquiterpenes has also been described by other authors (Baritaux *et al.*, 1992) and has been observed during other herb drying. However, drying at room temperature is difficult to standardise and requires long drying times, so that it is not recommendable on an industrial scale.

Oven drying at 45°C and freeze-drying are rapid and completely controllable methods whose results, in terms of product quality, have proved entirely similar. Thus the high cost of freeze-drying could not be justified.

Sensory analysis revealed that the dried samples lose to a certain degree the typical aroma of the fresh herb, acquiring other characteristics such as hay-like and earth odors together with a mentholated aroma, and are spicier and sweeter than fresh basil.

Microwaves cause a rapid evaporation of water from the vegetable tissue limiting oxidation thus substantially preserving color, flavor, and sensory qualities of products.

In particular, Di Cesare *et al.* (2003) compared the impact of drying basil leaves by microwave oven at atmospheric pressure with the effect of two traditional methods, i.e., air-drying at 50°C and freeze-drying. The MD was carried out at different microwave powers and times on raw basil leaves, while for air and freeze-drying techniques, both raw and blanched leaves were used. In MD samples, the percentage retentions of the characteristic volatile compounds (eucalyptol, linalool, eugenol, and methyl eugenol) were generally higher than in the samples dried by traditional methods, with the exception of freeze-dried unblanched basil.

A further improvement of dried basil leaves by combining microwave and low pressure has been reported (Calín-Sánchez *et al.*, 2012). The main objective of the study of Calín-Sánchez and co-authors was indeed to evaluate the effects of two drying methods (CD, convective drying; and VMD, vacuum-microwave drying) and their combination (CPD Convective pre-drying and VMFD vacuum-microwave finishing drying) on the quality and volatile composition of sweet basil. According to the authors, this last combination (e.g., CPD at 40°C before VMFD at 360W connected to a vacuum system) led to the highest quantity of volatiles retained in the final products. An increase in the temperature of the hot air used for convective drying resulted in higher reductions of total volatiles, while an increase in the intensity of the treatment (higher microwave power) involved small decreases of total volatiles concentration.

The combination proposed by Calín-Sánchez and co-authors (Calín-Sánchez *et al.*, 2012) lead to high-quality dried products with high concentration of total volatiles in a relatively short time (about 250min).

In 2004 Barbieri and co-authors (Barbieri *et al.*, 2004) proposed to dry basil leaves by low pressure superheated steam (LPSS) instead of by hot air (50, 60, and 70°C), studying the impact of the two methods on the basil volatiles. According to the authors the original aroma profile of the fresh vegetable was kept almost constant when LPSS drying technique was applied, while air dried product showed a significant variation in the relative proportions of aroma compounds. Moreover, regarding air-drying, not only a significant reduction in the content of most of the volatiles was observed, but also a significant variation in their relative proportions according to changes in air temperature. Studying hot-air-drying, Baritaux and co-authors (1992) observed an increase in the concentration of several flavor volatiles of basil, including linalool. These authors concluded that the increase in oxygenated monoterpenes was likely due to hydrolysis of glycosides and to conversion of linalyl acetate into linalool.

Recently OD of basil leaves has been reported in literature (Boggia *et al.*, 2012) optimizing the experimental conditions of an industrial patent. The contemporary optimization of both process variables and mixture composition by DOE (Figure 78.2) allowed enhancing the quality of the dried product. Sensorial techniques, head space analysis, and Scanning Electron Microscopy (SEM) (Figure 78.3) were applied to monitor changes in appearance, color, texture, and flavor, respectively. The best appearance and aroma of dried leaves were obtained

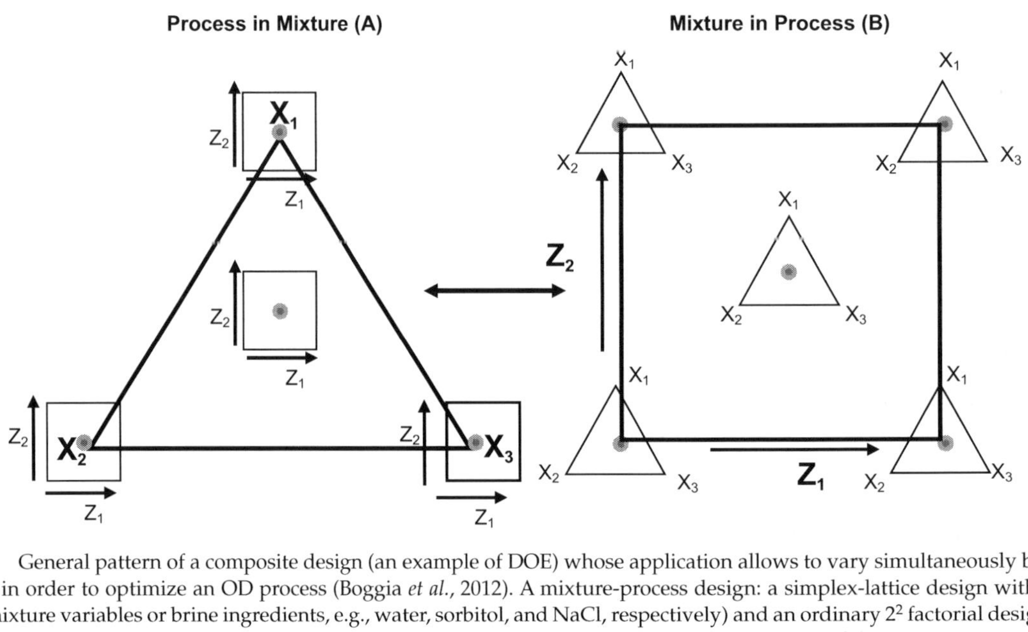

FIGURE 78.2 General pattern of a composite design (an example of DOE) whose application allows to vary simultaneously both process and mixture factors in order to optimize an OD process (Boggia *et al.*, 2012). A mixture-process design: a simplex-lattice design with a central point (X_1, X_2, X_3, the mixture variables or brine ingredients, e.g., water, sorbitol, and NaCl, respectively) and an ordinary 2^2 factorial design with a central point are combined (Z_1 and Z_2, the process variables, e.g., brine temperature and immersion time, respectively).

at 67.5°C with a dipping time of 55 s in a brine containing 65% water; 20% sorbitol, and 15% NaCl. The authors compared the impact of OD with the effect of air-drying at room temperature The OD products showed a higher quality than those treated with the traditional method.

Texture Impact

The impact of several drying methods on leaf structure has been observed by Scanning Electron Microscopy (SEM) and reported in literature (Yousif *et al.*, 1999). According to the authors, after air-drying (CD), basil leaves commonly exhibit severe shrinkage of their cuticle layer. Internally, the epidermis and mesophyllic cells of the palisade layer are extremely affected and look collapsed.

Conversely, when the samples are dried by VMD, the structure of cuticle is less affected by heat, as judged from the extent of shrinkage to that layer (e.g., the cells of the epidermis and palisade layers look "stretched" or "puffed"). This also results in a color improvement since air-dried samples have a darker appearance with respect to vacuum-microwave-dried basil leaves.

OTHER WAYS IN WHICH COMPOSITION IS ALTERED

Besides the decreased flavor, shrinkage, and unappetizing texture, other evident drawbacks of dried products are color damage, browning effects, and decreasing of nutrients.

Color Impact

During drying, as a function of drying conditions, a development of brown pigments is observed. The major green pigment, the blue-green colored chlorophyll a (Chl a) is less stable than the yellow-green chlorophyll b (Chl b), so in heat damaged products, the Chl a/Chl b ratio decreases and the color shifts from green-blue green to green-yellow green. In fact, the determination of the Chl a/Chl b ratio could be commonly used as a marker of the extent of color damage in herbs during processing.

Furthermore chlorophyllase may catalyze the chlorophyll breakdown with the loss of phytol and formation of chlorophyllides, which are more sensitive than chlorophylls to the loss of Mg^{2+}. As the leaf is heated, the intercellular spaces collapse, liberating plant acid compounds and releasing chlorophylls from the protein complex. These events promote the change of chlorophylls into pheophytins, brownish pigments (Di Cesare *et al.*, 2003).

Furthermore, the prevailing conditions of heat and atmospheric oxygen in air-drying facilitate the high enzymatic activity of polyphenol oxidase (PPO), which results in the browning effect that characterizes many air-dried food materials.

According to Di Cesare and co-authors, MD gave the best retention of both aroma compounds and chlorophylls and the lowest color differences between raw and dried basil leaves followed by freeze-drying. Moreover, it requires a shorter process with respect to traditional methods, and if coupled with simultaneous blanching,

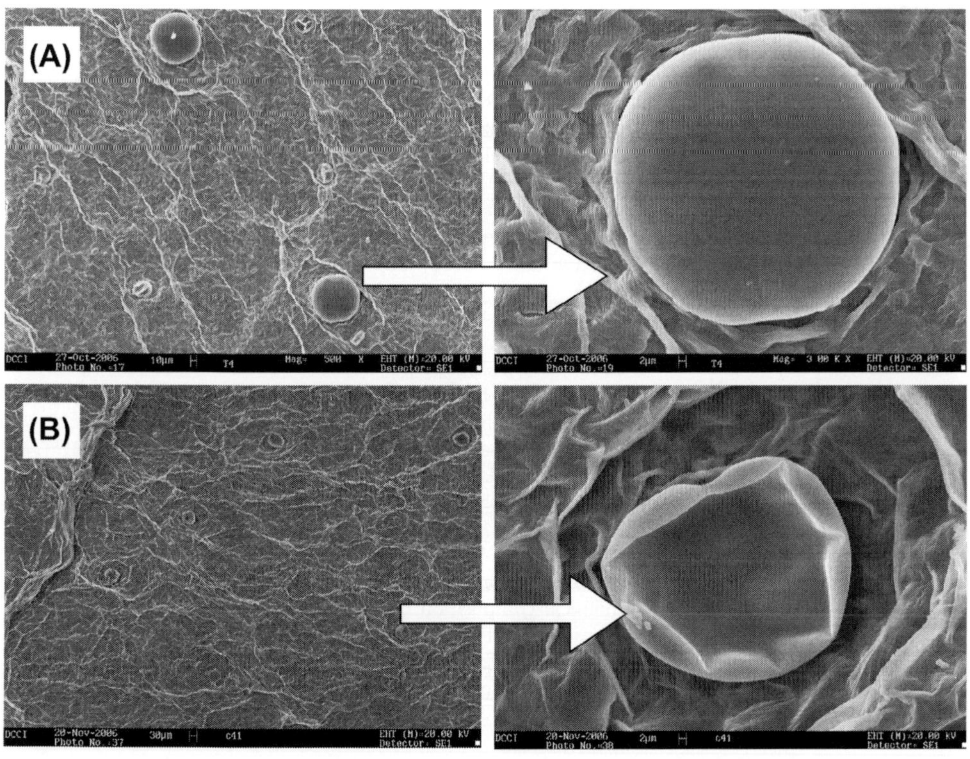

FIGURE 78.3 Scanning electron micrographs of peltate glandular trichomes on the adaxial surface of *Ocimum basilicum* L. leaves (on the left side) and enlargement of a peltate gland (on the right side). The impact of OD respectively in optimized (A) and non-optimized conditions (B) is evident, showing a whole peltate gland in A and a burst one in B.

reduces the loss of salts, acids, volatile compounds, and chlorophyll.

Usually OD and freeze drying of herbs minimize the thermal damage to color by preventing enzymatic browning. Nevertheless, as far as basil leaves are concerned, a comparison among numerous drying methods on the same color evaluations is yet to be published.

Mineral Content

There are only a few studies on effects of drying processes on the mineral content of foods. As far as basil is concerned, in 2005, Ozcan and co-authors (Ozcan *et al.*, 2005) published a paper about the effects of oven and sun drying on the mineral composition of basil herbs.

The authors determined K, P, Ca, S, Mg, Fe, and Al contents as main minerals in both oven and sun dried basil. The effect of these drying methods on the mineral content of fresh, oven dried and sun dried basil were compared. The changes in the concentrations of minerals were dependent on the method and on the drying temperature. Oven drying resulted in increased drying rates and shorter drying time. The content of all the minerals analyzed in the basil oven-dried (at 50°C to 17.31% moisture content after 15 h) were higher than those of the sun-dried basil. Thus the oven drying method is recommended by the authors when the mineral composition should be preserved.

ANALYTICAL TECHNIQUES

Most of the published studies were conducted on the effects of dehydration on sensory qualities of basil leaves such as:

Texture, mainly described by sensory evaluation with trained panels and/or by SEM, Scanning Electron Microscopy. A trained panel is composed of expert assessors with a high degree of sensory sensitivity, experience of sensory methodology and ability to make consistent and repeatable sensory assessments of food samples. Leaf fragments are mounted onto aluminium disc mounts using colloidal silver paste, subsequently coated with gold, and finally stored under desiccation until SEM analysis.

Color, mainly evaluated by Colorimetric Analysis and/or by sensory evaluation. Usually color is studied both by measuring lightness (L), redness (R), and yellowness using a chromatometer and by sensorial analysis using panellists. Pigments (e.g., chlorophyll a and chlorophyll b) are commonly analyzed by HPLC.

Flavor, mainly studied by sensory evaluation and/or by GC/MS, Gas-Chromatography-Mass Spectrometry analysis. Hydrodistillation (HD), microwave-assisted hydrodistillation (MWHD), and supercritical fluid extraction (SFE) can be employed to isolate volatiles prior to GC/MS analysis. Alternative methods such as SPME (Solid Phase Micro Extraction) and SBSE (Stirbar Sorptive Extraction) are often used for sample preparation. These solvent-less techniques minimize the sample preparation reducing organic solvent consumption. Besides, on-line coupling of extraction and analysis results in a higher sensitivity, a reduced potential of analyte loss, and a reduction on the sample amount needed for analysis.

Taste, mainly evaluated by panel test.

There are only a few studies dealing with the effects of drying processes on the mineral content of basil leaves and they are performed using ICP-AES (Inductively Coupled Plasma Atomic Emission Spectrometer).

In conclusion, drying is a complex process which affects basil leaf characteristics in many ways. The adverse changes can be minimized by choosing among the mildest drying technologies and by appropriately designing the process conditions. Nevertheless it is difficult to compare the comprehensive quality of the dried products obtained with different methods and sometimes the best quality is reached using processes and conditions too expensive in terms of time and/or money.

Finally, it is important to underline that the design of food drying must be done in a comprehensive way, also considering pre-drying and post-drying processes (Lewicki, 2006). A thorough knowledge of the pros and cons of the whole drying process is needed in order to optimize it and to obtain high quality products.

References

Antonelli, A., Fabbri, C., Boselli, E., 1998. Modifications of dried basil (*Ocimum basilicum*) leaf oil by gamma and microwave irradiation. Food Chem. 63, 485–489.

Barbieri, S., Elustondo, M., Urbicain, M., 2004. Retention of aroma compounds in basil dried with low pressure superheated steam. J. Food Eng. 65, 109–115.

Baritaux, O., Richard, H., Touche, J., Derbesy, M., 1992. Effects of drying and storage of herbs and spices on the essential oil. Part I. Basil. Ocimum basilicum L. Flavor and Fragrance J. 7, 267–271.

Boggia, R., Leardi, R., Zunin, P., Bottino, A., Capannelli, G., 2013. Dehydration of PDO Genovese basil leaves (*Ocimum basilicum maximum* L. cv Genovese Gigante) by direct osmosis. J. Food Process. Preservation. 37 (5), 621–629.

Calín-Sánchez, A., Lech, K., Szumny, A., Figiel, A., Carbonell-Barrachina, A.A., 2012. Volatile composition of sweet basil essential oil (*Ocimum basilicum* L.) as affected by drying method. Food Res. Int. 48, 217–225.

Contreras, C., Martin, M.E., Martínez-Navarrete, N., Chiralt, A., 2005. Effect of vacuum impregnation and microwave application on structural changes which occurred durino air-drying of apple. LWT-Food Sci. Technol. 38, 471–477.

Di Cesare, L.F., Forni, E., Viscardi, D., Nani, R.C., 2003. Changes in the chemical composition of basil caused by different drying procedures. J. Agri. Food Chem. 51, 3575–3581.

Diaz-Maroto, M.C., Sanchez Palomo, E., Castro, L., Gonzales Vinas, M.A., Perez-Coello, M.S., 2004. Changes produced in the aroma compounds and structural integrity of basil (*Ocimum basilicum* L.) drying. J. Sci. Food Agri. 84, 2070–2076.

Fleisher, A., 1981. Essential oils from two varieties of *Ocimun basilicum* L. grown in Israel. J. Sci. Food Agri. 32, 1119–1122.

Hussain, A.I., Anwar, F., Hussain Scherazi, S.T., Przybylski, R., 2008. Chemical composition, antioxidant and antimicrobical activities of basil (*Ocimum basilicum*) essential oils depends on seasonal variations. Food Chem. 108, 986–995.

Javanmardi, J., Stushnoff, C., Locke, E., Vivanco, J.M., 2003. Antioxidant activity and total phenolic content of Iranian *Ocimum accessions*. Food Chem. 83, 547–550.

Jirovetz, L., Buchbauer, G., Shafi, M.P., Kaniampady, M.M., 2003. Chemotaxonomical analysis of the essential oil aroma compounds of four different *Ocimum* species from southern India. Eur. Food Res. Technol. 217, 120–124.

Karawya, M.S., Hashin, F.M., Hifnaway, M.S., 1974. Oil of *Ocimum basilicum* L. and *Ocimum rubrun* L. grown in Egypt. J. Agri. Food Chem. 22, 520–522.

Kwee, E.M., Niemeyer, E.D., 2011. Variationsin phenolic composition and antioxidant properties among fifteen basil (*Ocimum basilicum* L.) cultivars. Food Chem. 128, 1044–1050.

Leardi, R., 2009. Experimental design in chemistry: a tutorial. Anal. Chim. Acta 652, 161–172.

Lee, J., Scagel, C.F., 2009. Chicoric acid found in Basil (*Ocimun basilicum* L.) leaves. Food Chem. 115, 650–656.

Lee, S.J., Umano, K., Shibamoto, T., Lee, K.G., 2005. Identification of volatile components in basil (*Ocimum basilicum* L.) and thyme leaves (*Thymus vulgaris* L.) and their antioxidant properties. Food Chem. 91, 131–137.

Lewicki, P.P., 2006. Design of hot air drying for better foods. Trends Food Sci. Technol. 17, 153–163.

Makri, O., Kintzios, S., 2008. *Ocimum* sp. (basil): Botany, cultivation, pharmaceutical properties, and biotechnology. J. Herbs Spices Med. Plants 13, 123–150.

Miele, M., Dondero, R., Ciarallo, G., Mazzei, M., 2001. Methyleugenol in *Ocimum basilicum* L. cv. Genovese Gigante. J. Agri. Food Chem. 49, 517–521.

Ozcan, M., Arslan, D., Unver, A., 2005. Effect of drying methods on the mineral content of basil (*Ocimum basilicum* L.). J. Food Eng. 69, 375–379.

Raghavan, G.S.V., Silveira, A.M., 2001. Shrinkage characteristics of strawberries osmotically dehydrated in combination with microwave drying. Drying Technol. 19, 405–414.

Singh, A.P., Luthria, D.L., Wilson, T., Nicholi, V., Singh, V., Banuelos, G.S., Pasakdee, S., 2009. Polyphenols content and antioxidant capacity of eggplant pulp. Food Chem. 114, 955–961.

Surveswaran, S., Cai, Y.Z., Corke, H., Sun, M., 2007. Systematic evaluation of natural phenolic antioxidants from 133 Indian medicinal plants. Food Chem. 102, 938–953.

Tepe, B., Sokmen, M., Akpulat, H.A., Sokmen, A., 2006. Screening of the antioxidant potentials of six *Salvia* species from Turkey. Food Chem. 95, 200.

Tortoe, C., 2010. A review of osmodehydration for food industry. Afr. J. Food Sci. 4, 303–324.

Yousif, A.L., Scaman, C.H., Durance, T.D., Girard, B., 1999. Flavor volatiles and physical properties of vacuum-microwave and airdried sweet basil (*Ocimum basilicum* L.). J. Agri. Food Chem. 47, 4777–4781.

Zhang, M., Tang, J., Mujumdar, A.S., Wang, S., 2006. Trends in microwave-related drying of fruits and vegetables. Food Sci. Technol. 17, 524–534.

CHAPTER

79

Variations in Polyphenols and Lipid Soluble Vitamins in *Moringa oleifera*

Julia P. Coppin, H. Rodolfo Juliani, Qingli Wu, James E. Simon

New Use Agriculture and Natural Plant Products Program, Rutgers University, New Jersey, USA

CHAPTER POINTS

- Studies show that moringa leaves from trees grown and collected in the sub-Saharan African countries of Ghana, Senegal, and Zambia contained high levels of vitamins and polyphenols.
- Rapid profiling and quantitation of γ- and α-tocopherol and α- and β-carotene as well as chlorogenic acid analogs in moringa leaves from the same variety, when grown under varying environmental conditions, would indeed vary, yet still could contribute significantly to the improvement of health and nutrition.
- There were consistently higher amounts of α-tocopherol and β-carotene than their respective isomers in samples of the *M. oleifera* leaves collected from the different locations across Africa.
- The commercial sample of PKM-2 from the Mitengo Women's Cooperative gave the highest overall amounts of tocopherols, carotenes, or chlorogenic acid analogs in *M. oleifera* leaves. The leaves were from the PKM introduced Moringa varieties.
- Moringa leaves contained isomers and analogs of chlorogenic acid.
- Peak 6, the isomer of caffeoylquinic acid has the highest concentration amongst the six chlorogenic acid analogs found in *M. oleifera* leaves.
- The geographical locations of the leaves had little significant influence on the polyphenol and vitamin content.
- *M. oleifera* leaves are rich sources of vitamins and polyphenols.

INTRODUCTION

Moringa leaf, reported to be a good source of nutrients, is known to have extraordinary antioxidative properties (DanMalam *et al.*, 2001; McBurney *et al.*, 2004; Fahey, 2005). Consumed by indigenous people in many tropical areas, the leaves are eaten fresh or cooked in soups, salads, and porridges or used in various cultural cuisines (Locket *et al.*, 2000). The leaf of this multi-purpose tree is said to be rich in vitamins, and polyphenols including phenolic acids and flavonoids (Makkar and Becker, 1996; Fuglie, 1999; Prouilleta *et al.*, 2004; Lakshminarayana *et al.* 2005; Sánchez-Machado *et al.*, 2006; Amaglo *et al.*, 2010; Coppin *et al.*, 2013). For many African nations that suffer from high cases of malnutrition, and have limited healthcare access, the inclusion of plants like moringa for consumption can be significant as the leaves are nutritionally rich and contain a wide range of phytonutrients (Juliani *et al.*, 2009, Kunyanga *et al.*, 2012). With such a high nutritious content, pregnant women and lactating mothers with vitamin A and E deficiencies use the powdered leaves to enhance their children's nourishment (Makkar and Becker, 1996; Lockett *et al.*, 2000; Rodas *et al.*, 2003). Farmers also add the leaves to animal feed to maintain a healthy livestock and use moringa powder as fish food in aquacultural systems (Sarwatt *et al.*, 2002; Dongmeza *et al.*, 2006). The leaves, a significant source of proteins (10.2–17.0%), are used as a protein supplement for animals such as cows, to further fortify the nutrient content in milk (Juliani *et al.*, 2009; Mendieta-Araica *et al.*, 2011; Soliva *et al.*, 2005).

Isomers of vitamin A and E in moringa leaves contribute to its phytochemical content. Epidemiological reports suggest that the vitamins and polyphenols administered as supplements in foods improve the status of different oxidative biomarkers that possess free radical scavenging

Processing and Impact on Active Components in Food
http://dx.doi.org/10.1016/B978-0-12-404699-3.00079-2

capabilities (Bowman and Morbarhan, 1995; Bennett el al., 2003; Scalbert *et al.*, 2005). The antioxidative functionality of these phytochemicals, in part addresses the use of *M. oleifera* as a promising synergistic natural plant that can be applied to ethnic foods to improve human nutrition.

Though the plant is known to contain essential vitamins, traceable amounts of chlorogenic acid (CGA) were also detected within the leaves. Scientific validation suggested that analogs of chlorogenic acid were one of the major phenolic acids identified. Quinic acid and residues of certain *trans*-cinnamic acids, mostly caffeic acid and *p*-coumaric acid were the main constituents, which give structure to CGA isomers (Clifford *et al.*, 2006; Chan *et al.*, 2009). Predominantly 5-O-caffeoylquinic acid and 3-O-caffeoylquinic acid have been detected in moringa leaves collected in Ghana (Amaglo *et al.*, 2010). In this research, a total of six CGA analogs were identified based on the UV and MS data. These CGA isomers are known to function as dietary neurosignalling compounds at adenosine receptors and as antioxidants (Nakatani *et al.*, 2000).

According to Coppin (2008), this study was the first to show that analogs of chlorogenic acid, vitamins E (α–and γ–tocopherols) and A (α- and β-carotenes) were found within the leaves of *Moringa oleifera* from the sub-Saharan African nations of Ghana, Senegal, and Zambia. The identification and quantitation of these phytochemicals by HPLC/UV/MS provides both conformational and new scientific information on the nutritional applications of moringa.

HOW COMPOSITION IS ALTERED

Processed *Moringa oleifera* Leaves Collected from Ghana, Senegal, and Zambia

The first batch of dried leaves of *M. oleifera* were collected from two locations in Ghana (Volta Dan, Dodowa), one location in Senegal (Dakar), and three locations in Zambia (Kafue, Binga, Mungu) in the months of January, May, and August of 2006, respectively. The species were botanically authenticated as *M. oleifera* by local botanists, and herbarium voucher specimens were collected and deposited in ASNAPP offices in each country. A second batch of leaves was collected from Zambia (Lusaka) in April, June, and August 2007, respectively. The leaves were harvested from nurseries and orchards, shade dried, and packaged for experimental analysis. In Lusaka, the Mitengo Women's Cooperative processed Moringa leaves from the PKM-2 variety by drying the leaves under a shed, milling them to fine grains and selling locally. Three commercial samples of their finished products were included in this research. Prior to analysis, the dry *M. oleifera* leaves were ground to a fine powder by a Perten Laboratory Mill 3100 and stored at room temperature in the dark.

Variations of Tocopherols and Carotenes in *Moringa oleifera* Samples

Many studies optimize the extraction of α- and γ-tocopherols, and α- and β-carotenes from natural substances by saponification or lipid extraction. In this research, saponification gave higher concentrations of tocopherols and carotenes within the moringa leaves and hence this method was adopted. Seasonal collection of the leaves was sporadic and inconsistent, therefore seasonal analysis was not taken into consideration for this study. Leaves were collected from naturalized Moringa trees and the introduced PKM-1 and PKM-2 Moringa varieties, which were the improved varieties developed and released from India. This study showed that the PKM Moringa varieties adapted to several countries in sub-Saharan Africa were rich in nutrients and phytochemicals. The naturalized *M. oleifera* already introduced and growing in each country was used to assess which genetic sources were higher in nutrients and vitamins, and to assess whether Moringa regardless of source evaluated, would still contribute to improving health and nutrition.

Both α- and γ-tocopherols (Figure 79.1) and α- and β-carotenes (Figure 79.2) were identified by comparison of retention time with their corresponding standards. All moringa leaf samples analyzed had varying amounts of tocopherols and carotenes. However, results showed that *M. oleifera* leaves contained consistently higher amounts of β-carotene and α-tocopherol than their other respective isomers. A commercial sample (from PKM-2) tested from the Mitengo Women's Cooperative in Lusaka (Zambia) contained the highest overall amount of tocopherols [(116.79 mg/100 g dry weight (d.w.)] and carotenes (43.53 mg/100 g d.w.) within the leaves. This commercial sample also had the highest amounts of β-carotene and α-tocopherol of the entire sample collection (Table 79.1).

The lowest quantity of tocopherols and carotenes were determined to be from different locations in Africa and of different varieties. These were the sample collected from Volta Dam (Ghana) which had the lowest amount of total tocopherols (15.25 mg/100 g d.w.) amongst the naturalized variety, and the PKM-1 sample collected from Lusaka (Zambia) which had the lowest overall amount of total tocopherols (12.58 mg/100 g d.w.). The lowest amount of total carotenes was 4.49 mg/100 g d.w. detected in the naturalized sample received from Dodowa (Ghana). On analysis of the leaves, there was no significant influence on geographical location, or the variety, on the Vitamin A and E content.

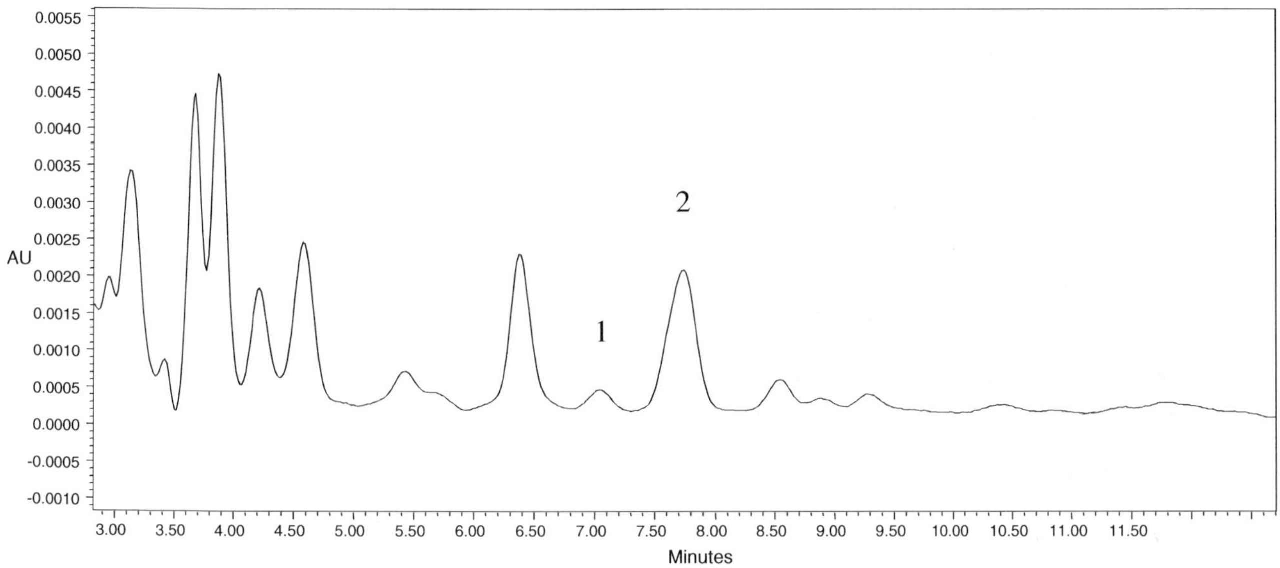

FIGURE 79.1 Chromatographic separation of γ- tocopherol and α- tocopherols in *Moringa oleifera* leaves. A representative UV chromatogram at 300 nm showing (1) γ- tocopherol and (2) α- tocopherols extracted from the leaves of *Moringa oleifera* grown and collected in sub-Saharan Africa. *Reproduced from Coppin (2008).*

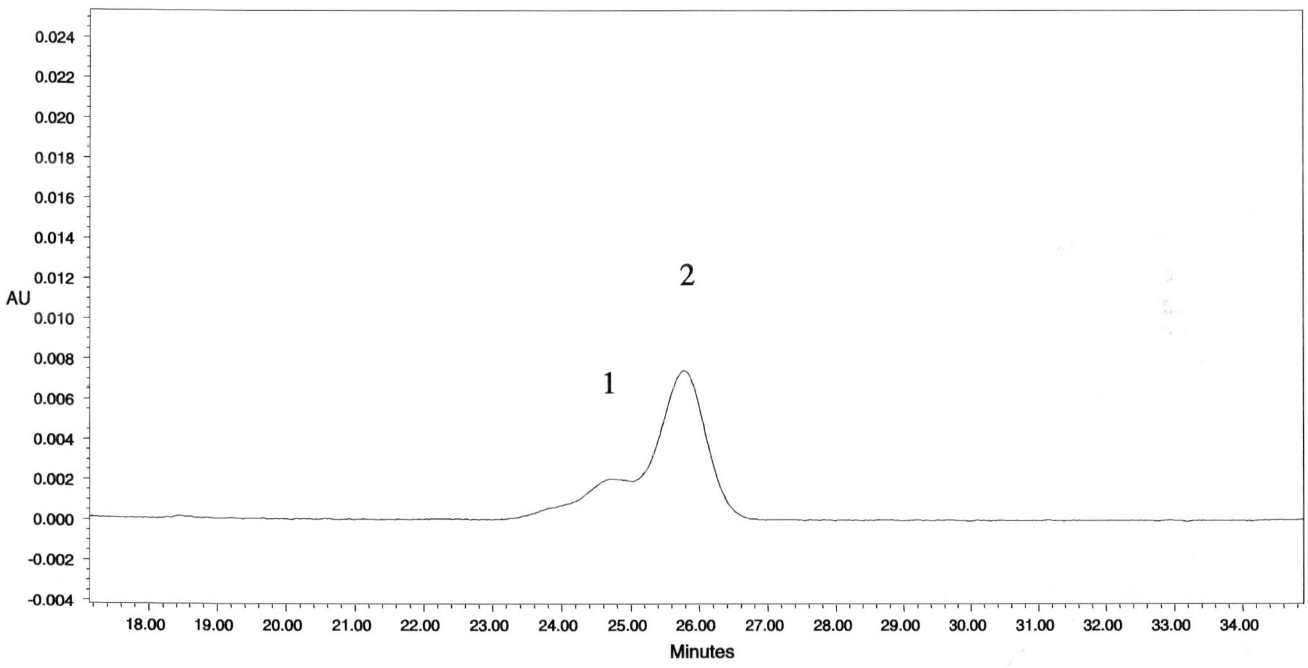

FIGURE 79.2 Chromatographic separation of α-carotene and (2) β-carotene in *M. oleifera* leaves. A representative UV chromatogram at 450 nm showing (1) α-carotene and (2) β-carotene extracted from the leaves of *M. oleifera* grown and collected in sub-Saharan Africa. *Figure taken from Coppin (2008).*

Out of the PKM variety, the PKM-1 leaves from Dakar (Senegal) showed highest levels of total tocopherols and carotenes of 104.71 and 16.32 mg/100 g d.w., respectively. The PKM samples collected in Zambia had lower amounts of tocopherols and carotenes with a slight variation between them; however they were still significantly lower than the sample obtained from Dakar Senegal (Table 79.1). The local PKM (1 and 2) varieties, the indigenous variety, the naturalized Moringa samples, and the finished commercial products (from the PKM-2 variety) showed consistent variations in the amount of carotenes and tocopherols. This suggests that each sample is independent of the other whether origin or variety (Table 79.1) and yet all were found to contain significant levels of these vitamins.

The recommended dietary allowances (RDAs) for Vitamin E (α-tocopherol) vary by gender and age. The

TABLE 79.1 Contents of Isomers of Tocopherols and Carotenes in *Moringa oleifera* Leaves

		Content (mg/100 g d.w.)					
Country	**Location**	**γ-Tocopherol**	**α-Tocopherol**	**Total Tocopherols[a]**	**α-Carotene**	**β-Carotene**	**Total Carotenes[b]**
Zambia	Commercial (PKM-2)	5.46	111.34	116.79	4.85	38.69	43.53
	Lusaka)	1.79	14.94	16.73	0.34	18.55	18.89
		1.56	33.43	35.00	0.64	19.60	20.24
	Lusaka (PKM-1)	0.66	11.91	12.58	0.16	6.34	6.50
	Kafue	7.40	51.35	58.76	0.22	13.48	13.70
	Lusaka (PKM-1)	0.91	11.86	12.77	0.14	13.54	13.69
	Mungu	1.60	18.23	19.83	0.20	12.29	12.49
	Binga	1.38	14.26	15.64	0.10	7.46	7.56
	Lusaka (PKM-2)	2.97	25.60	28.57	0.15	9.80	9.95
	Kafue	7.76	74.08	81.84	0.51	13.88	14.39
	Lusaka (PKM-1)	1.05	12.98	14.03	0.10	7.60	7.70
	Mungu	7.73	78.65	86.38	0.46	14.23	14.68
	Binga	10.54	101.51	112.04	0.35	9.75	10.10
	PKM-2	4.06	26.70	30.76	0.15	10.95	11.10
Ghana	Dodowa	1.73	17.47	19.20	0.31	4.17	4.49
	Volta Dan	2.10	13.14	15.25	0.16	5.66	5.82
Senegal	Dakar (Indigenous)	1.45	23.57	25.02	0.57	23.53	24.10
	Dakar (PKM-1)	6.60	98.11	104.71	0.59	15.73	16.32
	Dakar (PKM-1)	2.98	39.75	42.73	0.30	11.96	12.26
	Dakar (Indigenous)	3.55	104.67	108.22	0.68	19.23	19.91

PKM-1 and PKM-2 were improved varieties from India grown in different locations. Values represent means of duplicate determinations (n = 2).
[a] *Total tocopherols only referring to the additions of both α-tocopherol and γ-tocopherol.*
[b] *Total carotenes only referring to the additions of both α-carotene and β-carotene.*
Data taken from Coppin (2008).

highest RDA of vitamin E for infants is 4–5 mg/day, children, 6–7 mg/day, teenagers and adults, 15 mg/day, and pregnant and lactating women 15 and 19 mg/day, respectively (Food Nutrition Board, 2000). Therefore, using a simulated calculation while recognizing such RDA levels could be different across populations in other countries, 100 g d.w. of moringa leaves from the naturalized variety could supply or exceed the RDA of vitamin E (total tocopherols) for infants, children, teenagers, adults, and pregnant women (except for PKM-1 samples from Zambia) (Table 79.1). Lactating women would require 124.6 g d.w. of moringa leaves to obtain their RDA quota for vitamin E.

The RDA of vitamin A for adults 18 years and over is 0.7–0.9 mg/day (Food Nutrition Board, 2000). Thus, 100 g of moringa leaves from any of the locations included in this study would exceed RDA of vitamin A as the sample with the lowest amount of total carotenes contained 4.49 mg/100 g. This amount would also exceed the RDA of lactating and pregnant women whose intake is 1.2–1.3 mg/day and 0.75–0.77 mg/day, respectively. Infants (0.4–0.5 mg/day) and children (0.6 mg/day) would require 13.34 g or less of moringa leaves with the lowest vitamin A content to satisfy their dietary quota. This indicates moringa leaves provide significant amounts of these vitamins to the diet.

Qualitative Analysis of Chlorogenic Acid

Moringa extract was chemically profiled using LC/UV/MS in comparison with the chlorogenic acid standard, which led to the identification of three isomers (structural and/ or spatial, peaks 1, 2, and 6) and three analogs (dehydroxy derivatives, peaks 3 and 4; and methyl derivative, peak 5) of chlorogenic acid

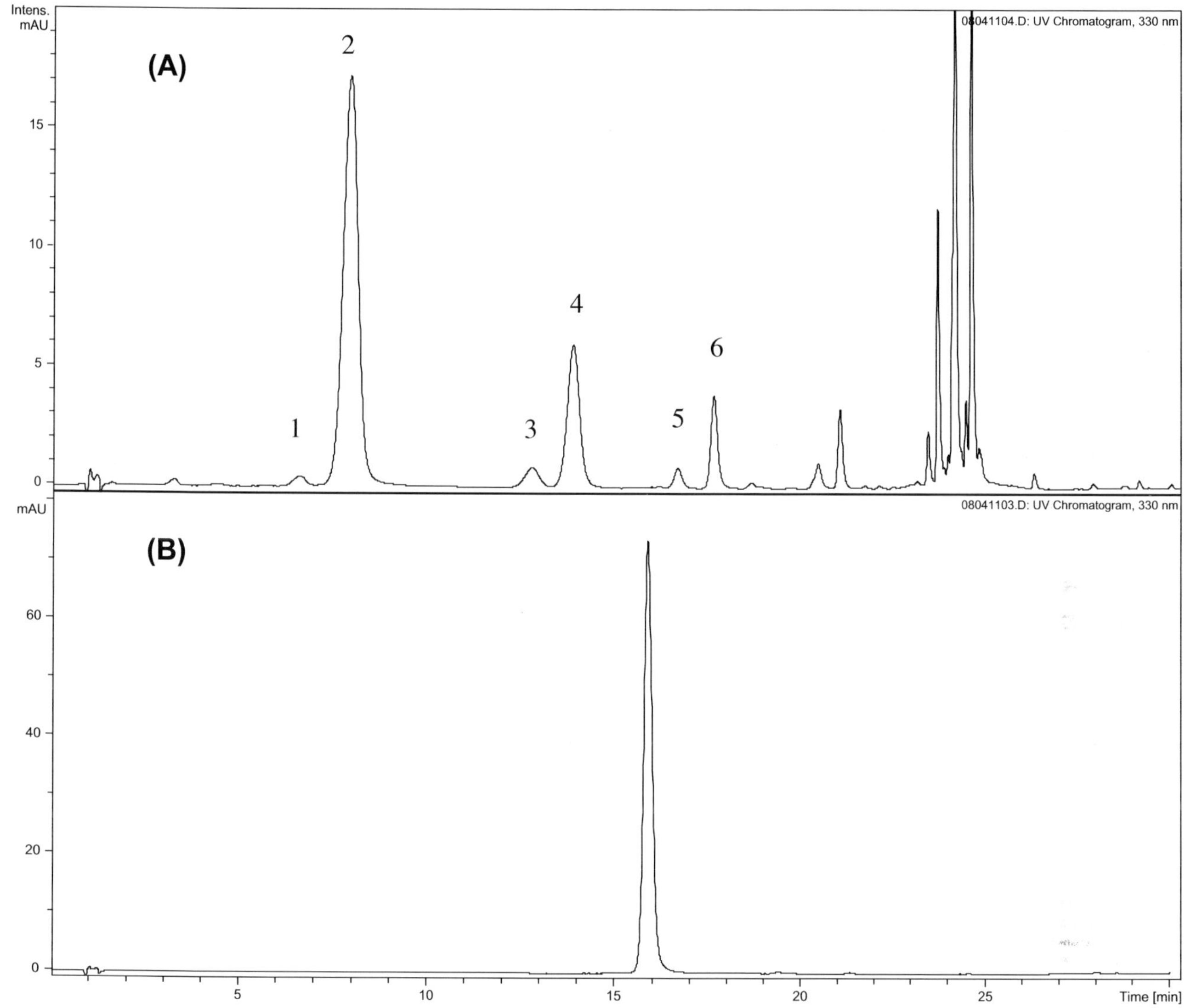

FIGURE 79.3 Chromatographic separation of six chlorogenic acid analogs in *M. oleifera* leaves and the standard chlorogenic acid. UV chromatograms at 330 nm of (A) chlorogenic acid analogs in moringa leaf extract and (B) standard compound of chlorogenic acid. *Data taken from Coppin (2008).*

based on analysis of the UV (λ_{max} 330 nm) and MS data (Figures 79.3 and 79.4). The retention time of the six identified peaks did not match that of the chlorogenic acid standard. However, peaks 1, 2, and 6 analyzed by mass spectrometry (MS), have the same molecular weight as that of chlorogenic acid, while peaks 3 and 4 differed by a molecular weight of −16, and peak 5 differed by a molecular weight of +14 (Figures 79.4 and 79.5). Due to lack of the authenticated pure standards for all the analytes, it was necessary to use chlorogenic acid to estimate the content of each peak for quantitative purposes using the correction factor of molecular weight ratio using UV detection. The MS spectra of the chlorogenic acid analogs detected in Moringa leaves shows the sodium adduct molecular ion and characteristic fragment ions (Figure 79.4).

The UV chromatograms showed that analogs of chlorogenic acid, or 3-caffeoylquinic acid, were the major phenolic acids found in the *M. oleifera* leaf (Figure 79.3). Chlorogenic acids characterized by the ester bond between quinic acid and one to four residues of specific *trans*-cinnamic acids, are most commonly caffeic, *p*-coumaric, and ferulic acid. The mass spectra in Figure 79.3 showed that peaks 1, 2, and 6 could potentially be compounds of CGA isomers, while peaks 3 and 4 are assume to be dehydroxyl derivatives, and peak 5 is assumed to be a methyl derivative of CGA (Figure 79.5).

Each CGA analog and total content of chlorogenic analogs varied between moringa varieties. There was no apparent relationship between the PKM and naturalized varieties. The varieties had values between 0.181–0.414% with the lowest amount of CGA found in

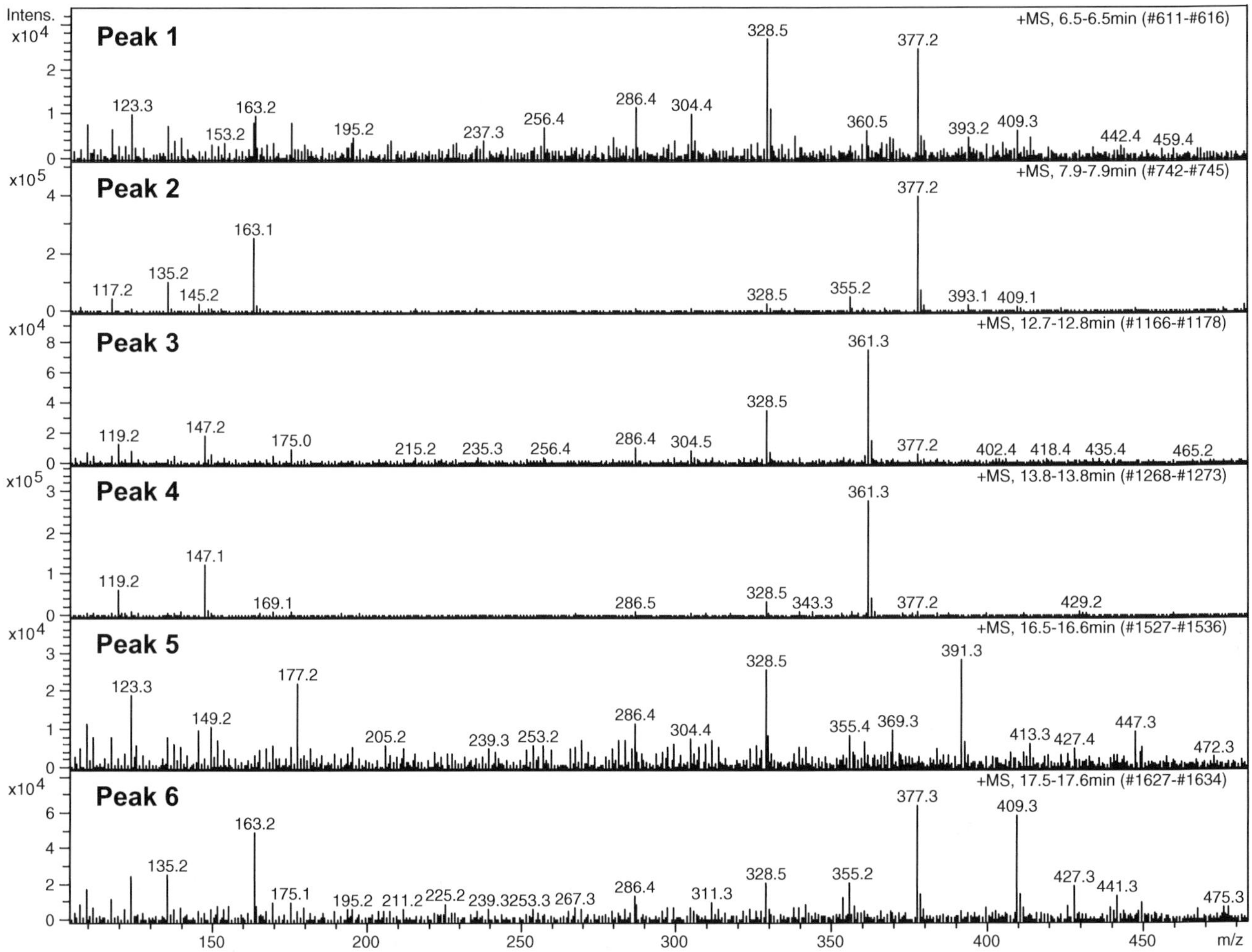

FIGURE 79.4 MS spectra of the major chlorogenic acid analogs in *M. oleifera* leaves. The six chlorogenic acid analogs with peaks (1–6) corresponding to those shown in the UV chromatogram in Figure 79.3. *Data taken from Coppin (2008).*

the sample from Mungu (Zambia) and the highest CGA content from the Mitengo Women's commercial products (from PKM-2) from Lusaka (Zambia) (Table 79.2). It was determined that peak 6, the isomer of caffeoylquinic acid (CQA) has the highest overall concentration amongst all six CGA analogs. This CGA analog was assumed to be an isomer of chlorogenic acid as it had the same molecular weight (Figure 79.5). However, further investigation would be necessary to determine the structure of this isomer.

The PKM-1 sample from Dakar, Senegal, exhibited the highest concentrations of 0.358% while the PKM-2 varieties were lower than the overall median of 0.298% with values of 0.228% and 0.248%. There were variations seen in the indigenous and naturalized varieties with a range of 0.388–0.181% with the highest collected from Dakar, Senegal and the lowest from Mungu, Zambia (Table 79.2).

The dietary intake of antioxidant polyphenols has been determined to be 1 g per day (Scalbert and Williamson, 2000). It has been shown that the moringa leaves can provide significant amount of antioxidant phenols as 100 g of the indigenous samples provide 0.18–0.39 g of polyphenols to the diet (18–39% of the dietary intake) (Juliani *et al.* 2009).

OTHER WAYS IN WHICH COMPOSITION IS ALTERED

Isomers of Vitamin A and E and of chlorogenic acid showed varying quantities in the *M. oleifera* leaves from the three locations collected. This variation could be attributed to many factors, including maturity level at the time of collection, agronomic practices, climatic and geographical locations. For samples prepared at the different locations, it is unknown as to whether there was a standard packaging procedure utilized throughout. Leaves can be eaten fresh, cooked, or stored as a dried powder for many months without loss of nutritional value (Arabshahi-D *et al*, 2007). The tree is also known to be adaptable to a wide range of

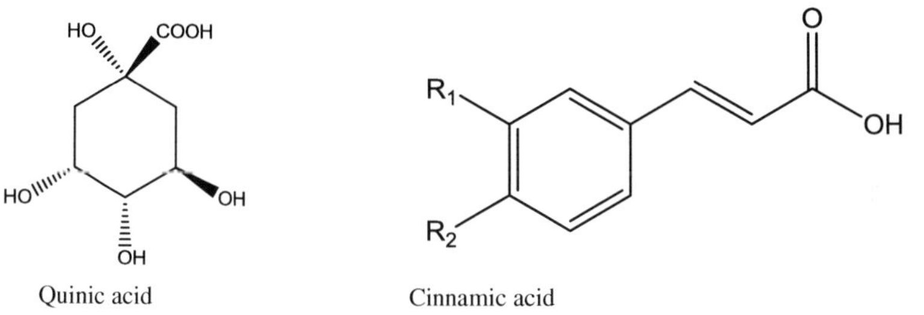

Peak No.	[M+Na]+ (m/z)	MS fragment ion (m/z)	Identity	R Groups
1	377	163	Ca-QA	R_1=R_2=OH
2	377	163	Ca-QA	R_1=R_2=OH
3	361	147	Co-QA	R_1=H, R_2=OH
4	361	147	Co-QA	R_1=H, R_2=OH
5	391	177	Me-Ca-QA	R_1/R_2=OH/OMe
6	377	163	Ca-QA	R_1=R_2=OH

FIGURE 79.5 Molecular structures of isomers and analogs of chlorogenic acid found in *M. oleifera* leaves. Peak No. refers to peaks in the UV chromatogram in Figure 79.3. Ca-QA: Caffeoylquinic acid; Co-QA: Coumaroylquinic acid; Me-Ca-QA: Methylcaffeoylquinic acid. *With permission from Coppin (2008).*

environmental conditions from hot and dry to hot, humid, wet conditions. Under chilling/freezing temperature the tree does not survive but can withstand light frost. To what extent these factors contributed to the phytochemical content within the leaves used is unknown. However, it has been determined that vitamins disintegrate when heated or exposed to light. To reduce this probability, the samples were stored in the dark, and minimal light exposure occurred during analysis of vitamins A and E.

ANALYTICAL TECHNIQUES

Simultaneous Determination of Tocopherols and Carotenes

The procedure for the determination of tocopherols and carotenes within moringa leaves was slightly modified from the method reported by Sánchez-Machado *et al.* (2006).

Determination of Chlorogenic Acid Analogs

For determination of the chlorogenic analogs, about 100 mg of the powdered moringa leaf was extracted in 25 ml of 70% methanol and sonicated for 30 min. Approximately 1 ml of the extract was filtered into light sensitive HPLC vial through a 0.45 μm filter and stored at −20°C in the dark (Coppin, 2008).

HPLC Conditions for Tocopherols and Carotenes

For identification of the tocopherols and carotenes, HPLC separation was performed on a Luna 5 μm Phenyl-Hexyl 150 × 4.6 mm, Phenomenex Co., Alliance HPLC equipped with Waters 2795 Separation module and Waters 2487 UV Dual λ Absorbance Detector, and Waters Millennium Software. The mobile phases consisted of both water (solvent A) and acetonitrile (ACN) (solvent B), however solvent C was 0.1% formic acid (v/v) in methanol (MeOH) for tocopherols, and MeOH for carotenes analysis. For chromographic separation the following gradients were based on reports made by Coppin (2008).

LC/MS Conditions for Analysis of Phenolic Acids

For the identification of chlorogenic acid analogs, HPLC separation was performed on a Prodigy ODS3 5 μm, 150 x 3.2 mm, 5 micro, Phenomenex Co., Hewlett Packard Agilent 1100 Series LC/MS equipped with autosampler, quaternary pump system, DAD detector, degasser, MSD trap with an electrospray ion source (ESI). The mobile phase for chromatographic separation consisted of solvent A (0.1% formic acid in water, v/v) and solvent B (0.1% formic acid (v/v) in ACN) with a linear gradient of 0–50 % B (Coppin, 2008).

TABLE 79.2 Contents of Chlorogenic Acid Analogs in *Moringa oleifera* Leaves from Ghana, Senegal and Zambia

		Content (% d.w.)						
	Location	**1**	**2**	**3**	**4**	**5**	**6**	**Total Chlorogenic Acid Derivatives***
Zambia	Commercial (PKM-2) Lusaka	T	0.017	0.021	0.048	0.033	0.095	0.214
		0.024	0.013	0.038	0.162	0.036	0.066	0.339
		0.026	0.015	0.043	0.209	0.032	0.089	0.414
	Lusaka (PKM-1)	0.021	0.019	0.020	0.051	0.036	0.093	0.240
	Mungu	0.021	0.013	T	0.025	0.028	0.095	0.181
	Lusaka (PKM-1)	0.021	0.014	0.022	0.046	0.032	0.090	0.223
	Mungu	0.022	0.015	0.023	0.070	0.030	0.091	0.252
	Binga	0.024	0.014	T	0.038	0.029	0.113	0.218
	Lusaka (PKM-2)	0.024	0.017	0.020	0.046	0.035	0.105	0.248
	Mungu	0.021	0.021	0.026	0.090	0.046	0.138	0.343
	Lusaka (PKM-1)	0.025	0.015	0.021	0.051	0.036	0.099	0.246
	Mungu	0.021	0.014	0.027	0.093	0.047	0.139	0.340
	Binga	0.024	0.013	0.020	0.05	0.040	0.126	0.273
	Lusaka (PKM-2)	0.021	0.015	0.018	0.042	0.035	0.097	0.228
Ghana	Dodowa	0.021	0.014	0.029	0.098	0.028	0.149	0.339
	Volta Dan	T	0.015	0.025	0.097	0.039	0.094	0.271
Senegal	Dakar (Indigenous)	0.027	0.019	0.032	0.086	0.042	0.181	0.388
	Dakar (PKM-1)	0.023	0.018	0.028	0.072	0.028	0.189	0.358
	Dakar (PKM-1)	0.022	0.017	0.023	0.048	0.027	0.160	0.296
	Dakar (Indigenous)	0.023	0.016	0.026	0.074	0.041	0.078	0.326

The values are expressed as average. Locations from Zambia: Lusaka, Mungu and Binga; Ghana: Dodowa, Volta Dan; Senegal: Dakar. T = trace amounts detected only. Values represent means of duplicate determinations (n = 2).

* *Total analogs of chlorogenic acid in the study refer only to the additions of the six analogs of chlorogenic acid.*

Data taken from Coppin (2008).

References

Amaglo, N.K., Bennett, R.N., Lo Curto, R.B., Rosa, E.A.S., Lo Turco, V., Giuffrida, A., Lo Curto, A., Crea, F., Timpo, G.M., 2010. Profiling selected phytochemical and nutrients in different tissues of the multipurpose tree *Moringa oleifera* L. grown in Ghana. Food Chem. 122, 1047–1054.

Arabshahi-D, S., Devi, D.V., Urooj, A., 2007. Evaluation of antioxidant activity of some plant extracts and their heat, pH and storage stability. Food Chem. 100, 1100–1105.

Bennett, R.N., Mellon, F.A., Foidl, N., Pratt, J.H., Dupoint, S.M., Perkins, L., Kroon, P.A., 2003. Profiling glucosinolates and phenolics in vegetative and reproductive tissues of the multi-purpose trees *M. oleifera* L. (Horseradish Tree) and *Moringa stenopetala* L. J. Agric. Food Chem. 51, 3546–3553.

Bowman, P.E., Morbarhan, S., 1995. Evidence from cancer intervention and biomarker studies and the development of biochemical markers. Am. J. Clin. Nutr. 62, 1403S–1409S.

Chan, E.W.C., Lim, Y.Y., Ling, S.K., Tan, S.P., Lim, K.K., Khoo, M.G.H., 2009. Caffeoylquinic acids from leaves of *Etlingera* species (Zingiberaceae). LWT Food Sci. Technol. 42, 1026–1030.

Clifford, M.N., Wu, W., Kuhnert, N., 2006. The chlorogenic acids of *Hemerocallis*. Food Chem. 95, 574–578.

Coppin, J., 2008. A study of the nutritional and medicinal values of *Moringa oleifera* leaves from sub-saharan Africa: Ghana, Rwanda, Senegal and Zambia. Masters thesis. Rutgers, the State University of New Jersey, New Brunswick, New Jersey, USA.

Coppin, J.P., Xu, Y., Chen, H., Pan, M.H., Ho, C.T., Juliani, R., Simon, J.E., Wu, Q.L., 2013. Determination of Flavonoids and Anti-inflammatory Activity in *Moringa oleifera* by LC/MS. J. Funct. Foods 4, 1892–1899.

DanMalam, H.U., Abubakar, Z., Katsayal, U.A., 2001. Pharmacognostic studies on the leaves of *Moringa oleifera*. Nig. J. Nat. Prod. Med. 5, 45–49.

Dongmeza, E., Siddhuraju, P., Francis, G., Becker, K., 2006. Effects of dehydrated methanol extracted Moringa (*Moringa oleifera* Lam.) leaves and three of its fractions on growth performance and feed nutrient assimilation in Nile tilapia (*Oreochromis niloticus* (L.). Aquaculture 261, 407–422.

Fahey, J.W., 2005. *Moringa oleifera*: A Review of the Medical Evidence for Its Nutritional, Therapeutic, and Prophylactic properties. Part 1. Trees for Life Journal 1:5.

Fuglie, L.J., 1999. The Miracle Tree: *Moringa oleifera*: Natural Nutrition for the Tropics. Church World Service, Dakar. pp. 68; revised in 2001 and published as The Miracle Tree: The Tree Multiple Attributes of Moringa, pp. 172. http://www.echotech.org/bookstore/advanced_search_result.php?keywords=Miracle+Tree%20.

Food and Nutrition Board, 2000. Dietary Reference Intakes. Vitamin C, Vitamin E, Selenium, and Carotenes. National Academy Press, Washington, DC.

Juliani, H.R., Fonseca, Y., Acquaye, D., Malumo, H., Malainy, D., Simon, J.E., 2009. Nutritional Assessment of Moringa (*Moringa* spp.) from Ghana, Senegal and Zambia. In: Juliani, H.R., Simon, J.E., Ho, C.T. (Eds.), African Natural Plant Products: Discoveries and Challenges in Quality Control. American Chemical Society Symposium Series 1021. American Chemical Society, Washington, D.C. USA, pp. 469–484.

Kunyanga, C.N., Imungi, J.K., Okoth, M.W., Biesalski, H.K., Vadivel, V., 2012. Total phenolic content, antioxidant and antidiabetic properties of methanolic extract of raw traditionally processed Kenyan indigenous food ingredients. LWT Food Sci. Technol. 45, 269–276.

Lakshminarayana, R., Raju, M., Krishnakantha, T.P., Baskaran, V., 2005. Determination of major carotenes in a few leafy vegetables by high performance chromotography. J. Agric. Food Chem. 53, 2838–2842.

Lockett, C.T., Calvert, C.C., Grivetti, L.E., 2000. Energy and micronutrient composition of dietary and medicinal wild plants consumed during drought. Study of rural Fulani, northeastern Nigeria. Int. J. Food Sci. Nutr. 51, 195–208.

Nakatani, N., Kayano, S., Kikuzaki, H., Sumino, K., Katagiri, K., Mitani, T., 2000. Identification, quantitative determination, and antioxidative activities of chlorogenic acid isomers in prune (*Prunus domestica* L.). J Agric. Food Chem. 48, 5512–5516.

Makkar, H.P.S., Becker, K., 1996. Nutritional value and whole and ethanol antinutritional components of extracted *Moringa oleifera* leaves. Animal Feed Sci. Technol. 63, 211–228.

McBurney, R.P.H., Griffin, C., Paul, A.A., Greenberg, D.C., 2004. The nutritional composition of African wild food plants: from compilation to utilization. J. Food Composition Anal. 17, 277–289.

Mendieta-Araica, B., Spörndly, R., Reyes-Sánchez, N., Spörndly, E., 2011. Moringa (*Moringa oleifera*) leaf meal as a source of protein in locally produced concentrates for dairy cows fed low protein diets in tropical areas. Livestock Sci. 137, 10–17.

Prouilleta, C., Mazie'reb, J., Mazie'reb, C., Wattela, A., Braziera, M., Kamela, S., 2004. Stimulatory effect of naturally occurring flavonols quercetin and kaempferol on alkaline phosphatase activity in MG-63 human osteoblasts through ERK and estrogen receptor pathway. Biochem. Pharmacol. 67, 1307–1313.

Rodas Mendoza, B., Morera Pons, S., Castellote Bargalló, A.I., López-Sabater, M.C., 2003. Rapid determination by reversed-phase high-performance liquid chromatography of Vitamins A and E in infant formulas. J. Chromatogr. A. 1018, 197–202.

Sarwatt, S.V., Kapange, S.S., Kakengi, A.M.V., 2002. Substituting sunflower seed-cake with *Moringa oleifera* leaves as a supplemental goat feed in Tanzania. Agroforestry Syst. 56, 241–247.

Sánchez-Machado, D.I., Lopez-Cervantes, J., Rios Vasquez, N.J., 2006. High performance liquid chromatography method to measure α- and γ-tocopherol in leaves, flowers and fresh beans from *M. oleifera*. J. Chromatogr. A. 1105, 111–114.

Scalbert, A., Williamson, G., 2000. Dietary Intake and Bioavailability of Polyphenols. J. Nutr. 130 (8), 2073S–2085S.

Scalbert, A., Johnson, I.T., Saltmarsh, M., 2005. Polyphenols: antioxidants and beyond. Am. J. Clin. Nutr. 8 (Suppl), 215S–2175S.

Soliva, C.R., Kreuzer, M., Foidl, N., Foidl, G., Machmüller, A., Hess, H.D., 2005. Feeding value of whole and extracted *Moringa oleifera* leaves for ruminants their effects on ruminal fermentation in vitro. Animal Feed Sci. Technol. 118, 47–62.

SECTION 10

CONFECTIONARY AND OTHER FOOD ITEMS

CHAPTER

80

Effect of Processing on Dark Chocolate Composition: A Focus on Allergens

Sefat E. Khuda, Kristina M. Williams

US Food and Drug Administration, Maryland, USA

CHAPTER POINTS

- Chocolate matrix components and the tempering process differentially affected the detection of peanut, egg, and milk allergens using commercial ELISA methods.
- The extraction of total proteins and allergenic proteins from a chocolate matrix depends on the extraction buffer components and the characteristics of the individual protein.

INTRODUCTION

Global consumption of chocolate and chocolate products is increasing due to the positive health effects of many of its constitutent components (Hii *et al.*, 2009; Rusconi and Conti, 2010; Waterhouse et al., 1996). The increase in chocolate consumption poses potential health risks to populations who are sensitive to allergens that may be present in chocolate products. This is evidenced by the number of chocolate-associated safety recalls by the US Food and Drug Administration (http://www.fda.gov/Safety/Recalls/ucm307962.htm; http://www.fda.gov/Safety/Recalls/ucm302056.htm). Allergy to chocolate components is rare, but amine in chocolate can trigger restaurant syndrome, resulting in symptoms like migraine or hypertension (Settipane, 1987). However, chocolate products may contain allergenic material as a part of the product formulation or as an adulterant. For example, milk is often added to chocolate to soften its texture. Unfortunately, milk allergy is the most common food allergy among children (Afoakwa, 2010). The majority of inadvertent allergen exposures occur from adulterated chocolate products containing one or more allergens that are not on the product label. Since strict avoidance is the only preventive measure for allergic consumers, the presence of allergens poses a consumer risk. Cross contact, incomplete cleaning of production lines, or improper package labeling are the biggest threats to allergic consumers, as determined by surveillance studies (Pele *et al.*, 2007; Spanjersberg *et al.*, 2010).

Detection methods for food allergens play a major role in developing appropriate labeling guidelines and assuring consumer safety. The most common detection methods are DNA-based [polymerase chain reaction (PCR)], and antibody-based [enzyme-linked immunosorbent assay (ELISA) and lateral flow] (Kirsch *et al.*, 2009). Other techniques, like liquid chromatography-mass spectrometric (LC-MS) are becoming more common, but LC-MS is more difficult to standardize and requires specialized equipment (Fæste et al., 2011). Regardless of the detection method, there are critical assay parameters that must be considered, such as sample preparation and food matrix.

Food matrices may adversely affect detection procedures. Chocolate is a difficult matrix because of its composition and physical characteristics. High fat, polyphenol, and tannin content of chocolate may directly interfere with extraction and detection procedures. Matrix components may also form covalent, ionic, and hydrogen bonds or hydrophobic interactions with target proteins. Physical and chemical properties and the solubility of target allergens also change in a chocolate matrix (Taylor *et al.*, 2009).

HOW COMPOSITION IS ALTERED

Food Processing Effects on Chocolate Components and Allergenic Proteins

Processed and mixed food products are more popular with consumers than unprocessed and single

Processing and Impact on Active Components in Food
http://dx.doi.org/10.1016/B978-0-12-404699-3.00080-9

ingredient food, increasing the challenge of protecting food-allergic consumers (Grimshaw *et al.*, 2003; Pele *et al.*, 2007). Chocolate and chocolate products have to undergo a tempering procedure to improve texture and appearance, the major determinants of chocolate quality and acceptance. During manufacturing processes, such as melting and tempering, the fat in chocolate influences the crystallization of end products. Food industries employ a variety of chocolate processing procedures including heating, chilling, high pressure treatment, ultra filtration, irradiation, hydrolysis, and fermentation (van Hengel, 2007). Each of these processes modifies chocolate components structurally and functionally, including allergenic material that may be present. During industrial processing of chocolate, chemical rearrangements including aldol condensation, polymerization, cyclization, and alkalization lead to formation of a variety of products and compounds that can enhance flavor, color, shelf-life, bioavailabilty, and nutritional value (Afoakwa, 2010).

Food processing also causes physical and chemical modifications to allergenic proteins such as unfolding, aggregation, fragmentation, hydrolysis, covalent modification, or total degradation. These structural changes may modify protein epitopes that are responsible for eliciting allergic reactions or are the target of immunochemical detection methods. One common chemical modification, usually requiring heat, is the Maillard reaction between the carbonyl group of a reducing sugar in chocolate and the amino group of a protein. (Davis *et al.*, 2001). Glycosylated allergenic proteins resulting from the Maillard reaction may have altered allergenicity or immunoreactivity either by the formation of new IgE binding epitopes or by the destruction of existing ones (Maleki *et al.*, 2000). Therefore, processing can result in increased or decreased protein allergenicity. During processing, new epitopes can also be revealed that were hidden within the three dimensional structure of the allergenic protein. Certain components of the food matrix may also interact with the allergenic proteins creating neoepitopes, or conversely, food matrix components may block IgE epitopes in processed food (Berrens, 1996; Clare Mills et al., 2009; Nowak-Wegrzyn and Fiocchi, 2009).

OTHER WAYS IN WHICH COMPOSITION IS ALTERED

Effect of Tempering on the Analysis of Food Allergens in a Dark Chocolate Matrix

This study has focused on detection of common allergens processed into a model food matrix. Dark chocolate was chosen since chocolate consumption and adverse reactions to allergens in this matrix are increasing. An understanding of factors complicating food allergen detection in this matrix may also have a wider application to other food matrices. The objectives of this work were to: (i) investigate the effect of tempering on simultaneous detection of multiple allergens (peanut, egg, and milk) in a dark chocolate matrix; (ii) determine the effect of tempering on recovery of proteins from dark chocolate using different extraction buffers; and (iii) evaluate tempering effects on the extractability of specific allergenic proteins from a dark chocolate matrix using allergen-specific antibodies.

ANALYTICAL TECHNIQUES

Detection of Peanut, Egg, and Milk Allergens in a Tempered Dark Chocolate Matrix using Commercial ELISA Kits

The following materials were used to produce the chocolate samples: semi-sweet 62% cacao chocolate, produced on a peanut-, egg-, and milk-free dedicated line (ScharffenBerger); nonfat dry milk-NIST SRM 1549 and spray-dried whole egg powder-NIST RM 8445 (National Institute of Standards and Technology) and peanut flour, 12% fat light roast (Golden Peanut Co.). The 500, 100, 25, 10, 5, and 2.5 ppm tempered samples were prepared by adding the required amounts of dry milk, egg powder, and peanut flour to melted chocolate and tempered for 4 h at 46°C. Untempered chocolate was prepared by grinding chocolate to a fine powder over dry ice followed by addition of the required amounts of allergens. Homogeneity of both samples was determined as previously described (Khuda *et al.*, 2012a).

Food processing effects on allergenic proteins can hamper critical procedural steps of ELISA methods. Matrix components of food may cross-react with the reagents of detection methods, causing false-positive results. Tannins and polyphenol compounds of chocolate have a high binding affinity to proteins and, occasionally, for the antibody components of the ELISA. Addition of fish gelatin and nonfat dry milk (NFDM) to the chocolate extraction can improve the extraction efficiency of peanut allergens and block non-specific binding of matrix components (Pomes *et al.*, 2004). In the present study, peanut, egg, and milk (casein and BLG) allergens were quantified in untempered and tempered dark chocolate samples using five commercial test kits: (i) RIDASCREEN FAST peanut, egg, and casein from R-Biopharm (RB); (ii) Veratox peanut, egg, and total milk allergen quantitative test kits from Neogen (NE) Corp.; (iii) Morinaga (MO) peanut, egg, and milk [casein and β-lactoglobulin (BLG)] protein ELISA kits (Crystal Chem); (iv) Tepnel (TE) BIOKIT peanut, egg, casein, and BLG assay kits (Neogen Corp.); and (v) ELISA Systems (ES) peanut, egg, casein, and BLG residue kits (BioMericux). For analytical purposes, four samples at each incurred level were extracted according to the kit instructions. Four aliquots of each sample were used to quantitate the allergen concentration (16 total aliquots per incurred level). The data obtained from the assays were analyzed based on calculated protein content of each allergen at each incurred level.

Peanut: Mean recovery at all incurred levels of peanut for untempered and tempered chocolate, respectively, were as follows: ES (1.8%, 1.6%), MO (17.1%, 15.2%), NE (35.1%, 26.1%), RB (92.6%, 66.9%), and TE (11.8%, 18.8%). Chocolate tempering did not significantly decrease the mean measured concentration or the percent recovery of peanut (Figure 80.1.A). For most kits, however, the recovery was well below the incurred levels for both untempered and tempered samples indicating that matrix components affected peanut recovery more than the tempering process.

Egg: Egg protein was close to or higher than the incurred levels in the untempered and tempered dark chocolate matrix. Mean recovery at all incurred levels of egg for untempered and tempered chocolate, respectively, were as follows: ES (78.5%, 67.9%), MO (110.4%, 88.6%), NE (253.1%, 285.4%), RB (264.1%, 256.8%), and TE (77.3%, 81%) (Figure 80.1.B), indicating that neither the tempering process nor matrix factors impacted egg protein determination.

Milk Casein and BLG: Casein and BLG recoveries were reduced following tempering, and in some cases the reduction was dramatic (Figure 80.1.C, D). The negative impact of processing on milk protein detection could be due to reduced solubility and heat sensitivity of the proteins being measured and/or a reduction in binding of detection antibody to the proteins, possibly due to conformational changes of milk proteins during tempering. The matrix itself did not interfere with detection methods for milk, since the recovery of untempered samples was similar to that of tempered samples.

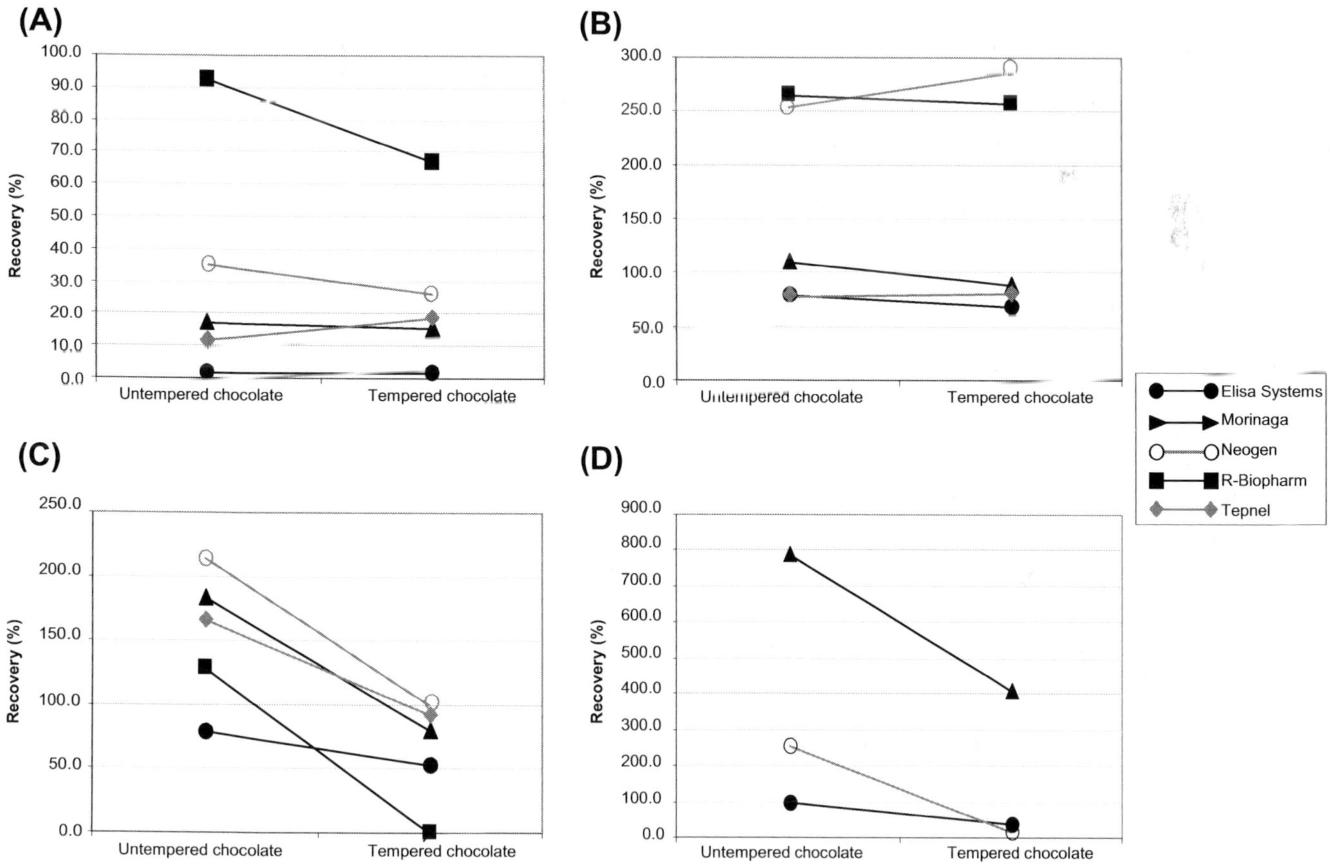

FIGURE 80.1 Mean per cent recovery of all incurred levels in untempered and tempered dark chocolate for (A) peanut, (B) egg, (C) casein, and (D) BLG using indicated commercial ELISA kits. Recovery of each test allergen was decreased following tempering.

Extraction of Proteins from Tempered Chocolate Samples using Different Buffer Solutions

The use of appropriate extraction procedures with processed food and difficult matrices is critical for the accurate quantification of allergenic proteins (Khuda *et al.*, 2012b). There is no universal extraction buffer that is optimal for all foods, because efficiency of allergen extraction depends on the matrix, extraction buffer components, food texture, and specific extraction procedures, such as time and temperature.

The condition of the allergenic material may also affect extraction efficiency. Studies with defatted and non-defatted peanut flour showed no difference in extraction efficiency of peanut proteins with various buffers, but there was poor extractability of peanut protein from light and dark roasted peanuts as compared to raw peanuts (Rudolf *et al.*, 2012). Previous evaluations of extraction solutions showed that buffers containing Tween 20, Triton X-100, sodium dodecyl sulfate (SDS), high NaCl and 2-mercaptoethanol (2-ME) can improve extraction of allergenic proteins from heat-processed foods (Steinhoff *et al.*, 2011; Westphal *et al.*, 2004; Watanabe *et al.*, 2005).

Protein extraction efficiency was evaluated quantitatively and qualitatively using five different extraction buffers: phosphate-buffered saline (PBS)-10 mM pH 7.4; high-salt buffer (HSB)-20 mM $NaH_2PO_4 \cdot H_2O$, 1 M NaCl pH 7.4; 0.1 M carbonate, pH 9.5; 0.05 M Tris-HCl, pH 8, plus 0.5 N NaCl (Cabanillas *et al.*, 2012); and HSB containing 1% SDS plus 7% 2-ME. A 1 g portion of the 0 ppm and 500 ppm untempered and 500 ppm tempered samples were extracted by vigorous mixing with 10 ml of buffer solution at room temperature, then centrifuged at 20,000 × g for 30 min at 4°C and clarified by filtration through a 0.45 μm PVDF filter. Protein levels of extracts were determined with a Pierce 660 nm protein assay kit with ionic detergent compatibility reagent (IDCR) (Piece Chemical Co.). Table 80.1 shows the protein yield from test samples using the different buffers. Significantly less total protein was extracted with HSB, PBS, and Tris buffers as compared to protein extracted with HSB/SDS/ME and carbonate buffers. Additionally, total protein extracted from tempered samples exceeded that of untempered samples using HSB/SDS/ME and carbonate buffers. The high pH of the carbonate buffer and the addition of denaturing or reducing agents, such as SDS and 2-ME, to high salt buffer apparently increased the solubility of the proteins from untempered and tempered samples. These results show that solubilization of protein from a processed chocolate matrix can be optimized by using different extraction buffer components.

TABLE 80.1 Extraction Yield of Total Protein in Untempered and Tempered Dark Chocolate with HSB, HSB/SDS/ME, Carbonate, Tris, and PBS

Chocolate Samples	Total Protein Concentration (mg/ml) at Indicated Extracts				
	HSB	HSB/SDS/ME	Carbonate	Tris	PBS
Untempered 0 ppm	0.115	2.75	1.63	0.178	0.041
Untempered 500 ppm	0.256	2.04	1.59	0.177	0.068
Tempered 500 ppm	0.249	2.55	3.7	0.25	0.191

Protein concentrations of different extracts were measured using Pierce 660 nm protein assay kit with IDCR reagent.

Detection of Specific Allergenic Proteins by Western Blot Analysis of Tempered Chocolate Extracts

Although certain buffers may be preferable for extraction of total protein from food samples, specific food allergenic proteins may not be efficiently extracted using the same buffers. Western blot analysis was employed for qualitative characterization of different extracts from untempered and tempered chocolate samples. Allergen specific immunochemical staining was performed using polyclonal antibodies and labeled proteins were detected with Odyssey Infrared Imaging System (Li-Cor Biosciences).

Peanut

Figure 80.2 shows the immunoblot using five different buffer extracts of chocolate samples stained with rabbit polyclonal antibodies raised against whole denatured peanut and individual peanut proteins Ara h 1, Ara h 2, Ara h 3, Ara h 6. Using Ara h 1-specific antibody no Ara h 1 protein was detected in any of the sample extracts. This was probably due to the effects of tempering on Ara h 1, which is known to become rapidly denatured and/or aggregated following various processing conditions, including heat. These structural changes result in reduced extraction efficiency of the protein and, possibly, reduced affinity of the anti-Ara h 1 antibody. Ara h 2 dimers were only observed in the blots of HSB/SDS/ME extracts, demonstrating that Ara h 2 proteins are optimally extracted using reducing and denaturing conditions. Faint Ara h 3 bands were detected in the carbonate and HSB/SDS/ME extracts. Ara h 6 and denatured peanut antibody recognized Ara h 6 and Ara h 1 (faint), respectively, only in the HSB/SDS/ME extract. Staining intensity was typically lower using tempered as compared to untempered samples when extracted with the same buffer. Ara h 2 and Ara h 6 bands are known to be more resistant to processing than other peanut allergens

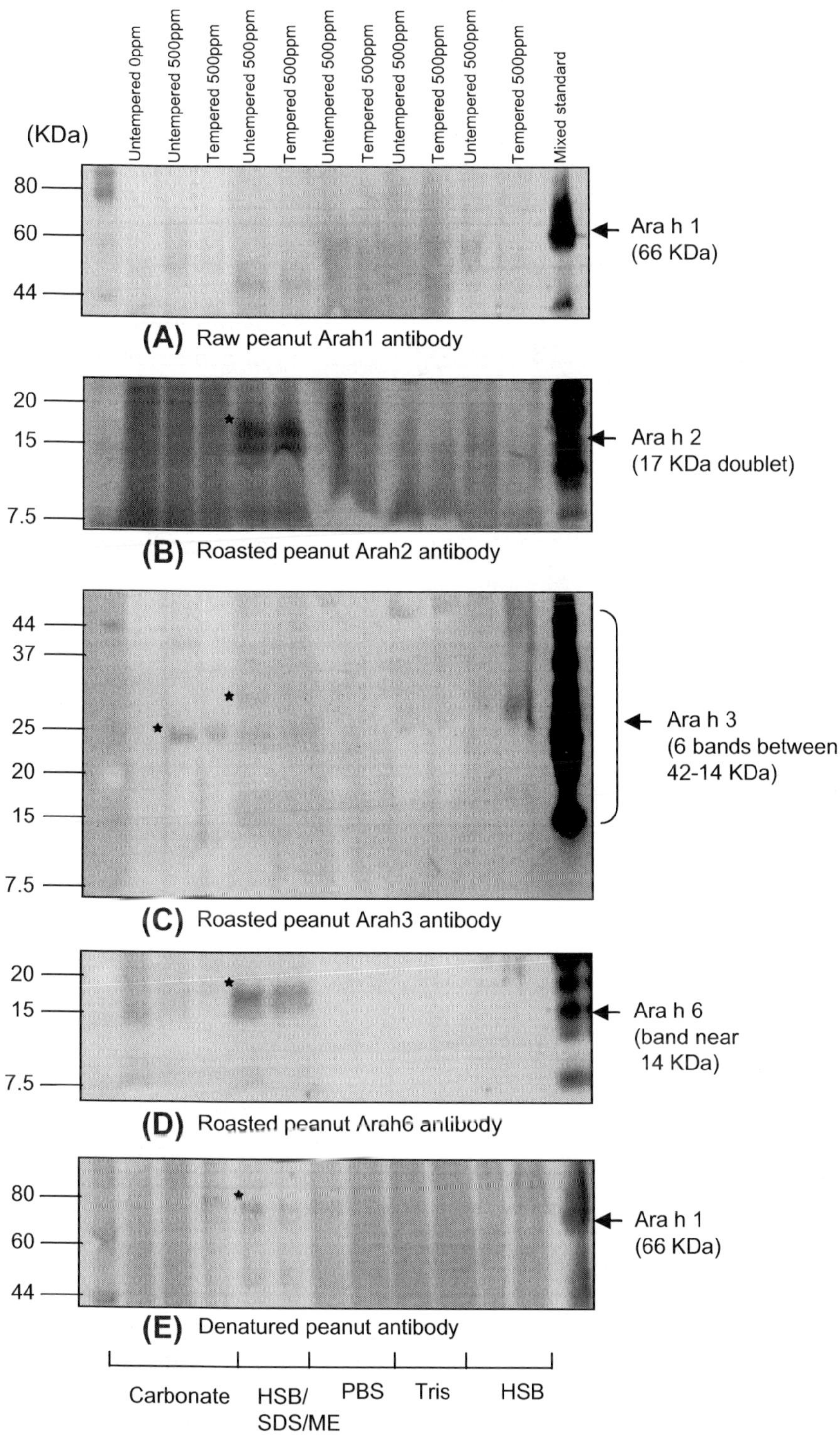

FIGURE 80.2 Immunoblot of five different extracts with polyclonal antibodies against raw peanut Ara h 1 (A); roasted peanut Ara h 2 (B); roasted peanut Ara h 3 (C); roasted peanut Ara h 6 (D); and denatured peanut protein (E). * Indicates the peanut allergen corresponding to protein specificity of the antibody. Ara h 1 purified from raw peanuts was used to generate polyclonal antibody because of the inability to purify this protein from roasted peanuts.

and are identified as the major peanut allergens (Zhuang and Dreskin, 2012; Chen *et al.*, 2012). These observations led to the conclusion that the optimum extraction of these major peanut allergenic proteins from tempered chocolate samples required a buffer with added denaturants and/or reducing agents.

Egg

To determine if rabbit polyclonal antibodies produced against denatured egg proteins could result in improved detection of egg allergens from processed food, antibodies were generated using total proteins prepared from raw egg white, boiled egg white, and

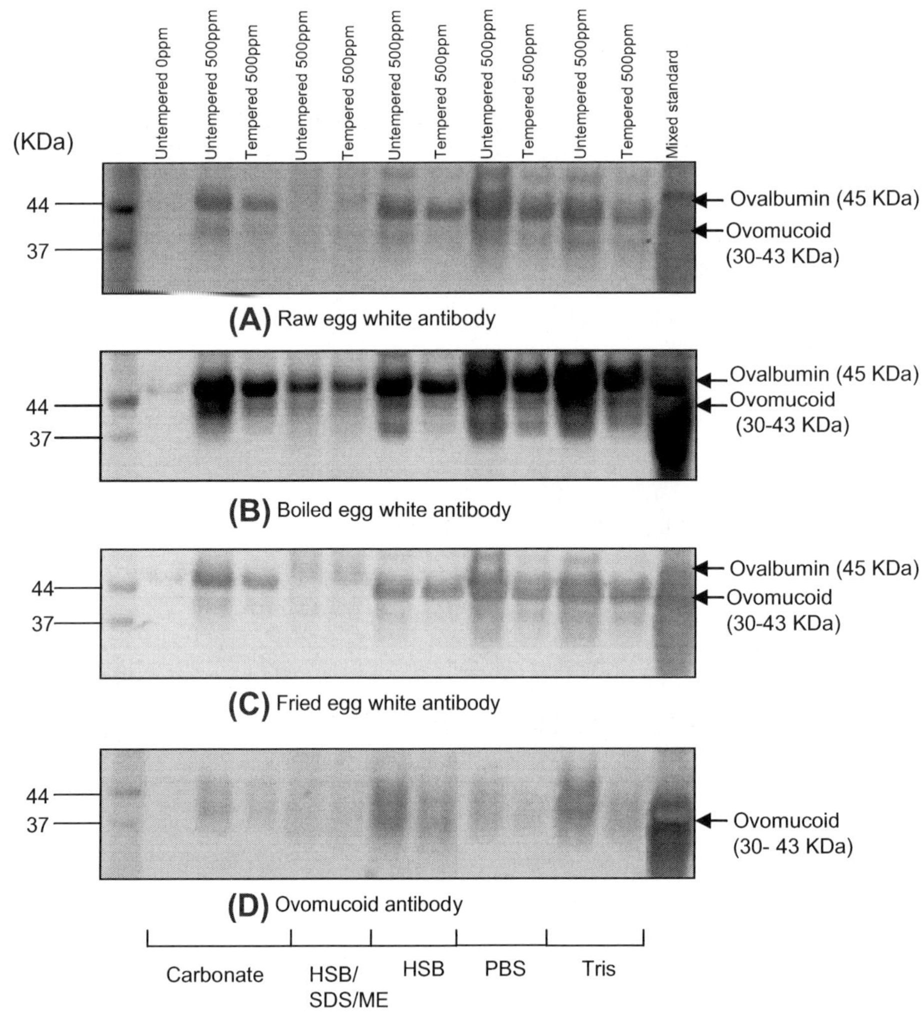

FIGURE 80.3 Immunoblots of egg allergenic proteins using antibodies against raw and processed egg. Antibodies produced against raw egg white (A), boiled egg white (B), fried egg white (C), and ovomucoid (D) were used to detect egg proteins in different buffer extracts of untempered and tempered chocolate samples.

fried egg white. Antibodies were also raised against purified ovomucoid, since this egg allergen is relatively heat resistant. Egg allergenic proteins in different buffer extracts were evaluated by immunoblot analysis (Figure 80.3). The analysis focused on the egg proteins ovalbumin and ovomucoid, which are the most common egg allergens. Antibody produced against raw egg, white, and fried egg white detected ovalbumin and ovomucoid in all buffer extracts except HSB/SDS/ME, but staining intensity was lower with tempered extracts. When using antibody generated against boiled egg white, proteins were detected in all extracts, but staining intensity was also lower using extracts from tempered samples. Ovomucoid-specific antibody detected proteins mostly in HSB and Tris extracts with reduced intensity in tempered samples. This indicates that antibody generated against denatured (boiled) egg proteins may be preferable for detection of egg allergens from processed samples, but recovery may still be reduced following tempering.

Milk casein and BLG

Casein and BLG were evaluated with commercial polyclonal antibodies generated against casein from milk extracts (Bioorbyt LLC) and a synthetic peptide generated from the BLG amino acid sequence (Genetex Inc.). These common milk allergenic proteins are heat sensitive and difficult to detect from processed samples (Figure 80.1). The effect of tempering on the detection of casein and BLG in buffer extracts was determined by immunoblot analysis (Figure 80.4). Carbonate, PBS, Tris and HSB buffers were optimum for casein extraction, but tempering decreased antibody binding to casein regardless of the buffer used for extraction. Comparable results were obtained using the anti-BLG antibody, although staining was slightly better using the HSB extracts.

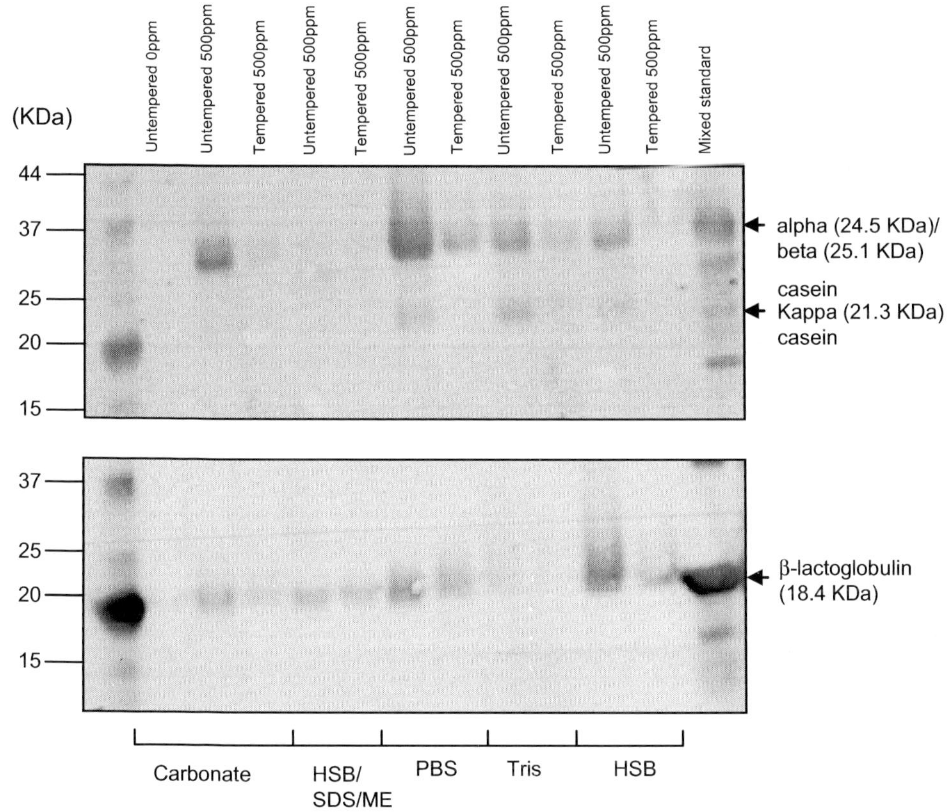

FIGURE 80.4 Evaluation of milk allergenic proteins in different buffer extracts of untempered and tempered chocolate samples. Casein and β-lactoglobulin polyclonal antibodies were used. Apparent molecular weight of casein proteins is higher than predicted due to casein phosphorylation.

Unlike the situation with peanut allergenic proteins, the addition of detergent or a reducing agent did not improve extraction efficiency of milk proteins.

References

Afoakwa, E.O., 2010. Chocolate Science and Technology (Wiley-Blackwell).

Berrens, L., 1996. Neoallergens in heated pecan nut: products of Maillard type degradation? Allergy 51, 277–278.

Cabanillas, B., Pedrosa, M.M., Rodríguez, J., Muzquiz, M., Maleki, S.J., Cuadrado, C., Burbano, C., Crespo, J.F., 2012. Influence of enzymatic hydrolysis on the allergenicity of roasted peanut protein extract. Int. Arch. Allergy Immunol. 157, 41–50.

Clare Mills, E.N., Sancho, A.I., Rigby, N.M., Jenkins, J.A., Mackie, A.R., 2009. Impact of food processing on the structural and allergenic properties of food allergens. Mol. Nutr. Food Res. 53, 963–969.

Chen, X., Wang, Q., El-Mezayen, R., Zhuang, Y., Dreskin, S.C., 2012. Ara h 2 and Ara h 6 Have Similar Allergenic Activity and Are Substantially Redundant. Int. Arch. Allergy Immunol. 160, 251–258.

Davis, P.J., Smales, C.M., James, D.C., 2001. How can thermal processing modify the antigenicity of proteins? Allergy 56, 56–60.

Fæste, C.K., Rønning, H.T., Christians, U., Granum, P.E., 2011. Liquid chromatography and mass spectrometry in food allergen detection. J. Food Prot. 74, 316–345.

Grimshaw, K.E., King, R.M., Nordlee, J.A., Hefle, S.L., Warner, J.O., Hourihane, J.O., 2003. Presentation of allergen in different food preparations affects the nature of the allergic reaction—a case series. Clin. Exp. Allergy 33, 1581–1585.

Hii, Law, C.L., Suzannah, S., Misnawi, Cloke As, M., 2009. Polyphenols in cocoa (Theobroma cacao L.). J. Food Ag-Ind. 2, 702–722.

http://www.fda.gov/Safety/Recalls/ucm307962.htm.

http://www.fda.gov/Safety/Recalls/ucm302056.htm.

Khuda, S., Slate, A., Pereira, M., Al-Taher, F., Jackson, L., Diaz-Amigo, C., Bigley 3rd, E.C., Whitaker, T., Williams, K., 2012a. Effect of Processing on Recovery and Variability Associated with Immunochemical Analytical Methods for Multiple Allergens in a Single Matrix: Dark Chocolate. J. Agric. Food Chem. 60, 4204–4211.

Kirsch, S., Fourdrilis, S., Dobson, R., Scippo, M.L., Maghuin-Rogister, G., De Pauw, E., 2009. Quantitative methods for food allergens: a review. Anal. Bioanal. Chem. 395, 57–67.

Maleki, S.J., Chung, S.Y., Champagne, E.T., Raufman, J.P., 2000. The effects of roasting on the allergenic properties of peanut proteins. J. Allergy Clin. Immunol. 106, 763–768.

Nowak-Wegrzyn, A., Fiocchi, A., 2009. Rare, medium, or well done? The effect of heating and food matrix on food protein allergenicity. Curr. Opin. Allergy Clin. Immunol. 9, 234–237.

Pele, M., Brohée, M., Anklam, E., Van Hengel, A.J., 2007. Peanut and hazelnut traces in cookies and chocolates: relationship between analytical results and declaration of food allergens on product labels. Food Addit. Contam. 24, 1334–1344.

Pomes, A., Vinton, R., Chapman, M.D., 2004. Peanut Allergen (Ara h 1) detection in foods containing chocolate. J. Food Prot. 67, 793–798.

Rudolf, J., Ansari, P., Kern, C., Ludwig, T., Baumgartner, S., 2012. Food Addit. Contam. Part A Chem. Anal. Control Expo. Risk Assess. 29, 1–11.

Rusconi, M., Conti, A., 2010. Theobroma cacao L., the Food of the Gods: a scientific approach beyond myths and claims. Pharmacol Res. 61, 5–13.

Settipane, Guy, A., 1987. The Restaurant Syndromes. N. Engl. Reg. Allergy Proc. 8, 39–46.

Spanjersberg, M.Q., Knulst, A.C., Kruizinga, A.G., Van Duijn, G., Houben, G.F., 2010. Concentrations of undeclared allergens in food products can reach levels that are relevant for public health. Food Addit. Contam. Part A Chem. Anal. Control Expo. Risk Assess. 27, 169–174.

Steinhoff, M., Fischer, M., Paschke-Kratzin, A., 2011. Comparison of extraction conditions for milk and hen's egg allergens. Food Addit. Contam. Part A Chem. Anal. Control Expo. Risk Assess. 28, 373–383.

Taylor, S.L., Nordlee, J.A., Niemann, L.M., Lambrecht, D.M., 2009. Allergen immunoassays—considerations for use of naturally incurred standards. Anal. Bioanal. Chem. 395, 83–92.

van Hengel, A.J., 2007. Food allergen detection methods and the challenge to protect food-allergic consumers. Anal. Bioanal. Chem. 389, 111–118.

Watanabe, Y., Aburatani, K., Mizumura, T., Sakai, M., Muraoka, S., Mamegosi, S., Honjoh, T., 2005. Novel ELISA for the detection of raw and processed egg using extraction buffer containing a surfactant and a reducing agent. J. Immunol. Methods. 300, 115–123.

Waterhouse, A.L., Shirley, J.R., Donovan, J.L., 1996. Antioxidants in chocolate. Lancet. 348, 834.

Westphal, C.D., Pereira, M.R., Raybourne, R.B., Williams, K.M., 2004. Evaluation of extraction buffers using the current approach of detecting multiple allergenic and nonallergenic proteins in food. J. AOAC Int. 87, 1458–1465.

Zhuang, Y., Dreskin, S.C., 2012. Redefining the major peanut allergens. Immunol. Res. [Epub ahead of print].

CHAPTER

81

Honey Production and Compositional Parameters

Jonathan M. Stephens, David R. Greenwood†, Liam Fearnley†, Jessie Bong†, Ralf C. Schlothauer*, Kerry M. Loomes†*

*Comvita Innovations, Institute for Innovation in Biotechnology, University of Auckland, Auckland, New Zealand, †School of Biological Sciences, University of Auckland, New Zealand

CHAPTER POINTS

- Honey contains a diverse array of compounds whose composition varies widely, depending on floral origin and storage characteristics.
- Floral origin has a significant effect on composition, as found with Manuka honey and clover honey, where the percentage of similar compounds was 35% in the m/z 100–499 mass range and only 2.4% with $m/z \geq 500$.
- Temperature and storage time account for most of the observed compositional differences between Manuka honeys of similar geographical origin.
- The conversion of dihydroxyacetone (DHA) to methylglyoxal (MGO), a principal component responsible for the unique Manuka factor (UMF) in Manuka honey, occurs non-enzymatically, is irreversible, and occurs progressively with honey aging.
- The conversion of the precursor DHA to the active compound MGO in maturing honey is dependent on the concentration of DHA present originally in the Manuka honey.
- As Manuka honey aged, the DHA concentration decreased exponentially while MGO concentration increased according to a parabolic curve and peaking at age 3 and 4 years.
- Concentrations of the compound 5-hydroxymethylfurfural (HMF) in Manuka honeys increased with aging and could be accelerated if the honey has been subjected to unduely high temperatures during processing or storage.
- The concentration of MGO as a measure of the UMF value is positively correlated with the concentration of certain phenolics in Manuka honey.

INTRODUCTION

Honey is a complex mixture made up of mainly sugars (70–80%) and water (10–20%) and is produced from the nectar of flowering plants. Honey is comprised principally of the monosaccharides fructose and glucose and it is utilized as a natural sweetener. Honey also possesses anti-infective properties due to the presence of hydrogen peroxide (Postmes *et al.*, 1993; Snowdon and Cliver, 1996; Willix *et al.*, 1992), acidity (Bognadov, 1997; Snowdon and Cliver, 1996), and to a certain extent, the high osmolarity and low water activity due to the high sugar content (Snowdon and Cliver, 1996; Willix *et al.*, 1992).

Depending on its floral origin and age, honey can exhibit varying degrees of therapeutic benefits above and beyond the peroxide activity. These high-value honeys have found clinical applications in the area of wound healing (Molan, 1992a,b) and are exemplified by the indigenous New Zealand Manuka (*Leptospermum scoparium*) and the related Kanuka (*Kunzea ericoides*) New Zealand honeys. Manuka honey displays a unique non-peroxide-dependent antibacterial activity largely attributable to the dicarbonyl compound methylglyoxal (MGO), which is derived from the precursor molecule dihydroxyacetone (DHA) in *L. scoparium* nectar. In addition, Manuka honey possesses an array

Processing and Impact on Active Components in Food
http://dx.doi.org/10.1016/B978-0-12-404699-3.00081-0

of phenolic compounds (Stephens *et al.*, 2010; Weston *et al.*, 1999) that may play antioxidative, antibacterial, or radical-scavenging roles (Gheldof *et al.*, 2002; Pyrzynska and Biesaga, 2009; Weston *et al.*, 1999). Recent evidence also indicates that there may be other components such as arabinogalactan proteins that contribute to these therapeutic activities (Gannabathula *et al.*, 2012).

The commercial value of Manuka honey has led to the establishment of an industry standard equivalent scale known as unique Manuka factor (UMF), which collectively represents this non-peroxide antibacterial activity. However, not all Manuka honey possesses appreciable levels of this antibacterial activity, and it has been reported that Manuka honeys harvested from different geographical origins exhibit a wide variation in MGO content and thus UMF activity, as well as other chemical constituents (Stephens *et al.*, 2010; Stephens, 2006). Therefore, it is of interest to understand how floral origin contributes to the molecular composition of these premium honeys. It is also of both scientific and commercial interest to know how the composition of the principal antibacterial component, MGO, and other compounds are affected by honey age and storage conditions.

HOW COMPOSITION IS ALTERED

Effects of Floral Origin and Storage Conditions on the Molecular Composition of Manuka Honey

Clover honey possesses peroxide antibacterial activity but lacks any measurable UMF activity. By contrast, Manuka honey is utilized medicinally as a wound-healing agent due to the presence of additional bioactive substances such as MGO, phenolic compounds, and potentially other compounds. We therefore hypothesized that clover honey would provide an informative comparison to gain an insight into how floral origin affects the compositional make-up of honey.

Using Fourier-transform ion cyclotron resonance (FT-ICR) mass spectrometry we first compared the non-carbohydrate components present in Manuka, Kanuka, and clover honeys, which were isolated using solid-phase extraction (Table 81.1). Analyses (<2 ppm mass accuracy) revealed similar mass range distributions between these honeys (Figure 81.1A). However, the compounds detected differed significantly in the number of similar compounds as defined by their atomic composition and consequent MS/MS fragmentation patterns. For

TABLE 81.1 Source, Age, and Storage Conditions for Honey Samples

Identifier	Description	Age	Date of Analysis
1A	Ph03—*L. scoparium* var. *incanum*, gumlands, Kaitaia, Northland	4–5 years at room temperature	March, 2009[1]
1B	Ph06—blend, approx. 50% *L. scoparium* var. *incanum*, native forest, Kaitaia, Northland	0.75 years at room temperature	March, 2009[1]
1C	Ph07—*L. scoparium* var. *linifolium*, wetlands, Waikato	<1 year at 4°C	March, 2009[1]
1D	Ph09—*L. scoparium* var. *linifolium*, wetlands, Waikato	4 years at room temperature	March, 2009[1]
1E	Ph16—*Kunzea ericoides*, Waikato	3 years at room temperature	March, 2009[1]
1F	Ph20—*Trifolium* spp., South Island	<0.75 years at room temperature	March, 2009[1]
2A	Ph01—*L. scoparium* var. *incanum*, gumlands, Kaitaia, Northland	<0.75 years at 4°C	March, 2010[2]
2B	Ph04—*L. scoparium* var. *incanum*, gumlands, Kaitaia, Northland	4–5 years at room temperature	March, 2010[2]
3A	PMHC1 4C—*L. scoparium* var. *incanum*, gumlands, Kaitaia, Northland	2 years at 4°C	April, 2010[2]
3B	PMHC1 RT—*L. scoparium* var. *incanum*, gumlands, Kaitaia, Northland	2 years at room temperature	April, 2010[2]

Honeys were prepared for mass spectrometry analyses using solid-phase extraction (Stephens *et al.*, 2010).
[1]*Samples prepared by Hill Laboratories (Hamilton, New Zealand).*
[2]*Samples prepared at the University of Auckland.*
Table taken with permission and adapted from Fearnley et al. *(2012).*

example, a comparison between a Manuka honey that had been stored for between 4 and 5 years at room temperature and a clover honey stored for less than 9 months at room temperature showed only 36% and 33% similar compounds in the *m/z* 100–299 and 300–499 mass ranges, respectively (Figure 81.1B). In the higher mass ranges the number of similar compounds decreased with only 5% in the *m/z* 500–699 mass range and no similar compounds with masses greater than *m/z* 700.

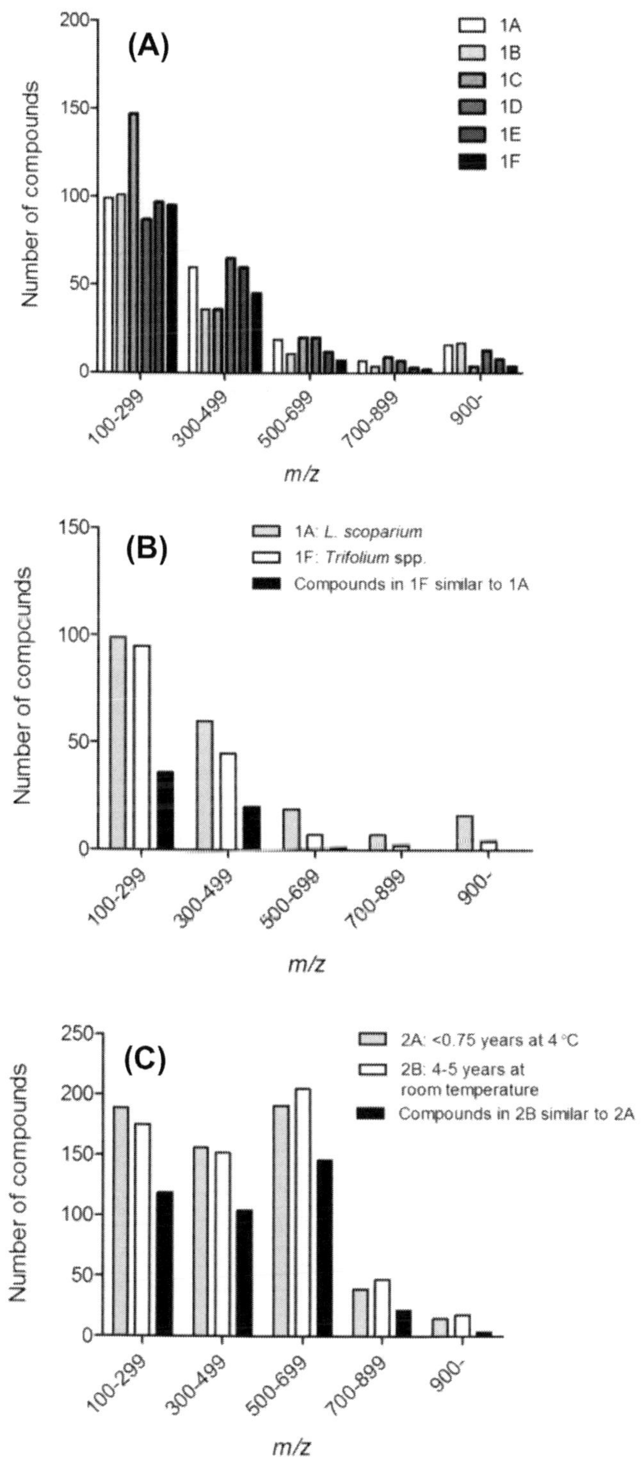

These findings confirmed that there are intrinsic compositional differences between Manuka honey and clover honey. However, as compositional differences could also arise due to storage conditions we next analyzed two distinct Manuka honey batches from the same geographical region but with different storage conditions (Figure 81.1C). Here we found that these two samples shared a higher and more uniform compound similarity across all mass ranges shown in the distribution (shown by black shaded region). However, in the higher (*m/z* 900+) region there was a noticeable reduction in similarity (Fearnley *et al.*, 2012).

Effects of Storage Temperature on the Molecular Composition of Manuka Honey

To examine more closely the specific effects of storage temperature on compound distribution, we analyzed two Manuka honey samples from the same source (Figure 81.2). These two samples differed only in the conditions at which they were stored, with the first sample stored under refrigeration at 4°C and the second at room temperature for approximately 18 months. Mass spectrometry analyses revealed that for the Manuka honey sample stored at 4°C, 69% of the total compounds analyzed across the *m/z* 100–699 mass range (Figure 81.2) were similar to those found in the room-temperature Manuka honey. This level of similarity was comparable to the compound similarities observed between the Manuka honeys from the same geographical region (Figure 81.1C).

These findings showed that honey contains a diverse array of compounds whose composition varies widely, depending on floral origin, geographical region, and storage characteristics. The presence of new ions in the

FIGURE 81.1 Mass range distributions of parent ions grouped by *m/z* values from representative honeys. (A) *m/z* Distributions of ions analyzed in six honeys of differing floral source, age, geographical origin, and storage conditions. (B) Comparison of honey sourced from *L. scoparium* stored for 4–5 years at room temperature (sample 1A), and clover honey sourced from *Trifolium* spp. stored <9 months at room temperature (sample 1F). The black shaded region indicates the number of ions within the *m/z* window detected for *L. scoparium* that were similar to *Trifolium* spp. (C) *m/z* Distributions of ions analyzed from separate honey batches sourced from *L. scoparium* from the same variety and geographic region. One of the honeys had been stored <9 months at 4°C (sample 2A) and the other for 4–5 years at room temperature (sample 2B). The black shaded region indicates the number of ions within the *m/z* window detected for honey sample 2A (stored <9 months at 4°C) that were similar to honey sample 2B (stored for 4–5 years at room temperature). Ion similarity assignments were based on monoisotope mass and fragmentation data as analyzed by the Mass Frontier software package. *Figures taken with permission and adapted from Fearnley* et al. *(2012).*

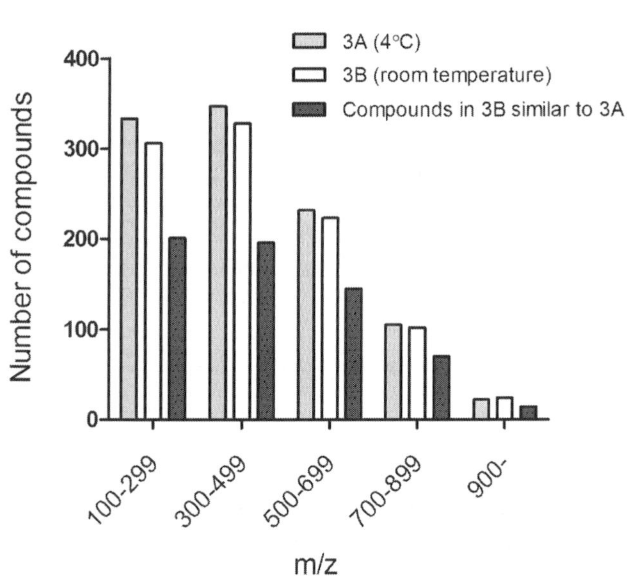

FIGURE 81.2 Effects of differing storage conditions on molecular composition. A Manuka honey was divided into separate samples, one of which was stored for 2 years at 4°C (sample 3A) and the other stored for 2 years at room temperature (sample 3B). The black bar indicates the number of ions within the *m/z* window detected for honey sample 3B that were similar to honey sample 3A. *Figure taken with permission from Fearnley* et al. *(2012).*

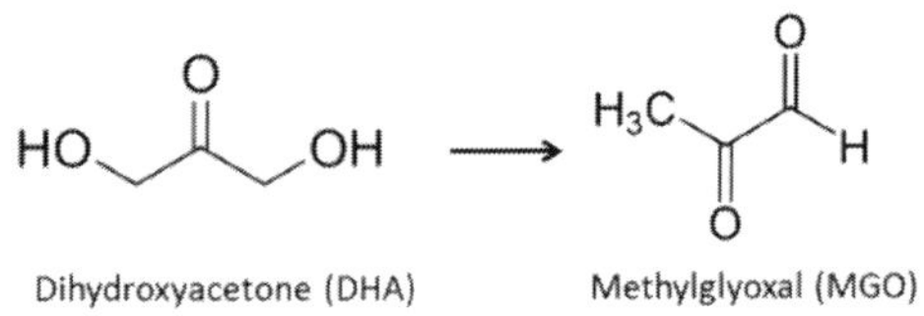

FIGURE 81.3 Conversion of DHA to MGO.

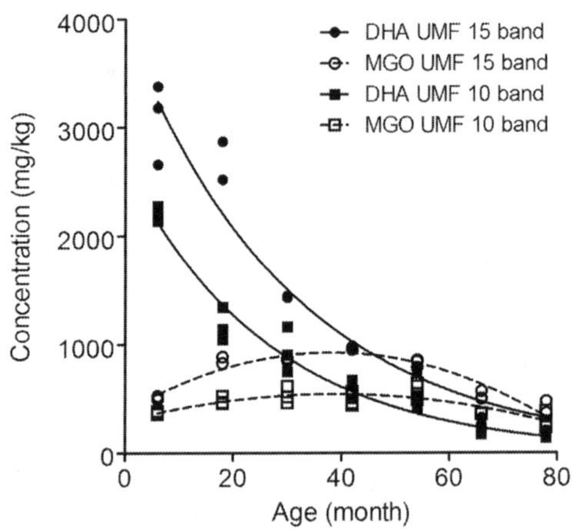

FIGURE 81.4 The concentration of MGO and DHA in aging Manuka honeys from 6 to 78 months, representing different UMF levels at commencement of storage under constant conditions. Each circle and square represents a single analysis of honey samples taken from UMF 15 and UMF 10 activity bands, respectively. *Data taken from Stephens* et al. *(unpublished).*

room-temperature Manuka sample (Figure 81.2) suggests that temperature and storage time are the main factors responsible for the observed compositional differences between Manuka honeys of similar geographical origin. Compared to the 4°C sample, the room temperature-stored Manuka honey also displayed significant increases in the quantities of a large number of ions, both major (*m/z* 582–625 range) and minor (*m/z* 300–450 range), without a significant fall in the amounts of others (data not shown). This observation implies that some of the observed mass changes were probably attributable to breakdown products.

Methylglyoxal and Dihydroxyacetone: Compositional Changes in Manuka Honey

MGO in Manuka honeys forms spontaneously from the precursor compound, DHA (Figure 81.3). This conversion occurs non-enzymatically, is irreversible, and occurs progressively with honey aging (Adams *et al.*, 2009; Stephens *et al.*, unpublished). MGO is not present in the nectar of *Leptospermum scoparium*, however DHA is present at elevated levels in the nectar (Adams *et al.*, 2009; Stephens *et al.*, unpublished). Therefore, the amount of MGO that forms in Manuka honey, and consequently the UMF activity, is dependent on the amount of DHA present originally in the *L. scoparium* nectar that was incorporated into the Manuka honey. The concentration of MGO is therefore typically low in fresh Manuka honeys. The presence of DHA in nectar appears to be unique to the *Leptospermum* genus with similar observations reported in a number of Australian honeys harvested from various *Leptospermum* species (Windsor *et al.*, 2012).

Figure 81.4 demonstrates the relationship between DHA and MGO concentrations over time from two UMF activity bands of Manuka honey samples (Stephens *et al.*, unpublished). For both sample groups, the DHA concentration decreased exponentially over time whilst the MGO concentration followed a parabolic curve, peaking in honeys aged between 3 and 4 years. There was a subsequent decline in the MGO concentration after approximately 4 years, presumably due to the low concentration of the residual precursor molecule DHA (Stephens *et al.*, unpublished).

Because the increase of MGO concentration in a Manuka honey is dependent on DHA concentration, the relationship between these components can be defined in terms of the DHA:MGO ratio (Stephens *et al.*, unpublished). This ratio follows an exponential decline as the honey matures, and honeys with different UMF activities, as shown in Figure 81.4, nevertheless follow the same ratio decline curve (Figure 81.5). This observation shows that the increase in MGO content within a honey is directly dependent on the initial DHA concentration. However, it

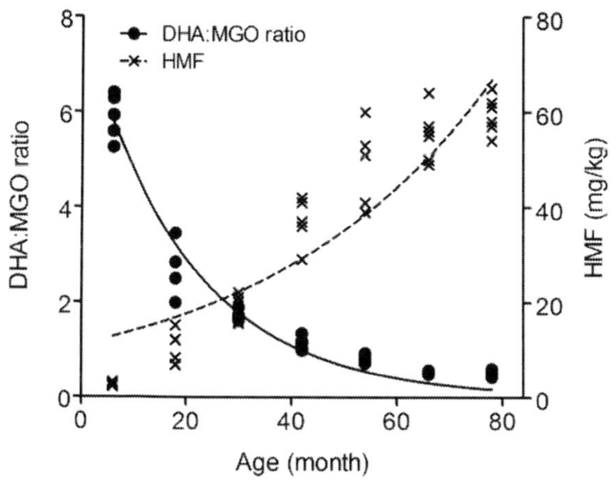

FIGURE 81.5 DHA:MGO ratio and the concentration of HMF in aging Manuka honeys from 6 to 78 months, illustrating the similarity of ratio regardless of initial DHA concentration and the exponential increase of HMF. Symbols represent single analyses. *Data taken from Stephens* et al. *(unpublished).*

is likely that Manuka honey stored at elevated temperatures would not display the same ratio relative to honey age. Indeed, elevated storage temperatures accelerate the conversion of DHA to MGO and the addition of amino acids also has a similar effect (Adams *et al.*, 2009).

The increase of 5-hydroxymethylfurfural (HMF) in honeys during storage has been well documented (White *et al.*, 1964) and maturing Manuka honey stored at constant temperature conditions exhibited the same effect (Figure 81.5). The increase of HMF followed an exponential curve; however, this curve can be accelerated if the honey has been subjected to unduly high temperatures during extraction, processing, or storage.

The close relationship between these three sugar-derived compounds, MGO, DHA, and HMF, in Manuka honey is illustrated by the exponential manner in which the DHA:MGO ratio declines whereas the HMF concentration increases with aging (Figure 81.5). Manuka honey that has been processed according to normal industry standards, and not been exposed to storage at elevated temperatures, is expected to display the respective DHA:MGO ratio and concentration of HMF during the honey's shelf-life.

Several compounds have been proposed as floral markers of Manuka honey, including phenyllactic acid, 4-methoxyphenyllactic acid, methyl syringate, 2-methoxyacetophenone, and 2-methoxybenzoic acid (Stephens *et al.*, 2010; Tan *et al.*, 1988; Wilkins *et al.*, 1993). These compounds are also the predominant constituents in Kanuka (*Kunzea ericoides*) honeys, but in different relative proportions. The concentration of MGO is positively correlated with the concentration of certain phenolics (Oelschlaegel *et al.*, 2012; Senanayake, 2006; Stephens *et al.*, 2010). For example, Manuka honey less than one year of age exhibited a positive relationship between the 2-methoxybenzoic acid and MGO concentrations (Stephens *et al.*, 2010). 4-methoxyphenyllactic acid, on the other hand, was negatively correlated with MGO concentration. It is likely that this compound exists at elevated levels in honey harvested from the related Kanuka, a plant species that often co-inhabits environments with *L. scoparium* (Stephens *et al.*, 2010).

OTHER WAYS IN WHICH COMPOSITION IS ALTERED

Effect of Heating on Methylglyoxal Content

Industry anecdotal evidence strongly suggests that heating Manuka honey during processing, or storage at elevated temperatures, will promote the formation of MGO and HMF. However, two published studies confirmed that prolonged storage of Manuka honey at 50°C and above led to the degradation of both DHA and MGO, but up to a 10-fold increase in the concentration of HMF (Adams *et al.*, 2009; Atrott *et al.*, 2012). Yet, a 70°C 10-min pasteurization-type heat exposure alone did not reduce the MGO concentration.

ANALYTICAL TECHNIQUES

Compositional Changes in Honey

Honey samples were analyzed by mass spectrometry following solid-phase extraction (Fearnley *et al.*, 2012). Briefly, honey samples (0.1 g) were weighed and dissolved in a 5% methanol and 0.1% formic acid mixture (4 ml). Samples were then extracted using pre-conditioned C18 SPE cartridges eluting with an 80% methanol mixture. Aliquots of these eluants were then injected on to a 2 × 150 mm Phenomenex Luna PFP chromatographic column and the column developed with a linear gradient of 5–95% acetonitrile in water containing 0.1% formic acid. The PFP column was attached to a ThermoFinnigan LTQ-FT electrospray ion trap Fourier transform ion cyclotron resonance mass spectrometer operating in positive ESI mode.

A triple top 3 tandem mass experiment was performed in parallel where a full scan in the ICR cell (accurate mass to < 2 ppm) was followed coincidently by MS^2 in the ion trap on the top three most intense parent ions, and then MS^3 performed on the top three daughter ions from each MS^2 scan. MassFrontier 5.1 (HighChem Ltd) was used for qualitative analysis and provisional identification of peaks. Spectral trees consisting of MS^n data were assembled and loaded into databases for comparisons between samples, which were performed using library search algorithms to produce score values indicative of match quality (Fearnley *et al.*, 2012).

Measurements of MGO and DHA Content

MGO and DHA were derivatized to allow quantification for the following analyses. Briefly, MGO analysis can be performed routinely either with 1,2-diaminobenzene followed by headspace SPME-GC-NPD (Stephens *et al.*, 2010), *o*-phenylenediamine derivatization and RP-HPLC (Adams *et al.*, 2008), or it can be derivatized with pentafluorobenzylhydroxylamine and detected using RP-HPLC (Windsor *et al.*, 2012). DHA is most commonly derivatized with pentafluorobenzylhydroxylamine and detected using RP-HPLC (Windsor *et al.*, 2012) or LC-MS/MS (Stephens *et al.*, unpublished). HMF analysis follows the industry standard method (Driffield *et al.*, 2005).

References

Adams, C.J., Boult, C.H., Deadman, B.J., Farr, J.M., Grainger, M.N.C., Manley-Harris, M., Snow, M.J., 2008. Isolation by HPLC and characterization of the bioactive fraction of New Zealand manuka (*Leptospermum scoparium*) honey. Carbohydr. Res. 343, 651–659.

Adams, C.J., Manley-Harris, M., Molan, P.C., 2009. The origin of methylglyoxal in New Zealand manuka (*Leptospermum scoparium*) honey. Carbohydr. Res. 344, 1050–1053.

Atrott, J., Haberlau, S., Henle, T., 2012. Studies on the formation of methylglyoxal from dihydroxyacetone in Manuka (*Leptospermum scoparium*) honey. Carbohydr. Res. 361, 7–11.

Bognadov, S., 1997. Nature and origin of the antibacterial substances in honey. LWT Food Sci. Technol. 30, 748–753.

Driffield, M., Chan, D., Macarthur, R., MacDonald, S., Brereton, P., Wood, R., 2005. Single laboratory validation of a method for the determination of hydroxymethylfurfural in honey by using solid-phase extraction cleanup and liquid chromatography. J. AOAC Int. 88, 121–127.

Fearnley, L., Greenwood, D.R., Schmitz, M., Stephens, J.M., Schlothauer, R.C., Loomes, K.M., 2012. Compositional analysis of manuka honeys by high-resolution mass spectrometry: Identification of a manuka-enriched archetypal molecule. Food Chem. 132, 948–953.

Gannabathula, S., Skinner, M.A., Rosendale, D., Greenwood, J.M., Mutukumira, A.N., Steinhorn, G., Stephens, J., Krissansen, G.W., Schlothauer, R.C., 2012. Arabinogalactan proteins contribute to the immunostimulatory properties of New Zealand honeys. Immunopharmacol. Immunotoxicol. 34, 598–607.

Gheldof, N., Wang, X.-H., Engeseth, N., 2002. Identification and quantification of antioxidant components of honeys from various floral sources. J. Agric. Food Chem. 50, 5870–5877.

Molan, P.C., 1992a. The antibacterial activity of honey. 1. The nature of antibacterial activity. Bee World 73, 5–28.

Molan, P.C., 1992b. The antibacterial activity of honey: 2. Variation in the potency of antibacterial activity. Bee World 73, 59–76.

Oelschlaegel, S., Gruner, M., Wang, P.N., Boettcher, A., Koelling-Speer, I., Speer, K., 2012. Classification and Characterization of Manuka Honeys Based on Phenolic Compounds and Methylglyoxal. J. Agric. Food Chem. 60, 7229–7237.

Postmes, T., Van den Bogaard, A.E., Hazen, M., 1993. Honey for wounds, ulcers, and skin graft preservation [8]. Lancet 341, 756–757.

Pyrzynska, K., Biesaga, M., 2009. Analysis of phenolic acids and flavonoids in honey. Trends Anal. Chem. 28, 893–902.

Senanayake, M.J., 2006. A chemical investigation of New Zealand unifloral honeys. PhD thesis. University of Waikato, Hamilton, New Zealand.

Snowdon, J.A., Cliver, D.O., 1996. Microorganisms in honey. Int. J. Food Microbiol. 31, 1–26.

Stephens, J.M., Schlothauer, R.C., Morris, B.D., Yang, D., Fearnley, L., Greenwood, D.R., Loomes, K.M., 2010. Phenolic compounds and methylglyoxal in some New Zealand manuka and kanuka honeys. Food Chem. 120, 78–86.

Stephens, J.M., Schlothauer, R.C., Morris, B.M., Yang, D. Examination of dihydroxyacetone and methylglyoxal in New Zealand manuka (*Leptospermum scoparium*) honeys. *Unpublished*.

Stephens, J.M.C., 2006. The factors responsible for the varying levels of UMF in manuka (*Leptospermum scoparium*) honey. PhD thesis. University of Waikato, Hamilton, New Zealand.

Tan, S.-T., Holland, P.T., Wilkins, A.L., Molan, P.C., 1988. Extractives from New Zealand Honeys. 1. White clover, manuka and kanuka unifloral honeys. J. Agric. Food Chem. 36, 453–460.

Weston, R.J., Mitchell, K.R., Allen, K.L., 1999. Antibacterial phenolic components of New Zealand manuka honey. Food Chem. 64, 295–301.

White, J., Kushnir, I., Subers, M.H., 1964. Effect of storage and processing temperatures on honey quality. Food Technol. 18, 153–156.

Wilkins, A.L., Lu, Y., Molan, P.C., 1993. Extractable organic substances from New Zealand unifloral manuka (*Leptospermum scoparium*) honeys. J. Agric. Food Chem. 32, 3–9.

Willix, D.J., Molan, P.C., Harfoot, C.G., 1992. A comparison of the sensitivity of wound-infecting species of bacteria to the antibacterial activity of manuka honey and other honey. J. Applied Bacteriol. 73, 388–394.

Windsor, S., Pappalardo, M., Brooks, P., Williams, S., Manley-Harris, M., 2012. A convenient new analysis of dihydroxyacetone and methylglyoxal applied to Australian *Leptospermum* honeys. J. Pharmacog. Phytotherapy 4, 6–11.

CHAPTER

82

Jam Processing and Impact on Composition of Active Compounds

Taha M. Rababah[*], *Muhammad H. Al-u'datt*[*], *Susan Brewer*[†]

[*]Department of Nutrition and Food Technology, Jordan University of Science and Technology, Irbid, Jordan,
[†]Department of Food Science and Human Nutrition, University of Illinois at Urbana-Champaign, Urbana, USA

CHAPTER POINTS

- Jam is a good processing method for fruit preservation over seasonal variation.
- Jam processing occurs by addition of pectin, citric acid, and sugar to a fruit slurry at high temperature to produce a viscous semi-solid texture.
- High temperatures achieved during processing lead to changes in water content, fiber content, water-soluble vitamins, fat-soluble vitamins, and minerals.
- Vitamins, minerals, phenolic compounds, antioxidants, and flavonoids are affected by light, temperature, and type of pectin added during processing.
- The storage of jam at room temperature for more than 6 months results in significant alterations in the active components of jam.
- Anthocyanins lost during storage of processed jams are accompanied by increases in the polymeric color values, indicating that anthocyanins are extensively polymerized.
- Jam texture depends on the type of pectin, sugar, and citric acid added.

INTRODUCTION

As many studies have revealed, fruits and their products are rich in phenolic and antioxidant compounds. Consumption of fruits and their products may improve health outcomes related to a number of diseases. Fruits and fruit jams have been shown to be a rich source of bioavailable phenolics and antioxidants (Sun *et al.*, 2005). There is a considerable demand for fresh fruits as well as their products. Since many types of fruit are seasonal and their shelf-life is limited, they must be processed to maintain quality (Scibisz and Mitek, 2009). Jam processing is considered one of the most common preservation methods for fruit. It is usually produced as a result of cooking fruits with sugar and other ingredients such as pectin, and citric acid (Abers and Wrolstad, 1979). A good-quality jam usually has a bright color, typical sweet–sour flavor, and a pleasing consistency which is neither too liquid nor too stiff. However, jam quality is affected by many factors including fruit cultivar, degree of fruit ripeness, jam formulation, processing procedure, jar types, and storage conditions (Haffner *et al.*, 2003). Anthocyanins are a group of polyphenolics that are responsible for the color of many fruits, vegetables, and flowers, such as orange, red, and blue. Their stability depends on various factors such as pH-value, presence of oxygen, enzyme activity, temperature, sugar and ascorbic acid contents. Jam structure is determined by the equilibrium obtained between the pectin, sugar, and acid contents in the fruit jams structure, and it is fundamentally a property of the soluble solids contents. Jam firmness occurs through combination of hydrogen bonds and hydrophobic interactions which are responsible for the gel formation of high methoxy (HM) pectin. The total phenolics, antioxidant activity, anthocyanins, and some physiochemical properties of fruits and fruit jams, including pH, soluble solids, and firmness are discussed in this chapter.

JAM PROCESSING

Fruit jams have been one of the most common methods used to preserve fruits for decades. Fruits that have fiber content or are high in pectin need less pectin in

Processing and Impact on Active Components in Food
http://dx.doi.org/10.1016/B978-0-12-404699-3.00082-2

their jam formulation. Appropriate acidity, total soluble solid substances, and calcium content are all imperative for pectin gelation (Javanmard and Endan, 2010). Jam formulation varies due to the composition of the matrix ingredients which have an impact on the rheology of the jam produced. Any small alteration in the jam matrix, for example replacing part of the sugar with other sweeteners or using different pectins, can create changes in the food matrix constituents. As a consequence of changes in the interactions, jam quality will be dramatically affected (Javanmard and Endan, 2010). Fruit jams are usually prepared as described by Downing (1996): the primary ingredients are 40% fruit, 51% sugar, 8.6% water, and 0.4% high methoxy pectin. Fruits are sorted, washed, and ground, then heated in a double-jacketed steam kettle at 80°C to inactivate enzymes. Pectin is added and the mixture is heated for 2 min to allow proper pectin hydration. Acidity is checked and adjusted if necessary by addition of 0.3% of citric acid solution to achieve a pH value between 3.0 and 3.2. Sugar is added and the mixture is boiled to a final soluble solids concentration of 65%. The jam is hot-packed at 85°C into 100-ml glass jars, immediately sealed with metal covers, and inverted for 5 min to sterilize the glass containers. The jars are then returned to the upright position and held at 50°C. Jam quality such as total phenolics, antioxidant activity, anthocyanins, color, acidity, soluble solid, texture, and nutraceaticals can be affected during storage (Amakura *et al.*, 2000).

HOW COMPOSITION IS ALTERED

Phenolic compounds, antiradical and anthocyanin pigments can be altered during jam cooking possibly due to the reduction of disrupted cell structures during fruit processing and the fruits become more prone to non-enzymatic oxidation. The decrease in phenolics, especially total ellagic acid, could be due to jam processing. The degradation rate of anthocyanins was significantly lower at pH 3.0. Hydrolysis of the glycoside linkage is known to be the first step of anthocyanin degradation, and this proposed mechanism could be due to the fact that elevated temperature during jam processing shifts the anthocyanins equilibrium toward the colorless chalcones, which are unstable especially if lacking the sugars. The chalcones may be subjected to degradation via oxidation reactions during food processing, producing brown compounds or pigments with high molecular weight, or the reduction of anthocyanin pigments could be due to the flavylium salts which are stable only in highly acidic conditions. These salts lose the proton at higher pH and transform into quinoidal base, which is an unstable pigment, and immediately bond to water and form a colorless compound called chromenol. The reduction of antioxidant activity of orange juices was based on the scavenging of free radicals (hydroxyl and 1,1-diphenyl-2-picrylhydrazyl radicals) that are attributed to the degradation of ascorbic acid by heating. Soluble solids can have an effect on anthocyanin pigments which could be because the higher soluble solids content of jam processing increases the degradation of anthocyanin.

ACTIVE COMPOUNDS IN FRUITS AND THEIR JAMS

Water

Fruits and vegetables generally contain 90–96% water. Fresh or pre-cooked fruit is boiled with a solution of cane or beet sugar until sufficient water is evaporated to produce a mixture which will gel on cooling and contain about 32–34% water. The moisture content is maintained in the recommended range for a stable jam formulation because it has affects the length of the storage period (Ndabikunze *et al.*, 2011).

Fiber and Pectin

Fibers are non-starch polysaccharides that are poorly digested by humans, such as cellulose and lignin. They include both water-soluble and water-insoluble fractions of plant components that are indigestible by the small intestine. They can be lost during separation steps such as peeling, filtration, or stem removal, and during processing as a result of changes in moisture content (Rickman *et al.*, 2007). In the food industry, pectin is used as a thickening agent in jams and jellies for texture formulation. When water is added to pectin, it will give the solution a viscous-like behavior. It can be dispersed in sugar-concentrated syrups when the soluble solids concentration is less than 30% with heating and smooth mixing (Javanmard and Endan, 2010). In addition, acidity, total soluble solid concentration, and calcium content all influence pectin gelation (Mizrahi, 1979); increasing the pH and ionic strength can decrease the viscosity of the pectin solution.

Vitamins and minerals

Vitamins are conveniently divided into two major groups: fat-soluble and water-soluble. Fat-soluble vitamins are A, D, E, and K. Their absorption by the body depends upon the normal absorption of fat from the diet. Water-soluble vitamins include vitamin C and several members of the vitamin B complex (Dauthy, 1995). Fruits are rich in vitamins and minerals such as vitamin C and B-complex vitamins. Vitamin C is present in all fresh fruit; strawberry, citrus fruit, apple, and peach provide moderate amounts of vitamin C and can contribute substantially to the diet when consumed in sufficient quantity. Vitamin C, which is also known as ascorbic acid, is easily destroyed by oxidation, especially at high temperatures, and most easily lost during processing, storage, and cooking. (Dauthy, 1995). For example, guava jam retains only 37.5% of ascorbic acid found in the fruit compared to 79.6% in juice (Jawaheer *et al.*, 2003). The stability of ascorbic acid might also be further enhanced by the presence of a high concentration of citric acid which has been reported to have a protective effect on L-ascorbic acid and tends to slow its oxidation due to its metal sequestering properties (Langdon, 1987). Many factors influence vitamin C stability including temperature, presence of oxygen in the headspace, and light; so jam stored at ambient temperature (22–25°C) has a relatively high rate of degradation of ascorbic acid. However, glass containers are impermeable to oxygen which causes low ascorbic acid destruction due to oxidation (Jawaheer *et al.*, 2003). Most fruits also supply varying amounts of water-soluble vitamins, such as B-complex, and fat-soluble vitamins, such as β-carotene. Sodium, magnesium, and potassium account for the alkaline ash content of fruit. Vitamin E and carotenoids in carrots, spinach, tomatoes, and apricots are susceptible to oxidation and can be lost during cooking by leaching into the cooking liquid (Rickman *et al.*, 2007).

Phenolic and Antioxidant Compounds

Phenolic compounds are a group of chemical compounds that are classified as simple phenolics or complex phenolic compounds (Shahidi and Naczk, 2007). Fruits and vegetables have high amounts of simple phenolics (volatile compounds), but most plant leaves and roots, have polymerized compounds such as condensed tannins. Simple phenolics such as *o*-cresol, 4-ethylphenol, guaiacol, eugenol, and 4-vinyl guaiacol are major flavor components of citrus fruits and orange juice (Buttery *et al.*, 1976). Complex phenolic compounds, hydroxycinnamic and hydroxybenzoic, are also found in plants. The primary hydroxycinnamic acids found in fruits are sinapic, ferulic and caffeic acids, found in apples, apricots, blueberries, and tomatoes representing more than 75% of hydroxycinnamic acid (Kono *et al.*, 1995).

Phenolic compounds in fruits and jams are affected by light, pH, pectin, and temperature (Laleh *et al.*, 2006); the impact of thermal processing on the total phenol content was assessed on the basis of that recorded in the fruit and jam. It was observed that the thermal processing of wild bilberries resulted in major alterations in the total phenol content. Previous studies reported losses in total phenolics as an effect of thermal processing of various berries, but the degree of alteration of total phenolics can be of different magnitudes (Rickman *et al.*, 2007). A significant decrease in total phenol contents (between 42–51%) was observed as a result of processing. The largest loss in total phenol content was noticed after processing in jam with 0.3% pectin. In addition, the storage period affects the total phenol content. After one month of storage at 20°C, total phenol content decreased by 5–12% from the value recorded one day after processing. By extending the storage time to 7 months, the losses reached 41–47% of that recorded one day after processing. Antioxidant activities in fruits have been investigated by many researchers. Epidemiological studies have found strong evidence that antioxidants in fruits and vegetables can help the human body to prevent, or affect the development of diseases, such as cardiovascular disease, cancer, neurological conditions, and type 2 diabetes (Shahidi and Naczk, 2007). Researchers have focused on fruit juices having high levels of antioxidants such as pomegranate, cranberry, and blueberry as well as other dark fruits such as cherry and blackcurrant which contain high levels of polyphenolic compounds as flavanoids and tannins. In addition, primary research focus has been on orange, grape, and apple juices containing hydroxycinnamic acids and vitamins primarily (Lim *et al.*, 2007). Capecka *et al.* (2005) evaluated the antioxidant effects of other phytochemicals of lemon balm herbs, oregano, and peppermint before and after drying and concluded that the drying process adversely affected antioxidant activity. Tomczak (2007) studied the effect of antioxidants in pasteurized juice, during storage at different temperatures as well as oxygen availability under many conditions. He showed that antioxidant activity decreased in all samples, both after pasteurization and after storage. Rababah *et al.* (2011a) studied the total phenolics, antioxidant activity, and anthocyanins of five fresh fruits (strawberry, cherry, apricot, fig, and orange) and their jams. The results showed that fresh strawberry had the highest content of total phenolics (8503.1 mg GAE/kg) followed by cherry, apricot, fig, and orange, respectively (Table 82.1). Jam processing decreased the total phenolics, antioxidant activity, and anthocyanins of all fruits. Total phenolics of jam decreased during storage only in apricot, fig, and orange; also fresh strawberry had the highest antioxidant activity (54.88% inhibition) followed by the other fruits. Antioxidant activity did not change in strawberry during jam storage, while reductions in the other fruits were observed. Fresh strawberries had the highest anthocyanin content (2323.8mg cya-3-glu/kg), followed by cherries and the other fruits, respectively. Anthocyanin content decreased in apricot and fig jams during 5 months of storage. Rababah *et al.* (2012b) compared grape jam, raisin, and raisin jams for phenolic

TABLE 82.1 Total Phenolics, Antioxidant Activity, Anthocyanins, and pH of Different Fruits

Fruit	Total Phenols (mg GAE/kg)	DPPH Inhibition (%)	Anthocyanins (mg cya-3-glu/kg)	pH
Strawberry	8503.13±287.0[a]	54.88±0.2[b]	2323.80±72.0[a]	3.50±3.50[c]
Cherry	4512.90±57.0[b]	31.64±5.7[c]	215.25±1.1[b]	3.66±0.01[b]
Apricot	1859.68±10.7[c]	29.20±2.0[c]	25.40±4.7[c]	4.03±0.01[a]
Fig	1390.35±34.2[d]	25.42±1.4[c]	41.50±1.3[c]	3.57±0.01[c]
Orange	1224.24±21.2[d]	30.22±2.9[c]	27.75±1.2[c]	3.66±0.01[b]
BHT	—	94.71±0.3[a]	—	—

All values are calcuated as dry basis and means of three replicates.
Means±SD in the same column with the same letters (a–d) are not significantly different ($P \leq 0.05$).
From Rababah et al. *(2011a).*

TABLE 82.2 Phenolics, Antioxidant Activity, and Anthocyanins of Grape Fruit, Raisin, and their Jams

Treatment	Phenolics (mg GAE/100g)	Antioxidant Inhibition (%)	Anthocyanin (mg cya3-glu/100g)
Grape	616.1±16.5[a]	36.2±1.7[a]	64.7±2.8[a]
Raisin	331. 6±12.3[c]	26.5±1.3[b]	34.5±1.6[c]
Grape jam	449.2±14.9[b]	28.1±1.5[b]	40.2±2.1[b]
Whole raisin jam	281.2±11.9[d]	19.9±1.2[c]	23.5±1.5[d]
Ground raisin jam	274.9±10.2[d]	19.3±1.1[c]	22.8±1.3[d]

All values are calculated on dry basis and means of three replicates.
Means±SD in the same column with the same letters are not significantly different ($P \leq 0.05$).
From Rababah et al. *(2012b).*

TABLE 82.3 Lightness (L^*), Redness (a^*), and Yellowness (b^*), ΔE, and Chroma Values of Grape and Raisin Jams

Treatment	L^*	a^*	b^*	ΔE	Chroma
Grape jam	17.2±1.2[c]	2.5±0.2[a]	4.9±0.3[b]	18.0±0.9[c]	5.5±0.3[b]
Whole raisin jam	23.1±1.4[a]	1.8±0.1[b]	7.1±0.6[a]	24.2±1.5[a]	7.3±0.7[a]
Ground raisin jam	19.8±1.7[b]	2.2±0.1[ab]	5.4±1.1[b]	20.6±2.3[b]	5.8±1.3[b]

All values are calculated on dry basis and means of three replicates.
Means±SD in the same column with the same letters are not significantly different ($P \leq 0.05$).
From Rababah et al. *(2012b).*

content, antioxidant activity, and anthocyanin content. Fresh grapes had the highest antioxidant activity, total phenolic content and anthocyanin content followed by grape jam, raisins, and raisin jam, respectively (Table 82.2). There were no differences in color parameters (Table 82.3). Rababah *et al.* (2011b) and Rababah *et al.* (2012a) also studied the phenolic content, antioxidant activity, and anthocyanin content of strawberries and cherries and their jams during storage at different temperatures. Results showed that fresh strawberries had the highest concentration of total phenolics, antioxidant compounds, and anthocyanins (1693.55 mg GAE/100 g, 84.91%, and 18.59 mg cya-3-glu, respectively). The antioxidant activity and anthocyanins of jam decreased quickly at storage at temperatures of 35–55°C. In spite of the significant reduction in these nutraceutical compounds, processed jams retained substantial amounts during storage at different temperatures (Table 82.4). Q10 was calculated for cherry jam as an indication of the dependency of a quality degradation reaction on temperature expressed by temperature coefficient (Q10), which is defined as the ratio between the reaction rate constant k at a temperature T+10°C and the one at T (Table 82.5).

FLAVONOIDS

Flavonoid compounds are widely distributed in nature as pigments in flowers, fruit, vegetables, and tree barks. Chemically, they are glycosides of flavones forming flavonols, flavones, flavanones, isoflavones, catechins, anthocyanidins, and chalcones. Flavonoid compounds contribute to the color of fruits which is one of the most important characteristics of food from the consumer's point of view. Anthocyanins are the most common class of flavonoids and they are commonly orange, red, or blue in color. Anthocyanins are present in flowers, fruits, and vegetables. In industrial food production, during processing and storage there is a degradation of food color. Color deterioration is generally caused by loss of red anthocyanin pigment, alteration of the chemical configuration of brown pigments, and discoloration (Abers and Wrolstad, 1979). These changes might be due to polymerization which usually occurs as a result of temperature, pH, and oxygen and water activity (Tsai and Huang, 2004). Anthocyanin stability in jam processing was evaluated under different conditions by Ozela *et al.* (2007), who concluded that degradation of the anthocyanins pigment increased with increased temperature and in the presence of light at 25°C, 40°C, and 60°C, respectively, and more stable at pH5.0 and 6.0 than at pH4.0 both in the presence and absence of light. They also found that

TABLE 82.4 Total Phenolics, Antioxidant Activity, Anthocyanins, and pH of Strawberry Jam during Storage at 25°C

Jam	Storage Period (Month)	Total Phenols (mg GAE/kg)	Dpph Inhibition (%)	Anthocyanins (mg cya-3-glu/kg)	pH
Strawberry	Fruit	8503.13 ± 287.3^{a}	54.88 ± 0.2^{b}	2323.80 ± 72.1^{a}	3.50 ± 0.00^{a}
Strawberry jam	0	578.26 ± 82.0^{b}	42.50 ± 0.9^{c}	78.85 ± 0.3^{b}	2.71 ± 0.01^{b}
Strawberry jam	1	507.61 ± 23.1^{b}	33.08 ± 0.7^{d}	75.65 ± 7.9^{b}	2.77 ± 0.03^{b}
Strawberry jam	2	487.68 ± 30.7^{b}	33.53 ± 0.4^{d}	73.75 ± 7.9^{b}	2.77 ± 0.01^{b}
Strawberry jam	3	476.81 ± 41.0^{b}	33.60 ± 0.3^{d}	78.70 ± 7.3^{b}	2.80 ± 0.01^{b}
Strawberry jam	4	467.75 ± 30.7^{b}	33.05 ± 0.2^{d}	73.00 ± 6.2^{b}	2.86 ± 0.01^{b}
Strawberry jam	5	455.07 ± 43.6^{b}	28.49 ± 2.5^{d}	69.80 ± 1.9^{b}	2.89 ± 0.01^{b}
BHT	—	—	94.71 ± 0.3^{a}	—	—

All values are calcuated as dry basis and means of three replicates.
Means ± SD in the same column with the same letters (a–d) are not significantly different ($P \leq 0.05$).
From Rababah et al. *(2011a).*

TABLE 82.5 Q10 Values over Three Temperature Ranges for Phenolics, Antioxidant Inhibition, and Anthocyanins

	Q10		
Temperature Range (°C)	Total Phenolics	Antioxidant Inhibition	Anthocyanins
25–35	1.40	1.27	1.78
35–45	1.20	2.72	1.48
45–55	1.14	1.34	1.37
25–55	1.24	1.78	1.54

From Rababah et al. *(2012a).*

jam processing caused a decrease in total anthocyanin content by 81–84% compared with fresh fruit (Rababah *et al.*, 2012a). Rababah *et al.* (2011a) studied the effect of processing of strawberry and cherry jams on the anthocyanin and color parameters. They showed that the lightness and yellowness color parameters of strawberry and cherry jams increased, while the redness decreased during the storage. Also they found that Q10 values of phytochemicals were between 1 and 2 and were less sensitive to temperature changes compared with most chemical and biochemical reactions. The loss of monomeric anthocyanins is most likely due to the formation of anthocyanin polymers or condensation reactions among anthocyanins and procyanidins during the processing steps and storage period of jams (Brownmiller *et al.*, 2008).

LIGHT, TIME, AND TEMPERATURE

Temperature is one of the most important parameters that influences rheological behavior of food (Javanmard and Endan, 2010). It can destabilize flavonoid molecular structure (Laleh *et al.*, 2006). Light exposure promotes pigment destruction while reducing water activity enhances stability (Skrede *et al.*, 2000). Fruit and vegetable rheological properties are of interest to food scientists and food engineers because of their impact on the quality (and acceptability) of the final product. Throughout the heating process of fruit purees and jams, there is a significant change in apparent viscosity (Fugel *et al.*, 2005). The time parameter that affects jam rheology during the cooking process is a major factor. It has been shown that some jams, such as pineapple jam, exhibit thixotropic behavior (Javanmard and Endan, 2010). The temperature parameter also adversely affects flavonoids, especially the anthocyanin molecule; at higher temperatures hydrolyzation of 3-glycoside structure can occur, which normally has a protective effect on the unstable anthocyanin or the pyrilium ring can be hydrolyzed resulting in production of chalkones, which are responsible for the brown color that developes in foods containing anthocyanins (Laleh *et al.*, 2006).

ACIDITY

Acidity is another factor affecting fruit jam stability. Citric acid is considered necessary to correct the pH balance which is needed for jam production. Lime and lemon juice are high in citric acid, therefore, they are often used in jam manufacturing. Increasing the pH and decreasing the dextrose equivalent of high methyl ester pectin increases the viscous shear. Comparing the gel made from pectin that is extracted from apples to that from citrus pectin gels, the apple pectin gel shows higher viscous shear whereas citrus pectin gels have more elastic shears and are less viscous. Products that have elastic shear are less spreadable and exhibit more brittle behavior; moreover they have a higher tendency to exhibit syneresis. Gel formation is dependent on the presence of the carbohydrate pectin, which at a pH of 3.2–3.4 in the presence of a high concentration of sugar, forms a viscous semi-solid (Dauthy, 1995).

ANALYTICAL TECHNIQUES

Total phenolics content in methanolic extracts can be determined according to the Folin–Ciocalteu procedure as decribed by Rababah *et al.* (2011a,b) and the results are calculated as gallic acid equivalents. Antioxidant activities are determined based on determination of radical DPPH-scavenging activity according to the method of Rababah *et al.* (2011a,b). The radical scavenging activity is expressed as percentage of inhibition. Total anthocyanins content in the samples is determined according to the following procedure: after extraction of anthocyanins with acidified methanol, the absorbance of the extracts is measured at 530 and 657 nm. The anthocyanins content is expressed as milligrams of cya-3-glucoside equivalent per 100 g of dry sample weight. The firmness measurement is performed in the samples with a texture analyzer. The firmness value (N) is recorded as the value of peak force. The color of the samples is measured by a colorimeter and recorded in the L^*, a^*, b^* color system.

CONCLUSION

It can be concluded that many changes occur in physiochemical and nutraceautical properties in fruits and jams during processing, storage, and cooking. Fiber appears to exhibit minimal change during processing. Antioxidant activity, phenolic compounds, and anthocyanin content in jams are affected by thermal processing. Low methoxy pectin concentration and storage time affect the antioxidant properties as well as the color characteristics of jams. Thus, by increasing the proportion of low methoxy pectin in a jam formulation, there is an increase in bioactive compounds and antioxidant values. However, color quality, polymeric color is higher. The storage of jam at room temperature for 1–3 months results in mild or moderate alterations of investigated parameters, while more than 6 months of storage induced statistically significant alterations. The losses of anthocyanins during storage of processed jams were accompanied by increases in the polymeric color values, indicating that anthocyanins were extensively polymerized. Despite significant losses of total anthocyanins, the color density values changed little during jam storage at room temperature. Taking into consider all these factors, and in spite of the fact that some losses occur in the content of total phenolics, anthocyanins, L-ascorbic acid, as well as antioxidant activity, jams are still excellent sources of nutritional substances with antioxidant potential. Fruit jams have different rheological characteristics when temperature is increased. Furthermore, increase in temperature results in a decrease of apparent viscosity.

References

Abers, J., Wrolstad, R., 1979. Causative factors of color deterioration in strawberry preserves during processing and storage. J. Food Sci. 44, 75–79.

Amakura, Y., Umino, Y., Tsuji, S., Tonogai, Y., 2000. National Institute of Health Sciences. J. Agric. Food Chem. 48, 6292–6297.

Brownmiller, C., Howard, L.R., Prior, R.L., 2008. Processing and storage effects on monomeric anthocyanins, percent polymeric color, and antioxidant capacity of processed blueberry products. J. Food Sci. 73, H72–79.

Buttery, G., Guadgni, D., Ling, C., Seifert, M., Lipton, W., 1976. Additional volatile component of cabbage, broccoli and cauliflower. J. Agric. Food Chem. 24, 223–238.

Capecka, E., Mareczek, A., Leja, M., 2005. Antioxidant activity of fresh and dry herbs of some Lamiaceae species. J. Agric. Food Chem. 93, 223–226.

Dauthy, M.E., 1995. Fruit and vegetable processing, Food and Agriculture Organization of the United Nations, FAO Agricultural Services Bulletin. No.119.

Downing, D.L. (Ed.), 1996. A complete course in canning and related processes—book III processing procedures for canned food products. CTI Publications, Inc, Baltimore, Maryland.

Fugel, R., Carle, R., Schieber, A., 2005. Quantity and authenticity control of fruit puree, friut preparation and jams—A Review. Trends Food Sci. Technol. 16, 433–441.

Haffner, K., Finstad, M., Rosenfeld, H., Skrede, G., 2003. Color of raspberry jam as influenced by cultivar, temperature and light during storage. Acta Horticulturae 628, 829–834.

Javanmard, M., Endan, J., 2010. A Survey on Rheological Properties of Fruit Jams. Int. J. Chem. Eng. Appl. 1, 31–37.

Jawaheer, B., Goburdhun, D., Ruggoo, A., 2003. Plant foods for human nutrition. Kluwer Academic Publishers. Printed in the Netherlands. 58, 1–12.

Kono, Y., Shibata, H., Kodama, Y., Ueda, A., Sawa, Y., 1995. Chlorogenic acid as a natural scavenger for hypochlorous acid. Biochem. Biophys. Res. Commu. 217, 972–978.

Laleh, G., Frydoonfar, H., Heidary, R., Jameeil, R., Zare1, S., 2006. The effect of light, temperature, pH and species on stability of anthocyanins pigments in four berberis species. Pak. J. Nutr. 5 (1), 90–92.

Langdon, T.T., 1987. Preventing of browning in fresh prepared potatoes without the use of sulfiting agents. J. Food Technol. 41, 64–67.

Lim, Y.Y., Lim, T.T., Tee, J.J., 2007. Antioxidant properties of several tropical fruits: A comparative study. Food Chem. 103 (3), 1003–1008.

Mizrahi, S., 1979. A review of physicochemical approach to the analysis of the structural viscosity fluid fruit products. J. Texture Stud. 10, 67–82.

Ndabikunze, B.K., Masambu, B.N., Tiisekwa, B.P.M., Issa-Zacharia, A., 2011. The production of jam from Indigenous fruits using baobab (*Adansonia digitata L.*) powder as a substitute from commercial pectin. Afr. J. Food Sci. 5 (3), 168–175.

Ozela, E., Stringheta, P., Chauca, M., 2007. Stability of anthocyanins in spinach vine (*Basella rubra*) fruits. Cien. Cien. Inv. Agr. 34 (2), 115–120.

Rababah, T.,A., Al-Mahasneh, M., Kilani, I., Yang, W., Alhamad, M., Ereifej, K., Al-u'datt, M., 2011a. Effect of jam processing and storage on total phenolics, antioxidant activity, and anthocyanins of different fruits. J. Sci. Food Agr. 91, 1096–1102.

Rababah, T., Al-u'datt, M., Al-Mahasneh, M., Feng, H., Alothman, A., Almajwal, A., Yang, W., Kilani, I., Alhamad, M., Ereifej, K., Abu-Darwish, M., 2011b. Effect of storage on the physicochemical properties, total phenolic, anthocyanin, and antioxidant capacity of strawberry jam. J. Food Agri. Environ. 9 (2), 101–105.

Rababah, T., Al-u'datt, M., Al-Mahasneh, M., Yang, W., Feng, H., Kilani, I., Abu Ishmais, M., 2012a. Effect of jam processing and storage on phytochemicals and physiochemical properties of cherry at different temperatures. J. Food Process. Pres. Doi: 10.1111/j.1745-4549.2012.00770.x.

Rababah, T., Al-u'datt, M., Almajwal, A., Brewer, S., Feng, H., Al-Mahasneh, M., Ereifej, K., Yang, W., 2012b. Evaluation of the Nutraceutical, Physiochemical and Sensory Properties of Raisin Jam. J. Food Sci. 77 (6), c609–c613.

Rickman, J., Bruhn, C., Barrett, D., 2007. A review of Nutritional comparison of fresh, frozen, and canned fruits and vegetables II. Vitamin A and carotenoids, vitamin E, minerals and fiber. J. Sci. Food Agric. 87, 1185–1196.

Scibisz, I., Mitek, M., 2009. Effect of processing and storage conditions on phenolic compounds and antioxidant capacity of high bush blueberry jams. Polish J. Food Nutri. Sci. 59, 45–52.

Shahidi, F., Naczk, M., 2007. Food phenolics: An overview. In: Shahidi, F., Naczk, M. (Eds.), Food phenolics: sources, chemistry, effects, applications, 1995. Technomic Publishing Company Inc, Lancaster, PA, pp. 1–5.

Skrede, G., Wrolstad, R.E., Durst, R.W., 2000. Changes in anthocyanins and polyphenolics during juice processing of highbush blueberries (*Vaccinium corymbosum L.*). J. Food Sci. 65, 357–364.

Sun, T., Tang, J., Powers, J.R., 2005. Effect of pectolytic enzyme preparations on the phenolic composition and antioxidant activity of asparagus juice. J. Agric. Food Chem. 53 (1), 42–48.

Tomczak, W., 2007. Changes in antioxidant activity of black chokeberry juice concentrate solutions during storage. ACTA Sci. Polonorum Technol. Alimentaria. 6 (2), 49–55.

Tsai, P.J., Huang, H.P., 2004. Relationship between anthocyanin paterns and antioxidant capacity in mulberry wine during storage. J. Food Quality 27, 497–505.

Index

Note: Page numbers with "*f*", denote figures; "*t*" tables.

B

C

D

CPI Antony Rowe
Eastbourne, UK
October 16, 2014